BIOLOGY
PRINCIPLES & EXPLORATIONS

George B. Johnson
Peter H. Raven

HOLT, RINEHART AND WINSTON

Harcourt Brace & Company

Austin • New York • Orlando • Atlanta • San Francisco
Boston • Dallas • Toronto • London

Requests for permission to make copies of any part of the work should be mailed to:
Permissions Department, Holt, Rinehart and Winston, Inc.,
6277 Sea Harbor Drive, Orlando, Florida 32887-6777.

For permission to reprint copyrighted material,
grateful acknowledgment is made to the following sources:

Princeton University Press: "Out of Africa" map.

Dorion Sagan and Lynn Margulis: From *Garden of Microbial Delights: A Practical Guide to
the Subvisible World* by Dorion Sagan and Lynn Margulis.
Copyright ©1988 by Dorion Sagan and Lynn Margulis

Printed in the United States of America.

ISBN 0-03-072457-0

2 3 4 5 6 7 041 00 99 98 97 96 95

Biology: Principles and Explorations

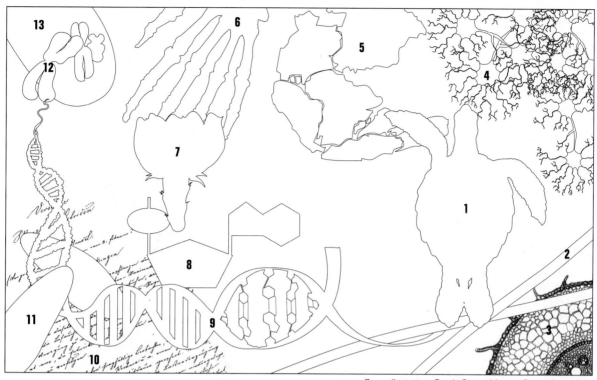

Cover illustration: Randy Gates, Morgan-Cain & Associates

1. A young green sea turtle (*Chelonia mydas*)
2. *Spirogyra*
3. Cross section of monocot root
4. Human neurons
5. Pangaea
6. Left hand of human skeleton
7. Prickly pear cactus flower
8. Nucleotide
9. DNA
10. Title page of Mendel's 1866 treatise on heredity
11. Paramecium
12. Human chromosomes
13. *Chilodonella*, a eukaryotic cell

Editorial Director of Math and Science
Richard Monnard

Executive Editor
Ellen Standafer

Senior Editor
Susan Feldkamp

Project Editors
Carolyn Biegert
John Gallo
Mitchell Leslie
Jennifer Linn

Production
Beth Prevelige
Simira Davis
George Prevelige
Rose Degollado
Nancy Hargis
Susan Mussey

Editorial Staff
Jane Martin
Steve Oelenberger
Tanu'e White

Design and Production
Morgan-Cain & Associates

Acknowledgments

Contributing Writers

Tracey Cohen
Science Writer
Highland Park, NJ

Thomas R. Koballa, Jr.
Professor
Science Education Department
University of Georgia
Athens, GA

George Nassis
Kenneth Rainis
WARD'S Natural Science Establishment
Rochester, NY

Lab Reviewers

Alex Molinich
George Nassis
Laboratory Investigations
WARD'S Natural Science Establishment
Rochester, NY

Kenneth Rainis
Safety
WARD'S Natural Science Establishment
Rochester, NY

Reviewers

Hugh Clement Allen
Adjunct Professor of Natural Sciences
Miami-Dade Community College
Miami, FL

David Armstrong, Ph.D.
University of Colorado
Boulder, CO

Carol C. Baskin
Adjunct Professor
School of Biological Sciences
University of Kentucky
Lexington, KY

Jerry M. Baskin, Ph.D.
Professor
School of Biological Sciences
University of Kentucky
Lexington, KY

Barry Bogin, Ph.D.
Professor of Anthropology
Department of Behavioral Sciences
University of Michigan
Dearborn, MI

Linda Butler, Ph.D.
Lecturer
University of Texas at Austin
Austin, TX

Mark Coyne
Assistant Professor
Department of Agronomy
University of Kentucky
Lexington, KY

Joe Crim, Ph.D.
Professor of Zoology
University of Georgia
Athens, GA

Susan Chattan Dabb, Ed. D.
Chair, Science Department
King Senior High School
Tampa, FL

Mary Pitt Davis
Science Teacher
Glenelg High School
Glenwood, MD

Andrew Dewees, Ph.D.
Chair, Department of Biological Sciences
Sam Houston State University
Huntsville, TX

William J. Ehmann, Ph.D.
Chair, Department of Environmental Science
Trinity College of Washington, DC
Washington, DC

William Forward
Biology Teacher/Science Chair
Rio Linda Senior High School
Rio Linda, CA
Science Methods Professor
California State University
Sacramento, CA

Bill Gasper
Science Department Chair
Clearwater Catholic Central High School
Clearwater, FL

Gary W. Goodnight
Biology Teacher
Denver South High School
Denver, CO

Jerald Halpern
Editor
Brookline, MA

Philip Hastings, Ph.D.
Assistant Research Scientist
Department of Ecology and
Evolutionary Biology
University of Arizona
Tucson, AZ

Andrea L. Huvard, Ph.D.
Department of Biology
California Lutheran University
Thousand Oaks, CA

Wojciech Kedzierski, Ph.D.
Assistant Professor
University of Texas
Southwestern Medical Center
Dallas, TX

Jo Ann D. Lane
Science Department Chair
St. Ignatius High School
Cleveland, OH

Glenn E. Mitchell
Science Department Chair, Mentor Teacher
Coalinga High School
Coalinga, CA

Martin Nickels, Ph.D.
Professor of Physical Anthropology
Illinois State University
Normal, IL

Nancy R. Parker
Associate Professor of Biological Sciences
Southern Illinois University
Edwardsville, IL

Sharon Perlman
Biology Teacher
Coral Park Senior High School
Miami, FL

Celia T. Rainwater
Biology Teacher
Tom C. Clark High School NISD
San Antonio, TX

Irving Rashkover
Assistant Principal for Curriculum
Kinloch Park Middle School
Miami, FL

Marian Smith, Ph.D.
Associate Professor of Biology
Southern Illinois University
Edwardsville, IL

Gerald Summers
Associate Professor of Biological Sciences
University of Missouri
Columbia, MO

Susan Green Talkmitt
Biology Teacher
Monterey High School
Lubbock, TX

William Thwaites, Ph.D
Biology Department
San Diego State University
San Diego, CA

Betty H. Tumminello
Assistant Principal/Curriculum Coordinator
Pineville High School
Pineville, LA

Jerry Warren
Science Department Chair
Elk Grove High School
Elk Grove, CA

BIOLOGICAL PRINCIPLES

UNIT 4 PRINCIPLES OF ECOLOGY

PART 2   BIOLOGICAL EXPLORATIONS

BIOLOGICAL EXPLORATIONS

BIOLOGICAL EXPLORATIONS

LABORATORY Investigations

FEATURES

Highlights..

UP CLOSE

SCIENCE➡TECHNOLOGY➡SOCIETY

PART 1

BIOLOGICAL PRINCIPLES

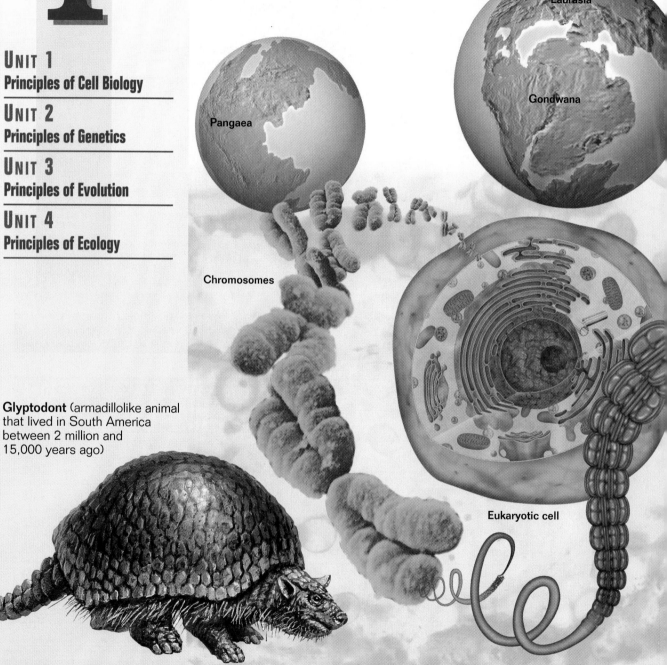

Laurasia

Gondwana

Pangaea

Chromosomes

Eukaryotic cell

Glyptodont (armadillolike animal that lived in South America between 2 million and 15,000 years ago)

> *During the voyage of the Beagle I had been deeply impressed by discovering great fossil animals covered with armour like that on the existing armadillos; secondly, by the manner in which closely allied animals replace one another in proceeding southward over the [South American] continent; and thirdly, by the South American character of most of the productions of the Galapagos archipelago, and more especially by the manner in which they differ slightly on each island of the group . . . It was evident that such facts as these, as well as many others, could be explained on the supposition that species gradually became modified; and the subject haunted me.*
>
> — Charles Darwin

Africa

DNA

Pygmy armadillo, or pichis (armadillo found in South America)

Visiting the Galapagos Islands off the coast of Ecuador in 1832, the young English biologist Charles Darwin recognized that each island was populated by different species of animals, yet the animals of each island resembled animals on the South American mainland. Darwin's observations led him to propose that organisms evolved, or changed, over time. Today we know more about the structure and function of living things than scientists in Darwin's day could have ever imagined. Biologists have learned how cells are structured, how they obtain and release energy, how they pass information from one generation to the next, and even how individual organisms work together to form a living community. The findings of modern scientists reveal how species change over time, supporting Darwin's observations in a multitude of ways.

CHAPTER 1

BIOLOGY AND YOU

Biology students
in Tucson, Arizona

1-1 Biology at Work Today

*T*he science of biology developed from scientists sharing their observations and studies with each other. Their fascination with and investigation of the real world has led to a vast body of knowledge that is continuously growing. As you read this section, think about the changes that might be taking place in your surroundings and how they might eventually affect you. What type of changes, for example, have you observed in your neighborhood? What effects might those changes have on you, your family, and your neighbors?

What Is Killing the Frogs?

Sometimes important things happen, right under our eyes, without anyone noticing. That thought occurred to David Bradford as he stood looking at a quiet lake high in the Sierras of California in the summer of 1988. Bradford, a biologist, had hiked all day to get to the lake, and when he got there his worst fears were confirmed. The lake was on a list of mountain lakes that Bradford had been visiting that summer in Sequoia–Kings Canyon National Parks while looking for a little frog with yellow legs, shown in Figure 1-1. The frog's scientific name was *Rana muscosa*, and it had lived in the lakes of the park for as long as anyone had kept records. But this silent summer evening the little frog was gone. The last major census of the frog's populations within the park had been taken in the mid-1970s, and *R. muscosa* had been everywhere, a common inhabitant of the many freshwater ponds and lakes within the park. Now, for some reason that Bradford did not understand, the frogs had disappeared from 98 percent of the ponds that had been their homes.

After Bradford reported this puzzling disappearance to other biologists, an alarming pattern soon became evident as other scientists reported similar findings. Throughout the American West, local populations of amphibians (frogs, toads, and salamanders) were becoming extinct. In Oregon, for example, 80 percent of the 30 populations of the Cascades frog, *Rana cascadae*, which had been studied in the mid-1970s, were now gone. Something was killing local populations of amphibians.

Amphibians have been around for 370 million years, since long before the dinosaurs. Their sudden disappearance from so many of their natural homes sounded an alarm among biologists. It was not just a concern for frogs.

Figure 1-1 Sequoia–Kings Canyon National Parks are outlined in the map of California, *top*. The many lakes and rivers of these parks are home to the mountain yellow-legged frog, *Rana muscosa*, *bottom*.

Scientists were afraid that the vanishing amphibians might be an example of "miner's canary" syndrome. In the nineteenth century, coal miners took canaries down into the mines as an early warning system to signal the presence of poisonous carbon monoxide gas. Canaries, which are very sensitive to the deadly gas, would die long before people would be affected. Amphibians are very sensitive to their environment, too, because their moist skins absorb chemicals from pond water. Biologists worried that the frogs' disappearance was an early warning that something very damaging was happening to the environment.

Searching for a Culprit

Important questions often do not have simple answers, and so it proved with this problem. Humans are putting stress on the environment in so many ways that many factors might be contributing to the decline of isolated amphibian populations. To understand what is going on, a scientist must focus on a particular situation, analyze it carefully, and try to learn the source of the problem for that particular amphibian population. Only after many populations are studied in this way can a general conclusion be drawn about the cause of the population decline.

One of the first amphibian populations to be analyzed in detail was that of a tiger salamander, *Ambystoma tigrinum*, shown in Figure 1-2. This salamander lives in a cluster of ponds located near Galena Lake, high on the western mountain slopes in west-central Colorado. John Harte, a biologist studying this population, had seen its numbers decline by 65 percent over a seven-year period.

Figure 1-2 Biologist John Harte observed the decline in numbers of the tiger salamander, *Ambystoma tigrinum*, beginning in the early 1980s.

There were many factors in the environment that might have been responsible for the salamander population's decline. However, Harte suspected acid rain. **Acid rain** results when sulfur in smoke produced by the burning of coal and oil reacts with water in the air to form sulfuric acid, which falls back to Earth in rain or snow. The acidity of a solution is described by **pH**—a number that represents the hydrogen ion concentration in a solution. Solutions with a low pH (between 0 and 7) are acidic. A solution with a pH of 2 is far more acidic than a solution with a pH of 6.

Colorado's Rocky Mountains, shown in Figure 1-3, get their moisture from winds that blow from the west, winds that carry acid from industry and urban smog. Over 90 percent of the annual moisture on the mountains falls as snow. Every spring when the snow melts, the acid accumulation of many months is released all at once into the high mountain ponds. Perhaps that pulse of acid was killing the salamanders.

Testing an Idea

To test his idea, Harte first measured the amount of acid released into the ponds at spring snowmelt for the years 1984 through 1988. In all the ponds, the amount of acid shot up to very high levels when the snow melted in early to mid-June. After a few weeks, the added acid was neutralized by minerals in the rocks lining the ponds. Acid levels then fell back to normal. Harte concluded that a pulse of acid was indeed being released into the ponds at snowmelt.

Harte then asked himself whether the acid released by snowmelt could be killing the salamanders. To answer this question, he placed salamander eggs in pond water, added acid, and examined the eggs to see how many hatched. What happened depended on how much acid he added and when he added it. Many of the eggs were killed by the level of acid released during snowmelt—but only if the eggs were exposed to the acid between the fifth and tenth days of egg development. This is the time when organs develop within the growing embryo. Exposure to acid then is fatal.

So the question for Harte boiled down to this: Were the salamander eggs in the ponds at a critical stage of their development when snowmelt delivered the jolt of acid to the ponds? For every year but one, the answer proved to be "yes." The acid arrived from snowmelt just at the time the salamander eggs were sensitive to it. The one exception was 1988, a year of exceptionally early snowmelt. In that year egg-laying occurred after acid levels had fallen back to their normal values, and, therefore, the eggs in the pond survived. It was the only year in Harte's study in which the number of salamanders increased.

Thus, Harte concluded that the steady decline he observed in the tiger salamander population during the 1980s was probably caused by acid rain. In this one case, the most likely culprit responsible for a declining amphibian population was tentatively identified. ◻

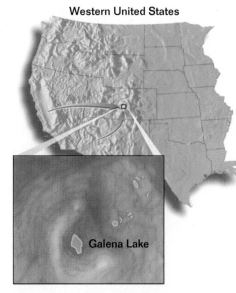

Western United States

Galena Lake

Figure 1-3 Windblown acidic compounds from Arizona smelters and California smog may be responsible for the decline of the tiger salamander in the Colorado Rockies. The location of Harte's study is shown in the inset.

◻ CAPSULE SUMMARY

Scientists expand scientific knowledge by making observations and posing questions about what they observe. To test an idea, scientists gather and record specific information. It is very important in science to pose questions that can be examined through experimentation.

Section Review

1. *Identify one observation that made David Bradford aware that* Rana muscosa *had disappeared from the lakes in Sequoia–Kings Canyon National Park.*

2. *Why was communication so important among the biologists who studied amphibians?*

Critical Thinking

3. *Why do you think waterfowl in Harte's study area were not affected by acid rain?*

4. *Amphibians living in the lakes along the coast of California were not affected by acid rain. Offer an explanation for this observation.*

1-2 The Scientific Process

Section Objectives

- Describe the six stages of a scientific investigation.
- Distinguish between hypothesizing, predicting, and experimenting.
- Define the elements of a control in an experiment.
- Define theory and explain why theories form the framework of science.

*J*ohn Harte is a biologist, and what he was doing in the Colorado mountains was science. Science is a particular way of investigating the world, of forming general rules about why things happen by observing particular situations. A scientist like Harte is an observer, someone who looks at the world in order to understand how it works. Stated briefly, a scientist determines principles from observation.

A Scientific Investigation Has Six Stages

It was once fashionable to claim that scientific investigations always progress by a rigid series of "either/or" steps called "the scientific method." In each of these steps, one of two incompatible alternatives is rejected. It is as if trial-and-error testing inevitably leads through the maze of uncertainty that always slows scientific progress. If this were true, a computer could be programmed to be a good scientist. But science is not done this way.

If you ask successful scientists like John Harte how they do their work, you will discover that without exception they design their experiments with a pretty good idea of the outcome. A scientist integrates all that he or she knows and allows his or her imagination full play in an attempt to get a sense of what might be true.

Although there is no single scientific method, all scientific investigations can be said to have six stages: collecting observations, forming hypotheses, making predictions, verifying predictions, performing control experiments, and forming a theory. Figure 1-4 summarizes these six stages.

Figure 1-4 Most scientists use variations of the scientific process when conducting their scientific investigations. Here, Asa Bradman, a student of John Harte, collects water samples from a Colorado pond.

Scientific Process
- Collecting observations
- Forming hypotheses
- Making predictions
- Verifying predictions
- Performing control experiments
- Forming a theory

MC-242   MC-243   MC-244

As you will see, insight and imagination play a large role in scientific progress. It is for this reason that some scientists are better at science than others—in the same way that Beethoven and Mozart stand out above most other composers.

Collecting Observations

The heart of any scientific investigation is careful observation. **Observation** is the act of noting or perceiving objects or events by using one or more of the five senses. Harte had studied the Colorado salamander population for many years. He had noted a thousand details of how the salamanders lived and what their ponds, such as the one shown in Figure 1-5, were like. If he had not kept careful records of what he saw, he might not have noticed that the salamander population was slowly declining over the years.

Forming Hypotheses

Observing the decline, Harte made a guess about why the salamanders were dying—perhaps they were being killed by acid rain. We call such a guess a hypothesis. A **hypothesis** is a proposed explanation that might be true. Hypotheses must be able to be tested by additional observations or experimentation. It is important to note that a hypothesis is not just any guess. A scientist makes an educated or informed guess based on everything he or she already knows. Using what he knew, Harte guessed that acids created in the upper atmosphere by industry were falling onto the mountains in the winter snows. By making the ponds acidic, the melted snow was killing the salamander embryos.

Making Predictions

Harte knew that if his hypothesis was correct, he could reasonably expect several consequences. We call these expected consequences predictions. A **prediction** is what you expect to happen if a hypothesis is accurate. Harte predicted that if acid precipitation was killing the salamanders, then it should be possible to detect the acid entering the ponds when the snow melts. Moreover, if acid precipitation was the culprit, then the amount of acid entering the ponds should be enough to kill salamander embryos.

Verifying Predictions

Harte set out to test his hypothesis by trying to verify its predictions. We call the test of a hypothesis an **experiment**. Harte did two series of experiments. In the first, he sought to determine if a rise in acid levels in the ponds occurred at the time of snowmelt. He took water samples from both deep and shallow parts of several ponds at frequent intervals, starting before snowmelt began and continuing until well after snowmelt concluded. He found that a large amount of acid was indeed introduced into the ponds by the melting snow.

Salamanders lay their eggs only once a year, as soon as the pond ice melts. Any salamander eggs developing in the

Figure 1-5 Natural changes in a habitat can be mistaken as trends if detailed, accurate, long-term records are not kept. Biologists of the Rocky Mountain Biological Lab are performing long-term studies that carefully monitor conditions in lakes such as this one in Mexican Cut, Colorado.

Figure 1-6 Deformities like the curved spine and stunted gills of this salamander occur when animals develop in water of pH 5, common for lakes and ponds contaminated by acid rain.

☐ *CAPSULE SUMMARY*

Control experiments lend validity to conclusions by allowing a single variable to be held constant, sorting out other factors that could affect results.

ponds would thus be exposed to the acid from the melting snow. But was it enough to do real harm? To test his prediction that the melting snow would release enough acid to damage the eggs, Harte carried out a second series of experiments in the laboratory. He allowed captive salamanders to lay eggs in water from the ponds. Then he added acid to the water at different times, to correspond with the acid levels observed in the pond during snowmelt, and looked to see if the eggs developed properly into adult salamanders. Harte found that the amount of acid being delivered to the ponds by snowmelt was indeed enough to deform and kill developing salamander embryos, particularly when introduced from five to ten days after the eggs were laid. Figure 1-6 shows an example of a deformed salamander.

Performing Control Experiments

Harte was now able to conclude that salamander eggs were being exposed to lethal amounts of acid. He decided to test his hypothesis further by using a control. A **control** is that part of an experiment in which the key factor is not allowed to change. In this experiment, the key factor was the exposure of early embryos to lethal amounts of acid in the ponds. As a control, Harte used the data gathered in 1988, a year when the snowmelt was unusually early. Just as in other years of Harte's study, acid was released into the ponds. The only difference between 1988 and the other years was that the acid was released earlier in the year. It was released so early, in fact, that by the time the salamanders laid their eggs, the acid in the pond had already been neutralized by chemicals in the rocks. Had the salamanders continued to decrease in number in the absence of acid, then acid could not have been responsible for killing the salamanders, and Harte's hypothesis would have been rejected. Through careful monitoring of the number of salamanders in the population in that unusual year, however, Harte observed a dramatic increase in their numbers! The result of the control was thus consistent with Harte's prediction. ☐

Forming a Theory

The essence of science is to reject hypotheses that are not supported by observations. The predicted consequences of a hypothesis are tested by experiments. If the experimental results do not agree with the predictions, the hypothesis is rejected. If they do agree, the hypothesis is not rejected—yet.

A scientist works systematically by attempting to show that certain hypotheses are not valid, that they are not consistent with the results of experiments. A successful experiment is one in which one or more of the alternative hypotheses are shown to be inconsistent with observations and are thus rejected. Harte, for example, was able to show that enough acid was being introduced into the ponds to kill the salamander embryos; he could therefore retain the hypothesis that acid in the snow was killing the salamanders.

Scientific progress is made the same way a marble statue is, by chipping away the unwanted bits.

A collection of related hypotheses that have been tested many times is called a theory. A **theory** is a unifying explanation for a broad range of observations. The hypothesis that acid rain is contributing to the decline of isolated amphibian populations will require a great deal more evidence than Harte's before becoming generally accepted. Many other factors may play important roles, such as increased ultraviolet (UV) radiation due to ozone depletion, and it is important not to be misled by what happens in a single instance. Only after many studies like Harte's will scientists be able to assemble a picture that accurately reveals what is harming the amphibians.

Constructing a theory often involves contrasting ideas and conflicting hypotheses. For example, Harte's conclusions have been questioned by other scientists who suggest that the correlation he indicates may be a coincidence. Maybe in heavy snowpack years it isn't extra acid that kills the larvae, they suggest, but rather increased shading of the ponds (which retards photosynthesis, leading to oxygen depletion and larval death). Further experiments will tell who is right. Argument, disagreement, and unresolved questions are a healthy part of science, a true reflection of how science is done, and the speediest path to increased knowledge. ◻

The word *theory* is used very differently by scientists and by the general public. To a scientist a theory represents that of which he or she is most certain. To the general public, *theory* implies a lack of knowledge, a guess. How often have you heard someone say "It's only a theory" to imply lack of certainty? As you can imagine, confusion often results. In this text, the word *theory* will always be used in its scientific sense, as a generally accepted scientific principle.

Theories are the solid ground of science, that of which scientists are most certain. There is, however, no absolute certainty or scientific "truth" in a theory. The possibility always remains that future evidence will cause a theory to be revised or rejected. A scientist's acceptance of a theory is always provisional. ◻

Section Review

1. *Name the stages of a scientific investigation.*
2. *How do scientists use the following: hypotheses, predictions, and experiments?*
3. *What is the purpose of a control experiment?*

Critical Thinking

4. *Explain why there is no absolute certainty or scientific "truth" in a theory.*

1-3 Properties of Life

What does the word alive *mean? Most dictionaries will define* alive *as "living or having life." Stop for a moment and write a short definition of* life. *You will not find it an easy task. The problem is not a deficiency on your part but rather the loose way in which the word* life *is used in everyday speaking.*

Section Objectives

■ Explain why life is difficult to define in terms of visually observable properties.

■ Identify the five properties of life.

Figure 1-7 Sensitivity to external stimuli is one of the key properties of life.

❑ *CAPSULE SUMMARY*

Some of the most obvious properties of life cannot be used alone to decide whether something is alive.

Living Things Share Certain Properties

Biology is defined as the study of life. The science of biology stems from knowledge acquired by scientists systematically studying living things, just as Harte did while investigating the effects of acid on salamander embryos. While most people are capable of distinguishing between living and nonliving, actually defining life can be quite difficult. When writing your definition of life, what sort of lifelike qualities did you mention? Perhaps you considered some of the most obvious examples: movement, sensitivity, the ability to respond to stimuli, change over time in the form of development, complexity, and even death. While unique to living things, these examples are terribly inadequate criteria of life. You can probably think of something that is not alive that has some of the qualities that you listed.

Clouds, for example, move when stimulated by the wind and develop from moisture that is suspended in the atmosphere. Clouds grow and change their shapes. Clouds are complex because of their chemical and structural makeup, and some might view their dissipation as being similar to death. However, death is not the same as disorder. All living things die, while inanimate objects do not. Clouds may break up and vanish, but they do not die.

So you see that movement, sensitivity, development, complexity, and death are properties you might expect to see in living things, but none of these properties individually is adequate to define life. While nonliving things may exhibit some of these properties, only living things exhibit all of them. Thus, you could argue that collectively these properties do define life. But is that enough?

Biologists recognize that all organisms with which we are familiar share certain general properties, as shown in Figure 1-8. It is by these properties that you can differentiate living things from nonliving things. ❑

Cellular Organization

All living things are composed of one or more cells—tiny chambers with thin coverings called membranes. A **cell** is

Properties of Life
- Cellular organization
- Metabolism
- Homeostasis
- Reproduction
- Heredity

Figure 1-8 Organisms like the soybean seedlings and the sand dollar have the same properties of life as you.

the smallest unit capable of all life functions. Some cells have more complex interiors than others, but all are able to grow and reproduce. The simplest organisms, such as the paramecium shown in Figure 1-9, have only a single cell. As a **multicellular** (composed of many cells) individual, your body contains more than 100 trillion cells.

Metabolism

All living things use energy to grow, to move, and to process information. That is why living things must eat to continue living, as shown in Figure 1-9. All the energy you use is captured from sunlight by plants and algae. To get your energy, you extract it from plants or plant-eating animals in a process called metabolism. **Metabolism** is the sum of all chemical reactions that an organism carries out.

Figure 1-9 Living things like the paramecium, *below,* exhibit organization. Humans, *below left,* and other organisms extract energy from food sources to perform the many chemical activities that are essential to life.

Homeostasis

All living things maintain relatively stable internal conditions, often quite different from their surroundings, by a process called **homeostasis.** Your body, for example, attempts to maintain a temperature of about 37°C (98°F) regardless of how cold or warm the weather might be.

Figure 1-10 If they live to maturity, these hatchling snakes may someday produce their own offspring. Reproduction ensures the ongoing success of a species.

Reproduction

The ability to reproduce from one generation to the next is characteristic of all species of living things. Rapidly growing bacteria divide into daughter cells every 15 minutes, and bristlecone pine trees that are 5,000 years old still produce seedlings. Since no organism lives forever, reproduction, as shown in Figure 1-10, is an essential part of living.

Heredity

All living things have DNA molecules inside their cells that encode information to direct growth and development—a set of blueprints, called genes, that determines what the organism will be like. In reproduction this set of instructions is passed on to the offspring in a process called **heredity**, resulting in family resemblances like those shown in Figure 1-11.

Figure 1-11 All living things pass on genetic information that makes their offspring similar yet unique.

Section Review

1. *Why would it be difficult to use development as a single criterion to define life?*
2. *What are the properties of life?*

Critical Thinking

3. *Suppose you are a biologist who has found an object that looks like an organism. What steps might you take to determine if your discovery is indeed alive?*

1-4 Addressing Real-World Problems

N o one can read a newspaper or magazine today without noticing, time and again, issues involving biology. We learn that tropical rain forests are being destroyed, that more Americans have died of AIDS than were killed in the Vietnam War, and that genetic engineering may offer a cure for cystic fibrosis. In this text you will encounter many areas in which biologists are actively working to solve many of today's problems.

Section Objectives

- Identify some major environmental concerns caused by the growing human population.
- Describe efforts being made to increase the world's food production.
- List three diseases that are the subject of current scientific research and describe efforts to combat them.

Figure 1-12 The ever-increasing bulk of solid waste produced by modern lifestyles poses a troublesome challenge. Garbage in this dump site in Point Barrow, Alaska, pollutes ground water and despoils the tundra.

Protecting Our Earth

As the world's population approaches 6 billion, human needs are placing serious stress on the planet. Some of the problems described below may seem easy to solve, but in reality their solutions are anything but simple.

What to Do With the Garbage

What should we do with the garbage? Our growing population makes more and more of it, as shown in Figure 1-12, and disposal of this waste has become a global problem. Have you ever stopped to think about how much plastic you personally have used in your life? The National Solid Waste Management Association in Washington, D.C., estimated in 1990 that one person disposes of 113.4 g (0.4 lb.) of plastic per day. All that plastic is still around somewhere. Plastic does not break down in nature. Imagine if it were stacked beside you!

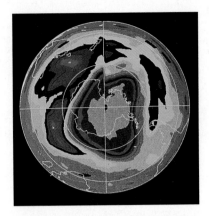

Figure 1-13 The pink area in the center of this satellite image represents the loss of ozone over the Antarctic. Industrial chemicals known as CFCs are responsible for this destruction of ozone.

Figure 1-14 The clearing of this rain forest will have disastrous environmental consequences for the many living organisms that depend on it for their survival.

Polluting the Atmosphere

Increasingly in the last century, humanity has taken to treating the atmosphere as if it were a garbage dump, one of limitless size into which chemicals could be "dumped" to be carried away by the wind. Only now are we beginning to realize that the Earth's atmosphere is far from limitless and that we cannot continue to treat it as a chemical sewer without paying a serious price. The tall smokestacks of power plants that burn high-sulfur coal release their smoke high in the sky, and it eventually falls back to Earth in rain or snow. This acid precipitation kills not only frogs, but also trees and many other creatures. The supposedly harmless aerosols and coolants made of chlorofluorocarbons (CFCs) have been released without concern into the atmosphere for decades. Only within the last few years have we learned that all those CFCs are still in the air, slowly destroying Earth's protective layer of ozone, as shown in Figure 1-13. Yet these compounds are still being produced and used. The ozone layer blocks harmful amounts of ultraviolet radiation from striking our skin. Its destruction would ultimately lead to ours.

Extinction

Earth's growing human population has also begun to seriously harm the other creatures with whom we share the planet. The world's tropical forests, home to one-fifth of the world's species of animals and plants, are being destroyed at the rate of more than an acre per second, as shown in Figure 1-14. At today's rate of destruction, all the rain forests will be gone in less than 30 years! With them will be lost over a million species, the greatest extinction event since the disappearance of the dinosaurs 65 million years ago. It is wise to remember that extinction is forever and that our children and their children and their children's children will never see any of these animals and plants alive again. Who knows what medicines and food plants we have forever discarded without knowing about them? Like burning a library without reading the books, extinction caused by humans is a tragedy beyond measure.

Feeding a Growing Population

Approximately 1.2 billion people live in a condition of extreme poverty, many of them malnourished. The demand for food is only going to increase.

To produce more food, we are going to have to develop new crops. Agricultural scientists are vigorously seeking new crops that will grow more efficiently in tropical soils, and crops that will grow without intensive use of fertilizers and pesticides. Also, genetic engineers are transplanting genes from one crop plant to another to create crop plants that are more resistant to insects. The success or failure of these efforts will have an enormous impact on the world's future. ☐

Combating Disease

Perhaps one of the most exciting ways that biology is affecting modern life is in the battle against disease. New technologies have enabled biological scientists to combat disease in ways scarcely imagined only a few years ago. Among the many diseases that you will encounter in this text, consider the following.

AIDS

It is estimated that by the end of this century 110 million people worldwide will be infected with HIV. **HIV** (human immunodeficiency virus), shown in Figure 1-15, is a virus that destroys the immune system, causing acquired immune deficiency syndrome, or AIDS. AIDS is fatal. All of the people infected with HIV are expected to develop AIDS. Over 300,000 Americans have already died, with 187,000 new cases reported in 1993–1994 alone. AIDS is now the leading cause of death among American males between the ages of 24 and 44. As the number of people with the HIV infection and AIDS grows, there is an increasing demand for more research to find a way to halt the spread of the disease and to help those already suffering from it.

Figure 1-15 HIV antibody testing and counseling are offered by public health clinics, *left,* to people who are at risk for infection by HIV, *above,* or who experience AIDS symptoms.

Cancer

At current rates, over one-third of the students who read this text will die of cancer someday. **Cancer** is a disorder of cells in which the normal controls on growth have been damaged and the cells divide unchecked within the body. Although the number of people with cancer is increasing, scientists studying this

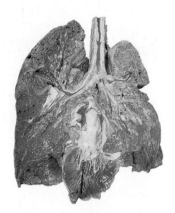

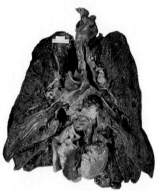

Figure 1-16 Cigarette smoking is the major cause of lung cancer. Its devastating effects on lungs are shown here, *bottom*.

disease have found that many cancer deaths are preventable. Almost all lung cancers and mouth cancers, for example, are due to the use of tobacco. Eliminating smoking and other uses of tobacco would prevent most lung and mouth cancers. Figure 1-16 shows how the use of tobacco can cause serious damage to your body. Your diet and the chemicals to which you expose yourself can affect the incidences of some major kinds of cancer. Research into everyday enemies, such as tobacco, could greatly lower the number of deaths caused by cancer in the future.

Cystic Fibrosis

Cystic fibrosis is an often fatal genetic disorder, passed from parent to child, that is caused by a defective gene encoding a cell membrane protein. Without a healthy copy of the gene, the child's cells cannot pump chloride ions in and out of cells correctly, causing thick mucus to build up in the lungs and other organs. In the 1990s biologists isolated a healthy version of the cystic fibrosis gene, and they are learning how to transfer it into cystic fibrosis patients. In 1993 the gene was first introduced into a human patient, carried into the cells of the lung aboard a cold virus. Researchers are hopeful that within a few years it will prove possible to cure humans by transferring into their lung cells, healthy copies of their defective gene. Other serious genetic disorders, such as muscular dystrophy, are also prime candidates for this gene-transfer therapy. The future for this approach looks very bright. ☐

What You Can Contribute .... It is clear that a scientific education has become necessary for everyone. The study of biology must play a major part in that education if you are to play any role in improving the standard of living for yourself and for future generations. Biological literacy is no longer a luxury for those who want to play a constructive role in improving the world; it is now a necessity.

Section Review

1. *Explain why saving the rain forests is such a pressing issue.*
2. *How is genetic engineering used in crop development?*

Critical Thinking

3. *Why is biological literacy essential in the battle against disease?*

CHAPTER REVIEW

Vocabulary

acid rain (6)
biology (12)
cancer (17)
cell (12)
control (10)
cystic fibrosis (18)
experiment (9)
heredity (14)
HIV (17)

homeostasis (13)
hypothesis (9)
metabolism (13)
multicellular (13)
observation (9)
pH (6)
prediction (9)
theory (11)

Concept Mapping

Construct a concept map that shows the relationships among scientific processes employed in biology. Use as many terms as needed from the vocabulary list. Try to include the following terms in your map: biology, observation, communication, hypotheses, predictions, experiments, and theories. Include additional terms in your map as needed.

Review

Multiple Choice

1. Scientific knowledge is ultimately based on
   a. beliefs.
   b. observations.
   c. models.
   d. experiments.

2. Biologists communicating about their observations of amphibian populations led to
   a. a decrease in the number of *Rana cascadae* in Oregon and California.
   b. the extinction of frogs in Kings Canyon National Park and other parks.
   c. a further decline in the number of amphibians in United States wetlands.
   d. the recognition of a widespread problem related to amphibian survival.

3. Harte suspected that acid rain was responsible for the tiger salamander population's decline because
   a. acid-laden winds provide the moisture received by the region.
   b. coal and oil are used to heat homes in nearby Colorado.
   c. the pulse of acid was detected long before the snow melted.
   d. local factories produce industrial acids which accumulate in the ponds inhabited by the tiger salamander.

4. A hypothesis is a testable statement that
   a. is considered unscientific because it is not a theory.
   b. can be supported in all circumstances.
   c. if false, can be demonstrated false through experimentation.
   d. is the final step in every research project.

5. Which of the following is *not* a stage of a scientific investigation, as described in the chapter?
   a. theorizing
   b. observing
   c. experimenting
   d. classifying

6. A girl wants to determine which of two foods produces the greatest weight gain in gerbils. She takes two gerbils of identical weight from the same litter and feeds one gerbil 20 g of food A each day and feeds the second gerbil 20 g of food B each day. After 30 days, she weighs the two gerbils and finds that the one fed food A has gained 50 g and the one fed food B has gained 80 g. A control in this experiment is
   a. gerbils of identical weight.
   b. the food fed the gerbils.
   c. the gerbils' weight gains.
   d. the color of the gerbils' fur.

7. Problems of waste management and air pollution are due to
   a. the spread of disease.
   b. growth of the human population.
   c. uncontrolled scientific experimentation.
   d. genetic engineering and global warming.

8. Using the criteria of movement and complexity to define life
   a. makes distinguishing living things from nonliving things a difficult task.
   b. is acceptable for children and adolescents.
   c. means that cars are alive but trucks and buses are not.
   d. means that single-celled organisms are alive whereas multicellular organisms are not.

9. To test the effectiveness of a new drug against human immunodeficiency virus (HIV), a biologist would
   a. take 100 cultures of HIV and put the new drug in 50 of them.
   b. take 100 cultures of HIV and put the new drug in all of them.
   c. put the new drug in none of the 100 cultures.
   d. avoid using cultures consisting of animal cells.

Completion

10. By one estimate, the tropical forests will be gone in less than _____ years.

11. Agricultural scientists seek new crops that will grow without intensive use of _____ and pesticides.

12. A scientist using a model to weigh the benefits and risks of a course of action is engaged in _____ .

13. Biology is defined as the study of _____ .

14. Before a hypothesis is proposed, _____ must be made. A test of a hypothesis is called a(n) _____ .

15. Harte hypothesized that _____ was the cause of the decline of the salamander population.

Short Answer

16. What characteristic of frogs makes them the "miner's canary" of wetlands?

17. A virus consists of DNA or RNA in a protein coat. It can reproduce and is capable of movement from one host to another. Is a virus alive? Explain.

18. Explain why news accounts are much more influential than scientific papers in educating the public about scientific issues.

19. "In science all hypotheses and theories are tentative." How does this statement apply to the work of John Harte?

20. Compare scientists' use of the term *theory* with that of the general public.

21. Scientists often try to repeat, or replicate, the results of other scientists' experiments. Why do you think this is an important part of scientific investigation?

Critical Thinking

22. **Making Inferences** One of the most important parts of any scientific paper is the part called "Methods and Materials," in which the scientist describes the procedures used in the experiment. Why do you think such details are so important?

23. **Making Predictions** An experiment was conducted to test the effect of a new drug on the growth of human cancer cells. In the experimental group, the drug was added to cancer cells growing in 100 test tubes. In the control group, genetically similar cancer cells were grown under exactly the same conditions as the experimental group but without the drug. The results of the experiment showed slower growth of cancer cells in the experimental group. What would be the next logical step in testing the effectiveness of the new drug to slow the growth of human cancer cells?

24. **Making Inferences** Explain why it is inappropriate to say that a hypothesis has been proven.

25. **Making Inferences** Scientific knowledge is constructed like a tower built from blocks rather than discovered like a gold nugget in a stream. Explain how the report of John Harte's work with the tiger salamander provides support for this statement.

Activities and Projects

26. **Cooperative Group Project** Design an experiment to find out how the pH of the water used to water bean plants affects their growth. First, identify the variables that should be controlled and describe the control and experimental groups. Explain how the results from the control and experimental groups should be compared. Finally, write the components of your design on a poster and display it for other students to see.

27. **Cooperative Group Project** Create a television commercial intended to convince high school students to become biologically literate. To make your commercial as persuasive as possible, survey students about their opinions on the advantages and disadvantages of studying biology in high school. Address their responses in your commercial.

28. **Science-Technology-Society** Learn what you can about integrated pest management (IPM) by reading and talking with farmers and agricultural extension agents. What are some of its advantages and disadvantages? Why are agricultural companies that produce pesticides less than enthusiastic about IPM?

29. **Multicultural Perspective** Tropical rain forests are located mainly within the borders of impoverished nations in Central America, South America, and Africa and are home to at least one-third of all animal and plant species. It is predicted that in less than 100 years, all the world's rain forests will have been eliminated due to lumbering and clearing for farms. People in the United States and other wealthy nations argue that assisting the economic growth of the nations containing the rain forests is the best way to save the rain forests. Explore both sides of this argument. Report what you find to your class.

30. **Multicultural Perspective** Do library research to discover examples of ways that some cultures around the world speak of fire as a living being (many cultures, for example, refer to fire as "grandmother" or "grandfather"). Which of the five characteristics often associated with living things (movement, sensitivity, development, complexity, and death) do you think fire might possess? What about the five general properties of living things (cellular organization, metabolism, reproduction, homeostasis, and heredity)? Defend your answer.

Readings

31. Read the article "Silence of the Frogs," in *The New York Times Magazine*, December 13, 1992. Describe Cynthia Carey's research. What does her work seek to discover about the causes of red-leg disease in the western frog? Name three factors that scientists think could possibly be responsible for the disappearance of frog populations in various locations around the world.

32. Read the article "The Long Shot," in *Discover*, August 1993. Explain why scientific techniques using killed or weakened whole viruses have thus far failed to produce a successful vaccine for AIDS. Why is it difficult for scientists attempting to develop such a vaccine to perform experiments and thus verify their predictions regarding the success of any prospective vaccine?

Observing Effects of Acid Rain

OBJECTIVE

Observe the effects of acid rain on seeds from three different species of plants.

PROCESS SKILLS

- measuring pH
- predicting effects of acid rain
- organizing and graphing class data

MATERIALS

- safety goggles
- lab apron
- bean, pea, or corn seeds
- metric ruler
- 250 mL beakers
- mold inhibitor
- wax pencil or marker
- plastic bags (zipper type)
- paper towels
- 2 water solutions with different pH values
- graph paper

BACKGROUND

1. What threat does acid rain pose to the environment?
2. What are major sources of the pollutants that cause acid rain?
3. Write your own **Focus Question** on your Vee Form.
4. **Knowing Side of the Vee** List the **Concepts** and new **Vocabulary Words** on your Vee Form. In the **Concept Statements** section of the Vee, use these words in sentences that define and explain them.

TECHNIQUE

Doing Side of the Vee

Part A: Measuring and Sowing Seeds

For this procedure, you will work with one water solution while your partner works with a different water solution. Before beginning the lab, put on safety goggles and a lab apron. You and your partner will use the same kind of seed.

1. Measure the length in millimeters of each of 20 seeds. Determine the average length. In the **Records** section of your Vee, make a table similar to the one shown on the following page. Record the seed type, your solution's pH, and the average seed length (Day 0) in the table.

2. Place the 20 seeds into a beaker, 10 seeds for you and 10 seeds for your partner. Slowly add mold inhibitor until the seeds are covered. Soak the seeds for 10 minutes. While the seeds soak, label a plastic bag with your name and the pH of the water solution assigned to you. Moisten three layers of paper towels with your assigned water solution. Your partner will also label a plastic bag and use a different assigned solution. Fold the moistened towels in half and put them inside the bag. Drain the mold inhibitor from the beaker and gently rinse the seeds with water. Place the seeds on clean paper towels.

3. Place your 10 seeds between the layers of the moistened paper towels. Slide the towels and seeds into your plastic bag. Close the bag. Your partner should do the same with the other 10 seeds.

Part B: Germination and Growth

4. After 2 days, examine your seeds. Note the changes in appearance, length, average seed size, and the number of germinated seeds. Record your observations in the **Records** Section of your Vee Form.

Effects of Acid Rain on Seed Growth

	You	Your Partner
Seed Type		
pH of Solution		
Average Seed Length (Day 0)		
First Observation		
Second Observation		
Last Observation		
Number of Seeds That Germinated		
Overall Average Seed Growth		

5. Add your data to the class chart on the chalkboard. Compare your results with those of your classmates.

6. Moisten the paper towels with your assigned solution and return the towels and seeds to the plastic bag.

7. Observe your seeds again after 2–3 days. Add more of your solution when necessary. In the **Records** section of your Vee Form, record the average length and number of germinated seeds.

8. Make your final observations and measurements 7–10 days after you began this investigation. In the **Records** section, note any overall changes in appearance you observe. Check for changes in length. Note average seed size and the number of germinated seeds. Add your data to the class data.

9. Subtract the Day Zero average seed length from the Last Observation average seed length. This is the overall average seed growth. Record this in the **Records** section of your Vee Form and the class chart.

10. Make a graph to organize the class data on the effects of pH on the germination and growth of each type of seed. Label the vertical axis with the number of millimeters of overall seed growth. Along the horizontal axis, show the range of pH values for the entire class. Copy this graph onto the back of your Vee Form in **Additional Records and Observations**. In the **Procedure** section of the Vee, briefly summarize the procedure you followed.

INQUIRY

1. Develop a hypothesis to explain the possible effects of acid on the germination of your seeds.

2. How does the pH value appear to affect seed germination?

3. Use the information on the **Knowing Side** of the Vee to interpret your results from the **Doing Side**, and then write your **Knowledge Claim**. Write a **Value Claim** for this lab.

ANALYSIS

1. In which solution did each type of seed germinate first?

2. What pH appears to be best suited for both successful germination and continued seed growth? Which is the least beneficial?

FURTHER INQUIRY

Write a **New Focus Question** that could be the point of a new investigation. The following is an example:

Are certain stages in the plant life cycle more susceptible to acid rain than others?

CHAPTER 2

NATURE OF CELLS

REVIEW
- characteristics of life (Section 1-3)

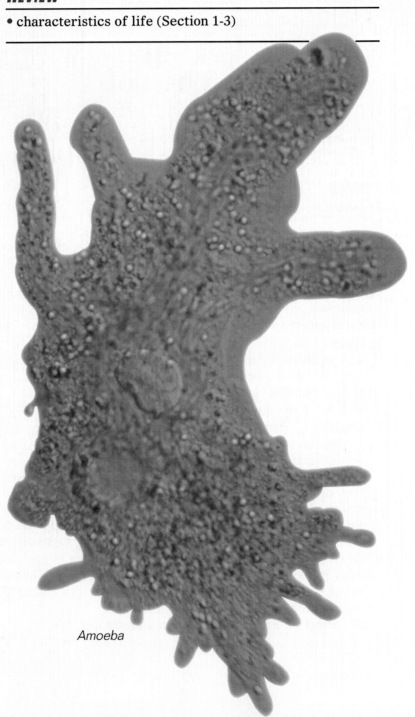

Amoeba

2-1 Cells: The Smallest Units of Life

Organisms are more than collections of chemicals. They are chemicals organized to carry out the functions of living. The organization of all living things begins with the cell, the smallest unit capable of carrying out the functions of life. All living things are composed of one or more cells.

How Cells Were Discovered

Most cells are too small to see with the naked eye. A typical human body cell is many times smaller than a grain of sand! Scientists became aware of cells only after microscopes were invented in the 1600s. When the English scientist Robert Hooke used one of the first microscopes to observe a thin slice of cork in 1665, he saw "a lot of little boxes," shown in Figure 2-1. These little boxes reminded him of the small rooms in which monks live, so he called them cells. Later, Hooke observed the same pattern in the stems and roots of carrots and other plants. What Hooke still did not know, however, was that cells are the basic unit of living things.

Ten years later, the Dutch scientist Anton van Leeuwenhoek focused a microscope on what seemed to be clear pond water and discovered a wondrous world of living creatures! He named them "animalcules," or tiny animals. Today we know that they were not animals, but single-celled protists, among the most diverse of all living things.

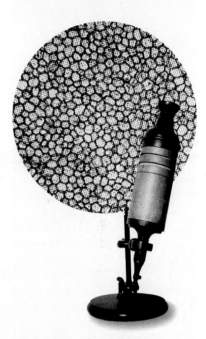

Figure 2-1 It was the English scientist Robert Hooke who first coined the term *cell*. Hooke used a simple microscope to view a slice of cork. The "little boxes" that he saw reminded him of the monastery cells that served as individual living quarters for monks, and a new biological term arose as a result. Although Hooke did not know it, all living things are made of one or more cells.

Formation of the Cell Theory

It took scientists more than 150 years to fully appreciate the discoveries of Hooke and Leeuwenhoek. In 1838, the German botanist Matthias Schleiden concluded that cells compose not only the stems and roots but every part of a plant. A year later, the German zoologist Theodor Schwann made the same claim about animals. And in 1858, a German physician, Rudolph Virchow, observed that cells come only from other cells. The observations of Schleiden, Schwann, and Virchow form what is known today as the cell theory. The **cell theory** is usually stated in three parts:

1. All living things are composed of one or more cells.

2. In organisms, cells are the basic units of structure and function.

3. Cells are produced only from existing cells.

All Cells Share Certain Characteristics

All cells, even those that are very simple, share certain structural characteristics. All cells have a **cell membrane** or **plasma membrane** that separates the cell's contents from materials outside the cell. The cell membrane also regulates what moves in and out of a cell, helping it to maintain homeostasis. Without the cell membrane to contain the substances the cell needs for life, the cell would die. All cells have **cytoplasm** *(SYT uh plaz uhm)*, which is everything inside the cell membrane except the cell's genetic material. The fluid portion of cytoplasm is called cytosol *(SYT uh sol)*.

The cytosol is packed full of free-floating **ribosomes** *(RY buh sohmz)*, the structures on which proteins are made. You will learn how ribosomes make proteins in Chapter 9.

All cells have the ability to reproduce themselves, and they all possess genetic material, which contains the instructions for making proteins and carrying out the cell's day-to-day activities.

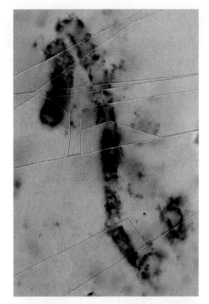

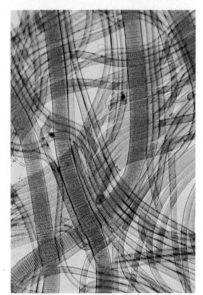

Figure 2-2 Fossils of prokaryotic microorganisms, *top*, have been found in Western Australia in rock formations over 3.4 billion years old. Modern cyanobacteria such as *Oscillatoria*, *bottom*, are little changed from their primitive ancestors. Like all prokaryotes, cyanobacteria lack nuclei and other membrane-bound organelles.

A Short History of Cells

The oldest fossils we have of cells are those of tiny cyanobacteria, shown in Figure 2-2. These **prokaryotic** *(pro KAR ee AHT ik)* cells lived at least 3.5 billion years ago. Prokaryotes are single-celled organisms that lack internal membrane-bound compartments. The term *prokaryote* is from the Greek *pro,* meaning "before," and *karyote,* meaning "kernel." Early cells were simple and small (1–2 μm in diameter). Like their fossil ancestors, modern prokaryotes are very small (1–15 μm) and do not have internal compartments. Without separate compartments that isolate materials, cells cannot carry out many specialized functions. In prokaryotes, the genetic material is a single, circular molecule that is not enclosed in a membrane-bound compartment. For nearly 2 billion years—half the age of Earth—prokaryotes were the only organisms that existed.

The First Cells With Internal Compartments Were Eukaryotes

The first cells with internal compartments evolved about 1.5 billion years ago. Much larger than any bacteria, these cells range from 2 to 1,000 μm in size. Such cells are **eukaryotic** *(yoo KAR ee AHT ik)*. The term *eukaryotic* comes from the Greek words *eu,* meaning "true," and *karyote,* meaning "kernel" or "nucleus." Eukaryotes have a **nucleus**, a membrane-bound compartment that houses the cell's DNA. Eukaryotes possess other small, specific membrane-bound internal compartments called **organelles** that carry out specific functions. Such organization allows eukaryotic cells to function in more complex ways than do prokaryotic cells.

How Eukaryotes Evolved

Most biologists who study eukaryotic cell structure think that eukaryotes evolved from prokaryotes. Many of the organelles of eukaryotes resemble bacteria, perhaps engulfed long ago by much larger cells. Scientists hypothesize that bacterial "trespassers" remained inside these cells, gradually losing their ability to live independently. These invading bacteria became organelles, and eukaryotic cells were the result. The fact that some organelles have their own distinctive DNA provides additional evidence for this hypothesis.

All living cells that are not bacteria are eukaryotes. Your cells are eukaryotic, as are tree cells and elephant cells. The "animalcules" seen by van Leeuwenhoek also were eukaryotic.

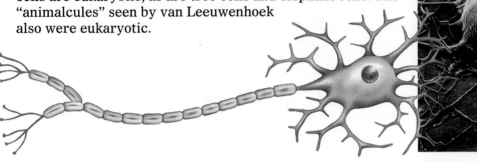

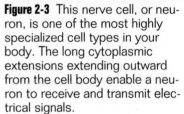

Figure 2-3 This nerve cell, or neuron, is one of the most highly specialized cell types in your body. The long cytoplasmic extensions extending outward from the cell body enable a neuron to receive and transmit electrical signals.

Multicellularity

Early eukaryotes were single-celled, but eventually many of them became aggregated (clustered) into multicellular organisms. Multicellular organisms are those that are composed of more than one cell. Being multicellular was a great evolutionary advance because it enabled particular cells to specialize in certain activities. For example, nerve cells, such as the one shown in Figure 2-3, are highly specialized cells that conduct messages in the form of nerve impulses from one part of the body to another. You are a multicellular individual. Your body is composed of trillions of cells whose specialized activities are coordinated with one another.

Not all eukaryotes are multicellular. In fact, if you were to survey all living organisms on Earth today, you would find that most living eukaryotes are unicellular protists, single-celled organisms. Whether single-celled or multicellular, the cells of all eukaryotes are similar in design, more similar to each other than to the prokaryotes. The basic design of eukaryotic cells will be explored in Section 2-4. ❏

❏ CAPSULE SUMMARY

Except for bacteria, organisms consist of one or more eukaryotic cells that contain membrane-bound organelles. Prokaryotes and eukaryotes share several characteristics: a cell membrane, cytoplasm, ribosomes, and DNA.

Cells Must Be Small
....

There are some 100 trillion cells in the human body, typically ranging from 5 to 20 μm in diameter.

Why is your body made of many tiny cells instead of a few large cells? There are two limits that affect how efficiently cells work and that govern cell size. One limit is related to the exchange of materials between the

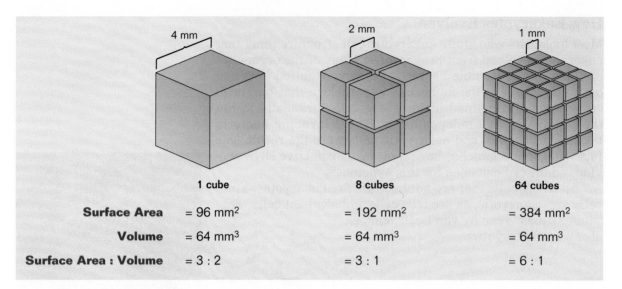

	1 cube	8 cubes	64 cubes
Surface Area	$= 96 \text{ mm}^2$	$= 192 \text{ mm}^2$	$= 384 \text{ mm}^2$
Volume	$= 64 \text{ mm}^3$	$= 64 \text{ mm}^3$	$= 64 \text{ mm}^3$
Surface Area : Volume	$= 3 : 2$	$= 3 : 1$	$= 6 : 1$

Figure 2-4 As a cell gets larger, its volume increases at a faster rate than its surface area. A cell's surface area must be large enough to meet the needs of its volume.

◻ *CAPSULE SUMMARY*

Smaller cells are more efficient than larger cells because it takes less time to transport information and materials across the surface to the inside where they are needed.

inside and the outside of the cell, and the other limit is related to the distribution of materials within the cell.

A cell's surface provides the only opportunity for interaction with its environment. All information and materials entering or leaving the cell, including wastes, must pass through "doors" in the cell's membrane. The efficiency of such an exchange depends on the ratio of the cell's surface area to its volume. The surface area of a cell is the measurement of the exterior of the cell. Large cells have far less membrane surface per unit of volume with which to supply materials to the cell's interior and to rid it of wastes. You can understand surface-area-to-volume ratios by looking at the imaginary cube-shaped cell in Figure 2-4. A cell with six times as much surface area per unit as volume per unit can move materials in and out of a cell more efficiently than a cell with a smaller surface-area-to-volume ratio.

A second limit governing the size of cells has to do with transport within the cell itself. As cell size increases, it takes longer for information and materials to reach their destination. Small cells, therefore, function more efficiently than larger cells. ◻

Section Review

1. *What observations were used to develop the cell theory?*
2. *What function do organelles serve?*
3. *Explain the importance of surface-area-to-volume ratio in a cell.*

Critical Thinking

4. *In prokaryotes, DNA exists as a single molecule. Why is it not considered an organelle?*

2-2 The Chemistry of Living Cells

*L*iving cells carry on many complicated chemical processes. In order for you to understand living things, some knowledge of basic chemistry is necessary. The chemistry that you will encounter in this chapter will be useful in helping you understand biology, for all organisms are chemical machines.

Atoms Are the Cell's Smallest Components

All living and nonliving things are composed of **atoms.** Every atom consists of a cloud of tiny particles called electrons that spin in undefined paths around a small, very dense core called a nucleus, as shown in the model in Figure 2-5. The nucleus is a cluster of two kinds of particles, protons and neutrons. To understand atoms you must remember that **electrons** carry a negative charge and protons a positive charge (neutrons are not charged). It is the attraction between positive and negative charges that keeps the electrons spinning about the nucleus.

Kinds of Atoms

When the atoms in a sample of matter are all alike, the sample represents an element. An **element** is a substance that cannot be broken down to any other substance by ordinary chemical means. There are currently more than 100 known elements. Each is denoted with a one-, two-, or three-letter symbol. For example, carbon is represented by C, oxygen by O, and hydrogen by H.

Atoms as a whole have no electrical charge. Some atoms can react with other atoms to form particles with unequal numbers of electrons and protons. These kinds of atoms are called **ions.**

The Nature of Atoms

The chemical behavior of an atom is determined by the electron cloud that surrounds its nucleus. Almost all the volume of an atom is empty space. When two atoms meet, the electron clouds can overlap, but the nuclei never come into contact with each other.

Atoms Have Energy

Energy is the ability to do work. Electrons in atoms have energy; it takes energy to keep a negative electron from crashing into the positive nucleus of an atom.

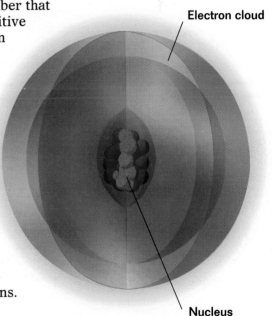

Figure 2-5 The volume of an atom is mostly empty space. The cloud portions of the atom represent regions where moving electrons are most likely to be found.

Electron cloud

Nucleus

☐ CAPSULE SUMMARY

All matter is composed of atoms. Atoms consist of electrons, protons, and neutrons. Electrons determine the nature of atoms and store considerable amounts of energy. Electrons that are close to the nucleus have lower energy levels than those that are farther away.

Electrons are organized outside the nucleus by the amount of energy they possess. Electrons close to the nucleus are at low energy levels. Electrons farther from the nucleus are at higher energy levels. Because an electron is so small and moves so fast, we do not know its energy or position in space with a high degree of certainty. The locations of electrons, therefore, are described in terms of probability. We can mathematically determine how likely it is that an electron will be at a certain location. The pathways an electron takes in moving about the nucleus, however, are unknown. It was once thought that electrons orbited the nucleus like the planets orbit the sun. We now know that this model for the atom is inaccurate. ☐

Atoms Chemically React to Form Compounds

A **compound** is a group of atoms held together by chemical bonds. Compounds are represented by chemical formulas like NaCl (sodium chloride, or table salt) and H_2O (water). The formula identifies the elements in the compound as well as their proportions. The force that links the atoms of compounds is called a chemical bond. There are three kinds of bonds that are important to biological systems: covalent bonds, ionic bonds, and hydrogen bonds.

Covalent Bonds

Covalent bonds form when two atoms share electrons. Like the rivets and welds that link the steel girders in a skyscraper, covalent bonds are the strong links that hold together the atoms of most of the compounds in your body. The chemistry of living cells is based on the element carbon, which accounts for more than one-half the dry weight of cells. The ability of carbon atoms to form very stable carbon-carbon bonds by bonding covalently is of great significance in biology. A carbon atom has four outer electrons and can form four covalent bonds with another carbon atom or with a different kind of atom. A group of atoms held together by covalent bonds is called a **molecule.**

Stable atoms have filled outer energy levels. All atoms in living things (except hydrogen and helium) have outer levels that hold eight electrons. If an atom has only seven electrons in its outer energy level, it will react readily with an atom that has a single electron in its outer level. Water (H_2O) is a molecule made of oxygen (six outer electrons) that forms covalent bonds with two hydrogen atoms (one outer electron each).

There are two key properties of covalent bonds that make them ideal for their role in living systems. They are very strong and very directional, meaning they can form bonds in one or more specific directions. Thus, bonds can form between two or more specific atoms sharing electrons.

1. Electron energy level model

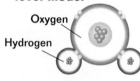

2. Ball and stick model

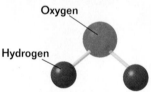

3. Space filling model

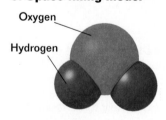

Figure 2-6 Water is held together by the sharing of electrons (covalent bonds) between the hydrogen atoms and oxygen. Three different ways of representing the water molecule are shown.

Ionic Bonds

Ionic bonds form between two atoms of opposite charge. An actual exchange of an electron occurs to form the ions that form the ionic bond. The force of attraction between a positive and negative ion is an **ionic bond.** For example, sodium atoms become positively charged ions (Na^+) by losing their electrons to chlorine atoms to form chloride ions (Cl^-). Each chloride ion has an extra electron that attracts it electrically to surrounding sodium ions of opposite charge. Substances that form ionic bonds break apart when placed in water, producing free ions. Many such ions perform essential roles in biological activities. Sodium, for example, is essential for the functioning of nerve cells.

Hydrogen Bonds

Hydrogen bonds, which are weak bonds of a very special sort, play a key role in living systems. They differ from ionic and covalent bonds in that they link *molecules* together rather than atoms. Look at the water molecules in Figure 2-6. Oxygen forms covalent bonds with two hydrogen atoms. The shared electrons in water are more strongly attracted by the oxygen nucleus than by the hydrogen nuclei. Water molecules act like a molecular magnet, with positive and negative ends, or "poles." Molecules that have unequal areas of charge, like water, are **polar molecules.** A **hydrogen bond** is a weak chemical bond that forms between two polar molecules. The positive end of one polar molecule is attracted to the negative end of another, as shown in the model in Figure 2-7.

Hydrogen bonds are weak, so they do not form if there are long distances between molecules. Also, hydrogen bonds play critical roles in determining the shapes of many important biological molecules, such as DNA and proteins. ▢

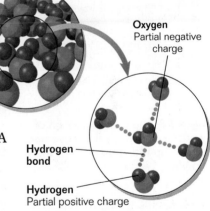

Figure 2-7 Ionic bonds in sodium chloride result from an electron transfer between a sodium atom and chlorine atom. The transfer produces ions that arrange in a cubic structure, *above.* Water molecules have slight positive and negative charges that cause the molecules to attract each other forming hydrogen bonds, *below.*

Sodium ion

Electron transfer

Positive (sodium) ion

Negative (chloride) ion

Chloride ion

Oxygen
Partial negative charge

Hydrogen bond

Hydrogen
Partial positive charge

Water Is a Major Component of Cells

When life on Earth was beginning between 3 and 4 billion years ago, water provided a medium in which other molecules could interact. Life as we know it could not have evolved without these interactions. Today, three-fourths of Earth's surface is covered by water. Every cell in your body contains water; in most cells there is an abundance of it. About two-thirds of the molecules in your body are water molecules.

Water Stores Heat

Water heats more slowly than most other substances. It also retains its temperature longer than other substances when

Water   Dye

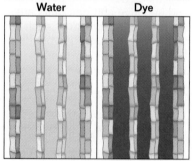

Liquid movement up a stem through capillary action

Figure 2-8 The capillary action of water is shown by the movement of dye up the stem to color the flower.

its surrounding environment cools—it stores heat well. Many organisms dispose of excess heat through water evaporation. For example, humans cool down by sweating.

Water Clings to Itself and Other Molecules

Water molecules readily form hydrogen bonds with one another, so water clings to itself in an attraction called cohesion *(koh HEE zhuhn)*. It is because of cohesion that water is a liquid and not a gas at room temperature. Hydrogen bonds link many individual water molecules together at the water's surface, like a crowd of people linked by holding hands. Surface tension forms across the surface of water because of the cohesive attraction between individual water molecules.

The attraction of water to a substance other than water is called adhesion *(ad HEE zhuhn)*. Water adheres to any substance that it can form hydrogen bonds with. That is why some things get "wet" and others, such as waxy substances, that are composed of nonpolar molecules, do not. The adhesion of water to substances with surface charges causes capillary action. Capillary action and cohesion are responsible for the upward movement of water as shown in Figure 2-8.

Water Ionizes

When the covalent bonds of water break, a hydrogen ion (H^+) and a hydroxide ion (OH^-) are produced.

$$H_2O \rightarrow H^+ + OH^-$$

This ionization process goes on continuously in water. As a result, pure water always has a low concentration of hydrogen and hydroxide ions; roughly 1 out of every 550 million water molecules exists as ions at any instant.

Water Is a Powerful Solvent

When a solvent, such as water, dissolves a solute, such as table sugar, a solution is formed. Many compounds dissolve in water because of water's polar nature. When covalent compounds dissolve in water, molecules are evenly dispersed in the solution. When ionic compounds are dissolved

Figure 2-9 Ionic substances dissolve in water to form charged particles (ions), *left.* Covalent substances dissolve in water to form molecules that may have very slight charges, *right.* Many substances found in the body are dissolved ions.

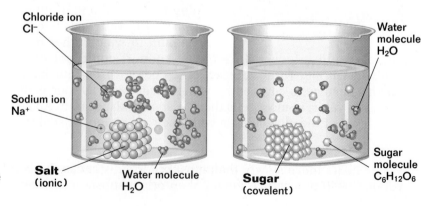

Chloride ion
Cl^-

Sodium ion
Na^+

Salt
(ionic)

Water molecule
H_2O

Water molecule
H_2O

Sugar
(covalent)

Sugar molecule
$C_6H_{12}O_6$

in water, ions are evenly dispersed in the solution. Elements in the body are present as compounds or dissolved ions in various fluids.

When nonpolar molecules (which do not form hydrogen bonds) are placed in water, the water molecules crowd the nonpolar molecules together. That is why oil and water do not mix. Nonpolar molecules, which are repelled by water, play many important roles in living things. They are responsible for fine-tuning the shapes of proteins and for maintaining the structure of cell membranes that surround every cell.

Acids and Bases Affect the Cell's Environment

Any compound that forms hydrogen ions when dissolved in water is called an **acid**. When an acid is added to water, the hydrogen ion concentration is increased. A convenient way of relating the amount of hydrogen ion from one solution to another is the **pH scale**. A simplified version of the pH scale is shown in Figure 2-10. The pH of most solutions falls within a numerical range from zero to fourteen. The pH of any solution can be determined using a pH meter or indicator papers.

Any substance that ionizes to form hydroxide ions when dissolved in water is called a **base**. Bases lower the hydrogen ion concentration of water because hydroxide ions react with hydrogen ions to form water molecules.

$$H^+ + OH^- \rightarrow H_2O$$

Bases have hydrogen ion concentrations that are lower than that of pure water. Bases thus have pH values above 7. Household ammonia has a pH of 11, and intestinal fluid has a pH of about 8. Your body's fluids are constantly monitored in a series of complex processes to keep the pH of these fluids within acceptable levels.

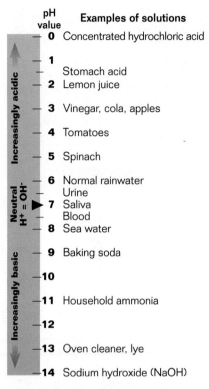

Figure 2-10 The pH scale enables you to compare the acidity of various materials. H^+ concentration increases in the range from pH 7 to 1. OH^- concentration increases in the range from pH 7 to 14.

Section Review

1. *How does an element differ from a compound?*
2. *How do covalent bonds differ from ionic bonds?*
3. *Explain why water is such an excellent solvent.*

Critical Thinking
4. *Antacids are often taken to relieve symptoms of "heartburn." How might an antacid work to relieve such stomach conditions?*
5. *Explain why you think melted candle wax will not mix with water.*

2-3 Chemical Building Blocks of Cells

Section Objectives

■ Compare and contrast the four principal kinds of macromolecules found in living organisms.

■ List the subunits that make up macromolecules.

■ Describe the function of nucleic acids and ATP.

*T*he basic chemical building blocks of your body are the same as those in all other organisms. Most of your body's molecules are organic compounds, which refers to a class of compounds containing carbon.

Macromolecules are built from small organic compounds the same way a railroad train is built, by linking a lot of units together into long chains. There are four principal kinds of macromolecules found in living organisms: carbohydrates, lipids, proteins, and nucleic acids.

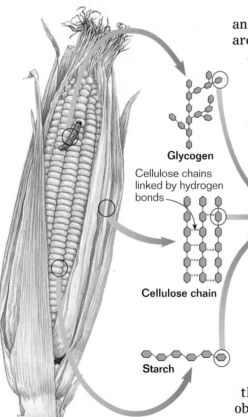

Figure 2-11 Starch, glycogen, and cellulose are made of glucose units organized in different structural arrangements.

Glycogen

Cellulose chains linked by hydrogen bonds

Cellulose chain

Glucose molecule

Starch

Cells Use Carbohydrates to Store Energy and Provide Support

Carbohydrates are composed of carbon, hydrogen, and oxygen atoms in the proportion of 1:2:1. A general formula for the carbohydrate class of compounds is $(CH_2O)_n$, where n is the number of carbon atoms. The sugar glucose is a small carbohydrate; its n equals 6. Its chemical formula is $C_6H_{12}O_6$.

Carbohydrates like glucose play a key role in the storing and transporting of energy in your body. **Polysaccharides** are carbohydrates that are made by linking individual sugars together to form long chains. Organisms store the energy contained in sugars like glucose by converting glucose into an insoluble form for future use. **Starch,** a common storage form of glucose, is composed of long chains made of hundreds of glucose molecules, as shown in Figure 2-11. When your body digests starch, the long starch chains are broken into short fragments. Your body then stores the glucose-containing fragments in longer chains called **glycogen**.

Many organisms use polysaccharides as structural material. In plants, for example, glucose molecules are joined together in long chains forming cellulose. **Cellulose,** a major component of the cell wall of plants, provides structural support for plants. Though they have similar components, cellulose and glycogen have different structural formulas. Your body is not able to break the links joining the glucose subunits in cellulose chains, so you cannot obtain energy from eating grass.

Lipids Store Energy and Are a Component of Cell Membranes

Lipids are a class of organic macromolecules that differ from other macromolecules in that they do not dissolve in water. Olive oil and vegetable oil are lipids, and so are waxes such as beeswax and earwax. Though lipids are a diverse class of compounds, most of their functions can be placed in one of three categories: energy storage, structural support in cell membranes, and specific reactants for metabolic reactions.

Fats are energy-storage lipid molecules that have more hydrogens bonded to their carbon chains than do carbohydrates. The structure of a fat molecule is shown in Figure 2-12. The fatty acid chains are usually 14 to 20 $-CH_2^-$ units long. This structure enables fat to supply more energy than carbohydrates. A gram of fat provides nine calories; a gram of carbohydrate provides only four calories.

When all the carbon atoms on the fatty acid chains are bonded to hydrogen atoms ($-CH_2^-$ units), these fats are called saturated fats. **Saturated fats** are called saturated because they contain the maximum number of C—H bonds possible. **Unsaturated fats** have carbon-carbon double bonds at various points along the fatty acid chain. They are called unsaturated because fewer hydrogen atoms can bond to the carbon chain when there are double bonds between carbon atoms.

Lipids also provide structural support in cell membranes. The membranes that surround the cells of your body are composed of lipids to which phosphorus molecules are attached. Such molecules are called phospholipids. Animal cell membranes also contain cholesterol, which is yet another kind of lipid called a **steroid.** Many of the hormones that your body uses to control its activities are steroids. There are many other kinds of lipids, including important pigments, light-absorbing substances such as the chlorophyll of green plants, and the pigment retinal found in your eyes. ◻

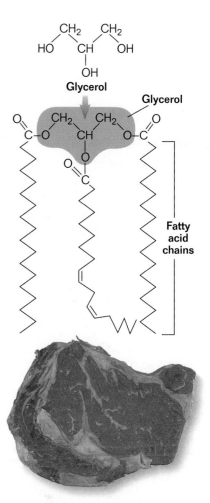

Figure 2-12 Fats, found in foods, consist of three long chains of fatty acids bonded to a glycerol backbone.

◻ **CAPSULE SUMMARY**

Lipids are not soluble in water. Fats store more energy per gram than carbohydrates because they have more carbon-hydrogen bonds.

Many Cellular Activities Involve Proteins

Proteins are the third major group of macromolecules. Proteins have many important structural functions. Your hair and muscles are made of protein, and so are a spider's web, a peacock feather, and the fibers of a blood clot. The most abundant protein in your body is **collagen,** a fibrous protein that forms the matrix of your skin, ligaments, tendons, and bones. Proteins also play a vital role in the metabolic (chemical and physical) activities of all living things. Proteins called **enzymes** assist the chemical reactions of metabolism. As you will learn in Chapter 4, few of the chemical reactions that take place in your body can proceed quickly without enzymes.

Enzymes Control Chemical Reactions

An enzyme is a catalyst because it increases the rate of a chemical reaction without the enzyme itself being destroyed in the process. Organisms maintain internal balance because enzymes control chemical reactions. The long developmental process that turns you from a fertilized egg into an adult human is controlled by the proper starting and stopping of production of particular enzymes at the appropriate time.

Amino Acids

Amino acids are the building blocks of proteins. There are 20 different kinds of amino acids that humans use, as shown in Figure 2-13. Because amino acids differ in chemical character, it is not likely that any two proteins with different amino acids will be alike chemically. Long chains of amino acids are called **polypeptides**. A protein is composed of one or more polypeptides.

CONTENT LINK

How the shape of an enzyme enables it to catalyze a particular chemical reaction is explained in more detail in **Chapter 4.**

Figure 2-13 The formation of different proteins from the combinations of just 20 different amino acids is much like the formation of all the words in the English language from just 26 different letters.

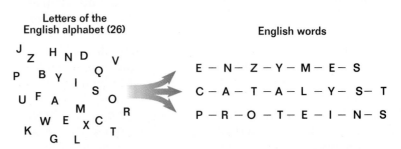

Letters of the English alphabet (26)

English words

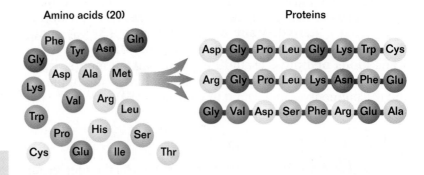

Amino acids (20)

Proteins

The amino acid chains of enzymes tend to fold into compact three-dimensional shapes, as shown in Figure 2-14. It is the precise shape of an enzyme that enables it to catalyze a particular chemical reaction.

Proteins also function as hormones and neurotransmitters. In these functions, proteins serve as signaling devices that are involved in regulating the activities of the cells of organisms. ■

□ CAPSULE SUMMARY

Proteins are part of an organism's structure. As enzymes, proteins increase the rate of chemical reactions within cells. The sequence of amino acids in a particular protein determines its shape, chemical properties, and function.

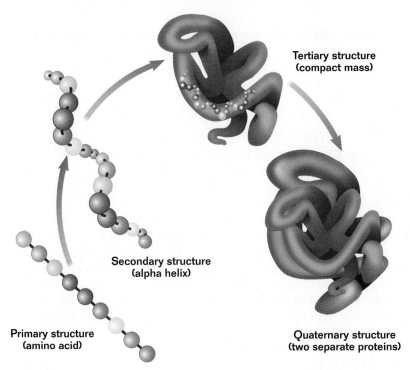

Tertiary structure
(compact mass)

Secondary structure
(alpha helix)

Primary structure
(amino acid)

Quaternary structure
(two separate proteins)

Figure 2-14 The biological activity of a protein depends on its structure. Four levels are used to describe protein structure. Primary structure shows amino acids linked by covalent bonds. Secondary structure can be a helix that shows recurring arrangements of amino acids. Tertiary structure shows the folding of that helix to make a polypeptide. Quaternary structure shows the arrangement of multiple tertiary proteins.

Nucleic Acids Contain the Cell's Hereditary Information

Nucleic acids are the fourth group of macromolecules found in living things. Nucleic acids are long chains of small repeating subunits, called **nucleotides**. A nucleotide consists of a five-carbon sugar with a phosphate (PO_4^-) group attached to one side of the sugar ring and an organic base attached to the other, as shown in Figure 2-15. The major function of nucleic acids is to store hereditary information that can be later translated to form new proteins. The nucleic acid that stores hereditary information in your cells is **DNA** (deoxyribonucleic acid).

CONTENT LINK

You will learn more about the structure and function of DNA in *Chapter 8*.

Figure 2-15 This portion of a DNA strand shows the arrangement of the sugar, base, and phosphate subunits in a nucleotide.

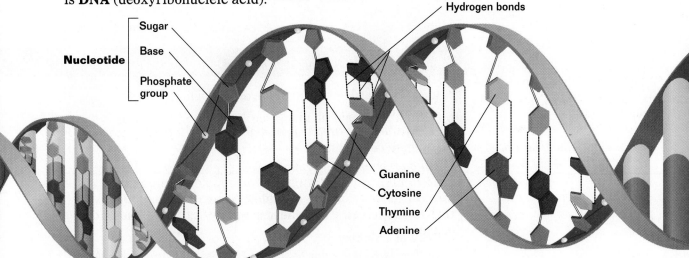

Hydrogen bonds

Nucleotide
  Sugar
  Base
  Phosphate group

Guanine
Cytosine
Thymine
Adenine

CONTENT LINK
....
You will encounter RNA molecules in **Chapter 9** when you
study how genes are expressed
—how particular sequences of
nucleotide bases are used by
the cell to assemble particular
proteins.

CONTENT LINK
....
You will learn more about ATP in
Chapter 4.

A second nucleic acid found in organisms, called **RNA** (ribonucleic acid), has a slightly different structure. RNA plays a variety of roles in the process of making proteins.

ATP Is the Cell's Fuel
....

There is one additional biological molecule that should be mentioned because of its importance in living systems and its structural similarity to nucleic acids—ATP, adenosine triphosphate. ATP contains an organic base, a sugar, and three phosphate groups. The term *triphosphate* means the molecule has three phosphate groups.

ATP is the energy currency of the cell. Its phosphate groups store energy like a coiled spring, their negative charges repelling one another. It takes energy to put the terminal phosphate in place, and energy is released when it is removed. When living cells break down food molecules containing carbohydrates and fats, part of the energy from those reactions is stored temporarily in ATP. A steady supply of ATP is necessary to ensure that a cell can perform all the tasks essential for life.

Figure 2-16 ATP is structurally similar to the nucleotides of DNA, except it includes three phosphate groups instead of one.

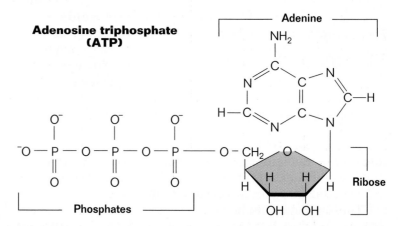

Section Review

1. Name the elements found in all biological macro-molecules.
2. Why do some athletes eat high-carbohydrate diets the day before competition?
3. List the subunit(s) that makes up each of the following:
   a. cellulose   d. fats   g. enzymes
   b. proteins   e. nucleic acids   h. starch
   c. polypeptides   f. glycogen

Critical Thinking
4. How is ATP similar to sugar? How is ATP similar to DNA?

2-4 The Interior of the Cell

I *magine yourself inside the cell. Scientists once thought that the cytoplasm was a structureless gel. Using microscopes, scientists have discovered that the cell's interior is filled with membranes and structures. If you journeyed into the cytoplasm, you would pass many structures, called organelles. In this section you will learn more about these principal structures of the eukaryotic cell.*

Microscopes Reveal Cell Structure

Even before Robert Hooke first glimpsed cells of cork in 1665, scientists realized that they needed more than the human eye to study objects and living things. As microscopy has evolved, scientists have learned more about plant and animal life than Hooke and van Leeuwenhoek could have ever imagined. Modern microscopes serve as passports into the unseen world, enabling biologists to observe cellular processes and to see details of cell structure. Microscopes continue to provide scientists with new insight into how cells work—and ultimately how whole organisms function.

Two important concepts relating to microscopes are magnification and resolution. **Magnification** is the ability of a microscope to make an image appear larger. **Resolution** is the ability to distinguish small, close objects. The resolution of a microscope refers to its ability to show details clearly. Resolution and magnification are equally important. If an image appears larger but its details are unclear, the user of the microscope will see only a fuzzy blur.

There are several basic types of microscopes. Each type of microscope has its own strengths and limitations. Scientists have learned which microscope is the most appropriate for the organisms they wish to study.

Light Microscopes

Light microscopes use a beam of light passing through one or more lenses to produce an enlarged image of the object or specimen being viewed. Microscopes that use two sets of lenses are called compound microscopes. The ocular (*AHK yoo luhr*) lens is positioned near the viewer's eye. The objective lens set is positioned near the specimen. A typical compound microscope, such as the one shown in Figure 2-17, has a light bulb or mirror in the base that sends light upward through the specimen. Light rays pass through the objective and then through the ocular. The image you see is magnified by both sets of lenses. The total magnification is determined by multiplying the magnifications of the two lenses. If your

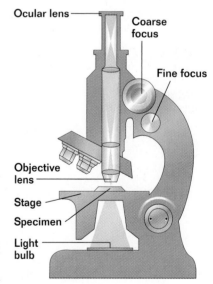

Figure 2-17 The compound light microscope illuminates a specimen with a beam of light. Two sets of lenses, the ocular (located in the eyepiece) and the objective (located just above the specimen), magnify the specimen, which is mounted on a glass slide.

microscope has a 10× ocular, and the 40× objective is in place, the object you are looking at will appear 400 times larger than it actually is.

A biologist can use a compound microscope to study living cells. Cells appear to be essentially transparent, although there are small variations in thickness and density. As a result, the cell and some of its internal structures are visible, but the image is not very distinct. More details of the structures inside cells can be seen by slicing cells thinly and dyeing them with stains. Looking at a cell this way has obvious disadvantages—only one thin slice of cell is seen, and, of course, the cell is dead. However, sectioning and staining cells enables biologists to see many structures not visible in living cells.

Light microscopes are important tools, but they have one important shortcoming. As magnifications increase, the resolution decreases and the details of the object viewed appear fuzzy. The most powerful light microscopes can magnify an object 2,000 times. Practically speaking, bacteria with a diameter of 0.5 µm are about the smallest living things that can be distinguished using a good mass-produced light microscope.

Electron Microscopes

Microscopes using electrons instead of light to form images can magnify images at least 100 times as much as the light microscope. Because electrons would bounce off the gas molecules in air, the stream of electrons and the specimen to be viewed must be placed in a vacuum chamber. Therefore, living cells cannot be viewed with electron microscopes.

The **transmission electron microscope (TEM)** produces a stream of electrons that passes through a specimen and strikes a fluorescent screen. By replacing the fluorescent screen with a piece of photographic film, a photograph called a transmission electron micrograph can be made. Figure 2-18 shows a transmission electron micrograph of a *Giardia* cell. Sections of specimens that are to be viewed with a TEM are sliced much more thinly than sections prepared for a light microscope. These sections are treated with stains that block electrons, causing details to become visible.

The **scanning electron microscope (SEM)** enables biologists to see detailed three-dimensional images of cell surfaces, such as the image of the cells of *Giardia* shown in Figure 2-18. Specimens are not sliced but are placed on a small metal cylinder and coated with a very thin layer of metal. Like the picture on a television set, the image is formed one line at a time as the beam of electrons scans the specimen from side to side. The electrons that bounce off the specimen form an image that can be viewed on a video screen, or a scanning electron micrograph can be made.

The micrographs made with electron microscopes are always black and white—never in color. However, electron micrographs often have color added in the darkroom to make certain structures stand out in the micrograph.

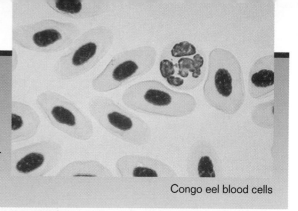

Light Microscopes

- Living specimens can be viewed.
- Selective stains enable specific organelles to be seen.
- Light microscopes are comparatively affordable, hence more available than electron microscopes.
- Light microscopes magnify up to 2,000✕.

Congo eel blood cells

Electron Microscopes

- Specimens must be dead.
- Electron microscopes reveal details not visible with light microscopes.
- Electron microscopes are expensive, hence less available than light microscopes.

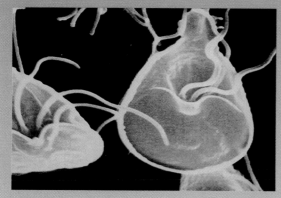

Giardia lamblia

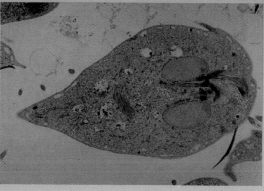

Giardia lamblia

Scanning (SEM)

- Specimens are coated with a thin layer of metal.
- A beam of electrons reveals surface details of specimens.
- The SEM magnifies up to 100,000✕.

Transmission (TEM)

- Specimens are thinly sliced, then stained.
- A beam of electrons passes through the specimen, revealing its internal structure.
- The TEM magnifies up to 200,000✕.

The use of each type of microscope—light microscope, transmission electron microscope, and scanning electron microscope—has advantages and disadvantages. Figure 2-18 summarizes the differences among these three kinds of microscopes.

Figure 2-18 Light microscopes and electron microscopes each reveal different aspects of an organism's structure.

Scanning Tunneling Microscope

New video and computer techniques are extending the resolution and level of detail that can be detected by microscopes. The scanning tunneling electron microscope (STM) uses a needle-like probe to measure differences in voltage due to electrons that leak, or tunnel, from the surface of the object being viewed. A computer tracks the movement of the probe across the object, creating a three-dimensional image of the specimen's surface. The STM can be used to study living organisms.

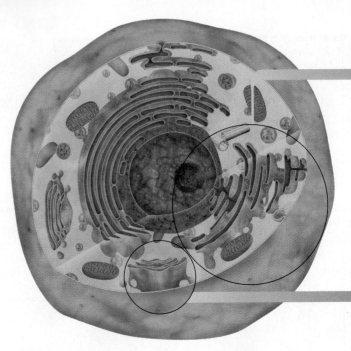

Internal Membranes Transport Materials Within the Cell

In order for a eukaryotic cell to maintain homeostasis, it is necessary for supplies to be moved from one part of the cell to another. In prokaryotic cells, a molecule can go from one place to another fairly quickly. In eukaryotic cells, molecular traffic is directed more precisely by an extensive system of internal membranes called the **endoplasmic reticulum** *(ehn doh PLAZ mihk rih TIHK yuh luhm)*, or **ER**. Like the plasma membrane that surrounds the cell, the ER is composed of a lipid bilayer with embedded proteins. Weaving in sheets through the cell's interior, the ER creates a series of channels between the membranes that isolates some spaces as membrane-enclosed sacs called **vesicles**. This system of internal compartments is a fundamental distinction between eukaryotes and their prokaryote ancestors.

Manufacturing Centers

The cell manufactures many proteins and lipids on the ER's surface. Some proteins and lipids are used within the cell, for example, to replace damaged or worn parts of the plasma membrane. Other proteins and lipids, such as digestive enzymes or hormones, are exported from the cell.

Proteins that are exported from the cell are manufactured by ribosomes on the surface of portions of the ER. Ribosomes are complex molecules composed of dozens of different proteins and RNA. The endoplasmic reticulum that is densely studded with ribosomes is called rough ER, as shown in Figure 2-19. The endoplasmic reticulum of cells specialized in making lipids, such as many brain and intestinal cells, has relatively few or no ribosomes and is called smooth ER. Cells can have both kinds of ER.

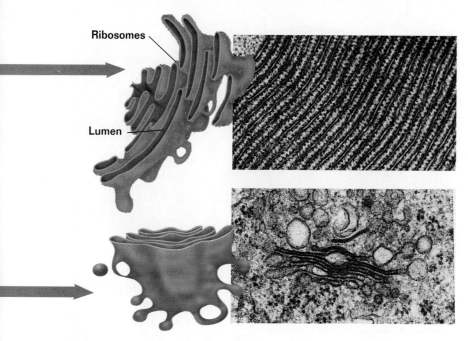

Figure 2-19 The pebbly appearance of the cell's rough endoplasmic reticulum (rough ER) is due to the presence of ribosomes. It is here that the cell manufactures proteins intended for export.

Figure 2-20 The Golgi apparatus packages and distributes proteins and lipids.

Packaging and Distribution Centers

Proteins and lipids destined for export are passed across the ER membrane as they are made. They then pass into an interior compartment of the ER called the lumen. These proteins and lipids move through the lumen to an area of smooth ER. The protein is then enclosed in a vesicle that buds off from the ER surface. The vesicle migrates across the cytoplasm to an adjacent organelle called a Golgi apparatus. The **Golgi** *(GOHL jee)* **apparatus,** shown in Figure 2-20, is the packaging and distribution center of the cell.

The Golgi apparatus contains a variety of enzymes that act on proteins and lipids and that serve as molecular address labels to determine where the protein or lipid will go. The newly made molecules are transported to different compartments of the cell or to the plasma membrane where they are exported from the cell. You can see this process in Figure 2-21.

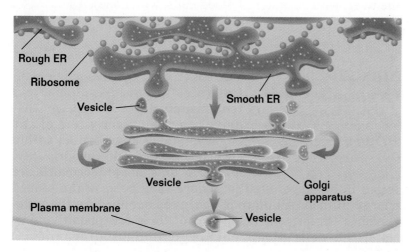

Figure 2-21 Proteins made on the endoplasmic reticulum travel through the cytoplasm to the Golgi apparatus, where they are packaged in vesicles and sent to different cellular compartments or exported from the cell.

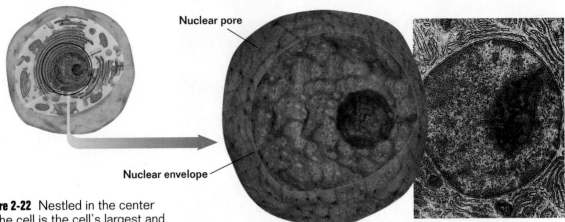

Nuclear pore

Nuclear envelope

Figure 2-22 Nestled in the center of the cell is the cell's largest and most easily seen organelle, the nucleus. First described by the English botanist Robert Brown in 1831, the nucleus directs all cell activities. It also contains the cell's hereditary information, the DNA instructions for making the next generation.

The Nucleus Directs Cell Activities and Stores DNA

Nestled in the center of the cell is the cell's largest and most easily seen organelle, the nucleus. The nucleus, shown in Figure 2-22, directs all cell activities and serves as the storage center for a eukaryotic cell's DNA.

Much like wearing two layers of clothing, the surface of the nucleus is bound by a complex double membrane (formed of a lipid bilayer) called the **nuclear envelope**. The cell's many activities are regulated by protein and RNA molecules that pass in and out of the nucleus across this nuclear envelope. How do they get across? Scattered over the surface of the nuclear envelope like craters on the moon are shallow depressions called nuclear pores that provide passageways. These pores contain many embedded proteins that act as molecular channels permitting certain molecules to pass into and out of the nucleus.

Inside the nucleus, DNA and proteins associated with DNA are organized into rod-shaped structures called **chromosomes.** All eukaryotic species have a characteristic number of chromosomes. The cells of your body each contain 46 chromosomes (except for egg or sperm cells, which have 23 chromosomes, and a few specialized tissues such as the liver, with 92, and red blood cells, with none). ❑

❑ *CAPSULE SUMMARY*

The endoplasmic reticulum (ER) transports the cell's proteins and lipid vesicles. The Golgi apparatus packages and distributes proteins and lipids for export. The nucleus directs all of the cell's activities and houses the genetic material, DNA.

Specialized Organelles Act as Cellular Powerhouses

Two kinds of organelles play essential roles in energy release and food manufacture. The energy that drives the many activities of the eukaryotic cell is generated within organelles called **mitochondria** *(myt uh KAHN dree uh)*. Mitochondria are found in eukaryotic cells, where they release the stored energy in food. All of your energy is supplied by the mitochondria within your cells. Except for a few primitive protists, the cells of all eukaryotes have mitochondria.

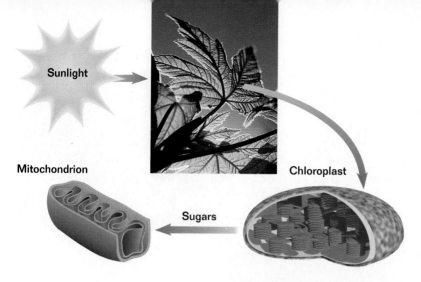

Sunlight

Mitochondrion

Chloroplast

Sugars

Figure 2-23 Chloroplasts in green plants capture sunlight, which enables them to make sugars. These sugars are the ultimate source of energy for all living things. A eukaryotic cell's mitochondria continuously release this energy, providing the cell with the energy necessary for life.

Chloroplasts are organelles that make food in the form of sugars, using water, carbon dioxide in the air, and energy from sunlight. This process is called photosynthesis. Chloroplasts are found only in algae, such as seaweed, and plants, and they are the only places within these organisms where photosynthesis occurs. Organelles like mitochondria and chloroplasts are thought to have evolved from separate organisms, such as bacteria, that were once ingested by a cell.

All the chemical energy that mitochondria extract from the food you eat originated from the photosynthetic process carried out by chloroplasts. Figure 2-23 shows how energy flows in connected pathways throughout the living world. ☐

CONTENT LINK

You will learn more about photosynthesis and cellular respiration in Chapter 5.

☐ *CAPSULE SUMMARY*

Mitochondria are the energy-releasing organelles of all eukaryotic cells. Chloroplasts enable algae and plants to capture energy during photosynthesis.

Biochemical Factories of the Cell

Eukaryotic cells contain a variety of organelles that isolate specialized biochemical activities. For the same reason that a kitchen, bathroom, and bedroom of a house are separated, so a cell isolates certain activities.

Chemical Specialty Shops

Almost every eukaryotic cell contains small vesicles called peroxisomes *(puhr AHKS ih sohms)*, which are derived from the smooth ER. **Peroxisomes** contain several kinds of enzymes. Some peroxisome enzymes convert fats to carbohydrates. Others alter potentially harmful molecules within the cell by forming hydrogen peroxide, H_2O_2, which is converted to water.

Recycling Centers

Another group of spherical organelles the same size and appearance as peroxisomes are lysosomes *(LY seh sohms)*. **Lysosomes** are vesicles that contain the cell's digestive enzymes. The enzymes within lysosomes cause the rapid breakdown of proteins, nucleic acids, lipids, and carbohydrates. Lysosomes digest and recycle the cell's used components. ☐

☐ *CAPSULE SUMMARY*

Peroxisomes neutralize potentially dangerous molecules within the cell. Lysosomes digest and recycle the cell's used components.

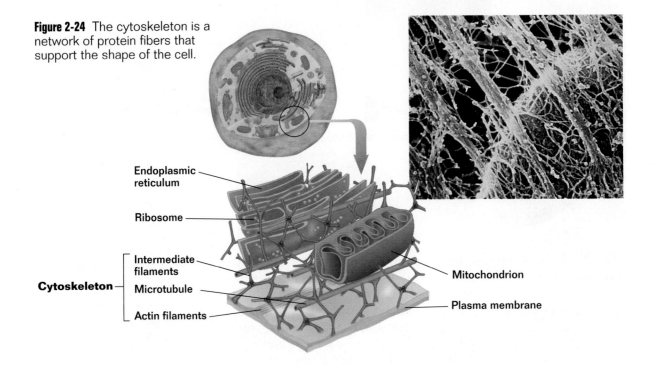

Figure 2-24 The cytoskeleton is a network of protein fibers that support the shape of the cell.

Endoplasmic reticulum

Ribosome

Cytoskeleton ⎧ Intermediate filaments
⎨ Microtubule
⎩ Actin filaments

Mitochondrion

Plasma membrane

Protein Fibers Provide an Internal Framework

The cytoplasm of eukaryotic cells is crisscrossed by a network of several kinds of protein fibers that supports the shape of the cell and anchors its organelles. This meshlike network is called the **cytoskeleton** and is shown in Figure 2-24. Within the cytoskeleton, hollow protein fibers called **microtubules** aid in moving chromosomes during cell division by forming cylindrical organelles called **centrioles**. The cells of plants and fungi lack centrioles. Two additional kinds of protein fibers, actin filaments and intermediate filaments, provide the cell with mechanical support and help to determine the shape of the cell.

Protruding from the surfaces of many eukaryotic cells are long threadlike organelles called **flagella,** which are used in locomotion. Flagella are actually complex cables of microtubules. Each flagellum consists of a circle of nine microtubules surrounding two central microtubules. This 9 + 2 arrangement is a fundamental feature of the flagella of eukaryotes, such as those on fast-swimming protists and those on the cells of sensory hairs inside the human ear. If flagella are numerous, shorter, and organized in tightly packed rows, they are called **cilia**. An example of cells that have cilia is shown in Figure 2-25. The 9 + 2 arrangement is also shown in Figure 2-25. ▫

Interior Spaces Are Storage Areas

Plant cells, unlike animal and most fungal cells, store waste products in a large internal space called a **central vacuole**. Vacuoles store large amounts of water and nutrients, as well

 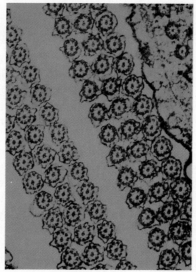 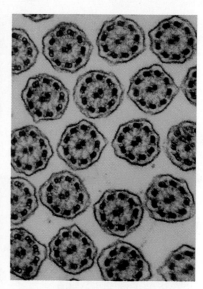

as wastes. The pressure exerted by this stored water acts as a sort of "skeleton" for the plant, keeping it rigid. This rigidity allows a plant to stand upright; when a plant's vacuoles lack water, it will become limp and will wilt.

Figure 2-25 The continuous beating of the hairlike cilia lining the mammalian windpipe, or trachea, *left*, helps the body remove foreign particles. All cilia exhibit a 9 + 2 internal arrangement of microtubules. The cilia of a single-celled microorganism, *center*, appear strikingly similar to the cilia that line the human oviducts (the tubes that carry eggs from the ovary to the uterus), *right*.

Diversity and Themes in Biology

As you have learned in this chapter, eukaryotic cells share certain similarities. Yet they also differ in some characteristic ways. Plant cells, for example, have a rigid cell wall made of the polysaccharide cellulose, which lies just outside the plasma membrane. Plant cells also have central vacuoles and sometimes chloroplasts. Cells of fungi *(FUHN jeye)*, a group of organisms that includes mushrooms and molds, also have a rigid cell wall, but it is made of a different polysaccharide called chitin *(KEYE tihn)*. Animal cells lack a cell wall. An animal cell is surrounded only by its plasma membrane, making it possible for the cell—and ultimately the animal itself—to move. Within the three cellular "floor plans" seen in plants, fungi, and animals, an enormous variety exists.

Just as there are many kinds of cells, there is so much diversity among living things that it is easy to overlook the fact that all living things have much in common. As you study biology, you will discover that certain basic principles or themes tend to recur. When viewed together, these themes will make the underlying unity of life increasingly clear. The first of these themes—"Evolution"—is described on the following page. This theme will help you recognize how an understanding of cells fits into the "big picture" of biology. In the next four chapters, you will be introduced to four additional themes.

Evolution

Evolution is the change of organisms over time. Although biologists had long considered that the diversity of life on Earth was the result of evolution, it was not until 1859 that a plausible mechanism for evolution was suggested. In that year, the English naturalist Charles Darwin published his book *On the Origin of Species,* in which he suggested that organisms change over time in response to changes in their environment.

Darwin called this process natural selection. Simply stated, Darwin's theory of evolution by natural selection says: Those organisms that are better able to successfully respond to the challenges of their environment are more likely to survive to reproduce; thus, the characteristics of those organisms become more common.

Today biologists recognize that evolution provides a unifying thread that runs throughout the living world. In this chapter, for example, you learned that all living things are made of one or more cells, each surrounded by a plasma membrane. Additions to this basic cell design—for example, organelles such as the nucleus and mitochondria—represent evolutionary embellishments upon this simple cellular "floor plan." Throughout your study of biology, you will see recurring examples of ways that species have changed over time in response to their environments. No other concept serves to unify the study of biology as powerfully as evolution.

At a glance, the *Euglena, Vorticella,* and the in-line skater would seem to have little in common. Yet the cells of each contain energy-releasing mitochondria, an evolutionary inheritance found among all eukaryotes.

Section Review

1. *A biologist wants to study how substances enter and exit a certain type of cell. Describe the benefits and disadvantages of using an electron microscope for this purpose.*

2. *Name five organelles. Describe how each enables the cell to display the properties of life.*

Critical Thinking

3. *Are lysosomes in animal cells similar to central vacuoles in plant cells? Explain your answer.*

4. *What does the presence of mitochondria in eukaryotic cells suggest about the evolution of eukoryotes?*

Vocabulary

acid (33)	enzyme (35)	peroxisome (45)
amino acid (36)	eukaryotic (26)	pH scale (33)
atom (29)	evolution (48)	plasma membrane (26)
base (33)	flagella (46)	polar molecule (31)
carbohydrate (34)	glycogen (34)	polypeptide (36)
cell membrane (26)	Golgi apparatus (43)	polysaccharide (34)
cell theory (25)	hydrogen bond (31)	prokaryotic (26)
cellulose (34)	ion (29)	protein (35)
central vacuole (46)	ionic bond (31)	resolution (39)
centriole (46)	light microscope (39)	ribosome (26)
chromosome (44)	lysosome (45)	RNA (38)
cilia (46)	magnification (39)	saturated fat (35)
collagen (35)	microtubule (46)	scanning electron microscope (SEM) (40)
compound (30)	mitochondrion (44)	starch (34)
cytoplasm (26)	molecule (30)	steroid (35)
cytoskeleton (46)	nuclear envelope (44)	transmission electron micro-scope (TEM) (40)
DNA (37)	nucleic acid (37)	unsaturated fat (35)
electron (29)	nucleotide (37)	vesicle (42)
element (29)	nucleus (26)	
endoplasmic reticulum (ER) (42)	organelle (26)	

Concept Mapping

Construct a concept map that shows the similarities and differences between prokaryotic cells and eukaryotic cells. Use as many terms as needed from the vocabulary list. Try to include the following items in your map: organelles, DNA, cell membrane, cytoplasm, mitochondria, nucleus, endoplasmic reticulum, ribosomes, chloroplasts, and energy. Include additional terms in your map as needed.

Review

Multiple Choice

1. Organelles that are present in plant cells but absent from animal cells include the
   a. chloroplasts and central vacuole.
   b. flagellum and cell wall.
   c. mitochondria.
   d. endoplasmic reticulum, cell wall, and lysosomes.

2. What evidence supports the hypothesis that eukaryotic cells evolved from prokaryotic cells?
   a. Fossils have been found of eukaryotic cells that look like prokaryotic cells.
   b. Organelles of eukaryotic cells are similar to bacteria.
   c. Eukaryotic cells have DNA identical to that found in their organelles.
   d. Most living eukaryotes are unicellular protists.

3. The growth of a cell is limited by the ratio between its
   a. surface area and volume.
   b. organelles and surface area.
   c. organelles and cytoplasm.
   d. nucleus and cytoplasm.

4. Eukaryotic cells differ from prokaryotic cells in that eukaryotic cells
   a. lack organelles.
   b. have DNA, but not ribosomes.
   c. are single-celled.
   d. have a nuclear membrane.

5. A substance that forms hydrogen ions when dissolved in water is called a(n)
 a. acid.
 b. hydrophilic solute.
 c. base.
 d. lipid.

6. An ionic bond forms when
 a. one atom transfers its electrons to another atom.
 b. two atoms share electrons.
 c. atoms act like molecular magnets.
 d. the neutrons of one atom fill the outer energy level of another atom.

7. Macromolecules that enable corn stalks to stand tall are
 a. lipids.
 b. proteins.
 c. carbohydrates.
 d. amino acids.

8. Water is a good solvent because it
 a. is a nonpolar molecule.
 b. has a pH of 7.
 c. has charged regions that interact with polar substances.
 d. gains electrons that produce oppositely charged ions.

9. Which of the following is the appropriate pairing of structure and function?
 a. smooth endoplasmic reticulum : protein manufacture
 b. Golgi apparatus : protein packaging and distribution
 c. mitochondrion : photosynthesis
 d. lysosomes : conversion of carbohydrates to fat

Completion

10. An atom with unequal numbers of electrons and protons is called a(n) _____ .

11. In water, one oxygen atom is _____ bonded to two hydrogen atoms. The oppositely charged regions of nearby water molecules are held together by weak _____ bonds.

12. Enzymes act as _____ that change the rate of a chemical reaction without being destroyed in the process.

13. DNA is an example of a(n) _____ . The role of DNA in the cell is to carry _____ information.

14. Endoplasmic reticulum that has ribosomes attached to its surface is called _____ . This type of endoplasmic reticulum is associated with the production of _____ .

15. The network of protein fibers that crisscrosses the cytoplasm of eukaryotic cells is called the _____ . This network provides _____ for the cell and holds organelles in place.

Short Answer

16. Describe how proteins, polypeptides, and amino acids are related to one another.

17. How are the functions performed by ATP and fat similar yet different?

18. Multicellular organisms are those composed of more than one cell. In Section 2-1, several advantages of multicellularity are described. What are two disadvantages of multicellularity?

19. What kind of microscope was used to produce the photograph of the insect shown below? State your reasons for your choice.

Themes Review

20. Evolution Name two structures found in all cells. How do these structures underscore the role of evolution in biology?

Critical Thinking

21. Making Inferences Look at the water strider in the photograph below. What property of water is responsible for the insect's ability to stand on the water's surface?

22. Making Inferences Animal fats usually exist as solid fats, and those in plants exist as oils. However, in many animals of the Arctic and Antarctic, animal fats are mostly oils. What adaptive advantage would the storage of body fat as oil instead of a solid be to animals that live in freezing climates?

23. Making Predictions The chemical erythromycin is particularly effective in inhibiting protein synthesis by the ribosomes in bacteria cells, but it does not inhibit protein synthesis in human cells. Suggest a possible use for erythromycin.

Activities and Projects

24. Cooperative Group Project Write and perform a series of skits to demonstrate the unique properties of water that make it essential for living things.

25. Multicultural Perspective Some of the world's earliest cultural symbols depict human relationships to the four elements of water, earth, fire, and air. Look for examples of these symbols in your library's art, anthropology, psychology, or science sections. What similarities do you find between cultures? What colors and shapes would you use to draw symbols for the four elements?

26. Researching and Writing Learn more about the work of Schleiden, Schwann, or Virchow. Write a brief report summarizing the processes that the researcher used to arrive at his conclusions about cells.

27. Cooperative Group Project Construct a cell organelle using modeling clay. After constructing your model, present it to the class. In your presentation, describe the organelle's structure, function, and relationship to other organelles.

28. Research and Writing Find out how the laboratory techniques of fractionation, centrifugation, and electrophoresis enable biologists to experiment with cells and analyze the substances that cells produce. In your report, describe the equipment used for each technique and what can be accomplished when the technique is employed.

Readings

29. Read the article "Revenge of the Killer Microbes" in *Time*, September 12, 1994. How do disease-causing bacteria become resistant to antibiotics?

30. Read the article "No-'Stick' Tips for Heart-Healthy Diets," in *Science News*, March 6, 1993, page 150. What percentage of calories in the experimental diet described in the article came from fat? What differences did the researchers observe between test subjects who used corn oil and those who used corn-oil margarine?

Animal and Plant Cells

OBJECTIVE

Differentiate between animal and plant cells by observing organelles and cellular structures of epithelial cells and *Elodea* cells.

PROCESS SKILLS

- observing cell organelles
- comparing and contrasting cell types

MATERIALS

- prepared slides of human epithelial cells from the skin lining the mouth
- compound light microscope
- sprigs of *Elodea*
- forceps
- microscope slides and coverslips
- safety goggles
- lab apron
- dropper bottle of Lugol's iodine solution
- prepared slides of three unknowns
- prepared slides of human blood

BACKGROUND

1. What characteristics distinguish plant cells from animal cells?
2. Write your own **Focus Question** on your Vee Form.
3. **Knowing Side of the Vee** List the **Concepts** and new **Vocabulary Words** on your Vee Form. In the **Concept Statements** section of the Vee, use these words in sentences that define and explain them.

TECHNIQUE

Doing Side of the Vee

Part A: Epithelial Cells

1. Examine a prepared slide of epithelial cells under low power. Locate cells that are separate from each other and place them in the central field of view. Examine the cells under high power. Adjust the diaphragm to reduce the light intensity for greater clarity.

2. In the **Records** section of your Vee Form, make a drawing of two or three cells as they appear under high power. Identify and label the cell membrane, the cytoplasm, the nuclear envelope, and the nucleus of at least one of the cells.

3. In the **Records** section of your Vee, make a drawing of these cells as you imagine they might look in the lining of your mouth.

Part B: Plant Cells

4. Carefully tear off a small leaf near the top of an *Elodea* sprig. Using forceps, place the whole leaf in a drop of water on a slide, and add a cover slip.

5. Observe the leaf under low power. The outermost part of the cell is the cell wall. The many small, green organelles in the cells are the chloroplasts. Look for a cell you can see clearly, and move the slide so that it is in the center of the field of view. Examine this cell under high power using the fine adjustment to bring it into focus.

6. Find an *Elodea* cell that is large enough to allow you to see the cell wall and the chloroplasts clearly. Draw this cell in the **Additional Records and Observations** section on the back of your Vee Form.

7. The chloroplasts may be moving in some of the cells. If no movement is observed, warm the slide in your hand or shine a bright lamp on it for a minute or two. Reexamine the slide under high power. Look for movement of the cell contents. Such movement is called cytoplasmic streaming.

8. Put on safety goggles and a lab apron. Because the cell membrane is pressed against the cell wall you may not see it. Due to the abundance of chloroplasts,

other organelles might also be hidden. To see the vacuole, nucleus, and nucleolus, prepare another wet-mount slide substituting Lugol's iodine solution for the water. Allow the iodine to diffuse throughout the cell. Then observe the stained cells under low and high power.

9. In the **Additional Records and Observations** section of your Vee Form, draw a stained *Elodea* cell. Label the central vacuole, nucleus, nucleolus, chloroplasts, cell wall, and cell membrane if they are visible.

Part C: Blood Cells

10. Obtain a prepared slide of human blood and examine it under low and high power. The stain used on the red blood cells causes the endoplasmic reticulum to resemble a blue mesh. Other blood cells may be blue or purple.

11. Obtain prepared slides of three unknown specimens from your teacher, and examine each under low and high power. In the **Additional Records and Observations** section on the back of your Vee, make a table like the one below. Record the code number assigned to each unknown its classification (plant or animal), and your reasons for its classification.

Classification of Unknowns

Unknown (code number)	Classification (plant or animal)	Reasons for Classification

INQUIRY

1. Describe the shape of epithelial cells.
2. Describe the appearance of the cytoplasm.
3. What shape are the *Elodea* cells?

4. Describe the shape of the chloroplasts. Where are they located in the cell?
5. Describe the movement of the chloroplasts in the *Elodea* cells.
6. After staining with iodine, list the organelles that are visible in the *Elodea* cell.
7. Use the information on the **Knowing Side** of the Vee to interpret your results from the **Doing Side,** and then write your **Knowledge Claim.** Write a **Value Claim** for this lab.

ANALYSIS

1. In what observable ways are animal and plant cells structurally similar? In what ways are they different?
2. How does the structure of each of the three kinds of cells that you observed relate to the function of each type of cell? Based on what you know about the organism each cell is from, how might each structure help that organism survive?
3. If some of the epithelial cells were folded over on themselves but were still transparent, what could you conclude about their thickness?
4. Lugol's iodine solution causes the movement of chloroplasts to stop. Why?

FURTHER INQUIRY

Write a **New Focus Question** that could be the point of a new investigation. The following are examples:

Obtain other specimens from different plants and from different parts of plants, such as the roots or stems. Try to relate any differences you observe to special functions of each type of cell.

List cell structures that you were unable to see with light microscopy. Use library resources to locate electron micrographs of these structures.

CHAPTER 3

CELLS AND THEIR ENVIRONMENT

REVIEW

- polar and nonpolar particles (Section 2-2)
- hydrogen bonding (Section 2-2)
- organic compounds (Section 2-3)
- ADP and ATP (Section 2-3)

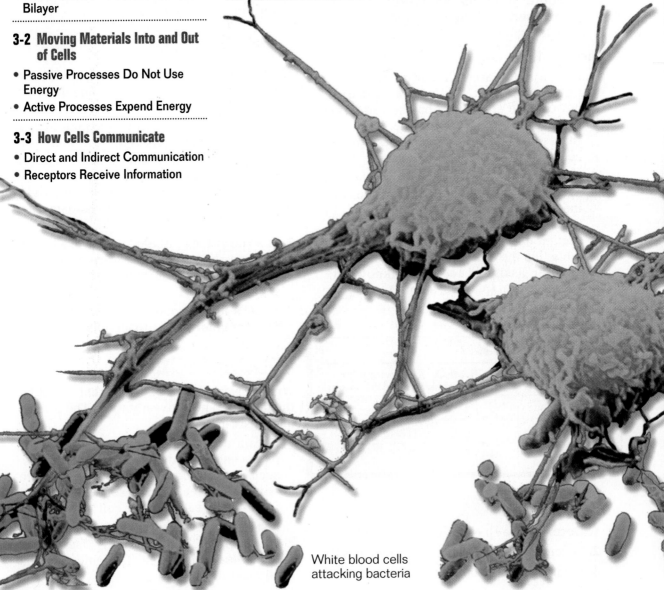

White blood cells attacking bacteria

3-1 Structure of the Plasma Membrane

The envelope that makes up the outer surface of a cell is called the plasma membrane. If you were small enough to stand on the outer surface of a cell, you would float, because most of the cell's plasma membrane is liquid. Floating alongside you would be proteins shaped like boulders, while towering overhead like huge trees would be a forest of other proteins. A plasma membrane is a complex and dynamic structure. It is the source of a cell's identity and the site of much of its activity.

Section Objectives

- Describe the arrangement of phospholipid molecules in a plasma membrane.
- Recognize the importance of cell surface proteins.
- Identify three main types of cell surface proteins.

A Lipid Bilayer Is the Foundation of the Membrane

The plasma membrane is composed of two main groups of organic molecules—phospholipids and proteins. A **phospholipid** is a molecule shaped like a head with two tails, as shown in Figure 3-1. The "head" is polar and the two "tails" are nonpolar. The polar head attracts water molecules, which are also polar. These attractions are known as hydrogen bonds. Since the polar water molecules cannot form hydrogen bonds with the nonpolar tails, the water tends to push them away. All this pushing and pulling of phospholipids by water results in the formation of a double layer of phospholipids called a **lipid bilayer**.

The lipid bilayer is not strong and firm like a hard shell, but is fluid like a soap bubble. The individual phospholipid molecules, arranged side by side, float within the bilayer. The most important feature of the lipid bilayer is its nonpolar interior zone. This zone is the true barrier that separates the cell from its surroundings. Most polar particles

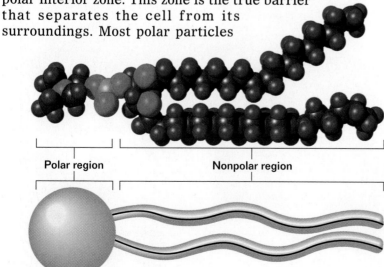

Polar region   Nonpolar region

Figure 3-1 The polar "head" of a phospholipid molecule contains a polar phosphorus group, and the two "tails" are long, nonpolar carbon chains.

cannot go across this zone, because they are repelled by the nonpolar interior of the bilayer. Sugars and many other food molecules are polar particles, as are proteins, ions, and most cell wastes. The nonpolar zone of the bilayer is such an excellent barrier that if the plasma membrane were made only of lipid bilayer, most food molecules and other polar substances would be unable to pass into and out of the cell.

CAPSULE SUMMARY

The foundation of a cell's plasma membrane is a fluid double layer of phospholipid molecules called the lipid bilayer.

The solution to this problem is to have passageways through the barrier that enable the cell to regulate precisely which substances go in and out. The passageways through plasma membranes are membrane proteins that traverse the lipid bilayer. ❑

Proteins Are Embedded Within the Bilayer

Every cell is isolated from its environment by its lipid bilayer, able to communicate with the outside world only by means of the proteins embedded within its lipid shell. The proteins within the plasma membrane of cells are often called **cell surface proteins.**

Figure 3-2 shows that a protein, which is typically polar, can extend all the way through a membrane's nonpolar bilayer zone. The protein is anchored into the membrane by its nonpolar middle region, unable to sink inside the cell or float up onto the surface. Notice that Figure 3-2 also indicates that there are different kinds of cell surface proteins. Each type plays a vital role in the life of a cell.

Figure 3-2 In a plasma membrane, phospholipid molecules align themselves in two layers to form a lipid bilayer. The polar heads of the phospholipids point toward water, intracellular water on one side and extracellular water on the other side. The lipid tails point toward each other. Various proteins penetrate the lipid bilayer. Indeed, from 50 to 70 percent of a typical body cell's membrane is protein.

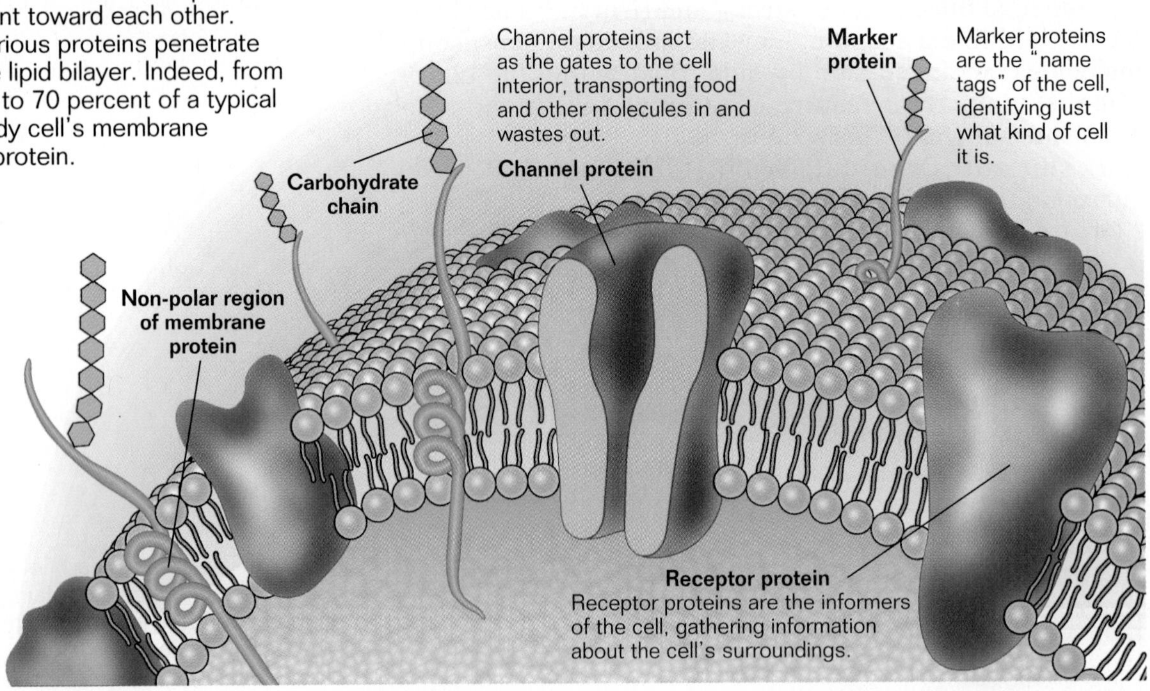
Channel proteins act as the gates to the cell interior, transporting food and other molecules in and wastes out.
Marker protein
Marker proteins are the "name tags" of the cell, identifying just what kind of cell it is.
Carbohydrate chain
Channel protein
Non-polar region of membrane protein
Receptor protein
Receptor proteins are the informers of the cell, gathering information about the cell's surroundings.

Channels Pass Through the Membrane

Cell surface proteins called **channels** have a series of non-polar amino acid sequences, causing them to loop back and forth through the membrane bilayer many times. This forms a doughnut-shaped channel through the bilayer. Polar sugars, amino acids, ions, and many other particles cross the membrane through these channels. However, particular channels fit only certain particles. The different cells of your body differ from one another largely by the kinds of channels they have. Nerve cells, for example, have a large number of channels specialized to transport sodium ions.

Receptors Transfer Information

Receptor proteins are specialized to transmit information from the world outside the cell to its interior. The end of the receptor that sticks out from the cell surface has a special shape that fits only a particular type of molecule. When a molecule of the right shape fits into the receptor—like a hand into a glove—it causes a change at the opposite end of the receptor protein, the end that protrudes into the cell interior. This change, in turn, triggers responses within the cell. Many of the hormones within your body act by binding to specific receptors on the surfaces of your cells. **Hormones** are chemicals that are secreted by tissues called glands.

Markers Identify Cells

Many cell surface proteins have long exterior arms, often with carbohydrates attached to them, called **cell surface markers**. Every cell of your body has markers on its surface saying that it is part of you and not of some other individual. Other markers say whether it is a liver cell or a heart cell or a brain cell. Still others convey additional information. Your body's immune defenses rely on cell surface markers to tell one cell from another. Cell surface markers will play an important role in Unit 9, where the immune system is discussed. ❑

❑ CAPSULE SUMMARY

Channels, receptors, and cell surface markers are proteins that are embedded within the lipid bilayer of plasma membranes. Channels serve as passageways, receptors serve as information receivers, and markers serve as cell identifiers.

Section Review

1. *Explain why the plasma membrane is an effective barrier to many molecules.*
2. *Draw a simplified diagram of a plasma membrane.*
3. *Describe the functions of the three types of cell surface proteins found in plasma membranes.*

Critical Thinking
4. *If the water molecules inside and outside a cell suddenly became nonpolar, would the transport of materials into and out of the cell be affected? Explain your answer.*

3-2 Moving Materials Into and Out of Cells

I f a cell were simply a bag, with nothing but a phospholipid bilayer for a plasma membrane, its interior would be an inactive place. Practically nothing would be able to enter or leave the cell. Sitting inside such a cell would be like sitting inside a room with no windows or doors. No cell is like that, of course, because every cell has an assortment of proteins extending through its plasma membrane that act as its windows and doors.

Passive Processes Do Not Use Energy

Particles move into or out of a cell by passing through the cell's plasma membrane. **Passive transport** is the movement of a substance through a cell's membrane without the expenditure of cellular energy. Substances pass into or out of your cells by way of several different passive transport processes.

Diffusion

If you were submerged within a drop of water, with your body no bigger than a protein, what would you see? Water molecules as big as basketballs would be zipping around, many of them bashing into you! Molecules do not stand still; they are in constant motion. Water molecules and particles dissolved in water move randomly. Their path is not predictable.

This random movement of individual dissolved particles within water has an important consequence that *is* predictable, however. Since the movement is random, a particle is more likely to move from an area where there are a lot of them (an area of high concentration) to an area where there are fewer of them (an area of lower concentration). This net movement of particles from an area of high concentration to an area of lower concentration is called **diffusion** *(dif FYOO zhuhn)*. In your lungs, oxygen diffuses into the bloodstream because there is a higher concentration of oxygen molecules in the lung's air sacs than there is in the blood. Eventually, dissolved particles diffuse within a liquid until they fill the volume uniformly, as shown in Figure 3-3. At this point the system can be described as being in **equilibrium**, the situation that exists when the concentration of a substance is the same throughout a space.

A substance that dissolves in another is called a **solute** *(SAHL yoot)*. Sugars, amino acids, and ions are all solutes in cells. The more plentiful substance that dissolves the solute is called the **solvent** *(SAHL vuhnt)*. In cells, the solvent is water. The mixture of solutes and solvent is called a **solution**.

Lump of sugar

Osmosis

Solute and solvent particles tend to diffuse from areas where their concentration is high to areas where their concentration is lower. Now imagine that a membrane separates two regions of a liquid. As long as solute particles and solvent (water) molecules can pass freely through the membrane, diffusion will soon equalize the amount of solute and solvent on the two sides. Equilibrium will be reached.

But what if a polar solute is added to one side and it cannot pass through the membrane? This situation arises in cells all the time. An amino acid cannot cross a lipid bilayer, and neither can an ion or a sugar molecule. What happens? Unable to cross the membrane, the polar solute particles form hydrogen bonds with the water molecules surrounding them, as shown in Figure 3-4. These "bound" water molecules are no longer free to diffuse through the membrane. In effect, the polar solute has reduced the number of free water molecules on that side of the membrane. This means the opposite side of the membrane (without solute) has more free water molecules than the side with the polar solute. As a result, water molecules move by diffusion from the opposite side toward the side with the polar solute.

Figure 3-3 If you drop a lump of sugar into a beaker of water, the sugar particles will diffuse and become evenly distributed throughout the water.

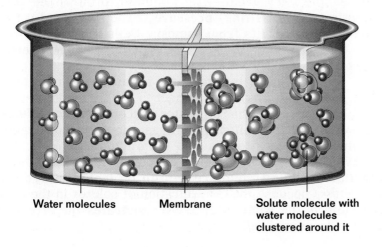

Water molecules   Membrane   Solute molecule with water molecules clustered around it

Figure 3-4 The addition of solutes to one side of a membrane reduces the number of water molecules that can move freely on that side. This is because the water molecules become bound to solute molecules. Water then moves by osmosis from the side where water molecule concentration is higher to the side where their concentration is lower.

		Conditions	Environment Solution Is	Cell Solution Is	Water Will Move
Hypotonic solution		Solute concentration in the environment is lower than in the cell.	Hypotonic	Hypertonic	Into the cell, and cell will burst
Isotonic solution		Solute concentration in the environment is equal to that in the cell.	Isotonic	Isotonic	Equal amounts will move into and out of the cell, and cell volume is maintained
Hypertonic solution		Solute concentration in the environment is higher than that in the cell.	Hypertonic	Hypotonic	Out of the cell, and cell will shrivel

Figure 3-5 If almost any animal cell is suspended in a hypotonic solution, it will burst because of an increase in internal osmotic pressure. In an isotonic solution, a cell's volume will be maintained. In a hypertonic solution, a cell will shrivel.

Eventually, the concentration of free water molecules will equalize on both sides of the membrane. At this point, however, there are more water molecules (bound and unbound) on the side of the membrane with the polar solute. Net water movement through a membrane in response to the concentration of a solute is called **osmosis** *(ahz MOH sihs)*. Stated another way, osmosis is the diffusion of water molecules through a membrane in the direction of higher solute concentration.

As a result of osmosis, extra water molecules accumulate on one side of the membrane. If these water molecules accumulate inside the cell, they will exert a pressure that can become very great—great enough to burst the cell! **Osmotic** *(ahz MAH tihk)* **pressure** is the increased water pressure that results from osmosis. Cells with strong cell walls (like plant and fungi cells) can withstand high internal osmotic pressures.

A cell immersed in pure water is said to be **hypertonic** *(heye puhr TAHN ihk)* with respect to the surrounding solution because the cell has a greater concentration of solutes. The surrounding solution, which has a lower concentration of solutes than the cell, is said to be **hypotonic** *(heye poh TAHN ihk)*. Most of your body cells and the tissue fluid that circulates around them are said to be **isotonic** *(eye soh TAHN ihk)* because the concentration of solutes in the cells and fluid is the same. Figure 3-5 illustrates the effect of osmosis on animal cells.

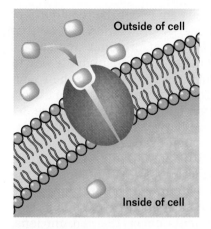 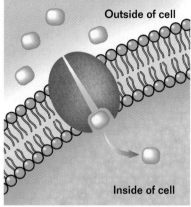

Figure 3-6 Facilitated diffusion is the transport of specific particles through a membrane by a channel protein. A molecule that is outside of the cell locks into the channel protein, *left*. The molecule is then transported through the channel to the inside of the cell, *right*.

Facilitated Diffusion

A cell is able to function because it is able to control what enters and leaves. Like a home, a cell has "doors with locks" that only certain particles can pass through. The doors of the eukaryotic cell are the protein channels in the plasma membrane. The structure of each protein channel is such that it can accept only particular particles, much as a lock is shaped to accept a particular key. The plasma membrane is said to be **selectively permeable** *(PUR mee uh buhl)* because it allows the passage of some solutes but not others. Selective permeability is one of the most important properties of the plasma membrane of every cell.

Many of the selective protein channels through the plasma membrane are "two-way" channels called pores. Pores transport molecules or ions through the membrane in either direction, much as an open door will allow you to pass into or out of a room. When a solute that fits a pore is more plentiful on one side of the membrane than the other, solute particles will pass by diffusion toward the side of lower concentration. Eventually, the concentration of the solute will become the same on both sides of the membrane. Different solute particles fit different pores. Their ability to fit depends on the size, polarity, and shape of the solute. Diffusion through selective pores is called **facilitated** *(fah SIHL uh tayt ehd)* **diffusion** and is shown in Figure 3-6. The particles move from higher to lower particle concentration. Like diffusion and osmosis, facilitated diffusion is a passive process, meaning that no energy is expended by the cell. Glucose is a sugar that moves into most cells by facilitated diffusion. ◻

◻ *CAPSULE SUMMARY*

Diffusion, osmosis, and facilitated diffusion are particle transport processes that do not require an expenditure of energy by a cell. Of these three processes, only facilitated diffusion involves the movement of particles through membrane protein channels.

Active Processes Expend Energy
....

If facilitated diffusion were the only tool cells had to harvest particles from the environment, they would have serious problems. Many amino acids and sugars are even more scarce outside cells than inside. Facilitated diffusion of these particles would

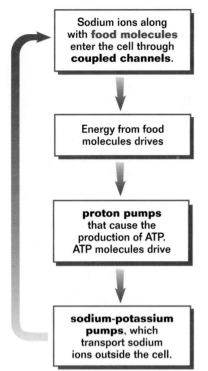

Sodium ions along with **food molecules** enter the cell through **coupled channels**.

↓

Energy from food molecules drives

↓

proton pumps that cause the production of ATP. ATP molecules drive

↓

sodium-potassium pumps, which transport sodium ions outside the cell.

Figure 3-7 The interplay between coupled channels and two kinds of energy-driven pumps enables cells to maintain a steady inward flow of food molecules.

result in their pouring out of cells, an event that could affect cellular protein synthesis and energy production. Recall that amino acids are the building blocks of proteins, and sugars are a major energy source for cells. Cells must have a way to beat the diffusion game so that they can maintain their concentration of these important food molecules at a level different from the concentration level outside the cell. Figure 3-7 demonstrates how the interplay between several important transport mechanisms in cell membranes enables cells to maintain a high internal level of amino acids and sugars. **Proton pumps** cause the production of ATP molecules, the energy currency of the cell. **Sodium-potassium pumps** use some of this ATP to accumulate an abundance of sodium ions outside the cell. **Coupled channels** carry the sodium ions, along with food molecules, back inside the cell. Proton pumps and sodium-potassium pumps are highly specialized protein channels. As you will see, both use energy to transport ions (charged particles) against a concentration gradient (toward the side of higher particle concentration). Using energy to transport a particle through a membrane against a concentration gradient is called **active transport**.

Proton Pumps

A proton pump, illustrated in Figure 3-8, is one type of active transport channel. Proton pumps actively transport protons through the internal plasma membranes of mitochondria and chloroplasts. A proton is a hydrogen atom that is missing its electron. Proton pump channels are used to make ATP from ADP. ATP is the cell's key energy-storing molecule. This active transport of protons to make ATP is called **chemiosmosis** *(kehm ee ahz MOH sihs)*. All the energy harvested by plants in photosynthesis and practically all the energy you get from the food you eat is derived from chemiosmosis.

Figure 3-8 Inside a cell, proton pump channels actively transport protons through the internal membranes of chloroplasts and mitochondria. In a chloroplast, *right,* protons that build up on one side of the membrane diffuse back to the other side through protein channels by facilitated diffusion. These channels use the force of the proton pushing through to power the manufacture of ATP.

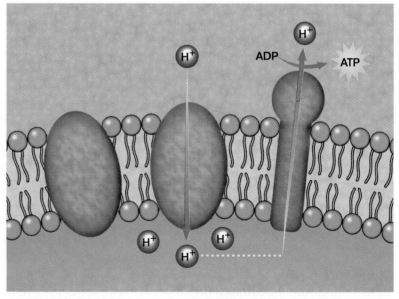

Sodium-Potassium Pumps and Coupled Channels

A second kind of active transport channel is called a sodium-potassium pump. A sodium-potassium pump uses energy stored in the form of ATP to power the active transport of sodium ions (Na^+) out through a cell's membrane. The action of the sodium-potassium pump is the most important energy-using process in your body. More than one-third of all the energy expended by a human cell that is not actively dividing is used to transport sodium ions in this way!

Why do your cells use so much of their energy pumping sodium and potassium ions? For one thing, nerve cells use the differences in sodium and potassium ion concentrations produced by sodium-potassium pumps to send signals throughout the body, like electrical signals passing over wires. The details of how nerve signals are created and transmitted will be discussed in Chapter 40. Sodium-potassium pumps also help to transport food particles into cells.

The transport of many food particles and a variety of other particles into your cells involves two different kinds of channels, as shown in Figure 3-9. One of the channels is the sodium-potassium pump. The other channels are coupled channels. In the first step, the active transport of sodium ions out of the cell by the sodium-potassium pump increases the sodium ion concentration outside the cell. Each channel is capable of transporting as many as 300 sodium ions per second when working full tilt. The fact that there are so many sodium ions outside the cell due to the action of the

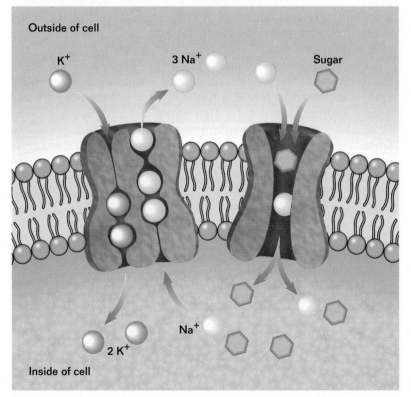

Figure 3-9 In a sodium-potassium pump, *left,* for every three sodium ions, Na^+, pumped out, the energy from a molecule of ATP is used, and two potassium ions, K^+, are pumped in. As a result of the pumping, most of your cells have a high concentration of sodium outside their membrane. As a sodium ion diffuses back into the cell through a coupled channel, *right,* it brings along with it a sugar molecule.

Proton pumps cause the production of ATP molecules in chloroplasts and mitochondria. Sodium-potassium pumps use ATP molecules to create a high concentration of sodium ions outside the cell. Many cells maintain a high internal concentration of food molecules by coupling the inward transport of food molecules with the inward migration of sodium ions.

sodium-potassium pump leads to the second step: sodium ions move back into the cell by means of coupled channels that also carry sugar molecules. The coupled channel has two passageways through the membrane, both of which must be used for the channel to work. Imagine the two passageways as two turnstiles attached to the same gear. One passageway fits a particular molecule, such as a sugar molecule. The other fits sodium ions. Because there are so many sodium ions outside the cell due to the action of the sodium-potassium pump, sodium ions will diffuse back into the cell. The force of their entry is so great that it pulls sugar molecules into the cell too, even though the cell interior already has a generous supply of sugar. Just as it takes two hands to clap, so it takes two membrane channels to actively transport many food particles into the cells of your body. This two-step transport process is one of the most fundamental and important activities of any cell. ❏

Endocytosis and Exocytosis

Although, as you have read, individual food molecules are routinely imported through protein channels, other food particles are too large for that method of transport. In order to consume these larger meals, the cell literally engulfs the particle. This process, shown in Figure 3-10, is called **endocytosis** *(ehn doh seye TOH sihs)*. The cellular movements of endocytosis require an expenditure of energy by the cell. If the material brought into the cell is liquid and contains dissolved molecules, the endocytosis is referred to as **pinocytosis** *(peyen oh seye TOH sihs)*. Maturing human egg cells obtain dissolved nutrients secreted by the surrounding "nurse" cells of the ovary by pinocytosis. If the material brought into the cell is another cell or other fragment of organic matter, the process is known as **phagocytosis** *(fag oh seye TOH sihs)*. Phagocytosis is very common among unicellular eukaryotes. Amoebas, for example, commonly devour their prey in this fashion. Some animal cells also carry out phagocytosis. The white blood cells of your immune system use phagocytosis to protect you from infection, engulfing invading bacteria. Some types of white blood cells may take in as much as 25 percent of their cell volume each hour.

The reverse of endocytosis is exocytosis *(ek soh seye TOH sihs)*. **Exocytosis** is the dumping of excretions or waste

Figure 3-10 Cells trap extracellular fluid and particles within membrane vesicles by endocytosis, *left.* Exocytosis is the dumping of excretions or waste materials outside a cell by discharging them from waste vacuoles that fuse with the plasma membrane, *right.*

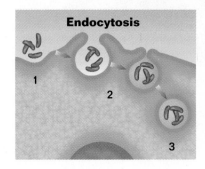
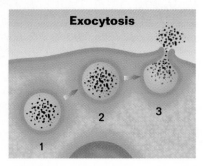

materials outside a cell by discharging them from waste vacuoles that fuse with the plasma membrane. Exocytosis also occurs in cells that secrete important chemical products. Gland cells, for example, secrete hormones into the bloodstream by exocytosis.

One of the major differences between a house and a tent is that a house does a better job of maintaining a constant environment by keeping out the cold, rain, and unwanted visitors. The same is true of your body and of every organism. On a cellular level, the plasma membrane of every cell determines what gets in and out. Cells use their membranes to maintain conditions inside that are far more constant than those on the outside. Without these constant internal conditions, many of the complex interactions that take place within cells would be impossible. Similarly, your body maintains constant internal conditions. For example, it attempts to maintain a temperature of 37°C, no matter how cold or warm the weather might be. The levels of salt and glucose in your blood are maintained at nearly the same level at all times. The maintaining of constant internal conditions is called **homeostasis.** Control of internal conditions is an essential characteristic of all living things.

How do the emperor penguin and the athlete maintain constant internal body temperatures?

Section Review

1. *How does osmosis differ from diffusion?*
2. *If a living cell is placed in a hypertonic solution, in which direction will water molecules move?*
3. *Identify one similarity and two differences between facilitated diffusion and active transport.*
4. *Explain the role the sodium-potassium pump plays in the transport of glucose into some cells.*
5. *What is endocytosis?*

Critical Thinking

6. *A living cell is placed into a hypotonic solution. Compare the solute and solvent concentration inside the cell with the solute and solvent concentration in the surrounding solution.*

3-3 How Cells Communicate

Bringing the "groceries" in and taking the "garbage" out are not the only things that cells do. Just as in your house you answer the phone and watch television, so the cells of your body communicate with the world outside. Your cells communicate with each other in order to coordinate the body's growth, development, and other activities.

Section Objectives

- Identify three ways cells communicate with each other.
- Describe how receptors can influence the cytoplasm of a cell.
- Distinguish between chemically gated and voltage-gated channels.

Direct and Indirect Communication

Some of your cells are in physical contact with each other, as shown in Figure 3-11. However, most cells of your body communicate less directly. The two systems involved in indirect communication are the endocrine (ductless gland) system and the nervous system. The endocrine system communicates with body cells by using chemical signals called hormones. A nerve cell communicates with another nerve cell or a muscle cell with chemicals called **neurotransmitters** *(noo roh trans MIHT uhrs)*.

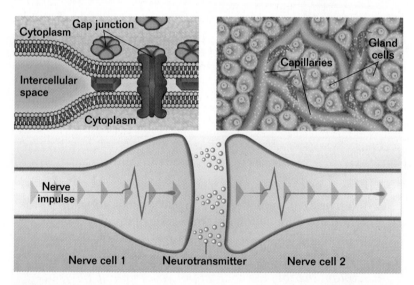

Figure 3-11 Cells communicate in several ways. Some cells have tiny openings called gap junctions between them that directly join their cytoplasms, enabling exchange of small particles, *top left.* Gland cells, *top right,* release hormones into the bloodstream. Hormones interact with specific cells elsewhere in the body. Communication between nerve cells, *bottom,* requires chemical communication across the short gap between adjacent cells. These chemicals are called neurotransmitters.

Receptors Receive Information

The key to understanding cell communication is to focus on the cell's communication "machinery," special proteins on the cell surface. Called **receptors**, these proteins are specialized antennae that signal the cytoplasm when a particular particle has just bumped into the cell surface. They are able to do this because each kind of receptor has a particular shape that fits only a particular signal particle. When that signal particle is encountered by the cell, it binds to the receptor specialized to fit it, like two pieces of a jigsaw puzzle.

The binding of a signal particle to its receptor can cause the receptor to influence the cytoplasm in one of three ways, as shown in Figure 3-12. First, the receptor can act as an enzyme, chemically changing molecules in the cytoplasm. Second, the receptor can cause the formation of another signal, called a second messenger. These second messengers will have an effect elsewhere in the cytoplasm. A modified form of ATP called cyclic AMP is a very common second messenger in eukaryotic cells. Third, the receptor can open a channel through the membrane. This is how neurotransmitters work.

Chemically Gated Channels

In order to better understand how receptors work, we will focus for a moment on a particular receptor, the receptor in muscle cells. The muscle receives the signal to contract from nerve cells. The chemical signal that the nerve releases into the gap between it and the muscle is a small neurotransmitter molecule called **acetylcholine** *(uh seet uhl KOH leen)*. This molecule binds to a receptor in the plasma membrane of the muscle cell, as shown in Figure 3-13. This acetylcholine (ACh) receptor contains a channel. Such a channel is called a **gated channel** because it may be opened or shut, like a gate in a fence.

The binding of acetylcholine to the receptor causes the pore to open up like the shutter of a camera. It doesn't stay open long, only a microsecond or so. That's how long the acetylcholine molecule stays bound to the receptor. During this brief period, sodium ions flood in, setting off the muscle contraction. The ACh receptor is said to be **chemically gated**, because its opening and closing depends on the binding of a chemical, the neurotransmitter acetylcholine. ❏

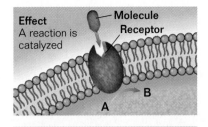
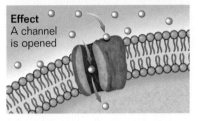

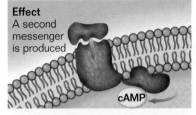

Figure 3-12 When the proper molecule locks into the membrane receptor of a cell, a change will occur within the membrane or cytoplasm of that cell. There may be chemical changes in cytoplasmic molecules, *top;* the production of cytoplasmic second messenger molecules, *center;* or the opening of channels within the plasma membrane, *bottom.*

❏ *CAPSULE SUMMARY*

The binding of an acetylcholine molecule to a receptor in the membrane of a muscle cell causes a sodium channel in the receptor to open. Sodium ions rush into the muscle cell, and the cell contracts.

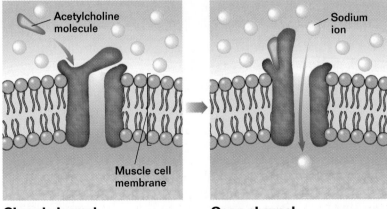

Closed channel   **Open channel**

Figure 3-13 An acetylcholine molecule released from a nerve cell binds to an acetylcholine receptor in the plasma membrane of the muscle cell. The receptor is a sodium channel that is closed; the binding of acetylcholine opens it.

Voltage-Gated Channels

Not all gated channels are open and shut by chemicals. In nerve cells, another type of gated channel is found, as shown in Figure 3-14. Nerve cells are specialized to transmit electrical signals. The signals travel down long cytoplasmic extensions as electrical disturbances on the surface of the plasma membrane. What makes the signal move along the membrane? The nerve signal moves by opening a series of gated sodium channels one after another, like a line of falling dominoes. In this case, however, the "open" signal is electrical rather than chemical. These sodium channels are said to be **voltage-gated**. Like the chemically gated ACh receptor, voltage-gated sodium channels are composed of proteins that weave in and out of the membrane.

How does voltage act to open the channel? Some of the amino acids that make up the channel have ionized and become positively charged. This makes these charged areas of the channel very sensitive to voltage. When the immediate surroundings of the channel experience a change in voltage, the charged areas rotate out of the center of the channel, opening it to enable the passage of sodium ions. ◻

Normal state
(closed channel)

Nerve signal formation
(open channel)

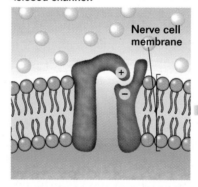 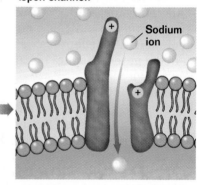

Nerve cell membrane

Sodium ion

Figure 3-14 When a voltage change occurs in the vicinity of a voltage-gated receptor in the plasma membrane of a nerve cell, the gate of this channel protein opens.

Section Review

1. *Compare and contrast hormones with neurotransmitters.*
2. *Define the term* receptor.
3. *What is acetylcholine, and what is its effect on the plasma membrane of a muscle cell?*

Critical Thinking

4. *Adrenaline is a hormone that binds with liver cells and causes them to release glucose by way of a cyclic AMP mechanism. Draw and label a cell diagram that demonstrates the action of an adrenaline molecule on a liver cell.*

CHAPTER REVIEW 3

Vocabulary

acetylcholine (67)
active transport (62)
cell surface marker (57)
cell surface protein (56)
channel (57)
chemically gated (67)
chemiosmosis (62)
coupled channel (62)
diffusion (58)
endocytosis (64)
equilibrium (58)
exocytosis (64)
facilitated diffusion (61)

gated channel (67)
hormone (57)
hypertonic (60)
hypotonic (60)
isotonic (60)
lipid bilayer (55)
neurotransmitter (66)
osmosis (60)
osmotic pressure (60)
passive transport (58)
phagocytosis (64)
phospholipid (55)
pinocytosis (64)

proton pumps (62)
receptor (66)
receptor protein (57)
selectively permeable (61)
sodium-potassium pump (62)
solute (58)
solution (58)
solvent (58)
voltage-gated (68)

Concept Mapping

Construct a concept map that shows how cells maintain homeostasis. Use as many terms as needed from the vocabulary list. Try to include the following terms in your map: passive process, energy, concentration gradient, active process, diffusion, osmosis, facilitated diffusion, proton pump, sodium-potassium pump, endocytosis, and exocytosis. Include additional concepts in your map as needed.

Review

Multiple Choice

1. Cell membrane functions such as serving as receptor sites and surface markers are accomplished by
   a. phospholipid molecules.
   b. ions.
   c. proteins.
   d. lipids.

2. The rate of diffusion of ink in a container of water can be increased by
   a. increasing the volume of the water.
   b. increasing the concentration of the ink.
   c. decreasing the permeability of the ink.
   d. decreasing the mass of the water.

3. If frog eggs are placed in a saltwater aquarium,
   a. water will move from inside the eggs to outside the eggs.
   b. water will move from outside the eggs to inside the eggs.
   c. the water balance will be maintained.
   d. the eggs will burst.

4. Facilitated diffusion in cells differs from ordinary diffusion in
   a. its use of energy.
   b. the direction of particle movement.
   c. the use of selective pores.
   d. the conversion of ATP to ADP.

5. The secretion of thyroid-stimulating hormone by cells of the pituitary gland is an example of
   a. pinocytosis.   c. chemiosmosis.
   b. diffusion.   d. exocytosis.

6. A two-step process involving the sodium-potassium pump and facilitated diffusion of sodium into the cell is required for
   a. gated channels to operate.
   b. glucose to enter cells.
   c. the functioning of cell surface markers.
   d. phagocytosis.

Cells and Their Environment 69

7. Researchers observed that electrically charged ions and fluorescent-tagged tracer molecules of different sizes moved from cell to cell without appearing outside the cells. These observations support the existence of
   a. receptors.
   b. neurotransmitters.
   c. ductless glands.
   d. gap junctions.

8. Production of cytoplasmic second messengers and opening of plasma membrane channels are possible results of
   a. signals received by membrane receptors.
   b. the actions of amino acids.
   c. the actions of sugar molecules.
   d. the opening of gated channels.

9. Signals for muscle contraction are sent from nerve cells by way of
   a. voltage-gated receptors.
   b. peptide hormones.
   c. chemically gated channels.
   d. phospholipid molecules.

10. A white blood cell engulfs, digests, and destroys an invading bacterium through the process of
    a. osmosis.   c. pinocytosis.
    b. phagocytosis.   d. facilitated diffusion.

Completion

11. The _____ molecules in the plasma membrane arrange themselves so that the carbon chain ends of the molecules point toward each other and the phosphate ends point away from each other.

12. ATP molecules are _____ by proton pumps, and ATP molecules are _____ by sodium-potassium pumps.

13. A cheerleader experiences a burst of energy just before running onto the field. It is likely that the cheerleader's liver received a chemical signal called a _____ from the adrenal gland causing it to release glucose.

14. The sodium channels involved in the movement of nerve impulses are _____ gated channels, and the acetylcholine receptors associated with muscle contractions are _____ gated channels.

15. When a cell needs to accumulate particles in greater concentration than is found outside the plasma membrane, _____ transport is likely.

16. Particles too big to move through protein channels are able to enter the cell by _____.

17. The part of the cell that functions to maintain homeostasis relative to the cell's environment is the _____.

Short Answer

18. How are the functions of channel proteins and receptor proteins alike and different?

19. Inflated Mylar balloons and rubber balloons can be purchased at most any party store. Mylar balloons won't deflate unless punctured, but rubber balloons deflate within days without being punctured. Which balloon is analogous to a plasma membrane? Explain your answer.

20. Most boats of any size are equipped with a bilge pump to remove water that has leaked into the boat. Explain how the movement of water by a bilge pump is similar to the transport of particles through a membrane by active transport.

21. Use your understanding of osmosis to describe why putting salt on a pork chop before cooking it on a grill is likely to result in a dry and tough piece of meat.

Themes Review

22. **Homeostasis** A sodium-potassium pump is one kind of active transport channel. Describe how the function performed by this transport channel contributes to homeostasis in a human.

23. Evolution Explain why there are no cells with non-permeable membranes.

24. Making Inferences A gelatin block is prepared with some phenolphthalein solution added to it. (Phenolphthalein is an acid-base indicator that is clear in the presence of an acid, but turns pink in a base.) The block is enclosed in a membrane. The membrane-enclosed block is then suspended 25 cm above a beaker containing ammonium hydroxide. After half an hour, the block begins to turn pink. Account for the gelatin's pink color.

25. Evaluating Data An apparatus called an osmometer was used to determine the rate of osmosis during an experiment. Line A on the graph reflects the data collected by one student group. Line B reflects the class average, which was calculated using the data collected by the first student group and seven other groups. How do the data collected by the first group compare with the class average? What might account for any discrepancies identified between the two data sets?

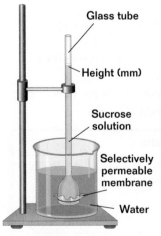

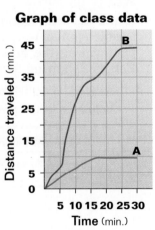

Graph of class data

26. Multicultural Perspective Research the attitudes of Japanese people toward eating the meat of the Japanese puffer fish. Find out about the relationship between the tetrodotoxin sometimes present in the meat of the puffer fish and human fatalities.

27. History Find out about the model of the plasma membrane proposed by cell biologist J. D. Robertson in the 1950s. How does it compare with the "fluid mosaic model" described in this chapter and proposed by Singer and Nicholson in 1972?

28. Research and Writing Find out how the membrane of a bacterium compares with the membrane of a eukaryotic cell. In your report, include labeled cross-sectional sketches showing the similarities and differences between the two membrane types.

29. Read "Toward a Future With Memory: Researchers look high and low for the essence of Alzheimer's," in *Science News*, February 24, 1990, pages 120–123. Why are scientists studying the cell membranes of brain cells to discover the cause of Alzheimer's disease?

30. Read "Fiddling With Salt Intake," in *Bioscience*, June 1992, page 411. What is the scientific evidence that suggests diet may affect the development of sodium channels?

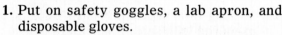

Diffusion and Cell Size

OBJECTIVES

- Calculate ratios of surface area to volume.
- Recognize the relationship between rate of diffusion and cell size.

PROCESS SKILLS

- analyzing models
- predicting outcomes
- measuring rates of diffusion
- applying mathematical formulas

MATERIALS

- safety goggles
- lab apron
- disposable gloves
- block of phenolphthalein agar
  (3 cm × 3 cm × 6 cm)
- plastic knife
- metric ruler
- 250 mL beaker
- 150 mL of vinegar
- plastic spoon
- paper towel

BACKGROUND

1. What factors affect the rate of diffusion across a cell membrane?
2. What is the relationship between surface area, volume, and diffusion in cells?
3. Write your own **Focus Question** on your Vee Form.
4. **Knowing Side of the Vee** List the **Concepts** and new **Vocabulary Words** on your Vee Form. In the **Concept Statements** section of the Vee, use these words in sentences that define and explain them.

TECHNIQUE

Doing Side of the Vee

1. Put on safety goggles, a lab apron, and disposable gloves.
2. **CAUTION: Use the knife carefully to avoid injury.** Trim the agar block with your knife to make three cubes: 3 cm, 2 cm, and 1 cm on a side, respectively. Each cube will represent a cell.

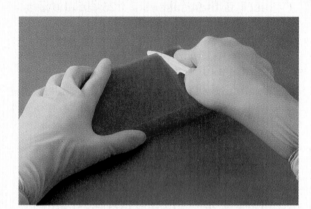

3. Place the three cubes in the beaker, and add vinegar until the cubes are submerged. Keep the cubes submerged 10 minutes, using the plastic spoon to turn the cubes frequently. Be careful not to scratch the surfaces of the cubes.
4. In the **Records** section of your Vee, make a table similar to Table 1. Complete it by performing the necessary calculations. Use the following formulas:

Surface area = length × width × number of sides

Volume = length × width × height

Table 1 Agar Cube Comparisons

Cube Dimension	Surface Area (cm²)	Volume (cm³)	Ratio of Surface Area/Volume*
3 cm			
2 cm			
1 cm			

*(reduce to simplest fraction)

5. After 10 minutes, use the spoon to remove the agar cubes, and then blot them dry on a paper towel.

6. Cut the first cube in half. Measure the distance of diffusion of the vinegar in millimeters. Record the distance in a table similar to Table 2 below in the **Additional Records and Observations** section on the back of the Vee Form. A color change from red or pink to clear indicates diffusion. Be sure to thoroughly rinse and dry the knife before the next cut.

Table 2 Rate of Diffusion

Cube Dimension	Depth of Diffusion (mm)	Time (min.)	Rate (mm/min.)
3 cm			
2 cm			
1 cm			

7. Repeat step 6 with the other two cubes.

8. Record the total time in minutes in Table 2. Calculate the rate of diffusion as millimeters per minute, and record. In the **Procedure** section of the Vee, briefly summarize the procedure you followed.

9. Clean up your materials and wash your hands before leaving the lab.

INQUIRY

1. Why is it important not to scratch the surface of a cube when it is submerged in vinegar?

2. Propose a reason for rinsing and drying the knife thoroughly each time it is used to cut the cubes in half.

3. Cite evidence that vinegar diffuses into an agar cube.

4. List the agar cubes in order of size from largest to smallest. Then list them in order, from largest to smallest, according to their ratios of surface area to volume. How do the lists compare?

5. Compare the surface-area-to-volume ratio with the rate of diffusion for each cube.

6. Now compare the the surface-area-to-volume ratio with the extent of diffusion for each cube.

7. Use the information on the **Knowing Side** of the Vee to interpret your results from the **Doing Side,** and then write your **Knowledge Claim**. Write a **Value Claim** for this lab.

ANALYSIS

1. The size of some living cells is 0.01 cm. Using the formulas in the lab, calculate the surface area and volume of a cell that is 0.01 cm. Which cell has the greater surface area in proportion to its volume, the living cell or the largest agar block "cell"?

2. As cells increase in size, what happens to the surface-area-to-volume ratio?

3. Write a generalized statement about the relationship between the extent of diffusion and cell size.

4. In what ways do your agar models simplify or ignore the features of real cells?

FURTHER INQUIRY

Write a **New Focus Question** that could be the point of a new investigation. The following is an example:

How have the cell membranes of some cells of living organisms become modified to be more efficient?

ENERGY AND METABOLISM

REVIEW

- chemical bonds (Section 2-2)
- carbohydrates, lipids, proteins, and nucleic acids (Section 2-3)
- structure of proteins (Section 2-3)
- structure of ATP (Section 2-3)

Male African lion

4-1 Energy and Living Things

*E*nergy may sound like a topic that belongs in a physics course, but it is vital to living things. Organisms use energy to carry out the many tasks of living. The life processes of every cell—growth, reproduction, movement, and transport of molecules and ions across cell membranes—are all driven by energy. Plants make life on Earth possible by capturing energy from the sun. The capacity to use energy for life processes, as the insect in Figure 4-1 is doing, keeps us alive. In fact, biologists often view life as a constant flow of energy channeled by organisms to do the work of living.

Section Objectives

- Distinguish between potential and kinetic energy.
- Explain what happens when energy is transformed.
- Describe how energy flows through living systems.
- Identify what happens in an oxidation-reduction reaction.

Figure 4-1 This click beetle, *Pyrophorus noctilucus,* produces light by a chemical process that requires energy. Bioluminescence, an important form of communication that helps this animal locate a mate, is only one of the many ways in which organisms expend energy.

Energy Takes Many Forms

Energy is a quantity that is quite familiar to us all. You know that it is found in the motion of a falling boulder, in the sound of a guitar, in the blast of an explosion, and in the welcome warmth of a blazing fire. The food you eat also contains energy. But what, exactly, is energy? Although difficult to define, **energy** can be most easily understood as the ability to cause matter to move or change. Kick a football, for example, and the energy of your kick makes the football move from one place to another. Fry an egg, and the clear, liquid egg white changes color and solidifies as energy causes a change in the arrangement of its atoms and molecules. Because there are so many different ways that matter can move and change, energy appears to exist in many different forms. These forms include mechanical energy, light energy, sound energy, thermal energy (which is related to heat), and electric energy.

The two most basic forms of energy are kinetic *(kih NEHT ihk)* energy and potential energy. **Kinetic energy** is the energy an object has because of its motion. All moving objects have kinetic energy. A cheetah racing across a savannah has a lot of kinetic energy, while a worm inching along the ground has much less. **Potential energy**, on the other hand, is the energy that is stored in an object because of its position. It is called potential energy because it is ready to make matter move or change when it is released. A boulder atop a hill, for example, has energy because of its position— *the energy that was required to lift the boulder against the force of gravity is stored as potential energy.* The higher the hill, the more potential energy the boulder has. ◻

◻ *CAPSULE SUMMARY*

Energy has many different forms. Kinetic energy is work being done. Potential energy is the ability to do work.

Food	Calories per 100 g
Granola	482
Strawberries	31
Kidney beans	91
Cheddar cheese	375
Broccoli	28

Figure 4-2 Foods, *top,* contain stored energy. The amount of energy food contains is measured in calories, *bottom.* One food calorie is the amount of energy needed to raise the temperature of one kilogram of water one degree Celsius.

◻ *CAPSULE SUMMARY*

Energy can be changed from one of its forms to another. Energy cannot be created or destroyed.

Energy Can Be Transformed

A boulder can also be used to illustrate a fundamental characteristic of energy: energy can be transformed (changed from one form to another). When a boulder rolls downhill, energy stored by virtue of its position is released, becoming kinetic energy. In fact, most of the energy we use every day comes from converting one form of energy to another. An electric fan converts electric energy to mechanical energy. As gasoline burns in a car's engine, chemical energy (potential energy a substance has because of the position of the atoms in its molecules) is converted to mechanical energy and heat.

All energy conversions are governed by the laws of **thermodynamics** *(THUR moh deye NAM ihks),* the study of energy transformations. One of these laws describes the observation that the total amount of energy in a closed system never changes. In science, a **system** is a collection of related objects (matter) that can be studied. For example, an organism is a system, as is the planet Earth and the universe. A **closed system** is one like the universe, into which no energy or matter can enter from the outside. A system that exchanges matter and energy with its surroundings (like Earth) is called an **open system**. The fact that the total amount of energy in a closed system remains constant is stated as the **first law of thermodynamics:** *Energy cannot be created or destroyed; it can only be converted from one form to another.*

Much of the work done by living things involves energy transformation. When you use food as a source of energy for movement, you are converting chemical energy into mechanical energy. When you use food to help maintain your body temperature, you are converting chemical energy into thermal energy (the random movement of all particles of matter). Figure 4-2 shows how much energy some foods contain. Just as potential energy is stored in a boulder's position at the top of a hill, chemical energy is stored in the arrangement of atoms in the molecules that make up food. ◻

Energy Tends to Become Disorganized

Not all of the potential energy in food is recovered when it is transformed into kinetic energy. Almost half escapes into the environment as heat (thermal energy transferred by moving particles). Thermal energy is generally not a useful form of energy. It can do work only when it is concentrated and can flow from one object to another. Many systems, including living systems, cannot concentrate thermal energy enough to make it do work. A burning match can ignite wood, for example, but the thermal energy of air molecules cannot.

The reason some thermal energy is lost as heat when energy is transformed is stated in another law that also governs energy conversions. The basis of this law is that all systems tend to change in ways that make them more stable. Again, think of the boulder sitting on top of a hill. This is not a stable system, because the boulder may roll downhill. Once the boulder rolls to the bottom of the hill, it can go no farther and thus the system is more stable. To better understand this law, look at the soft-drink cans in Figure 4-3. Because thermal energy is the *random* motion of particles, it is a disorganized and stable form of energy. The amount of disorder in a system, or its amount of unavailable energy, is called **entropy**. The tendency of systems to become more stable by gaining entropy is stated in the **second law of thermodynamics:** *Disorder (entropy) in the universe constantly increases; in a closed system, energy tends to be converted to less organized (more stable) forms.* ◻

Figure 4-3 If soft-drink cans are organized into a stack, *top*, the stack may tumble over. Therefore, the system is unstable. When the cans fall over, *bottom*, the system becomes less organized but more stable. To organize the system again (restack the cans), energy must be put into it.

◻ **CAPSULE SUMMARY**

In every energy transformation, some energy is converted to heat, increasing the disorder and the stability of a system.

Energy Flows Through Living Systems

All organisms need a constant supply of energy to fuel the activities of life. Directly or indirectly, almost all of the energy for living systems comes from the sun. Energy from the sun enters living systems when sunlight is absorbed by molecules found in plants, algae, and certain bacteria. In the process, electrons in these molecules are boosted to higher energy states. Like a boulder on top of a hill, an electron that has been boosted to a higher energy state has additional energy because it is farther away from the nucleus of an atom. This additional energy is released by dropping the electron back to a lower energy state. The energy is then used to do work, or it is captured and stored.

Organisms that harvest energy by boosting electrons use it to produce energy-storing macromolecules. Most of these organisms boost electrons with energy from sunlight and convert light energy to chemical energy. The process that converts light energy to chemical energy is called **photosynthesis** *(foht oh SIHN thuh sihs)*. Instead of using sunlight, certain bacteria obtain energy for boosting electrons from

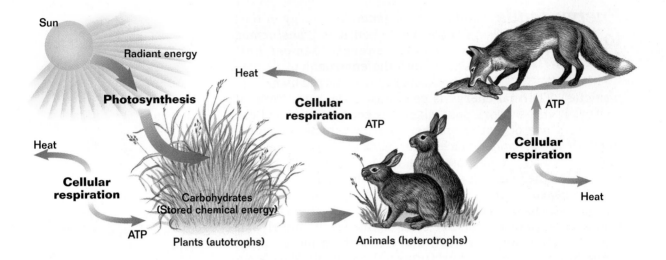

Figure 4-4 Energy flows through an ecosystem from sunlight or inorganic chemicals to autotrophs and then to heterotrophs. Because some heat escapes with every energy conversion, much of the energy captured by autotrophs eventually flows out of living systems.

❏ CAPSULE SUMMARY

Living systems are open systems. Energy from sunlight flows through living systems, from autotrophs to heterotrophs. Energy is captured during photosynthesis and made available for use by cells during cellular respiration.

inorganic molecules and convert it to organic chemical energy. The process that converts inorganic chemical energy to organic chemical energy is called **chemosynthesis** *(kee moh SIHN thuh sihs)*. Organisms that harvest energy from either sunlight or chemicals are called **autotrophs** *(AWT oh trohfs)*. The word *autotroph* comes from the Greek words *autos*, meaning "self," and *trophikos*, meaning "to feed."

Organisms that cannot harvest energy directly from sunlight or inorganic molecules but consume food instead are called **heterotrophs** *(HEHT uhr oh trohfs)*. The word *heterotroph* comes from the Greek words *heteros*, meaning "other," and *trophikos*, "to feed." Heterotrophs obtain their energy from other organisms. Energy stored in food is released by a process that is similar to burning. This process, which is called **cellular respiration**, is a series of chemical reactions that converts energy stored in food to a more useful form. While burning converts almost all of the energy in a fuel to heat, much of the energy released during cellular respiration is used to make ATP, which fuels the activities of living. The flow of energy through the living world is illustrated in Figure 4-4. ◼

Energy Is Carried by Electrons
. . . .

Energy can be extracted from chemical bonds in food molecules. As you learned in Chapter 2, the chemical bonds that hold atoms together in food molecules are composed of electrons that are shared by two atoms. Like rubber bands, shared electrons keep two atoms from pulling apart. The energy in food is actually stored in the shared electrons of carbon-to-hydrogen bonds. The energy of these electrons is portable; it can be transferred to new chemical bonds by transferring the electrons. That energy is stored and released again in processes that involve chains of chemical reactions.

Flow of Energy

As you might imagine, there are many ways biological systems can be studied with an eye to understanding how they work. One of the best ways is to follow the energy needed for life. The energy-capturing process of photosynthesis and the energy-releasing process of cellular respiration are two of the fundamental pathways by which energy travels.

Taken together, the processes within cells that capture, transfer, and use energy are called metabolism.

All living things are united in a network in which each organism is linked to another by the common need for energy. Almost all the energy that powers life on Earth comes from the sun, captured by green plants, algae, and certain bacteria during photosynthesis. These organisms serve as sources of energy for other organisms, the energy passing from one organism to another. Ultimately, energy flow is a key factor in shaping ecosystems. ■

Energy that was stored as a result of photosynthesis is transferred to this broad-billed hummingbird when it drinks sugary nectar from the flowers of a thistle plant.

Chemical reactions that pass electrons from one atom or molecule to another are **oxidation-reduction reactions. Oxidation** *(AWKS ih DAY shun)* is the loss of electrons. **Reduction** is the gain of electrons. In organisms, however, electrons usually are not transferred alone. Instead, each electron moves with a proton as part of a hydrogen atom, H. Thus, as Figure 4-5 illustrates, the transfer of energy in organisms usually involves the removal of hydrogen atoms from one molecule and the addition of hydrogen atoms to another molecule. Oxidation-reduction reactions play key roles in organisms because they enable energy (carried by hydrogen atoms) to pass from one molecule to another.

Oxidation

$$C_6H_{12}O_6 + 6O_2 \longrightarrow 6CO_2 + 6H_2O$$

Reduction

Figure 4-5 $C_6H_{12}O_6$ is oxidized when it loses hydrogen atoms. O_2 is reduced when it gains hydrogen atoms. Oxidation and reduction always take place together because every electron that is lost by one atom or molecule is gained by another.

Section Review

1. *Define kinetic energy and potential energy.*
2. *Summarize the first and second laws of thermodynamics.*
3. *What roles do photosynthesis and cellular respiration play in the flow of energy through living systems?*
4. *How are oxidation-reduction reactions important to cells?*

Critical Thinking

5. *Use the laws of thermodynamics to explain why a rock that tumbles off a cliff eventually comes to a stop.*

4-2 Energy and Chemical Reactions

In a sense, organisms can be compared to race cars; both are complex "machines" powered by chemical energy. Chemical reactions power the movement of mechanical parts, creating the force that moves a race car. Chemical reactions also power most cellular activities. Just as a successful race car driver must "look under the hood" to learn how a race car's engine works, you must look at the chemical machinery of cells, such as the ones in Figure 4-6, to understand how organisms function. Understanding cell chemistry, however, requires some knowledge of chemical reactions.

Section Objectives

- Distinguish between exergonic and endergonic chemical reactions.
- Explain why activation energy and catalysts help chemical reactions occur.
- Describe the role of enzymes in cell chemistry.

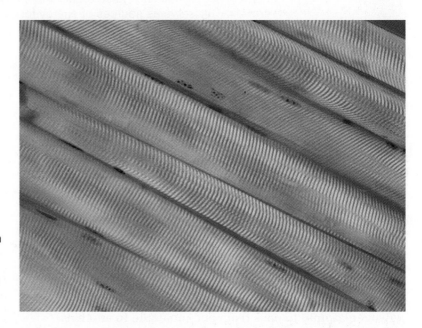

Figure 4-6 Chemical reactions power cellular activities that result in the movement, growth, response, and reproduction of an organism. For example, the contraction and relaxation of cells, like these human striated muscle cells, are translated into the movements of your body.

Chemical Reactions Absorb or Release Energy

A **chemical reaction** occurs when chemical bonds between atoms are broken or formed, resulting in the formation of one or more different substances. The substances that are combined or broken apart during chemical reactions are called **reactants**. The new substances that form are called **products**. You can think of the chemical reactions in cells as either gluing atoms or molecules together or tearing molecules apart. Extracting energy from sugar, for example, involves ripping apart the carbon backbones of sugar molecules. Making a protein is a matter of sticking amino acid molecules together to form long chains. Like most chemical reactions, the reactions that make proteins are reversible, meaning that they can also occur in the opposite direction. As you digest food, for instance, proteins that were assembled by other organisms are torn apart into individual amino acids.

The energy from chemical reactions that drives cell activities is called **free energy**. While most biochemical reactions release free energy into their surroundings, others absorb it. A reaction that releases free energy is called an **exergonic** (*ehks uhr GAHN ihk*) **reaction**. As the top graph in Figure 4-7 indicates, the products of exergonic reactions contain less energy than their reactants. Cellular respiration is an exergonic process. In contrast, a reaction that absorbs free energy is an **endergonic** (*ehn duhr GAHN ihk*) **reaction**. As the bottom graph in Figure 4-7 indicates, the products of endergonic reactions contain more energy than their reactants. Photosynthesis is an endergonic process. ◻

◻ CAPSULE SUMMARY

Chemical reactions produce new substances by breaking or forming chemical bonds between atoms. Exergonic reactions release free energy into their surroundings. Endergonic reactions absorb free energy from their surroundings.

Starting Chemical Reactions Requires Energy

As the second law of thermodynamics predicts, exergonic reactions readily occur because systems tend to become less organized and therefore more stable. If this is so, why haven't all exergonic reactions already occurred? For example, the burning of gasoline is an exergonic reaction. Therefore, you can think of a tank full of gasoline as crammed with chemical reactions waiting to happen. So why doesn't all the gasoline in the world burn up right now? It doesn't because gasoline needs a "kick in the pants" to burn. Energy from a spark or a match is needed to start the combustion reaction that burns gasoline. In order for the reaction's products (carbon dioxide and water) to form, existing chemical bonds in the reactants (gasoline and oxygen) must first be broken, and this takes energy.

The extra energy required to break existing chemical bonds and to initiate a chemical reaction is called **activation energy**. If you look at Figure 4-7 again, you will notice that the energy curve for the exergonic reaction initially rises. This "uphill" curve represents the reaction's activation energy. The heat from a spark or a match supplies the sizable amount of activation energy that is needed to start the chemical reactions waiting to happen in gasoline. How does heat become the activation energy for a chemical reaction? The heat from a spark or a match transfers free energy to the reactants. This extra energy causes the reactant particles to move and vibrate faster—they have more kinetic energy. In other words, the reactant particles are higher on the activation energy "hill." If the kinetic energy of reactant particles is the same as the required activation energy, it can be used to break chemical bonds in the reactants so that new bonds and products can form.

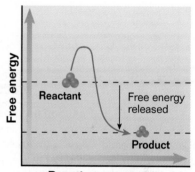

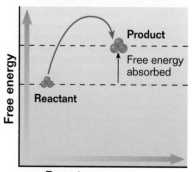

Figure 4-7 Exergonic reactions, *top*, release free energy, leaving their products with less total energy than their reactants. Endergonic reactions, *bottom*, absorb free energy, giving their products more total energy than their reactants.

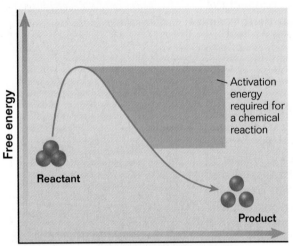

CAPSULE SUMMARY

Chemical reactions require an input of energy to get started. The energy required to get a reaction started is called activation energy. A catalyst lowers the amount of activation energy needed to start a reaction.

A chemical reaction—exergonic or endergonic—will proceed at a much faster rate than it normally would if the necessary activation energy can be lowered. A substance that alters a chemical reaction so that it can proceed at a lower activation energy is called a **catalyst** (*KAT uh lihst*). The process of increasing the rate of a chemical reaction through the use of a catalyst is called **catalysis** (*kuh TAL uh sihs*). Substances that act as catalysts are not changed by the reactions they assist. Catalysts do not make chemical reactions begin spontaneously; that is, they do not reduce the activation energy to zero. The need for activation energy cannot be avoided, as you can see from the graphs in Figure 4-8. ◻

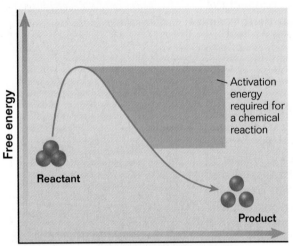

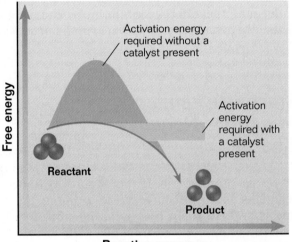

Figure 4-8 Chemical reactions are more likely to happen when the large amount of activation energy required, *left,* can be lowered. Catalysts reduce the amount of energy needed to start a reaction, *right,* but do not change the amount of energy contained in either the reactants or the products.

Enzymes Catalyze Chemical Reactions in Cells

Like a tank of gasoline, cells are crammed with chemical reactions waiting to happen. Each reaction waits for the nudge that will give it enough activation energy to start. Just as the heat from a flame supplies the activation energy needed for gasoline to burn, heat (molecular motion) can supply activation energy for chemical reactions that occur in cells. However, the amount of heat living things can tolerate is not enough to make these reactions occur as quickly as they must. As you might suspect, cells need catalysts to help start chemical reactions. And as you will learn later, cells control chemical reactions by controlling the activity of these catalysts.

Enzymes (*EHN zeyemz*) are the catalysts used by cells to trigger and control particular chemical reactions. An enzyme is a protein, often a rather large one, that works by binding to specific reactant molecules. Like other catalysts, *enzymes are not permanently changed by the reactions they catalyze.* The reactant molecules to which an enzyme binds are called **substrates.** Amylase (*AM uh lays*), for example, is the enzyme that catalyzes the breakdown of starch into glucose. Starch is this enzyme's substrate. Starch synthetase *(SIHN thuh tays)* is the enzyme that catalyzes the reverse reaction, the formation of starch from glucose. Glucose is this enzyme's substrate. Enzymes lower the activation energy required to start biochemical reactions, allowing them to occur more readily. *Without enzymes, chemical reactions necessary for life would not occur at a rate sufficient for sustaining life.*

Because of their shapes, enzymes bind to specific substrates. In Figure 4-9, you can see that an enzyme's surface provides a mold that fits the molecules of a certain substrate almost exactly. Other molecules that fit less perfectly simply don't adhere to the enzyme's surface. The site where a substrate binds to an enzyme is called the **active site**. At the active site, the enzyme and the substrate molecule interact in such a way that the activation energy for a particular reaction is lowered, making the substrate more likely to react.

Enzymes are very effective catalysts. As an example, consider a reaction that takes place in your bloodstream.

$$\text{carbon dioxide} + \text{water} \xrightarrow{\text{carbonic anhydrase}} \text{carbonic acid}$$

By itself, the reaction is very slow. In a cell, perhaps 200 molecules of carbonic acid form in an hour. However, in the presence of the right enzyme—carbonic anhydrase *(an HEYED rays)*—600,000 molecules of carbonic acid form every second! Thus, the enzyme speeds up the reaction rate about 10 million times. Carbonic anhydrase plays an important role in your blood. By allowing carbon dioxide and water in blood to combine and form carbonic acid so quickly, the enzyme prevents your blood from becoming saturated with carbon dioxide, a waste product. In your lungs, the reverse reaction occurs, and CO_2 is eliminated from your body when you exhale (breathe out). This enables your blood to continue removing carbon dioxide from your body efficiently.

Figure 4-9 One way that enzymes lower activation energy is by weakening chemical bonds, *left,* so that they will break more easily. For example, amylase speeds the breakdown of starch into glucose by weakening the bonds linking glucose subunits. Enzymes also bring molecules together in a certain order, *right,* by serving as templates (patterns). This is how starch synthetase works.

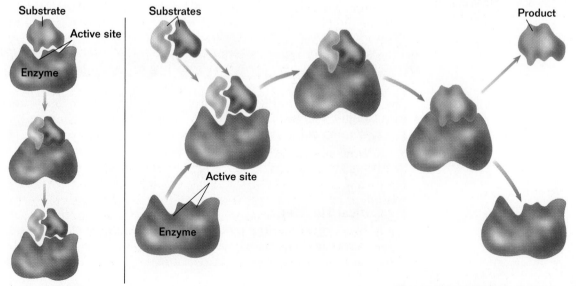

Factors that change the shape of an enzyme affect the enzyme's activity. One of these factors is temperature. For each enzyme, there is a range of temperatures within which the enzyme operates efficiently, as indicated by Figure 4-10. Temperatures outside of this range either cause some of the hydrogen bonds that determine an enzyme's shape to be broken or cause other hydrogen bonds to form, also changing the shape of the enzyme. Another factor that affects enzyme activity is pH (the relative acidity or alkalinity of a solution). Again, each enzyme has an optimal range of pH within which it can operate. A pH outside of this range can cause the breaking of certain bonds that determine an enzyme's shape. Figure 4-11 illustrates the relationship between the activity of two digestive enzymes and pH. ◻

Figure 4-10 The darker fur of this Himalayan cat indicates cooler regions of its body. These regions have a high concentration of a dark pigment that is produced with the aid of an enzyme that becomes inactive at warmer temperatures.

Figure 4-11 Pepsin and trypsin are digestive enzymes whose activities are affected by pH. Pepsin's activity peaks at a much lower pH than trypsin's. Pepsin works in the very acidic environment of the stomach, while trypsin works in the less acidic small intestine.

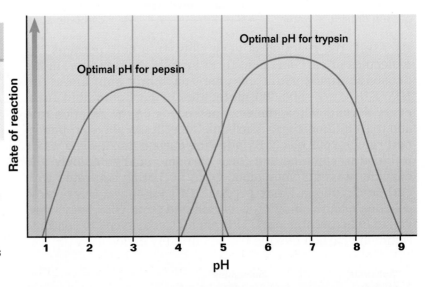

Section Review

1. *What is the difference between exergonic chemical reactions and endergonic chemical reactions?*
2. *What is activation energy?*
3. *How do catalysts help to start a chemical reaction?*
4. *Why are enzymes needed for chemical reactions in cells?*

Critical Thinking

5. *What effect might a molecule that changes the shape of carbonic anhydrase have on your body?*

4-3 Metabolism

E ach of the significant properties we use to define
life—growth, movement, reproduction, heredity—
uses energy. To sustain life, energy must be continually supplied (like putting more logs on a fire) and converted
into useful forms (like boiling water for cooking). But unlike
burning wood in a fire, energy must be released slowly to
prevent living cells from burning up. The energy must also
be released in such a way that it can be put to use doing the
work of cells. Thus, energy passes along a network of controlled chemical reactions that acts as a highway for energy
transportation within a cell. The chemistry of living, this
symphony of energy-transferring reactions within cells, is
called *metabolism* (muh TAB uh lihz uhm).

Life Processes Are Driven by Energy

Cells use energy to do those things that require work. One of the most obvious of these is movement. Some cells swim through water, propelling themselves by rapidly spinning a long tail-like flagellum. During development, an embryo's cells use energy to crawl over one another and reach new positions in the embryo. Cells also expend energy to change their shape. When one of your white blood cells engulfs an invading bacterium, it accomplishes the neces-sary change in shape by extending its cytoskeleton. Movement also occurs within cells. Mitochondria and cellu-lar materials are passed several feet along the narrow extensions of nerve cells that connect your feet to your spine. All of these movements use energy.

A second major way cells use energy is to build new molecules. Very few biological molecules are created by a single chemical reaction. Instead, most are manufactured by a long series of enzyme-catalyzed chemical reactions in which each reaction modifies the product of the previous reaction. An ordered series of enzyme-catalyzed chemical reactions that forms a product in a step-by-step manner, as illustrated in Figure 4-12, is called a **biochemical pathway**. Because building molecules takes energy, synthetic bio-chemical pathways (which make the specific molecules cells need) almost always involve at least one endergonic chemical reaction. Remember, these reactions consume energy because their products contain more energy than their reactants did. And nothing can happen until extra energy from an external source triggers the reaction.

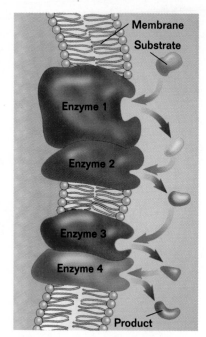

Figure 4-12 A biochemical pathway consists of a series of enzyme-catalyzed chemical reactions. The product of each reaction in a pathway becomes the reactant of the next reaction in the pathway.

Energy and Metabolism **85**

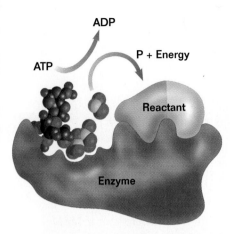

ADP

ATP

P + Energy

Reactant

Enzyme

Figure 4-13 Coupling energy-requiring reactions with the splitting of ATP molecules is one of the key processes that help cells manage energy. The two parts of a coupled reaction (ATP-splitting and the endergonic reaction) take place in concert, separate parts of a single process.

ATP Supplies Energy for Metabolism
....

ATP supplies most of the energy that drives metabolism. In a sense, it is a cell's "energy currency"—the "money" it uses to "pay for" endergonic processes. As you learned in Chapter 2, each ATP molecule is made of three parts: ribose, a sugar; adenine, a nitrogen-containing base; and a chain of three phosphate groups. Energy that was temporarily stored in an ATP molecule is made available for use by a cell when the end phosphate group is transferred to another molecule. In this exergonic reaction, a sizable packet of energy is transferred along with the phosphate group. This energy then activates a reaction involving the molecule that received the phosphate group. Almost all of the endergonic reactions in cells require less activation energy than that transferred with a phosphate group from ATP. The breakdown of ATP is thus able to power many of a cell's endergonic activities.

An enzyme that catalyzes an endergonic reaction in cells has *two* active sites. As you can see in Figure 4-13, one is for the reaction's reactant and the other is for ATP. The ATP site splits the end phosphate group from an ATP molecule, releasing energy. The phosphate group (P) and some of the energy released are received by the reactant molecule attached to the second active site. This transfer of a phosphate group and its energy drives the reaction at the second active site. Since both reactions occur on the surface of the same enzyme, they are physically linked, or coupled. An endergonic reaction that is driven by the splitting of ATP molecules is therefore called a **coupled reaction**. In a similar way, you can make water in a swimming pool leap straight up in the air, despite the fact that gravity prevents water from rising spontaneously—just jump in the pool! The energy you add going in more than compensates for the force of gravity holding the water down. ■

Coenzymes Help Transport Energy
....

In many metabolic pathways, energy released from one set of reactions is transferred to another set of reactions. Remember that in living systems, electrons carry energy from one atom or molecule to another in oxidation-reduction reactions. Often, high-energy electrons are passed from the active site of the enzyme that is catalyzing a reaction to an organic molecule called a **coenzyme** *(koh EHN zeym)*. The coenzyme then carries the electrons to another enzyme that is catalyzing a different reaction, as you can see in Figure 4-14. Just as tanker trucks transport energy (in the form of gasoline) from storage tanks to gas stations, coenzymes shuttle energy (in the form of hydrogen atoms with high-energy electrons) from one place to another in a cell.

One of the most important coenzymes in cell metabolism is nicotinamide *(nihk uh TIHN uh meyed)* adenine

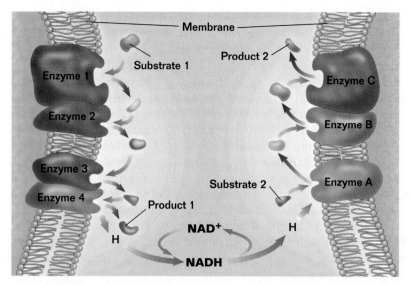

Figure 4-14 NAD⁺ accepts high-energy electrons (paired with protons as hydrogen atoms) and becomes NADH, which transfers the electrons from one set of enzyme-catalyzed reactions to another. When NADH reaches an enzyme in a second set of reactions, it releases the electrons (and their energy) to the reaction that this enzyme is catalyzing. NAD⁺ can then return to the enzyme in the first set of reactions and get more electrons.

dinucleotide, usually referred to by the abbreviation **NAD⁺**. When NAD⁺ acquires a hydrogen atom from the active site of an enzyme, it becomes **NADH** (and has been reduced). In cells, hydrogen atoms with high-energy electrons are stripped from food molecules and donated to NAD⁺, forming NADH. The NADH molecules then carry high-energy electrons. Like carrying money in a wallet, cells carry hydrogen atoms with high-energy electrons (one form of their energy "money") in molecules like NADH. And just as you can exchange one type of money for another, cells can exchange one type of energy carrier for another. The energy in NADH, as you will learn in Chapter 5, can be converted to the cell's main energy currency—ATP.

> **□ CAPSULE SUMMARY**
>
> NAD⁺ is a coenzyme that, when reduced to NADH, carries energy from one reaction to another in cells.

Enzymes Direct Metabolism in Cells

A cell may contain thousands of different kinds of enzymes, each specific to a particular substrate and each promoting a different chemical reaction. The enzymes that are active in a cell at any one time determine what happens in that cell, much as traffic lights determine the flow of traffic in a city. An increase in a particular enzyme's concentration accelerates the rate of the reaction it catalyzes. Conversely, a decrease in an enzyme's concentration slows the rate of the reaction it catalyzes. Furthermore, not all cells contain the same enzymes. The chemical reactions occurring in a nerve cell are very different from those occurring in a red blood cell because the two kinds of cells have a different array of enzymes. By controlling the concentration of enzymes and when they are active, a cell is able to control its chemical reactions, just as a conductor controls the music an orchestra produces by dictating the tempo and volume at which the music is played.

> **CONTENT LINK**
>
> More information about how cells regulate biochemical pathways is found in **Chapters 5, 35, and 41**.

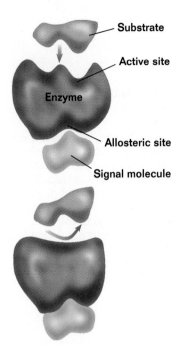

Substrate

Active site

Enzyme

Allosteric site

Signal molecule

Figure 4-15 An allosteric enzyme has an active site and an allosteric site, *top*. When a signal molecule binds to the allosteric site, *bottom*, the enzyme and its active site change shape.

☐ *CAPSULE SUMMARY*

Metabolism is regulated by controlling the kinds of enzymes present in a cell, their concentration, and their activity.

Because an enzyme must have a precise shape to work correctly, it is possible for a cell to control an enzyme's activity by altering the enzyme's shape. Many enzymes have shapes that can be altered by the binding of "signal" molecules to their surfaces. Such an enzyme is called an **allosteric** *(al oh STEHR ihk)* **enzyme**. *Allosteric*, which means "other shape," is derived from the Greek words *allos*, meaning "other," and *stereos*, meaning "solid." If an allosteric enzyme is unable to bind to a substrate because of the new shape produced by the binding of a signal molecule, the signal molecule is said to *repress* the enzyme's activity. If an allosteric enzyme is unable to bind to a substrate unless a signal molecule is bound to it, the signal molecule is said to *activate* the enzyme. The site where the signal molecule binds to an allosteric enzyme's surface is called the **allosteric site**. Figure 4-15 illustrates how a signal molecule may affect an allosteric enzyme.

When a cell already has an adequate amount of the chemical produced by a biochemical pathway, the pathway will often shut down. Thus, wasteful overproduction is avoided. How does a pathway "know" to shut itself down? The first enzyme in the pathway is an allosteric enzyme with a site that has the shape of the pathway's product. The binding of the product molecule to the allosteric enzyme *inhibits* the enzyme's activity so that when the concentration of the product is high, the first step in the pathway is effectively turned off. The shutting down of a biochemical pathway caused by a key enzyme's sensitivity to the level of the pathway's product is called **feedback inhibition**. When the product concentration drops, the pathway is reactivated. Feedback inhibition is another way that cells simply and effectively regulate their biochemical activities. ☐

Section Review

1. *Name three types of processes that require a cell to expend energy.*
2. *What role does ATP play in coupled reactions?*
3. *Name two molecules that carry energy in biochemical pathways, and describe how each transfers its energy.*
4. *How is metabolism controlled in cells?*

Critical Thinking

5. *Excess glucose, the primary fuel for cellular respiration, is stored in your liver as the complex molecule glycogen. How do you suppose your body knows when to convert glucose to glycogen for storage and when to convert glycogen back to glucose for release from the liver?*

Vocabulary

activation energy (81)
active site (83)
allosteric enzyme (88)
allosteric site (88)
autotroph (78)
biochemical pathway (85)
catalysis (82)
catalyst (82)
cellular respiration (78)
chemical reaction (80)
chemosynthesis (78)
closed system (76)
coenzyme (86)

coupled reaction (86)
endergonic reaction (81)
energy (75)
entropy (77)
enzyme (82)
exergonic reaction (81)
feedback inhibition (88)
first law of thermodynamics (76)
free energy (81)
heterotroph (78)
kinetic energy (76)
metabolism (85)
NAD⁺ (87)

NADH (87)
open system (76)
oxidation (79)
oxidation-reduction reaction (79)
photosynthesis (77)
potential energy (76)
product (80)
reactant (80)
reduction (79)
second law of thermodynamics (77)
substrate (82)
system (76)
thermodynamics (76)

Concept Mapping

Construct a concept map that shows the function of enzymes in cells. Use as many terms as needed from the vocabulary list. Try to include the following terms in your map: metabolism, chemical reaction, activation energy, catalyst, substrate, active site, and biochemical pathway. Include additional terms in your map as needed.

Review

Multiple Choice

1. Which of the following is an example of potential energy?
   a. water flowing over a dam
   b. a rock rolling downhill
   c. a burning match
   d. a can full of gasoline

2. Energy flows through living systems from
   a. the sun, to heterotrophs, and then to autotrophs.
   b. autotrophs, to the environment, and then to heterotrophs.
   c. the sun, to autotrophs, and then to heterotrophs.
   d. the environment, to heterotrophs, and then to autotrophs.

3. When a molecule loses an electron, the molecule has been
   a. reduced.   c. activated.
   b. metabolized.   d. oxidized.

4. If the beaker in which two chemicals are mixed becomes hot, the reaction that occurred is probably a(n)
   a. energetically neutral reaction.
   b. exergonic reaction.
   c. endergonic reaction.
   d. coupled reaction.

5. Which statement is correct for the following chemical reaction?

$$CO_2 + H_2O \xrightarrow{\text{carbonic anhydrase}} H_2CO_3$$

   a. H_2O is a product of the reaction.
   b. Carbonic anhydrase is the energy source for the reaction.
   c. H_2CO_3 is a catalyst for the reaction.
   d. CO_2 is a reactant of the reaction.

6. Enzymes speed up chemical reactions in cells by
   a. lowering the activation energy.
   b. remaining unaltered by the reaction.
   c. releasing the products of the reaction.
   d. taking the place of activation energy.

7. Energy is transferred from one place to another in a biochemical pathway by
   a. enzymes.   c. catalysts.
   b. electrons.   d. ATP.

8. Metabolism
   a. includes processes that capture, transfer, and use energy.
   b. is a series of exergonic reactions that provide energy to cells.
   c. functions only in closed systems.
   d. enables photosynthesis to occur without energy from the sun.

9. A coupled reaction
   a. increases the amount of ATP in a cell.
   b. includes the transfer of energy from ATP.
   c. has two parts that are endergonic.
   d. enables a cell to produce energy.

10. When a product molecule attaches to the allosteric site of an enzyme, the
    a. active site disappears.
    b. formation of product increases.
    c. enzyme's shape changes.
    d. substrate concentration decreases.

Completion

11. The energy of position is called _____ energy, while the energy of motion is called _____ energy.

12. Energy transformations tend to make _____ become more disorganized. _____ is a disorganized and less useful form of energy.

13. Simple organization and random molecular motion are characteristics of a system with a high degree of _____ .

14. Energy needed for life is captured during _____ and used by cells during cellular respiration. Much of the energy released during cellular respiration is used to make _____ .

15. All chemical reactions require _____ , which is lowered by the process of

    _____ .

16. A substance that speeds the progress of a chemical reaction is called a(n) _____.

17. In a biochemical pathway, the _____ of one enzyme-catalyzed reaction becomes the _____ for the enzyme catalyzing the next reaction.

18. When an enzyme's _____ changes, its activity also changes.

19. Cells use _____ to move, to change their shape, and to build new molecules.

20. A series of enzyme-catalyzed reactions makes up a(n) _____ .

21. Coupled reactions transfer energy from _____ to power _____ reactions.

22. NAD^+ is a _____ that functions in cell metabolism. Your cells extract the electrons and protons needed to convert NAD^+ to NADH from _____ .

23. The binding of a product molecule to a key enzyme that shuts down a biochemical pathway is called _____ .

Short Answer

24. Living organisms, such as animals and plants, appear to violate the second law of thermodynamics because of their high degree of organization. Explain why these observations are misleading.

25. What are oxidation-reduction reactions? Describe how these reactions are used by cells.

26. Explain how the presence of a catalyst affects the rate of a chemical reaction.

27. An enzyme's efficiency increases with greater substrate concentration, but only to a point. Why?

28. Explain how energy is stored in and released from ATP molecules.

29. Describe two ways in which cells are able to control the activity of enzymes that are present in the cytoplasm.

Themes Review

30. **Flow of Energy** Describe the flow of energy that results in your being able to obtain energy for metabolism by eating a T-bone steak.

31. **Homeostasis** Explain why every reduction reaction is accompanied by an oxidation reaction.

32. Levels of Organization A tennis trainer may treat a player's painful but minor elbow injury by spraying it with a surface anesthetic containing ethyl chloride. The anesthetic rapidly evaporates, cooling the skin and reducing the pain. Realizing that pain is caused by chemical reactions, explain how you think this treatment works.

Critical Thinking

33. Using Analogies Using an analogy, develop a model to describe the concept of entropy.

34. Making Inferences For every 100 units of energy available from alfalfa, cattle that eat the alfalfa capture only 10 units of that energy, and humans that eat the cattle as hamburger capture only 1 unit of that same energy. How do you account for this inefficient transfer of energy?

35. Interpreting Data How do the two reactions represented below compare in terms of energy?

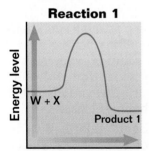
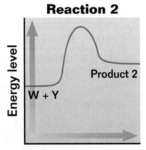

Reaction 1

Energy level

W + X

Product 1

Progress of reaction

Reaction 2

Energy level

W + Y

Product 2

Progress of reaction

36. Making Predictions The figure below depicts a metabolic pathway. Predict how a high concentration of Substrate C would affect the activity of Enzyme 1 if the shape of Enzyme 1 is changed when it binds with Substrate C.

Substrate A $\longrightarrow$ Substrate B $\longrightarrow$ Substrate C
　　　　Enzyme 1　　　　　　Enzyme 2

Activities and Projects

37. Cooperative Group Project Using modeling clay, build a model of an enzyme and its substrate. In your model, show the enzyme's active site, how the enzyme changes shape when it binds to its substrate, and how the enzyme returns to its original shape when the products of the reaction are released.

38. Research and Writing Find out about the relationship between vitamins and coenzymes. In a written report, identify several coenzymes and the vitamins with which they are associated.

39. Research and Writing Investigate the lethal effect of hydrogen cyanide on the human body. In a written report, discuss the chemical structure of hydrogen cyanide, and explain its effects in terms of its competitive inhibition of the enzyme cytochrome oxidase.

40. Multicultural Perspective Interview an expert on, or do research on, the following martial arts: tai chi, tae kwan do, and aikido. Find out what these forms of exercise have in common. How do they relate to the flow of energy? Where and how did each originate? Relate your findings in a report to your class.

Readings

41. Read Elizabeth Pennisi's article "Juiced-up fruit: Unbelievably flavorful," in *Science News*, September 14, 1991, page 173. How did Berger and his associates make use of natural biosynthetic pathways to enhance the flavor of fruits?

42. Read David Freedman's article "Life's Off-Switch," in *Discover*, July 1991, pages 61–67. Arthur Kornberg believes that "life is chemistry." What is his latest discovery? Why did Kornberg once receive a telephone call from President Lyndon Johnson?

Observing Enzyme Activity

OBJECTIVE

Recognize the presence and function of enzymes in commercial products.

PROCESS SKILLS

- organizing data
- predicting results

MATERIALS

- detergents (various brands)
- safety goggles
- lab apron
- 18 g of regular instant gelatin or 1.8 g of sugar-free instant gelatin
- 0.7 g washing soda (Na_2CO_3)
- 150 mL beaker
- glass stirring rod
- boiling water
- 6 test tubes
- test-tube rack
- graduated cylinder
- Pasteur pipette
- plastic wrap
- tape
- wax pencil
- 50 mL beakers
- pH paper
- balance

BACKGROUND

1. What is an enzyme? What is a substrate?
2. What environmental factors influence enzyme activity?
3. Many enzymes are named after their substrates. What kind of substrate might the enzymes known as *proteases* act upon?
4. Write your own **Focus Question** on your Vee Form.
5. **Knowing Side of the Vee** List the **Concepts** and new **Vocabulary Words** on your Vee Form. In the **Concept Statements** section of the Vee, use these words in sentences that define and explain them.

TECHNIQUE

Doing Side of the Vee

Day 1

1. Bring a sample of laundry detergent from home in a plastic bag. Be sure to write down the ingredients of the product as they are listed on the label. Note any mention of the word *enzyme* either in the ingredients list or on the label.
2. Put on safety goggles and a lab apron.
3. Place 18 g of regular instant gelatin or 1.8 g of sugar-free instant gelatin in a beaker. Slowly add 50 mL of boiling water and mix, using a stirring rod. **CAUTION: Be careful handling boiling water.** To this hot gelatin solution, very slowly add 0.7 g of washing soda (Na_2CO_3) while stirring. Record your observations. What gas do you think is being released?
4. Pour 5 mL of the liquid gelatin–Na_2CO_3 mixture into each of 6 test tubes placed in a test-tube rack. Remove the bubbles on the top of each tube with a Pasteur pipette. Cover the tubes tightly with plastic wrap and secure with tape. Allow these to cool at room temperature or in a refrigerator until Day 2.
5. Clean up your materials and wash your hands before leaving the lab.

Day 2

6. Put on safety goggles and a lab apron.
7. Use a wax pencil to mark the test tubes at the upper level of the cooled gelatin in

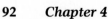

each tube. It is from this point that you will measure the breakdown, or hydrolysis, of the gelatin from day to day. Label the test tubes 1–6.

8. Prepare a 10 percent solution of each of five different detergents brought in by you and your classmates. Test the pH of each with your pH paper. Record the pH for each numbered detergent sample in the **Records** section of your Vee.

Hydrolysis of Gelatin

Detergent	pH	Amount of Hydrolysis (mm)
1		
2		
3		
4		
5		

9. Add 15 drops (1 mL) of the first detergent solution to the gelatin surface of the first test tube. Repeat for each of the other samples. Reseal the tubes and place in a test-tube rack for each observation. To the sixth tube add only 15 drops (1 mL) of water to the gelatin surface. Put the tubes aside for 24 hours at room temperature.

10. Clean up your materials and wash your hands before leaving the lab.

Day 3

11. After 24 hours at room temperature, draw another wax pencil line at the top of the gelatin layer. Then measure the distance in millimeters between the first line and the second line. This indicates the amount of hydrolysis of the protein in the liquid gelatin by the enzymes in the detergent. Record your data in the chart in the **Records** section of your Vee Form.

12. Clean up your materials and wash your hands before leaving the lab.

INQUIRY

1. Why was it important to include a sixth test tube to which you added water?

2. Why did you observe a difference in gelatin levels among the different test tubes from Day 2 to Day 3?

3. Suggest a reason for adding washing soda (Na_2Co_3) to the gelatin solution.

4. Does the rate of hydrolysis in sugar-free gelatin differ from that in regular gelatin?

5. Use the information on the **Knowing Side** of the Vee to interpret your results from the **Doing Side,** and then write your **Knowledge Claim.** Write a **Value Claim** for this lab.

ANALYSIS

1. Why are enzymes added to many laundry detergents?

2. Do you think detergent enzymes are stable at temperatures found during a hot-water cycle in a washing machine? Are enzymes in detergent stable in the presence of bleach?

3. What kind of stains do you think detergent enzymes could most effectively remove? Offer an explanation for your answer.

4. How do you think changes in room temperature could affect your results?

FURTHER INQUIRY

Write a **New Focus Question** that could be the point of a new investigation. The following is an example.

Are there proteases in other products in the supermarket? Check the labels on items such as meat tenderizer, papaya shampoo, and contact-lens cleaner.

CHAPTER 5

PHOTOSYNTHESIS AND CELLULAR RESPIRATION

REVIEW

- proton pumps and chemiosmosis (Section 3-2)
- autotrophs and heterotrophs (Section 4-1)
- laws of thermodynamics (Section 4-1)
- oxidation-reduction reactions (Section 4-1)
- coenzymes and coupled reactions (Section 4-3)

Mule deer eating yellow fawn lilies

5-1 How Energy Cycles

*L*ike a rechargeable battery, your body eventually runs low on energy and needs to be supplied with more. You get this energy from food. The energy in your food was first captured from sunlight by photosynthesis. You extract that energy by the process of cellular respiration. Together, photosynthesis and cellular respiration form a cycle, seen in Figure 5-1, that links organisms to each other and to the environment.

Section Objectives

■ Explain why photosynthesis and cellular respiration form a continuous cycle.

■ Describe what happens to the sugars produced during photosynthesis.

■ Discuss the importance of food chains.

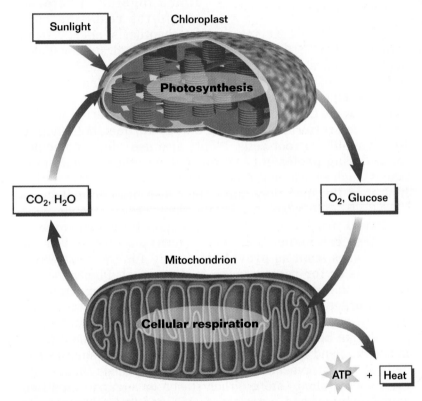

Figure 5-1 Photosynthesis and cellular respiration form a continuous cycle because the products of one process are the starting materials for the other.

Photosynthesis and Cellular Respiration Form a Cycle

The carbon atoms needed to make all the organic molecules of living things ultimately come from a nonliving part of the environment. During photosynthesis, carbon atoms are pulled from the carbon dioxide gas, CO_2, in air and used to form carbohydrates and other organic compounds. These materials are made, step by step, in biochemical pathways that are powered by transferring hydrogen atoms from one reaction to the next. The hydrogen atoms needed for these pathways are extracted from water molecules, H_2O. Leftover oxygen atoms combine to form oxygen gas, O_2, as a byproduct.

Photosynthesis uses carbon dioxide, water, and light energy and produces carbohydrates and oxygen. Cellular respiration uses carbohydrates and oxygen and produces carbon dioxide, water, and energy for cell activities.

Cellular respiration is essentially the reverse of photosynthesis. During cellular respiration, hydrogen atoms are pulled from carbohydrates and joined with oxygen atoms from oxygen gas, forming water. In biochemical pathways, the carbon chains of carbohydrates and other organic molecules are systematically dismantled. Energy stored in the molecules is then made available for the activities of cells. Carbon atoms that are split from carbohydrates are linked to oxygen gas, forming carbon dioxide gas as a byproduct. ■

Energy Is Stored in Carbon Compounds

Some of the energy that is required to link carbon atoms together is stored within the resulting molecules. In plants carrying out photosynthesis, energy from sunlight is stored within newly made carbohydrate molecules such as sugars. Sugars can be transported from leaves to other parts of a plant. In that way, parts of a plant that are not exposed to light, such as roots, are supplied with carbohydrates. Later, the energy in sugars can be converted to ATP by cellular respiration, making energy for metabolic activities available to root cells. Plants also use sugar molecules made during photosynthesis for making other organic molecules, such as proteins, lipids, nucleic acids, and other carbohydrates. Because they make their own organic molecules directly from inorganic materials, plants are autotrophs.

Plants usually produce more sugars by photosynthesis than they can immediately use. Excess sugar is stored for future uses such as providing energy for the next day's activities or for rapid growth in the spring. Plants mainly convert excess sugar to starch for both short-term and long-term storage. As you will recall from Chapter 2, starch is a carbohydrate made of large numbers of glucose molecules joined, like beads in a necklace, by chemical bonds. Starch molecules tend to cluster in certain plant cells, such as the cells of fruits (avocados and bananas) and seeds (beans and corn). Some plants store either starch or sugar in modified stems (potatoes and sugar cane) or in modified roots (carrots and yams). Excess sugar may also be converted to oils and stored within seeds (peanuts and safflower seeds).

Because they are rich in organic molecules that store energy, plants often serve as food for animals. Of the more than 250,000 known species of plants, about 150 are cultivated (grown) extensively as food. A type of plant that is cultivated for use by humans is called a **crop**. The major food crops grown in the world today have been cultivated for thousands of years, selected by early farmers for their food value and ease of cultivation. For example, just three species of grasses—rice, wheat, and corn—directly or indirectly supply more than half of all human energy needs. ■

CONTENT LINK

The origin and importance of several crop plants are discussed in Chapter 26.

Energy is stored in organic molecules made by linking carbon atoms together. Excess carbohydrates produced by plants provide food for animals.

Energy Flows Through Food Chains

In a sense, energy moves through the living world on the "shoulders" of carbon atoms. The journey begins when carbon atoms captured from carbon dioxide gas are incorporated into organic molecules. In the process, energy is stored in organic molecules that can serve as food for other organisms. Autotrophs (plants, algae, and certain bacteria), which make food molecules by capturing energy and carbon atoms from their environment, are called **producers**. Once captured by producers, energy stored in organic molecules can be passed to heterotrophs, which are not able to capture energy and carbon directly from their environment. Because they must obtain energy and organic molecules from other organisms, heterotrophs are also called **consumers**. Some consumers obtain most of their food by eating producers. Some eat other consumers. When you eat beef, for example, you are a consumer eating another consumer (a cow) that ate a producer (grass). A series of organisms through which energy flows from a producer to one or more consumers, as illustrated in Figure 5-2, is called a **food chain**. ◻

◻ **CAPSULE SUMMARY**

Energy stored in organic molecules is passed from producers to consumers through food chains.

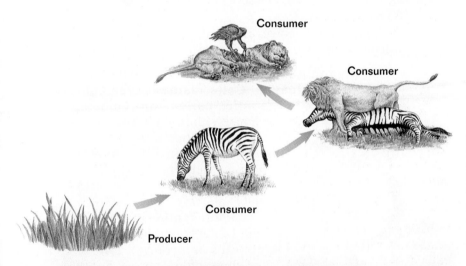

Consumer

Consumer

Consumer

Producer

Figure 5-2 A food chain consists of a series of organisms through which energy flows. Because some energy is lost every time it is transferred from one organism to another (second law of thermodynamics), most food chains consist of no more than three or four organisms.

Section Review

1. *How do photosynthesis and cellular respiration form a continuous cycle?*

2. *How do plants use the products of photosynthesis?*

3. *How are the organic molecules produced as a result of photosynthesis important to you and other heterotrophs?*

Critical Thinking

4. *Because of the second law of thermodynamics, food chains normally involve no more than three or four organisms. Why do you think this is so?*

5-2 How Photosynthesis Works

E arth is bathed in energy streaming from the sun. Each day, our planet is bombarded with an amount of energy equal to the energy of about 1 million atomic bombs the size of the one dropped on Hiroshima, Japan, in 1945. Approximately 1 percent of this energy is captured by photosynthesis, providing the energy upon which almost all life on Earth depends.

Section Objectives

■ *Summarize the overall process of photosynthesis.*

■ *Identify the roles of photons, pigments, electron transport chains, and proton pumps in photosynthesis.*

■ *Explain how oxygen gas is produced during photosynthesis.*

■ *Discuss how the products of the light reactions of photosynthesis are used in the Calvin cycle.*

■ *Recognize how various environmental factors affect the rate of photosynthesis.*

Overview PHOTOSYNTHESIS

Photosynthesis takes place in three stages.

1. Energy is captured from sunlight.
2. Light energy is converted to chemical energy (ATP and NADPH).
3. ATP and NADPH power the synthesis of organic molecules, using carbon from carbon dioxide.

In plants, photosynthesis occurs primarily within leaves. Cells inside leaves contain organelles called chloroplasts, which house chlorophyll, the light-absorbing substance needed for photosynthesis. In all plants and photosynthetic protists, photosynthesis occurs within chloroplasts.

Photosynthesis also occurs in certain bacteria that contain chlorophyll but do not have chloroplasts. Some of the organic molecules produced during photosynthesis store energy for later needs. Others are used as building blocks for the macromolecules that make up a cell.

The overall process of photosynthesis can be summarized by the following chemical equation:

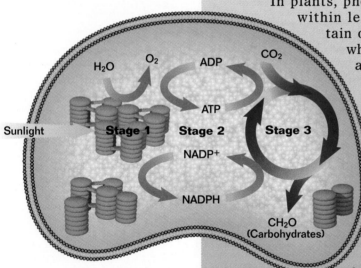

Figure 5-3 Photosynthesis occurs in three distinct phases: energy is captured from sunlight, converted to chemical energy, and stored in organic molecules.

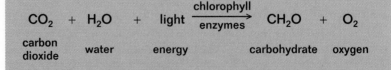

$$CO_2 + H_2O + light \xrightarrow[\text{enzymes}]{\text{chlorophyll}} CH_2O + O_2$$

carbon dioxide   water   energy   carbohydrate   oxygen

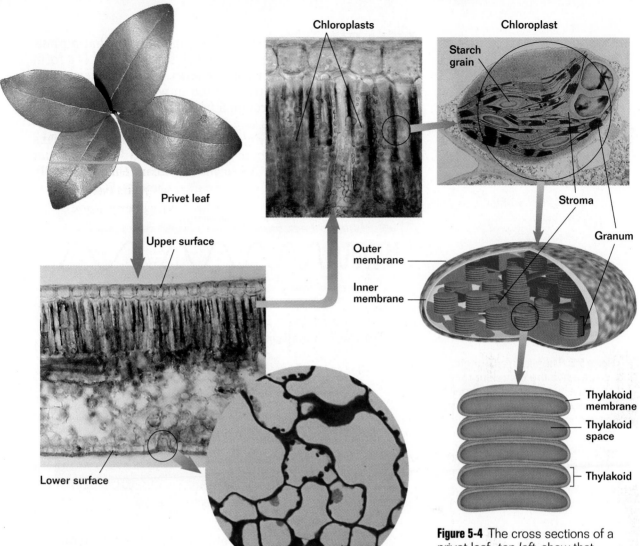

Chloroplasts

Chloroplast

Starch grain

Stroma

Granum

Privet leaf

Upper surface

Outer membrane

Inner membrane

Lower surface

Opening

Thylakoid membrane

Thylakoid space

Thylakoid

Figure 5-4 The cross sections of a privet leaf, *top left,* show that most of its chloroplasts are located in the upper part of the leaf. Tiny openings in the bottom surface of the leaf, *bottom center,* enable carbon dioxide to enter and oxygen and water vapor to exit. A photomicrograph of a chloroplast, *top right,* reveals its internal structure. The stroma surrounding the grana contains the enzymes needed for making carbohydrates. The cross section of a single granum, *bottom right,* shows that each thylakoid is a closed compartment. Chlorophyll molecules are embedded in the thylakoid membranes.

Plants Are Specialized for Photosynthesis

You may not think of plants as being exceptional. However, even the simplest leaf is a wonderfully complex and sophisticated photosynthetic machine, in which structure and function form a near-perfect union. Figure 5-4 shows the structure of one type of leaf and how it is adapted for photosynthesis. In plants and most other photosynthetic organisms, photosynthesis occurs inside chloroplasts, which are tiny green organelles found only in eukaryotic cells. If you could cut into a chloroplast, you would find that its interior is surprisingly complex. As Figure 5-4 also shows, the interior of a chloroplast is filled with many flattened membrane-bound sacs called **thylakoids** *(THEYE lah koydz).* Thylakoids are often stacked like coins, in columns called **grana** *(GRAN uh).* A chloroplast usually contains dozens of grana and individual thylakoids, which are all suspended in a fluid matrix called the **stroma** *(STROH muh).*

Light Energy Is Packaged in Photons

What is there in sunlight that a plant can use to make sugars? Light is a form of **radiant energy**, which is energy that is transmitted in waves that can travel through a vacuum. The complete range of radiant energy forms is called the **electromagnetic spectrum**. As you can see in Figure 5-5, the many different forms of radiant energy differ both in wavelength and in the amount of energy they transmit. All forms of radiant energy actually consist of tiny packets of energy called **photons** (FOH tahnz).

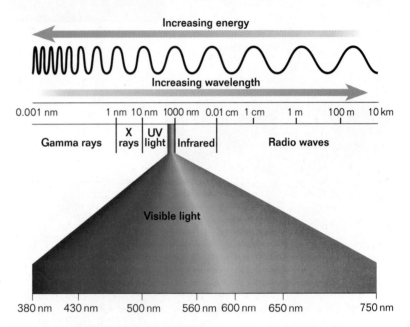

Figure 5-5 The electromagnetic spectrum is the complete range of all the types of radiant energy. The wavelengths of some electromagnetic waves are measured in millionths of a millimeter, or nanometers (nm). Photons of visible light (wavelengths of 380 nm to 750 nm), which includes all the colors of light in sunlight, have intermediate amounts of energy.

As sunlight shines on an object, the object's surface is bombarded by energetic photons. The photons striking an object are either reflected or absorbed by the object, or they are transmitted (passed) through it. An object's color results from the visible light photons it reflects. Human eyes see visible light because they have light receptors that can absorb those kinds of photons. Leaves absorb mostly photons of blue and red light and reflect mostly photons of green light, making them appear green to us. A plant uses some of the energy in the photons it absorbs to make sugars.

How do human eyes and plants "choose" which photons to absorb? The answer to this question is in an atom's structure. Recall that electrons spin in energy-specific regions near an atom's nucleus. When a photon strikes an atom, energy in the photon may boost one of the atom's electrons to a higher energy state. Just as you must lift your foot only so far to step to the next rung of a ladder, so boosting an electron to a particular energy state requires a precise amount of energy. Thus, a particular atom absorbs only photons of certain kinds of light—those with just the right amount of energy. ■

☐ CAPSULE SUMMARY

When a photon of light with the right amount of energy is absorbed by an atom, one of the atom's electrons is raised to a higher energy state.

Photosynthetic Pigments Absorb Photons

A molecule containing atoms that enable it to absorb light is called a **pigment. Chlorophyll,** which is a green pigment, is the primary light-absorbing agent for photosynthesis. Most plants contain two types of chlorophyll, chlorophyll *a* and chlorophyll *b*, that both play important roles in photosynthesis. The yellow and orange plant pigments that produce fall colors, and the colors of many fruits, vegetables, and flowers, are **carotenoids** *(kuh RAH tuh noydz)*. Carotenoids assist in photosynthesis by capturing energy from light of different wavelengths than those absorbed by chlorophyll. Figure 5-6 shows the wavelengths of light absorbed by chlorophylls and carotenoids as well as how these pigments affect photosynthesis.

The pigments used in photosynthesis are arranged in molecule clusters, each of which is called a **photosystem.** These clusters, which contain both chlorophylls and carotenoids, are embedded in the thylakoid membranes inside chloroplasts. In photosynthetic bacteria, which lack chloroplasts, photosystems are located in membranes inside the bacterial cell. Most photosynthetic organisms have two kinds of photosystems. **Photosystem I** clusters boost electrons to a higher energy state by absorbing light with a wavelength of 700 nm. **Photosystem II** clusters boost electrons by absorbing more energetic light with the slightly shorter wavelength of 680 nm. Energy from a photon strike anywhere in a photosystem is funneled to particular chlorophyll *a* molecules that are located in the **reaction center** of the photosystem. Upon receiving the right amount of energy from a photon strike, electrons from the reaction center's chlorophyll *a* are boosted to higher energy states.

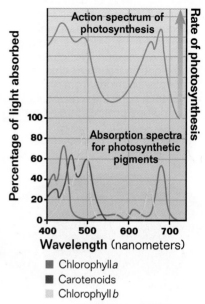

■ Chlorophyll *a*
■ Carotenoids
▨ Chlorophyll *b*

Figure 5-6 The absorption spectra, *bottom*, show the wavelengths of light absorbed by photosynthetic pigments. Chlorophylls absorb mostly red, blue, and violet light. Carotenoids absorb mostly blue and green light. The action spectrum, *top*, shows the combined effect of all photosynthetic pigments on the rate of photosynthesis.

Light Energy Is Converted Into Chemical Energy

In the second stage of photosynthesis, light energy is converted into chemical energy by a series of reactions that is initiated when light is absorbed by the photosystems. These reactions are commonly called the "light reactions" or the "light-dependent reactions." The two types of photosystems work in partnership, one passing electrons to the other like runners passing a baton in a relay race. The easiest way to understand how the two photosystems work together is to follow the "baton"—an electron.

Action of Photosystem II

In a sense, photosystem II clusters are the first to act during the light reactions. The absorption of light by photosystem II pigments initiates a series of events that results in the production of ATP. First, photons of light striking a chloroplast

Figure 5-7 Photosystems and electron transport chain molecules are embedded in the thylakoid membrane. Notice that the electron transport chain that accepts electrons from photosystem II contains a molecule that pumps protons into a thylakoid. As protons leave a thylakoid via ATP synthetase, ATP is generated.

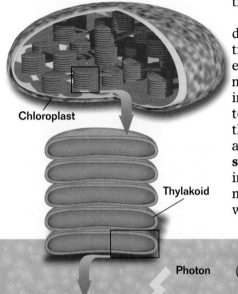

Chloroplast

Thylakoid

are absorbed by photosystem II pigment molecules, causing electrons from the reaction center's chlorophyll *a* to be boosted to a higher energy state. Traveling as part of a hydrogen atom, each excited electron leaves its chlorophyll molecule and jumps to a nearby membrane protein. Then, like the baton in a relay race, each excited electron is passed through a series of membrane-bound protein and pigment molecules called an **electron transport chain**. As excited electrons pass through this electron transport chain, some of their energy is used to power ATP production. Eventually, the electrons are accepted by photosystem I clusters.

To see how an electron's energy is used to generate ATP during photosynthesis, look at Figure 5-7. One of the electron transport chain molecules acts as a proton pump, using the energy of the excited electrons to "pump" protons (hydrogen nuclei) across the thylakoid membrane, into the thylakoid's interior. Because a thylakoid is a sealed compartment, protons cannot diffuse back out. Instead, they build up until there is enough pressure to drive them out through the only available exit—the ATP-generating protein channel **ATP synthetase**. Recall from Chapter 3 that making ATP by forcing protons through a membrane channel is called chemiosmosis. The ATP is released into the stroma of a chloroplast, where it is used in the production of organic molecules.

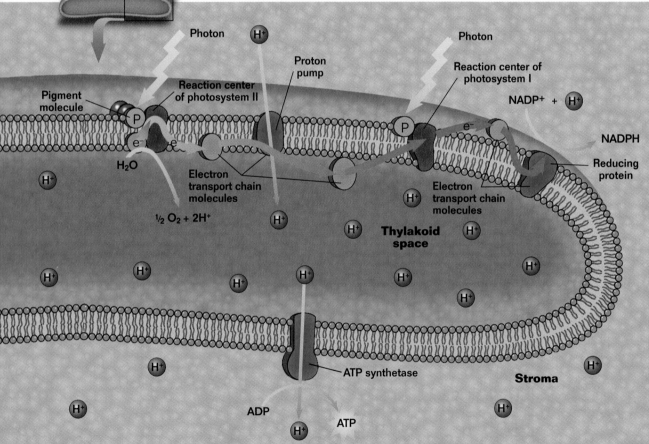

Action of Photosystem I

The point of photosynthesis, however, is not simply to make ATP. Rather, it is to extract carbon atoms from carbon dioxide and use them to make new organic molecules, such as sugars, lipids, proteins, and nucleic acids. Synthesizing such molecules takes a lot of energy. ATP provides this energy, but something else is needed as well. Organic molecules contain hydrogen as well as carbon and oxygen. Because carbon dioxide has no hydrogen atoms, a source of attachable hydrogen atoms is needed for synthesizing organic molecules. Recall that the addition of hydrogen atoms to a molecule is called reduction. Biologists therefore refer to a ready supply of attachable hydrogen atoms as **reducing power**.

The absorption of light by photosystem I pigments initiates a series of events that results in the generation of reducing power, which is in the form of NADPH molecules. **NADPH** is the reduced form of NADP⁺ (nicotinamide adenine dinucleotide phosphate). Like NAD⁺, **NADP⁺** is a coenzyme that carries hydrogen atoms (and energy) from one place to another in a cell. The NADPH is released into the stroma of a chloroplast, where both the hydrogen atoms and the energy NADPH carries are used in the production of organic molecules. Remember, NADPH does not result from the action of photosystem I alone. Figure 5-8 illustrates how the two types of photosystems work together to convert light energy into chemical energy.

Figure 5-8 The two types of photosystems work together simultaneously to convert light energy to chemical energy in the form of ATP and NADPH. Notice that electrons transferred to NADPH still contain excess energy.

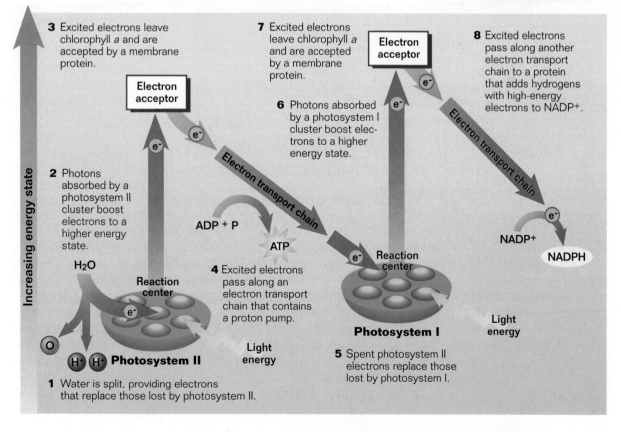

3 Excited electrons leave chlorophyll *a* and are accepted by a membrane protein.

7 Excited electrons leave chlorophyll *a* and are accepted by a membrane protein.

8 Excited electrons pass along another electron transport chain to a protein that adds hydrogens with high-energy electrons to NADP⁺.

6 Photons absorbed by a photosystem I cluster boost electrons to a higher energy state.

2 Photons absorbed by a photosystem II cluster boost electrons to a higher energy state.

Electron acceptor

Electron acceptor

Electron transport chain

Electron transport chain

Electron transport chain

ADP + P

ATP

H₂O

Reaction center

4 Excited electrons pass along an electron transport chain that contains a proton pump.

Reaction center

NADP⁺

NADPH

Increasing energy state

O

H⁺ H⁺ **Photosystem II**

Light energy

Photosystem I

Light energy

1 Water is split, providing electrons that replace those lost by photosystem II.

5 Spent photosystem II electrons replace those lost by photosystem I.

Levels of Organization

Organization is characteristic of all living things. At their most fundamental level of organization, living things are composed of atoms and molecules. These atoms and molecules interact in systems of chemical reactions (biochemical pathways) that direct the basic processes of life. Biochemical pathways that carry out particular processes, such as photosynthesis and cellular respiration, are organized within organelles in most cells.

Cells are the next level of organization. They are the smallest level of organization that can be considered alive. Multicellular organisms such as yourself consist of many different cell types that are each specialized to do different things. Cells of the same type form tissues such as muscles and nerves. Finally, the different tissues of your body can be found working together in organs, biological machines (such as the heart and the liver) that carry out particular jobs.

No organism lives in isolation, however. At the highest levels of organization, living things interact with each other. Organisms live together in populations, which are groups of individuals of a particular species. The different kinds of populations that live in a place make up a community. A community and its physical environment make up an ecological system, or ecosystem. Taken together, all of Earth's ecosystems make up the biosphere, the highest level of life's organization.

Atoms make up the chlorophyll molecules that are part of many plant cells. Different types of cells, organized into tissues, make up the leaves of a plant.

Source of Oxygen

Research indicates that the reactions that convert light energy to ATP and NADPH are also the source of the oxygen gas produced during photosynthesis. Oxygen gas results from the splitting of water molecules. Why does this occur? Look again at Figures 5-7 and 5-8. Each photosystem II cluster continually passes electrons to a photosystem I cluster, which passes electrons to NADPH. The loss of electrons leaves the photosystem II cluster unable to participate in photosynthesis until the electrons it has contributed are replaced. Therefore, each photosystem II must have a source of electrons. The electrons (and protons) released when water molecules are split replace those lost by a photosystem II cluster. The oxygen atoms that are left combine to form oxygen gas. Virtually every oxygen molecule in the air you breathe was once split from a water molecule by one of the light reactions of photosynthesis. ◻

◻ CAPSULE SUMMARY

Excited electrons from each photosystem II aid in the production of ATP. Excited electrons from each photosystem I aid in the production of NADPH. Water molecules are split, releasing the electrons needed to replace those lost by each photosystem II and producing oxygen gas as a result.

Chemical Energy Is Stored in Organic Molecules

In the final stage of photosynthesis, the chemical energy of ATP and NADPH is used by enzymes. These enzymes incorporate carbon atoms from carbon dioxide into organic molecules, a process called **carbon fixation.** The reactions that "fix" carbon to build organic molecules are sometimes called the "dark reactions" or the "light-independent reactions." Actually, they are neither. Although carbon fixation *can* occur in the dark, it only occurs if the products of the light reactions, ATP and NADPH, are present. Thus, while the reactions do not use light *directly*, they are not "light-independent." Furthermore, carbon dioxide enters the leaves of most plants through openings that close at night. In most plants, therefore, the reactions that fix carbon *do not* occur in the dark.

Among photosynthetic organisms, there are actually several biochemical pathways in which carbon is fixed. The most common carbon-fixing pathway is called the **Calvin cycle**, honoring Melvin Calvin. During the 1940s and 1950s, Calvin and his associates at the University of California at Berkeley worked out the cycle's reactions. The Calvin cycle, which occurs in all plants and algae, employs a complex battery of enzymes that are found in the stroma of a chloroplast. These enzymes form a cycle because they regenerate the starting material for further reactions. A total of six carbon dioxide molecules must enter the Calvin cycle to produce one six-carbon sugar molecule. Figure 5-9 illustrates the important events of the Calvin cycle. ◘

CONTENT LINK
Alternate carbon-fixing pathways for photosynthesis are discussed in **Chapters 24** and **25.**

CAPSULE SUMMARY

ATP and NADPH are used to fix carbon in the Calvin cycle. For every six carbon dioxide molecules that enter the Calvin cycle, one six-carbon sugar molecule can be produced.

Figure 5-9 The Calvin cycle is a series of reactions that attaches carbon atoms from carbon dioxide molecules to growing carbon chains while regenerating the starting material to which the carbon atoms are initially attached.

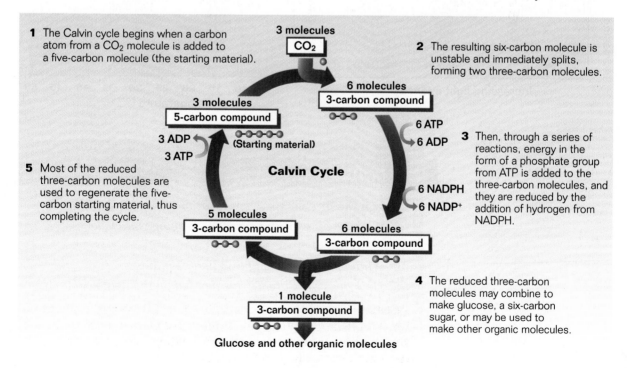

1 The Calvin cycle begins when a carbon atom from a CO_2 molecule is added to a five-carbon molecule (the starting material).

3 molecules
CO_2

2 The resulting six-carbon molecule is unstable and immediately splits, forming two three-carbon molecules.

6 molecules
3-carbon compound

3 molecules
5-carbon compound

3 ADP
3 ATP
(Starting material)

6 ATP
6 ADP

Calvin Cycle

3 Then, through a series of reactions, energy in the form of a phosphate group from ATP is added to the three-carbon molecules, and they are reduced by the addition of hydrogen from NADPH.

6 NADPH
6 NADP+

5 Most of the reduced three-carbon molecules are used to regenerate the five-carbon starting material, thus completing the cycle.

5 molecules
3-carbon compound

6 molecules
3-carbon compound

4 The reduced three-carbon molecules may combine to make glucose, a six-carbon sugar, or may be used to make other organic molecules.

1 molecule
3-carbon compound

Glucose and other organic molecules

Environment Affects the Rate of Photosynthesis

Photosynthesis is directly affected by environmental factors. The most obvious of these factors is light. In general, the rate of photosynthesis increases as light intensity increases, until a point called the light saturation point is reached. At that point, the rate of photosynthesis levels off, as the graph on the left in Figure 5-10 indicates. Carbon dioxide concentration affects the rate of photosynthesis in a similar manner. Once a certain concentration of carbon dioxide is present, the reactions of photosynthesis cannot proceed any faster. The temperature graph on the right in Figure 5-10 shows that photosynthesis also operates best within a certain range of temperatures. Remember that like all biochemical pathways, photosynthesis is a series of enzyme-catalyzed reactions. As you learned in Chapter 4, enzymes operate properly only within certain temperature ranges. Light, carbon dioxide concentration, temperature, and other environmental factors interact with one another to determine the optimum level of photosynthesis for a particular plant in its environment.

Figure 5-10 A graph of the effect of light intensity on photosynthesis, *left,* shows that the rate of photosynthesis increases as light intensity increases, but only to a certain point. A graph of the effect of temperature on photosynthesis, *right,* shows that as temperature increases, the rate of photosynthesis increases to a certain point and then decreases.

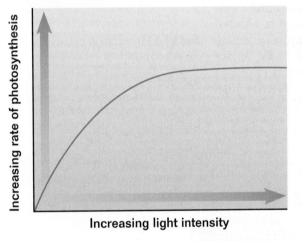

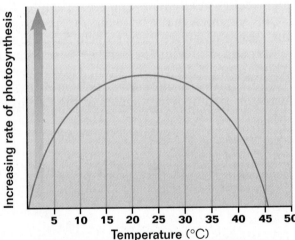

Section Review

1. *List the three main events that make up photosynthesis.*
2. *Why are pigments necessary for photosynthesis?*
3. *What is a photosystem?*
4. *What is the role of water in photosynthesis?*
5. *How does the Calvin cycle depend on the light reactions?*

Critical Thinking
6. *What combination of environmental factors would produce an optimum rate of photosynthesis for most plants?*

5-3 How Cellular Respiration Works

Almost all organisms, autotrophs and heterotrophs, obtain energy for their activities from organic molecules assembled during photosynthesis. Energy that was invested in building these molecules is retrieved by stripping electrons from them and using these electrons to make ATP. This process, called cellular respiration, is possible because the electrons have extra energy obtained from a chlorophyll molecule's encounter with a photon of light.

Overview   CELLULAR RESPIRATION

Cellular respiration takes place in two stages.

1. Glucose is converted to **pyruvate** *(peye ROO vayt)*, producing a small amount of ATP and NADH.

2. When oxygen is present, pyruvate and NADH are used to produce a large amount of ATP. When oxygen is absent, pyruvate is converted to lactic acid or ethyl alcohol.

The breakdown of glucose into pyruvate, which does not require oxygen, occurs in the cytosol of all cells. The conversion of pyruvate and NADH to ATP occurs in the mitochondria of all eukaryotic cells. In bacterial cells, which lack mitochondria, the conversion of pyruvate and NADH to ATP occurs in the cell membrane. Many of the intermediate products of cellular respiration are used to produce organic molecules needed for building and maintaining cells.

Metabolic processes that require oxygen are termed **aerobic** *(air OH bihk)*, from the Greek word *aer*, meaning "air." Metabolic processes that do not require oxygen are termed **anaerobic** *(AN air oh bihk)*, meaning "without air." Because far more ATP per food molecule results when oxygen is present, the aerobic pathways of cellular respiration are the primary source of energy for most cells.

The overall process of cellular respiration can be summarized by the following equation:

$$\text{CH}_2\text{O} \; + \; \text{O}_2 \; \xrightarrow{\text{enzymes}} \; \text{CO}_2 \; + \; \text{H}_2\text{O} \; + \; \text{36ATP}$$

carbohydrate   oxygen   carbon   water   energy
                                      dioxide

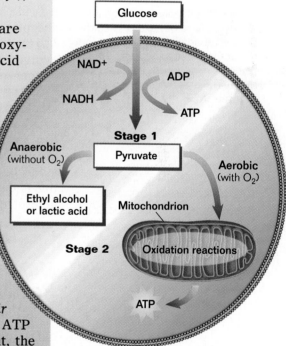

Figure 5-11 Cellular respiration occurs in two distinct stages. First, glucose is converted to pyruvate. The outcome of the second stage depends on the presence of oxygen.

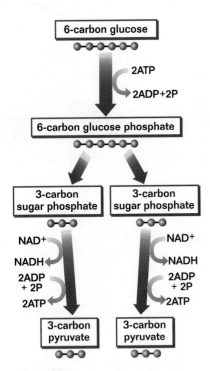

6-carbon glucose

2ATP

2ADP+2P

6-carbon glucose phosphate

3-carbon sugar phosphate

3-carbon sugar phosphate

NAD+

NADH

2ADP + 2P

2ATP

NAD+

NADH

2ADP + 2P

2ATP

3-carbon pyruvate

3-carbon pyruvate

Figure 5-12 Glycolysis, which is the first stage of cellular respiration, is an anaerobic process that occurs in a cell's cytosol. During glycolysis, coupled reactions produce ATP by adding a phosphate group to ADP.

☐ *CAPSULE SUMMARY*

Cellular respiration begins with glycolysis. A cell gains two pyruvate molecules, two ATP molecules, and two NADH molecules for every glucose molecule entering this anaerobic pathway.

Glucose Is Split During Glycolysis

Although other organic molecules can be used to produce ATP for a cell's energy needs, glucose is usually thought of as the starting material for cellular respiration. During the first stage of cellular respiration, glucose is split into smaller molecules in a biochemical pathway called **glycolysis** (*gleye KAHL uh sihs*). The word *glycolysis* is derived from the Greek words *glykys*, meaning "sweet," and *lysis*, meaning "to dissolve." Glycolysis is a sequence of enzyme-catalyzed reactions that converts a six-carbon glucose molecule into two three-carbon molecules of pyruvate, as shown in Figure 5-12.

In addition to pyruvate, glycolysis produces a small amount of ATP and NADH. Coupled reactions that occur during the formation of each molecule of pyruvate produce a total of four ATPs. But because glycolysis also uses two ATPs to get started, there is a net gain of only two ATPs for each molecule of glucose entering the pathway. The formation of each pyruvate molecule also results in the removal of a hydride ion with high-energy electrons. Each hydride ion is donated to a molecule of NAD+, forming two NADH molecules for every glucose molecule that is split. Once a hydride ion (with its electrons) is passed to another hydrogen acceptor, NAD+ can return to glycolysis for more electrons.

The anaerobic reactions of glycolysis are thought to have evolved more than 3 billion years ago, when there was no oxygen gas in Earth's atmosphere. At that time, most life-forms on Earth were probably single-celled heterotrophs that "fed" on the ocean's rich supply of organic molecules. The small amount of ATP produced by glycolysis was sufficient for their energy needs. Although more efficient pathways for producing ATP evolved later, all organisms still use glycolysis to begin cellular respiration. In the absence of oxygen, this ancient energy-extracting pathway is the only way that a heterotrophic cell can harvest energy from food. ☐

Making More ATP Requires Oxygen

When oxygen is available, a series of reactions called **oxidative respiration** follows glycolysis. The reactions of oxidative respiration drive the production of large amounts of ATP. The pathways of oxidative respiration are thought to have appeared after photosynthetic bacteria began to fill the atmosphere with oxygen gas. The presence of this oxygen gas enabled the development of a far more efficient way of using glucose as an energy source.

In preparation for the ATP-producing part of oxidative respiration, three-carbon pyruvate molecules from glycolysis are first converted to a two-carbon fragment. As this takes place, a carbon dioxide molecule and a hydride ion with high-energy electrons are extracted from each pyruvate molecule.

The carbon dioxide, a byproduct of cellular respiration, leaves the mitochondrion and then the cell. The hydride ions are donated to an NAD^+ molecule, forming NADH that is used later in oxidative respiration. The remaining two-carbon fragment is called an acetyl *(uh SEET uhl)* group. As Figure 5-13 shows, this group is attached to a coenzyme (coenzyme A), forming a compound called **acetyl-CoA** *(uh SEET uhl-koh ay)*. If a cell already has a plentiful supply of ATP, acetyl-CoA is funneled into fat synthesis. Thus, high-energy electrons are stored for later needs. If the cell needs ATP immediately, acetyl-CoA is directed to the next stage of oxidative respiration.

The Krebs Cycle

The next phase of oxidative respiration is the **Krebs cycle,** which is a repeating series of reactions that produces ATP, electron carriers, and carbon dioxide. The cycle is named for the biochemist Hans Krebs, who first proposed the cycle in 1937. The cycle's reactions, however, were not worked out until the 1950s. The cycle, which is summarized in Figure 5-13, begins when the two-carbon fragment of acetyl-CoA is attached to a four-carbon molecule found in mitochondria. Once the Krebs cycle is completed, the same four-carbon molecule that began the cycle has been regenerated. High-energy electrons harvested by the cycle are taken away by hydride ions that are attached to NADH and $FADH_2$ (another electron carrier). These electron carriers now hold most of the chemical energy that was previously stored in glucose.

Figure 5-13 During oxidative respiration, pyruvate is first converted to acetyl-CoA, which enters the Krebs cycle. In the Krebs cycle, the acetyl group is added to a four-carbon compound, the starting material. In an additional series of reactions, two carbon atoms are expelled in carbon dioxide, one ATP is generated, high-energy electrons are harvested, and the four-carbon starting material is regenerated.

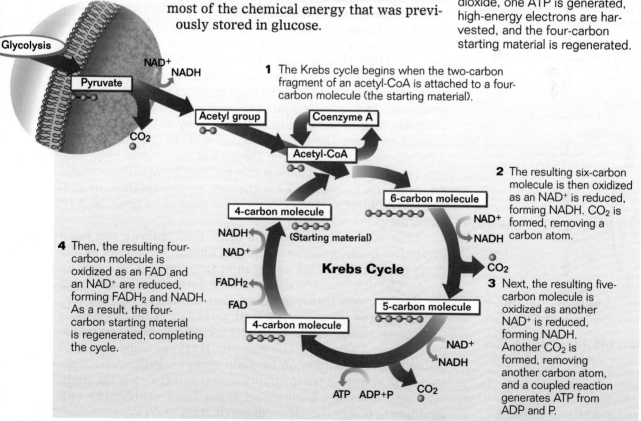

1 The Krebs cycle begins when the two-carbon fragment of an acetyl-CoA is attached to a four-carbon molecule (the starting material).

2 The resulting six-carbon molecule is then oxidized as an NAD^+ is reduced, forming NADH. CO_2 is formed, removing a carbon atom.

3 Next, the resulting five-carbon molecule is oxidized as another NAD^+ is reduced, forming NADH. Another CO_2 is formed, removing another carbon atom, and a coupled reaction generates ATP from ADP and P.

4 Then, the resulting four-carbon molecule is oxidized as an FAD and an NAD^+ are reduced, forming $FADH_2$ and NADH. As a result, the four-carbon starting material is regenerated, completing the cycle.

Glycolysis
Pyruvate
NAD^+ NADH
CO_2
Acetyl group
Coenzyme A
Acetyl-CoA
6-carbon molecule
NAD^+
NADH
CO_2
4-carbon molecule (Starting material)
NADH
NAD^+
FADH2
FAD
4-carbon molecule
Krebs Cycle
5-carbon molecule
NAD^+
NADH
ATP ADP+P
CO_2

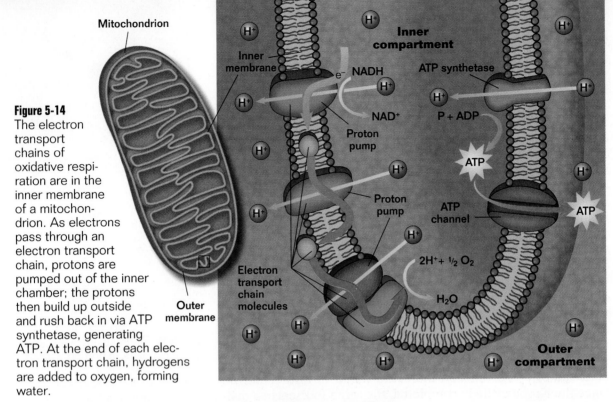

Figure 5-14
The electron transport chains of oxidative respiration are in the inner membrane of a mitochondrion. As electrons pass through an electron transport chain, protons are pumped out of the inner chamber; the protons then build up outside and rush back in via ATP synthetase, generating ATP. At the end of each electron transport chain, hydrogens are added to oxygen, forming water.

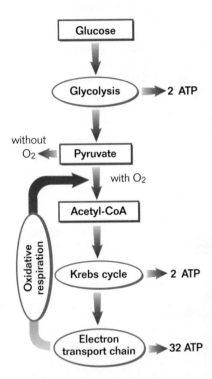

Figure 5-15 The complete oxidation of glucose by cellular respiration begins with glycolysis, which yields a net of only 2 ATPs. The pathways of oxidative respiration yield an additional 34 ATPs.

Electron Transport Chain

The NADH and FADH$_2$ made during the Krebs cycle, and the reactions that precede it, are used to generate the bulk of the ATP made by oxidative respiration. The electrons these molecules carry are passed through a series of membrane-bound proteins—an electron transport chain—by oxidation-reduction reactions. The electron transport chains for oxidative respiration, illustrated in Figure 5-14, are embedded in the inner membranes of mitochondria. In bacteria, they are located in the cell membrane. Each electron transport chain passes high-energy electrons to proton-pumping membrane channels. The energy of these electrons is used to drive the generation of ATP by chemiosmosis. ATP leaves a mitochondrion through other protein channels, entering the cytosol where it can be used by other reactions. As indicated in Figure 5-15, the complete oxidation of glucose by cellular respiration yields a total of at least 36 ATP molecules.

What happens to electrons after their energy has been expended in an electron transport chain? Hydrogen atoms carrying the electrons are joined to oxygen gas, forming water. Oxygen, therefore, acts as the final electron acceptor for the electron transport chains of oxidative respiration. Energy cannot be extracted from pyruvate unless oxygen is present to siphon off the electrons entering electron transport chains. Otherwise, the electron-ferrying components of mitochondria would soon become clogged with spent electrons. As long as a fresh supply of oxygen is available, high-energy electrons harvested from food molecules can continue to power the efficient production of ATP.

Fermentation Occurs in the Absence of Oxygen

If there is no oxygen for oxidative respiration, the pyruvate produced by glycolysis has a different fate. As you read earlier, electrons extracted during glycolysis are carried away by NAD+. In the absence of oxygen, these electrons do not enter electron transport chains but remain attached to their carriers. Soon a cell's NAD+ becomes saturated with electrons. With no more NAD+ available to carry away electrons, the pathways of oxidative respiration "back up," and glycolysis cannot proceed. To continue obtaining energy from food when oxygen is absent, another acceptor must be found for the electrons. Under anaerobic conditions, the electrons from glycolysis are added to organic molecules, a process that is called **fermentation**.

As Figure 5-16 indicates, there are two different products of fermentation among eukaryotes. In animals, such as yourself, lactic acid is produced simply by adding the electrons from glycolysis back to the pyruvate produced by glycolysis. If you lift a heavy weight up and down rapidly 100 times, for example, your muscle cells use up all the available oxygen. Consequently, the muscle cells have only the ATP made by glycolysis for further energy needs. This situation immediately produces a tired feeling in your muscles. Later, lactic acid in your muscle cells causes your muscles to feel sore. However, if oxygen becomes available, lactic acid is converted back to pyruvate, which then can enter oxidative respiration.

Fungi and plants have a different pathway for fermentation. They first convert pyruvate to two-carbon molecules by removing carbon dioxide. The electrons from glycolysis are added to these molecules, producing ethyl alcohol. For centuries, humans have used fungi (yeast) in the preparation of foods and beverages. Wine and beer contain ethyl alcohol made by yeast performing fermentation. Carbon dioxide made by yeast during fermentation causes bread to rise. ∎

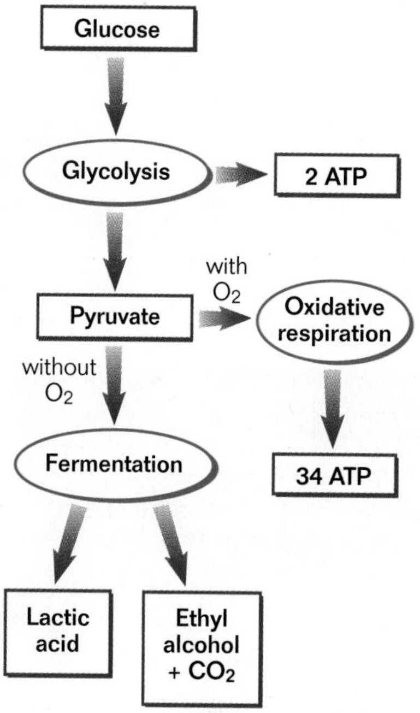

Figure 5-16 Under anaerobic conditions, the pyruvate generated by glycolysis undergoes fermentation, enabling cellular respiration to yield only 2 ATPs.

❏ CAPSULE SUMMARY

When oxygen is present, oxidative respiration follows glycolysis and produces a large amount of ATP. Fermentation follows glycolysis when no oxygen is available for oxidative respiration.

Many Fuels Are Used for Cellular Respiration

Thus far, only glucose has been discussed as a fuel for cellular respiration. The glucose for cellular respiration is obtained by eating carbohydrates such as starch and sugar. But if there are not enough carbohydrates in an organism's diet to supply its energy needs, other molecules must be oxidized to obtain energy. The long carbon chains of fatty acids have many hydrogen atoms, and thus fats provide a rich harvest of energy. In fact, humans obtain more energy from fats than they do from glucose (a gram of fat contains more than twice the energy of a gram of glucose). Although their subunits are normally used for building important cell components, proteins and nucleic acids also can be oxidized by cellular respiration to make ATP. ∎

❏ CAPSULE SUMMARY

Fats, proteins, and nucleic acids, in addition to glucose, can be oxidized by cellular respiration to make ATP.

Cells Control the Rate of Cellular Respiration

A cell's energy-producing machinery operates only when there is a need for energy. The rate of cellular respiration slows down when a cell already has ample supplies of ATP. This is very sensible, but how does a mitochondrion "know" whether to speed up or slow down cellular respiration? The control of cellular respiration works through the system of feedback inhibition illustrated in Figure 5-17. Key reactions early in glycolysis and the Krebs cycle are catalyzed by enzymes that have a regulatory allosteric site that is the same shape as ATP. When ATP levels in a cell are high, it is very likely that ATP molecules will bind to these allosteric sites, causing the enzymes to change shape. The new shape, however, deactivates the enzymes. Thus, high levels of ATP act to shut down the processes a cell uses to make ATP. ▢

□ **CAPSULE SUMMARY**

A system of feedback inhibition enables cells to control the rate of cellular respiration.

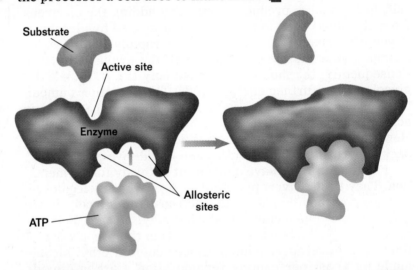

Figure 5-17 Feedback inhibition controls the production of ATP in cells. Excess ATP shuts off a pathway when ATP binds to an allosteric site on a key enzyme, stopping ATP production. As ATP is used by a cell, the enzyme releases the ATP, and ATP production begins again.

Substrate
Active site
Enzyme
Allosteric sites
ATP

Section Review

1. *How does oxygen affect the efficiency of cellular respiration?*
2. *What is the starting material for glycolysis? What is the product?*
3. *What are the three main events of oxidative respiration?*
4. *Using a concept map, indicate how ATP, lactic acid, and ethyl alcohol result from cellular respiration.*

Critical Thinking

5. *What happens when a cell's supply of ATP gets low? Why do you think this happens?*
6. *Considering what you have learned about cellular respiration, what are several things you could do to reduce your amount of body fat?*

Vocabulary

acetyl-CoA (109)
aerobic (107)
anaerobic (107)
ATP synthetase (102)
Calvin cycle (105)
carbon fixation (105)
carotenoids (101)
chlorophyll (101)
consumer (97)
crop (96)
electromagnetic spectrum (100)

electron transport chain (102)
fermentation (111)
food chain (97)
glycolysis (108)
granum (99)
Krebs cycle (109)
$NADP^+$ (103)
NADPH (103)
oxidative respiration (108)
photon (100)
photosystem (101)

photosystem I (101)
photosystem II (101)
pigment (101)
producer (97)
pyruvate (107)
radiant energy (100)
reaction center (101)
reducing power (103)
stroma (99)
thylakoid (99)

Concept Mapping

Construct a concept map that shows how photosynthesis and cellular respiration are related. Use as many terms as needed from the vocabulary list. Try to include the following terms in your map: ATP, photosynthesis, cellular respiration, light reactions, carbohydrates, glycolysis, Krebs cycle, electron transport chain, photosystems, and Calvin cycle. Include additional terms in your map as needed.

Review

Multiple Choice

1. The products that result from photosynthesis and serve as the starting materials for cellular respiration are
   a. carbohydrates and oxygen.
   b. carbon dioxide and water.
   c. NADP and hydrogen.
   d. ATP and water.

2. Carotenoids
   a. cause plants to look green.
   b. are found in the reaction centers of photosystems.
   c. trap light energy that chlorophyll cannot absorb.
   d. do not play a role in photosynthesis in most plants.

3. The thylakoid membranes of a plant cell are the sites where
   a. carbohydrates are formed.
   b. the light reactions occur.
   c. light energy is packaged into photons.
   d. ATP is used to produce NADPH.

4. The end product(s) of the electron transport chains of photosynthesis is (are)
   a. electrons.
   b. water.
   c. glucose.
   d. ATP and NADPH.

5. The oxygen that results from photosynthesis comes directly from the
   a. splitting of carbon dioxide molecules by the Calvin cycle.
   b. splitting of water molecules to provide electrons for photosystem II.
   c. action of proton pumps in the electron transport chains.
   d. absorption of photons by carotenoids in the photosystems.

6. Which of the following is the correct pairing of a metabolic process and its need for oxygen?
   a. conversion of glucose to pyruvate: no oxygen required
   b. fermentation: oxygen required
   c. conversion of pyruvate and NADH to ATP: no oxygen required
   d. synthesis of organic molecules from carbon dioxide: oxygen required

7. Oxidative respiration involves all of the following *except*
   a. conversion of pyruvate to acetyl-CoA.
   b. the Krebs cycle.
   c. the Calvin cycle.
   d. electron transport.

8. The final electron acceptor in cellular respiration is
   a. carbon dioxide.
   b. NADH.
   c. glucose.
   d. oxygen.

9. In cellular respiration, the greatest number of ATPs is generated by
   a. glycolysis.
   b. the Krebs cycle.
   c. the electron transport chain.
   d. fermentation.

10. When a cell's supply of ATP is more than is needed,
   a. the rate of cellular respiration slows because ATP binds to an enzyme's allosteric site.
   b. the rate of cellular respiration increases in order to maintain homeostasis.
   c. the excess ATP is converted to ADP by the Krebs cycle.
   d. ATP enters glycolysis and binds with coenzyme A to make acetyl-CoA.

Completion

11. Both plants and animals use the biochemical pathways of _____ to break down carbohydrates formed during _____ .

12. _____ are plants that are grown for use by humans. As producers, these plants are the first step in many food chains, through which _____ from the sun is distributed to other organisms.

13. Photosynthesis begins when light is absorbed by _____ , which are arranged in clusters called photosystems. The principal light-absorbing compound in green plants is _____ .

14. In photosynthesis, the Calvin cycle uses _____ and _____ from the light reactions and produces _____ .

15. Light is a form of radiant energy that is packaged in units called _____ .

16. The product of fermentation in animals is _____ , and the product in plants is _____ and _____ .

17. Besides glucose, the possible energy sources for cellular respiration include _____ and _____ .

Short Answer

18. How are food chains important to the organisms that are consumers?

19. What happens in the Calvin cycle? Why are these reactions neither dark reactions nor light-independent reactions?

20. How could the availability of water affect the rate of photosynthesis?

21. Contrast the aerobic and anaerobic pathways of cellular respiration.

22. How are the processes of photosynthesis and cellular respiration alike?

Themes Review

23. **Flow of Energy** Why is it said that all life depends on autotrophs?

24. **Levels of Organization** Trace the movement of light energy through the structures of a leaf to the point where electrons are excited for photosynthesis.

25. **Evolution** The electron transport chains of oxidative respiration are located in the inner membranes of mitochondria in protists, plants, and animals. They are located in the cell membrane of bacteria. How might this difference indicate that mitochondria might have originated from bacteria?

Critical Thinking

26. **Making Predictions** In an experiment, several plants were placed in each of two growth chambers, one with red light and one with green light. Which plants should grow better? Why?

27. Interpreting Data In an experiment conducted to test the effect of light intensity on the rate of photosynthesis, light intensity was varied by moving a light source to different distances from a water plant. The rate of photosynthesis was estimated by counting the number of oxygen bubbles generated per minute. Data collected during the experiment are presented in the graph below. What is the relationship between light intensity and the amount of oxygen produced? If light intensity were increased, would oxygen production continue to increase?

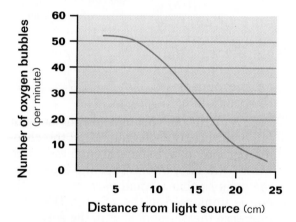

Distance from light source (cm)

28. Making Inferences Evidence indicates that Earth's early atmosphere contained no oxygen. What types of metabolism would have been possible? What is the source of the oxygen present in Earth's atmosphere today?

29. Making Inferences Bread rises as a result of yeast fermentation. How might fermentation cause bread to rise? Why is sugar often added to bread dough?

Activities and Projects

30. Cooperative Group Project Design and produce a mural that shows how the processes of photosynthesis and cellular respiration link all organisms.

31. Cooperative Group Project Prepare a script for a three-act play that shows what takes place in the three stages of photosynthesis. With the other members of your group, perform the play for your class.

32. Multicultural Perspective For thousands of years, the seeds of the buffalo gourd, *Cucurbita foetidissima*, have been a source of oil for Native Americans. Research the characteristics of this plant, its uses in ancient Native American cultures, and its potential importance to modern society. In a written report or an oral report, relate your findings and explain why this plant is such a rich source of oil and other useful products.

33. Research and Writing Research several other ways, besides bread making, that fermentation is used in food preparation. Relate your findings in a written report.

Readings

34. Read Arthur W. Galston's article "Photosynthesis as a Basis for Life Support on Earth and in Space," in *BioScience*, July/August 1992, pages 490–493. What are the components of a CELSS system? How might the NASA-sponsored CELSS program help solve the problems humans face because of overpopulation?

35. Read James Utley's article "Chemistry that comes naturally," in *New Scientist*, July 31, 1993, pages 24–28. How do electrochemical reactions resemble the reactions that occur during cellular respiration and photosynthesis? How is electrochemistry being used to more safely make industrial chemicals and destroy toxic wastes?

Chromatography of Plant Pigments

Doing Side of the Vee

1. Put on a lab apron. Use scissors to cut the bottom end of a chromatography paper strip to a tapered end. **CAUTION: Be careful when using the scissors.**

OBJECTIVES

- Separate and observe the pigments that give a leaf its color.
- Describe the function of plant pigments during photosynthesis.

PROCESS SKILLS

- collecting data
- calculating ratios

2. Draw a faint pencil line 1 cm above the pointed end of the paper strip (now called a chromatogram). Use a capillary tube to apply a tiny drop of the simulated plant pigments extract on the center of the pencil line.

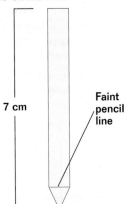

7 cm

Faint pencil line

MATERIALS

- lab apron
- Ward's Chromatography of Simulated Plant Pigments Kit:
  1 vial of simulated plant pigments extract
  1 bottle of chromatography solvent
  capillary tubes
  chromatography paper strips
  chromatography reaction chambers
- metric ruler
- scissors
- 10 mL graduated cylinder
- tape or glue

3. Pour 5 mL of the chromatography solvent into your reaction chamber. Pull the chromatography paper strip through the opening of the cap, and adjust the length of the strip so that a small portion of the tip end is immersed in the solvent. **DO NOT** immerse the pigment spot in the solvent.

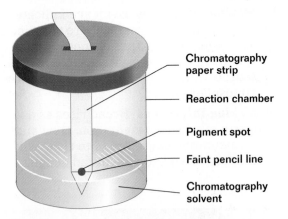

Chromatography paper strip

Reaction chamber

Pigment spot

Faint pencil line

Chromatography solvent

BACKGROUND

1. What is a pigment? What pigments are found in the leaves of plants?
2. What color are chlorophylls? carotenes? xanthophylls?
3. Write your own **Focus Question** on your Vee Form.
4. **Knowing Side of the Vee** List the **Concepts** and new **Vocabulary Words** on your Vee Form. In the **Concept Statements** section of the Vee, use these words in sentences that define and explain them.

4. Place the cap over the reaction chamber, and carefully bend the end of the strip over the cap. Be sure that the chromatogram is level and does not touch the walls of the reaction chamber.

5. Within 5–7 minutes you will notice bands of different colors—orange, yellow, and two shades of green—traveling up the strip. Remove the chromatogram from the solvent when the solvent nears the top of the reaction chamber.

6. With a pencil, mark the position of the uppermost end of the solvent. Then mark the farthest distance that each of the separated pigments moved. Measure the distances that the solvent and each of the pigments moved. In the **Records** section of your Vee Form, make a table similar to the one below, and record your measurements. Tape or glue your chromatogram to the **Records** section. Label the pigment colors.

7. Clean up your materials and wash your hands before leaving the lab.

Chromatography of Plant Pigments

Band no.	Pigment color	Migration (distance in mm)	R_f value
1 (top)			
2			
3			
4			
Solvent			

8. Use the formula below to calculate the R_f value.

$$R_f = \frac{\text{Distance substance (pigment) traveled}}{\text{Distance solvent traveled}}$$

The R_f value is a number that represents the ratio of the distance that the pigment moved relative to the distance that the solvent moved. Since the R_f value is constant for each pigment type, it is used to help scientists identify compounds. In the **Records** section of your Vee Form, record a decimal fraction for the R_f of each pigment. In the **Procedure** section of the Vee, briefly summarize the procedure you followed.

INQUIRY

1. What is a chromatogram?

2. The word *chromatography* comes from the Greek words *chromat*, which means "color," and *graphon*, which means "to write." Write a descriptive sentence that explains what happened to your plant pigments during the lab.

3. How do your R_f values compare with those of your classmates?

4. Which color moved the farthest distance? the shortest distance?

5. Use the information on the **Knowing Side** of the Vee to interpret your results from the **Doing Side,** and then write your **Knowledge Claim.** Write a **Value Claim** for this lab.

ANALYSIS

1. Using the data from your lab, what pigments do you think would be indicated by an R_f value of .92? of .42?

2. What is the name of the main pigment used during photosynthesis?

3. At what time during the year do the other pigments become visible? Why?

4. List other uses for chromatography.

FURTHER INQUIRY

Write a **New Focus Question** that could be the point of a new investigation. The following is an example:

Design an experiment that uses various solvents to separate pigments found in substances such as fruit juices, writing ink, and dyes. Then calculate the R_f values.

CHAPTER 6

CELL REPRODUCTION

REVIEW

- characteristics of cells (Section 2-1)
- structure of proteins (Section 2-3)
- DNA and chromosomes (Section 2-4)

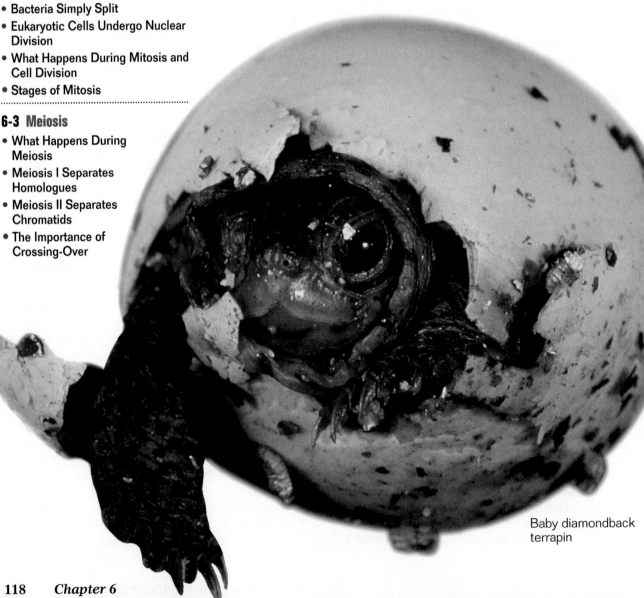

Baby diamondback terrapin

6-1 Chromosomes

Your body is composed of approximately 100 trillion cells. The intricate organism that you are today is the result of millions of cell divisions. Each cell division is a carefully orchestrated performance carried out by many kinds of structures. In this section you will meet one of these structures, the chromosome, and learn how it is transmitted from one generation to the next.

Chromosome Structure

DNA (deoxyribonucleic acid) is a long, thin molecule that contains the information needed to direct a cell's activities and to determine a cell's characteristics. This vast amount of vital information encoded in DNA is organized into genes. A **gene** is a segment of DNA that transmits information from parent to offspring. A single molecule of DNA has thousands of genes, which are lined up like the railroad cars of a train. When genes are being used, the strand of DNA is extended, enabling other molecules to retrieve its information. However, when a cell prepares to divide, the DNA molecule coils and twists into a dense structure called a chromosome. A **chromosome** is a rod-shaped structure that forms when a single DNA molecule and its associated proteins coil tightly before cell division.

If DNA did not coil and form a chromosome, a single DNA strand would be about 5 cm (approximately 2 in.) long. This is too long to fit inside a cell. But when the thread of DNA is coiled around a protein scaffold, it can be compacted into a smaller, more manageable structure, as shown in Figure 6-1. Your chromosomes are approximately 40 percent DNA and 60 percent protein.

Figure 6-1 A chromosome's scaffold is made of proteins called histones. Histones are rich in positively charged amino acids. Because DNA is a negatively charged molecule, it wraps tightly around the histones, causing the molecule to coil.

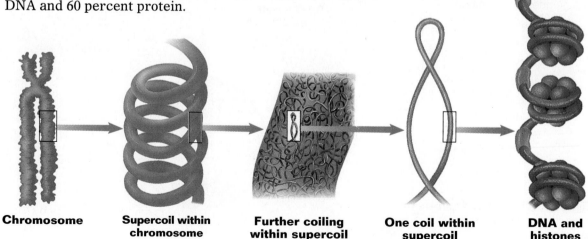

Histones

DNA

Chromosome   Supercoil within chromosome   Further coiling within supercoil   One coil within supercoil   DNA and histones

Cell Reproduction **119**

Chromosomes become visible through a microscope only after they have condensed prior to cell division. By this time, each chromosome has formed a copy of itself. The two copies of each chromosome are called **chromatids** (*KROH muh tihdz*). Chromatids form prior to cell division when the DNA molecule duplicates itself, ensuring that each new cell will have the same genetic information as the old cell. The two chromatids are attached by a protein disk at a point called a **centromere** (*SEHN troh mihr*), shown in Figure 6-2. □

Human cells have 23 different chromosomes. Your body cells (also called somatic cells) contain two copies of each chromosome, for a total of 46 chromosomes. The two copies of each chromosome are called **homologous** (*hoh MAHL uh gus*) **chromosomes**, or homologues. Homologous chromosomes are similar in shape and size and have similar genetic information. You received one homologue of each chromosome from your mother and the other homologue from your father. When a cell contains two homologues of each chromosome, it is termed **diploid** (*DIHP loyd*). Biologists use the symbol $2n$ to represent the diploid number of chromosomes in a cell. For humans, $2n = 46$.

Not all cells are diploid. In the life cycle of animals, including humans, **gametes** (*GAH meets*)—egg cells and sperm cells—are haploid. A **haploid** (*HAP loyd*) cell contains only one homologue of each chromosome. The fusion of two haploid gametes forms a diploid zygote, as shown in Figure 6-3. A **zygote** (*ZY goht*) is a fertilized egg cell, the first cell of a new

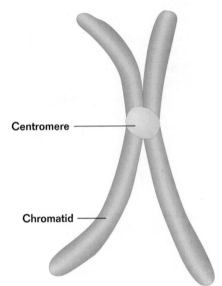

Centromere

Chromatid

Chromosome

Figure 6-2 An easy way to determine the number of chromosomes in a cell is to count the number of centromeres. The number of centromeres will equal the number of chromosomes.

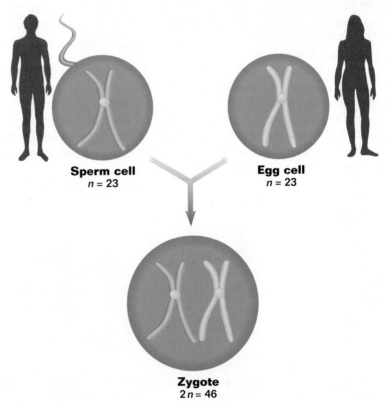

Sperm cell
$n = 23$

Egg cell
$n = 23$

Zygote
$2n = 46$

Figure 6-3 When haploid gametes fuse, they produce a diploid zygote.

Table 6-1 Chromosome Number of Various Organisms

Organism	Number
Fungi (haploid)	
Penicillium	1–4
Saccharomyces	18
Insect (diploid)	
Mosquito	6
Drosophila	8
Housefly	12
Plants (number of chromosome sets per cell varies)	
Garden pea	14
Corn	20
Sugarcane	80
Adder's tongue fern	1,262
Vertebrates (diploid)	
Frog	26
Mouse	40
Human	46
Chimpanzee	48
Orangutan	48
Gorilla	48
Horse	64
Dog	78
Duck	80

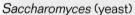

Saccharomyces (yeast)

Corn plant

Drosophila (fruit fly)

Orangutan

individual. Being haploid ensures that when an egg and a sperm fuse, the resulting zygote will contain the characteristic diploid number of chromosomes for that organism. Biologists use the symbol n to represent the haploid number of chromosomes. For humans, $n = 23$. Table 6-1 lists the characteristic chromosome number of a variety of organisms.

☐ CAPSULE SUMMARY

The number of chromosomes in cells is constant within a species. Haploid cells contain one homologue of each chromosome. Diploid cells contain two homologues of each chromosome.

Chromosomes Affect Development

Each of the 46 human chromosomes has thousands of genes that play important roles in determining how a person's body develops and functions. All these genes must be present in an individual's cells for the same reason that a car must have an engine, a transmission, and wheels—to function properly. Therefore, a person must have the characteristic number of chromosomes in his or her cells. In most cases, humans who are missing even one chromosome do not

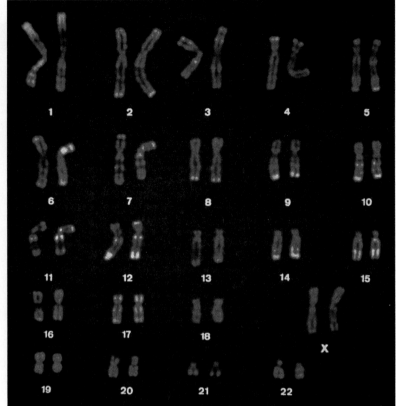

Figure 6-4 To examine an individual's karyotype, investigators chemically treat and stain the chromosomes in cells from a blood sample. Afterward, the chromosomes are usually photographed, cut out, arranged in pairs from largest to smallest, and numbered, *above right*. People with Down syndrome, *bottom left*, have three copies of chromosome 21 in their karyotypes, *top left*.

survive embryonic development. The condition in which a diploid cell is missing a chromosome is called monosomy *(MAHN uh soh mee)*. And just as a car will not function correctly with two engines stuffed under the hood, a human embryo will not develop properly with more than two copies of most chromosomes. The condition in which a diploid cell has an extra chromosome is called **trisomy** *(try SOH mee)*.

Deviations in chromosome number can be detected by analyzing a **karyotype** *(KAR ee uh typ)*, the collection of chromosomes found in an individual's cells. Figure 6-4 shows a typical karyotype.

Figure 6-4 also shows a karyotype from an individual with an extra copy of chromosome 21. The traits produced by having an extra copy of chromosome 21 were first described in 1866 by the British physician J. Langdon Down and are collectively called **Down syndrome**, or trisomy 21 syndrome. The features that characterize Down syndrome include a short stature, a round face with upper eyelids that cover the inner corners of the eyes, and, most significantly, varying degrees of mental retardation.

Down syndrome occurs in all racial groups with the same frequency, approximately 1 in 1,000 children. It is much more common in children of older mothers. In mothers younger than 30 years old, the incidence is only about 1 in 1,500 births, while in mothers 30 to 35 years old, the incidence doubles to 1 in 750 births. In mothers older than 45, the risk is as high as 1 in 16 births. The reason that more babies with Down syndrome are born to older mothers is that all the eggs a female will ever produce are present in her ovaries when she is born. As the female ages, the eggs can accumulate an increasing amount of damage; males, in contrast, produce new sperm throughout adult life.

What events cause an individual to have an extra copy of a chromosome? When a cell divides normally, each chromosome and its homologue separate, an event called disjunction. When normal disjunction does not occur, one or more chromosomes may fail to separate properly. This accident in chromosome separation is called **nondisjunction** *(nahn dihs JUHNK shuhn)*, which results in one new cell receiving both chromosomes and the other new cell receiving none. Trisomics arise as a result of nondisjunction. In the case of Down syndrome, nondisjunction occurs with chromosome 21.

Prenatal Testing

Because of the risk of Down syndrome, a pregnant woman over the age of 35 may be advised to have prenatal testing to check for an extra copy of chromosome 21 in the fetus. This can be done through procedures such as amniocentesis and chorionic villi sampling, which are illustrated in Figure 6-5.

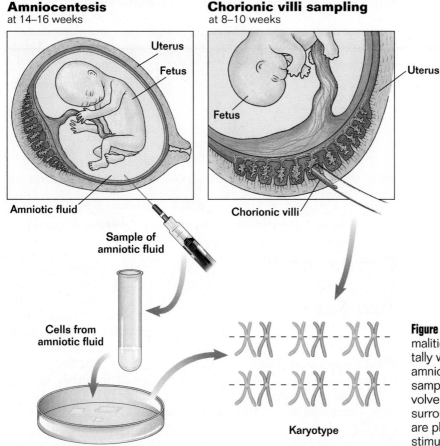

Amniocentesis
at 14–16 weeks

Uterus

Fetus

Amniotic fluid

Sample of amniotic fluid

Cells from amniotic fluid

Cell culture

Chorionic villi sampling
at 8–10 weeks

Fetus

Uterus

Chorionic villi

Karyotype

Figure 6-5 Chromosome abnormalities can be diagnosed prenatally with procedures such as amniocentesis and chorionic villi sampling. Both procedures involve taking cells from tissues surrounding the fetus. The cells are placed in a culture medium to stimulate growth, and then karyotypes are prepared.

Deviations in a chromosome number can cause abnormal development. In humans, Down syndrome is caused by an extra copy of chromosome 21, which is a result of nondisjunction.

In **amniocentesis** *(am nee oh sehn TEE sihs)*, a physician uses a needle and syringe to remove a small amount of the fluid from the amnion, the sac that surrounds the fetus. Fetal cells in the amniotic fluid are used to make a karyotype, which can then be analyzed. In **chorionic villi sampling**, a physician analyzes a karyotype made using cells grown from a sample of the chorionic villi, fingerlike extensions of the placenta that grow into the mother's uterus. Since the villi have the same genetic makeup as the fetus, the physician is able to detect abnormalities in the chromosome number. □

Alterations in Chromosome Structure

Although rare, changes in an organism's chromosome structure do occur. Some alterations cause **mutations**, changes in an organism's genetic material. Some of these mutations are illustrated in Figure 6-6. When a fragment of a chromosome breaks off, it can be lost when the cell divides, causing a mutation called a **deletion**. As a result of deletion, a new cell will lack a certain set of genes. In a mutation called a **duplication**, the chromosome fragment attaches to its homologous chromosome, which will then carry two copies of a certain set of genes. Sometimes the fragment reattaches to the original chromosome in the reverse orientation, producing a mutation called an **inversion**. Or, the fragment may join a nonhomologous chromosome, in an event called **translocation**.

Figure 6-6 When a chromosome breaks, its genes can become rearranged in different ways. Notice how each kind of mutation affects the shaded genes shown below.

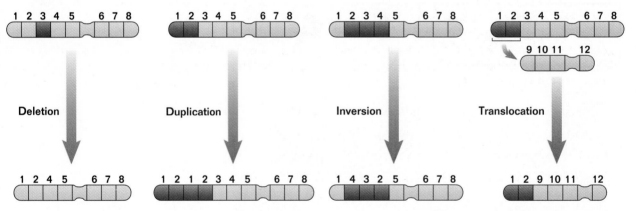

Chromosomes Determine Your Sex

Of the 23 pairs of chromosomes in human somatic cells, 22 pairs are the same in males and females. These chromosomes are called **autosomes** *(AWT uh sohmz)*. The chromosomes that differ between males and females are called the **sex chromosomes** because they carry the genes that determine an individual's sex. Sex chromosomes exist in either of two forms—as an X chromosome or as a shorter Y chromosome. In humans and many other organisms, the genes that cause a fertilized egg to develop into a male are located on the Y chromosome. Thus, any individual with a Y chromosome is a male, and any individual without a Y chromosome is a female.

Table 6-2 Sex Chromosome Abnormalities

Sex	Abnormality	Effects	Occurrence
Male	XXY	Klinefelter's syndrome: longer than average limbs, sparse body hair, underdeveloped genitalia, slight breast development; sterile	1 per 800–1,000 male births
	XYY	Some studies suggest increased risk of antisocial behavior; fertile	1 per 1,000 male births
Female	XO	Turner's syndrome: short stature, pronounced webbing of the neck, undeveloped secondary sexual characteristics; sterile	1 per 2,000 female births
	XXX	Triplo-X syndrome: most are of normal intelligence and are fertile	1 per 1,200 female births

In these cases, females are designated XX because they have two X chromosomes, and males are designated XY because they have one X chromosome and one Y chromosome. Because a female can donate only an X chromosome to her offspring, the sex of an offspring is determined by the male, who can donate either an X or a Y. Table 6-2 explains the abnormalities that arise when a person has an abnormal number of sex chromosomes.

In some insects, such as grasshoppers, there is no Y chromosome. In such cases, the females are characterized as XX and the males as XO (the O indicates the absence of a chromosome). In birds, moths, and butterflies, the male has two X chromosomes and the female only one. ❏

❏ *CAPSULE SUMMARY*

Sex chromosomes carry information that determines an organism's sex. In humans and many other organisms, females are designated XX, and males are designated XY.

Section Review

1. *When are chromosomes visible in a cell?*
2. *If a cell that is about to divide has 22 chromosomes, how many chromatids does it have? How many centromeres does it have?*
3. *Are homologous chromosomes normally found in gametes? Explain why or why not.*
4. *What is trisomy? How does trisomy arise in a cell?*
5. *Describe the karyotype of an individual with Down syndrome. What are the effects of this deviation?*
6. *What are sex chromosomes? In what combination do they exist in human females? in human males?*

Critical Thinking

7. *Why might two organisms have the same number of chromosomes but not the same traits?*

6-2 Mitosis and Cell Division

Section Objectives

- Discuss the two stages of cell division in bacteria.
- Define cell cycle, and explain what occurs during each of its phases.
- Explain the events of mitosis, and describe the structures involved in them.

*F*or cells, reproducing by cell division is essential for life. What triggers a cell to divide? Cell biologists have observed that the size of a cell seems to be the signal that stimulates it to divide. Recall that as a cell grows, its surface area becomes too small to enable the necessary amounts of nutrients to enter the cell and to allow wastes to pass out. Therefore, as the surface-area-to-volume ratio becomes relatively small, the cell may divide. In this section, you will explore cell division by first looking at how the simplest kinds of cells—bacteria—divide.

Bacteria Simply Split

A bacterium is a single, prokaryotic cell. It has a cell wall but lacks a nucleus and membrane-bound organelles. A bacterium's single DNA molecule is not coiled around proteins to form chromosomes. Instead, its DNA is a circular chromosome, attached to the inner surface of the plasma membrane like a rope attached to the inner wall of a tent. For these tiny organisms, cell division takes place in two stages: first the DNA is copied, and then the cell splits.

Before a bacterium can divide, it must have two copies of its DNA so that each of the two new cells will have a complete copy of the genetic information of that bacterium. To make a copy of itself, the DNA molecule begins to "unzip" lengthwise, exposing its two strands. As this happens, each strand is made into a complete DNA molecule. Shortly after the entire DNA circle has unzipped, the cell has two identical copies of its hereditary information, attached side by side to the interior plasma membrane, as illustrated in Figure 6-7.

Once the DNA has been copied and the cell has grown to an appropriate size, the bacterium splits into two equal halves through a process called binary fission. **Binary fission** is a form of asexual reproduction that produces identical offspring. First, a new plasma membrane is added at a point on the membrane between the two DNA copies. As new material is added in this zone, the growing plasma membrane pushes inward and the cell is constricted in two, like a long balloon being squeezed around its middle. A new cell wall forms around the new membrane. Eventually the dividing bacterium is pinched into two independent cells. Each cell contains one of the circles of DNA and is a complete, functioning bacterium.

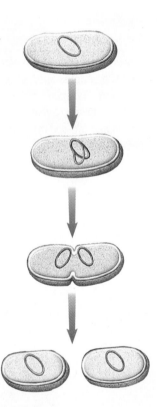

Figure 6-7 During binary fission, a bacterium makes a copy of its DNA and forms a new membrane and cell wall. Afterward, the cell splits into two cells.

Eukaryotic Cells Undergo Nuclear Division

The cells of eukaryotic organisms (protists, fungi, plants, and animals) have a far more complex internal structure than that of the simple cells of bacteria. Cell division in eukaryotic cells must take into account the nucleus with the chromosomes inside and the many other internal organelles, all of which must be strategically maneuvered before the cell can properly divide.

Cell Cycle

The life of a eukaryotic cell is traditionally diagrammed as a **cell cycle,** a repeating sequence of growth and division through which many kinds of eukaryotic cells pass. The five phases of the cell cycle are shown in Figure 6-8. The cell cycle may be summarized as follows:

$$G_1 \rightarrow S \rightarrow G_2 \rightarrow M \rightarrow C$$

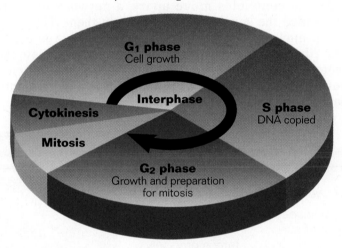

Figure 6-8 In the cell cycle, growth phases, *shaded blue,* alternate with a cell division phase, *shaded brown.*

1. The G_1 phase is the growth phase of a cell. During this phase, a cell grows rapidly and carries out its routine functions. For most organisms, this phase occupies the major portion of the cell's life between cell divisions.

2. The S phase is when the DNA is copied. At the end of this phase, an individual chromosome consists of two chromatids attached at the centromere.

3. In the G_2 phase, preparations are made for nuclear division. Mitochondria and other organelles replicate. Microtubules are reassembled; they will be used to form the spindle apparatus that moves the chromosomes.

4. The M phase is the phase in which mitosis occurs. **Mitosis** *(my TOH sihs)* is the process by which the nucleus of a cell is divided into two nuclei, each with the same number and kinds of chromosomes.

5. The C phase is when the cytoplasm divides during a process called **cytokinesis** *(syt oh kih NEE sihs).*

What Happens During Mitosis and Cell Division

A eukaryotic cell spends most of its life in the G_1, S, and G_2 phases, which are collectively called **interphase.** During interphase, a cell does a great deal of growing. Chromosomes are loosely wound, and genes are being used to make the enzymes that direct the activities of the cell.

Forming Spindle Fibers

As interphase ends and mitosis begins, the chromosomes begin to condense. Enzymes in the cell begin to break down the nuclear envelope. The cell starts building the equipment that will move copies of the chromosomes to opposite ends, or poles, of the cell. In the center of an animal cell, a pair of cylindrical structures called centrioles *(SEHN tree ohlz)* start to separate, each moving toward opposite poles of the cell. As the centrioles move apart, a network of protein cables, called the spindle, forms between them. The spindle will help move chromosomes apart. Each cable is called a **spindle fiber** and is made of microtubules, long hollow tubes of protein. Recall that plant cells do not have centrioles. However, plant cells form a spindle that is almost identical to that of an animal cell, as shown in Figure 6-9.

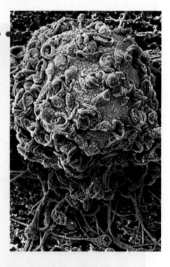

Because evolution has shaped life to meet the challenges of survival, it should come as no surprise that biological structures are very well suited to their functions. The excellent correspondence between structure and function will be seen time and again in your journey through biology. You have already seen how the shape of an enzyme is intimately related to the chemical reaction it carries out, and that the shapes of the proteins protruding from a cell's surface determine in large measure what goes on inside the cell. You have also seen how mechanisms within the cell precisely sort and distribute the cell's genetic materials at the time of cell division.

As you learn about the diversity of life on Earth, you will encounter a parade of structures, some bizarre and some common- place. Each structure is precisely adapted to carry out its particular function, from the harpoon-like nematocysts that jellyfish use to capture prey, to the waterproof coverings that enable birds and reptiles to lay their eggs on dry land. Of all the lessons of biology, the relationship between structure and function is one of the most fundamental.

When a cell divides, top left, or an egg and sperm fuse, above right, chromosomes, bottom left, ensure that an organism's genetic information will be safely transmitted to new cells.

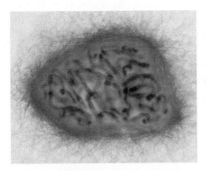
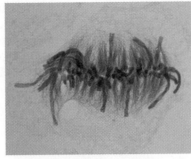

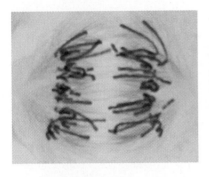
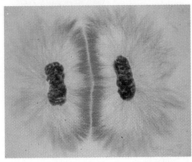

Attaching Spindle Fibers to Centromeres

As the chromosomes continue to condense, a second group of microtubules extends out from a region of the centromere of each chromosome called the **kinetochore** (*kuh NEHT uh kawr*). The kinetochore is a disk of protein that serves as a platform for assembling the microtubules. The two sets of microtubules extend out toward opposite poles of the cell. Each set of microtubules continues to grow longer until it makes contact with the pole of the spindle. When the process is complete, as shown in Figure 6-9, one chromatid is attached by a set of microtubules to one pole, and the other chromatid is attached to the other pole.

Separating Chromatids

Once the microtubules are attached to the centromeres, the centromeres split, freeing the chromatids from each other. Mitosis is now simply a matter of reeling in the microtubules and dragging the chromatids, each now considered a chromosome, to the poles, as shown in Figure 6-9. At the poles, the ends of the spindle fibers are dismantled bit by bit. As the fibers become shorter and shorter, the attached chromosomes move closer and closer to the poles. When they finally arrive, each pole has one complete set of chromosomes.

Dividing the Cell

In the final step of cell division, a new nuclear envelope forms around each pole, forming two nuclei, and the chromosomes within uncoil. Cytokinesis then takes place; the cytoplasm of the cell divides in half, as shown in Figure 6-9. Each half includes one of the two new nuclei and an assortment of organelles, which replicated earlier in the cell cycle. ◻

Figure 6-9 In preparation for division, this cell from an African blood lily, *top left*, has formed a spindle. Then, spindle fibers attach to centromeres, *top center*. Next, the chromatids in this cell begin to separate, *bottom left*. Cell division produces two new cells with identical genetic information, *bottom center*.

◻ *CAPSULE SUMMARY*

The life of a eukaryotic cell includes periods of growth, replication, and division. During interphase, the DNA and the cellular organelles are copied, During mitosis, the nucleus of the cell divides, forming two nuclei with identical genetic information. The cytoplasm in the cell splits during cytokinesis.

Stages of Mitosis

Biologists traditionally divide the process of mitosis into four general stages: prophase, metaphase, anaphase, and telophase. These four stages are illustrated in Figure 6-10, which shows mitosis and cytokinesis in a cell from a whitefish.

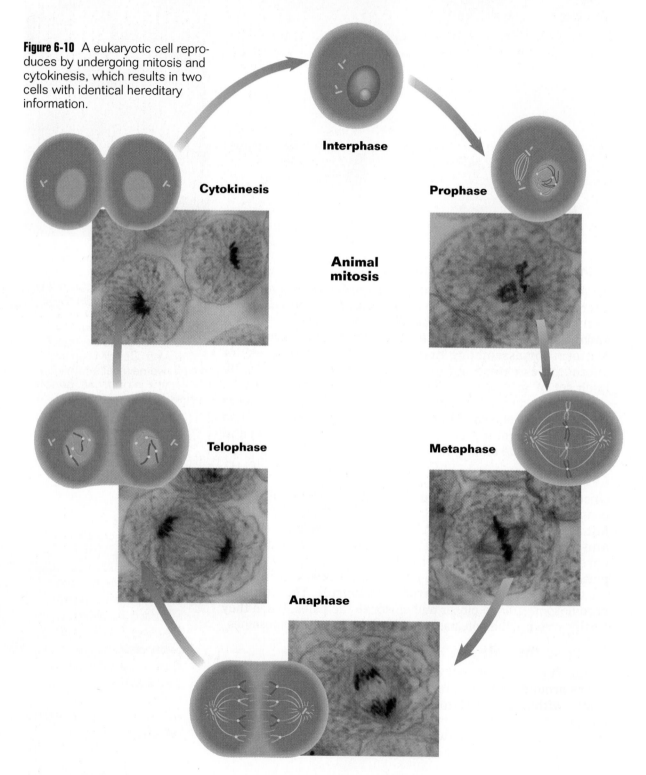

Figure 6-10 A eukaryotic cell reproduces by undergoing mitosis and cytokinesis, which results in two cells with identical hereditary information.

Interphase

Cytokinesis

Prophase

Animal mitosis

Telophase

Metaphase

Anaphase

Prophase

During prophase, chromosomes begin condensing and become visible. The nuclear envelope surrounding the nucleus begins to break down, and the network of spindle fibers becomes visible.

Metaphase

In metaphase, the chromosomes move to the center of the cell and line up along the "equator." Once in place at the equator of the cell, each chromosome is held in place by the microtubules attached to the kinetochore.

Anaphase

In anaphase, the two chromatids physically separate when the centromere divides. The chromatids, each which now may be called a chromosome, move toward opposite poles of the cell as the fibers attached to them shorten.

Telophase

The chromosomes, now at opposite ends of the cell, uncoil. A new nuclear envelope forms. The spindle fibers break down and disappear, and mitosis is complete.

As mitosis ends, cytokinesis begins. During cytokinesis, the cytoplasm of the cell is cleaved in half, and the cell membrane grows to enclose both cells. Animal cells and other cells that lack cell walls are pinched in half by a belt of protein threads. Plant cells have a rigid cell wall and a different strategy of cell division. In plant cells, vesicles formed by the Golgi bodies fuse at the equator of the cell and form the cell plate, a membrane across the middle of the cell. A new cell wall then forms on both sides of the cell plate. Once the cell wall is complete, there are two new plant cells. ❑

❑ *CAPSULE SUMMARY*

Mitosis can be divided into four stages: prophase, metaphase, anaphase, and telophase. Cytokinesis is the division of the cytoplasm.

Section Review

1. *How does the structure of bacterial DNA differ from the structure of your DNA?*

2. *Describe the process in which bacterial DNA copies itself.*

3. *During which phase of the cell cycle does DNA replicate?*

4. *During which stage of mitosis do spindle fibers first appear? How do they function during metaphase? during anaphase?*

Critical Thinking

5. *What is cytokinesis? How does cytokinesis differ between plant cells and animal cells?*

6. *Why does it take more time for a cell to complete mitosis than to complete cytokinesis?*

6-3 *Meiosis*

You have just learned that bacteria reproduce asexually by binary fission. Humans, like most animals and plants, reproduce sexually. In sexual reproduction, gametes from the opposite sexes unite to form a zygote. Because the zygote is diploid, the number of chromosomes in the gametes must be halved. Otherwise, the number of chromosomes in a zygote would be twice the diploid number for the species. In this section, you will learn how the correct number of chromosomes is maintained from generation to generation.

Section Objectives

- *Define sexual reproduction.*
- *Summarize the events that happen during meiosis I and meiosis II.*
- *State the outcome of meiosis.*
- *Define crossing-over and explain how it affects evolution.*

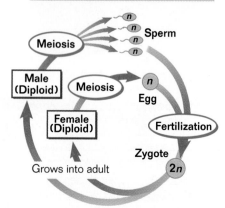

Figure 6-11 The figure above shows the pattern of alternation between the diploid and haploid chromosome number in a sexually reproducing organism such as yourself.

What Happens During Meiosis

The mechanism that halves the number of chromosomes in cells is a form of cell division called **meiosis** (*my OH sihs*). Meiosis consists of two successive nuclear divisions. Before the first division, the DNA is copied, just as it is before mitosis. In the first division, called meiosis I, homologous chromosomes separate into two cells. In the second division, called meiosis II, the two chromatids of each chromosome separate into two haploid cells. Thus, one diploid cell that undergoes meiosis produces four haploid cells. In animals, meiosis often results in haploid gametes, as shown in Figure 6-11. In plants, meiosis often leads to **spores**, haploid cells that later lead to the production of gametes.

During meiosis, two unique events occur:

1. **Crossing-over** In the beginning of meiosis I, homologous chromosomes pair up next to each other. While paired,

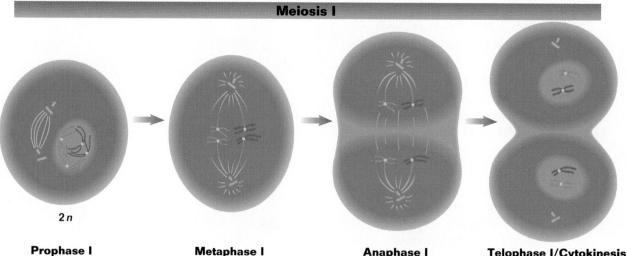

Meiosis I

2*n*

Prophase I   Metaphase I   Anaphase I   Telophase I/Cytokinesis

the arms of the chromosomes exchange reciprocal segments of DNA in a process called **crossing-over.**

2. **Skipping replication** Because there is only one replication of DNA but *two* divisions, meiosis halves the number of chromosomes.

Meiosis I Separates Homologues

The two nuclear divisions of meiosis are traditionally divided into eight stages, as illustrated in Figure 6-12. Although these stages have the same names as those of mitosis, the events differ in significant ways.

Prophase I

As in prophase of mitosis, the chromosomes condense, and the nuclear envelope breaks down. However, a unique event not seen in mitosis occurs: first the homologous chromosomes pair up, and then crossing-over occurs.

Metaphase I

In metaphase I, the pairs of homologous chromosomes are moved by spindle fibers to the equator of the cell. The homologues, each made up of two chromatids, remain together.

Anaphase I

In anaphase I, the homologues separate. As in mitosis, the chromosomes of each pair are pulled by action of the spindle fibers to opposite poles of the cell. The difference between anaphase of mitosis and anaphase of meiosis I is that in the latter, the chromatids do not separate at their centromeres. Each chromosome is still composed of two chromatids joined by a centromere. During this phase of meiosis, nondisjunction can occur.

Figure 6-12 Meiosis produces four cells, each with half as much genetic material as the original cell, *below across.*

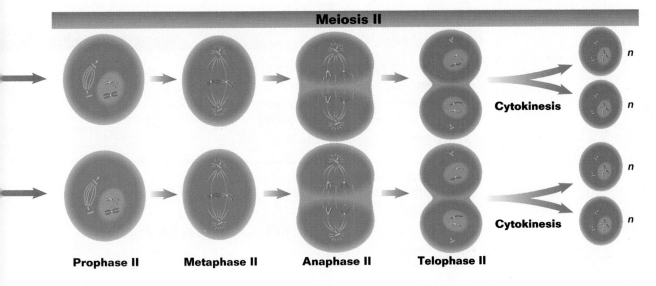

Meiosis II

Cytokinesis n

Cytokinesis n

n

n

Prophase II   Metaphase II   Anaphase II   Telophase II

Telophase I

In telophase I, individual chromosomes gather at each of the two poles. In most organisms, the cytoplasm divides, forming two new cells. Note that each of the cells produced now contains half the number of chromosomes of the original cell. For this reason, meiosis I is often called **"reduction division."**

Meiosis II Separates Chromatids

The stages of meiosis I are similar to those of mitosis, with some important differences. Meiosis II is identical to mitosis except that the chromosomes do not replicate before they divide at their centromeres. In anaphase II, the centromeres divide, and the chromatids, now called chromosomes, move to opposite poles of the cell. Meiosis II is followed by cytokinesis, in which new membranes are formed around the four products of meiosis to create four haploid cells. ◻

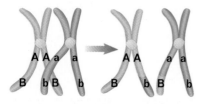

Figure 6-13 At sites of crossing-over, portions of a chromatid on one homologous chromosome are broken and exchanged with the corresponding portions of one of the chromatids on the other homologous chromosome.

The Importance of Crossing-Over

Crossing-over is an efficient way to produce **genetic recombination**, the formation of new combinations of genes. As a result of crossing-over, the two chromatids of a chromosome no longer contain identical genetic material, as shown in Figure 6-13. Crossing-over thus provides a source of genetic variation. Since the speed at which a species can change is often limited by the amount of genetic variation available, crossing-over has an enormous impact on how rapidly organisms evolve.

Section Review

1. *Why is meiosis necessary in organisms that reproduce sexually?*
2. *What happens to homologous chromosomes in meiosis I? Why is meiosis I often called "reduction division"?*
3. *How is meiosis II similar to mitosis?*
4. *If a cell in a mosquito undergoes meiosis, how many chromosomes will each resulting cell contain? what if the cell is that of a dog?*

Critical Thinking

5. *What is crossing-over? How does it affect the rate of evolution?*

CHAPTER REVIEW

6

Vocabulary

amniocentesis (124)
anaphase (131)
autosome (124)
binary fission (126)
cell cycle (127)
centromere (120)
chorionic villi sampling (124)
chromatid (120)
chromosome (119)
crossing-over (133)
cytokinesis (127)
deletion (124)
diploid (120)

Down syndrome (122)
duplication (124)
gamete (120)
gene (119)
genetic recombination (134)
haploid (120)
homologous chromosome (120)
interphase (128)
inversion (124)
karyotype (122)
kinetochore (129)
meiosis (132)
metaphase (131)

mitosis (127)
mutation (124)
nondisjunction (123)
prophase (131)
reduction division (134)
sex chromosome (124)
sexual reproduction (132)
spindle fiber (128)
spore (132)
telophase (131)
translocation (124)
trisomy (122)
zygote (120)

Concept Mapping

Construct a concept map that shows the sequence of major events of the cell cycle. Use as many terms as needed from the vocabulary list. Try to include the following items in your map: cell cycle, mitosis, cytokinesis, interphase, prophase, metaphase, anaphase, telophase, chromosomes, DNA, and protein. Include additional concepts in your map as needed.

Review

Multiple Choice

1. In humans, a zygote will develop into a male if the
   a. sperm and egg both contribute an X chromosome.
   b. egg contributes an X chromosome and the sperm contributes a Y chromosome.
   c. sperm and egg both contribute a Y chromosome.
   d. egg contributes a Y chromosome and the sperm contributes an X chromosome.

2. The haploid number of chromosomes in a human gamete is
   a. 23.   c. 46.
   b. 22.   d. $2n$.

3. A human cell may have more or less than 46 chromosomes as a result of nondisjunction, a process in which
   a. chromosomes fail to separate during cell division.
   b. a fragment of a chromosome breaks off and is lost.
   c. homologous chromosomes pair up and exchange segments of DNA.
   d. genes are combined in new ways.

4. Meiosis is important for sexually reproducing organisms because it
   a. reduces the chromosomes in a cell from the diploid to the haploid number.
   b. ensures that each new cell will contain identical genetic information.
   c. prevents alterations in chromosome structure.
   d. ensures that the chromosome number is increased from one generation to the next.

5. Prokaryotes divide by binary fission, a form of asexual reproduction in which
   a. the nucleus divides into two nuclei.
   b. the number of chromosomes in the cell is reduced.
   c. a bacterium splits into two equal halves with identical genetic information.
   d. spindle fibers attach to the poles of the cell.

6. How does cell division differ between animal and plant cells?
   a. Plant cells do not have centrioles; animal cells do.
   b. Animal cells form a cell plate during cytokinesis.
   c. Animal cells are always haploid.
   d. Plant cells are pinched in half by a belt of protein threads.

7. During the metaphase stage of mitosis,
   a. the cell membrane begins to fold inward.
   b. chromosomes line up at the cell's equator.
   c. spindle fibers shorten, pulling chromosomes to the poles of the cell.
   d. chromosomes are at opposite ends of the cell.

8. During the C phase of the cell cycle the
   a. nucleus begins division.
   b. cytoplasm divides.
   c. chromosomes replicate.
   d. centromeres replicate.

9. Homologous chromosomes pair and undergo crossing-over during
   a. interphase.
   b. prophase of meiosis I.
   c. metaphase of mitosis.
   d. telophase of meiosis II.

Completion

10. The information that determines a cell's characteristics and directs its activities is organized into segments of DNA called _____.

11. Two methods of prenatal diagnosis are _____ and chorionic villi sampling.

12. Using a photograph of chromosomes from a cell, a _____ can be constructed by sorting and arranging the chromosomes by size and centromere position.

13. Meiosis results in new cells with the _____ number of chromosomes of the parent cell. When the _____ from a mother and a father fuse in the process of fertilization, the normal chromosome number is restored.

14. Humans have 23 pairs of chromosomes, 22 pairs of _____ , and one pair of _____ chromosomes.

15. Chromosomes that are similar in shape and size and carry similar genetic information are called _____ chromosomes.

16. A chromosome is composed of _____ wrapped around a protein scaffold. Before mitosis takes place, a chromosome is copied, resulting in two unseparated copies called _____ .

17. Mitosis is the process by which the _____ of a cell divides, while cytokinesis is the process by which the cytoplasm of a cell divides.

Short Answer

18. Under what conditions would a doctor recommend that a prospective mother have an amniocentesis?

19. Why is meiosis I often called the reduction division?

20. What is mutation? In what ways can a mutation alter the structure of a chromosome?

21. Why is DNA replication (forming a chromosome consisting of two chromatids) essential before mitosis begins?

22. What is the fundamental characteristic of sexual reproduction?

Themes Review

23. **Evolution** Explain how meiosis leads to genetic variation.

24. **Evolution** What is crossing-over? When does it occur? How does crossing-over help a species survive?

25. **Homeostasis** For a cell to function efficiently, the magnitude of its surface area must greatly exceed that of its volume. Explain how cell division functions to maintain this relationship between surface area and volume and in doing so maintains cell homeostasis.

26. **Structure and Function** The events of mitosis in plants and animals are very similar with the exception of the absence of centrioles in plants. How has the absence of centrioles in plant cells influenced scientists' thinking about the function of centrioles in mitosis?

27. **Structure and Function** Binary fission and mitosis are successful evolutionary adaptations that ensure that DNA and cytoplasm are equally distributed between two daughter cells. How do binary fission and mitosis differ in the ways that this outcome is achieved?

Critical Thinking

28. **Making Inferences** Why is there a concern that more infants with Down syndrome will be born in the United States as more women delay having children?

29. **Communicating Effectively** Construct a poster comparing mitosis and meiosis in a specific organism. (Choose one with a small number of chromosomes.) First, depict a cell undergoing mitosis. Include diagrams of the cell and its chromosomes in all of the mitotic phases described in the text. Second, depict a cell undergoing meiosis, again including diagrams of the cell and its chromosomes in all of the meiotic phases described in the text. Indicate the number of chromosomes contained within each nucleus after the two processes are completed. Which process is likely to produce chromosomes genetically different from those of the original cell?

Activities and Projects

30. **Research and Writing** Do library research and talk with medical professionals to learn how cancer cells differ from normal cells in relation to the cell cycle. Share the results of your research with your class.

31. **Research and Writing** Find out what biologists observing nuclear division in primitive eukaryotes such as *Amoeba* have learned about the evolution of mitosis.

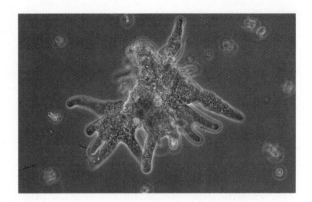

32. **Multicultural Perspective** Many cultures have stories and tales about dwarfs or giants who have special powers. Look for examples of cultural heroes and heroines who are chosen as leaders and protectors because of their physical differences. What attitudes do other characters in these tales demonstrate toward these individuals?

Readings

33. Read the article "An Introduction to DNA Fingerprinting," in *The American Biology Teacher*, April 1993, pages 216–221. What is DNA fingerprinting? What are possible sources for the DNA used in this technique? Describe some applications of this technique

34. Read the article "Cells That Reach Out For the Light," in *Discover*, June 1993, page 64. According to Albrecht-Buehler, how do cells know where to go as they differentiate? Explain the experiment he conducted that led to his hypothesis. What do other scientists think of this hypothesis?

Comparing Mitosis and Meiosis

OBJECTIVES

- Understand the differences between mitosis and meiosis.
- Recognize the relationships between mitosis and genetic continuity, and between meiosis and genetic variation, among living things.
- Use models to understand such concepts as nondisjunction and trisomy.

PROCESS SKILLS

- comparing and contrasting structures
- relating structural features to functions
- comparing and contrasting processes

MATERIALS

- Colored pipe cleaners
- Yarn
- Wooden beads
- Small white labels for "genes"

BACKGROUND

1. How do cells make new cells identical to themselves?
2. What cellular procedures ensure that gametes contain one-half as many chromosomes as body cells?
3. Where in the human body does mitosis occur? Where does meiosis occur?
4. What is nondisjunction? What is trisomy?
5. Write your own **Focus Question** on your Vee Form.
6. **Knowing Side of the Vee** List the **Concepts** and new **Vocabulary Words** on your Vee Form. In the Concept Statements section of the Vee, use these words in sentences that define and explain them.

TECHNIQUE

Doing Side of the Vee

Part A: Mitosis

As you go through the phases of mitosis and meiosis, record your drawings of the setup for each phase in the **Records** section of your Vee Form.

1. Select four pieces of equal-sized pipe cleaner in one color and four of equal size in a second color. Obtain four wooden beads and a piece of yarn approximately 90 cm long.

2. With the yarn, construct a "cell membrane" circle on your lab desk. When the cell is about to divide, the chromatin shortens and thickens into chromosomes inside the cell. Remember, the nuclear envelope has already disappeared.

3. The end of interphase and the beginning of prophase are indistinguishable and blend into one another, hence the use of the word *phase* instead of *stage*. Recall that nuclear division really begins with prophase, when chromosomes become distinct and visible under a microscope. Show your cell with two pairs of replicated chromosomes. Use the wooden bead as the centromere connecting the pipe cleaner chromosome and its double. Add labels for "genes."

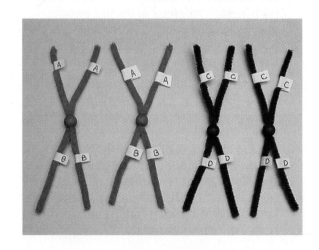

4. Now move on to metaphase. Where are the chromosomes at metaphase? What part of your chromosome model attaches to the spindle fibers that extend from the centrioles?

5. Now show your cell at anaphase. What is happening to the spindle fibers? What is happening to the chromatids? What are they now called?

6. Show your cell at telophase. What is happening to the cell membrane? Show that with your yarn. What other events occur?

Part B: Meiosis

7. Repeat steps 1–3, doing the same things you did for mitosis. At prophase I of meiosis, the homologous chromosomes pair up, forming tetrads. Show this with your materials.

8. In metaphase I, each chromosome attaches to a separate spindle fiber, rather than each chromatid, as during mitosis. Show this with your materials.

9. Does each lab student have the chromosome pairs in the same order (look at the "genes" or letters)? What are some of the gene combinations in your chromosome pairs?

10. During anaphase I, the spindle fibers contract. A pair of chromatids of each tetrad is pulled toward each pole. Telophase I is next.

11. Meiosis II follows the same procedure as mitosis, dividing the two cells resulting from meiosis I. The chromatids will separate. Four cells will result, so stretch out the yarn to make four cells. These cells represent gametes.

12. Take one of your gametes and combine it with the gamete belonging to another student. This is comparable to fertilization. What combinations of genes did you make?

13. Repeat step 10, only this time use your materials to show what would happen if a chromosome failed to separate from its homologue during anaphase.

INQUIRY

1. At metaphase during mitosis, do you and the students to either side of you have the chromosomes lined up in the same order? Does it matter?

2. After telophase of mitosis, how does the size of each new cell compare with that of the original cell?

3. What "genes" are in each of your gametes? When a parent has two different forms of a gene for a trait, how many forms of the gene for that trait are in each gamete produced by that parent?

4. Use the information on the **Knowing Side of the Vee** to interpret your results from the **Doing Side**, and then write your **Knowledge Claim**. Write a **Value Claim** for this lab.

ANALYSIS

1. Does a new cell forming as a result of mitosis have the same genes as the original cell? Explain why or why not.

2. How is mitosis similar to meiosis? How does it differ?

3. Give some examples of variety in a population. Why do you think variety is important to populations?

4. Describe some consequences that might arise if a gamete received an extra copy of a chromosome.

FURTHER INQUIRY

Write a **New Focus Question** that could be the point of a new investigation. The following are examples:

How often do different cells of the body undergo mitosis? Are there any body cells that do not divide?

How can certain chemicals disrupt the events of mitosis and meiosis?

MENDEL AND HEREDITY

REVIEW

- scientific method (Section 1-2)
- gene (Section 6-1)
- meiosis (Section 6-3)

Similar traits in three generations of a family

7-1 Fundamentals of Genetics

Many of your traits—the color and shape of your eyes, the texture of your hair, even your height and weight—resemble the characteristics of your parents. The transmission of traits from parent to offspring is called **heredity**. Humans have long been interested in heredity. From the beginning of recorded history, we have attempted to develop improved varieties of crop plants, like rice, and domestic animals, such as the horse. And almost certainly, parents of long ago wondered why a particular trait appeared in family members. Before DNA and chromosomes were discovered and understood, heredity was one of the greatest mysteries of science. In this section you will learn how this mystery was solved.

Figure 7-1 Breeding the garden pea *Pisum sativum*, Gregor Mendel formulated the basis of modern genetics.

The Beginnings of Genetics

The key to understanding heredity was discovered over a century ago by an Austrian monk named Gregor Mendel. In the garden of a monastery, Mendel bred the garden pea *Pisum sativum*, shown in Figure 7-1. From these experiments, he developed a simple set of rules that accurately predicted patterns of heredity. The patterns he found form the basis of **genetics,** the branch of biology that studies heredity. When Mendel's rules became widely known, scientists all over the world set out to discover the physical mechanism behind these patterns. Eventually, their studies taught them that traits are

CONTENT LINK

The molecular basis of heredity is discussed in **Chapter 8.**

Mendel and Heredity **141**

Genetics is the branch of biology that studies heredity, the transmission of traits from parents to offspring. Early investigators such as T. A. Knight and Gregor Mendel bred varieties of the garden pea Pisum sativum *in attempts to understand heredity.*

determined by genes, the instructions encoded in the DNA of the chromosomes an individual receives from each parent.

Mendel was not the first person to try to understand heredity by studying pea plants. Over 200 years ago, British farmers performed similar experiments and even obtained results comparable to Mendel's. In the 1790s, for example, the British farmer T. A. Knight bred a variety of the garden pea that had purple flowers with a variety that had white flowers. All the offspring had purple flowers. When two of the offspring were bred, however, some of their offspring had purple flowers and some had white. Knight's explanation of this phenomenon noted only that purple flowers had a "stronger tendency" to appear than white flowers. ▢

Mendel Studied the Garden Pea

Gregor Mendel was born in 1822 in a region of Austria that is now part of the Czech Republic. As a son of peasants, Mendel learned much about agriculture, knowledge that became invaluable later in his life. In 1843 Mendel entered an Augustinian monastery to study theology. Three years later, after he failed a teacher's examination, Mendel went to the University of Vienna to study science and mathematics. There, he learned how to study science through experimentation and how to use mathematics to explain natural phenomena.

After attending the university, Mendel returned to the monastery and joined a local science research society. Each member undertook a scientific investigation that could be discussed at meetings and published in the society's journal. Mendel repeated Knight's experiments with pea plants. However, his approach to the experiments differed substantially from that of his predecessor; Mendel attacked the problem in a mathematical fashion and *counted* the number of each kind of offspring. Quantitative approaches to science—those that include measuring and counting—were just becoming popular in Europe, so Mendel's method was on the cutting edge of research at the time. Now, most research in the natural sciences must be quantified.

Why Mendel Chose Peas

For his experiments, Mendel chose to study the garden pea. The garden pea is a good subject for genetic study for several reasons.

1. Many varieties of *P. sativum* exist. Mendel initially examined 32 varieties. From these he selected seven pairs of varieties that differed in easily distinguishable forms of various traits, such as flower color, seed color, and seed shape.

Figure 7-2 In 1857, Gregor Mendel began breeding garden peas in the garden of a monastery.

2. Mendel knew from earlier experiments that he could expect one of the two forms of each trait to disappear in one generation and then reappear in the next. This gave him something to count.

3. *P. sativum* is a small, easy-to-grow plant that matures quickly and produces a large number of offspring. Mendel would be able to conduct many experiments and obtain results quickly.

4. The male and female reproductive parts of *P. sativum* are enclosed within the same flower, as shown in Figure 7-3. When left undisturbed, the flower does not open fully; it simply fertilizes itself through a process called **self-pollination**. As a result, one individual plant can produce offspring. To cross two pea plants, Mendel first had to remove the anthers (the pollen-producing organs) from a flower of one plant. He could then dust the pistil (the egg-producing organ) with pollen from a flower of a different pea plant. Transferring the pollen from the flower of one plant to the flower of a different plant is called **cross-pollination**. Scientists use the term *cross* to refer to the breeding between two flowers from separate plants. ❑

❑ *CAPSULE SUMMARY*

The garden pea is a good subject for genetic study because it has many varieties that grow quickly and are able to self-pollinate, important characteristics for Mendel's experimental design.

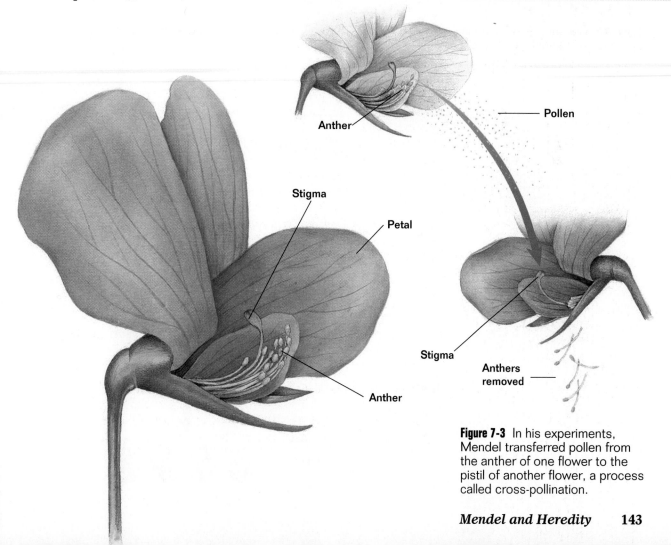

Figure 7-3 In his experiments, Mendel transferred pollen from the anther of one flower to the pistil of another flower, a process called cross-pollination.

Mendel's Experimental Design

Mendel carried out his experiments with garden peas in three steps.

1. **Step 1** Mendel began his experiments by allowing each variety of garden pea to self-pollinate for several generations. This method ensured that each variety was **true-breeding** for a particular trait, which means that all the offspring would display only one form of a particular trait. For example, a true-breeding, purple-flowering plant produced only plants with purple flowers in subsequent generations. Mendel called these plants the parental generation, or **P generation.**

2. **Step 2** Mendel then cross-pollinated two varieties from the P generation that exhibited contrasting traits, such as purple flowers and white flowers. He called the offspring of these plants the first filial generation, or **F_1 generation.**

3. **Step 3** Finally, Mendel allowed the F_1 generation to self-pollinate. He called the offspring of these plants the second filial generation, or **F_2 generation.** These were the plants that he counted. Figure 7-4 summarizes Mendel's experimental design.

Figure 7-4 Mendel studied traits in three generations of garden peas.

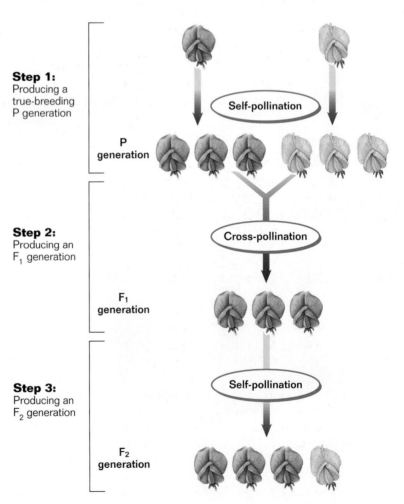

Step 1:
Producing a true-breeding P generation

Self-pollination

P generation

Step 2:
Producing an F_1 generation

Cross-pollination

F_1 generation

Step 3:
Producing an F_2 generation

Self-pollination

F_2 generation

Mendel Observed Two Ratios

For each cross, Mendel obtained F_1 generation plants that had only one form of the crossed traits. The contrasting trait had disappeared. Mendel described the remaining, or expressed trait, as **dominant.** The trait that was not expressed in the F_1 generation was described as **recessive.** Table 7-1 identifies the dominant and recessive forms of the seven traits that Mendel studied.

Table 7-1 Mendel's Crosses and Their Results

Trait	Dominant vs. Recessive		F_2 Generation Results Dominant Form	Recessive Form	Ratio
Flower color	Purple	White	705	224	3.15:1
Seed color	Yellow	Green	6,022	2,001	3.01:1
Seed shape	Round	Wrinkled	5,474	1,850	2.96:1
Pod color	Green	Yellow	428	152	2.82:1
Pod shape	Round	Constricted	882	299	2.95:1
Flower position	Axial	Top	651	207	3.14:1
Plant height	Tall	Dwarf	787	277	2.84:1

When the F_1 generation was allowed to self-pollinate, the recessive trait reappeared in some of the plants in the F_2 generation. At this point Mendel *counted* each type of plant in the F_2 generation. For example, he counted 705 plants with purple flowers and 224 plants with white flowers. From these data, Mendel calculated a ratio of approximately 3 purple-flowering plants to every 1 white-flowering plant (3:1). For

Mendel and Heredity 145

each cross, Mendel obtained the same 3:1 ratio of plants expressing the dominant trait to plants expressing the recessive trait. Table 7-1 lists the numbers of F₂ individuals and the ratios Mendel obtained for each cross.

Mendel's next question was, Will the 3:1 ratios continue in subsequent generations? He found that plants showing the recessive traits were true-breeding when they were allowed to self-pollinate. When plants with the dominant trait self-pollinated, Mendel found that only one-third of them were true-breeding, whereas two-thirds were not. For these plants, Mendel observed a 3:1 ratio of dominant to recessive traits. These results suggested that the 3:1 ratio in the F₂ generation was really a disguised 1:2:1 ratio: 1 true-breeding dominant plant to 2 not-true-breeding dominant plants to 1 true-breeding recessive plant. ❑

❑ **CAPSULE SUMMARY**

In Mendel's experiments, only the dominant traits were expressed in the F_1 generation. The recessive traits reappeared in the F_2 generation, in which Mendel calculated a 3:1 ratio of dominant to recessive traits.

Mendel Proposed a Theory of Heredity

To explain his results, Mendel proposed a theory that has become the foundation of the science of genetics. His theory has five elements.

1. Parents do not transmit traits directly to their offspring. Rather, they pass on units of information that operate in the offspring to produce the trait. Mendel called these units of information "factors." In modern terminology, Mendel's factors are called genes. As you learned in Chapter 6, a gene is a segment of a DNA molecule that transmits hereditary information.

2. For each trait, an individual has two factors: one from its mother and one from its father. The two factors may or may not have the same information. If the factors have the same information (for example, if both factors have information for purple flowers), the individual is said to be **homozygous** (hoh moh ZY guhs). If the factors are different (for example, one factor has information for purple flowers and the other has information for white flowers), the individual is said to be **heterozygous** (heht uhr oh ZY guhs). Each copy of a factor, or gene, is called an **allele** (uh LEEL).

3. In modern terms, the physical appearance, or **phenotype** (FEE noh typ), of an individual is determined by the alleles that code for traits. The set of alleles that an individual has is called its **genotype** (JEE noh typ). See Figure 7-5.

4. An individual receives one allele from one parent and the other allele from the other parent. Each allele can be passed on when the individual matures and reproduces.

5. The presence of an allele does not guarantee that a trait will be expressed in the individual that carries it. In heterozygous individuals, only the dominant allele is expressed; the recessive allele is present but unexpressed. ❑

Figure 7-5 The physical appearance of an organism is its phenotype. These two rabbits have different phenotypes because their collection of genes, or genotypes, varies.

❑ **CAPSULE SUMMARY**

Mendel theorized that for each trait, an individual has two "factors," or genes, one from each parent. Each copy of a factor is called an allele. The two alleles may or may not contain the same information. When they differ, the allele with the dominant information will be expressed.

Mendel's Theory Became Laws of Heredity

Mendel's theory brilliantly predicts the results of his crosses and also accounts for the ratios he observed. Similar patterns of heredity have since been observed in countless other organisms. Because of its overwhelming importance, Mendel's theory is often referred to as the law of segregation. In modern terms, the **law of segregation** states that the members of each pair of alleles separate when gametes are formed.

Mendel went on to study how different pairs of genes are inherited, such as the genes for flower color and plant height. He found that for the pairs of traits he studied, the inheritance of one trait did not influence the inheritance of any other trait. This observation eventually became known as the law of independent assortment. The **law of independent assortment** states that pairs of alleles separate independently of one another during gamete formation. We now know that this principle applies only to genes located on different chromosomes or far apart on the same chromosome.

Mendel's paper describing his results was published in 1866. Unfortunately, it failed to arouse much interest and was forgotten. In 1900, sixteen years after Mendel's death, several scientists independently rediscovered the pioneering paper. They had been searching the literature in preparation for publishing their own findings, which were similar to those Mendel had quietly presented more than three decades earlier. ❑

❑ **CAPSULE SUMMARY**

The law of segregation states that the two alleles for a trait separate when gametes are formed. The law of independent assortment states that when two or more pairs of alleles are located on different chromosomes or far apart on the same chromosome, they separate independently of one another during gamete formation.

Section Review

1. *List four reasons why Mendel used* Pisum sativum *as a subject for genetic experiments.*

2. *Explain how Mendel designed the pea plant experiments that led to his theory of heredity.*

3. *At what point in his experiments did Mendel count the individual pea plants? How did he derive the 3:1 ratio?*

4. *At what point in his experiments did Mendel derive the 1:2:1 ratio? What hereditary characteristics in pea plants does this ratio represent?*

5. *What is the difference between the terms* dominant *and* recessive? heterozygous *and* homozygous? genotype *and* phenotype?

Critical Thinking

6. *How are Mendel's two laws explained in terms of meiosis?*

7. *Would Mendel's results have been different if he had experimented with squash plants, which usually do not self-pollinate? Why or why not?*

7-2 *Analyzing Heredity*

Section Objectives

- *Define* probability *and explain how it is used to predict the results of genetic crosses.*
- *Use a Punnett square to predict the results of monohybrid and dihybrid genetic crosses.*
- *Identify five factors that influence patterns of heredity.*

☐ CAPSULE SUMMARY

Geneticists use capital and lowercase letters to represent alleles. For each trait in an organism, three possible genotypes exist: TT, tt, *or* Tt.

Stripped of ratios and symbols, Mendel's work simply describes how units of information are passed from parents to offspring. The search for the physical nature of the units of information dominated the science of biology for more than half a century after Mendel's work was rediscovered in 1900. We now know that the units of heredity are genes, which are found on the chromosomes that an individual inherits from its parents.

Interpreting Mendel's Model

Geneticists still rely on Mendel's model to predict the likely outcome of genetic crosses. They use letters to represent the alleles of an organism. *Capital letters refer to dominant alleles, and lowercase letters refer to recessive alleles.* Note that capital and lowercase forms of the same letter must be used to designate the two forms of one gene. For example, the allele for the dominant trait of tallness in pea plants is represented by *T*, and the allele for the recessive trait of shortness by *t*. Since there are two alleles for each trait, the genotype of a pea plant that is homozygous dominant for tallness is *TT*. A pea plant that is homozygous recessive for shortness has the genotype *tt*. If these two plants are crossed with each other, the offspring will be heterozygous for the trait and will be designated *Tt*. ☐

Probability

Mendel's crosses can be interpreted according to rules of probability because these rules can predict how genes will be distributed among the offspring of two parents. **Probability** is the likelihood that a specific event will occur. For example, when you toss a coin, there's a chance that it will land with the "heads" side up. There's also a chance that it will land with the "tails" side up. The probability of either event happening can be determined by the following formula:

$$\text{Probability} = \frac{\text{number of one kind of possible outcome}}{\text{total number of all possible outcomes}}$$

Thus, when you toss the coin, the chance of its landing heads up is 1 out of 2 possibilities, or 1/2. The same formula can be used to predict the outcome of a genetic cross. For example, consider Mendel's crosses that studied seed color. Out of 8,023 F_2 pea plants, 6,022 had the dominant yellow seed color and 2,001 had the recessive green seed color.

Using the formula, the probability that the yellow seed color will appear in such a cross is 6,022/8,023, or 0.75 (75 percent). Expressed as a reduced fraction, the probability is 3/4. The probability that the green seed color will appear in the F_2 generation is 2,001/8,023, or 0.25 (25 percent). Expressed as a reduced fraction, the probability is 1/4. In other words, probability tells us that there are three chances in four that an offspring of two heterozygous individuals will have the dominant trait and one chance in four that it will have the recessive trait.

Monohybrid Crosses

A cross that provides data about one pair of contrasting traits is called a **monohybrid cross.** A cross between a pea plant that is true-breeding for tallness and one that is true-breeding for shortness is an example of a monohybrid cross. Biologists can also predict the probable outcome of a cross by using a diagram called a **Punnett square,** named for its inventor, Reginald Punnett. In the Punnett square in Figure 7-6, the genotype of the tall plant and the alleles (*TT*) it can contribute to its offspring are written on the top left side of the square. The genotype of the short plant and the alleles (*tt*) it can contribute to its offspring are written on the bottom left of the square. The interior of the square is a grid of boxes. Each box is filled with two letters—one letter from the left side of the square and one letter from the top of the square. These letters indicate the possible genotypes of the offspring.

In the case of the monohybrid cross in Figure 7-6, 100 percent of the offspring are expected to be heterozygous (*Tt*), expressing the dominant trait of tallness. Note that by convention, the dominant form of the trait is written first, followed by the lowercase letter for the recessive form of the trait. ◼

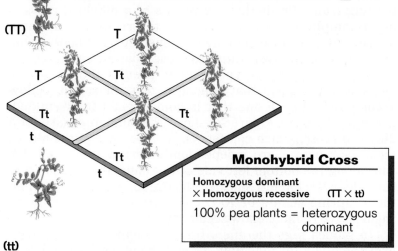

Monohybrid Cross

Homozygous dominant
× Homozygous recessive   (*TT* × *tt*)

100% pea plants = heterozygous dominant

Figure 7-6 A cross between a tall homozygous pea plant and a short homozygous pea plant will produce only tall heterozygous offspring.

Punnett squares can also be used to predict the outcome of a heterozygous cross. For example, in rabbits the allele for a black coat (*B*) is dominant over the allele for a brown coat (*b*). Figure 7-7 shows a Punnett square that predicts the results of a monohybrid cross between two rabbits that are both heterozygous (*Bb*) for coat color. As you can see, one-fourth of the offspring would be expected to have the genotype *BB*, two-fourths (or one-half) would be expected to have the genotype *Bb*, and one-fourth would be expected to have the genotype *bb*. Since *B* is dominant over *b*, three-fourths of the offspring would have a black coat, and one-fourth would have a brown coat. Here you can see the two ratios that Mendel observed in his experiments— 1*BB* : 2*Bb* : 1*bb* (genotype) and 3 black : 1 brown (phenotype). ◼

□ CAPSULE SUMMARY

A monohybrid cross between two individuals that are heterozygous for a trait that has dominant and recessive forms is expected to yield offspring with a 1:2:1 genotypic ratio and a 3:1 phenotypic ratio.

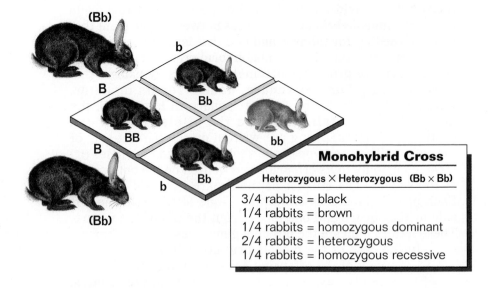

Figure 7-7 Crossing two rabbits that are both heterozygous for black coats will produce two heterozygous black offspring, plus one true-breeding black and one true-breeding brown offspring.

Monohybrid Cross

Heterozygous × Heterozygous (Bb × Bb)

3/4 rabbits = black
1/4 rabbits = brown
1/4 rabbits = homozygous dominant
2/4 rabbits = heterozygous
1/4 rabbits = homozygous recessive

Dihybrid Crosses

A **dihybrid cross** is a cross that involves two pairs of contrasting traits. Predicting the results of a dihybrid cross is more complicated than predicting the results of a monohybrid cross because you have to consider how the two alleles of each of the two traits from each parent can combine. For example, suppose you want to predict the results of crossing a pea plant that is homozygous for round, yellow seeds (*RRYY*) with one that is homozygous for wrinkled, green seeds (*rryy*). Figure 7-8 shows the Punnett square for a dihybrid cross, which consists of 16 boxes. When the alleles from each parent are independently sorted and listed, *RY* runs along the bottom left side of the Punnett square and *ry* runs along the top left side. As illustrated, the genotype of all the offspring should be *RrYy*. Therefore, all the offspring should have round, yellow seeds. ◼

In guinea pigs the allele for short hair (*S*) is dominant over the allele for long hair (*s*), and the allele for black hair (*B*) is dominant over the allele for brown hair (*b*). The Punnett square in Figure 7-9 predicts the probable offspring of a cross between two individuals heterozygous for both

□ CAPSULE SUMMARY

A dihybrid cross between two homozygous individuals with contrasting traits yields heterozygous individuals expressing the dominant phenotype.

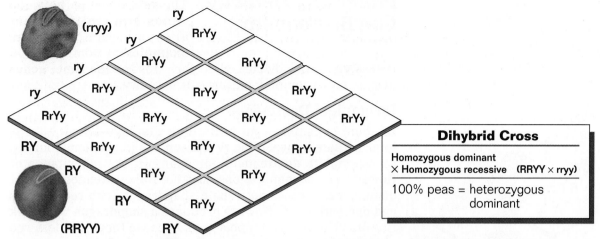

Dihybrid Cross

Homozygous dominant
× Homozygous recessive   (RRYY × rryy)

100% peas = heterozygous dominant

Figure 7-8 This dihybrid cross produces only one type of offspring.

characteristics (*SsBb*). The offspring are likely to have nine different genotypes that will result in the following four phenotypes:

- Nine-sixteenths (9/16) of the guinea pigs will have short, black hair. These include individuals with the genotypes *SSBB*, *SsBB*, *SSBb*, and *SsBb*.

- Three-sixteenths (3/16) will have short, brown hair. These include individuals with genotypes *SSbb* and *Ssbb*.

- Three-sixteenths (3/16) will have long, black hair. These include individuals with the genotypes *ssBB* and *ssBb*.

- One-sixteenth (1/16) will have long, brown hair. These include individuals with the genotype *ssbb*.

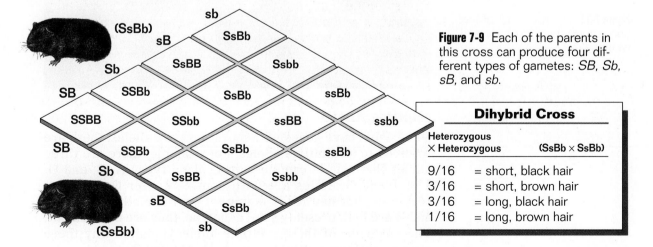

Figure 7-9 Each of the parents in this cross can produce four different types of gametes: *SB, Sb, sB,* and *sb*.

Dihybrid Cross

Heterozygous
× Heterozygous   (SsBb × SsBb)

9/16	= short, black hair
3/16	= short, brown hair
3/16	= long, black hair
1/16	= long, brown hair

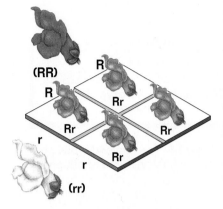

Patterns of Heredity Can Be Complex

The relationships between genes and traits are not always as simple as the examples of dominant and recessive alleles discussed so far. Most of the time, genes display more complex patterns of heredity.

Incomplete Dominance

In some organisms, an individual displays a trait that is intermediate between the two parents, a phenomenon known as **incomplete dominance.** For example, the inheritance of flower color in snapdragons does not follow Mendel's idea of dominance. A cross between a snapdragon with red flowers and one with white flowers produces a snapdragon with pink flowers. The flowers appear pink because they have less red pigment than the red flowers. See Figure 7-10.

Figure 7-10 These snapdragons appear pink because they have less red pigment than the red flowers.

Codominance

In some cases, two dominant alleles are expressed at the same time, a phenomenon called **codominance.** Codominance is different from incomplete dominance because both traits are displayed. An example of codominance is the roan coat in horses. A cross between a homozygous red horse and a homozygous white horse results in heterozygous offspring with a roan coat, which consists of red hairs and white hairs, as seen in Figure 7-11.

Figure 7-11 The roan coat of this horse consists of red hairs and white hairs.

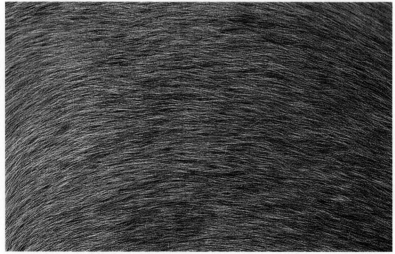

Multiple Alleles

Some traits have genes with more than two alleles; these are referred to as **multiple alleles.** For example, there are three alleles that can determine human blood type—A, B, and O. The A and B alleles are both dominant over O, which is recessive, but neither is dominant over the other. When A and B are both present in the genotype, they are codominant. The existence of these multiple alleles explains why there are four different blood types—A, B, AB, and O.

Continuous Variation

When several genes influence a trait, such as height or weight, determining the effect of one of these genes is difficult, just as it is difficult to follow the flight of one bee within a swarm. Because the genes that determine a phenotype such as height or weight may segregate independently of one another, slight differences in phenotypes are expressed when many individuals are compared. These traits are said to be exhibiting **continuous variation** because you see a variety of phenotypes on a continuum from one extreme to the other.

Environmental Influences

An individual's phenotype often depends on conditions in the environment. For example, during the winter, the pigment-producing genes of the arctic fox shown in Figure 7-12 do not function due to the cold temperature. As a result, the coat of the fox is white, and the animal blends into the snowy background. In summer, the genes function to produce pigments and the coat darkens to a reddish brown, resembling the color of the tundra where the fox lives. ☐

☐ CAPSULE SUMMARY

Patterns of inheritance are more complex than those explained in Mendel's model. Heredity is affected by phenomena such as incomplete dominance, codominance, multiple alleles, continuous variation, and environmental influences.

Figure 7-12 Many arctic mammals develop white fur during the cold winter and dark fur during the warm summer.

Section Review

1. *Write the formula used to determine the likelihood that an event will occur. How is probability used in genetics?*

2. *Construct a Punnett square to predict the outcome of a monohybrid cross between two heterozygous tall pea plants. What are the expected phenotypic and genotypic ratios?*

3. *What is a dihybrid cross? What is the expected phenotypic ratio for a heterozygous dihybrid cross?*

4. *Describe three phenomena that can affect patterns of heredity. Give an example for each phenomenon.*

Critical Thinking

5. *Is it possible to have the blood type ABO? Why or why not?*

7-3 Human Genetics

Just as flower color in Mendel's peas was determined by the particular chromosome the plant received, the degree to which you resemble your mother or father was established before your birth by the particular chromosomes you received. Many of the alleles present in human populations demand more serious consideration than the color of a flower. Alleles that code for defective forms of proteins can cause debilitating and even deadly disorders.

Section Objectives

■ *Explain how mutations can cause genetic disorders.*

■ *List two genetic disorders, and describe their causes and symptoms.*

■ *Explain what sex-linked traits are, and give an example.*

■ *Describe how a pedigree is constructed and analyzed.*

■ *Define genetic counseling, and list three techniques it employs.*

CONTENT LINK
You will learn more about mutations in **Chapter 9**.

Mutations Cause Genetic Disorders

In order for a person to develop and function normally, the proteins encoded by his or her genes must function in a very precise manner. Unfortunately, sometimes genes become damaged or are copied incorrectly, resulting in faulty proteins. Changes in genetic material are called **mutations.** Most of the time, mutations are rare because cells have efficient systems for correcting errors. Still, some mutations occur, and they usually have harmful effects.

The harmful effects produced by mutated genes are called **genetic disorders.** Many mutations that cause genetic disorders are carried by recessive genes in heterozygous individuals. This means that two phenotypically normal people who are heterozygous carriers of a recessive mutation can produce children who are homozygous for the recessive gene. In such cases, the effects of the mutated genes cannot be avoided.

Sickle Cell Anemia

An example of a recessive genetic disorder is **sickle cell anemia,** a condition caused by a mutated allele that produces a defective form of the protein **hemoglobin** (*HEE moh gloh bihn*). Hemoglobin is found within red blood cells, where it binds with oxygen and transports it through the body. In sickle cell anemia, the defective form of hemoglobin causes many red blood cells to bend into a sickle shape, as seen in the micrograph in Figure 7-13. The sickle-shaped cells rupture easily, impairing the oxygen-carrying capability of the blood. They also tend to get stuck in blood vessels, which can cut off blood supply to an organ altogether.

The recessive allele responsible for causing sickle-shaped red blood cells also helps protect the cells of heterozygous individuals from the effects of malaria, a disease caused by protozoans of the genus *Plasmodium*. (*Plasmodium* is a parasite that invades red blood cells.) People who are heterozygous have a mixture of sickled red blood cells and normal red blood cells. The sickled cells cannot

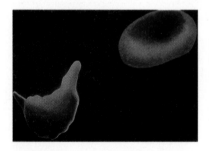

Figure 7-13 Sickle cell anemia is caused by a gene mutation that produces a defective form of hemoglobin. The cell bends, or sickles, in the absence of oxygen.

readily be invaded by the parasite, and the normal cells support adequate oxygenation of tissues. Therefore, these people are protected from the effects of malaria that threaten the lives of individuals who are homozygous dominant for the gene. The defective hemoglobin allele apparently arose in populations that lived in malaria-prone areas of central Africa, where 1 person in 100 is homozygous for the defective allele and thus has the disease. In the United States, however, only 1 African American out of 500 has sickle cell anemia. This genetic disorder is almost unknown in other races.

Hemophilia

Another recessive genetic disorder is **hemophilia** *(hee moh FIHL ee uh)*, a condition that impairs the blood's ability to clot and can cause excessive and prolonged bleeding. A dozen genes code for the proteins involved in blood clotting, and mutations that cause hemophilia can occur in any of them. Two of these genes are found only on the X chromosome, a situation that has an important consequence for males. If a mutation appears in one of these genes in a male, he doesn't have a normal gene on the Y chromosome to compensate. Therefore, he will develop hemophilia. A trait that is determined by a gene found only on the X chromosome is said to be a **sex-linked trait.** Hemophilia and several other important genetic disorders are described in Table 7-2. ❑

CAPSULE SUMMARY

Mutations in genetic material can cause genetic disorders such as sickle cell anemia and hemophilia. Both disorders are caused by faulty proteins. The alleles for both disorders are recessive.

Table 7-2 Some Important Genetic Disorders

Disorder	Symptom	Defect	Dominant or Recessive	Frequency Among Human Births
Cystic fibrosis	Mucus clogs lungs, liver, and pancreas; usually don't survive to adulthood	Failure of chloride ion transport mechanism	Recessive	1/2,080 (whites)
Sickle cell anemia	Poor blood circulation	Abnormal hemoglobin molecules	Recessive	1/500 (African-Americans)
Tay-Sachs disease	Deterioration of central nervous system in infancy; usually don't survive to adulthood	Defective form of enzyme hexosamin-idase A	Recessive	1/1,600 (Jews)
Phenyl-ketonuria	Failure of brain to develop in infancy; usually don't survive to adulthood (if untreated)	Defective form of enzyme phenylalanine hydroxylase	Recessive	1/18,000
Hemophilia	Failure of blood to clot	Defective form of blood clotting factor VIII	Sex-linked recessive	1/7,000
Huntington's disease	Gradual deterioration of brain tissue in middle age; shortened life expectancy	Production of an inhibitor of brain cell metabolism	Dominant	1/10,000
Muscular dystrophy	Wasting away of muscles; shortened life expectancy	Muscle fibers degenerate and atrophy	Sex-linked recessive	1/10,000

Tracking Traits in Families

Imagine that you want to learn about an inherited trait present in your family. How would you find out whether the trait is dominant or recessive and what the chances are of transmitting it to your children? If you wanted to study a trait in organisms such as garden peas or fruit flies, you would conduct genetic crosses. However, studying human heredity requires a different approach.

To study human heredity, scientists look at family histories called **pedigrees.** By identifying which relatives exhibit a trait, scientists can determine whether the gene producing the trait is dominant or recessive and whether it is sex-linked or autosomal (located on one of the other 22 pairs of chromosomes). It is also possible to infer whether a family member is homozygous or heterozygous for a particular trait.

Analyzing a Pedigree for Albinism

How are pedigrees of a trait constructed? Consider the genetic disorder albinism, a condition in which the body is unable to synthesize pigment. As a result of albinism, the skin and hair appear white. In the pedigree of the family presented in Figure 7-14, each symbol represents one individual in the family history. The circles represent females and the squares represent males. Individuals who exhibit the trait being studied, in this case albinism, are indicated by yellow symbols. Marriages are represented by horizontal lines connecting a

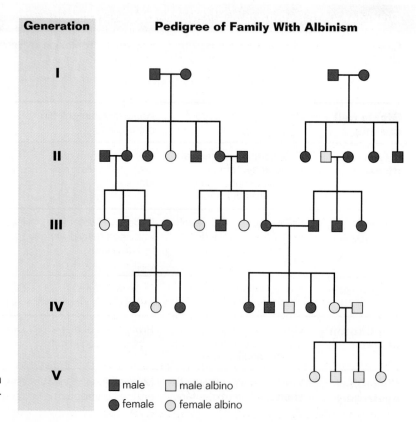

Figure 7-14 This pedigree shows how the genetic disorder albinism is distributed throughout five generations in a family.

circle and a square. Vertical lines indicate children, arranged from left to right in order of their birth.

How do you analyze the pedigree?

1. First, determine whether albinism is sex-linked or autosomal. If the trait is sex-linked, it is usually seen only in males because they have only one X chromosome. If the trait is autosomal, it will appear in both sexes equally. In Figure 7-14, the proportion of males with albinism (5/13, or 39 percent) is similar to the proportion of females with albinism (8/21, or 37 percent). So you can conclude that the trait is probably carried on an autosomal chromosome rather than on a sex chromosome.

2. Second, determine whether albinism is inherited in a dominant or recessive manner. If the trait is dominant, every individual with albinism will have a parent with albinism. If the trait is recessive, an individual with albinism can have heterozygous parents who appear normal. In Figure 7-14, most individuals with albinism have parents who do not exhibit the trait, which indicates that albinism is recessive.

3. Third, find out whether the trait is determined by a single gene or by several. If the trait is determined by a single recessive gene, then normal parents should produce affected children in a 3:1 ratio in the family. This means that approximately 25 percent (the *1* of 3:1) of the children should express the trait. If the trait were determined by several genes, the proportion of affected individuals would be much lower. In this case, 9/34, or 27 percent, of the individuals have albinism, strongly suggesting that only one gene determines the trait.

This example shows how a family pedigree can be analyzed to understand the inheritance pattern of a particular trait. By looking at the pattern of inheritance of albinism in a single family, a scientist can predict that the albinism affecting these individuals is an autosomal recessive trait controlled by a single gene. ◻

❏ *CAPSULE SUMMARY*

By constructing and analyzing a pedigree, a scientist can determine the pattern of inheritance of a trait within a family. It may also be possible to determine whether an individual is heterozygous or homozygous.

Identifying and Treating People at Risk

Most genetic disorders cannot be cured, although progress is being made in many cases. If a person has a disorder for which there is no treatment, he or she may choose to avoid having children. How can a person find out if he or she is a potential carrier for a genetic disorder? People at risk of having children with a genetic disorder can often be identified through genetic counseling. In genetic counseling, a trained genetic counselor helps an individual or family understand the nature of the hereditary disorder and the risks of passing it on to children. Genetic counseling can involve analyzing a family pedigree. It can also include

CONTENT LINK

You will learn more about gene technology in **Chapter 10**.

examining the genetic makeup of an embryo using amniocentesis or chorionic villi sampling, two forms of prenatal testing that you learned about in Chapter 6.

In some cases, therapy is available to treat a genetic disorder if it is diagnosed early enough. For example, **phenylketonuria** (PKU) is a genetic disorder in which an individual lacks an enzyme that converts the amino acid phenylalanine into the amino acid tyrosine. As a result, phenylalanine builds up in the body, a condition that causes severe mental retardation. If PKU is diagnosed soon after birth, the newborn can be placed on a low-phenylalanine diet. Such a diet ensures that the baby gets enough phenylalanine to make proteins, but not so much that it causes any damage. The child maintains the low-phenylalanine diet until approximately 10 years of age, when the brain is fully developed and PKU is usually no longer a problem. Because this disease can be easily diagnosed after birth by inexpensive laboratory tests, many states require the testing of all newborns for PKU.

Gene technology is also making it possible for scientists to correct certain genetic disorders by replacing defective genes with copies of healthy ones. In 1990, this approach was tried for the first time on human patients. In one case a young girl was homozygous for a defective gene that normally encoded an enzyme necessary for a properly functioning immune system. Doctors were successful in transferring a healthy copy of the gene into the girl's cells, and she is considered to be cured of her genetic disorder. In another case, the transferred gene was a potent cancer-fighting gene. Currently, gene transfer therapy is being tested by scientists seeking cures for cystic fibrosis and muscular dystrophy. Scientists are hopeful that they will soon find cures for these genetic disorders. ◻

☐ CAPSULE SUMMARY

The techniques involved in genetic counseling can help identify parents at risk of having children with a genetic disorder. Current treatments for genetic disorders are beginning to include gene transfer therapy.

Section Review

1. *How does a mutation give rise to a genetic disorder?*
2. *Identify two genetic disorders. Explain the defect in the protein that causes each disorder.*
3. *Explain how males inherit hemophilia.*
4. *What is phenylketonuria? How is this disorder treated?*

Critical Thinking

5. *When analyzing a pedigree, how can you determine if an individual is heterozygous for the trait being studied?*
6. *What is genetic counseling? Why might a couple undergo genetic counseling?*

Vocabulary references: allele (146), codominance (152), continuous variation (153), etc.

Vocabulary

allele (146)
codominance (152)
continuous variation (153)
cross-pollination (143)
dihybrid cross (150)
dominant (145)
F₁ generation (144)
F₂ generation (144)
genetic disorder (154)
genetics (141)
genotype (146)
hemoglobin (154)

hemophilia (155)
heredity (141)
heterozygous (146)
homozygous (146)
incomplete dominance (152)
law of independent
   assortment (147)
law of segregation (147)
monohybrid cross (149)
multiple alleles (152)
mutation (154)
pedigree (156)

P generation (144)
phenotype (146)
phenylketonuria (158)
probability (148)
Punnett square (149)
recessive (145)
self-pollination (143)
sex-linked trait (155)
sickle cell anemia (154)
true-breeding (144)

Concept Mapping

Construct a concept map that shows the nature and results of Mendel's experiments and also the myths surrounding them. In constructing your map, use the following terms: *P. sativum,* varieties, segregation, independent assortment, P generation, F₁ generation, F₂ generation, genes, dominant trait, and recessive trait. Include additional concepts in your map as needed.

Review

Multiple Choice

1. Which of the following is not a good reason that *Pisum sativum* makes an excellent subject for genetic study?
   a. It has many varieties.
   b. It requires cross-pollination.
   c. It grows quickly.
   d. It demonstrates complete dominance.

2. Offspring that are the product of true-breeding parents are called the
   a. F₁ generation.
   b. F₂ generation.
   c. dominant offspring.
   d. phenotypic expression.

3. If two parents with dominant phenotypes have an offspring with a recessive phenotype, then
   a. the parents are heterozygous.
   b. other offspring must be homozygous dominant.
   c. the parents are homozygous.
   d. the mother carries a lethal gene.

4. The law of segregation
   a. is demonstrated by the phenotypic ratio of 9:3:3:1.
   b. states that pairs of alleles separate when gametes form.
   c. does not apply to the phenomenon of codominance.
   d. predicts the probability of a penny landing "tails" side up when tossed in the air 50 times.

5. Whenever a 9:3:3:1 ratio appears among the offspring of two parents who are both heterozygous for two different traits, it means that
   a. continuous variation is at work.
   b. the traits are expressions of incomplete dominance.
   c. the two sets of alleles probably sorted independently.
   d. mutations have likely occurred in both F₁ and F₂ generations.

6. An example of continuous variation is
   a. a pink snapdragon produced by crossing red and white snapdragons.
   b. phenylketonuria.
   c. the height of people.
   d. the changing color of the Arctic fox's coat.

7. Sickle cell anemia is an autosomal recessive disorder that occurs once in every 500 African American births. If the first child of two healthy African American parents is affected by the disorder, what is the probability that their second child will also have sickle cell anemia?
   a. 1 in 500
   b. 1 in 250
   c. 1 in 4
   d. 0

8. Genetic counseling involves
   a. separating pairs of alleles when gametes form.
   b. educating parents about hereditary disorders.
   c. replacing defective genes with healthy ones.
   d. supplementing an infant's diet with the amino acid tyrosine.

Completion

9. When Mendel crossed F_1 generation individuals, the phenotypic ratio he observed in their offspring was _____, while the genotypic ratio he inferred was _____ .

10. The color of a dog's coat can be described as the dog's _____ , which is determined by sets of alleles, or its _____ .

11. A recessive genetic disorder that impairs the blood's ability to clot is _____ . It is said to be a _____ trait because it is carried on the X chromosome.

12. Tay-Sachs disease and phenylketonuria are examples of _____ . They are caused by _____ in genetic material.

Short Answer

13. In science, it has often been said that one scientist's discovery is built on work of others that came before him. How does this saying apply to the work of Mendel?

14. Most of the members of one animal species are brown, but occasionally a black individual appears. Assuming that the trait for color shows complete dominance, how could two brown parents produce a black offspring? What is the genotype of the black offspring? What are the possible genotypes of the black offspring's siblings?

15. The law of independent assortment applies not only to genetics but also to many lottery games. Suppose that ten ping-pong balls are placed in a machine, each marked from 0 to 9. If they are mixed thoroughly and one is taken out, the chances are 1 in 10 that a particular number, such as 3, will be drawn. If two such machines are used, the chances of a 3 being drawn from both machines is 1 in 100.

$$1/10 \times 1/10 = 1/100$$

(1 in 10 for each jar, or a total of 1 in 100 for drawing 3 and 3)

Suppose you have chosen the number 35097 to win $1 million in a state lottery. What is the chance of your number being drawn? Should you count on winning the lottery?

16. Why did Mendel begin his experiments by allowing pea plants to self-pollinate for several generations?

17. Why are calculated ratios in a Punnett square probable rather than certain?

18. What is the difference between self-pollination and cross-pollination?

19. What would be the result of a cross between two pea plants that were heterozygous for both yellow seed color and axial flower position. Show the Punnett square, and state the genotypic and phenotypic ratios.

20. **Structure and Function** Describe the organization of information in a pedigree chart. How is a pedigree chart used by a genetic counselor?

21. **Structure and Function** Many scientists claim that Mendel was very lucky because he studied traits determined by genes located on different chromosomes or located on the same chromosome, but some distance apart. How might Mendel's conclusions have differed if he had studied two traits determined by genes located close to one another on the same chromosome?

22. **Levels of Organization** How is the genotype for a particular trait related to the phenotype for that same trait?

23. **Making Inferences** Mendel based his conclusions about inheritance patterns on studies of large samples. Why do you think the use of large samples is advantageous when studying inheritance patterns?

24. **Interpreting Graphics** The partial pedigree below shows a family with a genetic disorder. Is the trait sex-linked? Is it inherited in a dominant or recessive manner?

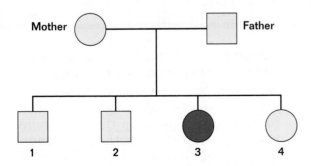

25. **Communicating Effectively** A 20-year-old man diagnosed with muscular dystrophy has a sister who is soon to be married. If you were the man, what would you tell your sister?

26. **Research and Writing** Interview health care workers or genetic counselors to learn why testing newborns for PKU is required by law and how the mental retardation associated with PKU can be avoided.

27. **Cooperative Group Project** Raise the fruit fly *Drosophila melanogaster* to investigate the results of different crosses.

28. **Multicultural Perspective** Hopi Native Americans have high numbers of albinos in their tribe. Research Dr. Frank C. Dukepoo's idea of cultural selection that explains this phenomenon.

29. Read the article "Cystic Fibrosis: Tests, Treatments Improve Survival," in *FDA Consumer*, June 1993. What are the symptoms associated with cystic fibrosis? How is CF inherited? How is the discovery of the CF gene enabling reseachers to develop new diagnostic tests and treatments for the fatal disease?

30. Read the article "Origin of the Punnett Square," in *The American Biology Teacher*, April 1993, pages 204–208. Why did Punnett originally devise his "chessboard" approach to genetics?

Monohybrid Crosses

OBJECTIVE

Using a Punnett square, predict the genotypic and phenotypic ratios of offspring resulting from the random pairing of gametes.

PROCESS SKILLS

- predicting results of monohybrid crosses
- organizing data using Punnett squares

MATERIALS

- lentils
- green peas
- petri dishes

BACKGROUND

1. How many traits are involved in a monohybrid cross? How many alleles?
2. What prevents the expression of a recessive allele?
3. When gametes form, what happens to the alleles for each trait?
4. Write your own **Focus Question** on your Vee Form.
5. **Knowing Side of the Vee** List the **Concepts** and new **Vocabulary Words** on your Vee Form. In the **Concept Statements** section of the Vee, use these words in sentences that define and explain them.

TECHNIQUE

Doing Side of the Vee

You will model the random pairing of alleles by choosing lentils and peas from petri dishes. These dried seeds will represent the traits for seed color. A green pea will represent **G**, the dominant allele for green seeds; a lentil will represent **g**, the recessive allele for yellow seeds.

1. Count 25 of each dried seed and place them in two petri dishes. Mark one petri dish "female gametes" and the other dish "male gametes."
2. Since each parent contributes one allele at random to each offspring, model a cross between these two parents by choosing 10 random pairings of the dried seeds from the two containers. Do this by simultaneously picking one seed from each container, without looking. Place the pair together on the lab table. The pair of seeds represents one offspring. Record the results in the **Records** section of your Vee in a table like the one below.

Table 1 Gamete Pairings

Trial	Offspring Genotype	Offspring Phenotype
1		
2		
3		
4		
5		
6		
7		
8		
9		
10		

3. Return the seeds to their original dishes and repeat step 2 nine more times.
4. Determine the genotypic and phenotypic ratios among the offspring. Using a data table like the one atop the next page, record your ratios in the **Additional Records and Observations** section on the back of your Vee Form.
5. Compare your ratios with those of your classmates. Now pool the data for the whole class and record it in the table

Table 2 Offspring Ratios

Genotypes:	Total	Genotypic Ratio
Homozygous dominant (GG)		
Heterozygous (Gg)		___ : ___ : ___
Homozygous recessive (gg)		
Phenotypes:		**Phenotypic Ratio**
Green fruit		___ : ___
Yellow fruit		

below. Compare the greater numbers with your small sample of 10. Calculate the genotypic and phenotypic ratios for the class data, and record them in the table.

Table 3 Offspring Ratios (Entire Class)

Genotypes:	Total	Genotypic Ratio
Homozygous dominant (GG)		
Heterozygous (Gg)		___ : ___ : ___
Homozygous recessive (gg)		
Phenotypes:		**Phenotypic Ratio**
Green fruit		___ : ___
Yellow fruit		

6. Construct a Punnett square showing the parents and their offspring in the **Additional Records and Observations** section. In the **Procedure** section of the Vee, briefly summarize the procedure you followed.

INQUIRY

1. What are the genotypes and phenotypes of the parents?

2. What does each seed represent in the petri dish?

3. When the seeds were selected and paired, what did pairs represent?

4. Describe the genotypes of both parents using the terms *homozygous* or *heterozygous*, or both.

5. Did tables 2 and 3 reflect a classic mono-hybrid cross phenotypic ratio of 3:1?

6. What trait is being studied in this investigation?

7. Use the information on the **Knowing Side** of the Vee to interpret your results from the **Doing Side**, and then write your **Knowledge Claim**. Write a **Value Claim** for this lab.

ANALYSIS

1. When the class data were tabulated, did a classic monohybrid cross ratio of a phenotype of 3:1 result?

2. If a genotypic ratio of 1:2:1 is observed, what must the genotypes of both parents be?

3. Show what the genotypes of the parents would be if 50 percent of the offspring were green and 50 percent of the offspring were yellow.

4. Diagram the cross of a heterozygous black guinea pig and an unknown guinea pig whose offspring include a recessive white-furred individual. What are the possible genotypes of the unknown parent?

FURTHER INQUIRY

Write a **New Focus Question** that could be the point of a new investigation. The following is an example:

How could you model a dihybrid cross of two parents that are heterozygous for two traits? Construct and complete a Punnett square for this cross.

CHAPTER 8

DNA: THE GENETIC MATERIAL

REVIEW

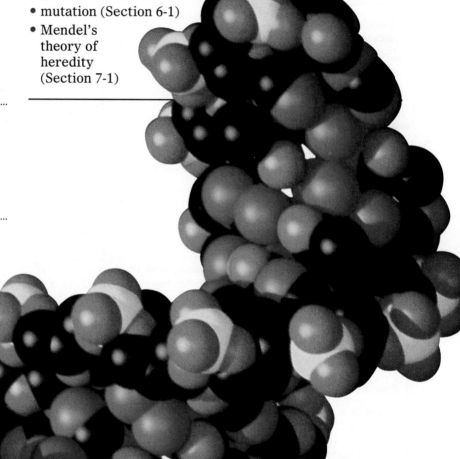

Computer-generated model of DNA

8-1 Identifying the Genetic Material

Mendel's experiments and results answered the question of why you resemble your parents. You resemble your parents because you have copies of their chromosomes, which contain sets of instructions called genes. Mendel's work, however, created more questions, such as, What are genes made of? Scientists believed that if they could answer this question they would understand how chromosomes function as the bearers of heredity. In this section, you will read about the events that led to the identification of the genetic material.

Section Objectives

- Describe what Frederick Griffith observed during his transformation experiments.
- Summarize the steps involved in Oswald Avery's transformation experiments, and state the results.
- Outline the experiment of Alfred Hershey and Martha Chase, and describe its significance in identifying the genetic material.

The Genetic Material: Protein or DNA?

By the 1940s, scientists knew that chromosomes consisted of deoxyribonucleic acid (DNA) and protein. While both were candidates for the role of the genetic material, proteins seemed the more likely choice because they were thought to be more chemically complex than DNA. However, little was known about DNA. It seemed far too simple a molecule to carry all the information needed for heredity. This view began to change when experiments yielded unexpected results.

Griffith Demonstrates Transformation

In 1928, an experiment completely unrelated to the field of genetics led to an astounding discovery about DNA. Frederick Griffith, a public health bacteriologist in England, was trying to prepare a vaccine against the pneumonia-causing bacterium *Streptococcus pneumoniae*. A **vaccine** *(vahk SEEN)* is a substance prepared from killed or weakened microorganisms that is introduced into the body to produce immunity. Griffith worked with two types, or strains, of *S. pneumoniae*. The first strain is enclosed in a capsule made of polysaccharides. The capsule protects the bacterium from the body's defense systems, which helps make the microorganism **virulent** *(VIHR yoo luhnt)*, or able to cause disease. The capsule also causes the bacteria to grow smooth-edged *(S)* colonies when grown in a Petri dish. The second strain of *S. pneumoniae* lacks the polysaccharide capsule and does not cause disease. When grown in a Petri dish, the second strain forms rough-edged *(R)* colonies.

Griffith knew that mice infected with the *S* bacteria grew sick and died. He also knew that mice infected with the *R* bacteria were not harmed and remained alive. To prepare a vaccine, Griffith weakened *S* bacteria by raising their temperature to a point at which the bacteria were "heat-killed," meaning that they could no longer divide. When Griffith injected the mice with heat-killed *S* bacteria, the mice lived. He then mixed the harmless live *R* bacteria and the harmless heat-killed *S* bacteria. Mice injected with this mixture of two harmless substances died. When Griffith examined the blood of the dead mice, he found that the live *R* bacteria had polysaccharide capsules. Somehow, the harmless *R* bacteria underwent a change, or **transformation** *(trans fuhr MAY shuhn)*, and became the virulent *S* bacteria. Griffith's transformation experiments are summarized in Figure 8-1.

Figure 8-1 Griffith discovered transformation when he showed that harmless bacteria could turn virulent when they were mixed with bacteria that caused disease.

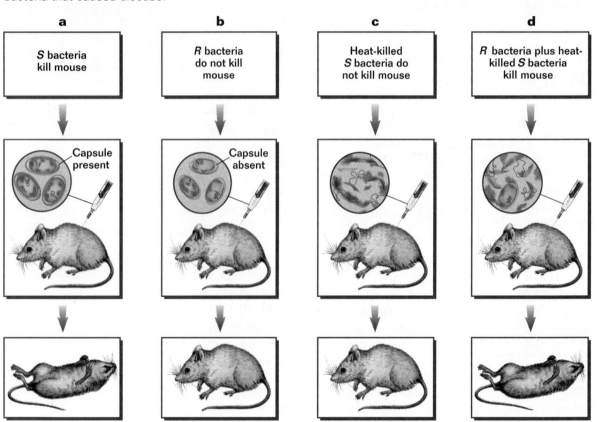

The search for the material responsible for transformation continued until 1944. In that year, Oswald Avery and his co-workers, biologists at the Rockefeller Institute in New York City, demonstrated that DNA was responsible for transformation. In an elegant series of experiments, they showed that the activity of the material responsible for transformation was not affected by protein-destroying enzymes, but was destroyed when a DNA-destroying enzyme was present. Avery and his colleagues made the announcement that the genetic material was DNA. ◻

◻ CAPSULE SUMMARY

The transformation experiments of Griffith and of Avery yielded results that suggested DNA was the genetic material.

Hershey and Chase Show That Genes Are Made of DNA

Even though Avery's experiments concluded that the genetic material was composed of DNA, many scientists remained skeptical. They thought that DNA was relevant only to certain kinds of bacteria and preferred to think that protein was the genetic material. In 1952, however, Alfred Hershey and Martha Chase, two scientists at Cold Spring Harbor Laboratory on Long Island, New York, performed an experiment that settled the controversy. In this experiment, Hershey and Chase used a type of bacteriophage *(bak TIR ee uh fayj)*. A **bacteriophage** is a virus that infects bacteria. Hershey and Chase used the bacteriophage T_2, or just the T_2 phage for short, shown in Figure 8-2. The T_2 phage attaches to the surface of a bacterium and injects its hereditary information into the cell, functioning like a tiny hypodermic needle. The protein coat of the phage, including the tail, remains outside of the bacterium. Once inside the bacterium, this hereditary information replicates and directs the production of hundreds of new phages. When the new phages are mature, they burst out of the infected bacterium and attack new cells.

The only molecule in the phage that contains phosphorus is the DNA. Likewise, the only molecules that contain sulfur are the proteins in the phage's coat. The Hershey and Chase experiment, summarized in Figure 8-3, took advantage of this difference in chemical composition. Hershey and Chase grew one batch of phage in a nutrient medium that contained radioactive phosphorus (^{32}P). They grew another batch of phage in another nutrient medium that contained radioactive sulfur (^{35}S). The radioactive phosphorus became part of the phages' DNA, and the radioactive sulfur became part of the phages' protein coat. Radioactive elements were used because they can be followed, or traced, in a reaction or process. If these two kinds of labeled phages were used to infect a bacterial host, Hershey and Chase could use these labels to locate the genetic material of the phages after they infected bacteria.

In their experiments, Hershey and Chase first infected *Escherichia coli* bacteria with ^{35}S-labeled phages and allowed them to inject their genetic material. After a few minutes, the scientists mixed the infected bacteria in an ordinary kitchen blender. The violent agitation tore the phages off the surfaces of the bacteria. The investigators then used a rapidly spinning centrifuge to separate the bacteria and phages into two separate layers. When the layers were examined, Hershey and Chase found that most of the ^{35}S label was still part of the phage, meaning the protein was not injected into the bacteria. A very different result was observed in the second phase of the experiment when bacteria were infected with ^{32}P-labeled phages. When Hershey and Chase examined the layers, they found that most of the ^{32}P was now part of the bacteria, meaning the DNA had been

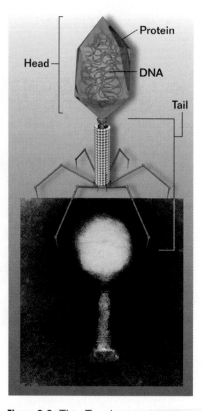

Figure 8-2 This T_2 phage, magnified 95,700X, is one of many phages that infect *E. coli*, a bacterium that normally lives in the intestines of mammals. These phages have a very simple structure—a core of DNA surrounded by a coat of protein.

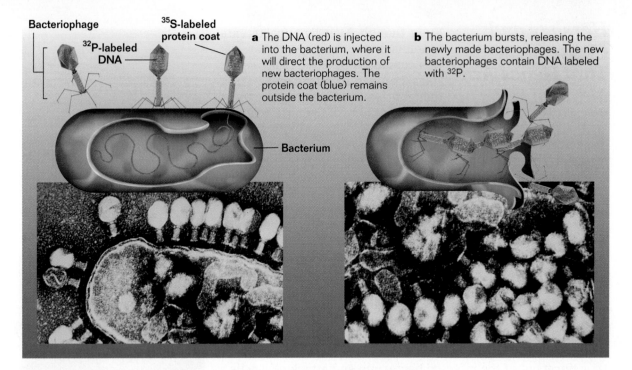

Bacteriophage

^{32}P-labeled DNA

^{35}S-labeled protein coat

a The DNA (red) is injected into the bacterium, where it will direct the production of new bacteriophages. The protein coat (blue) remains outside the bacterium.

b The bacterium bursts, releasing the newly made bacteriophages. The new bacteriophages contain DNA labeled with ^{32}P.

Bacterium

Figure 8-3 Hershey and Chase used bacteriophages to show that DNA, not protein, is the genetic material of viruses.

injected into the hosts. The new generation of phages that were produced by these bacteria also contained the radioactive DNA. The conclusion of the Hershey and Chase experiment was clear and indisputable—the genetic material is made of DNA and not protein.

These important experiments and many others have shown that DNA is the molecule that stores genetic information in living cells. As you will see in the next section, DNA is particularly well suited to this function. ☐

Section Review

1. *What did Frederick Griffith observe when he injected mice with living R bacteria and heat-killed S bacteria?*

2. *How did Oswald Avery's experiment supply evidence that DNA, and not protein, is the genetic material?*

3. *How did Alfred Hershey and Martha Chase alter the molecular structure of bacteriophage T$_2$ in their experiments? For what purpose did they use a kitchen blender? How did they conclude that DNA is the genetic material?*

Critical Thinking

4. *Why were many scientists reluctant to believe that DNA was the genetic material?*

8-2 The Structure of DNA

By the early 1950s, most scientists were convinced that genes were made of DNA. They began studying DNA in earnest, hoping that the mystery of heredity could be solved by understanding the structure of the molecule.

Nucleotides Are the Building Blocks of DNA

DNA is an extraordinarily long, thin molecule made of subunits called **nucleotides** (NOO klee uh tydz) that are linked together like a chain. Each nucleotide is constructed of three parts: a phosphate group, a five-carbon sugar molecule, and a nitrogen base. Figure 8-4 shows how these three parts are arranged to form a nucleotide. The five-carbon sugar is called **deoxyribose** (dee ahk see RY bohs), from which DNA gets its full name, deoxyribonucleic acid.

While the sugar molecule and the phosphate group are the same for each nucleotide in a molecule of DNA, the nitrogen base may be any one of four different kinds. Figure 8-5 illustrates the molecular configurations of the four nitrogen bases: **adenine** (AHD uh neen), **guanine** (GWAH neen), **thymine** (THY meen), and **cytosine** (SYT oh seen). Adenine and guanine belong to a class of organic molecules called **purines** (PYUR eenz). Purines are large molecules, each with a double ring of carbon and nitrogen atoms. Thymine and cytosine are **pyrimidines** (py RIHM uh deenz). Pyrimidines have a single ring of carbon and nitrogen atoms.

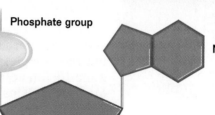

Phosphate group

Nitrogen base

Sugar (deoxyribose)

Figure 8-4 Each nucleotide is made of the sugar deoxyribose, a phosphate group, and a nitrogen base.

Figure 8-5 The nitrogen base in a nucleotide can be either a purine, which has a double ring of carbon and nitrogen atoms, or a pyrimidine, which has a single ring of carbon and nitrogen atoms.

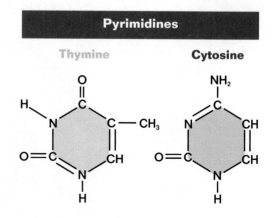

Purines	Pyrimidines
Adenine Guanine	Thymine Cytosine

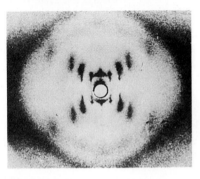

CAPSULE SUMMARY

The DNA molecule is a long chain of subunits called nucleotides. Each nucleotide is made of a phosphate group, the five-carbon sugar deoxyribose, and one of four different nitrogen bases.

In 1949, Erwin Chargaff, an American biochemist working at Columbia University in New York City, made an interesting observation about DNA. Chargaff's research data showed that for the DNA in each organism, the amount of adenine always equals the amount of thymine. Likewise, the amount of guanine always equals the amount of cytosine. However, the amount of adenine and thymine and of guanine and cytosine varied between different organisms. These findings, known as Chargaff's rules, or more commonly as the **base-pairing rules,** suggested that the precise arrangement of nucleotides within a DNA molecule specifies genes. ☐

The DNA Molecule Is a Double Helix

The significance of Chargaff's rules became clear in the 1950s when scientists began using X-ray diffraction to study the structures of molecules. In X-ray diffraction, a beam of X rays is focused at an object. The X rays bounce off the object and are scattered in a pattern onto a piece of film. By analyzing the complex patterns on the film, scientists can determine the structure of the molecule. In the winter of 1952, Maurice Wilkins and Rosalind Franklin, two scientists working at King's College in London, developed some high-quality X-ray diffraction photographs of the DNA molecule. These photographs, such as the one in Figure 8-6, suggested that the DNA molecule resembled a tightly coiled helix and was composed of two or three chains of nucleotides.

Figure 8-6 This X-ray diffraction photo of DNA fibers taken by Rosalind Franklin revealed the X pattern characteristic of a helix; the dimensions of the pattern indicated that the DNA helix had a diameter of about 2 nanometers and made a complete spiral turn every 3.4 nanometers.

The problem now was to discover the three-dimensional structure of the DNA molecule. The model had to take into account Chargaff's rules and the X-ray diffraction data. In 1953, James Watson and Francis Crick, two scientists at Cambridge University, used this information along with their knowledge of chemical bonding to come up with a solution. Using tin-and-wire models of molecules, they built a model of DNA with the configuration of a **double helix,** a "spiral staircase" of two strands of nucleotides twisting around a central axis. Figure 8-7 shows Watson and Crick next to their tin-and-wire model of DNA.

Figure 8-7 A breakthrough in genetics came in 1953 when James Watson and Francis Crick deduced the structure of DNA. They first built models of nucleotides and then assembled the nucleotides into a configuration that fit information from Chargaff's data and Franklin's photos.

As you can see in Figure 8-8, the double helix looks something like a twisted ladder. The sides of the ladder are constructed of alternating sugar and phosphate units, and each rung is a purine and a pyrimidine held together by hydrogen bonds. Why is a purine always paired with a pyrimidine? These base pairings are the only possible arrangement because adenine (A) can form hydrogen bonds only with thymine (T), and cytosine (C) can form hydrogen bonds only with guanine (G). Notice that this arrangement of the nitrogen bases explains Chargaff's observations. The strictness of base pairing results in two strands that are **complementary** to each

Figure 8-8 Watson and Crick's model of DNA is a double helix, a spiral staircase of two nucleotide chains that are hydrogen-bonded to each other and twisted around a central axis.

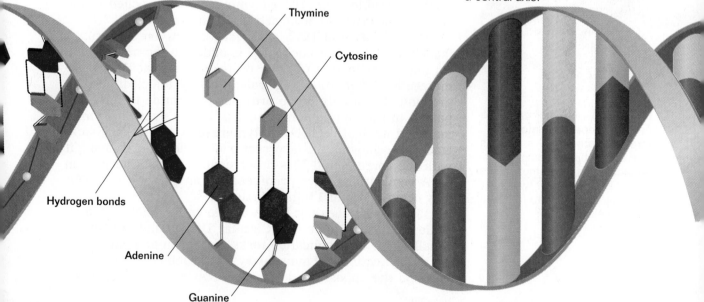

Thymine

Cytosine

Hydrogen bonds

Adenine

Guanine

The two strands of the double helix are complementary because a purine can form hydrogen bonds only with a pyrimidine. Specifically, only adenine and thymine can form hydrogen bonds together, and only cytosine and guanine can form hydrogen bonds together.

other; that is, the sequence of bases on one strand determines the sequence of bases on the other strand. For example, if the sequence of one strand of a DNA molecule is TCGAACT, the sequence on the other strand must be AGCTTGA. □

How DNA Is Copied

When the double helix was first discovered, scientists were very excited about the complementary relationship between the sequences of bases. They predicted that the complementary structure was used as a basis to make new DNA. Watson and Crick proposed that one strand could serve as a template, or surface, upon which the other strand is built. Within five years of the discovery of DNA's structure, scientists had firm evidence, from a number of different kinds of experiments, that the complementary strands of the double helix did indeed separate and serve as templates for building new DNA.

The process of synthesizing a new strand of DNA is called **replication.** Before replication can begin, the double helix must be unwound. This is accomplished by enzymes called **helicases,** which open up the double helix by breaking the hydrogen bonds that link the complementary bases. Once the two strands are separated, additional enzymes and proteins attach to the individual strands and hold them apart, preventing them from twisting.

The point at which the double helix separates is called the **replication fork** because of its Y shape. At the replication fork, enzymes known as **DNA polymerases** move along each of the DNA strands, adding nucleotides to the exposed bases according to the base-pairing rules. As the DNA polymerases move along, two new double helixes are formed, as shown in Figure 8-9. Once a DNA polymerase has begun adding nucleotides to a growing double helix, the enzyme remains attached to the strand until it reaches a signal that tells it to detach.

In the course of DNA synthesis, errors are sometimes made and the wrong nucleotide is added to the new strand. An important feature of DNA replication is that DNA polymerase has a "proofreading" role; it can add nucleotides to a growing strand only if the previous nucleotide is correctly paired to its complementary base. In the event of a mismatched nucleotide, DNA polymerase is capable of backtracking, removing the incorrect nucleotide, and replacing it with the correct one. This proofreading prevents errors in DNA replication. After proofreading, an error in the DNA may occur once per 1 billion nucleotides.

Replication does not begin at one end of the DNA molecule and end at the other. Circular DNA found in bacteria usually have two replication forks that begin at a single origin of replication and move away from each other until they meet on the opposite side of the DNA circle. Linear DNA

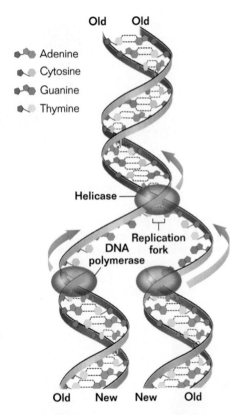

Figure 8-9 A DNA molecule replicates by separating into two strands. Each strand is used as a template to build a complementary strand. When the complementary strand is complete, it twists with the template strand to form a double helix.

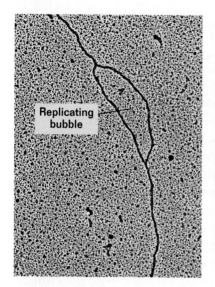

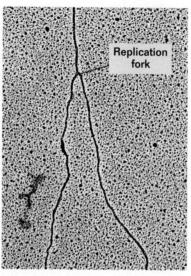

Figure 8-10 A replicating bubble enlarged 20,000X, *left,* has formed in the DNA of a mammalian cell. On the right, the bubble has enlarged into a replication fork with two double-stranded arms made of old and new DNA.

molecules found in eukaryotic organisms usually have many replication forks that begin in the middle and move in both directions, creating replicating "bubbles" along the molecule, as shown in Figure 8-10. If replication did not occur this way, it would take 16 days to copy just one DNA molecule of a fruit fly. But because approximately 6,000 replication forks exist simultaneously, replication of the fruit fly DNA takes only three minutes. Human DNA is also copied in segments, with a replication fork approximately every 100,000 nucleotides. ❑

❑ CAPSULE SUMMARY

In DNA replication, enzymes work to unwind and separate the double helix and add complementary bases to the exposed strands. Each new double helix is composed of one old DNA strand and one new DNA strand.

Section Review

1. List the three parts of a nucleotide. Which part can be one of four molecules? Name these four different molecules.

2. Describe the structure of the DNA molecule. What kind of chemical bond holds its two strands together?

3. What two pieces of information enabled Watson and Crick to discover the structure of DNA?

4. Why are the two strands of the double helix described as "complementary"?

5. Define replication. Explain the two roles that DNA polymerases play in replication.

Critical Thinking

6. Suppose a strand of DNA has the base sequence CCAGATTG. What is the base sequence of the complementary strand?

8-3 The Structure of a Gene

Ever since it became clear that DNA is the hereditary material of the cell, scientists have peered ever more closely at DNA in an attempt to learn more about genes. New questions revolved around the nature of the information held in the DNA molecule. Scientists soon learned that genes hold information specifying how to build particular proteins. Remember that a protein is a string of amino acids. Each amino acid is coded for in the DNA. A gene affects the phenotype of an individual because of the activity of the protein that it specifies. If the protein is an enzyme that makes brown pigment, then the gene may affect hair color. Simply stated, genes are the DNA-encoded information that specifies particular proteins; each gene is made of a specific sequence of nucleotides.

Section Objectives

■ Describe how eukaryotic genes are organized, and explain why this organization is beneficial.

■ Explain the significance of a multigene family.

■ Define transposons, and identify the scientist responsible for discovering them.

■ Recognize the relationship between the p53 gene and cancer.

CONTENT LINK

You will learn how a gene is used to make a protein in **Chapter 9**.

Genes in Eukaryotes Are Often Interrupted

While it is tempting to think of a gene as an unbroken stretch of nucleotides that codes for a protein, this actually occurs only in the genes of bacteria. In all other organisms, much of the DNA does not code for protein. Most genes are frequently interrupted by long segments of nucleotides that have no coding information. These noncoding sequences are called intervening sequences, or **introns** (IHN trahns). The nucleotide segments that code for amino acids are called **exons** (EHK sahns) because they are expressed.

How does the cell make a protein from the fragments of genes? Inside the cell nucleus, both the introns and exons are used to make a very large molecule that is complementary to the DNA strand. Then, enzymes chop out all the introns from the molecule. The exons are "stitched" back together to form a smaller molecule that will be used to make a protein. These steps are summarized in Figure 8-11.

Why are so many genes of eukaryotic cells split up with noncoding sequences? Do these interruptions in genes fulfill some special purpose? Many biologists think that this organization of genes adds evolutionary flexibility. Each exon encodes a different part of a protein. One exon may influence which molecules an enzyme is able to recognize, while another may determine whether a protein will respond to particular signal molecules. By having introns and exons, cells can occasionally shuffle exons between genes and make new genes over time. Natural selection probably favored the intron-exon system of organization because it

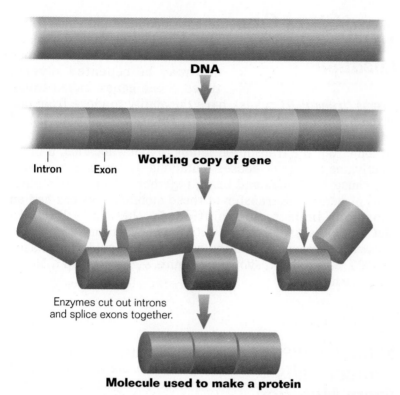

DNA

Intron | Exon

Working copy of gene

Enzymes cut out introns
and splice exons together.

Molecule used to make a protein

Figure 8-11 The genes of eukaryotic organisms contain noncoding information that must be cut out. The remaining fragments are then stitched together and used to make a protein.

☐ CAPSULE SUMMARY

Many eukaryotic genes are interrupted by segments of DNA that do not code for proteins. Coding regions of DNA are called exons, and noncoding regions are called introns.

enables cells to manufacture many different proteins by juggling exons between genes. The many thousands of proteins that occur in human cells appear to have arisen from only a few thousand exons. ☐

Some Genes Exist in Many Copies

Introns were not the only surprise that scientists found in eukaryotic genes. At first they assumed chromosomes carried one copy of each kind of gene—for example, one to make the enzyme that breaks down the sugar lactose, and another one to encode the protein hemoglobin that carries oxygen in your blood. They soon found out that this is not always true. Some genes in your cells exist in multiple copies, clusters of almost identical sequences called **multigene families.** Multigene families may contain as few as three or as many as several hundred versions of a gene. For example, your cells each contain 12 different hemoglobin genes in two families. Three genes are "silent," meaning they do not make proteins; five others are active only during embryonic or fetal development. Only four of the genes encode the two chains of human hemoglobins in adults. The nucleotide sequences of each of the 12 hemoglobin genes are clearly related to each other. Scientists theorize that the 12 different sequences arose from one ancestral gene, which duplicated and evolved into two gene families.

Some Genes Can Jump to New Locations

In both bacteria and eukaryotes, individual genes are scattered randomly about on chromosomes and may be repeated several times. Some genes, called **transposons** (*trans POH zahns*), have the ability to move from one chromosomal location to another. Once every few thousand cell divisions, a transposon jumps to a new location on a chromosome. When a transposon jumps to a new location, it often inactivates a gene or causes mutations. Because transposons can cause mutations and bring together different combinations of genes, the transfer of these mobile genes has had an enormous impact on evolution. In the 1950s Barbara McClintock, a geneticist at Cold Spring Harbor Laboratory, discovered transposons in the course of her studies of maize. In 1983, McClintock, shown in Figure 8-12, received a Nobel prize for her work.

Figure 8-12 The geneticist Barbara McClintock did research that revealed that transposons were the cause of the pigmented, spotted, and colorless kernels found in maize.

Some Genes Protect Cells From Mutation

Recently, scientists have discovered that some genes are able to prevent the ill effects caused by mutation. One such gene, which was the focus of intense research in 1993, is called p53 (short for protein 53). In normal cells, p53 acts as a tumor suppressor—its protein product helps coordinate a complex system of responses to the presence of damaged DNA that might otherwise lead to cancer. The p53 gene works by shutting off cell growth and division in damaged cells or, in some cases, by causing cells to self-destruct. However, when the p53 gene loses its activity by mutation, the cells of the body lose the protection it provides and, as a result, can begin to grow unchecked, signalling the development of cancer. Since 1989, researchers have found mutated forms of the p53 gene in more than 51 kinds of human tumors. As scientists learn more about the molecules that interact with p53, they will uncover new possibilities for drug development.

Section Review

1. *What is the difference between exons and introns? How are the noncoding segments of DNA extracted?*
2. *What is a multigene family? Give an example.*

Critical Thinking

3. *How do transposons and exons differ in the way they affect genes?*
4. *Why is p53 called a tumor suppressor gene?*

adenine (169)
bacteriophage (167)
base-pairing rules (170)
complementary (171)
cytosine (169)
deoxyribose (169)
DNA polymerase (172)
double helix (170)

exon (174)
helicase (172)
intron (174)
guanine (169)
multigene family (175)
nucleotide (169)
purine (169)
pyrimidine (169)

replication (172)
replication fork (172)
thymine (169)
transformation (166)
transposon (176)
vaccine (165)
virulent (165)

Concept Mapping

Construct a concept map that shows the structure of DNA and how it is copied. In constructing your map, use the following terms: nucleotides, phosphate group, five-carbon sugar, nitrogen base, purine, pyrimidine, double helix, replication, DNA polymerases, and gene. Include additional concepts in your map as needed.

Review

Multiple Choice

1. A DNA nucleotide does *not* include a
   a. five-carbon sugar.
   b. phosphate group.
   c. double helix.
   d. nitrogen base.

2. James Watson and Francis Crick
   a. built a structural model of DNA.
   b. determined the matches between the four nitrogen bases.
   c. studied the structure of molecules using X-ray diffraction.
   d. discovered transposons.

3. The experiment of Hershey and Chase showed that
   a. bacteriophages can be injected into human cells.
   b. DNA controls heredity.
   c. bacteria undergo transformation.
   d. a vaccine for pneumonia could be produced.

4. If the sequence of bases in one strand of a DNA molecule is GCCATTG, what is its complementary sequence?
   a. CGGTAAC   c. ATTGCCA
   b. GCCATTG   d. TAAGCCG

5. In his experiments with *Streptococcus pneumoniae,* Griffith found that
   a. the DNA of the heat-killed smooth cells entered some of the rough cells.
   b. the S bacteria were transformed.
   c. the capsule does not protect the bacterium.
   d. mice infected with the R bacteria always died.

6. Multiple replication forks along the DNA molecule
   a. proofread and correct replication errors.
   b. reduce the time required for DNA replication.
   c. ensure that the new and old DNA strands are complementary.
   d. signals DNA polymerase to detach from the strand.

7. The results of Barbara McClintock's experiments with corn led her to conclude that
   a. transposons are inactive.
   b. some genes move about on a chromosome.
   c. chromosomes may contain many versions of a gene.
   d. genes are composed of sequences of nucleotides.

8. Multigene families
   a. exist in prokaryotic cells.
   b. show that chromosomes carry one copy of a single gene.
   c. demonstrate convergent evolution.
   d. are nearly identical copies of the same gene.

9. What properties of the bacteriophage T_2 enabled Hershey and Chase to convincingly show that the genetic material is DNA?
   a. The protein shell of the T_2 phage attaches to a bacterium, and its DNA is injected inside.
   b. Phages containing radioactive sulfur and phosphorus are found in nature.
   c. The T_2 phage lacks protein.
   d. The T_2 phage is a eukaryote, whereas a bacterium is a prokaryote.

10. Eukaryotic genes generally differ from prokaryotic genes in that eukaryotic genes
    a. have no more than one copy of each kind of gene.
    b. are interrupted by noncoding segments of nucleotides.
    c. consist of nucleotides that code for proteins.
    d. never include exons.

Completion

11. According to _____ rules, the amount of thymine equals the amount of adenine, and the amounts of these nitrogen bases vary between different organisms.

12. Thymine and cytosine belong to a group of nitrogen bases known as _____ , while guanine and adenine belong to a group of nitrogen bases known as _____ .

13. A _____ consists of a phosphate group, a nitrogen base, and a five-carbon sugar. Many nucleotides joined together form a _____ molecule.

14. In eukaryotic cells, nucleotide sequences that code for amino acids are _____ ; those that are removed from the gene are _____ .

15. In 1983, Barbara McClintock received a Nobel Prize for her discovery of _____, which are sometimes called "jumping genes."

16. Using protein- and DNA-destroying enzymes, _____ demonstrated that the transforming factor discovered by Griffith is DNA.

17. During the process of _____ , the two strands of a double helix separate and are used as templates to build _____ strands of DNA.

Short Answer

18. What was the nature of the research done by Erwin Chargaff? What conclusions did Chargaff draw from his work?

19. Would Watson and Crick have been able to build their double helix model of DNA in 1953 without the photographs they obtained from Maurice Wilkins and Rosalind Franklin? Explain.

20. The majority of scientists were not convinced by the findings of Oswald Avery and his colleagues that DNA is the primary heredity mechanism in all cells. However, they were convinced by the findings reported by Hershey and Chase. Explain.

21. Why are the experiments of Griffith and Avery and his colleagues considered important but not conclusive steps in showing that DNA is the genetic material?

22. How is a DNA molecule copied in the process of replication?

23. Why is damage to exons very likely to affect the synthesis of a protein, while damage to introns is not?

24. A scientist observed that corn kernels that usually have no pigment were spotted with pigment. He knew that the spots were not due to the normal expression of the gene for kernel pigment. What genetic phenomenon could explain his observations?

25. **Evolution** What does the presence of DNA in all living organisms and its function as the genetic material suggest about how life evolved on Earth?

26. **Evolution** Why do scientists believe that evolutionary flexibility is increased by having exons separated by introns?

27. **Structure and Function** What are some ways in which the structure of a DNA molecule is related to its function?

28. **Levels of Organization** A molecule of DNA consists of two long strands of nucleotides bonded together. What molecules make up a nucleotide?

29. **Making Inferences** In what ways was the process of communication vital to the work of Watson and Crick?

30. **Making Inferences** Rosalind Franklin died of cancer at an early age. How might her work with X-ray diffraction have contributed to her death?

31. **Interpreting Graphics** The data in the table below were collected by using thin-layer chromatography. They show the amount of each type of base, by percentage, in several DNA samples. Do the data support Chargaff's rules? Explain.

	G	C	A	T
mold	15.2	34.3	14.9	35.1
plant	19.7	41.2	19.5	42.1
mollusk	17.4	32.3	17.9	34.7
reptile	12.9	35.6	13.2	36.7
mammal	14.6	39.5	13.8	37.6

32. **Making Predictions** A scientist extracted 4.6 picograms (or 10^{-12} grams) of DNA from mouse muscle cells. She also extracted 4.6 picograms of DNA from an equal amount of mouse liver cells. How much DNA could be extracted from the same amount of mouse kidney cells? how much from the same amount of mouse sperm? Explain.

33. **Research and Writing** Find out about DNA fingerprinting and how a DNA fingerprint can be used by police to solve a crime.

34. **Research and Writing** Shortly before Watson and Crick presented their model of DNA to the scientific community, Linus Pauling proposed a three-stranded model of DNA. Find out how the work of Linus Pauling influenced Watson and Crick's thinking about the structure of DNA. What led Watson and Crick to propose a double helix model of DNA rather than to accept Pauling's model?

35. **Multicultural Perspective** Discuss the kinds of obstacles that you think women scientists such as Rosalind Franklin and Barbara McClintock faced in the past. What kind of discrimination might they have had to overcome?

36. Read the article "DNA's Stroke of Genius," in *New Scientist*, volume 138, pages 21–23. What did reviewers of *The Double Helix* say about the uniqueness of Watson and Crick's contribution?

37. Read the article "Happy Birthday, Double Helix," in *Time*, March 15, 1993. Who gathered at Cold Spring Harbor Laboratory and what were they celebrating? Describe three ways their discovery has impacted the world.

DNA and Its Structure

OBJECTIVE

Construct and analyze a model of DNA.

PROCESS SKILLS

- constructing, identifying, and manipulating a model

MATERIALS

- plastic soda straws, 3 cm sections (48)
- metric ruler
- scissors
- permanent marker
- 48 pushpins (12 red, 12 blue, 12 yellow, and 12 green)
- 48 paper clips

BACKGROUND

1. What provides the instructions that direct the activities in a cell?
2. What is the structure of DNA?
3. How can can you use models of DNA to demonstrate the process of DNA replication?
4. Write your own **Focus Question** on your Vee Form.
5. **Knowing Side of the Vee** List the **Concepts** and new **Vocabulary Words** on your Vee Form. In the **Concept Statements** section of the Vee, use these words in sentences that define and explain them.

TECHNIQUE

Doing Side of the Vee

1. Work with a classmate to complete the investigation.
2. **CAUTION: Pointed objects can cause injury if not properly used**. Cut the soda straws into 3-cm pieces to make 48 segments.

3. Insert a pushpin midway along the length of each straw segment. Push a paper clip into one end of each straw segment until it touches the pin.

4. Keep the pins in a straight line, and insert the paper clip of a blue pushpin segment into the open end of a red pin segment. Add segments of straw to the red segment end in the following order: green, yellow, blue, yellow, blue, yellow, green, red, red, and green. Use the permanent marker to label the blue segment at the end "top." This strand of segments is one-half of your first model.

5. Construct the other half of your first model beginning with a yellow pin segment. Keep the pins in a straight line.

Link segments together in the following order: green, red, blue, yellow, blue, yellow, blue, red, green, green, and red. Label the yellow segment at the end "top."

6. Place the strands parallel to each other on the table with the "top" blue pin of the first strand facing the "top" yellow pin of the second strand.

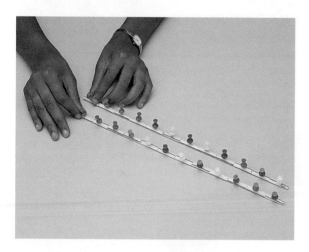

7. Demonstrate replication by simulating a replication fork at the top pair of pins. Add the remaining straw segments to complete a new DNA model. Be sure to use the base-pairing rules.

8. Assign the following bases to the pushpin colors: red = adenine, blue = guanine, yellow = cytosine, and green = thymine. Make a sketch of the original DNA model in the **Records** section of your Vee Form. Sketch the process of DNA replication in the **Records** section of your Vee Form. In the **Procedure** section of the Vee, briefly summarize the procedure you followed.

INQUIRY

1. How many nucleotides did the original DNA model contain?

2. What color pin is always across from a blue pin?

3. What color pin is always across from a green pin?

4. Write the base-pair order for the DNA molecule you created, using the following code: red = adenine, blue = guanine, yellow = cytosine, and green = thymine.

5. How does the replicated model of DNA compare to the original model of DNA?

6. Use the information on the **Knowing Side** of the Vee to interpret your results from the **Doing Side,** and then write your **Knowledge Claim.** Write a **Value Claim** for this lab.

ANALYSIS

1. What is the name given to the point at which replication starts on a DNA molecule?

2. What would the complementary bases be if one side of a DNA molecule had the bases adenine, cytosine, cytosine, thymine, thymine, and adenine?

3. Speculate what would happen if the base pairs in the replicated model were not in the same sequence as the original model.

4. Of what benefit is DNA replication to humans?

5. What are the advantages of having DNA remain in the nucleus?

FURTHER INQUIRY

Write a **New Focus Question** that could be the point of a new investigation. The following are examples:

How do DNA molecules differ among various species of animals and plants? How are they similar?

How might transposing genes produce cancer?

CHAPTER 9

GENE EXPRESSION

REVIEW

- mutation (Section 6-1)
- cell division (Section 6-2)
- nucleotide (Section 8-1)
- structure of DNA (Section 8-2)
- DNA replication (Section 8-2)

Active genes, appearing as bright spots, on fruit fly chromosomes

9-1 From Genes to Proteins

*T*he discovery that genes are made of DNA led to more questions about heredity. How is the information in DNA used to determine an organism's characteristics? Investigators soon found a partial answer: an organism's traits are determined by proteins that are built according to the plans specified in its DNA. The next question was, How does DNA determine the nature of a protein?

Section Objectives

■ Describe three roles of RNA in gene expression.

■ Summarize the process of transcription.

■ Recognize the evolutionary significance of the genetic code.

■ Outline the major steps of translation.

The Path of Genetic Information

Proteins are not built directly from genes. Your cells preserve hereditary information by transferring the information in genes into sets of working instructions for use in building proteins. The working instructions of the genes are made of molecules of **ribonucleic** *(ry boh noo KLAY ihk)* **acid,** or RNA. RNA, like DNA, is a nucleic acid. However, it differs from DNA in three ways. First, RNA consists of a single strand of nucleotides instead of the two strands that form the DNA double helix. Second, RNA contains the five-carbon sugar ribose *(RY bohs)* rather than the sugar deoxyribose. And third, RNA has a nitrogen-containing base called **uracil** *(YUR uh sihl)*—abbreviated as U—instead of the base thymine found in DNA. Like thymine, uracil is complementary to adenine.

RNA is present in cells in three different forms, each of which has a different function: messenger RNA (mRNA), ribosomal RNA (rRNA), and transfer RNA (tRNA). All three types of RNA are essential for processing the information from DNA into proteins, a process called **gene expression**. Gene expression occurs in two stages—transcription and translation. In **transcription**, the information in DNA is transferred to mRNA. In **translation**, the information in mRNA is used to make a protein. The path of genetic information is summarized in Figure 9-1.

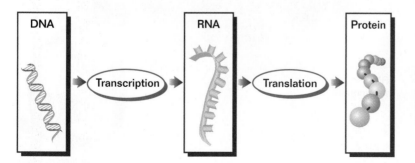

Figure 9-1 During gene expression, the information in DNA is first "rewritten" (transcribed) as a molecule of mRNA and then "deciphered" (translated) and used to build a protein.

Transcription: Making RNA

The first step in using DNA to direct the making of a protein is transcription, the process that "rewrites" the information in a gene in DNA into a molecule of mRNA. In eukaryotic organisms, transcription occurs inside the nucleus; in prokaryotic organisms, it takes place in the cytoplasm. Transcription begins when an enzyme called **RNA polymerase** binds to the beginning of a gene on a region of DNA called a promoter. A **promoter** is a specific sequence of DNA that acts as a "start" signal for transcription. After RNA polymerase binds to a promoter, the enzyme starts to unwind and separate the double helix's two strands, exposing the DNA's nitrogen-containing bases. Like DNA replication, transcription uses DNA bases as a template for making a new molecule (RNA). In transcription, however, only one of the two strands of DNA serves as a template.

Once a portion of the DNA double helix has separated, RNA polymerase moves along the bases of the template strand like a train on a track, always in the same direction. The enzyme reads each nucleotide and pairs it with a complementary RNA nucleotide, as shown in Figure 9-2. In eukaryotic cells, the RNA nucleotides are found in the nucleus; in prokaryotic cells, they are in the cytoplasm. Transcription follows the same base-pairing rules as DNA replication except that uracil, rather than thymine, pairs with adenine. When the RNA nucleotides are added, they are linked together with sugar-to-phosphate covalent bonds.

As RNA polymerase works its way down the strand, a single strand of RNA grows and

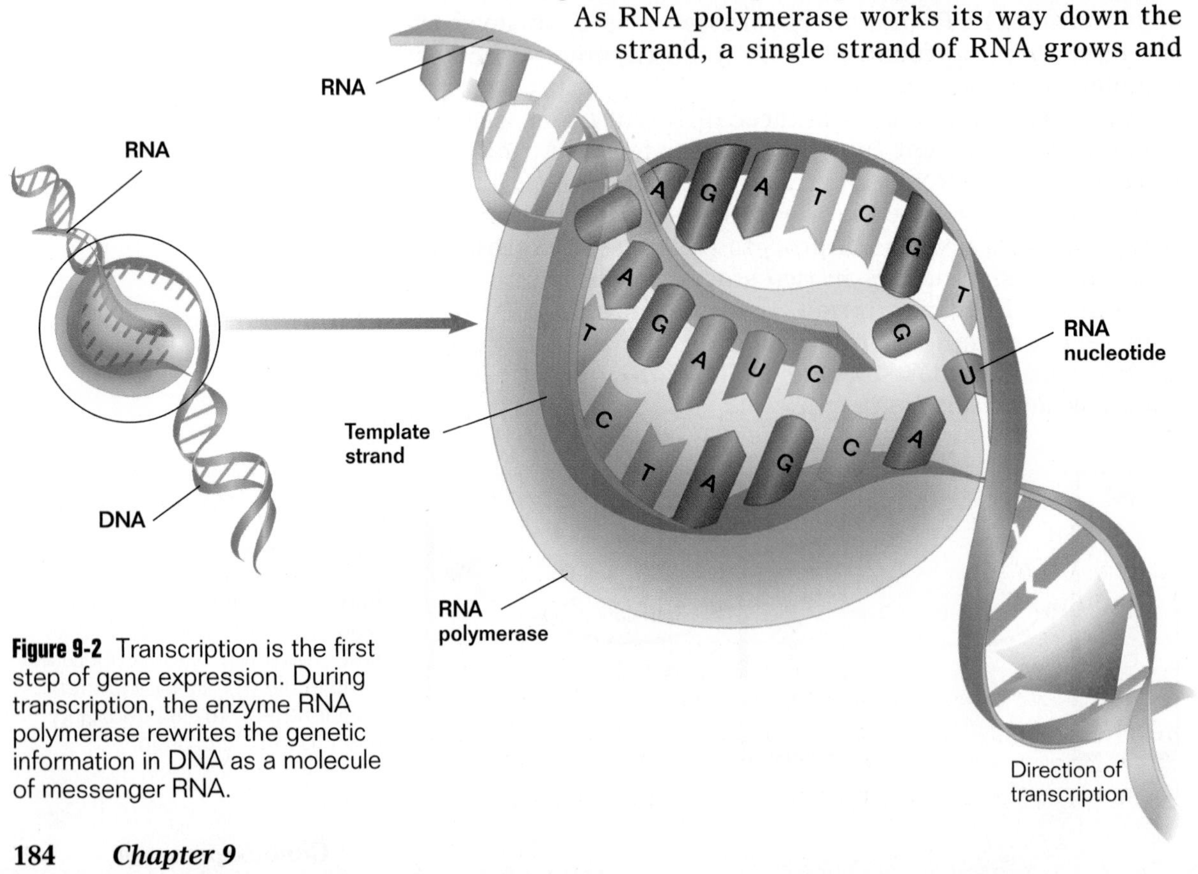

Figure 9-2 Transcription is the first step of gene expression. During transcription, the enzyme RNA polymerase rewrites the genetic information in DNA as a molecule of messenger RNA.

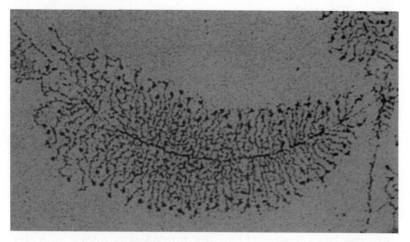

Figure 9-3 A cell can produce large amounts of a particular protein by transcribing a single gene with several molecules of RNA polymerase. The multiple copies of RNA being made in this electron micrograph give the DNA a feathery appearance.

dangles off the enzyme like a tail, as shown in Figure 9-3. Behind RNA polymerase, the two strands of DNA close up by forming hydrogen bonds between complementary bases, re-forming the double helix.

Transcription proceeds at a rate of about 60 nucleotides per second until the RNA polymerase reaches a stop signal on the DNA called a terminator. A **terminator** is a sequence of bases that tells the RNA polymerase to stop adding nucleotides. At this point the enzyme detaches from the DNA and releases the RNA molecule into the cell for the next stage of gene expression.

Three Types of RNA

Transcription manufactures three types of RNA: messenger RNA (mRNA), transfer RNA (tRNA), and ribosomal RNA (rRNA). **Messenger RNA** is an RNA copy of a gene used as a blueprint for a protein. When a cell needs a particular protein, a specific mRNA is made. Messenger RNA is appropriately named because it *carries* hereditary information from DNA and *delivers* it to the site of translation. During translation, mRNA serves as a template for the assembly of amino acids.

The functions of tRNA and rRNA differ from that of mRNA. **Transfer RNA** acts as an interpreter molecule, translating mRNA sequences into amino acid sequences; **ribosomal RNA** plays a structural role in ribosomes, the organelles that function as the sites of translation. You will learn more about these two molecules when you learn about translation.

Processing mRNA in Eukaryotic Cells

After transcription is completed in eukaryotes, a messenger RNA molecule must be processed before it can serve its role in building a protein. In Chapter 8 you learned that many eukaryotic genes are split into coding regions called exons, which are interrupted by noncoding regions called introns. Both exons and introns are transcribed into mRNA. Before mRNA leaves the nucleus, the introns are cut out. The exons are joined to form a single molecule of mRNA, which leaves the nucleus through a pore, and enters the cytoplasm. ❑

❑ CAPSULE SUMMARY

During transcription, RNA polymerase assembles a molecule of RNA on a DNA template. Messenger RNA, one of three types of RNA, carries genetic information from the DNA of genes to the site of translation.

The Genetic Code

....

After transcription, the genetic message is ready to be translated from the language of RNA to the language of proteins. The instructions for building a protein are written as a series of three-nucleotide sequences called **codons** (*KOH dahnz*). Each codon along the mRNA strand either corresponds to an amino acid or signifies a stop signal. Recall from Chapter 2 that amino acids bond together to form a protein chain. You will learn more about the role of "stop" signals in translation later in this section.

From trial-and-error experiments, biologists worked out which codons correspond to which amino acids. In 1961, Marshall Nirenberg, an American biochemist, deciphered the first codon by making artificial mRNA that contained only the base uracil. When the mRNA was added to a test tube containing amino acids, it was translated into a polypeptide made of the amino acid phenylalanine. From this, Nirenberg learned that the codon UUU is the instruction for the amino acid phenylalanine. Soon, more elaborate techniques enabled scientists to decipher codons consisting of more than one kind of base. Table 9-1 presents the entire **genetic code**, the amino acids and stop signals that are coded for by each of the possible mRNA codons.

The genetic code is nearly universal. With few exceptions, it is the same in all organisms. For example, the codon GUC codes for the amino acid valine in bacteria, in eagles, in dogs, and in your own cells. Thus, it appears that all life forms had a common evolutionary ancestor with a single genetic code. The only exceptions biologists have found to this rule are in the ways cell organelles (such as mitochondria and chloroplasts) and a few microscopic protists (ciliates such as *Paramecium*) read stop codons. ▢

▢ CAPSULE SUMMARY

All organisms have a genetic code made of three-nucleotide sequences called codons. Codons correspond to particular amino acids and to stop signals. The genetic code is nearly universal.

	U	C	A	G	
U	Phe	Ser	Tyr	Cys	**U**
	Phe	Ser	Tyr	Cys	**C**
	Leu	Ser	stop	stop	**A**
	Leu	Ser	stop	Trp	**G**
C	Leu	Pro	His	Arg	**U**
	Leu	Pro	His	Arg	**C**
	Leu	Pro	Gln	Arg	**A**
	Leu	Pro	Gln	Arg	**G**
A	Ile	Thr	Asn	Ser	**U**
	Ile	Thr	Asn	Ser	**C**
	Ile	Thr	Lys	Arg	**A**
	Met	Thr	Lys	Arg	**G**
G	Val	Ala	Asp	Gly	**U**
	Val	Ala	Asp	Gly	**C**
	Val	Ala	Glu	Gly	**A**
	Val	Ala	Glu	Gly	**G**

Table 9-1 In the genetic code, each amino acid is coded for by a three-nucleotide sequence called a codon. The first base in a codon is found along the left side of the chart, the second base is at the top, and the third base is found along the right side of the chart. Because any of four different nucleotides may be used at each of the three positions, there are 64 different possible codons ($4 \times 4 \times 4 = 64$) in the genetic code.

Translation: Making Proteins

The equipment for translation is located in the cytoplasm, where a cell keeps its supply of transfer RNA (tRNA). A tRNA molecule is a single strand of RNA folded into a compact shape with three loops, illustrated in Figure 9-4. One of the loops has a three-nucleotide sequence called an **anticodon** *(an tee KOH dahn)*. It is called an anticodon because the three-nucleotide sequence is complementary to one of the 64 codons of the genetic code. This enables tRNA to bind to mRNA through hydrogen bonding. In most organisms, there is no tRNA molecule with an anticodon complementary to the codons UAG, UAA, or UGA, which is why these codons act as stop codons. Opposite the anticodon on a tRNA molecule is a site at which the molecule carries an amino acid. The amino acid that a tRNA molecule can carry corresponds to a particular codon.

A cell's cytoplasm also contains thousands of ribosomes, the protein-making factories of the cell. A ribosome, illustrated in Figure 9-5, is composed of two subunits, which are bound together only when they are involved in translation.

A ribosome has three binding sites that play important roles in translation. One binding site holds mRNA so that its codons are accessible to tRNA molecules. The other two binding sites recognize tRNA. The **A site** holds a tRNA molecule that is carrying its specific amino acid. The **P site** holds a tRNA molecule that is carrying its specific amino acid attached to the growing protein chain. As you can see in Figure 9-5, these two binding sites are next to each other on the ribosome.

Figure 9-4 The unique shape of tRNA enables it to act as an interpreter molecule. Its anticodon complements a specific codon and corresponds to a specific amino acid.

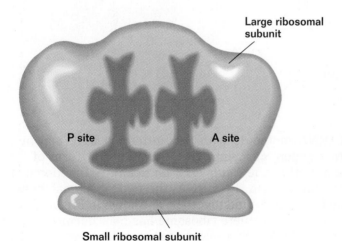

Figure 9-5 Each ribosome is made of over 50 different proteins and several segments of ribosomal RNA.

Assembling the Protein

Translation begins when mRNA binds to the smaller ribosomal subunit in the cytoplasm. The mRNA is now oriented so that the "start" codon, a codon that signals the beginning of a protein chain, is sitting in the P site. Research has shown that in most cases, the start codon has the sequence AUG.

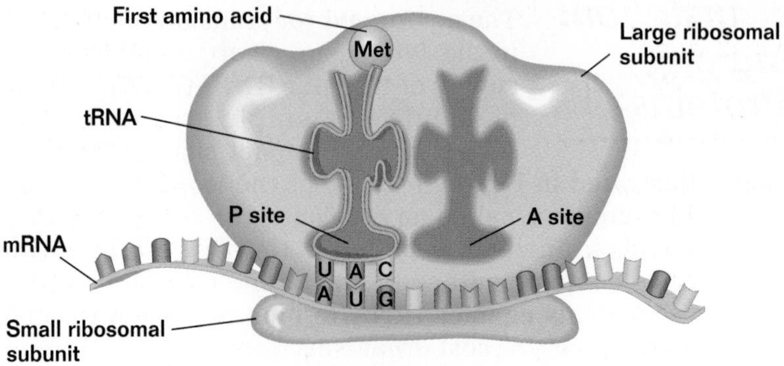

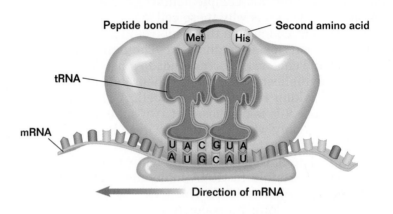

First amino acid — Met

Large ribosomal subunit

tRNA

P site   A site

mRNA

U A C
A U G

Small ribosomal subunit

Figure 9-6 When translation begins, mRNA is bound to a complete ribosome so that the start codon is positioned in the P site, ready for the first amino acid of the protein chain.

Peptide bond — Met   His — Second amino acid

tRNA

mRNA

U A C G U A
A U G C A U

Direction of mRNA

Figure 9-7 When both sites on the ribosome are filled, a peptide bond can form and link the amino acids.

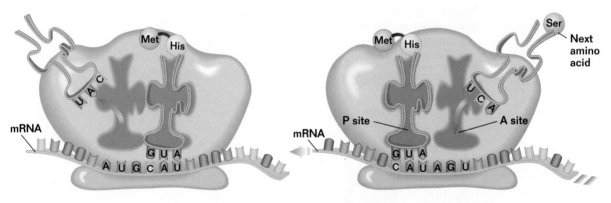

Met His

U A C

mRNA

G U A
A U G C A U

Met His   Ser — Next amino acid

U C A

P site   A site

mRNA

G U A
C A U A G U

Figure 9-8 After a peptide bond is formed, mRNA shifts on the ribosome so that a new codon is present in the A site.

A tRNA molecule with the anticodon UAC can bind to the start codon, carrying with it a modified form of the amino acid methionine (*muh THY uh neen*). A functional ribosome is formed when the mRNA, the two ribosomal subunits, and the first tRNA bind together, as shown in Figure 9-6.

Once a complete ribosome has been formed, the codon in the vacant A site is ready to receive the next tRNA. A tRNA molecule with the complementary anticodon arrives and binds to the codon, carrying its specific amino acid with it. Now, both the A site and the P site are holding amino acids. Next, an enzyme helps form a peptide bond between the adjacent amino acids, forming the first link of the protein chain, shown in Figure 9-7. Afterward, the tRNA in the P site detaches and moves away from the ribosome, leaving behind its amino acid. The tRNA in the A site moves over to fill the vacant P site, with the protein chain in tow. Because the anti-codon remains attached to the codon, the tRNA molecule and mRNA molecule move across the ribosome into the P site as a unit. As a result, a new codon is present in the A site, ready to receive the next tRNA and its amino acid, as shown in Figure 9-8. Amino acids are carried to the A site and bonded to the protein chain until the end of the mRNA sequence is reached. At this point, a stop codon is encountered for which, as you know, there is no anticodon. With nothing to fit into the empty A site in the ribosome, the ribosome complex falls apart and the newly made protein is released into the cell, as shown in Figure 9-9. ☐

☐ **CAPSULE SUMMARY**

After mRNA has bound to a ribosome, tRNA carries amino acids to the ribosome according to the three-base codons. The amino acids are joined to form a protein chain.

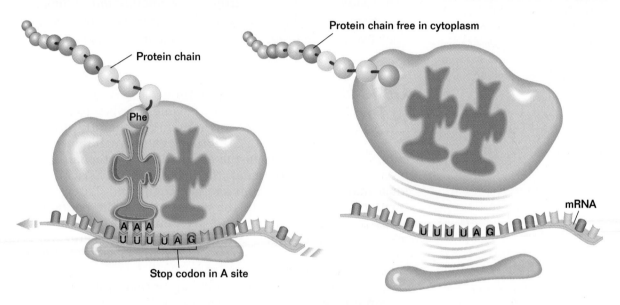

Protein chain free in cytoplasm

Protein chain

Phe

A A A
U U U U A G

Stop codon in A site

mRNA

U U U U A G

Figure 9-9 The sequence of amino acids in a gene determines the primary structure of a protein. After a protein is synthesized, it begins to fold into a molecule with secondary and tertiary structures.

Section Review

1. *Define the two stages of gene expression. Could translation take place if transcription did not occur? Explain why or why not.*

2. *What is the evolutionary significance in describing the genetic code as "nearly universal"?*

Critical Thinking

3. *Why is the term* transcription *appropriate for describing the process of making RNA?*

4. *Why is the term* translation *appropriate for describing the synthesis of proteins?*

5. *How does the structure of tRNA enable it to act as an interpreter molecule?*

6. *How do a ribosome's A site and P site enable it to function as a protein-making factory?*

9-2 Regulating Gene Expression

Section Objectives

- Discuss why the regulation of gene expression in living organisms is important.

- Explain how the lac operon is affected by the presence and absence of lactose.

- Explain how the steroid hormone estrogen acts to regulate gene expression.

*B*eing able to translate a gene into a protein is only part of gene expression. Every cell must also be able to regulate when particular genes are used. Imagine if every instrument in an orchestra played at full volume constantly. All you would hear is noise! No orchestra plays that way, because music is not noise—it is the controlled expression of sound. Similarly, every function that a living organism carries out is the controlled expression of genes, each used at the proper moment to achieve precise effects.

Gene Regulation in Prokaryotes

In order to survive, bacteria must be able to adjust to changes in their environment, such as fluctuations in available nutrients. For example, when the amino acid tryptophan is not present in the environment, the bacterium *Escherichia coli* must manufacture it from another compound. Later, when tryptophan is present in the environment, the bacterium stops making the amino acid. By being able to adjust its metabolism to changes in its environment, a bacterium is saved from wasting its energy and resources on producing a substance that is readily available.

Scientists first studied gene expression in bacteria. An example of gene regulation that is well understood at the molecular level is found in *E. coli*. When you consume a milk product, the disaccharide lactose (milk sugar) is soon present in your intestinal tract and available to the *E. coli* living there. Before a bacterium can absorb the lactose, it must first make beta-galactosidase, an enzyme that cleaves lactose into glucose and galactose.

As with tryptophan, it is in a bacterium's best interest to focus its energy on using available nutrients. Therefore, *E. coli* should make beta-galactosidase only when lactose is present. To understand how this happens, you must first understand the basic mechanism that controls gene regulation in prokaryotes—the operon. An **operon** *(AHP uhr ahn)* is a cluster of genes that codes for proteins with related functions. The gene for beta-galactosidase is part of a group of genes called the *lac* **operon.** The *lac* operon is divided into several different regions, which are shown in Figure 9-10. One region is the promoter, the RNA polymerase binding site that signals the beginning of a gene. Another region is made of structural genes, the genes that code for polypeptides. Between the promoter and the structural genes is a region of DNA called the **operator**. Because of its position in

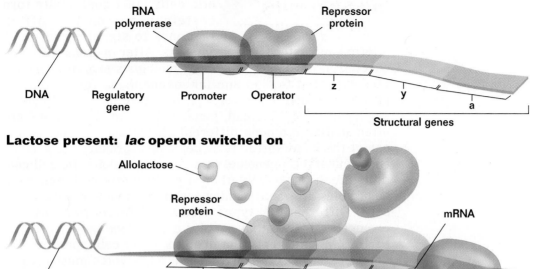

Lactose absent: *lac* operon switched off

RNA polymerase

Repressor protein

DNA

Regulatory gene

Promoter

Operator

z

y

a

Structural genes

Lactose present: *lac* operon switched on

Allolactose

Repressor protein

mRNA

DNA

Promoter

Operator

z

y

a

Direction of transcription

Figure 9-10 The *lac* operon has three structural genes that code for proteins used to metabolize lactose. When lactose is absent from a bacterium's environment, the repressor protein is bound to the operator and the *lac* operon is switched off. When lactose is present, the repressor protein detaches from the operator and the *lac* operon is switched on.

the operon, the operator is able to control RNA polymerase's access to the structural genes; it acts like a switch, turning the operon on or off.

What determines whether the *lac* operon is in the "on" or "off" mode? When no molecules are bound to the operator, the *lac* operon is switched on; RNA polymerase can bind to the promoter, move across the operator, and transcribe the structural genes. The *lac* operon is switched off when a protein called a **repressor** is bound to the operator. A bound repressor creates a barrier, preventing RNA polymerase from transcribing the structural genes (although RNA polymerase can still bind to the promoter).

Transcription can resume when the repressor is removed by a molecule called an **inducer**. In the case of the *lac* operon, the inducer is allolactose, a molecule that is made from lactose when it enters the cell. When allolactose is present in the cell, it binds to the repressor and changes the shape of this protein. As a result, the repressor falls off the operator, as shown in Figure 9-10. Now, RNA polymerase has access to the structural genes. The bacterial cell can begin building the proteins it needs to metabolize lactose.

When lactose is not present in the intestinal tract, allolactose is not produced. Therefore, there is nothing to change the shape of the repressor, which remains bound to the operator. As a result, transcription of the *lac* operon is blocked and the structural genes remain unexpressed. By producing the enzymes only when the nutrient is available, the bacterium avoids wasting its energy making proteins it does not need. ◻

◻ CAPSULE SUMMARY

The lac *operon is a well-understood mechanism of gene regulation in* E. coli. *It is a cluster of genes that enables a bacterium to build the proteins needed for lactose metabolism only when lactose is present.*

Gene Regulation in Eukaryotes

Like prokaryotic cells, eukaryotic cells must continually turn certain genes on and off in response to signals from their internal and external environments. After operons were discovered in the 1960s, molecular biologists studying eukaryotes expected to find similar mechanisms for regulating gene expression. However, operons have not been found in eukaryotic cells. Instead, genes with related functions are often scattered among different chromosomes. This is just one of the many differences between a prokaryotic genome and a eukaryotic genome. In addition, a eukaryotic cell contains much more DNA than a prokaryotic cell, and most eukaryotes are multicellular organisms made of specialized cells. Different cell types produce different proteins. For example, the gene that enables a red blood cell to produce hemoglobin is also present in all the other cell types, but it is unexpressed. Gene expression in eukaryotes must involve mechanisms that account for differences such as these.

Much of eukaryotic gene expression is still a mystery. What scientists do know is that gene expression is affected by the way the DNA in a chromosome is physically arranged. Transcription takes place in the regions of a chromosome where the DNA has uncoiled. Biologists are able to see evidence of this process in polytene (*PAHL ih teen*) chromosomes. Polytene chromosomes are giant chromosomes found in salivary gland cells of flies such as *Drosophila*. These giant chromosomes consist of a large number of partially replicated chromosomes stacked neatly side by side. Figure 9-11 shows a polytene chromosome from the fruit fly *Drosophila* in which DNA has uncoiled to form a chromosome puff. A **chromosome puff** is a region of intense transcription that forms when DNA loops out from the chromosome, perhaps making the genes in that region more accessible to RNA polymerase. Chromosome puffs along a chromosome change locations as the insect develops.

Figure 9-11 When the genes in a particular region of a polytene chromosome are activated, biologists are able to see obvious changes in the shape of the chromosome.

Once a eukaryotic gene is available to be expressed, what kinds of mechanisms regulate its expression? As with prokaryotes, gene expression can be regulated before, during, or after transcription. However, because a eukaryotic cell has a nuclear envelope that physically separates transcription from translation, there are more opportunities for regulating gene expression. Control mechanisms have been found to occur after the mRNA leaves the nucleus, before translation, and even after the protein is functional. ❏

Enhancer Control

Like prokaryotes, eukaryotes regulate gene expression primarily by controlling when RNA polymerase binds to the beginning of a gene. In eukaryotes this binding cannot take place without the aid of a cluster of proteins called transcription factors. Sometimes one or more of the necessary transcription factors are located at a site called an **enhancer,** which is far distant from the gene being regulated. Because each such factor can be sensitive to different aspects of the cell, enhancers make eukaryotic gene regulation very flexible.

To understand how an enhancer stimulates transcription, look at the action of steroid hormones in the cells of vertebrates. A steroid hormone is a molecule made of lipids that acts as a chemical signal. Estrogen is a steroid hormone responsible for secondary sex characteristics in females. When estrogen passes through the cell membrane of specific cells it binds to a receptor protein in the nuclear envelope, forming a hormone-receptor complex. The hormone-receptor complex has the right shape to bind to a specific protein called an acceptor protein, which in turn binds to enhancer regions in the DNA. Binding of the acceptor protein to the enhancer region stimulates RNA polymerase to begin transcription. ❏

❏ **CAPSULE SUMMARY**

Gene expression in eukaryotes involves mechanisms that must uncoil the appropriate regions of DNA in specific cells at specific times.

❏ **CAPSULE SUMMARY**

Enhancers are regions of DNA that stimulate transcription of certain genes in eukaryotic organisms.

Section Review

1. *Why is gene regulation necessary in a living organism?*
2. *What effect does a repressor have on the* lac *operon when lactose is not present? when lactose is present?*
3. *How does the steroid hormone estrogen trigger transcription in a eukaryotic cell?*

Critical Thinking
4. *How would the* lac *operon in E. coli living in your intestinal tract be affected if you ate a bowl of ice cream?*
5. *Why are there more opportunities for regulating gene expression in eukaryotic cells than in prokaryotic cells?*

Genes, Mutation, and Cancer

- Describe three ways point mutations alter genetic material.
- Discuss how the process of cell division can be affected by mutations.
- Give an example of an oncogene, and describe how it leads to cancer.

*I*f all the DNA in your cells could be pulled out, uncoiled, and connected end to end, it would be about 200 billion kilometers long. When all this DNA is being made, errors in replication could cause fatal mistakes. Fortunately, cells are equipped with mechanisms that work to preserve the integrity of your hereditary information. These mechanisms do not always perform perfectly, however, and errors do occur.

Mutations Are Changes in DNA

Although rare, changes in an organism's hereditary information can occur. As you learned in Chapter 6, a change in the DNA of a gene is called a mutation. The effects of a mutation vary, depending on whether it occurs in a gamete or in a body cell. Mutations in gametes can be passed on to offspring of the affected individual, but mutations in body cells affect only the individual in which they occur. Mutations are an important basis of evolution. Some mutations alter the structure of the chromosome itself. Other mutations called **point mutations,** illustrated in Figure 9-12, change one nucleotide or just a few

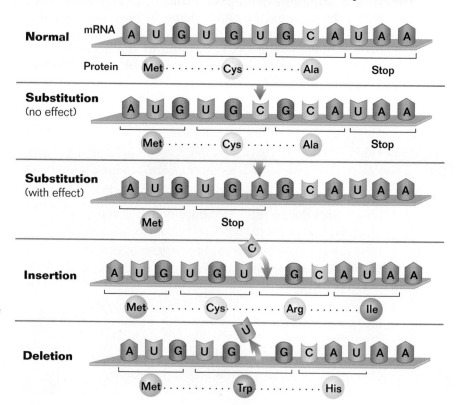

Figure 9-12 Point mutations arise when a chromosome breaks. The product of a normal gene is shown in the top row of this figure. Four effects of point mutations are shown below it.

nucleotides in a gene. There are two general types of point mutations. In point mutations called **substitutions,** one nucleotide in a gene is replaced with a different nucleotide. For instance, the codon UGU becomes UGC. Sometimes substitutions have little or no effect. In this case, the correct codon, UGU, and the mutated codon, UGC, both translate into the same amino acid—cysteine. However, if UGU changed into UGA, the codon would become a stop codon, and translation would end prematurely. As a result, the protein that the gene codes for would be shortened and incomplete.

In point mutations called **insertions** and **deletions,** one or more nucleotides are added to or deleted from a gene. Because the genetic message is read as a series of triplet codons, insertions and deletions can upset the triplet groupings. Imagine deleting the letter C from the sentence "THE CAT ATE THE RAT." Keeping the triplet groupings, the message would read "THE ATA TET HER AT," which is meaningless. A mutation that causes a gene to be read in the wrong three-nucleotide sequence is called a frameshift mutation because the reading pattern is displaced one or two positions. If a reading is displaced three positions, there may be no effect at all on the protein's function.

What Causes Mutations?

Some mutations are chemical mishaps that arise spontaneously. Other mutations, however, are induced by exposure to environmental agents called **mutagens** (*MYOOT uh jehnz*). Mutagens include ionizing radiation (such as X rays and gamma rays), ultraviolet light, and certain chemicals or irritants referred to as carcinogens (*kahr SIHN uh jihns*). **Carcinogens** are cancer-causing agents such as asbestos (fibers used in insulation), benzene (a liquid used in making detergents, plastics, and paints), and other industrial pollutants. ❑

❑ *CAPSULE SUMMARY*

Mutations are changes in the genetic message in an organism. A mutation may change the structure of a chromosome or may just alter a single nucleotide.

Mutations Can Cause Cancer

Of all the health problems that can afflict humans, few are as mysterious as cancer. **Cancer** is a term used to indicate a disease characterized by abnormal cell growth. Normally, cell growth is a highly regulated process that is controlled by chemical and physical signals that inhibit or stimulate cell division. A cell that has become cancerous does not respond to the signals that stop cell division. As a result, cancerous cells, such as the ones shown in Figure 9-13, divide without stopping.

Health problems begin when a cell that has turned cancerous evades the body's immune system, which normally destroys cancerous cells. The cancerous cell then proliferates, forming a mass of cells called a **tumor.** Tumors are classified as either benign or malignant. A benign tumor does not

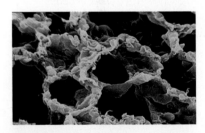

Figure 9-13 In the healthy lung tissue, *top,* cell division is highly regulated. When cells escape regulation and continue to grow, *bottom,* a tumor forms.

invade surrounding tissues. They usually cause few problems and can be surgically removed. Tumors described as malignant, on the other hand, are very harmful. A malignant tumor spreads into other tissues and interferes with organ functions. The most devastating property of a malignant tumor is that its cells are able to break free of the tumor and enter blood and lymph vessels. These cancerous cells are then carried to new locations in the body and form new growths. The spread of malignant cells beyond their original site is called **metastasis** *(muh TAHS tuh sihs)*.

Cancer Genes

Through many years of intense work, researchers have learned that a cell becomes cancerous when mutations occur in genes that regulate cell growth. An example of growth-regulating genes is a class of genes called the *ras* genes (because they were first discovered in viruses that cause tumors known as sarcomas in rats). The *ras* genes code for proteins that help prevent uncontrolled cell division. When there is a mutation in a *ras* gene, a faulty *ras* protein is built. As a result, the cell divides more rapidly than normal, a sign that it has become cancerous. The *ras* genes are examples of oncogenes *(ang KOH jeens)*. An **oncogene** is a gene that, when mutated, can cause a cell to become cancerous.

Researchers have found that a mutated *ras* gene usually contains a point mutation. For example, in a form of human bladder cancer caused by a mutated *ras* gene, a single G nucleotide has been replaced with a T, transforming the normal amino acid glycine into a valine. However, the *ras* gene is only one of several controls that the cell normally exercises over unwanted growth. All of these controls must be inactivated before cancer results. This is why most cancers occur in people over 40 years old—it takes time for an individual cell to accumulate the necessary mutations. ∎

☐ *CAPSULE SUMMARY*

Cancer is the uncontrolled growth of cells. It begins when the genes that control cell growth become mutated. An example of such a mechanism is a class of genes called the ras genes.

Section Review

1. *Which could have a more serious effect on a protein: a base-pair substitution or a frameshift mutation? Why?*
2. *What is cancer? How do the terms* tumor *and* metastasis *relate to this health problem?*
3. *What is the function of the protein encoded by the* ras *gene? How can a cell be affected when the* ras *gene is mutated?*

Critical Thinking

4. *Why is a malignant tumor considered more harmful than a benign tumor?*

Vocabulary

anticodon (187)
A site (187)
cancer (195)
carcinogen (195)
chromosome puff (192)
codon (186)
deletion (195)
enhancer (193)
gene expression (183)
genetic code (186)
inducer (191)

insertion (195)
lac operon (190)
messenger RNA (185)
metastasis (196)
mutagen (195)
operator (190)
operon (190)
oncogene (196)
point mutation (194)
promoter (184)
P site (187)

repressor (191)
ribonucleic acid (183)
ribosomal RNA (185)
RNA polymerase (184)
substitution (195)
terminator (185)
transcription (183)
transfer RNA (185)
translation (183)
tumor (195)
uracil (183)

Concept Mapping

Construct a concept map that shows the role of RNA in gene expression. In constructing your map, use the following terms: transcription, translation, mRNA, tRNA, rRNA, gene, promoter, codons, anticodons, proteins, amino acids, ribosome, and cytoplasm. Include additional concepts in your map as needed.

Review

Multiple Choice

1. RNA differs from DNA in that RNA
   a. includes thymine rather than uracil.
   b. consists of a single chain of nucleotides.
   c. contains the sugar deoxyribose.
   d. has four nitrogen bases.

2. A short chain of DNA has the nucleotide sequence ATA CCG GAC ATC. What is its complementary mRNA nucleotide sequence?
   a. TAT GCC CTG TAG
   b. AUA CCG GAC AUG
   c. UAU GGC CUG UAG
   d. ATA CCG GAC ATC

3. During translation, a codon pairs with a(n)
   a. complementary sequence of mRNA.
   b. three-nucleotide sequence of rRNA.
   c. complementary sequence of tRNA.
   d. specific amino acid.

4. What happens when RNA polymerase reaches a terminator on a strand of DNA?
   a. It kills the DNA.
   b. Transcription stops.
   c. Anticodons are no longer added to the growing strand of RNA.
   d. The promoter takes over, and translation starts.

5. In prokaryotes, transcription is regulated by
   a. the presence of enhancers.
   b. metastasis.
   c. the operon model.
   d. frameshift mutations.

6. Which of the following blocks transcription of lactose metabolizing genes?
   a. the absence of lactose
   b. the binding of lactose to the repressor
   c. the binding of lactose to the operator
   d. the presence of beta-galactosidase

7. What protein is able to stop RNA polymerase from transcribing structural genes in DNA?
   a. operator   c. promoter
   b. repressor   d. enhancer

8. When polytene chromosomes puff, they are
   a. undergoing transcription.
   b. stimulating cancerous transformations.
   c. initiating estrogen production.
   d. keeping mRNA from leaving the cell nucleus.

9. Cancer is caused by
   a. tumors.
   b. mutations in growth regulating genes.
   c. *ras* genes.
   d. radiotherapy and chemotherapy.

Completion

10. The making of RNA from DNA is called _____ , while the construction of proteins from the information carried by mRNA is called _____ .

11. The genetic code is made of three-nucleotide sequences called _____ . The number of codons possible in the genetic code is _____ .

12. Amino acids of a protein chain are held together by _____ bonds. These bonds are formed as _____ is moved from the A site to the P site on the ribosome.

13. A(n) _____ is a group of structural genes that includes the promoter and operator. The _____ is located between the promoter and the structural genes.

14. Gene expression in prokaryotes is regulated by _____ . However, gene expression in eukaryotic cells is regulated by the _____ of DNA in a chromosome.

15. Mutations in _____ are likely to be genetically significant, while mutations in _____ affect only the organism in which they occur.

16. A cancerous cell can proliferate, forming a mass of cells called a _____ .

Short Answer

17. Mutations may occur in gametes or in body cells. In which cell type is a mutation likely to be a source of genetic variation for evolution? Why?

18. Three types of RNA are made from DNA. What are the three types of RNA? How does each kind function in protein synthesis?

19. What is the role of RNA polymerase in transcription?

20. What is the condition of the operator on the *lac* operon when transcription is occurring? Would this condition be a result of a high or a low concentration of lactose?

21. What is the function of allolactose in the operation of the *lac* operon?

22. Two strains of *E. coli* bacteria, A and B, are placed in a medium containing the amino acid tryptophan. Strain A is able to turn off the genes that code for the enzymes that produce tryptophan. Strain B is unable to do this. What is likely to happen to the two strains after several days in the medium? Why?

23. Could gene expression occur without RNA? Explain.

24. Students studying together for a test reached the conclusion that a deletion mutation that occurs at the end of a gene has a greater chance of being lethal than one that occurs near the middle of a gene. Do you agree with their conclusion? Explain.

25. A ribosome has three binding sites that function during translation. What is the function of each site?

26. During translation, a protein is made by linking amino acids. How are amino acids linked to form proteins?

27. How is translation initiated?

Themes Review

28. **Structure and Function** Chromosome puffs have been observed in the polytene chromosomes of *Drosophila* and other flies. How are chromosome puffs thought to promote transcription in eukaryotic cells?

29. Homeostasis How does gene regulation of the *lac* operon promote homeostasis in the *E. coli* bacteria that live in the human intestine?

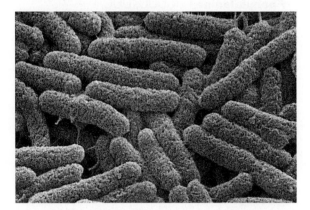

30. Evolution Genetic engineering involves inserting segments of DNA taken from one organism into the DNA of another organism. What would be the likely result of an experiment in which a scientist replaces a faulty stop codon in the DNA of mouse cells with the codon UAG taken from the DNA of a frog, a pine tree, or a clam? What do the results of this experiment suggest about the evolutionary ancestry of these organisms?

31. Levels of Organization Write the mRNA transcription of the DNA sequence presented below.

CTG TTC ATA ATT

Next, write the tRNA anticodons that would pair with the mRNA transcription. Finally, write the abbreviations for the amino acids coded for by the mRNA transcription using Table 9-2.

Critical Thinking

32. Making Inferences In eukaryotic cells, there is a lag time between the beginning of transcription and the beginning of translation. This does not occur in prokaryotic cells. What reason can you give for this difference?

Activities and Projects

33. Research and Writing Do library research to find the latest information on oncogenes and their protein products. Present an oral report to the class in which you show how your findings support the theory that cancer arises from out-of-control cell division.

34. Cooperative Group Project Construct models showing gene expression in prokaryotes and eukaryotes. Your model of the prokaryote should include promoter, operator, and structural genes; repressor; RNA polymerase; inducer; mRNA; and ribosome. Your model of the eukaryote should show DNA, enhancer, receptor protein, acceptor protein, RNA polymerase, and ribosomes.

35. Research and Writing Tetracyclines are antibiotics that are commonly used as an antibacterial drug. Do library research and talk with health care professionals to learn how tetracyclines fight bacterial infection.

36. Multicultural Perspective Research the various types of cancer that seem to have a higher rate of incidence among particular cultures. For example, the Japanese have a very low rate of breast cancer compared with Americans, but they have a higher rate of other types of cancer. What possible reasons could explain these phenomena?

Readings

37. Read the article "How Does a Cell Become a Whole Body?" in *Discover*, November, 1992. What two discoveries helped scientists begin unraveling the mysteries of development? How does the homeobox differ from other gene clusters? What are the two sources for the signals that determine a cell's fate?

DNA and Proteins

OBJECTIVE

Demonstrate how the proteins made by a cell are determined by the cell's DNA.

PROCESS SKILLS

- analyzing models of protein synthesis
- recognizing the relationship between DNA structure and its function

MATERIALS

- paper and pencil

BACKGROUND

1. What is the basic structure of a DNA molecule?
2. How does DNA code for proteins, which are made in the cytoplasm, if it does not leave the nucleus?
3. What are the building blocks of protein?
4. What can result if only one amino acid is coded incorrectly?
5. Why are proteins important in cellular functions?
6. What is a mutation?
7. What conditions can cause mutations?
8. How can a mutation affect protein structure?
9. Write your own **Focus Question** on your Vee Form.
10. **Knowing Side of the Vee** List the **Concepts** and new **Vocabulary Words** on your Vee Form. In the **Concept Statements** section of the Vee, use these words in sentences that define and explain them.

Amino Acids	
Phenylalanine	UUU, UUC
Leucine	UUA, UUG, CUU, CUC, CUA, CUG
Isoleucine	AUU, AUC, AUA
Methionine or START	AUG
Valine	GUU, GUC, GUA, GUG
Cysteine	UGU, UGC
Tryptophan	UGG
Arginine	CGU, CGC, CGA, CGG, AGA, AGG
Glycine	GGU, GGC, GGA, GGG
Serine	UCU, UCC, UCA, UCG, AGU, AGC
Proline	CCC, CCA, CCU, CCG
Threonine	ACU, ACC, ACA, ACG
Alanine	GCU, GCC, GCA, GCG
Tyrosine	UAU, UAC
STOP	UAA, UAG, UGA
Histidine	CAU, CAC
Glutamine	CAA, CAG
Asparagine	AAU, AAC
Lysine	AAA, AAG
Aspartic acid	GAU, GAC
Glutamic acid	GAA, GAG

TECHNIQUE

Doing Side of the Vee

1. Assume that the base sequence on one strand of a DNA molecule is:

   C A C G C T T G G T G A C C G T A A

2. List the base sequences of the complementary DNA strand that would form during replication. Place the information in the **Records** section of your Vee Form.

3. List what the mRNA base sequences would be if the strand shown in step 1 were being read by mRNA.

4. Using the table on the preceding page, determine the amino acid sequence this mRNA codes for. List these amino acids in the **Records** section of the Vee Form.

5. If the fifth nucleotide in the DNA strand shown in step 1 were changed from cytosine to adenine, what would the resulting mRNA be? Record your answer in the **Records** section of the Vee Form.

6. List the sequence of amino acids that would reflect the change described in step 5.

7. Here is the base sequence of a strand of DNA:

   C T C C T C A G G A G T C A G C G T G C A A C A

   a. List the bases in the mRNA strand coded from this DNA strand.

   b. For which amino acids would this mRNA code?

   c. If the twelfth base in the original DNA strand were changed to cytosine, for what mRNA would the new DNA code?

   d. How are the proteins in b and c different?

8. In the **Procedure** section of the Vee, briefly summarize the procedure you followed.

INQUIRY

1. How many mRNA bases code for an amino acid?

2. What is the significance of the start and stop codons?

3. If one base is changed, what happens to the protein that is formed?

4. Using the table on the preceding page as a guide, write a statement relating mRNA codons and amino acids.

5. Use the information on the **Knowing Side** of the Vee to interpret your results from the **Doing Side,** and then write your **Knowledge Claim.** Write a **Value Claim** for this lab.

ANALYSIS

1. Study the table shown on the preceding page. If the base G in the mRNA codon GAU (aspartic acid) were changed, what possible codons (and their corresponding amino acids) could result?

2. If one base is changed in the mRNA code, a mutation may result. Describe the effects of a mutation by contrasting normal red blood cells with "sickle-cell anemia" blood cells.

3. Write a generalized statement about mutations.

FURTHER INQUIRY

Write a **New Focus Question** that could be the point of a new investigation. The following is an example:

How can changes in DNA that yield incorrect proteins cause genetic diseases such as sickle cell anemia, hemophilia, PKU, or cystic fibrosis?

CHAPTER 10

GENE TECHNOLOGY

REVIEW

- allele (Section 7-1)
- vaccine (Section 8-1)
- structure of DNA (Section 8-2)
- transcription and translation (Section 9-1)
- ribosomal RNA (Section 9-1)

Dr. Paul Umbeck and his genetically engineered cotton with longer, stronger fibers.

10-1 What Is Genetic Engineering?

*I*n 1973, biochemists Stanley Cohen and Herbert Boyer constructed a creature that was part bacterium and part frog. How did these scientists construct such an organism? Using organisms such as those shown in Figure 10-1, they first isolated the gene that codes for ribosomal RNA from the DNA of a frog. Then they inserted the frog gene into the DNA of the bacterium Escherichia coli. During transcription, the bacterium busily produced frog rRNA. Never before had such a genetically altered organism existed.

Section Objectives

- List the four steps involved in a genetic engineering experiment.
- Explain how recombinant DNA is produced.
- Describe how restriction enzymes are used in genetic engineering.

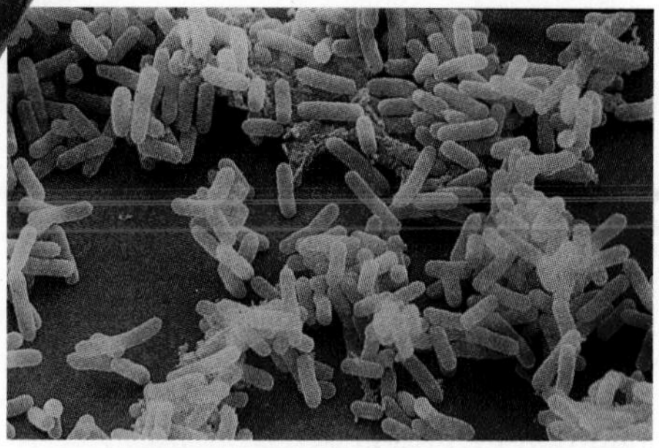

Figure 10-1 In their laboratories in San Francisco, Stanley Cohen and Herbert Boyer successfully put together the first genetically engineered organism. They used a gene from the frog *Xenopus laevis* and a chromosome from the bacterium *Escherichia coli*.

The Basics of Genetic Engineering

The Cohen and Boyer experiment ushered in a scientific revolution in biology. It yielded laboratory techniques that enable researchers to locate, isolate, and study small segments of DNA obtained from much larger chromosomes. These methods are changing basic research in agriculture, medicine, and many other fields. Today, certain human genes can be transferred into bacteria to produce enormous amounts of the protein encoded by the human gene.

The process used to isolate a gene from the DNA of one organism and transfer the gene into the DNA of another is called **genetic engineering**. Genetic engineering involves building **recombinant DNA**, a molecule made from pieces of DNA from separate organisms. Every genetic engineering

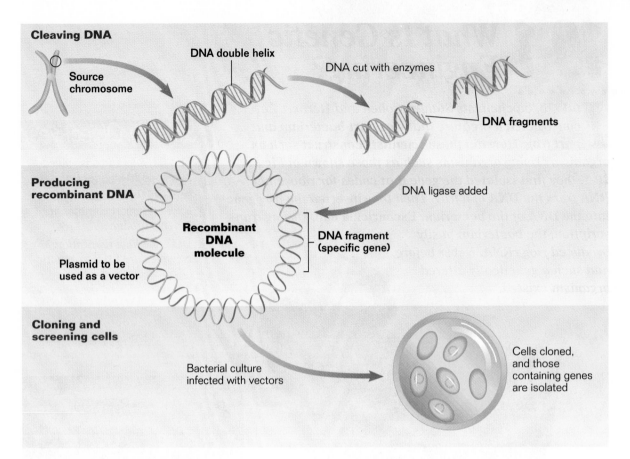

Cleaving DNA

Source chromosome

DNA double helix

DNA cut with enzymes

DNA fragments

Producing recombinant DNA

DNA ligase added

Recombinant DNA molecule

Plasmid to be used as a vector

DNA fragment (specific gene)

Cloning and screening cells

Bacterial culture infected with vectors

Cells cloned, and those containing genes are isolated

Figure 10-2 A genetic engineering experiment has four basic steps: cutting the gene of interest out of its source chromosome; inserting that gene into a vector that will carry the gene into bacterial cells; cloning the infected bacteria; and finding the cells that have taken up the vector with the gene.

experiment presents unique problems, but all share four distinct steps, which are illustrated in Figure 10-2.

1. **Cleaving DNA**   The DNA containing the gene of interest (the gene to be transferred) is cut into fragments using special enzymes that cleave, or separate, sequences of nucleotides.

2. **Producing recombinant DNA**   In genetic engineering, recombinant DNA is made when a DNA fragment is put into the DNA of a **vector,** an agent that is used to carry the fragment into another cell. Commonly used vectors include viruses and plasmids *(PLAZ mihds)*. A **plasmid** is a circular DNA molecule, usually found in bacteria, that can replicate independently from the main chromosome.

3. **Cloning cells**   A culture of bacteria is infected with the fragment-containing vectors. Some of the bacteria will take in the vectors. These cells are isolated and allowed to reproduce. Growing a large number of genetically identical cells from a single cell is called **cloning.**

4. **Screening cells**   Bacterial cells that have received the particular gene of interest are identified and isolated.

Perhaps the best way to understand the steps in a genetic engineering experiment is to follow one from the beginning to the end. Let's go back and learn how Cohen and Boyer transferred the frog rRNA gene into the DNA of a bacterium.

Cleaving DNA

In their experiment, Cohen and Boyer sought to transfer a gene that codes for ribosomal RNA in the African clawed frog, *Xenopus laevis*. To cut this gene out of the chromosome in which it was found, the investigators performed a type of molecular surgery using bacterial enzymes called restriction enzymes. **Restriction enzymes** cleave DNA at specific sequences, generating a set of small fragments of DNA. These sequences are unusual because they are made of two strands of DNA that have the same nucleotides running in opposite directions. For example, Cohen and Boyer used a restriction enzyme called *Eco*R1, which recognizes the sequence GAATTC. Try writing the sequence of the complementary strand; it is CTTAAG, the same as the original sequence written backward, as shown on the left side of Figure 10-3.

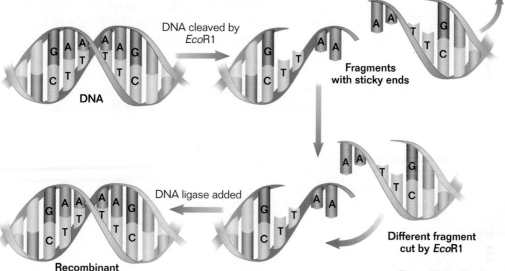

Figure 10-3 *Eco*R1 recognizes the nucleotide sequence GAATTC and makes its cut after the G. Any fragments cut by *Eco*R1 have the same sticky ends and therefore can be joined with DNA ligase to make recombinant DNA.

Many restriction enzymes do not make their incision in the center of the sequence; rather, the cut is staggered, made to one side of the sequence. For example, in the sequence GAATTC, *Eco*R1 makes its cut after the nucleotide G, as shown in Figure 10-3. The cuts produce fragments of DNA with short single strands dangling from each end. Because these dangling tails are complementary to each other, they are called cohesive ends, or "sticky ends." These ends can pair with each other, and the cuts can be sealed with the aid of an enzyme called DNA ligase. Or, the sticky ends can pair with *any other DNA fragment cut by the same restriction enzyme* because these would have the same complementary sticky ends. Any two fragments of DNA cut by the same restriction enzyme can be joined together, as shown in Figure 10-3. Fragments of elephant and ostrich DNA cleaved by the same enzyme can be joined just as readily as two fragments of bacterial DNA, because they have the same complementary sequences at their ends. ❑

❑ CAPSULE SUMMARY

Restriction enzymes recognize specific nucleotide sequences and cleave DNA into fragments with short sticky ends. The sticky ends enable DNA fragments from different organisms to join together.

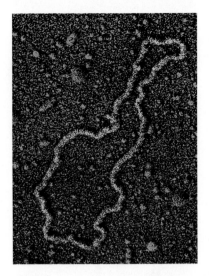

Figure 10-4 This circular molecule of DNA, pSC101, was the first plasmid to successfully carry a gene from a frog into a bacterium.

Producing Recombinant DNA

To make a molecule of recombinant DNA, Cohen and Boyer used the restriction enzyme *Eco*R1 to cut apart a large plasmid, which would be used as the vector to carry the ribosomal RNA frog gene. From this plasmid DNA they isolated a fragment that included two important genes: the gene for plasmid DNA replication and the gene that makes the cell carrying this plasmid resistant to the antibiotic tetracycline *(teh trah SY klihn)*. These two genes would be important later in the experiment, when they would be used to identify cells that took in the frog gene.

Because both ends of this fragment from the large plasmid were cut by the same restriction enzyme, they could be joined together to form a circular molecule of DNA. This smaller plasmid, shown in Figure 10-4, was called pSC101 (because it was the 101st plasmid isolated by Stanley Cohen). Cohen and Boyer produced recombinant DNA by mixing the frog's DNA fragments and pSC101. This process is illustrated in Figure 10-5.

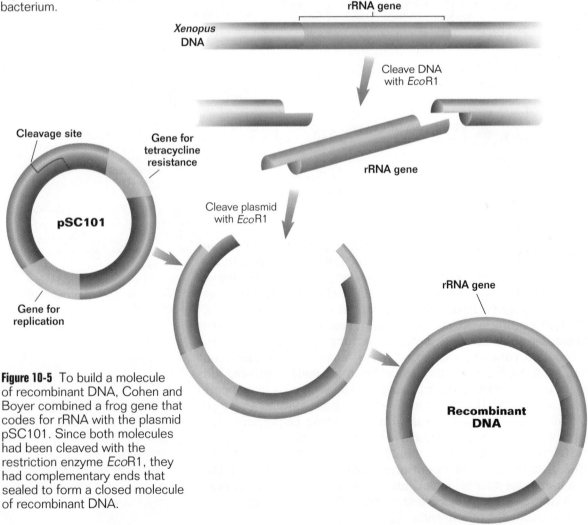

Figure 10-5 To build a molecule of recombinant DNA, Cohen and Boyer combined a frog gene that codes for rRNA with the plasmid pSC101. Since both molecules had been cleaved with the restriction enzyme *Eco*R1, they had complementary ends that sealed to form a closed molecule of recombinant DNA.

Cloning Cells

Once the recombinant DNA that would serve as a vector was made, Cohen and Boyer were ready to introduce it into *E. coli* bacteria. They treated growing cultures of bacteria so that the cells would take up the recombinant DNA. Because only a few cells would take up the vectors, a method was needed to identify them. This is when the gene for tetracycline resistance became important. Cohen and Boyer added the antibiotic tetracycline to the bacterial cultures. As shown in Figure 10-6, the only cells that were not killed by the antibiotic were those that had become resistant to tetracycline because they had taken up the vector. All bacterial cells without vectors were eliminated. Every surviving cell was then allowed to reproduce, forming a colony of identical cells, or clones, all of which had the recombinant DNA. Cohen and Boyer obtained thousands of these clones. ◻

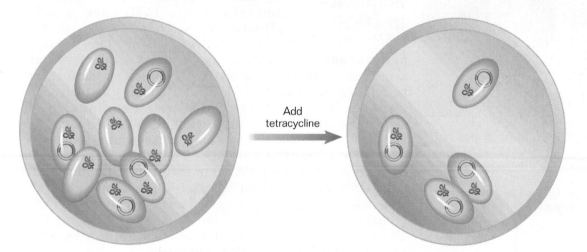

Add tetracycline

Screening Cells

Once Cohen and Boyer eliminated the bacteria that did not contain the recombinant DNA, they needed to test each clone to see if the frog rRNA gene was indeed present. A useful procedure for identifying a specific gene is known as the Southern blot. The **Southern blot** is a technique that uses radioactively labeled RNA or single-stranded DNA as a "probe" to identify a specific gene. The probes have specifically ordered nucleotide bases that are complementary to the sequence of bases in the gene being sought. In a Southern blot, the cloned DNA is first cleaved into fragments by restriction enzymes. The fragments are separated by **gel electrophoresis** *(ee lehk troh fuh REE sihs)*, a technique that uses an electrical field within a gel to separate molecules in a mixture. Because DNA is negatively charged, the various fragments move through the gel according to their size, forming a pattern of bands. The DNA fragments are then split into single-stranded DNA, which is then blotted onto filter paper. Afterward, the filter paper is moistened with a solution

Figure 10-6 To identify bacterial cells that contained the vector, Cohen and Boyer added the antibiotic tetracycline to the bacterial cultures. Only the cells that had taken up the vectors were resistant to tetracycline and survived.

containing the radioactively labeled probes. Among the thousands of DNA fragments, only the fragments that contain the gene of interest will bind with the probes, because of the complementary nucleotide sequence. Cohen and Boyer searched patiently and eventually uncovered cells containing the DNA that paired with the ribosomal RNA probe. These are the cells Cohen and Boyer sought—bacterial cells that contained the frog rRNA gene.

While the details vary from one experiment to another, all genetic engineering experiments employ the same basic strategy used in this first successful gene transfer. ■

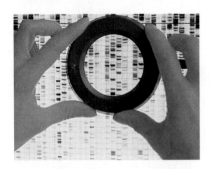

Figure 10-7 When fragments of DNA are separated by gel electrophoresis, they form a pattern of bands called a DNA fingerprint, which can be analyzed and used to establish identity.

Identifying Sequences in DNA

The knowledge that restriction enzymes cut DNA at specific sites has produced powerful tools that have greatly affected the biological sciences. One such tool is **restriction fragment length polymorphism** (RFLP) **analysis**, used to identify base sequences in DNA. RFLP analysis is based on the fact that restriction enzymes cut DNA into fragments that have specific lengths. Because of differences in nucleotide sequences, the number of cutting sites for a restriction enzyme varies among alleles and, consequently, among the DNA of different individuals. Therefore, a restriction enzyme will cut the DNA from different individuals into fragments with different lengths. When samples of these fragments are separated by gel electrophoresis, each forms a characteristic pattern of bands, shown in Figure 10-7. These bands represent a unique array of RFLP sites, which serve as an individual's **DNA fingerprint.**

RFLP analysis can be used to establish the identity of a person. Because it can determine a sequence of nucleotides from a sample of DNA found in blood, semen, bone, and hair, RFLP analysis has been useful in forensics. RFLP analysis is also valuable for identifying the genes that cause genetic disorders, which sometimes have unique restriction sites. Alleles for Huntington's disease, sickle cell anemia, and a number of other genetic disorders are currently detected in this way.

In 1985, Kary B. Mullis, a biochemist working for a biotechnology corporation in California, discovered a simple way to make unlimited copies of a gene, a process now known as **polymerase chain reaction** (PCR). The starting materials for PCR include a single molecule of DNA (even a short fragment will do), DNA polymerase (the enzyme responsible for DNA replication), and a supply of all four nucleotides. When all these ingredients are incubated in a test tube, millions of copies of a DNA segment can be made in a few hours, as shown in Figure 10-8. PCR has revolutionized research in molecular biology, giving biologists as much of the particular DNA as they need. Prior to this technique, a piece of DNA was

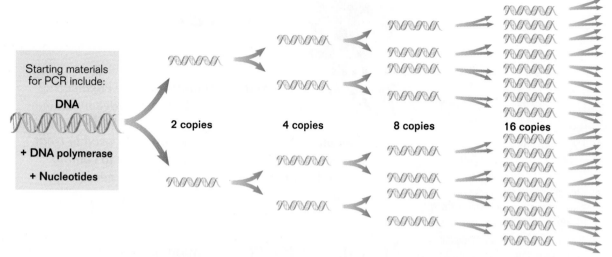

Starting materials for PCR include:

DNA

+ DNA polymerase

+ Nucleotides

2 copies   4 copies   8 copies   16 copies

cloned by inserting it into a plasmid or a virus, a process that usually took weeks. In the past few years, PCR has been used to duplicate DNA from drops of blood and semen found at crime scenes, from the cells of an embryo for prenatal diagnosis, and even from a 40,000-year-old woolly mammoth frozen in a glacier. Many more applications are sure to follow.

Figure 10-8 Using DNA polymerase and nucleotides, the process of polymerase chain reaction produces millions of copies of DNA in just a few hours.

Mapping the Human Genome

The potential ability of gene technology to aid the fight against disease is great, and many efforts are underway to increase its usefulness. One of the most significant efforts is the **Human Genome Project,** a research effort to identify and locate the entire collection of genes in a human cell. Goals of this project include improving existing human genetic maps, constructing physical maps of entire chromosomes, and ultimately determining the complete DNA sequence in the human genome. In 1993, a preliminary map of the entire human genome was completed using RFLP analysis. One major goal for the future of the project is to develop new methods for mapping and sequencing DNA.

CONTENT LINK

Read about the social and ethical issues surrounding the Human Genome Project on **page 222.**

Section Review

1. *Describe the four steps of the first experiment that successfully transferred a gene from a frog to a bacterium.*

2. *What is recombinant DNA? What role does it play in genetic engineering?*

3. *What is unique about the cuts that restriction enzymes make to DNA?*

Critical Thinking

4. *What safety and ethical issues do you think might arise over the use of genetic engineering?*

10-2 The New Medicine

Much of the excitement about genetic engineering has focused on its potential to improve medicine by curing and preventing illnesses. Major advances have been made in the production of proteins used to treat illnesses, in the development of new vaccines to combat diseases, and in the replacement of defective genes with healthy ones.

Section Objectives

- Explain how genetic engineering can benefit human health.
- Describe two diseases that can be treated with products of genetic engineering.
- Explain how to genetically engineer a vaccine.

Figure 10-9 Large amounts of insulin can be obtained through genetic engineering. The gene for insulin is cut out of its chromosome and inserted into a bacterial plasmid. The bacteria containing the plasmids produce insulin, which can be collected to treat disorders such as diabetes.

Making Genetically Engineered Drugs

Many genetic disorders and other human illnesses occur when the body fails to make critical proteins that are essential for proper functioning. For example, **diabetes mellitus** *(muh LY tuhs)* **type I**, also called insulin-dependent diabetes, is an illness that occurs when the body cannot make sufficient amounts of the protein insulin. Diabetes mellitus type I can be treated by regular injections of insulin or by an insulin pump. However, insulin is typically present in the body in very low amounts, making the large quantities needed for pharmaceuticals difficult and expensive to obtain. With the availability of genetic engineering techniques, this problem has been largely overcome. The genes encoding the protein insulin can now be inserted into bacteria, which then produce insulin. Because the host bacteria can be grown cheaply in bulk, large amounts of insulin can be easily obtained, as shown in Figure 10-9.

In 1982, the U.S. Food and Drug Administration approved the use of the first commercial product of genetic engineering—human insulin. Today, hundreds of pharmaceutical

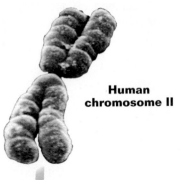

Human chromosome II

Insulin gene transferred

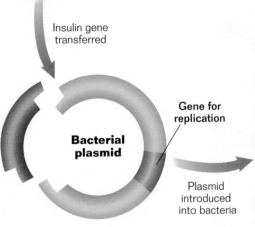

Gene for replication

Bacterial plasmid

Plasmid introduced into bacteria

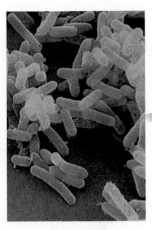

Bacteria culture

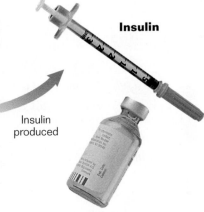

Insulin

Insulin produced

companies around the world are busy producing other medically important proteins using the genetic engineering techniques you learned about in the last section. These products include **anticoagulants** (proteins involved in dissolving blood clots), which are effective in treating heart attack patients, and factor VIII, a protein that promotes blood clotting. A deficiency in factor VIII leads to hemophilia, an inherited disorder characterized by prolonged bleeding. For a long time, hemophiliacs received blood factors that had been isolated from donated blood. Unfortunately, some of the donated blood was infected with viruses such as HIV and the hepatitis B virus, which were then unknowingly transmitted to those people who received blood transfusions. Today, the use of genetically engineered factor VIII eliminates the risks associated with blood products obtained from other individuals. Other genetically engineered pharmaceutical products are listed in Table 10-1.

Table 10-1 Genetically Engineered Medicines

Product	Examples and Uses
Colony-stimulating factors	Growth factors that stimulate white blood cell production; used to treat immune system deficiencies and to fight infections
Erythropoetin	Stimulates red blood cell production; used to treat anemia in individuals with kidney diseases
Growth factors	Stimulate differentiation and growth of various cell types; used to promote wound healing
Human growth hormone	Used as a treatment for dwarfism
Interferons	Interfere with reproduction of viruses; also used to treat some cancers
Interleukins	Activate and stimulate different classes of white blood cells; can be used in treating wounds, HIV infections, cancer, immune deficiencies

Making Genetically Engineered Vaccines

A vaccine is a solution containing a harmless version of a pathogen (disease-causing microorganism) or its toxins. When a vaccine is injected, the recipient's immune system will recognize the pathogen's surface proteins. The immune system will then respond by making defensive proteins called **antibodies**, which will later combat the pathogen.

CONTENT LINK

You will learn more about the immune system in Chapter 39.

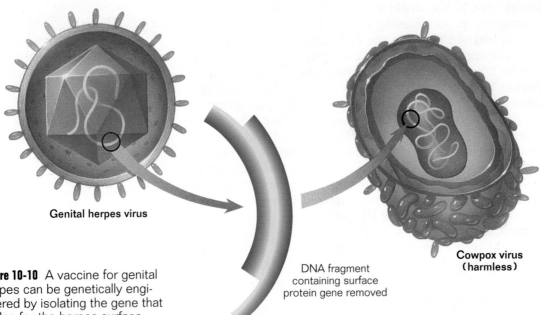

Genital herpes virus

Cowpox virus
(harmless)

DNA fragment
containing surface
protein gene removed

Figure 10-10 A vaccine for genital herpes can be genetically engineered by isolating the gene that codes for the herpes surface protein and inserting it into a harmless cowpox virus. The virus will now manufacture the herpes surface proteins. A person vaccinated with the engineered virus will make the antibodies against the virus.

Antibodies will stop the growth of the pathogen before the disease it causes can develop.

Traditionally, vaccines have been prepared either by killing a specific pathogenic microbe or by making it unable to grow. This ensures that the vaccine will not cause the disease. The problem with this approach is that any failure in the process to kill or weaken a pathogen will result in the transmission of the disease to the very patients seeking protection. While the majority of vaccines are safe, a fraction of a percentage of vaccines cause the treated individuals to contract the disease. This small, but real, danger is one of the reasons why rabies vaccines are administered only when a person has actually been bitten by an animal suspected of carrying rabies.

Today there is a new and much safer method of making vaccines, as illustrated in Figure 10-10. Using genetic engineering techniques, the genes that encode the pathogen's surface proteins can be inserted into the DNA of harmless bacteria (or viruses). The modified, but still quite harmless, bacteria become an effective and safe vaccine. These harmless bacteria can be used to stimulate the body to make the antibodies that will attack the pathogen. As a result, the body is protected against infection.

Among the vaccines now being manufactured in this way are ones directed against the herpes II virus (which produces small blisters on the genitals) and hepatitis B virus (which causes a sometimes fatal inflammation of the liver). A major effort is underway to produce a vaccine that will protect people against malaria, a disease caused by a protozoan for which there is currently no effective protection against infection. ◻

◻ CAPSULE SUMMARY

Genetic engineering techniques are being used to manufacture proteins such as insulin and factor VIII, as well as vaccines.

Curing Genetic Disorders

In 1990 the first attempts were made to use genetic engineering to combat genetic disorders. Many genetic disorders arise when an individual lacks a normally functioning copy of a particular gene. One obvious way to cure such disorders is to give the person a working copy of the gene. Until recently this approach was not practical for three reasons. First, the defective gene was difficult to identify and isolate. Second, it was hard to transfer a healthy copy of such a gene into the cells of body tissues that use it. Finally, it was necessary to find a way to keep the altered cells or their offspring alive in the body for a long time. With genetic engineering, it is now possible to overcome these difficulties.

One of the first gene therapy attempts involved two young girls, shown in Figure 10-11, who suffered from an immune system disorder caused by a defective gene. Doctors extracted bone marrow cells from the girls and replaced the defective gene, which failed to produce an important immune-system enzyme, with a normal gene. These cells were returned to the girls' bones and began to produce the missing enzyme. Because this kind of bone marrow cell actively divides, researchers hope that offspring of the genetically engineered cells will continue to secrete the enzyme into their blood for a long time.

Genetic engineering is also providing a new and powerful weapon in the battle against cancer. All humans have white blood cells that secrete a protein called tumor necrosis factor (TNF). TNF attacks and kills cancer cells. Unfortunately this does not happen often. Genetic engineers recently developed a method of adding the TNF gene to a kind of white blood cell that is very effective at locating cancer cells but not very effective at harming them. Once armed with this TNF gene, however, these white blood cells will secrete TNF and kill cancer cells. Genetic engineering has enabled these cells to become like cruise missiles with a deadly payload zeroing in on cancer cells. ◻

Figure 10-11 Ashanthi DeSilva, *top,* and Cynthia Cutshall, *bottom,* were among the first patients to receive gene therapy. Three years after their treatment, they both appear to be thriving.

☐ CAPSULE SUMMARY

Some human genetic disorders are being treated and "corrected" by inserting copies of the corresponding normal gene into individuals whose copy of the gene is defective.

Section Review

1. *Discuss two ways human health has benefited from genetic engineering.*

2. *Explain how genetic engineering can be useful in the treatment of human illnesses such as diabetes mellitus type 1 and hemophilia.*

3. *Explain how vaccines can be genetically engineered.*

Critical Thinking

4. *What other illnesses or disorders can you think of whose treatment might benefit from genetic engineering?*

10-3 The New Agriculture

*O*ne of genetic engineering's greatest successes has been the manipulation of genes in crop plants and livestock. Gene transfers have resulted in crop plants that are more resistant to plant diseases and pests, as well as tomatoes that have a longer shelf life. Genetic engineering has also been used to boost milk production in dairy cows and increase the growth rate of certain livestock.

Section Objectives

- Explain the role of the Ti plasmid in agricultural research.
- Describe how herbicide-resistant genes in crop plants can benefit the environment.
- Explain how genetic engineering techniques have been used to improve crop yields and make livestock more productive.

Transporting Genes Into Plants

The key to the great progress in genetic engineering of plants in recent years was the discovery of a suitable vector to transport a gene from one plant to another. For years, genetic engineering in plants was not possible because, unlike bacteria, plants have few viruses or plasmids that can perform this critical role. Said simply, genetic engineers lacked a suitable vector to carry the gene into plant cells.

The breakthrough came in the form of an unusual bacterial plasmid responsible for crown gall, a disease characterized by large bulbous tumors. This plasmid is called the Ti plasmid ("Ti" stands for tumor-inducing). The Ti plasmid easily infects broadleaf crop plants such as tomatoes, tobacco, and soybeans. When it has infected a plant cell, it proceeds to insert itself into the plant cell's chromosome. To make a genetic engineering vehicle, scientists removed the tumor-causing genes from the Ti plasmid. The vacant space in the now-harmless plasmid was then filled with DNA introduced by the scientists, as shown in Figure 10-12. This DNA could then be carried into the chromosomes of a target plant.

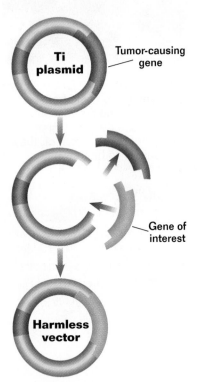

Figure 10-12 Genetic engineering can turn a tumor-causing Ti plasmid into a suitable vector for broadleaf crops. The tumor-causing gene is removed and replaced with a gene of interest, which can then be carried into plant cells by bacteria.

Making Crops Resistant to Herbicides and Insects

A recent improvement in agriculture is the development of crop plants that are resistant to the herbicide **glyphosate**, a powerful, biodegradable weedkiller. Glyphosate kills most actively growing plants by destroying an enzyme needed to make certain kinds of protein. Genetic engineers found a bacterial strain that could make the enzyme despite the presence of glyphosate. Then they isolated the gene encoding the enzyme. Using the Ti plasmid as a vector, genetic engineers then successfully introduced the resistance gene into non-cereal crop plants. More recently, genetic engineers successfully "shot" the Ti plasmid into cells of wheat plants using a gene gun.

Herbicide-resistant crops are advantageous; they lower the cost of producing a crop because a field of crops resistant to glyphosate does not need to be weeded. The farmer simply treats the field by dragging a rope soaked in glyphosate across the field. All growing plants die except the crop, which is resistant to the herbicide, as shown in Figure 10-13.

The development of glyphosate-resistant crops is also beneficial to the environment. Glyphosate is quickly broken down in the environment, which makes its use a great improvement over most herbicides. Perhaps even more important, the tragic loss of fertile topsoil to erosion, one of the greatest environmental challenges facing our country today, would be greatly reduced if cropland were not intensively cultivated to remove weeds.

Another important advance in agriculture that has resulted from genetic engineering is the development of crops resistant to insect pests. Such crops do not need to be sprayed with pesticides; this would be a great benefit to the environment.

Consider cotton. Its fibers are a major source of raw material for clothing throughout the world, yet the plant itself can hardly survive in a field because of the many insects that attack it. Over 40 percent of the chemical pesticides used today are employed to kill insects that harm cotton plants. The world's environment would be better off if these thousands of tons of pesticides were not needed. Biologists are now in the process of producing cotton plants that are resistant to insects, so that pesticides will not be needed.

One successful approach uses a kind of soil bacterium that produces enzymes that attack and kill the larvae of moths and butterflies, important pests of some crops. When the genes producing these enzymes were inserted into the chromosomes of tomato plants, the plants began to manufacture the new enzyme. The enzyme made the tomato plants highly toxic to tomato hornworms, a pest that seriously damages tomato crops. ▪

Figure 10-13 The effects of genetic engineering can be seen in this field of soybeans. The plants that have been genetically engineered to resist glyphosate continue to thrive after treatment with the herbicide.

□ *CAPSULE SUMMARY*

Genetic engineers have manipulated the genes of certain kinds of crop plants to make them resistant to herbicides and destructive pests.

Developing Crops That Need No Fertilizer

Nitrogen is an element that all plants must have in order to make proteins and DNA. The most abundant source of nitrogen in the environment is atmospheric nitrogen, N_2. However, plants cannot obtain nitrogen from the air. All of the nitrogen that plants need must be obtained from the soil. How does this nitrogen get in soil? Bacteria living within the roots of soybeans, peanuts, and clover "fix" nitrogen by converting N_2 gas from the atmosphere into nitrates, nitrites, and ammonia, which are forms of nitrogen that plants can use.

Because crops rapidly consume nitrogen, farmers replenish the soil by adding high-nitrogen fertilizers. Farming would be much cheaper and far more productive if major crops such as wheat, rice, and corn could be grown without such massive applications of fertilizer—or without any fertilizer at all. The task of genetically engineering major crops to carry out nitrogen fixation has become the focus of many researchers. The problem has not been an inability to discover and isolate the necessary genes or to get them into crop plants. The problem is that when genes are introduced into plants, they do not seem to function properly in their new host. Researchers all over the world are working to find a way around this difficulty. ❑

❑ **CAPSULE SUMMARY**

Genetic engineers are looking for ways to transfer genes for nitrogen fixation from bacteria into crop plants.

Improving Livestock Production

A very interesting advance in agriculture has been the introduction of growth hormone into the diet of dairy cows, which greatly improves milk production. Instead of extracting this hormone at great expense from the brains of dead cows, the relevant gene has been introduced into bacteria. The bacteria then produce the hormone so cheaply that it is practical to add it as a supplement to the cows' diet.

Extra copies of the gene encoding the same growth hormone have been introduced directly into the chromosomes of both cattle and hogs to increase their weight. Though still underway, these attempts promise to create new breeds of very large and fast-growing cattle and hogs. The human version of this same growth hormone is now being tested as a potential treatment for dwarfism, a disorder in which the pituitary gland fails to make adequate amounts of growth hormone. ❑

❑ **CAPSULE SUMMARY**

The addition of genetically engineered growth hormone to the diets of livestock increases milk production in dairy cows and weight gain in cattle and hogs.

Section Review

1. *How is the Ti plasmid used to insert genes into plant cells?*

2. *Explain how genetic engineering techniques can make crop plants resistant to weedkillers such as glyphosate.*

3. *How can genetic engineering reduce the amount of insecticides used in agriculture?*

Critical Thinking

4. *In spite of FDA approval, the use of genetically engineered growth hormone in dairy cows has been controversial. Would you have concerns about consuming milk from cows treated with growth hormone? Why or why not?*

Vocabulary

antibody (211)
anticoagulant (211)
cloning (204)
diabetes mellitus type I (210)
DNA fingerprint (208)
Human Genome Project (209)

gel electrophoresis (207)
genetic engineering (203)
glyphosate (214)
plasmid (204)
polymerase chain reaction (208)
recombinant DNA (203)

restriction enzyme (205)
restriction fragment length polymorphism analysis (208)
Southern blot (207)
vector (204)

Concept Mapping

Construct a concept map that shows the process and outcomes associated with genetic engineering. In constructing your map, use the following terms: restriction enzymes, DNA ligase, sticky ends, plasmids, DNA from host cells, recombinant DNA, agriculture, and medicine. Include additional concepts in your map as needed.

Review

Multiple Choice

1. Broken pieces of DNA can be joined together by
   a. restriction enzymes.
   b. DNA ligase.
   c. recombinant DNA.
   d. RFLP analysis.

2. Which is a characteristic of the fragments of DNA generated by using restriction enzymes?
   a. They are between four to six base pairs long.
   b. Each fragment has short, single-stranded "sticky ends."
   c. They contain identical nucleotides running in the same direction.
   d. The fragments unzip completely, forming single-stranded DNA.

3. What can scientists engaged in genetic engineering experiments do to prevent potential hazards?
   a. Use organisms unable to survive outside the laboratory.
   b. Conduct only moderately dangerous experiments.
   c. Restrict their use of the Southern blot technique.
   d. Immediately call the local police when an accident occurs.

4. Which human illness can be treated using a product of genetic engineering?
   a. malaria   c. flu
   b. hemophilia   d. a sinus cold

5. The product of a technique in which the genes that code for a pathogen's surface proteins are inserted into the DNA of harmless bacteria is
   a. a genetically engineered vaccine.
   b. the human growth hormone.
   c. factor VIII.
   d. the Ti plasmid.

6. The genetic engineering of crop plants that are resistant to the herbicide glyphosate is significant because
   a. its use reduces the erosion of topsoil.
   b. glyphosate is stable in the environment for many years.
   c. it also kills moth and butterfly larvae.
   d. glyphosate use increases the need for insecticides.

10 CHAPTER REVIEW

7. A legal issue likely to be raised by the Human Genome Project is
a. how methods used in the project can be applied to map plant genomes.
b. if doctors should use the project data to improve health care.
c. whether the genomes of animals should be mapped.
d. whether employers should know if a genetic disorder has been inherited.

8. The process used to make unlimited copies of a particular gene is called
a. RFLP analysis.
b. Southern blot.
c. DNA fingerprinting.
d. polymerase chain reaction.

Completion

9. Human _____ was the first product made by genetic engineering to be marketed.

10. Genetic engineering depends on the ability of _____ to cut DNA at specific sites along its length. When DNA from two different organisms is cut by the same restriction enzyme and mixed together, the _____ can undergo complementary base pairing.

11. The process of moving genetic material from the DNA of one organism to the DNA of another organism is called _____ .

12. The first experiment that successfully demonstrated genetic engineering was conducted by _____ and _____ . In the experiment, a plasmid from *E. coli* served as the _____ .

13. The genetically engineered pharmaceutical _____ is used to treat anemia in patients with kidney problems. Another product called _____ is used to treat HIV infection and cancer.

14. Genes have been introduced into soybean plants using _____ as a vector and "shot" into wheat plants using a _____ .

15. The gene for the protein _____ has been genetically engineered into white blood cells that locate cancer cells.

Short Answer

16. A scientist is conducting a genetic engineering experiment. What four steps will the scientist follow?

17. What role do restriction enzymes play in the production of recombinant DNA?

18. Why do you suppose a high priority has been given to developing new methods for mapping and sequencing DNA during the first 10 years of the Human Genome Project?

19. Vaccines for several human infectious diseases have traditionally been made from weakened or killed bacteria and viruses. The presence of specific surface proteins that trigger the production of antibodies is the key to a vaccine's success. Genetically engineered vaccines are now used in place of several traditionally made vaccines. Why are these vaccines preferred?

20. Describe two advances in genetic engineering that have made large scale tomato farming more profitable than ever before.

21. Restriction fragment length polymorphism analysis and polymerase chain reaction are processes used by genetic engineers. Explain how each process is used in gene technology.

Themes Review

22. **Evolution** Natural selection is a mechanism of evolution whereby the members of a population who are best able to adapt to their environment survive and produce offspring. How is natural selection affected by genetic engineering?

23. **Homeostasis** Homeostasis may be viewed as the tendency of organisms to remain relatively stable within a range of conditions. How does an insect-resistant crop that is the product of genetic engineering destabilize a population of insects that depends on the crop for food?

24. Structure and Function When restriction enzymes are used to cleave specific nucleotide sequences of DNA, the process results in two "sticky ends." What are sticky ends? How do they function in the preparation of recombinant DNA?

Critical Thinking

25. Making Inferences The United States government has stringent regulations requiring researchers to confine genetically engineered organisms that are considered high risk to the laboratory. What concerns might have led to the enactment of these regulations?

26. Interpreting Data Examine the restriction map of the *E. coli* plasmid pBR 322, shown below. This is a commonly used plasmid composed of 4,363 base pairs. The map shows sites where certain restriction enzymes cut the DNA of the plasmid. For example, *Sph* 1 cuts the plasmid at base pair 566. Suppose you want to isolate from the plasmid the gene that codes for resistance to the antibiotic tetracycline, which is indicated at Tcr. What restriction enzymes would you use? How many base pairs long is the Tcr gene? How might you check to be certain that your procedure was successful?

Restriction Map of pBR 322 DNA

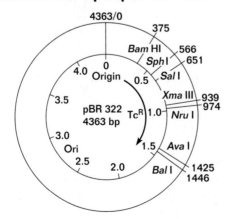

Activities & Projects

27. Research and Writing The question of awarding patents on genetically engineered organisms first arose when microbiologist Ananda Chakrabarty filed for a patent on a bacterium capable of digesting the components of crude oil. Find out about the bacterium that Chakrabarty engineered and about the court battle he waged for the right to obtain a patent for a genetically engineered organism.

28. Research and Writing The polymerase chain reaction (or PCR) is used by scientists to make multiple copies of single genes. Having multiple copies makes it easier to determine if the gene of interest is present in a sample. Find out how the PCR has been automated and how the automation has affected its use in detecting the virus responsible for AIDS.

29. Multicultural Perspective Find out which countries in Africa are suffering from drought. What kind of crops are grown in these countries? How could genetic engineering be used to help the people in these countries grow crops in spite of the recurrence of severe drought?

Readings

30. Read the article "The First Kids With New Genes," in *Time*, June 7, 1993. Describe the genetic disease that afflicts Ashanthi DeSilva and Cynthia Cutshall. What events were involved in developing and administering gene therapy to these two young girls?

31. Read the article "Whose Genome Is It, Anyway?" *Discover*, May 1992. How long is the Human Genome Project expected to take? How much is it expected to cost? Describe the characteristics of the hypothetical human whose genome will be the first to be sequenced. What are the reasons for this choice?

Genetic Engineering Model

OBJECTIVE

Construct and analyze a model representing genetic engineering.

PROCESS SKILLS

• manipulating models
• simulating a scientific process

MATERIALS

• plastic soda straws, 3 cm sections (56)
• metric ruler
• scissors
• permanent marker
• 56 pushpins (15 red, 15 green, 13 blue, and 13 yellow)
• 56 paper clips

BACKGROUND

1. What is the name of the process that isolates a gene from the DNA of one organism and transfers the gene into the DNA of another organism?

2. How can you demonstrate the manipulation of genetic material to produce new combinations of traits?

3. Write your own **Focus Question** on your Vee Form.

4. **Knowing Side of the Vee** List the **Concepts** and new **Vocabulary Words** on your Vee Form. In the **Concept Statements** section of the Vee, use these words in sentences that define and explain them.

TECHNIQUE

Doing Side of the Vee

1. Work in cooperative groups of four students and divide into two-person teams.

2. **CAUTION: Pointed objects can cause injury if not properly used.** Cut the soda straws into 3 cm pieces to make 56 segments.

3. One two-person team should complete steps 4 and 5 while the other team completes steps 6–8. Work with your entire group to complete steps 9–12.

4. Make a model of a bacterial DNA molecule by arranging the nucleotides of the master strand in the following order: blue, red, green, yellow, red, red, blue, blue, green, red, blue, green, red, blue, blue, green, yellow, and red. Create a double-stranded DNA model by constructing a second strand starting with yellow and proceeding with the correct color pushpin to complement the original strand.

5. With your double-stranded DNA model lying on the table, form a circular molecule by carefully joining the opposite ends of each strand. In the **Records** section of your Vee Form, make a sketch of the molecule that shows the arrangement of the bases. Use the abbreviations B, Y, G, and R for the colors of pushpins.

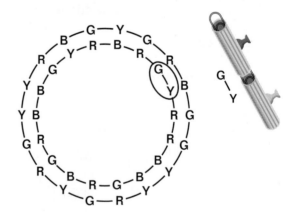

6. Make one strand of a donor, human DNA molecule with the following sequence: blue, blue, red, red, yellow, green, green, blue, red, and yellow.

7. Make a second strand of donor, human DNA having the following sequence: blue, red, red, yellow, green, blue, yellow, yellow, green, and green.

8. Match the complementary portions of the two strands of your human DNA fragment. Make a sketch of the donor molecule in the **Additional Records and Observations** section on the back side of your Vee.

9. Imagine that an enzyme moves around the circular molecule of bacterial DNA until it finds the sequence red-red-blue-blue and its complementary sequence, green-green-yellow-yellow. Find this sequence in the model of a bacterial DNA molecule that you drew in step 5.

10. Simulate the action of the enzyme by splitting the circular molecule at the sequence you identified in step 9. Separate the yellow nucleotide from the blue at one end of the sequence, and the green from the red at the opposite end of the sequence on the complementary strand. Make a sketch of the split molecule in the **Additional Records and Observations** section of your Vee Form.

11. Move the double-stranded donor, human DNA fragment into the break in the bacterial DNA molecule.

12. Imagine that a second enzyme joins the ends of the donor and recipient DNA, creating a new DNA molecule. Make a sketch of the final bacterial DNA molecule in the **Additional Records and Observations** section of your Vee Form. In the **Procedure** section of the Vee, briefly summarize the procedure you followed.

INQUIRY

1. How many nucleotides composed the bacterial DNA molecule?

2. What color pin is always across from a blue pin?

3. What color pin is always across from a red pin?

4. Use the following information to interpret the base-pair order for the DNA molecule you made in step 4, then write the code for the first DNA molecule you created: red = adenine, blue = guanine, yellow = cytosine, green = thymine. You may abbreviate the bases, using the letters A, G, C, and T.

5. Look at the model you created of donor, human DNA. What is unusual about the structure of this DNA fragment? (Hint: make sure to connect the bases to the correct complementary color.)

6. Use the information on the **Knowing Side** of the Vee to interpret your results from the **Doing Side,** and then write your **Knowledge Claim.** Write a **Value Claim** for this lab.

ANALYSIS

1. Compare and contrast the models of bacterial DNA and human DNA.

2. What knowledge did you have to possess in order to construct the donor, human DNA?

3. How does the original bacterial DNA molecule differ from the final DNA molecule?

4. Of what possible benefit could this process be to humans?

FURTHER INQUIRY

Write a **New Focus Question** that could be the point of a new investigation. The following is an example:

Which bacteria are the best hosts for human gene insertion?

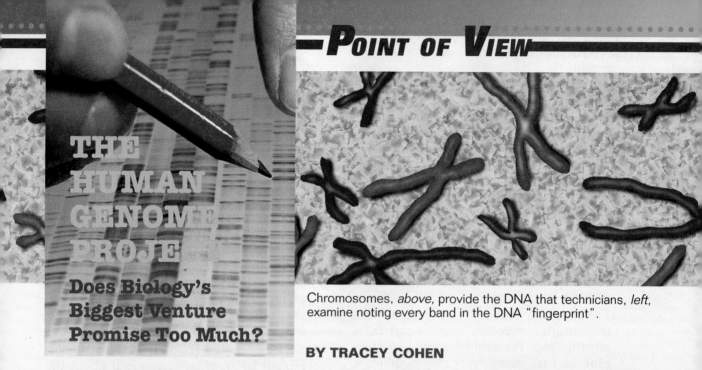

THE HUMAN GENOME PROJE

Does Biology's Biggest Venture Promise Too Much?

Chromosomes, *above,* provide the DNA that technicians, *left,* examine noting every band in the DNA "fingerprint".

BY TRACEY COHEN

*T*he biggest research effort in the history of biology— the Human Genome Project has as its goal to map and sequence the tens of thousands of genes that make up the human genome. The maps will show where on the chromosomes each gene is located. Research is going on in national laboratories, scientific institutes, universities, and private companies around the world and will take about 15 years to finish. In the United States, the project is sponsored jointly by the National Institutes of Health and the Department of Energy and is expected to cost $3 billion.

Supporters of the Human Genome Project claim it will revolutionize medicine. They say that medical scientists will be able to diagnose and eventually cure large numbers of diseases. Many promises and predictions have been made to create support for the project. James Watson, former director for the project, has said that "never will a more important set of instruction books be made available to human beings." The Office of Technology Assessment wrote in a report that "sequencing the human genome will provide one of the most powerful tools humankind has ever had for deciphering the mysteries of its own existence."

Such statements make the project sound exciting and heroic. But are they true? Not everyone thinks so. In the scientific community, the project has caused many bitter arguments. Researchers critical of the project claim that a lot of money will be wasted on something that is poor science. They say that those scientists trying to get money for the project have made unrealistic claims to gain public support.

Will the Project Improve People's Health?

The Human Genome Project makes genetic diseases the focus of medical research. Much of the drive behind the Project is the claim that it will greatly improve the practice of medicine, and thus people's health. This statement might be true if diseases were mainly genetic in origin. However, even though researchers have identified more than 3,000 genetic disorders, most of these are rare. For example, the most common lethal genetic disorder among Caucasians, cystic fibrosis, afflicts about 25,000 Americans. About 30,000 people of various races suffer from Huntington's disease, another deadly hereditary disease. Another 150,000 are thought to be at risk.

In contrast, over 900,000 people died of heart disease in 1990 alone according to the National Center for Health Statistics. Another 500,000 died of cancer. In the United States, heart disease, cancer, stroke, and accidental injury are the leading causes of sickness and death. For the last 20 years, nearly three-quarters of all deaths in this country have been the result of these four killers. These are not, strictly speaking, genetic diseases. Although people with rare disorders might benefit, the majority of society will not.

What Social and Ethical Problems Are Raised by the Project?

Genetic Discrimination Some evidence links genes with certain diseases. However, the actual role of these genes in disease is still unclear. They appear to make people susceptible to getting illnesses like cancer or heart disease. The

Human Genome Project will make it easier to find out if a person has susceptibility genes.

But having a susceptibility gene does not mean the person will always get the disease. Nor does the presence of the gene give any clue about the severity of the disease. The gene does not show how the person will respond to treatment either. And, even without the gene, a person might still get that disease.

What is certain about susceptibility genes is that they present enormous opportunity for discrimination. A survey by researchers at Harvard Medical School has found about 30 cases of genetic discrimination in which people were denied jobs or insurance because they carried a gene linked to a particular disease. Although these people had no symptoms, employers and insurance companies saw them as either sick or likely to become sick.

Impact on Social Responsibilities When scientists talk about genes as "causes," they make it seem as if an individual's genes are responsible for many health conditions that are really social in origin. Poverty, for instance, is not a genetic condition. Yet poverty is clearly a crucial factor in health and disease.

For example, a study by the United States Department of Health and Human Services reported a clear relationship between family income and cancer rate. As income decreases, cancer rates increase. Survival rates are lower for low-income cancer patients as well. People with low incomes have less money for good nutrition and less access to good health care. Their diseases are discovered in later stages when the cure rates are lower.

Who Benefits From the Project?

There is money to be made in every step of the Human Genome Project. Companies that make automatic DNA sequencing equipment will benefit, as will companies that make computer equipment and software.

The creation of new genetic screening tests will also be very profitable. As more genes thought to be linked with diseases are identified, more people will be tested. New screening tests could become a billion-dollar industry. It is estimated that almost 3 million people a year could be tested to determine if they have genes for cystic fibrosis, sickle cell anemia, hemophilia, and muscular dystrophy. Companies that make screening tests, doctors, employers, and insurance companies will likely create social pressure for widespread use of these tests.

Scientists, too, have economic interests in genetic information. For instance, more than 30 leading genome researchers have made deals with investors and new companies to market the results of their work.

Unfortunately, the Human Genome Project is the wrong approach both to understanding human biology and to improving medicine and public health. Before any more work is done, there should be public debate about how this project is likely to affect everyone. Decisions about the Human Genome Project should not be left just in the hands of those who have their own narrow interests at stake.

Tracey Cohen is a freelance writer specializing in science and environmental issues.

Analyzing the Issue

1. **Detecting Bias** How does the writer feel about the Human Genome Project? List the key facts the writer gives to support that opinion. List instances where the writer uses words or phrases to sway your opinion without presenting the facts.

2. **Formulating an Opinion** Obtain at least three books and articles about the genome project. Several possible references are listed below. Write a persuasive essay supporting your own view of the genome project.

Beckwith, Jon. "A Historical View of Social Responsibility in Genetics." *BioScience,* Vol. 43, No. 5, 1993, pp. 327–333.

Hubbard, Ruth and Elijah Wald. *Exploding the Gene Myth.* Boston: Beacon Press, 1993.

Kevles, Daniel J. and Leroy Hood, eds. *The Code of Codes: Scientific and Social Issues in the Human Genome Project.* Cambridge: Harvard University Press, 1992.

Lewontin, R.C. "The Dream of the Human Genome." *The New York Review of Books,* Vol. 39, No. 10, May 28, 1992.

Lee, Thomas F. *The Human Genome Project: Cracking the Code of Life.* New York and London: Plenum Press, 1991.

3. **Take Action** What legislation exists to protect people against genetic discrimination? Who is allowed to know the results of genetic screening tests? How can this information be used? Write to the following for information.

   The Council for Responsible Genetics
   19 Garden Street
   Cambridge, MA 02138

   National Center for Human Genome Research
   Cold Spring Harbor, NY 10098

4. **Examining Social Consequences** Are scientists responsible for the ways in which information from the Human Genome Project is used? Explain. Compare modern research in human genetics with the eugenics movement of the early 20th century. You may need to do library research. How is the comparison accurate? How is it false? Discuss the role of scientists in each case.

CHAPTER
11

THE ORIGIN OF LIFE

REVIEW

- polar nature of water (Section 2-2)
- structure of proteins and nucleic acids (Section 2-3)
- second law of thermodynamics (Section 4-1)
- role of enzymes in catalyzing chemical reactions (Section 4-2)
- natural selection (Section 5-3)
- roles of RNA and DNA in heredity (Sections 9-1 and 9-3)

Artist's view of the early Earth

11-1 The Mystery of Life's Origin

Everywhere you look there is life. It is found not only in fields, forests, and ponds but also in deserts, on polar icecaps, atop high mountains, and deep beneath the sea. Drops of water teem with life, mostly creatures too tiny to be seen without magnification as they dash about, bashing into each other. With life all around, it is hard to imagine there was ever a time when Earth was barren of life, but indeed there was. Studies of radioactive elements in rocks indicate that Earth is about 4.5 billion years old, or 1 billion years older than the fossil seen in Figure 11-1. Where did such life-forms come from? How did they arise? How did a barren mass of rock and water become a home for deer, butterflies, sea otters, and humans?

Section Objectives

- Identify three possible origins of life on Earth.
- Compare and contrast the primordial soup model with the bubble model.
- Explain how the first proteins might have formed.
- Describe how cellular organization might have begun.
- Recognize the importance of the development of heredity.

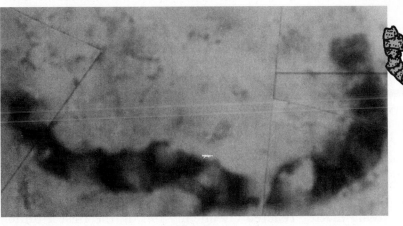

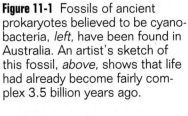

Figure 11-1 Fossils of ancient prokaryotes believed to be cyanobacteria, *left*, have been found in Australia. An artist's sketch of this fossil, *above*, shows that life had already become fairly complex 3.5 billion years ago.

There Are Several Ideas About the Origin of Life

Trying to explain how life might have originated on Earth is a difficult but fascinating quest and one that has been pursued by philosophers, theologians, and scientists alike. Still, we may never know *exactly* how life got here. No one was there to record what happened. What was it really like on the early Earth? Were the forces that caused life to begin ones that can be explained by science or ones that cannot be explained by science? A few scientists even question whether life originated on Earth. They have suggested instead that life on Earth had an **extraterrestrial** (*EHKS truh tuh REHS tree uhl*) **origin**—an origin outside of Earth. They hypothesize that life was carried here by an asteroid or by a meteorite like the one in Figure 11-2. Although there may always be questions about

Figure 11-2 Meteorites like this one might have brought life to Earth. Organic compounds make up about 2 percent of its weight.

The Origin of Life **225**

exactly how life arose on Earth, there are many possibilities. The two principal means by which life might have originated are discussed below.

Divine Creation

Traditionally, many cultures have believed that life was put on Earth by divine (relating to a god or gods) forces, as the act of a creator or creators. Belief in divine creation is common to many of the world's major religions, though the accounts of creation vary from one religion to another. By all accounts of divine creation, the process that gave rise to life on Earth was driven by forces that cannot be explained by science. Philosophers have debated the essence of these forces for centuries. It is important to understand, however, that a belief is not the same thing as a scientific hypothesis. The essence of any scientific hypothesis is that the proposed idea is subject to test—that the idea could, in principle at least, be proven false. As you learned in Chapter 1, science is a way of investigating the natural world (through observation and experimentation) and forming general rules about how things happen. A belief in divine creation, however, is not a scientific hypothesis that can be tested. Try to imagine an observation that would disprove divine creation. Whatever you propose, it is always possible to argue that a divine agent simply made things appear the way they do. Because the idea that life originated through divine creation cannot be tested by scientific methods, it falls outside the realm of science. This is not to say that the belief is wrong, but rather that science can never test it.

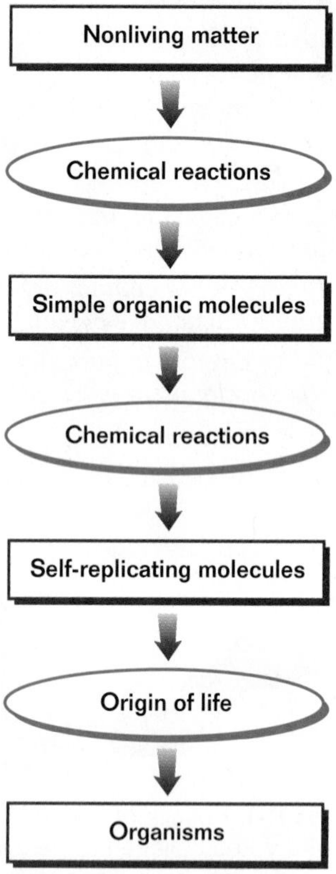

Spontaneous Origin

Most scientists think that life on Earth had a **spontaneous origin,** developing by itself through natural chemical and physical processes. They hypothesize that molecules of non-living matter reacted chemically during the first 1 billion years of Earth's history, forming a variety of simple organic molecules. They further hypothesize that complex organic molecules, some of which were capable of replicating themselves and other molecules, formed associations that became increasingly complex. In this view, outlined in Figure 11-3, the process by which life arose was driven by the natural force of selection. Changes that increased the stability of certain molecules would have allowed (selected) those molecules to persist for a longer time. Molecules that could be replicated would have become more common than those that could not be replicated. In laboratory experiments, many of the organic building blocks of life have been made from molecules of nonliving matter. Thus, the assumption that life began spontaneously can be tested by scientific methods. Because it is the only assumption that can be tested by scientific methods, this chapter deals with hypotheses based on the idea that life on Earth originated through natural chemical and physical processes. ∎

```
Nonliving matter
      ↓
Chemical reactions
      ↓
Simple organic molecules
      ↓
Chemical reactions
      ↓
Self-replicating molecules
      ↓
Origin of life
      ↓
Organisms
```

Figure 11-3 Most scientists think that life began spontaneously as the result of the process summarized here. A series of chemical reactions could have produced simple organic molecules, then more complex organic molecules, and eventually self-replicating, living matter.

❑ CAPSULE SUMMARY

Many possible origins of life have been proposed. Presently, only hypotheses based on the assumption that life arose naturally and spontaneously on Earth can be tested by scientific methods.

Life's Basic Chemicals Can Form Spontaneously

For life to have formed naturally on Earth, the materials that make up living things must have been present. As you learned in Chapter 2, all organisms are built from the same chemicals, just as all cars are assembled from the same materials. Cars are made from steel, glass, plastic, and rubber. Cells are made from proteins, lipids, carbohydrates, and nucleic acids. If life originated spontaneously, then the first question that must be answered is, Where did these molecules—the building blocks of life—come from? In their attempts to answer this question, scientists have focused on Earth's early oceans and atmosphere. Most scientists hypothesize that the basic chemicals of life formed during chemical reactions that occurred there. To find out if these reactions could have taken place, you must take an experimental journey more than 4.5 billion years back in time. Scientists do this in their laboratories by analyzing data from newly forming stars and by re-creating conditions they suspect existed on the early Earth.

The Primordial Soup Model

Charles Darwin, who developed the theory that life evolves through natural selection, once speculated that life began in "a warm little pond." In the 1920s, the Russian scientist A. I. Oparin proposed a hypothesis that extended Darwin's idea. He suggested that Earth's oceans were once a vast primordial *(preye MAWR dee uhl)* soup containing large amounts of organic molecules. Oparin envisioned these molecules forming spontaneously in chemical reactions activated by energy from solar radiation, volcanic eruptions, and lightning. Oparin thought that over millions of years, these molecules had gradually come together to form living matter.

Oparin, Harold Urey of the University of Chicago, and other investigators of the solar system proposed that Earth's early atmosphere lacked oxygen. They hypothesized that the early atmosphere was instead rich in nitrogen (N_2) and hydrogen-containing gases such as hydrogen (H_2), water vapor (H_2O), methane (CH_4), and ammonia (NH_3). Electrons in these gases would have been frequently pushed to higher energy levels by photons crashing into them from the sun or by electrical energy in lightning. Today, high-energy electrons are quickly soaked up by the oxygen in Earth's atmosphere (air is 21 percent oxygen) because oxygen atoms have a great "thirst" for such electrons. But in the absence of oxygen, high-energy electrons would have been free to do other things.

In 1953, Oparin's hypothesis was tested by Stanley Miller, who was then a graduate student working with Urey. To find out what sort of chemical "mischief" these electrons might have done, Miller placed the proposed gases into an apparatus like the one seen in Figure 11-4. Then, to simulate lightning, he zapped the mixture with electrical sparks. After a

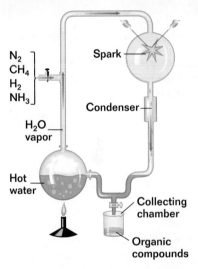

Figure 11-4 Lightning and heat from volcanic eruptions, *top*, might have been energy sources for the chemical reactions that led to the origin of life. In an apparatus similar to the one shown, *bottom*, Miller exposed a mixture of nitrogen and hydrogen-rich gases to electrical sparks simulating lightning. This produced high-energy electrons that jumped from molecule to molecule, causing the formation of chemical bonds and many new kinds of molecules.

The Origin of Life   227

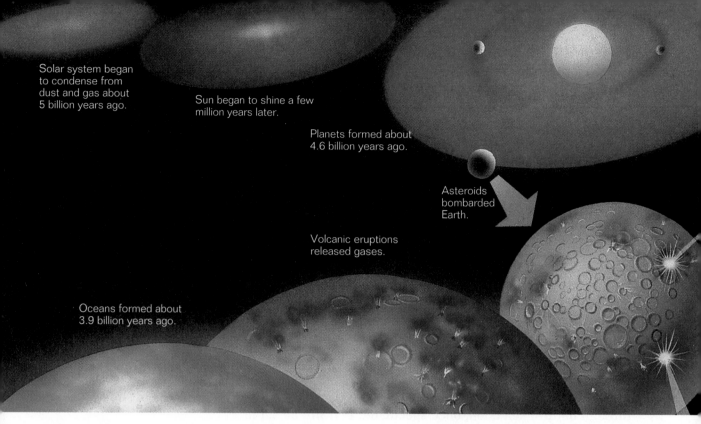

Solar system began to condense from dust and gas about 5 billion years ago.

Sun began to shine a few million years later.

Planets formed about 4.6 billion years ago.

Asteroids bombarded Earth.

Volcanic eruptions released gases.

Oceans formed about 3.9 billion years ago.

Figure 11-5 According to the current scientific model of the formation of the solar system, Earth would have been uninhabitable until about 3.8 billion years ago.

□ *CAPSULE*
SUMMARY

Miller demonstrated that basic organic molecules of life could have formed spontaneously from materials present on the early Earth. The primordial soup model is being reevaluated for two reasons: (1) life might have originated more quickly than previously assumed, and (2) methane and ammonia might not have been present in Earth's early atmosphere, as was assumed in Miller's experiment.

few days, Miller found a complex "chemical zoo" in the collecting chamber of his apparatus. Within this smelly mixture, he found some of life's basic building blocks: amino acids, fatty acids, and hydrocarbons (molecules made of carbon and hydrogen). These results demonstrated that some basic chemicals of life could have formed spontaneously on the early Earth under conditions like those in the experiment.

Recent discoveries have caused scientists to reevaluate Oparin's primordial soup model. At the time of Miller's experiment, scientists thought that life had taken more than a billion years to begin. However, today's model of Earth's formation, seen in Figure 11-5, and discoveries of 3.5-billion-year-old fossils indicate that the time available for life to begin was much shorter. Thus, it appears that life would have had to originate much faster than previously assumed. Another problem with the primordial soup model concerns the mixture of gases Miller used to simulate conditions on the early Earth. If the atmosphere had no oxygen 4 billion years ago, Earth would not have had a protective layer of ozone gas, O_3. Today, ozone shields our planet's surface from most of the sun's damaging ultraviolet light. Scientists think that without an ozone layer, ultraviolet light would have destroyed any ammonia and methane present in the atmosphere. When these gases are missing from experiments similar to Miller's, key biological molecules such as amino acids are not produced. This raises a very pointed question: If the necessary ammonia and methane were not in the atmosphere, where did they come from? □

The Bubble Model

In 1986, the geophysicist Louis Lerman suggested that problems with Oparin's model could be solved if the model were "stirred up" a bit. Lerman suggested that the key chemical processes took place not in a primordial soup but within bubbles on the ocean's surface. Bubbles produced by wind, wave action, the impact of raindrops, and the eruption of volcanoes cover about 5 percent of the ocean's surface at any given time. As you learned in Chapter 2, water molecules are polar. Therefore, water bubbles tend to attract other polar molecules. Lerman's bubble model, illustrated in Figure 11-6, proposes that chemicals in the early ocean collected within bubbles and reacted to form the key biological molecules.

Lerman's bubble model solves two key problems with Oparin's primordial soup model. First, chemical reactions would proceed much faster in bubbles (where reactants would be concentrated) than in Oparin's stagnant primordial soup. Thus, life could have originated in a much shorter period of time than it could have according to Oparin's model. Second, inside the bubbles, the methane and ammonia required to produce amino acids would have been protected from destruction by ultraviolet light.

Figure 11-6 Louis Lerman's bubble model suggests that chemicals reacted within bubbles on the early ocean, forming simple organic molecules. After countless generations of bubbles within which constantly changing mixtures of these molecules reacted, very complex biological molecules eventually resulted.

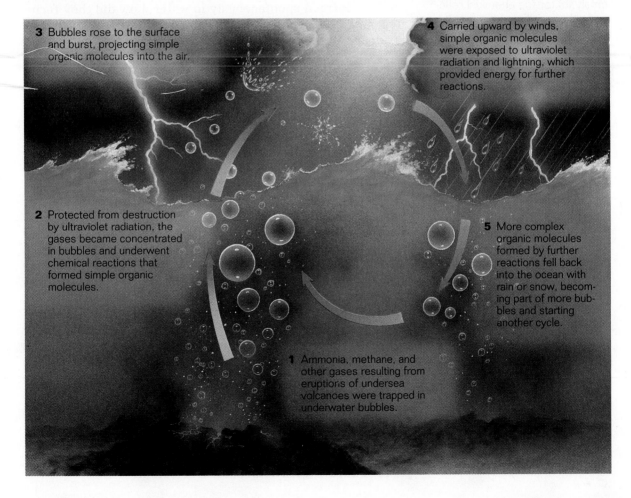

3 Bubbles rose to the surface and burst, projecting simple organic molecules into the air.

4 Carried upward by winds, simple organic molecules were exposed to ultraviolet radiation and lightning, which provided energy for further reactions.

2 Protected from destruction by ultraviolet radiation, the gases became concentrated in bubbles and underwent chemical reactions that formed simple organic molecules.

5 More complex organic molecules formed by further reactions fell back into the ocean with rain or snow, becoming part of more bubbles and starting another cycle.

1 Ammonia, methane, and other gases resulting from eruptions of undersea volcanoes were trapped in underwater bubbles.

Figure 11-7 Iron pyrite crystals have complex three-dimensional surfaces. Materials with such surfaces might have served as catalysts for the chemical reactions that produced the first building blocks of life.

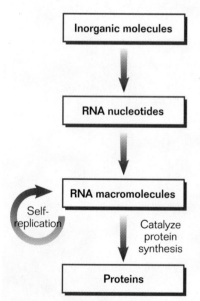

Figure 11-8 If RNA nucleotides from chemical reactions among inorganic molecules assembled into RNA molecules, these molecules might have been able to self-replicate and to catalyze the formation of proteins.

Proteins Can Be Assembled by RNA

Whatever the details of the process, most scientists accept that with the input of energy, the basic molecules of life could have formed spontaneously through simple chemistry. But knowing how the basic building blocks of life might have formed is like finding iron ore, sand, oil, and rubber trees—far from the steel, glass, plastic, and rubber needed to make a car. In other words, there is a long way to go from organic molecules to living cells. How did amino acids link together to form proteins? And how did nucleotides join to form long chains of DNA that store the instructions for assembling proteins? In the laboratory, scientists have not been able to make either of these macromolecules form spontaneously in water. Each link in an organic macromolecule is forged by a chemical reaction that also produces a water molecule. These reactions simply do not occur when their reactants are surrounded by a dense crowd of water molecules—that is, when they are dissolved in water. However, short chains of RNA, the nucleic acid that works with DNA to carry out DNA's instructions, can (with difficulty) be made to form spontaneously in water.

If the proteins and DNA that are necessary for life could not have formed spontaneously in water, how did they form? Some investigators speculate that early life could have developed on a solid surface rather than in water. The surfaces of clay minerals and iron pyrite *(PEYE reyet)* crystals have been suggested as possibilities. As you can see in Figure 11-7, materials such as these offer complex three-dimensional surfaces that might have acted as catalysts by serving as templates (patterns) for the formation of protein and DNA molecules. But while this might be possible in theory, researchers have not yet been able to make either protein or DNA molecules in this way, and most scientists are skeptical of the idea.

In the 1980s, Thomas Cech and his colleagues at the University of Colorado made a key discovery that may answer one question about life's mysterious origin. They found that certain RNA molecules can act like enzymes. RNA's three-dimensional structure provides surfaces with specific shapes for catalyzing reactions, much as protein shapes do. Like DNA, RNA acts as an information-storing molecule. Recall from Chapter 9 that messenger RNA molecules temporarily store the instructions for making proteins in their nucleotide sequences. As a result of Cech's work and experiments demonstrating that RNA molecules can form spontaneously in water, a very simple and attractive hypothesis has emerged: Perhaps RNA was the first self-replicating information-storage molecule. After it had formed, such a molecule could also have catalyzed the assembly of the first proteins, as suggested in Figure 11-8. But more important, such a molecule would have been capable of evolving through natural selection.

Microspheres Might Have Led to Cells

When you consider how some of the basic molecules of life behave in water, it is not difficult to imagine how the first cells might have formed. Remember that every cell is surrounded by a plasma membrane. Observations show that the basic molecules of plasma membranes—proteins and lipids—tend to aggregate (gather together) in water. By shaking up a bottle of oil-and-vinegar salad dressing, you can see a similar thing happening; the small spherical globs of oil formed by the shaking action attract one another and grow in size by fusing with other globs. Phospholipids, which form the bilayer of a plasma membrane, do the same. Similarly, short chains of amino acids produced abiotically (without life) in a laboratory aggregate into tiny vesicles called **microspheres**.

Scientists think that microspheres, similar to those shown in Figure 11-9, might have been the first step toward cellular organization. Once the basic molecules were present, the early oceans would have contained untold numbers of microspheres—billions in each spoonful of sea water. At first, microspheres would have formed spontaneously, persisted for a while, and then dispersed. Over millions of years, those microspheres that could survive longer by more efficiently incorporating molecules and energy would have become more common than here-today-gone-tomorrow kinds. Still, microspheres could not be considered alive unless they had acquired the capacity to transfer their abilities to offspring. ❑

❑ **CAPSULE SUMMARY**

Lipids and proteins tend to aggregate in water, forming microspheres. These tiny spheres could have been the first step in the organization of cells.

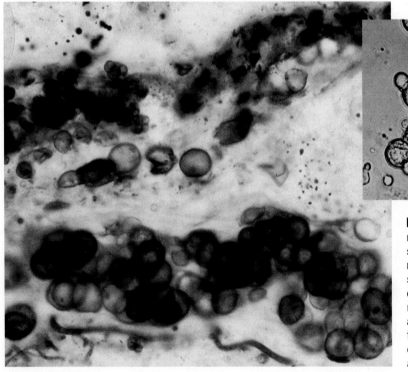

Figure 11-9 Ancient cells found in microfossils, *left*, resemble microspheres, *above*. Though they are not alive, microspheres share several characteristics with living cells. Both have a selectively permeable membrane, take in materials, grow in size, use energy to facilitate chemical reactions, and divide to form new individuals after reaching a certain size.

Origin of Heredity Remains a Mystery

There is considerable discussion among scientists about how hereditary mechanisms might have evolved. Most researchers now suspect that RNA was the first information-storing molecule to form and that RNA "enzymes" catalyzed the assembly of the earliest proteins. Scientists think that double-stranded DNA probably evolved later, as a way of ensuring the safety of hereditary information by storing it in a protected central location. However, scientists do not agree about whether RNA molecules first formed inside or outside of microspheres. Perhaps among microspheres that contained RNA, some might have developed a means of transferring their abilities to offspring (heredity). Once the mechanism of heredity developed, life as we know it began.

As you can see, the scientific vision of life's origin is at best a hazy outline viewed from a long distance through dark glasses. While scientists cannot disprove the hypothesis that life originated naturally and spontaneously, little is known about what actually happened. Many different scenarios—some of them quite imaginative—seem possible, and some have solid support from experiments. But because researchers do not yet understand how DNA, RNA, and hereditary mechanisms first developed, science is currently unable to resolve disputes concerning the origin of life. How life might have originated naturally and spontaneously remains a subject of intense interest, research, and discussion among scientists. ◻

◻ CAPSULE SUMMARY

Scientists now hypothesize that RNA was the first self-replicating information-storage molecule, but they cannot explain how heredity developed.

Section Review

1. *List three ideas about how life could have originated on Earth.*
2. *Why can't hypotheses that assume life had a divine origin be tested scientifically?*
3. *How are the primordial soup model and the bubble model similar? How are they different?*
4. *What properties of RNA make it the most likely candidate for the first information-storage molecule?*
5. *What important process had to develop before life could begin?*

Critical Thinking

6. *Why are microspheres a logical first step toward cellular organization?*
7. *Besides DNA, what other important feature that is a characteristic of all cells had to develop before living cells could exist?*

11-2 Evaluating the Spontaneous Origin Hypothesis

Most scientists accept spontaneous origin as the best available explanation of life's origin. However, like any scientific hypothesis, this explanation is subject to future modification or rejection based on new evidence. Indeed, strong objections to the spontaneous origin hypothesis have been voiced by people outside the scientific community. Evaluating these objections will provide you with a better understanding of the hypothesis.

Section Objectives

- Explain how radioisotopes can be used in determining Earth's age.
- Identify the biological processes that account for the diversity and complexity of life on Earth.
- Explain why life processes do not violate the second law of thermodynamics.
- Explain why proteins do not assemble spontaneously in water, and describe how they are assembled.

Is Earth Old Enough?

There is no doubt among scientists that our planet is very old. However, some people disagree, believing that Earth is no more than 10,000 to 20,000 years old—not nearly old enough for life to have arisen and then evolved as most scientists infer. The accepted scientific estimate of Earth's age is 4.5 billion years. This estimate is derived from **radiometric dating**, which involves the measurement of radioactive isotopes *(EYE soh tohps)* of certain elements found in rocks. Isotopes are forms of an element that differ in atomic mass. Radioactive isotopes, or **radioisotopes**, gradually change into other, more stable isotopes through a process called radioactive decay. For example, certain rocks contain minute traces of potassium-40, a radioisotope of the element potassium, K. As Figure 11-10 shows, it takes about 1.3 billion years for one-half of the potassium-40 in a rock to decay into other isotopes. The period of time it takes for one-half of a radioisotope to decay is called its **half-life**. By estimating how many half-lives have passed since a rock was formed, scientists can approximate the rock's age. Other long-lived radioisotopes are also used for radiometric dating.

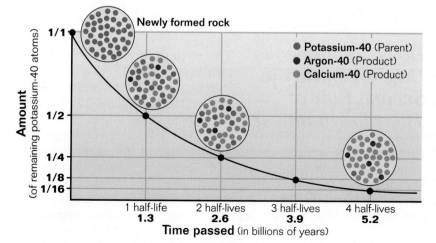

Figure 11-10 This graph shows the rate of decay for the radioisotope potassium-40. Rocks are dated by measuring the ratio of a radioisotope to its decay product. The radiometric clock starts "ticking" when rock minerals first crystallize from magma. After one *half-life* has passed, half of the original amount of a radioisotope remains, and the other half has been converted to decay product. The rest of the radioisotope continues to decay at a constant rate until all of it becomes the decay product.

Is radiometric dating reliable? The regular and measurable rate at which radioisotopes decay is the basis for radiometric dating. The constancy of radioactive decay is an elementary principle of physics that springs from the fundamental forces that hold all atoms together. The behavior of these forces has been verified many times. Though the rate of radioactive decay does vary slightly for certain radioisotopes under specific conditions, any possible variations are taken into account during the dating process. These variations are not significant enough to invalidate dating techniques. For a radioisotope's decay rate to fluctuate significantly, the fundamental nature of matter would have to change. Still, every scientist knows that *no* technique is 100 percent accurate. Determining the age of a rock, therefore, involves multiple measurements, using several different radioisotopes if possible. When it is done properly, radiometric dating is one of the most reliable and reproducible measurements scientists can make. ❑

❑ *CAPSULE SUMMARY*

Scientific estimates of Earth's age are derived from radiometric dating, which is based on the regular rate at which radioisotopes decay. For these rates to change, the fundamental nature of matter would have to change as well.

Is Life Too Complex to Have Arisen Naturally?

It may seem that organisms are too complex to have originated by random natural processes. To some, this implies that the complexity of life must be the result of an intelligent design. Most biologists, however, think that the complexity and diversity of life can result from the enormous amount of variation that can be encoded in DNA, coupled with the power of natural selection when applied over long periods of time. Natural selection—the very powerful agent of change that guides evolution—is neither random nor directionless. Instead, it produces very specific changes that are determined by the environment. Each favorable change is built upon a previous one, and, therefore, favorable changes are cumulative. With variation and natural selection working together over billions of years, very complex life-forms could have evolved naturally. ❑

The diversity and complexity of organisms can be explained as the result of genetic variation and the power of natural selection.

Does Spontaneous Origin Violate the Second Law of Thermodynamics?

The second law of thermodynamics (disorder tends to increase in the universe) is often cited as a reason to reject the spontaneous origin hypothesis. For life to have begun spontaneously, simple chemicals present on the early Earth must have become more ordered. Recall, however, that the second law of thermodynamics applies only to closed systems, while Earth and its organisms are open systems. In Chapter 4, you learned that increasing the order in an open system simply requires an input of energy. And as you have also learned, radiant energy from the sun continually enters Earth's living systems through photosynthesis, fueling the processes that organize life from metabolism to evolution.

To better understand why the second law of thermodynamics does not apply to processes involving organisms, consider the changes that occurred as your body developed. Like every human being, you began life as a single fertilized egg cell, similar to the one in Figure 11-11. You are now a highly organized creature, far more complex than you were when you started life. You have not violated the second law of thermodynamics because you are an open system. Any open system is capable of increasing in complexity as it absorbs energy from its surroundings. Therefore, even life's first chemicals could have organized spontaneously if enough energy sources were present on the early Earth. As Miller demonstrated in his experiment, electrical energy from lightning could have been one of those energy sources.

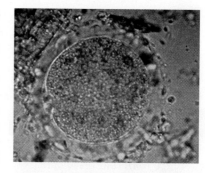

Figure 11-11 A fertilized human egg cell, like this one, grows and increases in complexity as it absorbs energy-containing nutrients from its surroundings.

Can Proteins Assemble Spontaneously?

Scientists have not been able to cause amino acids dissolved in water to join together to form proteins. The energy-requiring chemical reactions that join amino acids are freely reversible and do not occur spontaneously in water. However, most scientists no longer argue that the first proteins assembled spontaneously. Instead, they now propose that the initial macromolecules were composed of RNA, and that RNA later catalyzed the formation of proteins. The assembly of RNA from nucleotides is *not* reversible in water and could easily have occurred spontaneously. As scientists also point out, energy for such reactions could have come from many sources, including ultraviolet light, lightning, and the extreme heat of volcanic events. ◼

❑ CAPSULE SUMMARY

Proteins cannot assemble spontaneously in water, but they can be assembled with the aid of a catalyst and an input of energy.

Section Review

1. *How does radiometric dating indicate a rock's age?*
2. *What biological processes account for the diversity and complexity of life on Earth?*
3. *Does the second law of thermodynamics apply to the organization of life? Why or why not?*

Critical Thinking
4. *The cytosol of a cell is mostly water. Why, then, can proteins be assembled there?*

11-3 Is There Life on Other Worlds?

*O*n a dark, clear night you can look up and see countless stars in the sky. Have you ever wondered if we are alone in the universe? Does "someone" somewhere out there look up and wonder if we exist? For centuries, people have speculated about whether life exists elsewhere in the universe. Jules Verne wrote novels about people on the Moon. Reports of canals on Mars, like those seen in Figure 11-12, led to stories about beings from Mars. Scientists now know there is no evidence of life on either the Moon or Mars. But what about all the places humans have not yet explored?

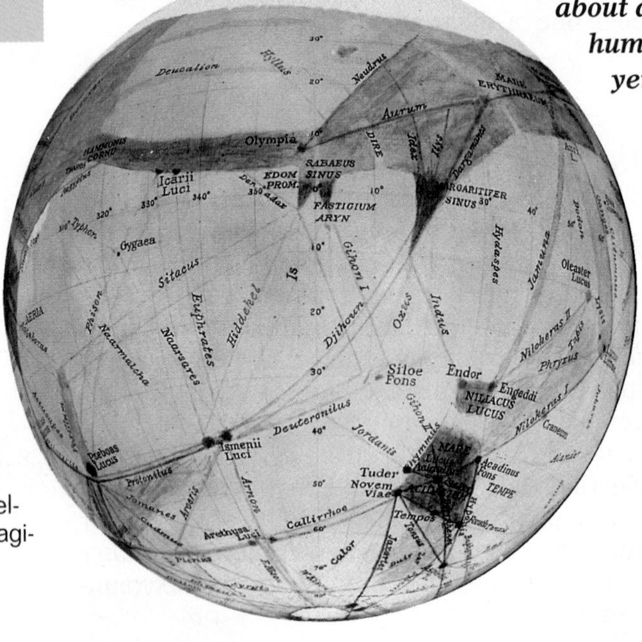

Section Objectives

- Relate Earth's size and distance from the sun to the life-forms it supports.
- Explain why life probably exists elsewhere in the universe.
- Describe how scientists are looking for intelligent life in other parts of our galaxy.

Figure 11-12 The American astronomer Percival Lowell, who believed that distinct lines seen on the surface of Mars were canals built by intelligent life-forms, drew this imaginative map of Mars in 1903.

Life on Earth Reflects the Nature of the Planet

Life as it has evolved on Earth closely reflects the nature of the planet and its history. For one thing, life as we recognize it exists mostly within a narrow range of temperatures—approximately –18° to 38°C (0° to 100°F). The spontaneous evolution of carbon-based life is probably possible only within the narrow range of temperatures that exists on Earth, a range directly related to its distance from the sun. If Earth were farther from the sun, it would be colder, and chemical processes would be greatly slowed down. Water, for example, would be a solid, and many carbon compounds would be brittle. If Earth were closer to the sun, it would be warmer, chemical bonds would be less stable, and few carbon compounds would be stable enough to persist. For life as we know it to exist on another planet, that planet would have to orbit its sun at just the right distance to have the necessary range of temperatures.

Earth's mass (which is a function of its size and density) is also just right for life to exist as we know it. Because of its mass, Earth has the right amount of gravitational pull to hold the gases found in our atmosphere. Earth's atmosphere helps insulate it from temperature extremes and radiation that is harmful to life, while allowing enough energy to reach its surface for fueling life's activities. If Earth were less massive (smaller or less dense), its gravitational pull would not be great enough to hold an atmosphere. If Earth were more massive (larger or more dense), it might hold such a dense atmosphere that all solar radiation would be absorbed before it reached the planet's surface. So for life as we know it to exist on another planet, that planet would have to have just the right mass (size and density) and gravitational pull to hold a suitable atmosphere. ◻

◻ **CAPSULE SUMMARY**

Earth has just the right mass and is just the right distance from the sun for life as we know it to exist.

Life Probably Exists Elsewhere
....

The universe as a whole is awash with places where life might have arisen. Within our solar system, the tiny moon of Jupiter seen in Figure 11-13 is the place most likely to support extraterrestrial life. In fact, conditions there would be far less hostile to life than the conditions that are thought to have existed in Earth's primordial oceans. Our own Milky Way galaxy and the nearby Andromeda galaxy each contain more than 100 billion stars. And the universe holds more than a billion galaxies. Astronomers estimate that the universe contains some 10^{20} (100,000,000,000,000,000,000) stars with physical characteristics that resemble those of our sun. At least 10 percent of these stars are thought to have planetary systems. New telescopes, such as the Hubble Space Telescope, reveal that several nearby stars seem to have planets orbiting them. If only 1 in 10,000 of the planets in the universe has the right combination of mass and distance from its sun to duplicate Earth's development, life could have arisen 10^{15} (a million billion) times. Undoubtedly, many other worlds have physical characteristics resembling those of Earth. Therefore, we might not be alone.

Life processes also might have arisen and evolved differently on other planets. A functional genetic system that is capable of accumulating and replicating changes is the basis of the evolution of life on Earth. But heredity does not require DNA—only a way to preserve and pass on information. Under different conditions, such a system theoretically could form from substances other than the carbon-based compounds and water that make up life on Earth. Silicon and ammonia are the most likely possibilities. Like

Figure 11-13 Europa, one of the moons of Jupiter, is the place in our solar system most likely to harbor extraterrestrial life. In 1995 scientists learned that its thin atmosphere contains oxygen gas. Beneath Europa's thick skin of ice, pressure due to gravity might create enough heat for water to exist in a liquid form. If that is so, life might have arisen and flourished there, hidden from view.

carbon, silicon needs four electrons to fill its outer energy level. And ammonia is even more polar than water. Perhaps under radically different temperatures and pressures, these substances might have formed complex molecules as diverse and flexible as the carbon-based ones on Earth. ■

Scientists Are Listening for Other Life-Forms

Serious attempts are now being made to look for messages from intelligent life-forms that might exist on planets circling distant stars. These efforts are referred to as the Search for Extraterrestrial Intelligence (SETI). The first SETI program was carried out by astronomer Frank Drake in 1960. He listened to two nearby stars for two weeks at one particular radio frequency but heard only sounds from military aircraft on Earth. Since then, other attempts (usually lasting several months or more) have been made to detect signals from intelligent life on other worlds. Although unusual signals have been detected, none have been found to repeat in a manner that could be detected again later.

In 1992, NASA stepped up the efforts to locate extraterrestrial life by aiming the world's largest radio telescope, seen in Figure 11-14, at approximately 1,000 of the stars nearest Earth. Using a new computer technology, NASA simultaneously monitored 8.4 million radio channels for signs of intelligent life. About 25 signals interesting enough to require further analysis were detected in the first year, although none were proved to be from alien life-forms. In 1993, however, Congress cut the program from the NASA budget, calling it "a great Martian chase" and "a waste of time." Astronomers are seeking ways to continue the search.

Figure 11-14 The world's largest radio telescope, located in the hills above Arecibo, Puerto Rico, is a huge disk more than 305 m (1,000 ft.) in diameter.

Section Review

1. *What evidence would you cite to support the hypothesis that life could exist elsewhere in the universe?*

Critical Thinking

2. *What do you think would happen to life on Earth if the planet's orbit suddenly shifted so that it was twice as far from the sun?*

3. *All stars give off radio waves, which result from the nuclear reactions that make stars shine. What, then, would you look for in extraterrestrial radio signals that might indicate they were produced by an intelligent life-form?*

CHAPTER REVIEW 11

Vocabulary

extraterrestrial origin (225)
half-life (233)
microsphere (231)
radioisotope (233)
radiometric dating (233)
spontaneous origin (226)

Concept Mapping

Construct a concept map that shows how life might have originated by natural forces. Use as many terms as needed from the vocabulary list. Try to include the following items in your map: spontaneous origin, primordial soup model, bubble model, RNA, proteins, microspheres, and radioisotopes. Include additional terms in your map as needed.

Review

Multiple Choice

1. The divine creation explanation for the origin of life
   a. can be scientifically tested.
   b. is identical to the extraterrestrial origin explanation.
   c. is based on forces that cannot be explained by science.
   d. is virtually the same among the many religions of the world.

2. A reevaluation of the primordial soup model was prompted by
   a. the hypothesis that ammonia and methane may have been missing from Earth's early atmosphere.
   b. Miller's finding that amino acids can be formed from ammonia and other chemicals.
   c. evidence that 4.5 billion years were available for life to begin.
   d. the discovery that ozone gas was present in Earth's atmosphere 4 billion years ago.

3. Lerman's bubble model, which is a modification of the primordial soup model,
   a. accounts for the formation of bubbles in Earth's early oceans.
   b. describes how life could have formed in a short time in the presence of ultraviolet light.
   c. proposes how polar molecules and bubbles gave rise to cells.
   d. explains how ammonia, nitrogen, and oxygen combined to form amino acids.

4. RNA is thought to have been the first information-storage molecule because it
   a. can replicate on solid surfaces.
   b. can form spontaneously without water.
   c. can store information and catalyze reactions.
   d. can assist in the assembly of DNA.

5. Which of these sequences of chemical reactions is the most likely to have led up to the formation of proteins?
   a. inorganic molecules, RNA nucleotides, RNA macromolecules, proteins
   b. inorganic molecules, amino acids, microspheres, proteins
   c. silicon and ammonia, RNA nucleotides, microspheres, proteins
   d. methane and ammonia, amino acids, DNA, proteins

6. Cells are different from microspheres because cells
   a. contain amino acids.
   b. have an outer boundary made of two layers.
   c. grow by taking in molecules from their surroundings.
   d. transfer instructions for heredity.

7. The reliability of radiometric dating is based on the assumption that
   a. the rates of radioactive decay have not changed over time.
   b. radioisotopes gradually change to more stable isotopes.
   c. carbon isotopes do not decay.
   d. all rocks are less than 50,000 years old.

8. Scientists feel comfortable offering an explanation for the diversity and complexity of the biological world based on what is known about
   a. radioisotopes.
   b. supernatural forces.
   c. genetic variation and natural selection.
   d. the forces that hold all atoms together.

9. If Earth's mass were less than it is, life would not exist because
   a. Earth would be too close to the sun.
   b. the gases that form Earth's atmosphere would have dissipated into space.
   c. the atmosphere would be too dense for solar radiation to reach the surface.
   d. Earth's gravitational pull would make carbon compounds brittle.

10. What characteristics of radio waves might suggest that they were produced in another part of our galaxy by intelligent life-forms?
    a. They would blend with background noise.
    b. They would repeat and be louder and different from background noise.
    c. They would be transmitted at or above 2 million Hz.
    d. They would be transmitted at less than 2 million Hz.

Completion

11. Some scientists propose that life had a(n) _____ origin and was later transported to Earth by _____.

12. Stanley Miller's experiment tested the _____ model for the origin of life and showed that _____ might have served as an energy source for the formation of organic molecules from hydrogen-rich gases in Earth's early atmosphere.

13. Tiny vesicles called _____, which tend to form when lipids and proteins aggregate, are thought to be the first steps toward _____ organization.

14. Using a procedure called _____, scientists estimate that Earth's age is about _____ years.

15. The processes by which life is organized do not violate the second law of thermodynamics because Earth and each of its organisms are _____ systems.

16. Because the chemical reactions that form them are reversible, _____ cannot assemble spontaneously in water, but they can be assembled with the aid of a(n) _____.

17. Two critical factors that have affected the way in which life developed on Earth are the planet's mass and its _____.

18. Because of the tremendous number of _____ in the universe, it is _____ that life exists somewhere besides Earth.

Short Answer

19. How does Lerman's bubble model of the origin of life differ from Oparin's primordial soup model?

20. According to the spontaneous origin theory, how was the development of heredity important?

21. Sedimentary rocks are made up of the shells of dead organisms or pieces of older rocks. Why, then, is radiometric dating *not* used to determine the age of sedimentary rocks?

22. Why do scientists think that life as we know it is more likely to exist on Jupiter's moon Europa than on Mars?

Themes Review

23. **Evolution** The phrase *chemical evolution* is sometimes applied to the series of events that might have resulted in the spontaneous origin of life. How do the events that are thought to have produced living organic matter from nonliving inorganic matter represent a process of evolution?

24. **Levels of Organization** How do the steps in a spontaneous origin of life reflect the organization of organisms?

25. **Flow of Energy** How do the primordial soup model and the bubble model account for the input of energy needed to organize life? Where did the energy come from, how was it incorporated, and where did it go?

26. **Structure and Function** How does RNA's structure enable it to function as a catalyst, and how might RNA's structure have enabled it to evolve?

Critical Thinking

27. **Designing Experiments** Suppose that you are asked to design an experiment to test whether organic matter might have formed from inorganic matter on the early Earth. What variables would you test? What type of control would you use for each experiment?

28. **Interpreting Data** A sample of a certain type of igneous rock contains 25 g of a radioisotope. The same amount of this type of rock has 200 g of the radioisotope when it forms. Using the graph showing the radioisotope's decay rate, determine (a) about how long ago the rock sample was formed and (b) the radioisotope's approximate half-life.

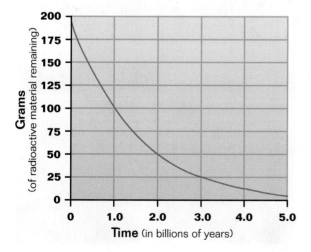

29. **Making Inferences** Suppose that while listening to recordings of radio waves coming from a solar system located in the Milky Way galaxy, you repeatedly hear strong radio signals that have patterns similar to music. What might you infer about the origin of such signals?

Activities and Projects

30. **Research and Writing** Thomas Cech and Sidney Altman shared a Nobel prize in 1989 for their work on RNA. Research their work and the rewards associated with winning a Nobel prize. Relate your findings in a written report.

31. **Cooperative Group Project** Collect information about the radio telescope at Arecibo, Puerto Rico, and about why the United States Congress decided in 1993 to stop funding NASA's use of the telescope to listen for signs of intelligent life. Then, debate the benefits and drawbacks of funding the listening project.

32. **Multicultural Perspective** Research stories from other cultures about visits by intelligent extraterrestrial beings. Relate your findings in an oral report.

Readings

33. Read R. Cowen's article "Taking a chemical look at the early Earth," in *Science News*, April 2, 1992, page 214. What radioisotope decayed to produce neodymium-142? What do Carlson and Jacobsen think that their findings may indicate about the formation of Earth's first crust and of the Moon?

34. Read Mitchell Waldrop's article "Finding RNA Makes Proteins Gives 'RNA' World a Big Boost," in *Science*, June 2, 1992, pages 1396–1397. How has Noller's research changed biologists' view of RNA's role in protein synthesis? How do Noller's findings support the hypothesis that RNA catalyzed the formation of the first proteins?

Making Microspheres

Make microspheres from amino acids, and compare their structure with that of living cells.

PROCESS SKILLS

- observing
- comparing and contrasting

MATERIALS

- safety goggles
- lab apron
- 500 mL beaker
- hot plate
- two 125 mL Erlenmeyer flasks
- ring stand with clamp
- balance
- aspartic acid
- glutamic acid
- glycine
- glass stirring rod
- tongs
- clock or timer
- 1% NaCl solution
- 50 mL graduated cylinder
- dropper
- microscope slides
- coverslips
- compound light microscope
- 1% NaOH solution

BACKGROUND

1. What are microspheres?
2. Offer a scientific explanation for how life could have begun from something similar to simple microspheres of heated amino acids.
3. How would microspheres be similar to living cells? How would they differ?
4. Write your own **Focus Question** on your Vee Form.
5. **Knowing Side of the Vee** List the **Concepts** and new **Vocabulary Words** on your Vee Form. In the **Concept Statements** section of the Vee, use these words in sentences that define and explain them.

TECHNIQUE

Doing Side of the Vee

1. Put on safety goggles and a lab apron.
2. Fill a 500 mL beaker half full with water and heat it on a hot plate. **CAUTION: Use care to avoid burns when working with the hot plate.** You will use the beaker as a hot-water bath. Leave space on the hot plate for a 125 mL Erlenmeyer flask to be added later.
3. While waiting for the water to boil, clamp a 125 mL Erlenmeyer flask to a ring stand. Add 1 g each of aspartic acid, glutamic acid, and glycine to the flask, and combine these dry powders with a stirring rod.
4. When the water in the beaker begins to boil, move the ring stand carefully so the flask of amino acids sits in the hot-water bath.

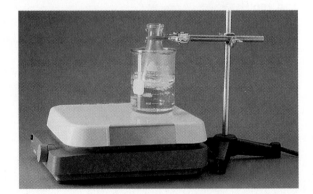

5. When the amino acids have heated for 20 minutes, measure 10 mL of 1% NaCl solution in a graduated cylinder and pour the solution into a second Erlenmeyer flask. Place the second flask on the hot plate beside the hot-water bath.

6. When the NaCl solution begins to boil, use tongs to remove the flask containing the NaCl solution from the hot plate. Then, still holding the flask with tongs, slowly add the NaCl solution to the hot amino acids while stirring.

7. Let the NaCl–amino acid solution boil for 30 seconds.

8. Remove the NaCl–amino acid solution from the water bath, and allow it to cool for 10 minutes.

9. Use a dropper to place a drop of the solution on a microscope slide, and cover the drop with a coverslip.

10. Place the slide on the microscope stage. Examine the slide under low power for tiny spherical structures. Then examine the structures under high power. These tiny sphere-shaped objects are microspheres. In the **Records** section of your Vee Form, draw what you see.

11. Place a drop of a 1% NaOH solution at the edge of the coverslip to raise the pH as you observe the microspheres. What happens?

12. Make a chart similar to the one shown below in the **Records** section of your Vee Form.

Characteristics of Microspheres

Cell-like characteristics	Non-cell-like characteristics

13. Clean up your materials and wash your hands before leaving the lab.

INQUIRY

1. Why was the NaCl–amino acid solution heated in step 7?

2. Suggest how microspheres were formed.

3. What did you observe when the pH was raised in step 11?

4. Use the information on the **Knowing Side** of the Vee to interpret your results from the **Doing Side,** and then write your **Knowledge Claim.** Write a **Value Claim** for this lab.

ANALYSIS

1. Compare and contrast microspheres with living cells.

2. What characteristics would microspheres have to exhibit before they could be considered living?

3. How might the conditions you created in the lab be similar to those that are thought to have existed when life first evolved on land?

4. Predict what would happen to microspheres if they were placed in hypotonic or hypertonic solutions.

FURTHER INQUIRY

Write a **New Focus Question** that could be the point of a new investigation. The following are examples:

What do you think would happen if you added too much or too little heat? How can you test for the right amount of heat to use?

Do you think your microsphere experiment would have worked if you had substituted other amino acids? How can you test your hypothesis?

CHAPTER 12

THEORY OF EVOLUTION

REVIEW

- structure of proteins (Section 2-3)
- mutations (Section 7-3)
- structure of genes (Section 8-1)
- gene sequencing (Section 10-1)
- radiometric dating (Section 11-2)

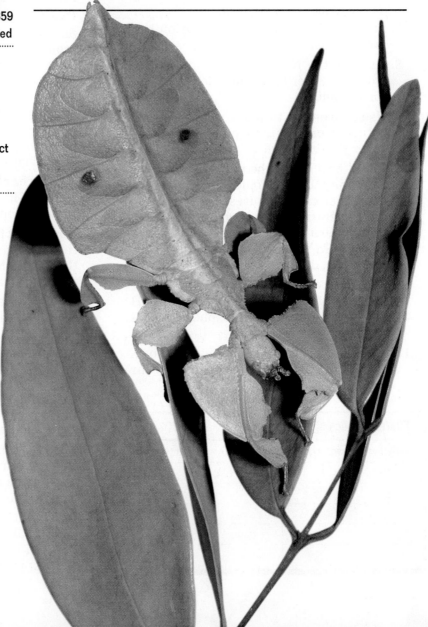

Phyllium pulchrifolium
(beautiful moving leaf)

12-1 The Work of Charles Darwin

Of all the major themes of biology, evolution is perhaps the best known but least understood by the general public. Scientists have concluded that the great diversity of life on Earth is the result of more than 3.5 billion years of evolution, during which species were replaced over time by other species. The idea that life evolves was expressed by several eighteenth- and nineteenth-century scientists. Lacking a reasonable explanation for evolution, other scientists of the time adamantly opposed the idea. Then, in 1859, the English naturalist Charles Darwin provided convincing evidence that species do evolve and suggested a mechanism (means) by which evolution occurs.

Section Objectives

■ Summarize the modern theory of evolution.

■ Identify several observations that led Darwin to conclude that species evolve.

■ Describe the process of natural selection and its outcome.

■ Discuss each of the main points of Darwin's theory of evolution by natural selection as it is stated today.

Overview THEORY OF EVOLUTION

The theory of evolution consists of the following four major points.

1. Variation exists within the genes of every species (the result of random mutation).

2. In a particular environment, some individuals of a species are better suited for survival and so leave more offspring (natural selection).

3. Over time, change within species leads to the replacement of old species by new species as less successful species become extinct.

4. There is clear evidence from fossils and many other sources that the species now on Earth have evolved (descended) from ancestral forms that are extinct (evolution).

Scientific theories are unifying explanations for many related observations. Like all scientific theories, the theory of evolution was developed through decades of scientific observation and experimentation. The modern theory of evolution began to take shape as a result of the work of Charles Darwin, seen in Figure 12-1. Today, virtually all scientists recognize evolution as the basis for both the diversity and relatedness of life on Earth.

Figure 12-1 This portrait of Charles Darwin was painted shortly after he returned from an around-the-world voyage. Observations made during the voyage led Darwin to look for a mechanism by which evolution occurs.

Darwin Became a Naturalist

Even though Charles Darwin became one of the greatest scientists of all time, he struggled in school. Darwin's father was a wealthy doctor who wanted Darwin to become either a doctor or a minister. Not inspired by the subjects his father urged him to study, Darwin frequently spent more time outdoors than in class. At the age of 16, Darwin was sent to Edinburgh, Scotland, to study medicine. Repelled by surgery, which at the time was done without anesthetics, Darwin repeatedly skipped lectures to collect biological specimens. In 1827, Darwin's father sent him to Cambridge University to prepare for the ministry. Although he did complete a degree in theology, Darwin again spent much of his time with friends who were interested in natural science.

In 1831, one of his professors at Cambridge recommended Darwin for a post as the unpaid naturalist on a naval voyage of the HMS *Beagle*. The ship and its route can be seen in Figure 12-2. Darwin was offered the position but regretfully declined when his father refused to let him go. Fortunately, Darwin's uncle interceded for him at the last moment. At the age of 22, Darwin was off on a journey that would change his life and forever change how we think of ourselves.

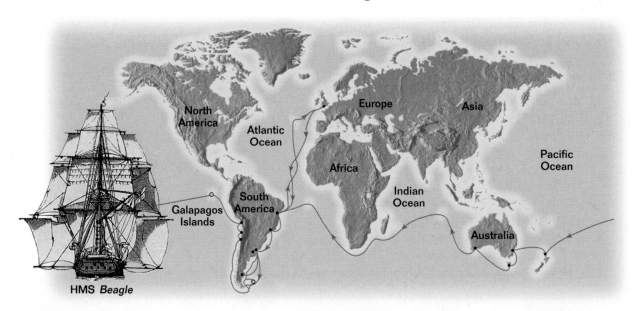

HMS *Beagle*

Figure 12-2 The HMS *Beagle* sailed around the world along the route shown on this map. The purpose of the ship's five-year voyage was to survey the coast of South America.

Darwin's Views Changed During the Beagle's Voyage

When the *Beagle* sailed on December 27, 1831, young Darwin accepted the prevailing view that each species was a divine creation, unchanging and existing as it was originally created. This view of divine creation was thought to explain why species are often uniquely adapted to their environments. However, scientists had begun to see that traditional views of divine creation could not explain the kinds and distribution of fossils they had found. Some scientists tried to explain their observations by modifying traditional accounts of creation, while others (including Darwin's own grandfather) proposed that the diversity of life resulted from evolution. In 1809, the French scientist Jean Baptiste Lamarck proposed a mechanism to explain how evolution occurs. Lamarck's hypothesis (now known to be incorrect) is illustrated in Figure 12-3.

While on the *Beagle,* Darwin's belief that species are unchanging began to weaken. During the voyage, Darwin read Charles Lyell's book *Principles of Geology,* which contains a detailed account of Lamarck's theory of evolution. As he visited different places, Darwin also saw things that he thought could only be attributed to a process of gradual change (evolution). For example, in South America, Darwin found fossils of extinct armadillos. These fossilized animals closely resembled, but were not identical to, the armadillos currently living in the area. Darwin wondered why living and fossilized species found in the same place would be similar and yet different. The most probable explanation, he reasoned, was that one species had given rise to the other.

On the Galapagos Islands, located about 1,000 km off the coast of Ecuador, Darwin found his most convincing evidence that species evolve. Darwin was struck by the fact that the plants and animals of the Galapagos Islands resemble those of the nearby coast of South America, as illustrated in Figure 12-4. If each species had been created independently and placed on the Galapagos Islands, why would they resemble the plants and animals of the adjacent South American coast? Why did they not instead resemble the plants and animals of similar islands, such as those near Africa, for instance? Darwin felt that the simplest explanation was that the ancestors of Galapagos species must have migrated there from South America long ago and changed after they arrived. Darwin referred to such change as "descent with modification"—evolution.

Figure 12-3 Lamarck believed that giraffes' necks became longer from stretching to reach higher for food and that the longer neck could be inherited by offspring. According to Lamarck, evolution occurs as structures develop through use, or disappear because of disuse, and as these "acquired characteristics" are passed to offspring.

Figure 12-4 On the Galapagos Islands, Darwin collected more than a dozen species of finches, each with a specialized diet and way of obtaining food. Close examination of these finches showed that all of the species closely resembled the blue-black grassquit, a South American finch species.

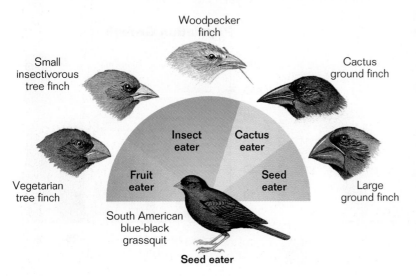

Woodpecker finch

Small insectivorous tree finch

Cactus ground finch

Insect eater

Cactus eater

Vegetarian tree finch

Fruit eater

Seed eater

Large ground finch

South American blue-black grassquit

Seed eater

Darwin Sought an Explanation for Evolution

When Darwin returned from his voyage at the age of 27, he continued his lifelong study of plants and animals. However, he did not publish his ideas about evolution until many years later. Shortly after his return, Darwin married and started a family. He and his wife, Emma, had 10 children. Darwin formulated his ideas on evolution over several years. During these full and active years, he painstakingly analyzed the specimens he had collected during his voyage and recorded his thoughts in several notebooks. As Darwin studied his data, his conviction that organisms had evolved grew ever stronger. However, he remained deeply puzzled about the most crucial question: *How* does evolution occur?

Malthus's Contribution

The key that unlocked Darwin's thinking about how evolution occurs was an essay written by the English economist Thomas Malthus. In his "Essay on the Principle of Population" (1798), Malthus stated that human populations have the potential to increase faster than the available food supply. As Malthus pointed out, a population grows in size by a geometric progression, such as the one illustrated in Figure 12-5. Food supply, however, increases at best by an arithmetic progression, which is also illustrated in Figure 12-5. According to Malthus, the human population would cover Earth's entire surface within a very short period of time if it could reproduce unchecked. This does not occur, Malthus pointed out, because death caused by disease, war, and famine intervenes.

The term *population*, as it is used today, does not refer only to the number of people living in a particular area. In biology, a **population** is a group of individuals that belong to the same species, live in a defined area, and breed with others in the group. For example, the alligators in the Everglades are a population because they live in a particular area and breed primarily with each other.

Figure 12-5 The "Food Supply Increase" graph, *left,* shows an arithmetic progression, in which the numbers increase by an added constant. The "Population Growth" graph, *right,* shows a geometric progression, in which the numbers increase by a multiplied constant.

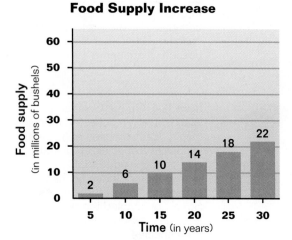

Food Supply Increase

Food supply (in millions of bushels) vs. Time (in years)

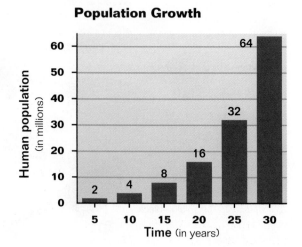

Population Growth

Human population (in millions) vs. Time (in years)

Natural Selection

Darwin realized that Malthus's principle of population applies to all species. Every organism has the potential to produce many offspring during its lifetime. In most cases, however, only a limited number actually survive to reproduce. Combining this observation with what he had seen on his voyage and with his own experiences in breeding domestic animals, Darwin made a key association: *Individuals that possess superior physical or behavioral attributes are more likely to survive than those that are not so well endowed.* Darwin saw that by surviving, individuals have the opportunity to reproduce and pass on their favorable characteristics to offspring. Thus, these characteristics will increase in a population, and the nature of the population will gradually change. Darwin called the process by which populations change in response to their environment **natural selection.**

Evolution by natural selection is a simple and logical explanation for the biological diversity and similarity Darwin saw on his voyage. As he saw it, organisms differed from place to place because their habitats presented different challenges to, and opportunities for, survival. Each species had evolved in response to its environment. The changing of a species that results in its being better suited to its environment is called **adaptation** *(ad uhp TAY shun)*. Darwin also noticed that organisms more closely resemble those living in nearby geographic locations than those living in similar yet widely separated parts of the world. The idea that species were created individually does not explain this pattern of distribution. Thus, Darwin concluded that the species of a particular place evolved from species that previously lived in the same area or that migrated from areas nearby. ◻

◻ **CAPSULE SUMMARY**

Darwin developed the theory of evolution by natural selection by applying Malthus's ideas on population to the observations he had made during the voyage of the Beagle.

Darwin Published His Ideas in 1859

In 1844, Darwin finally wrote down his ideas about evolution and natural selection in a preliminary manuscript that he showed only to a few scientists he knew and trusted. In that year, however, a burst of public criticism was directed against a book called *Vestiges of the Natural History of Creation.* The book claimed that evolution had occurred in Earth's past. Although it was a bestseller at the time, the book was criticized vehemently by both the government and the church. Lamarck's theory of evolution by the inheritance of acquired characteristics had also received severe criticism. Shrinking from such controversy, Darwin put aside his manuscript. For the next 14 years, he went about his life, enlarging and refining his notes on evolution but saying nothing about his ideas in public. While he published several books on other scientific subjects, his manuscript on evolution remained in a drawer in his study.

Figure 12-6 This cartoon (from 1874) of Darwin with a monkey-like "ancestor" shows how some people ridiculed Darwin because of his work.

❑ CAPSULE SUMMARY

Darwin did not publish his ideas about natural selection until similar ideas were presented to him by Alfred Russel Wallace. Today, Darwin's theory is widely accepted.

The stimulus that finally brought Darwin's work into print was a letter he received in June of 1858. The letter and a short essay were sent from Malaysia by the young English naturalist Alfred Russel Wallace. Wallace's essay concisely described the idea of evolution by natural selection. In his letter, he asked if Darwin would help get the essay published. Darwin's scientific friends urged him to get his own work ready. They arranged for an abstract of Darwin's manuscript to be presented with Wallace's paper at a public scientific meeting. Neither Darwin nor Wallace actually attended the meeting on July 1, 1858. Wallace was still in Malaysia, and one of Darwin's children had died of scarlet fever two days prior to the meeting. At the time, little notice was taken of the two papers, but Darwin had finally begun to write what he considered to be a "short abstract of his work."

When Darwin's book *On the Origin of Species by Means of Natural Selection* appeared in November of 1859, it caused an immediate sensation. Many people were deeply disturbed by certain aspects of Darwin's theory, such as the implication that humans are related to apes. While recognizing that humans closely resemble apes, they found the possibility of a direct evolutionary link unacceptable, as Figure 12-6 suggests. Darwin's arguments and evidence were so compelling, however, that his views were soon widely accepted by biologists around the world. Today, after more than a century of scientific progress, the evidence supporting evolution and natural selection is even more convincing. Although some people still find Darwin's ideas unacceptable, virtually all scientists agree that they are correct. *On the Origin of Species* has had a strong and lasting impact for two reasons: it presented a vast body of evidence that evolution has occurred, and, more important, it presented a reasonable hypothesis explaining how evolution takes place. ❑

Darwin's Ideas Have Been Updated

Since Darwin's work was published, his hypothesis—that natural selection is the mechanism by which evolution occurs—has been carefully examined by biologists. New discoveries, particularly in the area of genetics, have given scientists new insight into how natural selection brings about the evolution of species. Darwin's ideas, restated in modern terms, are summarized below.

Natural Selection Causes Change *Within* Populations

Darwin's key inference was that within any population, those individuals best suited to survive and prosper in their environment will leave the most offspring. Thus, the traits of those individuals will become more common in successive generations. As a result, a population will adapt to its environment.

Scientists now know that genes are responsible for inherited traits. Therefore, certain traits become more common in a population because more individuals in the population carry the genes for those traits. In other words, natural selection causes the *frequency* of certain genes in a population to vary over time. Mutations and the recombination of genes that occurs during sexual reproduction are constant sources of new variations for natural selection to act upon. Today, biologists use the term **microevolution** to refer to change that occurs within a species over time. Figure 12-7 illustrates the extent to which microevolution can change a species.

Isolation Leads to Species Formation

The natural environment is not uniform; it differs from place to place. Therefore, populations of the same species living in different locations tend to evolve in different directions. The condition in which two populations of the same species are separated from one another is called **isolation.** As two isolated populations of the same species become increasingly different over time, the populations may no longer be able to interbreed (breed with one another). When the individuals of two populations can no longer interbreed, the two populations are considered to be different **species.**

Extinction Leads to Species Replacement

Over long periods of time, events such as climatic changes and natural disasters result in some species becoming **extinct**, which means that they disappear permanently. Species that are better suited to the new conditions may replace those that become extinct. Therefore, the organisms alive today are but the latest members of a long parade of life. Today, biologists use the term **macroevolution** to refer to change among species over time. The replacement of the dinosaurs by mammals is an example of macroevolution. ❏

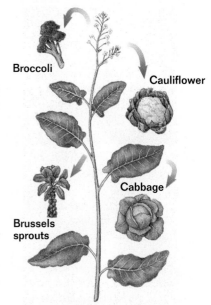

Broccoli

Cauliflower

Cabbage

Brussels sprouts

Wild *Brassica oleracea*

Figure 12-7 These familiar vegetables all belong to the same species, *Brassica oleracea*. Each was developed through selective breeding. Although the mechanism is *artificial* selection instead of *natural* selection, the breeding of crops is an example of microevolution.

❏ *CAPSULE SUMMARY*

Microevolution is change within a species. Macroevolution is change that leads to replacement of species.

Section Review

1. *List three observations made by Charles Darwin during his five-year voyage that led him to conclude that living species have evolved from species that are now extinct.*

2. *Briefly describe how natural selection occurs.*

3. *Using the terms* adaptation, microevolution, *and* macroevolution, *summarize the modern theory of evolution in as few words as you can.*

Critical Thinking

4. *Why are scientists' ideas about evolution stated as a theory? How is a scientific theory different from a scientific hypothesis?*

12-2 Evidence of Macroevolution

Since Darwin's death in 1882, scientists have learned a great deal about how evolution occurs. Discoveries of microscopic fossils have extended the known history of life on Earth to more than 3.5 billion years. Today, Darwin's theory is almost universally accepted by scientists as the best available explanation for the biological diversity on Earth. Based on a large body of supporting evidence, most scientists agree on the following three major points: (1) Earth is about 4.5 billion years old; (2) organisms have inhabited Earth for most of that time; and (3) all organisms living today evolved from earlier, simpler life-forms.

Section Objectives

- Describe how the fossil record supports evolution.
- Explain how biological molecules such as proteins and DNA show evidence of macroevolution.
- Explain how comparing the anatomy and development of living species provides evidence of macroevolution.
- Differentiate between the gradualism and punctuated equilibrium models of evolution.

Fossils Provide a Record of Macroevolution

The most direct evidence that macroevolution has occurred comes from fossils. A **fossil** is the preserved or mineralized remains (bone, tooth, or shell) or traces (footprint, burrow, or imprint) of an organism that lived long ago. Fossils, therefore, provide an actual record of Earth's past life-forms. Change over time (evolution) can be observed in this fossil record. For instance, fossilized species found in older rocks are different from those found in newer rocks, as you can see in Figure 12-8. After observing such differences, Darwin predicted that "missing links" (intermediate forms) between the great groups of organisms would eventually be found. Many of these links have been found. For example, fossil links have been found between fishes and amphibians, between reptiles and birds, and between reptiles

Figure 12-8 Fossils of simple unicellular life-forms, such as the coccoid cyanobacteria, *left*, characterize 700 million-year-old rocks found in western Canada. Fossils of more complex, multicellular life-forms, such as the crinoid, *center*, occur in 300 million-year-old rocks found in Indiana. Fossils of highly complex life-forms, such as the pterodactyl, *right*, occur in rocks formed during the Jurassic period 210 million to 140 million years ago.

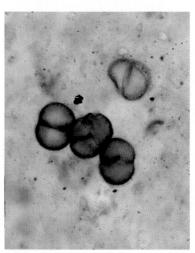
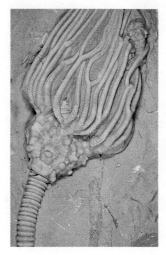
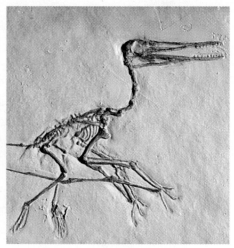

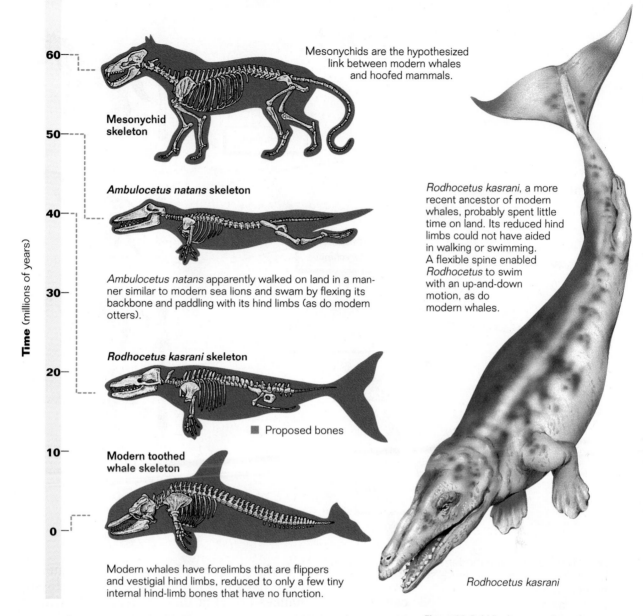

Mesonychids are the hypothesized link between modern whales and hoofed mammals.

Mesonychid skeleton

Time (millions of years)

60 —

50 —

40 —

30 —

20 —

10 —

0 —

Ambulocetus natans **skeleton**

Ambulocetus natans apparently walked on land in a manner similar to modern sea lions and swam by flexing its backbone and paddling with its hind limbs (as do modern otters).

Rodhocetus kasrani **skeleton**

■ Proposed bones

Modern toothed whale skeleton

Modern whales have forelimbs that are flippers and vestigial hind limbs, reduced to only a few tiny internal hind-limb bones that have no function.

Rodhocetus kasrani, a more recent ancestor of modern whales, probably spent little time on land. Its reduced hind limbs could not have aided in walking or swimming. A flexible spine enabled *Rodhocetus* to swim with an up-and-down motion, as do modern whales.

Rodhocetus kasrani

and mammals, making the fossil history of the vertebrates remarkably complete. In the early 1990s, the discoveries of the fossilized remains of two whale ancestors, illustrated in Figure 12-9, provided new links in the evolution of whales from four-legged land mammals.

However, the fossil record, and thus the record of the evolution of life on Earth, is not complete. Many species live in environments where fossils do not form. Most fossils form when organisms and traces of organisms are rapidly buried in fine sediments deposited by water, wind, or volcanic eruptions. Thus, the environments that are most conducive to fossil formation are wet lowlands, slow-moving streams, lakes, shallow seas, and areas near volcanoes that spew out volcanic ash. What are the chances that organisms living in upland forests, mountains, grasslands, or deserts

Figure 12-9 Whales are thought to have evolved from four-legged land mammals that are also the ancestors of modern hoofed mammals. *Ambulocetus natans* and *Rodhocetus kasrani* are recently discovered transitional (intermediate) forms, or "missing links," in whale evolution.

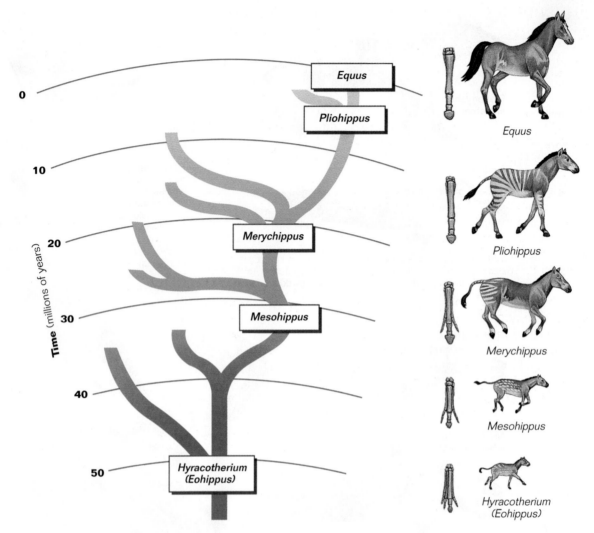

Figure 12-10 *Equus,* the modern horse, evolved from the dog-sized *Hyracotherium.* Notice the transition, *right,* from *Hyracotherium's* four-toed front foot to the one-toed front foot of the modern horse. By including more of the modern horse's ancestors and their descendants, a branching diagram called a family tree, *left,* can be constructed.

will die in just the right place to be buried in sediments? Even if an organism lives in an environment where fossils can form, the chances are slim that its dead body will be buried in sediment before it decays or is eaten and scattered by scavengers.

Although the fossil record will never be complete, it contains striking evidence that macroevolution has occurred. **Paleontologists,** scientists who study fossils, can determine the age of fossils fairly accurately by using radiometric dating, as you learned in Chapter 11. Radiometric dating enables paleontologists to arrange fossils in sequence from oldest to youngest. When this is done, patterns of successive change over time become evident, as you can see from the series of horses illustrated in Figure 12-10. The diagram on the left in Figure 12-10 is a **family tree,** which shows how organisms are related through evolution. Each branch point in a family tree indicates a **common ancestor,** which is a species from which *two or more* species diverged (separated).

Molecules Contain a Record of Macroevolution

The picture of successive change seen in the fossil record enables scientists to make a prediction that can be tested. If species have changed over time, then the genes that determine their characteristics also should have changed. As species evolved, one change after another should have become part of their genetic instructions through mutation. Therefore, more and more changes in a gene's nucleotide sequence should accumulate over time.

This prediction was first tested by analyzing the amino acid sequences of proteins found in several species. Recall that the nucleotide sequence of the gene coding for a protein determines the amino acid sequence of that protein. If evolution has occurred, species that diverged from a common ancestor in the distant past should have more amino acid sequence differences between their proteins than do species that recently diverged. Comparing the beta chain of human hemoglobin (146 amino acids) with that of several species reveals these differences. The results of such a comparison are given in Table 12-1. Notice that species that recently shared a common ancestor (humans and gorillas) have few amino acid sequence differences, while species that shared a common ancestor in the distant past (humans and frogs) have many amino acid sequence differences.

Nucleotide substitutions (mutations) are responsible for changes in the amino acid sequence of a protein. Scientists estimate the number of nucleotide substitutions that have occurred in a gene since two species diverged by counting the resulting amino acid sequence differences. Today, gene sequencing also enables scientists to determine the exact nucleotide sequence of a gene. Using the data obtained from the study of this "molecular record," scientists produce family trees like the one in Figure 12-11. These family trees provide very strong evidence supporting evolution because

Table 12-1 Hemoglobin Comparison

Species	Amino Acid Differences
	Compared with human hemoglobin
Gorilla	1
Rhesus monkey	8
Mouse	27
Chicken	45
Frog	67
Lamprey	125

Hemoglobin Family Tree

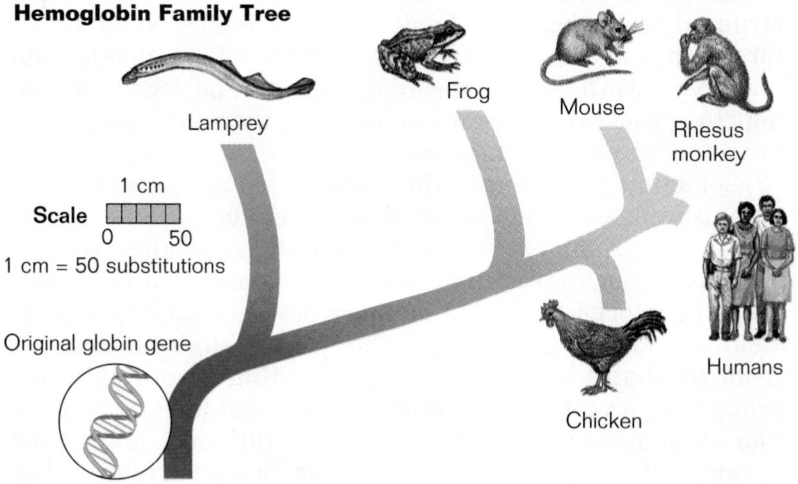

Scale

1 cm

0   50

1 cm = 50 substitutions

Original globin gene

Lamprey

Frog

Mouse

Rhesus monkey

Humans

Chicken

Figure 12-11 The length of the branches on this molecular family tree for the hemoglobin gene indicates the number of nucleotide substitutions that have occurred in the original "globin" gene's nucleotide sequence. The relationships indicated by this family tree are not a guess, but a direct observation of evolution, recorded in the DNA of many species.

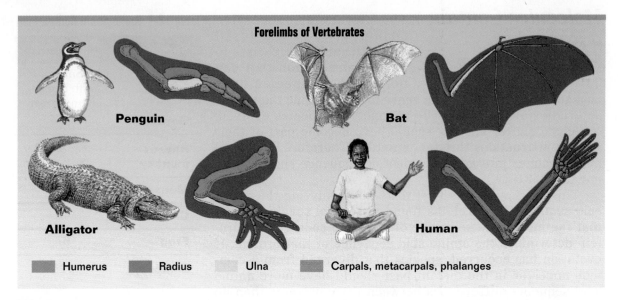

Forelimbs of Vertebrates

Penguin

Bat

Alligator

Human

| | Humerus | | Radius | | Ulna | | Carpals, metacarpals, phalanges |

Figure 12-12 Though they are modified for different functions, the forelimbs of vertebrates contain the same kinds of bones, which form in the same way during embryological development. Such structures are called homologous structures.

they show the same relationships indicated by the fossil record. Time after time, comparisons of amino acid and nucleotide sequences among organisms have provided a wealth of direct evidence of successive change over time.

Anatomy and Development Reflect Macroevolution
....

Comparisons of the anatomies (structures) of different types of organisms often reveal basic similarities in body structures, even though their functions may be very different. As vertebrates evolved, for example, particular sets of bones were sometimes put to different uses. And yet, the similarity in the structure of these bones can still be seen, suggesting that all vertebrates share a common ancestor. As you can see in Figure 12-12, the forelimbs of all vertebrates are constructed from the same basic array of bones. Such structures are said to be homologous *(hoh MAHL uh guhs)*, from the Greek words *homo*, meaning "same," and *legein*, meaning "to speak." **Homologous structures,** therefore, are structures that share a common ancestry.

Sometimes, bones (or other structures) are present in an organism but are reduced in size and either have no use or have a less prominent function than they do in other, related organisms. Such structures, which are considered to be evidence of an organism's evolutionary past, are called **vestigial** *(vehs TIJ ee uhl)* **structures.** The word *vestigial* is derived from the Latin word *vestigium*, meaning "footprint." An example of a vestigial structure can be seen in Figure 12-13. The human coccyx *(KAHK sihks)*, or tailbone, is another example of a vestigial structure. The fused vertebrae that make up the coccyx are homologous to those of other

Figure 12-13 The vestigial wings of this flightless cormorant, which lives in the Galapagos Islands, are too small to enable it to fly.

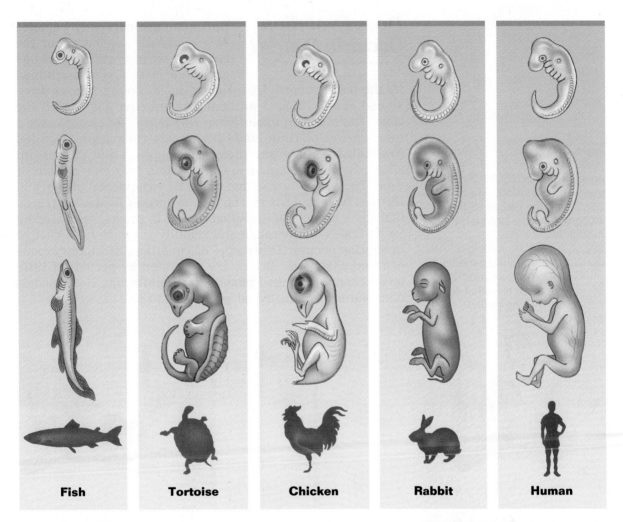

| Fish | Tortoise | Chicken | Rabbit | Human |

vertebrate tails. Although the coccyx serves as a point of attachment for certain muscles, this small internal "tail" does not resemble the long external tail of most other vertebrates. Furthermore, the human coccyx has no function in locomotion, balance, or behavioral displays as the tails of other vertebrates do. Thus, the coccyx is a vestigial vertebrate tail.

The evolutionary history of organisms is also evident in their embryonic development. Compare the development of a human embryo with that of the other vertebrate embryos shown in Figure 12-14. Notice that in addition to their early similarity, each embryo develops a tail, buds that become limbs, and pharyngeal pouches (which house the gills of fish and amphibians). The tail remains in most adult vertebrates. Although vertebrate bones are homologous, limbs develop somewhat differently in each group of vertebrates. Only adult fish and immature amphibians retain pharyngeal pouches. In humans, the tail and pharyngeal pouches are relics (forms from the past) that disappear by the time of birth. These relics and the similarities among the embryos strongly suggest that the development of all vertebrates evolved as new genetic instructions were layered on top of older ones. ◻

Figure 12-14 Early in development, all vertebrate embryos are remarkably similar. As development continues, various structures are modified until they take on their characteristic adult forms.

◻ **CAPSULE SUMMARY**

Fossils and DNA molecules contain a record of evolution. Homologous structures and vestigial structures provide evidence of common ancestry. Vertebrate embryonic development indicates that new genetic instructions have been layered on top of older ones.

Does Evolution Occur in Spurts?

Biologists are currently engaged in a lively discussion about the rate at which evolution has occurred in the past, as judged by the fossil record. Historically, most biologists have envisioned evolution as a gradual process that goes on all the time. The model of evolution in which gradual change over a long period of time leads to species formation is called **gradualism.** But some biologists suggest that successful species would remain virtually the same over long periods of time. They hypothesize that major environmental changes in the past have had a major impact on species formation. These biologists argue that evolution occurs in spurts, separated by long periods of environmental stability in which little change in species occurs. This model of evolution, in which periods of rapid change in species are separated by periods of little or no change, is called **punctuated equilibrium.** Both models of evolution are illustrated in Figure 12-15.

Figure 12-15 These hoofed mammals known as titanotheres, which lived from 50 to 35 million years ago, illustrate two models describing the rate of evolution. According to the gradualism model, *left,* change occurs gradually over time. According to the punctuated equilibrium model, *right,* change occurs rapidly in short periods of time that are separated by long periods of little or no change.

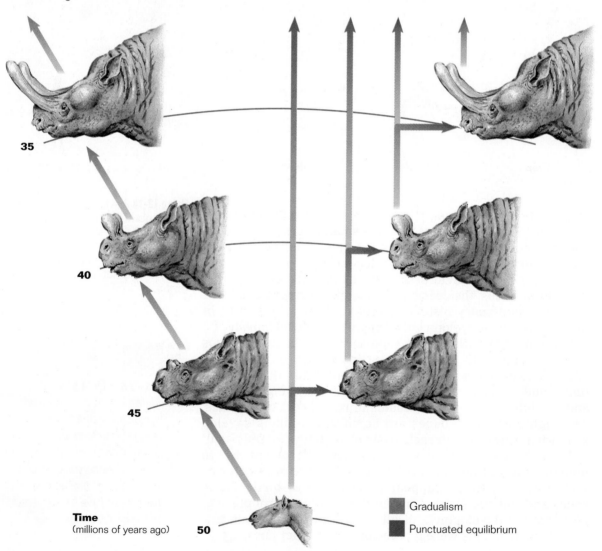

35

40

45

Time
(millions of years ago)   50

■ Gradualism
■ Punctuated equilibrium

What could cause major environmental changes, and why would they lead to spurts in evolution? The fossil record shows that drastic environmental changes have occurred very infrequently, separated by quiet periods that often last tens of millions of years. Events such as volcanic eruptions, asteroid impacts, and ice ages have been linked to sudden and drastic changes in climates, both locally and across the entire planet. Such changes have also been linked to the extinction of many groups of organisms. As a result, habitats that were once occupied became vacant and provided opportunities for colonization by species that could rapidly adapt to the new conditions through natural selection.

A prediction that can be made using the punctuated equilibrium model is that the fossil record should be very discontinuous. Is this prediction supported by evidence? There is considerable disagreement among biologists on this point, and the discussion continues. Of course, large gaps in the fossil record exist as a result of erosion and other destructive geologic processes. However, the fossil record seems to provide evidence of both types of evolution. Many groups of organisms appear suddenly in the fossil record. Some of these groups remain virtually unchanged for millions of years, while other groups disappear as suddenly as they appear. Still other groups of organisms change gradually through time, as predicted by the gradualism model. More careful study of the fossil record may reveal additional examples of both types of evolution. ◻

◻ CAPSULE SUMMARY

Gradualism is the model of evolution in which change occurs gradually over time. Punctuated equilibrium is the model of evolution in which change occurs in spurts separated by long periods of equilibrium.

Section Review

1. *How does the fossil record indicate that macroevolution has occurred?*

2. *What is a common ancestor?*

3. *What do the similarities in the development of vertebrate embryos indicate about vertebrates?*

4. *How does the punctuated equilibrium model of evolution differ from the gradualism model?*

Critical Thinking

5. *Why do fishes and mammals have more nucleotide sequence differences in their genes for a particular protein than do reptiles and mammals?*

6. *How are the wings of the flightless cormorant seen in Figure 12-13 homologous to the limbs of other vertebrates? Why are they considered vestigial structures?*

7. *Does acceptance of the punctuated equilibrium model mean that the gradualism model of evolution must be totally rejected? Why or why not?*

12-3 *Evidence of Microevolution*

*T*he heart of Darwin's theory of evolution is its assertion that natural selection is the mechanism responsible for evolution. Darwin wrote: "Can we doubt . . . that individuals having any advantage, however slight, over others, would have the best chance of surviving and of procreating their kind? On the other hand, we may feel sure that any variation in the least degree injurious would be rigidly destroyed. This preservation of favorable variations, I call Natural Selection." In his writings, Darwin offered examples of how natural selection has shaped life on Earth. There are now many well-documented examples of how natural selection has acted to change the genetic makeup of species.

Section Objectives

- Identify five steps in the process of natural selection.
- Describe how natural selection has affected the European peppered moth.
- Explain the role of natural selection in the distribution of sickle cell anemia.
- Describe the process of species formation.

Overview   NATURAL SELECTION

The process of natural selection depends on five main elements.

1. All species have genetic variation.
2. The environment presents many different challenges to an individual's survival.
3. Organisms tend to produce more offspring than their environment can support; thus, individuals of a species often compete with one another to survive (struggle for survival).
4. Individuals that are better able to cope with the challenges of their environment tend to leave more offspring than those less suited to the environment (survival of the fittest).
5. The characteristics of the individuals best suited to a particular environment tend to increase in a population over time.

The key lesson scientists have learned about evolution is that the environment dictates the direction and extent of change, as illustrated in Figure 12-16. Just as success determines which plays a football coach keeps in his team's game plan, so it determines which changes in a species are "kept" through natural selection.

Figure 12-16 Natural selection and the environment played important roles in the evolution of the polar bear *(Ursus maritimus), right,* and the grizzly bear *(Ursus horribilis), above.* The polar bear's white fur enables it to hunt more successfully in its snowy environment. The grizzly bear's brown fur enables it to hunt more successfully in its forested environment.

Industrial Melanism Is Natural Selection at Work

A particularly well studied example of natural selection in action is **industrial melanism,** the darkening of populations of organisms over time in response to industrial pollution. Although it occurs in many insect species, the best-known case of industrial melanism involves the European peppered moth, *Biston betularia.* Among the members of this species, there are two color variations, shown in Figure 12-17. The darker moths have genes for increased production of melanin (a black pigment). Once treasured by butterfly collectors, the dark variety of *Biston betularia* was extremely rare until the 1850s. Starting around 1850, however, dark peppered moths began to appear more often, usually in heavily industrialized areas. Every year, more dark moths were seen. After 100 years, almost all of the peppered-moth populations near industrial centers were composed of dark individuals.

The Concealment Hypothesis

Using Darwin's theory of evolution by natural selection, a hypothesis explaining the replacement of light moths by dark moths can be formed. Dark peppered moths are common in industrial regions where tree trunks are darkened by the soot of pollution. Perhaps dark moths are camouflaged against a background of soot-darkened bark and thereby escape being eaten by birds. Light moths, on the other hand, would stand out against a dark background and would be easy prey for hungry birds. A prediction can be made that the dark peppered moths would be favored in industrial areas because their dark color conceals them from birds that eat moths. This hypothesis is known as the concealment hypothesis.

Testing the Hypothesis

To see if natural selection could have caused the color change in the peppered-moth populations, the British ecologist H.B.D. Kettlewell performed an experiment during the late 1950s. Kettlewell raised populations of light and dark peppered moths in a laboratory. He then marked the underside of their wings with a dot of paint so they could be recognized later. Next, he released equal numbers of light and dark moths in two separate wooded areas of England. One of the wooded areas, near the city of Birmingham, was heavily polluted. The other wooded area, in the rural county of Dorset, was unpolluted. Finally, Kettlewell set rings of traps around the woods to recapture the moths and see which ones survived. As the graph in Figure 12-17 indicates, more of the moths matching the color of the tree trunks in each location survived. Many subsequent experiments confirmed these results. Hidden observers even saw birds passing by dark moths on polluted tree trunks and attacking the more conspicuous light moths. Kettlewell concluded that natural selection indeed causes industrial melanism in peppered-moth populations. ▢

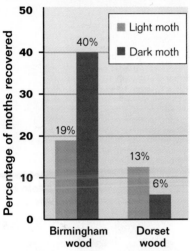

Figure 12-17 Two color variations, *top,* occur among European peppered moths, *Biston betularia.* A graph shows the results of Kettlewell's experiments, *bottom.* In the polluted woods near Birmingham, two-thirds of the surviving moths were dark. In rural Dorset, two-thirds of the surviving moths were light.

▢ **CAPSULE SUMMARY**

Experiments show that microevolution has occurred within populations of the European peppered moth.

Sickle Cell Anemia Reveals Evidence of Microevolution

One of the best-studied examples of the effect of natural selection involves hemoglobin proteins in humans. As you learned in Chapter 8, sickle cell anemia is a hereditary disease that affects hemoglobin molecules in human red blood cells. It arises from a single nucleotide change in the gene that codes for beta-hemoglobin and is one of the best understood of all human genetic disorders. In Figure 12-18, you can see how the sickle cell allele affects red blood cells. Because sickled red blood cells tend to clog tiny arteries, sickle cell anemia is usually lethal (deadly). Heterozygous individuals, who have both a defective hemoglobin gene and a normal hemoglobin gene, make enough functional hemoglobin to keep their red blood cells healthy.

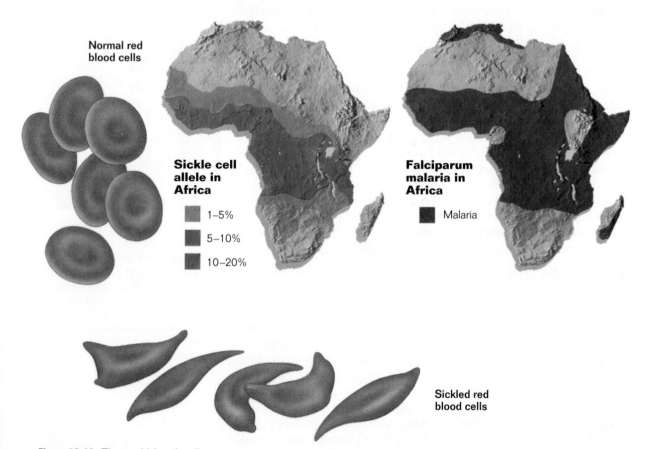

Normal red blood cells

Sickle cell allele in Africa

1–5%

5–10%

10–20%

Falciparum malaria in Africa

Malaria

Sickled red blood cells

Figure 12-18 The red blood cells of people who are homozygous for the sickle cell allele collapse into sickled shapes, *bottom left,* when the oxygen level in the blood is low. The distribution of the sickle cell allele in Africa, *center,* coincides closely with that of falciparum malaria, *right.*

First detected in 1910 in Chicago, sickle cell anemia affects roughly 2 out of every 1,000 African Americans in the United States, but it is almost unknown within other ethnic groups. In central Africa where the sickle cell allele is thought to have originated, 1 in every 100 people is homozygous for the defective allele and develops the fatal disorder. Why has natural selection not eliminated the defective sickle cell allele from human populations? Why instead is this potentially fatal allele so common in Africa?

The sickle cell allele persists in human populations because having a single copy of the defective allele provides a definite advantage in certain environments. People who are heterozygous for the sickle cell allele are far more resistant to malaria (a leading cause of death in central Africa) than are people who are homozygous for normal hemoglobin. While one in a hundred individuals is homozygous for the sickle cell allele and dies of anemia, one in five individuals is heterozygous for the sickle cell allele and survives malaria. Thus, even though people who are homozygous for the sickle cell allele die, a population that lives where malaria is prevalent experiences far fewer deaths from sickle cell anemia than would occur from malaria if the allele were not present in the population. As Figure 12-18 also illustrates, the distribution of the sickle cell allele in Africa closely coincides with the occurrence of falciparum malaria, a particularly devastating type of malaria.

Balancing Selection

In central Africa, natural selection affects the sickle cell allele in opposing ways: (1) it tends to eliminate the sickle cell allele because it is lethal to homozygous individuals, and (2) it tends to preserve the sickle cell allele because heterozygous individuals are resistant to malaria. Selection that acts in opposite directions is called **balancing selection.** When malaria is present, *both* homozygous genotypes are selected against (neither homozygous genotype has a selective advantage). Thus, heterozygous individuals are the most "fit." Balancing selection, therefore, maintains both the normal hemoglobin allele and the sickle cell allele in the population of central Africa. Even defective alleles such as the sickle cell allele will remain in a population as long as the benefit balances the cost.

Directional Selection

While the sickle cell allele remains common in central Africa, it is gradually being eliminated from human populations in other parts of the world. In the United States, for example, the environment does not favor the sickle cell allele because its lethal potential is not balanced by any advantage. Because malaria no longer occurs in the United States, African Americans gain no advantage from being heterozygous for the sickle cell allele. As a result, the sickle cell allele is becoming less common among African Americans. When natural selection is unopposed, the frequency of a particular allele tends to move in one direction—in this case, toward elimination. Biologists call such unopposed selection **directional selection.** As it does with the peppered moth, the environment determines the direction and extent of evolutionary change. ◻

◻ CAPSULE SUMMARY

Balancing selection maintains an allele in a population when the environment acts on the allele in opposing ways. Directional selection causes an allele to become either more common or less common.

Species Formation Begins With Microevolution

Species formation occurs in a series of successive stages. Because natural selection favors changes that increase reproductive success, microevolution continually molds and shapes a species to improve the "fit" between a species and its environment. Over time, separate populations of the same species may become very different from one another as different kinds of changes accumulate in each isolated group. The accumulation of differences between groups (populations, species, genera, etc.) is called **divergence.** Within populations, divergence leads to the formation of new species. Biologists call the process by which new species form **speciation** (*spee see AY shun*).

Forming Ecological Races

Figure 12-19 The formation of these ecological races of seaside sparrows may indicate that speciation has begun. If the populations of these ecological races remain isolated, they will become increasingly specialized as they adapt to their local environments.

Figure 12-19 illustrates what may be an early stage in the process of speciation. A species, such as the seaside sparrow, often occurs in several different kinds of environments. In each environment, natural selection acts to make the local population of a species better adapted to that environment. If their environments differ enough, local populations can become quite distinct. Over time, populations of the same

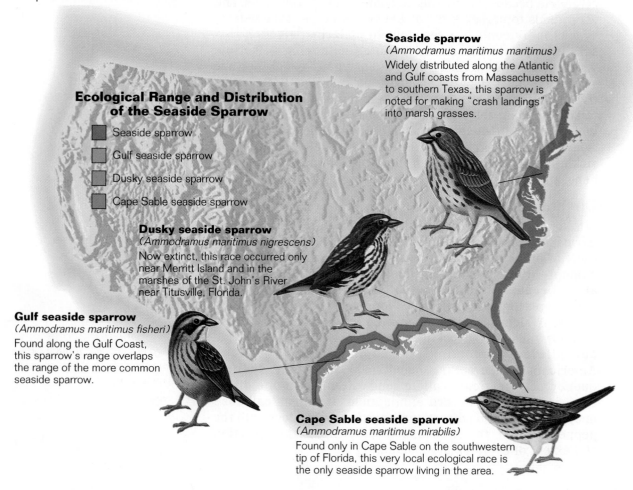

Ecological Range and Distribution of the Seaside Sparrow

- Seaside sparrow
- Gulf seaside sparrow
- Dusky seaside sparrow
- Cape Sable seaside sparrow

Seaside sparrow
(*Ammodramus maritimus maritimus*)
Widely distributed along the Atlantic and Gulf coasts from Massachusetts to southern Texas, this sparrow is noted for making "crash landings" into marsh grasses.

Dusky seaside sparrow
(*Ammodramus maritimus nigrescens*)
Now extinct, this race occurred only near Merritt Island and in the marshes of the St. John's River near Titusville, Florida.

Gulf seaside sparrow
(*Ammodramus maritimus fisheri*)
Found along the Gulf Coast, this sparrow's range overlaps the range of the more common seaside sparrow.

Cape Sable seaside sparrow
(*Ammodramus maritimus mirabilis*)
Found only in Cape Sable on the southwestern tip of Florida, this very local ecological race is the only seaside sparrow living in the area.

species that differ genetically because of adaptations to different living conditions become what biologists call **ecological races.** Although the members of different ecological races are not yet different enough for the groups to be considered different species, they have taken the first step on that road. Ecological races may continue to diverge and become more and more different from each other as natural selection favors different survival strategies in different environments. Eventually, the races may become so different that they can no longer interbreed successfully. Biologists then consider them separate species.

Maintaining New Species

What keeps new species that diverge from a common ancestor separate? Why are even closely related species usually unable to interbreed with one another? A variety of isolating mechanisms are responsible for preventing interbreeding between closely related species. Once ecological races become different enough, some type of barrier to reproduction, like the one illustrated in Figure 12-20, usually prevents different groups from breeding with each other. The prevention of mating between formerly interbreeding

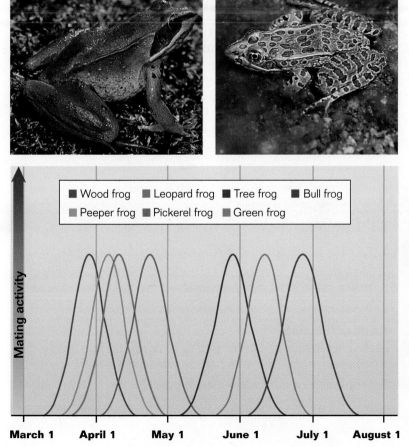

Figure 12-20 Though they appear to be very similar aside from their coloration, wood frogs (*Rana sylvatica*), *top left,* and leopard frogs (*Rana pipiens*), *top right,* are different species. The graph, *bottom,* shows that the time of peak mating activity varies among several species of frogs in the genus *Rana.* Such differences in mating activity represent one type of barrier to reproduction that promotes speciation.

Legend (graph): ■ Wood frog ■ Leopard frog ■ Tree frog ■ Bull frog ■ Peeper frog ■ Pickerel frog ■ Green frog

Y-axis: Mating activity

X-axis: March 1, April 1, May 1, June 1, July 1, August 1

Table 12-2 Mechanisms for Reproductive Isolation

Mechanism	Description
Geographical isolation	Groups are physically separated.
Ecological isolation	Groups occupy different habitats, and their hybrids are not suited to either environment.
Temporal isolation	Groups reproduce at different times of the day or year.
Behavioral isolation	Groups are not attracted to each other for mating.
Mechanical isolation	Structural differences prevent mating between individuals of different groups.
Reproductive failure	Matings between groups do not produce fertile offspring.

groups or the inability of these groups to produce fertile offspring is called **reproductive isolation.** Several types of barriers that may isolate two or more closely related groups are listed in Table 12-2.

Microevolution Leads to Macroevolution

Biologists have observed the stages of speciation in many different organisms. Thus, the way that natural selection leads to the formation of new species has been thoroughly documented. As changes continue to accumulate over time, living species may become very different from their ancestors and from other species that evolved from the same common ancestor. Biologists therefore agree that change within species caused by natural selection (microevolution) eventually leads to the appearance of new species and even new kingdoms of organisms (macroevolution). ◻

◻ CAPSULE SUMMARY

Speciation begins as a population adapts to its specific environment. Reproductive isolation keeps newly forming species from interbreeding. Over time, change within species caused by natural selection leads to the rise of new kinds of organisms.

Section Review

1. *List the five steps in the process of natural selection.*
2. *How does the distribution of sickle cell anemia suggest that natural selection has acted on the sickle cell allele?*
3. *What role does isolation play in speciation?*

Critical Thinking

4. *Why were peppered moths able to adapt to their environment whether it was polluted or unpolluted?*
5. *What is the difference between balancing selection and directional selection? Which is likely to lead more quickly to evolution? Why?*

Vocabulary

adaptation (249)
balancing selection (263)
common ancestor (254)
directional selection (263)
divergence (264)
ecological race (265)
extinct (251)
family tree (254)

fossil (252)
gradualism (258)
homologous structures (256)
industrial melanism (261)
isolation (251)
macroevolution (251)
microevolution (251)
natural selection (249)

paleontologist (254)
population (248)
punctuated equilibrium (258)
reproductive isolation (266)
speciation (264)
species (251)
vestigial structure (256)

Concept Mapping

Construct a concept map that shows how microevolution leads to macroevolution. Use as many terms as needed from the vocabulary list. Try to include the following terms in your map: microevolution, macroevolution, natural selection, genetic variation, environment, speciation, divergence, and extinction. Include additional terms in your map as needed.

Review

Multiple Choice

1. According to the modern theory of evolution,
   a. extinct species are always replaced by gradualism.
   b. random gene mutation is a component of evolution.
   c. punctuated equilibrium has replaced natural selection.
   d. the diversity of life resulted from the inheritance of acquired characteristics.

2. Natural selection
   a. leads to a reduction in the diversity of life.
   b. involves the inheritance of acquired characteristics.
   c. allows organisms to live in environments for which they are best suited.
   d. causes species to become better suited to their environment.

3. Which of the following is *not* an element in the process of natural selection?
   a. Genetic variation exists within a species.
   b. Individuals struggle for survival.
   c. Organisms cause environmental change.
   d. Fit with the environment leads to a population increase.

4. According to the modern version of Darwin's theory of evolution by natural selection,
   a. the frequency of the genes in a population does not change over time.
   b. populations that are isolated from one another remain the same over time.
   c. microevolution results from species extinction.
   d. the individuals of two populations become distinct species when they can no longer interbreed.

5. Fossil evidence of all past life on Earth has not been found because
   a. fossils could not form in all environments in which organisms live.
   b. paleontologists need to perfect the techniques of radiometric dating.
   c. some organisms do not have the gene that codes for fossil formation.
   d. common ancestors first must be identified using family trees.

6. Anatomical structures that share a common ancestry are called
   a. vestigial structures.
   b. homologous structures.
   c. analogous structures.
   d. evolutionary structures.

7. The punctuated equilibrium model differs from the gradualism model by stating that
   a. species formation occurs slowly and continuously.
   b. Darwin's theory of natural selection is inaccurate.
   c. periods of rapid species formation alternate with periods of little or no change.
   d. the fossil record should be continuous.

8. Kettlewell's experiments with peppered moths in polluted and unpolluted areas
   a. provided support for the concealment hypothesis.
   b. showed that light-colored moths are favored in industrial areas.
   c. suggested that birds prefer the taste of dark-colored moths.
   d. disproved the theory of evolution by natural selection.

9. The effect of the sickle cell allele on central Africa's population is an example of
   a. industrial melanism.
   b. punctuated equilibrium.
   c. balancing selection.
   d. directional selection.

10. The process by which isolated populations of the same species become new species is called
    a. speciation.
    b. microevolution.
    c. macroevolution.
    d. natural selection.

Completion

11. Darwin observed that Galapagos species were more similar to the species of _____ than to the species of other similar but more distant _____ .

12. Evolution within populations of a species is called _____ , while the evolution of new species is called _____ .

13. _____ are scientists who study fossils. By arranging fossils from oldest to youngest, these scientists are able to construct _____ , which show how organisms are related through evolution.

14. Evidence indicating that evolution has occurred is found in the _____ sequences of proteins and in the _____ sequences of DNA.

15. The _____ model of evolution proposes that new species form during brief spurts of rapid change, while the _____ model proposes that evolution occurs slowly over long periods of time.

16. The allele for sickle cell anemia exhibits _____ selection within the population of central Africa and _____ selection among African Americans.

17. _____ are populations of the same species that live in different environments and differ genetically. Over time, these groups may become new _____ .

Short Answer

18. What is the theory of evolution?

19. Charles Darwin was not the first person to suggest that organisms evolved. Yet his name is often the only one associated with the theory of evolution. Why?

20. How are the structures of proteins and DNA used to support fossil evidence that evolution has occurred?

21. How do structures such as the vestigial legs of whales and the human coccyx indicate that evolution has occurred?

22. At the time that dinosaurs became extinct, mammals became more numerous and diverse. Explain how such events might result from natural selection.

Themes Review

23. **Structure and Function** The wings of butterflies and the wings of bats serve similar functions but differ in structure. They are called analogous structures. Do analogous structures provide evidence of evolution? Explain.

24. **Homeostasis** How does natural selection help populations maintain homeostasis in changing environments?

25. Levels of Organization Look at the photos below. The larvae of barnacles and lobsters are virtually identical. What does this indicate about the evolutionary history of these organisms?

Critical Thinking

26. Making Inferences About 40 years after the publication of Darwin's book *On the Origin of Species*, genetics was recognized as a science. At this time, support for Darwin's theory of evolution by natural selection began to grow among scientists. How might these two events be related?

27. Using Analogies Darwin used the metaphor of an evolutionary tree to illustrate the diversity and relatedness of species. What do an evolutionary (family) tree's trunk, limbs, and twigs represent?

28. Interpreting Data Use the data in the table to answer the following questions. Which species is most closely related to humans? Which species is most distantly related to species B?

Species	Number of Amino Acid Differences Compared With Human Hemoglobin
Humans	0
Species A	8
Species B	17
Species C	39

29. Identifying Variables Why did H.B.D. Kettlewell mark the wings of the moths he released with a dot of paint? Name at least four variables that existed in Kettlewell's experiments.

Activities and Projects

30. Cooperative Group Project Locate and examine photographs and drawings of the tortoises Darwin observed on the Galapagos Islands. Plan and produce a mural showing the tortoises in their natural environments. Display the mural in your school.

31. Research and Writing Research Alfred Russel Wallace's contributions to the science of biology. Relate your findings in a written report.

32. Multicultural Perspective Most of Africa lies within the tropics, and more than one-half of the land is a desert. As a result, many African species have unique structural and behavioral adaptations for surviving in the desert. Use books, magazines, and films to research ways that African species, including humans, are adapted to life in the desert. Relate your findings to your class.

Readings

33. Read Jeremy Greenwood's article "Theory fits the bill in the Galápagos Islands," in *Nature*, April 22, 1993, page 699. How did *Geospiza fortis* respond to a change in its environment? How is the Grants' work important to the study of evolution and ecology?

34. Read David Raup's article "Extinction: bad genes or bad luck?" in *New Scientist*, September 14, 1991, pages 46–49. Mass extinctions affected species in many different habitats over wide geographic areas. What does Raup think this suggests about the process of extinction? How does he think most extinctions occur?

Recognizing Patterns of Variation

OBJECTIVES

- Observe genetic variation in particular traits among members of a population.
- Infer how individual variation can potentially contribute to the continuing evolution of a species.

PROCESS SKILLS

- measuring
- calculating means
- organizing data in tables and graphs

MATERIALS

- metric ruler
- mung bean sprouts
- graph paper (optional)
- peanuts in the shell
- tape measure
- calculator
- tree leaves

BACKGROUND

1. How much variation in genetic expression exists within a population?

2. How might variation within a species be useful? How might variation be harmful?

3. How can variation within plant and animal populations be measured?

4. How is variation within a species the basis for natural selection?

5. Write your own **Focus Question** on your Vee Form.

6. **Knowing Side of the Vee** List the **Concepts** and new **Vocabulary Words** on your Vee Form. In the **Concept Statements** section of the Vee, use these words in sentences that define and explain them.

TECHNIQUE

Doing Side of the Vee

Part A: Bean Sprouts and Peanuts

You will need 50–55 peanuts in the shell and 20 mung bean sprouts.

1. Measure the lengths of 20 mung bean sprouts to the nearest millimeter from the tip to the site where the sprout first emerged from the seed.

2. Record your measurements in the **Records** section of your Vee Form. Organize your measurement values on a single line from the lowest to highest value. Beneath each value, write the number of sprouts of that length. On your Vee form or on graph paper, make a graph showing the distribution curve .

3. Crack open the peanut shells and measure the lengths of 100 peanuts to the nearest millimeter. Record your results in a table in the **Records** section of your Vee, giving the lengths and the number of peanuts of each length.

4. Make a bar graph using your data.

Part B: Humans

5. Using the tape measure, measure your lab partner's left forearm from the wrist bone to the elbow. Record your measurements in the **Additional Records** section of your Vee. Pool your data with those collected by others in your class. In the **Additional Records** section of your Vee, make a bar graph listing the measurements on the horizontal axis and the number of forearms of each length on the vertical axis.

6. Calculate the mean forearm length. Record your results in the **Additional Records** section of your Vee.

Part C: Tree Leaves

You will need 10 tree leaves collected from one tree species.

7. Measure the lengths of each of the 10 leaf blades. Record your results in the **Additional Records** section of your Vee.

8. Measure the petiole (leaf stem) of each leaf and record the data in the **Additional Records** section of your Vee.

9. Using the data from all class teams, record the lengths for the blades and petioles, beginning with the shortest length and ending with the longest length. Record your results in the **Additional Records** section of your Vee. In the **Procedure** section of the Vee, briefly summarize the procedure you followed.

10. Clean up your materials and wash your hands before leaving the lab.

INQUIRY

1. What is the range of measurements for your mung bean sprouts?

2. The mode is the most frequently occurring value. What is the mode for your bean sprouts?

3. List the advantages and disadvantages for survival of the longest sprouts.

4. Analyze the bar graph showing the lengths of peanuts. Compare the numbers of peanuts of average length with the numbers of very large and very small peanuts.

5. Since the peanut supplies the seed embryo with energy, what is the relationship of peanut size to the amount of energy available to the peanut embryo?

6. For Part B, calculate the difference between the lowest measurement and the mode and the difference between the highest measurement and the mode. How does the mode differ from the mean?

7. Use the information on the **Knowing Side** of the Vee to interpret your results from the **Doing Side**, and then write your **Knowledge Claim**. Write a **Value Claim** for this lab.

ANALYSIS

1. What is variation?

2. Explain why it is advantageous for a species to show variation among individuals.

3. Describe the relationship between natural selection and variation.

4. If the environment changed so that a very large peanut was an advantageous variation, what would happen to the peanut population over time in response to such a change?

5. Describe an environmental change that would favor a much larger peanut. Can you think of a change that would favor a much smaller peanut?

6. How might leaf size be important to the success of a plant?

7. How might you determine whether a trait such as those measured in this lab investigation is a result of genetic or environmental factors?

FURTHER INQUIRY

Write a **New Focus Question** that could be the point of a new investigation. The following is an example:

What are some examples of variations found in other species? For example, measure adult human height, pine needle length, acorn weight, or the length of the hind legs of grasshoppers. For each characteristic measured, hypothesize about environmental changes that could lead to the selection of an extremely small or large size.

CHAPTER 13

HISTORY OF LIFE ON EARTH

REVIEW

- prokaryotes and eukaryotes (Section 2-1)
- chloroplasts and mitochondria (Section 2-4)
- photosynthesis (Section 4-1)
- origin of the Earth (Section 11-1)
- radiometric dating (Section 11-2)

Earth as viewed from the *Apollo 17* spacecraft

13-1 Life in the Ancient Seas

Several methods of radiometric dating have determined that Earth is approximately 4.5 billion years old. For most of the first several hundred million years, our planet was a fiery ball of molten rock, but eventually the surface cooled and formed a rocky crust over the hot molten interior. When Earth's surface cooled, water vapor in the air condensed to form great oceans. It was within these seas that life first evolved.

Section Objectives

- Contrast the Earth's age with the age of Earth's oldest fossil organisms.
- Describe the evolution of bacteria.
- Contrast bacteria with protists.
- Recognize an evolutionary advancement first seen in protists.
- Describe how mass extinction events have affected the evolution of life on Earth.

Bacteria Were the First Cells to Evolve

The earliest traces of life are found as tiny fossils in 3.5-billion-year-old rocks from the ancient seas. Earth's first cells were bacteria. Like bacteria that exist today, these ancient bacteria were prokaryotes. Unlike today's animal and plant cells, the first cells did not have internal compartments that could perform special functions. Instead, the interior of one of these early cells was like a warehouse, an open space within which all the cell's contents were free to move about.

Among the first of the bacteria to appear were the **cyanobacteria** (SY ahn oh bak TIHR ee ah). Cyanobacteria, shown in Figure 13-1, are photosynthetic. As ancient cyanobacteria carried out photosynthesis, they released oxygen gas into the oceans. This was something new,

Figure 13-1 The explosive growth of a particular kind of cyanobacteria in this lake, *left*, has caused what is known as a bloom. Cyanobacteria, *above*, were the first cells to use chlorophyll to capture light. Chlorophyll is the same pigment that green plants use today for photosynthesis.

Figure 13-2 Archaebacteria are rare today. They are found mainly in unfavorable environments where conditions resemble those of early Earth, such as in this hot spring in Yellowstone National Park. The densely growing bacteria stain the runoff channels orange.

CONTENT LINK
. . . .
More information about the evolution of mitochondria and chloroplasts can be found in **Chapter 20.**

Figure 13-3 Some of the major events that have occurred during the evolution of life on Earth are shown in this continuous time line.

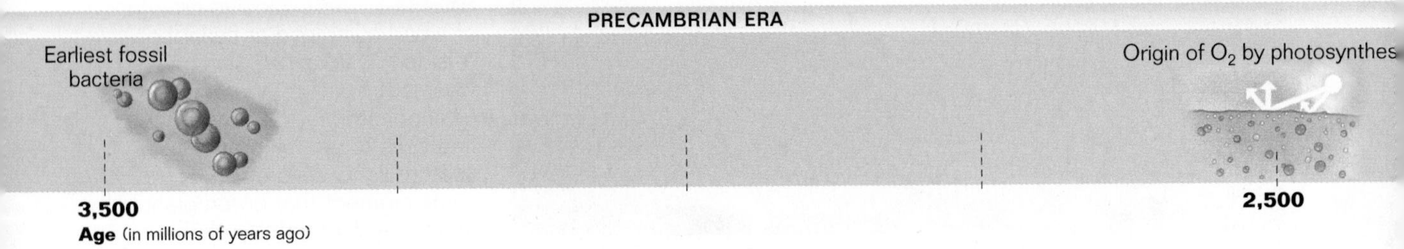

because oxygen gas was rare on Earth until then. After hundreds of millions of years, when the waters of the ancient oceans had become saturated with oxygen, the oxygen produced by cyanobacteria began to escape into the air. Over the billions of years that followed, more and more oxygen was added to the air. Today, oxygen gas is 21 percent of the Earth's atmosphere.

Research has shown that some of the early cyanobacteria may be ancestors of chloroplasts and that other kinds of early bacteria may have given rise to mitochondria. Remember that mitochondria and chloroplasts are organelles found only in eukaryotic cells. Scientists propose that bacteria may have first entered larger cells as parasites or undigested prey and then gradually evolved into the mitochondria and chloroplasts that characterize eukaryotic cells today.

Two Kingdoms of Bacteria

Fossils and other evidence indicate that ancient cyanobacteria are among the direct ancestors of the most common group of bacteria that exist today, the **eubacteria** *(yoo bak TIHR ee ah)*. Eubacteria include those bacteria that cause disease and decay. Other ancient bacteria were very different from eubacteria and formed a separate branch early during the evolution of the first cells. These bacteria are more closely related to a second group, or kingdom, of modern bacteria, the **archaebacteria** *(AHR kee bak TIHR ee ah)*, shown in Figure 13-2. The cell walls and membranes of archaebacteria are quite different from those of the eubacteria, and the way the archaebacteria produce protein from

PRECAMBRIAN ERA

Earliest fossil bacteria

Origin of O₂ by photosynthes

3,500
Age (in millions of years ago)

2,500

their DNA is also different. Since archaebacteria evolved before there was oxygen in the atmosphere, they had to develop ways of producing energy without using oxygen. Some modern archaebacteria get their energy not by photosynthesis but by combining hydrogen, H_2, with carbon dioxide, CO_2, to form methane, CH_4. Other kinds of archaebacteria produce energy using other chemical pathways. Chemical evidence suggests that the first eukaryotic cells, as shown in Figure 13-3, were more likely to have evolved from archaebacteria than eubacteria. Indeed, archaebacteria appear to be the ancestors of all eukaryotic cells, including yours! ■

❑ CAPSULE SUMMARY

The earliest fossil cells closely resemble modern-day bacteria. Primitive cyanobacteria produced the first oxygen in Earth's atmosphere and are believed to be the ancestors of chloroplasts. Archaebacteria probably gave rise to the first eukaryotic cells.

Dawn of the Eukaryotes

For over a billion years bacteria were the only living things on Earth. Then, starting about 1.5 billion years ago, a new kind of organism evolved from the archaebacteria—the **protist.** Protists were the first eukaryotes. They have DNA that is enclosed within a nucleus and complex systems of internal membranes.

Most biologists group all living things into six great categories called kingdoms, as shown in Figure 13-4. The two

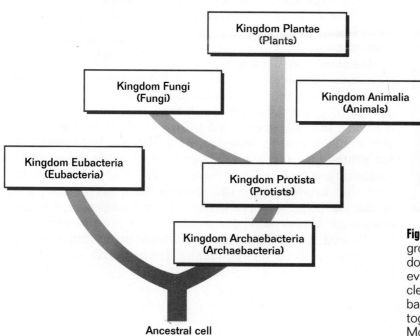

Kingdom Plantae (Plants)

Kingdom Fungi (Fungi)

Kingdom Animalia (Animals)

Kingdom Eubacteria (Eubacteria)

Kingdom Protista (Protists)

Kingdom Archaebacteria (Archaebacteria)

Ancestral cell

Figure 13-4 Living things are grouped into six separate kingdoms. Until recent molecular evidence made the relationships clearer, eubacteria and archaebacteria were traditionally placed together into one kingdom called Monera.

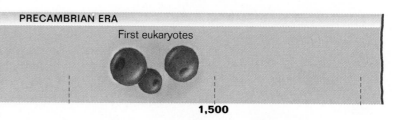

PRECAMBRIAN ERA

First eukaryotes

1,500

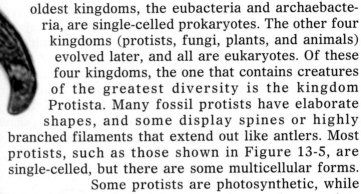

oldest kingdoms, the eubacteria and archaebacteria, are single-celled prokaryotes. The other four kingdoms (protists, fungi, plants, and animals) evolved later, and all are eukaryotes. Of these four kingdoms, the one that contains creatures of the greatest diversity is the kingdom Protista. Many fossil protists have elaborate shapes, and some display spines or highly branched filaments that extend out like antlers. Most protists, such as those shown in Figure 13-5, are single-celled, but there are some multicellular forms. Some protists are photosynthetic, while others hunt bacteria or other protists for food.

Figure 13-5 Protists, a very diverse group, are much larger than bacteria. Some protists, like these *Euglena*, have both plant and animal characteristics. What are these characteristics?

Multicellularity Evolves Many Times in Protists

The development of multicellular organisms marks a great step in the evolution of life on Earth. It allowed "division of labor" among cells and allowed cell specialization that led to organism complexity. Almost every creature large enough to see with the naked eye is multicellular. The first known fossils of multicellular organisms were found in 630-million-year-old rocks from southern Australia.

Multicellular organisms apparently evolved independently many different times among the protists. Some of the multicellular lines that resulted did not produce diverse groups of organisms, although they still survive. Among them are the red, green, and brown algae often seen swept onto the seashore as seaweed. Algae, shown in Figure 13-6, are protists, (some biologists used to classify these photosynthetic organisms with plants). Three of the multicellular lines that evolved from the protists were very successful, producing large, diverse groups. Each group was assigned a separate kingdom. These three groups are the fungi, the plants, and the animals. Each of these three multicellular kingdoms evolved independently from a different kind of protist.

Figure 13-6 Brown algae, called kelps, are multicellular protists that form vast underwater "forests" in some coastal waters.

PRECAMBRIAN ERA

Early eukaryotes

Diverse protists

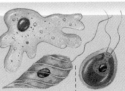

1,500

1,000

Age (in millions of years ago)

Today's Organisms Have Their Origins in the Cambrian Period

··

All major groups of organisms that survive today, except plants, originated sometime during the first hundred million years of multicellular life. This time period, called the Cambrian period, lasted from just less than 600 million years ago to about 500 million years ago. The Cambrian period was a time of great evolutionary experimentation. Many very unusual animals appeared at this time, animals for which there are no close living relatives. A particularly rich collection of Cambrian fossils, the Burgess Shale, has been uncovered on a rocky mountainside in Canada. The remains of bizarre creatures in the Burgess Shale, as shown in Figure 13-7, are unlike anything alive today. There appear to be more kinds of extinct animal phyla in the Burgess Shale than there are

Figure 13-7 By studying fossils from the Burgess Shale, artists were able to re-create a scene from the shallow seas of the Cambrian period. Animals of this environment included those with odd structural adaptations such as torpedo-shaped bodies, tripod-like tails, and elongated grasping organs.

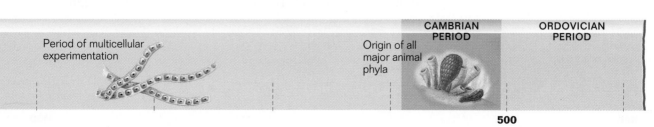

Period of multicellular experimentation

Origin of all major animal phyla

CAMBRIAN PERIOD

ORDOVICIAN PERIOD

500

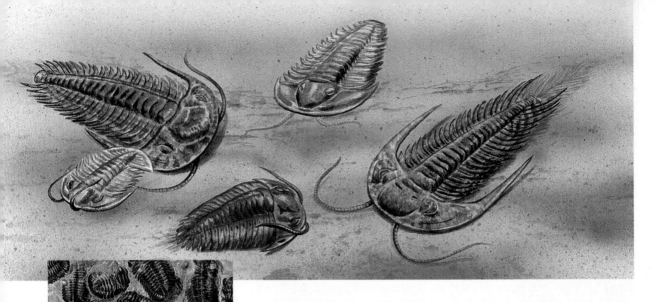

Figure 13-8 The fossil record contains nearly 4,000 trilobite species. These marine arthropods were extremely common 500–600 million years ago. By about 250 million years ago, they were all extinct.

kinds of living animal phyla today! Life was more diverse in the Cambrian seas than it has ever been since. In the period following the Cambrian, the Ordovician period, diverse animals continued to abound in the seas. Among them were the trilobites shown in Figure 13-8.

Mass Extinctions Have a Major Impact

The end of the Ordovician period is marked by a drastic change in the fossil record. A large portion of the organisms on Earth suddenly became extinct about 440 million years ago. This was the first of five major **mass extinctions** that have occurred during the history of life on Earth.

Another mass extinction of similar magnitude happened abruptly about 360 million years ago. Then, about a hundred million years later, the third and greatest of all mass extinctions literally devastated our planet. It happened at the end of what is called the Permian period, some 250 million years ago. About 96 percent of all species of animals living at the time became extinct. Approximately 35 million years later, a fourth less devastating mass extinction occurred. Although the specific causes for this and the other extinction events are not clear, evidence suggests that massive geological or climatic changes were likely contributing factors.

The last mass extinction event will be discussed in more detail in Chapter 33. It occurred 65 million years ago and

ORDOVICIAN PERIOD

Animal diversity abounds, early vertebrates

Jawless fishes

500
Age (in millions of years ago)

450

Figure 13-9 Although tropical rain forests cover only 7 percent of the Earth's land surface, they contain more than one-half of all the world's animal and plant species. But rain forests are being destroyed at an alarming rate. This rain forest in Brazil is being burned for farmland.

brought about the extinction of about two-thirds of all land species, including the dinosaurs.

There is good reason to believe that another mass extinction may be occurring on Earth today. In contrast to previous mass extinctions, this extinction event has a known cause. It is occurring because the Earth's ecosystems, especially tropical rain forests, are being destroyed by human activity, as shown in Figure 13-9. ☐

Section Review

1. *What is the age of the Earth's oldest fossil organisms, and what present-day organisms do they most closely resemble?*

2. *Explain why this statement is true: A eukaryotic cell is descended from both eubacteria and archaebacteria.*

3. *How is a protist different from a bacterium?*

4. *Define multicellularity, and identify the kingdom in which it first occurred.*

Critical Thinking

5. *Predict how human evolution would have been affected if mass extinctions had never occurred.*

First mass extinction	SILURIAN PERIOD	DEVONIAN PERIOD

Plants, arthropods, and fungi invade land; jawed fishes appear

400

13-2 Four Invasions of the Land

*L*ife has existed on Earth for at least 3.5 billion years, and for 90 percent of that time living things remained in water. Only recently, in geologic terms, has life emerged from the seas to live on land.

The Importance of Ozone

During the Cambrian period and for millions of years afterward, while the seas teemed with life, there was no living thing on the dry, rocky surface of the land. Exposed to the sun's harsh rays, nothing could live there. During all this time, however, a subtle change was occurring. Photosynthesis by cyanobacteria was adding oxygen gas to Earth's atmosphere. As significant amounts of oxygen began to reach the upper atmosphere, the sun's rays caused the atoms of oxygen gas, O_2, to bond and form a new compound. This reaction of oxygen atoms resulted in the formation of a gas called **ozone,** O_3. In the upper atmosphere, the ozone acted like a great shield to block the harsh ultraviolet radiation of the sun from reaching the surface. For the first time, Earth's surface was a safe place to live.

Plants and Fungi Cooperate to Invade Land

The first multicellular creatures to populate the land were plants and fungi. Cooperating with each other, they solved a particularly difficult challenge—how to survive on bare rock. Each brought to the task a unique talent.

Plants, which evolved from photosynthetic protists, could carry out photosynthesis. Recall from Chapter 5 that in photosynthesis plants use the energy from sunlight to make their own food. They could not, however, extract needed minerals from bare rock. Fungi, which evidently evolved from a hunting kind of protist, could not produce food from sunlight but were able to absorb minerals, even from bare rock.

DEVONIAN PERIOD

Bony fishes become abundant

Second mass extinction

Early amphibians

400
Age (in millions of years ago)

350

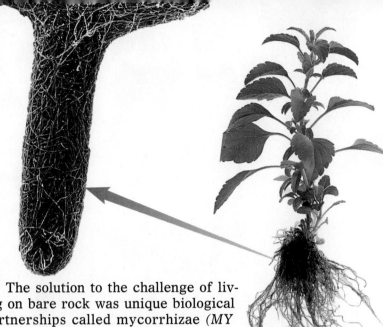

Figure 13-10 Mycorrhizae, the partnership between plants and fungi, have been very successful. Indeed, 80 percent of all living plants have mycorrhizae occurring within their roots. Fungi account for 15 percent of the total weight of the world's plant roots!

The solution to the challenge of living on bare rock was unique biological partnerships called mycorrhizae *(MY koh REYE zee)*. **Mycorrhizae** are associations between fungi and the roots of plants, as shown in Figure 13-10. In mycorrhizae one creature lives inside the other, and each helps the other. The fungi provide minerals to the plant, and the plant provides the food to the fungi. This kind of "you-help-me-and-I-help-you" partnership is called **symbiosis** *(sihm beye OH sihs)*. Both plants and fungi invaded the surface of the land at the same time, approximately 430 million years ago.

Arthropods Crawl out of the Sea

Within 100 million years of their initial invasion, plants covered the surface of the Earth and formed extensive forests. The landscape must have been eerily quiet when plants first invaded land, for no animals had yet left the sea. But the picture soon changed. Two groups of animals had been particularly successful in the oceans, and both groups were soon to populate the land. The first animals to venture forth were the **arthropods,** a kind of animal with a hard outer skeleton and jointed appendages. Crabs are arthropods, as are lobsters, insects, and spiders. As near as biologists are able to determine, the first arthropods to live on land were scorpions, like the one shown in Figure 13-11.

Figure 13-11 This fossil scorpion, a carnivorous relative of the spider, has two great pincers in front and a venomous stinger at the end of its tail.

CARBONIFEROUS PERIOD

Amphibians dominate the land

Early reptiles

300

Figure 13-12 Swamp forests 320 million years ago were dominated by tall, seedless canopy trees and shorter tree ferns. Dragonflies with wingspans of more than 1 m (3 ft.) hovered over these ancient swamps.

Figure 13-13 This blister beetle, an insect, is eating pollen produced by a flowering plant in India. Insects are important to plants because they help transfer pollen from one plant to another.

The arthropod invasion of the land is one of nature's great success stories. From the initial land invaders soon evolved a unique kind of terrestrial arthropod—the insect. Insects have become the most abundant and diverse group of animals ever. Today there are more than 200 million insects alive for each person on Earth! The special characteristic of insects that opened up the world to their invasion was the ability to fly. Insects were the first animals to evolve wings. Early forms, like the dragonfly shown in Figure 13-12, had two pairs of wings. Some insects that evolved later, like flies, had one pair.

Flying enabled individual insects to patrol the landscape in search of food, mates, and nesting sites. It also led directly to the great partnership between insects and flowering plants, as is shown in Figure 13-13. The oldest fossils of flowering plants are from about 127 million years ago, but the group may be much older than that. The association between flowering plants and insects has led to the current dominance of both these groups.

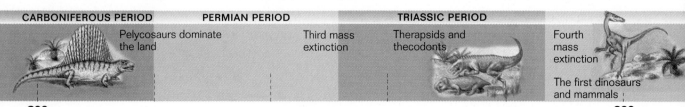

CARBONIFEROUS PERIOD **PERMIAN PERIOD** **TRIASSIC PERIOD**

Pelycosaurs dominate the land

Third mass extinction

Therapsids and thecodonts

Fourth mass extinction

The first dinosaurs and mammals

300

Age (in millions of years ago)

200

Vertebrates Follow Onto Land

While plants and arthropods were dominant on land, vertebrates were widespread in the sea. **Vertebrates** are animals with backbones. An example is shown in Figure 13-14. The first vertebrates were fishes that evolved in the oceans 550 million years ago. Fishes soon came to dominate the seas, and for hundreds of millions of years that is where vertebrates stayed. The first vertebrates to inhabit the land did not venture out of the sea until 350 million years ago. Those first land vertebrates were amphibians, ancestors of today's frogs, toads, and salamanders. Because of their strong, flexible internal skeleton, vertebrates can be far larger than insects, and they soon dominated the landscape. ◻

❑ **CAPSULE SUMMARY**

Ancient cyanobacteria produced the oxygen that was converted into ozone, which forms Earth's protective ozone layer. Once the Earth's surface was safe for habitation, fungi, plants, arthropods, and vertebrates were able to live on land.

Figure 13-14 This fish skeleton clearly shows the backbone, the structure that is characteristic of all vertebrate animals.

Section Review

1. *What are mycorrhizae, and what role did they play in the evolution of life on land?*
2. *What role did arthropods play in the evolution of life on land?*
3. *When did vertebrates first invade land?*
4. *List in order, from first to most recent, the four groups of organisms to invade land.*

Critical Thinking
5. *Identify and explain the relationship between the evolution of eubacteria and the evolution of the first land organisms.*

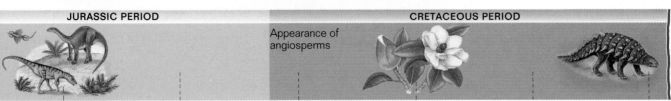

JURASSIC PERIOD

CRETACEOUS PERIOD

Appearance of angiosperms

100

13-3 Vertebrate Evolution

Section Objectives

- Recognize the advantages that sharks had over jawless fishes in the ancient seas.
- Describe the function of a swim bladder.
- Explain why reptiles are more completely adapted to life on land than amphibians.
- Explain how the Permian mass extinction turned out to be an opportunity for reptiles.
- Identify the dominant vertebrates on land today.

*O*f all animals, vertebrates are the most familiar to us, both because we are vertebrates and because almost all other land animals bigger than our fist are vertebrates, too.

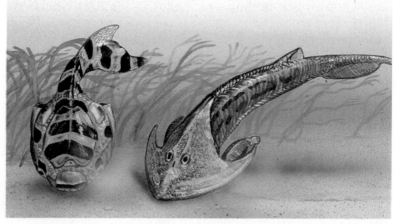

Figure 13-15 Early jawless fishes appear to have fed in a head-down position, their fins acting as stabilizers while their small mouths vacuumed organic particles from the bottom.

Fishes Are the First Vertebrates

The first vertebrates, shown in Figure 13-15, were small jawless fishes. Although many species of jawless fishes filled the seas 450 million years ago, only two types, hagfishes and lampreys, survive today.

Jaws Evolve

Fishes with jaws first appeared around 430 million years ago and rapidly replaced jawless fishes in the seas of the world. Jaws allowed fishes to bite and chew their food instead of sucking it up. As a result, jawed fishes became efficient predators. The earliest kinds of jawed fishes had bulky, armored bodies, but they were soon replaced by jawed fishes with flexible and agile bodies. These new species consisted of sharks and bony fishes.

Sharks

From the heavily armored jawed fishes evolved a very efficient predator—the shark. Sharks lack the bulky armor plating of their ancestors. Their skeletons are made of cartilage

---

CRETACEOUS PERIOD

Fifth mass extinction

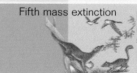

Birds and mammals spread

100   80   60

Age (in millions of years ago)

rather than bone, making them lighter and more buoyant. They also have large, strong mobile fins, which allow them to swim fast and to quickly adjust their direction through the water. A Caribbean reef shark is shown in Figure 13-16.

Bony Fishes

Many hundreds of kinds of sharks evolved. About 250 species survive. Sharks were largely replaced by bony fishes, particularly versatile newcomers. The bony fishes have become an extremely diverse group. A unique way to regulate depth in the water evolved among bony fishes, shown in Figure 13-17. A gas-filled sac called a swim bladder enables bony fishes to change their depth with little effort. By letting gas into or out of the sac, bony fishes can sink, rise, or remain motionless at any depth without having to rapidly beat their fins.

Figure 13-16 Sleek and fast, sharks could swim much more skillfully than their armored relatives.

Amphibians Venture Onto Land
....

When you look at a frog, it is difficult to appreciate the major evolutionary advances that took place as the first amphibians evolved from bony fishes. After 200 million years of success in the sea, fishes were uniquely adapted for success in water. Major changes had to occur in a fish's body design to evolve into an animal capable of living on land.

Structural Innovations

Amphibians were able to adapt to land because of the development of several structural innovations. Early amphibians had moist breathing sacs called lungs, which they used to absorb oxygen from air. Although amphibians could also absorb oxygen through their skin, lungs eventually proved to be far more efficient. The lungs of amphibians required associated changes in the circulatory system, including the development of a new blood vessel and a modified heart. The development of limbs was another structural adaptation for life on land. The limbs of amphibians are derived from the bones of fish fins, but imagine trying to walk on the tips of flippers! What made walking possible was the evolution of a strong support system of bones in the region just behind the head. This platform of bone provided a rigid base for the limbs to work against.

Figure 13-17 Many bony fish species, like these herring, swim in large schools. About half of all species of vertebrates on Earth today are bony fishes.

TERTIARY PERIOD		
Major mammal groups evolve	Mammals dominate the land	

40   20   10

Figure 13-18 An amphibian, like this salamander, has a body that stays close to the ground. In contrast, the body of a reptile, such as a lizard, is held higher off the ground, enabling its legs to function very efficiently.

Amphibians, such as the salamander shown in Figure 13-18, seem primitive compared with the reptiles that eventually replaced them as the dominant vertebrates on land. But amphibians are in fact a very successful group. Having survived for 350 million years, the amphibians evolved long before the dinosaurs and have thus far outlasted them by 65 million years.

Amphibians are an imperfect solution to the challenge of living on land. They must return to water to reproduce, and they must live in moist places because their bodies are in constant danger of losing too much water through their skin by evaporation. A more complete solution to the challenges of terrestrial living is seen in the reptiles. Reptiles evolved from amphibians while amphibians were still dominant on the land, some 300 million years ago. ◻

◻ CAPSULE SUMMARY

Amphibians evolved from primitive bony fishes. The evolution of lungs, a more efficient circulatory system, and legs in amphibians enables them to spend part of their life on land. Skin and eggs that easily dry out require amphibians to remain near water.

Reptiles Break the Ties to Water

....

Besides improving upon the land-adapted features of amphibians, reptiles exhibit two key adaptations in vertebrate body design. Reptiles evolved a watertight skin, which meant that they would not dehydrate by losing moisture to the atmosphere. They also developed a watertight egg. Unlike amphibians, reptiles can lay their eggs on dry land because the eggs are surrounded by a shell that prevents water loss.

The next 50 million years after the reptiles evolved was a period of widespread drought. Reptiles, better adapted for dry times, had an apparent advantage over amphibians. Gradually, reptiles became the dominant group on Earth. Then came the great Permian mass extinction.

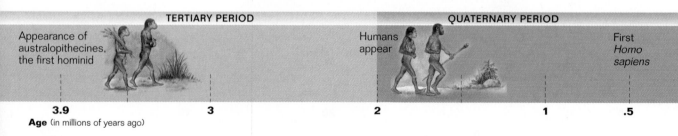

TERTIARY PERIOD

QUATERNARY PERIOD

Appearance of australopithecines, the first hominid

Humans appear

First *Homo sapiens*

3.9   3   2   1   .5

Age (in millions of years ago)

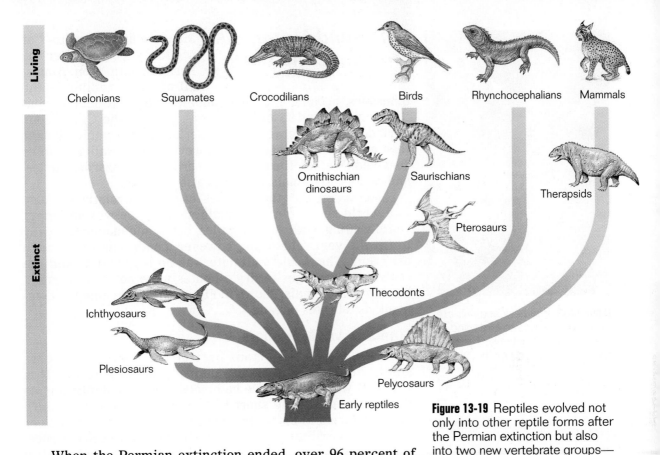

Living: Chelonians, Squamates, Crocodilians, Birds, Rhynchocephalians, Mammals

Extinct: Ornithischian dinosaurs, Saurischians, Therapsids, Pterosaurs, Ichthyosaurs, Thecodonts, Plesiosaurs, Pelycosaurs, Early reptiles

Figure 13-19 Reptiles evolved not only into other reptile forms after the Permian extinction but also into two new vertebrate groups—the birds and mammals.

When the Permian extinction ended, over 96 percent of all species on Earth had disappeared, including many reptiles. But not all of them were gone.

Reptiles Branch Out

Following the Permian extinction, many new reptile groups evolved, as is shown by the evolutionary tree in Figure 13-19. One hundred million years after the Permian mass extinction, a vast assortment of reptile species dominated the land, water, and air. At about the same time, a geological event of massive proportion was just beginning. **Continental drift,** the movement of Earth's giant landmasses, has resulted in the present-day position of the continents. Understanding the movement of the continents provided scientists with an explanation for many formerly confusing geographical distributions. For example, there are a large number of marsupial (pouched) mammal species found in Australia and South America, continents that were formerly connected via Antarctica.

THE FUTURE

Homo nextus???

NOW

Figure 13-20 Feathers and hair are structures unique to birds and mammals, respectively. These two vertebrate groups have successfully adapted to most of Earth's environments.

Birds and Mammals Dominate the Earth

Many evolutionary lines of reptiles appeared after the Permian extinction. Almost all of them, including the dinosaurs, became extinct in another mass extinction event that took place 185 million years later in the Cretaceous period. In the Cretaceous extinction, most species larger than a small dog disappeared forever. The smaller reptiles survived. We see them today as crocodiles, lizards, turtles, and snakes. Mammals and birds also survived.

Once again, the stage was set for a great evolutionary play in a world swept clean by extinction. There was an empty world waiting, with small reptiles, birds, and mammals ready to fill the void. This time, however, conditions were much different. The world's climate was no longer dry, and the reptiles' advantages in dry climates were not so important. Birds and mammals, shown in Figure 13-20, became the dominant vertebrates on land.

Humans Are Mammals

Humans are an important part of the mammal success story, the latest editions of dominant mammal forms. It is true that we humans have had many great accomplishments during our relatively short history, from creating magnificent art forms to discovering antibiotics. But it is also clear that our contributions have not always been positive ones. We have polluted the waters where fish still swim little changed from when vertebrates first ventured onto land. Humans are the first species to do major harm to the Earth, and it seems that the fate of our ancient home may ultimately lie in our hands. □

Section Review

1. *Identify the first vertebrates, and describe their lifestyle.*
2. *Explain why sharks have been a more successful group than lampreys.*
3. *Identify the fish group that has a swim bladder, and describe the function of this structure.*
4. *If a snake and a frog both lived in the same drought-stricken area, which animal would be more likely to survive? Why?*
5. *What effects did the Permian and Cretaceous mass extinctions have on the evolution of vertebrates?*

Critical Thinking
6. *Predict the impact humans will have on the Earth in the future.*

CHAPTER REVIEW 13

Vocabulary

archaebacteria (274)
arthropod (281)
continental drift (287)
cyanobacteria (273)

eubacteria (274)
mass extinction (278)
mycorrhizae (281)
ozone (280)

protist (275)
symbiosis (281)
vertebrate (283)

Concept Mapping

Construct a concept map that shows the different types of life on Earth and the paths by which they evolved. Use as many terms as needed from the vocabulary list. Try to include the following items in your map: prokaryotes, eukaryotes, eubacteria, archaebacteria, protists, plants, fungi, animals, invertebrates, vertebrates, mitochondria, and chloroplasts. Include additional terms in your map as needed.

Review

Multiple Choice

1. It has been determined that the age of the Earth's oldest fossil organism is
   a. 4.5 billion years.
   b. 3.5 billion years.
   c. 2.0 billion years.
   d. 1.5 billion years.

2. Protists differ from bacteria in that protists
   a. contain DNA.
   b. are photosynthetic.
   c. have a nucleus.
   d. are the ancestors of mitochondria and chlororplasts.

3. Which of the following was not a characteristic of the first prokaryotic cells?
   a. The nucleus was absent.
   b. The flagella were absent.
   c. Special functions occurred in internal compartments.
   d. They were unicellular.

4. The formation of ozone made it less dangerous for organisms to invade the land because ozone
   a. blocks visible light.
   b. blocks ultraviolet radiation.
   c. blocks water loss from the land.
   d. encourages photosynthesis in cyanobacteria.

5. Which of the following is most likely to have been a problem for plants when invading the land?
   a. availability of light
   b. sources of water
   c. accessible minerals
   d. gravity

6. Which list reflects the correct order, from first to most recent, of the four groups of organisms to invade the land?
   a. fungi and plants, arthropods, amphibians
   b. amphibians, insects, fungi, plants
   c. plants, fungi, vertebrates and amphibians
   d. archaebacteria, fungi, arthropods, amphibians

7. Sharks were better adapted than jawless fishes to survive in the ancient seas because sharks
   a. had skeletons made of bone.
   b. were armor plated and had fins.
   c. used swim bladders.
   d. had jaws and agile, flexible bodies.

8. The gas-filled structure that allows bony fishes to remain motionless at any depth in the water is called a
   a. lung.   c. urinary bladder.
   b. swim bladder.   d. fin.

9. What structural features of reptiles make them better suited for life on land than amphibians?
   a. lungs and walking legs
   b. the ability to absorb oxygen through their skin
   c. watertight skin and eggs
   d. an internal skeleton and sharp teeth

10. Which list reflects the correct sequence of evolution for groups of vertebrates?
    a. jawless fishes, bony fishes, amphibians, reptiles
    b. sharks, reptiles, amphibians, birds
    c. sharks, jawless fishes, reptiles, mammals
    d. bony fishes, jawless fishes, birds and mammals

11. Which of the following groups of vertebrates had a reptilian ancestor that existed prior to the Permian extinction?
    a. modern reptiles   c. mammals
    b. birds   d. all of the above

Completion

12. The fossil records suggest that _____ mass extinction events have occurred during the history of life on Earth. The most recent occurred about _____ years ago and brought about extinction of the dinosaurs.

13. One group of common bacteria that exists on Earth today, the _____ , are different in enough ways from another, less common group of bacteria, the _____ , that the two groups are considered to form two separate kingdoms of bacteria.

14. Most plants are able to absorb minerals from the soil because they have _____ in their roots.

15. The first multicellular organisms to live on land were _____ and plants. They were followed by animals—first by _____ and then by amphibians.

16. The first vertebrates, the _____ , evolved _____ million years ago.

17. Jawed fishes gave rise to _____ and bony fishes.

18. Structural innovations that enabled early amphibians to adapt to life on land included lungs and _____ . But their advances on land were hampered by _____ and eggs that easily dry out.

19. Understanding _____ has enabled scientists to explain the abundance of marsupial mammal species in Australia and South America.

20. Reptiles that survived the Permian extinction gave rise to many new species of reptiles as well as two new groups of vertebrates, the _____ and the _____ .

Short Answer

21. Oxygen has not always been a part of Earth's atmosphere. Where did Earth's first oxygen come from? Why was it necessary for oxygen to accumulate in Earth's atmosphere before life could inhabit land?

22. Scientists predict that a sixth mass extinction is imminent. What actions can be taken to prevent this mass extinction?

23. Why is the Burgess shale significant in terms of our understanding of the history of life on Earth?

24. Multicellularity first evolved in protists. What advantages do multicellular organisms have over single-celled organisms?

25. The invasion of the land by plants and fungi was possible because of a symbiotic partnership. What is the name given to this partnership? What are the contributions of plants and fungi to this partnership?

26. A partnership exists between flowering plants and bees. In the partnership, bees get nectar that they use as food from the flowers and the flowers are pollinated by the bees. Is this a symbiotic partnership? Explain.

27. What structural feature of vertebrates enabled them to become the most dominant life-forms on the land?

28. **Structure and Function** Amphibians evolved from bony fishes and were the first vertebrates to invade the land. What structural features not found in bony fishes did early amphibians have that enabled them to spend part of their life on land?

29. **Evolution** What are the dominant vertebrates on land today? How did a change in the Earth's environment favor the dominance of these vertebrates?

30. **Making Inferences** Cyanobacteria, a type of eubacteria, were once classified by scientists as members of the plant kingdom. Why do you think scientists did this?

31. **Making Inferences** Continental drift involves the movement of the Earth's great landmasses. How might the movement of the Earth's great landmasses have brought about mass extinction?

32. **Interpreting Graphics** The following diagram compresses the history of the Earth into a 12-hour clock to help you understand the relative time of different events. Note that the formation of the Earth occurred at midnight on the clock. The oceans formed at about 1:30 A.M., while the oldest human fossils date from just before noon. Based on fossil evidence, about what time on the clock did the first prokaryotes appear? About what time did the first eukaryotes appear?

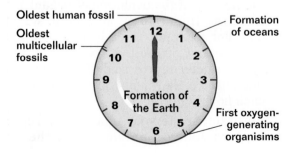

33. **Research and Writing** Lampreys are one of the two remaining kinds of jawless fishes. Examine biology books and science journals to learn more about the feeding habits of lampreys and how lampreys have affected commercial fishing in the Great Lakes.

34. **Cooperative Group Project** Demonstrate the operation of a swim bladder using 2-liter soft drink bottles. Fill several bottles with different amounts of water and place them in an almost full aquarium. Observe how the bottles float at different levels depending on the amount of water they contain. Relate the way in which you are able to regulate the depth at which the bottles float to the operation of the swim bladder in bony fishes. Have different group members gather materials, fill the aquarium with water, manipulate the soft-drink bottles, and share the findings with your class.

35. **Cooperative Group Project** Examine bacteria and protists using a microscope. Have each group member collect a different bacterium or protist and prepare a slide of the organism to be viewed by all group members. Exchange information about each organism viewed.

36. Read the article "Oldest Living Bacteria Tell All," in *Discover*, January 1992, pages 30–31. What two important discoveries did scientists make from the mastodon that was unearthed in an Ohio golf course?

37. Read the article "Life Beyond Boiling," in *Discover*, May 1993, pages 87–91. Why are scientists interested in studying the bacteria that grow in deep-sea thermal vents? What do scientists hypothesize about the structure of enzymes in ancient bacteria? Explain the scientific evidence that supports their hypothesis.

LABORATORY Investigation Chapter 13

Overview of Life on Earth

OBJECTIVES

- Compare and contrast the distinguishing characteristics of representative organisms of the six kingdoms.
- Organize the appearance of life on Earth in a time line.

PROCESS SKILLS

- observing and inferring relationships
- organizing data

MATERIALS

You will need a variety of specimens, photographs, and other materials representing the six kingdoms. Some of these specimens will require viewing with a compound light microscope.

Examples could include:

- cyanobacteria
- archaebacteria
- eubacteria
- protists
- fungi
- plants
- arthropods
- lampreys
- sharks
- bony fishes
- amphibians
- reptiles, birds, mammals

BACKGROUND

1. What are some examples of archaebacteria?
2. What characteristics enabled amphibians to live on land?

3. What traits enabled birds and mammals to be successful on land?
4. Write your own **Focus Question** on your Vee Form.
5. **Knowing Side of the Vee** List the **Concepts** and new **Vocabulary Words** on your Vee Form. In the **Concept Statements** section of the Vee, use these words in sentences that define and explain them.

TECHNIQUE

Doing Side of the Vee

1. To observe organisms representative of the six kingdoms, first complete a table similar to the one below in the **Additional Records and Observations** section of your Vee Form. Then observe each specimen and record your data in the table.

Representative Organisms

Organism name	Kingdom	Characteristics/ adaptations for life on Earth	Sketch

2. Use care when handling the jars of preserved animals. The liquid preservative can leak if the jars are tilted. In the **Procedure** section of the Vee, briefly summarize the procedure you followed.
3. Work in your lab group and use your book as a guide to create a time line of the history of the evolution of organisms. Base the time line on an 8-hour school day. For example, the 8-hour school day represents 3,500 million years. You will have to determine a scale for your time line based on how many millions of years each hour represents.
4. Add sketches to your time line to represent various organisms from the six

kingdoms. Label your sketches. Be creative in the drawings to represent the various organisms that have existed on Earth. Also, account for the disappearance of organisms.

5. In the **Procedure** Section of the Vee, briefly summarize the procedure you followed.

INQUIRY

1. Cyanobacteria are green. What did these organisms contribute to the environment 2.5 billion years ago?

2. Look at your example of an archaebacterium.
   a. What type of environment does the archaebacterium live in?
   b. Why did early archaebacteria have to produce energy without oxygen?

3. Refer to your specimens from the kingdom Protista to answer the following.
   a. What structures are visible in the protists?
   b. If the protist is heterotrophic, what might it eat?
   c. Some protists that were green algae evolved into plants. What process enabled them to contribute oxygen to the environment?

4. Why is it an advantage for a plant to have a great amount of leaf surface area?

5. Look at the examples of fungi.
   a. How are fungi similar to plants? How are they different?
   b. What enabled fungi to live on land?

6. Examine your arthropod examples. What structures enabled them to live on land?

7. Compare the lamprey and the shark. What are several adaptations that allowed for the shark's success as a species?

8. What adaptation did bony fishes develop that enabled them to surpass the shark?

9. Compare the skins of the frog and the lizard. Describe the type of habitat each must live in because of their skin.

10. What enabled frogs to live on land?

11. What adaptations did reptiles develop to completely break their ties to water?

12. Birds and mammals are both endotherms. How did this adaptation allow for their success on land?

13. Use the information on the **Knowing Side** of the Vee Form to interpret your results from the **Doing Side,** then write your **Knowledge Claim**. Write a **Value Claim** for this lab.

ANALYSIS

1. What organelle in your cells might be descended from early cyanobacteria?

2. Offer an explanation for the fact that multicellularity has evolved several times.

3. According to your time line, what period in the day did the mammals arise on Earth?

4. In terms of your time line, during what period of the 8-hour day did the first plants arise on Earth?

5. Write a sentence to describe, in terms of time, how long humans (mammals) have existed compared to all other organisms.

6. Speculate on what you think might have happened in the Earth's history if mass extinctions had not occurred.

FURTHER INQUIRY

Write a **New Focus Question** that could be the point of a new investigation. The following is an example:

What characteristics might be used to determine which kinds of plants are the most successful? Which characteristics are valid indicators of success? Which characteristics are not?

CHAPTER 14

HUMAN EVOLUTION

REVIEW

- DNA structure (Section 2-3)
- protein structure (Section 2-3)
- genes (Section 6-1)
- gametes (Section 6-1)
- natural selection (Section 12-1)

Early hominid

14-1 The Evolution of Primates

*I*n Charles Darwin's book **The Descent of Man,** *he proposed that humans, gorillas, and chimpanzees all evolved from a common ancestor. When the book was published in 1871, there was little fossil evidence to support Darwin's case. Although the fossil record of the origin of humans is still incomplete today, numerous fossil finds made since Darwin's death have added substantial validity to his hypothesis.*

Section Objectives

- Describe the mammal that gave rise to the first primates.
- List two distinctive characteristics of primates.
- Contrast prosimians with monkeys.
- Contrast monkeys with apes.
- Describe the evolutionary relationship between humans and apes.

Primates Have Unique Characteristics

The story of primate evolution begins about 80 million years ago, during the age of dinosaurs. Scurrying about in the trees at this time was an inconspicuous, insect-eating mammal the size of your fist. As is shown in Figure 14-1, this creature had big eyes and tiny, sharp teeth. Biologists think these ancient mammals were the ancestors of the first **primates,** the mammalian group that includes prosimians, monkeys, apes, and humans.

The first primates evolved about 60 million years ago. Evolution favored two distinct anatomical changes that made primates better than their ancestors at stalking and capturing insect prey in the branches of trees.

1. One change was the development of grasping fingers and toes. These fingers and toes are tipped with nails, not claws, as shown in Figure 14-2. Unlike the clawed, unbendable toes of their ancestors, primates have

Figure 14-1 The ancestor of primates closely resembled a modern-day mammal called a shrew, except that this early mammal lived in trees—a kind of "tree shrew."

Figure 14-2 This tarsier from the Philippines is a primate. Its bendable, clawed fingers and toes and forward-facing eyes are key adaptations for life in the trees.

grasping hands and feet that let them grip limbs, hang from branches, seize food, cling to their mothers when they are young, and, significantly, use tools.

2. The second change was in the position of the eyes in primates. The eyes of their ancestors were located on the sides of the head so that their two fields of vision did not overlap. The eyes of primates are shifted forward to the front of the face. This forward placement of the eyes produces overlapping "binocular vision" that enables the primate brain to judge distance more precisely.

Judging distance accurately is a very important ability for an animal that leaps from branch to branch high above the ground. Other mammals have binocular vision, but only primates have both binocular vision and grasping hands. Three-dimensional sight and the ability to manipulate objects have played central roles in directing the evolution of increased intelligence in primates. ◘

❏ *CAPSULE SUMMARY*

Primates evolved about 60 million years ago from small, insect-eating mammals that lived in trees. Primates have grasping fingers and toes as well as binocular vision.

Figure 14-3 Like other prosimians, many lemur species are nocturnal. Nocturnal lemurs, like this sportive lemur, hunt for insects at night and sleep during the day. All 24 surviving species of lemurs live on Madagascar, an island about twice the size of the state of Arizona that is located off the southeast coast of Africa.

The First Primates Were Prosimians

The term **prosimian** is from the Latin *pro*, meaning "before," and *simia*, meaning "ape." Fossils indicate that 38 million years ago prosimians were common in North America, Europe, Asia, and Africa. Only a few kinds of prosimians survive today, and their present range is severely limited. Modern prosimians include lorises, tarsiers, and lemurs. A lemur, shown in Figure 14-3, is about the size of a house cat and has a long tail used for balancing as it climbs through the trees. But the lemurs are in great danger today, and they may soon become extinct in the wild. The forest homes of these animals are being rapidly destroyed by the activities associated with an expanding human population. And as the lemurs' forest homes disappear, so does one of the oldest living links to our past.

Monkeys Adapt to Daylight Activity

About 36 million years ago, a revolution occurred in how primates lived. They became diurnal *(deye UR nuhl)*. **Diurnal** primates are active during the day and sleep at night. This change had far-reaching effects. Because vision is far more important for daytime hunting, evolution favored many improvements in eye design. One of these was the development of specialized cells, called cone cells, in the sensory tissue that lines the back of the eye. Cone cells enable the eye to see in color, thus giving primates color vision. This improved sense of sight was accompanied by the development of a larger, more complex brain. These new day-active primates are called monkeys.

Monkeys rapidly replaced most of the diurnal prosimians; only nocturnal prosimians survive today in areas where monkeys or apes have lived. Feeding mainly on fruits and leaves rather than insects, monkeys were among the first primates with fully developed opposable thumbs. An **opposable thumb** stands out at an angle from the other fingers and can be bent toward them to grasp an object. This provides a hand with a much improved level of dexterity.

Monkeys, shown in Figure 14-4, appear to have evolved first in central Africa and quickly spread to Asia. Modern African and Asian species are commonly referred to as Old World monkeys. Some monkeys migrated to Central and South America, where they developed in isolation. These are the New World monkeys. ◻

◻ *CAPSULE SUMMARY*

Monkeys evolved from prosimian ancestors about 36 million years ago. The development of color vision and opposable thumbs is associated with a more developed brain in these diurnal primates.

Figure 14-4 Baboons, like this mandrill from Africa, *below*, are Old World monkeys that spend most of their time not in the trees but on the ground. Unlike their Old World relatives, many New World monkey species, like this wooly spider monkey from Brazil, *left*, grasp objects with their long, flexible tails.

The Path to Humans

Fossil evidence indicates that humans evolved from the evolutionary line that gave rise to apes. Apes evolved independently from Old World prosimian ancestors about 30 million years ago. Modern apes include gibbons, orangutans, gorillas, chimpanzees, and bonobos (pygmy chimpanzees). Apes have even larger, more developed brains than monkeys, and none of the apes have tails. Once common, apes are rare today. Modern apes are confined to relatively small areas in Africa and Asia. Apes never reached North or South America.

Studies of their DNA tell us a great deal about how apes evolved. The evolution of apes and humans is diagrammed in Figure 14-5. Because the split between the human line and the line leading to the gorillas and chimpanzees was so recent, the genes of humans and chimpanzees have not had time to evolve many differences. Human and chimpanzee DNA nucleotide sequences differ by only 1.6 percent. Your hemoglobin—a protein composed of 573 amino acids—and the hemoglobin of a chimpanzee differ in only a single amino acid. ■

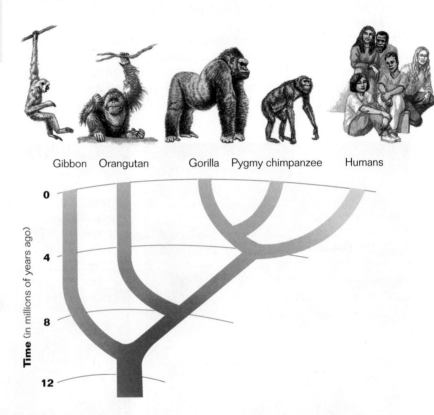

Figure 14-5 This is a phylogenetic tree of apes and humans. The most primitive apes, those found in the evolutionary line leading to gibbons, diverged from other apes about 10 million years ago. Orangutans split off about 8 million years ago. The key split between the gorillas and chimpanzees and the line leading to humans occurred about 4 million years ago.

Section Review

1. *Describe the ancestors of primates, and state when they lived.*
2. *List two primate characteristics that lemurs, chimpanzees, and humans have in common.*
3. *Describe the evolution of apes.*
4. *Identify and explain the evidence that closely links humans to chimpanzees.*

Critical Thinking
5. *Many trees produce brightly colored fruit on high branches. What advantages would monkeys have in finding and picking this fruit?*

14-2 Early Hominids

Fifteen million years ago the world's climate began to get cooler, and the great forests of Africa were largely replaced with open savannas. In response to these environmental changes, natural selection resulted in the divergence of the human line from the gorilla and chimpanzee line about 4 million years ago.

Australopithecines Were the First Hominids

Our earliest known direct ancestors belong to the genus *Australopithecus*. Australopithecines (*aw stray loh PIHTH uh seenz*), along with humans, are classified as **hominids** (*HAHM ih nihds*), belonging to the human line.

Australopithecines exhibited two characteristics that were early milestones on the path leading to the evolution of humans. First, they were **bipedal,** meaning they were able to walk upright on two legs. Table 14-1 compares the skeleton of a gorilla with that of an australopithecine. The structure of the ape skeleton makes it difficult for apes to stand and walk upright. In contrast, australopithecines had a skeletal structure that enabled them to be bipedal.

Second, most australopithecines had large brains—with a greater volume, relative to body weight, than apes had. Some australopithecines weighed about 18 kg (40 lb.) and were approximately 1.1 m (3.5 ft.) tall, about the size of a small

Section Objectives

- State the genus and place of origin of the first hominids.
- Contrast apes with australopithecines.
- Describe several australopithecine species.
- Describe the evidence that indicates human ancestors walked upright before their brains enlarged.

Table 14-1 Comparison of Gorilla and Australopithecine Skeletons

Gorilla	Australopithecine
Skull atop C-shaped spine; spinal cord exits near rear of skull	Skull atop S-shaped spine; spinal cord exits at bottom of skull
Arms longer than legs; arms and legs used for walking	Arms shorter than legs; only legs used for walking
Tall and narrow pelvis, allowing the body weight to shift forward	Bowl-shaped pelvis, centering the body weight over the legs
Femurs (thighbones) angled away from pelvis when walking upright	Femurs angled inward, directly below body to carry its weight

■ Skull ■ Spine ■ Arms ■ Pelvis ■ Femurs

Gorilla

Australopithecine

A. afarensis

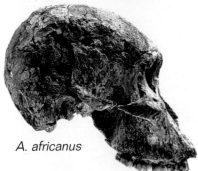
A. africanus

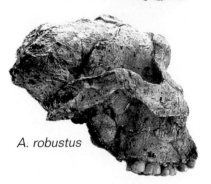
A. robustus

A. boisei

Figure 14-6 Most species of *Australopithecus* had brains quite a bit larger than those of apes, and their teeth were more like those of humans than of apes.

chimpanzee. Other australopithecine species were larger—more than 45 kg (100 lb.) and over 1.5 m (5 ft.) tall. Their brains were typically larger than a chimp's and in some species even larger than a gorilla's, occupying a volume of up to 550 cu. cm (34 cu. in.). While some australopithecine brains were bigger than a gorilla's brain, they were still much smaller than your brain, which is about 1,350 cu. cm (83 cu. in.).

The Discovery of Australopithecus

The first australopithecine fossil, a skull, was discovered in 1924 by Raymond Dart, an anatomy professor in Johannesburg, South Africa. Beautifully preserved, the skull was that of an individual who had died at about five years old. This skull has a rounded jaw, unlike the pointed jaw of apes, and the brain case is far larger than the brain case of an ape of similar size. What really attracted Dart's attention was the age of the skull. The rock in which the skull was embedded was from a geological formation that contained other kinds of fossils estimated to be several million years old. At that time, the oldest reported fossils of hominids were less than 500,000 years old, so the extreme age of this skull was unexpected and exciting. Scientists now believe the skull to be 2.8 million years old.

Other Australopithecine Finds

Dart named his discovery *Australopithecus africanus*. The name *Australopithecus africanus* is from the Latin *australis*, meaning "southern," the Greek *pithekos*, meaning "ape," and the Latinized *africanus*, meaning "African." Dart argued from the start that *Australopithecus* was a bipedal "manlike ape," the long-sought evolutionary link between humans and apes. At first, few thought he was right, but soon the evidence began to mount, and eventually the scientific community was persuaded. In 1938 a stockier species of *Australopithecus* was unearthed in South Africa. Called *A. robustus*, it had massive teeth and jaws. In 1959 in East Africa, Mary Leakey discovered a third species, *A. boisei*, which was even more stockily built. Nicknamed "nutcracker man," *A. boisei* had a great bony ridge on the crest of the head that anchored immense jaw muscles. Like the other australopithecines, *A. boisei* was very old—almost 2 million years old. In 1989 yet another species was discovered, a massively boned ancestor of *A. boisei*.

Figure 14-6 shows skulls of several different australopithecine species.

Lucy: The Oldest Hominid?

In 1974 the anthropologist Donald Johanson went to a remote desert region of northern Ethiopia in search of early human fossils, and he hit the jackpot. Johanson found the most complete and best-preserved skeleton of a prehuman hominid ever discovered. Nicknamed "Lucy," the skeleton was 40 percent complete. It was also nearly 3.2 million years old, the oldest australopithecine fossil then known. The skeleton was assigned to a new species and named *Australopithecus afarensis*; it is shown in Figure 14-7.

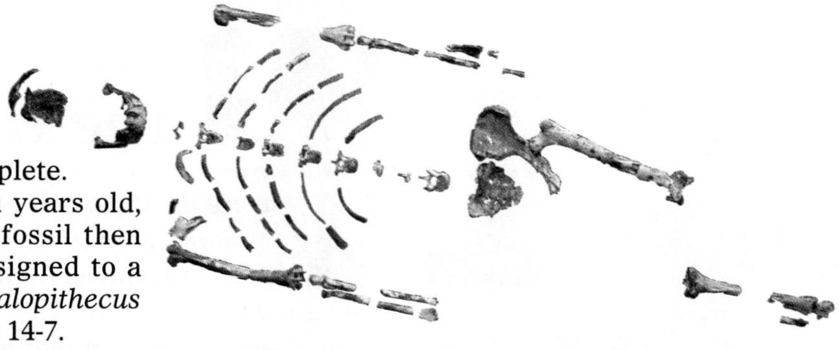

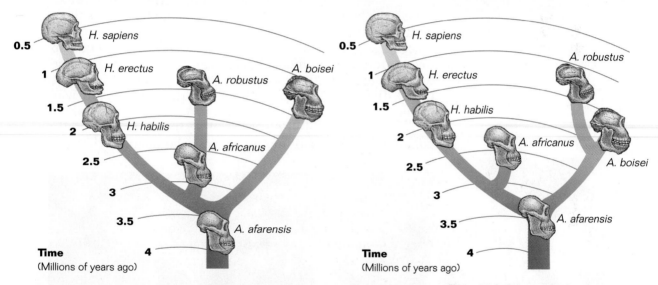

Figure 14-7 The shape of Lucy's leg bones indicates that she must have walked upright. Her teeth are distinctly humanlike, but her head resembles that of an ape. Lucy's brain was no larger than an ape's, about 400 cu. cm.

Since Johanson's discovery, many other specimens of *A. afarensis* have been unearthed, some of which are estimated to be 3.9 million years old. Most researchers agree that these smaller *A. afarensis* individuals represent the true base of the human family tree. They appear to be the first members of the genus *Australopithecus*. As shown in Figure 14-8, they are probably ancestors of all the other australopithecines as well as of the genus *Homo*, which includes our species.

Figure 14-8 Because of an incomplete fossil record, scientists differ in their interpretations of how australopithecines evolved. Some researchers would put *A. robustus* on one branch, while others would put the same species on another branch.

Section Review

1. Where did the first hominids evolve, and what is their genus name?
2. How do the skeletal features of an ape and an australopithecine differ?
3. Explain why the discovery of "Lucy" was a significant scientific accomplishment.

Critical Thinking

4. All the australopithecine species were bipedal. Develop a scenario that explains why and how bipedalism evolved. Be sure to think in terms of selective advantages.

14-3 The Human Line

We are the third and only surviving species of humans, members of the genus **Homo**. The first humans evolved from australopithecine ancestors about 2 million years ago. They were replaced in turn by a second species of human that moved out of Africa and spread across the Earth.

Section Objectives

- Compare and contrast Homo habilis *with australo-pithecines.*
- Describe the characteristics and evolution of Homo erectus.
- Contrast the two theories of the origin of Homo sapiens.
- Compare and contrast Neanderthals with modern humans.

Homo habilis *Was the First Member of Our Genus*

In the early 1960s more hominid bones were discovered close to the site where *Australopithecus boisei* had been unearthed. Scattered among the bones were stone tools. Although the fossils were badly crushed, painstaking reconstruction of the many pieces suggested a skull with a volume of about 640 cu. cm (39 cu. in.). This is much larger than the australopithecine brain volume of 400–550 cu. cm (24–34 cu. in.). There was much discussion at first about whether this fossil was human or australopithecine. Then, in 1972, Richard Leakey, shown in Figure 14-9, discovered a similar and virtually complete skull. Many of the critics who doubted that the

Figure 14-9 Persistent effort by Richard Leakey, far right, and his associates was rewarded. In 1972 Leakey's team found the most complete *Homo habilis* skull to date, *above*. About 1.6 million years old, it has a brain volume of 775 cu. cm (47 cu. in.) and many of the characteristics of modern human skulls.

first skull was human were silenced by Leakey's discovery. Because of its association with tools, this early human was named *Homo habilis*. The name *Homo habilis* is from the Latin *homo*, meaning "man," and *habilis*, meaning "handy." Skeletons discovered in 1987 indicate that *H. habilis* was short in stature, about 1.2 m (4 ft.) tall. *H. habilis* lived in Africa for 500,000 years and then became extinct, replaced by a new species of human with an even larger brain. ◻

◻ **CAPSULE SUMMARY**

Homo habilis, the first hominid assigned to our own genus, was known to make and use stone tools. It evolved from an australopithecine ancestor about 2 million years ago and survived in Africa for about 500,000 years.

Homo erectus Evolved Next

Our picture of *Homo habilis* lacks detail because it is based on only a few specimens. Because *H. habilis* had not evolved far from its australopithecine roots, some scientists still dispute the classification of *H. habilis* as a true human. There is no such doubt, however, about the species that replaced it, *Homo erectus*. Many specimens have been found, and they all indicate that *H. erectus* was distinctly human. The story of how this second species was discovered is a fascinating one.

The Discovery of Java Man

After the publication of Darwin's book *The Origin of Species* in 1859, there was much public discussion about the fossil ancestor common to both humans and apes. Puzzling over this, the Dutch physician and anatomist Eugene Dubois took a very simple approach to the problem—he went to the zoo. Looking at the apes there, he was most drawn to the orangutans. Dubois thought the orangutans resembled what a missing link should look like, so he decided to seek fossil evidence of the missing link in the area where orangutans live, in Indonesia on the island of Java.

Dubois set up a medical practice in a river village in eastern Java. Digging into a hill that villagers claimed had "dragon bones," he unearthed the top of a skull and a thighbone in 1891. He was very excited by his find for three reasons. First, the structure of the thighbone clearly indicated that the individual had long, straight legs and was an excellent, upright walker. Second, the size of the skull cap suggested a *very* large brain, about 900 cu. cm (54.9 cu. in.)—a great deal larger than any ape brain. Most surprising to Dubois, the bones seemed to be as much as 500,000 years old, judging by other fossils he unearthed with them.

The fossil hominid bones that Dubois had found were far older than any fossil hominid discovered up to that time, and at first few scientists were willing to accept that it was an ancient species of human. After years of trying to convince a doubting audience that the bones were human, Dubois became disgusted. He is said to have buried the "Java man" skull cap and thighbone under the floorboards of his dining room and for 30 years refused to let anyone see them.

The Discovery of Peking Man

It was a generation before scientists were forced to admit that Dubois was right all along. In the 1920s a skull was discovered near Peking (now called Beijing), in China. This skull closely resembled that of Java man. Continued excavation at the site eventually uncovered 14 similar skulls. Many of the skulls, along with lower jaws and other bones, were excellently preserved. Crude tools were also found, as were the ashes of campfires. Casts made from the fossils were distributed to laboratories around the world for study. The originals were loaded onto a truck and evacuated from Peking at the beginning of World War II, only to disappear during the confusion of the time. No one knows what happened to the truck or its priceless cargo.

African Origins

Java man and Peking man are now recognized as belonging to the same species, *Homo erectus*. *Homo erectus*, depicted in Figure 14-10, was larger than *H. habilis*—about 1.5 m (5 ft.) tall. It had a large brain of about 1,000 cu. cm (60 cu. in.) and clearly walked erect. The shape of the skull interior suggests that *H. erectus* may have been capable of speech.

Where did *H. erectus* originate? It should come as no surprise to you that it came from Africa. In 1976 a complete *H. erectus* skull that is 1.5 million years old, a million years older than the Java and Peking finds, was discovered in East Africa. *Homo erectus* marked the beginning of a great human migration from Africa. Far more numerous than *H. habilis*, *H. erectus* quickly became widespread and

Figure 14-10 The skull of *H. erectus* had prominent brow ridges and, like modern humans, *H. erectus* had smaller teeth and a less protruding face than *Australopithecus* or *H. habilis*.

abundant in Africa and eventually migrated into Asia and Europe. A social species, *H. erectus* lived in groups of 20 to 50 people. They often lived in caves, but there is evidence that they also built crude wooden shelters. They successfully hunted large animals, butchered them using flint and bone tools, and cooked the meat over fires. The Peking site contains the remains of horses, bears, elephants, deer, and rhinoceroses.

Homo erectus survived for over 1 million years, longer than any other species of our genus. These very adaptable humans disappeared in Africa and Europe only about 500,000 years ago, as early modern humans were emerging. Interestingly, they survived much longer in Asia, until about 250,000 years ago. *Homo erectus* was without serious doubt the direct ancestor of our species, *Homo sapiens*. ◻

◻ *CAPSULE SUMMARY*

Homo erectus *was the second species of human to live on Earth. It evolved in Africa 1.5 million years ago, and by 500,000 years ago it had migrated to Europe and Asia. Homo erectus lived in groups, produced efficient stone tools, and was the first hominid to use fire.*

Modern Humans Are Homo sapiens
....

The human evolutionary journey approaches present times with the appearance of *Homo sapiens*, our species, about 500,000 years ago. The name *H. sapiens* is from the Latin *homo*, meaning "man," and *sapiens*, meaning "wise." *Homo sapiens* is a newcomer to the human family. It has not been around nearly as long as *H. erectus* was. Early *H. sapiens* left behind many fossils and artifacts, including the first known paintings by humans.

African Origins Again?

The geographic origin of our species is a much-debated topic among researchers studying human evolution. Many scientists have argued that independent *H. erectus* lines living in Africa, Europe, and Asia interbred as they evolved and that *H. sapiens* thus arose as a new species more or less simultaneously all over the globe. Others think it unlikely that *H. erectus* evolved into *H. sapiens* more than once. They argue that *H. sapiens* appeared in one place, then spread over the world, replacing *H. erectus* as they went. Recently, scientists have added fuel to the fire of this controversy by studying the DNA within modern human mitochondria.

The reason these scientists looked at mitochondrial DNA to study evolution is that the DNA within mitochondria is transmitted only by females. It is possible to trace genetic variations within a mitochondrial gene back through a family tree, from mother to grandmother to great-grandmother. Since DNA accumulates mutations (changes in genes) over time, the oldest mitochondrial DNA should show the largest number of mutations.

It turns out that the greatest number of different mitochondrial DNA sequences occurs among modern Africans. The results from the DNA studies indicate that *H. sapiens* have been living in Africa longer than on any

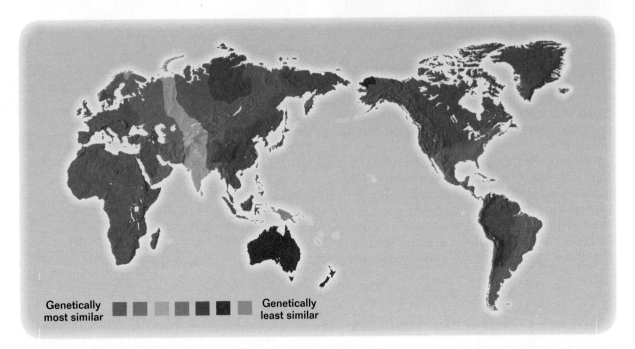

Genetically most similar ■ ■ ■ ■ ■ ■ ■ Genetically least similar

Figure 14-11 By analyzing geographic patterns of variations in nuclear genes and their relative predominance, a map showing the migration of *Homo sapiens* from Africa throughout the world can be derived.

❏ *CAPSULE SUMMARY*

Results from mitochondrial DNA studies suggest that Homo sapiens, *our species, evolved from* Homo erectus *ancestors in Africa about 500,000 years ago.* Homo sapiens *then migrated throughout the world, evolving into the different human varieties.*

other continent. While there is not yet universal agreement among researchers, this evidence supports the hypothesis that *H. sapiens* evolved first in Africa. If this is true, the varieties of living humans evolved after that, and not independently from separate populations of *H. erectus*. This would mean our species was born in Africa and spread from there to all parts of the world. Results from nuclear gene research agree with the mitochondrial study findings, as indicated by the map in Figure 14-11. Indeed, it seems that *H. sapiens* retraced the path taken by *H. erectus* half a million years earlier. ❏

Homo sapiens Migrates Into Europe

Homo sapiens first appeared in Europe about 130,000 years ago. The first *H. sapiens* fossils were found in 1856 in the Neander Valley of Germany. These early European humans, called Neanderthals (*nee AN dur THALZ*), are shown in Figure 14-12. The name *Neanderthal* is from the German *Neander*, the name of a river, and *thal*, meaning "valley." Compared with ourselves, the European Neanderthals were powerfully built, short, and stocky. Their skulls were massive, with protruding faces and heavy, bony ridges over the brows. Their brains were even larger than those of modern humans.

Rare at first outside of Africa, Neanderthals became progressively more abundant in Europe and Asia, and by 70,000 years ago they had become common. Neanderthals took care of their injured and sick and commonly buried their dead, often placing food, weapons, and even flowers with the bodies. Such attention to the dead suggests that they believed in a life after death. The Neanderthals were the first hominids to show evidence of the abstract thinking characteristic of modern humans.

Modern *Homo sapiens*

About 34,000 years ago the European Neanderthals were abruptly replaced by *Homo sapiens* of essentially modern appearance. Early modern humans lived by hunting animals like the bison shown in Figure 14-13. It was the time of the

Figure 14-12 The Neanderthals lived in huts or caves and made diverse tools, including scrapers, spearheads, and hand axes.

Figure 14-13 Early modern humans used intricate tools and weapons made out of stone, horn, and bone. The animals they hunted can be seen in elaborate and often beautiful cave paintings throughout Europe. These paintings were discovered inside Altamira Cave, located near the north coast of Spain.

Neanderthals were powerfully built, early Homo sapiens that lived in Europe and Asia 70,000 years ago. By 34,000 years ago they had been replaced by early modern humans, also H. sapiens whose physical features were very similar to ours.

last great ice age, and Europe was covered with grasslands inhabited by large herds of grazing game. Early modern humans had complex patterns of social organization and are believed to have had sophisticated language capabilities. They eventually spread across Siberia and reached North America at least 13,000 years ago. They made this journey after the great fields of ice had begun to retreat and while a land bridge still connected Siberia and Alaska. There were no more than several million people living in the entire world 10,000 years ago, compared with more than 5.4 billion living today. ◻

Figure 14-14 While not the only animal capable of conceptual thought, human beings have refined and extended this ability until it has become the hallmark of our species.

Humans Are Unique

Like all living things, the humans shown in Figure 14-14 are the product of evolution. Our evolution has been marked by a regular increase in brain size. Our ability to make and use tools effectively is a capability that has, more than any other, been responsible for our dominant position in the animal kingdom. Humans use symbolic language and can shape concepts out of experience. Language has allowed us to transmit accumulated experience from one generation to another. Thus, humans have what no other animal has ever had—cultural evolution. Through culture, we have found ways to change and mold our environment to our needs, rather than changing ourselves in response to the demands of the environment. The human species controls its future in a way never before possible. This is both an exciting potential and an enormous responsibility.

Section Review

1. *State the genus and species name of the first humans, and describe when and how they lived.*
2. *Describe the evidence that identifies* Homo erectus *as the first human species to leave Africa.*
3. *Describe the appearance and lifestyle of* Homo erectus.
4. *Explain the scientific evidence that supports the hypothesis that* Homo sapiens *evolved in Africa.*
5. *When did* H. sapiens *with physical features like yours first appear?*

Critical Thinking

6. *The cause of the extinction of the Neanderthals is not known with certainty. Develop a hypothesis that explains the disappearance of these early* Homo sapiens.

bipedal (299)
diurnal (296)
hominid (299)
opposable thumb (297)
primate (295)
prosimian (296)

Concept Mapping

Construct a concept map that shows the hominids and their relationships to each other. Use as many terms as needed from the vocabulary list. Try to include the following items in your map: australopithecines, humans, Africa, *A. boisei*, *A. africanus*, *A. afarensis*, *H. habilis*, *H. erectus*, *H. sapiens*, bipedal, brain, Mary Leakey, Dart, Johanson, Lucy, Java man, 3.9 million years ago, 2 million years ago. Include additional terms in your map as needed.

Review

Multiple Choice

1. Which anatomical feature distinguishes primates from their ancestors?
   a. grasping hands
   b. canine teeth
   c. body hair
   d. monocular vision

2. Monkeys differ from most prosimians in that monkeys
   a. are color blind.
   b. have thumbs.
   c. sleep at night.
   d. live alone.

3. Among the apes, the most distant relatives to humans are the
   a. chimpanzees.
   b. orangutans.
   c. gorillas.
   d. gibbons.

4. The discovery of Lucy by Donald Johanson is most important because it shows that
   a. hominids inhabited the Earth 3 million years ago.
   b. our ancestors developed bigger brains after becoming bipedal.
   c. *A. africanus* is the base of the human family tree.
   d. Dart's theory about the evolution of man from apes was wrong.

5. Which pair reflects a correct match between hominid and its discoverer?
   a. nutcracker man : Mary Leakey
   b. *Australopithecus afarensis* : Raymond Dart
   c. *Homo erectus* : Donald Johanson
   d. *A. africanus* : Richard Leakey

6. Which of these is *not* a characteristic of an australopithecine skeleton?
   a. arms shorter than legs
   b. spinal cord that exits at bottom of skull
   c. tall and narrow pelvis
   d. skull atop S-shaped spine

7. Compared with australopithecines, *Homo habilis*
   a. used a more sophisticated language.
   b. developed better hand dexterity due to opposable thumbs.
   c. was more apelike.
   d. had greater brain volume.

8. The first hominid known to use fire and tools was
   a. *Homo erectus*.
   b. *A. africanus*.
   c. "handy man."
   d. *Homo sapiens*.

9. Studies of mitochondrial DNA from *Homo sapiens* suggest that
   a. egg cells carry mitochondria.
   b. modern man first evolved in Africa.
   c. populations of *Homo sapiens* evolved independently.
   d. Neanderthals lived in present-day Germany.

10. Neanderthals differ most from modern humans in their
    a. tooth structure.
    b. walking gait.
    c. foot structure.
    d. forehead shape.

Completion

11. The mammal that gave rise to the first primates looked a lot like a modern-day _____ . The first primates evolved about _____ years ago and had forward facing _____ and grasping fingers and toes.

12. Lorises and lemurs are _____ . They live on the island of Madagascar, and many hunt for food at _____ .

13. Orangutans, gorillas, gibbons, and chimpanzees are not monkeys, but _____ . They are larger than monkeys, and none have a _____ .

14. The first hominids were members of the genus _____ . They evolved in _____ and lived on Earth between _____ and 1.5 million years ago.

15. Modern humans and "handy man" are included in the genus _____ . The first members of this genus are thought to have evolved from _____ about 2 million years ago.

16. *Homo erectus* came from Africa and was the first human species to migrate into _____ and Asia. *Homo erectus* used _____ to cook meat and lived in _____ and wooden shelters.

17. Neanderthals and modern people are *Homo* _____ . Neanderthals became _____ about 34,000 years ago, about the same time that modern people appeared.

Short Answer

18. Describe the positioning of the eyes in the first primates. How did this positioning of the eyes improve the primates' ability to survive?

19. In areas inhabited by monkeys today, very few diurnal prosimians are found. How can you explain this?

20. Differences in the nucleotide sequence of DNA from the hemoglobin gene provide evidence for the evolutionary relationship between human and ape species. What assumptions about the nucleotide sequence of DNA support this evidence?

21. Why do scientists dispute the classification of *Homo habilis* as a true human?

22. Two theories about the evolution of *Homo sapiens* have been proposed. What are they? Which theory is best supported by existing evidence?

Themes Review

23. **Levels of Organization** In trying to determine where and when modern humans evolved, scientists have made use of mitochondrial DNA. Why have they chosen to use mitochondrial DNA rather than DNA in the nuclei of cells?

24. **Structure and Function** Explain how you can tell that the skeleton shown below is from a primate? How do you know it is not from a hominid?

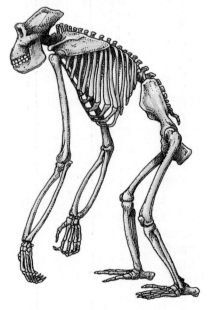

25. Making Predictions The reasons why our ancestors first stood on two feet are not known with certainty. But it has been proposed that bipedalism evolved as a mechanism to keep the brain cool on the open, equatorial savanna. Develop a hypothesis that explains how being bipedal and being hairless except for the top of the head function as inter-related adaptations to protect the brain from overheating.

26. Making Inferences Thermoluminescence and electron spin resonance are new dating techniques used by anthropologists. Using these new techniques, flint tools used by modern humans were determined to be about 92,000 years old. Flint tools used by Neanderthals were determined to be much younger—about 60,000 years old. What do these data suggest about the assumption that Neanderthals lived before modern humans?

27. Making Predictions The life expectancy of men and women today is about 70 years. By comparison, analyses of the bones of Neanderthals indicate that both males and females died at the end of the female reproductive cycle, at roughly 40 years of age. This suggests that grandparents did not exist in Neanderthal society. How do you think the absence of grandparents impacted the Neanderthal society as a whole?

28. Research and Writing In the past few years two new techniques, thermoluminescence (TL) and electron spin resonance (ESR), have been used by anthropologists to date fossils. Find out how these two techniques work and on what kinds of fossils the techniques can be used.

29. Cooperative Group Project Select a hominid described in this chapter. Then prepare a group report on the hominid. Have each member of your group choose one of the following tasks to complete. (1) Sketch a map of where the remains were found. (2) Write about who found the remains and when. (3) Write a description of the remains. (4) Draw an evolutionary tree that shows how the hominid is related to modern man. Consult scientific journals for the latest information on hominids. Finally, as a group, decide how the report will be presented to the class.

30. Multicultural Perspective Imagine that you are a biological anthropologist living in Africa 1.5 million years ago. You study human skeletons, both prehistoric and modern. In your imaginary visit to Africa, describe your encounter with a male and female *Homo erectus*. Do they behave like modern *Homo sapiens*? What similarities might you find? What differences? What technologies would *Homo erectus* have? Language? Fire-making skills? Can the *Homo erectus* couple tan animal hides?

31. Read the article "Climate and the Rise of Man," in *U.S. News and World Report*, June 8, 1992, pages 60–67. Explain how changes in climate might have caused bipedalism to evolve in the ancestors of humans. Why do some scientists regard the development of agriculture by humans as a last-ditch effort to survive?

32. Read the article "Women Leave Indelible Mark on Evolution," in *New Scientist*, July 24, 1993, page 15. Why do some scientists consider women rather than men to be the cornerstone of human evolution?

Human Evolution

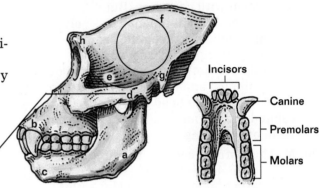

Ape

OBJECTIVES

- Identify anatomical differences and similarities between apes and humans.
- Categorize fossil forms of hominids by examining their anatomical features.

PROCESS SKILLS

- measuring and comparing anatomical features
- organizing and classifying data

MATERIALS

- metric ruler
- protractor

Human

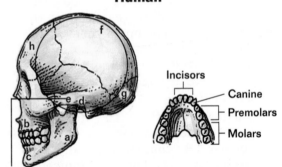

BACKGROUND

1. How do scientists use comparative anatomy to infer evolutionary relationships among species?
2. Write your own **Focus Question** on your Vee Form.
3. **Knowing Side of the Vee** List the **Concepts** and new **Vocabulary Words** on your Vee Form. In the **Concept Statements** section of the Vee, use these words in sentences that define and explain them.

TECHNIQUE

Doing Side of the Vee

Part A: Ape and Human Skulls

1. Refer to the drawings of the ape and human skulls and the checklist of anatomical features to identify similarities and differences between the brain areas, faces, teeth, and jaws of apes and humans. In the **Records** section of your Vee Form, make a table to record your observations and measurements for each of the features in the checklist.

Checklist of Anatomical Features

Brain Capacity: The circle drawn on each skull represents the brain capacity. Measure the radius of each circle in centimeters, cube this number, and multiply by 1,000 to approximate the life-size brain capacity in cubic centimeters.

Lower Face Area: Measure *a* to *b* and *c* to *d* in centimeters for each skull. Multiply these two numbers together, and multiply the product by 40 to approximate the life-size lower face area in square centimeters.

Brain Area: Measure *e* to *f* and *g* to *h* in centimeters for each skull. Multiply these two numbers together, and multiply the product by 40 to approximate the life-size brain area in square centimeters.

Jaw Angle: Note the two lines in the nose area of each skull. Measure the inside angle with your protractor to determine how far outward the jaw projects.

Brow Ridge: Note the presence or absence of a bony ridge above the eye sockets.

Teeth: Note the number and kind of teeth in the lower jaw.

Part B: Fossil Hominids

2. Refer again to the checklist of anatomical features. Measure any features possible for the four fossil hominid skulls. Record your measurements in a table in the **Additional Records and Observations** section on the back of your Vee Form.

3. Classify each feature of the hominid skulls as being apelike, humanlike, or intermediate by writing an A, H, or I next to the feature. Note that a large brain capacity and the absence of a prominent brow ridge are characteristic of humans. Also, less lower face area and more brain area is typical of modern humans. Refer to your data for the ape and human skulls to help you classify the features of the hominid skulls. In the **Procedure** section of the Vee, briefly summarize the procedure you followed.

Fossil Hominids

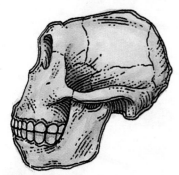

Australopithecus robustus

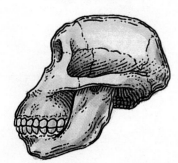

Australopithecus africanus

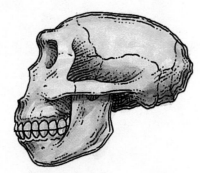

Homo erectus

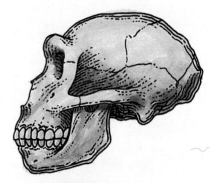

Neanderthal

INQUIRY

1. What is the relationship between the skull features and the brain size of humans compared with that of apes?

2. Based on your observations and measurements, which of the fossil hominid skulls is most apelike? Most humanlike?

3. Use the information on the **Knowing Side** of the Vee to interpret your results from the **Doing Side**, and then write your **Knowledge Claim**. Write a **Value Claim** for this lab.

ANALYSIS

1. What can you infer about the diets of apes and humans from the shapes of their teeth and jaws?

2. How can the hominid fossils be related to both humans and apes without actually being either?

3. Describe the overall changes in skull structure in the evolution of humans.

4. Most skeletons of fossil hominids are not complete. Why do you think this is so?

FURTHER INQUIRY

Write a **New Focus Question** that could be the point of a new investigation. The following is an example:

What was the probable diet of fossil hominids, based on their facial features?

Human Evolution 313

CHAPTER 15

POPULATIONS

REVIEW

- Mendel's laws (Section 7-1)
- phenotype and genotype (Section 7-1)
- mutation (Section 9-3)
- natural selection (Section 12-1)

A population of flamingos.

15-1 How Populations Grow

The commuters crowded into the New York subway shown in Figure 15-1 are members of a population of about 16 million people living in the greater New York City area. On a larger scale, they belong to the rapidly growing human population of the world. Since 1930, the world's human population has nearly tripled. What causes populations to grow? What determines how fast they grow? Is there anything that can slow their growth? You will be able to answer these questions after reading this section.

Figure 15-1 The world's human population has reached an unprecedented size of nearly 6 billion. Today, more people, including these rush-hour commuters, live in the New York City area than lived on Earth 10,000 years ago.

What Is a Population?

A **population** consists of all of the individuals of a species that live together in one place at one time. This is a flexible but useful definition; it allows scientists to use similar terms when speaking of the world's human population, the population of bacteria that live in your intestine, and the population of Devil's Hole pupfish that swims in the tiny pool shown in Figure 15-2.

Figure 15-2 A population can be widely distributed or confined to a small area. An extreme example is the Devil's Hole pupfish, *right*. This species consists of one population of several hundred individuals, which inhabit this small pool, *left*, in Death Valley, Nevada.

Populations   315

Every population tends to grow in size for the simple reason that individuals tend to have more than one offspring. The environment's capacity to support the population limits its size. The statistical study of populations is called **demography** *(duh MAW gruh fee)*. This term is derived from two Greek words: *demos*, meaning "people" (the same root word as in "democracy"), and *graphos*, meaning "measurement." Demography helps to predict how the size of a population will change.

Three Key Features of Populations

To predict how a population will grow, it is first necessary to look at it closely. Every population has a set of key features that play a large role in determining its future. One of the most important features of any population is its current size. **Population size,** the number of individuals in a population, has an important effect on the ability of the population to survive. Many studies have shown that very small populations are most likely to become extinct. Random events or natural disturbances may threaten the continued survival of a small population containing only a few individuals more than they would endanger a large population. Inbreeding (breeding with relatives) may become common in small populations because only relatives are available as mates. Inbreeding produces a more genetically uniform population in which recessive traits, many of them unfavorable, are more likely to be homozygous and therefore expressed. Also, the reduced level of variability that results from inbreeding is likely to detract from the population's ability to adjust to changing conditions.

A second important feature of any population is its **population density.** Density refers to the number of individuals found in a given area, as shown in Figure 15-3. If the individuals of a population are widely spaced, they may rarely, if ever, encounter one another, making reproduction rare. This can happen even if the number of individuals over a wide area is relatively high.

Figure 15-3 Both islands contain the same number of tortoises. However, because the upper island is twice as large as the lower island, it has half the population density of tortoises.

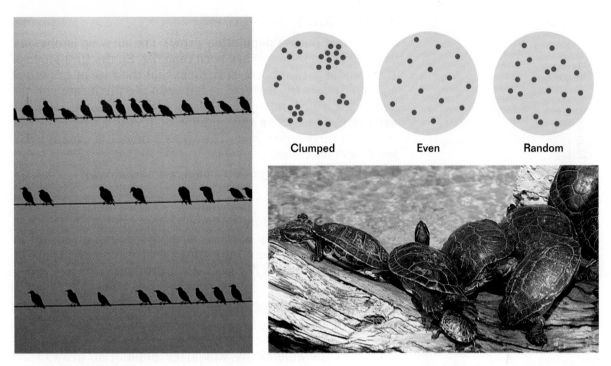

Clumped   Even   Random

A third important feature of any population is the way in which the individuals of the population are arranged, a measure called **dispersion**. The three main patterns of dispersion possible within a population are illustrated in Figure 15-4. If individuals are randomly spaced, the location of each individual is determined by chance. If individuals are evenly spaced, they are located at regular intervals. In a clumped distribution, individuals are bunched together in clusters. Each of these patterns reflects the interactions between the population and its environment. Clumped distributions are the most common type of distribution in nature. One reason they are common is that individual organisms tend to seek out particular sets of conditions (called microhabitats) that may occur only in certain spots—combinations of soil type, moisture, and host trees, for example.

Figure 15-4 This panel, *top right,* shows the three patterns of dispersion possible in populations. Turtles seeking the warmest basking sites form a clumped distribution, *bottom right.* Starlings arrange themselves evenly along telephone lines, *left.*

When Do Populations Grow?

When trying to predict how a population will grow, demographers construct a model of the population. A **model** of a population is a hypothetical population that has key characteristics of the real population being studied. By making a change in the model and observing what happens, it is possible to learn something about what might take place in nature if similar changes occur.

To learn how demographers study a population, you will construct a simple model of population growth. It is not difficult to do. You will develop your model in three stages by asking some basic questions.

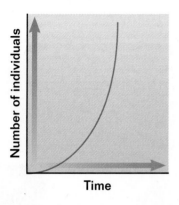

Figure 15-5 This J-shaped curve is characteristic of exponential growth.

Stage I Model

First, when does a population grow? The answer is obvious: when more individuals are born than die. So the first stage of your model is the simple statement that the rate of growth of a population is the difference between the birth rate and the death rate. For human populations, birth and death rates are usually expressed as the number of births and deaths per thousand people per year. The stage I model is usually written as a brief equation.

$$r \text{ (rate of growth)} = \text{birth rate} - \text{death rate}$$

Stage II Model

If you want to calculate the number of individuals that will be added to the population as it grows, symbolized by ΔN (read as "delta N"), multiply the size of the current population, symbolized by N, by the rate of growth.

$$\Delta N = r N$$

When population size is plotted against time, this stage II model produces what is called an **exponential growth curve.** In exponential growth, the rate of increase (r) remains constant, but the amount by which the population grows (ΔN) rises quickly as the size of the population increases. Note that a key assumption of this model is that the birth and death rates (which determine the rate of growth) do not change. Figure 15-5 illustrates exponential growth. In an exponentially growing population that initially contained just a single bacterial cell that divided every 30 minutes, there would be over a million bacteria after only 10 hours.

In fact, no population exhibits exponential growth for very long. Within a year, a population that began as a single bacterium and reproduced at its maximal rate could cover the Earth in a layer over a kilometer thick. The reason populations do not continue to grow unchecked is that they begin to run out of resources and to accumulate wastes. Eventually growth slows and the size of the population stabilizes. The population size that an environment can sustain is called the **carrying capacity,** symbolized by K.

Stage III Model

Your population growth model can be adjusted to compensate for diminishing resources by multiplying it by the fraction of resources still available. This fraction is $(K-N)/K$. As a population grows, smaller and smaller amounts of resources remain. The final model thus becomes:

$$\Delta N = r N [(K-N)/K]$$

Because this model accounts for the declining resources available to populations as they grow, it is called the **logistic model** of population growth. The word *logistics* means procuring, maintaining, and transporting materials—solving the day-to-day problems of living. The logistic model assumes that birth and death rates are not constant, but vary with population size. As the population grows in size, birth rates decline and death rates rise, and consequently the rate of growth falls. Eventually, as N approaches K, the population ceases to grow (that is, $\Delta N = 0$) because the birth rate equals the death rate. Logistic growth is illustrated in Figure 15-6.

The logistic model of population growth, though simple, provides an excellent description of how populations grow in nature. Competition for food, shelter, mating sites, and other resources, as well as the accumulation of toxic wastes, tends to increase as a population approaches its carrying capacity. ◻

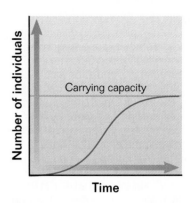

Figure 15-6 The curve of logistic growth looks like a stretched-out letter *s*.

◻ *CAPSULE SUMMARY*

In the exponential model of population growth, the growth rate remains constant. Rapid, continuous growth is the result. In the logistic model of population growth, the growth rate declines as population size rises. The population stabilizes at the carrying capacity, the maximal population size the environment can sustain.

Two Strategies of Population Growth

The population growth curve described by the logistic model illustrates two very different ways that a population can prosper in a competitive world. One approach to population growth depends largely on the r term (rate of growth) of the logistic equation, while the other is influenced predominantly by the K term (carrying capacity). In nature, most organisms employ one strategy or the other (not consciously, of course—natural selection simply favors those taking particular approaches). Some organisms, however, employ a strategy somewhere between the two extremes or change from one to the other as their environment changes.

Populations of *r*-strategists Can Grow Rapidly

Many species, including bacteria, some annual plants (those with only one growing season), and a number of species of insects, are r-strategists. Populations of species that are **r-strategists** are characterized by exponential growth—which results in temporarily large populations—followed by sudden crashes in population size. These species tend to live in unpredictable and rapidly changing environments, where it pays to be able to reproduce quickly when conditions are favorable. In general, r-strategists reproduce early in life and have many offspring each time they reproduce. Their offspring are small, mature rapidly, and receive little or no parental care. The parent thus spreads its reproductive investment among many offspring, investing little in each.

Populations of *K*-strategists Tend to Grow Slowly

Other species, such as redwood trees, whales, and rhinoceroses, have small population sizes and slow population growth. Populations of these species, called **K-strategists,** are characterized by a high degree of specialization. K-strategists tend to live in environments that are stable and predictable, where it pays to be able to compete effectively. K-strategists tend to reproduce late in life and to have few offspring each time they reproduce. These offspring are large, mature slowly, and often receive extensive parental care. Many of the plants and animals that are in danger of extinction today, such as tigers, rhinoceroses, and gorillas, are K-strategists.

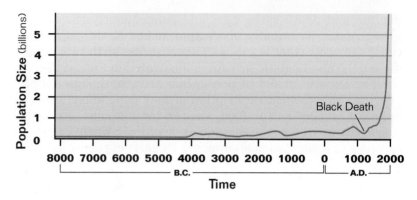

Figure 15-7 Compare this graph of human population growth with the graph of exponential growth shown in Figure 15-5. The decrease in population size in the fourteenth century was caused by the Black Death, which devastated Europe and parts of Asia. This disease, which is transmitted by fleas, may have killed up to 50 percent of Europe's population between 1347 and 1352. Notice that World War I (1914–1918), in which almost 10 million died, and World War II (1939–1945), in which more than 50 million died, did not cause declines in the world's population.

Population Size (billions) — 5, 4, 3, 2, 1, 0

8000 7000 6000 5000 4000 3000 2000 1000 0 1000 2000
B.C. ——————————— A.D.
Black Death
Time

The Rapidly Growing Human Population

Human populations have evolved with many of the characteristics of *K*-strategists, including small families, extended parental care, and a large investment in each offspring. However, in recent times humans have greatly expanded the carrying capacity of their environment. The development of agriculture and technology has enabled humans to increase their food supply, combat pests, and cure diseases. This increase in the value of *K* has led to a rapid increase in ΔN. In other words, the human population is growing explosively, as shown in Figure 15-7. By 1995 there were more than 5.6 billion people on Earth, and the human population was growing at a rate of about 1.7 percent annually. Nearly 94 million people are added to the population every year. At this rate, the human population will double in just over 40 years. No one knows if the Earth can continue to support such a large population, although it seems clear that the human population cannot expand at this rate for long without seriously damaging the environment. Damage to the planet will eventually reduce the carrying capacity for humanity and therefore slow the growth of the human population.

Section Review

1. *Are fans attending a basketball game randomly dispersed, evenly dispersed, or clumped? Explain your answer.*

2. *The logistic model of population growth assumes that as population size increases, birth rates fall and death rates rise. Does this assumption reflect what is likely to happen in nature? Justify your answer.*

3. *Explain why an r-strategist might be better suited for an unpredictable environment than a K-strategist would be.*

Critical Thinking

4. *Suppose that a population is shrinking. What must be true of its birth and death rates?*

15-2 How Populations Evolve

D arwin proposed that biological diversity is the result of natural selection. In the modern statement of Darwin's theory, every natural population contains individuals with alternative versions of genes, known as alleles. Those individuals having alleles that improve the chances of survival and reproduction are favored, and so these alleles become more common. In Darwin's day, no one knew what caused genetic variation. Biologists now know it is produced by mutations, changes in DNA.

In the century since Darwin's death, the science of genetics has revolutionized how biologists think of heredity and has allowed them to construct detailed models of how natural selection alters the proportions of alleles within populations. Models of allele changes within populations have proven very useful, often allowing biologists to predict how a particular population will respond to a change in its habitat. Before you can understand how populations change in response to evolutionary forces, you need to understand how they behave in the absence of these forces.

Unchanging Populations

When Mendel's work was rediscovered in the early 1900s, biologists began to investigate how alleles might increase or decline in abundance. Specifically, they wondered if dominant alleles, which are usually more common than recessive alleles, would replace recessive alleles simply by sheer weight of numbers. In 1908, the English mathematician G. H. Hardy and the German physician Wilhelm Weinberg independently demonstrated that dominant alleles do not in fact replace recessive ones. With simple algebra and probability, they showed that the frequencies of alleles and proportions of heterozygous and homozygous individuals remain constant from generation to generation. Their demonstration, called the **Hardy-Weinberg principle,** states that populations do not change unless evolutionary forces act upon them. The Hardy-Weinberg principle holds true only for large populations in which individuals mate randomly and in which the forces that change the proportions of alleles are not acting.

The Hardy-Weinberg principle is usually stated as an equation. For a gene with two alternative alleles, A and a, the equation looks like the following.

Figure 15-8 In Chapter 12, you learned about the evolution of coloration in peppered moths like these. In this chapter you will learn how alleles responsible for traits such as coloration change in abundance under natural selection.

$$p^2 \quad + \quad 2pq \quad + \quad q^2 \quad = \quad 1$$

frequency of individuals homozygous for allele A	frequency of heterozygous individuals, with alleles A and a	frequency of individuals homozygous for allele a	

A frequency is the proportion of a group that is of one type. For example, the frequency of boys in your class is the number of boys divided by the total number of students in the class. In a population of 100 tigers containing 84 striped tigers and 16 white (albino) tigers, the frequency of striped tigers is 84/100, or 0.84; the frequency of white tigers is 16/100, or 0.16. The p and q symbols in the Hardy-Weinberg equation are **allele frequencies.** The frequency of allele A is the proportion of all alleles for this gene in the population that are A. Similarly, the frequency of allele a is the proportion of alleles that are a. By convention, the frequency of the more common of the two alleles is designated p, and the frequency of the rarer allele is designated q. Because there are only two alleles, $p + q$ must always equal 1.

Figure 15-9 White coloration in tigers is caused by a rare recessive allele. Do you think this allele is likely to become more abundant?

Applying the Hardy-Weinberg Principle

You can use the Hardy-Weinberg principle to predict genotype frequencies (frequencies of homozygotes and heterozygotes) from allele frequencies. How can you do that? Remember that p represents the frequency of allele A, and q represents the frequency of allele a. Individuals homozygous for allele A have two copies of this allele, so they occur at a frequency of p times p, or p^2. Individuals homozygous for allele a have two copies of a and occur at the frequency of q times q, or q^2. Heterozygous individuals have one copy of A and one copy of a, but heterozygotes can occur in *two* ways—A from the father and a from the mother, or a from the father and A from the mother. Therefore, the frequency of heterozygotes is $2pq$. Every individual in the population must be either AA, Aa, or aa, so the sum of the three frequencies must equal 1.

$$p^2 + 2pq + q^2 = 1.$$

The importance of the Hardy-Weinberg equation is that it allows you to predict the frequency of each genotype in a population if you know the allele frequencies. The Hardy-Weinberg principle has proven very useful in assessing real-life situations. For example, as you learned in Chapter 7, the often-fatal human disorder cystic fibrosis is caused by a recessive allele. This disorder affects Caucasian North Americans at a frequency of about 1 in every 2,080 individuals, or 0.00048. What proportion of Caucasian North Americans are expected to be heterozygous for this allele? You can easily figure this out.

1. **Calculate the frequency of the recessive allele.** You know that q^2, the frequency of homozygous recessive individuals, is 0.00048. Therefore q, the frequency of the cystic fibrosis allele, is the square root of 0.00048, or 0.022.

2. **Calculate the frequency of the dominant allele.** You can calculate p, the frequency of the dominant allele, by subtraction. Because $p + q = 1$, $p = 1 - q$. So $p = 1 - 0.022$, or 0.978.

3. **Determine the frequency of heterozygotes.** The frequency of heterozygous individuals is expected to be *2pq*, or 2 times 0.978 times 0.022, or 0.043. This means that 43 of every 1,000 Caucasian North Americans are predicted to carry the cystic fibrosis allele unexpressed. Figure 15-10 summarizes the genotype frequencies for this example.

Genotype	Phenotype	Frequency
CC	Normal	$p^2 = 0.96$
Cc	Normal	$2pq = 0.043$
cc	Cystic fibrosis	$q^2 = 0.0048$

Figure 15-10 This chart summarizes the calculations of genotype and phenotype frequencies made in the text.

How valid are the estimates of genotype frequencies that are made using the Hardy-Weinberg equation? For all but small, isolated human populations, they prove to be very accurate. Most human populations are large and not inbred and therefore resemble the ideal population assumed by Hardy and Weinberg. Remember that the frequencies of genotypes deviate from those predicted by the Hardy-Weinberg equation only when there are evolutionary forces acting on the population that tend to favor one allele over another. *The Hardy-Weinberg equation is valid only when nature is blind to which allele an individual carries.* When the identity of the allele does matter, the equation no longer applies. ◧

CAPSULE SUMMARY

The Hardy-Weinberg principle states that the frequencies of alleles and genotypes remain constant in populations in which evolutionary forces are not acting.

Five Forces Cause Populations to Evolve

In large, randomly mating (that is, not inbred) populations, such as most human populations, the frequencies of alleles and genotypes remain constant from generation to generation unless evolutionary forces act on the population. There are five evolutionary forces that can cause the proportions of homozygotes and heterozygotes in a population to differ significantly from those predicted by the Hardy-Weinberg equation.

Mutation

Although mutation from one allele to another can obviously alter allele frequencies, mutation rates in nature are so low

Figure 15-11 Normal fruit flies have two wings. This mutant has four. This rare mutation, like most mutations, is harmful. Beneficial mutations are the raw material for natural selection.

that proportions of common alleles are not measurably affected. Most genes mutate only about 1 to 10 times per 100,000 cell divisions, so mutation itself does not significantly change allele frequencies. It is, however, the ultimate source of all variation and thus makes evolution possible. Figure 15-11 illustrates an example of mutation.

Migration

Migration is the movement of individuals from one population to another. It can be a powerful force for genetic change. Migration of individuals to or from a population creates **gene flow,** the movement of alleles into or out of a population. Gene flow occurs because new individuals add alleles to the population, and departing individuals take alleles away.

Nonrandom Mating

Sometimes individuals prefer to mate with others of their own genotype, a situation called **nonrandom mating.** Mating with relatives, which is known as inbreeding, is a type of nonrandom mating that causes the frequency of heterozygotes to be much less than that predicted by the Hardy-Weinberg equation. Inbreeding does not change the frequencies of alleles; it increases the proportion of homozygotes in a population. For example, populations of self-fertilizing plants consist primarily of homozygous individuals. Nonrandom mating also results when organisms choose their mates. In animals, females often select males based on their size, coloration, ability to gather food, or other characteristics, as shown in Figure 15-12.

Figure 15-12 What characteristics does a female widowbird find attractive? As Malte Andersson of Oxford University discovered, she prefers long tails like the one on this male. Moreover, the longer the tail is, the more attractive it is. Andersson showed that females preferred to nest in areas controlled by males whose tails had been artificially extended. Darwin recognized that female preference for extreme male traits could explain the evolution of elaborate male ornamentation that seems to be detrimental to survival.

Genetic Drift

In small populations, the frequency of an allele can be changed drastically by a chance event, such as a fire, landslide, or lightning strike. When an allele is found in only a few individuals, the loss of even one individual from the population can have

Figure 15-13 Cheetahs are an endangered species, and efforts to save them are complicated by their history. Cheetahs have gone through at least two drastic declines in population size. As a result, the surviving cheetahs are descendants of only a few individuals, and the species is genetically uniform. One consequence of this genetic uniformity is reduced disease resistance—cheetah cubs are more likely to die from disease than are the cubs of lions or leopards. These chance reductions in the cheetah's population size, examples of genetic drift, may result in the extinction of this species.

major effects on its frequency. Since this sort of change in allele frequency appears to occur randomly, as if the frequencies were drifting, it is called **genetic drift.** Small populations that are isolated from each other can differ greatly as a result of genetic drift. Because we humans lived in small groups for much of our history, genetic drift must have been a particularly important factor in the evolution of our species. Cheetahs are another species whose evolution has been profoundly affected by genetic drift, as described in Figure 15-13.

Natural Selection

Natural selection causes deviations from the Hardy-Weinberg proportions by directly changing the frequencies of alleles. An allele can increase or decrease in frequency, depending on its effects on survival and reproduction. For example, the allele for sickle cell anemia is slowly declining in frequency in the United States because individuals homozygous for this allele rarely produce children. Selection is one of the most powerful agents of genetic change. You will learn more about how natural selection alters allele frequencies in the next section. ◻

Section Review

1. *The frequency of striped tigers in a population is 0.99. From this information, can you calculate the frequencies of alleles and genotypes? If not, what additional information is needed?*

2. *Distinguish between natural selection and genetic drift.*

Critical Thinking

3. *Recall from Chapter 12 that in Africa the frequency of the sickle cell allele is not changing. Is this an example of the Hardy-Weinberg principle? Explain your answer.*

How Natural Selection Shapes Populations

Section Objectives

- *Explain why selection against unfavorable recessive alleles is slow.*
- *Distinguish between polygenic and single-gene traits.*
- *Contrast directional, stabilizing, and disruptive selection.*

D*arwin argued, and more than a century of research has demonstrated, that natural selection is a powerful force in nature, constantly adjusting populations to accommodate an ever-changing environment. However, although selection is a powerful agent of genetic change, there are limits to what it can accomplish. These limits arise because selection does not act directly on genes. It does not favor certain individuals just because they have particular genes, but because of the **consequences** of possessing those genes—the individuals are bigger, or smarter, or have other favorable characteristics. Stated differently, natural selection acts on phenotype, not genotype.*

Selection Against Unfavorable Recessive Alleles Is Slow

When a mutation creates a new allele by altering the DNA sequence of a gene, the effect of the new allele on the phenotype is almost always recessive. Remember that genes code for proteins that influence the phenotype (what the individual is like) through their activity as enzymes, regulators, or structural elements. Most random changes in a gene are destructive, resulting in a protein whose shape is no longer well suited to its function. In a heterozygous individual, both altered and normal forms of the protein are made. As long as enough of the normal form is produced to perform its function, the phenotype of the heterozygous individual is normal.

Now think carefully about how natural selection might operate on a mutant allele. Only characteristics that are expressed can be targets of natural selection. It follows that selection cannot operate efficiently against rare recessive alleles, even if they are unfavorable. Only when the allele becomes common enough that individuals carrying it come together and produce homozygous offspring does natural selection have an opportunity to act.

To better understand this limitation on natural selection, go back for a moment to the Hardy-Weinberg equation. When a recessive allele *(a)* is present at a frequency equal to 0.1, only 1 out of 100 individuals will be homozygous recessive *(aa)* and display the phenotype associated with this allele. But 18 out of 100 individuals will be heterozygotes *(Aa)* and carry the allele unexpressed. So natural

Figure 15-14 Alexis, *far right*, the son of Nicholas II, last Tsar of Russia, suffered from severe hemophilia, a disease caused by a recessive allele. For Alexis, even a fall or cut could lead to uncontrollable and potenially dangerous bleeding. Why hasn't natural selection eliminated such a harmful allele? You will learn the answer in this section.

selection can act on only 1 out of every 19 individuals that carry the allele. If the frequency in the population of the recessive allele is 0.01, the frequency of homozygous recessive individuals in that population will be only 1 in 10,000. For a slightly higher frequency of 0.02, the frequency of homozygous recessive individuals is still only 1 in 2,500. Many human diseases caused by recessive alleles have frequencies similar to this. These genetic disorders are not eliminated by natural selection, because very few of the individuals bearing the alleles express them. ▢

▢ CAPSULE SUMMARY

Natural selection reduces the frequency of a harmful recessive allele slowly because very few individuals are homozygous recessive and express the allele.

Many Genes Have More Than One Common Allele

Selection is a very powerful force for genetic change in biological communities, for the simple reason that not all alleles are rare. When biologists first learned how to measure the amount of genetic variation in natural populations, they found that a minimum of 15 percent of the genes of an average insect are heterozygous. Most of the variation is due to a small number of common alleles. When a gene in a population has more than one allele appearing at a significant frequency, the population is said to be polymorphic *(pah lee MAWR fihk)* for that gene. The genetic variation that results is called **genetic polymorphism.** Vertebrates are a little less polymorphic than insects. Humans, for example, are heterozygous for a minimum of 5 percent of their genes. This is an enormous pool of variation on which natural selection can act. Figure 15-15 illustrates one example of polymorphism in vertebrates.

Figure 15-15 All of the geese in this photo are lesser white snow geese. However, the two forms were once considered separate species. Now scientists know that one polymorphic gene with two alleles is responsible for the color difference. White coloration is recessive. Because geese prefer mates of their own color (an instance of nonrandom mating), both colors are maintained in the species.

Selection on Traits Controlled by More Than One Gene

A second reason that natural selection acts powerfully to shape populations is that many important characteristics of plants, animals, and other organisms are affected by a large number of genes. A characteristic influenced by several genes is called a **polygenic trait** (*pah lee JEHN ihk*). Human height, for example, is determined by dozens of genes, each adding a little bit to the overall effect. Natural selection thus can alter the allele frequencies of many different genes governing a trait, influencing most strongly those genes that make the greatest contribution to the phenotype. It is therefore difficult to keep track of any one particular gene—it is like trying to follow one duck in a flock. Instead, what biologists do is more like keeping track of the entire flock. They measure the polygenic trait in each individual in the population and calculate the average. If selection is favoring increased height, for instance, then over time the average height within the population should increase.

Because of genetic polymorphism, polygenic traits tend to exhibit a range of phenotypes clustered around an average value. If you were to plot the height of everyone in your class on a graph, the values probably would form a hill-shaped curve called a **normal distribution,** with the average value at the summit, as illustrated in Figure 15-16.

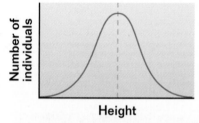

Figure 15-16 This hill-shaped curve, *top,* is a normal distribution. The dotted line indicates the average height. In this photo taken around the turn of the century, cadets of the Connecticut Agricultural College, *bottom,* showed that height was distributed normally in their class.

Directional Selection

When selection acts to eliminate one extreme from a range of phenotypes, the genes promoting this extreme become less common in the population. This can be demonstrated in a simple experiment. If fruit flies are raised in the dark and then given a chance to fly toward light, some will fly toward light and some will not. In generation after generation, investigators selected those flies that had the strongest tendency to fly toward light as parents for the next generation (thus selecting *against* the light-avoiders).

After 20 generations, the average tendency to fly toward light had increased 50 percent within the population. Elimination of flies that failed to move toward light caused the population to contain fewer individuals with alleles that resulted in light-avoidance.

As a consequence, a fly picked at random from the new fly population would be more likely to move spontaneously toward light than a fly selected from the original population. The population has been changed by selection in the direction of greater attraction to light.

In selection against one extreme form of a polygenic trait, the average value is shifted toward the other extreme. For this reason, this form of selection is called **directional selection.** It is illustrated in the upper panel of Figure 15-17.

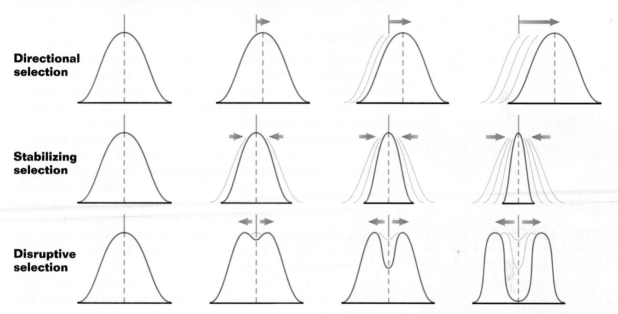

Directional selection

Stabilizing selection

Disruptive selection

Figure 15-17 This figure shows the three kinds of selection on polygenic traits. As in Figure 15-16, the dotted line marks the average value of the trait.

Stabilizing Selection

When selection acts to eliminate extremes at both ends of a range of phenotypes, the frequencies of the intermediate phenotypes increase. When selection acts in this way, the population will contain fewer individuals with alleles promoting extreme types. This type of selection is very common in nature. In humans, for example, babies with intermediate weight at birth have the highest survival rate. In ducks and chickens, eggs of intermediate weight have the highest hatching success.

In selection against both extremes of a polygenic trait, the average value (the peak of the normal distribution) does not change. Instead, as you can see in the middle panel of Figure 15-17, the distribution becomes narrower, tending to "stabilize" the average by increasing the proportion of individuals that are similar. Therefore, this form of selection on polygenic traits is called **stabilizing selection.**

Figure 15-18 In Africa, the swallowtail butterfly mimics several species of butterflies. The left column in the photo shows two individual swallowtail butterflies, and the right column shows the species each individual mimics.

Disruptive Selection

Sometimes selection acts to eliminate rather than favor the intermediate type. When selection acts in this way, the population will contain more individuals exhibiting the two extremes. For example, some butterflies that birds find palatable mimic the coloration and patterns of species that birds find distasteful and thus gain protection from being eaten. In different localities, a palatable butterfly species may closely resemble different distasteful species, as illustrated in Figure 15-18. However, any individuals that do not closely resemble one of the distasteful species are readily detected and eaten by birds, so intermediate patterns are selected against.

In selection against the intermediate types of a polygenic trait, the average value does not change. Instead, the distribution becomes separated or "disrupted" into extreme classes, as you can see in the lower panel of Figure 15-17. This form of selection on polygenic traits is called **disruptive selection.** ❑

❑ *CAPSULE SUMMARY*

There are three types of selection on polygenic traits. In directional selection, the range of phenotypes shifts toward one extreme. In stabilizing selection, the range of phenotypes narrows. In disruptive selection, the range of phenotypes "splits" and moves toward both extremes.

Section Review

1. *Explain why you would expect natural selection against a harmful dominant trait to proceed rapidly.*

2. *In Mendel's peas, height was controlled by a single gene with two alleles. How would a graph of height versus frequency for peas differ from the graph for humans shown in Figure 15-16?*

3. *In disruptive and stabilizing selection, the average value of the trait does not change. Explain how these two forms of natural selection differ.*

Critical Thinking

4. *What role might disruptive selection play in speciation?*

Vocabulary

allele frequency (322)
carrying capacity (318)
demography (316)
directional selection (329)
dispersion (317)
disruptive selection (330)
exponential growth curve (318)
gene flow (324)

genetic drift (325)
genetic polymorphism (327)
Hardy-Weinberg principle (321)
K-strategist (319)
logistic model (318)
model (317)
nonrandom mating (324)
normal distribution (328)

polygenic trait (328)
population (315)
population density (316)
population size (316)
r-strategist (319)
stabilizing selection (329)

Concept Mapping

Construct a concept map that shows how the forces of genetic change cause evolution. Use as many terms as needed from the vocabulary list. Try to include the following items in your map: Hardy-Weinberg principle, genetic drift, nonrandom mating, natural selection, mutation, migration. Include additional terms in your map as needed.

Review

Multiple Choice

1. Fish commonly called king mackerel follow schools of menhaden, their primary food source. The distribution of king mackerel in the ocean is most likely
   a. randomly spaced.
   b. evenly spaced.
   c. clumped.
   d. normal.

2. The growth exhibited by a colony of bacteria that has a limited food supply will most likely be
   a. exponential.   c. natural.
   b. logistic.   d. random.

3. Which of the following reflects a *correct* match between a species and the population growth strategy it employs?
   a. rhinoceros: *r*-strategist
   b. human: *K*-strategist
   c. bacterium: *K*-strategist
   d. cockroach: *K*-strategist

4. According to the Hardy-Weinberg principle, allele frequencies in randomly mating populations
   a. change when birth rate exceeds death rate.
   b. increase and then decrease.
   c. achieve disequilibrium.
   d. do not change.

5. The frequency of homozygous recessive albino rats in a population is 0.01. What is the expected frequency of the dominant allele in this population?
   a. 0.9   b. 0.09   c. 0.18   d. 1.8

6. Which of the following is *not* a cause of genetic change?
   a. genetic drift   c. natural selection
   b. random mating   d. mutation

7. Why is it unlikely that the frequency of muscular dystrophy in the human population will be reduced quickly by means of natural selection?
   a. Natural selection acts only on homozygous recessive individuals.
   b. Muscular dystrophy is not a genetic disorder.
   c. The frequency of homozygous recessive individuals is too great.
   d. Homozygous dominant individuals can have affected children.

8. In an effort to develop the best-tasting apple, a grower eliminated all trees that produced apples considered too sweet or too sour. What type of selection is being carried out by the grower?
   a. directional   c. disruptive
   b. stabilizing   d. choice

Completion

9. Population growth can be predicted by knowing something about a population's size, its _____ , and its dispersion.

10. A population will _____ if the number of deaths exceeds the number of births. But, if the number of births exceeds the number of deaths, the population will _____ .

11. Populations of *r*-strategists tend to be _____ in size and grow _____ , while populations of *K*-strategists tend to _____ in size and grow _____ .

12. According to the Hardy-Weinberg principle, the frequency of dominant and recessive traits within a population will remain _____ unless the population is evolving.

13. In the Hardy-Weinberg equation, q^2 represents homozygous _____ individuals, p^2 represents homozygous _____ individuals, and $2pq$ represents heterozygous individuals.

14. One of the evolutionary forces that disrupts the constant state of allele frequency as predicted by the Hardy-Weinberg principle is _____ , which may involve inbreeding. Another is _____ , which results from random environmental events such as landslides.

15. Selection against an unfavorable recessive allele is slow because natural selection can act only on _____ individuals.

16. Human characteristics such as height are _____ traits because they are influenced by several genes. A graph of the distribution of these types of traits within a population usually shows a _____ distribution.

Short Answer

17. Distinguish between the size of a population and its density.

18. Explain why genetic drift is more likely to occur in a small population than in a large one.

19. Many of our most serious pests, including rats, cockroaches, and mosquitoes, are *r*-strategists. What features of *r*-strategists make them particularly difficult to control?

20. Is the evolution of the peppered moth discussed in Chapter 12 a case of directional, disruptive, or stabilizing selection? Explain your answer.

Themes Review

21. **Homeostasis** Homeostasis may be viewed as a state of biological equilibrium. According to the Hardy-Weinberg principle, allele frequencies remain in a state of equilibrium from one generation to the next unless certain forces act to upset the equilibrium. What forces can upset the equilibrium and bring about evolution?

22. **Levels of Organization** What is a population? What is the relationship between the biological concepts of population and species?

23. **Structure and Function** Draw a graph that depicts the distribution of a population that has undergone disruptive selection. How does disruptive selection affect the population?

24. **Evolution** After a forest fire, plants quickly recolonize the burned area. Are these "pioneer" plants more likely to be *r*-strategists or *K*-strategists? Explain your answer.

25. Interpreting Data Biologists introduced pheasants onto an island in Washington State in the 1930s. Using the data shown on the graph below, answer the following questions. What is the island's carrying capacity? Describe how you reached your answer. If the island's area is 25 sq. km, what was the population density at the beginning of the fourth year?

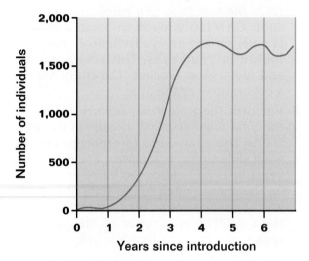

26. Responding Critically The frequency of sickle cell anemia among the African Americans is 1 in 500 individuals. A reporter writing about the disease has asked you to check the accuracy of the data, which show (1) frequency of the sickle cell allele as 0.25; (2) frequency of homozygous recessive individuals as 0.065; (3) frequency of the dominant allele as 0.75; (4) frequency of homozygous dominant individuals as 0.563; and (5) frequency of heterozygotes as 0.375. What advice would you give the reporter?

27. Evaluating Data Would you rather get $1 million a day for 30 days or receive a nickel the first day, two nickels the second day, four nickels the third day, eight nickels the fourth day, and so on, for the same 30-day period? Carry out the calculations to check your decision.

28. Research and Writing China and Thailand greatly reduced their birth rates through government-sponsored population control programs. Research the population control measures taken in both countries, and write a report that summarizes what you have learned. In your report you should evaluate each country's program and explain whether it could be applied to other countries.

29. Cooperative Group Project Research human population growth in selected countries on the continents of North America, South America, Asia, Europe, and Africa. Make a presentation to your class describing population trends around the world. In your presentation, be sure to indicate where population growth is most rapid and if population growth is leveling off in any countries. Different group members can take responsibility for researching the population growth in different countries, preparing graphics for the presentation, and presenting the group's findings to the class.

30. Read the article "Sweet Death," in *Natural History*, February 1992, pages 2–6. According to Jared Diamond, what advantage might the genes that cause diabetes have once provided? What lifestyles changes have made these genes harmful?

31. Read the article "The Allure of Symmetry," in *Natural History*, September, 1993, pages 30–36. How did Dr. Thornhill demonstrate that female scorpion flies prefer the most symmetrical males? Suggest a hypothesis that explains why the females have this preference.

A Population Study

Observe the growth and decline in a population of yeast cells and apply the underlying principles to changes in human populations.

PROCESS SKILLS

- sampling population density
- graphing changes in populations
- interpreting data

MATERIALS

- yeast culture
- 1 mL pipette
- test tube
- iodine in dropper bottle
- ruled microscope slide (2 mm × 2 mm)
- safety goggles
- coverslip
- compound microscope

BACKGROUND

1. What is a population?
2. What is carrying capacity?
3. What are some common limiting factors that prevent populations from exceeding their carrying capacity?
4. How do populations of *r*-strategists differ from populations of *K*-strategists?
5. How can populations be sampled to achieve an accurate count?
6. Write your own **Focus Question** on your Vee form.
7. **Knowing Side of the Vee** List the **Concepts** and new **Vocabulary Words** on your Vee Form. In the **Concept Statements** section of the Vee, use these words in sentences that define and explain them.

TECHNIQUE

Doing Side of the Vee

1. Your teacher will transfer approximately 1 mL of the yeast culture to a test tube, and then add 2 drops of iodine to the test tube.

2. Put on safety goggles. Use the 1 mL pipette to transfer 0.1 mL (one drop) from the test tube to a ruled microscope slide. **CAUTION: Work with care—iodine is a poison and an eye irritant.** If you get iodine in your eyes, flush with water and alert your teacher. Carefully lower a coverslip over the drop.

3. Perform the following steps to estimate the total number of yeast cells in 0.1 mL. Using the compound microscope, view your slide under low power. Focus on the yeast cells and notice the black grid lines on the slide. Switch to the 400× objective and align the slide so that you can just see the top left-hand corner of one square, area 1, shown in the figure below.

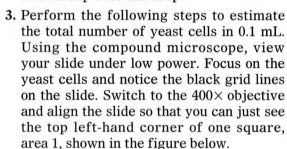

4. Count all the yeast cells in area 1 and record the number. Then move the slide to area 2. Continue counting cells and recording data until you have counted the cells in each of the four areas that make up one square. Add the total number of cells in the square and record this number in the **Records** section of your Vee Form.

5. Switch the microscope to low power and move the slide one square to the left. Under high power, count the cells and record the number.

6. Repeat steps 4 and 5 to count the cells in a total of six squares. Add the total number of cells you counted in the six squares. Calculate the average number of yeast cells in a 2 mm square by dividing the total by 6. Record this initial number in the **Records** section of your Vee Form.

7. Clean up your materials and wash your hands before leaving the lab.

8. Repeat steps 1 through 6 each day for four more days. Record your data for 24, 48, 72, and 96 hours. If possible, repeat steps 1 through 6 again after the weekend.

9. To find the total population of yeast cells in 1 mL (the amount in the test tube), multiply the average number of cells counted in a 2 mm square by 2,500.

10. Graph your data using the values you calculated for the entire test tube population over the five-day period.

11. In the **Procedure** section of the Vee, briefly summarize the procedure you followed.

INQUIRY

1. What effect does iodine have on yeast cells?

2. Why were several areas counted and then averaged each day?

3. From your graphed data, what was the change in population size each day?

4. When was the most rapid growth? the slowest growth?

5. When was the growth peak reached?

6. To find the total number of cells in 1 mL, you multiplied the number of cells you counted in a 2 mm square by 2,500. Show the equation that gives the ratio of the number of cells counted in one square to the total population in 1 mL. (Hint: The cells and liquid formed a layer about 0.1 mm thick on the microscope slide. As a result, each small square held about 0.4 mm^3 of liquid. To find the number of cells in 0.1 mL, you must convert mm^3 to mL. Remember that 1 mL = 1 cm^3 and that 1 cm^3 = 1,000 mm^3.)

7. Use the information on the **Knowing Side** of the Vee to interpret your results from the **Doing Side**, and then write your **Knowledge Claim**. Write a **Value Claim** for this lab.

ANALYSIS

1. How did the yeast population change over time?

2. Why did the population grow?

3. What limiting factors probably caused the yeast population to decline?

4. If you kept counting for several more days, what do you think would happen to the yeast population?

5. What limiting factors that affect yeast also affect human populations?

6. Is yeast an *r*-strategist or a *K*-strategist? Defend your answer.

7. In what ways do population growth and decline in a yeast population resemble growth and decline in a human population? In what ways do they differ?

FURTHER INQUIRY

Write a **New Focus Question** that could be the point of a new investigation. The following is an example:

Is it possible to set up a population of yeast that continues to grow, without declining, for a week? Would it be possible to keep the population size growing indefinitely?

CHAPTER
16

ECOSYSTEMS

REVIEW

- first and second laws of thermodynamics (Section 4-1)
- photosynthesis (Section 5-2)
- cellular respiration (Section 5-3)
- six kingdoms of life (Section 13-1)

One of the last dusky seaside sparrows

16-1 An Introduction to Ecosystems

The coasts of the United States are dotted with salt marshes similar to the one shown in Figure 16-1. Several years ago in one such marsh, a team of biologists captured a little sparrow like the one shown on the opposite page. This bird was unusual because he was a dusky seaside sparrow, the last member of this subspecies alive in nature. He was a ragged bird, no longer in his prime; the biologists who caught him estimated his age as over 10 years. They had caught him, and earlier two other elderly males, in an attempt to preserve some of the genes of this almost extinct subspecies through crossbreeding with another subspecies of seaside sparrow. The other two males died in 1986, and so for a year this sparrow was alone, the last of his kind. With his death on July 14, 1987, the dusky seaside sparrow disappeared forever.

Living Things Are Interdependent

Why be concerned about the loss of a sparrow? There are, after all, lots of kinds of sparrows. Why mourn this one? Few of us would have been able to recognize a dusky seaside sparrow if we had been lucky enough to see one. And what does an extinct sparrow have to do with biology anyway?

The answers to these questions lie at the heart of biology. Although it is easy to think of the environment as external to us—something we always use, sometimes enjoy, and occasionally pollute—this viewpoint ignores all that biologists have learned about life on Earth. In fact, we share the world with all of Earth's other organisms. Together, we and the Earth's other inhabitants weave an intricate tapestry of interactions that defines what our environment is like. Just as removing one resistor from an enormous mainframe computer can alter the interactions of its many components in ways that influence the computer's operation, so removing something as simple as a sparrow from our environment can have many diverse consequences, not all of them easily predictable.

In 1866, the German biologist Ernst Haeckel gave a name to the study of how organisms fit into their environment. He called it *ecology*, from the Greek words *oikos*, meaning "house, or place where one lives," and *logos*, meaning "study of." **Ecology** is the study of the interactions of living organisms with one another and with their physical environment (soil, water, weather, and so on). Our study of

Figure 16-1 A salt marsh in Florida was the home of the last dusky seaside sparrow until his capture. This subspecies of sparrow was driven to extinction by pesticides used to control mosquitoes.

ecology, then, is the study of the house in which we live. The place where a particular population of a species lives is its **habitat.** The many different species that live together in a habitat are called a **community.** A community and all the physical aspects of its habitat—the soil, water, and weather—are called an **ecosystem,** or ecological system.

The Inhabitants of an Ecosystem

Most of us know that tropical rain forests teem with life; an area the size of a football field may contain hundreds of species of plants and thousands of species of insects. But almost any ecosystem is rich with species. **Species diversity,** or **diversity,** is the number of species living within an ecosystem.

To gain some idea of the diversity of ecosystems, consider for a moment a pine forest in the southeastern United States, such as the one illustrated in Figure 16-2. If you could fence in a square kilometer of this forest and then go in and collect every animal, what would you expect to get? Starting with the largest animals, you might find a black bear and almost certainly would find a white-tailed deer. In times past, before human activities drove them away, you might have found a cougar or a red wolf as well. The woods also contain smaller mammals—raccoons, foxes, gray squirrels, rabbits, and chipmunks. Lizards dart among the leaves, while snakes and toads often remain hidden. Beneath the ground, moles tunnel through the soil. A host of large birds can be found, including red-tailed hawks, turkey vultures, turkeys, and quail. The trees are alive with smaller birds: warblers, sparrows, cardinals, and wrens. If the square kilometer included a lake, you might find catfish, bass, perch, a variety of turtles, and perhaps an alligator.

This collection includes only a few of the members of the animal kingdom to be found in this forest ecosystem. The soil contains immense numbers of earthworms and flatworms.

Figure 16-2 Pine forests like this one are common in the southeastern United States.

Figure 16-3 Unlike most other spiders, jumping spiders, *left,* do not build webs. Instead, they lie in wait and pounce on passing insects. Male stag beetles, *right,* use their large jaws to compete for mates.

Hidden under the bark of trees and beneath the pine needles and leaves covering the ground are many different species of arthropods, two of which are shown in Figure 16-3.

Now stop and consider this: the animals you have been collecting belong to only one of the six kingdoms of life. Among the members of the plant kingdom, the pine trees are obvious, but they are by no means the only kind of plant present. There are also a variety of smaller trees and shrubs, and beneath them are vines, ferns, and mosses. Trees in part of the forest may be sheathed in kudzu vine, an introduced species from Asia that covers shrubs and trees with a dense, leafy blanket. Grasses and many kinds of flowers grow on the forest floor.

The fungi constitute a separate kingdom. Most fungi are multicellular organisms. They obtain their food by secreting materials that digest organic material and then absorbing the nutrients. You would find many kinds of fungi growing on fallen trees and spreading as fine threads through the decaying material on the forest floor. Figure 16-4 shows two examples of fungi. The roots of nearly all the kinds of trees will be surrounded by fungi growing in intimate associations called mycorrhizae. Other fungi might be present on the surface of trees or rocks as lichens, which are associations between fungi and algae or cyanobacteria.

A fourth kingdom of life contains the protists, most of which are single-celled and too small to see without a microscope. Protists in a pine forest include the algae and related organisms that grow in fresh water, and the many different kinds of microscopic creatures that inhabit any drop of water.

The fifth and sixth kingdoms of organisms contain bacteria—simple, single-celled organisms that occur everywhere in the forest, although mainly in the soil. A single teaspoon of soil is home to billions of individual bacteria. They play many critical roles in the life of the forest, including changing atmospheric nitrogen gas into a form that organisms can use to build their tissues (you will learn more about this function later in the chapter). In addition, bacteria and fungi are responsible for the decomposition of dead organisms. Without bacteria and fungi, organisms would not decompose, and the

Figure 16-4 Among the fungal residents of pine forests are these mushrooms, *top,* and these shelf fungi, *bottom,* which are digesting the tree on which they are growing.

nutrients contained within their bodies would be unavailable to other organisms. No ecosystem can persist without bacteria and fungi to break down the bodies of its dead members.

If you were to collect and remove every single organism of each of the six kingdoms from your square kilometer, it would be bare, stripped of life. Only the nonliving surroundings would remain—the soil, rocks, and water; the wind that blows over it; and the rain and sunlight that fall upon it. This portion of the forest ecosystem is the physical habitat. The minerals and water that a physical habitat contains and the weather to which it is exposed determine in large measure what kinds of organisms are able to live there.

CONTENT LINK

In **Chapter 19**, you will learn more about the characteristics of the six kingdoms.

Figure 16-5 No ecosystem is completely isolated. This coconut, which may have drifted thousands of kilometers, is sprouting on the beach of a small Pacific island.

What Determines the Boundaries of an Ecosystem?

The physical boundaries of an ecosystem are not always obvious and depend in large measure on how the ecosystem is being studied. For example, a scientist might consider a single rotting log on the forest floor to be an ecosystem if he or she is interested in the fungi and insects living there. Quite often, individual fields, forests, or lakes are studied as isolated ecosystems. In truth, of course, no one location is ever totally isolated from other places. Even distant oceanic islands get occasional migrant visitors—birds blown off course, seeds carried by the wind or waves, and lizards that arrive on floating logs or mats of vegetation.

Natural Changes in Ecosystems

Natural changes in the physical environment of ecosystems happen all the time. When a volcano forms a new island, a glacier recedes and exposes bare soil, or a fire burns all of the vegetation in an area, new habitat is created. This change sets off a process of colonization and ecosystem development. Small, fast-growing plants quickly invade this new habitat. They do not persist, however, because their pioneering efforts make the ground more hospitable for other species. As a result, later waves of plant immigrants soon outcompete and replace the original inhabitants, only to be replaced in turn by still other species that are better able to compete in the new environment.

This regular progression of species replacement is called **succession.** When succession occurs on land where nothing has grown before, it is called **primary succession.** When succession occurs in areas where there has been previous growth, such as in abandoned fields or forest clearings, it is called **secondary succession.** It used to be thought that the stages of succession were predictable and that they always led to the same final community of organisms within any particular ecosystem, a group called a climax community. Ecologists now recognize that chance plays a major role in the competitive interactions that govern each successional change. For this reason, no two successions are exactly alike, although the progression of changes will tend toward similar communities in similar physical conditions. That is why characteristic communities such as tropical rain forest, desert, and tundra occur over large areas of the planet. ◼

❑ *CAPSULE SUMMARY*

Ecology is the study of the interactions of organisms with their living and nonliving environment. An interacting group of organisms and their nonliving environment constitute an ecosystem. Ecosystems change through the process of succession. Succession occurring on newly formed habitat is primary succession. Secondary succession occurs on habitat that has previously supported growth.

An Example of Ecological Succession: Glacier Bay
....

You can better understand succession by looking at a real example. A good choice is a receding glacier, because land is continually being exposed as the face of the glacier moves back. So imagine that you have traveled to the glacier that dominates the head of Glacier Bay, Alaska. The face of the glacier has been melting and receding for the last 200 years, and has moved back some 100 km (62 mi.). A walk across the ground exposed by the receding glacier is in a sense a walk through time—the farther one walks away from the glacier face, the longer the land has been exposed. In Figure 16-6, you can follow the changes that take place as time passes.

The most recently exposed areas, at the face of the receding wall of ice, are piles of bare rock and gravel that lack the usable nitrogen essential to plant and animal life. For about 10 years the piles remain devoid of life. The seeds and spores of the first pioneering plants to colonize this barren landscape—mosses, fireweed, willows, cottonwood, and *Dryas*, a sturdy plant with clumps about 30 cm (1 ft.) across —are carried in by the wind. At first all of these plants grow close to the ground, severely stunted by mineral deficiency. But *Dryas* has mycorrhizae on its roots that supply the nutrients it needs. After a few years, *Dryas* uses this competitive advantage to crowd out the other plants. What remains is a dense mat of *Dryas*.

After about 10 years, seeds of alder arrive, blown in from distant sites. It is a matter of chance when they come, but inevitably some finally make their way to the site, landing on the mat of *Dryas*. Alder roots have nitrogen-fixing nodules, so they are able to grow even more rapidly than *Dryas*. Dead leaves and fallen branches from the alder trees add even more usable nitrogen to the soil, in time allowing the willows and cottonwoods to invade and grow with vigor. After about 30 years, there are dense thickets of alder, willow, and cottonwood that shade and eventually kill the pioneering *Dryas*.

The alder thicket matures until, some 80 years after the glacier first exposed the land, Sitka spruce invades the

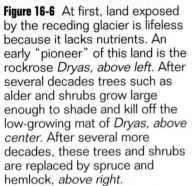

Figure 16-6 At first, land exposed by the receding glacier is lifeless because it lacks nutrients. An early "pioneer" of this land is the rockrose *Dryas, above left.* After several decades trees such as alder and shrubs grow large enough to shade and kill off the low-growing mat of *Dryas, above center.* After several more decades, these trees and shrubs are replaced by spruce and hemlock, *above right.*

thickets. Spruce trees cannot fix nitrogen, but the nitrogen released by the alders enables them to form a dense forest that shades out the alders and willows. In the competition for light needed to carry out photosynthesis, the alders lose, just as *Dryas* did before them.

After the spruce forest is established, hemlock trees arrive at the site. They are very shade-tolerant and have a root system that competes well against spruce for soil nitrogen. Hemlock trees soon become dominant in the forest. This last community of spruce and hemlock proves to be a very stable ecosystem—a climax community. Similar communities are found over broad areas of Alaska, Canada, and Russia.

Climax communities are not permanent creations of nature, fixed and unchangeable. As local climates change, the distribution of particular species within the forest ecosystem may change too. Whatever these fluctuations, it remains true that climax communities are among the most stable of ecosystems, remaining relatively unchanged for long periods of time.

Section Review

1. *What components of an ecosystem are not part of a community?*
2. *What equipment would help an ecologist to obtain a more accurate count of the number of species in the pine forest ecosystem described previously?*
3. *A lawn is not a climax community. Explain why it usually does not go through succession.*

Critical Thinking
4. *Consider a pond ecosystem. In what ways does the surface of the water constitute a boundary for this ecosystem? Describe some ways in which the surface is not a boundary.*

16-2 Energy Flows Through Ecosystems

*E*verything that organisms do in ecosystems—running, breathing, burrowing, growing—requires energy. Of all the factors that organize an ecosystem, determining what organisms and how many of each it will contain, none is more important than the flow of energy. In this section you will learn where organisms get their energy and how this energy moves within an ecosystem.

The Path of Energy: Who Eats Whom in Ecosystems

Energy flows into the biological world from the sun. Life exists on Earth because photosynthesis makes it possible to capture some of the light energy from the sun and transform it into the chemical energy of organic molecules. These organic compounds compose what we call food. The amount of organic material that the photosynthetic organisms of an ecosystem produce is called **primary productivity.**

Primary productivity determines the energy budget of an ecosystem. All of the organisms in an ecosystem are chemical machines driven by the energy captured in photosynthesis. The organisms that first capture energy, the **producers,** include plants, some kinds of bacteria, and algae. Producers make energy-storing molecules. All other organisms in an ecosystem are **consumers,** which obtain the energy to build their molecules by consuming plants or other organisms.

Ecologists assign every organism in an ecosystem to a **trophic level,** which is determined by the organism's source of energy. Energy moves from one trophic level to another, as illustrated in Figure 16-7. The lowest trophic level of any ecosystem is occupied by the producers: plants in most terrestrial ecosystems and algae and bacteria in aquatic ones. Producers not only use the energy of the sun to build energy-rich sugar molecules, but also absorb nitrogen and other key substances from the environment and incorporate them into biological molecules. It is important to realize that plants respire as well as produce. The roots of a plant, for example, do not carry out photosynthesis, because there is no light underground. Roots obtain their energy the same way that you do—by using energy-storing molecules produced elsewhere (in this case, in the leaves of the plant).

At the second trophic level are **herbivores,** animals that eat plants. They are the primary consumers in ecosystems. Cows and horses are herbivores, as are caterpillars and ducks. A herbivore must be able to break down the plant's molecules. Simple sugars and starches present no problem,

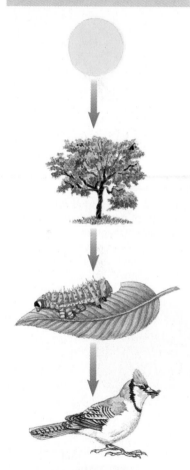

Figure 16-7 The sun is the ultimate source of energy for organisms. Producers, the first trophic level, trap solar energy in organic molecules. Animals at the second trophic level get energy by eating producers, and animals at the third level get energy by feeding on other animals.

Ecosystems **343**

but the digestion of cellulose, a molecule made of sugar units linked together, is a chemical feat that only a few organisms have evolved the ability to perform. Most herbivores rely on helpers. A cow, for instance, has a thriving colony of bacteria in its gut that digests cellulose. Humans cannot digest cellulose because we lack these bacteria. This is why a cow can live on a diet of grass but you cannot.

At the third trophic level are secondary consumers, animals that eat herbivores. Such flesh-eating animals are called **carnivores.** Tigers, wolves, and snakes are carnivores. Some animals, such as bears, are both herbivores and carnivores, eating both plants and animals. They use the simple sugars and starches stored in plants as food, but they cannot digest cellulose. Such animals are called **omnivores.** Humans are omnivores. Many ecosystems contain a fourth trophic level made up of carnivores that consume other carnivores. They are called tertiary consumers, or top carnivores. A hawk that eats a snake is a tertiary consumer. Very rarely do ecosystems contain more than four trophic levels, for reasons that will become clear in a moment.

In every ecosystem there is a special class of consumers called **detritivores,** which include fungal and bacterial decomposers, vultures, and worms. Detritivores obtain their energy from the organic wastes and dead bodies that are produced at all trophic levels. Bacteria and fungi are known as **decomposers** because they cause decay. Decomposition of bodies and wastes releases nutrients back into the environment to be used again by other organisms.

A path of energy through the trophic levels of an ecosystem is called a **food chain,** shown in Figure 16-8. In most ecosystems, energy does not follow simple linear paths because animals feed at several trophic levels. This creates a complicated, interconnected path of energy called a **food web,** illustrated in Figure 16-9.

Figure 16-8 This food chain shows one path of energy flow in an Antarctic ecosystem. The producers in this ecosystem are algae that live near the surface of the ocean.

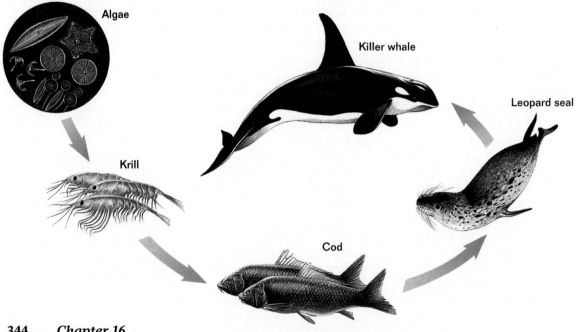

Algae

Killer whale

Leopard seal

Krill

Cod

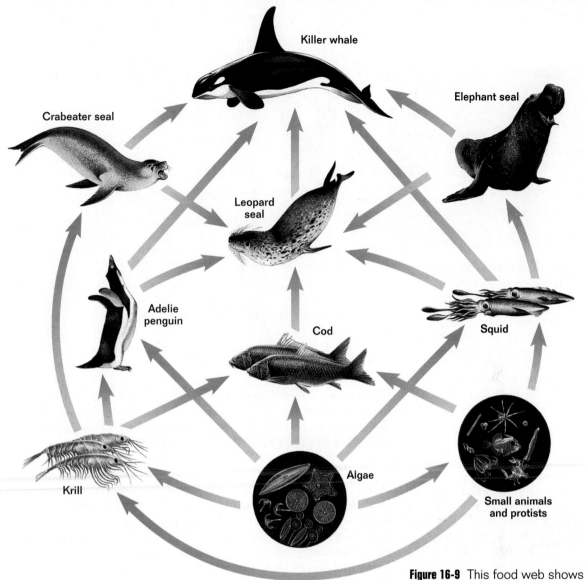

Killer whale

Elephant seal

Crabeater seal

Leopard seal

Adelie penguin

Cod

Squid

Krill

Algae

Small animals and protists

Figure 16-9 This food web shows a more complete picture of the feeding relationships in an Antarctic ecosystem. Notice that the food chain shown in Figure 16-8 is just one strand of this complex food web.

Energy Transfers Between Trophic Levels Are Inefficient

The deer that you see browsing on leaves in Figure 16-10 is busy acquiring energy. There is potential energy in the leaves, stored in the chemical bonds within their molecules. What happens to a leaf's energy after the deer consumes it? You can see what happens to the energy in Figure 16-10. Some of the energy is transformed to other forms of potential energy, such as fat. Another portion accomplishes mechanical work such as running, breathing, and eating more leaves. However, almost half is dissipated to the environment as heat.

Every transfer of energy within an ecosystem dissipates energy as heat. Although heat can be harnessed to do work (as in a steam engine), it is generally not a useful source of

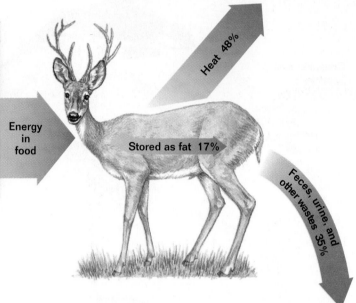

Energy
in
food

Heat 48%

Stored as fat 17%

Feces, urine, and other wastes 35%

Figure 16-10 A deer acquires energy from energy-rich molecules in the vegetation it eats. Some of that energy is stored as fat, and some is lost in wastes—urine and feces. However, most of the energy escapes as heat, which is inevitably produced when energy is transformed from one form to another.

□ *CAPSULE SUMMARY*

Energy transfers between trophic levels are very inefficient. On average, only 10 percent of the energy in any trophic level will be incorporated into the next level.

energy for biological systems. Thus, from a biological point of view, the amount of useful energy available to do work decreases as energy passes through an ecosystem. The loss of useful energy to heat limits how many trophic levels an ecosystem can support. When a plant harvests energy from sunlight to make structural molecules such as cellulose, it stores in chemical bonds only about one-half of the energy it is able to capture, losing the rest as heat. This is the first of many such losses as the energy passes through the ecosystem. When a herbivore uses plant molecules to make its own molecules, only about 10 percent of the energy present in the plant molecules ends up in the herbivore's molecules; 90 percent of the energy is lost. And when a carnivore eats the herbivore, 90 percent of what little energy remains is then lost in making carnivore molecules. At each trophic level, the energy stored by the organisms is about one-tenth of that stored by the organisms in the level below. □

Ecologists often portray the flow of energy through ecosystems by means of a block diagram called a pyramid of energy. In this diagram, each trophic level is represented by a block, and the blocks are stacked atop one another, with the lowest trophic level on the bottom. The width of each block is determined by the amount of energy stored in the organisms at that trophic level. Because the energy stored by the organisms at each trophic level is about one-tenth of that stored by the organisms in the level below, the diagram is a pyramid. In the river ecosystem shown in Figure 16-11, eelgrass and algae are producers; turtles, snails, and caddis flies are herbivores; and beetles, bass, and sunfish are carnivores. The pyramid of energy for this ecosystem is shown in Figure 16-11.

Trophic level

4 — Top carnivore

3 — Carnivores

2 — Herbivores

1 — Producers

Energy Loss Limits the Number of Trophic Levels in an Ecosystem

From what you have learned, you should not be surprised that there is a limit to the number of trophic levels in an ecosystem. Most terrestrial ecosystems involve only three or, rarely, four levels. Too much energy is lost at each level to allow more levels. For example, a large human population could not survive by eating lions captured on the Serengeti Plain of Africa; there are simply too few lions to make this possible. The amount of grass in that ecosystem cannot support enough zebras to maintain a large enough population of lions to feed lion-eating humans. The ecological complexity of the world is thus fixed in a fundamental way by the loss of potential energy that occurs at each trophic level. This loss is a consequence of the second law of thermodynamics, which you studied in Chapter 4.

The loss of useful energy to heat as energy passes from one trophic level to another is a fact of nature that we cannot change. There is an important lesson in the fact that it takes a very large population of zebras and wildebeest to support a small population of lions on the Serengeti Plain (lions are outnumbered by their prey by about 1,000 to 1). As we seek

Figure 16-11 This simple aquatic ecosystem contains four trophic levels. Each trophic level contains about 90 percent less energy than the level below it. Because of space considerations, the trophic levels are not shown to scale here.

Figure 16-12 Because energy transfer between trophic levels is inefficient, about 10 times more grain is required to feed carnivorous humans like the boy eating the steak than to feed herbivorous humans like the girl eating a slice of bread.

ways to maintain ever-larger human populations, we should remember the lion's situation. Humans are omnivores and, unlike the lion, can choose to eat either meat or plants. The choice makes a difference for our future because we eat steak and hamburger only at a great cost in energy. As explained in Figure 16-12, 10 kg of grain are needed to build 1 kg of human tissue if we eat the grain, but 100 kg of grain are needed to build 1 kg of human tissue if a cow eats the grain and we eat the cow.

There are more zebras and wildebeest than lions on the Serengeti Plain because the amount of available energy determines the numbers of animals living in an ecosystem at any one time. However, because some organisms are much bigger than others and therefore use more energy, the numbers of organisms often do not form a pyramid when one compares different trophic levels. In other words, there may be more individuals in a particular trophic level than in the level below it. For instance, caterpillars and other insect herbivores greatly outnumber the trees they feed on. The number of individuals in a trophic level can be an inaccurate indicator of the amount of energy in that level. To better assess the amount of energy present in trophic levels, ecologists usually measure **biomass,** the dry weight of tissue and other organic matter. By collecting, drying, and weighing all of the organisms in each trophic level of an ecosystem, ecologists obtain a pyramid of biomass, with each trophic level containing only about 10 percent of the biomass of the level below it.

Section Review

1. *Suppose you want to create an ecosystem in an aquarium. Explain why your ecosystem would require producers.*
2. *Draw a food web that reflects the feeding relationships in the ecosystem illustrated in Figure 16-11. What information not included in Figure 16-11 would help you to draw a more accurate food web?*
3. *Explain why a given area of land could support more vegetarian humans than omnivorous humans.*

Critical Thinking
4. *How would you modify the food web in Figure 16-9 to include decomposers? Explain your answer.*
5. *In which trophic level would you place humans? Justify your answer.*
6. *Nearly all the mammals that humans eat, including cows, sheep, and goats, are herbivores, not carnivores or omnivores. Explain why a herbivore is a more efficient meat producer from an energetic point of view.*

16-3 Materials Cycle Within Ecosystems

*U*nlike energy, which flows through the Earth's ecosystems in one direction (from sun to producers to consumers), the physical components of ecosystems often cycle constantly. These components include all of the inorganic (noncarbon) substances that make up the soil, water, and air. All materials that cycle through living organisms are important in maintaining the health of ecosystems, but four substances are particularly important: water, carbon, nitrogen, and phosphorus.

Materials Cycle Between Organisms and the Nonliving Environment

The paths of water, carbon, nitrogen, and phosphorus, as they pass from the nonliving environment to living organisms and then back to the environment, form closed circles, or cycles, called biogeochemical cycles. In each biogeochemical cycle, a substance enters living organisms from the atmosphere, water, or soil, resides for a time in the organisms, then returns to the nonliving environment. Ecologists often speak of such substances as cycling within an ecosystem between a living reservoir (organisms that live in the ecosystem) and a nonliving reservoir. In almost all cases, there is much less of the substance in the living reservoir than in the nonliving reservoir.

Figure 16-13 By taking in carbon dioxide during photosynthesis, these trees are participating in the carbon cycle.

The Water Cycle Is Driven by the Sun

Of all the nonliving components of an ecosystem, water has the greatest influence on the ecosystem's inhabitants. To a great degree, the availability of water determines the diversity of ecosystems, as you will see repeatedly in the detailed descriptions of the world's principal ecosystems presented in the next chapter.

Water cycles within ecosystems in two ways, each of which is driven by the sun. In the nonliving portion of the water cycle, water vapor in the atmosphere condenses and falls to the Earth's surface as rain or snow. A portion of this water seeps into the soil and becomes, for a time, part of the **ground water,** which is water retained beneath the surface of the Earth. Most of the remaining water that has fallen to the Earth does not stay long at the surface. Instead, heated

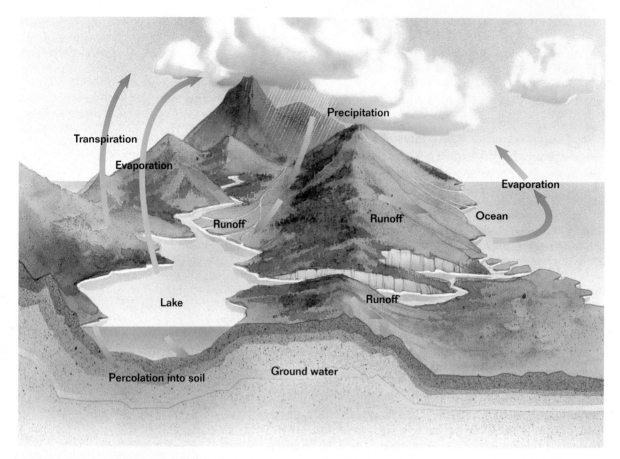

Figure 16-14 This diagram shows the major steps in the water cycle.

by the sun, it reenters the atmosphere by evaporation from lakes, rivers, and oceans. The path of water within an ecosystem is illustrated in Figure 16-14.

In the living portion of the water cycle, water is taken up by the roots of plants. After passing through a plant, the water moves into the atmosphere by evaporating from the leaves, a process called transpiration. Transpiration is also a sun-driven process: the sun heats the Earth's atmosphere, creating wind currents that draw moisture from the tiny openings in the leaves of plants.

In aquatic ecosystems (lakes, rivers, and oceans), the nonliving portion of the water cycle predominates; almost all of the water that falls to the Earth returns to the atmosphere by evaporation. In terrestrial ecosystems, the nonliving and living parts of the water cycle play important roles. In thickly vegetated ecosystems, such as tropical rain forests, more than 90 percent of the moisture in the ecosystem passes through plants and is transpired from their leaves. In a very real sense, these plants create their own rain. Moisture travels from plants to the atmosphere and falls back to the Earth as rain. ◻

When forests are cut down, the water cycle is disrupted and less moisture is returned to the atmosphere. Some of the water drains into rivers and ultimately into the sea instead of

◻ **CAPSULE SUMMARY**

In the water cycle, water falls to the surface as precipitation. Some water returns to the atmosphere through evaporation. Water also collects in lakes, rivers, and oceans. Some water seeps into the soil and joins the ground water.

rising to the clouds and falling again on the land. Because less water cycles within an ecosystem after the trees are cut down, extensive cutting can convert lush forests into wasteland with too little rain to support the return of the forest, as shown in Figure 16-15. It is a tragedy of our time that just such a transformation is occurring in many tropical areas.

Figure 16-15 Removal of the trees from this area in Brazil disrupted the water cycle, reduced rainfall, and left the soil without the protection normally provided by roots. The rain that did fall carried off the thin layer of topsoil, further diminishing the chances that vegetation could return.

The Carbon Cycle Is Linked to the Flow of Energy

Carbon also cycles between the nonliving environment and living organisms. You can follow the carbon cycle in Figure 16-16. The Earth's atmosphere is about 0.035 percent carbon dioxide. Carbon dioxide in the air or dissolved in water is used by photosynthesizing plants, algae, and bacteria as a raw material to build organic molecules. In effect, they trap the carbon atoms of carbon dioxide within the living world. Carbon atoms return to the pool of carbon dioxide in the air and water in three ways.

One way is through cellular respiration. Nearly all living organisms, including plants, perform cellular respiration. They use oxygen to oxidize organic molecules during cellular respiration, and carbon dioxide is a byproduct of this reaction.

Carbon also returns to the atmosphere through combustion, or burning. Much carbon is contained in wood and may stay there for many years, returning to the atmosphere only when the wood is burned. Sometimes carbon can be locked away beneath the Earth for a very long time—thousands or

Figure 16-16 This diagram shows the major steps of the carbon cycle. Humans are disrupting the carbon cycle by burning fossil fuels and by destroying vegetation that would have absorbed carbon dioxide.

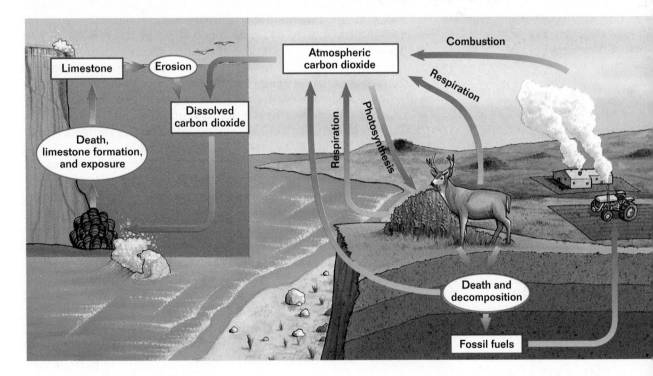

*Carbon enters the living portion of
the carbon cycle through photosynthesis. Organisms release carbon through cellular respiration.
Carbon long trapped in rock and
fossil fuels is freed by erosion and
burning, respectively.*

CONTENT LINK

*Over the last 150 years, the burn-
ing of fossil fuels has led to a
30 percent increase in the concentration of carbon dioxide in the
atmosphere. You will find out
about the possible consequences
of this atmospheric change in*
Chapter 18.

even millions of years. The remains of organisms that
become buried in sediments may be gradually transformed
by heat and pressure into fossil fuels—coal, oil, and natural
gas. The carbon contained in these remains is released back
into the atmosphere only when the fossil fuels are burned.

Erosion releases carbon as well. Marine organisms
extract a substantial amount of the carbon dioxide dissolved
in sea water and use it to build their calcium carbonate
shells. When these marine organisms die, their shells sink to
the ocean floor, become covered with sediments, and form
limestone. Eventually, as the limestone becomes exposed
and erodes, the carbon in it is returned to the pool of available carbon. This process takes millions of years. ◾

The Nitrogen and Phosphorus Cycles

Organisms contain large amounts
of nitrogen because proteins and
nucleic acids are both nitrogen-
rich. The atmosphere is 79 percent nitrogen gas, N_2. However,
organisms obtain that nitrogen only with great difficulty;
most organisms are unable to use the nitrogen gas that is so
plentiful in the atmosphere. The two nitrogen atoms in a molecule of nitrogen gas are connected by a particularly strong
triple covalent bond that is very difficult to break. The pro-
fusion of life that blankets our globe is possible only because
a few bacteria have enzymes that can break this triple bond
and bind the nitrogen atoms to hydrogen, forming ammonia,
NH_3. The process of combining nitrogen gas with hydrogen
to form ammonia is called **nitrogen fixation.** Bacteria
evolved the ability to fix nitrogen early in the history of life,
before photosynthesis had introduced oxygen gas into the
Earth's atmosphere. Today, nitrogen fixation occurs only in
the absence of oxygen; even a trace of oxygen poisons the
process. In today's world, awash with oxygen, nitrogen-
fixing bacteria live in the soil within capsules that admit no
oxygen or within swellings, or nodules, on the roots of beans,
alder trees, and a few other kinds of plants.

The nitrogen cycle, diagramed in Figure 16-17 on the
next page, is a complex process with four important stages.

1. **Assimilation**   The ammonia produced by nitrogen-fixing
   bacteria spreads through the soil and is picked up by
   plants, which use the nitrogen atoms to build proteins,
   nucleic acids, and a variety of other nitrogen-containing
   molecules. When animals eat these plants, they use the
   nitrogen to build their own molecules. Nitrogen assimilation is the absorption and incorporation of nitrogen into
   plant and animal compounds.

2. **Ammonification**   Many animals excrete excess nitrogen
   in their urine as urea or uric acid, which a second kind of
   soil bacterium converts back into ammonia. When an

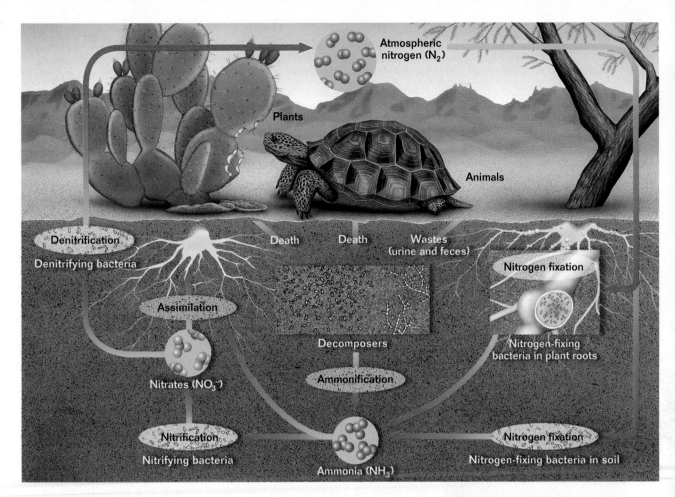

Atmospheric nitrogen (N₂)

Plants

Animals

Denitrification
Denitrifying bacteria

Assimilation

Nitrates (NO₃⁻)

Nitrification
Nitrifying bacteria

Death   Death   Wastes
(urine and feces)

Decomposers

Ammonification

Ammonia (NH₃)

Nitrogen fixation

Nitrogen-fixing
bacteria in plant roots

Nitrogen fixation
Nitrogen-fixing bacteria in soil

organism dies, these bacteria also convert the nitrogen compounds in the dead tissue to ammonia. Some of this ammonia is then reabsorbed by plants. Ammonification is the production of ammonia by bacteria during the decay of nitrogen-containing organic matter.

3. **Nitrification**   Some of the ammonia in the soil is converted by several kinds of bacteria to nitrate, NO_3^-. Some nitrate is then absorbed by plants, which can assimilate nitrate as well as ammonia. Nitrification is the production of nitrate from ammonia.

4. **Denitrification**   Another kind of bacterium acts on the remaining nitrate, converting it back into nitrogen gas. The nitrogen gas is then released into the atmosphere, completing the nitrogen cycle. Denitrification is the conversion of nitrate to nitrogen gas. ❏

The growth of plants in ecosystems is often limited by the availability of nitrate and ammonia in the soil, which is one of the reasons farmers fertilize fields. Today most of the ammonia and nitrate added to the soil by farmers is not organic. It is produced in factories by industrial, rather than bacterial, nitrogen fixation. The amount of nitrogen gas converted to ammonia or nitrates in chemical factories is immense, amounting to perhaps 30 percent of the total

Figure 16-17 Bacteria carry out many of the important steps in the nitrogen cycle, including the conversion of atmospheric nitrogen into a usable form, ammonia.

❏ **CAPSULE SUMMARY**

Nitrogen-fixing bacteria living in the soil or in plant roots transform nitrogen gas into ammonia. Plants lacking these bacteria take up ammonia and nitrate from the soil. Animals get nitrogen by eating plants or other animals.

nitrogen that cycles in biological systems worldwide. As you learned in Chapter 10, scientists are very interested in using genetic engineering to place the nitrogen-fixing genes of bacteria into the chromosomes of crop plants. If these attempts succeed, the plants themselves will be able to fix nitrogen, eliminating the need for nitrogen-supplying fertilizers.

The Phosphorus Cycle

Phosphorus is an essential element of all living organisms. It is a key part of both ATP and DNA. Phosphorus is usually present in soil and rock as calcium phosphate, which dissolves in water to form phosphate ions, PO_4^{3-}. This phosphate is absorbed by the roots of plants and used to build organic molecules such as ATP and DNA. Animals that eat the plants reuse the organic phosphorus. When the plants and animals die and decay, bacteria in the soil convert the phosphorus in organic molecules back into PO_4^{3-}.

Phosphorus is sometimes transferred from one ecosystem to another. For example, phosphorus from a forest ecosystem might be carried down a river to the sea, where it would be trapped in sediments that would eventually turn into rock. Millions of years later the sea floor might rise and expose the rock, and the phosphorus would be released by weathering and made available for living organisms once again.

The phosphorus level in freshwater lake ecosystems is often low, which prevents much growth of photosynthetic algae in these ecosystems. Phosphorus added to lakes as a result of human activities (some detergents and fertilizers are rich in phosphorus) can have disastrous consequences. Fertilization of a lake by the continual, inadvertent addition of phosphorus to its waters produces a green scum of algal growth on the surface of the lake that will eventually kill most of the organisms in the lake. After an initial "bloom" of rapid algal growth, the algae begin to die. Bacteria that feed on the dead algal cells use up so much of the lake's dissolved oxygen that fishes and invertebrate animals suffocate, as shown in Figure 16-18.

Figure 16-18 These fish suffocated after fertilizer, washed into the lake by runoff from farmland, caused a population explosion of algae. Bacteria that decomposed the algae used up the oxygen in the lake.

Section Review

1. *Explain how cutting down a rain forest disrupts the water cycle.*
2. *The burning of rain forests leads to increased atmospheric carbon dioxide levels in two ways. Explain how.*
3. *Describe three functions performed by bacteria in the nitrogen cycle.*

Critical Thinking
4. *Nutrients can be reused, but energy cannot. Explain why.*

Vocabulary

biomass (348)
carnivore (344)
community (338)
consumer (343)
decomposer (344)
detritivore (344)
diversity (338)
ecology (337)

ecosystem (338)
food chain (344)
food web (344)
ground water (349)
habitat (338)
herbivore (343)
nitrogen fixation (352)
omnivore (344)

primary productivity (343)
primary succession (340)
producer (343)
secondary succession (340)
species diversity (338)
succession (340)
trophic level (343)

Concept Mapping

Construct a concept map that describes the role of organisms in the flow of energy through an ecosystem. Use as many terms as needed from the vocabulary list. Try to include the following items in your map: trophic level, food web, food chain, producer, consumer, primary productivity, carnivore, detritivore, herbivore. Include additional terms in your map as needed.

Review

Multiple Choice

1. Extinction of the dusky seaside sparrow is noteworthy because
   a. it was the last remaining subspecies of seaside sparrow.
   b. of its possible effects on other organisms in the same ecosystem.
   c. bird-watchers will never again see the dusky seaside sparrow.
   d. other organisms will fill the habitat that it left vacant.

2. What critical role is played by fungi and bacteria in any ecosystem?
   a. primary production
   b. decomposition
   c. boundary setting
   d. physical weathering

3. A mountain lion is a(n)
   a. omnivore.   c. detritivore.
   b. herbivore.   d. carnivore.

4. Plants are called producers, rather than consumers, because they
   a. produce carbon dioxide when they respire.
   b. capture energy from the sun and build carbohydrates.
   c. produce leaves and roots in the spring.
   d. use the energy they produce in photosynthesis.

5. Which sequence shows the correct order of succession at Glacier Bay, Alaska?
   a. *Dryas*, alder, Sitka spruce
   b. mosses, hemlock, Sitka spruce
   c. *Dryas*, hemlock, alder
   d. alder, *Dryas*, hemlock

6. Assume that the energy stored by organisms at one trophic level is about 10 percent of that stored by organisms in the level below. How much energy would be available to organisms at the third trophic level of an energy pyramid if the energy stored by organisms at the first trophic level were 1,000 kcal?
   a. 1,000 kcal   c. 10 kcal
   b. 100 kcal   d. 1 kcal

7. In terrestrial ecosystems, plants return water to the atmosphere by
   a. condensation.   c. assimilation.
   b. transpiration.   d. desiccation.

8. Humans are adversely affecting the carbon cycle by
   a. planting trees.
   b. burning fossil fuels.
   c. mining limestone.
   d. breathing.

9. Which role is not performed by bacteria in the nitrogen cycle?
   a. converting nitrogen gas to ammonia
   b. changing urea to ammonia
   c. turning nitrates into nitrogen gas
   d. converting nitrates to uric acid

Completion

10. A(n) _____ includes the interactions among living organisms and everything that affects their lives. The study of these interactions was named _____ by Ernst Haeckel.

11. Weeds growing in a recently burned patch of forest represent the first stage in _____ , while lichens that colonize a newly formed volcanic island are the first stage of _____ .

12. Cows, which eat only plants, are called _____ , but humans, who eat both plants and animals, are called _____ .

13. A food chain is used to describe the passing of _____ from one organism to another. An organism's position in the food chain is called its _____ level.

14. A linear pathway that describes what organisms eat is called a food _____ . When all of the linear pathways in an ecosystem are linked together, they form a food _____ .

15. Carbon enters living things as _____ during photosynthesis. Carbon is returned to the nonliving environment by the burning of _____ .

16. Phosphorus is an essential element in many biologically important chemicals, such as _____ and DNA.

Short Answer

17. Picture a southeastern pine forest. How would it change if all bacteria and fungi were removed?

18. Explain why alders, willows, and cottonwoods replace *Dryas* at Glacier Bay, Alaska.

19. Food chains usually consist of no more than three or four trophic levels. Explain why.

20. Ecological pyramids can be constructed using biomass, numbers of individuals, and energy. Typically, the greatest amount of biomass, number of individuals, or amount of energy is found in the lowest trophic level, that of the producers. However, there are occasionally fewer individuals in the lowest trophic level than in the level above. Explain how this can occur.

21. Name two major processes involved in the carbon cycle.

22. Name the four main stages of the nitrogen cycle, and describe what occurs at each stage.

Themes Review

23. **Flow of Energy** Reptiles, such as the Komodo dragon shown below, require about one-tenth as much food as mammals, which have much faster metabolic rates. On the Serengeti Plain of Africa, the ratio of lions to their prey is about 1 to 1,000. If Komodo dragons instead of lions were the top predators on the Serengeti Plain, would the ratio of predators to prey be higher, lower, or the same? Explain your answer.

24. **Evolution** Humans, raccoons, and bears are omnivores. What adaptive advantage might this feeding strategy provide?

25. **Structure and Function** Explain why the nitrogen-fixing bacteria associated with plants live in airtight nodules on the plants' roots.

Critical Thinking

26. **Explaining Observations** A scientist measuring atmospheric carbon dioxide levels in a Michigan deciduous forest (deciduous trees, such as elms and maples, shed their leaves in the fall) finds that the carbon dioxide levels fluctuate during the year, as shown in the graph below. Using your knowledge of the carbon cycle, explain the cause of these fluctuations.

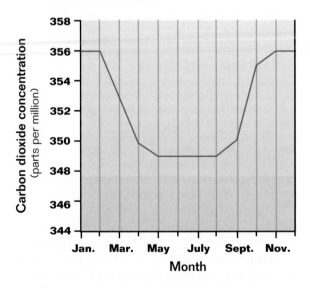

27. **Making Predictions** How would the water cycle change if all living organisms became extinct?

28. **Making Inferences** Energy flow in an ecosystem is described in terms of the activities of producers, consumers, and decomposers. What is the role of each of these in the carbon cycle?

Activities and Projects

29. **Research and Writing** Write a report that describes how ecologists are using stable-isotope tracing to construct food webs without observing feeding behavior. In your report, discuss the nature of the sample needed for analysis, and discuss the long-standing assumption that was questioned by ecologists as a result of data collected using stable-isotope tracing.

30. **Cooperative Group Project** Build your own ecosystem in an aquarium or terrarium. Decide as a group what kind of ecosystem you will build. Then assign each group member the task of collecting representative producers, consumers, or decomposers for the ecosystem. Finally, as a group, construct a food web for the ecosystem.

Readings

31. Read the article "The Recovery of Spirit Lake," in *American Scientist*, March–April 1993, pages 166–177. What was the condition of Spirit Lake before and soon after the eruption of Mount St. Helens in 1980? How has Spirit Lake changed since the eruption?

32. Read the article "Ecosystems For Industry," in *New Scientist*, February 5, 1994, pages 21–22. How are current industries and ecosystems different? How would we all benefit if industries functioned more like ecosystems?

33. Read the article "Why American Songbirds Are Vanishing," in *Scientific American*, May 1992, pages 98–104. What types of ecological damage explain the population decline of many bird species? What can be done to increase the numbers of songbirds in North America?

LABORATORY Investigation   Chapter 16

Ecosystem in a Jar

OBJECTIVE

Observe the interaction of organisms in a closed ecosystem, and compare this ecosystem with others observed in nature.

PROCESS SKILLS

- observing a model ecosystem
- recognizing relationships among components of an ecosystem

MATERIALS

- large glass jar with lid
- pond water or dechlorinated tap water
- gravel, rocks, and soil

Ecosystem 1
- pinch of grass seeds
- pinch of clover seeds
- 10 mung bean seeds
- 3 earthworms
- 4–6 isopods
- 6 mealworms
- 6 crickets

Ecosystem 2
- Strands of *Anacharis, Fontinallis,* and foxtail
- duckweed
- *Chlamydomonas* culture
- black ram's horn snail
- 4 guppies or platys

Ecosystem 3
- pinch of grass seeds
- pinch of clover seeds
- pond snails
- 2 *Anacharis* strands
- 10 *Daphnia*
- 3 *Fontinallis* strands
- *Chlamydomonas*
- 4 guppies or platys
- graph paper
- 8½ × 11 in. acetate sheets
- several colors of pens for overhead transparencies

BACKGROUND

1. How are living things affected by nonliving things in the environment?
2. How do different types of organisms interact with one another?
3. Write your own **Focus Question** on your Vee Form.
4. **Knowing Side of the Vee** List the **Concepts** and new **Vocabulary Words** on your Vee Form. In the **Concept Statements** section of the Vee, use these words in sentences that define and explain them.

TECHNIQUE

Doing Side of the Vee

Part A: Experimental Setup

1. In this investigation, you will be observing organisms in one of three different ecosystems. Look at the organisms listed in the materials list for each of the three systems, and choose the system you would like to observe. Think about the organisms that will live in your ecosystem. Which ones might be the most numerous? Which ones might decrease in number? Hypothesize how the organisms will interact.

2. Form a group with classmates who have chosen the same ecosystem. As a group, prepare the environment in the jar with a chosen substrate. The substrate for Ecosystem 1, a land environment, is rocks and soil. Ecosystem 2, a water environment, requires gravel and water. Ecosystem III requires both land and water environments. You will need to set up the water environment in a small dish that fits inside the larger container.

3. Plant the seeds and/or add the plants and/or algae to your ecosystem. Put a lid on the container and let the ecosystem remain undisturbed in indirect sunlight for a week.

4. Clean up your materials and wash your hands before leaving the lab.

Part B: Observing Your Ecosystem

5. Place the chosen animals into the jar and lightly close it.

6. Observe the jar for a few minutes daily.

7. Make a chart in the **Records** section of your Vee Form to record the number of each species in your original ecosystem. Record daily any changes you observe.

8. Make a graph for each species in your chart, plotting the number of organisms as a function of time. Place a clear acetate sheet over each graph. Using a pen for overhead transparencies, trace each graph onto an acetate sheet. Use a different color and a different acetate sheet for each organism.

9. Compare the acetate sheets of two organisms that you hypothesized would interact—a predator and its prey, for example. Hold one sheet on top of the other and analyze both graphs. Record the results in the **Records** section of your Vee Form. In the **Procedure** section of the Vee, briefly summarize the procedure you followed.

10. Clean up your materials and wash your hands before leaving the lab.

INQUIRY

1. What happened to the organisms in your ecosystem?

2. What are some possible causes of the changes in the populations you observed?

3. Construct a food chain for the ecosystem you observed.

4. What could be learned if more than one jar was set up in an identical manner?

5. How does your ecosystem resemble a natural ecosystem? How does it differ?

6. Use the information on the **Knowing Side** of the Vee to interpret your results from the **Doing Side,** and then write your **Knowledge Claim.** Write a **Value Claim** for this lab.

ANALYSIS

1. How did your observation compare with your hypothesis? If the results differed from what you expected, explain what might have caused the difference.

2. Looking at your graphs, what kind of relationship can you find between predator and prey populations?

3. How would you modify the ecosystem if you were to repeat this investigation?

FURTHER INQUIRY

Write a **New Focus Question** that could be the point of a new investigation. The following are examples:

What are the effects of certain abiotic factors—including temperature, light, and moisture—on the organisms in an ecosystem?

How could a scientist set up an experiment to find out how certain pesticides or fertilizers might affect an ecosystem?

CHAPTER 17

BIOLOGICAL COMMUNITIES

REVIEW

• community (Section 16-1)
• habitat (Section 16-1)
• ecosystems (Section 16-1)
• primary productivity (Section 16-2)
• energy flow in ecosystems (Section 16-2)

A desert community in Arizona

17-1 How Organisms Interact in Communities

You can view the community of organisms that inhabits an ecosystem as a web of interactions. Each species competes with other species for resources. Many species cooperate with other species, and every animal species feeds on some species and is fed on by others. As you will read in this chapter, these interactions between species play major roles in shaping ecosystems.

Section Objectives

■ Describe coevolution of predators and prey.
■ Identify one example of competition.
■ Contrast mutualism with commensalism.

Interacting Species Evolve in Response to One Another

The interactions between species in ecosystems today are the result of a long evolutionary history in which the participants have adjusted to one another. Thus, plants evolved flowers that promoted efficient dispersal of their pollen by insects and other animals, and pollinators evolved a number of traits that enabled them to obtain food or other resources efficiently from the flowers of the plants they pollinate. Natural selection has often led to a close match between the characteristics of flowers and their pollinators, as you can see in Figure 17-1. Back-and-forth evolutionary adjustments between interacting members of an ecosystem are called **coevolution.**

Figure 17-1 With its long tongue, the hawk moth is able to reach the nectar deep within these flowers.

Coevolving in Opposition: Predators and Prey

Predation occurs when one organism feeds on another, as the shark shown in Figure 17-2 is doing. Examples of predation include familiar situations such as lions eating zebras and snakes eating mice. A special case of predation is **parasitism,** in which one organism feeds on and usually lives on or in another, typically larger, organism. Parasites often do not kill their prey (known as the "host"), since they depend on it for food and a place to live, and as a means to transmit their offspring to new hosts. Among the parasites that may have fed on you at some time are ticks, mosquitoes, and fleas.

How would you expect predators and prey to coevolve? To answer this question, you must consider the effect of predation on both participants. The predator benefits from

Figure 17-2 This blacktip shark is devouring a tuna.

feeding on its prey. The prey, in contrast, is always harmed. At best, it only loses some tissue, which requires energy to replace. At worst, it is killed. Therefore, you would expect the prey to evolve ways to elude, avoid, or fight off predators. In response, the predator should evolve countermeasures to the defenses of its prey. Next, you will follow a real example of predator-prey evolution, the coevolution of plants and herbivores (plant-eating animals).

How Plants Defend Themselves From Herbivores

The most obvious ways that plants protect themselves from herbivores are with thorns, spines, and prickles. But many chemical compounds that occur in plants are much more widely distributed and even more important in defending the plants in which they occur. Virtually all plants contain defensive chemicals called **secondary compounds**. Estimates suggest that perhaps 50,000 to 100,000 different secondary compounds may exist; the chemical structures of some 15,000 of these compounds already have been characterized.

As a rule, each group of plants makes its own special kind of defensive chemical. For example, the mustard plant family produces a characteristic group of chemicals known as mustard oils. These oils give pungent aromas and tastes to such plants as mustard, cabbage, radish, and horseradish. The same tastes that we enjoy signal the presence of chemicals that are toxic to many groups of insects.

Figure 17-3 Defensive chemicals of the poison ivy plant, *right*, cause an itchy rash, *above*.

Some of the best-known plants with defensive chemicals are the species of the genus *Toxicodendron*, which includes poison ivy and poison oak. These plants produce a gummy oil called urushiol (*OO roo shee awl*) that causes a severe, itchy rash in susceptible people, as shown in Figure 17-3. This irritating substance almost certainly functions to protect the plants that produce it.

How Herbivores Overcome Plant Defenses

For each group of plants that is protected by a particular kind of defensive chemical, there are certain groups of herbivores that are able to feed on these plants, often as their exclusive food source. For example, the larvae of cabbage butterflies feed almost exclusively on plants of the mustard and caper families, which are defended from other herbivores by mustard oils. How do these animals manage to avoid the chemical defenses of the plants? Cabbage butterflies have evolved the ability to break down mustard oils and thus feed on mustards and capers without harm. This is one example of predator-prey coevolution. ▢

▢ CAPSULE SUMMARY

In a predator-prey interaction, the prey is harmed while the predator benefits. Thus, prey often evolve ways to escape being eaten, and predators evolve ways to overcome the defenses of the prey. For instance, many species of plants produce chemicals to protect them from herbivores. Many species of herbivores have evolved ways to break down these chemicals.

Competition: Common Use of Scarce Resources

When two species use the same resource, they are said to compete, and their interaction is called **competition.** Resources for which species compete include food, nesting sites, living space, light, mineral nutrients, and water. For competition to occur, however, the resource must be in short supply. In Africa, for example, lions and hyenas compete for prey. Fierce rivalry between these species can lead to battles that cause injuries to both sides. Though the term *competition* brings such battles to mind, most competitive interactions do not involve fighting. In fact, some competing species never encounter one another. They interact only by means of their effects on the abundance of resources. In the next section, you will learn how competition affects the nature of communities.

Coevolving in Cooperation

In **symbiosis** (*sihm beye OH sihs*), two or more species live together in a close, long-term association. The two types of symbiotic relationships are **mutualism,** in which both participating species benefit, and **commensalism,** in which one species benefits and the other is neither harmed nor helped. Cooperation is the hallmark of symbiosis. Examples of symbiosis include lichens, which are associations of certain fungi with green algae or cyanobacteria, and mycorrhizae, which are associations of fungi and the roots of plants.

One well-known instance of mutualism involves ants and aphids, as illustrated in Figure 17-4. Aphids are small insects that use their piercing mouthparts to suck fluids from the sugar-conducting vessels of plants. They extract a certain amount of the sucrose and other nutrients from this fluid. However, much of the fluid—so-called honeydew—runs out

Figure 17-4 The small green insects on this plant stem are aphids. They are pampered and protected by their ant guards, which feed on the sugary fluid the aphids secrete.

Figure 17-5 The clown fish can survive the stings of the sea anemone. Because it can live among the anemone's tentacles, the clown fish is protected from predators and can feed on the leftovers of its host's meal. The anemone is apparently unaffected by the fish.

in an altered form through their anus. Certain ants have taken advantage of this habit, in effect domesticating the aphids. The ants move the aphids from plant to plant, like a herd of dairy cows, "milking" them for the honeydew, which the ants use as food.

Among the best-known examples of commensalism are the relationships between certain small tropical fishes and sea anemones, marine animals that have stinging tentacles. These fishes, such as the clown fish shown in Figure 17-5, have evolved the ability to live among the tentacles of the sea anemones, even though these tentacles would quickly paralyze other fishes that touched them. The clown fishes feed on the leftovers from the meals of the host anemone.

Section Review

1. *Predator-prey coevolution has been described as an "arms race." Explain why this is an apt description.*

2. *Describe some ways that humans compete with other organisms.*

3. *Would you expect the unaffected member of a commensal relationship to evolve in response to the other member? Explain your answer.*

Critical Thinking

4. *The relationship between a hawk and the tree in which it nests may be a case of commensalism. However, the tree may benefit from the hawk's presence, perhaps gaining protection from herbivores. Describe an experiment to determine whether the hawk-tree relationship is mutualism or commensalism.*

17-2 How Competition Shapes Communities

Every biological community is part of an ecosystem, which encompasses not only living organisms but also the physical environment in which they live. In the previous section, you learned that a biological community is composed of organisms that interact in ways shaped by a long history of coevolution. Interactions among species have molded ecosystems over long periods of time, creating the world you see today. What determines which species live where? Why, for example, does the jaguar shown in Figure 17-6 live in Central America but not in Canada? To some degree, accident and history play a role. To a greater degree, however, biological interactions and the characteristics of the physical environment determine what animals and plants are able to live in a particular ecosystem. This section examines the role of biological interactions in shaping the nature of communities; the role of the physical environment is investigated in the next section.

Section Objectives

- Understand what a niche is.
- Distinguish between fundamental and realized niches.
- Contrast character displacement with competitive exclusion.
- Explain why competitive exclusion is not inevitable.

Figure 17-6 The jaguar is the largest cat native to the Americas. It once ranged from Argentina through Central America to central Texas and southern Arizona. Humans have eliminated the jaguar from much of its former range through over-hunting and habitat destruction. It is no longer found in the United States and is rare in Central America.

Each Species in a Community Plays a Role

To understand how biological interactions influence the makeup of communities, you must focus on the day-to-day events within the community—how organisms get food, where they live, and so on. The functional role of a particular species in an ecosystem is called its **niche.** A niche is how an organism lives, the "job" it performs within the ecosystem. A niche is often described in terms of how the organism affects energy flow within the ecosystem. Viewed in this way, the niche of a shrub growing in a forest

Biological Communities 365

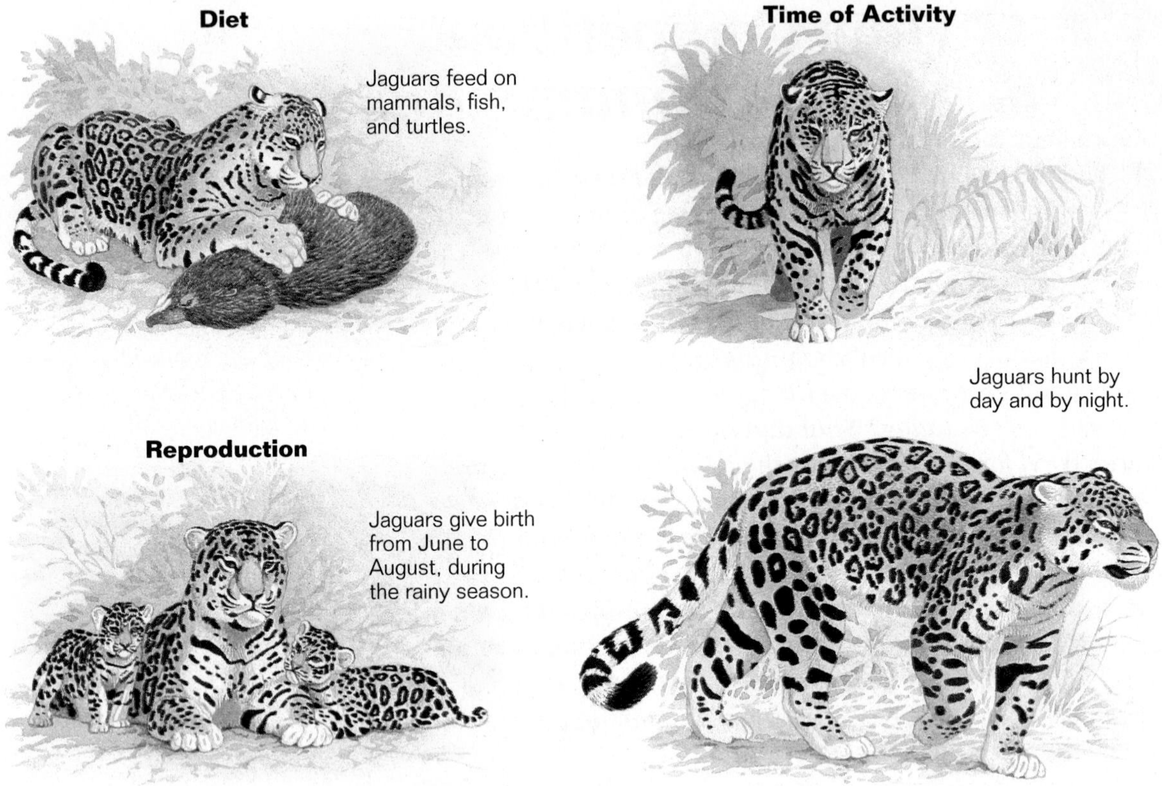

Diet

Jaguars feed on mammals, fish, and turtles.

Time of Activity

Jaguars hunt by day and by night.

Reproduction

Jaguars give birth from June to August, during the rainy season.

Figure 17-7 All the ways in which the jaguar interacts with its environment—including where it lives, what it eats, when it is active, and when it reproduces—constitute its niche.

meadow is that of a producer, while the niche of a deer that eats the shrub is that of a herbivore, and the niche of a cougar that eats the deer is that of a carnivore. Niches can also be described in terms of how organisms use space. For instance, arthropod carnivores feed at different heights from the ground: wolf spiders hunt their insect prey within leaf litter on the forest floor, ladybird beetles hunt on the stems of plants, and dragonflies hunt in the air. Within every ecosystem, each organism has its own niche, its particular way of making a living and influencing its surroundings. Figure 17-7 summarizes some aspects of the jaguar's niche in the Central American rain forest.

A Closer Look at Niches

To gain a better understanding of what a niche is, you must look more closely at a particular species, say a Cape May warbler (a small, insect-eating songbird) flying in a forest and landing to search for dinner in a spruce tree. In a broad sense, the niche of this bird is the temperature and humidity it prefers (this species summers almost exclusively in the northeastern United States and Canada), the time of year it nests (mid-summer), what it likes to eat (small insects), and where on the spruce tree it finds its food (high on the tree at the tips of the branches). The total niche an organism is potentially able to occupy within an ecosystem, the entire range of conditions it can tolerate, is its **fundamental niche.**

Now reconsider what the warbler is doing. It feeds mainly at the very top of the spruce tree. Why is this? Insects that the warbler could eat are located all over the tree, but this bird, like most other Cape May warblers, centers its attention on insects in the upper branches. Seeming to defy good sense, Cape May warblers occupy only a portion of their fundamental niche.

Closer study of the Cape May warbler and its potential competitors reveals that this surprising behavior is part of a larger pattern of niche restriction. In the late 1950s, Robert MacArthur, an ecologist from Princeton University, carried out a classic investigation of the feeding habits of five warbler species. MacArthur found that all five species fed on insects in the same spruce trees at the same time but that each species concentrated on a different part of the tree, as shown in Figure 17-8. Some of the species of warblers fed on insects near the ends of the branches, while others regularly penetrated well into the foliage. Some, such as the Cape May warbler, stayed high on the trees, while others, such as the myrtle warbler, hunted on the lower branches. Thus, although

Figure 17-8 In the forests of the northeastern United States, these five warblers are potential competitors because they feed on insects in the same trees. However, each species hunts only in a portion of the tree. Ecologist Robert MacArthur proposed that this feeding strategy reduced competition among these species of warblers.

Cape May warbler

Blackburnian warbler

Black-throated green warbler

Bay-breasted warbler

Myrtle warbler

all five species of warbler had very similar fundamental niches, they did not actually use the same resources. In effect, they divided the range of resources among them, each taking a different portion.

The part of its fundamental niche that a species actually occupies is called its **realized niche.** Stated in these terms, the realized niche of the Cape May warbler is only a small portion of its fundamental niche. What is the advantage in not foraging throughout the tree? MacArthur suggested that this feeding pattern reduces competition. Because each of the five warbler species uses a different set of resources by occupying a different realized niche, the species are not in competition with one another. In general, ecologists agree with MacArthur's conclusion that natural selection has favored a range of preferences and behaviors among the five species that "carves up" the available resources. In nature, similar, potentially competing species often differ more where their ranges overlap than where each occurs alone. This increased difference when living together is called **character displacement.** Many ecologists think character displacement is the result of evolution to reduce competition, although this conclusion remains controversial. ◻

◻ *CAPSULE SUMMARY*

An organism's niche is its "profession," its way of life. The entire range of conditions it can tolerate is its fundamental niche. In the face of competition, an organism may occupy only part of its fundamental niche. That part is called its realized niche.

Demonstrating Competition in Nature

It is one thing to propose that one species can limit the realized niche of another, as MacArthur did based only on observed patterns of resource use by warblers. But it is quite a different thing to demonstrate such an effect experimentally. A very clear case of competition was provided by experiments carried out in the early 1960s by Joseph Connell of the University of California. Connell worked with two species of barnacles that grow on the same rocks along the coast of Scotland. Barnacles are marine animals that are related to crabs, lobsters, and shrimp. They have free-swimming, microscopic larvae that cement themselves to rocks and remain attached there for the rest of their lives. As you can see in Figure 17-9, one species, *Chthamalus stellatus*, lives in shallow water, where it is often exposed to air by receding tides. A second species, *Balanus balanoides*, lives lower down on the rocks, where it is rarely exposed to the atmosphere.

When Connell removed *Balanus* from the deeper zone, *Chthamalus* was easily able to occupy the vacant surfaces, indicating that intolerance of the physical environment did not prevent *Chthamalus* from becoming established there; its fundamental niche clearly includes the deeper zone. However, when *Balanus* was reintroduced, it could always outcompete *Chthamalus* by crowding it off the rocks. In contrast, *Balanus* could not survive when placed in the

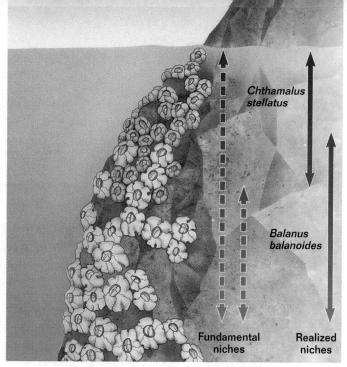

Figure 17-9 On the rocky coast of Scotland, *Balanus,* the larger species, and *Chthamalus* live on the same rocks, *above left.* *Balanus*'s superiority in the competition for space on the rocks stems from its faster growth rate: a *Balanus* individual can actually force a *Chthamalus* individual off a rock. Though their fundamental niches overlap considerably, *above right,* the realized niches of these two barnacles overlap little because of competition.

shallow-water habitats where *Chthamalus* normally occurs. *Balanus* apparently lacks the adaptations that permit *Chthamalus* to survive long periods of exposure to air. Connell's experiments show that *Chthamalus* occupies only a small portion of its large fundamental niche; the rest is unavailable because of competition with *Balanus.* As MacArthur suggested, competition can limit how species use resources.

The Outcome of Competition
....

Competition among species does not occur just because they use the same resource. They compete only when that resource is in limited supply. In nature, shortage is the rule, and species that use the same resource are almost sure to compete with each other. Such competitive interactions can have a profound effect on the structure of an ecosystem by stimulating character displacement. Alternatively, the species that is the better competitor may drive out the other, as occurred with Connell's barnacles (*Balanus* excluded *Chthamalus* from the rock surfaces). Local elimination of one competing species is known as **competitive exclusion.**

Darwin noted that competition should be most acute between very similar kinds of organisms because they tend to use the same resources in the same way. Does it follow, then, that when very similar species compete, one will always become extinct locally? In a series of carefully controlled laboratory experiments performed in the 1930s, the Russian biologist G. F. Gause looked into this question. In his initial

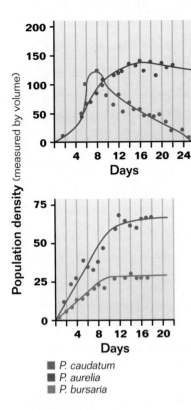

Figure 17-10 These graphs summarize Gause's experiments on competition. The lower graph also demonstrates the negative effect of competition on the participants. Both species reach about twice the density when grown without a competitor as when grown with one.

☐ CAPSULE SUMMARY

Competition can cause changes in a community. In the short term, the species that is the superior competitor may drive inferior competitors to extinction. Over the long term, competing species may evolve differences that reduce their use of common resources.

experiments, Gause grew two species of the protist *Paramecium* in the same culture tubes, where they had to compete for the same food (bacteria). Invariably, the smaller of the two species, which was more resistant to bacterial waste products, drove the larger one to extinction. From these results Gause formulated what is now called the **principle of competitive exclusion.** This principle states that if two species are competing, the species that uses the resource more efficiently will eventually eliminate the other locally—no two species can have the same niche.

When Can Competitors Coexist?

Is competitive exclusion the inevitable outcome of competition for limited resources, as Gause's principle states? No. The outcome depends on the fierceness of the competition and on the degree of similarity between the fundamental niches of the competing species. If it is possible for the species to avoid competing, they may coexist.

In a revealing experiment, Gause challenged *Paramecium caudatum*—the defeated species in his earlier experiments—with a third species, *P. bursaria*. Since these two species were also expected to compete for the limited bacterial food supply, Gause thought one would win out, as had happened in his previous experiments. But that's not what happened. Both species survived in the culture tubes. Like MacArthur's warblers, the paramecia found a way to divide the food resources. How did they do it? In the upper part of the culture tubes, where oxygen concentration and bacterial density were high, *P. caudatum* was dominant because it was better able to feed on bacteria than was *P. bursaria*. However, in the lower part of the tubes, the lower oxygen concentration favored the growth of a different potential food, yeast, which *P. bursaria* was better able to eat. The fundamental niche of each species was the whole culture tube, but the realized niche of each species was only a portion of the tube. Because the niches of the two species did not overlap too much, both species were able to survive. Figure 17-10 summarizes Gause's experiments. ☐

Predation Can Lessen Competition Among Prey
....

Many studies of natural ecosystems have shown that predation lessens the effects of competition. A very clear example is provided by the studies of Robert Paine of the University of Washington. Paine examined how sea stars affect the numbers and types of species within marine intertidal communities. Sea stars are fierce predators of marine animals such as clams and mussels. When sea stars were kept out of experimental plots in a rocky intertidal zone, the number of prey species fell from 15 to 8. As you can see in Figure 17-11, the eliminated

species were crowded out by mussels, the chief prey of sea stars. By preying on mussels, sea stars keep the mussel populations low, neutralizing the mussels' ability to outcompete the other animals for space on the rocks. When the sea stars were removed, the mussels were free to assert their competitive dominance. As a direct result, the number of species in the ecosystem declined. Thus, predation can have a very healthy effect on ecosystems: it can promote diversity by minimizing competition.

Figure 17-11 The sea star *Pisaster, above left,* altered the outcome of competition for space in the coastal ecosystem studied by Robert Paine. When freed of the control exerted by sea stars, mussels—which are the superior competitors—crowded seven other species out of the ecosystem and overgrew the habitat, *above right.*

Section Review

1. *Explain the difference between* niche *and* habitat.
2. *Can an organism's realized niche be larger than its fundamental niche? Justify your answer.*
3. *Why can character displacement occur only if competitive exclusion has not occurred?*
4. *How did diversity of habitat promote coexistence of the two species of barnacles studied by Connell?*

Critical Thinking

5. *A scientist finds no evidence that species in a community are competing and concludes that competition never played a role in the development of this community. Is this conclusion valid? Justify your answer.*

17-3 The Influence of the Physical Environment

Section Objectives

■ Explain the influence of the Earth's shape on climate.

■ Describe the cause of the rain-shadow effect.

■ Explain how ocean currents affect terrestrial climates.

*I*f you traveled across the country by car you would notice dramatic changes in the plants and animals outside your window. For example, the drought-tolerant cactuses of Arizona deserts do not live in the forests of Maine or the swamps of Florida. Although it is not surprising that different environments harbor different sets of organisms, it is surprising that there are relatively few sets. Certain characteristic groups of animals and plants occur together over very broad areas. Why is this? The physical environment, together with biological interactions, determines which animals and plants are able to live in a particular ecosystem. Climate has a particularly important influence on the character of ecosystems.

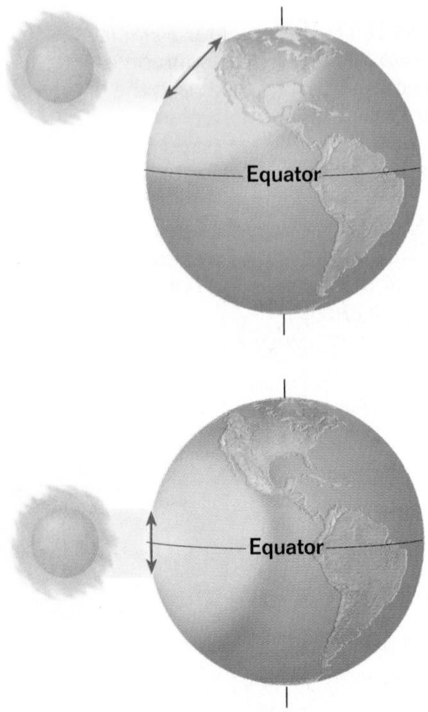

Figure 17-12 Near the equator, direct sunlight falls on the Earth's surface, *bottom*. Therefore, solar energy is concentrated and produces greater warming than at higher latitudes, *top*. There, the curvature of the Earth spreads the solar energy over a greater area.

What Causes Climate?

The distribution of the Earth's ecosystems results from the interaction of the physical features of the Earth (such as the different soil types and the occurrence of mountains, valleys, lakes, and oceans) with two key factors: (1) the amount of energy from the sun that reaches different parts of the Earth and the seasonal variation in that energy, and (2) the global pattern of atmospheric and oceanic circulation created by the unequal global distribution of solar energy. Together, these factors determine local climates, particularly the amount and distribution of precipitation.

The great diversity of ecosystems on Earth is due in part to variations in climate from place to place. Miami, Florida, and Bangor, Maine, for example, are likely to have very different weather on every day of the year. There is no mystery about the causes of climatic differences. Because the Earth is nearly spherical, different places on the surface receive different amounts of solar energy, as illustrated in Figure 17-12. These differences are directly responsible for many of the major climatic differences that occur over the Earth's surface and indirectly responsible for much of the variety of ecosystems. At regions near the equator, the sun's rays arrive almost perpendicularly, making the tropics warmer than the temperate regions. Nearer the poles, the angle at which the sun's rays hit the Earth spreads them out over a much greater area, providing less energy per unit of area.

The Earth's daily rotation on its axis and its revolution around the sun are also important in determining world

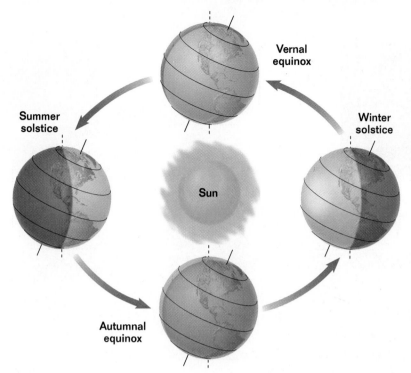

Vernal
equinox

Summer
solstice

Winter
solstice

Sun

Autumnal
equinox

Figure 17-13 Because the Earth is tilted 23.5°, the relationship shown in Figure 17-12 changes as the Earth orbits the sun. Between the vernal equinox (March 21) and the autumnal equinox (September 23), the Northern Hemisphere tilts toward the sun and therefore receives more direct sunlight than does the Southern Hemisphere, which tilts away from the sun. The result is spring and summer in the Northern Hemisphere and fall and winter in the Southern Hemisphere. After the autumnal equinox, the pattern reverses. For the next six months, the Southern Hemisphere tilts toward the sun and experiences spring and summer, while the Northern Hemisphere tilts away and has fall and winter.

climate. Because of the Earth's daily rotation, periods of cooling (night) and warming (day) alternate, preventing the buildup of extreme temperatures. The axis of the Earth is tilted 23.5° off vertical, so the revolution of the Earth around the sun produces a progression of seasons at higher latitudes (areas far from the equator), as diagramed in Figure 17-13.

Solar Heating Drives the Circulation of the Earth's Atmosphere

Since hot air rises, air near the equator, which is warmed by the sun, also rises. This air cools and loses most of its moisture, producing heavy rains in the tropics. Because this region of rising air is one of low pressure, it draws air from both north and south of the equator. As the rising air cools, it moves toward the poles. When the rising air masses reach about 20° to 30° north and south latitude, the air, now cooler and drier, sinks and heats, producing a zone of low precipitation; the sinking air acts like a sponge, because warm air holds more moisture than cooler air. Since air at these latitudes is still warmer than air at the polar regions, it continues to flow toward the poles, rising again at about 60° north and 60° south latitude. When this air mass descends near the poles, it produces a zone of very low temperature and precipitation. These movements of air result in the major pattern of atmospheric circulation that is illustrated in Figure 17-14.

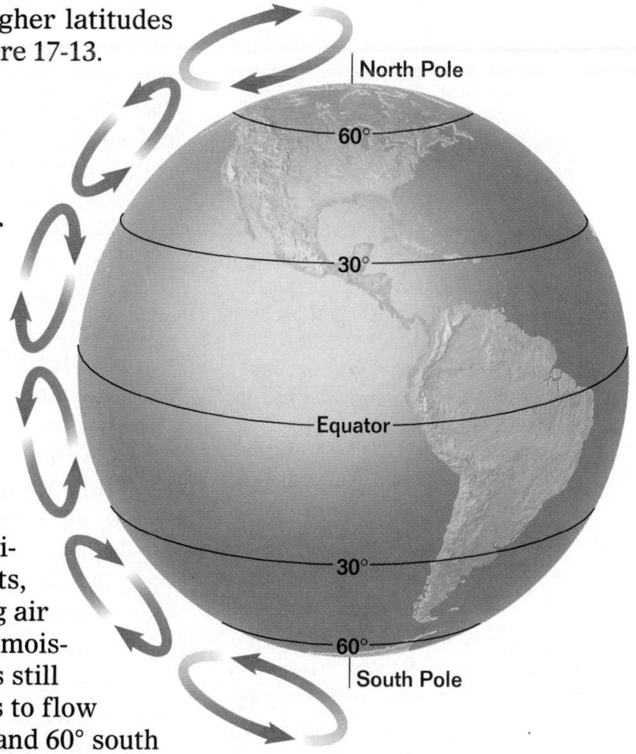

North Pole

60°

30°

Equator

30°

60°

South Pole

Figure 17-14 This diagram shows the global pattern of atmospheric circulation caused by the sun's heating of the Earth.

Coast ranges

Sierra Nevada range

Figure 17-15 Winds blowing in from the Pacific Ocean must ascend to cross the Sierra Nevada range, causing the air to cool and release its moisture. Abundant precipitation on the western side of the range supports dense forests, *top left*. Having already lost most of their moisture, the winds drop little precipitation on the eastern side of the Sierra Nevada range. The arid conditions there permit only sparse growth of drought-tolerant plants, *top right*. A similar but smaller rain shadow is caused by the lower Coast ranges.

Why It Rains Where It Does

The moisture-holding capacity of air increases when it is warmed and decreases when it is cooled. Consequently, precipitation is generally low near 30° north and 30° south latitude, where cool air is falling and warming, and relatively high near 60° north and 60° south latitude, where air is rising and cooling. Partly as a result of these factors, most of the great deserts of the world, such as the Sahara, lie near 30° north or 30° south latitude. And some of the great temperate forests are located near 60° north latitude (there is virtually no land at 60° south latitude).

Other major deserts, such as the Gobi Desert of Asia, are formed in the interiors of continents. Continental interiors receive limited precipitation because of their distance from the sea, and sometimes because mountain ranges intercept the moisture-laden winds from the sea, as shown in Figure 17-15. When mountains force incoming air upward, the air's moisture-holding capacity decreases as the air cools, resulting in increased precipitation on the windward side (from which the wind is blowing) of the mountains. As the air descends the other side of the mountains, known as the leeward side, it is warmed, and its moisture-holding capacity increases. Therefore, it tends to draw up moisture from the surface, rather than releasing the little moisture it contains. The leeward sides of mountains are often much drier than their windward sides, and the vegetation is often very different. This phenomenon is called the **rain-shadow effect.** In the United States, the Mojave and Great Basin Deserts lie in the rain shadow of the Sierra Nevada range, and the Great Plains lie in the rain shadow of the Rocky Mountains.

Patterns of Circulation in the Ocean

The patterns of circulation in the ocean, shown in Figure 17-16, are determined by the major patterns of atmospheric circulation discussed previously, but they are modified by the location of the landmasses around and against which the ocean currents must flow. Oceanic circulation is dominated by huge, circling surface currents that move around the subtropical oceans at about 30° north and 30° south latitude. These currents move clockwise in the Northern Hemisphere and counterclockwise in the Southern Hemisphere. By redistributing heat, oceanic currents profoundly affect life not only in the oceans but also on coastal lands. For example, in the Atlantic Ocean the warm Gulf Stream swings away from North America near Cape Hatteras, North Carolina, and reaches Europe near the southern British Isles. Because of the Gulf Stream, western Europe is much warmer than eastern North America at the same latitudes. ▢

❑ CAPSULE SUMMARY

Latitude, or distance from the equator, is the most important factor determining the climate of an area. Other important factors are distance from the ocean, elevation, and position relative to mountain ranges that can intercept moisture-carrying winds.

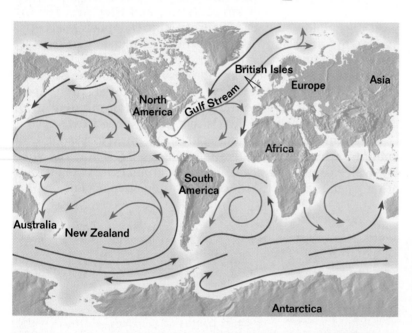

Figure 17-16 This diagram shows the major oceanic currents. Warm currents, which begin in tropical waters, are indicated in red. Cold currents are shown in blue.

Section Review

1. *Why are the tropics warmer than the temperate regions?*
2. *Describe two geographical factors that promote the development of deserts.*
3. *How does the Gulf Stream moderate the climate of western Europe?*

Critical Thinking
4. *Would there still be seasons if the Earth were not tilted in its orbit? Explain your answer.*

17-4 Major Biological Communities

Section Objectives

- Contrast the three major kinds of marine habitats.
- Describe the Earth's major biomes.

If you were to tour the world and look at biological communities in the oceans and on land, you would soon learn one of the great generalizations of ecology: very similar communities occur in many different places that have similar climates and geographies. In this section you will take just such a tour and visit the major types of biological communities found on Earth.

Marine Communities

Nearly three-fourths of the Earth's surface is covered by ocean. The seas have an average depth of more than 3 km (1.9 mi.), and they are, for the most part, cold and dark. Heterotrophic (nonphotosynthetic) organisms are found even at the greatest ocean depths, which reach nearly 11 km (6 mi.) in the Marianas Trench of the western Pacific Ocean. Photosynthetic organisms are confined to the upper few hundred meters of water. Organisms that live below this level obtain almost all of their food from organic debris that drifts downward.

The marine environment consists of three major kinds of habitats, which are illustrated in Figure 17-17: (1) the shallow waters along the coasts of the continents, (2) the surface of the open sea, and (3) the depths of the open ocean.

Figure 17-17
The shallow waters along the shores of the continents receive nutrients that have washed from the land. This habitat is the most diverse in the ocean. Farther out, in the surface layers of the open ocean, nutrients are often scarce, and the producers are floating photosynthetic algae and bacteria. In the cold, dark, deep layers of the ocean, where light cannot penetrate, nearly all organisms depend on organic material that drifts down from waters nearer the surface.

Figure 17-18 Coral reefs occur in shallow ocean waters in tropical regions. In numbers of species, coral reefs rival tropical rain forests.

Shallow Ocean Waters

The zone of shallow water is small in area, but compared with other parts of the ocean, it is inhabited by large numbers of species. The world's great fisheries are located on banks in the coastal zones, where nutrients, washed from the land, are often more abundant than in the open ocean.

Open Sea Surface

Drifting freely in the upper waters of the ocean is a diverse biological community called **plankton,** which consists mostly of microscopic organisms. Plankton includes bacteria, algae, fish larvae, and other small animals. Some of the members of the plankton, including algae and some bacteria, are photosynthetic. Collectively these organisms account for about 40 percent of all the photosynthesis that takes place on Earth.

Figure 17-19 These schooling mackerel are speedy inhabitants of the open ocean surface.

Deep Sea Waters

In the deep waters of the sea, below the depth of 300 m (1,000 ft.), live some of the most bizarre organisms found on Earth. Many of these animals can produce their own light, which they use to communicate with one another or to attract prey.

Figure 17-20 Surprisingly diverse communities of worms, clams, fishes, crabs, and bacteria, *left,* flourish near cracks in the sea floor through which extremely hot water escapes. The producers in this community are bacteria that use hydrogen sulfide gas ("rotten egg" gas) leaking from the cracks as an energy source. Another deep-sea inhabitant, the deep-sea swallower, *right,* seems to be all mouth and stomach.

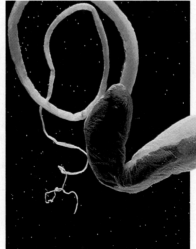

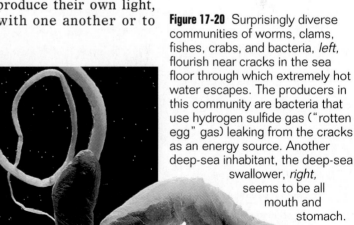

Figure 17-21 This pond in Pennsylvania is a typical freshwater ecosystem.

Figure 17-22 The seven biomes discussed in this chapter cover most of the Earth's land surface. Areas that cannot be assigned to one of these biomes are shown in gray. Because climate changes with elevation, mountains contain a variety of communities and so do not belong to any one biome. Mountainous areas are indicated on the map. Also, Antarctica is not shown because it has no biomes.

Freshwater Communities

Freshwater habitats (lakes, ponds, streams, and rivers) are distinct from both marine and terrestrial habitats, and they are very limited in area. Lakes cover about 1.8 percent of the Earth's surface, and rivers and streams cover about 0.3 percent. All freshwater habitats are strongly connected to terrestrial ones, with freshwater marshes and swamps constituting intermediate habitats. In addition, a large amount of organic and inorganic material continuously enters bodies of fresh water from communities on the land.

Like the oceans, large lakes have three zones in which organisms live: a shallow zone near the shore, an open-surface zone inhabited by plankton, and a deep-water zone below the limits of effective light penetration.

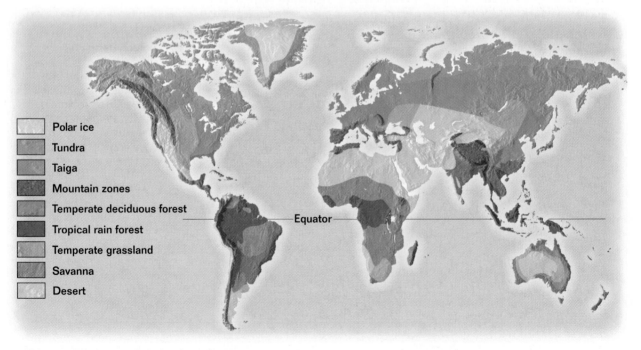

- Polar ice
- Tundra
- Taiga
- Mountain zones
- Temperate deciduous forest
- Tropical rain forest
- Temperate grassland
- Savanna
- Desert

Equator

Terrestrial Communities

A major terrestrial community that is found in different areas with similar climates is called a **biome.** A biome's structure and appearance are similar throughout its distribution. Biomes can be classified in a number of ways; the classification used here is chosen merely as a convenient means of discussing the properties of organisms from an ecological perspective. This classification recognizes seven biomes: (1) tropical rain forest, (2) savanna, (3) desert, (4) temperate grassland, (5) temperate deciduous forest, (6) taiga, and (7) tundra. These biomes differ greatly from one another because they have developed in regions with very different climates. The global distribution of these biomes is shown in Figure 17-22.

Tropical Rain Forests: Lush Equatorial Forests

The rainfall in tropical rain forests is generally 200–450 cm (80–180 in.) per year, with little difference in distribution from season to season. The tropical rain forest is the richest biome in terms of number of species, probably containing at least half of the Earth's species of terrestrial organisms— more than 2 million species. In a single square mile (just over 2.5 sq. km) of tropical forest in Peru or Brazil, there may be 1,200 or more species of butterflies, twice the total number of species found in the United States and Canada combined. However, although the communities that make up tropical forests are rich in species, each kind of animal, plant, or microorganism in a given area is often represented by very few individuals.

Tropical rain forests have a high primary productivity, even though they exist mainly on quite infertile soils. Most of the nutrients are held within the plants themselves and are rapidly recycled when the plants die or when parts, such as leaves, fall off and decompose. Most of the roots of the tall trees spread through the top 1 or 2 cm (less than 1 in.) of the soil and extract the nutrients from decomposing leaves and other plant parts.

Figure 17-23 Tropical rain forests, such as this one in Costa Rica, have very high species diversity.

Figure 17-24 Massive herds of wildebeest and zebras migrate across the Serengeti Plain, part of East Africa's savanna. Carnivores such as lions, leopards, African wild dogs, and hyenas attack the herds.

Figure 17-25 In deserts, conservation of water is essential. Spines on these cactuses help protect their moist tissues from thirsty herbivores.

Figure 17-26 Before European settlers arrived in the New World, much of the central United States and southern Canada was covered by prairie, or temperate grassland. As many as 60 million bison grazed on these prairies. Today, only fragments of undisturbed prairie and fewer than 100,000 bison remain.

Savannas: Dry Tropical Grasslands

Dry climates often favor the development of grassland. The world's great dry grasslands, called savannas, are found in tropical areas that have relatively low annual precipitation or prolonged annual dry seasons. On a global scale, the savanna biome is found in a transitional zone located between tropical rain forest and desert. Annual rainfall is generally 90–150 cm (35–60 in.) in savannas. There is a wider fluctuation in temperature during the year than in the tropical rain forests, and there is seasonal drought. These factors have led to the evolution of an open landscape with widely spaced trees. Many of the animals are active only during the rainy season. The huge herds of grazing mammals and their associated predators that inhabit the savannas of Africa are well known and spectacular.

Deserts: Arid Lands

Typically, fewer than 25 cm (10 in.) of precipitation falls annually in the world's desert areas. The scarcity of water is the overriding factor influencing most biological processes in the desert. In desert regions, the vegetation is characteristically sparse. Deserts are most extensive in the interiors of continents, especially in Africa (the Sahara), Asia, and Australia. Less than 5 percent of North America is desert. Lands in which water is not so scarce are called semideserts. These two dry biomes are combined in Figure 17-22 on page 378. In deserts and semideserts, the amount of water that actually falls at a particular place can vary greatly, both during a given year and between years.

Temperate Grasslands: Seas of Grass

Moderate climates promote the growth of rich temperate grasslands. Temperate grasslands once covered much of the interior of North America and were widespread in Eurasia and South America. Such grasslands are often highly productive when converted to agriculture, and much of the rich agricultural land in the United States and southern Canada was originally occupied by prairie, another name for

Figure 17-27 Temperate deciduous forests, such as this one in Pennsylvania, are dominated by hardwood trees that lose their leaves in autumn. In North America, the area occupied by temperate deciduous forests has shrunk considerably since the arrival of European settlers.

temperate grassland. The roots of grasses characteristically penetrate far into the soil, which tends to be deep and fertile.

Temperate Deciduous Forests: Rich Hardwood Forests

Relatively mild climates and plentiful rain promote the growth of forests. Temperate deciduous forests (deciduous trees shed their leaves all at once in the fall) grow in areas with relatively warm summers, cold winters, and sufficient precipitation. The annual precipitation generally ranges from 75 to 250 cm (30 to 100 in.) and is well distributed throughout the year. Moisture is generally unavailable to animals and plants in the winter, because it is usually frozen.

This biome covers very large areas, including much of the eastern United States and southeastern Canada and extensive areas of Europe and Asia. In North America, deciduous forests are home to deer, bears, beavers, raccoons, and the other familiar animals of the temperate regions. The trees are hardwoods (oak, hickory, and beech), and shrubs and herbs grow on the forest floor.

Figure 17-28 Coniferous trees predominate on the taiga. This photo of taiga was taken near Churchill, Manitoba, in north-central Canada.

Taiga: Great Coniferous Forests of the North

Cold, wet climates promote the growth of coniferous forests. A great ring of northern forests of coniferous trees, primarily spruce and fir, extends across vast areas of Eurasia and North America. This biome, one of the largest on Earth, is called by its Russian name, taiga *(TEYE guh)*. Winters in the taiga are long and cold, and most of the precipitation falls in the summer. At the northern

Figure 17-29 The scarcity of trees on the tundra is apparent in this view of Denali National Park in Alaska.

latitude where taiga occurs, the days are short in winter (as little as six hours) and long in summer. During the summer, plants may grow rapidly. Marshes, lakes, and ponds are common and are often fringed by willows or birches. Most of the trees in the taiga tend to occur in dense stands of one or only a few species. Many large mammals live in the taiga, including herbivores such as elk, moose, and deer, and carnivores such as wolves, bears, lynxes, and wolverines.

Tundra: Cold Plains of the Far North

In the Far North, there are few trees. Between the taiga and the permanent ice surrounding the North Pole is the open, sometimes boggy, biome known as the tundra. This enormous biome covers one-fifth of the Earth's land surface. Annual precipitation in the tundra is very low, usually less than 25 cm (10 in.), and water is unavailable for most of the year because it is frozen. During the brief Arctic summers, when water is not frozen, the surface of the tundra is often extremely boggy because permafrost, or permanent ice, usually exists within 1 m (about 3 ft.) of the surface.

A land of grasses, sedges, dwarf willows, and mosses, the tundra is open and windswept. Foxes, lemmings, owls, and caribou are among the vertebrate inhabitants. Trees are small and usually confined to the edges of streams and lakes. The tundra's appearance is extremely uniform.

Section Review

1. *Why can't photosynthesis occur in the deepest parts of the ocean?*
2. *What common environmental hardship confronts organisms in deserts and tundra?*

Critical Thinking
3. *Explain why there are no biomes in Antarctica.*

Vocabulary

biome (379)
character displacement (368)
coevolution (361)
commensalism (363)
competition (363)
competitive exclusion (369)

fundamental niche (367)
mutualism (363)
niche (365)
parasitism (361)
plankton (377)
predation (361)

principle of competitive
  exclusion (370)
rain-shadow effect (374)
realized niche (368)
secondary compound (362)
symbiosis (363)

Concept Mapping

Construct a concept map that shows how the biomes can be classified based on precipitation, temperature, and geographical location. Use as many terms as needed from the vocabulary list. Try to include the following terms in your map: tropical rain forest, savanna, desert, temperate deciduous forest, temperate grassland, taiga, and tundra. Include additional terms in your map as needed.

Review

Multiple Choice

1. In predator-prey coevolution, if the prey evolves a defense to stop predation then the predator likely will evolve
   a. in a way that enables it to overcome the prey's defense.
   b. so that it can parasitize the prey.
   c. countermeasures to secondary compounds.
   d. into the prey.

2. The interaction between a spruce tree and a hemlock tree, both of which are trying to get nitrogen from the soil, is an example of
   a. mutualism.
   b. commensalism.
   c. competition.
   d. succession.

3. The ways in which an organism interacts with its environment make up its
   a. niche.　　c. habitat.
   b. space.　　d. ecosystem.

4. Connell's experiment with barnacles demonstrated
   a. that barnacles can be grown in captivity.
   b. how barnacles extend their realized niche.
   c. character displacement in the ocean.
   d. competitive exclusion in nature.

5. The principle of competitive exclusion indicates that
   a. a niche can be shared by two species.
   b. niche subdivision may occur.
   c. one species will give way to another if their niches are very similar.
   d. competition ends in worldwide elimination of a species.

6. Compared with the sun's rays falling on a location at the equator, the sun's rays striking a location at 30 degrees north latitude would be
   a. spread out over a greater area.
   b. focused on a smaller area.
   c. almost perpendicular.
   d. tilted at 23.5 degrees off vertical.

7. The Great Plains of the United States get little rain because
   a. of the rain-shadow effect of the Sierra Nevada range.
   b. of the dry winds that blow across the western deserts.
   c. they lie on the leeward side of the Rocky Mountains.
   d. they lie on the windward side of the Cascade range.

8. The highest species diversity is found in which biome?
   a. taiga　　　　c. deciduous forest
   b. grassland　　d. tropical rain forest

9. Marine life is most diverse in the
   a. deep sea waters.
   b. open sea surface.
   c. shallow ocean waters.
   d. polar oceans.

10. The Gulf Stream affects the climate of coastal lands by
    a. lowering air temperature.
    b. raising air temperature.
    c. lowering the freezing point.
    d. causing beach erosion.

11. Cold and long winters, very few trees, and little precipitation describe the
    a. tundra.   c. deciduous forest.
    b. taiga.   d. grasslands.

Completion

12. An interaction in which one species benefits and the other is not affected is _____ . The other type of symbiosis is _____ , in which both species benefit. An example of this interaction is the relationship between _____ and aphids.

13. A niche is how an organism _____ in an ecosystem. It may be described in terms of _____ flow or how organisms use space.

14. Robert MacArthur studied niche restriction among five species of _____ . His work showed that a species' _____ niche may be a small portion of its fundamental niche.

15. The rain-shadow effect predicts that the _____ side of mountains will get less rain than the _____ side.

16. In marine environments, species diversity is highest in the _____ , and photosynthetic organisms are least abundant in the _____ .

17. The two terrestrial biomes that receive less than 25 cm of precipitation annually are desert and _____ . The biome that receives the most precipitation is _____ .

Short Answer

18. Briefly summarize Joseph Connell's barnacle experiment. What did he conclude from his work?

19. What are two possible outcomes of competition between two species of squirrel for the same type of nut?

20. What observations led to the formulation of the competitive exclusion principle by the Russian biologist G. F. Gause?

21. Identify two factors that affect the temperature patterns in a region. How are the temperature patterns affected by each factor?

22. List the major kinds of marine habitats and describe the characteristics of each.

23. What is a biome? What climatic factors determine a biome's characteristics? List the seven major biomes and describe the characteristics of each.

Themes Review

24. **Homeostasis** The blacktail jackrabbit, *Lepus californicus*, shown below, lives in the deserts of the southwestern United States. How do its large ears, which are richly supplied with blood vessels, and its tendency to rest in deep shade on hot days help the rabbit maintain a relatively stable body temperature?

25. **Energy Flow** Describe the niches of a lion, a zebra, and the grass that grows on the African plain in terms of how each species affects energy flow in the ecosystem.

26. **Evolution** Explain the role of coevolution in the coexistence of potential competitors.

Critical Thinking

27. **Evaluating Data** Do MacArthur's observations of warblers furnish enough evidence to conclude that character displacement had occurred in these birds? If not, specify the observations required to make a stronger case for character displacement.

28. **Making Predictions** In Gause's experiments, *Paramecium caudatum* could coexist with *P. bursaria* but not with *P. aurelia*. Predict what would happen if *P. aurelia* and *P. bursaria* were grown together.

29. **Making Predictions** Tillamook, Oregon, is on the windward side of the Pacific coastal mountains, and Portland, Oregon, is on the leeward side. In which city would you expect the least precipitation? Explain why.

30. **Designing an Experiment** Suggest an experiment to test the hypothesis that the extinction of the bluebirds native to New York City's Central Park was due to their being outcompeted by starlings released in the park by humans.

31. **Making Inferences** A scientist captures two species of *Paramecium* in a pond. When the two species are grown together in cultures in the laboratory, one species always outcompetes and eliminates the other. In the pond, however, the two species coexist. Suggest a hypothesis to account for this difference in outcomes. How would you test your hypothesis?

32. **Making Inferences** Early human settlers of Hawaii introduced predators—including cats, dogs, rats, and pigs—that the native animals had never encountered. These introduced predators proved devastating to the native animals, many of which became extinct. Using your knowledge of coevolution, explain why prey should be more vulnerable to introduced predators than to native predators.

Activities and Projects

33. **Research and Writing** What is dendrochronology? Find out how the work of dendrochronologists provides insights into the past physical environment of an area.

34. **Multicultural Perspective** The Australian Aborigines inhabited all of Australia, even the arid central region. Research the Aborigines' way of life to learn how they were able to live in such a hostile environment with such simple technology.

Readings

35. Read the article "Singing Caterpillars, Ants and Symbiosis," in *Scientific American*, October 1992, pages 76–82. In the mutualistic relationship between ants and caterpillars, what does each species gain? How do the sounds produced by the caterpillars affect the ants?

36. Read the article "Living Together," in *Scientific American*, January 1992, pages 122–133. Describe the traditional view of how coevolution between host and parasite should proceed. According to Dr. Paul Ewald, some parasites should not evolve in this way. Explain his reasoning.

LABORATORY Investigation

Habitat Selection

OBJECTIVE

Assess the effect of environmental variables, including temperature and light, on habitat selection by brine shrimp.

PROCESS SKILLS

- identifying and controlling variables
- graphing and interpreting data

MATERIALS

- marking pen
- clear, flexible plastic tubing
- meter stick
- 4 test tubes with stoppers and test tube rack
- 2 corks to fit tubing
- graduated cylinder or beaker
- funnel
- brine shrimp culture
- aluminum foil
- 3 screw clamps
- 1 pipette
- petri dish
- Detain™ or methyl cellulose
- tape
- ice bag
- hot water bag
- fluorescent lamp or grow light
- 14 pieces of screening
- calculator

BACKGROUND

1. What variables are involved for habitat selection? What is a niche?
2. Write your own **Focus Question** on your Vee Form.

3. **Knowing Side of the Vee** List the **Concepts** and new **Vocabulary Words** on your Vee Form. In the **Concept Statements** section of the Vee, use these words in sentences that define and explain them.

TECHNIQUE

Doing Side of the Vee

You and your partner will complete Part A, B, or C and then share your results with the other students on your team.

Part A: Control Setup

1. Use a marking pen to mark the plastic tubing at 12 cm, 22 cm, and 32 cm from one end. You have now divided the tube into 4 sections. Start at one end and label the sections 1, 2, 3, and 4. Label 4 test tubes 1, 2, 3, and 4.

2. Place a cork in one end of the tubing. Use a graduated cylinder or beaker and a funnel to transfer 50 mL of brine shrimp culture to the tubing. Cork the open end and lay the tubing on a desk top.

3. Cover the tubing with aluminum foil and let it remain undisturbed for 30 minutes.

4. After the time has passed, attach screw clamps to each spot that you marked on the tubing. While your partner holds the corks firmly in place, tighten the middle clamp first and then the outer clamps.

5. Immediately pour the contents of each section of tubing into the test tube labeled with the same number.

6. Stopper test tube 1 and invert it gently to distribute the shrimp. Use a pipette to draw a 1 mL sample of shrimp culture, and transfer it to the petri dish. Add a few drops of Detain™ to the petri dish to slow down the shrimp, and then count the live shrimp. Record the count in the **Records** section of your Vee Form. Dispose of the shrimp as your teacher directs, and repeat this procedure four more times for a total of five counts.

Calculate the average number of shrimp in test tube 1 and record the result on your Vee Form.

7. Repeat step 6 for the contents of each of the remaining three test tubes.

8. Clean up your materials and wash your hands before leaving the lab.

9. Make a histogram showing the number of brine shrimp in each section of tubing.

10. In the **Procedure** section of the Vee, briefly summarize the procedure you followed.

Part B: Temperature Gradient

1. Repeat steps 1 and 2 of Part A.

2. Tape the tubing to the desk top and cover it with aluminum foil. Mark the aluminum foil to show the approximate positions of sections 1 and 4.

3. Place a bag of ice over section 1 of the tubing. Place a hot-water bag over section 4 of the tubing. **CAUTION: Handle the hot-water bag carefully.** Do *not* use water over 70°C, which can burn you.

4. After 10 minutes, replace the hot water bag with a fresh bag. Replace the bag again after 20 minutes. After 30 minutes, quickly complete Steps 4 and 5 of Part A.

5. *Immediately* read and record the temperature of the test tube contents.

6. Make a histogram showing the number of shrimp in each section of tubing. Identify each section with its temperature.

7. Complete steps 6 through 10 of Part A.

Part C: Light Gradient

1. Repeat steps 1 and 2 of Part A.

2. Set a light source about 20 cm away from the tubing. Use a low-wattage light bulb or soft white fluorescent tubes to give light without much heat. **CAUTION: Light bulbs get very hot and can burn your skin. Do not touch the bulb.**

3. Cover section 1 of the tubing with eight layers of screening. Place four layers on section 2, two layers on section 3, and leave section 4 uncovered. Wait 30 minutes.

4. Make a histogram showing the number of shrimp in each section. Identify each section with the amount of screening.

5. Complete steps 4 through 10 of Part A.

INQUIRY

1. How did the brine shrimp react to changes in temperature and light?

2. Why were five counts taken in each test tube?

3. Why was a control (Part A) necessary?

4. Use the information on the **Knowing Side** of the Vee to interpret your results from the **Doing Side**, and then write your **Knowledge Claim**. Write a **Value Claim** for this lab.

ANALYSIS

1. After examining the histograms made by other teams, describe the niche of brine shrimp.

2. Brine shrimp cannot migrate, or move, from one body of water to another. How is it helpful for shrimp to react to changes in their environment?

3. How might a positive phototactic response (moving toward light) be advantageous for brine shrimp? How could a negative response be helpful?

FURTHER INQUIRY

Write a **New Focus Question** that could be the point of a new investigation. The following is an example:

How do brine shrimp react to a gradient of motion? How would this relate to turbidity?

CHAPTER 18

HUMAN IMPACT ON THE ENVIRONMENT

REVIEW

- pH (Section 1-1)
- ozone layer (Section 13-2)
- population growth (Section 15-1)
- carrying capacity (Section 15-1)
- diversity (Section 16-1)
- ground water (Section 16-3)
- carbon cycle (Section 16-3)

One of the 7,500 remaining tigers

18-1 Global Change

As you read this, the Earth faces unprecedented environmental problems, ranging from global changes in the atmosphere to the loss of topsoil. This section will focus on a collection of environmental issues that affect the entire world—changes in the atmosphere that influence every ecosystem on Earth. Human-induced environmental changes that affect ecosystems worldwide are referred to as "global change." This section focuses on three such human-induced environmental changes: acid rain, ozone destruction, and global warming. Although other global environmental changes are also occurring, these three serve to clearly illustrate the general nature of the changes.

Section Objectives

- *Recognize the causes and effects of acid rain.*
- *Describe the cause and effects of ozone depletion.*
- *Explain how the burning of fossil fuels has changed the atmosphere, and describe the controversy over global warming.*

Burning High-Sulfur Coal Creates Acidic Precipitation

The power plant you see in Figure 18-1 burns coal and sends the smoke high into the atmosphere through stacks more than 65 m (210 ft.) tall. This smoke contains high concentrations of sulfur because the coal that the plant burns is rich in sulfur. Burning high-sulfur coal causes several unpleasant effects. The sulfur-rich smoke smells like rotten eggs, and it combines chemically with the limestone in many public buildings, producing unsightly black surfaces. The intent of those who designed the power plant was to release the sulfur-rich smoke high into the atmosphere where winds would disperse and dilute it. Tall smokestacks were first introduced in Britain in the mid-1950s, and they rapidly became popular in

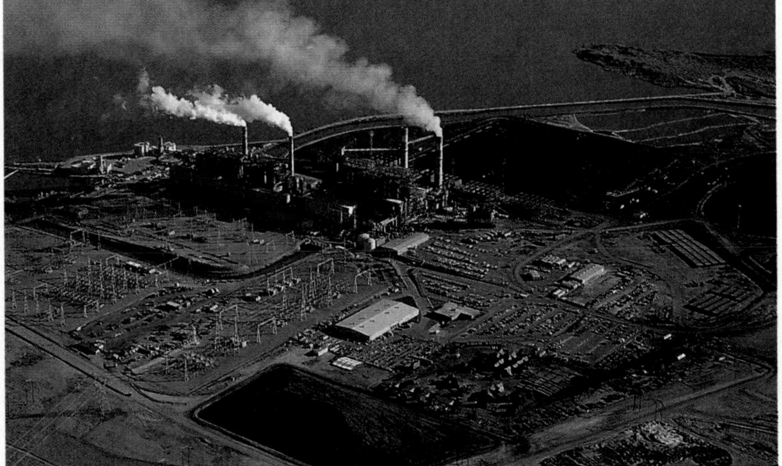

Figure 18-1 The Four Corners Power Plant in New Mexico burns high-sulfur coal and releases sulfur dioxide, which can combine with water in the atmosphere to produce acid rain.

Figure 18-2 The bare trees in this North Carolina forest died as a result of acid rain. Acid rain can change the acidity of the soil and interfere with nutrient absorption by trees and other plants.

the United States and Europe—there are now over 800 of them in the United States alone.

In the 1970s, ecologists began to report evidence that the tall stacks were not eliminating the problems of high-sulfur coal, just exporting the ill effects elsewhere. Throughout northern Europe, lakes were reported to have suffered drastic drops in species diversity; some even became lifeless. The trees of the great Black Forest of southern Germany seemed to be dying as well. But the ill effects were not limited to Europe. Forests in the eastern United States and southern Canada were also being damaged, as you can see in Figure 18-2. It is now estimated that at least 570,000 hectares (1.4 million acres) of forest in the Northern Hemisphere have been adversely affected. Scientists discovered that the sulfur introduced into the atmosphere by smokestacks combines with water vapor to produce sulfuric acid. Rain and snow carry the sulfuric acid back to the surface. This acidified precipitation is called **acid rain.** In North America, acid rain is most severe in the northeastern United States and southeastern Canada, downwind from the coal-burning plants in the Midwest. In 1989, when schoolchildren in a nationwide project measured the pH of rainwater, locations around the United States rarely had a pH lower than 5.6—except in the northeastern United States, where rain and snow had a pH of about 3.8, almost 100 times as acidic as the typical values for the rest of the country.

Acid rain destroys many forms of life. Thousands of lakes in Sweden and Norway no longer support fish, and in the United States and Canada, tens of thousands of lakes are dying as their pH levels fall below 5.0. Throughout the United States and Europe, local populations of frogs and other amphibians are becoming extinct, at least partly because their young cannot develop properly in acidic water. Destruction of the ozone layer, which will be discussed next, may also be a factor in these extinctions.

The solution to acid rain at first seems obvious: capture the sulfur-rich emissions instead of releasing them into the atmosphere. But there have been serious problems with implementing this solution. For one thing, it is expensive. In the United States, it is estimated that installation of the necessary smokestack "scrubbers," which would remove most of the sulfur from the emissions, would cost about $5 billion. An additional difficulty is that the polluter and the recipient of the pollution often are far from one another, and neither wants to pay for what is viewed as the other's problem. Revisions of the Clean Air Act passed by Congress in 1990 have begun to address this problem by requiring some cleaning of emissions in the United States. Much still remains to be done worldwide, however. ◻

❑ **CAPSULE SUMMARY**

Acid rain is any precipitation with higher than normal acidity. Acid rain is caused by the interaction between pollutants and water in the atmosphere.

Destruction of the Ozone Layer

As you learned in Chapter 13, life was trapped in the oceans for 3 billion years because ultraviolet radiation from the sun reached the Earth's surface unchecked, and nothing could survive that bombardment of destructive energy. Living things were able to leave the oceans and colonize the surface of the Earth only after a protective shield of ozone, O_3, had developed in the upper atmosphere. Imagine if that shield were taken away. Alarmingly, it appears that the ozone layer is disappearing and that we are destroying it ourselves.

In 1985, a British researcher in Antarctica noticed that ozone levels in the atmosphere seemed to be as much as 30 percent lower than they had been 10 years earlier. Satellite images taken over the South Pole revealed that the ozone concentration was unexpectedly lower over Antarctica than elsewhere in the Earth's atmosphere, as if some "ozone-eater" were chewing up the ozone and leaving a mysterious zone of below-normal concentration, an ozone "hole." Alarmed, scientists examined satellite images taken in previous years. They found that the disintegration of the Earth's ozone shield was evident as far back as 1978. Every year since then, more ozone has disappeared, and the ozone hole has grown larger. Moreover, a smaller hole has appeared over the Arctic, as you can see in Figure 18-3. Currently, the global concentration of ozone in the upper atmosphere has fallen by more than 3 percent. Because more ultraviolet radiation is reaching the Earth's surface, scientists expect more cases of diseases caused by exposure to ultraviolet radiation: skin cancer, cataracts (a disorder in which the lens of the eye becomes cloudy), and cancer of the retina, the light-sensitive part of the eye. For example, in the United States, the number of cases of malignant melanoma, a potentially lethal form of skin cancer, has almost doubled since 1980.

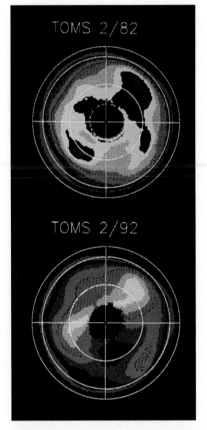

Figure 18-3 Ozone is invisible, but it can be detected by instruments on the ground or carried in satellites. In these computer-enhanced satellite images of the Northern Hemisphere, dark colors indicate zones with the lowest ozone concentration.

Figure 18-4 Because of the thinning ozone layer, it is wise to minimize your exposure to the sun. When out in the sun, you should cover as much of your skin as possible and apply sunscreen to any exposed areas.

What Is Destroying the Ozone?

The major culprit of ozone destruction is a class of chemicals called **chlorofluorocarbons** (CFCs). High over the South and North Poles, where it is very cold, CFCs stick to frozen water vapor and catalyze the conversion of ozone, O_3, into molecular oxygen, O_2. Invented in the 1920s, CFCs were considered miracle chemicals because they were stable, supposedly harmless, and a nearly ideal heat exchanger. Throughout the world, CFCs were used in large amounts as the coolant in refrigerators and air conditioners, as the propellant in aerosol dispensers, and as the foaming agent in the production of plastic foam cups and containers. Though CFCs were escaping into the atmosphere, at first no one worried because CFCs were thought to be chemically inert and because the atmosphere was thought of as limitless. CFCs are very stable chemicals, so over many years they accumulated in the atmosphere. In the early 1970s, however, chemists Sherwood Rowland and Mario Molina warned that CFCs reacted with ozone and therefore could damage the ozone layer. As a result, CFCs were banned as aerosol propellants in spray cans in the United States. Discovery of the ozone hole in 1985 substantiated Rowland and Molina's warning and stimulated the international community to control CFCs.

International agreements to end CFC production by 1996 were signed by the United States and 92 other countries in the early 1990s. Although this development is important and encouraging, the vast majority of the CFCs that have already been manufactured have not yet reached the upper atmosphere. Also, because of their great stability, CFCs currently in the atmosphere will remain there and continue to destroy ozone for more than a century. Moreover, several other kinds of chemicals destroy ozone. Agreements to ban some of these chemicals have been signed, but other ozone-destroying chemicals remain unregulated. Table 18-1 summarizes the uses and regulatory status of some of the major ozone-destroying chemicals. ◻

◻ CAPSULE SUMMARY

Destruction of the ozone layer is caused by chlorofluorocarbons (CFCs) and several other kinds of chemicals. Many, but not all, of these ozone-destroying chemicals are controlled by international treaties.

Table 18-1 Uses and Status of Some Known Ozone-Depleting Chemicals

Chemical	Use	Status
CFCs (Chlorofluorocarbons)	Air conditioners, refrigerators, several industrial processes	Production to cease in 1996
Halon	Fire extinguishers	Production ceased in 1994
Carbon tetrachloride	Industrial solvent	Production to cease in 1996
HCFCs (Hydrochloro-fluorocarbons)	Temporary replacements for CFCs	Production to cease in 2030
Methyl bromide	Pesticide	Unregulated globally

Is the Climate Warming?

For over 150 years the growth of our industrial society has been fueled by cheap energy, most of it obtained by burning fossil fuels—coal, oil, and natural gas. Fossil fuels are the remains of ancient organisms that have been transformed by pressure and heat into carbon-rich substances. When fossil fuels are burned, carbon atoms combine with oxygen atoms, yielding carbon dioxide, CO_2. Industrial society's burning of fossil fuels has released huge amounts of carbon dioxide into the atmosphere. Additional carbon dioxide has been added to the atmosphere by the burning of vegetation to clear land for agriculture. As with CFCs, no one worried because carbon dioxide was thought to be harmless, and the atmosphere was thought to be a limitless reservoir that was able to absorb and disperse any amount of carbon dioxide. That, as it turns out, is not true.

Carbon dioxide in the atmosphere influences global temperatures. The chemical bonds in carbon dioxide molecules absorb solar energy, trapping heat within the atmosphere like glass traps heat within a greenhouse. The warming of the atmosphere that results from the heat-trapping ability of carbon dioxide and several other gases (including methane, CH_4, and nitrous oxide, N_2O) is known as the **greenhouse effect.** Earth's moderate climate results, at least in part, from the insulating effect of carbon dioxide and these other gases.

However, human activities are greatly increasing the concentrations of the greenhouse gases in the atmosphere. Since studies show that the world's temperatures are directly correlated with the concentration of greenhouse gases in the atmosphere, many scientists fear that the rising

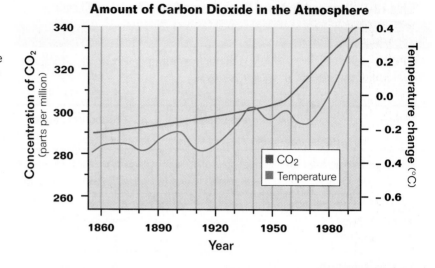

Amount of Carbon Dioxide in the Atmosphere

Figure 18-5 This graph shows the average atmospheric CO_2 concentration and the average change in global temperature over the last 130 years. Currently, average global temperatures are 0.3°C–0.6°C (0.5°F–1.1°F) higher than in 1860. The concentrations of other greenhouse gases, such as methane, nitrous oxide, and CFCs (which also trap heat in the atmosphere) have been increasing as well.

☐ **CAPSULE SUMMARY**

Humans are increasing the concentrations of greenhouse gases such as carbon dioxide in the atmosphere. Since these gases trap heat in the atmosphere, scientists worry that global temperatures may increase.

levels of greenhouse gases are causing **global warming,** an increase in global temperatures. Temperature records of the last 50 years seem to support this concern; global levels of greenhouse gases are rising, and so are global temperatures, as shown in Figure 18-5. In fact, eight of the warmest years on record occurred between 1980 and 1992, and 1990 was the hottest ever measured. Some scientists do not agree that these high temperatures can be blamed on rising carbon dioxide levels. They argue that these years of warm weather are nothing unusual and that such fluctuations have often occurred in the past. As you might expect, there is considerable debate among governments about what ought to be done about global warming. Some countries, such as Denmark and Netherlands, are already reducing their production of carbon dioxide. In late 1993, the United States announced a plan to hold production of greenhouse gases at 1990 levels—although compliance by industry is largely voluntary. ☐

Section Review

1. *Describe two harmful effects of acid rain.*
2. *How will the depletion of the ozone layer affect human health?*
3. *Seven of the eight warmest years on record occurred in the 1980s. Explain why this is not conclusive evidence of global warming.*

Critical Thinking

4. *Some of the pollution from power plants in the United States causes acid rain in Canada. Should the United States compensate Canada for damage to its ecosystems? Justify your answer.*

18-2 Ecosystem Damage

Many of the most serious aspects of the environmental crisis concern changes that do not occur on the same scale as ozone depletion or global warming. This section focuses on the two most important examples of local environmental problems—pollution and consumption of nonreplaceable resources. You will also learn about the core environmental problem that drives local and global pollution and consumption: human population growth.

Section Objectives

- Identify three nonreplaceable resources.
- Describe the history of human population growth.
- Contrast population growth in developing and industrialized countries.

Pollution

Since the industrial revolution, society has had a tendency to assume that the environment can absorb any amount of pollution. Lake Erie, shown in Figure 18-6, and other large lakes became polluted because of the unthinking assumption that they could absorb unlimited amounts of industrial chemicals. Because of overly casual attitudes and poor regulation in industrialized and developing countries, the problem of pollution has grown very serious in recent years.

Many of the most disastrous incidents of pollution involve industrial chemicals that are toxic or carcinogenic (cancer-causing). Until recently, there has been relatively little regulation of the manufacture, transportation, storage, and destruction of such chemicals. A particularly clear example of the casual attitude that has prevailed until recently occurred in Basel, Switzerland, in 1986. Firefighters putting out a warehouse fire inadvertently washed 27,000 kg (30 tons) of mercury and pesticides that were stored in the warehouse into the Rhine River. These poisons flowed down the Rhine, through Germany and Netherlands, and into the North Sea, killing fish and other aquatic animals and plants. Today, the river is recovering, but its species diversity remains far lower than before the disaster. In response, European governments have recently passed laws prohibiting the location of dangerous chemicals so close to major rivers. In the United States, few such laws exist.

In an example of pollution closer to home, an oil tanker named the *Exxon Valdez* ran aground on the coast of Alaska in 1989, as shown in Figure 18-7. Oil that gushed from a tear in the tanker's hull heavily polluted many kilometers of coastline and killed thousands of marine animals. If the tanker had been loaded no higher than the waterline, little oil would have been lost. But it was loaded far higher than that, and the weight of the oil above the waterline forced 42 million L (11 million gal.) of oil out the hole in the ship's hull. Today, years after the spill, cleanup efforts continue, and the evidence of damage to local wildlife continues to mount.

Figure 18-6 In the early 1970s, Lake Erie was so polluted that few fish could survive there. This aerial photo shows a slick of raw sewage spreading into the lake from a damaged sewage treatment plant in Cleveland, Ohio.

Figure 18-7 The *Exxon Valdez* ran aground on the Alaskan coast in 1989. The spilled oil, *above right*, polluted 1,600 km (1,000 mi.) of coastline. About 6,000 sea otters, *above left*, were killed by the spill. By 1994, Exxon, owner of the tanker, had spent over $3 billion in cleanup costs and fines.

Obviously, a "back to nature" approach—one that ignores the benefits of using chemicals intelligently—will not allow us to meet the needs of a very crowded world or to feed the additional billions of people who will join us during the next few decades. But it is essential that we use our technology as intelligently as possible, with due attention to the protection of the productive capacity of all parts of the Earth, on which we all depend.

Consuming Nonreplaceable Resources

Among the many ways that ecosystems are being damaged, one class of problems stands out as potentially more serious than the rest: the consumption or destruction of resources that we cannot replace. Though a polluted stream can be cleaned up, no one can restore an extinct species. Three kinds of nonreplaceable resources are being consumed or destroyed at alarming rates: topsoil, ground water, and species.

Loss of Topsoil

The United States is one of the most productive agricultural countries on Earth, largely because of its particularly fertile soils. These soils have accumulated over tens of thousands of years. The Midwestern farm belt sits astride what was once a great prairie. The topsoil of that ecosystem accumulated bit by bit as the remains of countless animals and plants decayed. By the time humans came to plow the prairie, the topsoil extended down more than a meter.

We can never replace this rich topsoil, and yet we are allowing it to be lost at a rate of several centimeters each decade. By repeatedly turning over the soil to eliminate weeds, by allowing animals to overgraze ranges and pastures, and by practicing poor land management, we permit wind and rain to remove more and more of the topsoil, as shown in Figure 18-8. Our country has lost one-fourth of its topsoil since 1950.

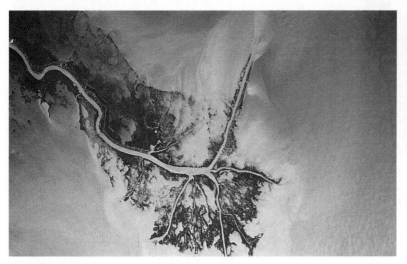

Pollution and Depletion of Ground Water

A second resource that we cannot replace is ground water—water trapped beneath the soil, much of it within porous rock reservoirs called **aquifers** *(AWK wuh furz)*. Water seeps into aquifers at too slow a rate to replace the large amount of water now being withdrawn. In most areas of the United States, there is relatively little control over the use of ground water, with the unfortunate consequence that a very large portion is wasted watering lawns, washing cars, and running fountains. A great deal more is being polluted by poor disposal of chemical wastes. Once pollution enters the ground water, there is no effective means of removing it.

Extinction of Species

Over the last 50 years, about half of the world's tropical rain forests have been destroyed, burned to make pasture and farmland or cut for timber. About 150,000 sq. km (58,000 sq. mi., about the area of the state of Georgia) will be destroyed this year. At this rate, all of the rain forests of the world will be gone within your lifetime. As the rain forests disappear, so do their inhabitants, and tropical rain forests have the highest species diversity of any biome. It is estimated that at least one-fifth of the world's species of animals and plants—about 1 million species—will become extinct during the next 50 years. This is an extinction event unparalleled for at least 65 million years, since the end of the age of the dinosaurs. And the number of species in danger of extinction during your lifetime is far greater than the number that became extinct at the end of the Cretaceous period. ◻

This disastrous loss of species is important to every one of us, because as species disappear, so do our chances to learn about them and their possible benefits. The fact that much of our supply of food is based on only 100 species of plants, out of the tens of thousands of species that have been used at least occasionally as food, should give us pause. Like burning a library without reading the books, we don't know what it is we

Figure 18-8 The Mississippi River, shown here where it flows into the Gulf of Mexico, *above left*, dumps the fertile soil of the Midwest into the ocean. Runoff carries this topsoil into the river from farmland. Loss of topsoil is a worldwide problem. For example, this photo shows Melbourne, Australia, *above right*, during a choking dust storm. Winds pick up topsoil from overgrazed rangeland and abandoned farms in the interior of Australia and carry it to the coast.

◻ *CAPSULE SUMMARY*

Two serious environmental problems are pollution and destruction of nonreplaceable resources. Three nonreplaceable resources in particular danger are topsoil, ground water, and species.

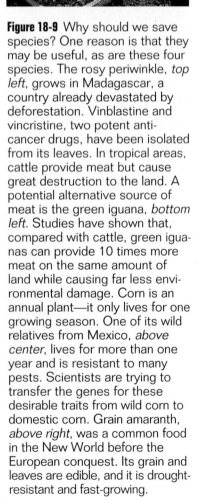

Figure 18-9 Why should we save species? One reason is that they may be useful, as are these four species. The rosy periwinkle, *top left*, grows in Madagascar, a country already devastated by deforestation. Vinblastine and vincristine, two potent anti-cancer drugs, have been isolated from its leaves. In tropical areas, cattle provide meat but cause great destruction to the land. A potential alternative source of meat is the green iguana, *bottom left*. Studies have shown that, compared with cattle, green iguanas can provide 10 times more meat on the same amount of land while causing far less environmental damage. Corn is an annual plant—it only lives for one growing season. One of its wild relatives from Mexico, *above center*, lives for more than one year and is resistant to many pests. Scientists are trying to transfer the genes for these desirable traits from wild corn to domestic corn. Grain amaranth, *above right*, was a common food in the New World before the European conquest. Its grain and leaves are edible, and it is drought-resistant and fast-growing.

waste. All we can be sure of is that we cannot retrieve the lost knowledge. Figure 18-9 shows four tropical species with actual or potential benefits.

You should not be lulled into thinking that extinction is a problem limited to the tropics. The ancient forests of the northwestern United States are being cut swiftly, largely to supply jobs, and much of the cost of logging is subsidized by our government (the U.S. Forest Service builds the necessary access roads, for example). At the current rate, very little of these ancient forests will remain in a decade. It is hypocritical of us to scold tropical nations for destroying their rain forests when we do such a poor job of preserving our own country's species.

The Core Problem: Population Growth

If we were to solve all of the problems of pollution and consumption mentioned in this section, all we would gain is more time to address a more fundamental problem. The human population is growing extremely rapidly, as illustrated in Figure 18-10.

Human beings first reached North America at least 12,000 years ago, by crossing the narrow strait between Siberia and Alaska, and then spread throughout North America and South America. Ten thousand years ago the continental ice sheets that covered northern Europe and North America withdrew, and agriculture first developed. There were only about 5 million people on Earth then, distributed over all of the continents except Antarctica. With the new and much more dependable sources of food that became available as a result of agriculture, the human population began to grow more rapidly. By about 2,000 years ago, there were an estimated 130 million people on Earth. By 1650, the world's population had nearly quadrupled, reaching 500 million.

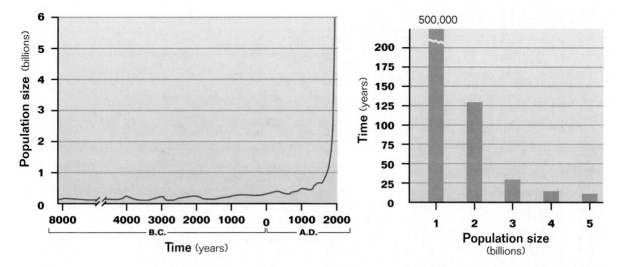

Since at least 1650, and probably for much longer, the human birth rate (as a global average) has remained fairly constant, near 30 births per 1,000 people per year. However, with the spread of better sanitation and improved medical techniques, the death rate has fallen steadily, to an estimated 1994 level of about 9 deaths per 1,000 people per year. The difference between the current annual birth rate (now estimated to be 26 births per 1,000 people) and death rate amounts to an annual worldwide increase in the human population of approximately 1.7 percent. This number may seem small, but don't be deceived. The world's population will double in just over 40 years *if it continues to grow at this rate* (but most scientists think it will not).

The world's population exceeded 5 billion in early 1987, and the annual increase now amounts to about 94 million people; about 260,000 people are added to the world population each day, or about 180 every minute. The world's population is expected to exceed 6 billion by the year 2000. Population growth is fastest in the developing countries of Asia, Africa, and Latin America and slowest in the industrialized countries of North America, Europe, Japan, New Zealand, and Australia. For example, the growth rate of the American population is only 0.8 percent, less than half of the global rate. Most European countries are growing even more slowly, and the populations of Germany and Russia are actually shrinking. By contrast, Kenya's population is increasing by about 3.7 percent per year.

Many countries are devoting considerable attention to slowing the growth rate of their populations, and there are genuine signs of progress. For example, by encouraging families to have only two children, Thailand reduced its growth rate from 3.2 percent to 2.4 percent between 1960 and 1994. In Mexico over the last 30 years, the average number of children per family has decreased from five to less than three. Though the global rate of population growth has been declining, experts project that the world's population will increase

Figure 18-10 The graph on the left shows the soaring human population. Another way to visualize the speed of human population growth is to plot the time required for the population to increase by 1 billion, as done in the graph on the right. It took over 500,000 years—from the origin of our species until about the year 1800—for the human population to reach 1 billion. The population grew from 1 billion to 2 billion in only 130 years, and from 4 billion to 5 billion in only 12 years.

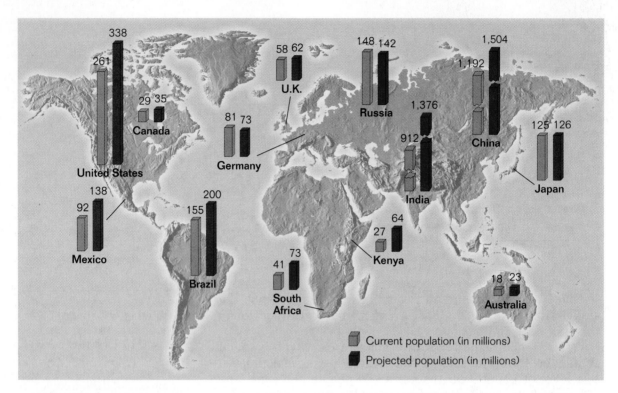

Figure 18-11 This map shows the current populations of several countries (figure on the left in each entry) and their projected populations in 2025 (figure on the right in each entry). Data were obtained from the Population Reference Bureau. These estimates assume that growth rates will continue to slowly decline.

to 8.5 billion by the year 2025. Figure 18-11 shows the current populations of several countries and their projected populations for the year 2025. If the world's growth rate continues to decrease, it is estimated by the United Nations that the population may stabilize by the end of the next century at about 10 billion to 15 billion people. No one knows whether the world can support so many people indefinitely. Finding a way for it to do so is the greatest task facing humanity. The quality of life that will be available to your children will depend to a large extent on our success.

Section Review

1. *Explain why supplies of ground water and soil are dwindling even though these resources are replenished by natural processes.*
2. *How does the growth of the human population affect the populations of other species?*
3. *Describe two instances in which technology has caused the growth rate of the human population to increase.*

Critical Thinking
4. *Unlike most other organisms, humans often use more resources than are needed for survival. How does our standard of living affect the Earth's human carrying capacity?*

18-3 Solving Environmental Problems

The pattern of global change and ecosystem damage that is overtaking our world is very disturbing. Human activities are placing severe stresses on ecosystems worldwide, and we must quickly find ways to reduce their harmful impact. Worldwide attention is focused on solving these problems, and a great deal of progress is being made. In this section you will learn how environmental problems can be solved.

Section Objectives

- Explain how our economic system can make it profitable to pollute.
- Identify the five major steps necessary to solve environmental problems.

Reducing Pollution

One of the most encouraging developments of the early 1990s was the worldwide increase in efforts to reduce pollution. International agreements to stop CFC production are but one example. The release of many dangerous industrial and agricultural chemicals—notably the insecticide DDT and the carcinogens asbestos and dioxin—has been restricted in the United States. A great deal of progress has also been made in reducing air pollution. Emissions of sulfur dioxide, carbon monoxide, and soot—three pollutants produced by the burning of coal—have been cut by over 30 percent in 10 years. The number of secondary sewage treatment facilities, which remove chemicals as well as bacteria from sewage, has increased 72 percent in the last 10 years. The Environmental Protection Agency estimates that businesses and private agencies are spending about $100 billion a year on pollution control,

Figure 18-12 This power plant is equipped with scrubbers that capture pollutants before they are released into the environment.

Figure 18-13 Pollution sometimes kills. In 1952, unusual weather conditions trapped London's pollution, generated mainly by the burning of coal, over the city for five days. This photo, *right,* was taken at noon during the "killer smog." In this case, at least some of the costs of pollution are obvious: 4,200 people died, most from respiratory illnesses aggravated by the choking smog. Pollution also erodes buildings, cars, and statues like this one in England, *above.*

twice the amount spent 10 years ago and five times the amount spent in 1970. However encouraging, this progress represents only a beginning. A serious attempt to address the overall problem of pollution requires more fundamental changes in how the economy of our society operates.

Why Pollution Is Profitable

To learn what needs to be done, it is first necessary to understand the ultimate cause of pollution. In essence, it is a failure by our economy to set an appropriate value on environmental health. To understand how this happens, you must think for a moment about money. The economy of the United States (and much of the rest of the industrialized world) is based on a simple feedback system of supply and demand. As something gets scarce, its price increases. This added profit acts as an incentive for the production of more of the item. If too much of the item is available, the price falls, and because it is no longer so profitable to produce the item, less of it is made.

This system works very well and is responsible for the economic strength of our nation. But it has one great weakness: if demand is set by price, then it is very important that all of the costs be included in the price. Imagine if the person selling the item is able to pass off part of the production cost to a third person; the seller would be able to set a lower price and sell more of the item. And, stimulated by the lower price, the buyer would purchase more than if all the costs had been added into the price.

That sort of pricing error is what has driven industry's pollution of the environment over the last century. The true costs of energy and of the many things made by industry are composed of direct production costs, such as materials and wages, and indirect costs, such as pollution and the risk of unanticipated ill effects to the environment. Imagine if all the medical costs associated with a 20 percent worldwide increase in skin cancer were factored into the price of a refrigerator or air conditioner (because of the ozone-destroying CFCs they contain). Indirect costs from the use of fossil fuels include reduced harvests of fish and shellfish due to oil spills, and crop and timber losses caused by air pollution. As shown in Figure 18-13, pollution from fossil fuels also damages buildings and causes illness and death. Since the indirect costs are not included in the price that the consumer pays, far more is consumed than if they had been included. By not adding the indirect costs to the price of energy and manufactured goods, our society has made it profitable to pollute. The indirect costs do not disappear because we ignore them; they are simply passed on to future generations, who must pay the bill in terms of damage to their own health and to the ecosystems on which they depend. ◾

Paying for Environmental Damage

Two effective approaches have been taken to reduce pollution in this country. The first approach has been to pass laws forbidding it. In the last 20 years, laws have begun to significantly slow the spread of pollution by imposing stiff standards for what can be released into the environment. All cars are required to have effective catalytic converters to reduce emissions, for example. Similarly, the Clean Air Act of 1990 requires that power plants install "scrubbers" on their smokestacks to reduce sulfur emissions. Catalytic converters make cars more expensive, and the installation and maintenance of smokestack scrubbers increase the price of energy. The overall effect is that the consumer pays to avoid polluting the environment. The new higher prices are closer to the true costs, and they lower consumption to more appropriate levels.

A second approach to reducing pollution has been to increase the consumer costs directly by placing a tax on pollution, such as the sulfur emitted by smokestacks. The problem with this approach is that the taxes have never been high enough to reflect the true costs of the pollution. If they were, the consumer costs might be so high as to inhibit industrial growth and development, thus eliminating jobs and depressing economic growth. In recent years, some economists have instead advocated an artificial price hike imposed by the government in the form of a tax added to the cost of goods and energy. This added cost would lower consumption, too, but by adjusting the tax, the government could attempt to balance the conflicting demands of environmental safety and economic growth. Such taxes, often imposed as "pollution permits," are becoming an increasingly important part of laws regulating pollution. They are a key part of the Clean Air Act of 1990. An example of taxes that help control pollution is shown in Figure 18-14.

Figure 18-14 One way to reduce pollution is to tax products whose use or production creates pollution. For example, though the costs of producing gasoline are about the same in the United States and Great Britain, the British pay more than twice as much for gasoline because of higher taxes. At mid-1994 exchange rates, the price of gasoline at this British station is equivalent to $3.20 per gallon. Such high taxes, though primarily designed to increase government revenues, also reduce the use of gasoline and therefore decrease the amount of pollution.

□ **CAPSULE SUMMARY**

There are two ways to factor the costs of environmental damage into the prices of goods and services. One way is to require pollution control devices. The other way is to tax products or services that create pollution.

Environmentalists have objected to this added-cost approach, pointing out that true environmental costs are very hard to calculate. How do you put a price on the possibility that sulfur emissions might drive a distant frog population to extinction? More important, in most cases we do not know enough about ecosystems to predict the effects of pollution with any certainty. What if the long-range consequences of pollution turn out to be disastrous? Pollution permits do not take into account these very real biological risks. □

How to Solve Environmental Problems

It is easy to get discouraged when considering the world's many serious environmental problems. But do not lose track of the single most important conclusion that emerges from our examination of these environmental problems—each of the world's many problems is solvable. A polluted lake can be cleaned up, a dirty smokestack can be altered to remove noxious gases, and the waste of key resources can be stopped. What is required is a clear understanding of the problem and a commitment to do something about it. The extent to which American families recycle aluminum cans and newspapers is evidence of the degree to which people want to become part of the solution, rather than part of the problem.

If one looks at how success was achieved in cases where environmental problems have been overcome, a clear pattern emerges. Viewed simply, there are five components to successfully solving any environmental problem.

1. **Assessment** The first stage of addressing any environmental problem is scientific analysis, the gathering of information about what is happening, as the scientist shown in

Figure 18-15 An environmental problem must be documented and understood before it can be solved. Here, biologist David Mech is fitting a radio-tracking collar to a rare gray wolf so that its movements can be tracked. Such information is necessary before a plan to conserve gray wolves can be created.

Figure 18-15 is doing. Data must be collected and experiments must be performed to construct a model of the ecosystem that describes how the ecosystem is responding to the situation. Such a model can then be used to make predictions about the future course of events in the ecosystem.

2. **Risk analysis** Using the information obtained by scientific analysis, it is possible to predict the consequences of environmental intervention—what could be expected to happen if a particular course of action were followed. It is necessary to evaluate not only the potential for solving the environmental problem, but also any adverse effects that a plan of action might create.

3. **Public education** When a clear choice can be made among alternative courses of action, the public must be informed. This involves explaining the problem in understandable terms, presenting the alternative actions available, and explaining the probable costs and results of the different choices.

4. **Political action** The public, through its elected officials, selects and implements a course of action. Individuals can have a major impact at this stage by exercising their right to vote and by contacting their elected officials. Many voters do not realize how much they can achieve by writing letters and supporting special interest groups.

5. **Follow-through** The results of any action should be carefully monitored to see if the environmental problem is being solved. The results can also be used to evaluate and improve the initial assessment and model of the problem. We learn by doing.

Figure 18-16 Your lifestyle affects the environment. Recycling cans, bottles, and newspapers, *above left*, saves energy and resources. An automobile produces its own mass in pollutants in a year. You can help diminish pollution by riding a bike, walking, or taking public transportation, such as buses and subways. You can find many other ways to help the Earth in books and magazines.

What You Can Contribute

You cannot hope to preserve what you do not understand. Although solving the world's environmental problems will take the efforts of many people, including politicians, economists, and engineers, the issues are largely biological. When all is said and done, your knowledge of ecology is the essential tool that you will need to contribute to the effort. If economists and politicians understood as much ecology as you have learned in these chapters, the world might not be as polluted as it is today. Figure 18-16 shows two simple ways you can participate in solving the problems described in this chapter. Humans rely on the Earth's ecosystems for food and for all of the other materials on which our civilization depends. It has been said that we do not inherit the Earth from our parents but borrow it from our children. We must preserve for them a world in which they can live.

Section Review

1. *Explain two ways to factor the costs of environmental damage into the price of products.*
2. *At which step in the solution of an environmental problem could you have the greatest influence? Explain your answer.*

Critical Thinking

3. *Of the five steps for solving environmental problems, which do you think might be the most likely to fail? Explain your answer.*

Vocabulary

acid rain (390)
aquifer (397)
chlorofluorocarbon (392)
global warming (394)
greenhouse effect (393)

Concept Mapping

Construct a concept map that shows how human activities are disrupting the atmosphere and that describes the effects of these disruptions. Use as many terms as needed from the vocabulary list. Try to include the following terms in your map: greenhouse effect, carbon dioxide, greenhouse gases, global warming, CFCs, ozone layer, acid rain, and high-sulfur coal. Include additional terms in your map as needed.

Review

Multiple Choice

1. Which of the following causes acid rain?
   a. releasing chlorofluorocarbons
   b. burning high-sulfur coal
   c. polluting ground water
   d. scrubbing smokestack emissions

2. An expected effect of ozone depletion is
   a. increased incidence of skin cancer.
   b. depletion of water stored in aquifers.
   c. increased availability of CFCs.
   d. a rise in pH levels in lakes and streams.

3. Burning of fossil fuels has changed the atmosphere by
   a. increasing the global concentration of ozone in the upper atmosphere.
   b. reducing the amount of CFCs.
   c. producing an ozone hole.
   d. increasing the concentration of carbon dioxide.

4. Which of the following is *not* true of the drastic extinction of species that is now occurring?
   a. It is the largest extinction event since the dinosaurs disappeared.
   b. One of its causes is the destruction of tropical rain forests.
   c. It is confined to tropical countries.
   d. Potentially useful species are becoming extinct.

5. In the last 100 years the human population has grown more rapidly than ever before. What factor has contributed the most to this growth?
   a. destruction of the rain forest
   b. the end of global wars
   c. improved sanitation
   d. discovery of new species

6. During the last 200 years, the human death rate has
   a. increased.
   b. decreased.
   c. remained stable.
   d. decreased then increased.

7. Which of these countries has the largest population growth rate?
   a. Germany   c. Kenya
   b. United States   d. Russia

8. The fact that the pollutants produced by industries can be dumped into the environment at no cost has contributed to
   a. goods being sold at prices below their true cost.
   b. bargain prices that weaken consumer confidence.
   c. an increase in the costs of cars since 1970.
   d. government reluctance to tax industries.

9. Of the five major steps to solving environmental problems, which involves determining the potential outcomes of an environmental plan before it is tried?
   a. assessment   c. follow-through
   b. risk analysis   d. political action

Completion

10. Acid rain is created when sulfur combines with _____ , forming sulfuric acid. The accumulation of acid rain has _____ the pH in many lakes in the United States and Canada, often killing the organisms that live there.

11. Scientists suspect that _____ are primarily responsible for the thinning of the Earth's ozone layer. A decreased ozone concentration will allow higher levels of _____ radiation to reach the Earth's surface and will _____ the number of cases of skin cancer.

12. The _____ is due to the accumulation of carbon dioxide and other gases in the atmosphere.

13. Ground water, which collected in _____ , is one nonreplaceable resource. Ways to _____ ground water include controlling its use and disposing of chemical wastes properly.

14. Ten thousand years ago, about _____ people inhabited the Earth. About 2,000 years ago, the human population was about _____ . In the year 2000, the world's population will exceed _____ people.

15. The price paid for most goods is _____ the actual cost because the cost of _____ and the unexpected harmful effects on the environment are not included.

Short Answer

16. Tall smokestacks are a part of many coal-burning power plants. What is the purpose of these tall smokestacks?

17. The stability of CFCs, long considered one of their virtues, is now recognized to be a serious drawback. Explain how the stability of these chemicals multiplies their destructive effects on the ozone layer.

18. What is the difference between the greenhouse effect and global warming?

19. The human population began to grow rapidly about 300 years ago. What caused this increase in the population growth rate?

Themes Review

20. **Structure and Function** A number of land animals have one or more layers of nonliving material covering their living tissue. The exoskeleton (external skeleton) of insects and the feathers of birds are two examples. How might such an adaptation improve an animal's chances for survival and reproduction in an environment with high levels of ultraviolet radiation?

21. **Evolution** The Earth's species diversity has declined drastically several times in the past, during so-called mass extinction events. Each time, however, new species evolved, and species diversity rebounded. Some critics of efforts to save species have argued that we need not worry about the loss of species, since new ones will evolve to take their place. Using your knowledge of evolution, identify a flaw in this argument.

Critical Thinking

22. **Making Inferences** Why is it wise to wear a hat and apply sunscreen to exposed areas of your body when you plan to be outdoors for an extended period of time?

23. **Interpreting Data** It is a widely held belief that the birth rate will decline when countries make progress toward industrialization. However, data show that birth rates have declined in China, Sri Lanka, and Costa Rica, countries with minimal industrial development, and have remained constant in Brazil, which has made progress toward industrialization. How might these data be explained?

24. Making Predictions Acid rain can damage or even kill many plants and animals. The photo shows a situation that contributes to the production of acid rain. What methods could be used to reduce the pollutants that cause acid rain?

25. Making Predictions Harmful ultraviolet radiation kills photosynthetic plankton, which absorb large amounts of carbon dioxide from the air. What effect would ozone depletion have on the levels of atmospheric gases such as carbon dioxide?

26. Making Predictions To prevent starvation, food production must increase at least as fast as population size. How will loss of topsoil and extinction of species hamper efforts to increase the production of food?

Activities and Projects

27. Research and Writing Contact local or state wildlife officials to obtain a list of endangered species that live in your state or area. Select one of these species and research the efforts being made to protect it. Write a report that explains and evaluates these efforts. In your report you should also describe what environmental changes caused the species to become endangered. Share your findings with your class.

28. Research and Writing CFCs deplete ozone in the upper atmosphere and contribute to the greenhouse effect. For these reasons, the chemical industry is now developing a range of alternatives to CFCs. Write a report that describes the requirements established for CFC replacements and that identifies several alternatives to CFCs.

29. Multicultural Perspective Many medicines are derived from chemicals produced by plants. One such plant is the neem tree of Asia. Research the medicines and other products obtained from this tree in India, where it is known as the "village pharmacy." Share your findings with your class.

Readings

30. Read *The Population Explosion* by Paul and Anne Ehrlich. Explain the logic behind the Ehrlichs' assertion that the United States is overpopulated. What does each term in the equation I=PAT represent? How does the magnitude of each term in the equation differ between developed and developing countries?

31. Read the article "The Scandal of Siberia," in *New Scientist*, November 27, 1993, pages 28–33. What kinds of damage have been done to the forest and swamp ecosystems of Siberia as a result of oil exploration?

32. Read the article "Plight of the Plover," in *Science News*, December 7, 1991, pages 382–383. How has the presence of humans on the beaches in the eastern United States affected the natural habitat of piping plovers? What actions have been taken to ensure that the plovers have peaceful nesting sites? What can the average citizen do to help save these birds from extinction?

Human Impact on the Environment **409**

Effects of Thermal Pollution

OBJECTIVE

Model the effects of thermal pollution on organisms in the laboratory, and apply the underlying scientific principles to current environmental issues.

PROCESS SKILLS

- forming a hypothesis
- designing an experiment
- observing
- analyzing data

MATERIALS

- two 400 mL beakers
- ice
- hot water
- thermometer
- U-shaped glass tube, 30 cm long
- two corks to fit ends of tubing
- 125 mL beaker
- graduated cylinder
- water
- *Paramecium* culture
- hand lens
- clock or watch
- glass-marking pen or wax pencil

BACKGROUND

1. What is a pollutant?
2. How can heat be considered a pollutant?
3. How can power plants pollute water, even when they have added no chemicals or waste products to the water?

4. How can power plants release nonharmful water?
5. Write your own **Focus Question** on your Vee Form.
6. **Knowing Side of the Vee** List the **Concepts** and new **Vocabulary Words** on your Vee Form. In the **Concept Statements** section of the Vee, use these words in sentences that define and explain them.

TECHNIQUE

Doing Side of the Vee

1. Discuss the objective of this investigation with your partners and develop a hypothesis concerning the effect of temperature on *Paramecium*.
2. Design an experiment to test your hypothesis. In your experiment, *Paramecium* will be contained in the U-shaped tube. One large beaker will be filled with ice, and the other large beaker will be filled with hot water. Other available supplies are listed in your materials list.
3. In designing your experiment, decide which factor will be an independent variable. Plan how you will vary the independent variable.
4. Decide which factor will be the dependent variable. Plan how you will measure changes in the dependent variable.
5. Remember that in most experiments a control is necessary. Plan a control for this investigation.
6. Discuss your planned experiment with your teacher. Proceed with the experiment once you have received approval from your teacher.
7. Once you begin the investigation, proceed until it is completed. Remember to record your results, including the numbers of individual *Paramecium* and the time involved.

8. Fill a 400 mL beaker with ice and water. Make sure ice remains in the beaker for the entire experiment. Fill another 400 mL beaker with 60°C tap water. **CAUTION: Be careful handling hot water.** Water more than 60°C can scald.

9. In a 125 mL beaker, gently swirl 20 mL of water and 20 mL of *Paramecium* culture. **Note:** Your teacher will supply you with aged tap water or spring water. (Chlorinated water would kill *Paramecium*.)

10. While your partner holds the tube steady, carefully pour the *Paramecium* and water mixture into the U-shaped tube. Fill the tube completely, leaving just enough room for a cork. Make sure there are no large bubbles of air in the tube. Place a cork in each end of the tube.

11. Proceed with your experiment using the tubing, ice water, hot water, and hand lens to observe any response of *Paramecium* to their environment.

12. Record the data you collect in the **Records** section of your Vee Form. In the **Procedure** section of the Vee, briefly summarize the procedure you followed.

13. Clean up your materials and wash your hands before leaving the lab.

INQUIRY

1. What effect did heat have on *Paramecium* in your experiment? What was the effect of cold?

2. What evidence did you have that *Paramecium* preferred one temperature range to another?

3. What was the independent variable? How did you vary it?

4. What was the dependent variable? How did you measure changes in it?

5. What controls did you use?

6. Why are the length of time and the temperature range important factors?

7. Use the information on the **Knowing Side** of the Vee to interpret your results from the **Doing Side**, and then write your **Knowledge Claim.** Write a **Value Claim** for this lab.

ANALYSIS

1. Did the results of your experiment support your hypothesis? Explain.

2. What are some possible sources of error in your experiment?

3. How could a pollutant cause an increase in the number of organisms? Explain.

4. Judging from your experiment, how do you think other organisms might react to a change in water temperature?

5. How could a power plant change the type of organisms that live in the water where it releases its cooling water?

FURTHER INQUIRY

Write a **New Focus Question** that could be the point of a new investigation. The following is an example:

How can an experiment be designed to test how acid rain affects *Paramecium*?

BIOLOGICAL DIVERSITY

How Are We Changing Life on Earth?

The extinction of the tiger means the elimination of a single species. The destruction of rain forests results in the loss of hundreds of species.

BY TRACEY COHEN

Between 1989 and 1992, 18 tigers were killed by poachers in India's Ranthambhore National Park. Although the country is dedicated to protecting the last big carnivores in its parks, India's tiger populations have fallen about 35 percent over the last five years. Elsewhere in Asia, tiger populations are also endangered. Probably no more than 5,000 to 7,000 of the great striped cats are left in the wild. The situation is so dire that many conservationists fear that wild tigers will be extinct in only a few years. Sadly, if the tiger becomes extinct, it will be just one of thousands of species driven to extinction by human activities.

Extinction is actually a natural process. Scientists estimate that every million years for the last 200 million years about 900,000 vertebrate species have become extinct. For higher plants, the extinction rate has been about 37,000 species every million years for the past 400 million years.

If extinction is part of the course of natural events, why does it matter if tigers and other organisms die out?

Why Are Different Species Important?

From a human perspective, natural ecosystems and the organisms in them are important for a number of reasons. The most obvious is that plants and animals provide us with food. For example, throughout history people have eaten about 7,000 different kinds of plants. At least 75,000, however, are known to be edible. Plants and animals also provide raw materials for clothing, fuels, oils, dyes, industrial chemicals, and building materials. Over 40 percent of the medicines used in the industrialized world come directly from organisms or were synthesized based on chemical compounds taken from them: 25 percent from plants, 13 percent from microorganisms, and 3 percent from animals. Aspirin, antibiotics, and the anticancer drugs vincristine and vinblastine are just a few examples. The possibilities for new foods, drugs, and other substances from the world's genetic library are enormous.

Biological diversity is crucial for other less visible, but no less important, reasons. The vast web of living things provides a variety of "life-support" services on a scale that would be impossible to replace. Plants and algae, for instance, capture the sun's energy and make it available to other organisms through photosynthesis. They also purify the planet's air by taking in carbon dioxide and releasing oxygen. The roots of plants break apart rocks and, with the help of specific microbes, form soil and keep it fertile. The global cycling of nutrients and the decomposition of wastes depend on countless organisms in a variety of ecosystems, as do climate patterns and water circulation and purification.

What Threatens Biological Resources?

According to some estimates, human activities have increased the natural rate of extinction by as much as 10,000 times. For example, 11 percent of the world's bird species are endangered. About 20 percent of the world's freshwater fish species are extinct or nearly so. In the United States, over 200 plant species are extinct and another 680 species may be gone by the end of the decade.

The single greatest threat to biodiversity is the destruction of habitat. According to the International Union for the Conservation of Nature and Natural Resources, 73 percent of those species with declining populations are suffering from habitat destruction. Loss of habitat occurs when people convert ecosystems from one type to another, say from forest to farmland or from wetlands to suburban housing developments or shopping malls. Habitat destruction also happens as a result of pollution. Such changes destroy hundreds of thousands of square kilometers of habitat. According to biologist E. O. Wilson, of the many habitats in the world that cover at least 1 sq. km, not many have fewer than a thousand species of plants and animals. Patches of rain forest and coral reefs—the most productive habitats—have tens of thousands of species.

Introduction of foreign species also seriously threatens biodiversity. Plants, animals, and other organisms have been spread by humans to new locations both accidentally and on purpose. Foreign species displace or destroy native ones. For example, in 1959 British colonists released Nile perch into Lake Victoria, in Africa. This predator fish has greatly reduced the native fish population and caused the extinction of some species in the lake. In the United States, the most serious threat to ecosystems in the national park system are new species of plants introduced into a park.

What Can Be Done to Preserve Biological Diversity?

Scientists agree that the first step is to collect as much information as possible. One approach, pioneered by Conservation International, is the Rapid Assessment Program (RAP). RAP is designed to quickly survey possible ecosystems that are "hot spots"—areas with the largest number of

endangered species. Emergency recommendations can then be made about conservation strategies for these areas.

At the same time, there are some immediate actions that can help. Zoos, for instance, are breeding endangered and threatened animals so that they can provide a reservoir for restoring wild populations—assuming, of course, that natural habitats remain.

Botanical gardens and seed banks are helping to preserve plant species. Employees of botanical gardens grow plants and carry out long-term research of their biology, reproduction, and roles in ecosystems. Seed banks conserve plant genetic material through seed storage. There are currently more than 50 seed banks worldwide. Many are devoted to preserving the numerous local varieties of food crops like corn, wheat, and potatoes.

Efforts to restore destroyed or degraded ecosystems are also underway. Projects range in size from small local streams to the Florida Everglades.

The Limits of Science and Technology: Can Biological Diversity Really Be Saved?

People tend to think that the crisis of biodiversity, as well as other environmental problems

created by human activities, can be corrected with things like better technology. Science and technology can provide crucial information and strategies, but there are limits to their usefulness. Captive breeding programs, for instance, can help save no more than a few hundred of the 1 million known animal species. On a larger scale, too many interdependent species are being lost, often before their ecological roles are fully known. And it would be impossible to create artificially all the conditions necessary for complex, healthy ecosystems.

The true problem of preserving life on Earth is ethical. It means recognizing that other species have a right to continue living. This recognition involves drastic changes in the ways we see and use natural resources. It also means changing the way we see ourselves in the larger community of living things. We need what environmental science professor David Orr called a "biophilia revolution." Biophilia is, in essence, a love of life and things that are alive. A biophilia revolution would combine a reverence for life with efficient use of resources and simpler lifestyles. Until that happens, the plight of biodiversity will not be solved.

Tracey Cohen is a freelance writer specializing in science and environmental issues.

Analyzing the Issue

1. **Reading Critically** What is the thesis, or main point, of the article? List the facts that the writer uses to support the thesis.

2. **Formulating an Opinion** Many scientists believe that humans have an ethical responsibility to preserve biodiversity. Yet, according to biologists Paul and Anne Ehrlich, "One cannot assert this ethical responsibility on scientific grounds."

   a. What do you think the Ehrlichs mean by this state-

   ment? Do you agree? Explain your answer.

   b. What role do scientists have in helping societies make ethical or moral decisions related to nature?

3. **Take Action** Find out what legislation exists at the local, state, and federal level to protect species and ecosystems. How does this legislation work? Has there been criticism of it? Why? By whom?

PART 2

BIOLOGICAL EXPLORATIONS

Protist

South African
bullfrog

Nile crocodile

Catfish

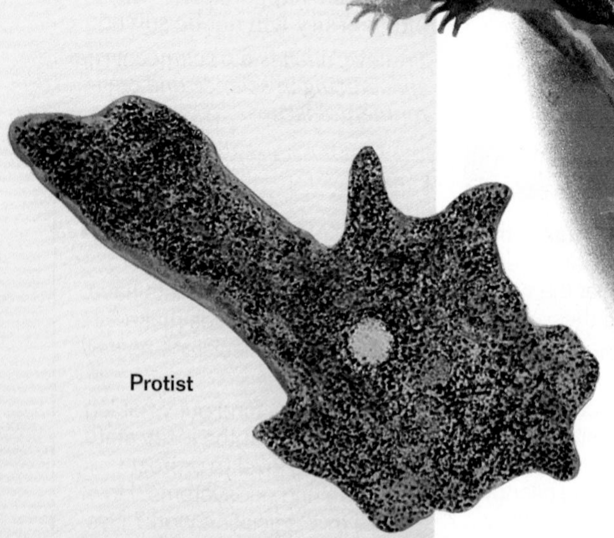

> *Since our genes stretch back in time in an unbroken line to the first lifeforms—just as the genes of mushrooms, bacteria, and rabbits do—it is questionable to assume that we and our domesticated mammals are higher forms of life. Certainly more recent, in some ways more complex, our intelligence is, nonetheless, simply another successful strategy for survival. . . . We simply exist today because our particular collections of genes, like stacks of chips growing on a roulette table, have not yet exhausted their winning streaks.*
>
> — Dorion Sagan and Lynn Margulis
> *Garden of Microbial Delights*

Mountain biker

Lappet-faced vulture

African elephant

Although they appear dissimilar, the African elephants, catfish, and underwater plants (and the unseen bacteria and fungi also present in this scene) have much in common. Living things have evolved surprisingly similar strategies for obtaining energy, retaining water, reproducing, and a host of other tasks essential for life. These strategies have been modified through time, fine-tuning each organism to conquer the environmental challenges posed by life on Earth. Whether a bacterium, a protist, a plant, a crocodile, or a human, each living thing is the result of a long, successful evolutionary journey.

Biological Explorations   **415**

CHAPTER 19

OVERVIEW OF DIVERSITY

REVIEW

- characteristics of prokaryotes and eukaryotes (Section 2-1)
- binary fission and mitosis (Section 6-2)
- crossing-over and meiosis (Section 6-3)
- introns (Section 8-3)
- homologous characters (Section 12-2)
- reproductive isolation (Section 12-3)
- six kingdoms of life (Section 13-1)
- symbiotic relationships (Section 13-2)

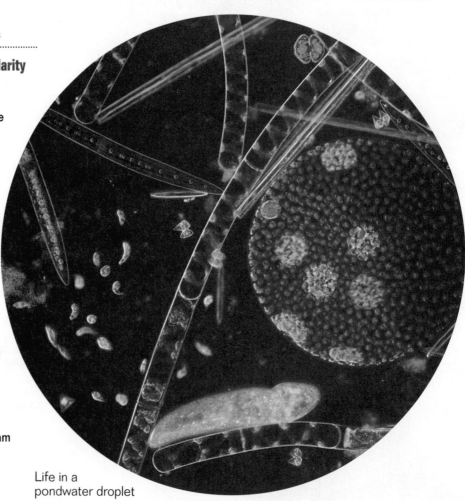

Life in a pondwater droplet

19-1 Origin of Eukaryotes

*L*ife on Earth is over 3.6 billion years old. For the first 2 billion years, the biggest organisms were prokaryotes no larger than 6 µm thick. Evidence of different life-forms is found in 1.5-billion-year-old fossils of larger cells (some as large as 60 µm in diameter) that appear to have had internal membranes and small, membrane-bound structures. These larger organisms mark one of the most important events in the evolution of life—the appearance of the eukaryotic cell. Highly adaptable, eukaryotes evolved rapidly, populating the Earth with a diversity of larger, more complex organisms such as yourself.

Prokaryotes: The Oldest Organisms

Prokaryotes are the planet's most abundant inhabitants; one teaspoon of soil can be home for over 1 billion bacteria. Although microscopic and structurally simple, bacteria are the most diverse of all living organisms. They are found in the widest possible range of habitats, playing important roles in almost all of them. The reason that bacteria are able to play such varied roles is their metabolic diversity. For example, some bacteria obtain energy from inorganic sources such as sulfur or ammonia; others can metabolize petroleum. Such metabolic diversity is what has enabled bacteria to adapt to environmental niches too harsh to support other forms of life.

Early in the history of life, two structurally and metabolically different groups of bacteria evolved—archaebacteria and eubacteria. Figure 19-1 illustrates their evolution from an ancestral prokaryote.

Even though both archaebacteria and eubacteria are prokaryotes, they are so different from each other that biologists assign them to two separate kingdoms—kingdom Archaebacteria and kingdom Eubacteria, which are the subjects of the *Highlights* features on pages 443 and 444. The characteristics that

Figure 19-1 An ancestral prokaryote gave rise to two different kinds of bacteria—archaebacteria and eubacteria. *Sulfolobus* is an archaebacterium that thrives in hot, sulfur springs. *E. coli* is a eubacterium that lives inside your intestines.

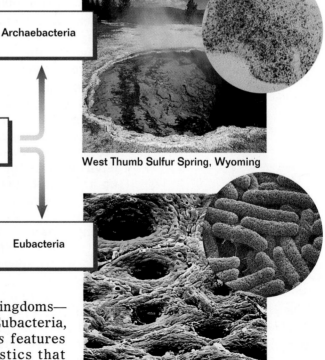

Archaebacteria

Ancestral prokaryote

West Thumb Sulfur Spring, Wyoming

Eubacteria

Lining of human intestine

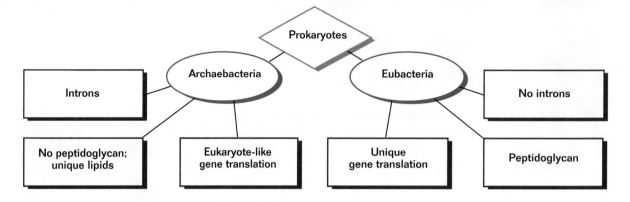

Figure 19-2 Archaebacteria and eubacteria are so different from each other that biologists assign them to separate kingdoms.

CONTENT LINK

The kingdom Eubacteria will be discussed in more detail in **Chapter 20.**

□ CAPSULE SUMMARY

Bacteria, the oldest and most abundant form of life, evolved into two branches: the archaebacteria and the eubacteria.

differentiate these two kingdoms of bacteria are summarized in Figure 19-2.

1. **Cell wall**   All bacteria have a plasma membrane surrounded by a cell wall. The cell walls of eubacteria are made of **peptidoglycan** *(pep tih doh GLY kan)*, a carbohydrate-protein compound that forms a strong mesh that strengthens the cell wall. Peptidoglycan is not found in the cell walls of archaebacteria.

2. **Plasma membrane**   In all bacteria, the plasma membrane is made of a phospholipid bilayer. The plasma membranes of archaebacteria, however, contain unique lipids that are not found in any other organism.

3. **Gene translation machinery**   Archaebacteria have a ribosomal protein and an RNA polymerase that are very similar to those found in eukaryotic cells. The gene translation machinery of eubacteria, however, is distinctly different from that of both archaebacteria and eukaryotes.

4. **Gene architecture**   The genes of eubacteria are not interrupted by introns as are those of eukaryotes. However, some of the genes of archaebacteria have been found to have introns.

Most of the archaebacteria that survive today are methanogens *(muh THAN uh jehnz)*, bacteria that use hydrogen gas (H_2) to reduce carbon dioxide (CO_2) to methane (CH_4). Methanogens are **obligate anaerobes,** organisms that are poisoned by oxygen. Swamps and marshes are perfect environments for them; the methane they produce bubbles up as "marsh gas." Methanogens also live in the digestive tracts of cows and other herbivores that consume a diet rich in cellulose, converting carbon dioxide to methane gas. An essential part of the sewage treatment process is encouraging the growth of these organisms in anaerobic digestion tanks, where sewage sludge is converted into methane gas.

Two other types of archaebacteria live in unusually harsh environments. The archaebacteria called extreme halophiles live in very salty places such as the Dead Sea (an inland body of salt water on the Israel-Jordan border) and the Great Salt Lake (a shallow saltwater lake in Utah). The thermoacidophiles grow in hot, acidic environments such as the sulfur springs of Yellowstone National Park. ■

The First Eukaryotes: A New Kind of Cell

One and a half billion years ago the first eukaryotes appeared. Recall that a eukaryotic cell has a complex system of internal membranes and DNA that is enclosed within a nucleus. What were the first eukaryotes like? Although biologists cannot be sure, an organism that may resemble a very early stage in the evolution of the eukaryotic cell is *Pelomyxa palustris*, shown in Figure 19-3.

Pelomyxa is a single-celled, nonphotosynthetic organism found on the bottom of freshwater ponds. Like some of the early eukaryotes, *Pelomyxa* is much larger than a prokaryotic cell and contains a complex system of internal membranes. It also resembles some of the fossils of early eukaryotes. However, *Pelomyxa* still resembles bacteria in two ways. First, it does not have mitochondria, the sites of cellular respiration in eukaryotic cells. However, it does contain two kinds of bacteria that may play the same role that mitochondria do in all other eukaryotes. Second, *Pelomyxa* does not undergo mitosis like eukaryotic cells. Instead, its nuclei divide in a manner similar to binary fission in bacteria: they divide by pinching apart into two new nuclei around which membranes form. Many of the fundamental characteristics of *Pelomyxa* resemble those of the archaebacteria far more than those of the eubacteria.

The Evolution of Mitochondria

If one of the main differences between prokaryotes and eukaryotes is the presence of mitochondria, how did these organelles arise in eukaryotic cells? To explain the origins of mitochondria, most biologists accept the theory of endosymbiosis. The **theory of endosymbiosis,** presented in the 1970s by Lynn Margulis, a biologist at the University of Massachusetts at Amherst, proposes that mitochondria are the descendants of symbiotic, aerobic (oxygen-requiring) eubacteria. The symbiotic bacteria may have first entered large cells (similar to *Pelomyxa*) as parasites or undigested prey and eventually taken up residence there, as illustrated in Figure 19-4. The eubacteria most similar to mitochondria are the nonsulfur purple bacteria, which are able to carry out the key metabolic process of oxidative respiration. Before they had acquired these bacteria, the larger host cells were unable to perform oxidative respiration, a process essential for living in an atmosphere that contained increasing amounts of oxygen.

Transitional Prokaryote

Resembles prokaryotes
- Lacks mitochondria
- Reproduces in a manner similar to binary fission

Resembles eukaryotes
- Larger than prokaryotic cells
- Contains complete system of internal membranes

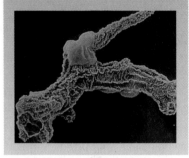

Figure 19-3 The protist *Pelomyxa palustris* is a eukaryotic cell with prokaryotic features. It may resemble some of the first eukaryotic cells that appeared on Earth.

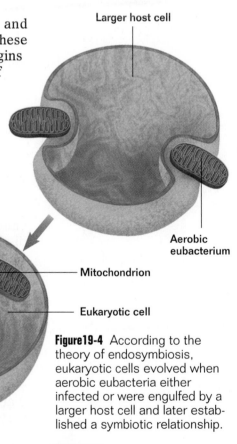

Larger host cell

Nucleus

Aerobic eubacterium

Mitochondrion

Eukaryotic cell

Figure 19-4 According to the theory of endosymbiosis, eukaryotic cells evolved when aerobic eubacteria either infected or were engulfed by a larger host cell and later established a symbiotic relationship.

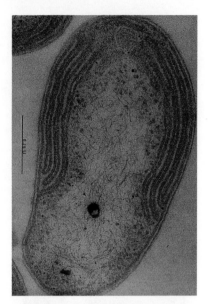

Figure 19-5 The theory of endosymbiosis is supported by the observation that the plasma membrane infoldings of aerobic eubacteria, such as *Nitrobacter,* function in cellular respiration and resemble the membranous layer found in mitochondria.

What observations support the idea that mitochondria are descended from bacteria? There are several key lines of evidence.

1. **Size and structure** Mitochondria are sausage-shaped organelles 1–3 μm long, about the same size as most eubacteria. They are bound by two membranes. The smooth outer membrane is thought to be derived from the endoplasmic reticulum of the larger host cell. The inner membrane is folded into numerous layers, resembling the plasma membranes of aerobic eubacteria, shown in Figure 19-5. Embedded within this membrane are proteins that carry out oxidative respiration.

2. **Genome** Mitochondria have DNA in the form of a circular, closed molecule similar to the chromosomes found in bacteria. This DNA molecule contains the genes encoding the essential proteins of oxidative respiration. However, during the billion and a half years in which mitochondria have existed within eukaryotic cells, most of their genes have been transferred to the chromosomes of the host cells.

3. **Gene translation machinery** Genes within mitochondrial DNA are expressed within mitochondria. Mitochondrial ribosomes resemble bacterial ribosomes in size and structure.

4. **Reproduction** Mitochondria, like bacteria, reproduce by simple fission. However, genes in the cell nucleus direct mitochondrial reproduction.

Other Examples of Endosymbiosis

In addition to mitochondria, many eukaryotic cells contain other organelles thought to be descended from bacteria. Plants and photosynthetic protists like algae contain chloroplasts. According to the theory of endosymbiosis, chloroplasts in the cells of plants and algae are descendants of photosynthetic prokaryotes that were incorporated into a larger host cell. The chloroplast is bound by two membranes that resemble those of mitochondria and were apparently derived in the same fashion. As in mitochondria, the gene translation machinery of the chloroplast closely resembles that of bacteria.

While all mitochondria almost certainly arose from a single symbiotic event, it has only recently become clear that chloroplasts also have a single origin. Three biochemically distinct classes of chloroplasts exist, each resembling a different bacterial ancestor. Red algae have chloroplasts containing pigments similar to those of cyanobacteria. The chloroplasts of plants and green algae, however, contain pigments that more closely resemble those of the photosynthetic bacterium *Prochloron*. Brown algae contain pigments resembling yet a third group of bacteria. This diversity of chloroplasts suggested that eukaryotic cells acquired chloroplasts by symbiosis at least three different times. However, recent comparisons of chloroplast DNA sequences indicate that there was a single origin of chloroplasts, followed by very different evolutionary histories. In each of the three main lines, different genes became relocated to the nucleus, lost, or modified. Chloroplasts seem to have had a rich evolutionary history after their entry into eukaryotic cells.

Centrioles, organelles associated with the assembly of microtubules, also appear to have had an endosymbiotic origin. They resemble spiral-shaped bacteria (spirochetes) and contain bacteria-like DNA involved in the production of their structural proteins. The body of evidence supporting endosymbiotic spirochetes as the evolutionary forerunner of centrioles is still incomplete and is an area of active research. ◻

◻ CAPSULE
SUMMARY

The evolution of organelles such as mitochondria and chloroplasts is explained by the theory of endosymbiosis. This theory proposes that eukaryotic cells evolved when certain prokaryotes established symbiotic relationships within larger cells.

The Origin of Mitosis Is Unknown

Another characteristic that distinguishes eukaryotes from prokaryotes is mitosis, the process of nuclear division. How did this intricate mechanism in a eukaryotic cell arise? Mitosis did not evolve all at once. There are traces of very different, and possibly intermediate, mechanisms in some of the eukaryotes surviving today. In fungi and some groups of protists, for example, the nuclear envelope does not disintegrate. Only after the chromosomes have been replicated and sorted is the nuclear envelope constricted to form two new nuclei. This is followed by cytokinesis. This separate nuclear division phase of mitosis does not occur in most protists or in plants or animals. Biologists are not certain whether it represents an intermediate step on the evolutionary journey to the form of mitosis that is characteristic of most eukaryotes today, or whether it is simply a different way of preserving the number of chromosomes in a cell. Scientists have yet to find fossils in which the interiors of dividing cells can be seen well enough to trace the history of mitosis.

Section Review

1. List four differences between Sulfolobus and E. coli. Do you agree or disagree with biologists who place the two bacteria in separate kingdoms? Explain.

2. Why is Pelomyxa palustris thought to resemble the first eukaryotes?

3. Explain the theory of endosymbiosis. What evidence would you present to support this theory?

Critical Thinking

4. How does mitosis in a mushroom differ from mitosis in your body? Does the form of mitosis in the mushroom have any evolutionary significance? Why or why not?

19-2 Evolution of Sexual Reproduction

O *f all the differences between prokaryotes and eukaryotes, the most essential is found in their mode of reproduction. Prokaryotes reproduce by binary fission, a type of asexual reproduction involving one parent that produces genetically identical offspring without forming gametes. Eukaryotes are able to undergo sexual reproduction. In sexual reproduction two parents contribute genetic material that recombines during the formation of offspring.*

Section Objectives

- *Differentiate between asexual and sexual reproduction. Explain the benefits of sexual reproduction.*
- *Discuss a current theory about the evolution of sexual reproduction.*
- *Describe three major sexual life cycles found in eukaryotes.*

Asexual Reproduction Is Primitive and Prevalent

Asexual reproduction is the simplest and undoubtedly the most primitive method of reproduction. It is advantageous for organisms in a stable environment, enabling them to produce many well-adapted offspring in a short period of time. Also, an organism that reproduces asexually does not need to expend energy producing gametes or finding a mate.

The oldest groups of eukaryotes are all protists that are able to reproduce asexually. Some protists are also able to reproduce sexually. For example, the flagellates, a group of single-celled eukaryotic organisms that propel themselves with whiplike flagella, are predominantly asexual. Other protists, such as the green algae, exhibit a true sexual cycle, but only occasionally. Indeed, many protists reproduce asexually almost all the time. The fusion of two haploid cells to create a diploid zygote, the essential act of sexual reproduction, usually occurs in the presence of stressful conditions in the environment, such as a shortage of nutrients.

In some forms of asexual reproduction a single parent fragments or forms buds to give rise to offspring. These methods are found in multicellular eukaryotes. For example, consider a sponge, a representative of a group of animals living in the sea. Frequently, a sponge reproduces by fragmentation, the breaking of the body into several small pieces. Each small portion is able to grow into a whole new individual. On the other hand, some organisms, like the tiny, cylindrical *Hydra* shown in Figure 19-6, reproduce by budding. In budding, a small part of the parent's body grows into a new individual. The bud may break from the parent and become an independent organism, or it may remain attached to the parent and eventually give rise to a colony composed of many other individuals.

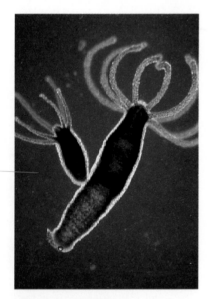

Figure 19-6 Some organisms, such as this *Hydra,* are able to reproduce by budding. In this form of asexual reproduction, a new individual grows out of the body of the original.

Sexual Reproduction Has Advantages

One of the most important evolutionary innovations of eukaryotes was the ability to reproduce sexually. Sexual reproduction provides a powerful means of shuffling genetic material, quickly generating different combinations of genes among individuals. The rapid generation of genetic diversity produced by sexual reproduction has been a principal factor in the evolutionary success of eukaryotes. Genetic diversity is the raw material for evolution. In many cases, the pace of evolution appears to be related to the level of genetic diversity available for selection to act upon: the greater the genetic diversity, the more rapid the evolutionary pace. The processes that occur during sexual reproduction and give rise to genetic diversity are shown in Figure 19-7.

However, as you learned in Chapter 12, evolution is the result of changes that occur at the level of *individual* survival and reproduction rather than at a species level. It is not immediately obvious what advantage is gained by the offspring of an individual that reproduces sexually. The segregation of chromosomes that occurs in meiosis tends to disrupt advantageous combinations of genes more often than it assembles new, better-adapted ones. As a result, some of the offspring produced by sexual reproduction are not as well adapted as their parents. Because all of the offspring could maintain a parent's successful gene combinations if the parent reproduced asexually, the appearance of sexual reproduction among eukaryotes raises a question: What are the benefits that promoted its evolution?

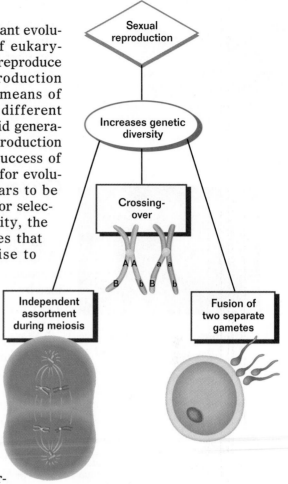

Figure 19-7 Sexual reproduction represents an advance in the ability of an organism to generate genetic variability.

How Sexual Reproduction Evolved

In attempting to answer this question, biologists have looked more carefully at the origins of sexual reproduction among the protists. Many species of protists form a diploid cell in response to stress in the environment. Why? Biologists think this occurs because only in a diploid cell can certain kinds of chromosome damage be effectively repaired, particularly breaks in both strands of DNA. As organisms became larger and lived longer, it must have become increasingly important for them to be able to repair such damage. The early stages of meiosis, in which the two copies of each chromosome line up and pair with each other, may have originally been a mechanism for repairing damage to DNA. Perhaps the undamaged version of the chromosome was used as a template to fix the damaged one. Indeed, the molecular events that occur during crossing-over in meiosis involve many of the enzymes that repair DNA damage. In yeasts, mutations that inactivate the system that repairs double-stranded breaks in chromosomes also prevent crossing-over. Thus it seems likely that sexual reproduction and the close association between pairs of meiotic chromosomes first evolved as mechanisms to repair chromosomal damage. ▢

Eukaryotes Have Life Cycles

Many protists are haploid all their lives. On the other hand, animals and most plants are diploid for most of their life cycle. When organisms are diploid, their cells contain two sets of chromosomes, one set derived from the male parent and one set derived from the female parent. The production of haploid gametes by meiosis, followed by the union of two gametes in sexual reproduction, is a **sexual life cycle.** Eukaryotes are characterized by three types of sexual life cycles.

1. **Zygotic meiosis** In the simplest of sexual life cycles, the zygote is the only diploid cell. This sort of life cycle is said to exhibit **zygotic meiosis** because the zygote undergoes meiosis immediately after it is formed. Thus, haploid cells occupy the major portion of the life cycle. Zygotic meiosis is found in many algae, such as the unicellular *Chlamydomonas* shown in Figure 19-8.

2. **Gametic meiosis** In almost all kinds of animals, including humans, the gametes are the only haploid cells; all of the other cells of the individuals in the life cycle are diploid. This sort of sexual life cycle is said to exhibit **gametic meiosis** because meiosis produces gametes. In this type of life cycle, illustrated in Figure 19-9, the diploid zygote occupies the major portion of the life cycle.

3. **Sporic meiosis** Plants have a life cycle that regularly alternates between a haploid phase and a diploid phase.

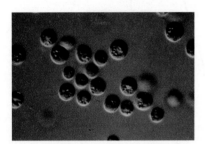

Figure 19-8 Some organisms, such as the green alga *Chlamydomonas,* have a life cycle characterized by zygotic meiosis. In zygotic meiosis a diploid zygote undergoes meiosis and develops into haploid individuals.

Figure 19-9 Some organisms, such as humans, have a life cycle characterized by gametic meiosis. In gametic meiosis a diploid reproductive cell undergoes meiosis and gives rise to haploid gametes.

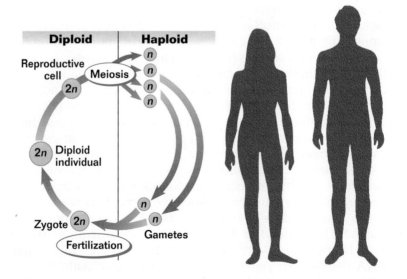

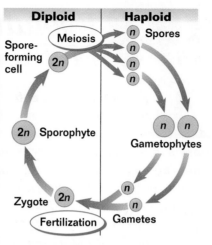

Figure 19-10 Some organisms, such as tree ferns, have a life cycle characterized by sporic meiosis. In sporic meiosis a diploid spore-forming cell undergoes meiosis and gives rise to haploid spores.

The diploid phase, called a **sporophyte** *(SPOHR uh fyt)*, produces spores. The spores give rise to the haploid phase, called a **gametophyte** *(guh MEET uh fyt)*, which produces gametes that fuse to give rise to the diploid phase. Plants are said to exhibit **sporic meiosis.** In this form of meiosis, illustrated in Figure 19-10, certain diploid cells of the sporophyte undergo meiosis to form haploid spores. ◻

☐ *CAPSULE SUMMARY*

Eukaryotic organisms can have one of three kinds of sexual life cycles—zygotic meiosis, gametic meiosis, or sporic meiosis—depending on when meiosis occurs and what kind of cell it produces.

Section Review

1. *What advantage does an asexually reproducing organism have over one that reproduces sexually?*

2. *How have eukaryotes benefited from sexual reproduction?*

3. *What kind of sexual life cycle is found in the green alga* Chlamydomonas? *When does meiosis take place in these organisms?*

4. *What kind of organisms exhibit gametic meiosis? What is the function of meiosis in organisms that have gametic meiosis?*

5. *What kind of organisms display sporic meiosis? What are the names of each of the phases in this kind of life cycle? What kind of cells are produced during each phase?*

Critical Thinking

6. *What is the mechanism in ancestral cells that may have served as the basis for the evolution of sexual reproduction? In what form is this mechanism still present in cells involved in sexual reproduction?*

19-3 *Evolution of Multicellularity*

*A*rchaebacteria and eubacteria are unicellular organisms (individuals composed of a single cell), as are most protists. For these organisms, unicellularity has been tremendously successful; single-celled organisms make up more than half the biomass on Earth. Animals, plants, and fungi, however, are not only numerous, but also extraordinarily diverse. Their success can be attributed to a very different way of life that evolved among the protists some 1.5 billion years ago—multicellularity.

Multicellularity Allows for Specialization

An organism that is unicellular, such as *Vorticella* shown in Figure 19-11, has size limits. It can only grow to a certain size before encountering serious surface-area-to-volume problems; as cell size increases, there is soon too little surface area to meet the needs of the cell volume. The evolution of multicellularity solved this dilemma. A **multicellular organism** is one composed of many cells that are permanently associated with one another and that integrate their activities, such as the green algae shown in Figure 19-11. The main advantage of multicellularity is that it allows for specialization. Distinct types of cells form **tissues,** groups of cells with a common structure and function. Different tissues are organized into **organs,** specialized structures with specific functions. With such functional "division of labor" within its body, a multicellular organism can have cells devoted specifically to protecting the body, others devoted to moving it about, and still others devoted to seeking mates and pursuing prey. In short, multicellularity enables an organism to carry on a host of activities of a complexity that would have been impossible

Figure 19-11 Multicellularity evolved among the protists. Unicellular protists such as *Vorticella, left,* are more complex than prokaryotes, but multicellular protists such as green algae, *right,* have specialized cells with distinct functions.

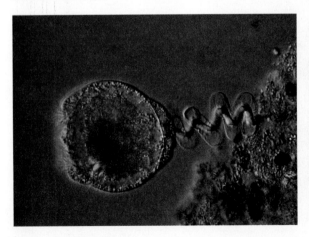

for its unicellular ancestors. In just this way, a small city of 50,000 people is vastly more complex and organized than a crowd of 50,000 people in a football stadium; each inhabitant plays a role that keeps the city functioning, rather than just being another body in a mob. ☐

☐ **CAPSULE SUMMARY**

The evolution of multicellularity enabled organisms to grow more complex.

Colonies and Aggregates Are Not Truly Multicellular
....

True multicellularity, in which the activities of the individual cells are coordinated and the cells themselves are in contact, occurs only in eukaryotes; it is one of their major characteristics. Occasionally, the cell walls of bacteria adhere to one another, and bacterial cells may even be held together within a common sheath. In fact, some bacteria, such as the cyanobacteria in Figure 19-12, form filaments, sheets, or three-dimensional formations of cells. However, these formations cannot be considered truly multicellular because little integration of cell activities occurs. Such bacteria may properly be considered colonial (living together). A **colonial organism** is a collection of cells that are permanently associated, but in which little or no integration of cell activities occurs. Many protists form colonies, consisting of many cells with little differentiation or integration.

An **aggregation** is a more temporary collection of cells that come together for a period of time and then separate. For example, a plasmodial slime mold (a member of the kingdom Protista) is a unicellular organism that spends most of its life moving about and feeding as a single-celled amoeba. When starved, however, these cells aggregate into a large colony, shown in Figure 19-13. This weblike mass produces spores, which can travel to distant locations where there may be more food.

In some protists, the distinction between being a colonial organism and one that is multicellular is blurry. For example,

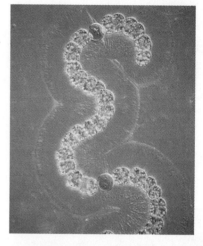

Figure 19-12 Although the cyanobacterium *Anabaena* can appear as a filament made of many cells, it is not considered a multicellular organism.

Figure 19-13 Plasmodial slime molds are brightly colored protists that feed as a large mass, which may grow to a diameter of several centimeters. These masses are considered to be aggregations, not multicellular.

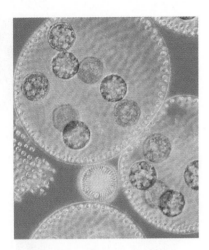

Figure 19-14 *Volvox* is a colonial organism shaped like a hollow ball. Its wall is composed of hundreds or thousands of flagellated cells embedded in a jellylike layer.

the green alga *Volvox*, shown in Figure 19-14, is a colonial protist composed of hundreds or thousands of individual cells aggregated into a hollow ball. *Volvox* moves by a coordinated beating of the flagella of individual cells—like scores of rowers all pulling their oars in concert. A few cells near the rear of the moving colony are reproductive cells, but most are relatively undifferentiated. Some species have cytoplasmic connections between the cells that might permit coordination of some activities. Does this make *Volvox* a multicellular organism? Is coordination of flagella evidence of integration? To return to the analogy comparing the inhabitants of a small city with a crowd in a football stadium, imagine that the stadium crowd is carrying out a "wave," in which sections of the crowd stand and raise their arms one after another like falling dominoes. Does this indicate that the football watchers are truly integrated? No. In this sense, *Volvox* is more properly considered a colonial organism, although the distinction is a difficult one. ◻

◻ **CAPSULE SUMMARY**

Colonial organisms and aggregations are not considered to be multicellular, but rather a permanent or temporary collection of cells.

Multicellularity Appears in Four Kingdoms

Multicellularity has evolved many times among the protists. Three groups of photosynthetic protists that contain species with independently attained true multicellularity are red algae (phylum Rhodophyta), brown algae (phylum Phaeophyta), and green algae (phylum Chlorophyta), all shown in Figure 19-15. In these algae, individuals are composed of many cells that interact with one another and coordinate their activities, and individual cell specialization is fairly complex. Multicellularity does not imply small size or limited adaptability. Some marine algae grow to be enormous. An individual kelp, a type of brown algae, may grow to dozens of meters in length—some taller than a redwood tree! Red algae grow at great depths in the sea, far below where kelp or other algae are found. Nevertheless, not all algae are multicellular. Green algae, for example, include many kinds of multicellular organisms, but an even larger number of unicellular ones.

Figure 19-15 Red algae, brown algae, and green algae are multicellular protists that can grow to be many meters in length.

Plants, animals, and fungi are composed of many kinds of highly specialized cells that coordinate their activities. Protists were the ancestors of the three complex multicellular kingdoms whose representatives are shown in Figure 19-16.

Plantae Plants, such as the cactus shown in Figure 19-16, are descendants of multicellular green algae. Most green algae are aquatic and have simpler reproductive and vegetative structures than plants.

Animalia Animals, such as the meerkat in Figure 19-16, arose from a unicellular protist ancestor. Formerly, several phyla of heterotrophic protists were grouped with the animals, but today, scientists include only multicellular organisms in this kingdom. The simplest and seemingly most primitive animals today, the sponges, seem to have evolved from a kind of flagellated protist.

Fungi Fungi, such as the mushrooms shown in Figure 19-16, also arose from a different heterotrophic unicellular protist than the one that gave rise to animals. Certain protists, including slime molds and water molds, have been considered to be fungi, although they are in fact protists.

Figure 19-16 The three kingdoms of multicellular eukaryotes arose from protists. Cactuses, *above left,* and all other plants are the descendants of a multicellular green alga. Mushrooms, *above center,* and other fungi evolved from a unicellular protist. A different type of unicellular protist gave rise to animals, such as meerkats, *above right.*

Multicellularity Requires Cell Specialization and Coordination

The main difference between the more complex multicellular organisms and multicellular protists like marine algae is **cell specialization.** In an animal, plant, or fungus, the body of an individual has different kinds of cells with very different structures and functions, which you can learn about in *Highlights: Six Kingdoms of Life* on pages 446–448. Even the simple sponges, for example, which lack specialized tissues and organs, have a variety of different kinds of cells. The outer body of the sponge is lined with flattened cells that function as a protective coat of skin. The inner layer, or body cavity, is lined by special collar cells that catch food particles in the water and move water through the sponge by beating the flagella back and forth. Between the two layers of cells, several kinds of amoeboid cells circulate, some of which produce a tough protein skeleton of spongin (familiar as the bathtub sponge). Each of these kinds of cells has a different structure and a different job to do.

Multicellularity implies something very important about the genes of an individual: *different cells are using different genes!* For example, the genes encoding the flagellar protein of collar cells are not being expressed in amoeboid cells—they are "turned off," which is why the amoeboid cells do not have flagella. The process whereby a single cell (in humans a fertilized egg) becomes a multicellular individual with many different kinds of cells is called **differentiation.** The cell specialization that is the hallmark of complex multicellular life is the result of differential development—different cells developing in different ways by activating different genes.

A second characteristic of complex multicellular organisms is **intercellular coordination,** the adjustment of a cell's activity in response to what other cells are doing. Consider a fir tree. The cells growing at the very top of the tree secrete a chemical called auxin, which diffuses downward through the plant and suppresses the growth of side branches. Farther down from the top of the tree, less auxin is present so the side branches grow longer. The triangular appearance of the fir tree is thus a direct consequence of the chemical signal passed down from the topmost cells. Auxin is an example of a hormone, a chemical signal used to communicate between the cells of complex multicellular organisms. The cells of all complex multicellular organisms communicate with one another with hormones. In some organisms, like sponges, there is relatively little coordination between the cells; in other organisms, like humans, almost every cell is under complex coordination. ◼

Section Review

1. *What advantage does a multicellular organism have over one that is unicellular?*

2. *Describe* Anabaena, *plasmodial slime molds, and* Volvox. *Is each considered to be a multicellular organism? Explain.*

3. *What three kingdoms arose from protists? Did they all arise from the same kind of protist? Explain.*

Critical Thinking

4. *In terms of multicellularity, why can a fir tree grow on land, whereas a sponge must live in the sea?*

5. *Name three species that are complex multicellular organisms. What can you infer about the genes of these organisms?*

19-4 Classification of Living Things

Our world is populated by at least 10 million different kinds of organisms. In order to study and discuss these organisms, scientists give each kind they discover a name. It would be impossible to remember the name of every kind of organism, just as it would be impossible for a postal worker to sort mail bearing only the addressee's first name. To make the process easier, the postal worker sorts mail first by zip code, then by street name and house number. By categorizing individuals according to their addresses, the post office can accurately locate a specific individual. Similarly, to help identify an organism, a biologist groups specific kinds into categories, which are assigned to larger and more inclusive categories.

Section Objectives

■ *Describe Linnaeus's role in developing the modern system of naming organisms.*

■ *Explain the scientific system for naming a species.*

■ *Distinguish scientific naming from biological classification.*

■ *Define the term* species.

■ *Describe how classification reflects evolutionary history.*

Linnaeus Assigned Organisms Two-Word Names

Humans have been naming and describing organisms since the beginning of language. Over 2,000 years ago, the Greek philosopher and naturalist Aristotle (384–322 B.C.) grouped plants and animals on the basis of their structural similarities. This simple classification system was expanded by later Greeks and Romans, who grouped plants and animals into basic categories such as oaks, dogs, and horses. Eventually each unit of classification was called a **genus** *(JEE nuhs)*, the Latin word for "group." The plural of genus is genera *(JEHN uhr uh)*. Starting in the Middle Ages, a genus was given a name in Latin, the language of scholars. Thus, cats were assigned to the genus *Felis*, dogs to *Canis*, and horses to *Equus*. The science of naming and classifying organisms is called **taxonomy.**

Until the mid-1700s, biologists referred to a particular species by adding a series of descriptive terms to the name of the genus. These phrases, sometimes consisting of 12 or more Latin words, were called polynomials (from *poly*, meaning "many," and *nomen*, meaning "name"). As you can see in Figure 19-17, polynomials became quite unwieldy and awkward. Polynomials were sometimes altered by various biologists so that a given organism rarely had a universal name. A simplified system for naming organisms came from the work of the Swedish biologist Carl Linnaeus (1707–1778), whose ambition was to catalog all the known kinds of organisms. In the 1750s he published several books that employed the well-established polynomial system. But as a kind of shorthand, Linnaeus included a two-word Latin name for each species.

Figure 19-17 The polynomial system of naming organisms was cumbersome. For example, under this system the European honeybee had a 12-part name: *Apis pubescens, thorace subgriseo, abdomine fusco, pedibus posticis glabis, untrinque margine ciliatus.*

For example, the European honeybee became *Apis mellifera*. Linnaeus's system for naming organisms is called **binomial nomenclature** because each name is composed of two words.

Scientific Names Are Universal

For nearly 250 years, Linnaeus's binomial nomenclature has remained the standard way of identifying a species. The unique two-word name for a species is its **scientific name.** The first word in a scientific name is the genus to which the organism belongs. An organism is assigned to a genus based on its major characteristics. For example, oak trees, all of which produce acorns, are placed in the genus *Quercus*. The second word in a scientific name identifies one particular kind of organism within the genus. Table 19-1 lists and describes two species of oak in the genus *Quercus*.

As you can see, when you write a scientific name, the first letter of the genus name is always capitalized and the first letter of the second word is always lowercased. As with all foreign words, scientific names are italicized or underlined.

Table 19-1 Two Species of Oaks

	Common Name	Genus	Scientific Name	Traits
	Red oak	*Quercus*	*Quercus rubra*	Lobed leaves; produces acorns approximately 25 mm (1 in.) long
	Willow oak	*Quercus*	*Quercus phellos*	Unlobed leaves; produces acorns approximately 15 mm (0.6 in.) long

Figure 19-18 Common names often differ from place to place. A "robin" in Great Britain is *Erithacus rubicula*, *left*, but in North America it is *Turdus migratorius*, *right*.

The scientific name of an organism is the same throughout the world, providing a standard for communication among biologists, regardless of their native language. This system of classification is a great improvement over the use of common names, which often vary from place to place. For example, the birds shown in Figure 19-18 are both commonly called robins, one in Great Britain, the other in North America. However, they belong to different species, genera, and families. In order to avoid ambiguity in published research, scientists refer to them by their scientific names, *Erithacus rubicula* and *Turdus migratorius*.

The name given to a species must conform to rules formulated by an international commission of scientists. All scientific names must be Latin words or terms constructed according to the rules of Latin grammar. Two different organisms cannot be assigned the same scientific name. Since all the members of a genus will share their genus name, the second word in the name of each member of that genus must be different. For example, only one species of the genus *Homo* can be given the name *sapiens*. Organisms in different genera cannot have the same genus name but can share the second word of their scientific names. For example, the green anole lizard, *Anolis carolinensis*, and the Carolina chickadee, *Parus carolinensis*, share the name *carolinensis*, meaning they were described from specimens collected in North Carolina and South Carolina. ◻

◻ CAPSULE SUMMARY

To ensure accurate communication about organisms, biologists use binomial nomenclature to name species. Binomial nomenclature is a system of scientific names developed by the eighteenth-century Swedish biologist Carl Linnaeus.

All Organisms Are Classified in a Hierarchy

Linnaeus worked out a fairly extensive system of classification for both plants and animals, emphasizing an organism's form and structure as the basis for arranging specimens in a collection. Subsequently, the genera and species that he described were organized into a hierarchical system of groups that increase in inclusiveness. The different groups into which organisms are classified have

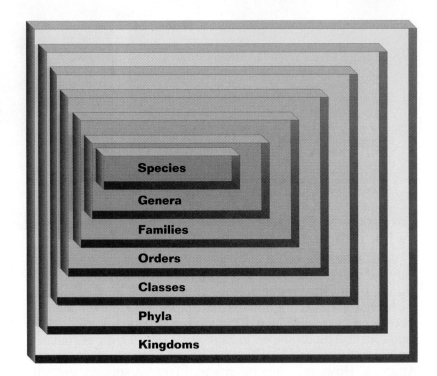

Figure 19-19 Each living thing is assigned to a series of groups that increase in inclusiveness, beginning with species (least inclusive) and ending with kingdom (most inclusive).

Within the image, from innermost to outermost:
Species
Genera
Families
Orders
Classes
Phyla
Kingdoms

☐ **CAPSULE SUMMARY**

Scientists classify organisms into a hierarchical system of groups. Today, there are seven basic levels in the hierarchy: kingdom, phylum, class, order, family, genus, and species.

expanded since Linnaeus's time and now consist of seven levels, as illustrated in Figure 19-19. Genera with similar properties are clustered into a **family.** Similar families are combined into an **order.** Orders with common properties are united in a **class.** Classes with similar characteristics are assigned to a **phylum** (*FY luhm*). Finally, similar phyla are collected into a **kingdom.** The term **division** is an alternative term for phylum in the classification of bacteria, fungi, and plants. As you learned in Chapter 13, living things are grouped into six kingdoms—Archaebacteria, Eubacteria, Protista, Fungi, Plantae, and Animalia. The seven-level hierarchy can be subdivided into more specific categories, such as superclass, subclass, superorder, and suborder. In all, more than 30 taxonomic levels are recognized.

Each category at every level of classification is based on characteristics shared by all the organisms it contains. Consider again the classification of the honeybee, illustrated in Figure 19-20. Its scientific name, *Apis mellifera*, indicates that it belongs to the genus *Apis*, which is classified in the family Apidae. All members of the family Apidae are bees that either live alone or in hives, as does *A. mellifera*. The order to which the honeybee belongs, Hymenoptera, includes ants, bees, and wasps, which usually have two pairs of wings and are likely to be able to sting. At the next level of classification, *A. mellifera* belongs to the class Insecta, meaning it is an insect with three major body segments and three pairs of legs. Its phylum, Arthropoda, indicates that the honeybee is an arthropod, an organism with a hard cuticle of chitin and jointed appendages. Its kingdom, Animalia, tells you that *A. mellifera* is a multicellular heterotroph whose cells lack walls. ☐

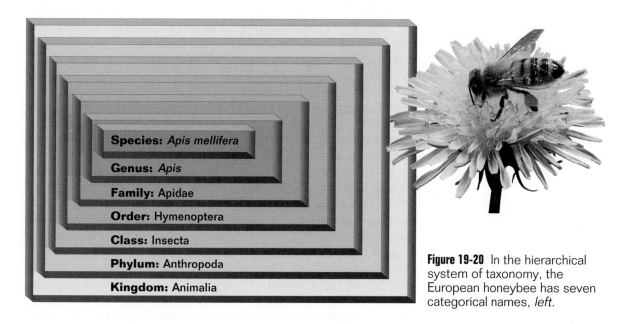

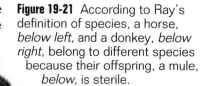

Species: *Apis mellifera*

Genus: *Apis*

Family: Apidae

Order: Hymenoptera

Class: Insecta

Phylum: Anthropoda

Kingdom: Animalia

Figure 19-20 In the hierarchical system of taxonomy, the European honeybee has seven categorical names, *left.*

Recognizing a Species

The basic biological unit in the Linnaean system of classification is the species. A definition of species was proposed by John Ray (1627–1705), an English clergyman and scientist. A species, Ray suggested, is a group of individuals that can breed with one another and produce fertile offspring. Even offspring that looked very different were considered the same species. By Ray's definition, all dogs are one species, all pigeons are one species, and so on. However, carp and goldfish are not the same species because they cannot interbreed. And although a horse and a donkey can mate, they are not the same species because the offspring, a mule, shown in Figure 19-21, is sterile.

With Ray's definition, the species became regarded as an important biological unit that could be cataloged—the task that Linnaeus undertook a generation later. The

Figure 19-21 According to Ray's definition of species, a horse, *below left,* and a donkey, *below right,* belong to different species because their offspring, a mule, *below,* is sterile.

genus was still the fundamental unit of classification, but more emphasis came to be placed on the distinctions between the individual species within genera. When Darwin's ideas about evolution were joined with Mendel's ideas about genetics to form the area of study known as population genetics, biologists found that Ray's definition of species had its shortcomings. For example, sometimes it is difficult to determine whether a set of organisms represents several species or just one species composed of dissimilar individuals.

With the emergence of population genetics, biologists sought a more precise definition of species. A result of this effort was the **biological species concept,** a definition of species first stated in 1942 by the biologist Ernst Mayr of Harvard University. Mayr explained that a biological species is a group of actually or potentially interbreeding natural populations that are reproductively isolated from other such groups. As you learned in Chapter 12, reproductive isolation arises when a barrier isolates two or more groups of organisms and prevents them from interbreeding. The biological species concept emphasizes interfertility between organisms. However, reproductive barriers between sexually reproducing species are not always effective. Sometimes, individuals of different species interbreed and produce offspring called **hybrids.** For example, wolves and dogs are members of separate species in the genus *Canis*. However, interbreeding between wolves and dogs produces fertile offspring, such as the hybrid shown in Figure 19-22. Coyotes, members of another separate species in the genus *Canis*, also can interbreed with dogs to produce fertile hybrids.

The biological species concept works fairly well for members of the kingdom Animalia, in which strong barriers to hybridization usually exist. It accurately defines many species in which members reproduce sexually. However, the biological species concept fails to describe species that are predominantly asexual in

Figure 19-22 Even though a wolf, *below left,* and a dog, *below right,* belong to separate species, they can interbreed and produce fertile offspring such as this dog-wolf hybrid, *below.*

their reproduction, including all bacteria, as well as some protists, fungi, plants, and even a few animals. Most of the time, such asexually reproducing organisms give rise to clones that are genetically identical to a single parent. These species clearly cannot be defined in the same way as sexually reproducing animals and plants that derive genetic information from two parents.

Essentially, there are no barriers to interbreeding between the species within many groups of organisms. Most species of plants, some mammals, and many fishes are able to form fertile hybrids with one another, even though they may not do so in nature. In practice, biologists today recognize species much the way their predecessors did—by studying an organism's visible features. ❏

❏ **CAPSULE SUMMARY**

Today, biologists define a species as a group of organisms that share many characteristics and, in nature, interbreed with each other and not with members of other species.

How Many Species Exist?

Since the time of Linnaeus, approximately 1.5 million species have been named, a far greater number of organisms than Linnaeus ever thought existed. But the actual number of species in the world is undoubtedly much greater, judging from the large numbers still being discovered, such as *Lecythis prancei* shown in Figure 19-23. Scientists estimate that there are at least 10 million different kinds of organisms on Earth and that many of them (more than 5 million species) live in the tropics. Considering that no more than half a million tropical species have been named, our knowledge of these organisms is obviously very limited.

Figure 19-23 *Lecythis prancei* is a species of a nut tree found in rain forests along the Amazon River in Brazil. First described in 1990, virtually nothing is known about the biology of this spectacular new species.

After naming and classifying approximately 1.5 million species of organisms, what have biologists learned? Identifying particular species has enabled biologists to understand which species are unique sources of food and medicine. For example, being able to distinguish the fungus *Penicillium* from the fungus *Aspergillus* is necessary for the production of the antibiotic penicillin. Taxonomy has also given biologists the ability to catch a glimpse of the evolutionary history of life on Earth.

Reconstructing Evolutionary Histories
....

Linnaeus's classification system was based on the fact that organisms exhibit different degrees of similarity. For instance, tigers resemble gorillas more closely than they resemble lampreys. According to Darwin's views, organisms that are similar descended from a common ancestor; therefore, classification provided strong evidence supporting evolution. However, making evolutionary connections based on similar traits can be misleading because not all traits are inherited from a common ancestor. Consider the wings of a bird and the wings of an insect. Both equip the respective organism for flight, but the two kinds of wings are built differently and evolved independently of each other. Similar traits such as wings are the result of convergent evolution. In **convergent evolution,** organisms evolve similar features independently, often because they live in similar habitats. Similar features that evolved through convergent evolution are called **analogous characters.** Figure 19-24 illustrates an example of convergent evolution found in two different plant families—the cactus family and the spurge family.

Figure 19-24 Thick water-storing stems are not only found in the Cactaceae (cactus family) of North and South American deserts, *above left,* but also in members of the Euphorbiaceae (spurge family) native to the African deserts, *above right.*

Biologists must be able to distinguish homologous traits from analogous ones in order to reconstruct evolutionary history. The evolutionary history of a species is called its **phylogeny** *(fy LAHJ uh nee).* How do taxonomists determine the phylogeny of a species? In general, this is determined by the overall similarity between the characteristics of different kinds of organisms. In one approach, called **cladistics** *(kluh DIHS tihks),* biologists reconstruct a phylogeny in which relationships are inferred based on similarities derived from a common ancestor. Cladistics is used to determine the sequence in which different groups of organisms evolved. To do this, cladistics focuses on a set of unique characteristics found in a particular group of organisms. These unique characteristics are called **derived traits.** Using patterns of shared derived traits, a biologist using cladistics constructs a branching diagram called a **cladogram,** which shows the evolutionary relationships among groups of organisms. The key to cladistics is identifying morphological, physiological, or behavioral traits that differ among the organisms being studied and that can be attributed to a common ancestor.

In practice, a biologist constructing a cladogram is interested in studying the evolutionary relationships of certain groups of organisms, such as species within a genus or genera within a family. Cladograms do not convey direct information about ancestors and descendants, showing who came from whom. Instead, cladograms convey comparative information about relationships. Organisms that are grouped more closely on a cladogram share a more recent common ancestor than those farther apart. Because the analysis is comparative, a cladogram deliberately includes an organism that is only distantly related to the other organisms. This distantly related organism is called an **out-group.** The out-group serves as a baseline for comparisons among the other organisms being evaluated, the **in-group.**

How to Construct a Cladogram

To see how a biologist might go about reconstructing a phylogeny using a cladogram, consider a collection of seven vertebrates: a lamprey, a lizard, a tiger, a salamander, a shark, a gorilla, and a human. For each of these animals, seven different traits have been recorded in Table 19-2; a plus sign (+) indicates the presence of the trait, and a minus sign (–) indicates its absence.

Table 19-2 Traits Identified in Collection of Vertebrates

Organism Traits:	Jaws	Dry Skin	Hair	Lungs	Tail	Pectoral Fins	Bipedal
Lamprey	–	–	–	–	+	–	–
Lizard	+	+	–	+	+	–	–
Tiger	+	+	+	+	+	–	–
Salamander	+	–	–	+	+	–	–
Shark	+	–	–	–	+	+	–
Gorilla	+	+	+	+	–	–	–
Human	+	+	+	+	–	–	+

As you learned in Chapter 13, the lamprey is only distantly related to the other six vertebrates. It is an agnathan (jawless fish) and was among the first vertebrates to appear before jaws or paired appendages evolved. Compared with the other groups being considered, the lamprey is the out-group, the one with the fewest traits in common with the other vertebrates. Any deviation from the basic characteristics that are

Table 19-3 Identification of Derived Traits

Organism Traits:	Jaws	Dry Skin	Hair	Lungs	Tail	Pectoral Fins	Bipedal
Lamprey	0	0	0	0	0	0	0
Lizard	1	1	0	1	0	0	0
Tiger	1	1	1	1	0	0	0
Salamander	1	0	0	1	0	0	0
Shark	1	0	0	0	0	1	0
Gorilla	1	1	1	1	1	0	0
Human	1	1	1	1	1	0	1
Total Number of Taxa	6	4	3	5	2	1	1

present in the out-group is considered an evolutionary change, or a derived trait.

The first step in constructing a cladogram is identifying the derived traits. In Table 19-3, the traits in the row for the out-group are marked with a zero (0). When a vertebrate has a trait not found in the out-group, the trait is considered a derived trait and is marked with a one (1). Next, the numbers of shared derived traits are calculated. This is done by adding the numbers of derived traits in each column.

It is obvious that two traits, pectoral fins and bipedalism, are each present in only one group of organisms; they are not shared with any of the other animals under consideration. Such traits are not useful in reconstructing the phylogeny because they do not allow the elimination of any potential relationships. Now you can begin sketching the cladogram.

1. Starting with a diagonal line, the out-group is drawn as the first branch of the cladogram and is labeled *Lamprey*. Just past this first branch, the most common derived trait present in the organisms is listed. In this example, jaws are present in six organisms (all but the out-group lampreys), so the derived trait *Jaws* is written on the diagonal line just past the branch representing lampreys, as shown in the cladogram in Figure 19-25.

2. Next, the second most common derived trait is determined, which in this case is lungs, found in five of the seven organisms. The two organisms that lack lungs are lampreys and sharks. Since lampreys are already in the cladogram, a second branch labeled *Shark* is now drawn above the first branch on the cladogram. The derived trait *Lungs* is listed on the main line past this second branch point.

3. The third most common derived trait is dry skin, present in four organisms. All the organisms except lampreys, sharks, and salamanders have dry skin. A third branch labeled *Salamander* is drawn above the second branch of

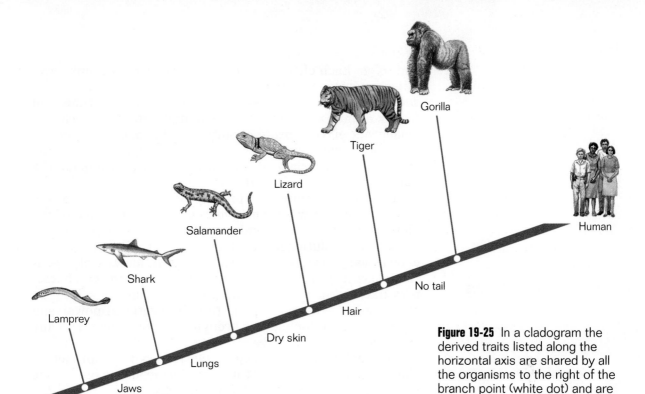

Figure 19-25 In a cladogram the derived traits listed along the horizontal axis are shared by all the organisms to the right of the branch point (white dot) and are not present in any organisms to the left.

the cladogram with the derived trait *Dry skin* indicated past the branch.

4. The fourth most common derived trait is hair. Of the four kinds of organisms remaining, all have hair except lizards, which are labeled on a fourth branch of the cladogram with the derived trait *Hair* indicated past the branch.

5. The fifth most common derived trait is a tail. Of the three kinds of organisms remaining, all lack a tail except tigers, which are labeled on the fifth branch of the cladogram with the trait *No tail* indicated just past it.

6. The other two traits, each present in only one taxon, are of no use in constructing the cladogram. While it is true that gorillas do not walk upright and humans do, a branch off the main axis with gorilla on it and humans as the remaining group exhibits exactly the same information as a branch off the main axis with human on it and gorilla as the remaining group. ◻

☐ *Capsule Summary*

An organism's evolutionary history is its phylogeny. Cladistics is a method of taxonomy that uses information about the presence of traits among a group of organisms to provide information about the relative relationships among organisms.

Considering the Weight of a Character

The great strength of a cladogram is its objectivity. A computer fed the data will generate exactly the same cladogram time and again. The great disadvantage of a cladogram is that it ignores too much information. It simply indicates that a character does or does not exist. A cladogram cannot take into account variations in the "strength" of a character, such as the size or location of a fin, the effectiveness of a lung, and so on. Each character is treated the same. Because evolutionary success depends so much on high-impact events, such as the evolution of feathers, cladograms sometimes fail to look at information of great potential importance. Thus, a cladogram of vertebrate evolution will group birds among the reptiles with crocodiles, accurately reflecting their true ancestry but ignoring the immense evolutionary impact of a derived character like feathers.

In order to avoid this pitfall, most practicing taxonomists attempt to weigh the evolutionary significance of the characters they study and to produce a more subjective analysis of evolutionary relationships. This approach, called **evolutionary systematics,** places birds in an entirely separate class from reptiles, giving extra weight to the characters like feathers that made powered flight possible. In evolutionary systematics, the full observational power and judgment of the biologist is brought to bear—along with any biases he or she may have.

In practice, evolutionary systematics is the approach of choice when a great deal of information about the organisms is available to consider. You cannot give a character due evolutionary weight without having enough information to make an accurate judgment. When little information is available about how the character affects the life of the organism, cladistics is the approach of choice. ◻

CAPSULE SUMMARY

Evolutionary systematics is a subjective method of taxonomy in which a biologist uses judgment to consider the importance of the characters among a group of organisms.

Section Review

1. Why is Linnaeus sometimes called "the father of modern taxonomy"?
2. Why do biologists use both words of a scientific name to correctly identify an organism?
3. From least to most inclusive, what are the names of the seven levels of the hierarchical system of classification?
4. What is cladistics? What kind of information does it reveal about evolutionary lines?
5. What is the biological species concept? Why does it fail to define many species that reproduce asexually?

Critical Thinking
6. What is the relationship between an organism's evolutionary history and its classification?
7. Why is evolutionary systematics considered more subjective than cladistics?

Kingdom Archaebacteria

Animalia | Plantae | Fungi
Eubacteria | Protista
Archaebacteria

Characteristics of Archaebacteria

- Unicellular prokaryotes
- Cell walls lack peptidoglycan
- Genes have introns
- Unique lipids in plasma membranes
- Either heterotrophic or autotrophic
- May be ancestors of eukaryotic cells
- Three groups that live in extreme environments—methanogens, extreme halophiles, and thermo-acidophiles

Morning glory pool

Thermoacidophiles

As their name implies, thermoacidophiles are archaebacteria that thrive in environments that are both **acidic** and **hot**. These bacteria flourish in temperatures between 60°C and 80°C (140°F and 176°F) and in pH between 2 and 4. The hot sulfur springs in Yellowstone National Park are inhabited by the thermoacidophile *Sulfolobus,* which obtains its energy by oxidizing sulfur.

Methanogens

Methanogens make up the largest group of archaebacteria identified so far. They are among the most strictly **anaerobic** of all organisms, poisoned by even traces of oxygen. Methanogens convert carbon dioxide into methane. Their metabolism is ideally suited to the kind of atmosphere thought to have existed on the primitive Earth. This methanogen, *Methanobacterium formicum,* has a cell wall made mostly of protein.

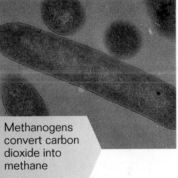

Methanogens convert carbon dioxide into methane

In ancient times methanogens could have lived anywhere, but today they live only where oxygen has been excluded and hydrogen and carbon dioxide are available. Methanogens are found in stagnant water, in sewage treatment plants, and in the intestinal tracts of animals. They can be found living on the ocean bottom and in hot springs. In spite of their intolerance to oxygen, they are obviously distributed throughout the world.

Sewage treatment plant

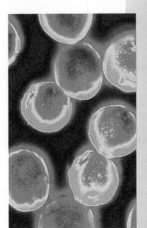

Halobacterium

Extreme Halophiles

The extreme halophiles are archaebacteria that require high concentrations of salt in order to survive. They grow in salty habitats along shorelines and in inland waters such as the Great Salt Lake and the Dead Sea. Some species require an environment 10 times saltier than sea water to grow. Colonies of halophiles often form in seawater evaporating ponds used in commercial salt production; the halophilic bacteria are harmless.

Kingdom Eubacteria

Animalia | Plantae | Fungi

Eubacteria | Protista

Archaebacteria

Characteristics of Eubacteria

- Unicellular prokaryotes
- Cell walls contain peptidoglycan
- Genes lack introns
- Reproduce asexually by binary fission
- Either autotrophic or heterotrophic
- Extremely diversified groups
- Certain types may be ancestors of mitochondria

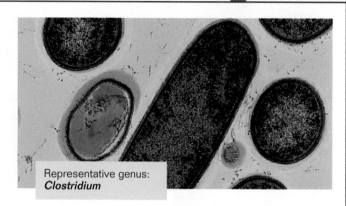

Representative genus:
Clostridium

Clostridium is a typical eubacterium that thrives in the soil. It is an **obligate anaerobe** (poisoned by oxygen) because it lacks the enzymes that break down the toxic peroxides produced by aerobic metabolism. *Clostridium* lives in tiny oxygen-free pockets in the soil, where it acts to decompose cellulose plant matter.

Forming Endospores

Clostridium is one of the few eubacteria able to form **endospores,** heat-resistant packages containing a copy of the DNA. The endospores can persist in the environment for as long as 50 years and then reinitiate growth.

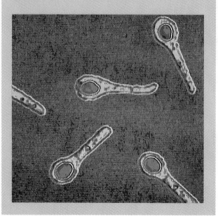

Causing Diseases

Several species of *Clostridium* can cause disease in humans. For example, *Clostridium botulinum* causes a form of food poisoning called **botulism.** *C. botulinum* produces an extremely potent toxin that affects nerve activity. Additional diseases caused by other species of *Clostridium* include tetanus and gas gangrene.

Botulism-tainted peppers

Fixing Nitrogen

A large portion of the Earth's **nitrogen fixation** is carried out by *Clostridium pasteurianum,* along with other eubacteria such as *Azotobacter* and *Rhizobium*. These eubacteria are able to convert nitrogen gas (N_2) into ammonia (NH_3).

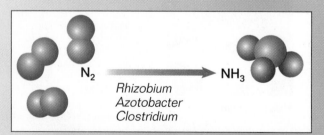

$N_2 \rightarrow NH_3$

Rhizobium
Azotobacter
Clostridium

Kingdom Protista

Characteristics of Protists

- Eukaryotic organisms
- Mostly unicellular; some multicellular
- Autotrophic, heterotrophic, or both
- Reproduce asexually; some also sexually

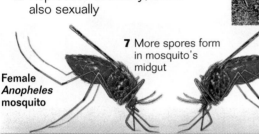

Female Anopheles mosquito

7 More spores form in mosquito's midgut

6 Mosquito picks up *Plasmodium* from infected individual

5 Spores released as red blood cells burst

4 Spores reproduce in host's red blood cells

1 Mosquito infects host with *Plasmodium*

2 Asexual spores form in host's liver cells

3 Spores burst out of liver cells

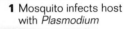

Representative genus:
Plasmodium

The protist *Plasmodium* is a **spore-forming parasite** with a complex life cycle that alternates between two hosts. *Plasmodium* causes malaria, a serious disease characterized by severe chills, fever, sweating, an enlarged spleen, confusion, and great thirst.

Life Cycle of *Plasmodium*

Plasmodium spends part of its life cycle growing in mosquitoes of the genus *Anopheles,* and another part of its life cycle in humans. A female mosquito acquires *Plasmodium* when she bites an infected human. Later, when she feeds off a different human, she injects saliva into the wound to prevent her victim's blood from clotting. Her saliva transfers *Plasmodium* into the bloodstream.

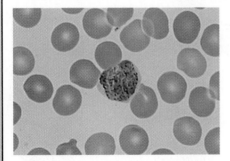

In humans, *Plasmodium* parasites are carried to the liver, where they reproduce. They reenter the bloodstream and invade blood cells. *Plasmodium* divides so rapidly that by 48 hours the cells rupture, releasing toxic substances into the bloodstream and causing a high fever. The cycle repeats itself every 48 hours until the infection is brought under control by the person's immune system, or until the person dies.

Affected Areas

Over 100 million people are infected with *Plasmodium* at any one time, and every year about 1 million of them (mostly children under the age of 5) die of malaria. Since *Anopheles* mosquitoes are abundant in the tropics, nearly all of malaria's victims live in Africa, Asia, and South America.

Kingdom Fungi

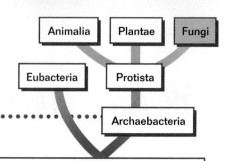

Characteristics of Fungi

- Eukaryotic organisms
- Mostly multicellular; some unicellular
- Cell walls made of chitin
- Filamentous bodies
- External heterotrophs; principal decomposers
- Reproduce sexually or asexually

Representative organism: **Mushroom**

Fungi, such as this mushroom *Amanita muscaria*, are external heterotrophs that acquire their nutrition by **absorption.** They digest food outside their bodies by secreting an enzyme that breaks organic materials into molecules fungi can absorb.

Mushroom

Hyphae

Mycelium

Filamentous Bodies

Most fungi are composed of filaments called **hyphae.** Hyphae are woven together to form a dense mat known as a **mycelium.** The mycelium of a fungus is usually hidden within the tissues of its food source.

Forming Spores

Mushrooms are reproductive structures. The underside of the cap is lined with clublike structures that give rise to haploid spores. One mushroom with a cap 3 in. across can produce as many as 40 million sexual spores per hour! Carried by wind or water, a spore can germinate if it lands in a moist environment.

Spores are haploid. When they germinate, they give rise to haploid hyphae. When two such haploid hyphae touch, they fuse together, forming secondary hyphae with paired but separate nuclei. It is from these filaments that mushroom caps form.

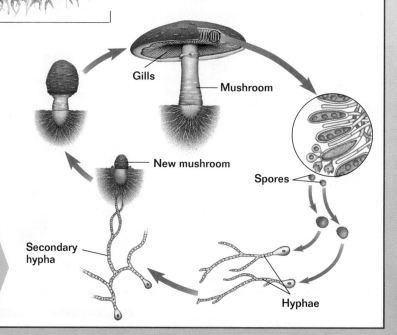

Gills

Mushroom

New mushroom

Spores

Secondary hypha

Hyphae

Kingdom Plantae

Characteristics of Plants

- ■ Eukaryotic organisms
- ■ All multicellular
- ■ Cell walls made of cellulose
- ■ Most consist of roots and shoots
- ■ Mostly autotrophic and terrestrial
- ■ Highly specialized structures for reproduction and survival on land
- ■ Reproduce sexually; some capable of reproducing asexually

Representative organism:
Cherry tree

Flowering plants such as a cherry tree are called **angiosperms,** a name derived from the Greek words *angion,* meaning "vessel" and *sperma,* meaning "seed." A flower is a reproductive structure that encourages cross-pollination by means of animals. Cherry flowers have nectar, which attracts bees. In seeking nectar within a flower, a bee is covered with pollen, which will produce male gametes; it carries this pollen to ovules, which will produce female gametes in other flowers as it seeks more nectar.

Forming Seeds

Fertilization in angiosperms is a unique process called **double fertilization,** in which sperm possess two nuclei. One sperm nucleus fertilizes the egg to form the zygote, which produces the embryo. The other sperm nucleus fuses with other nuclei to form nutritive tissue called **endosperm.** The embryo and the endosperm are enclosed within the seed coat. **Seeds** are usually contained within fruits, structures that aid in their dispersal.

Cross section of cherry flower ovary

Transporting Substances

Like most plants, the cherry tree is a vascular plant. Vascular plants have tissues specialized for conducting water **(xylem)** and transporting carbohydrates **(phloem)** throughout the plant. The veins seen in leaves are bundles of vascular tissues.

Fruits with fleshy coverings, such as cherries, are eaten by birds and other vertebrates. The fruit's bright red color signals an abundant food supply. By feeding on these fruits, the animals may carry seeds to suitable habitats where new trees can grow.

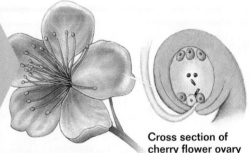

Kingdom Animalia

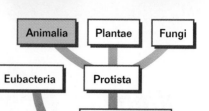

Animalia Plantae Fungi
Eubacteria Protista
Archaebacteria

Characteristics of Animals

- Eukaryotic organisms
- Multicellular; no cell walls
- Most reproduce sexually
- Interior heterotrophs
- Specialized tissues for impulses and movement
- Inhabit nearly all environments in biosphere

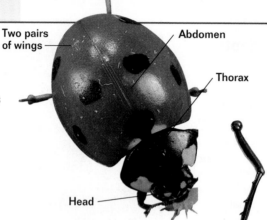

Two pairs of wings

Abdomen

Thorax

Head

Jointed appendage

Representative organism: *Ladybird beetle*

The ladybird beetle, like all arthropods, has a rigid **exoskeleton,** in which chitin is an important element. The exoskeleton provides places for muscle attachment, protects the animal from injury, and prevents water loss. Insects have three body sections —the head, thorax, and abdomen—and are noted for possessing jointed appendages, such as legs and wings.

Fungi
Flowering plants
Beetles
Protists
Other
Bees, wasps
Mollusks
Vertebrates
Butterflies, moths
Spiders
Other insects
Flies, mosquitoes
Millipedes, centipedes

Arthropod Diversity

Insects such as the ladybird beetle are **arthropods,** the most successful group of animals that has ever lived. Two-thirds of all named species of organisms on Earth are arthropods. Of the 770,000 named species of insects, 350,000 are beetles.

Increasing in Size

As an individual arthropod grows larger, its exoskeleton splits and is shed, allowing the animal to increase in size. The eggs of arthropods develop into immature forms that may bear little or no resemblance to the adult. Most arthropods change their characteristics as they develop, a process called **metamorphosis.**

Although many of the millions of insects are harmful to humans, many others are helpful. Ladybird beetles are fierce predators, feeding on plant-eating insects.

A hallmark of insects such as the ladybird beetle is the diversity of the specialized appendages attached to their heads. Insects have **antennae** specialized for sensing the environment. Insects also have jaws, or **mandibles,** and other mouthparts that have evolved into different shapes for grinding, scraping, piercing, and sucking.

CHAPTER REVIEW 19

Vocabulary

aggregation (427)
analogous character (438)
binomial nomenclature (432)
biological species concept (436)
cell specialization (429)
cladistics (438)
cladogram (438)
class (434)
colonial organism (427)
convergent evolution (438)
derived traits (438)
differentiation (430)
division (434)

evolutionary systematics (442)
family (434)
gametic meiosis (424)
gametophyte (425)
genus (431)
hybrid (436)
in-group (439)
intercellular coordination (430)
kingdom (434)
multicellular organism (426)
obligate anaerobe (418)
order (434)
organ (426)

out-group (439)
peptidoglycan (418)
phylogeny (438)
phylum (434)
scientific name (432)
sexual life cycle (424)
sporic meiosis (425)
sporophyte (425)
taxonomy (431)
theory of endosymbiosis (419)
tissue (426)
zygotic meiosis (424)

Concept Mapping

Construct a concept map that shows the evolutionary events leading to the origin and development of eukaryotes. In constructing your map, use the following terms: archaebacteria, cell organelles, mitochondria, chloroplasts, eubacteria, sexual reproduction, asexual reproduction, DNA repair mechanism, and genetic diversity. Include additional concepts in your map as needed.

Review

Multiple Choice

1. Archaebacteria differ from eubacteria because archaebacteria
   a. have cell walls made of peptidoglycan.
   b. have plasma membranes made of phospholipids.
   c. contain genes interrupted by introns.
   d. cannot live in harsh environments.

2. According to the theory of endosymbiosis, mitochondria evolved from
   a. eubacteria.
   b. archaebacteria.
   c. chloroplasts.
   d. protists.

3. Which of the following does *not* relate to the subject of asexual reproduction?
   a. binary fission   c. budding
   b. fragmentation   d. zygotic meiosis

4. Sexual reproduction is believed to have evolved originally as a way for cells to
   a. shuffle genetic material.
   b. repair damaged DNA.
   c. produce diploid individuals.
   d. increase their population growth at a maximum rate.

5. Sporic meiosis differs from zygotic meiosis and gametic meiosis because
   a. sporic meiosis produces a gametophyte stage.
   b. sporic meiosis produces a sporophyte stage.
   c. gametes do not fuse in sporic meiosis life cycles.
   d. meiosis in the sporic meiosis life cycle occurs immediately after the formation of the zygote.

6. The main advantage of multicellularity is that
   a. it allows the most rapid population growth.
   b. it tends to regulate population growth more efficiently.
   c. it allows division of labor.
   d. it prevents cells from specializing in only one or a few functions.

7. A colonial organism differs from an aggregation because a colonial organism
   a. is a temporary collection of cells.
   b. is a permanent collection of cells.
   c. is truly multicellular.
   d. forms only upon starvation or stress of its component cells.

8. A difference between the scientific name of an organism and the classification of that organism is that
   a. the scientific name includes the family and class of the organism.
   b. the scientific name always contains three words (trinomial nomenclature).
   c. the classification includes more categories than the scientific name.
   d. classification can vary from place to place.

Completion

9. Prokaryotes belong to kingdom _____ or kingdom Eubacteria. Members of these two kingdoms have different cell walls, plasma membranes, gene architectures, and _____ .

10. According to the text, biological classification now uses a total of _____ groups or levels. The names of the _____ group and the _____ group of an organism compose its scientific name according to the modern system of _____ nomenclature.

11. Prokaryotes reproduce without producing gametes, while many _____ are able to undergo sexual reproduction. An advantage of sexual reproduction is that it can _____ genetic diversity.

12. *Volvox* is a(n) _____ organism; its cells are permanently associated. However, individual cells of *Dictyostelium discoideum* come together as a(n) _____ only when food is scarce.

13. *Pelomyxa palustris* represents an early stage in the evolution of _____ . Like archaebacteria, *Pelomyxa palustris* lacks _____ and does not undergo mitosis.

14. Sexually reproducing plants undergo _____ meiosis. During the sexual life cycle, the diploid sporophyte generation produces _____ spores through meiosis.

Short Answer

15. How do representatives of the two kingdoms of bacteria differ in cell wall and plasma membrane composition? How do they differ in gene architecture and translation machinery?

16. Name three organelles found in eukaryotic cells that have endosymbiotic origins. Briefly describe the likely prokaryotic ancestors of these organelles.

17. Name and describe two forms of asexual reproduction, and identify an organism that uses each form.

18. Examine the life cycles of *Clamydomonas* and humans in Figures 19-8 and 19-9. How are these two life cycles similar? How are they different?

Highlights Review

19. Until recently, all bacteria were classified in the kingdom Monera. What observations led scientists to classify bacteria into the two kingdoms Archaebacteria and Eubacteria?

20. Name two species of *Clostridium*, and describe their harmful and beneficial effects on humans.

21. What criteria are used to classify organisms as members of the kingdom Protista?

22. Name several external heterotrophs and internal heterotrophs. In what kingdom are they found? How do such organisms differ in the way they obtain nutrients?

23. Describe the process of fertilization found in angiosperms. How does this process differ from that found in animals?

24. Complete the chart at the top of next page.

Kingdom	Prokaryote or Eukaryote	Cellular Organization	Mode of Nutrition	Representative Organism
Archaebacteria	_____	Unicellular	Heterotroph	*Sulfolobus*
_____	Prokaryote	_____	Heterotroph or autotroph	_____
_____	Eukaryote	Unicellular; multicellular	_____	*Plasmodium*
Fungi	_____	Mostly multicellular	_____	*Amanita*
Plantae	Eukaryote	_____	_____	_____
Animalia	_____	_____	Heterotroph	_____

Themes Review

25. **Evolution** Compare a unicellular alga, such as *Vorticella*, with a green plant. What distinguishing characteristics of multicellular organisms are present in the green plant, but not in the alga?

26. **Levels of Organization** Rank the seven major taxonomic groups from the group at the top of the hierarchy to the group at the bottom of the hierarchy. Which two taxonomic groups provide the scientific name of an organism? Where are they located in the hierarchy?

27. **Evolution** Describe the evidence that supports the theory that mitochondria evolved from eubacteria.

Critical Thinking

28. **Making Inferences** How does the cladistic approach to the classification of birds differ from the method used in evolutionary systematics? Which method best reflects the ancestry of the birds? Defend your answer.

29. **Making Inferences** Many organisms are repulsed by the odor of decaying organic matter, which often contains infectious bacteria. What adaptive advantage might this reaction give an organism?

Activities and Projects

30. **Research and Writing** The endosymbiotic hypothesis is one of two hypotheses that explain the origin of eukaryotic cells. The second hypothesis is the autogenous hypothesis. Research the autogenous hypothesis. In your report, present the highlights of both hypotheses and explain why most scientists favor the endosymbiotic hypothesis.

31. **Research and Writing** The bacterium *Escherichia coli* usually reproduces by binary fission, but it can also transfer genetic material through a sex pilus. Research how this transfer of genetic material takes place. Summarize this process and speculate how the transfer of genetic material might impact the use of antibiotics.

Readings

32. Read the article "Marriage of Convenience" in *Sciences*, September–October 1990. To whom did Lynn Margulis and Mark McMenamin give credit for first proposing endosymbiotic origins for chloroplasts, mitochondria, and centrioles? What 1960s technological advancement enabled Margulis to provide evidence for the endosymbiotic origins of eukaryotic organelles?

Dichotomous Keys

OBJECTIVES

- Use a dichotomous key to identify leaves.
- Construct a dichotomous identification key.

PROCESS SKILLS

- identifying and comparing characteristics of objects
- organizing data

MATERIALS

- shoes
- masking tape
- marker

BACKGROUND

1. Who first developed the classification system used by scientists today? What is taxonomy?
2. How do scientists classify organisms? Why is classification an essential tool of biology?
3. What is a dichotomous key?
4. Write your own **Focus Question** on your Vee Form.
5. **Knowing Side of the Vee** List the **Concepts** and new **Vocabulary Words** on your Vee Form. In the **Concept Statements** section of the Vee, use these words in sentences that define and explain them.

TECHNIQUE

Doing Side of the Vee

Part A: Using a Dichotomous Key

1. Field guides often use dichotomous keys to identify organisms. Use the dichotomous key shown here to identify the tree leaves below. Begin with descriptions 1a and 1b, and follow the directions. Proceed through the list of paired descriptions until you identify the leaf in question. In the **Records** section of your Vee Form, write the names of the leaves.

1a. If the edge of the leaf has no teeth, waves, or lobes, go to 2 in the key.

1b. If the edge of the leaf has teeth, waves, or lobes, go to 3 in the key.

2a. If the leaf has a single bristle at its tip, it is a shingle oak.

2b. If the leaf has no single bristle at its tip, go to 4 in the key.

3a. If the leaf edge is toothed, it is a lombardy poplar.

3b. If the leaf edge has waves or lobes, go to 5 in the key.

4a. If the leaf is heart-shaped with veins branching from the base, it is a redbud.

4b. If the leaf is not heart-shaped, it is a live oak.

5a. If the leaf edge has lobes, it is an English oak.

5b. If the leaf edge has waves, it is a chestnut oak.

Part B: Making a Dichotomous Key

2. Ask 10 or more student volunteers to remove their shoes and use masking tape and a marker to label the sole of one of their shoes with their name. The labeled shoes should then be placed on a single table in the classroom.

3. Form into small groups. Discuss the appearance of the shoes. In the **Additional Records and Observations** section on the back of the Vee Form, make a table that lists some general characteristics of the shoes, such as the type and size. Also list the names of the students that own the shoes. Complete the chart by describing the characteristics of each person's shoe.

4. Use the information in your table to make a dichotomous key that can be used to identify the owner of each shoe. Remember that a dichotomous key includes pairs of opposing descriptions. At the end of each description, the key should either identify an object or give directions to go to another specific pair of descriptions. Write your dichotomous key in the **Additional Records and Observations** section of your Vee.

5. After all groups have completed their key, exchange keys with a member of another group. Use the key to identify the owner of each shoe, and then verify the accuracy of your identification by reading the label on the shoe. If the key has led you to an inaccurate identification, return the key so that corrections can be made.

6. In the **Procedure** section of the Vee, briefly summarize the procedure you followed.

INQUIRY

1. What other characteristics might be used to identify leaves with a dichotomous key?

2. How was the shoe identification key that your group designed dichotomous?

3. Were you able to identify the shoes using another group's key? If not, describe the problems you encountered.

4. Use the information on the **Knowing Side** of the Vee to interpret your results from the **Doing Side,** and then write your **Knowledge Claim.** Write a **Value Claim** for this lab.

ANALYSIS

1. How was it helpful to list the characteristics of the shoes before making the key?

2. Does a dichotomous key begin with general descriptions and then proceed to more specific descriptions, or vice versa? Explain your answer, giving an example from the key you made.

FURTHER INQUIRY

Write a **New Focus Question** that could be the point of a new investigation. The following is an example:

What characteristics might be used to identify birds or other animals using a dichotomous key?

CHAPTER 2

VIRUSES AND BACTERIA

REVIEW

- prokaryotes (Section 2-1)
- autotroph (Section 4-1)
- photosynthesis (Section 5-2)
- bacteriophage (Section 8-1)
- vaccine (Section 8-1)
- structure of DNA (Section 8-2)
- transcription and translation (Section 9-1)
- differences between archaebacteria and eubacteria (Section 19-1)

The oil-eating bacterium *Psuedo-monas putida* can be used to help clean up oil spills, such as the one that occurred when the oil tanker *Braer* ran aground in Quendale Bay in the Shetland Islands in January 1994.

20-1 Viruses

*I*n Chapter 1 you learned about the properties of life. All living things are cellular, able to grow and reproduce, and are guided by information encoded in the nucleic acid DNA. The smallest organisms that have these properties are bacteria, single cells that were the earliest forms of life. Even smaller than bacteria are viruses, mere segments of nucleic acids wrapped in a protein coat. Because viruses depend on the cells of other living organisms in order to reproduce, biologists do not consider viruses to be alive. However, when they are able to reproduce, viruses often cause damage to the host organism. For this reason, viruses have had, and continue to have, a major impact on the living world, as you can see in Figure 20-1.

Section Objectives

- Discuss the events that led to the discovery and understanding of tobacco mosaic virus.
- Define virus, and explain why a virus is not considered a living organism.
- Describe the basic structure of a virus.
- Describe the structure of HIV and explain how it reproduces.

Is a Virus a Living Organism?

Biologists first suspected the existence of viruses near the end of the nineteenth century. At that time, European scientists were seeking to identify the infectious agent responsible for tobacco mosaic disease, a disease that stunts the growth of tobacco plants and makes their leaves blotchy. One experiment used fine-pored porcelain filters to strain bacteria from sap extracted from infected tobacco plants. The pores in these filters were so small that bacteria could not pass through them. However, the infectious agent passed through the filters without difficulty. Scientists concluded that the infectious agent must be smaller than a typical bacterial cell. After further investigation, they found that the agent could reproduce only inside the living cells it infected. They called the agent a *virus*, a Latin word meaning "poison."

For many years after this discovery, viruses were erroneously regarded as primitive forms of life, tiny cells that were perhaps the ancestors of bacteria. The true nature of viruses was discovered in 1933 when the biologist Wendell Stanley of the Rockefeller Institute tried to purify an extract of the tobacco mosaic virus (TMV). To Stanley's great surprise, the purified TMV extract formed crystals, a property of chemicals. The crystals retained the ability to infect healthy tobacco plants and were therefore the virus itself. Stanley concluded that TMV is chemical matter rather than a living organism.

Within a few years of Stanley's findings, scientists were able to disassemble TMV and confirm Stanley's conclusion. TMV is a chemical, not a cell. In fact, each particle of TMV is made of only two kinds of molecules: the nucleic acid RNA

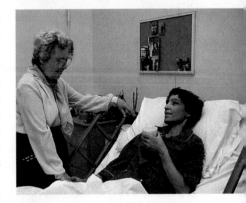

Figure 20-1 This AIDS patient is receiving treatment in a clinic at San Francisco General Hospital. In 1994, nearly 81,000 new AIDS cases were reported.

and protein. As you can see in Figure 20-2, TMV consists of a core of RNA surrounded by a coat of protein. Later scientists were able to separate the RNA from the protein. When they reassembled the two components, the reconstructed TMV particles were fully able to infect healthy tobacco plants. Clearly, the chemicals were the virus itself, not merely derivatives of it. From these experiments, and a host of others carried out using other viruses, a general picture of virus structure and function has emerged. A **virus** is a strand of nucleic acid encased in a protein coat that can infect cells and replicate within them. Biologists do not consider viruses to be living organisms.

Figure 20-2 Tobacco mosaic virus is rod-shaped and has a coat of proteins spiraling around a single strand of RNA. TMV causes a disease that stunts the growth of tobacco plants and discolors their leaves.

Figure 20-3 Influenza virus is made of a coil of RNA surrounded by a lipid-rich envelope studded with protein spikes.

Viruses Are Made of Protein and Nucleic Acid

Like TMV, most viruses have a protein sheath, or **capsid,** surrounding a core of nucleic acid. Many plant viruses, as well as some animal viruses such as the human immunodeficiency virus (HIV) that causes AIDS, contain RNA. However, the nucleic acid found in most viruses is DNA. Many viruses found in animals, such as the influenza virus shown in Figure 20-3, have a membranous **envelope** surrounding the capsid. The envelope helps the virus gain entry into cells. It contains proteins, lipids, and **glycoproteins** (proteins with carbohydrate molecules attached) derived from the host cell.

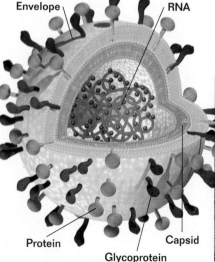

Envelope

RNA

Protein

Capsid

Glycoprotein

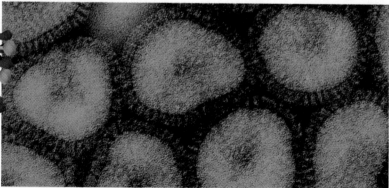

Viruses differ greatly in appearance. The simplest viruses consist of a single molecule of a nucleic acid and a capsid made of a single protein or a few different protein molecules repeated several times. The more complex viruses may consist of several different segments of DNA or RNA contained within a capsid made of several different kinds of protein. Most viruses have an overall structure that is either helical or polyhedral. A helical virus, like the tobacco mosaic virus, is rodlike in appearance, with capsid proteins winding around the core in a helix. A polyhedral virus has many sides and is roughly spherical. The capsid of most polyhedral viruses is in the shape of an **icosahedron** (*eye koh suh HEE druhn*), a shape with 20 triangular faces and 12 corners. By arranging themselves as an icosahedron, the proteins form an external shell that has more volume than would be possible with other shapes. Figure 20-4 shows the polyhedral shape of an adenovirus, which can cause upper respiratory infections in humans.

Some viruses, particularly bacterial viruses called **bacteriophages,** have very complicated structures. By taking a close look at Figure 20-5, you can see that the capsid is a polyhedral head attached to a helical tail. A long DNA molecule is coiled within the head.

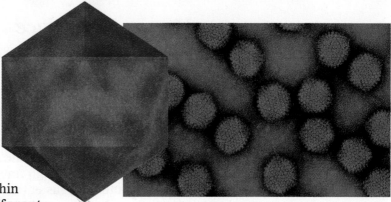

Figure 20-4 Adenovirus, *right,* is in the shape of an icosahedron, *left.* It has a capsid with a protein spike at each of its 12 corners.

Figure 20-5 Many bacteriophages (viruses that infect bacteria) have a complex capsid made of a polyhedral head attached to a tail. Such viruses usually inject their genetic material into a cell like a hypodermic needle.

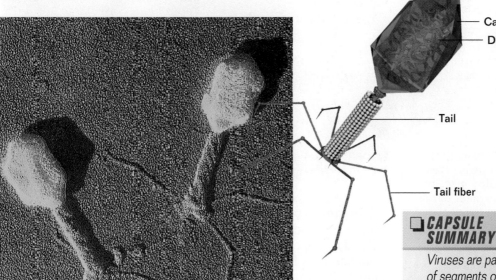

Capsid
DNA

Tail

Tail fiber

The smallest viruses are only about 17 nm in diameter. The largest ones may be up to 100 nm, barely big enough to be visible with a light microscope. Most viruses can be detected only by using the higher resolution of an electron microscope. ◻

Viruses and Bacteria 457

Viruses Reproduce Inside Living Cells

Viruses lack the enzymes for metabolism and have no ribosomes or other equipment for protein synthesis. Therefore they must rely on living cells for reproduction. Before a virus can reproduce, it must first infect a living cell. How does a virus get into a cell? A bacterial virus, like bacteriophage T4, punches a hole in the bacterial cell wall and injects its DNA into the cell like a hypodermic needle. A plant virus, like TMV, enters a plant cell through tiny rips in the cell wall at points of injury. An animal virus enters its host cell by endocytosis, the process by which the cell engulfs materials that are too large to enter through channels in the cell membrane. Once they are inside a cell, many viruses are **pathogens** *(PATH uh jehnz)*, agents that cause disease.

Before any virus can be engulfed by a cell, the virus must first bind to the cell membrane. What does the virus bind to? The envelope of an animal virus has spikes of glycoproteins and lipids that are able to bind to specific receptor molecules on the cell membrane. An animal virus is able to infect only cells with surface receptor proteins to which the virus's envelope molecules can attach. This is why human immunodeficiency virus (HIV), shown in Figure 20-6, infects only human white blood cells. It is also why the polio virus enters only certain spinal nerve cells and the hepatitis virus enters only liver cells.

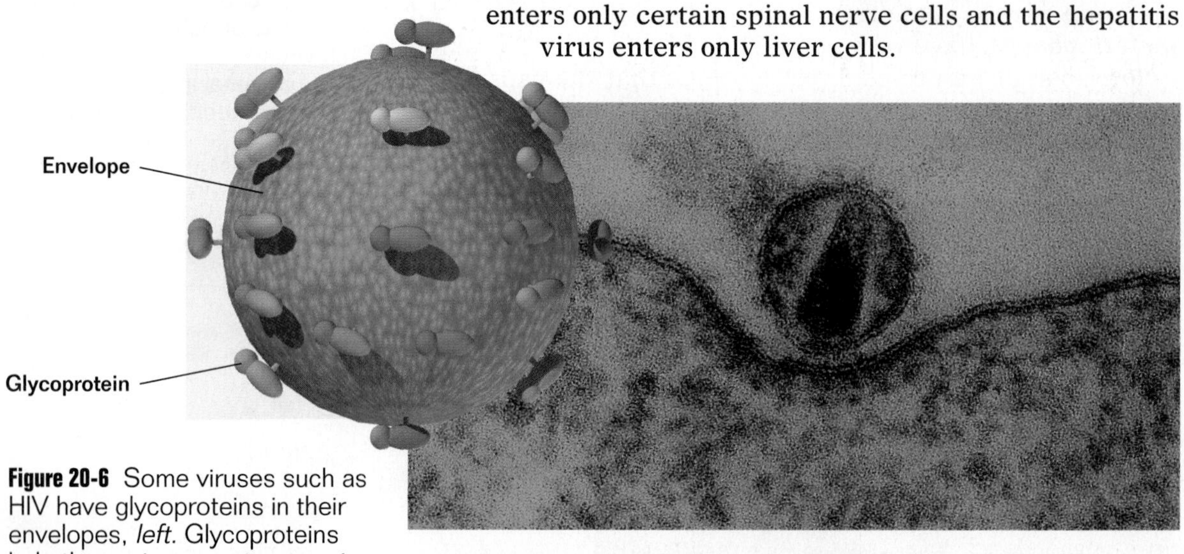

Envelope

Glycoprotein

Figure 20-6 Some viruses such as HIV have glycoproteins in their envelopes, *left.* Glycoproteins help these viruses enter an animal cell by binding to specific receptors on the cell surface, *right.*

Mammals protect themselves from viral infections by producing antibodies to the virus envelope's glycoproteins. An **antibody** is a protein secreted by cells in the immune system in response to a foreign substance in the body. The antibodies enable mammalian immune systems to identify and destroy viruses. However, mutations in viruses often change their glycoproteins and therefore make it difficult for the antibodies to recognize the virus. That is why new strains of influenza (flu) to which people are not resistant continually arise. As time passes, the envelope of the flu virus keeps changing. Like a fugitive changing disguises, the altered virus avoids detection.

Mutations in the virus genes that encode the structure of its glycoproteins may also enable the virus to bind to a receptor protein that it failed to recognize earlier. Many scientists think that HIV became a human pathogen when a mutation occurred in the HIV gene that encodes a glycoprotein (gp 160) in the envelope. The mutation altered the glycoprotein so that it was able to recognize and bind to the human white blood cell protein called CD4. The cells with the CD4 protein play a key role in the production of antibodies. Once able to bind to these cells, HIV enters and destroys them, disrupting the human immune system and eventually causing the disease AIDS. ☐

☐ **CAPSULE SUMMARY**

Viruses reproduce inside living cells. They can enter a cell by injecting their genetic material into the cell, slipping through tears in the cell wall, or binding to molecules on the cell surface and triggering endocytosis.

Viruses Take Over a Cell's Machinery

For an example of how a virus is able to reproduce within a host cell, study Figure 20-7, which illustrates how HIV infects human white blood cells. HIV gains access to a white blood cell first by binding to the cell membrane. The binding triggers endocytosis, and the cell membrane folds inward and surrounds the virus. The virus enters the cell within a membrane-bound vesicle, which soon releases the virus into the cell cytoplasm.

Once within the host cell, HIV sheds its envelope and capsid, leaving two strands of the virus's RNA floating in the cytoplasm. HIV also contains an enzyme called **reverse transcriptase,** which manufactures DNA that is complementary to the virus's RNA. A virus that transcribes DNA from an RNA template is called a **retrovirus.** Reverse transcriptase and other structural features of HIV are shown in *Up*

CONTENT LINK

*HIV's devastating effect on the human body's immune system will be discussed in **Chapter 39**.*

Figure 20-7 After HIV enters a cell by endocytosis, its nucleic acid (RNA) is used to make viral proteins and RNA.

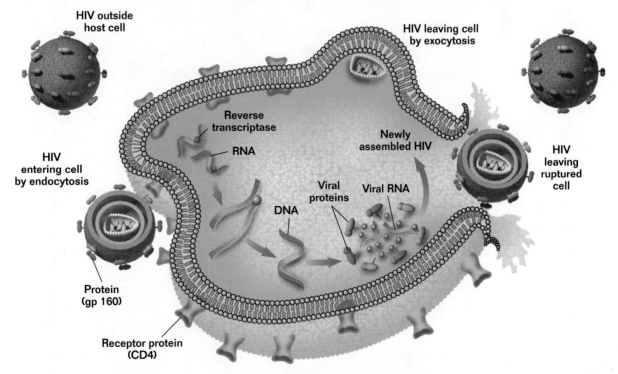

HIV outside host cell

HIV leaving cell by exocytosis

HIV entering cell by endocytosis

Reverse transcriptase

RNA

Newly assembled HIV

HIV leaving ruptured cell

Viral proteins

Viral RNA

DNA

Protein (gp 160)

Receptor protein (CD4)

UP CLOSE AIDS VIRUS

- **Name:** Human immunodeficiency virus (HIV)
- **Habitat:** Interior of human white blood cells
- **Size:** 125 nm

Characteristics

Envelope The outer envelope of HIV is composed of a lipid bilayer derived from the membrane of the host cell. Beneath the envelope is a protein core.

Viral proteins Embedded within the HIV envelope are glycoproteins such as gp 120 (used in preparing the first experimental vaccines) and gp 160, which enables the virus to recognize the human white blood cell surface protein CD4 and thus enter the cell.

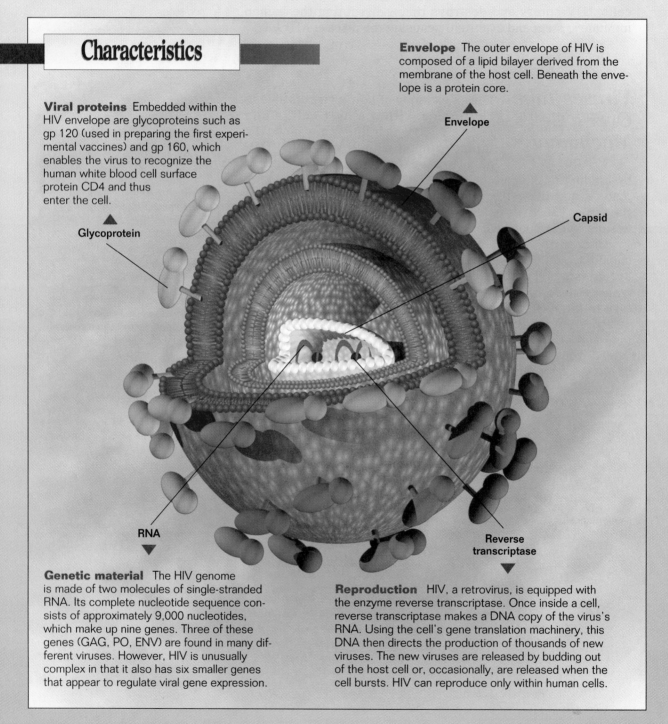

Envelope

Glycoprotein

Capsid

RNA

Reverse transcriptase

Genetic material The HIV genome is made of two molecules of single-stranded RNA. Its complete nucleotide sequence consists of approximately 9,000 nucleotides, which make up nine genes. Three of these genes (GAG, PO, ENV) are found in many different viruses. However, HIV is unusually complex in that it also has six smaller genes that appear to regulate viral gene expression.

Reproduction HIV, a retrovirus, is equipped with the enzyme reverse transcriptase. Once inside a cell, reverse transcriptase makes a DNA copy of the virus's RNA. Using the cell's gene translation machinery, this DNA then directs the production of thousands of new viruses. The new viruses are released by budding out of the host cell or, occasionally, are released when the cell bursts. HIV can reproduce only within human cells.

Close: AIDS Virus on page 460. In some viral infections, such as herpes, the DNA then inserts itself into the host cell's chromosome, where it remains inactive until some future event causes it to be removed from the chromosome and to resume activity. In other viral infections, the DNA does not insert itself into the host chromosome. Whether HIV inserts itself into the chromosomes of the white blood cells it infects is currently being disputed. Individuals infected with HIV typically do not exhibit high levels of the virus for an average of eight years. There is some evidence that the immune system succeeds in suppressing an ongoing infection over this long period, until its resources eventually become depleted and the infection escapes control.

After the viral RNA is transcribed into DNA, the genes are translated into HIV proteins. The host cell's machinery is then used to produce and assemble many copies of the virus. Some of the newly assembled virus particles leave the cell by exocytosis (budding out). Eventually the host cell ruptures, releasing thousands of additional virus particles. These newly released virus particles are then free to infect other white blood cells and continue the cycle of infection. Most other animal viruses follow a similar course of infection, although details may differ in individual cases. ❑

❑ CAPSULE SUMMARY

Viruses are able to reproduce by taking over a host cell's machinery. Retroviruses, such as HIV, are equipped with reverse transcriptase, a unique enzyme that can transcribe DNA from an RNA template.

Origins of Viruses

Viruses are considered to be escaped fragments of host genomes. This is why viruses are almost always highly specific to the hosts they infect. Because viruses originated as fragments of bacterial and eukaryotic genomes, their great diversity is not surprising. Biologists think there are at least as many kinds of viruses as there are kinds of organisms. Because a given organism often is susceptible to many kinds of viruses, the actual number of kinds of viruses may be much greater than the number of organisms.

Section Review

1. *How did Stanley's experiment with tobacco mosaic virus help reveal the nature of viruses?*
2. *What is a virus? Why don't biologists consider viruses to be living organisms?*
3. *Why do viruses rely on living cells for reproduction?*
4. *Why are some viruses able to avoid detection by the immune system?*

Critical Thinking

5. *Based on your knowledge of HIV structure and reproduction, describe one way to interrupt its life cycle.*

20-2 Bacteria

Section Objectives

- Differentiate between gram-positive and gram-negative bacteria.
- List seven differences between bacteria and eukaryotic cells.
- Describe the external and internal structure of Escherichia coli.
- Describe three different ways bacteria can obtain energy.

Bacteria are the oldest, simplest, and most abundant form of life on Earth. In a single gram of soil, there may be 2.5 billion bacteria. It is not surprising, then, that bacteria play a very important part in the web of life on Earth. They play a key role in recycling minerals within the Earth's ecosystems. In fact, photosynthetic bacteria were responsible in large measure for the introduction of oxygen into the Earth's atmosphere. Bacteria cause some of the most deadly diseases that cause injury to plants and animals, including humans. Our constant companions, bacteria are present in everything we eat and on everything we touch.

Bacteria Differ in Cell Wall Structure

Bacteria are small, single cells and are the only ones characterized by prokaryotic organization. Early in the evolution of life, bacteria split into the two branches archaebacteria and eubacteria, which are now classified in separate kingdoms. Too tiny to see with the naked eye, a bacterial cell is usually one of three basic shapes, as shown in Figure 20-8: **bacillus** (buh SIHL uhs), a rod-shaped cell; **coccus** (KAHK us), a spherical cell; or **spirillum** (spy RIHL uhm), a spiral cell. A few kinds of bacteria aggregate into stalked structures or filaments.

A bacterium's plasma membrane is encased within a cell wall. Members of the kingdom Eubacteria have a cell wall made of peptidoglycan, a network of polysaccharide molecules linked together with chains of amino acids. Some eubacteria have a cell wall covered with an outer membrane layer made of large molecules called lipopolysaccharides.

Figure 20-8 Approximately 4,800 different kinds of bacteria are currently recognized, but undoubtedly thousands more await discovery and description.

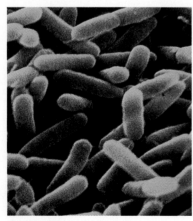

Pseudomonas

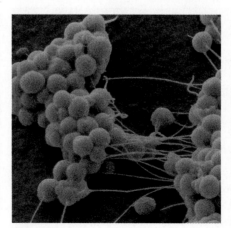

Staphylococcus

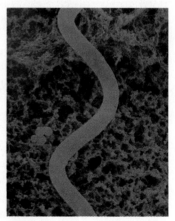

Spirillum

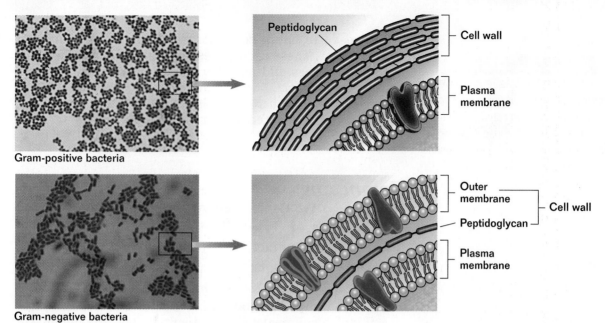

Gram-positive bacteria

Gram-negative bacteria

Peptidoglycan

Cell wall

Plasma membrane

Outer membrane

Cell wall

Peptidoglycan

Plasma membrane

Figure 20-9 The peptidoglycan layer is much thicker in gram-positive bacteria than in gram-negative bacteria, causing them to retain violet dye and appear purple.

(A lipopolysaccharide is a chain of sugar molecules with a lipid attached to one end.) Outside of the cell wall and membrane, many bacteria have a gelatinous layer called a **capsule.**

Eubacteria are commonly classified by differences in their cell walls. A bacterium with a cell wall containing a large amount of peptidoglycan is classified as **gram-positive.** A bacterium with a cell wall containing a thin layer of peptidoglycan covered by an outer membrane is classified as **gram-negative.** These terms refer to a bacterium's reaction to a staining procedure developed by the Danish microbiologist Hans Gram. These reactions are illustrated in Figure 20-9. In this procedure, a sample of bacteria is covered in a series of chemicals, beginning with a purple dye that stains the cell wall and ending with an alcohol rinse that breaks down cell membranes. Gram-positive bacteria retain the purple dye because their thick peptidoglycan layer remains intact and prevents the dye from leaving the cell. Gram-negative bacteria are not stained purple because the outer membrane is removed by the alcohol rinse and their thin peptidoglycan layer allows the dye to escape the cell. Gram-negative bacteria are usually identified by a pink stain that is applied after the alcohol wash.

Gram staining is an important technique in medicine. In many cases, the reaction to a Gram stain provides valuable information for the treatment of a bacterial disease. For example, gram-positive bacteria tend to be killed by penicillin, an antibiotic that prevents the proper formation of peptidoglycan in cell walls. Because of their outer membranes, gram-negative bacteria tend to be resistant to penicillin, but much more susceptible to the antibiotic tetracycline. Thus Gram stain identification of a bacterium can help determine which drug will be most effective against a disease.

Some bacteria form thick-walled **endospores** around their chromosomes and a small bit of cytoplasm when they

Bacteria can be classified into two groups according to cell wall structure. Gram staining can be used to identify bacteria because it distinguishes between two different kinds of bacterial cell walls.

are exposed to harsh conditions such as drought or high temperatures. These endospores are highly resistant to environmental stress and may germinate after years to form new, active bacteria. The formation of endospores in the bacterium *Clostridium botulinum* is responsible for botulism, a disease often considered to be a form of food poisoning. □

Comparing Bacteria With Eukaryotic Cells

Bacteria, which outnumber all eukaryotes combined, differ from eukaryotes in at least seven important respects. You can study the characteristics of *Escherichia coli*, a eubacterium that resides in your intestinal tract, in *Up Close: Escherichia coli* on page 465.

1. **Internal compartmentalization** Bacteria are prokaryotes and lack a cell nucleus, unlike eukaryotes. The cytoplasm of bacteria has very little internal organization. It contains no internal compartments or membrane systems.

2. **Cell size** Most bacterial cells are about 1 μm in diameter; most eukaryotic cells are well over 10 times that size.

3. **Multicellularity** All bacteria are single cells. Some bacteria may stick together in a matrix or may form filaments. However, these formations cannot truly be considered multicellular because the cytoplasm in the cells does not directly interconnect as is the case with many multicellular eukaryotes. Also, the activities of the cells are not specialized.

4. **Chromosomes** Bacterial chromosomes consist of a single circular piece of DNA. Eukaryotic chromosomes are linear pieces of DNA that are completely integrated with proteins.

5. **Cell division** Cell division in bacteria typically takes place by binary fission, a process in which one cell simply pinches into two cells. In eukaryotes, however, microtubules pull chromosomes to opposite poles of the cell during the nuclear division process called mitosis. Afterward the cytoplasm of the eukaryotic cell divides in half, forming two cells.

6. **Flagella** Bacterial flagella are simple structures composed of a single fiber of protein that spins like a corkscrew to move the cell. Eukaryotic flagella are more complex structures made of microtubules that whip back and forth rather than spin. Some bacteria also have shorter, thicker outgrowths called **pili** that act as docking cables, helping the cell to attach to surfaces or to other cells. Both of these bacterial appendages can be seen in Figure 20-10.

7. **Metabolic diversity** Bacteria have many metabolic abilities that eukaryotes lack. For example, bacteria perform several different kinds of anaerobic and aerobic respiration while eukaryotes are aerobic organisms. Unlike eukaryotes, certain bacteria can obtain their energy by oxidizing inorganic compounds such as sulfur. Other bacteria have the ability to fix atmospheric nitrogen. □

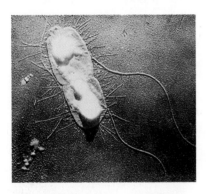

Figure 20-10 The short, hairlike fibers covering the *E. coli* above are pili, structures that help the bacterium adhere to surfaces. The two long strands are flagella, structures used for locomotion.

Bacteria and eukaryotes differ in their cellular organization, cell structures, and in metabolic diversity.

UP CLOSE *ESCHERICHIA COLI*

- ■ **Scientific Name:** *Escherichia coli*
- ■ **Habitat:** Inhabits the intestines of many mammals
- ■ **Size:** Up to 1 μm
- ■ **Mode of nutrition:** Heterotrophic

Characteristics

Cell structure *E. coli* is a prokaryotic cell; its genetic material and organelles are not enclosed within membranes. It has a rigid cell wall composed of a strong network of polysaccharides cross-linked by polypeptide chains. *E. coli* are gram-negative eubacteria—a lipopolysaccharide layer is deposited over the cell wall, forming an outer membrane.

Locomotion *E. coli* is motile, that is, it has the ability to move on its own. By rotating its slender whiplike flagella, *E. coli* propels itself through its environment.

Adherence Like many gram-negative bacteria, *E. coli* has pili—short, thin, hairlike appendages attached to the bacterial cell. Pili have two main functions. The first is to adhere to surfaces, including the surfaces of intestinal-lining cells. The second function of pili is to join bacterial cells prior to the transfer of DNA from one cell to another.

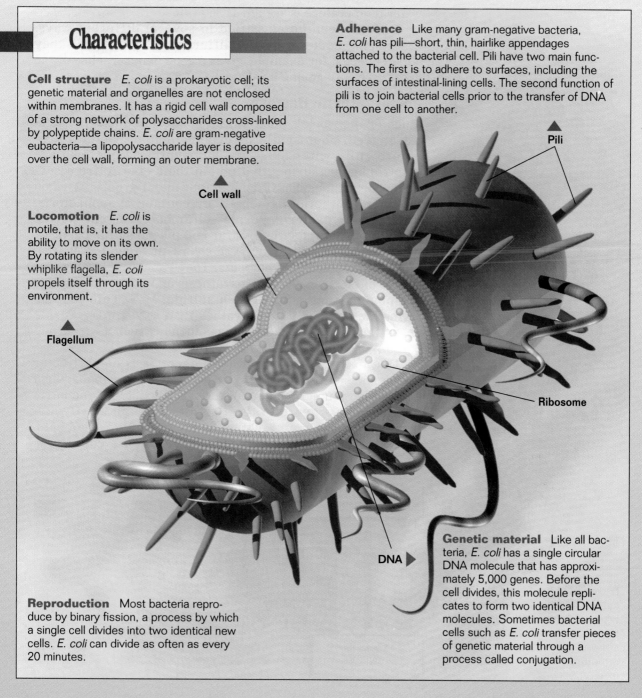

▲ Cell wall

▲ Flagellum

▲ Pili

Ribosome

DNA ▶

Reproduction Most bacteria reproduce by binary fission, a process by which a single cell divides into two identical new cells. *E. coli* can divide as often as every 20 minutes.

Genetic material Like all bacteria, *E. coli* has a single circular DNA molecule that has approximately 5,000 genes. Before the cell divides, this molecule replicates to form two identical DNA molecules. Sometimes bacterial cells such as *E. coli* transfer pieces of genetic material through a process called conjugation.

Grouping Bacteria According to How They Obtain Energy

Over 4,000 species of bacteria have been named, and undoubtedly many thousands more exist. Bacteria occur in the widest possible range of habitats and play key ecological roles in virtually all of them. They thrive in hot springs, where the usual temperature may range as high as 100°C (210°F); they have been found living beneath 430 m of ice in Antarctica. Bacteria are abundant in ground water and are even present at high pressures in the deep ocean, growing around deep sea vents where the water temperature is as high as 360°C (680 °F).

Bacteria can be classified in several different ways. Classifying bacteria by the different ways in which they obtain energy, for example, gives a good general sense of the great diversity among bacteria.

Photosynthetic Bacteria

Much of the world's photosynthesis is carried out by bacteria. Indeed, photosynthesis evolved among bacteria long before eukaryotes existed. Photosynthetic bacteria are autotrophs, organisms that obtain their energy from sunlight. They can be classified into four major groups based on the photosynthetic pigments they contain: cyanobacteria, green sulfur bacteria, purple sulfur bacteria, and purple nonsulfur bacteria. Green sulfur and purple sulfur bacteria, which are not necessarily these colors, are found in anaerobic (oxygen-free) environments. They cannot use water as a source of electrons for photosynthesis and instead use sulfur compounds such as hydrogen sulfide, H_2S. This process forms elemental sulfur, S, rather than oxygen, O_2. Purple nonsulfur bacteria use organic compounds such as acids and carbohydrates for photosynthesis.

One group of photosynthetic bacteria is of particular importance. In Chapter 13 you learned that cyanobacteria (also called blue-green algae) were responsible for the introduction of oxygen into the Earth's atmosphere. Cyanobacteria often clump together in large mats of filaments. Each filament is a chain of cells encased in a continuous

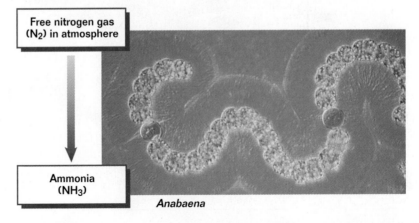

Figure 20-11 *Anabaena* is a photosynthetic cyanobacterium in which individual cells adhere in filaments. The two large orange-colored cells are heterocysts, structures in which nitrogen fixation occurs.

Free nitrogen gas (N_2) in atmosphere

Ammonia (NH_3)

Anabaena

jellylike capsule. A considerable number of cyanobacteria, such as *Anabaena,* shown in Figure 20-11, are capable of fixing nitrogen. Structures called **heterocysts** contain enzymes that fix nitrogen gas, N_2, into ammonia, NH_3, for use by the growing cell. Cyanobacteria such as *Anabaena* come the closest to multicellularity among the bacteria.

Chemoautotrophic Bacteria

Not all autotrophs obtain energy from sunlight. Bacteria called chemoautotrophs obtain energy by removing electrons from inorganic molecules such as ammonia, NH_3, methane, CH_4, or hydrogen sulfide, H_2S. In the presence of one of these hydrogen-rich chemicals, chemoautotrophic bacteria can manufacture all their own amino acids and proteins. Chemoautotrophic bacteria that live in the soil, such as *Nitrosomonas* and *Nitrobacter,* are of great importance to the environment and to agriculture. As depicted in Figure 20-12, they have a crucial role in a sequence of reactions in the nitrogen cycle called **nitrification,** a process that involves the oxidation of ammonia into nitrate. Nitrate is the form of nitrogen most commonly used by plants.

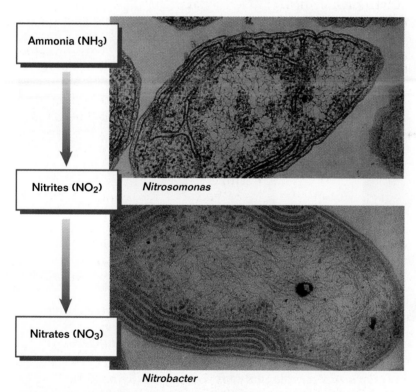

Ammonia (NH_3)

Nitrites (NO_2)

Nitrosomonas

Nitrates (NO_3)

Nitrobacter

Figure 20-12 *Nitrosomonas, above,* and *Nitrobacter, below,* are chemoautrophic bacteria that obtain their energy from inorganic molecules. They both play an important role in the nitrogen cycle. *Nitrosomonas* oxidizes ammonia into nitrites. *Nitrobacter* oxidizes nitrites into nitrates.

Heterotrophic Bacteria

Most bacteria are heterotrophs, feeding on organic material formed by other organisms. Together with fungi, heterotrophic bacteria are the principal decomposers of the living world; they break down the bodies of dead organisms and make the nutrients in these molecules available for recycling. Most of the odors associated with soil come from substances produced by heterotrophic bacteria. In recent decades, mutant strains of bacteria that break down synthetic products such as nylon and pesticides have been discovered.

Other activities of heterotrophic bacteria may be helpful or harmful to humans. For example, more than half of our antibiotics are produced by species of *Streptomyces*, a filamentous bacterium abundant in soil. On the other hand, one species of *Staphylococcus* secretes a toxin into food, causing severe reactions, including nausea, diarrhea, and vomiting, within a few hours after the food is eaten.

Species of the symbiotic bacteria *Rhizobium* are by far the most important of all nitrogen-fixing organisms. *Rhizobium* species are heterotrophic bacteria that usually live within nodules (lumps) on the roots of legumes (plants such as soybeans, beans, peas, peanuts, alfalfa, and clover), as shown in Figure 20-13. The plant furnishes anaerobic conditions and growth nutrients for the bacteria, and the bacteria fix nitrogen, which is incorporated into plant protein. Farmers take advantage of *Rhizobium*'s nitrogen-fixing abilities when they "rotate" their crops every few years and grow leguminous plants to replenish the soil with nitrogen.

Figure 20-13 Nitrogen fixation is carried out primarily by bacteria of the genus *Rhizobium*. This bacterium forms nodules in the roots of legumes.

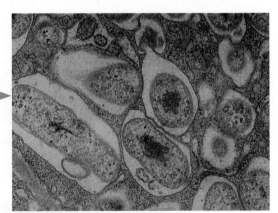

Rhizobium

Section Review

1. *Draw a cross section of the cell wall of a gram-positive bacterium and a gram-negative bacterium and explain the Gram staining procedure. How can the information obtained from a Gram stain be useful?*

2. *Construct a table that lists the seven ways bacteria differ from eukaryotic cells.*

3. *What two purposes do pili serve in* E. coli?

4. *In what three ways can bacteria obtain energy?*

Critical Thinking

5. *How would you classify bacteria used in cleaning up oil spills?*

20-3 Bacteria and Viruses as Pathogens

Bacteria are beneficial to humans in many ways. They recycle nutrients in ecosystems and play key roles in the manufacture of foods and drugs. However, bacteria have their most noticeable impact on humans as pathogens, agents of disease. At times, you may have blamed an illness on bacteria when in fact it was caused by a virus.

Section Objectives

- Describe the symptoms of tuberculosis.
- Explain antibiotic resistance.
- List five human diseases caused by bacteria.
- List five diseases that are caused by viruses.

Diseases Caused By Bacteria

Occasionally your body becomes a host for parasitic bacteria that may cause infection and disease. Pathogenic bacteria are harmful because they attack cells or secrete toxins. For example, tuberculosis is a bacterial disease that once ranked among the most common causes of death. **Tuberculosis** is a disease of the respiratory tract caused by the bacterium *Mycobacterium tuberculosis*, shown in Figure 20-14. In most instances, a person becomes infected with *M. tuberculosis* by inhaling tiny droplets of moisture that contain the bacteria. Some bacteria settle in the lungs. The body responds by forming small nodules called tubercles around the organisms. Bacteria can survive in tubercles for an indefinite amount of time. Eventually the tubercles may become scar tissue and trap the bacteria, rendering them harmless. Or, if the body's immune system becomes weakened, the bacteria may break out of the tubercles and enter a blood vessel. They then may spread to new sites. This condition leads to a progressive disease characterized by coughing up sputum and blood, chest pain, fever, fatigue, weight loss, and loss of appetite. Tuberculosis commonly occurs as a long-term, progressively worsening disease, although it can lead to a rapid death. Other bacterial diseases that have had notable impacts on human populations are described in Table 20-1.

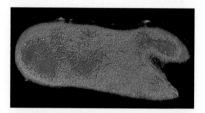

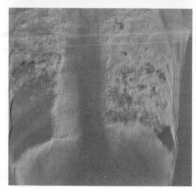

Figure 20-14 When inhaled, the bacterium *Mycobacterium tuberculosis, above,* forms small nodules called tubercles in lungs, *below.*

Treating Bacterial Diseases

In 1928, the British bacteriologist Alexander Fleming observed a mold of the genus *Penicillium* growing on a culture of bacteria. He noticed that bacteria did not grow near the mold. Apparently the mold was secreting a substance that killed the bacteria. Fleming isolated the substance and named it penicillin. In the early 1940s, scientists found that penicillin was very effective in treating many bacterial diseases, and its various forms have since been used in treating pneumonia, scarlet fever, rheumatic fever, and many other diseases.

Penicillin is an example of an **antibiotic**, a substance obtained from bacteria or fungi that is used as a drug to

Table 20-1 Important Bacterial Diseases

Disease	Symptoms	Infectious Agent	Mode of Transmission
Bubonic plague	Fever, bleeding lymph nodes form swellings called buboes; often fatal	*Yersinia pestis*	Carried from infected rodents by fleas
Cholera	Severe diarrhea and vomiting; often fatal	*Vibrio cholerae*	Contaminated water
Dental caries	Destruction of minerals in tooth enamel	species of *Streptococcus*	Dense collections of bacteria in mouth
Dysentery	Fever, diarrhea, vomiting	*Shigella dysenteriae*	Contaminated food or water
Lyme disease	Rash, pain, swelling in joints	*Borrelia burgdorferi*	Carried from infected animals by ticks
Typhoid fever	Headache, fever, diarrhea, rash; often fatal	*Salmonella typhi*	Contaminated water or food

fight pathogenic microorganisms. Antibiotics work by interfering with the microorganism's cellular processes. In most cases, this includes preventing cell wall formation, breaking up cell membranes, or disrupting chemical processes. Because these processes do not occur in viruses, antibiotics cannot be used for fighting viral diseases.

Since penicillin's discovery, other antibiotics such as tetracycline, streptomycin, and ampicillin have been discovered in nature or produced chemically. In the past, these drugs helped conquer diseases such as tuberculosis, bacterial pneumonia, syphilis, and gonorrhea. However, in recent years, strains of bacteria have become resistant to the antibiotics that once killed them. Antibiotic resistance arises when a few bacteria harbor mutant genes that make them immune to the drug. These bacteria survive and pass on their resistance genes to their offspring. Doctors have recently seen an increase in antibiotic-resistant strains of bacteria that cause diseases such as tuberculosis, pneumonia, meningitis and gonorrhea. New drugs, as well as new strategies, will be needed to tackle these invulnerable bacteria.

Diseases Caused By Viruses

Like pathogenic bacteria, viruses damage a body's cells and cause illness. For example, the common cold, which is usually spread by inhalation, is familiar to all of us. More than 200 kinds of viruses can cause a cold. Influenza, one of the most devastating diseases of all times, is caused by a virus found in wild ducks living in central and northern Asia.

Table 20-2 Important Viral Diseases

Disease	Symptoms	Mode of Transmission
Chickenpox	Blisters, painful rash, fever	Air currents
Measles	Blotchy rash, respiratory congestion, high fever	Air currents
Rubella	Rash, swollen glands	Air currents
Mumps	Painful swelling in salivary glands	Air currents
Smallpox	Blisters, lesions, fever, malaise; often fatal	Air currents
Infectious hepatitis	Fever, chills, nausea, swollen liver, jaundice, painful joints	Contaminated food or water
Polio	Headache, stiff neck, possible paralysis	Contaminated food or water
AIDS	Immune system failure; fatal	Sexual contact, contaminated-blood products or hypodermic needles

Commonly known as the flu, influenza is a viral disease of the respiratory tract characterized by chills, fever, and muscular aches. When the virus is inhaled, it comes in contact with the cells of the upper respiratory tract and penetrates them. There, the viruses reproduce and spread to new cells. People usually develop antibodies, substances that attach to the influenza viruses and prevent them from infecting new cells. However, the influenza virus mutates readily, constantly changing its surface proteins. Occasionally a new influenza variant is so different that the human immune system fails to recognize it. When this happens, the body has little defense against the infection, and influenza can be fatal. Other viral diseases that have had a serious effect on society are described in Table 20-2.

Section Review

1. *What causes tuberculosis? What are the symptoms of this disease?*
2. *How does antibiotic resistance in bacteria arise?*
3. *Describe five diseases caused by bacteria and five caused by viruses.*

Critical Thinking

4. *Why wouldn't a doctor prescribe an antibiotic to treat influenza?*

CHAPTER REVIEW

antibiotic (469)
antibody (458)
bacillus (462)
bacteriophage (457)
capsid (456)
capsule (463)
coccus (462)
endospore (463)

envelope (456)
glycoprotein (456)
gram-negative (463)
gram-positive (463)
heterocyst (467)
icosahedron (457)
nitrification (467)
pathogen (458)

pili (464)
retrovirus (461)
reverse transcriptase (459)
spirillum (462)
tuberculosis (469)
virus (456)

Review

Multiple Choice

1. What evidence led Stanley to conclude that tobacco mosaic virus (TMV) is not a living organism?
   a. The extract of TMV crystallized.
   b. TMV is made of RNA and protein.
   c. TMV reproduces only in cells.
   d. The virus poisons tobacco plants.

2. The basic components of all viruses are a nucleic acid and a(n)
   a. endospore.   c. protein coat.
   b. glycoprotein.   d. icosahedron.

3. What triggers the entry of HIV into human white blood cells?
   a. HIV glycoproteins bind to receptor proteins.
   b. The cells absorb genetic material.
   c. HIV injects its genetic material.
   d. The cell begins to divide.

4. Bacteria are different from eukaryotic cells in that
   a. only bacteria have flagella.
   b. eukaryotic cells are smaller than bacteria.
   c. bacteria perform aerobic and anaerobic respiration.
   d. eukaryotic cells lack a nucleus.

5. Bacteria that do not require sunlight but that obtain energy by removing electrons from hydrogen-rich chemicals are called
   a. heterotrophs.
   b. photosynthetic bacteria.
   c. cyanobacteria.
   d. chemoautotrophs.

6. Which of the following is *not* a mechanism by which antibiotics kill bacteria?
   a. preventing cell wall construction
   b. inhibiting mitosis
   c. breaking up cell membranes
   d. disrupting chemical processes

7. Which is the correct match of a viral disease and its effect on the human body?
   a. polio: paralysis
   b. influenza: skin blisters and lesions
   c. mumps: immune system failure
   d. dental caries: tooth decay

Completion

8. The shape of a virus is typically either _____ or polyhedral. The capsid of most polyhedrals is in the shape of a(n) _____ , which provides more volume than other shapes. Viruses are very small, ranging in size from about 17 nm in diameter to about _____ nm.

9. HIV is a _____ , which means that it transcribes DNA from RNA. It attacks human _____ cells and apparently causes the disease _____ , for which there is no known cure.

10. Bacteria with much peptidoglycan in their cell walls (and typically susceptible to penicillin) are called _____ bacteria, while those with less peptidoglycan (and typically susceptible to tetracycline) are called _____ bacteria.

11. Cyanobacteria introduced _____ into the Earth's atmosphere and are capable of fixing _____ .

12. Antibiotics are used to treat diseases caused by _____ , although they are usually ineffective against diseases caused by _____ .

13. A symptom of tuberculosis is the formation in the lungs of nodules called _____ .

14. **Structure and Function** The typical virus consists of either DNA or RNA encased in a protein coat and ranges in diameter from about 17 to 100 nm. What advantages do viruses derive from their relatively simple composition and extremely small size?

15. **Flow of Energy** Heterotrophic bacteria function as decomposers. Describe how these bacteria contribute to the flow of energy in an ecosystem.

16. **Evolution** Suppose cold viruses invade your body. Your body's immune system may destroy most but not all of these viruses. How does the response of your body's immune system affect the evolution of the cold viruses?

17. **Making Inferences** Some medical experts suggest that the drug AZT (azidothymidine) can help patients with AIDS. This drug blocks the action of the enzyme reverse transcriptase. Explain how AZT might help these patients.

18. **Making Inferences** In the 1520s, the Spanish explorer Cortes and his armies introduced smallpox to the Americas. The death rate among the native Aztecs ranged between 50 to 90 percent compared with a death rate of about 10 percent among people in Europe. What accounts for the difference in death rates?

19. **Making Predictions** Over the last 20 years, the number of antibiotic-resistant bacterial pathogens has steadily increased. This is thought be a result of antibiotic abuse by patients and doctors. Doctors tend to overprescribe antibiotics for patients who demand a quick fix for their illness. Just recently, the World Health Organization has established a global computer database so that doctors can report outbreaks of antibiotic resistance. What are the potential benefits of this database?

20. **Multicultural Perspective** AIDS, the disease caused by HIV, is a major health concern worldwide. Locate statistics on AIDS cases for as many countries as possible. Then, draw a world map on poster-size paper and devise a color legend for the map that shows the number of AIDS cases in countries for which data were located. Color the map to match the legend, and give it a title. Next, write a set of questions that can be answered using the map. Display the map and the questions where they can be seen by other students. Different students can take responsibility for gathering the AIDS data, drawing and coloring the map, creating the legend, and writing the questions.

21. **Research and Writing** Edward Jenner developed the first successful vaccine for smallpox in 1796. His vaccine was based on the fact that people who contracted the milder disease cowpox did not contract smallpox. Jenner inoculated people with the cowpox virus. Research the concerns people of Jenner's time had about his vaccine. Then, write a message to persuade them to get vaccinated.

Identifying and Staining Bacteria

OBJECTIVE

Identify types of bacteria by their shape.

PROCESS SKILLS

- comparing and contrasting different types of bacteria

MATERIALS

- prepared slides of coccus, bacillus, and spirillum bacteria
- compound light microscope
- safety goggles
- lab apron
- disposable gloves
- 3 culture tubes of bacteria, marked A, B, and C
- test tube rack
- sterile cotton swabs
- 3 microscope slides
- microscope slide forceps or wooden alligator-type clothespin
- 150 mL beaker
- methylene blue stain in dropper bottle
- paper towels

BACKGROUND

1. Are bacteria prokaryotes or eukaryotes?
2. How do prokaryotes and eukaryotes differ?
3. How do bacteria reproduce?
4. Write your own **Focus Question** on your Vee Form.

5. **Knowing Side of the Vee** List the **Concepts** and new **Vocabulary Words** on your Vee Form. In the **Concept Statements** section of the Vee, use these words in sentences that define and explain them.

TECHNIQUE

Doing Side of the Vee

Part A: Prepared Slides

1. Observe each of the three prepared slides of bacteria under low power and high power. Draw and label each type of bacterium in the **Records** section of your Vee Form. Compare your slides to the photographs shown below.

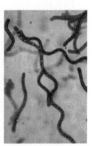

Part B: Live Bacteria

In this part of the investigation, you will transfer bacteria from each culture tube (A, B, and C) to three microscope slides for staining and observation. **CAUTION: Put on safety goggles, a lab apron, and disposable gloves.**

2. Have your partner remove the cap from the culture tube marked "A" with the tips of his or her fingers. The cap should not be placed on the table. Insert a sterile cotton swab into the test tube and transfer a very small amount of bacterial culture to a microscope slide. Spread the culture on the slide with the swab and allow it to dry. Dispose of the swab properly.

3. Use the microscope slide forceps or the wooden alligator-type clothespins to hold the slide. Place the slide across the mouth of a beaker half filled with water. Apply drops of methylene blue stain with a dropper. **CAUTION: Methylene blue is an eye irritant and stains skin and clothing. Avoid contact with your eyes and skin. Do not ingest.** Flood the slide with the stain, but do not allow the stain to spill into the beaker. Allow the stain to remain on the slide for two minutes.

4. Rinse the slide by dipping it into the water in the beaker several times. Blot the slide dry with a paper towel, being careful not to rub the slide. Allow the slide to dry.

5. When the slide is dry, observe it under the microscope and determine the type of bacteria on the slide—coccus, bacillus, or spirillum—based on your observations of the prepared slides. In a table similar to the one shown below, record the identity of the bacteria in the **Additional Records and Observations** section on the back of the Vee Form.

Types of Bacteria

Culture Tube	Type
Tube A	
Tube B	
Tube C	

6. Repeat steps 2 through 5 using the culture tubes marked "B" and "C." In the **Procedure** section of the Vee, briefly summarize the procedure you followed.

7. Clean up your materials and wash your hands before leaving the lab. Dispose of all materials according to instructions provided by your teacher.

INQUIRY

1. When observing the prepared slides, which power provided the clearest view of the bacteria?

2. Describe the shapes of coccus, bacillus, and spirillum bacteria.

3. Why should the test tube cap in Part B not be placed on the table?

4. Use the information on the **Knowing Side** of the Vee to interpret your results from the **Doing Side,** and then write your **Knowledge Claim.** Write a **Value Claim** for this lab.

ANALYSIS

1. What were some possible sources of contamination as you transferred the bacteria from the test tubes to the slides?

2. What is the advantage of staining bacteria before observing them under a microscope?

3. Why is caution necessary in handling bacteria, even if you are working with a bacterial species known to be harmless?

FURTHER INQUIRY

Write a **New Focus Question** that could be the point of a new investigation. The following is an example:

Are antiseptics equally effective against the three types of bacteria?

BIOLOGICAL WEAPONS

Should the Research Continue?

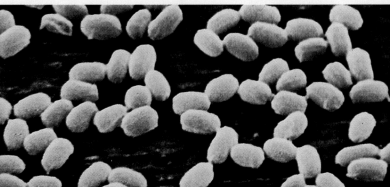

Can soldiers defend themselves from deadly biological weapons, like *Bacillus anthracis* bacteria?

BY TRACEY COHEN

The following article is an editorial that presents one person's viewpoint on the uses of scientific research and the social responsibility of researchers. Read the article carefully and see if you can determine the author's point of view and whether or not the evidence presented is balanced and fair.

I magine that you turn on the radio one morning and hear the following news.

"*The number of civilian casualties in the eastern European conflict continues to rise following the deliberate poisoning of a reservoir with a biological warfare agent last week. Another 32 people have died, bringing the death toll to 465, health officials reported. Hundreds more people have become ill, and hospitals are overrun with sick, frightened people. Medical experts have identified the pathogen as a new strain of intestinal bacteria that has been genetically altered to make it resistant to the three most widely used antibiotics.*"

Biological weapons use bacteria, viruses, fungi, and other organisms, or substances derived from them, to cause illness and death in humans, animals, or plants. Two well known agents developed for military use are anthrax and botulin toxin. Anthrax, a disease caused by the bacterium *Bacillus anthracis*, quickly kills between 95 and 100 percent of untreated infected victims. Botulin toxin, a bacterial toxin derived from *Clostridium botulinum*, is one of the most potent natural poisons known.

Warfare by contamination is actually not new. There are historical accounts dating back 2,000 years. But the development of biological weapons did not begin in earnest until the twentieth century. Though the full details were not known at the time, Japan started a large-scale biological weapons program in 1935. Fear that Nazi Germany was conducting a similar program led the United States to begin its own research on biological weapons in 1941. Today, at least 25 nations are developing biological weapons or already have them, according to the Pentagon.

Biological weapons are so terrible and uncontrollable that

most nations have decided to outlaw their use. In support of this view, 126 nations signed the Biological Weapons Convention (BWC) of 1972. This treaty bans the development, production, and stockpiling of biological weapons (BWs). Although the treaty prohibits research leading to the development of offensive weapons, it does permit countries to continue research aimed at defending themselves against attack by biological weapons.

Can We Really Defend Against BWs?

Many bacteria, viruses, fungi, and other agents known to cause disease can be adapted to serve as weapons. Genetic engineering and other biotechnologies can increase the diversity of potential pathogens further. It would be technically and economically impossible to vaccinate entire populations of people or livestock against all potential viruses or to develop antibiotics for all the pathogenic bacteria. Protecting plants would be no easier. Disease-resistant crops would have to be planted months ahead or enough of the right fungicide would have to be available at the right time. But, as in the case of human and animal pathogens, advance knowledge of exactly what weapons the enemy might use would be needed to enact defensive measures. This kind of information is not likely to be widely available.

Can Research Be Limited?

Given the horrible consequences of biological warfare, there is a growing concern that some defense research may apply to creating offensive weapons. In 1993, a study was issued by the Center for Public Integrity, an independent group that examines public services and ethic-related issues. They reported that approximately one-quarter of the current research could be converted to the development of offensive weapons. In their study, they cited several cases. In one instance, a biologist had created a new, highly infectious and antibiotic-resistant strain of anthrax. In another case, researchers had altered the deadly botulin neurotoxin so that it could not be treated by the usual antidote.

Who Will Decide the Future of Biological Weapons Research?

Creating a defense against biological weapons requires the expertise of many scientists from a variety of backgrounds. Specialists in public health, medicine, and biology all have a role to play in deciding the fate of future research. In a democracy, however, every citizen also has a view in these decisions–not just the scientists. It is in the public's interest that these issues be debated in the national forum where an informed citizen can be a part of the decision-making process.

What Steps Are Possible?

The United States has been a leader in the effort to eliminate the use of biological weapons. It may be possible, however, for the United States to take an even more active role in eliminating this threat. One possibility is refocusing its research efforts on work that does not involve producing pathogens or toxins. For instance, more research funds could be diverted to developing better protective clothing and equipment for military personnel who may be exposed to biological weapons in the field.

Another possibility is to transfer research on pathogens and toxins to civilian agencies such as the National Institutes of Health. This would increase public access to information and improve accountability. It would also help to counter another criticism made in the Center for Public Integrity's 1993 report: The study found that in 1989-90 (the most recent year for which complete data is available), just 3 percent of the $84.8 million spent on research went to institutions well known for biomedical research.

What Is the Role of Scientists in Shaping Research?

Through their work, scientists are able to cause broad and lasting changes in the world. This power carries with it a special obligation for scientists to consider the consequences of their work. It is not hard for example, for biomedical researchers to anticipate the impact of creating a deadly new virus.

Recently, several hundred scientists at universities and in industry signed a nationwide pledge not to accept research grants to develop biological warfare agents. These scientists have made a personal choice about the nature and extent of their involvement in research. Their actions, however, serve to remind all of us that every citizen has a stake in determining the future of biological weapons research.

Tracey Cohen is a freelance writer specializing in science and environmental issues.

Analyzing the Issue

1. **Detecting Bias** There are many different kinds of writing. For example, some articles attempt to describe events objectively, leaving readers to draw their own conclusions. Other articles present an analysis with an obvious point of view.

a. What kind of article is this? How do you know?

b. What is the writer's thesis, or main point?

c. Read through the article carefully. List all the opinions or unsupported judgments made in the article.

d. List the facts the writer uses to support the thesis.

e. Writers use facts selectively to support their arguments. What information not used in the article could challenge the writer's viewpoint?

2. **Researching the Issue** Research the history of the Manhattan Project. Find any accounts of how work on the project adversely affected the scientists who worked on it.

3. **Take Action** Read at least three articles concerning this issue. Your teacher will have a list of references that you can use. Based on your reading, list four reasons why the United States should continue defensive biological weapons research and four reasons why all such research should be abandoned.

CHAPTER 21

PROTISTS

• •

REVIEW

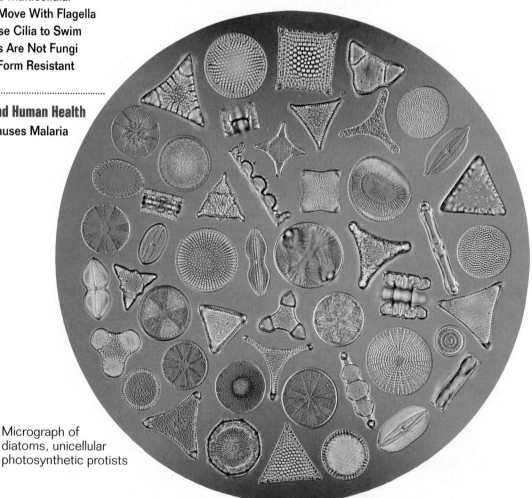

Micrograph of diatoms, unicellular photosynthetic protists

21-1 Characteristics of Protists

The most structurally diverse collection of organisms is the kingdom Protista. Most protists are unicellular, microscopic organisms, but a few are complex and multicellular. Protists are also the most diverse in terms of their life cycles, which can involve asexual or sexual reproduction. Composed of metabolically diverse organisms, the kingdom Protista contains the ancestral life-forms that gave rise to the three kingdoms of multicellular organisms—the fungi, the plants, and the animals.

Protists Were the First Eukaryotes

The first eukaryotic cells are thought to have evolved approximately 1.5 billion years ago, but there is a gap in the evolutionary record between prokaryotes and the first eukaryotes. And because eukaryotes differ from bacteria in so many ways, they could not have evolved quickly. Although no living organism has the characteristics associated with the first transitional eukaryotes, the amoeba *Pelomyxa palustris* exhibits many of them. *Pelomyxa* is a unique protist that lacks mitochondria and does not undergo mitosis; it may represent a very early stage in the evolution of eukaryotic cells.

Two important eukaryotic features that evolved among the protists are sexual reproduction and multicellularity. Many protists reproduce only asexually, by mitosis; some employ meiosis and sexual reproduction in times of environmental stress, and others reproduce sexually most of the time. Multicellularity, involving a small degree of coordination and interaction among specialized cells, also evolved independently in different groups of protists at different times. And, early during the evolution of protists, complex flagella and cilia appeared. These organelles are composed of nine pairs of microtubules surrounding two single microtubules in the center. Biologists refer to this as the "9 + 2" structure, which is present in all eukaryotes that have these organelles. Figure 21-1 lists the eukaryotic features that evolved in protists.

What Unites Members of the Kingdom Protista?

The kingdom Protista contains all eukaryotes that cannot be classified as animals, plants, or fungi. Protists are distinguished from these other organisms because they lack specialized features that characterize the three other multicellular kingdoms. For example, unlike plants and animals, protists do not reproduce by forming embryos, nor do they develop complex multicellular reproductive structures.

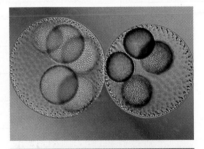

Eukaryotic Features That Evolved in Protists

Sexual reproduction

Meiosis

Mitosis

Multicellularity

Flagella and cilia with "9 + 2" structure

Figure 21-1 The protist *Volvox* is a colony of hundreds or thousands of flagellated cells. Like all other eukaryotes, *Volvox* exhibits key eukaryotic characteristics that first evolved among protists.

Instead of containing a well-defined group of similar organisms, the kingdom Protista consists of an unusual assortment of organisms that have different ways of meeting life's needs. For example, some protists have chloroplasts and manufacture their own food (like plants), some ingest their food (like animals), and some absorb their food (like fungi). The major phyla of protists are strikingly different from one another and, with a few exceptions, are only distantly related.

In older systems of classification, photosynthetic protists (algae) were classified in the plant kingdom, as were protists that absorbed their food (water molds and slime molds). Heterotrophic protists were thought to consume food "like animals" and were considered small, simple members of the animal kingdom. Today all groups of eukaryotic organisms that do not have the specialized characteristics of either animals, plants, or fungi are classified in the kingdom Protista, as shown in Table 21-1. ◻

Table 21-1 Members of Kingdom Protista

Common Name	Approximate Number of Species
Amoebas	300
Brown algae	1,500
Cellular slime molds	70
Chytrids	575
Ciliates	8,000
Diatoms	more than 11,500
Dinoflagellates	2,100
Euglenoids	1,000
Foraminiferans (Forams)	300
Green algae	more than 7,000
Plasmodial slime molds	500
Red algae	4,000
Sporozoans	3,900
Unicellular flagellates	3,000
Water molds	580

Protists Are Confined to Moist Environments

Protists are found almost anywhere there is water. Many live in lakes and oceans, floating as plankton or anchored to rocks. They are also common inhabitants of damp soil and sand, and thrive in other moist terrestrial environments such as leaf litter. Some species of protists are parasites. They live in the tissues and bloodstream of humans and other animals, where they can cause potentially fatal diseases. Other protists parasitize plants.

Many protists have mechanisms for monitoring and responding to stimuli in their environment. For example, some protists have **eyespots**, small organelles containing light-sensitive pigments that detect changes in the quality and intensity of light. Protists are also sensitive to touch and chemical changes in their environment. When certain protists encounter a noxious chemical, for example, they will back up and try to bypass it. ❑

❑ CAPSULE SUMMARY

Protists live in moist environments and can be either free-living or parasitic.

Some Protists Can Reproduce Sexually
....

Some kinds of protists reproduce only asexually, giving rise to new genetically identical individuals. Others can reproduce sexually or asexually. In these protists, sexual reproduction is often triggered by periods of environmental stress.

Reproduction Among Unicellular Protists

The unicellular green alga *Chlamydomonas* reproduces in a manner typical of unicellular protists. *Chlamydomonas* exhibits sexual as well as asexual reproduction, as outlined in Figure 21-2. As a mature organism, the single-cell protist is haploid. *Chlamydomonas* reproduces sexually in times of environmental stress such as a shortage of nutrients. Under such circumstances, mitosis produces haploid gametes. After they are released, gametes of opposite mating types fuse to form pairs. There are two different mating types (designated + or −) in *Chlamydomonas*. Each mating type has distinctive

Figure 21-2 The unicellular green alga *Chlamydomonas* reproduces sexually when gametes of opposite mating types fuse. It reproduces asexually when it divides by mitosis.

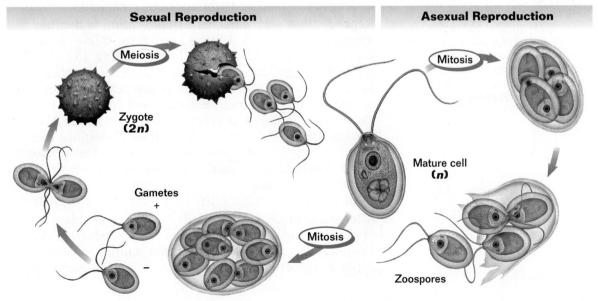

Sexual Reproduction   **Asexual Reproduction**

Meiosis

Mitosis

Zygote (2*n*)

Mature cell (*n*)

Gametes +

Mitosis

−

Mitosis

Zoospores

cell-surface proteins. These mating types ensure that only individuals with different genetic lineages will participate in sexual reproduction. When gametes of opposite mating types pair, they shed their cell walls and fuse into a diploid zygote with a thick protective wall called a **zygospore.** A zygospore can withstand unfavorable environmental conditions for long periods. When conditions become favorable again, meiosis within the zygospore produces four haploid individuals, which break out of the zygospore wall. These haploid cells, two of each mating type, grow into mature cells and complete the sexual life cycle.

When it reproduces asexually, *Chlamydomonas* first absorbs its flagella. The haploid cell then divides mitotically one to three times, producing two to eight haploid cells called **zoospores,** which remain within the wall of the parent cell. After developing flagella, mature zoospores break out of the parent cell and complete the asexual life cycle.

Sexual Reproduction Among Multicellular Protists

Sexual reproduction among multicellular protists occurs in many different ways, some of which are quite complex. Three modes of reproduction found among the green algae demonstrate the reproductive variation that exists among multicellular protists.

1. **Conjugation** *Spirogyra,* a filamentous green alga, reproduces sexually by a process called **conjugation.** Conjugation begins when two filaments align side by side. Portions of the walls between adjacent cells then dissolve and form a cytoplasmic bridge between the cells called a conjugation tube. As seen in Figure 21-3, the contents of one cell then pass through the tube into the cell of the adjacent filament, where the two nuclei fuse to form a diploid nucleus. The resulting zygote then develops a thick wall, falls from the parent filament, and becomes a resting spore. When conditions are favorable, the resting spore undergoes meiosis and produces a new haploid filament.

2. **Reproduction with gametes** *Oedogonium* is another filamentous green alga. Unlike *Spirogyra, Oedogonium* has specialized cells for producing and holding gametes. One type of cell produces flagellated sperm; the other type produces an egg. When the sperm are released, they swim to the egg-carrying cell, enter it, and fuse with the egg. The resulting zygote is released and forms a thick-walled resting spore. This diploid spore then undergoes meiosis, forming four haploid zoospores that are released into the water. Each zoospore settles and divides by mitosis. When a zoospore divides, one of the new cells becomes an anchor, while the others divide to form the new filament.

3. **Alternation of generations** *Ulva* is a very common marine green alga. Figure 21-4 shows that the reproductive cycle of *Ulva* is characterized by two distinct multicellular phases: a haploid, gamete-producing phase called the gametophyte

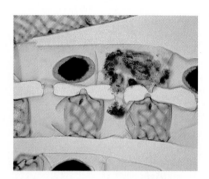

Figure 21-3 *Spirogyra* is a filamentous green alga that contains large spiral-shaped chloroplasts. During sexual reproduction, adjacent filaments join by conjugation tubes. Genetic material from the cell of one filament passes into the adjacent cell of the other filament.

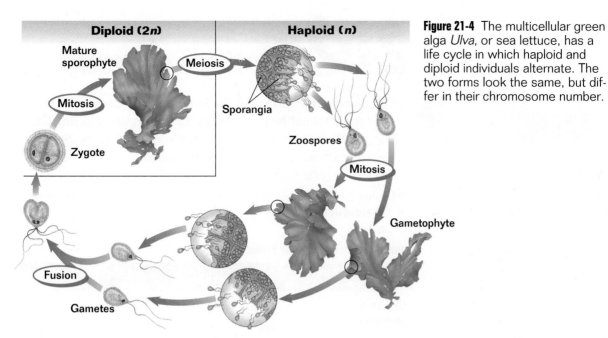

Figure 21-4 The multicellular green alga *Ulva*, or sea lettuce, has a life cycle in which haploid and diploid individuals alternate. The two forms look the same, but differ in their chromosome number.

generation, and a diploid, spore-producing phase called the sporophyte generation. The adult sporophyte alga, called sea lettuce, is large and distinctive in appearance. It has reproductive cells called **sporangia** (*spoh RAN jee uh*), which produce haploid zoospores by meiosis. The spores land on rocks and grow into multicellular haploid gametophytes. When mature, the gametophyte superficially resembles the sporophyte. The mature gametophyte eventually produces haploid gametes that unite and form zygotes. These diploid zygotes then complete the life cycle by dividing mitotically to form a new diploid sporophyte. ◻

❑ *CAPSULE SUMMARY*

Some protists are able to reproduce sexually in times of stressful environmental conditions.

Section Review

1. *What are eukaryotic features that first evolved in protists?*

2. *Why aren't protists classified in the other three multicellular kingdoms?*

3. *Describe three environments in which protists thrive.*

4. *Draw the asexual and sexual life cycles of the protist* Chlamydomonas. *What causes the protists to reproduce sexually?*

5. *Describe three ways sexual reproduction can occur in multicellular protists.*

Critical Thinking

6. *Why is kingdom Protista casually referred to as a "catchall" kingdom?*

21-2 *Protist Diversity*

Section Objectives

- Explain how amoebas and forams move.
- Describe the structure of diatoms.
- Compare and contrast the three kinds of algae.
- Name three different kinds of flagellates.
- Describe the general characteristics of Euglena and Paramecium.
- Discuss the unique features that distinguish protistan molds and sporozoans from other protists.

The diverse nature of the kingdom Protista can best be understood by considering 15 major phyla of protists. These phyla can be placed into seven groups distinguished from one another by features such as structure, means of locomotion, and formation of spores. Table 21-2 lists the major phyla of protists and the features that make them unique.

Table 21-2   15 Phyla of Protists

Distinguishing Features	Phylum	Mode of Nutrition
Move using pseudopodia	Rhizopoda (amoebas)	Heterotrophic
	Foraminifera (forams)	
Have double shells made of silica	Bacillariophyta (diatoms)	Photosynthetic
Photosynthetic protists; can be multicellular	Chlorophyta (green algae)	Photosynthetic
	Rhodophyta (red algae)	
	Phaeophyta (brown algae)	
Move using flagella	Dinoflagellata (dinoflagellates)	Photosynthetic
	Zoomastigina (unicellular flagellates)	Heterotrophic
	Euglenophyta (euglenoids)	Most are heterotrophic; some are photosynthetic
Move using cilia	Ciliophora (ciliates)	Heterotrophic
Funguslike protists	Acrasiomycota (cellular slime molds)	Heterotrophic
	Myxomycota (plasmodial slime molds)	
	Oomycota (oomycetes)	
	Chytridiomycota (chytrids)	
Form resistant spores	Sporozoa (sporozoans)	Heterotrophic

Protists That Move Using Cytoplasmic Extensions

An interesting group of protists consists of amoebas *(uh MEE buhs)* and forams. These protists are distinguished by their unique form of locomotion, in which they use extensions of cytoplasm. Amoebas and forams are unicellular heterotrophs.

Amoebas, members of the phylum Rhizopoda, are protists that live in fresh and salt waters and are especially abundant in soil. Because an amoeba has no cell walls or flagella, it is extremely flexible. It moves through its environment using extensions of cytoplasm called **pseudopodia** *(soo doh POH dee uh)*, from the Greek words *pseudo*, meaning "false," and *podium*, meaning "foot." A pseudopodium bulges from the cell surface, stretches outward, and anchors itself to a nearby surface. The cytoplasm from the rest of the amoeba then flows into the pseudopodium. Pseudopodia are also used to surround and engulf food particles in the process of endocytosis, as you can see in Figure 21-5. Meiosis and sexual reproduction do not occur in amoebas. They reproduce by fission, simply dividing into two new cells. The majority of amoebas are free-living, but some species are parasites, such as *Entamoeba histolytica,* the protist that causes amoebic dysentery in humans. These organisms are transmitted by contaminated food or water.

Forams, members of the phylum Foraminifera, are marine protists that live in sand or attached to other organisms or rocks. Forams are characterized by their porous shells, called tests. Tests usually have many chambers arrayed in a spiral shape resembling a tiny snail and consisting of organic material that contains grains of calcium carbonate. Long, thin projections of cytoplasm extend through the pores in the tests to aid in swimming and in capturing prey. For 200 million years, tests have accumulated in limestone deposits and are important components of land formations such as the white cliffs of Dover, the famous landmark in southern England, shown in Figure 21-6. The life cycle of forams is complex and involves alternation between haploid and diploid generations.

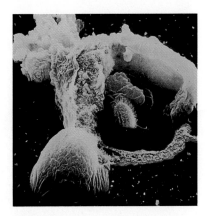

Figure 21-5 Members of the phylum Rhizopoda, such as this amoeba, ingest prey by endocytosis, capturing and engulfing prey with pseudopodia.

Figure 21-6 Forams live in snail-like shells made of calcium carbonate, *below right.* Fossils of these shells make up the sedimentary deposits in the white cliffs of Dover in England, *below left.*

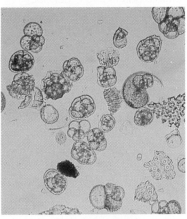

Diatoms Have Double Shells

Diatoms, members of the phylum Bacillariophyta, are photosynthetic, unicellular protists with unique double shells made of silica, which are often strikingly and characteristically marked, as shown in Figure 21-7. Their shells are like small boxes with lids, one half fitting inside the other. Abundant in oceans and lakes, diatoms are important food producers. Diatoms can have one of two types of symmetry: radial (like a wheel) or bilateral (two-sided). The empty shells of diatoms form thick deposits that are mined commercially as "diatomaceous earth," which is often used as an abrasive or to add the sparkling quality to paint used on roads. Diatoms are capable of a gliding movement made possible by chemicals secreted out of holes in their shells. Individuals are diploid and usually reproduce asexually. The two halves of the shell separate, and each half regenerates another matching half. As a consequence of this mode of reproduction, diatoms tend to get smaller and smaller with each generation. When an individual gets too small because of repeated division, it slips out of its shell, grows to full size, then regenerates a new shell. Sexual reproduction in diatoms is rare. ◾

CAPSULE SUMMARY

Amoebas and forams are protists that move using cytoplasmic extensions. Diatoms are protists with glasslike shells.

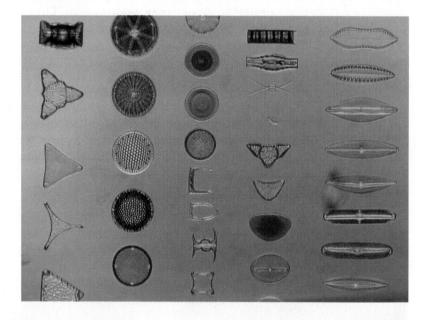

Figure 21-7 Members of the phylum Bacillariophyta, such as these diatoms, consist of two shells that fit together like a tiny box. Small pores in the shells enable gases and other substances to enter and leave the cell.

Some Algae Are Multicellular

The kingdom Protista also contains some of the fastest growing and most photosynthetically productive organisms—the algae. The three kinds of algae—green algae, red algae, and brown algae—are distinguished by the types of chlorophyll they contain.

Green algae, members of the phylum Chlorophyta, are an extremely varied group of protists. Most green algae are freshwater, unicellular organisms such as *Chlamydomonas*, but some are large multicellular marine organisms like *Ulva*, shown in Figure 21-8, a species found in marine intertidal zones. Other species are part of the marine plankton, inhabit damp soil, or even thrive within the cells of other organisms as photosynthetic symbionts. Green algae have chloroplasts that contain chlorophylls *a* and *b*, the same pigments found in the chloroplasts of plants. This is one reason why green algae are considered to be the ancestors of the plant kingdom. Most green algae have complex life cycles with sexual and asexual reproductive stages.

Red algae, members of the phylum Rhodophyta, are multicellular organisms found in warm ocean waters. Their color results from red pigments called phycobilins (*fy koh BYLIHNZ*), which are especially efficient at absorbing the green, violet, and blue light that penetrates into deep waters. For this reason, red algae can thrive at greater depths than other photosynthetic organisms. Red algae have complex bodies made up of interwoven filaments of cells. Some, such as the coralline algae shown in Figure 21-9, have calcium carbonate in their cell walls. Others have cell walls with a slippery outer layer that is used to make commercial products such as agar and carrageenan. Red algae have a complex life cycle, usually involving alternation of generations. None have flagella or centrioles, suggesting that red algae may be one of the most ancient groups of eukaryotes.

Brown algae, members of the phylum Phaeophyta, are all multicellular and almost exclusively marine. They are the most abundant seaweeds in many northern regions along rocky shores. The larger brown algae known as kelp grow in massive groves in relatively shallow water along coasts and provide food and shelter for many different kinds of organisms. Many have flattened blades, stalks, and anchoring bases and often contain complex internal conducting tissues like those of plants. Among the larger brown algae are genera such as *Macrocystis*, shown in Figure 21-10, whose blades float on the surface of the water while the base is anchored many meters below. The chloroplasts of brown algae resemble those of diatoms and dinoflagellates. Their life cycle involves an alternation of generations, with the sporophyte (diploid) generation consisting of the largest individuals.

Figure 21-8 *Ulva* is a green alga composed of a sheet of cells that is two cells thick. It grows attached to rocks in intertidal zones.

Figure 21-9 This species of coralline algae contributes to the great coral reefs.

Figure 21-10 Massive groves of the brown alga *Macrocystis* grow in temperate coastal waters throughout the world. These great kelp beds contain individuals that grow to a length of more than 60 m in a single season.

Some Protists Move With Flagella

Flagellates are protists that move using flagella. The three major phyla of flagellates are the dinoflagellates, the zoomastigotes, and the euglenoids.

Dinoflagellates (dy noh FLAJ uh layts), members of the phylum Dinoflagellata, are unicellular photosynthetic protists, most of which have two flagella. A few kinds of dinoflagellates are found in fresh waters, but the majority are marine, and they are often a component of plankton. Most dinoflagellates have a protective coat made of cellulose that is often encrusted with silica, giving them unusual shapes, as shown in Figure 21-11. Their flagella beat in two grooves—one encircling the body like a belt, the other perpendicular to it. As a result, dinoflagellates spin through the water like a top. Some species are luminous and produce a twinkling light that can be seen at night in tropical seas. A few dinoflagellates produce powerful toxins. The poisonous "red tides" that occur frequently in coastal areas are often associated with population explosions of dinoflagellates. Dinoflagellates reproduce asexually by mitosis.

Figure 21-11 This dinoflagellate, *Gonyaulax tamarensis*, is characterized by a pair of flagella set in perpendicular grooves.

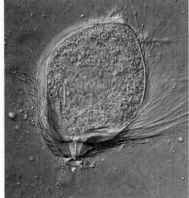

Figure 21-12 *Trichonympha* is a protist that inhabits the digestive tract of termites, where it aids in digesting cellulose. Rows of flagella protrude from one end of the cell.

Zoomastigotes (zoh oh MAS tih gohts), members of the phylum Zoomastigina, are unicellular, heterotrophic organisms that vary greatly in form. Each has at least one flagellum; some species have thousands. Most zoomastigotes reproduce only asexually, but some are known to produce gametes and reproduce sexually. Some zoomastigotes such as *Trichonympha*, shown in Figure 21-12, live symbiotically in the guts of termites, where they provide the enzymes that digest wood. Others, such as the trypanosomes, are dangerous pathogens in humans and domestic animals. The choanoflagellates (koh an o FLAG uh layts) are zoomastigotes that closely resemble collar cells in sponges and are thought to have given rise to the sponges and thus all other animals.

Euglenoids *(yoo GLEE noyds),* members of the phylum Euglenophyta, are freshwater protists with two flagella. They clearly illustrate the impossibility of classifying protists as animals or plants. About one-third of the 1,000 known species of euglenoids have chloroplasts and are photosynthetic; other species lack chloroplasts, ingest their food, and are heterotrophic. Some photosynthetic euglenoids may reduce the size of their chloroplasts and become heterotrophic if they are kept in a dark environment. If they are put back in the light, their chloroplasts return to normal size within a few hours, and photosynthesis resumes. Euglenoids are clearly related to zoomastigotes, and many taxonomists merge the two phyla and consider them to be one. *Euglena,* shown in Figure 21-13, has a protein scaffold called a **pellicle** *(PEHL ih kuhl)* inside the cell membrane. Since the pellicle is flexible, the euglenoid can change shapes. A light-sensitive organ called the eyespot helps orient the movements of these organisms toward light. Reproduction in this phylum occurs by mitosis; sexual reproduction has not been observed in this group.

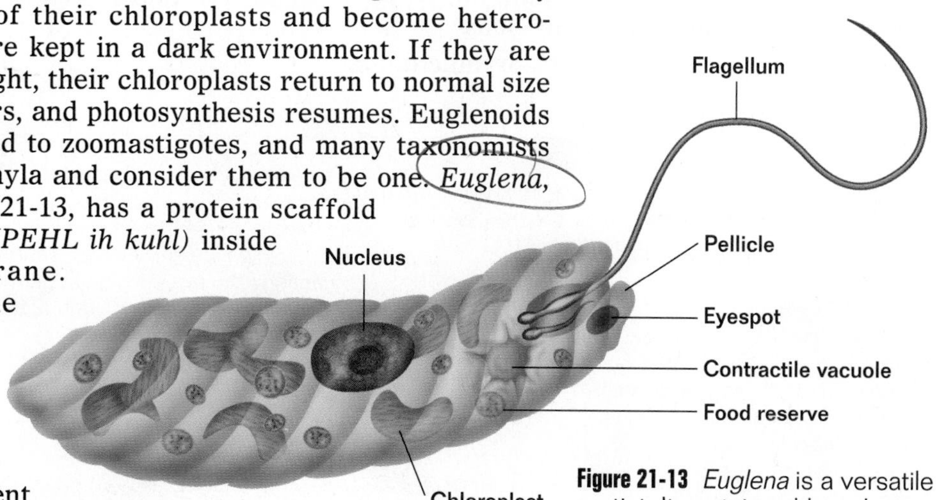

Flagellum

Nucleus

Pellicle

Eyespot

Contractile vacuole

Food reserve

Chloroplast

Figure 21-13 *Euglena* is a versatile protist. It contains chloroplasts and is photosynthetic, but it can also absorb organic nutrients and can live without light.

Protists That Use Cilia to Swim

As the name *ciliates* indicates, all members of the phylum Ciliophora have large numbers of cilia, usually arranged in long rows down the body or in spirals around it, as shown in Figure 21-14. Ciliates are complex unicellular heterotrophs. The body wall of ciliates is a tough but flexible outer pellicle that enables the organism to squeeze through or move around many obstacles. The pellicle consists of an outer membrane with numerous fluid-filled cavities beneath it. Ciliates, such as *Paramecium* shown in *Up Close: Paramecium* on page 490, form vacuoles for ingesting food and regulating their water balance. In addition to their characteristic cilia, most ciliates have two types of nuclei within their cells: small micronuclei and larger macronuclei. The micronuclei contain normal chromosomes that divide by mitosis. Macronuclei contain small pieces of DNA derived from micronuclei. Reproduction is usually by mitosis, with the body splitting in half. Cells divide asexually for about 700 generations and then die if sexual reproduction has not occurred. Most ciliates engage in the sexual process of conjugation, in which two cells unite and exchange genetic material.

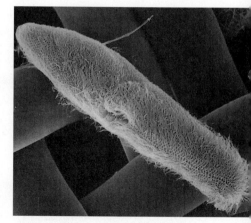

Figure 21-14 *Paramecium* is a fresh water protist that uses its thousands of cilia to move and feed.

UP CLOSE PARAMECIUM

- **Scientific name:** *Paramecium caudatum*
- **Habitat:** Lives in freshwater streams and ponds
- **Size:** Microscopic; up to 1 mm long
- **Diet:** Bacteria, small protists, organic debris

Characteristics

Surface *Paramecium*, a ciliate, is covered with thousands of cilia arranged in rows along the cell. Cilia beat in waves that move diagonally across the cell, causing the protist to spin through the water. *Paramecium* is surrounded by a rigid protein covering called a pellicle.

Maintaining Water Concentration Like other freshwater protists, *Paramecium* is constantly absorbing water by osmosis. Since these organisms need to maintain a relatively low concentration of water inside the cell in order to function normally, they must get rid of excess water. *Paramecium* does this with contractile vacuoles, saclike organelles that expand, collecting excess water, and then contract, squeezing water out of the cell.

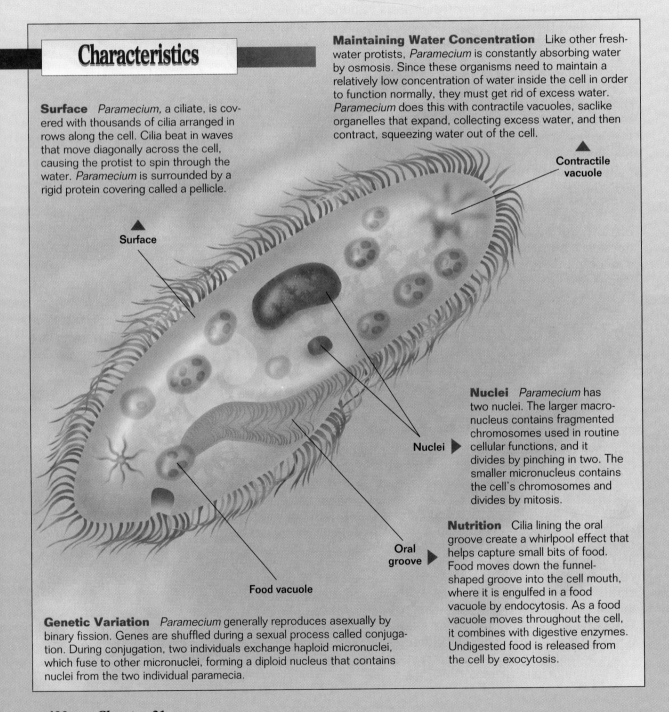

Contractile vacuole

Surface

Nuclei

Oral groove

Food vacuole

Nuclei *Paramecium* has two nuclei. The larger macronucleus contains fragmented chromosomes used in routine cellular functions, and it divides by pinching in two. The smaller micronucleus contains the cell's chromosomes and divides by mitosis.

Nutrition Cilia lining the oral groove create a whirlpool effect that helps capture small bits of food. Food moves down the funnel-shaped groove into the cell mouth, where it is engulfed in a food vacuole by endocytosis. As a food vacuole moves throughout the cell, it combines with digestive enzymes. Undigested food is released from the cell by exocytosis.

Genetic Variation *Paramecium* generally reproduces asexually by binary fission. Genes are shuffled during a sexual process called conjugation. During conjugation, two individuals exchange haploid micronuclei, which fuse to other micronuclei, forming a diploid nucleus that contains nuclei from the two individual paramecia.

Protistan Molds Are Not Fungi

Protistan molds are heterotrophs with restricted mobility. They were once thought to be related to fungi because they have a similar appearance and lifestyle. They are not fungi however; protistan molds have cell walls made of carbohydrates, like those found in other protists, whereas fungi have cell walls made of chitin. Also, protists carry out normal mitosis, whereas mitosis in fungi is unusual, as you will learn in Chapter 22. The four major phyla of protistan molds, each with different structures and life cycles, are not related to each other.

Cellular slime molds, members of the phylum Acrasiomycota, resemble amoebas in the phylum Rhizopoda but have many distinct features. The individual organisms behave as separate amoebas, moving through the soil and ingesting bacteria. In times of environmental stress, the individual amoebas aggregate and move toward a fixed center, as shown in Figure 21-15. There, they form multicellular colonies called slugs. Each slug develops a base, a stalk, and a swollen tip that develops spores. Each of these spores, when released, becomes a new amoeba, which begins to feed and repeat the life cycle. There are 70 known species of cellular slime molds, the best known of which is *Dictyostelium discoideum*.

Plasmodial slime molds, members of the phylum Myxomycota, are a group of bizarre organisms that stream along as a **plasmodium,** a mass of cytoplasm that looks like an oozing slime, as shown in Figure 21-16. As they move, they engulf and digest bacteria and other organic material. A plasmodial slime mold contains many nuclei, but these are not separated by cell walls. All nuclei undergo mitosis at the same time, in coordinated fashion. If the plasmodium begins to dry out or starve, it moves away rapidly, then stops, and often divides into many small mounds. Each mound produces a stalk tipped with a capsule in which haploid spores develop. The spores are highly resistant to unfavorable environmental conditions and can survive for years. When conditions are favorable, the spores germinate and become haploid cells that are either amoeboid or flagellated. These haploid cells are able to fuse into diploid zygotes, which undergo mitosis and form a new plasmodium.

Figure 21-15 When deprived of food, the individual amoebas of the cellular slime mold *Dictyostelium discoideum* aggregate, *top,* and form a moving mass called a slug, *center.* The slug will move to a new habitat and transform into a stalked structure that contains spores, *bottom.*

Figure 21-16 A plasmodial slime mold is a mass of cytoplasm containing many nuclei. Plasmodia can flow around objects and even pass through cloth.

Figure 21-17 Water molds usually help decompose dead animals in the water.

Oomycetes *(oh oh MY seets),* members of the phylum Oomycota, are the water molds, white rusts, and downy mildews that often grow on dead algae and dead animals in fresh water, as seen in Figure 21-17. All members of the group are either parasites or feed on dead organic matter. Oomycetes are unusual in that their spores have two flagella: one pointed forward, the other backward. Many oomycetes are plant pathogens, including *Phytophthora infestans,* which causes late blight in potatoes. This protist was responsible for the Irish potato famine of 1845–1847, during which about 400,000 people starved to death.

Finally, the Chytridiomycota, or chytrids, are a group of protists distinguished by their motile cells, which have a single, whiplike flagellum. There is some evidence that there may be an evolutionary relationship between chytrids and fungi.

□ CAPSULE SUMMARY

Protist diversity consists of organisms that vary in cellularity, in methods of locomotion and of reproduction, and in structure.

Some Protists Form Resistant Spores

Sporozoans, members of the phylum Sporozoa, are nonmotile, spore-forming, unicellular parasites. They infect animals with small spores that are transmitted from host to host. All sporozoans have a unique arrangement of microtubules and other organelles clustered at one end of the cell. Sporozoans have complex life cycles that involve both asexual and sexual reproduction. Sexual reproduction involves the fertilization of a large female gamete by a small, flagellated male gamete. The zygote that results soon becomes a thick-walled cyst called an **oocyte** *(OH oh syt),* which is highly resistant to drought and other unfavorable environmental conditions. The best known sporozoan is the malaria-causing parasite *Plasmodium,* which is shown in Figure 21-18. □

Section Review

1. *What are pseudopodia? How do amoebas and forams use them to move?*
2. *What are the shells of diatoms made of? What commercial uses are there for diatoms?*
3. *Construct a table that compares three kinds of multicellular algae.*
4. *How does* Euglena *differ from the other two phyla of flagellates?*
5. *Describe five general characteristics of* Paramecium.

Critical Thinking
6. *Why were protistan molds once classified as fungi?*

21-3 Protists and Human Health

One of the greatest impacts protists have on humans is as pathogens, agents of disease. In this section you will consider malaria, a significant human disease caused by a protist. Malaria is by no means the only disease caused by protists, but it will serve to acquaint you with the complex nature of the life cycles of pathogenic protists. Table 21-3 lists other diseases caused by protists.

Section Objectives

- Explain how malaria is transmitted.
- Describe the methods used to control malaria.
- Discuss three other human diseases caused by protists.

Plasmodium Causes Malaria

Malaria, caused by a sporozoan, is one of the most serious infectious diseases of recent history. Over 100 million people have malaria at any one time, and over a million, mostly children, die from it every year. The symptoms include severe chills, fever, sweating, confusion, and great thirst. Victims die of anemia, kidney failure, or brain damage unless the disease is brought under control by the person's immune system or by medical treatment.

The malarial sporozoan parasite is *Plasmodium*, shown in Figure 21-18. *Plasmodium* is spread from person to person by mosquitoes of the genus *Anopheles*. When an *Anopheles* mosquito "bites" a human to obtain blood, it injects saliva mixed with a substance that prevents the blood from clotting. If the mosquito is infected with *Plasmodium*, it will also inject about 1,000 elongated cells of this protist into the bloodstream of its victim. There are three stages in the *Plasmodium* life cycle, illustrated in Figure 21-18. The stage of *Plasmodium* while it

Figure 21-18 *Plasmodium* is a sporozoan that causes the disease malaria. *Plasmodium* has a complex life cycle that involves the mosquito *Anopheles* and human blood and liver cells.

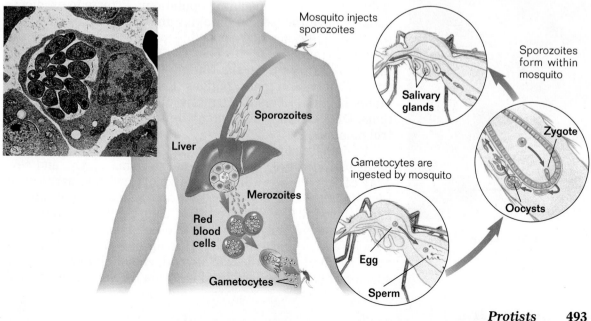

Mosquito injects sporozoites

Sporozoites form within mosquito

Salivary glands

Sporozoites

Liver

Zygote

Merozoites

Gametocytes are ingested by mosquito

Red blood cells

Oocysts

Egg

Gametocytes

Sperm

lives in mosquitoes and is injected into humans is called the **sporozoite.** Sporozoites make their way through the bloodstream to the human liver in about three minutes. In the liver, they rapidly divide and produce millions of cells of the second stage of the life cycle, called the **merozoite.** Merozoites re-enter the host's bloodstream, invade red blood cells, and divide rapidly. In about 48 hours the blood cells rupture, releasing merozoites and toxic substances throughout the host's body, initiating a cycle of fever and chills that characterizes malaria. The cycle repeats itself regularly every 48 hours as new waves of blood cells are infected.

Some of the merozoites in the human bloodstream undergo a sexual phase and develop into the third stage of the *Plasmodium* life cycle, called the **gametocyte.** In the human bloodstream, gametocytes are incapable of undergoing meiosis to form haploid gametes. However, if they are extracted from an infected person by a mosquito, they form sperm and egg cells within the gut of the mosquito. Gametes fuse to form a zygote, which develops in the wall of the mosquito's gut and produces large numbers of sporozoites. Sporozoites migrate to the salivary glands of the mosquito, where they may be injected by the mosquito into the bloodstream of a human, completing the life cycle. ❏

Treating and Preventing Malaria

Chemical treatments for malaria are hundreds of years old. In the middle of the seventeenth century, quinine, a chemical derived from the bark of the cinchona tree (*Cinchona officialis*, found in South America), was discovered to be a remedy for the disease. The native name for the tree, shown in Figure 21-19, was *quina*, hence quinine. Today, derivatives of quinine, such as chloroquine and primaquine, are used to treat infected individuals and prevent malaria in healthy individuals.

One way to reduce the number of cases of malaria is to reduce the size of mosquito populations. Efforts to eradicate malaria have focused on breaking the life cycle by eliminating *Anopheles* mosquitoes. Widespread application of the powerful insecticide DDT 30 to 50 years ago eliminated these mosquitoes from the United States, but DDT-resistant strains evolved in many regions. Long banned in the United States because of its devastating effects on natural ecosystems, DDT is still used in many developing countries to control mosquitoes.

Biologists continue to search for drugs that poison the parasites once they have entered the human body, and several effective agents are known. However, strains of *Plasmodium* have appeared that are resistant to all known antimalarial drugs. As a result of drug resistance, the number of new malaria cases has tripled since the mid-1970s. Recently, a potent antimalarial drug called artemisim has been isolated from the wormwood, *Artemisia annua.* This species of plant has been used in China for various medicinal

Figure 21-19 The bark of *Cinchona officinalis* yields quinine, a remedy for malaria.

purposes for over 2,000 years. Attempts to produce a vaccine using the techniques of genetic engineering are starting to produce promising results, and human trials are underway.

Table 21-3 Diseases Caused by Protists

Disease	Symptoms	Protist	Mode of Transmission
Amoebic dysentery	Amoebas feed on intestinal lining, causing bloody diarrhea	*Entamoeba histolytica*	Contaminated food or water
Giardiasis	Cramps, nausea, diarrhea, and vomiting	*Giardia lamblia*	Contaminated water
Leishmaniasis	Skin sores and deep, eroding lesions	*Trypanosoma*	Bites from sand flies
Sleeping sickness	Fever, weakness, lethargy	*Trypanosoma gambiense, Trypanosoma rhodesiense*	Bite from infected insects such as tsetse fly
Chagas' disease	Fever, severe heart damage	*Trypanosoma cruzi*	Bite from infected kissing bug
Toxoplasmosis	Primary danger is fetal infection; can cause convulsions, brain damage, blindness, and death in fetuses	*Toxoplasma gondii*	Contact with infected cats or improperly cooked meat
Late blight	Parasitizes and destroys potato plants	*Phytophthora*	Spores spread from diseased plants to healthy ones

Section Review

1. *What are the three stages in the life cycle of* Plasmodium? *Which one causes the characteristics of malaria?*
2. *Describe three methods used to treat and control malaria.*
3. *What are three additional human diseases caused by protists?*

Critical Thinking

4. *What might be a way to control sleeping sickness, a disease caused by a trypanosome and spread by the bite of the tsetse fly?*

CHAPTER REVIEW

amoeba (485)
conjugation (482)
diatom (486)
dinoflagellate (488)
euglenoid (489)
eyespot (481)

foram (485)
gametocyte (494)
merozoite (494)
oocyte (492)
pellicle (489)
plasmodium (491)

pseudopodia (485)
sporangia (483)
sporozoite (494)
zoomastigote (488)
zoospore (482)
zygospore (482)

Review

Multiple Choice

1. Two of the most important eukaryotic features that evolved in protists are
   a. photosynthesis and silica shells.
   b. forams and pseudopodia.
   c. multicellularity and sexual reproduction.
   d. spores and microtubules.

2. Which habitat is least likely to harbor any species of Protista?
   a. ocean waters
   b. the human liver
   c. a desert
   d. leaf litter

3. Which pair shows a correct match between a protist and its manner of reproduction?
   a. *Spirogyra:* sporangia
   b. *Ulva:* conjugation
   c. *Oedogonium:* gametes
   d. *Spirogyra:* alternation of generations

4. A photosynthetic single-celled protist that moves using flagella would likely be classified as a member of what phylum?
   a. Apicomplexa
   b. Oomycota
   c. Bacillariophyta
   d. Euglenophyta

5. Photosynthetic protists with boxlike shells are
   a. diatoms.
   b. plankton.
   c. zoomastigotes.
   d. euglenoids.

6. Red algae are different from green and brown algae because red algae
   a. are multicellular.
   b. have the pigment phycobilin.
   c. inhabit marine environments.
   d. display alternation of generations.

7. Why have efforts to control the spread of malaria not eradicated the disease?
   a. Resistant strains of mosquitoes and *Plasmodium* have evolved.
   b. The use of DDT has been erratic around the world.
   c. The symptoms of the disease are confused with those of the flu.
   d. The life cycle of *Plasmodium* is not well understood.

8. How is amoebic dysentery spread from person to person?
   a. by drinking unsanitary water
   b. from the bite of the conenose bug
   c. from the bite of the tsetse fly
   d. by eating overcooked pork

9. Which of the following identifies the correct mode of transmission of a human disease caused by a protist?
   a. Giardiasis is transmitted through the sting of a small wasp.
   b. Leishmaniasis is transmitted through the bite of a small bee.
   c. Giardiasis is transmitted through the bite of a sand fly.
   d. Leishmaniasis is transmitted through the bite of a sand fly.

Completion

10. Eukaryotes that do not belong to the fungus, plant, or animal kingdoms are classified as _____ .

11. *Chlamydomonas* produces haploid cells called _____ during the asexual phase of its life cycle and a protected diploid zygote called a _____ during the sexual phase of the life cycle.

12. Amoebas and forams are alike because they both use _____ to move around. Amoebas and forams are different because forams secrete external shells called _____ that are made of calcium carbonate.

13. The protist _____ is found in fresh water and exhibits both plantlike and animal-like characteristics. Inside the cell membrane of this protist is the flexible _____ , which allows the organism to change shape and move around obstacles.

14. The feeding stage of a plasmodial slime mold is a large mass of oozing cytoplasm called a _____ , whereas the feeding stage of a cellular slime mold is made up of separate _____ .

15. The _____ life cycle has three stages. The stage that lives in mosquitoes and is injected into humans is called the _____; the stage that invades human red blood cells is called the _____ ; and the stage that is extracted from a human before the formation of eggs and sperm in the gut of a mosquito is called _____ .

Themes Review

16. **Structure and Function** A zygospore is formed during the sexual phase of the life cycle of *Chlamydomonas*. Describe the structure of the zygospore and its function in the life cycle.

17. **Evolution** While red, brown, and green algae are alike in many ways, scientists hypothesize that green algae are the evolutionary ancestors of plants. What evidence supports this hypothesis?

18. **Homeostasis** Many freshwater protists that lack a cell wall have a contractile vacuole. How does the contractile vacuole maintain conditions inside the protist within the limits required by living cells?

Critical Thinking

19. **Interpreting Data** As you peer through your microscope, you see an organism that is single-celled, flexible, and that moves about with the aid of "false feet." You reason that it must be a protist. In which of the major protist phyla would you classify the organism? Explain your choice.

20. **Making Inferences** *Euglena* is often used in experiments in high school laboratories. In view of the fact that *Euglena* can be autotrophic, why is it a good choice for use in the lab?

Activities and Projects

21. **Multicultural Perspective** The Irish potato famine of 1845–1847 resulted in the deaths of many people and also the emigration of Irish people to other countries including the United States. Research this time period. Construct a graph that shows the number of Irish people who came to the United States between the years 1835 and 1857. Using the graph, write a paragraph that describes the impact of the Irish potato famine on the immigration of Irish people to the United States.

22. **Cooperative Group Project** Carrageenan and agar are extracts of red algae. Algin is an extract of brown algae that grow in ocean waters. Carrageenan, agar, and algin are added to many foods to make them thicker, smoother, and better-tasting. Examine food packages at home and in a grocery store to identify products that contain extracts of these algae. Make a poster-size chart that contains (1) the product names, (2) the algal extract added, and (3) how the product is improved by the addition of the extracts.

Comparative Protists

- Observe live protists under the microscope.
- Compare structural characteristics, methods of locomotion and feeding, and behavioral strategies among protists.
- Determine why protists are placed in a separate kingdom, based upon your observations.

- observing living organisms using a compound microscope
- comparing and contrasting organism traits and characteristics

- Detain™ (protist-slowing agent)
- microscope slides
- mixture of various protists, including *Amoeba, Paramecium, Euglena, Volvox, Spirogyra, Euplotes, Micrasterias, Vorticella, Stentor*
- toothpicks
- coverslips (22 × 22 mm)
- compound microscope
- references for identifying protists
- black construction paper
- paper punch
- scissors
- desk lamp or sunlit window
- white paper
- forceps
- plastic pipettes with bulbs
- vial of diatomaceous earth
- flashlight
- stereomicroscope
- culture tube of slime mold

1. What are some of the general characteristics of the kingdom Protista?
2. Are protists simple or complex organisms?
3. How are protists similar to plants? to animals? to fungi?
4. Write your own **Focus Question** on your Vee Form.
5. **Knowing Side of the Vee** List the **Concepts** and new **Vocabulary Words** on your Vee Form. In the **Concept Statements** section of the Vee, use these words in sentences that define and explain them.

Doing Side of the Vee

1. Create a table for the **Records** section of your Vee that will allow you to name each protist and to draw or describe its size, shape, and color; its type of movement, if any; how food is obtained; and its behavioral responses.

Part A: Protist Mixture

2. Place a drop of protist-slowing agent on the center of a clean microscope slide. Then add another drop from the bottom of the protist culture mixture. Mix the drops using a toothpick. Add a coverslip.

3. Observe your wet mount preparation under both low (100×) and high power (430×) of a compound microscope. Use references to identify at least five different organisms.

4. Using the drawing on the next page as a guide, prepare a protist "sun shade" by punching a hole in the center of a 40 × 20 mm piece of black construction paper that has a slight "curl" to it.

5. Prepare another wet mount without using any protist-slowing agent. View this under low power (100×) of a compound microscope, making note of the kind and

distribution of all protists you encounter. Record your findings in the **Records** section of your Vee.

6. Place the wet mount on top of a piece of white paper on a sunlit windowsill or under a table lamp if it is cloudy. Position the curled protist "sun shade" so that the hole is in the center of the coverslip.

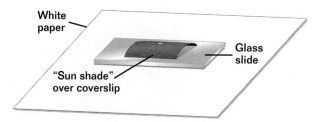

White paper

"Sun shade" over coverslip

Glass slide

7. After 10 minutes, gently pick up your slide, being careful not to disturb your protist "sun shade." Place the slide on the stage of your compound microscope. Use the low-power objective to focus in the center of the exposed opening. Record your observations. Have your partner use forceps to carefully remove the "sun shade" while you continue observing the protists. Record your observations in your data table.

Part B: Diatomaceous Earth

8. Using a pipette, place a drop of water in the center of a clean microscope slide.

9. Using a toothpick, mix a small amount of diatomaceous earth with the water drop. Add a coverslip and observe under both low and high power of a compound microscope. Use references to identify various diatom types—elongate (pennate) or circular (centric).

10. Observe the wet mount under low power as your partner shines a flashlight (at a 45° angle) onto the slide as it rests on the stage. Turn off your light source so that only the flashlight illuminates the diatomaceous earth. Record your observations in your data table.

Part C: Slime Mold Culture

11. Use a stereomicroscope to observe a sealed tube containing a slime mold culture, as your teacher directs. Identify the two primary stages of its life cycle.

12. In the **Procedure** section of your Vee, briefly summarize the procedures you followed in this investigation.

13. Clean up your materials and wash your hands before leaving the lab.

INQUIRY

1. Describe the different ways protists move. Give examples.

2. How do protists obtain food?

3. Which protists were affected by light? How? What structures were responsible?

4. Were all protists single cells?

5. Which observed protists had animal characteristics? plant characteristics? Which protist resembled a fungus?

6. Use the information on the **Knowing Side** of the Vee to interpret your results from the **Doing Side,** and then write your **Knowledge Claim.** Write a **Value Claim** for this lab.

ANALYSIS

1. Based on your observations in step 10, to what commercial use might the protists in diatomaceous earth be put?

2. How do the motions of *Vorticella* resemble the movement of muscle fibers?

FURTHER INQUIRY

Write a **New Focus Question** that could be the point of a new investigation. The following is an example:

Do samples from the middle and upper areas of the culture jar have different protist populations than those taken from the bottom?

CHAPTER 22

FUNGI

REVIEW

- surface-area-to-volume ratio (Section 2-1)
- stages of mitosis (Section 6-2)
- spore (Section 6-3)
- mycorrhizae (Section 13-2)
- lichen (Section 16-1)
- mating types (Section 21-1)

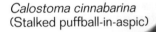

Calostoma cinnabarina
(Stalked puffball-in-aspic)

22-1 Characteristics of Fungi

Some of the most unusual and peculiar organisms on Earth are members of the kingdom Fungi. For example, mushrooms and molds are familiar fungi that grow so rapidly they sometimes appear overnight. Together with heterotrophic bacteria, fungi are the major decomposers of the biosphere, breaking down organic molecules and making them available for recycling.

Section Objectives

- Explain the role fungi play in ecosystems.
- List the characteristics of the kingdom Fungi.
- Describe the structure of a typical fungus body.
- Explain how fungi obtain nutrients.
- Describe how fungi reproduce.

Fungi Are Classified in Their Own Kingdom

The kingdom Fungi consists of eukaryotic, mostly multicellular organisms. They are an ancient group of organisms at least 400 million years old. Traditionally, biologists grouped fungi with plants, probably because fungi are immobile and appear "rooted" in the soil. However, fungi are as different from plants as they are from animals. Compare a familiar fungus, such as the mushrooms shown in Figure 22-1, with a plant you've recently studied, such as a pea plant. Upon careful examination, the following differences between fungi and plants become evident.

1. **Fungi are heterotrophic** Perhaps the most obvious difference between fungi and plants is their color: the stalk and cap of the mushroom are not green like the stem and leaves of the pea plant. Plants appear green because they contain chlorophyll, the pigment that enables them to make their own food using energy from the sun. Fungi do not contain chlorophyll; they are heterotrophs. They obtain energy by absorbing organic molecules from their surroundings.

2. **Fungi have filamentous bodies** Unlike the pea plant, which consists of a variety of cell and tissue types, the mushroom consists of long slender filaments. These filaments weave tightly together to form the fungus body and reproductive structures such as a mushroom.

3. **Fungal cells contain chitin** The cells of the mushroom, like the cells of all fungi, have walls made of **chitin** (KY tihn), the tough material found in the exoskeleton of insects and other arthropods. Cells of the pea plant, on the other hand, have walls made of cellulose, a different polysaccharide. Chitin is more resistant to bacterial decomposition than is cellulose.

4. **Fungi have nuclear mitosis** Mitosis in the mushroom is different from that in the pea plant and most other eukaryotic organisms, in which the nuclear envelope

Figure 22-1 These mushrooms are actually the reproductive structures of an extensive network of filaments that make up the body of a fungus.

disintegrates in prophase and re-forms in telophase. In dividing mushroom cells, by contrast, the nuclear envelope remains intact from prophase to anaphase. Consequently, spindle fibers form within the nucleus, dragging chromosomes to opposite poles of the nucleus, not opposite poles of the cell. Mitosis is completed when the nuclear envelope pinches in two.

The unique features of fungi strongly suggest that they are not closely related to any other group of organisms. The first fungi were probably unicellular eukaryotic organisms. The oldest fossils that have been identified as fungi are 450–500 million years old. ◻

◻ **CAPSULE SUMMARY**

Fungi are eukaryotic and heterotrophic. Their bodies are made up of slender woven filaments. Fungal cells contain chitin and have nuclear mitosis.

Fungi Are Well Suited for Absorbing Nutrients

In Figure 22-2, the fungus *Penicillium* is growing on an orange. The green and white fuzz you recognize as mold is actually the reproductive structures of the fungus. The body of the fungus lies woven within the tissues of the orange. All fungi except yeasts have bodies composed of slender filaments called **hyphae** (*HY fee*). When hyphae grow, they branch and form a tangled mass called a **mycelium** (*my SEE lee uhm*), shown in Figure 22-2. A mycelium can be made of many meters of individual hyphae. This body organization creates a high surface-area-to-volume ratio, which makes a fungus well suited for absorbing food from the environment.

Each hypha is a long string of cells divided by walls called **septa**. In most kinds of fungi, septa do not form a complete barrier between cells (another characteristic that makes fungi different from other eukaryotes). From one cell to the next, cytoplasm flows freely throughout

Figure 22-2 This orange is covered with the fungus *Penicillium*. The green and white fuzz growing on the orange's surface is the fungus's reproductive structures. Throughout the rest of the orange, the fungus grows as a mycelium, a mass of tangled filaments called hyphae, *below right.*

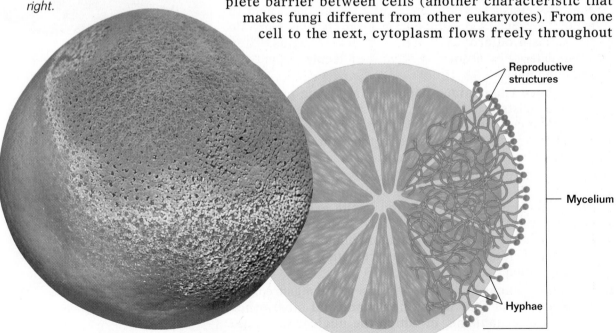

Reproductive structures

Mycelium

Hyphae

the hypha through perforations in the septa, as shown in Figure 22-3. Other organelles, such as ribosomes, mitochondria, and nuclei, also pass through these perforations. A typical fungal cell usually has many nuclei streaming in the cytoplasm.

How Fungi Obtain Food

All fungi digest food outside their bodies. Through the tips of their hyphae, they secrete powerful digestive enzymes that break down organic matter into small molecules. Fungi absorb these molecules and use them for energy. In their search for food sources, many fungi attack nonliving organic matter—leaves, branches, animal corpses, and waste—and decompose these materials. Fungi, such as the carbon fungus in Figure 22-4, are often found growing on dead trees, where they are the only major group of organisms capable of breaking down lignin, a major component of wood. Other fungi absorb nutrients from living hosts, which sometimes become weakened and succumb to infection or disease.

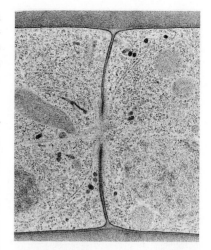

Figure 22-3 Fungal hyphae may be divided into cells by walls called septa. Septa rarely form complete barriers, so cytoplasm streams freely along the hypha, carrying proteins and nutrients to the rapidly growing tips.

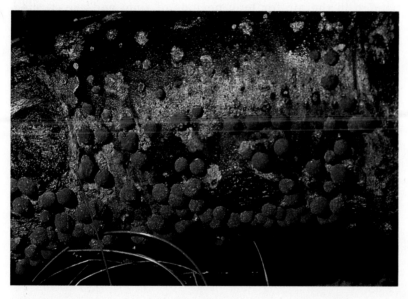

Figure 22-4 The carbon fungus *Hypoxylon fragiforme* appears in clusters on the wood it decomposes.

As decomposers, fungi often come into conflict with human interests. A fungus makes no distinction between a fallen log in the forest and bread, fruit, vegetables, meat or other items stored in a refrigerator. In fact, fungi have been known to attack materials humans consider inedible—paper, cardboard, cloth, paint, leather, waxes, fuel, and petroleum.

The success of fungi as commercial pests is due to their ability to grow under a wide range of conditions, a quality that also makes them commercially valuable. Fungi called yeasts produce substances such as carbon dioxide and ethanol, which are useful in baking, brewing, and wine-making. Other fungi provide the pungent flavors and aromas of specific kinds of cheese. Many kinds of antibiotics, such as penicillin, are produced by fungi. An extraordinary example of the use of compounds derived from fungi is cyclosporine, a drug derived from a fungus that dwells in the soil. Cyclosporine suppresses the reactions of the immune system that cause rejection of transplanted organs. This reduces the possibility of transplant rejection without the undesirable side effects caused by other drugs used for this purpose.

Figure 22-5 *Lycoperdon perlatum* is a kind of fungus called a puffball. Such fungi release hundreds of thousands of spores through a small opening in the top of the puffball.

How Fungi Reproduce

Fungi reproduce by releasing spores. Spores form sexually or asexually in reproductive structures at the tips of hyphae. When these reproductive structures form, they are cut off from the rest of the fungus by complete septa. Reproductive structures extend high above the food source so that air currents can carry the spores to a new habitat. Spores are well suited to the needs of an organism that is anchored to one place. As you can see in Figure 22-5, spores are so small and light that they remain suspended in the air for long periods of time and are carried great distances. When a spore lands in a suitable place, it begins to divide and soon gives rise to a new fungal hypha.

Fungal spores are haploid. Most spores are formed by mitosis during asexual reproduction. Like protists, fungi usually resort to sexual reproduction only in times of environmental stress. In sexual reproduction, hyphae from two mating types undergo cytoplasmic fusion. However, unlike sexual reproduction in plants and animals, in which nuclei fuse to form a diploid zygote, most fungal nuclei pair up but do not immediately fuse. Instead, the nuclei remain paired but separate in the same cytoplasm for most of the life of the fungus. A fungal hypha that has nuclei derived from two genetically different individuals is called a **dikaryotic** hypha. A mycelium with paired nuclei is called a **dikaryon.** ☐

Section Review

1. *What role do fungi play in the environment?*
2. *How does mitosis differ between fungi and plants?*
3. *What is a hypha? How does its structure enable fungi to obtain nutrients?*
4. *How do fungi reproduce? How does sexual reproduction in fungi differ from that in other eukaryotes?*

Critical Thinking
5. *Given fungi's ability to break down many kinds of substances, how might they be used by industries that produce waste?*

22-2 Fungal Diversity

*T*he one characteristic of fungi that distinguishes them from all other multicellular organisms is that their cells share nuclei. Three phyla of fungi—Zygomycota, Ascomycota, and Basidiomycota—are classified according to the way nuclei from different individuals are sorted in fused hyphae during sexual reproduction. These differences are summarized in Table 22-1. A fourth phylum, Deuteromycota, contains fungi in which sexual reproduction is unusual, rare, or has not been observed.

Section Objectives

- Explain how fungi are classified.
- Describe four phyla of fungi.
- Outline the life cycles of zygomycetes, ascomycetes, and basidiomycetes.
- Describe the mushroom Amanita muscaria.

Table 22-1 Three Sexually Reproducing Phyla of Fungi

Phylum	Distinctive Characteristics	Number of Species and Examples
Zygomycota	Haploid nuclei from different mating types fuse to form diploid zygotes, which develop into zygosporangia	665 species; black bread molds
Ascomycota	Formation of fine asexual spores; sexual spores in asci; hyphae divided by perforated septa; dikaryons	30,000 species; morels, truffles, yeasts, cup fungi
Basidiomycota	Formation of sexual spores in basidia; hyphae divided by perforated septa; dikaryons	16,000 species; mushrooms, puffballs rusts, smuts

Fungi That Form Zygosporangia

If you place an uncovered loaf of bread near a windowsill, a cottony mold will soon cover its surface. This common black bread mold, shown in Figure 22-6, is *Rhizopus stolonifer*, a member of the phylum Zygomycota *(zy goh my COHT uh).* The term *Zygomycota* refers to the thick-walled sexual structures called **zygosporangia** that characterize the members of this phylum. *Rhizopus* and other zygomycetes are fungi that live in the soil and feed on decaying plant and animal matter.

Asexual reproduction in zygomycetes occurs much more frequently than sexual reproduction. During asexual reproduction, haploid spores are produced in specialized hyphae that terminate with spore-producing structures called sporangia. When mature, spores

Figure 22-6 *Rhizopus stolonifer* is a familiar member of the phylum Zygomycota that is often found growing on bread.

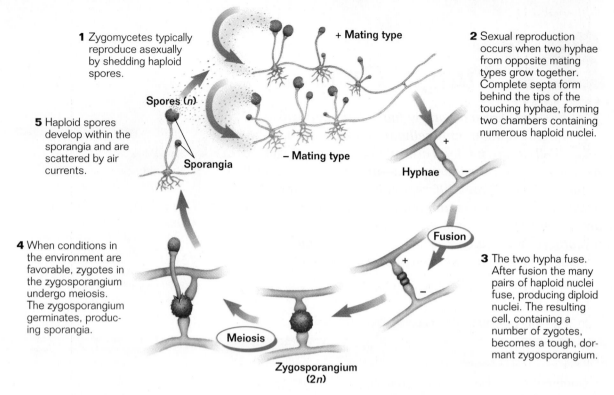

1 Zygomycetes typically reproduce asexually by shedding haploid spores.

+ Mating type

2 Sexual reproduction occurs when two hyphae from opposite mating types grow together. Complete septa form behind the tips of the touching hyphae, forming two chambers containing numerous haploid nuclei.

Spores (*n*)

5 Haploid spores develop within the sporangia and are scattered by air currents.

Sporangia

– Mating type

+

Hyphae

–

Fusion

4 When conditions in the environment are favorable, zygotes in the zygosporangium undergo meiosis. The zygosporangium germinates, producing sporangia.

+

–

3 The two hypha fuse. After fusion the many pairs of haploid nuclei fuse, producing diploid nuclei. The resulting cell, containing a number of zygotes, becomes a tough, dormant zygosporangium.

Meiosis

Zygosporangium
(2*n*)

Figure 22-7 In the sexual reproduction of a zygomycete, the fusion of hyphae from opposite mating strains produces a dormant zygosporangium. Meiosis and germination produce a sporangium, which bears haploid spores.

are shed and carried by the wind to new locations where they germinate and start new mycelia. Reproduction of *Rhizopus* is illustrated in Figure 22-7.

Fungi That Form Sacs of Spores

Prior to the twentieth century, the chestnut tree *Castanea dentata* was one of the dominant trees in forests in the eastern United States. Around 1890, a disease called chestnut blight wiped out virtually all the chestnut trees within a few years. Chestnut blight is caused by the fungus *Endothia parasitica,* a member of the phylum Ascomycota. The fungus that causes Dutch elm disease, another devastating plant disease, is also caused by a member of Ascomycota, *Ceratocystis ulmi.* Other ascomycetes are more familiar and economically beneficial fungi, such as yeasts used in baking and brewing, flavorful morels and truffles prized by gourmet chefs, and the salmon-colored bread mold *Neurospora,* which has played an important role in the development of modern genetics. The ascomycetes are named for their characteristic reproductive structure, the microscopic, club-shaped **ascus** *(AS kuhs),* a saclike structure in which haploid spores are formed. Asci usually form within the tightly interwoven hyphae of a complex structure called an **ascocarp.** In cup fungi and morels, the ascocarps are open, and the asci line the open cups. Other fungi have ascocarps that are closed or have a small opening at the top.

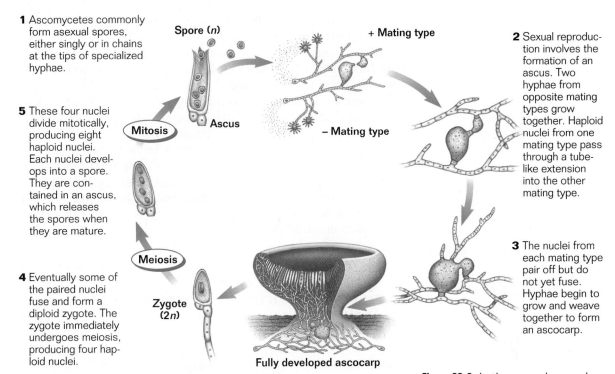

1 Ascomycetes commonly form asexual spores, either singly or in chains at the tips of specialized hyphae.

Spore (*n*)

+ Mating type

Mitosis

Ascus

− Mating type

2 Sexual reproduction involves the formation of an ascus. Two hyphae from opposite mating types grow together. Haploid nuclei from one mating type pass through a tube-like extension into the other mating type.

5 These four nuclei divide mitotically, producing eight haploid nuclei. Each nuclei develops into a spore. They are contained in an ascus, which releases the spores when they are mature.

Meiosis

4 Eventually some of the paired nuclei fuse and form a diploid zygote. The zygote immediately undergoes meiosis, producing four haploid nuclei.

Zygote (2*n*)

Fully developed ascocarp

3 The nuclei from each mating type pair off but do not yet fuse. Hyphae begin to grow and weave together to form an ascocarp.

Like zygomycetes, ascomycetes usually reproduce asexually. Asexual reproduction occurs when complete septa separate the tips of hyphae from the rest of the fungus. Mitosis in these tips gives rise to reproductive structures in which specialized spores called conidia form. When conidia are released, air currents carry them to other places, where they may germinate and form new mycelia. Reproduction in a typical ascomycete is illustrated in Figure 22-8.

Figure 22-8 In the sexual reproduction of an ascomycete, the fusion of hyphae from opposite mating strains produces structures called asci. Meiosis and mitosis within an ascus form haploid spores.

Yeasts

In general, **yeast** is the common name given to unicellular ascomycetes. There are about 350 named species of yeasts, including *Saccharomyces cerevisiae*, or baker's yeast, used for thousands of years in the production of baked goods, such as the bread shown in Figure 22-9, and many kinds of alcoholic beverages. Other yeasts, such as *Candida*, are human pathogens. *Candida* is a common source of thrush, a

Figure 22-9 The ability of yeast such as *Saccharomyces cerevisiae, below left,* to ferment carbohydrates by breaking down glucose to produce ethanol and carbon dioxide is fundamental to the production of breads and other baked goods.

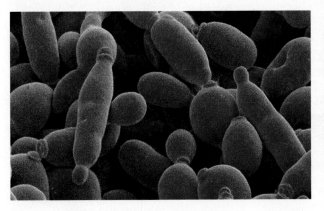

disease characterized by the formation of milk-white lesions on the mouth, lips, and throat.

Most yeasts reproduce asexually by fission or budding (the formation of a small cell from a portion of a larger one). Sexual reproduction among yeasts occurs when two cells fuse. One of these cells, containing two nuclei, functions as an ascus. Meiosis of the fused nuclei produces four spores that develop directly into new yeast cells.

Over the past few decades, yeasts have played an important role in genetic research. They were the first eukaryotic cells to be the subject of genetic engineering techniques. Because they reproduce rapidly, yeasts have become the eukaryotic cells of choice for many types of experiments in molecular and cellular biology.

Fungi That Form Clublike Structures

The fungi with which you are probably most familiar— mushrooms—are members of the third phylum of fungi, Basidiomycota. Other basidiomycetes include toadstools, puffballs, jelly fungi, and shelf fungi. Basidiomycetes are named for their characteristic club-shaped sexual reproductive structure, the **basidium** (buh SIHD ee um), which you can see in the *Up Close: Mushroom* feature on page 509. Unlike the other fungal phyla, asexual reproduction is rare among the basidiomycetes, except in some of the rusts and smuts, two important groups of plant pathogens that affect many crop plants. Sexual reproduction of a typical basidiomycete is illustrated in Figure 22-10.

Figure 22-10 In the sexual reproduction of a basidiomycete, structures called basidia form on the gills that line the mushroom cap. Meiosis within basidia produces spores.

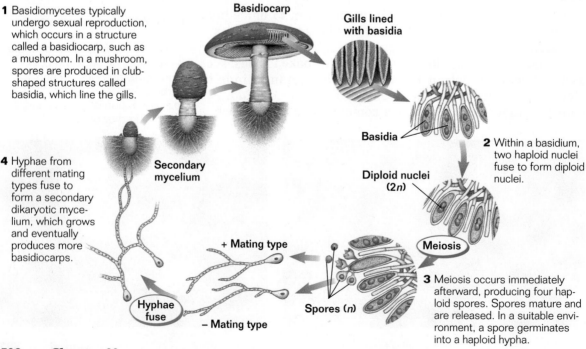

1 Basidiomycetes typically undergo sexual reproduction, which occurs in a structure called a basidiocarp, such as a mushroom. In a mushroom, spores are produced in club-shaped structures called basidia, which line the gills.

4 Hyphae from different mating types fuse to form a secondary dikaryotic mycelium, which grows and eventually produces more basidiocarps.

2 Within a basidium, two haploid nuclei fuse to form diploid nuclei.

3 Meiosis occurs immediately afterward, producing four haploid spores. Spores mature and are released. In a suitable environment, a spore germinates into a haploid hypha.

Basidiocarp

Gills lined with basidia

Basidia

Diploid nuclei (2*n*)

Meiosis

Secondary mycelium

+ Mating type

Spores (*n*)

Hyphae fuse

− Mating type

UP CLOSE MUSHROOM

- **Scientific Name:** *Amanita muscaria*
- **Habitat:** Moist organic soils
- **Size:** 10–15 cm
- **Nutrition:** Absorbs organic material in soil

Characteristics

Body structure The multicellular body of a fungus is basically filamentous, consisting of long strings of cells called hyphae. Hyphae are woven together to form a dense mat called a mycelium. Usually, the majority of a mycelium is hidden within a substrate such as soil.

Cell structure *Amanita* and other fungi have cell walls made of chitin, a complex polysaccharide also found in the external skeleton of insects and other arthropods. Some fungi have hyphae that are not divided into separate cells and have many nuclei in the same cytoplasm. Other fungi have hyphae that are divided into cells by perforated walls called septa.

Reproduction Under proper conditions, underground hyphae grow upward and weave together to produce a mushroom, the reproductive structure of fungi such as *Amanita*. A mushroom has a flattened cap attached to a stem called a stalk. The underside of the mushroom cap is lined with rows of gills. Thousands of club-shaped reproductive cells called basidia form on the gills. Through fusion and meiosis, each basidium produces spores that are released and germinate into new hyphae.

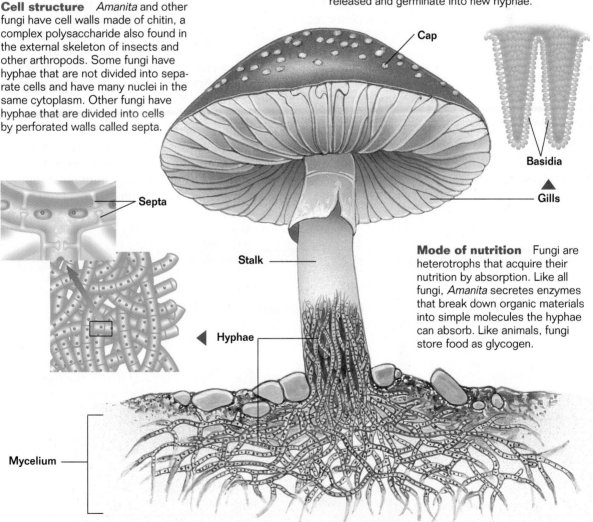

Cap

Basidia

Gills

Septa

Stalk

Hyphae

Mode of nutrition Fungi are heterotrophs that acquire their nutrition by absorption. Like all fungi, *Amanita* secretes enzymes that break down organic materials into simple molecules the hyphae can absorb. Like animals, fungi store food as glycogen.

Mycelium

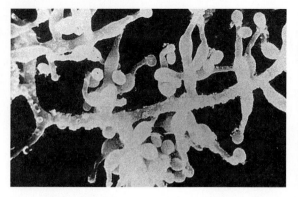
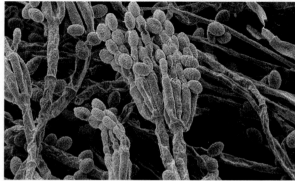

Figure 22-11 *Tolypocladium inflatum, above left,* a soil fungus, is one of the sources of cyclosporine, a drug that suppresses the immune system. Some species of *Penicillium, above right,* are the sources of the well known antibiotic penicillin.

Fungi in Which Sexual Reproduction Has Not Been Observed

As you have seen, the three phyla of fungi that have been discussed differ primarily in their mode of sexual reproduction. Not all fungi fit into one of these three phyla. In some 17,000 described species of fungi, sexual reproduction has not been observed. These fungi are placed in the phylum Deuteromycota, or **Fungi Imperfecti.** The Fungi Imperfecti are classified on the basis of their asexual reproductive structures.

Many of the Fungi Imperfecti, such as the two species in Figure 22-11, have great commercial importance. Some species of *Penicillium* are sources of the antibiotic penicillin, while other species of this genus give the characteristic flavors and aromas to cheeses such as Roquefort and Camembert. Species of *Aspergillus* are used for fermenting soy sauce and for the commercial production of citric acid. Most of the fungi that cause skin diseases in humans, including athlete's foot and ringworm, are also Fungi Imperfecti. □

Section Review

1. *On what basis are fungi classified into phyla? List the four phyla and give an example of each.*
2. *When does meiosis occur in the life cycle of a zygomycete? What is a zygospore? When do spores form?*
3. *What is an ascus? When does an ascus form in the life cycle of an ascomycete? What is contained within an ascus? When do spores form?*
4. *What is a basidium? Describe the activity within a basidium that leads to the formation of spores.*

Critical Thinking
5. *Why are yeasts gaining popularity in scientific research?*

22-3 Fungal Associations

Fungi are involved in a variety of intimate symbiotic associations with algae and plants that play very important roles in the biological world. Recall from Chapter 17 that mutualism is a form of symbiosis in which each partner benefits. These symbiotic associations typically involve a sharing of abilities between a heterotroph (a fungus) and a photosynthesizer (an alga or a plant). The fungus contributes the ability to absorb minerals and other nutrients efficiently from the environment; the photosynthesizer contributes the ability to use sunlight to power the building of organic molecules.

Lichens: Fungi and Photosynthetic Partners

A **lichen** is a symbiotic association between a fungus and a photosynthetic partner such as a green alga, a cyanobacterium, or both. The photosynthetic partner provides energy-rich compounds for both partners. It is protected from the environment by the fungal partner, which provides it with mineral nutrients. In most of the 15,000 described species of lichens, the fungal partners are ascomycetes. When you look at a lichen, such as the ones in Figure 22-12, you are seeing the fungus. The photosynthetic partner is hidden between the layers of hyphae, as shown in Figure 22-13. Sunlight penetrates the translucent layers of hyphae to fuel photosynthesis. Sometimes specialized fungal hypha penetrate the photosynthetic cells and serve as highways for transferring sugars and other organic molecules to the fungus. The fungus transmits biochemical signals that direct the cyanobacterium or the green alga to produce metabolic substances that it would not make if growing independently of the fungus. Many biologists characterize this particular symbiotic relationship

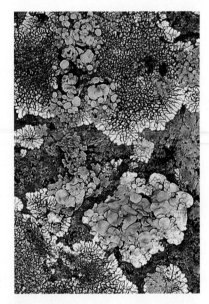

Figure 22-12 Lichens absorb some minerals from their substrate, but most of their minerals are derived from air and from rainfall.

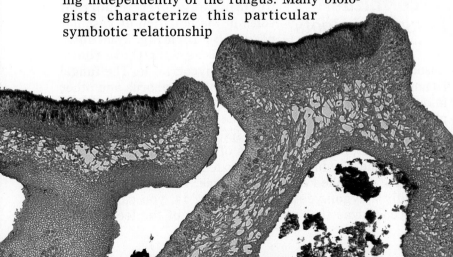

Figure 22-13 This cross section of a species of lichen called *Physica* shows algal cells, *colored magenta*, penetrating fungal cells, *colored green*.

as one of slavery rather than cooperation, a controlled parasitism of the photosynthetic organism by the fungal parasite.

The tough construction of the fungus, combined with the photosynthetic abilities of the alga, has enabled lichens to invade the harshest of habitats. They are extremely widespread in nature. Lichens have been found in arid desert regions, in the Arctic, growing on bare soil, on tree trunks, and on sunbaked rocks. In harsh, exposed areas, lichens are often the first colonists. They break down rocks and prepare the environment for other organisms. Lichens are a key component of primary succession because they are able to carry out nitrogen fixation and introduce useful forms of nitrogen into the environment.

Lichens are able to survive drought and freezing by becoming dormant. When moisture and warmth return, the lichen recovers quickly and resumes its normal activities, such as photosynthesis. In harsh environments, lichens may grow extremely slowly. Some lichens that grow high in the mountains appear to be thousands of years old and cover an area no larger than a fist. These lichens are among the oldest living organisms on Earth. ◻

Sensitivity to Pollution

Lichens are extremely sensitive to pollutants in the atmosphere because they readily absorb substances dissolved in rain and dew. Pollutants such as sulfur dioxide, a byproduct of automobile engine exhaust and industrial activity, quickly destroy a lichen's chlorophyll, decreasing its rate of photosynthesis. As a result, the physiological balance of the symbiotic relationship is upset. For this reason, lichens do not grow in or around cities. Biologists use the relative health of lichens and their chemical compositions as indicators of an environment's health. Recently, biologists have discovered that lichens are disappearing from national parks and other remote areas, which, despite their distance from the sources of pollution, are clearly being affected by the quality of air that reaches them.

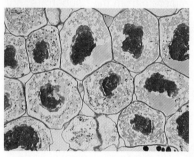

Figure 22-14 The roots of leek plants, *top*, have cells that contain endomycorrhizae with a zygomycete component, *bottom*.

Mycorrhizae: Fungi and Roots

Certain fungi play important roles in the nutrition of vascular plants by forming symbiotic associations with their roots. Associations of this kind are called mycorrhizae. The fungal filaments aid in the direct transfer of phosphorus and other minerals from the soil into the roots of the plant, while the plant supplies carbohydrates to the symbiotic fungus.

In the mycorrhizae of most species of plants, the fungal hyphae penetrate the outer cells of the root and form coils, swellings, and tiny branches that extend into the surrounding soil. These are called **endomycorrhizae.** The fungus involved is usually a zygomycete. In Figure 22-14, you can see the endomycorrhizae that grow in the roots of the leek *Allium* *porrum.* Fossils reveal that the rootlike appendages of the earliest plants often had endomycorrhizae, which may have played an important role in the invasion of land by plants. The soil of that time was completely lacking in organic matter, and mycorrhizal plants are particularly successful in infertile soil. Some archaic primitive vascular plants surviving today continue to depend strongly on endomycorrhizae.

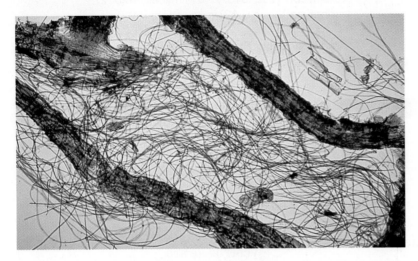

Figure 22-15 The hyphae of the fungus in ectomycorrhizae sometimes appear as a tangled mass around the root of the plant.

In at least 10,000 species of plants, the mycorrhizae do not physically penetrate the plant root but instead wrap around it. These are called **ectomycorrhizae.** In contrast to endomycorrhizae, the nonpenetrating ectomycorrhizae represent highly specialized relationships in which a particular species of plant has become associated with a particular fungus (usually a basidiomycete). These kinds of ectomycorrhizae are important because they involve many commercially significant trees that grow in temperate regions, such as pines, oaks, beeches, and willows. Figure 22-15 shows the roots of a pine tree surrounded by a tangled mass of ectomycorrhizae. At least 5,000 species of fungi have been identified in different ectomycorrhizae. They are mostly formed with basidiomycetes, but some involve associations with ascomycetes. ❑

□ *CAPSULE SUMMARY*

Mycorrhizae are symbiotic associations in which a fungus transfers minerals to a plant's roots, which in turn supply carbohydrates to the fungus. Endomycorrhizae penetrate root cells; ectomycorrhizae wrap around roots.

Section Review

1. *What is a lichen? What are the benefits for each partner in this symbiotic association?*
2. *Why are lichens so sensitive to pollution?*
3. *How do plants benefit from having mycorrhizae?*

Critical Thinking
4. *How might the Earth's landscape appear today if the earliest plants had not had mycorrhizae?*

22 CHAPTER REVIEW

Vocabulary

ascocarp (506)	dikaryotic (504)	lichen (511)
ascus (506)	ectomycorrhizae (513)	mycelium (502)
basidium (508)	endomycorrhizae (512)	septa (502)
chitin (501)	hypha (502)	yeast (507)
dikaryon (504)	Fungi Imperfecti (510)	zygosporangia (505)

Review

Multiple Choice

1. Fungi differ from plants in that fungi
   a. are multicellular.
   b. are immobile.
   c. have cell walls.
   d. are heterotrophic.

2. The cross walls that separate cells in hyphae are known as
   a. rhizoids.   c. asci.
   b. gills.   d. septa.

3. Which of the following characteristics is shared by all fungi and helps them obtain food?
   a. external digestion
   b. phagocytosis
   c. feed on nonliving matter
   d. anesthetize prey

4. Members of Deuteromycota are more difficult to classify than other fungi because
   a. they develop from zygosporangia.
   b. they are sexual and parasitic.
   c. they undergo meiosis.
   d. if sexual reproduction occurs, it is unknown.

5. The common edible mushroom is classified in the phylum
   a. Zygomycota.   c. Basidiomycota.
   b. Ascomycota.   d. Deuteromycota.

6. Mycorrhizae are symbiotic relationships of fungi and
   a. algae.   c. roots.
   b. lichens.   d. chloroplasts.

7. Some fungal associations no larger than a fist appear to be thousands of years old. These have been found
   a. in temperate forests.
   b. on well-irrigated alluvial plains.
   c. in fields of corn.
   d. in harsh environments, high in the mountains.

8. One might expect that plants without mycorrhizae are
   a. more likely to get fungal diseases.
   b. less successful in the transfer of minerals from the soil into the roots.
   c. best suited to poor soil conditions.
   d. primitive and might soon become extinct.

Completion

9. The kingdom _____ contains eukaryotic organisms that are immobile, lack chlorophyll, and produce chitin, hyphae, and spores.

10. Fungal _____ consists of a tangled network of hyphae. Each hypha is covered by a cell wall made of _____ .

11. Most fungi feed by secreting _____ into the organic matter that surrounds them and then _____ the digested food.

12. A mushroom develops from the _____ of different mating strains. On the hyphae lining the gills of mushrooms, haploid _____ are produced.

13. A mycelium with _____ but separate nuclei is called a _____ .

14. During the _____ phase of the life cycle of *Rhizopus*, hyphae from two mating types fuse and produce a cell that develops into a dormant _____ .

15. In _____ the fungal hyphae penetrate the outer cells of the plant root, but in _____ the fungal hyphae wrap around the plant root.

Themes Review

16. **Structure and Function** Describe the structure of the zygosporangium of *Rhizopus stolonifer*. How does the zygosporangium function to ensure survival of the species?

17. **Levels of Organization** Organisms are identified as producers, consumers, or decomposers based on the role they play in an ecosystem. What role do fungi play in most ecosystems? Give an example to support your answer.

18. **Levels of Organization** When observed, a member of the phylum Deuteromycota exhibits sexual reproduction involving the development of zygospores. Given this observation, should the fungus be reclassified? If so, in which phylum should it be placed?

Critical Thinking

19. **Making Inferences** The bacterium *E. coli* has long been a popular organism for use in studying the molecular biology of cells. But in recent years, the baker's yeast *Saccharomyces cerevisiae* has been used more often, especially when research questions relate to the molecular biology of human cells. Why do you suppose *Saccharomyces cerevisiae* is the preferred research organism for questions concerning the molecular biology of human cells?

20. **Interpreting Data** In the last 50 years, the truffle harvest in a certain area in Europe has drastically declined. During the same period, oak forests have been cut and the land cleared for farming and housing. Some truffle hunters in the area suspect that the decline in the truffle harvest is related to forest clearing. Based on the data presented in the graphs below, what is the relationship between the truffle harvest and forest clearing? What would you advise land-owners in the area to do if they would like to maintain the truffle harvest at its current level?

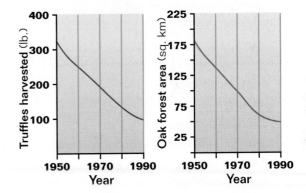

Activities and Projects

21. **History** Downy mildew caused by the fungus *Plasmopara viticola* devastated the crops of French grape growers during the 1870s and early 1880s. Then the Bordeaux mixture was invented by Pierre Millardet of the University of Bordeaux. Find out what chemicals are in the Bordeaux mixture and how the mixture prevents downy mildew. Also, learn how Millardet got the idea to develop the mixture. Write a paper that details what you learn, and share your results with the class.

22. **Research and Writing** Use the library to research ways in which medical science has been influenced by fungi. Investigate the discovery of penicillin and other drugs derived from fungi. Investigate the role played by fungi in various diseases. Present the findings of your research in a report.

LABORATORY Investigation — Chapter 22

Fungi on Food

OBJECTIVES

- Recognize fungal growth on food.
- Identify environmental conditions that favor the growth of fungi on food and those that inhibit it.

PROCESS SKILLS

- designing an experiment
- analyzing results

MATERIALS

- safety goggles
- lab apron
- disposable gloves
- 2 sterile petri dishes with nutrient medium
- 2 sterile petri dishes with nutrient medium and propionic acid
- fungal samples
- stereomicroscope
- toothpicks
- wax pencil
- masking tape

BACKGROUND

1. How do multicellular fungi, such as molds, obtain nutrients?
2. How do multicellular fungi reproduce and grow?
3. Write your own **Focus Question** on your Vee Form.
4. **Knowing Side of the Vee** List the **Concepts** and new **Vocabulary Words** on your Vee Form. In the **Concept Statements** section of the Vee, use these words in sentences that define and explain them.

TECHNIQUE

Doing Side of the Vee

Part A: Experimental Setup

1. **CAUTION:** Put on safety goggles, a lab apron, and disposable gloves. Obtain four sterile petri dishes, two with nutrient medium and two with nutrient medium plus propionic acid. Be sure the dishes are labeled for the presence or absence of propionic acid.

2. Examine the fungal samples through a dissecting microscope. Select a dense growth of a fungus, from which you will take samples.

3. Use a toothpick to scoop up a small sample of the fungus you selected. Gently touch the sample to the medium in four places in the two petri dishes without propionic acid. Raise the lids of the dishes as little as possible to do so. Do the same for the two petri dishes with propionic acid, using a clean toothpick and another small sample of the same fungus. Properly dispose of the toothpicks.

4. Place a piece of masking tape on opposite sides of each dish to hold the lid and bottom together. Label each petri dish with your name and the food source of the fungal sample you selected.

5. Design an experiment to determine which of two opposite environmental conditions is best for fungal growth. Some possible combinations are warm/cold, light/dark, and moist/dry. Label one of the two environmental conditions you selected on a dish with propionic acid and on a dish without propionic acid. Label the other environmental condition on the other two petri dishes (one with propionic acid and one without). Then incubate all four dishes under the appropriate conditions. Record the food source

of the fungal sample and the two environmental conditions that you are testing; use a table like the one below in the **Records** section of your Vee Form.

Fungal Growth in Different Environments				
Dish	Environmental Condition	Propionic Acid?	Source of Fungus	Growth
1				
2				
3				
4				

Part B: Comparing Amounts of Growth

6. After one week, examine each dish under the stereomicroscope without opening the dish.

7. Record your observations in the data table in the **Records** section, using the symbols given under the table for amount of growth observed.

8. Examine dishes belonging to other groups, especially those grown under different environmental conditions.

9. Add your observations of those dishes to the **Records** section of the Vee. In the **Procedure** section of the Vee, briefly summarize the procedure you followed.

10. Clean up your materials and wash your hands before leaving the lab. Dispose of all materials according to instructions provided by your teacher

INQUIRY

1. Besides the environmental conditions you chose, what additional factor was tested in your experiment?

2. What steps were taken in your experiment to avoid contamination of the plates?

3. How would contamination of the plates affect the results of your experiment?

4. What does extensive fungal growth on a plate indicate?

5. What effect does propionic acid have on fungal growth? How do you know?

6. Use the information on the **Knowing Side** of the Vee to interpret your results from the **Doing Side,** and then write your **Knowledge Claim.** Write a **Value Claim** for this lab.

ANALYSIS

1. Why do you think propionic acid is added to foods?

2. Which environmental conditions favor fungal growth? Which inhibit it?

3. Compare the results for the different kinds of fungi grown. Did fungi from certain food sources grow more rapidly than others?

4. Based on your conclusions, under what conditions would you keep a nonsterile food product if you wanted to prevent it from becoming moldy?

FURTHER INQUIRY

Write a **New Focus Question** that could be the point of a new investigation. The following is an example:

On what kinds of food do fungi grow best?

a. additive-free bread or bread containing chemical additives

b. natural cheese or processed cheese

c. regular strawberry preserves or low-sugar strawberry spread

OVERVIEW OF PLANTS

REVIEW

• mitosis (Section 6-2)
• meiosis (Section 6-3)
• mycorrhizae (Sections 13-2 and 22-3)
• life cycles, gametophyte, and sporophyte (Section 19-2)
• characteristics of the kingdom Plantae (*Highlights: Kingdom Plantae* on page 447)

Rose pogonia, an orchid native to North America

23-1 The Evolution of Plants

Plants (members of the kingdom Plantae) are complex multicellular organisms that are primarily terrestrial autotrophs—that is, they occur almost exclusively on land and produce their own organic molecules from inorganic materials by photosynthesis. Today, plants are one of the dominant groups of organisms on land, although some, like the tiny duckweeds seen in Figure 23-1, have returned to aquatic habitats. In this section, you will discover how plants began to adapt to life on land.

Section Objectives

- Identify several obstacles to living on land, and describe how plants overcame them.
- Distinguish nonvascular plants from vascular plants.
- Summarize alternation of generations in plants.
- Describe the moss life cycle.
- Describe the three basic features of vascular plants.

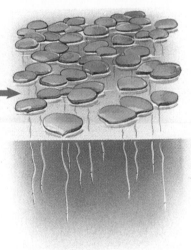

Figure 23-1 Duckweeds, the smallest flowering plants, live only in aquatic environments.

Plants Overcame Obstacles to Living on Land

The fossil record indicates that nothing lived on the land surfaces of our planet until about 440 million years ago. For the first 3 billion years of life's existence on Earth, life was confined to the sea. Scientists are not sure why it took so long for life to reach terrestrial habitats, but they suspect that intense solar radiation may have made the surface of the land uninhabitable. With the advent of photosynthesis in Earth's oceans, oxygen gas began to accumulate in the atmosphere. Some of this oxygen gas, O_2, was converted to ozone, O_3, leading to the development of an ozone layer high in the atmosphere. Then, as it does today, the ozone layer shielded Earth's surface from much of the harmful solar radiation.

Soon after the appearance of significant amounts of oxygen in Earth's atmosphere, plants and fungi invaded the land. Both plants and fungi probably evolved from multicellular protists. Multicellularity enabled plants to develop the complex structures and associations that have contributed to their success on land. However, the multicellular, aquatic green algae that were the ancestors of modern plants could not survive on land. Before the descendants of the earliest plants could thrive in terrestrial habitats, they had to overcome three obstacles: they had to be able to absorb minerals from the rocky surface; they had to be able to conserve water; and they had to have a way to reproduce on land.

Absorbing Minerals

Mutualistic associations similar to mycorrhizae may have played a key role in the initial occupation of land. As you learned in Chapter 13, mycorrhizae are symbiotic relationships between fungi and the roots of plants. The plants provide the fungi with carbohydrates and other organic molecules made during photosynthesis. The fungi absorb from the soil the phosphorus and other minerals that are needed by plants. Figure 23-2 shows how a plant benefits from mycorrhizae. Although the first plants lacked roots, fungi have been seen within and among the root cells of many fossilized early plants. Thus, some botanists (scientists who study plants) think that mutualistic associations similar to mycorrhizae may have enabled the first plants to absorb minerals from Earth's rocky surface. Today, approximately 80 percent of all living plant species form mycorrhizae.

Conserving Water

One of the key challenges to living on land is to avoid drying out. The first plants were very small and lived at the edges of oceans, where water was abundant. However, to occupy drier habitats, a means of conserving water was needed. For plants, the solution to this problem was the development of a watertight outer covering called a **cuticle**. Made of a waxy

Figure 23-2 The two pine seedlings, *below,* are the same age. The roots of the small pine seedling, *left,* have no mycorrhizal fungi. Mycorrhizal fungi growing within and between the root cells of the large pine seedling, *right,* have enhanced the seedling's growth.

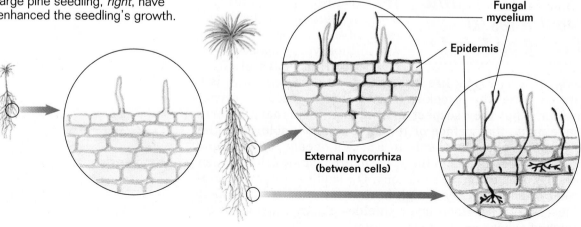

Fungal mycelium

Epidermis

External mycorrhiza (between cells)

Internal mycorrhiza (within cells)

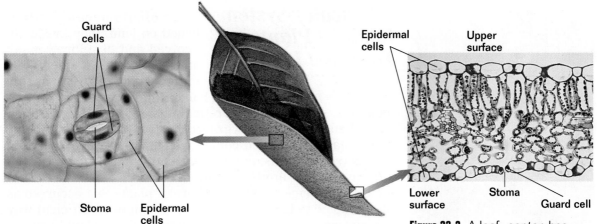

Guard
cells

Epidermal
cells

Upper
surface

Stoma

Epidermal
cells

Lower
surface

Stoma

Guard cell

Figure 23-3 A leaf, *center*, has numerous stomata. A pair of guard cells, *left*, surrounds each stoma. The cross section of the leaf, *right*, shows that a stoma is an opening through which air and water vapor can enter and exit the leaf.

substance, the cuticle covers the aboveground parts of a plant and prevents these tissues from losing water to the air. Like the wax on a shiny car, the cuticle is impermeable to water, but it is also impermeable to the gases required by plants for photosynthesis and cellular respiration.

Passages through the cuticle, in the form of specialized pores called **stomata** (*STOH muh tuh*), developed and enabled plants to exchange gases. The word *stoma* means "mouth" in Greek. Two views of a leaf's stomata are seen in Figure 23-3. Occurring on at least some portions of all plants except liverworts, stomata permit carbon dioxide gas to enter a plant body and permit water vapor and oxygen gas to exit. Because of the cuticle and stomata, water enters most plants primarily through their roots (which do not have a cuticle) and exits primarily through stomata. A pair of specialized cells called **guard cells** borders each stoma and controls its size by expanding and contracting. The timing of the opening and closing of stomata is critical to preventing excessive water loss while admitting the carbon dioxide required for photosynthesis.

Reproducing on Land

To reproduce sexually, an organism's male and female gametes must be able to reach one another. The male gametes of aquatic algae are able to swim through water to fertilize the female gametes. The gametes of plants that live on land, however, must be able to move in an environment where water is not abundant. Also, they must be protected from drying out while they are being transferred. The eggs of the first plants were surrounded by jackets of cells, and a film of water was required for a sperm to swim to an egg and fertilize it. Today, mosses, ferns, and several other groups of primitive plants still reproduce in this way. In more advanced plants, the sperm are enclosed in multicellular structures (pollen grains) that keep them from drying out. Such structures enable the male gametes of more advanced plants to be transmitted to female gametes by wind or animals rather than by water. ◻

CONTENT LINK

The location, structure, and function of stomata and guard cells are covered in more detail in **Chapter 24.**

◻ CAPSULE SUMMARY

Adaptations that have enabled plants to survive on land include mycorrhizae, a waxy cuticle, and structures that protect gametes.

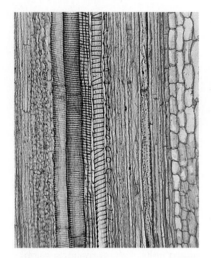

Figure 23-4 Thick-walled, tubular cells like these carry water from the tips of roots to the tips of leaves. Stacked end to end, these cells form tiny pipes called vessels.

A Vascular System Enabled Plants to Thrive on Land

Once plants became established on land, they gradually evolved and developed many other features that aided them in achieving success in this new and demanding habitat. For example, there was no fundamental structural difference between the aboveground and, when present, belowground parts of the earliest plants. Later, plants developed complex structures with specialized tissues (roots, stems, and leaves), each suited to its function and immediate environment. However, one of the most important changes in the structure of plants that occurred as they adapted to land was the development of an efficient way to move water and other materials through the plant body.

In order to survive in environments that have a limited water supply, most plants need an efficient "plumbing" system to carry water from their roots up to their leaves and to carry carbohydrates from their leaves down to their roots. These plumbing systems consist of specialized strands of hollow cells connected end to end like a pipeline, as you can see in Figure 23-4. The tissues that transport water and other materials within a plant make up the **vascular system.** The word *vascular* is derived from the Latin word *vasculum,* meaning "vessel" or "duct." In the dominant group of plants today, the cells of the vascular system run from near the tips of the roots to the tips of the stems and into the leaves. Not all plants have efficient vascular systems, however. Of the twelve phyla of living plants listed in Table 23-1, the three phyla that are referred to as nonvascular plants either have no vascular system or have only very simple vascular tissue. The nine remaining plant phyla have well-developed vascular systems and are referred to as vascular plants.

Table 23-1   Phyla of Living Plants

Phylum	Number of Species	Main Characteristics
Nonvascular Plants		
Hepatophyta Liverworts	6,000	Simplest plants; small, having a dominant gametophyte with a flattened or "leafy" body that lacks vascular tissue, a cuticle, stomata, roots, stems, and leaves
Anthocerophyta Hornworts	100	Small, with a flattened, dominant gametophyte that has stomata but lacks vascular tissue, roots, stems, and leaves
Bryophyta Mosses	10,000	Small; most have simple vascular tissue, a sporophyte consisting of a bare stalk and a spore capsule, and a dominant, "leafy" green gametophyte that lacks roots, stems, and leaves

Phylum		Number of Species	Main Characteristics
Vascular Plants			
Psilotophyta Whisk ferns		21	Seedless, with a small, independent gametophyte and a dominant sporophyte that is highly branched and has tiny leaves but is not differentiated into roots and stems
Sphenophyta Horsetails		15	Seedless, with a small, independent gametophyte and a dominant sporophyte consisting of roots and ribbed and jointed stems with soft needlelike leaves at the joints
Lycophyta Club mosses		1,000	Seedless, with a small, independent gametophyte and a dominant, mosslike sporophyte with roots, stems, and leaves
Pterophyta Ferns		12,000	Seedless, with a small, independent gametophyte and a dominant sporophyte consisting of roots, horizontal stems, and leaves called fronds; spores are produced in clusters of sporangia on lower surfaces of leaves
Coniferophyta Conifers		550	Gymnosperms (seed plants with tiny gametophytes, a large sporophyte, and ovules not enclosed by an ovary) that produce cones; mostly evergreen trees and shrubs with leaves modified as needles or scales
Cycadophyta Cycads		100	Gymnosperms with palmlike leaves; produce male and female cones on separate plants
Ginkgophyta Ginkgo		1	Gymnosperm; deciduous tree with fanlike leaves; produces conelike male reproductive structures and uncovered seeds on separate individuals
Gnetophyta Gnetophytes		70	Gymnosperms; diverse group of shrubs and vines
Anthophyta Flowering plants		250,000	Angiosperms (seed plants with tiny gametophytes, a large sporophyte, and ovules enclosed by an ovary); a very diverse group including trees, shrubs, vines, and herbs that produce flowers and fruits

Plants Evolved With Alternation of Generations

Plants evolved with a distinctive pattern of development in their life cycles. Among many algae, the zygote is often the only diploid (2*n*) cell, and it undergoes meiosis immediately after fertilization (zygotic meiosis) to form haploid cells. In early plants, however, meiosis was delayed. Recall from Chapter 19 that plant life cycles exhibit sporic meiosis, in which spores form by meiosis and gametes form by mitosis. The zygote divides by mitosis to produce many diploid cells that persist for a long portion of the life cycle. As a result, plants developed a life cycle in which a multicelled haploid individual that produces gametes, the gametophyte, alternates with a multicelled diploid individual that produces spores, the sporophyte. The pattern among life cycles in which a haploid individual alternates with a diploid individual is called **alternation of generations.**

Figure 23-5 The life cycles of all plants exhibit sporic meiosis and alternation of generations. The life cycle of a nonvascular plant, *left*, is characterized by a gametophyte that is larger than the sporophyte. The life cycle of a vascular plant, *right*, is characterized by a large sporophyte and a very small gametophyte.

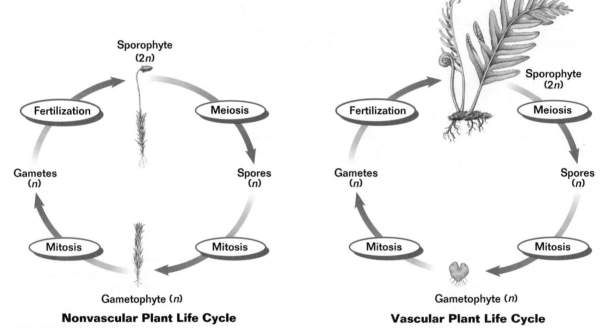

Nonvascular Plant Life Cycle

Vascular Plant Life Cycle

CAPSULE SUMMARY

Plant life cycles are characterized by alternation of generations in which a diploid sporophyte generation alternates with a haploid gametophyte generation. Over time, the plant life cycle evolved from one in which the gametophyte generation is dominant to one in which the sporophyte generation is dominant.

As plants evolved, however, a fundamental difference arose between the life cycles of the simpler, nonvascular plants and those of the more complex, vascular plants. As Figure 23-5 illustrates, the relative sizes of the gametophytes and sporophytes of nonvascular and vascular plants differ. In the nonvascular plants (mosses and liverworts), the gametophyte generation is the dominant (most noticeable) generation. In the vascular plants (ferns, gymnosperms, and angiosperms), the sporophyte generation is dominant. In fact, the very tiny gametophytes of most vascular plants grow *within* tissues of the sporophytes. ◻

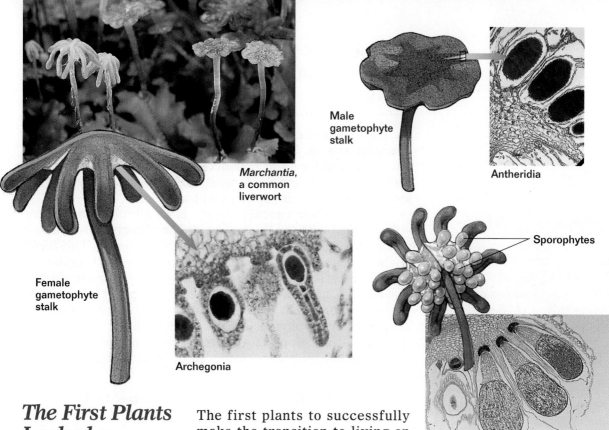

Marchantia,
a common
liverwort

Male
gametophyte
stalk

Antheridia

Female
gametophyte
stalk

Archegonia

Sporophytes

Sporophytes

The First Plants Lacked a Vascular System

The first plants to successfully make the transition to living on land probably had no vascular system for transporting materials throughout their bodies. All materials had to be transported by osmosis and diffusion, which greatly limited the maximum size of the plant body. Only two phyla of living plants, the liverworts (phylum Hepatophyta) and the hornworts (phylum Anthocerophyta), completely lack a vascular system. The members of these two groups have historically been grouped with the mosses and are still frequently referred to as "bryophytes." However, botanists no longer think that these three groups of relatively simple plants are directly related to one another. Liverworts and hornworts are usually inconspicuous and found growing in moist, shady places.

A common liverwort can be seen in Figure 23-6. The name *liverwort* dates back to the Middle Ages, when it was thought that plants resembling certain body parts might contain substances that could cure diseases of those body parts. The word *wort* meant "herb" in Old English. While the shape of some liverworts resembles a liver, the dominant gametophyte of most liverworts consists of simple leaflike and stemlike structures. Projections called rhizoids help anchor liverworts to the surfaces on which they grow. Gametes (eggs and sperm) are formed by mitosis in separate multicellular structures. **Archegonia** (ark uh GOHN ee uh) produce eggs, and **antheridia** (an thuhr IHD ee uh) produce sperm. When water is available, the sperm swim to a nearby archegonium and fertilize the egg within it. The resulting zygote grows into a very tiny diploid sporophyte. ❑

Figure 23-6 *Marchantia* is a common liverwort found living in moist, shady areas. The flattened, green gametophytes produce male and female gametes on separate stalks. Antheridia formed at the tip of a male stalk, *top right*, produce sperm. Archegonia formed at the tip of a female stalk, *bottom left*, produce eggs. The very tiny liverwort sporophytes, *bottom right*, grow from the archegonia under the caps of the female stalks.

❑ CAPSULE SUMMARY

Liverworts and hornworts are simple, nonvascular plants that lack roots, stems, and leaves. Water and other materials are distributed throughout their bodies by osmosis and diffusion.

Mosses Have Simple Vascular Tissue

The mosses (phylum Bryophyta) include many species in which a central strand of specialized conducting cells distributes water and carbohydrates throughout the plant. These conducting strands make up what is called **vascular tissue.** Simple in design, the vascular tissue in mosses is composed of conducting cells that lack thickened walls. To learn more about mosses and the other bryophytes, look at *Highlights: Bryophytes* on page 541.

Because their vascular tissue is so simple, mosses are still grouped with the liverworts and hornworts, and all three groups are considered to be nonvascular plants. The three phyla of nonvascular plants share an important similarity. Their life cycles, represented by the moss life cycle seen in Figure 23-7, are dominated by the gametophyte generation. The archegonia and antheridia of mosses are produced on separate gametophytes. The moss sporophyte consists of a bare stalk that supports a spore capsule, or **sporangium,** in which haploid spores are produced by meiosis. ■

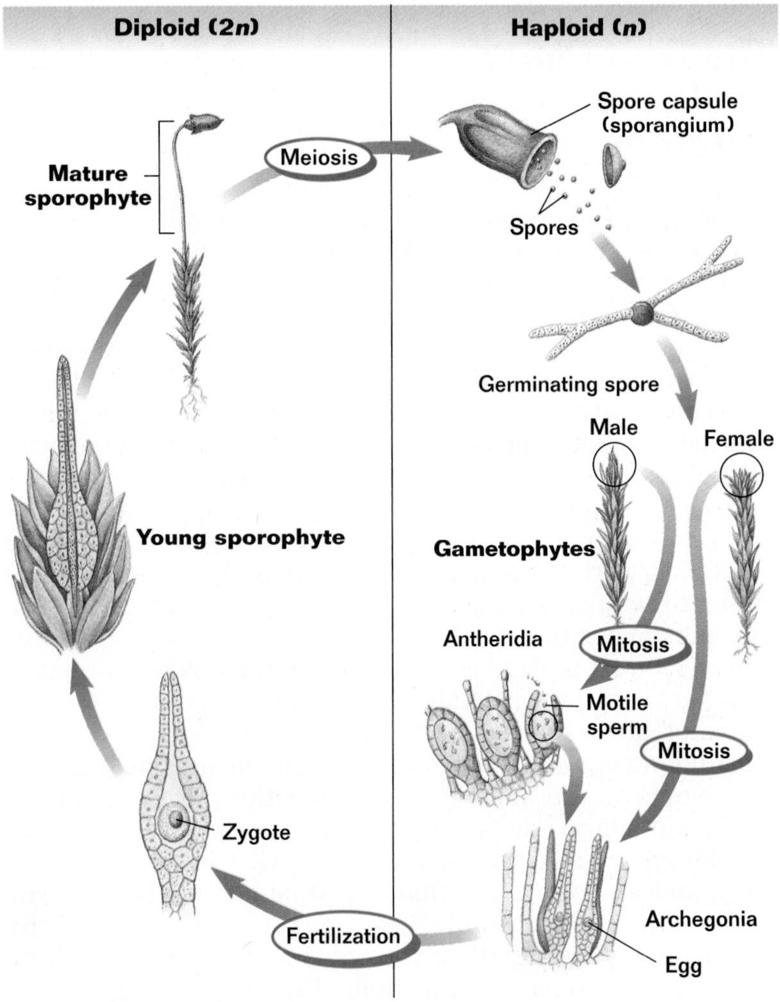

☐ **CAPSULE SUMMARY**

Mosses are considered nonvascular plants and are often grouped with liverworts and hornworts because their vascular tissue is very simple and because all three groups have similar life cycles.

Figure 23-7 The life cycle of a moss is characterized by a "leafy," green gametophyte that alternates with a smaller sporophyte, which consists of a bare stalk and a spore capsule. Motile sperm must swim through a film of water to fertilize an egg. The sporophyte grows on top of the gametophyte.

Vascular Plants Are Characterized by Several Features

The first vascular plants appeared approximately 430 million years ago, but only incomplete fossils of these plants have been found. The earliest known vascular plants for which there are relatively complete fossils, *Rhynia* and *Cooksonia,* are illustrated in Figure 23-8. Today, vascular plants occupy almost all terrestrial habitats except those perpetually covered by ice and snow. Unlike the nonvascular plants and the earliest vascular plants, many modern vascular plants grow very tall. For example, some trees are 50 m (163 ft.) or more in height and weigh many tons.

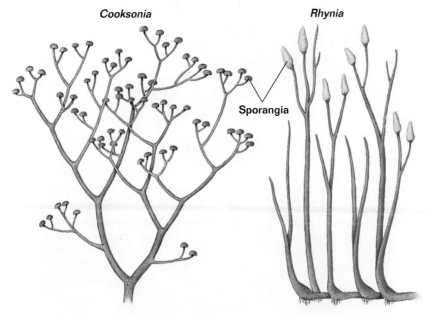

Cooksonia

Rhynia

Sporangia

Figure 23-8 Members of the extinct phylum Rhyniophyta, which appeared about 410 million years ago, included *Cooksonia, left,* the oldest known vascular plant, and *Rhynia, right.* These ancient plants had branched, leafless stems that were only a few centimeters long. Spore-forming sporangia were located at the tips of these stems. *Rhynia* also had horizontal underground stems, or rhizomes.

All vascular plants are distinguished by the following features: a dominant sporophyte, specialized conducting tissue, and a distinctive body form.

1. **Dominant sporophyte**   In contrast to nonvascular plants, the life cycles of vascular plants are dominated by a diploid sporophyte that is much larger than the gametophyte.

2. **Specialized conducting tissue**   The earliest vascular plants had two types of conducting tissue, still found largely unchanged in today's vascular plants. Both types of conducting tissue consist of strands of elongated cells that occur end to end like sections of pipe. Relatively soft-walled cells that conduct carbohydrates away from the areas where they are made, such as in leaves and stems, make up a kind of tissue called **phloem** (*FLOH uhm*). Hard-walled cells that transport water and dissolved minerals up from the roots make up a kind of tissue called **xylem** (*ZEYE luhm*). The reinforced walls of water-conducting cells can withstand considerable water pressure. Thus, water is able to rise to great heights in vascular plants.

CONTENT LINK

A more detailed discussion of the structure and function of xylem and phloem is contained in Chapter 24.

3. **Distinctive body form** All vascular plants have a body that consists of a vertical shaft from which specialized structures branch. This body form, which developed in the earliest plants, results as plants grow by adding new cells to the tips of their bodies. Zones of actively dividing cells, called **meristems** *(MEHR uh stehmz)*, produce plant growth. In the first vascular plants, there was no differentiation of the plant body into the aboveground structures, or **shoots**, and belowground structures, or **roots,** that characterize most plants. As the vascular plants became better adapted to living on land, however, they developed the familiar plant architecture—roots and a shoot consisting of stems and leaves—seen in Figure 23-9. ◻

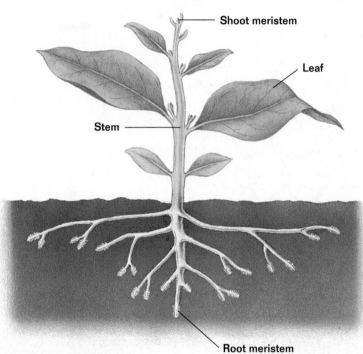

CONTENT LINK

Meristems and their role in plant growth are discussed in more detail in **Chapter 25**.

Figure 23-9 A typical vascular plant has an aboveground shoot with stems and leaves and a below-ground root. Plants grow from their tips as the cells of the meristems produce new cells by cell division.

Section Review

1. *What were three major obstacles that plants had to overcome to live successfully on land?*
2. *Describe three stages in the evolution of the plant vascular system. Use modern plants as examples of each stage.*
3. *How does alternation of generations occur in plants?*
4. *Summarize the events in the moss life cycle.*
5. *What are the three main characteristics of all vascular plants?*

Critical Thinking
6. *Why are vascular plants the most successful land plants?*

23-2 The Evolution of Seeds

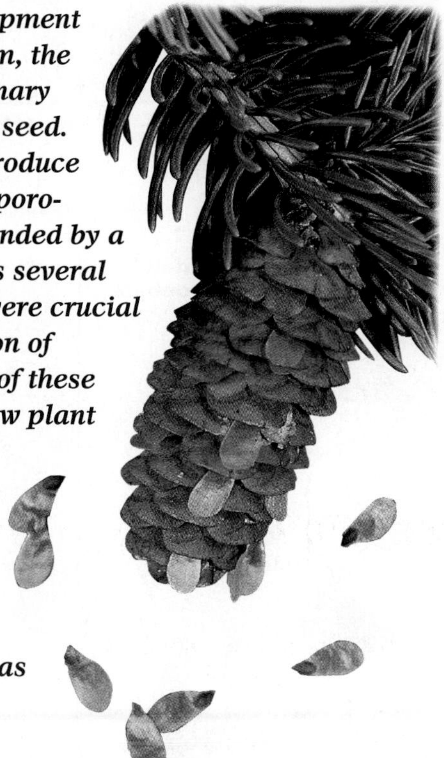

Following the development of a vascular system, the next great evolutionary advance in plants was the seed. Most plants living today produce seeds, which consist of a sporophyte plant embryo surrounded by a protective coat. A seed has several important functions that were crucial in the successful adaptation of plants to life on land. One of these functions is to protect a new plant during its most vulnerable stage—the embryo. Another function is to disperse (distribute) offspring to new locations away from their parents. Many kinds of seeds, such as the spruce seeds shown in Figure 23-10, have unique devices that are adaptations for carrying them farther away from their parent plants. Plants that produce seeds are called seed plants.

Section Objectives

- *Summarize the characteristics and the life cycle of ferns.*
- *Explain how gymnosperms differ from other seed plants.*
- *Compare and contrast the life cycles of the gymnosperms and the ferns.*
- *Describe a seed, and state its importance to the adaptation of plants to life on land.*

Figure 23-10 The seeds of a white spruce *(Picea glauca)* have a thin, papery "wing" that causes them to spin like a helicopter blade and keeps them airborne long enough to float some distance from their parent plant.

The First Vascular Plants Lacked Seeds

Although today's great forests are dominated by trees that produce seeds, the first forests were composed of vascular plants that did not produce seeds. Like the nonvascular plants, ferns and other seedless vascular plants have swimming sperm and require a film of water for fertilization. Forests of these plants flourished in the warm, humid climate of the late Paleozoic era. In these forests, plenty of water was available for successful reproduction. Ferns (phylum Pterophyta) are the most abundant and most familiar group of seedless vascular plants today. Though they are found throughout the world, ferns are most abundant in the tropics. Many ferns are small, measuring only a few centimeters in diameter. However, some of the largest living plants are tree ferns that can have trunks more than 24 m (79 ft.) tall and leaves up to 5 m (16 ft.) long. To learn more about ferns and other seedless vascular plants, look at *Highlights: Seedless Vascular Plants* on page 542.

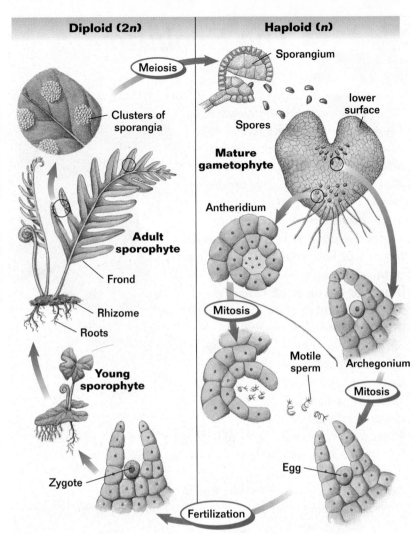

Figure 23-11 The life cycle of a fern is characterized by a large sporophyte (with leaves called fronds) that alternates with a small, heart-shaped gametophyte. Motile sperm must swim through a film of water to fertilize an egg.

(labels in figure: Clusters of sporangia, Adult sporophyte, Frond, Rhizome, Roots, Young sporophyte, Zygote, Meiosis, Sporangium, Spores, lower surface, Mature gametophyte, Antheridium, Mitosis, Motile sperm, Archegonium, Mitosis, Egg, Fertilization)

□ **CAPSULE SUMMARY**

Ferns are the most familiar seedless vascular plants. The fern life cycle has a dominant sporophyte generation and a smaller, but independent, gametophyte generation.

The fern life cycle, illustrated in Figure 23-11, represents an intermediate stage in the revolutionary change that took place in plant life cycles. Remember, the life cycles of nonvascular plants are dominated by a gametophyte that supports a smaller, dependent sporophyte. In ferns and other seedless vascular plants, however, the sporophyte is dominant, and the gametophyte is smaller, independent, and self-sufficient. The fern gametophyte is a thin, heart-shaped photosynthetic plant that lives in moist places and is usually no more than 1 cm in diameter. Fern gametophytes produce eggs in archegonia and sperm in antheridia, both of which are located on the lower surface of the plant. In ferns, the archegonia and antheridia are produced by the same individual. In other seedless vascular plants, the male and female structures are produced by separate gametophytes. When a film of water is available, sperm are able to swim to eggs and fertilize them. Fern sporophytes consist of roots, horizontal underground stems called rhizomes, and long, often highly divided leaves called **fronds.** Clusters of spore-producing sporangia form on the lower surfaces of fronds. □

The First Seed Plants Were Gymnosperms

Seeds apparently arose only once among the vascular plants, as the plant life cycle continued to shift toward a more dominant sporophyte generation and a more reduced gametophyte generation. Of the five phyla of living seed plants, four are collectively called **gymnosperms** *(JIHM noh spurmz).* The word *gymnosperm* comes from the Greek words *gymnos,* meaning "naked," and *sperma,* meaning "seed," and refers to the fact that gymnosperm seeds do not develop within a fruit (a mature ovary). First appearing about 380 million years ago, gymnosperms were the first seed plants. The flowering plants, or **angiosperms** *(AN jee oh spurmz),* evolved from gymnosperms and make up the fifth phylum of seed plants. The word *angiosperm* comes from the Greek words *angeion,* meaning "case," and *sperma,* meaning "seed," and refers to the fact that angiosperm seeds develop within a fruit. First appearing between 150 million and 200 million years ago, angiosperms are the most recently evolved of all plant phyla.

As Figure 23-12 illustrates, the gametophytes of seed plants have become highly reduced during the course of evolution. Developing from spores that are produced *within* the tissue of the sporophyte individuals, the gametophytes of seed plants are entirely dependent upon those sporophyte individuals for nutrients and water. Seed plants produce two kinds of gametophytes: a very tiny male gametophyte, or **microgametophyte,** that produces sperm, and a relatively large female gametophyte, or **megagametophyte,** that produces eggs. Thus, the spores that produce the microgametophytes are called **microspores,** and those that produce the megagametophytes are called **megaspores.** A **pollen grain,** which consists of only a few haploid cells surrounded by a thick protective wall, is a mature microspore that contains a microgametophyte. Each megagametophyte develops from a megaspore within an **ovule** *(AHV yool),* a multicellular structure that is part of the sporophyte. If the egg inside of an ovule is fertilized, the ovule and its contents become a seed.

Wind, insects, or other animals transport pollen grains to the female reproductive structures that contain ovules. The transportation of pollen grains from a plant's male reproductive structures to a female reproductive structure of a plant of the same species is called **pollination.** When a pollen grain reaches a female reproductive structure, the pollen grain cracks open. A **pollen tube** then grows from the pollen grain to an ovule and enables a sperm to pass directly to an egg. Thus, in seed plants there is no need for a film of water during the fertilization process. ◻

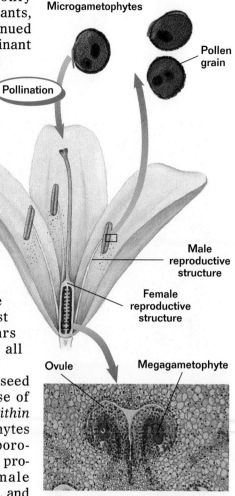

Figure 23-12 Reproductive structures found in cones or flowers produce the tiny gametophytes of seed plants. The transfer of pollen from a male reproductive structure to a female reproductive structure of the same plant species is called pollination.

◻ *CAPSULE SUMMARY*

The seed plants include gymnosperms, which produce seeds that do not develop within fruits, and angiosperms, which produce seeds that develop within fruits.

Most Living Gymnosperms Are Conifers
....

Members of the most familiar phylum of gymnosperms are trees that produce seeds in cones and thus are called **conifers** (*KAHN uh fuhrz*). Conifers (phylum Coniferophyta) include cedar, cypress, fir, hemlock, pine, redwood, spruce, and yew. The tallest living vascular plants, the giant redwoods of coastal California and Oregon, are conifers. One of the biggest redwoods, a giant sequoia (*Sequoia gigantea*) named after General Sherman of the Civil War, stands more than 80 m (262 ft.) tall and measures 20 m (66 ft.) around its base. Some individuals of another, much smaller species of conifer, the bristlecone pine (*Pinus longaeva*) that lives in the Rocky Mountains, are more than 5,000 years old—the oldest trees in the world. Most conifers have needle-like leaves that are an adaptation for limiting water loss. Conifers are often found growing in seasonally dry regions of the world, including the vast taiga forests of the northern latitudes. Many species of conifers are very important sources of timber and pulp. To learn more about conifers, look at *Highlights: Gymnosperms* on page 543.

CONTENT LINK
....

Plant products that are important to humans are discussed in **Chapter 26**.

Life Cycle of a Conifer

Conifers form two kinds of cones, which can be seen in Figure 23-13. Seed cones produce ovules on the surface of their scales. At the time of pollination, the scales of a seed cone are open, exposing the ovules. Pollen cones produce pollen grains within sacs that develop on the surface of their scales. The pollen grains of conifers are small and light, and they are carried by wind to seed cones. In pines and some other conifers, each pollen grain has a pair of air sacs that help to carry it in the wind. Because it is very unlikely that any particular pollen grain will be carried to a seed cone of the same species (the wind can take it anywhere), a great many pollen grains are needed to ensure that at least a few succeed in pollinating the species' seed cones. For this reason, pollen cones produce huge quantities of pollen grains. When pollen cones shed, the pollen grains often form a yellow layer on the surfaces of ponds, lakes, pavement, and car windshields.

Figure 23-13 This branch of an Austrian pine *(Pinus nigra)* has an immature seed cone and many pollen cones.

When a pollen grain lands near the ovule on a scale of a female cone, a slender pollen tube grows out of the pollen grain and into the ovule. Thus, the pollen tube delivers a sperm to the egg inside the ovule. Fertilization occurs when the sperm fuses with the egg, forming a zygote that is the beginning of a new sporophyte generation. Instead of growing directly into an adult sporophyte (a tree)—just as you grow directly into an adult from a zygote—the zygote first develops into a small embryo and then becomes dormant. While its further growth is postponed, the zygote and the sporophyte tissues that surround and protect it form a seed. Trace the stages in a conifer's life cycle in Figure 23-14. ■

☐ CAPSULE SUMMARY

Conifers produce massive amounts of pollen grains, which are carried to seed cones by wind.

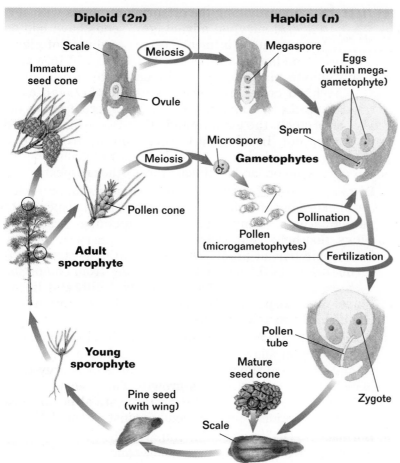

Diploid (2n)	Haploid (n)

Scale

Immature seed cone

Meiosis → Megaspore

Eggs (within mega-gametophyte)

Ovule

Meiosis → Microspore

Sperm

Gametophytes

Pollen cone

Pollination

Pollen (microgametophytes)

Adult sporophyte

Fertilization

Young sporophyte

Pine seed (with wing)

Pollen tube

Mature seed cone

Zygote

Scale

Figure 23-14 The life cycle of a conifer is characterized by a very large sporophyte (which produces cones) that alternates with tiny gametophytes (which form on the scales of the cones). Sperm are transported within pollen grains that are carried by wind to a female cone, where they then travel through a pollen tube to reach and fertilize an egg.

Other Gymnosperms

The other three gymnosperm phyla are much less common than the conifers. Living representatives of these groups can be seen in Figure 23-15. Cycads (phylum Cycadophyta), the dominant land plant during the Jurassic period, have short stems and palmlike leaves and are still widespread throughout the tropics. The only living species of ginkgo (phylum Ginkgophyta), the maidenhair tree (*Ginkgo biloba*), has fan-shaped leaves that are shed in autumn. The gnetophytes (phylum Gnetophyta) are all very unusual. *Welwitschia mirabilis*, perhaps the most bizarre of all plants, is a gnetophyte that grows in the harsh Namib Desert of southwestern Africa.

Figure 23-15 Cycads, *below left*, and the ginkgo, *below center*, are the only seed plants with motile sperm. One species of cycad is native to Florida. Because ginkgoes are resistant to air pollution, they are commonly planted along city streets. The two leathery beltlike leaves of *Welwitschia mirabilis*, *below right*, grow from a meristem at the base of the plant and split as they grow out over the desert sand.

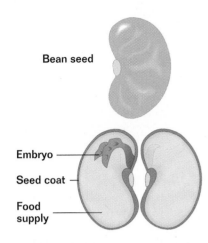

Bean seed

Embryo
Seed coat
Food supply

Figure 23-16 A bean, *top*, is a relatively large seed. It consists of an embryo and a stored food supply, both of which are surrounded by a seed coat, *bottom*.

□ *CAPSULE SUMMARY*

A seed is a sporophyte plant embryo surrounded by a protective coat. Seeds disperse a plant's offspring, nourish the embryo, and help control the time at which a new plant begins to grow.

What Is a Seed?
....

By providing the offspring of plants with several survival advantages, seeds have had an enormous influence on the evolution of plants on land. As you can see in Figure 23-16, a **seed** is a sporophyte plant embryo surrounded by a protective coat. The hard cover of a seed is called the **seed coat.** Formed from the sporophyte tissue of the parent plant, the seed coat protects the embryo and other tissues in the seed from drying out. In addition to their role in protecting a plant embryo, seeds have enabled plants to become better adapted to living on land in at least three other respects.

1. **Dispersal**   Seeds enable the offspring of plants, which are anchored in one place by their roots, to be dispersed to new locations. Many seeds have appendages, such as wings, that help wind, water, or animals carry them away from their parent plant. The dispersal of a plant's offspring prevents the parent and offspring from competing with each other for water, nutrients, light, and living space. Seed dispersal also facilitates the migration of a plant species to new habitats.

2. **Nourishment**   Most kinds of seeds have abundant food stored in them. Playing a role similar to that of the yolk in an egg, this food supply is a ready source of energy for a plant embryo as it starts its growth. Thus, seeds offer a young plant nourishment during the critical period just after germination when the seedling must establish itself.

3. **Dormancy**   Once a seed has fallen to the ground, it may lie dormant for many years. When conditions are favorable, particularly when moisture is present, the seed will begin to grow into a young plant. By remaining dormant until conditions improve, seeds enable plants to postpone development during unfavorable conditions such as a drought or a cold period. Thus, seeds aid in synchronizing the growth of a new plant with the season of the year. □

Section Review

1. *Briefly describe the fern life cycle.*
2. *What is a fundamental difference between gymnosperms and angiosperms?*
3. *How are the life cycles of ferns and gymnosperms similar? How are they different?*
4. *What is a seed? How are seeds an important adaptation to life on land?*

Critical Thinking
5. *Why is wind pollination more advantageous than the transfer of sperm to eggs through a film of water?*

23-3 The Evolution of Flowers

*T*he last difficult problem posed to plants by terrestrial living was resolved in angiosperms. Plants had previously been limited by a conflict between the need to obtain water and nutrients (solved by roots) and the need to find mates (solved by male gametes that can be carried to other plants). This problem was never completely solved in gymnosperms, whose lightweight pollen grains are carried passively by wind. Large numbers of pollen grains are therefore needed to improve the chances of a lucky encounter with a female cone—a very inefficient system. The pollen of many angiosperms, however, is delivered directly from one individual of a species to another. How? As Figure 23-17 indicates, animals carry the pollen for them! The innovation that made this great advantage possible is the flower.

Figure 23-17 Pollen covers the faces of four lesser long-nosed bats *(Leptonycteris curasoae)*, *above*, after a nighttime visit to organ pipe cactus flowers in the southwestern United States, *above left*. Worldwide, bats are important pollinators of cactuses and tropical fruit trees.

Angiosperms Achieved Evolutionary Success on Land

The most successful of all plants are angiosperms, seed plants that produce flowers. The remarkable evolutionary success of the angiosperms is the culmination of the plant kingdom's adaptation to life on land. Ninety percent of all living plants—more than 250,000 species of trees, shrubs, herbs, fruits, vegetables, and grains—are angiosperms. In short, nearly all of the plants that you see every day are angiosperms. Virtually all of your food comes directly or indirectly from angiosperms. In fact, more than half of the calories that humans consume come from just three species of angiosperms: rice, corn, and wheat.

What Is a Flower?

Flowers are the reproductive organs of angiosperms. Many flowers are sophisticated structures that are adapted to enable insect pollination. Bright colors attract the attention of insects. Nectar, which is a sugary secretion of many flowers, induces insects to enter a flower. Pollen-bearing structures coat the insects with pollen while they are visiting a flower. Then, when the insects visit another flower, they carry the pollen into that flower. The basic structure of a flower, seen in Figure 23-18, consists of the four concentric whorls (circles), of appendages described below.

1. **Calyx** The outermost whorl of a flower is called the **calyx** *(KAY liks)*, a name derived from the Greek word *kalyx*, meaning "cup." The calyx consists of one or more **sepals** *(SEE puhls)*, which are modified leaves, and protects a flower from physical damage while it is a bud.

2. **Corolla** The second whorl of a flower is called the **corolla** *(kuh ROHL uh)*, a name derived from the Latin word *corona*, meaning "crown." The corolla consists of one or more petals, which are also modified leaves. Often, the corolla produces vividly colored pigments or fragrances that attract particular pollinators to a flower.

3. **Androecium** The third whorl, called the **androecium** *(an DREE shee uhm)*, produces the microgametophytes, or pollen grains. The word *androecium* comes from the Greek words *andros*, meaning "male," and *oikos*, meaning "house." The androecium is made up of one or more **stamens** *(STAY muhnz)*, which consist of slender, thread-like **filaments** that are each topped by a pollen-containing sac called an **anther.**

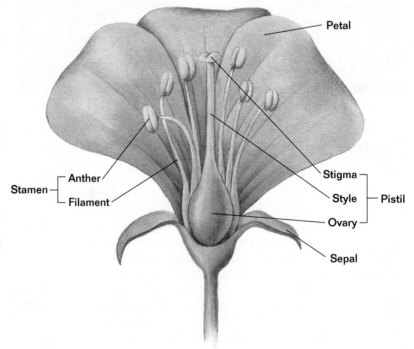

Figure 23-18 The parts of a flower are arranged in whorls—the calyx (all of the sepals), the corolla (all of the petals), the androecium (all of the stamens), and the gynoecium (all of the pistils). While some flowers have all four of the whorls shown here, many other flowers lack one or more of these whorls.

4. Gynoecium   The fourth and innermost whorl, called the **gynoecium** *(jeye NEE shee uhm)*, houses the ovules, in which the megagametophytes develop. The word *gynoecium* comes from the Greek words *gyne*, meaning "female," and *oikos*, meaning "house." The gynoecium consists of one or more **pistils** that are found in the center of a flower. Ovules develop in a pistil's swollen lower portion, which is called the **ovary.** An ovary may have one or more chambers. Usually, a slender stalk, called the **style,** rises from an ovary. The style has a swollen, sticky tip called the **stigma,** on which pollen lands and adheres. When a flower is pollinated (by pollen from the same species), a pollen tube emerges from each pollen grain and grows through the style and into the ovary.

Flowers may or may not have all four of these parts. A flower that has all four whorls of appendages is called a **complete flower.** If a flower has both a gynoecium and androecium, it is a **perfect flower.** Many flowers lack either a gynoecium or an androecium and are **imperfect flowers. Incomplete flowers** are those that lack any one of the four whorls. Therefore, flowers that lack *either* stamens or pistils, such as the squash flowers seen in Figure 23-19, are *both* incomplete and imperfect. ◻

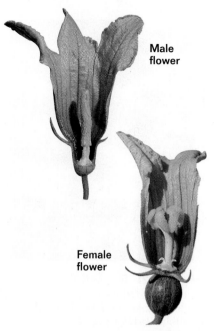

Figure 23-19 Squash and other members of the Cucurbitaceae family produce imperfect (and incomplete) flowers.

Flowering Plants Coevolved With Animals

If you were to watch insects visiting flowers, you would quickly discover that the visits are not random. Instead, certain insects are attracted by particular flowers. As you learned in Chapter 17, insects and plants have coevolved so that certain insects specialize in visiting particular kinds of flowers. An insect recognizes a particular color pattern and searches for flowers with that pattern. As a result, a particular insect carries pollen from the flowers of one individual to the flowers of another individual *of the same species.* This specificity is the key to successful insect pollination, making it much more effective than wind pollination.

Of all insect pollinators, the most numerous are bees. Bees evolved approximately 100 million years ago, about the time that flowering plants began to diversify greatly. Today, there are over 20,000 species of bees. Bees locate sources of nectar by odor at first, then by homing in on a flower's color and shape. Bee-pollinated flowers are usually yellow or blue and, as you can see in Figure 23-20, frequently have lines of dots, or "guides," to indicate the location of the nectar (usually in a flower's throat). Some of these markings are visible only to insects. While inside a flower, bees become coated with pollen. This coating is far from accidental. Most of the bees visiting flowers actively seek the pollen, which is a rich source of protein that they feed to their larvae.

◻ **CAPSULE SUMMARY**

Flowers are the reproductive structures of angiosperms. Most flowers consist of four whorls of appendages.

Figure 23-20 A yellow color, nectar guides, and a landing platform attract bees to the flowers of the unicorn plant, *Proboscidea altheifolia.*

Figure 23-21 This ruby-throated hummingbird transfers pollen while drinking nectar from the tubular red flowers of a salvia. While most insects cannot "see" red, it is a very distinct color to birds, as it is to humans.

Other insects also pollinate flowers. Butterflies tend to visit flowers that have "landing platforms" on which they can perch. These flowers are typically tube-shaped and filled with nectar that butterflies reach by uncoiling a long, hose-like mouthpart. Flies pollinate flowers that smell like rotting meat. Moths, which visit flowers at night, pollinate white, heavily scented flowers that are easy to locate in dim light.

Many angiosperms are pollinated by animals other than insects. Red flowers, for instance, are typically visited by hummingbirds, as you can see in Figure 23-21. Birds, which have keen vision, have a poor sense of smell. Knowing this, it is not surprising that red flowers usually lack a strong odor. Certain angiosperms have large, heavily scented, and pale-colored flowers that open at night. These flowers are pollinated by another nighttime visitor—bats.

Some angiosperms have reverted to wind pollination, a characteristic of their ancestors. Notable among the wind-pollinated angiosperms are oaks, birches, and grasses. To learn more about pollination in flowering plants, look at *Highlights: Angiosperms* on page 544. ◻

◻ *CAPSULE SUMMARY*

In many angiosperms, flower structure is an adaptation that enables pollination by animals, including insects, birds, and bats. Some flowers are wind pollinated.

Double Fertilization Provided Large Food Reserves in Seeds

Unlike the seeds of gymnosperms, the seeds of angiosperms develop a highly nutritious tissue, called **endosperm,** that originates at the same time that an egg is fertilized. In some angiosperms, such as corn and wheat, the endosperm is still

Table 23-2 Characteristics of Dicots and Monocots

Class	Seed	Flower	Leaf
Dicots	Two cotyledons	Flower parts in multiples of four or five	Veins usually netlike
	Bean	Evening primrose	Maple
Monocots	One cotyledon	Flower parts in multiples of three	Veins usually parallel
	Corn	Day lily	Corn

present in mature seeds. In other angiosperms, such as beans and peas, the endosperm is competely transferred into the embryo by the time a seed is mature. The food reserves are then stored in the embryo's fleshy, leaflike **cotyledons** *(kah tuh LEE duhnz)*, or seed leaves.

The angiosperms are divided into two classes based on the number of cotyledons in their seeds. Angiosperms that produce seeds with two cotyledons are called dicotyledons, or **dicots** (class Dicotyledones). Angiosperms that produce seeds with a single cotyledon are called monocotyledons, or **monocots** (class Monocotyledones). Monocots, which evolved somewhat later than dicots, differ from dicots in other important ways, as seen in Table 23-2.

You can see how the angiosperm embryo and endosperm originate at fertilization by studying the life cycle of a typical angiosperm, seen in Figure 23-22. The microgametophytes of seed plants contain two sperm cells. In most gymnosperms, one of these sperm dies. In angiosperms, however, *both* sperm fuse with certain cells of the megagametophyte. One sperm fuses with the egg, as in all sexually reproducing organisms, forming the zygote. The other sperm fuses with the haploid nuclei of two other cells produced by meiosis, forming a triploid (3*n*) cell that gives rise to endosperm. The term **double fertilization** is used to describe the process by which two sperm fuse with cells of the megagametophyte to produce both a zygote and endosperm. ◻

◻ *CAPSULE SUMMARY*

The endosperm produced during double fertilization stores nourishment within many angiosperm seeds. Angiosperms are classified as either dicots or monocots.

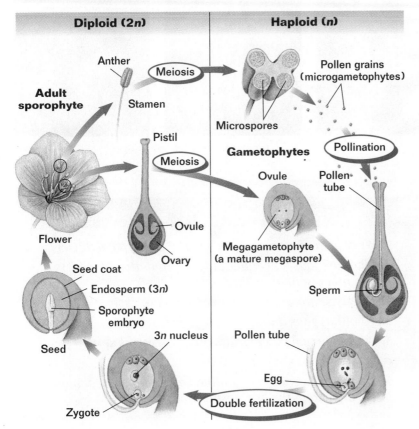

Figure 23-22 The life cycle of an angiosperm is characterized by a large sporophyte (which produces flowers) that alternates with tiny gametophytes (which form within the male and female structures of flowers). Sperm are transported within a pollen grain to a pistil, where they then pass through a pollen tube to reach and fertilize an egg.

Fruits Enabled Efficient Seed Dispersal

Among angiosperms, fruits evolved and enabled more efficient dispersal of seeds. A **fruit** consists of a mature ovary that contains one or more seeds and often includes other flower parts. Animals, which aid in pollination because they are attracted to and obtain nourishment from flowers, also aid in seed dispersal. Many mammals and birds are attracted to and eat fruits that are fleshy and tasty. You might have noticed that as fleshy fruits ripen, they often change from green and odorless to brightly colored and sweet smelling. The mature seeds within such ripe fruits are often resistant to chewing and digestion. Sometimes, virtually undamaged seeds pass out of an animal with its feces and are ready to germinate at a new location far from the parent plant. Most fruits, however, are not eaten by animals. Some are specialized for sticking to an animal's fur. Others are adapted for floating on wind currents or water. Figure 23-23 shows several types of angiosperm fruits that are specialized for different methods of seed dispersal. ☐

☐ **CAPSULE SUMMARY**

The fruits of angiosperms are specialized for seed dispersal by agents such as animals, wind, and water.

Figure 23-23 These fruits disperse seeds by different means. The fleshy fruits of raspberries, *left,* are eaten by birds and other animals. Cockleburs, *center left,* stick to animal fur. Wings enable silver maple fruits, *center right,* to fly. Coconuts, *right,* float on water.

Section Review

1. What is the function of each whorl in a typical flower?
2. What are four flower adaptations that attract insects and other animal pollinators?
3. Briefly summarize the angiosperm life cycle.
4. What happens during double fertilization?
5. What function do the fruits of angiosperms serve?

Critical Thinking

6. Why is the angiosperm method of seed dispersal more effective than the gymnosperm method?

Bryophytes

Becoming a Plant

The **nonvascular plants,** commonly referred to as **bryophytes,** represent the earliest stages in plant evolution. Arising from multicellular aquatic green algae, the first plants were small and were limited to habitats with abundant water, such as the edges of oceans or ponds. Like all bryophytes, they lacked roots, stems, and leaves. Included among the bryophytes are the two simplest groups of plants—liverworts (phylum Hepatophyta) and hornworts (phylum Anthocerophyta). Liverworts, such as this *Marchantia,* lack vessels that conduct water and other materials.

Hornworts also lack conducting tissue.

Marchantia

Developing Vascular Tissue

Mosses (phylum Bryophyta) are the most familiar bryophytes. Like this *Sphagnum* moss, most mosses have simple conducting tissue and thus may represent the next phase in plant evolution—the beginning of **vascular tissue.** As in all bryophytes, the moss body is not differentiated into roots, stems, and leaves, and the most conspicuous phase of the moss life cycle is the haploid gametophyte generation. Moss sporophytes consist of a bare stalk topped by a spore capsule and grow from the tips of the "leafy," green gametophytes.

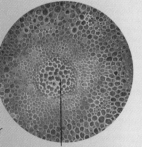

Sphagnum

Small Size

The sides of a moss's water-conducting cells are not thickened for added strength, as they are in true vascular plants. Individual mosses never become very large because they are able to conduct water only short distances.

Conducting strand

Tied to Water

Mosses almost always grow in mats in very moist places. When the mats are covered by dew or rain, motile moss sperm can easily swim to neighboring female gametophyte plants.

An Economically Important Moss

Sphagnum, or peat moss, is the most economically important moss. More than 290 species of *Sphagnum* mosses form extensive bogs in areas with cool climates and sufficient rainfall. The dense and deep mats formed by peat mosses are often cut into blocks, dried, and burned as fuel. Because of its moisture-holding capacity and its acidity, peat is also used extensively in gardening.

Seedless Vascular Plants

Cyathea sp.

Developing Shoots and Roots

Seedless vascular plants, such as this tree fern, *Cyathea,* were the first plants to develop a **vascular system.** A vascular system enabled the development of a body differentiated into **shoots** and **roots.** Because they had conducting cells with reinforced walls, vascular plants were able to grow much larger than nonvascular plants. The development of a waxy covering called a **cuticle** and spores that were resistant to drying aided in the survival of the seedless vascular plants on land. A life cycle dominated by the diploid sporophyte generation evolved with the seedless vascular plants. Today, ferns (phylum Pterophyta) are the only widely successful seedless vascular plants.

Reduced Gametophyte

Less than 1 cm (0.5 in.) across, this fern gametophyte is much smaller than the sporophyte but is independent of it.

Roots, Stems, and Leaves

Most fern sporophytes have horizontal stems, called **rhizomes,** that creep along below the ground and are anchored by **roots** that absorb water and minerals. The leaves, called **fronds,** stand vertically. In most ferns, the young leaves are called **fiddleheads.**

Frond

Fiddlehead

Rhizome

Roots

Drought-Resistant Spores . . .

Fern spores, which are highly resistant to drying, are produced in sporangia that grow in clumps on the lower side of the fronds.

. . . But Still Tied to Water

Ferns are most abundant in the tropics and prefer moist habitats. A film of water is required for the motile sperm to swim to eggs.

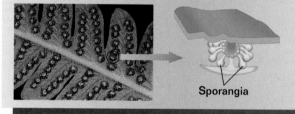

Sporangia

Other Seedless Vascular Plants

Three other phyla of seedless vascular plants that are well represented in the fossil record have only a few surviving species.

Phylum Sphenophyta (the horsetails)

Phylum Psilotophyta (the whisk ferns)

Phylum Lycophyta (the club mosses)

Gymnosperms

Reproducing With Seeds

Gymnosperms, represented by this lodgepole pine, *Pinus contorta*, were the first **seed plants**. A **seed** (which is a mature ovule) consists of a diploid sporophyte embryo surrounded by a protective coat. By postponing development until conditions are suitable for growth, seeds permit a plant embryo to survive long periods of unfavorable conditions. Seeds also serve to disperse an embryo far from its parent plant. With the evolution of gymnosperms, the sporophyte generation of the plant life cycle became completely dominant.

Pinus contorta

Greatly Reduced Gametophytes

The greatly reduced gameto- phytes of seed plants consist of the **microgameto- phytes**, encased in **pollen grains**, and the **megagameto- phytes**, housed in **ovules**. In most gymnosperms, the gametophytes are produced in cones, shown here on a lodgepole pine.

Seed cone   Pollen cones

Conserving Water

Like the lodgepole pine, most conifers have water- conserving needle-shaped leaves, an adaptation to dry environments. Some coni- fers, such as this juniper, have leaves that are reduced to tiny scales.

Scale-like leaves   Seed cone

Wind Pollination

The sperm of conifers do not swim but are carried within pollen grains that drift with the wind. Conifers and most other gymnosperms are wind pollinated. For this reason, they tend to occur in dense stands of one species. Extensive conifer forests are found in north- ern latitudes.

Pine pollen is equipped with two bladderlike wings that help to carry it on the wind.

Soft Wood

Secondary growth produces the woody stems of conifers. Many conifers grow very fast, so their wood is not as dense as other trees and is often referred to as "soft wood."

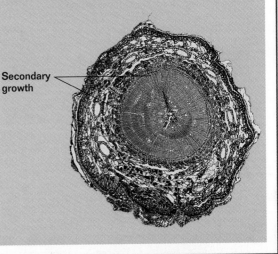

Secondary growth

Highlights
Plant Evolution

Angiosperms

Reproducing With Flowers

Angiosperms, such as the wild rose *(Rosa* sp.*)*, represent the culmination of the adaptation of plants to land. The great evolutionary advance of angiosperms is the **flower,** a reproductive structure that promotes pollination. The ovules of angiosperms are completely enclosed by the **ovary** of a flower. Mature ovaries form **fruits,** which greatly enhance seed dispersal among angiosperms. The evolution of the flower enabled the development of mechanisms for promoting **cross-pollination,** which tends to increase the genetic variation of a population and makes it better able to adapt via natural selection. For example, some flowers attract insects and other animals that carry pollen and thus increase the likelihood of cross-pollination.

Rosa sp.

Early angiosperms may have been pollinated by beetles, which were already abundant when angiosperms first evolved.

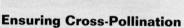

Self-Pollination

Many angiosperms with **perfect flowers** (both male and female parts), such as this garden pea, are adapted for **self-pollination,** which ensures pollination and fertilization (fusing of egg and sperm). Self-pollination occurs before a flower opens.

Ensuring Cross-Pollination

Some angiosperms, such as poplars, have two kinds of individuals: males (with flowers that produce only pollen) and females (with flowers that produce only ovules). The offspring of such plants result from **cross-fertilization** (the process in which a sperm from one individual fertilizes an egg of a different individual).

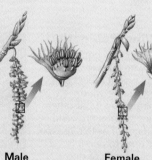

Male flower

Female flower

Preventing Self-Fertilization

In some angiosperms, such as oaks, **self-fertilization** (the fertilization of an egg by a sperm from the same individual) is unlikely because the male and female gametophytes form in separate flowers.

Female flower

Male flower

The male and female gametophytes of many angiosperms mature at different times, further preventing self-fertilization.

Wind Pollination

Some angiosperms, such as oaks, birches, and grasses, have reverted to wind pollination. Their small, greenish flowers do not attract pollinators. These plants grow in very large masses, making wind pollination particularly efficient and making cross-pollination more likely.

alternation of generations (524)
androecium (536)
angiosperm (531)
anther (536)
antheridium (525)
archegonium (525)
calyx (536)
complete flower (537)
conifer (532)
corolla (536)
cotyledon (539)
cuticle (520)
dicot (539)
double fertilization (539)
endosperm (538)
filament (536)

frond (530)
fruit (540)
guard cell (521)
gymnosperm (531)
gynoecium (537)
imperfect flower (537)
incomplete flower (537)
megagametophyte (531)
megaspore (531)
meristem (528)
microgametophyte (531)
microspore (531)
monocot (539)
ovary (537)
ovule (531)
perfect flower (537)
phloem (527)

pistil (537)
pollen grain (531)
pollen tube (531)
pollination (531)
root (528)
seed (534)
seed coat (534)
sepal (536)
shoot (528)
sporangium (526)
stamen (536)
stigma (537)
stomata (521)
style (537)
vascular system (522)
vascular tissue (526)
xylem (527)

Concept Mapping

Construct a concept map that shows how plants are classified. Use as many terms as needed from the vocabulary list. Try to include the following terms in your map: vascular plants, nonvascular plants, ferns, angiosperms, gymnosperms, mosses, cones, vascular tissue, seeds, and flowers. Include additional terms in your map as needed.

Review

Multiple Choice

1. What evolutionary development enabled plants to conserve water and, thus, helped them to adapt to life on land?
   a. mycorrhizae   c. pollen
   b. cuticle   d. seed

2. Because plants have an alternation of generations, their life cycles include a diploid individual that is called a(n)
   a. sporophyte.   c. zygophyte.
   b. gametophyte.   d. epiphyte.

3. Which of the following is *not* a characteristic of vascular plants?
   a. xylem and phloem
   b. a dominant haploid generation
   c. stems and leaves
   d. specialized underground structures

4. The first true seed plants were
   a. angiosperms.   c. ferns.
   b. gymnosperms.   d. mosses.

5. Gymnosperms and angiosperms are different in that gymnosperms
   a. are pollinated by wind.
   b. lack megagametophytes.
   c. have a diploid sporophyte generation.
   d. do not bear fruit.

6. The life cycle of a conifer differs from the life cycle of a fern in that
   a. alternation of generations does not occur in ferns.
   b. fertilization in conifers produces a haploid zygote.
   c. water is not required for fertilization in conifers.
   d. the sporophyte of a fern is larger than the gametophyte.

7. Seeds have helped plants adapt to life on land by
a. providing nourishment for embryos.
b. protecting embryos from air pollution.
c. allowing development to occur during unfavorable conditions.
d. limiting the dispersal of plant progeny.

8. Which of the following describes an imperfect flower?
a. petals, pistil, and stamens
b. anthers, ovary, and sepals
c. pistil, stamen, and corolla
d. calyx, corolla, and stamens

9. Which of the following flower characteristics would attract hummingbirds?
a. a large white corolla
b. a strong odor
c. a bright red, tubular corolla
d. heavily scented, blue petals

10. What is the function of the endosperm found in the seeds of an angiosperm?
a. to provide nourishment
b. to stimulate root growth
c. to inhibit embryonic development
d. to bring on a period of dormancy

Completion

11. Associations similar to _____ probably helped the first plants absorb water and minerals.

12. In the moss life cycle, the _____ grows on top of the leafy, green _____ . In mosses, eggs form in _____ , while sperm form in _____ .

13. The fern _____ is a heart-shaped plant that produces eggs and sperm. Fronds, rhizomes, and roots characterize the adult fern _____ .

14. The male reproductive part of a flower is the _____ ; it includes a filament and an anther. The female reproductive part of a flower is the _____ ; its enlarged base is called the _____ .

15. During double fertilization, one sperm fuses with the egg to form the _____ , while the other sperm fuses with the nuclei of other cells to form the _____ .

Short Answer

16. Why are mosses considered to be nonvascular plants even though they contain simple vascular tissue?

17. Suppose that the seeds of a particular plant species always germinate as soon as they land on soil. How might this affect the survival of the species?

18. How might imperfect flowers be an advantage to a plant species?

19. Describe two methods of seed dispersal in angiosperms. What plant adaptations facilitate the process of seed dispersal?

Highlights Review

20. What are the bryophytes?

21. Why can mosses such as *Sphagnum* never grow to be as large as oak trees?

22. Name four types of seedless vascular plants.

23. Explain how *Cyathea* is able to reproduce without seeds.

24. How did the process of plant reproduction change as plants evolved?

25. How does growth in stands of one species facilitate pollination in conifers?

26. Contrast male and female oak flowers.

27. Describe two ways that cross-pollination is promoted among angiosperms.

Themes Review

28. **Evolution** Alternation of generations occurs in all plant life cycles. Explain how this fact plus comparisons of plant structure and plant life cycles indicate that evolution has occurred in plants.

29. **Levels of Organization** Name the four phyla of living seed plants that are collectively called gymnosperms. What do these groups have in common? In addition, give an example of and a distinguishing characteristic of each group.

30. Structure and Function Name the four whorls of appendages that make up the basic structure of a flower. What is the function of each whorl?

Critical Thinking

31. Making Inferences Some liverworts have leaflike structures that absorb water rapidly. How might this feature be important to a nonvascular plant?

32. Making Predictions Suppose that the sporophyte of a type of fern has a diploid chromosome number of 14. What will be the chromosome number of the fern's gametophyte? Explain.

33. Communicating Effectively Write a paragraph that you could read to a classmate who cannot see Figure 23-22. Your paragraph should describe the important aspects of the angiosperm life cycle.

34. Making Inferences Look at the photograph of a cocklebur plant's seed. How might this seed be dispersed? Explain.

Activities and Projects

35. Research and Writing Identify a plant that grows well in your area. Make careful observations of the plant and do library research to determine what adaptations enable the plant to thrive in its environment. Describe what you learn in a written report.

36. Multicultural Perspective In Japan, gardening has long been a fine art. A well-planned Japanese garden combines both visual and musical art with botany, mythology, and the psychology of relaxation. Research the building of Japanese gardens and their importance to the Japanese people. Relate your findings in a written or an oral report that includes visuals such as slides or photographs from magazines and books.

37. Cooperative Group Project Investigate the plants in an area near your school. Then, draw a map of the area on poster-sized paper, and mark on the map the location of the different types of plants (liverworts, mosses, ferns, angiosperms, and gymnosperms) that you find. To identify some plants, you may need to use an identification key. Include on your map a legend that lists the groups of plants found and the symbol used to identify the plants on the map. Display your map in your classroom or school.

Readings

38. Read the article "Bogs: Thick Mats of Debris," in *American Horticulturist*, May 12, 1994, page 12. The author describes how *Sphagnum* mosses affect succession in wetlands. Why is a peat bog *not* an inviting place for most types of plants? What adaptations enable certain plants to live successfully in peat bogs?

39. Read the article "Angiosperm Endozoochory: Were Pterosaurs Cretaceous Seed Dispersers?" by Theodore Fleming and Karen Lips, in *The American Naturalist*, October 1991, pages 1058–1063. The authors hypothesize that flying reptiles of the Cretaceous period ate the fruit of angiosperms and contributed to the adaptive radiation of the angiosperms. What evidence supports their hypothesis?

Plant Diversity

OBJECTIVES

- Compare similarities and differences among phyla of living plants.
- Relate structural adaptations to the evolution of plants.

PROCESS SKILLS

- classifying plants
- relating the structure of plant parts to their function

MATERIALS

- live and preserved specimens representing four plant phyla
- stereomicroscope or hand lens
- compound microscope
- prepared slides of archegonia and antheridia of mosses and ferns

BACKGROUND

1. How do plants you commonly see compare with their ancestors, the green algae?
2. What is alternation of generations? Is it found in all plants?
3. What do you think was the evolutionary pressure for flowers to become colorful?
4. Write your own **Focus Question** on your Vee Form.
5. **Knowing Side of the Vee** List the **Concepts** and the new **Vocabulary Words** on your Vee Form. In the **Concept Statements** section of the Vee, use these words in sentences that define and explain them.

TECHNIQUE

Doing Side of the Vee

You will travel to four stations to observe plants that are representatives of four phyla of plants. Record the answers to the following questions in the **Records** section of your Vee Form.

Station 1: Mosses

1. Use a stereomicroscope or a hand lens to examine the samples of mosses. Which part of the moss is the gametophyte? Which part of the moss is the sporophyte? Make a sketch of your observations in the **Records** section of your Vee Form. In your drawing, label the gametophyte  and sporophyte portions of the moss plant and indicate whether each is haploid or diploid.

2. Use the compound microscope to look at the prepared slides of moss archegonia and antheridia. What kinds of reproductive cells are produced in each of these structures? Draw the archegonia and antheridia in the **Records** section of your Vee Form.

3. Do mosses have roots? How do mosses obtain water and nutrients from the soil?

Station 2: Ferns

4. Look at the examples of ferns at this station. The leafy green fern is called a frond.
   a. How does water travel throughout a fern? List observations supporting your answer.
   b. In the **Records** section of your Vee Form, make a drawing of the fern plant. Indicate whether the leafy, green frond in your drawing is haploid or diploid.
   c. Search the underside of the fern fronds for evidence of reproductive structures. In the **Records** section of your Vee Form, make a drawing of your findings. What kind of reproductive cells are produced by these structures?

5. Examine the examples of fern gameto-phytes.
   a. Locate and identify the reproductive organs found on the gametophytes. In the **Records** section of your Vee Form, sketch and label these organs and identify the reproductive cells produced by each.
   b. Are the gametophytes haploid or diploid?
   c. What portion of the fern life cycle immediately follows fertilization? What part of the fern appears after fertilization?
6. In what ways are ferns like bryophytes? In what ways are they different?

Station 3: Conifers

7. The gymnosperms most familiar to us are conifers. Look at the samples of conifers at this station.
   a. When you look at the limb of a pine tree, which portion (gametophyte or sporophyte) of the plant life cycle are you seeing?
   b. In what part of the conifer would you find reproductive structures?
8. Name an evolutionary advancement found in gymnosperms but lacking in ferns.

Station 4: Angiosperms

9. Draw one of the representative angiosperms at this station in the **Records** section of your Vee Form. In your drawing, label the sporophyte portion of the angiosperm's life cycle. Where are the sperm and eggs produced in an angiosperm?
10. What is an evolutionary development that is present in both gymnosperms and angiosperms but absent in bryophytes and ferns?
11. How do the seeds of angiosperms differ from those of gymnosperms?
12. Examine the fruits found at this station. How have fruits benefited angiosperms?
13. In the **Procedure** section of the Vee, briefly summarize the procedure you followed in this investigation.

INQUIRY

1. How do the sperm travel from the bryophyte antheridium to the archegonium?
2. In an angiosperm, how does the sperm get to the part of the flower containing the egg?
3. Which portion of the plant life cycle is dominant in bryophytes? Which portion is dominant in ferns, gymnosperms, and angiosperms?
4. Use the information on the **Knowing Side** of the Vee Form to interpret your results from the **Doing Side,** and then write your **Knowledge Claim.** Write a **Value Claim** for this lab.

ANALYSIS

1. What is a seed? Why is the seed a helpful adaptation to terrestrial life?
2. Why are gymnosperms referred to as "naked seed plants"?
3. Which type of plants are the most successful and diverse today? What are some adaptations found among members of this group?

FURTHER INQUIRY

Write a **New Focus Question** that could be the point of a new investigation. The following is an example:

How are the geographic distributions of the phyla of living plants related to their structures?

PLANT STRUCTURE AND FUNCTION

REVIEW

- adhesion and cohesion of water (Section 2-2)
- cell structure and organelles (Section 2-4)
- osmosis and active transport (Section 3-2)
- carbon fixation and the Calvin cycle (Section 5-2)
- stomata and guard cells (Section 23-1)
- roots, shoots, and meristems (Section 23-1)
- xylem and phloem (Section 23-1)

Flower of a widow's tears,
Commelinantia anomala

24-1 The Vascular Plant Body

Vascular plants are among the most successful groups of organisms on Earth, literally covering its surface with a carpet of life. Some vascular plants are smaller than your fingernail, while others are larger than a house. Yet, all share the same basic body plan. Like your body, a vascular plant's body is composed of tissues that form its organs—roots, stems, and leaves. In this section, you will look first at the cells and tissues that form a vascular plant's body and then at the structure of leaves, stems, and roots.

Section Objectives

■ Describe the three major tissue systems that compose the vascular plant body, and state the function of each.

■ Explain how leaves, stems, and roots are adapted for the functions they perform.

■ Describe several distinguishing features of sugar maple trees.

Three Tissue Systems Make Up the Vascular Plant Body

Like other complex multicellular organisms, the body of a vascular plant is made of different types of tissues. The fundamental plant tissues are organized into three basic functional units, or systems.

1. **Dermal tissue system** The **dermal tissue system** serves as a protective outer layer.

2. **Ground tissue system** The **ground tissue system** performs photosynthesis, stores water and carbohydrates, assists in transport, and surrounds and supports the conducting tissues.

3. **Vascular tissue system** The **vascular tissue system** conducts water, mineral nutrients, and carbohydrates made by photosynthesis.

The basic body plan of a vascular plant is seen in Figure 24-1, which shows how plant tissues are arranged.

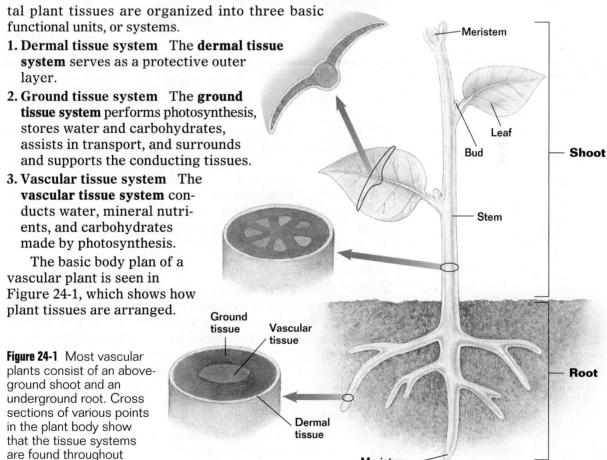

Figure 24-1 Most vascular plants consist of an above-ground shoot and an underground root. Cross sections of various points in the plant body show that the tissue systems are found throughout the plant.

Plant Structure and Function 551

Generalized Plant Cell

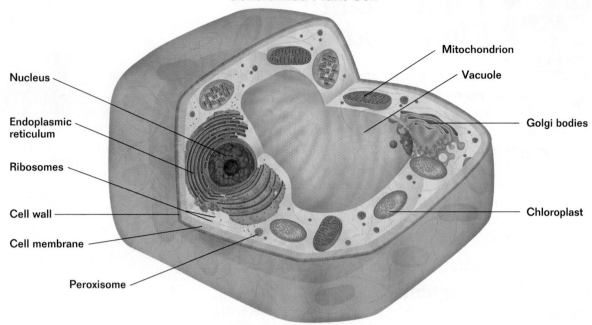

Nucleus

Endoplasmic reticulum

Ribosomes

Cell wall

Cell membrane

Peroxisome

Mitochondrion

Vacuole

Golgi bodies

Chloroplast

Figure 24-2 This diagram of a plant cell shows structures that occur in plant cells. However, each individual type of plant cell typically lacks one or more of these structures.

Each of the tissue systems is composed of one or more distinctive kinds of cells that are specialized for particular functions. The generalized plant cell in Figure 24-2 has the components typically found in plant cells. However, many specialized plant cells lack one or more of these components. Some plant cells perform their functions only after they have lost all of their contents. All plant tissues arise from the masses of actively dividing cells called meristems.

Dermal Tissue

Dermal tissue covers all parts of a plant's body. The outer covering, or "skin," of a plant is called the **epidermis** and is composed primarily of relatively flat cells. The word *epidermis* is derived from the Greek words *epi*, meaning "upon," and *derma*, meaning "skin." The epidermis of most above-ground plant parts is only one cell thick and is usually coated with a waxy cuticle that retards water loss.

Ground Tissue

Much of the body of a plant—including its roots, stems, and leaves—is made up of ground tissue. Ground tissue consists primarily of masses of thin-walled cells that are alive at maturity and have a functional nucleus. A variety of thick-walled cells that provide support are also found in ground tissue. In leaves, most of the ground tissue is packed with chloroplasts and is specialized for photosynthesis. In stems and roots, the ground tissue functions mainly to support the plant body and to store water and carbohydrates. Ground tissue also assists in the movement of water, minerals, and carbohydrates throughout the plant body by osmosis and active transport. ❑

Vascular Tissue

As you learned in Chapter 23, plants have two kinds of vascular tissue. Both xylem and phloem form tiny pipes that act like a plumbing system, carrying fluids up and down the plant body. However, the structure and function of the cells in each type of vascular tissue are very different. One major difference between the two types of vascular tissues is that mature xylem tissue consists of dead cells, while mature phloem tissue consists of living cells.

Xylem contains two principal types of conducting elements (cells), which can be seen in Figure 24-3. Both types of xylem cells lose their contents (plasma membrane, nucleus, and cytoplasm) at maturity and consist only of their cell walls. **Tracheids** (*TRAY kee ihdz*) are narrow, elongated, thick-walled cells that are tapered at each end. Water flows from one hollow tracheid to the next through pits in their cell walls. **Vessel elements,** the second type of xylem cells, are often wider than tracheids and develop large perforations in their ends. These perforations enable water to flow more quickly between the elements. Stacked end to end, a series of vessel elements forms a tube called a **vessel.**

Figure 24-3 Two types of conducting cells are found in xylem tissue. Tracheids are thin, elongated cells with tapered ends. Vessel elements are wider, and their ends have large perforations that enable water to flow more readily than it does through tracheids.

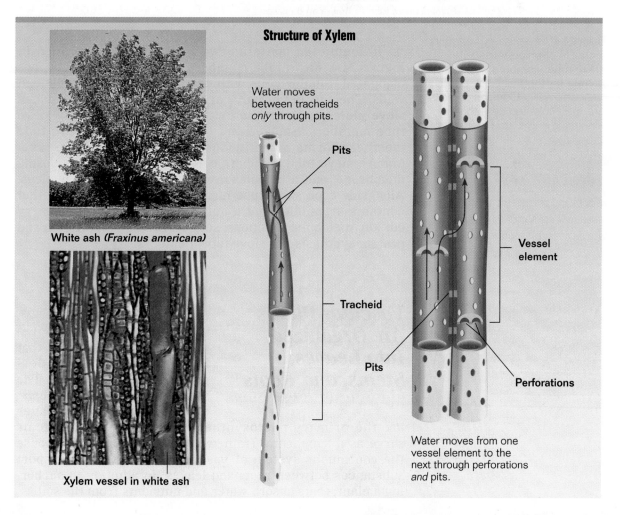

Structure of Xylem

White ash *(Fraxinus americana)*

Xylem vessel in white ash

Water moves between tracheids *only* through pits.

Pits

Tracheid

Pits

Vessel element

Perforations

Water moves from one vessel element to the next through perforations *and* pits.

Structure of Phloem

American basswood
(*Tilia americana*)

Both sieve tube members and companion cells are living at maturity, but sieve tube members lose their nucleus and organelles.

The cytoplasms of adjacent phloem cells are connected through the tiny pores in their cell walls.

Pores

Nucleus

Companion cell

Sieve plate

Sieve tube member

Phloem sieve tube in American basswood

Figure 24-4 The primary conducting cells of phloem tissue are the sieve tube members, which are named for the many tiny pores in their cell walls. Companion cells are the nucleated cells that lie alongside sieve tube members.

□ *CAPSULE SUMMARY*

Xylem and phloem are vascular tissues that consist of tubular cells. These cells transport materials throughout a plant's body.

Phloem also contains two kinds of cells, as seen in Figure 24-4. **Sieve tube members,** the conducting cells, are elongated, tubular cells with clusters of pores in their cell walls. The clusters of pores at the ends of these cells, called **sieve plates,** connect the cytoplasms of neighboring sieve tube members. Stacked end to end, a series of sieve tube members forms a continuous strand called a **sieve tube.** Mature sieve tube members consist of a cell wall, a plasma membrane, and a cytoplasm with no organelles or nucleus. Alongside each sieve tube member is a **companion cell** that contains organelles and a nucleus. Companion cells carry out key metabolic functions, such as cellular respiration and protein synthesis, for sieve tube members. □

Vascular Plants Are Organized Into Leaves, Stems, and Roots

As you learned in Chapter 23, most plants have an aboveground portion, the shoot, that is anchored by an underground portion, the root. The shoots of most plants consist of leaves and stems. Leaves are the primary photosynthetic organs of plants. Stems serve as a supporting framework for the leaves and house the continuous system of vascular strands that transport substances between roots and leaves. In addition to anchoring a plant, roots absorb water and nutrients from the soil.

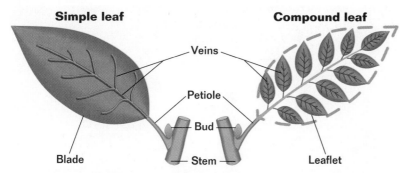

Figure 24-5 A simple leaf, *left,* consists of a continuous, flattened blade with many veins. The blade of a compound leaf, *right,* is divided into two or more leaflets. Most leaves are attached to a stem by a petiole.

Leaves

Most leaves have the basic structure seen in Figure 24-5. They consist of a flattened surface, the blade, that is often attached to the stem by a slender stalk, the **petiole** *(PEHT ee ohl).* Leaves with an undivided blade are called simple leaves. Those with a blade divided into two or more sections, or **leaflets,** are called compound leaves. Veins, which are bundles containing strands of both xylem and phloem tissue, are the plumbing system of a leaf. These veins are extensions of vascular bundles that run from the tips of roots to the edges of leaves. Recall from Chapter 23 that the veins in the leaves of most monocots run parallel to one another, while the veins in the leaves of most dicots branch and form a network. Many plants have highly modified leaves that are specialized for particular purposes, such as protection, water conservation, and climbing. Table 24-1 compares the structure and function of three types of modified leaves.

Table 24-1 Modified Leaves

Leaf	Name	Function
	Cactus spines	Protection, water conservation
	Garden pea tendrils	Climbing
	Venus' flytrap leaves	Photosynthesis, trapping insects to obtain nitrogen

Figure 24-6 This cross section of a leaf illustrates one type of structure that is characteristic of many leaves. Other types of leaves may have different internal arrangements.

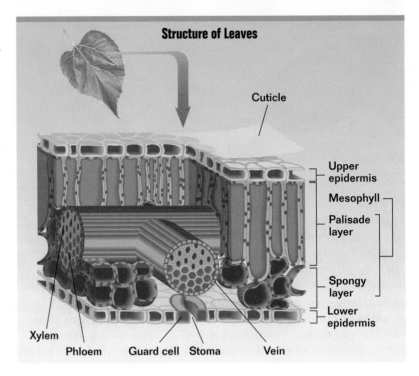

Structure of Leaves

Cuticle

Upper epidermis

Mesophyll

Palisade layer

Spongy layer

Lower epidermis

Xylem

Phloem   Guard cell   Stoma   Vein

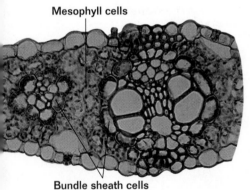

Mesophyll cells

Bundle sheath cells

Figure 24-7 The leaf of a corn plant, a C$_4$ plant, has a different internal structure than the leaf illustrated in Figure 24-6. The cells of the mesophyll are not differentiated into palisade and spongy layers. The mesophyll cells fix carbon by making a four-carbon compound. This compound is later transferred to the bundle sheath cells, where carbohydrates are made in the Calvin cycle.

A typical leaf is a mass of ground tissue that has veins running through it and that is encased in an envelope of epidermis. Figure 24-6 shows the internal structure of a typical leaf. The ground tissue in a leaf is called the **mesophyll** (MEHS oh fihl), which comes from the Greek words *mesos*, meaning "middle," and *phyllon*, meaning "leaf." Two kinds of mesophyll are found in most plants. Just beneath the upper epidermis of many kinds of leaves is the **palisade layer**, which consists of one or more rows of closely packed, columnar cells. The lower portion of the mesophyll usually consists of loosely packed, spherical cells and is called the **spongy layer**. The cells of the palisade and spongy layers are packed with chloroplasts, in which photosynthesis takes place. Scattered throughout the spongy layer are large air spaces, through which gases and water vapor travel. Stomata, the tiny holes that dot leaf surfaces, connect the air spaces of the mesophyll to the outside air.

The leaves of many plants have a modified internal structure that is an adaptation for a type of photosynthesis that operates very efficiently in hot climates. Recall from Chapter 5 that plants fix carbon (use carbon from carbon dioxide to make sugars) during photosynthesis by means of the Calvin cycle. Because the first detectable product of the Calvin cycle is a three-carbon compound, plants that fix carbon only with the Calvin cycle are called **C$_3$ plants.** In some plants, such as the plant shown in Figure 24-7, carbon is also fixed by an alternative pathway in which the first detectable product is a four-carbon compound. These plants are called **C$_4$ plants.** Because they fix carbon efficiently in high temperatures and intense light, C$_4$ plants are plentiful in the tropics.

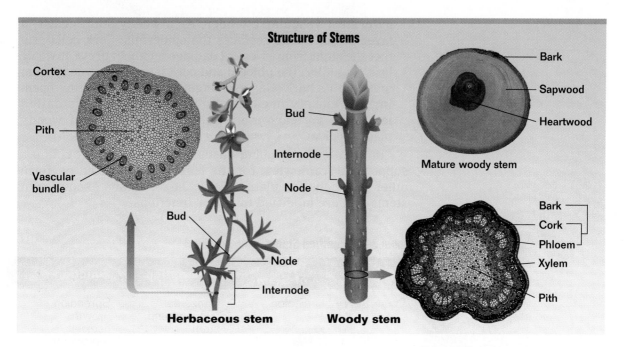

Structure of Stems

Cortex

Pith

Vascular bundle

Bud

Node

Internode

Herbaceous stem

Bud

Internode

Node

Woody stem

Bark

Sapwood

Heartwood

Mature woody stem

Bark

Cork

Phloem

Xylem

Pith

Stems

Figure 24-8 shows the external and internal structures of typical plant stems. The places where leaves attach to a stem are called **nodes. Internodes** are the areas of a stem between nodes. Also located at the nodes are **lateral buds** that grow into the branches of a stem. The bud at the tip of a stem is called the **terminal bud.** A typical stem consists of bundles of vascular tissues embedded in ground tissue. The outer layers of ground tissue in a stem are called the **cortex,** and the inner layers are called the **pith.** Other features of a stem vary, depending on whether the stem is woody or non-woody.

1. **Herbaceous stems** Flexible, relatively soft, and usually green stems, like those of violets and petunias, are called **herbaceous** *(huhr BAY shuhs)* **stems.** An epidermis forms the outermost layer of a herbaceous stem. Stomata in this epidermis enable the stem to exchange gases. The vascular tissues of a herbaceous (non-woody) stem are distributed within the ground tissue and arranged in **vascular bundles** that contain both xylem and phloem.

2. **Woody stems** Stiff, usually nongreen stems that contain layers of wood, like the trunks of trees, are called **woody stems.** In woody stems, the vascular tissues are arranged in solid cylinders. Xylem cells form the innermost cylinder. A new belt of xylem cells forms each year, adding to the width of the stem. Wood consists primarily of these xylem cells. The darker wood in the center of a tree trunk is called **heartwood.** Xylem cells in heartwood no longer conduct water because they have been filled with substances that help to strengthen the stem. The lighter wood in a tree trunk, which contains xylem cells that still conduct water, is called **sapwood.** Phloem cells form the

Figure 24-8 Compare two basic types of stems, *above.* Herbaceous stems, *left,* are typically soft and green. Woody stems, *right,* are typically stiff and nongreen with several layers of xylem cells. In herbaceous stems, the vascular bundles remain distinct. In woody stems, vascular bundles occur in young stems but fuse into solid cylinders in mature stems.

CONTENT LINK
....

The growth of woody and nonwoody plants is discussed in **Chapter 25.**

outermost vascular cylinder. As a woody stem grows, a layer of **cork cells** replaces the epidermis. Cork cells protect the stem from physical damage and help prevent water loss. Together, the phloem and cork layers of a woody stem make up its bark. Gas exchange occurs through tiny openings in loosely organized groups of cork cells called **lenticels.** To learn more about the structure of woody plants, look at *Up Close: Sugar Maple* on pages 560–561.

In addition to housing vascular tissues and providing a supporting framework for the leaves, stems often perform other functions for plants. Table 24-2 contains examples of stems that are modified for other functions.

Table 24-2 Modified Stems

Name	Type of Stem	Description	Function
Strawberry	Stolon	Horizontal, aboveground stem	Spreading growth, asexual reproduction
Potato	Tuber	Enlarged underground stem	Food storage
Cactus	Succulent	Fleshy, often leafless stem	Water storage

Roots

Roots have a simpler structure than stems. As Figure 24-9 shows, roots lack external features such as nodes, leaves, and buds. They also lack pith. Instead, the vascular bundles are at the center of a root. The vascular tissue of a root is surrounded by a thick layer of cortex that is covered by an outer sheath of dermal tissue. Epidermal cells cover the end of a root and are replaced by layers of cork cells as the root grows. The epidermal cells near the root tip produce slender projections called **root hairs.** By extending the cell membranes of epidermal cells, root hairs greatly increase the surface area of a root and play a critical role in the absorption of water and minerals. The actively growing tip of a root is covered by a protective layer of cells called the **root cap.**

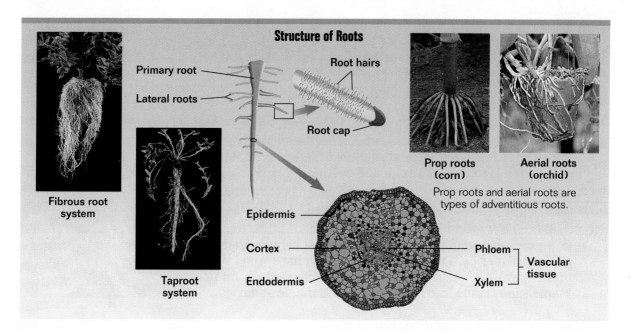

Structure of Roots

Primary root

Lateral roots

Root hairs

Root cap

Fibrous root system

Taproot system

Epidermis

Cortex

Endodermis

Phloem

Xylem

Vascular tissue

Prop roots (corn)

Aerial roots (orchid)

Prop roots and aerial roots are types of adventitious roots.

Based on branching patterns, there are two different types of root systems—taproot systems and fibrous root systems. Plants with a **taproot system,** such as carrots and radishes, have a large central root called a taproot. The lateral roots that branch from the taproot are usually much smaller than the taproot. Most dicots have a taproot system. The roots of a highly branched **fibrous root system** are all about the same size. Most monocots, such as grasses, have a fibrous root system. Many plants, such as orchids, also have **adventitious roots,** which grow from aboveground parts such as stems and leaves. Just like underground roots, adventitious roots provide support and absorb water and minerals. ❑

Figure 24-9 Most plants have either a taproot or fibrous root system. Some types of roots, such as prop roots and aerial roots, grow above ground and provide support as well as absorb water. Near a root tip, there are many root hairs, which enhance water absorption.

❑ *CAPSULE SUMMARY*

The leaves, stems, and roots of plants are made of the three types of plant tissues and are adapted to the particular functions they perform for the plant.

Section Review

1. *What are three functions of the ground tissue system?*
2. *Name and describe the two main types of conducting cells in xylem and phloem.*
3. *How does the structure of a leaf help it perform photosynthesis efficiently?*
4. *How does the structure of a root enable it to anchor a plant and to absorb water?*
5. *What structural features of the sugar maple make it economically important?*

Critical Thinking

6. *Why might a taproot system be an advantage to some plants, while a fibrous root system is an advantage to others?*

UP CLOSE SUGAR MAPLE

- **Scientific Name:** *Acer saccharum*
- **Range:** Northeastern United States and adjacent regions of southeastern Canada. Closely related species occur in the southeastern United States and in isolated areas throughout the Rocky Mountains, southwestern Oklahoma and Texas, and northern Mexico.
- **Habitat:** Northern temperate forests and canyons of southern mountains
- **Size:** Height—12 to 37 m (40 to 120 ft.), Canopy—up to 14 m (45 ft.) wide, Trunk—up to 1 m (3 ft.) in diameter
- **Importance:** Sugar maples are among the most commercially valuable trees in North America. They have high-quality, hard wood that is used to make furniture, musical instruments, kitchenware, and flooring. Maple sap is made into maple syrup and maple sugar.

External Structures

Seeds and fruits Seeds mature inside of winged fruits called samaras. The wings enable the fruits to carry seeds away from their parent tree on the wind.

Samaras

Summer

Flowers Maple flowers appear before the leaves in early spring. The flowers, which are arranged in small clusters, lack petals and thus are incomplete. Most are imperfect (either male or female) as well.

Spring

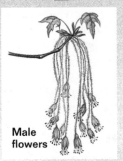

Male flowers

Winter

Leaves The blades of sugar maple leaves are palmately lobed (main veins branch from a single point, like fingers from the palm of a hand) and range from 7.5 to 15 cm (3 to 6 in.) across. Most leaves, which are opposite (attached in pairs), have five sharply toothed lobes. Spring and summer foliage (leaves) is light green. In fall, the foliage changes to bright yellow, orange, or red. Maples are deciduous, dropping their leaves in the fall.

Fall

Twigs and buds Sugar maple twigs are green at first, and then they change to reddish brown. They are also slender and glossy, with conspicuous, light-colored lenticels. The reddish brown buds are conical, pointed, and about 4 mm (0.2 in.) long. Maple twigs provide food for deer, moose, cottontail rabbits, and snowshoe hares.

Bark Young sugar maples have smooth, light-gray bark. The bark of older trees becomes dark gray or brown, rough, and deeply furrowed.

Internal Structures

Tissues All three tissue systems can be seen in a cross section of a maple leaf. Dermal tissue covers the leaf, while ground and vascular tissues are found inside the leaf. The ground tissue cells in a maple leaf contain chloroplasts, in which photosynthesis occurs. Xylem tissue delivers water and minerals from the roots to the leaves. Phloem tissue gathers sucrose (table sugar), the primary product of photosynthesis in maples, from surrounding leaf cells and transports it to other parts of the tree.

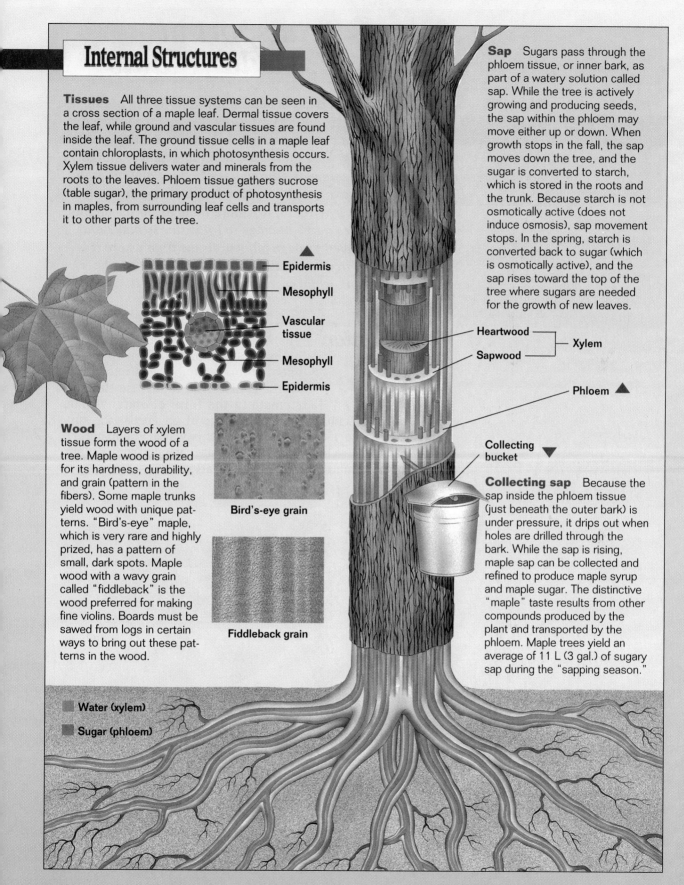

- Epidermis
- Mesophyll
- Vascular tissue
- Mesophyll
- Epidermis

Sap Sugars pass through the phloem tissue, or inner bark, as part of a watery solution called sap. While the tree is actively growing and producing seeds, the sap within the phloem may move either up or down. When growth stops in the fall, the sap moves down the tree, and the sugar is converted to starch, which is stored in the roots and the trunk. Because starch is not osmotically active (does not induce osmosis), sap movement stops. In the spring, starch is converted back to sugar (which is osmotically active), and the sap rises toward the top of the tree where sugars are needed for the growth of new leaves.

Heartwood ⎫
Sapwood ⎬ Xylem

Phloem ▲

Collecting bucket ▼

Wood Layers of xylem tissue form the wood of a tree. Maple wood is prized for its hardness, durability, and grain (pattern in the fibers). Some maple trunks yield wood with unique patterns. "Bird's-eye" maple, which is very rare and highly prized, has a pattern of small, dark spots. Maple wood with a wavy grain called "fiddleback" is the wood preferred for making fine violins. Boards must be sawed from logs in certain ways to bring out these patterns in the wood.

Bird's-eye grain

Fiddleback grain

Collecting sap Because the sap inside the phloem tissue (just beneath the outer bark) is under pressure, it drips out when holes are drilled through the bark. While the sap is rising, maple sap can be collected and refined to produce maple syrup and maple sugar. The distinctive "maple" taste results from other compounds produced by the plant and transported by the phloem. Maple trees yield an average of 11 L (3 gal.) of sugary sap during the "sapping season."

■ Water (xylem)

■ Sugar (phloem)

24-2 Transport in Vascular Plants

Section Objectives

- Explain how transpiration helps move water up a plant.
- Relate the cohesive and adhesive properties of water to its movement in a plant.
- Explain how guard cells regulate the rate of transpiration.
- Describe the process by which sugars are translocated throughout the body of a plant.

The body of a plant is beautifully adapted for obtaining and distributing materials. The sugar produced in the leaves travels through the phloem to other parts of the plant body, where it is used in metabolism or stored. Water and nutrients move through the body of a plant, from its roots to its leaves, within xylem tissue. However, many large trees have leaves that are more than 10 stories off the ground. How does a tree manage to lift water so high? And how is sugar moved through phloem tissue from where it is made or stored to where it is needed? Simply put, water is pulled up a plant, while sugar is pushed through it.

Transpiration Pulls Water Up a Plant

The leaves of plants have many tiny holes—the stomata. When they are open, stomata enable gas exchange. Water is also free to diffuse through stomata in the form of water vapor. As Figure 24-10 indicates, the passage of air across the surface of a leaf carries away much of this water vapor before it can reenter the leaf. The loss of water vapor from a plant through its stomata is called **transpiration**.

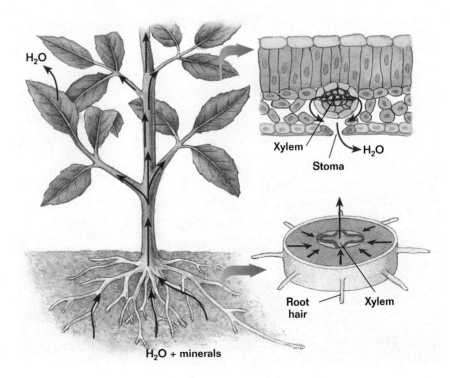

Xylem

H₂O

Stoma

Root hair   Xylem

H₂O + minerals

Figure 24-10 Transpiration is the loss of water from a plant. Most of this water loss is from the leaves. The water vapor that exits a leaf by transpiration first entered the plant through its roots and traveled to the leaves through the xylem.

Water is pulled up a plant because the loss of water by transpiration creates a suction that draws water out of the tracheids and vessels of the plant's xylem. This water extends in an unbroken column down through the stems and into the roots, where water is absorbed from the soil. As long as the column of water in the xylem remains unbroken, water will continue to move upward because of the pull of transpiration. More than 90 percent of the water taken in by the roots of a typical plant is ultimately lost through the plant's leaves in this way.

According to the **tension-cohesion theory,** two properties of water itself and a simple element of plant structure assist the pull of transpiration in moving water up a plant. Recall from Chapter 2 that water is a polar substance. As a result, water molecules readily form hydrogen bonds that enable them to stick to each other (cohere) and stick to other polar substances (adhere). The cohesion of water molecules, illustrated in Figure 24-11, gives a column of water great tensile strength. In other words, it can withstand a lot of tension (pull) without breaking. Thus, cohesion of water molecules helps to maintain an unbroken column of water in xylem tissue. Because of adhesion, water is able to move up the sides of a narrow tube by capillary action. Water absorbed by the roots of a plant moves through xylem cells that are elongated and very narrow, like straws. Thus, the adhesion of water molecules to the walls of the very narrow xylem cells helps to draw water to the top of a plant. ◻

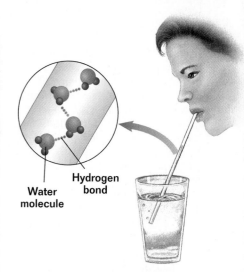

Figure 24-11 Water molecules are held together so strongly by hydrogen bonds that a column of water can withstand great tension without breaking.

◻ CAPSULE SUMMARY

Transpiration is the loss of water from a plant through its stomata. The cohesion and adhesion of water molecules help water to move up a plant in an unbroken column.

Guard Cells Regulate the Rate of Transpiration

The rate of transpiration must be regulated so that a plant does not lose too much water. Water loss by transpiration can be prevented only by the closing of a plant's stomata. However, stomata must be open at least part of the time so that the carbon dioxide needed for photosynthesis can enter the plant. Therefore, every plant must strike a balance between the conflicting demands of water conservation and photosynthesis.

A stoma is surrounded by a pair of guard cells that are shaped like two cupped hands. The stoma opens and closes because of changes in the water pressure within the guard cells. When guard cells take in water by osmosis, they become turgid (plump and swollen) and bowed in shape. The bowed shape results because the inner wall of a guard cell is thicker than the rest of the cell wall and cannot stretch when the guard cell swells. If you were to use a piece of tape to thicken one side of a long balloon, it would also bend when inflated. As the inner walls of a pair of guard cells separate, the stoma opens. When water leaves the guard cells, they lose turgor, their inner walls come back together, and the

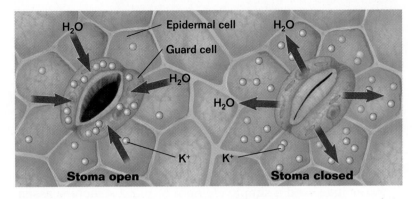

Figure 24-12 When potassium ions enter a pair of guard cells, the cells take in water by osmosis. The thickened inner walls of the guard cells enable a stoma to open when this water causes the cells to swell, *left*. When potassium ions leave the guard cells, water also leaves by osmosis, and the guard cells relax, closing the stoma, *right*. Blue arrows indicate water movement.

Epidermal cell

Guard cell

H₂O

H₂O

H₂O

H₂O

K⁺

K⁺

Stoma open

Stoma closed

stoma closes, as illustrated in Figure 24-12. Thus, loss of water from the guard cells for any reason causes stomata to close and stops further water loss. This is homeostasis in action.

Although the exact mechanism is not well understood, potassium ions, K⁺, play a critical role in opening and closing stomata. An active transport process that is triggered by light causes potassium ions to move into guard cells. The increased concentration of potassium ions inside the cells causes water to enter them by osmosis. When the movement of potassium ions is reversed and the potassium ion concentration becomes high in the surrounding cells, water diffuses out of the guard cells. The resulting loss of turgor by the guard cells causes the stoma to close. □

Sugar Is Pushed Through a Plant

In a plant, sugar moves from where it is made or stored to where it is needed through the sieve tubes of phloem. Botanists use the term **source** to refer to a part of a plant that provides sugar for other parts of the plant. For example, a leaf is a source because it makes sugar during photosynthesis. A root is also a source when sugar stored there is moved to other parts of the plant. Botanists use the term **sink** to refer to a part of a plant to which sugar is delivered. Areas of active growth where sugar is needed for metabolism, such as root tips and developing fruits, are examples of sinks. The movement of sugar within a plant from a source to a sink is called **translocation**. Look back at *Up Close: Sugar Maple* on pages 560–561 to learn more about how sugars move in a plant.

The movement of sugar in a plant is more complex than the movement of water. First, water flows freely through empty xylem elements, but sugar must pass through the cytoplasm of living cells. Second, water only moves upward within the xylem, while sugar moves both upward and downward in the same sieve tube, but at different times. Last, water diffuses freely through a plasma membrane, but sugar cannot. How, then, is sugar distributed throughout a plant? Many attempts have been made to answer this question. □

The model of translocation that most botanists favor was proposed in 1924 by the German botanist Ernst Münch. Münch's model was first tested by an experiment like the one seen in Figure 24-13. Using a concentrated sugar solution to represent the phloem near a source, water can be made to enter a tube and flow through it. The pressure created by water entering the tube pushes the water and some of the sugar in the solution to the other end of the tube. Therefore, Münch's model of translocation is often called the **pressure-flow model.** Once the tube is completely filled with water, sugar is also able to diffuse from one end to the other, as long as the sugar concentration remains higher at one end than the other. To see how Münch's model is thought to work in a plant, look at Figure 24-14.

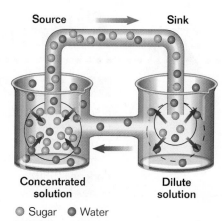

Source → Sink

Concentrated solution   Dilute solution

● Sugar ● Water

Figure 24-13 In a simple osmometer, *above,* the diffusion of water into a concentrated sugar solution pushes the sugar solution from the "source" to the "sink."

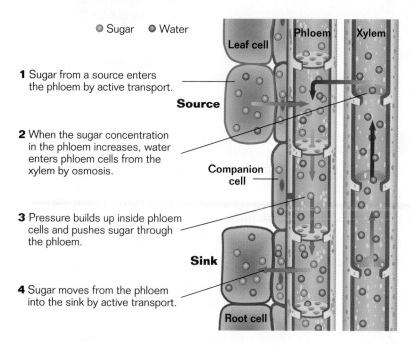

● Sugar ● Water

Leaf cell   Phloem   Xylem

1 Sugar from a source enters the phloem by active transport.

Source

2 When the sugar concentration in the phloem increases, water enters phloem cells from the xylem by osmosis.

Companion cell

3 Pressure builds up inside phloem cells and pushes sugar through the phloem.

Sink

4 Sugar moves from the phloem into the sink by active transport.

Root cell

Figure 24-14 According to Münch's hypothesis, sugars are pushed through the phloem by the pressure that results from the movement of water into the phloem by osmosis. Blue arrows indicate water movement, and red arrows indicate sugar movement.

Section Review

1. *What is transpiration, and how does it pull water up a plant?*
2. *What properties of water assist in its rise up a plant?*
3. *How do guard cells regulate the rate of transpiration?*
4. *What is translocation, and how does it occur?*

Critical Thinking

5. *Why might the lack of potassium, K, in the soil around a plant's roots interfere with the plant's ability to exchange gases and absorb water?*

adventitious root (559)
C_3 plant (556)
C_4 plant (556)
companion cell (554)
cork cell (558)
cortex (557)
dermal tissue system (551)
epidermis (552)
fibrous root system (559)
ground tissue system (551)
heartwood (557)
herbaceous stem (557)
lateral bud (557)
internode (557)

leaflet (555)
lenticel (558)
mesophyll (556)
node (557)
palisade layer (556)
petiole (555)
pith (557)
pressure-flow model (565)
root cap (558)
root hair (558)
sapwood (557)
sieve plate (554)
sieve tube (554)
sieve tube member (554)

sink (564)
source (564)
spongy layer (556)
taproot system (559)
tension-cohesion theory (563)
terminal bud (557)
tracheid (553)
translocation (564)
transpiration (562)
vascular bundle (557)
vascular tissue system (551)
vessel (553)
vessel element (553)
woody stem (557)

Review

Multiple Choice

1. Which of the following tissues transports water and carbohydrates through a plant?
   a. dermal
   b. cork
   c. vascular
   d. mesophyll

2. Which of the following is *not* a function of ground tissue in plants?
   a. storage
   b. support
   c. photosynthesis
   d. protection

3. The dermal tissue system functions in
   a. transport of water.
   b. transport of carbohydrates.
   c. protection.
   d. sexual reproduction.

4. Xylem cells that no longer conduct water make up the
   a. pith.
   b. sapwood.
   c. cortex.
   d. heartwood.

5. Which of the following is a modified stem?
   a. carrot
   b. Irish potato
   c. pine needle
   d. cactus spine

6. Transpiration rate is regulated by
   a. stomata.
   b. tracheids.
   c. pressure-flow.
   d. movement from source to sink.

7. Which of the following is *not* a characteristic of sugar maple trees?
   a. large, brightly colored flowers
   b. winged fruits called samaras
   c. five-pointed, palmately lobed leaves
   d. glossy, reddish brown twigs

8. Capillary action in the xylem of plants is made possible by
   a. pressure-flow.
   b. hydrogen bonding.
   c. low surface tension.
   d. high vapor pressure.

9. Guard cells become turgid and bow in shape when
   a. water moves out of the cells.
   b. potassium moves out of the cells.
   c. water moves into the cells.
   d. water levels are low in the cells.

10. According to the pressure-flow model, sugar moves through the phloem by
    a. osmosis, from areas of low sugar concentration to areas of high sugar concentration.
    b. pressure created when water moves by osmosis into areas of high sugar concentration.
    c. the diffusion of sugar molecules from sink to source.
    d. active transport of water from companion cells in the leaves.

Completion

11. In vascular tissue, _____ cells are dead at maturity, but _____ cells are living.

12. Phloem contains _____ members through which carbohydrates flow and _____ that control their activities.

13. In some leaves, the _____ includes two distinct layers of photosynthetic cells. The cells of the _____ layer are tightly packed and columnar. The _____ layer consists of loosely packed, spherical cells.

14. The movement of the products of photosynthesis within a plant is called _____. Water moves up a plant because of _____, the loss of water from leaves.

15. Because water is a _____ molecule, its movement in xylem is aided by adhesion and _____.

16. The guard cells regulate _____ loss in plants. The rate of transpiration increases when stomata are _____ and decreases when stomata are _____.

17. According to the pressure-flow model, sugars travel from a _____ to a _____.

Themes Review

18. **Structure and Function** Suppose that you are a water molecule passing through a plant. Trace your path from your entry into a root hair to your exit between two guard cells. Identify each structure that you encounter, and describe its function.

19. **Homeostasis** When a plant wilts, its stomata close. How, then, does wilting help a plant maintain homeostasis?

Critical Thinking

20. **Making Inferences** Compare taproot and fibrous root systems in Figure 24-9 on page 559. Would plants with taproot systems or plants with fibrous root systems be more likely to prevent erosion on a steep hillside? Explain.

21. **Interpreting Data** The rate of water movement in a plant is assumed to indicate the rate of transpiration. After measuring the rate of water movement in a plant during high and low humidity, the data was graphed, below. Which line, A or B, indicates a lower transpiration rate? Justify your answer.

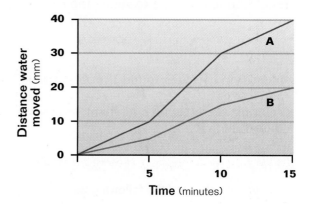

22. **Making Inferences** In the graph above, which line, A or B, indicates the rate of transpiration during high humidity? Which indicates the rate of transpiration during low humidity? Explain.

Activities and Projects

23. **Cooperative Group Project** Obtain a Venus' flytrap, a pitcher plant, and a sundew plant from a nursery or a biological supply company. Observe each plant to find out how its leaves trap insects. Make a videotape of the plants capturing insects. Report your findings to your class, using the videotape to enhance your report.

24. **Cooperative Group Project** Build clay models of cross sections of a C_3 and a C_4 leaf. Show the epidermis, mesophyll, veins, and guard cells of each. Using the models, explain to your class each leaf's structure and the function of its parts.

25. **Research and Writing** Research how sugar is obtained from plants. Relate your findings in a written report.

Roots, Stems, and Leaves

OBJECTIVE

Observe the tissues that make up roots, stems, and leaves, and examine their structure.

PROCESS SKILLS

- identifying tissues found in roots, stems, and leaves
- relating the structure of plant tissues to their function

MATERIALS

- Prepared slides of the following tissues:
  - *Allium* root tip
  - *Ranunculus* root cross section
  - *Zea mays* stem cross section
  - *Ranunculus* stem cross section
  - Lilac leaf cross section
- Compound light microscope

BACKGROUND

1. Which plant tissues are responsible for the absorption of water and dissolved minerals?

2. How is food, produced in the leaf, moved to other parts of the plant?

3. How do woody and herbaceous stems compare?

4. What tissues are continuous in the root, stem, and leaf?

5. How does the leaf conserve water?

6. Write your own **Focus Question** on your Vee Form.

7. **Knowing Side of the Vee** List the **Concepts** and new **Vocabulary Words** on your Vee Form. In the **Concept Statements** section of the Vee, use these words in sentences that define and explain them.

TECHNIQUE

Doing Side of the Vee

Part A: Roots

1. Study the cross section of the root tip shown in the photograph in Figure 24-9 on page 559.

2. Observe the prepared slide of the *Allium* root tip under low power. Locate the root cap and the root tip meristematic cells. Note the long root hairs in the area above the root tip.

3. Change slides to the *Ranunculus* cross section. The inner core is the vascular tissue, which is surrounded by the endodermis. This area is the tissue for transport of water, minerals, and food. Look for the star-shaped xylem and the smaller phloem cells surrounding the xylem. Draw what you see, and identify the tissues in the **Records** section of your Vee Form.

4. Locate the cortex, where starch is stored, surrounding the vascular cylinder. Outside the cortex you will find the epidermal cells and their root hairs. Draw a one-fourth section of the root tissues, as if you were cutting a pizza slice out of the cross section. Label all the tissues in the **Records** section of your Vee Form.

Part B: Stems

5. Observe a prepared slide of the stem of *Zea mays*, a monocot, and find the epidermis and the photosynthetic layer. In the center, look for the vascular bundles made up of xylem and phloem. Draw a diagram showing the location of the vascular bundles and the epidermis layer as they appear when viewed under low power.

6. Switch to high power and observe a vascular bundle. Draw the vascular bundle and label the tissues.

7. Observe a cross section of a herbaceous dicot stem such as that of *Ranunculus*. Look for the epidermis and cortex layers. Notice that the stem is more complex than a dicot root. Note the arrangement of the vascular bundles. In the **Records** section of the Vee Form, draw what you observed.

8. Now focus on a vascular bundle under high power. Draw and label a diagram in the **Records** section.

Part C: Leaves

9. Observe a prepared slide of a lilac leaf cross section under low power, and find the lower epidermis.

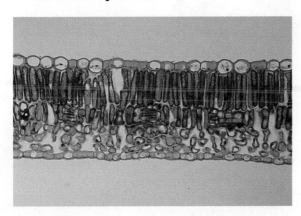

10. Identify the stomata on the lower epidermis. Find the guard cells that open and close a particular stoma. Locate an open stoma and a closed stoma. Draw and label diagrams of the stomata and guard cells in the **Additional Records** section of your Vee.

11. Note the spongy texture of mesophyll. Locate a vein containing xylem and phloem. Continue toward the palisade layer into the upper epidermis until you reach the clear continuous noncellular layer on top. This layer is called cutin. Draw and label a diagram of your observations in the **Additional Records** section of the Vee. In the **Procedure** section of the Vee, briefly summarize the procedure you followed.

INQUIRY

1. In the root, where are phloem and xylem located?

2. Where are the xylem and phloem found in the herbaceous stem?

3. How are the vascular bundles different in monocot and dicot stems?

4. Use the information on the **Knowing Side** of the Vee to interpret your results from the **Doing Side,** and then write your **Knowledge Claim.** Write a **Value Claim** for this lab.

ANALYSIS

1. How are the root cap cells different from the root tip meristematic cells?

2. What is the function of the root hairs?

3. How different is the arrangement of xylem and phloem in roots, stems, and leaves?

4. What is the function of a stoma?

5. What is the function of the air space in the mesophyll of the leaf?

6. Which leaf structures help to conserve water?

7. Which tissues of the leaf are continuous with the stem and root tissues? How is this functional?

FURTHER INQUIRY

Write a **New Focus Question** that could be the point of a new investigation. The following is an example:

How have the stems and roots of some plants become modified for storing sugars and starches?

CHAPTER 25

PLANT GROWTH AND DEVELOPMENT

REVIEW

- requirements for photosynthesis and cellular respiration (Section 5-1)
- meristem, vascular tissue, xylem, and phloem (Section 23-1)
- seeds (Section 23-2)
- endosperm and double fertilization (Section 23-3)
- cotyledons, monocots, and dicots (Section 23-3)
- stem and root structure (Section 24-1)

Bristlecone pine "flagged" by the wind

25-1 How Plants Grow and Develop

A *seed sprouts with a burst of growth in response to certain changes in the environment. These changes, such as warming temperatures and increasing soil moisture, signal the start of a season of favorable growing conditions. As a new plant grows (increases in size), the various tissues and organs of its body develop (differentiate and take form). All seed plants share certain fundamental patterns of growth and development.*

Growth and Development Begin as a Seed Forms

You learned in Chapter 23 that a new flowering plant begins its life after double fertilization occurs within the ovary of a flower. The resulting zygote and endosperm cell divide by mitosis, forming an embryo and endosperm. Then, a significant event occurs—the layers of protective tissue surrounding the embryo and endosperm toughen and become impermeable to both water and oxygen. Denied water and oxygen, the embryo stops growing, and a mature seed forms.

Compare the structures of two familiar types of seeds—a bean seed and a corn seed—illustrated in Figure 25-1. As you can see, most of the interior of a bean seed is occupied by two large, fleshy cotyledons. Before a bean seed fully matures, the endosperm is consumed by the embryo. The embryo's energy reserves are now stored in its cotyledons. Much of the interior of a corn seed, which has only one cotyledon, is occupied by endosperm. In both bean and corn seeds, the cotyledons are attached to the embryo, which has begun to develop specialized tissues. The embryo's shoot develops above the cotyledons and consists of an embryonic stem and tiny embryonic leaves. The embryo's root develops below the cotyledons.

Figure 25-1 A bean and a kernel of corn are two representative types of seeds that vary somewhat in structure.

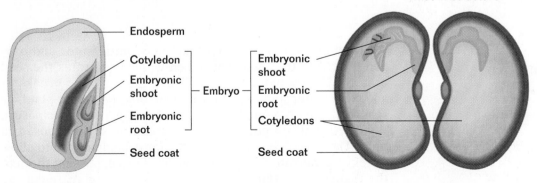

Corn Seed: Endosperm, Cotyledon, Embryonic shoot, Embryonic root, Seed coat, Embryo

Bean Seed: Embryonic shoot, Embryonic root, Cotyledons, Seed coat

Corn Seed

Bean Seed

Figure 25-2 The events in the germination of a corn seed, *below left,* differ from those of a bean seed, *below right.* As the corn shoot elongates and emerges from the soil, its leaves push through their protective cover, unfurl, and start to photosynthesize. Once the hook of a bean seedling breaks through the soil, it straightens out and pulls the cotyledons and leaves up into the air. As the food stored in a bean's cotyledons is used up, the cotyledons shrivel and eventually fall off.

Growth Continues When a Seed Germinates

The **germination** of a seed is the resumption of growth by a plant embryo. The first visible evidence that a seed is germinating is the emergence of the embryo's root. What happens next varies somewhat from one type of plant to another, as you can see in Figure 25-2. For example, the cotyledons of some plants emerge with the shoot, while the cotyledons of other plants remain underground. The embryonic shoots of many plants, such as the common bean *(Phaseolus vulgaris),* curve until they are shaped like a hook. This hook enables the shoots to push through the soil and prevents the tips of the shoots from being damaged. The embryonic shoots of other plants, such as corn *(Zea mays),* do not curve but are surrounded by a protective sheath.

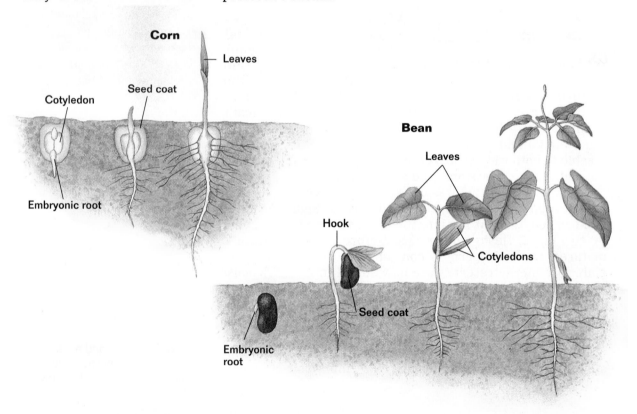

Corn

Leaves

Cotyledon

Seed coat

Embryonic root

Bean

Leaves

Hook

Cotyledons

Seed coat

Embryonic root

Seed germination cannot take place until water and oxygen penetrate the seed coat. When water penetrates the seed coat, the tissues in the seed swell, and the seed coat breaks. If adequate water and oxygen are available, the young plant, or **seedling,** will continue to grow. Many seeds require exposure to heat or cold to germinate, and some must be exposed to light. The seed coats of other seeds must be damaged to allow water to penetrate the seed. Exposure to fire, passing through the digestive system of an animal, and falling on rocks are several natural ways that seed coats are damaged. ◻

Apical Meristems Produce Primary Growth

All plants grow by cell division that occurs in meristems located at the tips of their shoots and roots. To better understand this pattern of growth, imagine a stack of dishes—the stack can get taller but not wider. Growth that occurs from the formation of new cells at the tips of a plant is called **primary growth.** The primary growth of a seedling occurs in **apical** *(AP ih kuhl)* **meristems.** As Figure 25-3 indicates, apical meristems are located at the tips of stems and at the tips of roots, just behind the root cap. The plant tissues that result from primary growth are called **primary tissues.**

Figure 25-3 A cross section of the shoot tip of a coleus plant, *below left,* shows the shoot apical meristem as well as young leaves and buds. A cross section of the root tip of a radish plant, *below,* shows the root apical meristem and the root cap.

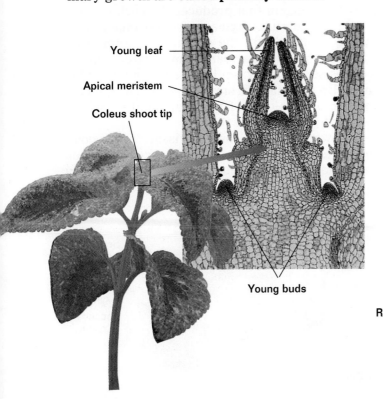

Young leaf

Apical meristem

Coleus shoot tip

Young buds

Radish root tip

Apical meristem

Root cap

During periods of growth, the cells of apical meristems divide and continually add more cells to the tips of a seedling's body. Thus, the seedling's stems and roots lengthen. After new cells are formed, they grow and undergo **differentiation,** the process by which cells become specialized in form and function. The new cells produced by the apical meristem of a stem become part of the primary dermal, ground, and vascular tissues of the lengthening stem and its young leaves. Some of the new cells produced by a root apical meristem elongate and become the primary dermal, ground, and vascular tissues of the lengthening root. Cells produced by the root apical meristem also become part of the root cap, replacing root cap cells that are constantly worn away as the root pushes through the soil. ◻

◻ CAPSULE SUMMARY

Apical meristems located at the tips of shoots and roots produce primary growth. Tissues that result from primary growth are called primary tissues.

Lateral Meristems Produce Secondary Growth

••••••••••••••••••••••••••••••••••

Many vascular plants also increase in width as they grow taller. Growth that causes a plant to increase in width is called **secondary growth.** The tissues that develop as a result of secondary growth are referred to as **secondary tissues.** Although secondary growth occurs in many non-woody plants, its effects are most dramatic in woody plants.

Woody plants have two cylinders of actively dividing cells called lateral meristems. Cell division in **lateral meristems** adds layers of new cells around the outside of a plant's body. Within the bark of a woody stem is the **cork cambium** (*KAM bee uhm*), a lateral meristem that produces the cork cells of the outer bark. Cork cambium originates from primary stem tissues that were produced by the apical meristem. Just beneath the bark is the **vascular cambium,** a lateral meristem that produces secondary vascular tissue. The vascular cambium, which also originates from primary tissues, forms between the xylem and phloem of vascular bundles, which are also the result of primary growth.

Figure 25-4 shows how the secondary tissues of a woody stem develop. Notice that the secondary vascular tissues form on opposite sides of the vascular cambium. Secondary

Figure 25-4 A woody stem first forms by primary growth, *below.* After the vascular cambium forms from primary stem tissues between the primary xylem and phloem, secondary growth begins, *below right.* Layers of secondary xylem and phloem form between the primary xylem and phloem, and the stem grows in diameter, *below center.* The cork cambium forms when the epidermis is stretched and broken as the stem thickens and produces cork cells that become part of the bark.

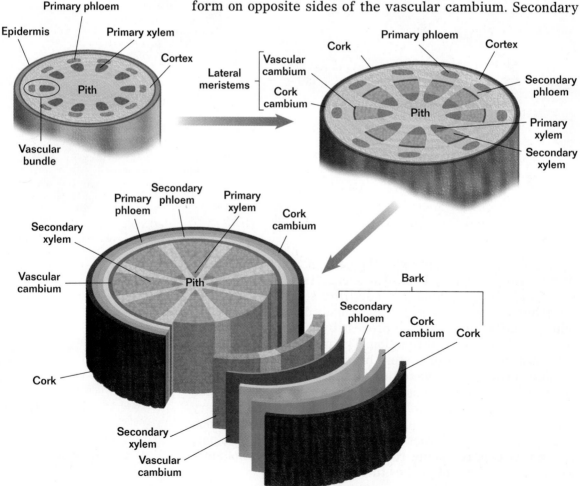

xylem forms in thick layers on the inner side of the vascular cambium and is the main component of wood. Secondary phloem forms on the outer side of the vascular cambium and becomes part of the inner bark. Thus, secondary phloem, which transports sugars from the leaves to the roots, is very close to the outer surface of a woody stem. Removing the bark of a tree damages the phloem and both cambia and may eventually kill the tree. ◻

Plant Development Is Continuous and "Reversible"

Although both plants and animals develop according to a genetic blueprint, their patterns of development are very different. The seedling that emerges from a germinating seed is a small version of an adult plant. As the plant grows, new cells are continuously produced in its apical and lateral meristems. These cells differentiate and replace or add to existing tissues, as seen in Figure 25-5. Thus, a plant continues to develop throughout its life. This developmental pattern is very different from that of animals, which develop until reaching a certain point at which they are considered to be adults.

Plant development and animal development also differ in how they are affected by the environment. The outcome of plant development is greatly influenced by a plant's immediate surroundings. A plant cannot move about, so it is critical that during development the seedling adapt to the particular conditions that the mature plant will have to face. Plants are able to change when the local environment changes because they develop continuously. For example, many trees and shrubs stop growing and drop their leaves during periods of cold weather or drought. When conditions improve, these trees and shrubs produce new leaves and resume growth. Animals, which can move to new environments when conditions change, develop without being as strongly influenced by environment.

Unlike most animal cells, many plant cells retain the functional genetic instructions necessary to produce *all* of the tissues of a mature plant. A different part of the complete set of genetic instructions for an organism is utilized in the cells of each of its tissues. During animal development, critical sets of genes that control development are lost or "turned off" and may not be reactivated. In many plant tissues, however, the unused sets of genes are preserved and can be activated in the future. When separated from the body of a mature plant, the cells of certain plant tissues can begin to undergo cell division and form masses of undifferentiated cells. Thus, in a sense, their development has been reversed. The masses of unspecialized plant cells can later resume the process of differentiation and form all of the tissues of a mature plant.

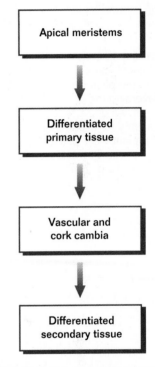

Figure 25-5 This graphic organizer shows the stages of plant cell differentiation. New cells form by cell division in the apical meristems. These cells differentiate to form primary tissues. In woody plants, the vascular and cork cambia differentiate from primary tissues and then give rise to secondary tissues.

The ability of specialized plant cells to reverse and resume development was first demonstrated in 1958 by the botanist F. C. Steward, whose famous experiment is illustrated in Figure 25-6. Techniques for growing pieces of living tissue in artificial media, such as those developed by Steward, are known as **tissue culture.** Today, tissue culture is an important tool in plant research and in the commercial propagation (reproduction) of plants such as orchids. Although many types of animal tissues are also grown in tissue cultures, most are unable to reverse and resume the process of development to produce a complete individual. To learn more about how plants can be propagated, look at *Up Close: Kalanchoë* on pages 578–579. ☐

Figure 25-6 F. C. Steward's experiments with carrots demonstrated that some plant cells are capable of reversing the process of development. All of the instructions necessary for producing a normal carrot plant were present in the phloem cells Steward extracted from a mature carrot.

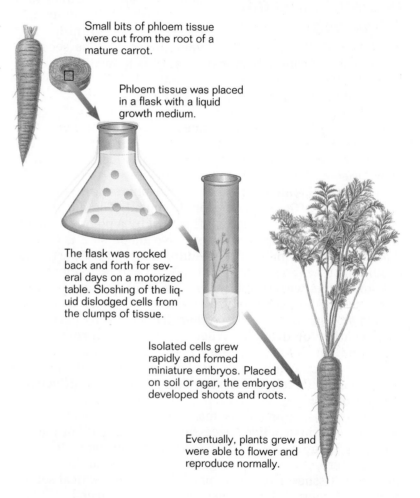

Small bits of phloem tissue were cut from the root of a mature carrot.

Phloem tissue was placed in a flask with a liquid growth medium.

The flask was rocked back and forth for several days on a motorized table. Sloshing of the liquid dislodged cells from the clumps of tissue.

Isolated cells grew rapidly and formed miniature embryos. Placed on soil or agar, the embryos developed shoots and roots.

Eventually, plants grew and were able to flower and reproduce normally.

How Long Does a Plant Live?

As you learned in Chapter 23, the oldest known trees are gymnosperms that are estimated to be about 5,000 years old. Some plants, however, live for only a few weeks. Depending on how long they live, all plants can be classified as one of three basic types: annuals, biennials, or perennials. Examples of each of these types of plants can be seen in Table 25-1.

1. **Annuals**   An **annual** plant completes its life cycle (grows, flowers, and produces fruits and seeds) within one growing season and then dies. Most annuals are non-woody, or herbaceous, plants that grow rapidly under favorable conditions and increase greatly in size if supplied with adequate water and nutrients. Plants that are annuals include sunflowers, dandelions, lupines, and many weeds.

2. **Biennials**   A **biennial** plant takes two years to complete its life cycle. During the first year, biennials produce a short stem and a rosette (circular cluster) of leaves. In the second year, the energy stored in the plant's roots and shoots is used to produce an elongated flowering stalk. After flowering and producing fruits and seeds, the plant dies. Biennial plants include yellow sweet clover, Queen Anne's lace (wild carrot), and parsley.

3. **Perennials**   A **perennial** plant lives for more than two years and may produce flowers, fruits, and seeds many times during its life. The majority of vascular plants are perennials, including many herbaceous plants and all woody plants. Herbaceous perennials include plants such as buttercups, morning glories, and evening primroses. The shoots of herbaceous perennials may die each year after a season of growth and food accumulation. Food for the next season's growth is stored in fleshy roots or underground stems. Woody perennials include trees (woody plants with one main stem), shrubs (woody plants with many stems), and many vines. Some woody perennials drop their leaves each year. Trees, shrubs, and woody vines that drop all of their leaves at the end of each growing season, such as elms, roses, and grapevines, are said to be **deciduous** (dee SIHJ oo uhs). Those that drop a few leaves at a time throughout the year, such as pines, junipers, and honeysuckles, are called **evergreens.** ❑

Table 25-1 Types of Plants

Type	Example
Annual	 Lupine
Biennial	 Yellow sweet clover
Herbaceous perennial	 Buttercup
Woody perennial	English ivy

❑ *CAPSULE SUMMARY*

Annuals complete their life cycle in one growing season. Biennials take two years to complete their life cycle. Perennials live for many years and may flower and produce seeds many times.

Section Review

1. *How does the germination and initial growth of a corn seedling differ from that of a bean seedling?*
2. *How is secondary growth produced?*
3. *How does plant development differ from animal development?*
4. *What is the difference between a biennial and a perennial?*
5. *Explain how a* Kalanchoë *can be propagated without seeds.*

Critical Thinking

6. *Why might being an annual be an advantage to a plant?*

UP CLOSE KALANCHOË

- **Scientific Name:** *Kalanchoë daigremontiana*
- **Range:** Native to southwestern Madagascar; cultivated worldwide as an indoor potted plant and as an outdoor perennial in warm climates
- **Habitat:** In nature, semiarid tropical grassland with well-drained, fertile soil and moist summers; in cultivation, warm, brightly lighted area with rich, well-drained soil
- **Size:** Height—30 cm to 1 m (1 ft. to 3 ft.)
- **Importance:** Kalanchoës *(kal an KOH eez)* are members of the Crassulaceae family, a group of succulent plants that are adapted to hot climates. They are valued as ornamentals for their colorful flowers, interesting foliage, and ease of cultivation and propagation.

External Structures

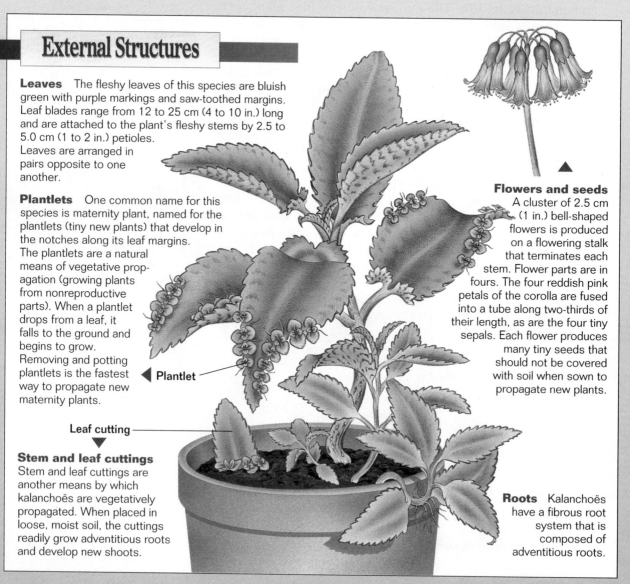

Leaves   The fleshy leaves of this species are bluish green with purple markings and saw-toothed margins. Leaf blades range from 12 to 25 cm (4 to 10 in.) long and are attached to the plant's fleshy stems by 2.5 to 5.0 cm (1 to 2 in.) petioles. Leaves are arranged in pairs opposite to one another.

Plantlets   One common name for this species is maternity plant, named for the plantlets (tiny new plants) that develop in the notches along its leaf margins. The plantlets are a natural means of vegetative propagation (growing plants from nonreproductive parts). When a plantlet drops from a leaf, it falls to the ground and begins to grow. Removing and potting plantlets is the fastest way to propagate new maternity plants.

◀ Plantlet

Leaf cutting
▼

Stem and leaf cuttings
Stem and leaf cuttings are another means by which kalanchoës are vegetatively propagated. When placed in loose, moist soil, the cuttings readily grow adventitious roots and develop new shoots.

Flowers and seeds
A cluster of 2.5 cm (1 in.) bell-shaped flowers is produced on a flowering stalk that terminates each stem. Flower parts are in fours. The four reddish pink petals of the corolla are fused into a tube along two-thirds of their length, as are the four tiny sepals. Each flower produces many tiny seeds that should not be covered with soil when sown to propagate new plants.

Roots   Kalanchoës have a fibrous root system that is composed of adventitious roots.

Internal Structures

Leaf structure Kalanchoës are succulents, which means that they have fleshy leaves and stems that store water. A look inside the leaf of a kalanchoë plant discloses how some succulents are adapted for conserving water. Notice that the epidermis consists of several layers of cells covered by a thick waxy cuticle. Relatively few, very small stomata dot the leaf surfaces. The leaf's mesophyll consists of uniformly large cells with little air space between them.

Vascular bundle

Stoma

Central vacuole

▼

Large central vacuole
A closer look at a mesophyll cell reveals that it has a large central vacuole that can hold a great deal of water.

Mesophyll

Epidermis

Cell wall

Organelles

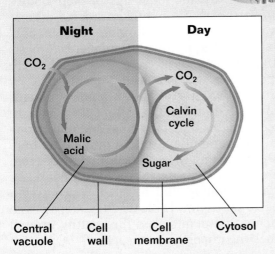

Night	Day

CO_2

CO_2

Calvin cycle

Malic acid

Sugar

Central vacuole

Cell wall

Cell membrane

Cytosol

CAM photosynthesis *K. daigremontiana*, like many other succulents, fixes carbon by a water-conserving pathway called crassulacean acid metabolism (CAM). Like C_4 plants, CAM plants fix carbon by first incorporating carbon dioxide into a compound called malic acid. Later, carbon dioxide is released from malic acid and used by the Calvin cycle to make sugar. Unlike C_4 plants, the two processes occur in the same cells but at different times of the day.

Unlike the stomata of other plants, the stomata of CAM plants open only at night. Thus, CAM plants make malic acid only at night. The malic acid is stored in the large central vacuoles of the mesophyll cells. A rising concentration of malic acid causes water to enter the vacuoles by osmosis, thus helping CAM plants to store water. During the day, malic acid diffuses out of the vacuoles, and the carbon dioxide released from it is fixed by the Calvin cycle.

25-2 Regulating Growth and Development

Section Objectives

■ *Identify basic requirements for healthy plant growth.*

■ *Explain why nutrients are important to a plant.*

■ *Name four types of plant hormones, and describe how each affects plant growth.*

■ *Describe how plant growth is affected by environment.*

Growing plants is more than just a popular hobby—it is essential to human survival. Plants provide food, medicines, clothing, and building materials, and they also brighten our lives, as Figure 25-7 illustrates. Success in growing plants—whether on an apartment windowsill, in a home garden, or in a wheat field—requires an understanding of the many factors that affect plant growth.

Figure 25-7 Many people enjoy gardening. Here, a gardener proudly displays a cabbage she has just harvested from her vegetable garden.

Figure 25-8 Three numbers on the front of a fertilizer bag indicate the percentage of the plant nutrients nitrogen, phosphorus, and potassium found in the fertilizer. This bag contains several other nutrients as well.

Nutrients Are Needed for Plant Growth

Multicelled organisms such as plants increase in size by adding new cells through cell division. Thus, to grow and develop, a plant must have a steady supply of the raw materials needed for building and operating new cells. Most plants are able to make all of the organic materials in their bodies because of photosynthesis. To photosynthesize, plants require light, water, and carbon dioxide. The carbon, hydrogen, and oxygen incorporated during photosynthesis make up more than 90 percent of the dry weight (weight after water has been removed) of a healthy plant.

However, the carbohydrates that a plant produces during photosynthesis do not satisfy all of its needs. In order to produce proteins, nucleic acids, and other molecules needed for metabolism, many other elements are required. Plants obtain these elements mostly from mineral **nutrients** that they extract from the soil. As Figure 25-8 shows, commercial fertilizers are also a source of mineral nutrients. Table 25-2 lists five of the many nutrients that plants need for healthy growth and describes the importance of each.

Table 25-2 Five Plant Nutrients and Their Importance

Nutrient	Importance
Nitrogen (N)	Part of all proteins, nucleic acids, chlorophylls, and coenzymes
Phosphorus (P)	Part of ATP, ADP, nucleic acids, phospholipids of plasma membranes, and some coenzymes
Potassium (K)	Needed to perform active transport, activate enzymes, regulate osmotic balance, and open stomata
Magnesium (Mg)	Part of chlorophyll; needed for photosynthesis and activation of enzymes
Sulfur (S)	Part of some proteins and coenzyme A; needed for cellular respiration

Plants also require oxygen for cellular respiration. While the green parts of a plant produce oxygen during photosynthesis, most of the oxygen needed by these parts comes from the atmosphere. A plant's roots, which do not usually conduct oxygen-producing photosynthesis, obtain oxygen from air in the spaces between soil particles. If the soil in which a plant is growing becomes compacted or saturated with water, there may not be enough oxygen available for its roots, and the plant could die. Figure 25-9 shows how one plant has adapted to living in places where little oxygen is present in the soil. ◻

◻ CAPSULE SUMMARY

Plants are able to produce all of their organic molecules as a result of photosynthesis, but they require several inorganic nutrients to convert carbohydrates into other important organic molecules.

Hormones Control Plant Growth and Development

In 1851, the great biologist Charles Darwin and his son Francis published a book called *The Power of Movement in Plants*. In the book, they reported the results of their experiments on how plants grow toward light. Young grass seedlings bend strongly toward a light source as their shoots elongate. The Darwins found that if they covered the tip of a seedling with material that prevented light from reaching the tip, the shoot would not bend. If they instead covered the tip of a seedling with a cap of gelatin that enabled light to reach the shoot, the shoot would bend. Charles and Francis Darwin concluded that some "influence" arose in the tip of the shoot and was transmitted downward, causing the shoot to bend.

With experiments conducted in the 1920s, the Dutch plant physiologist Frits Went showed that a chemical is the "influence" responsible for making plants bend toward light.

Figure 25-9 Mangroves, which grow in swamps, produce "air roots," which emerge from water and gather oxygen from the air.

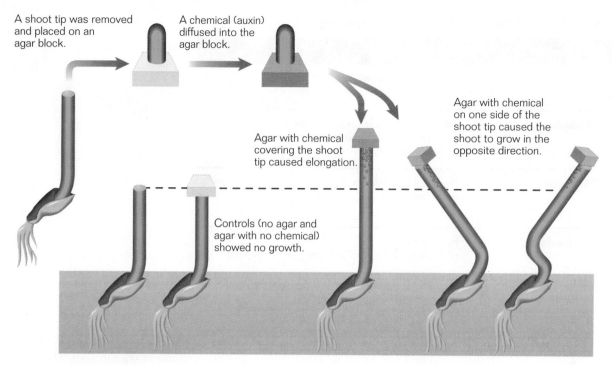

A shoot tip was removed and placed on an agar block.

A chemical (auxin) diffused into the agar block.

Agar with chemical on one side of the shoot tip caused the shoot to grow in the opposite direction.

Agar with chemical covering the shoot tip caused elongation.

Controls (no agar and agar with no chemical) showed no growth.

Figure 25-10 Frits Went's experiments, *above*, showed that a chemical (auxin) produced in the tips of oat (*Avena* sp.) seedlings causes shoots to grow toward light.

Figure 25-11 These coleus plants illustrate the effects of apical dominance and the removal of terminal buds. The terminal buds were not removed from the plant on the left, and thus it is taller and not as bushy as the plant on the right, from which the terminal buds were removed so that lateral buds could grow.

Went's experiments are described in Figure 25-10. Went named the growth-promoting chemical **auxin** *(AWK sihn)*. The word *auxin* comes from the Greek word *auxein,* meaning "to increase." Auxin causes the elongation of plant cells by increasing the plasticity of cell walls and enabling them to stretch during active cell growth. Influenced by light, auxin migrates from the lighted side of a stem to the darkened side. As a result of being exposed to a higher concentration of auxin, the stem cells on the darkened side of a seedling elongate faster than the cells on the lighted side. This differential growth causes the stem to grow toward light.

Auxin is one of many growth-regulating chemicals that act as hormones. A **hormone** is a chemical that is produced in one part of an organism and then transported to another part of the organism, where it brings about a response. The word *hormone* comes from the Greek word *horman,* meaning "to set in motion." Your body uses hormones to regulate growth and many other activities. Some of the hormones in your body are produced by specialized glands. In plants, however, all hormones are produced in various tissues throughout the plant. Four major types of plant hormones that have been identified by biologists are discussed below.

1. **Auxin** Auxin is produced in the tips of stems. In addition to promoting stem elongation, auxin stimulates fruit development, suppresses leaf and fruit drop, and inhibits lateral bud growth. The inhibition of lateral bud growth by auxin is called **apical dominance.** Removal of a terminal bud enables lateral buds to grow. Gardeners use this knowledge to shape the growth of their plants, as Figure 25-11 shows. Synthetic auxins are used in agriculture to control

leaf, flower, and fruit drop and are a component of some herbicides (chemicals that kill weeds).

2. **Cytokinins** The **cytokinins** *(seye toh KEYE nihnz)* are a group of chemicals that are produced in root tips. They stimulate cell division, promote lateral bud growth, and inhibit leaf drop. Synthetic cytokinins are used in tissue cultures to stimulate growth and in agriculture to break apical dominance.

3. **Ethylene** The simple gaseous compound **ethylene** is produced in most tissues of the plant body. It stimulates fruit ripening, promotes leaf, flower, and fruit drop, and retards lateral bud growth. Ethylene is used extensively in agriculture to ripen fruits that are harvested before they ripen naturally and to loosen fruits so that they drop more readily for machine harvest.

4. **Gibberellins** The **gibberellins** *(jihb uhr EHL ihnz)* are a group of chemicals that are produced in developing shoots and seeds. They were first discovered by Japanese scientists studying an abnormal elongation of rice seedlings, called "foolish seedling disease," and then were named for the fungus that causes the disease—*Gibberella.* Gibberellins cause stem elongation by stimulating cell division and elongation. They also induce seed germination and fruit development. Synthetic gibberellins are used in agriculture to produce seedless fruits, as seen in Figure 25-12, and to promote uniform seed germination. ◻

Figure 25-12 Seedless white table grapes are produced by treatment with gibberellin. Compare a cluster of treated grapes, *bottom right,* with a cluster of untreated grapes, *top left,* and you will notice another effect of gibberellins on plants.

◻ **CAPSULE SUMMARY**

Hormones, chemicals produced in one part of an organism and transported to another part of the organism, regulate plant growth and development.

Plant Growth Responds to Environmental Factors

Because plants are anchored in one spot, their growth must be adjusted continually in response to changes in the environment. In most cases, a plant's responses to environmental stimuli are triggered by the hormones that regulate plant growth.

Tropisms

A **tropism** *(TROH pihz uhm)* is a growth response in which the direction of growth is determined by the direction from which the stimulus comes. Auxins are responsible for producing tropisms. If the response is toward a stimulus, it is called a positive tropism. If the response is away from a stimulus, it is called a negative tropism. Three common stimuli to which plants respond are light, gravity, and touch. **Phototropisms** are growth responses to light. The growth of a plant's shoots toward light is a positive phototropism, while the growth of its roots away from light is a negative phototropism. **Gravitropisms** are growth responses to gravity. The upward growth of a germinating seed's shoot is a negative gravitropism, while the downward growth of its root is a

Figure 25-13 The coiling of grapevine tendrils, *above,* is a thigmotropism. The bending of an amaryllis toward a window, *above center,* is a positive photo-tropism. In germinating corn seedlings, *above right,* the upward growth of shoots is a negative gravitropism, and the downward growth of roots is a positive gravitropism.

positive gravitropism. **Thigmotropisms,** from the Greek word *thigma,* meaning "touch," are growth responses to touch. Effects of tropisms can be seen in Figure 25-13.

Photoperiodism

Have you ever wondered why certain plants bloom in the spring and others bloom in the summer or fall? Changes in the length of days (and nights) are responsible for the seasonal patterns of flowering in plants and also for many other seasonal patterns of plant growth and development. The response of a plant to the length of days and nights, or day length, is called **photoperiodism.** Plants can be categorized as one of three types, depending on how they respond to day length. If a response such as flowering occurs when days

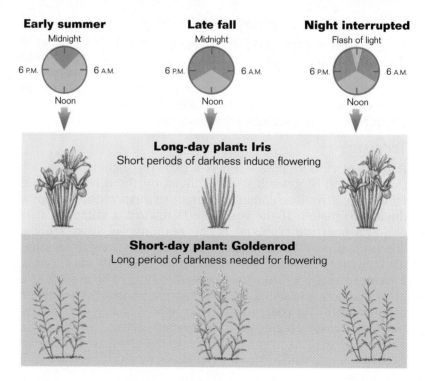

Figure 25-14 Day length (actually night length) controls flowering in many plants. Long-day plants flower when nights are short, and short-day plants flower when nights are long.

become shorter than a critical length, the plant is said to be a **short-day plant.** If the response occurs when days become longer than a critical length, the plant is said to be a **long-day plant. Day-neutral plants** are plants that are not affected by day length. However, as Figure 25-14 shows, it is not really the *day length* that initiates a photoperiodic response—it is the length of the nights. Knowledge of photoperiodism is very important to commercial flower growers. By artificially controlling the length of days and nights in greenhouses, plants such as poinsettias, Easter lilies, and chrysanthemums can be forced to flower when they ordinarily would not.

Dormancy

Dormancy is a condition in which a seed or a plant remains inactive for a period of time. It is thought that abscisic acid, another plant hormone, plays a role in initiating dormancy. Seeds often remain dormant for a period of time before germinating. Dormancy may be broken after the seeds have been exposed to a period of low temperatures. This prevents the seeds of plants that live in regions with distinct seasons from germinating before winter has ended. Many perennial plants temporarily stop growing and become dormant when the environment becomes unfavorable for continued growth. The buds of deciduous woody plants usually become dormant in late summer or fall when thick, protective scales form around each bud, as seen in Figure 25-15. Then, a drop in auxin concentration and a rise in ethylene production combine to initiate leaf drop. After a period of exposure to cold temperatures, bud dormancy is broken and plant growth resumes. The shoots of herbaceous perennials, on the other hand, typically die during cold or dry weather. Carbohydrates stored in the dormant underground stems or fleshy roots of these perennials serve as fuel for the growth of new shoots. ◻

Figure 25-15 Thick scales cover these dormant buds on this twig from an apple tree.

❏ CAPSULE SUMMARY

Plant hormones are produced in response to stimuli from the environment. Therefore, plant growth and development respond to the environment in many ways.

Section Review

1. *What could you do to promote healthy plant growth?*
2. *How are nutrients important to plant growth?*
3. *List four types of plant hormones, and briefly describe how each affects plant growth and development.*
4. *How do tropisms affect plant growth?*

Critical Thinking
5. *Why is it an advantage for plant growth and development to be regulated by stimuli received from the environment?*
6. *Why is apical dominance an advantage to a young plant?*

annual (577)
apical dominance (582)
apical meristem (573)
auxin (582)
biennial (577)
cork cambium (574)
cytokinin (583)
day-neutral plant (585)
deciduous (577)
differentiation (573)
dormancy (585)

ethylene (583)
evergreen (577)
germination (572)
gibberellin (583)
gravitropism (583)
hormone (582)
lateral meristem (574)
long-day plant (585)
nutrient (580)
perennial (577)
photoperiodism (584)

phototropism (583)
primary growth (573)
primary tissue (573)
secondary growth (574)
secondary tissue (574)
seedling (572)
short-day plant (585)
thigmotropism (584)
tissue culture (576)
tropism (583)
vascular cambium (574)

Review

Multiple Choice

1. The first sign of germination of a bean seed is the emergence of the embryo's
   a. root.
   b. hooked shoot.
   c. protected shoot.
   d. cotyledons.

2. Providing cells for growth at the tips of the plant is the primary function of the
   a. apical meristems.
   b. root cap.
   c. apical endosperm.
   d. dermal ground tissue.

3. Cell division that results in the increased width of a tree occurs in the
   a. apical meristems.
   b. primary tissues.
   c. lateral meristems.
   d. sieve tube vessels.

4. Plant and animal development differ in that plant development
   a. stops soon after the plant reaches maturity.
   b. is not affected by environmental factors.
   c. is controlled by genes that cannot be reactivated.
   d. is continuous and enables whole plants to develop in tissue cultures.

5. Plants that live three or more years are
   a. annuals.
   b. biennials.
   c. perennials.
   d. evergreens.

6. The fastest way to propagate kalanchoës is to plant
   a. leaf cuttings.
   b. plantlets.
   c. seeds.
   d. stem cuttings.

7. Which of the following is *not* a basic requirement for healthy plant growth?
   a. oxyen
   b. vitamins
   c. carbon dioxide
   d. water

8. Which nutrient is important to plants because of its role in active transport?
   a. nitrogen
   b. sulfur
   c. magnesium
   d. potassium

9. What hormone should be used if lateral bud growth is desired?
   a. ethylene
   b. cytokinin
   c. auxin
   d. gibberellin

10. The response of a plant to the length of days and nights is
    a. negative phototropism.
    b. positive phototropism.
    c. photoperiodism.
    d. dormancy.

Completion

11. In bean seeds, the embryo's energy reserves are stored in the _____ ; in corn seeds, the embryo's energy reserves are stored in the _____ .

12. Oaks and elms are examples of _____ perennials. Because these trees lose their leaves at the end of each growing season, they are said to be _____ .

13. Cork cambium produces _____ cells, and vascular cambium forms _____ and _____ .

14. Growing new plants from a plant's non-reproductive parts, such as stems and leaves, is called _____ .

15. Synthetic _____ are used to inhibit seed formation in grapes, while _____ is used to ripen apples and pears that are picked before they ripen naturally.

16. Plant growth *toward* light is called _____ . Plant growth *away* from gravity is called _____ .

17. Short-day plants tend to flower in the _____ , and long-day plants tend to flower in the _____ .

Themes Review

18. **Levels of Organization** Explain how a plant's tissues form through differentiation, beginning with the cells of the apical meristems and ending with secondary tissues.

19. **Flow of Energy** A carrot plant is a biennial that completes its life cycle in two years. How is the energy required for the plant's second year of growth stored at the end of the first year?

20. **Evolution** The pattern of plant growth seen in the photograph below is caused by apical dominance. What is the most obvious adaptive advantage of apical dominance to a young plant?

Critical Thinking

21. **Making Predictions** A student placed a green banana in each of several plastic bags. The student also placed a ripe pear in half of the bags and then sealed all of the bags. Which group of bananas (with or without pears) do you think will ripen sooner? Justify your answer.

22. **Making Inferences** Suppose that a friend who lives in North Dakota gives you some seeds from a plant that you admired when you saw it growing in your friend's yard. You plant the seeds at your home in Florida, but they fail to germinate. Based on your knowledge of seed germination, what might be preventing the germination of the seeds?

23. **Designing Experiments** Design an experiment that will test the inference you made for question number 22. Be sure to include a control for each variable in your experiment.

Activities and Projects

24. **Cooperative Group Project** Interview or write to people who work at commercial nurseries to find out how they produce the plants that are sold at garden centers and flower shops. Some specific plants that you might want to ask about include chrysanthemums, Easter lilies, fruit and nut trees, lawn grasses, orchids, pansies, and poinsettias. Summarize your findings in a report that can be shared with your class.

25. **Multicultural Perspective** Research the practice of and philosophical basis of bonsai, the Oriental art of growing miniature plants. Find out when and where bonsai originated and how bonsai plants are kept small. Relate your findings in a report that also explains how an understanding of plant growth and development is important to success in bonsai.

Seed Structure and Seedling Development

OBJECTIVE

Observe the structures of dicot and monocot seeds, and compare the development of their embryos.

PROCESS SKILLS

* relating structure to function
* comparing features of monocots and dicots

MATERIALS

* 1 pea seed soaked overnight
* 6 bean seeds soaked overnight
* 6 corn kernels soaked overnight
* stereomicroscope
* scalpel
* Lugol's iodine solution in dropper bottle
* paper towels
* 2 rubber bands
* 150 mL beakers (2)
* glass-marking pen
* metric ruler
* microscope slide
* medicine dropper
* compound light microscope

BACKGROUND

1. What are the parts of a seed?
2. In what ways are seeds like their parent plant?
3. How do monocotyledons and dicotyledons differ?
4. Write your own Focus Question on your Vee Form.

5. **Knowing Side of the Vee** List the **Concepts** and new **Vocabulary Words** on your Vee Form. In the **Concept Statements** section of the Vee, use these words in sentences that define and explain them.

TECHNIQUE

Doing Side of the Vee

Part A: Seed Structure

1. Obtain each of the seeds—pea, bean, and corn. Remove the seed coats of the pea and bean seeds. Open the seeds to reveal the two embryonic leaves.

2. Using the stereomicroscope, examine the embryos of the pea seed and the bean seed.

3. In the **Records** section of your Vee Form, draw the pea and bean embryos and label all of the parts that you can identify.

4. Examine a corn kernel and find a small, oval, light-colored area that shows through the seed coat. **CAUTION: Use the scalpel carefully to avoid injury.** Use the scalpel to cut the seed in half along the length of this area. Place a drop of iodine on the cut surface.

5. Use the stereomicroscope to examine the corn embryo. In the **Records** section of your Vee Form, sketch the embryo and label all the parts that you can identify.

Part B: Seedling Development

6. Set five corn kernels on a folded paper towel. Roll up the paper towel and put a

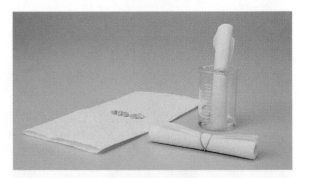

rubber band around the roll. Stand the roll in a beaker with 1 cm of water in the bottom. The paper towel will soak up water and moisten the corn. Keep water at the bottom of the beaker, but do not allow the corn kernels to be covered by water.

7. Repeat step 1 with five bean seeds.

8. After three days, unroll the paper towels and examine the corn and bean seedlings. Use a glass-marking pen to mark the roots and shoots of the developing seedlings. Starting at the seed, mark each 0.5 cm along the root of each seedling. And again starting at the seed, mark each 0.5 cm along the stem of each seedling.

9. Draw a corn seedling and a bean seedling on the **Records** section of the Vee and record the distance between the marks. Using a fresh paper towel, roll up the seeds, place the rolls in the beakers, and add fresh water to the beakers.

10. After two more days reexamine the seedlings. Measure the distance between the marks.

11. **CAUTION: Use the scalpel carefully to avoid injury.** Using the scalpel, make a cut about 2 cm from the tip of the root of a bean seedling. Place the root tip on a microscope slide and add a drop of water. Using a compound light microscope on low power, observe the root tip. In the **Records** section of your Vee, draw the root tip.

12. In the **Procedure** section of the Vee, briefly summarize the procedure you followed.

13. Clean up your materials and wash your hands before leaving the lab.

INQUIRY

1. What types of leaves first appear on the bean seedling?

2. What substance does the black color in the corn kernel indicate? Why might you expect to find this substance in the seed?

3. Has the distance between the marks changed? If it has, where has it changed?

4. Use the information on the **Knowing Side** of the Vee to interpret your results from the **Doing Side,** and then write your **Knowledge Claim.** Write a **Value Claim** for this lab.

ANALYSIS

1. What parts of the embryo were observed in all seeds on the third day?

2. How does the structure and development of the corn kernel differ from the structure and development of the pea and bean seeds?

3. What was the source of nutrients for each of the seed embryos? What is your evidence?

4. Describe the growth in the seedlings you observed.

5. What is the function of root hairs? How do they improve the function of the root?

6. Corn and beans are often cited as representative examples of monocots and dicots, respectively. Relate the seed structure of each to the terms *monocotyledon* and *dicotyledon.*

7. As they push through the soil, what protects the tips of corn shoots? of bean shoots?

FURTHER INQUIRY

Write a **New Focus Question** that could be the point of a new investigation. The following is an example:

How do monocots and dicots compare in general plant growth and in the structure of their leaves and flowers?

CHAPTER
26

PLANTS IN OUR LIVES

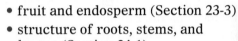

REVIEW

- amino acids and proteins (Section 2-3)
- cellulose (Section 2-3)
- crops (Section 5-1)
- nitrogen-fixation (Section 16-3)
- fruit and endosperm (Section 23-3)
- structure of roots, stems, and leaves (Section 24-1)
- seed structure (Section 25-1)

Pima Indian man harvesting squash

26-1 Plants as Food

Humans and other animals depend on the sugar that plants produce during photosynthesis. Certain plants produce far more sugar during photosynthesis than they can immediately use. Such plants store this extra sugar and use it later for activities such as spring growth. Plants with large energy stores are attractive to humans and other animals as food. In this section, you will examine some of the most important food plants.

Section Objectives

- Categorize familiar foods as roots, shoots, stems, leaves, flowers, seeds, or fruits.
- Identify the three most important cereal grains, and explain how each was derived from its wild ancestors.
- Describe the role of polyploidy in the origin of bread wheat.
- Identify nongrain foods derived from plants, and state their dietary importance.

All Plant Parts Serve as Food

All types of plant parts—roots, stems, leaves, flowers, fruits, and seeds—are eaten as food. For marketing purposes, each of the foods derived from plants is identified by an agricultural commodity term—cereal, fruit, or vegetable—that is registered in Washington, D.C. These terms have different meanings in botany than they do in agriculture, however, and this often causes confusion among the general public. For example, botanically, a fruit is the ripened ovary of a flower, and a **vegetative part** is any nonreproductive part of a plant. The foods that you think of as fruits—such as apples, bananas, and melons—are also fruits in the botanical sense. But the foods known agriculturally as cereals are derived from fruits as well. Vegetables, on the other hand, may be any botanical part of a plant, as you can see in Figure 26-1.

Figure 26-1 Familiar vegetables, *below*, have botanical functions. Broccoli, artichokes, and cauliflower are flowers. Cabbage, celery, and lettuce are leaves. Carrots, radishes, and turnips are roots. Pumpkins, squash, green beans, and tomatoes are fruits. Asparagus and alfalfa sprouts are shoots. Potatoes and onions are modified stems.

Flowers

Leaves

Roots

Fruits

Shoots

Storage structures

Stems   Stems and leaves

Plants in Our Lives **591**

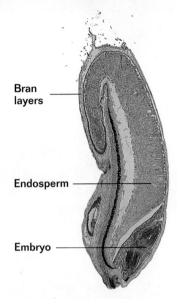

Bran
layers

Endosperm

Embryo

Figure 26-2 A wheat grain is a single-seeded dry fruit.

Figure 26-3 As this map illustrates, the United States is one of the world's major producers of rice, wheat, and corn. What other countries are major producers of these three important cereal crops?

Cereals Are the Most Important Sources of Food

Most of the foods that people eat come directly or indirectly from the fruits of **cereals,** which are grasses that are grown as food for humans and livestock. Cereal grasses produce large numbers of a type of edible, dry fruit called a **grain.** As you can see in Figure 26-2, a grain contains a single seed with a large supply of endosperm. A grain is covered by a dry, papery husk called the **bran,** which includes the wall of the ovary and the seed coat. Grains are rich in carbohydrates and also contain protein, vitamins, and dietary fiber. More than 70 percent of the world's cultivated farmland is used for growing cereal grains. The map in Figure 26-3 shows where the three most important cereal crops—rice, wheat, and corn—are grown.

Rice: An Important Source of Carbohydrates

For more than half of the people in the world, rice, *Oryza sativa*, is the main part of every meal. Although it is low in protein, rice is an excellent source of carbohydrates. While brown rice still has its vitamin-rich bran layers, white rice has been processed to remove the bran layers. This processing helps to prevent spoilage in stored rice. In societies where people eat mainly rice, vitamin-rich sauces such as soy sauce are added to white rice to make meals more nutritious. The white rice you buy at a grocery store is enriched

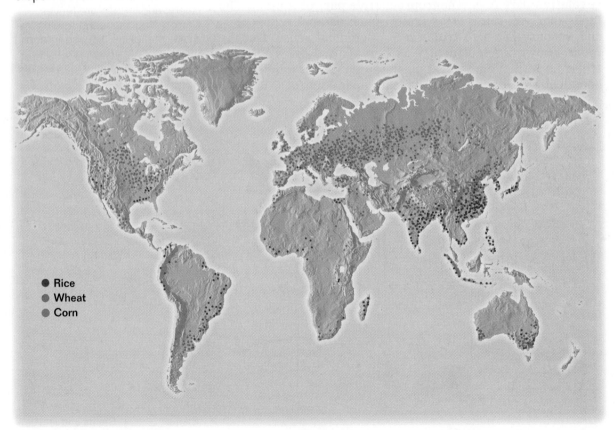

● Rice
● Wheat
● Corn

White rice   Brown rice

with added vitamins. Rice is often added to processed foods such as breakfast cereal, soup, baby food, and flour to increase their energy content.

Rice is native to tropical Asia, where it was gathered and eaten by people more than 10,000 years ago. Archaeologists have found evidence that people cultivated rice in southern China as early as 5000 B.C. From China, rice cultivation spread east to Japan and west to India. Rice thrives in areas with abundant rainfall, such as southeast Asia and the Gulf Coast of the United States, where it is grown in standing water. Rice is pictured in Figure 26-4.

Figure 26-4 Rice plants, *left,* are grown in standing water. White rice, *center,* lacks the vitamin-rich bran layers, but is still rich in carbohydrates. Brown rice, *right,* is more nutritious than white rice because it has not been processed to remove the bran layers.

Wheat: The World's Most Widely Grown Crop

For more than one-third of the world's population, wheat is the primary source of food. High in carbohydrates, the endosperm of wheat grains is commonly ground into white flour and used in breads and pasta. Vitamin-rich **wheat germ** consists of the embryos of wheat grains. **Whole wheat** consists of the endosperm plus the germ and bran layers. In the Middle East, wheat grains are often boiled or soaked, dried, and then pounded until they crack. The cracked grains, called bulgur *(BUL guhr),* are used in dishes such as tabbouleh *(tuh BOO lee)* and pilaf *(pih LAHF).* Most wheat is grown in temperate regions that have fertile soil and moderate rainfall. One of the world's best wheat-growing areas is the Great Plains region of the United States and Canada—a temperate grassland biome.

Even before the cultivation of wheat began in the Middle East about 11,000 years ago, people gathered and ate the grains of wild wheat plants. However, the grains of the first

Figure 26-5 The grains of modern bread wheat, *Triticum aestivum, top right* and *bottom left,* are much larger than the grains of *Triticum monococcum, top left,* one of three wild ancestors of modern bread wheat.

wild wheats that were cultivated bear little resemblance to the grains of modern wheat, as you can see in Figure 26-5. Early farmers selected grains from their best wheat plants—such as those with bigger and more numerous grains, those with stalks that did not break in the wind, and those with grains that fell off the stalk more easily during threshing—to use as seeds for planting. Over time, selection by farmers resulted in modern bread wheat, or *Triticum aestivum.* To learn more about the history of modern bread wheat, look at *Up Close: Bread Wheat* on pages 596–597.

Corn: The American Grain

Corn, *Zea mays,* is the most widely cultivated crop in the United States. American colonists of the 1600s and 1700s first learned how to grow corn from Native Americans. In the southeastern United States, corn was more widely grown than wheat, which does not grow as well in hot climates. Thus, corn-based foods—corn bread, corn pone, hominy, and grits—are a traditional component of the southeastern American diet. Corn is also one of the world's chief foods for farm animals. About 70 percent of the United States corn crop is consumed by livestock. Other uses for corn include the production of corn syrup, margarine, corn oil, cornstarch, and fuel-grade ethanol. Most of the corn grown in the United States today comes from a region known as the Corn Belt, which includes Iowa, Nebraska, Minnesota, Illinois, and Indiana.

Corn is the only major grain crop that originated in America. Early native farmers in Mexico are thought to have selected seeds from the largest flower spikes of teosinte *(tee oh SIHN tee), Zea mexicana,* a wild annual grass. Teosinte

Corn, *Zea mays*

Tassels (male flower spikes)

Immature ears (female flower spikes)

Teosinte, *Zea diploperennis*

Flower spike

Figure 26-6 Modern corn, *Zea mays, above left,* was derived from *Zea mexicana,* or teosinte, a native of Mexico. An ear of corn is a thick spike with many rows of grains. Teosintes such as *Zea diploperennis, above right,* a species that is almost identical to *Zea mexicana,* produce narrow spikes containing two rows of grains.

and modern corn are compared in Figure 26-6. Selective breeding by humans eventually produced plants that had flower spikes with many parallel rows of grain, resembling an ear of corn. Corn was later cultivated by the Aztecs of central Mexico, the Mayas of central Mexico and northern Central America, and the Incas of western South America. The cultivation of corn also spread among many other native North American cultures. Columbus and other European explorers introduced corn to many parts of the world. Now one of the principal crops grown in Africa, corn is the third most important source of food for humans. ◧

Many Other Plants Supply Important Foods

Although cereals provide most of the calories consumed by humans, they do not contain all of the nutrients needed for a healthy diet. Other significant foods derived from plants are seeds, stems, or roots that provide important vitamins, minerals, and amino acids. For example, your body produces only 11 of the 20 amino acids that are needed to make your proteins. The other 9 amino acids, called **essential amino acids,** must be obtained from the foods you eat. Meat provides all of the amino acids your body needs, and meat protein is therefore said to be a **complete protein.** Plant proteins, which lack one or more of the essential amino acids, are said to be **incomplete proteins.** In societies where there is little meat in the diet, it is particularly important that people eat combinations of grains and other foods that provide a complete protein.

❑ *CAPSULE SUMMARY*

Rice, wheat, and corn are the most important sources of food for humans. These crops, called cereals, produce dry fruits called cereal grains, which contain a single seed that is packed with energy-rich endosperm.

UP CLOSE BREAD WHEAT

- **Scientific Name:** *Triticum aestivum*
- **Range:** Agricultural regions worldwide
- **Habitat:** Cultivated fields in temperate and subtropical grasslands
- **Size:** 0.3 m (1 ft.) to 0.8 m (2.5 ft.) tall
- **Importance:** Bread wheat is the world's most agriculturally important annual grass. A staple food in temperate regions of the world, the grains of *Triticum aestivum* are high in gluten (a sticky mixture of proteins that makes dough elastic) and are usually ground into flour that is used to make bread.

External Structures

Leaves Wheat, which is a member of the Poaceae (grass) family, is a monocot. Its leaves are long and narrow (about 2 cm, or 0.8 in., wide) and have parallel veins. Wheat leaves do not have petioles. Instead, the leaf blades are attached to the stem by a sheath, which wraps around the stem.

Stems Wheat plants have hollow, jointed stems called culms. The culms of mature plants have from 3 to 6 nodes and internodes. Flower spikes top the uppermost internode of each culm. Given plenty of space, mature wheat plants may develop as many as 100 culms.

Roots Wheat plants have a fibrous root system, as do most grasses. The extensive root system, which may be more than 2.2 m (7 ft.) deep, is composed mainly of adventitious roots that develop from the base of the culms.

Leaf blade

Leaf sheath

Culms

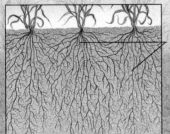
Adventitious roots

Awn

Spikelet

Floret

Palea

Anther

Stigma

Lemma

Glumes

Flower spike

Flowers The flowers of wheat plants occur in dense clusters called spikes, which are 5–13 cm (2–5 in.) long. The spikes of bread wheat consist of many spikelets, which are smaller clusters of up to 5 tiny individual flowers, or florets. Like all grass flowers, wheat flowers lack petals and sepals. Instead, two greenish bracts (modified leaves) called the palea and the lemma enclose the anthers and the pistil of each floret. The lemmas of some bread-wheat varieties have a long bristle called an awn. Two bracts called glumes occur at the base of each spikelet.

Fruit A kernel, or grain, of wheat is a one-seeded fruit called a caryopsis. Wheat kernels have a crease on one side and a brush of tiny hairs at one end.

Internal Structures

Grain structure A cross section of a wheat kernel shows that it is mainly (about 85 percent) starchy endosperm. The kernel's outer layers, consisting of the ovary wall, seed coat, and aleurone layer (a layer that contains protein and oils), make up the bran (about 12 percent of the kernel). The embryo, or wheat germ, is only a small portion (less than 3 percent) of the kernel.

Kernel

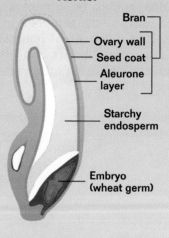

Bran —
- Ovary wall
- Seed coat
- Aleurone layer

Starchy endosperm

Embryo (wheat germ)

Apical meristems The apical meristems of a wheat plant are located within the culms at the point where they rise from the crown of the plant. This location is usually at, or just beneath, the soil line. As a result, the apical meristems of wheats and other grasses are protected from being eaten by grazing animals. Each new leaf grows up through the inside of a hollow culm and emerges from within the sheath of the culm's uppermost leaf.

Leaf sheath (oldest leaf)

Leaf sheath (younger leaf)

New leaf

▲ Apical meristem

Soil level

Chromosomes The cells of bread wheat have a chromosome number of $2n = 42$, or $n = 21$. Close examination of bread wheat's chromosomes reveals an interesting pattern. Among the 42 chromosomes in a bread-wheat cell, there are actually three distinct sets of 14 chromosomes (7 pairs). Because the chromosomes in each set are slightly different from those in the other two sets, botanists represent these sets of paired chromosomes with the letters AA BB DD. Thus, bread wheat is not a diploid plant but is instead a hexaploid ($6n$) plant—a polyploid plant with 6 of each kind of chromosome.

Wheat cell

■ A ■ B ■ D

Karyotype of wheat

Polyploidy Polyploidy, which means having many sets of chromosomes, is common among cultivated food and ornamental plants. Polyploids tend to be larger, or prettier, than their wild ancestors. Among plants, polyploidy is an important evolutionary tool that enables fertile hybrids to result from crosses among different species. Such is the case with bread wheat. Three different closely related species hybridized naturally to produce bread wheat. Polyploids that contain sets of chromosomes from two or more different species are called allopolyploids. The prefix *allo-* comes from the Greek word *allos*, meaning "other."

Legumes: An Important Source of Protein

Many members of the pea family, which are called **legumes,** produce protein-rich seeds in long pods. For example, approximately 45 percent of a soybean, the most important legume grown for food, is protein. Peas, peanuts, and the many different types of beans are the seeds of legumes. As Figure 26-7 illustrates, legumes are often eaten with cereal grains such as rice and corn. These combinations provide a complete protein. The genes that make essential amino acids in legumes may someday make the foods obtained from other plants more nutritious. Genetic engineers are currently attempting to transfer these genes into cereal grains.

As you learned in Chapter 16, many legumes form symbiotic associations with nitrogen-fixing bacteria. These associations, which appear as root nodules, enable the legumes to grow and form their protein-rich seeds in nitrogen-poor soil. Because some of the nitrogen fixed in the roots of legumes enriches the soil with nitrogen that other plants can use, legumes such as alfalfa are often rotated with other crop plants. Alfalfa is also fed to livestock.

Figure 26-7 This meal supplies a complete protein by combining foods made from two grains (rice and corn) with a legume (beans).

Tamales (made of corn and meat; supply essential amino acids)

Beans (legume that contains some essential amino acids)

Corn tortillas (made of grain; supply some essential amino acids)

Rice (grain rich in carbohydrates)

Root Crops: Dietary Staples for Many People

Potatoes, *Solanum tuberosum,* are an important food staple in many regions of the world. Although potatoes are thought of as a root crop because they grow underground, they are actually **tubers,** modified underground stems that store starch. Yams, an essential food crop in many tropical parts of the world, are also tubers. Native to the Andean region of

Figure 26-8 Cassava, *Manihot esculenta, left,* is a large, fleshy root that is rich in starch. Cassava, also known as Manioc, supplies more than one-third of the calories consumed in Africa. In Central America, cassava flour is used to make a type of bread. This native Panamanian woman, *far left,* is grinding cassava roots to release toxic chemicals found in the roots. Foods made from cassava must be cooked to completely destroy the roots' toxic components.

South America, potatoes quickly became an important crop in Europe after they were introduced there by the explorers who followed Columbus. Rich in calories and easy to cultivate, potatoes are an ideal crop for a small farm. A small plot can provide enough calories for an entire family.

Other important root crops include sweet potatoes, carrots, radishes, turnips, beets, and cassava *(kuh SAH vuh).* These vegetables are enlarged roots that store starch. Cassava, seen in Figure 26-8, is the staple food of more than 500 million people, making it the most important noncereal crop. Tapioca pudding is made from the starch extracted from cassava roots. Large quantities of sugar are extracted from sugar beets, which are large, fleshy roots that store sugar rather than starch. Much of the sugar you eat, however, comes from the stems of a grass—sugar cane. ■

❏ CAPSULE SUMMARY

Legumes such as beans and peas are important foods because they provide essential amino acids that grains lack. Root crops such as potatoes, yams, and cassava are a major source of calories in many cultures.

Section Review

1. *Name three common vegetables that are actually fruits and two vegetables that are actually flowers.*
2. *What are the three most important cereal grains, and where did each originate?*
3. *How does bread wheat differ from its wild ancestors?*
4. *How do potatoes and yams differ from other root crops?*

Critical Thinking
5. *How could transferring specific genes from legumes into rice plants help reduce malnutrition?*

26-2 *Other Uses of Plants*

Section Objectives

- Describe several ways that wood is used.
- Explain how plants are used to treat human ailments.
- Describe how plants are used to make rubber and cloth.

*P*lants are used by people for many purposes other than food. For example, if you were to look at this sheet of paper very closely through a magnifying glass, you would see that it is made of many fibers laid across one another in a thin mat. These fibers are strands of cellulose from the cell walls of plants. Plant fibers are also used to make cloth. Trees can be cut into lumber, burned as fuel for heating and cooking, and processed for use in paper, plastics, and rayon. Plants are also the original sources of many important medicines.

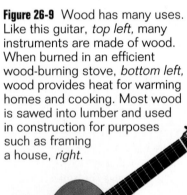

Figure 26-9 Wood has many uses. Like this guitar, *top left,* many instruments are made of wood. When burned in an efficient wood-burning stove, *bottom left,* wood provides heat for warming homes and cooking. Most wood is sawed into lumber and used in construction for purposes such as framing a house, *right.*

Wood Is the Most Valuable Plant Product Besides Food

After food, wood is the single most valuable resource obtained from plants. For more than a quarter of the world's people, wood is still the main source of fuel for heating and cooking. Thousands of products are also made from wood. The wood from trees that have been cut down and sawed into boards and planks is called **lumber.** Nearly 75 percent of the lumber cut in the United States is used for building construction, as Figure 26-9 illustrates. The rest is used to make products such as furniture, or it is ground and moistened to make **wood pulp.** Wood pulp is made into paper, rayon, and many other products. Although most of the cellulose fibers used to make paper are extracted from wood, paper-making fibers come from many different plants. For example, fibers from cotton, flax (from which linen is made), rice, papyrus *(puh PEYE ruhs),* and bamboo are found in many types of stationery.

Figure 26-10 A worker examines mountains of corrugated paper, *above left,* that will be used to make recycled paper, *above.*

Although trees are a renewable resource, they are currently being used faster than they can be replaced. As seen in Figure 26-10, recycling paper is one way that people are trying to limit their consumption of trees. Today, paper is often recycled by dissolving it in water, washing off the inks and dyes, and re-pressing the fibers into new paper. Unfortunately, the expense of ink and dye removal makes the cost of recycling paper exceed the cost of making new paper from wood pulp. As a result, commercial landfills contain more than 60 percent wastepaper. Efforts are now underway to recycle this wastepaper in a different way. By using genetically engineered bacteria that have the enzymes necessary to break down cellulose and ferment the resulting sugar, wastepaper is being used to produce fuel-grade ethanol that is added to gasoline. ◻

❏ CAPSULE SUMMARY

Wood is an important resource that is part of thousands of products. It is cut into lumber for use in building construction and ground into wood pulp for use in paper and rayon.

Many Medicines Are Obtained From Plants

People have always used substances obtained from plants to treat a variety of ailments. Rural and primitive cultures around the world still depend on native plants to ease pain and cure illnesses. By studying the plants traditionally used to treat human ailments, researchers have developed many "modern" medicines. For example, solutions made by soaking the bark of willow trees, *Salix,* were a traditional cure for aches and pains. The ingredient in willows that reduced pain was isolated in 1827 and named salicin (*SAL uh sihn*). Today, **aspirin** (acetylsalicylic acid), a derivative of salicin, is the most widely used drug in the world. Each of the

20 best-selling drugs in the world is either extracted directly from a plant or linked to a plant in another way.

In the 1960s, two cancer-treatment drugs were discovered in a familiar garden plant—the rosy periwinkle, *Catharanthus roseus*. The rosy periwinkle, seen in Figure 26-11, is the natural source of vinblastine *(vihn BLAS teen)* and vincristine *(vihn KRIHS teen)*. Vinblastine is used to treat Hodgkin's disease, a type of cancer that affects the lymph nodes. Vincristine is used to treat childhood leukemia. The use of these drugs, which are now made synthetically, has greatly increased the survival rate of patients with these cancers. For example, the chance that a child who contracts childhood leukemia will live past the age of five has increased from 1 chance in 20 to about 19 chances in 20.

Scientists have recently discovered that yew trees, *Taxus* sp., also produce a chemical with cancer-fighting properties. The chemical, called taxol, is found in several parts of yew trees, including the bark, as Figure 26-12 illustrates. However, only very tiny amounts of taxol are made by a single tree—4 tons of bark are needed to produce 1 lb. of taxol. Leaves may be a more promising source of taxol, a very complicated molecule that was produced synthetically for the first time in 1994.

Today, plants still directly provide many important medicines. For instance, sweet potatoes are the source of **cortisone** *(KAWRT uh sohn)*, a drug that is used to treat inflammation and allergies. The leaves of foxglove (an extremely poisonous European plant) yield **digitalis** *(dihj ih TAL ihs)*, a drug that is used to stabilize irregular heartbeats and to treat cardiac disorders. Reserpine *(rih SUR pihn)*, a drug obtained from the shrub *Rauwolfia serpentina*, is used to control high blood pressure. More than two-thirds of the people in the world obtain most of their medicinal drugs directly from plants. For example, more than 5,000 kinds of natural drugs obtained from plants are traded and sold in China. ◻

Figure 26-11 The rosy periwinkle, also known as vinca, is a popular bedding plant and the original source of two important drugs that are used to treat cancer.

◻ **CAPSULE SUMMARY**

Plants are the sources of many important medicines used by humans to treat diseases and other ailments.

Figure 26-12 These workers are stripping the bark from yew trees, a practice that may kill the trees. The bark is processed to obtain taxol, another drug that is used to treat cancer.

Figure 26-13 Cotton cloth is made from the fine fibers that are attached to the seeds in a cotton boll, *left,* which is the fruit of a cotton plant. After it is harvested, most cotton is sent to a gin, *right,* a machine that removes the cotton fibers from the seeds.

Fibers and Rubber Come From Plants

Plant fiber, in addition to its use in making paper, is also used to make cloth. Although synthetic fibers are now used in more than 30 percent of the world's clothing, natural plant fibers are still prized for their durability and comfort. Cloth made of cotton, the world's most important plant fiber, has been worn for centuries. Cotton thread is spun from the strong, fine fibers on cotton seeds, seen in Figure 26-13. The stems of flax, *Linum usitatissimum,* yield softer, more durable fibers that are used to make linen.

Rubber was first obtained from plants. Native Americans of Central and South America made rubber balls and waterproof shoes from **latex,** the milky white sap of tropical trees of the genus *Hevea.* Latex is extracted from "rubber" trees by the method seen in Figure 26-14. Today, natural rubber comes from rubber trees on plantations in Southeast Asia. Guayule (*gwah YOO lee*), a member of the sunflower family that is a native of the deserts in the southwestern United States, is a promising new source of natural rubber. Most of today's rubber, however, is synthesized from petroleum, a nonrenewable resource.

Figure 26-14 Latex, used to make natural rubber, is harvested from a rubber tree by cutting and removing strips of bark and collecting the milky white sap as it drips out.

Section Review

1. *What is the main way that people use wood?*
2. *What is wood pulp, and how is it used?*
3. *Name five medicines that are derived from plants, and state how they are used.*
4. *What is the most important plant fiber used in clothing?*

Critical Thinking

5. *Why might guayule become a valuable cultivated crop?*

Vocabulary

aspirin (601)
bran (592)
cereal (592)
complete protein (595)
cortisone (602)
digitalis (602)

essential amino acids (595)
grain (592)
incomplete protein (595)
latex (603)
legume (598)
lumber (600)

tuber (598)
vegetative part (591)
wheat germ (593)
whole wheat (593)
wood pulp (600)

Review

Multiple Choice

1. Which item correctly matches a food and the plant part from which it comes?
   a. yam : root
   b. sweet potato : leaves
   c. cassava : stem
   d. banana : fruit

2. Which one of the following cereal grains is eaten by more of the world's people?
   a. wheat   c. corn
   b. rice   d. oats

3. What trait did early farmers select when trying to increase the yield from wheat plants?
   a. fast growth in areas with much rainfall
   b. survival in hot climates
   c. large grains
   d. parallel rows of grain

4. Which of the following statements best describes the origin of bread wheat?
   a. It is a diploid plant essentially identical to the first wheat species cultivated about 11,000 years ago.
   b. It is a haploid plant.
   c. It is a tetraploid plant derived from the hybridization of two closely related species of wheat.
   d. It is a hexaploid plant derived from the hybridization of three closely related species of wheat.

5. A complete protein is provided by which of the following combinations of foods?
   a. beans and rice   c. potatoes and rice
   b. wheat and oats   d. rice and corn

6. An important nongrain food for many African people is
   a. sugarcane.   c. alfalfa.
   b. cassava.   d. corn.

7. What makes paper recycling so costly?
   a. collecting the paper to be recycled
   b. removing the inks and dyes from the paper
   c. growing the bacteria that digest cellulose
   d. paying taxes on recycled paper

8. Besides food, what is the most valuable plant product?
   a. medicine   c. wood
   b. oil   d. latex

9. Drugs derived from the rosy periwinkle are used in the treatment of
   a. heart disease.   c. arthritis.
   b. leukemia.   d. headaches.

10. Trees of the genus *Hevea* are a source of latex, which is used to make
    a. rubber.   c. petroleum.
    b. reserpine.   d. cloth.

Completion

11. As an agricultural commodity, tomatoes are classified as vegetables. But a tomato is actually a _____ because, like an apple, it is the ripened _____ of a flower.

12. The origin of _____ has been traced to the wild annual grass of Mexico, called _____ .

13. The dry fruit of a cereal plant is called a _____ ; it is covered by a dry husk called the _____ .

14. Plants that produce protein-rich seeds in long pods are called _____ . The most important of these plants is the _____ .

15. Wood is sawed into _____ and then used in construction. It is also processed into _____ and used to make paper.

16. The willow tree, *Salix*, was the orginal source of _____ . _____ is extracted from the leaves of the foxglove plant.

17. The world's most important plant fiber is _____ .

Themes Review

18. **Evolution** How did artificial selection by humans play a role in the origin of agricultural crops? How is artificial selection similar to natural selection?

19. **Levels of Organization** Enzymes are proteins (chains of amino acids) that catalyze chemical reactions in living cells. How might eating a diet that lacks some of the essential amino acids affect the functioning of living cells?

20. **Homeostasis** How do the drugs digitalis and reserpine help the human body maintain homeostasis?

Critical Thinking

21. **Communicating Effectively** Suppose that a friend asks you why corn, which he or she considers to be a vegetable, is listed as a cereal crop in the encyclopedia. To answer this question, write a paragraph that explains why corn is a cereal crop, agriculturally, and why it is also a fruit, botanically. Include other examples of foods that are classified as vegetables or grains but are also fruits.

22. **Making Inferences** Suppose that you discover a new strain of wheat that has much larger spikes and grains than bread wheat. How might this wheat differ from bread wheat? Explain.

23. **Interpreting Data** Recent medical reports suggest that a high-fiber diet may help to prevent colon cancer. The table below shows the approximate percentage of fiber in different parts of a wheat grain. Would you expect to find more cases of colon cancer among people who eat only whole wheat bread (made with endosperm, germ, and bran) or among people who eat only white bread (made with endosperm only)? Why?

Part of wheat grain	Percentage of fiber
Endosperm	4%
Germ	12%
Bran	40%

Activities and Projects

24. **Cooperative Group Project** Interview several people who are vegetarians. Find out what they eat and why they became vegetarians. Then, write an article about being a vegetarian. Include information from the interviews and from other sources such as books and magazines. Publish the article in your school newspaper.

25. **Multicultural Perspective** Native Americans of the Southwest and Great Plains refer to corn, beans, and squash as the "three sisters." Find out why Native Americans have traditionally grown these plants together and why they are important to human nutrition. Relate your findings in a written report.

26. **Research and Writing** Plywood is usually sold in sheets 4 ft. wide by 8 ft. long. Find out how plywood is made and why it is stronger than a regular wooden board of the same thickness. Summarize your findings in a written report.

Vegetative Propagation

OBJECTIVES

- Recognize the ability of some plants to reproduce vegetatively.
- Compare vegetative reproduction of duckweed in different nutrient solutions.

PROCESS SKILLS

- identifying the structures of duckweed
- observing vegetative propagation in duckweed
- hypothesizing on conditions that would increase growth of duckweed
- testing a hypothesis by performing experiments to observe optimum growth of duckweed

MATERIALS

- duckweed culture
- 5 petri dishes
- stereomicroscope
- glass-marking pen
- pond water
- Knop's solution
- 0.1% fertilizer solution
- distilled water
- safety goggles

BACKGROUND

1. What are vegetative plant parts?
2. How can a plant reproduce asexually by means of vegetative plant parts?
3. Write your own **Focus Question** on your Vee Form.
4. **Knowing Side of the Vee** List the Concepts and new **Vocabulary Words** on your Vee Form. In the **Concept Statements** section of the Vee, use these words in sentences that define and explain them.

TECHNIQUE

Doing Side of the Vee

1. Place a duckweed plant in a petri dish. Add a few drops of water. Observe the duckweed under a stereomicroscope. Record your observations in the **Records** Section of your Vee Form by sketching what you see.
2. Label four petri dishes as follows:

   A—pond water

   B—Knop's solution

   C—0.1% fertilizer solution

   D—distilled water
3. **CAUTION: Wear safety goggles for this step. Growth solutions are mild eye irritants. Avoid contact with your skin and eyes.** In case of contact, notify your teacher and flush the area with running water. Fill each petri dish three-quarters full with the solution identified on the label of the dish.
4. Place three duckweed plants in each Petri dish. Put the covers on the dishes and place the dishes in a well-lighted area.

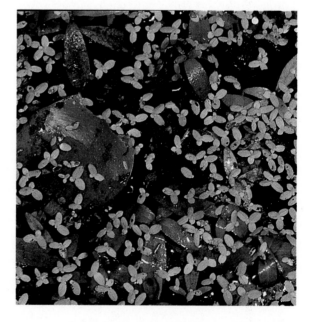

5. Hypothesize which solution should produce the largest number of duckweed plants. Record your hypothesis in the **Records** section of your Vee Form.

6. Observe the four petri dishes each day for two weeks. In the **Records** section of your Vee, make a chart to record the number of individual plants in each solution on each day, at the end of one week, and at the end of the two-week period. Record any other observations in the **Records** section of your Vee about the condition of the plants at those times.

7. In the **Additional Records** section of your Vee Form, make a bar graph with the Y-axis labeled "Number of plants" and the X-axis "Days." Make a key for the graph to indicate which lines represent the three solutions and the distilled water. Title your graph. In the **Procedure** section of the Vee, briefly summarize the procedure you followed.

 8. Clean up your materials and wash your hands before leaving the lab.

INQUIRY

1. In which dish did the greatest amount of growth take place?

2. In which dish did the least amount of growth take place?

3. Did the results you observed agree with your hypothesis? If not, how are they different?

4. As the number of new plants increases, what happens to the group of plants?

5. Describe the attachment of the new plants.

6. Use the information on the **Knowing Side** of the Vee to interpret your results from the **Doing Side,** and then write your **Knowledge Claim.** Write a **Value Claim** for this lab.

ANALYSIS

1. Describe the appearance of plants growing in the Knop's solution. Offer an explanation for your observations.

2. Describe the appearance of plants growing in distilled water. Explain your observations.

3. Based on this investigation, what is the connection between the general health of a plant and vegetative reproduction?

4. What factors regulate the rate of vegetative reproduction in duckweed?

5. Explain what factors accounted for the amount of growth in the dishes containing the fertilizer solution and the pond water.

6. Is there a control in this investigation. If so, what is the control? Give reasons for your answer.

7. Why is a new duckweed plant produced by vegetative reproduction genetically the same as the parent plant?

8. How does an organism benefit from receiving all its genetic information from and being identical to, a single parent? What disadvantages might there be?

FURTHER INQUIRY

Write a **New Focus Question** that could be the point of a new investigation. The following are examples:

How could you check for the effects of other environmental factors, such as light intensity or temperature, on the growth of duckweed?

How could your findings in this lab be applied to raising crop plants by means of vegetative propagation?

BIOTECHNOLOGY AND AGRICULTURE

What Will the Harvest Be?

The genetic engineering process for making BST starts with the use of a restriction enzyme to cleave DNA into fragments with sticky ends.

BY TRACEY COHEN

In February 1994 a genetically engineered drug called bovine somatotropin (BST) went on sale to dairy farmers in the United States. The drug, commonly known as recombinant bovine growth hormone (rBGH), increases milk production in cows by as much as 40 percent while increasing their food consumption by only 5 to 15 percent. Monsanto and over 400 other companies are collectively spending close to $2 billion a year on biotechnology research and development. Even federal and state governments are involved. In 1990, for example, federal funding for biotechnology was almost $3.5 billion.

Proponents of BST hailed its introduction as another means of reducing food costs and increasing food production. Not everyone greeted the news enthusiastically, however. Concerned about the possibility of hormone-tainted milk, several consumer groups called for a boycott of BST products and demanded that dairy products from BST cows be labeled as such.

Biotechnological innovations like BST are transforming agriculture. The list of traits that could potentially be genetically engineered into various agricultural organisms is endless. Researchers are working on tomatoes that ripen more slowly and keep their flavor longer, potatoes with less water and more flesh, and low-fat pigs that yield leaner cuts of pork. While these traits may be appealing to consumers, there are broad environmental, social, economic, and ethical issues of agricultural biotechnology that have yet to be adequately addressed.

What Effects Might Agricultural Biotechnology Have on the Environment?

Agricultural biotechnology poses environmental risks because many of the organisms being engineered will be released outdoors. A good example is the effort to develop herbicide-resistant crops (HRC's), which involves more than 30 companies, along with researchers at several universities and the U.S. Department of Agriculture. Scientists are trying to engineer corn, new varieties of wheat, and other major crops that will resist damage from high doses of specific herbicides. Farmers will

then be able to use heavier doses of herbicides to kill all the weeds in a field without harming their crops.

The use of HRCs, however, ignores complex ecological factors. For example, in any population of weeds, some plants will have a natural resistance to whatever herbicide is being applied. (It is the genes for such resistance that are being transferred into crop plants.) Those resistant weeds that are not affected by the herbicide will survive and reproduce, becoming the dominant population.

Eventually, farmers would have to try something new, creating a vicious cycle of applying more and more poison. This would result in the evolution of more resistant weeds. About 100 species of weeds are currently known to be resistant to one or more herbicides.

Will Farm Animals Really Be "Improved" by Genetic Engineering?

Many farm animals today are raised under conditions designed to produce the largest number of animals and animal products rapidly and at the lowest cost. This approach to animal husbandry, often justified as "progress," treats farm animals as if they were little more than machines for producing milk, eggs, or other foods. Biotechnological research directed at farm animals is likely to make this situation even worse.

For example, the product label for BST shows that it increases the risk of a painful udder infection in cows known as mastitis. Research also indicates that BST may cause cows to have more stillborn and deformed calves than do untreated cows. And the increased stress caused by excessive milk production may bring on early death in BST-treated animals.

Other work on animals involves inserting genes for desired traits directly into animal embryos. For example, the USDA developed transgenic pigs carrying the gene for human growth hormone. The idea was to create a variety of "superpig" that would grow rapidly to huge size. Only one of every 200 embryos survived the procedure. Those that survived had arthritis, vision problems, and an impaired immune system that left them with a tendency to contract pneumonia. There is no way to know for certain what effects genetic engineering will have on an animal and whether the animal will suffer as a result.

Do We Want More Milk?

The use of BST provides a good opportunity to examine the social costs of agricultural biotechnology. Though BST increases total milk production, its use comes at a time when there is already a milk surplus in the United States. A surplus of milk will drive milk prices down. A number of studies, including one by the U.S. Office of Technology Assessment, predict that use of BST will ultimately drive many small dairy farmers out of business. It is expected that the remaining dairy farms will become more efficient, but are society's interests really served by driving small farmers into bankruptcy?

The loss of a small farm has effects that ripple throughout the economy. Family farmers forced out of business often join the ranks of the poor and the unemployed. Rural businesses suffer as the demand for goods and services drops. It is estimated that for every farmer driven out of business, as many as 25 other dairy-related jobs are lost. Data from the USDA show that the unemployment rate is already much higher among agricultural workers than for all other areas of the civilian economy combined: 11.1 percent in agriculture versus 7.0 percent for all other areas in 1993. Poverty in rural America is also higher than in urban areas.

What Should Be Done?

When industry designs a new product, there are important questions to ask. What does society gain? Are real social needs being met? What are the risks? Are potential consequences shared equally or will some areas be threatened more than others? In the case of biotechnology, these questions are mostly unanswered.

It is also important to have a set of social standards for biotechnology products. For a product to be developed, it must be so needed by society that the benefits outweigh any risks. A product with minimal social value is not worth any risk at all.

Each biotechnological innovation must be thoroughly evaluated. It is not enough to look at only the usual business criteria of profit and loss. The broader ecological, social, cultural, and long-term economic implications must also be considered. Only then will this powerful new technology live up to its promises.

Tracey Cohen is a freelance writer specializing in science and environmental issues.

Analyzing the Issue

1. **Detecting Bias** This article is an argumentative essay. It attempts to convince a reader to agree with a certain viewpoint, take a specific action, or make a decision about a subject on which many possible opinions and actions are possible.

   a. What is the writer's point of view about biotechnology in agriculture?

   b. What action does the writer advocate?

   c. List the key factors the writer uses to support the argument.

   d. List the statements from the article that are opinions.

2. **Formulating an Opinion** Write an essay expressing your views about biotechnology in agriculture. You may need to do more library research. Before you begin writing, create an outline that includes the following:

   a. Main point or thesis to be proved

   b. Supporting facts, examples, and arguments

   c. Opposing facts and arguments

3. **Get the Facts** Investigate the movement for sustainable agriculture. How does this approach to farming compare with the most widely used current methods? What impact might biotechnology have on farmers' ability to practice alternative agriculture? Should research in biotechnology support the current system or help to change it? Explain your answer.

4. **Take Action** Find out what federal, state, and local laws regulate biotechnology. Contact the U.S. General Accounting Office and the Office of Technology Assessment for their evaluation of the regulatory framework. What changes, if any, need to be made?

OVERVIEW OF INVERTEBRATES

REVIEW

- diffusion and endocytosis (Section 3-2)
- haploid and diploid cells (Section 6-1)
- natural selection (Section 12-1)
- evolution of arthropods (Section 13-2)
- evolution of vertebrates (Section 13-3)

A jellyfish from the Palau Islands in the Pacific Ocean

27-1 Advent of Tissues and Symmetry

*I*n this chapter you will discover how a series of key evolutionary innovations has led to today's animal phyla. These phyla are shown in Figure 27-1 and are described in **Highlights: Invertebrate Evolution** *features that occur throughout the chapter. You will be provided with a broad overview of the body plan—the shape, symmetry, and internal organization—of different animals and learn how those body plans arose. An animal's body plan results from a pattern of development programmed into the animal's genes by natural selection.*

Section Objectives

- Describe the general characteristics of animals.
- Define radial symmetry.
- Compare and contrast sponges with cnidarians.

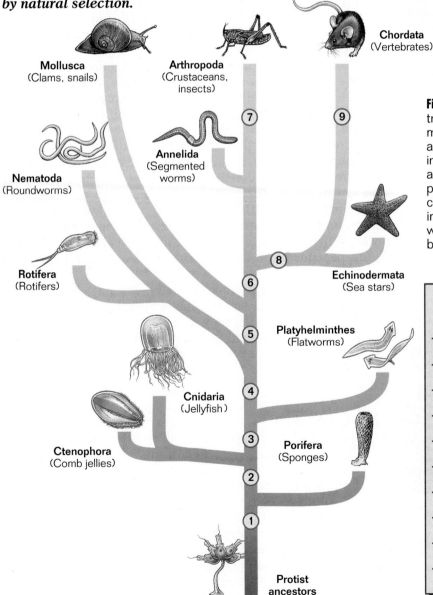

Figure 27-1 This phylogenetic tree identifies the major animal phyla. Although many animals live in water, some inhabit land. Most terrestrial animals are mollusks, arthropods, or chordates. The circled numbers indicate important evolutionary stages, which are listed in the table below.

Mollusca (Clams, snails)

Arthropoda (Crustaceans, insects)

Chordata (Vertebrates)

Nematoda (Roundworms)

Annelida (Segmented worms)

Rotifera (Rotifers)

Echinodermata (Sea stars)

Platyhelminthes (Flatworms)

Cnidaria (Jellyfish)

Ctenophora (Comb jellies)

Porifera (Sponges)

Protist ancestors

Evolutionary stages in the animal body	
Stage	**Milestone**
1	Multicellularity
2	Tissues
3	Bilateral symmetry
4	Body cavity
5	Coelom
6	Segmentation
7	Jointed appendages
8	Deuterostomes
9	Notochord

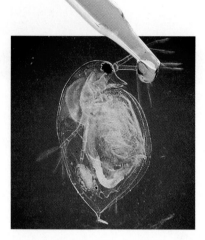

Figure 27-2 Some people may not realize that these tiny organisms, called *Daphnia*, are animals. Also known as water fleas, *Daphnia* make up part of the plankton in freshwater and marine ecosystems.

Figure 27-3 Sponges, like these purple tube sponges, are attached to the sea floor, unable to move around. While they may look inactive, they are actually very busy filtering food particles from sea water.

Some General Features of Animals

Animals are heterotrophs—they ingest their food before digesting it. Most animals move from place to place searching for food, which they then take into their body. In most animals, ingestion of food is followed by digestion in an internal cavity.

All animals are multicellular, and almost all (99 percent) are invertebrates (animals without backbones). Scientists estimate that there are between 5 million and 10 million living animal species. Only 42,500 of these species have a backbone. Animals range in size from organisms too small to see clearly with the naked eye, like the *Daphnia* in Figure 27-2, to enormous whales and giant squids. The animal kingdom includes about 35 phyla; most occur in the sea, with far fewer in fresh water and fewer still on land. Members of three phyla, Arthropoda (spiders, insects, and crustaceans), Mollusca (snails), and Chordata (vertebrates), dominate animal life on land.

Animal cells are distinct among multicellular organisms because animal cells lack rigid cell walls and are usually quite flexible. The cells of all animals except sponges are organized into structural and functional units called tissues.

The ability of animals to move, more rapidly and in more complex ways than the members of other kingdoms, is perhaps their most striking characteristic. Animals move by means of muscle cells, specialized cells able to contract with considerable force. A remarkable form of movement unique to animals is flying, an ability that is well developed among both insects and vertebrates. Among today's vertebrates, birds and bats are both strong flyers. At one time, flying reptiles called pterosaurs, now extinct, dominated the skies.

Most animals reproduce sexually. Animal egg cells are much larger than the small, usually flagellated sperm cells, and, unlike sperm cells, they do not swim. With few exceptions, animals are diploid and the gametes are the only haploid cells in their life cycles.

Sponges: The Simplest Animals

Sponges, shown in Figure 27-3, are the simplest animals. There are about 9,000 species of sponges, almost all of which live in the sea (a few species live in fresh water). The bodies of most sponges completely lack **symmetry**—they do not have body parts that grow around a central point or a central axis as do all other animals. The cells of sponges are not organized into tissues or organs. The bodies of sponges consist of little more than masses of specialized cells embedded in a gel-like substance, called matrix, like chopped fruit in gelatin. However, sponge cells do have a key property of all

Highlights
Invertebrate Evolution

Animals Without Tissues

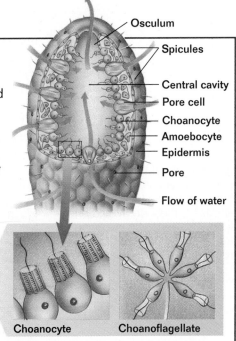

Osculum
Spicules
Central cavity
Pore cell
Choanocyte
Amoebocyte
Epidermis
Pore
Flow of water

Choanocyte

Choanoflagellate

Stage 1 Multicellularity

The bodies of *all* animals, including sponges (phylum Porifera), are **multicellular**—made of many cells. Sponges are composed of several different cell types whose activities are coordinated with each other.

The Body of a Sponge Is Built for Filter Feeding

The outside of a sponge is covered with epithelial cells that protect the sponge. Water, *blue arrows,* enters the sponge's central cavity by passing through many tiny pores that penetrate the skin. The beating action of many choanocyte flagella creates a water current inside the sponge. This current pulls a steady stream of water into the sponge. The choanocytes trap and ingest food particles that are suspended in the water, and the water eventually exits through the osculum.

Protist Ancestors

The choanocytes of sponges very closely resemble a kind of protist called a choanoflagellate, and ancient choanoflagellates are thought to have been the ancestors of sponges. Other, free-swimming colonial flagellates more closely resemble sponge larvae, however, and some believe them to be the true ancestors of sponges.

---

animal cells—cell recognition. The ability of a sponge cell to recognize another sponge cell can be demonstrated during lab experiments in which a sponge is passed through a fine silk mesh. Individual cells separate and then recombine on the other side of the mesh to re-form the sponge.

The body of an adult sponge is anchored in place on the sea floor and functions as a water-filtering machine. As sea water filters through the sponge, the sponge is able to trap protists and tiny animals that live in the water. The body of the sponge is perforated by tiny holes or pores. The name of the phylum, Porifera, refers to this system of pores. Inside the sponge is an internal cavity. Facing into the internal cavity are unique flagellated cells called **choanocytes** (*koh AN oh seyets*), also known as collar cells. The beating of the flagella of the many choanocytes that line the body cavity of the sponge draws water in through the pores and drives the water through the cavity. The water exits the sponge through one or more large openings in the sponge's body wall.

One cubic centimeter of a sponge can propel more than 20 L (5.3 gal.) of water a day into and out of the sponge's body. Why all this moving of water? The sponge is a filter feeder. The beating of each choanocyte's flagellum draws water down through its collar. The collar is made of small hairlike projections resembling a picket fence. Food particles in the water are trapped in the collar and later ingested by the choanocyte or by the sponge's other cells. ◼

CONTENT LINK

Information on sponge classification and reproduction can be found in **Chapter 28**.

❑ CAPSULE SUMMARY

Sponges are the simplest animals. Lacking tissues and body symmetry, sponges live attached to the sea floor, where they filter food particles from the water. The food is digested within the sponge's cells.

CONTENT LINK
....

More information on cnidarian
classification and life cycles can
be found in **Chapter 28**.

Cnidarians: Radially Symmetric Animals
....

All animals other than sponges have both tissues and symmetry and are called "true animals," or **eumetazoans** *(YOO meht uh zoh uhns)*. Early in their development, the embryos of most eumetazoans develop three distinct cell layers. These three embryonic tissues give rise to the many tissues of the adult body, as shown in Figure 27-4. An outer **ectoderm** layer gives rise to the outer coverings of the body and the nervous system; a middle **mesoderm** layer gives rise to the skeleton and muscles; and an inner **endoderm** layer gives rise to the digestive organs.

Figure 27-4 All of the organ systems of an animal develop from three embryonic tissue layers: ectoderm, mesoderm, and endoderm. The eyes you are using to read these words developed from embryonic ectoderm.

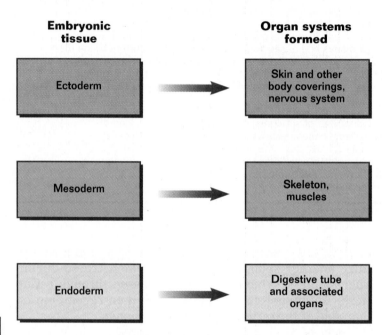

Embryonic tissue	Organ systems formed
Ectoderm	→ Skin and other body coverings, nervous system
Mesoderm	→ Skeleton, muscles
Endoderm	→ Digestive tube and associated organs

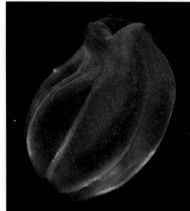

Figure 27-5 Ctenophores are marine animals that are found mostly in deep, open oceans. This particular species of ctenophore likes to feed on other ctenophores, so it is possible that the red object in its center is a relative.

The two most primitive eumetazoan phyla are Cnidaria *(nye DAIR ee uh)* and Ctenophora *(TEHN uh fawr ah)*. The phylum Cnidaria includes jellyfish, hydra, sea anemones, and corals. The phylum Ctenophora is a minor phylum that includes the comb jellies, like the one shown in Figure 27-5. Comb jellies are delicate marine animals that look like tiny jellyfish and swim by moving clusters of cilia. Both cnidarians and ctenophores are radially symmetric. **Radially symmetric** animals have body parts arranged around a central point, as shown in Figure 27-6.

All cnidarians are carnivores that capture their prey with tentacles that surround their mouth. These tentacles bear unique stinging cells called **cnidocytes** *(nih DOH seyets)*, which occur in no other organism and which give the phylum its name. Within each cnidocyte is a small barbed harpoon called a **nematocyst** *(NEHM uh toh sist)*, which cnidarians use to spear their prey. The captured prey is then drawn back to the tentacle containing the cnidocyte.

Tissues Lead to Greater Specialization

Stage 2 Tissues

The cnidarian body is more complex than that of sponges. Cnidarians (phylum Cnidaria) have specialized **tissues** that carry out particular functions. Cnidarians also exhibit **radial symmetry,** with their parts arranged around a central axis like the petals of a daisy. An interior digestive cavity is the site for **extracellular digestion**—digestion outside of cells.

Hydras Are Common Freshwater Cnidarians

The body wall of hydras, like those of other cnidarians, is composed of an outside ectodermal layer, an inside endodermal layer, and a middle, gel-like, mesogleal layer. The tentacles of cnidarians contain many stinging cells called cnidocytes, each of which houses a harpoon-like nematocyst. Cnidarians use their nematocysts to spear prey.

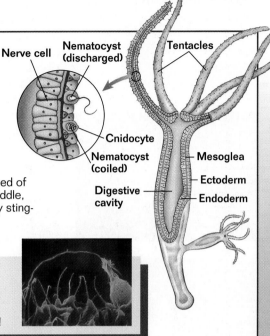

A Lethal Weapon

This scanning-electron micrograph shows a barbed nematocyst bursting out of its cnidocyte. The cnidocyte builds up a very high internal osmostic pressure and pushes the nematocyst outward so explosively that it can penetrate even the hard shell of a crab!

A major evolutionary innovation that occurred among the cnidarians is the extracellular digestion of food. Recall that food trapped by a sponge choanocyte is taken directly into that cell or another cell by endocytosis and digested intracellularly (within the cell). In a cnidarian, food is digested extracellularly (outside the cell), in a digestive cavity. Extracellular digestion is the same heterotrophic strategy used by fungi, except that fungi digest food outside their bodies. The extracellular

Figure 27-6 The body parts of this sea anemone, and of all other cnidarians, are arranged around a central point. This kind of body organization is called radial symmetry. Can you name another commonly known animal that has radial symmetry?

digestion of food has been retained by all of the more
advanced groups of animals.

Cnidarians have two basic body forms. **Medusae** are
free-floating, gelatinous, and often umbrella shaped. **Polyps**
are cylindrical, pipe-shaped animals that are usually
attached to a rock. Many cnidarians exist only as medusae,
others only as polyps, and still others alternate between
these two phases during the course of their life cycle, as
shown in Figure 27-7. ❑

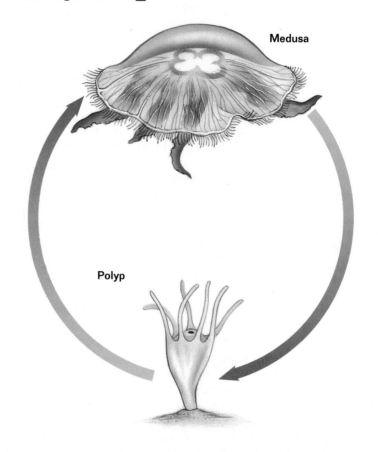

Medusa

Polyp

Figure 27-7 Many cnidarians, like
Aurelia, have life cycles that
include both a polyp and medusa
stage. Medusae, *top,* are free-
swimming. The black-and-white
photograph of the polyp,
bottom, shows that polyps
are stationary.

Section Review

1. *What characteristic of sponges could lead to their being
   mistakenly classified as plants instead of animals?*
2. *Describe the symmetry of a jellyfish.*
3. *Compare the way sponges capture and digest food with
   that of jellyfish.*

Critical Thinking

4. *Imagine a sponge that is living on the ocean floor and
   is covered by a watertight plastic bag. After a period
   of time, the ATP level in the sponge's cells begins to
   decrease. Explain why.*

27-2 Evolution of the Body Cavity

No process more clearly illustrates the way the animal body plan has evolved than the development of the body cavity. As you will see in this section, the fundamental architecture of the animal body is largely determined by the nature of the body cavity.

Section Objectives

■ *Define bilateral symmetry.*
■ *Describe the evolution of the body cavity in animals.*
■ *Compare and contrast flatworms with roundworms.*
■ *Describe the basic body plan of a mollusk.*

Flatworms: Worms With Solid Bodies

Although cnidarians and ctenophores have radial symmetry, all other eumetazoans have bilateral symmetry. **Bilaterally symmetric** animals have left and right halves that mirror each other when they are divided by an imaginary plane passing through their longitudinal center, as shown in Figure 27-8. In a bilaterally symmetric animal, the top surface of the animal is referred to as **dorsal** and the bottom surface as **ventral.** The front end of the animal is **anterior** and the back end is **posterior.** Bilateral symmetry was a major evolutionary advancement among animals because it enabled different parts of the body to become specialized in different ways. For example, most bilaterally symmetric animals have evolved a definite head end, a process called **cephalization.** Animals with heads are often active and mobile, moving through their environment headfirst, with sensory organs concentrated in front so the animal can sense food, danger, and potential mates as it enters new surroundings.

Figure 27-8 Most animals have bilateral symmetry, in which the right and left halves of the animal are mirror images of each other, *left.* Bilaterally symmetric animals have surfaces that are identified as anterior (head), posterior (rear), dorsal (back), and ventral (stomach), *right.*

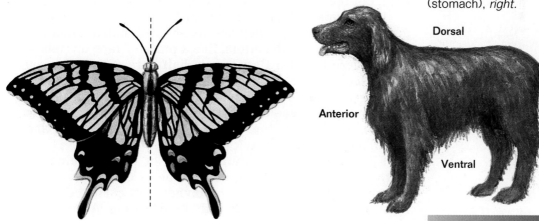

Bilateral symmetry

Dorsal

Posterior

Anterior

Ventral

CONTENT LINK

Learn more about flatworm classification and the life cycles of several parasitic flatworms in Chapter 28.

The simplest of all bilaterally symmetric animals are the solid worms. The largest phylum is Platyhelminthes, with about 20,000 species. The animals in this phylum are commonly called flatworms. Flatworms are the simplest animals

Highlights
Invertebrate Evolution

Organs and Bilateral Symmetry

Stage 3 Bilateral Symmetry

The evolution of a middle tissue layer called mesoderm in the solid-bodied flatworms (phylum Platyhelminthes) made possible the formation of organs such as testes and ovaries. Flatworms were the first animals to be **bilaterally symmetric**, with left and right halves that mirror each other. They were also the first animals to develop a distinct head region.

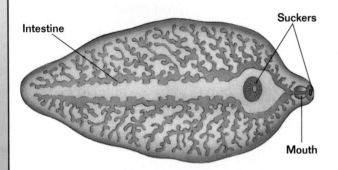

Intestine

Suckers

Mouth

A Parasitic Flatworm

Many flatworms, like these parasitic sheep **liver flukes,** have flat bodies with a mouth that is located on their undersurface. Adult flukes use two sucker-like mouths to attach themselves to the inside of a sheep's liver or gallbladder. The mouth of a flatworm is the only opening to their gut, and all food and wastes pass through it. Their intestine has many branches, ensuring that nutrients are brought close to all the worm's tissues.

Figure 27-9 This common freshwater flatworm is called a planarian. The two dark spots located on the animal's shovel-shaped anterior end are light-sensitive structures called eyespots. The darkly stained interior portion is the planarian's branched intestine.

that have organs. An **organ** is a collection of different tissues that work together as a unit to perform a particular function. The dark spots on the head of the flatworm shown in Figure 27-9 are sense organs called eyespots. Eyespots can detect light, but they cannot focus an image like your eyes can.

Flatworms lack any internal body cavity other than the gut (digestive tube). They are soft-bodied animals that are flattened from top to bottom, like a piece of tape or ribbon. If you were to cut a flatworm in half across its body, you would see that the gut is completely surrounded by tissues and organs. This solid body construction, shown in Figure 27-10, is called **acoelomate** *(ay SEEL oh mayt)*. The term *acoelomate* is from the Greek *a,* meaning "without," and *koilia,* meaning "body cavity."

Flatworms must be thin because of their acoelomate body design. Since flatworms have no circulatory system, dissolved substances such as oxygen and carbon dioxide must pass through the solid body by diffusion. A thin body shortens the distance that these substances must travel to reach each cell. The gut of a flatworm is highly branched. Portions of it run close to practically all of the flatworm's tissues, giving each cell access to food molecules. The gut has only one opening—the mouth. This means that material must move through the mouth in two directions: foods enter and wastes exit.

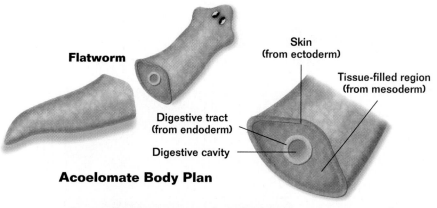

Flatworm

Skin
(from ectoderm)

Tissue-filled region
(from mesoderm)

Digestive tract
(from endoderm)

Digestive cavity

Acoelomate Body Plan

Figure 27-10 Flatworms are called solid-bodied worms because they have no body cavity between their digestive tract and skin. An animal without a body cavity has a limited amount of organ development.

Some flatworms are free-living, but many species are parasitic. Flatworms range in size from free-living forms less than 1 mm in length to parasitic intestinal tapeworms that grow to several meters long. ▢

Roundworms: Nematodes

All bilaterally symmetric animals other than the solid worms have an internal body cavity. A **pseudocoelom** (*SOO duh see luhm*) is a body cavity located between the endoderm and the mesoderm, as shown in Figure 27-11. The evolution of a body cavity was an important improvement in animal body design for several reasons.

CONTENT LINK
....

More information on roundworms and on rotifers, other pseudocoelomates, can be found in Chapter 28.

1. **Circulation**   Fluids that move within the body cavity can serve the function of a circulatory system, permitting the rapid passage of materials from one part of the body to another.

2. **Movement**   Fluid in the body cavity makes the animal's body rigid—like a balloon filled with water. A rigid body offers resistance to contracting muscle cells, enabling muscle-driven body movement.

3. **Organ function**   Body organs that are surrounded by a body cavity can function without being distorted by surrounding muscles. For example, food can pass freely through a gut suspended within a cavity. The rate at which the food moves through the digestive tube is not affected when the animal contracts its body muscles to move.

Figure 27-11 Roundworms are called pseudocoelomates because they have a pseudocoelom, a body cavity that develops between their digestive tract and their mesoderm layer. The fluid-filled pseudocoelom serves as a simple circulatory system in roundworms.

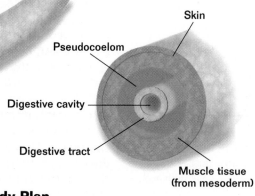

Skin

Pseudocoelom

Digestive cavity

Digestive tract

Muscle tissue
(from mesoderm)

Roundworm

Pseudocoelomate Body Plan

Highlights
Invertebrate Evolution

Advent of a Body Cavity
·····································

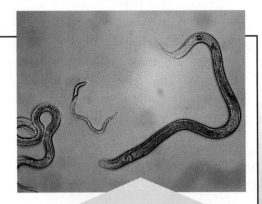

Stage 4 Body Cavity

Roundworms (phylum Nematoda) are bilaterally symmetric, cylindrical worms. The major innovation in their body design is a body cavity called a **pseudocoelom,** which forms between the gut and the body wall.

Most Roundworms Are Tiny

Some free-living roundworms are very small—less than 1 mm (0.04 in.) long. *Caenorhabditis elegans* is composed of only about 1,000 cells and is the only animal whose complete cellular anatomy is known.

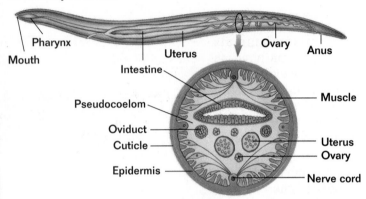

Pharynx
Mouth
Uterus
Intestine
Pseudocoelom
Oviduct
Cuticle
Epidermis
Ovary
Anus
Muscle
Uterus
Ovary
Nerve cord

Why Roundworms Wriggle

Roundworms, like these vinegar eels, have muscles that extend along the length of their bodies, rather than encircling them, enabling the worms to wriggle. The nematode body is covered with a flexible, thick cuticle that is shed as the worm grows. The nematode has a one-way digestive tube. Food enters the mouth at one end of the worm, and waste exits through the anus at the other end.

☐ CAPSULE SUMMARY

Roundworms were the first animals to have a body cavity. Called a pseudocoelom, this cavity is positioned between mesoderm and endoderm. Roundworms have muscle tissue and a gut that is open at both ends.

All pseudocoelomates (animals with a pseudocoelom) have a one-way gut through which food passes into the mouth and out the anus. This is a major improvement in body design, as it permits far greater specialization of the gut. A one-way digestive tube can function like an assembly line, with food being acted on in different ways in each section as it passes through.

Seven animal phyla have a pseudocoelom. Only one of the seven pseudocoelomate phyla contains a large number of species, the phylum Nematoda. Nematodes are commonly called roundworms. There are estimated to be between 500,000 and 600,000 nematode species with only about 13,000 species actually named, most of which are tiny. Most nematodes live in soil. It has been estimated that one spadeful of soil contains an average of 1 million nematodes! There are also some parasitic species, several of which infect humans. In nematodes, a layer of muscle extends along the length of the worm beneath a flexible, thick cover of epidermis and cuticle (protective layer). These long muscles push against the cuticle and the pseudocoelom, whipping the worm's body from side to side as the animal moves. ☐

Mollusks: Animals With Three-Part Bodies

CONTENT LINK

Learn more about mollusk organ systems and mollusk classification in **Chapter 29**.

Even though acoelomates and pseudocoelomates have flourished, a third way of organizing the animal body evolved among the mollusks. Mollusks and all animals that evolved after them, such as the earthworm shown in Figure 27-12, have a body cavity called a coelom and hence are called coelomates. A **coelom** *(SEE luhm)*, is a fluid-filled body cavity that develops not between endoderm and mesoderm, but entirely within the mesoderm.

What is the functional difference between a pseudocoelom and a coelom, and why has the latter type of body cavity been so successful? The answer has to do with the nature of animal embryonic development. In animals, the development of specialized tissues involves a process called primary induction. In **primary induction,** one of the three primary tissues (endoderm, mesoderm, or ectoderm) interacts with another. The interaction requires physical contact. A major advantage of the coelomate body plan is that it allows contact between mesoderm and endoderm, which enables primary induction to occur during development. For example, contact between mesoderm and endoderm permits localized portions of the digestive tract to develop into complex, highly specialized regions, like the stomach. In pseudocoelomates, mesoderm and endoderm are separated by the body cavity, limiting developmental interactions between these tissues.

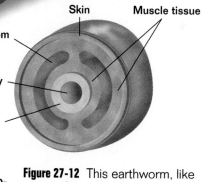

Segmented worm

Skin

Muscle tissue

Coelom

Digestive cavity

Digestive tract

Coelomate Body Plan

The formation of the coelom does re-create one old problem that was solved among pseudocoelomates—the circulation of nutrients, oxygen, and wastes. In coelomates, the digestive tube is again surrounded by tissue that presents a barrier to diffusion, just as in flatworms. In coelomates, however, the efficient circulation of nutrients, oxygen, and wastes is accomplished by a circulatory system. A **circulatory system** is a network of vessels that carries fluids to all parts of the body. The circulating fluid, called blood, carries nutrients and oxygen to the tissues and removes waste and carbon dioxide. Blood is usually pushed through the circulatory system by the contraction of one or more muscular hearts.

Figure 27-12 This earthworm, like most other animals, is called a coelomate because it has a coelom, a body cavity that develops within the mesoderm. The tissue interactions that occur in coelomates have enabled them to develop complex organ systems.

Highlights

Building a Better Body Cavity

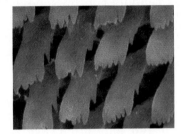

Stage 5 Coelom

The body cavity of a mollusk (phylum Mollusca) is called a coelom and is completely enclosed within mesoderm. The position of the coelom allows physical contact between mesoderm and endoderm, permitting tissue interactions that lead to the development of highly specialized organs in mollusks, such as a heart and stomach. All animal phyla that evolved after mollusks have a coelom.

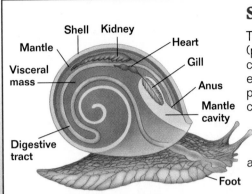

Shell Kidney

Mantle

Heart

Gill

Visceral mass

Anus

Mantle cavity

Digestive tract

Foot

Inside a Snail

The coelom of a snail is confined to a small zone around the heart. The **mantle** is a heavy fold of tissue that wraps around the snail's organs, or **visceral mass.** In aquatic snails, the cavity between the mantle and visceral mass contains the **gills.** The gills capture oxygen from water as it passes through the mantle cavity. In snails, the mantle secretes a hard outer shell. A simple kidney gathers wastes from the coelom and discharges them into the mantle cavity. Within the mouth of a snail is a unique rasping tongue called a **radula,** shown in the photo to the right. Snails creep along the ground on a muscular **foot.**

A radula's surface magnified many times

The least advanced of the coelomates, and thus probably the first to evolve, are the **mollusks.** Mollusks are the only major phylum of coelomates without segmented bodies. Mollusks, shown in Figure 27-13, are divided into three classes and, with over 100,000 named species, are the largest animal phylum except for the arthropods. Mollusks occur almost everywhere. Next to insects, they are the most successful land animals. There are more terrestrial mollusk species (35,000) than there are terrestrial vertebrate species (20,000).

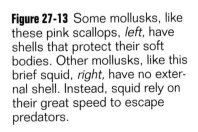

Figure 27-13 Some mollusks, like these pink scallops, *left,* have shells that protect their soft bodies. Other mollusks, like this brief squid, *right,* have no external shell. Instead, squid rely on their great speed to escape predators.

The bodies of all mollusks are composed of three distinct regions. There is a **visceral mass,** a central section that contains the body's organs. Wrapped around the visceral mass like a cape is a heavy fold of tissue called the **mantle.** In aquatic snails, gills are positioned on the inner surface of the mantle. **Gills** are filamentous projections of tissue, rich in blood vessels, that capture oxygen from the water. The water circulates between the mantle and visceral mass, and the oxygen diffuses into the gills. Carbon dioxide is released from the gills into the water. Terrestrial snails, which have no gills, breathe air with simple lungs. Finally, every mollusk has a muscular region called a foot.

The three classes of mollusks are different variations upon this same basic body design.

1. **Gastropods** Gastropods (snails and slugs) use their muscular foot to crawl, and their mantle often secretes a hard protective shell. All terrestrial mollusks are gastropods.

2. **Bivalves** Bivalves (clams, oysters, and scallops) secrete a two-part shell (also called a valve) with a hinge. They feed by drawing water into their shells and filtering out food particles.

3. **Cephalopods** Under their mantle, cephalopods (octopuses and squids) have a modified cavity that creates a jet propulsion system to propel them rapidly through the water.

A characteristic feature of mollusks is a structure called a **radula** (RAJ u luh), a rasping tongue-like organ. All mollusks except bivalves have a radula. With rows of pointed, backward-curving teeth, the radula is used by some snails to scrape algae off rocks. Cephalopods and many gastropods are active predators that use their radula as a weapon to puncture their prey. The small holes often seen in oyster shells are produced by the radula of certain gastropods. Once the hole is made, the snail extracts the oyster's body for food. ▢

❑ *CAPSULE SUMMARY*

The development of a coelom occurred first in the mollusks and is found in all animal phyla that evolved after this group. Embryos with a coelom have greater tissue interaction during development, enabling the formation of advanced body systems.

Section Review

1. *A dog is rolling on its back, wagging its tail. What kind of symmetry does the dog have? Name the body surface of the dog that is rolling on the ground, and identify the end with the wagging tail.*

2. *Compare and contrast the body plan of flatworms with that of roundworms.*

3. *Describe a coelom and identify the group of animals in which it first developed.*

4. *What are three major mollusk characteristics?*

Critical Thinking

5. *Explain why the wall of your digestive tube (intestine) contains muscle tissue but the wall of a roundworm's digestive tube does not.*

27-3 Four Innovations in Body Plan

*T*he mollusks represent one of the great success stories in animal evolution. Their coelomate body design enabled them to achieve a degree of complexity not seen in any of the previously mentioned groups. However, four fundamental changes in the evolution of the animal body have had an enormous impact, leading to the evolution of animals with even greater complexity.

CONTENT LINK

More information on annelid organ systems and annelid classification can be found in **Chapter 29**.

Annelids: Segmented Worms

One of the early key innovations in body plans to arise among the coelomates was segmentation, the building of a body from a series of similar segments. The first segmented animals to evolve were the annelid worms. Annelids, like the polychaete *(PAHL ih keet)* shown in Figure 27-14, are composed of a chain of nearly identical segments, like the boxcars that make up a train. The great advantage of such segmentation is the evolutionary flexibility it offers. A small change in an existing segment can produce a new kind of segment with a different function. Thus, in an annelid worm, some segments are modified for reproduction, some for feeding, and others for eliminating wastes.

Two-thirds of all annelids (about 10,000 species) live in the sea, and most others (about 5,000 species) are earthworms. The basic body plan of an annelid is often described as "a tube within a tube." The gut is a tube suspended within the coelom, which is itself a tube running from mouth to anus.

All annelids share three characteristics.

1. **Repeated segments** The body segments of an annelid are visible as a series of ringlike structures running the length of the body. The segments are divided from one another internally by partitions. In each of the cylindrical

Figure 27-14 Both of these animals are segmented worms. The marine polychaete, *left,* is called a fireworm. If you touch it, tiny hairlike structures break off in your skin and cause excruciating pain. An earthworm, *right,* is a terrestrial segmented worm. Earthworms are important because they aerate and fertilize the soil.

Highlights
The Rise of Segmentation
Invertebrate Evolution

Stage 6 Segmentation

The marine polychaetes, freshwater leeches, and terrestrial earthworms are all annelids (phylum Annelida). They were the first organisms to evolve a body plan based on repeated body segments. Most segments in an annelid look alike, and each segment is separated from the next by an internal partition.

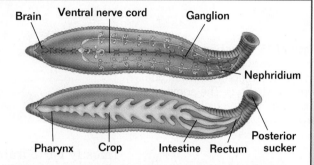

Brain Ventral nerve cord Ganglion Nephridium Pharynx Crop Intestine Rectum Posterior sucker

Leeches Have a Blood Diet

Many leeches are bloodsucking parasites. Like most annelids, they have a well-developed gut, *above,* that includes a muscular pharynx, a food storage area called a **crop,** and an intestine. Each segment of an annelid has a pair of excretory organs called **nephridia,** *top,* and a nerve center called a ganglion. A well-developed brain is located in an anterior segment.

segments, the digestive, excretory, and locomotor (movement) organs are repeated. The fluid within the coelom of each segment creates a hydrostatic (liquid-supported) skeleton that gives the segment rigidity. Muscles within each segment contract against the resistance offered by the fluid in the coelom. Because each segment is separate, each is able to expand or contract independently. This enables an annelid to move in ways that are quite complex. When an earthworm crawls on a flat surface, for example, it lengthens some parts of its body while shortening others.

2. **Specialized segments**   The anterior (front) segments of annelids are modified and contain the sensory organs of the worm. Some of these organs are sensitive to light; elaborate eyes with lenses and retinas have evolved in some annelids. A well-developed **cerebral ganglion,** or brain, is contained in one anterior segment.

3. **Connections between segments**   Although partitions separate the segments, materials and information must still pass between segments. A circulatory system carries blood from one segment to another, while a nerve cord connects the nerve centers, or ganglia, located in each segment with each other and with the brain. These nerve connections enable the brain to coordinate the worm's activities.

   Segmentation underlies the body organization of all advanced coelomate animals. Not only annelids but also arthropods (crustaceans, spiders, and insects) and chordates

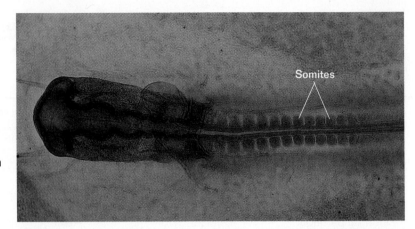

Figure 27-15 Your body's muscles formed from blocks of mesoderm tissue called somites when you were an embryo. The somites of this 33-hour-old chicken embryo are cleary visible.

(mostly vertebrates) exhibit some degree of body segmentation. Sometimes the segmentation is not obvious when viewing an adult arthropod or vertebrate. In many arthropods the segments are fused, and the underlying pattern is difficult to perceive. In the adult human, segments are not apparent, although they can be seen clearly during embryonic development. Vertebrate muscles develop from repeated blocks of tissue called somites, shown in Figure 27-15, that occur in the embryo. Another example of vertebrate segmentation is the vertebral column, which is a stack of very similar vertebrae. ■

Arthropods: Walking Animals

The first animals with jointed appendages were the arthropods, representatives of which are shown in Figure 27-16. Arthropods almost certainly evolved from the annelids. Arthropod bodies are segmented like those of annelids. Individual segments of arthropods often exist only during their larval stages, however. A **larva** is an immature organism that has a body form different from that of the adult. In the adult arthropod, the segments are fused into functional units. For example, the

Figure 27-16 The jointed appendages of this ghost crab, *left,* are obvious. Even its mouth parts are jointed. The Goliath beetle, *right,* an insect, has six jointed legs.

Highlights

Invertebrate Evolution

Stage 7 Jointed Appendages

Arthropods (phylum Arthropoda) have a coelom, segmented bodies, and **jointed appendages.** Each body region of an arthropod (head, thorax, and abdomen) is composed of a number of individual segments that fused during development. All arthropods have a strong **exoskeleton** made of **chitin.** Only one class of arthropods, the insects, has evolved wings, which permit them to fly rapidly through the air.

What an Aphid Left Behind

As arthropods grow, they shed their old exoskeletons in a process called **molting.** Here, insects called aphids voraciously feed on a plant. The white, paperlike objects are exoskeletons that the aphids have recently shed.

A Hovering Dragonfly

The jointed legs of insects attach to the central body region called the thorax. Insects have three pairs of legs and usually two pairs of wings (some insects, like flies, have retained only one wing pair).

---

larva of a butterfly, called a caterpillar, has many segments, while the adult butterfly has only three functional body units—a head, thorax, and abdomen. Each unit is composed of several fused segments. Having a head, thorax, and abdomen is characteristic of all adult insects.

All arthropods have jointed appendages. The name *arthropod* is from the Greek *arthron*, meaning "joint," and *podos*, meaning "foot." To gain some idea of the importance of jointed appendages, imagine yourself without them. Without jointed appendages—hips, knees, ankles, shoulders, elbows, wrists, knuckles—you could not walk or grasp any object. Arthropods use jointed appendages as legs for walking, as wings for flying, as antennae to sense their environment, and as mouthparts for sucking, ripping, and chewing their food. The scorpion, shown in Figure 27-17, seizes and tears apart its prey with jointed mouthparts that have been modified into large pincers. Jointed appendages have proven to be very successful—more than half of all named species on Earth are arthropods. Scientists estimate that 10^{18} insects are alive at any one moment!

CONTENT LINK

Learn more about arthropod body plans and classifications in **Chapter 30.**

Figure 27-17
The stinger of this desert hairy scorpion from Arizona is found at the tip of its tail. Scorpions, along with spiders and mites, belong to a group of arthropods called arachnids.

Another feature of the arthropod body plan is an **exoskeleton,** a rigid external shell made of a substance called chitin. In any animal, a skeleton provides places for muscle attachment. In arthropods, the muscles attach to the interior surfaces of their exoskeletons. An exoskeleton also protects an arthropod from predators and impedes water loss.

The arthropod exoskeleton does have one great limitation. Though chitin is tough, it is also brittle and cannot support great weight. To bear the pull of their more powerful muscles, the exoskeletons of large arthropods must be much thicker than the exoskeletons of small arthropods. The reason why you don't see beetles as big as birds or crabs the size of cows is that their exoskeletons would have to be so thick that the arthropod couldn't move its great weight. Because this size limitation is inherent in the body design of arthropods, no arthropods have ever grown to a great size. ◻

◻ **CAPSULE SUMMARY**

Jointed appendages and an exoskeleton characterize the phylum Arthropoda, which includes insects, spiders, scorpions, crabs, and lobsters.

Echinoderms: Animals With a Five-Part Body Plan

There are two major patterns of embryological development that occur among coelomate animals. The embryos of these animals begin as hollow balls of cells, as shown in Figure 27-18. The balls of cells then indent to form a ball two cell layers thick with an opening called a blastopore to the outside. In annelids, mollusks, and arthropods, the mouth develops from or near the blastopore. This same pattern of development is seen in all acoelomate animals. An animal whose mouth develops from the blastopore is called a **protostome.** The term *protostome* is from the Greek *protos,* meaning "first," and *stoma,* meaning "mouth."

Figure 27-18 Animal embryos develop in one of two ways. In protostome embryo development, cell divisions occur in a spiral pattern. The mouth of the animal forms from the blastopore, and the anus forms from a secondary opening, *left.* In deuterostome embryo development, cell divisions occur in a radial pattern. The anus forms from the blastopore, and the mouth forms from the secondary opening, *right.*

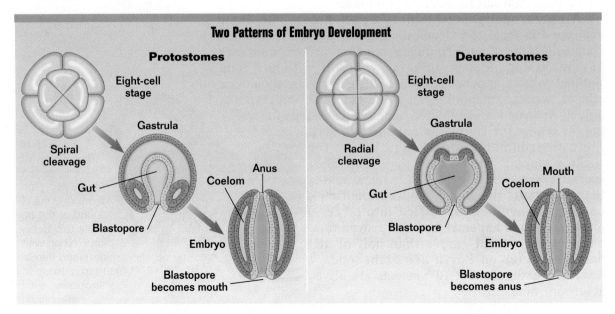

Two Patterns of Embryo Development

Protostomes

Eight-cell stage

Gastrula

Spiral cleavage

Gut

Blastopore

Coelom

Anus

Embryo

Blastopore becomes mouth

Deuterostomes

Eight-cell stage

Gastrula

Radial cleavage

Gut

Blastopore

Coelom

Mouth

Embryo

Blastopore becomes anus

A second kind of coelomate animal developed after the arthropods evolved. In these animals the basic pattern of embryological development differs from the pattern seen in protostome development. In the echinoderms, chordates, and a few other small phyla, the anus—not the mouth—develops from or near the blastopore. The mouth forms later, on another part of the embryo. These animals are called **deuterostomes** because the mouth forms second, from an opening other than the blastopore. The term *deuterostome* is from the Greek *deuteros*, meaning "second," and *stoma*, meaning "mouth." Echinoderms and chordates are clearly related to each other, as demonstrated by their shared pattern of embryonic development.

Deuterostomes represent a revolution in embryonic development. While the pattern of cell division in protostomes is spiral, it is radial in deuterostomes, as shown in Figure 27-18. Furthermore, in protostomes, the developmental fate of each embryonic cell is fixed, meaning that each cell is destined to develop into a particular part of the animal and no other. Because the chemicals that act as developmental signals to the cells are localized within the egg, there is no such developmental control programmed into each individual cell of the protostome embryo. However, if the cells of a deuterostome embryo are separated from each other at an early stage, each cell can develop into a complete organism. This is because the chemical developmental signals in deuterostome embryos are generated by the chromosomes contained in each individual cell. In protostomes, cell position is the key to the developmental fate of embryo cells, while in deuterostomes, whole groups of cells move around during the course of development to form new tissue associations. ▢

The first deuterostomes, marine animals called echinoderms, evolved more than 650 million years ago. Echinoderms, shown in Figure 27-19, were the first animals to develop an **endoskeleton,** an internal skeleton. The term

Figure 27-19 The vermillion biscuit sea star, *left,* and purple sea urchin, *right,* are echinoderms. Most sea stars have five arms, but there are some species that have many more. The long, movable spines of the sea urchin are used to push the animal along the ocean bottom.

Highlights

A New Pattern of Embryo Development
....

Invertebrate Evolution

Stage 8 Deuterostomes

Echinoderms (phylum Echinodermata), like sea stars, are coelomates that have a **deuterostome pattern of embryo development**. This same pattern of development occurs in the chordates. A delicate skin stretches over the internal endoskeleton of echinoderms. Their **endoskeleton** is composed of calcium-rich plates that often are fused together.

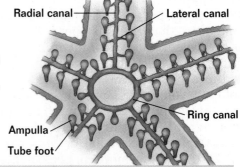

Radial canal — Lateral canal

Ampulla
Tube foot

Ring canal

How a Sea Star Crawls

Sea stars crawl by using thousands of tiny tube feet and a hydraulic **water vascular system** made up of water-filled canals. Each tube foot has a water-filled sac at its base called an ampulla. When the sac contracts, the tube foot is extended in the same way that a water-filled balloon extends when it is squeezed. Hundreds of tube feet extend from the bottom of each arm and attach to the sea floor. The sea star's body muscles can then pull against them, enabling the sea star to haul itself along.

CONTENT LINK
....
More information about echinoderm structure and classification can be found in **Chapter 30.**

❏ CAPSULE SUMMARY

The endoskeleton first appeared in the echinoderms. Radial symmetry, a simplified nervous system, and a water vascular system characterize these marine invertebrates.

echinoderm is from the Greek *echinos,* meaning "spiny," and *derma,* meaning "skin." This "spiny skin" refers to the echinoderm's endoskeleton, which is composed of hard calcium-rich plates called ossicles that are just beneath the delicate skin. When ossicles first form, they are enclosed in living tissue and so are truly an endoskeleton. In adults, the ossicles fuse, forming a hard shell. Today there are about 6,000 echinoderm species, almost all of which live on the ocean bottom. Many of the most familiar animals seen along the seashore are echinoderms, including sea stars (starfish), sea urchins, sand dollars, and sea cucumbers.

The body plan of echinoderms undergoes a fundamental shift during development. All echinoderms are bilaterally symmetric as larvae, but they become radially symmetric as adults. Adult echinoderms have a five-part body plan, easily seen in the five arms of a sea star. The nervous system consists of a central ring of nerves from which five branches arise. While the animal is capable of complex response patterns, there is no centralization of nerve function—no "brain." Apparently, the development of a central nervous system is not feasible in animals with radial symmetry.

A key evolutionary innovation of echinoderms is the development of a hydraulic system to aid movement. Called a **water vascular system,** this fluid-filled system is made up of interconnected internal canals and thousands of tiny hollow tube feet. ❏

Highlights
Invertebrate Evolution

Improving the Skeleton
. . . .

Stage 9 Notochord

Tunicates, lancelets, and all the vertebrates are chordates. Chordates (phylum Chordata) are coelomate animals that have a flexible dorsal rod called a **notochord**. Chordates also have **pharyngeal slits** and a **dorsal hollow nerve cord**. In the vertebrates, the notochord is replaced during embryonic development by a vertebral column.

How Lancelets Move

In a lancelet, a simple chordate, the flexible notochord persists throughout life. It aids swimming by giving the lancelet's muscles an anchor point to pull against. These muscles form a series of discrete blocks that can easily be seen.

How Lancelets Feed

Lancelets feed on microscopic protists that are caught as they filter through cilia and gills on the pharyngeal slits. The beating of cilia that line the front end of the digestive tract draws water, *blue arrow*, through the mouth, through the pharynx, and out the slits.

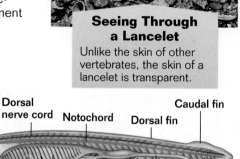

Seeing Through a Lancelet
Unlike the skin of other vertebrates, the skin of a lancelet is transparent.

Water → Mouth · Dorsal nerve cord · Notochord · Dorsal fin · Caudal fin
Water · Pharyngeal slits · Reproductive organs · Intestine · Anus · Muscle

Chordates: Animals With a Notochord
. . . .

Functionally, the endoskeleton of echinoderms is similar to the exoskeleton of arthropods. It is a hard shell that encases the body and serves as an attachment site for muscles. Members of the second major group of deuterostomes, the chordates, have a very different kind of endoskeleton. The chordate endoskeleton is completely internal and is characterized by a stiff rod that develops along the back of the organism during embryological development. Muscles attached to this rod allowed early chordates to swing their backs from side to side, enabling them to swim through the water. The attachment of muscles to an internal skeleton was an important evolutionary advancement. Endoskeletons were the first step along an evolutionary path that led to the vertebrates, and they made it possible for animals to grow large and move quickly.

The approximately 42,500 species of chordates share three characteristics.

1. **Nerve cord** Chordates have a single, hollow, dorsal nerve cord. Nerves attached to the nerve cord travel to different parts of the body.

2. **Notochord** Chordates have a long, stiff rod called a notochord that forms beneath the nerve cord. The notochord is positioned between the nerve cord and the developing gut in the early embryo.

CONTENT LINK
. . . .
*More detail about different kinds of invertebrate chordates is found in **Chapter 30**.*

Figure 27-20 This pine-barrens tree frog is found in New Jersey. Frogs belong to a group of vertebrates called amphibians. Amphibians were the first vertebrates to live successfully on land.

☐ *CAPSULE SUMMARY*

All chordates develop a hollow dorsal nerve cord, notochord, and pharyngeal slits sometime during their lifetime. Vertebrates are chordates that have a cartilaginous or bony endoskeleton that includes a vertebral column.

3. Pharyngeal slits Chordates have a series of slits that develop in the wall of their pharynx. The pharynx, located behind the mouth, is a muscular tube that connects the mouth to the digestive tract and windpipe.

All chordates have all three of these characteristics at some time in their life. For example, humans have a nerve cord, a notochord, and pharyngeal slits as embryos. As adults, humans retain only the nerve cord and the remnants of one pair of pharyngeal slits, which exist as the eustachian tubes that connect the throat to the middle ear.

With the exception of a group of animals called tunicates, which live attached to the ocean bottom, and a small group of fishlike marine animals called lancelets, all chordates are vertebrates. **Vertebrates,** like the frog shown in Figure 27-20, are chordates that have two important evolutionary advancements, a vertebral column and a head.

1. Vertebral column During embryological development, the notochord becomes replaced by a hollow, bony vertebral column, also called a backbone. The vertebral column is composed of many hollow bones called vertebrae that are stacked on top of one another. The vertebral column surrounds and protects the dorsal nerve cord.

2. Head In all vertebrates except the earliest fishes, there is a distinct and well-differentiated head containing a skull and a brain. For this reason, the vertebrates are sometimes called the craniate chordates. The term *craniate* is from the Greek *kranion*, meaning "skull."

All vertebrates have an internal skeleton made of bone or cartilage. The vertebrate skeleton is a solid framework against which attached muscles can contract. A strong supporting endoskeleton makes possible the great size and extraordinary powers of movement that characterize the vertebrates. ☐

Section Review

1. *Identify and explain the significance of a major evolutionary development that first evolved in annelids.*

2. *What limits the maximum size of an arthropod?*

3. *State the evidence that enables scientists to conclude that you are more closely related to echinoderms than to arthropods.*

Critical Thinking

4. *The* law of recapitulation *was proposed by the German scientist Ernst Haeckel in 1866. It states that a species' evolutionary history can be deciphered by studying the stages of its embryological development. Describe evidence that supports this law.*

Concept Mapping

Construct a concept map that shows the similarities and differences among flatworms, roundworms, and segmented worms. Use as many terms as needed from the vocabulary list. Try to include the following items in your map: bilateral symmetry, ectoderm, mesoderm, endoderm, acoelomates, pseudocoelomates, and coelomates. Include additional terms in your map as needed.

Review

Multiple Choice

1. Animals differ from other eukaryotes because they typically
a. reproduce sexually.
b. are organized into tissues.
c. use muscle cells to move.
d. are heterotrophs.

2. Which one of the following animals exhibits radial symmetry?
a. dog **c.** giant squid
b. sponge **d.** sea anemone

3. From what kind of embryonic tissue do adult digestive organs develop?
a. ectoderm **c.** protostome
b. endoderm **d.** deuterostome

4. Which of the following is a coelomate?
a. octopus **c.** sponge
b. nematode **d.** jellyfish

5. Flatworms and roundworms are different because flatworms
a. are bilaterally symmetric.
b. have a gut with one opening.
c. may be parasitic.
d. exhibit cephalization.

6. What specialized body parts evolved after animals developed segmentation?
a. tentacles **c.** wings
b. flagella **d.** blood vessels

7. Which is *not* a characteristic of annelids?
a. jointed appendages
b. repeated segmentation
c. a coelomate body plan
d. specialized segments

8. Which is a feature of the arthropod body plan?
a. a hydrostatic support system
b. pharyngeal slits
c. an exoskeleton
d. a nonsegmented body

9. Which is *not* a deuterostome and does not have an exoskeleton?
a. a butterfly
b. an earthworm
c. a crab
d. a starfish

10. Which of the following do adult chordates and adult echinoderms have in common?
    a. a nonsegmented body
    b. an internal skeleton
    c. a water vascular system
    d. bilateral symmetry

Completion

11. Crabs exhibit _____ symmetry, and hydras exhibit _____ symmetry.

12. Unlike sponges, cnidarians are _____, or true animals. This means that cnidarians have body symmetry and _____ .

13. Flukes and tapeworms are members of the phylum _____ . These animals have a digestive cavity with _____ for taking in food and eliminating waste.

14. The first animals to have body cavities were members of phylum _____ . Fluids moving in the body cavity function as a _____ system to transport nutrients, oxygen, and waste.

15. All mollusk bodies contain a muscular _____ that may be used for movement, a _____ that contains internal organs, and a fold of tissue called the _____ that may secrete a shell.

16. An earthworm is a terrestrial segmented worm, or _____ . Its _____ segments are modified and contain cerebral ganglia and other sensory organs.

17. A starfish has an _____ skeleton. It exhibits _____ symmetry as a larva and _____ symmetry as an adult.

Short Answer

18. Name and describe the two body forms of cnidarians.

19. What is cephalization? What is the main advantage of cephalization?

20. What is primary induction? Give an example of the outcome of primary induction.

21. Adult cats do not have a notochord. Does this mean that cats are not chordates?

Highlights Review

22. What is the evidence that suggests sponges evolved from protists?

23. Suppose you were a small floating animal that happened to brush against the tentacles of a hydra. Describe the action of the hydra that would result in your being pulled through its mouth and into its gut.

24. In animals, nutrients are transferred from the digestive tract to body cells. How does the digestive tract of a flatworm facilitate this process?

25. Compare the body plans and habitats of earthworms with those of roundworms.

26. Why might the mantle cavity of a snail be considered a "multifunctional" structure?

27. What is the evidence that suggests annelids are the ancestors of arthropods?

28. How would a sea star use its water vascular system to pry open the valves of a clam?

29. How is the food gathering mechanism used by lancelets similar to that used by sponges? How is it different?

Themes Review

30. **Evolution** What evolutionary trend do the body cavities of flatworms, roundworms, and segmented worms show?

31. **Levels of Organization** Describe the relationship between cells, tissues, and organs using an example from the basic flatworm body plan.

32. **Structure and Function** What is a radula? How do gastropods use the radula to obtain food?

Critical Thinking

33. **Making Inferences** How is bilateral symmetry better suited to terrestrial animals than radial symmetry?

34. Communicating Effectively The diagram below shows a cross section of a bilaterally symmetric animal. Name the primary germ layers and structures indicated by the lines labeled *a* through *d*. Is the animal an acoelomate, a pseudocoelomate, or a coelomate? Name the phylum whose representatives have this body plan.

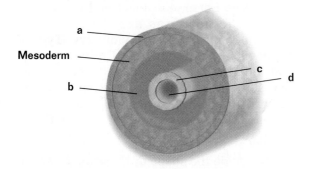

35. Interpreting Data The shell pictured to the right was found on a beach. To which class of mollusks did the organism that occupied the shell belong? What is the probable cause of the organism's death?

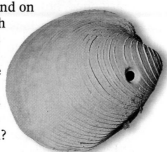

36. Responding Critically A boy who was hospitalized after being stung by bees was interviewed on the nightly news. During the interview, the boy said that the bees were as large as sparrows and seemed to fly at speeds nearing 30 mph. Do you think the boy's story is accurate or exaggerated? Explain.

Activities and Projects

37. Research and Writing Find out about vinegar eels and the conditions under which they thrive. Write a report detailing how to maintain a culture of vinegar eels and describing what we can learn from studying them.

38. Multicultural Perspective Working in groups of three to four students, give a class presentation on the ceremonial use of invertebrates by the indigenous people of another culture. You might consider types of musical instruments, art forms, diet, oral stories, and seasonal celebrations passed down over many generations.

39. Cooperative Group Project Collect clam chowder recipes from different cookbooks. After studying several of the recipes, write your own recipe for clam chowder. Then prepare a pot of clam chowder using your recipe. Allow members of the class to taste and critique your chowder. Different group members may take responsibility for collecting recipes, writing the group's recipe, preparing the chowder, and arranging the chowder tasting.

40. Research and Writing Use library resources and interview medical professionals to identify the species of the "medicinal leech." Find out how it was used in medicine during the eighteenth and nineteenth centuries and how it is used today. In your report, describe why medical professionals stopped using the leech at the turn of the century and why they have resumed medical use of the leech.

Readings

41. Read "Luck Be a Ladybug," in *National Wildlife,* June–July 1994, pages 30–32. Why are "bug-pickers" traveling to the foothills of California's Sierra range each fall season?

42. Read "Dusk and Dawn Are Rush Hours on the Coral Reef," in *Smithsonian,* October 1993, pages 104–115. Describe the daily activity of three different invertebrates that live in the reef along the Gulf of Aqaba.

Survey of Invertebrates

OBJECTIVES

- Compare similarities and differences among invertebrates.
- Relate structural adaptations to the evolution of invertebrates.

PROCESS SKILLS

- classifying invertebrates
- relating structural features to functions
- analyzing data for trends from simple to complex development

MATERIALS

- preserved or living specimens of invertebrate animals
- blunt probes
- hand lens or stereomicroscope

BACKGROUND

1. What type of body form does a sea anemone have? a jellyfish?
2. What term describes concentration of nervous tissue at the anterior end of an animal's body?
3. What are the three body regions of a mollusk?
4. Name the three body regions in adult insects.
5. What is the name of the system of movement in echinoderms?
6. Write your own **Focus Question** on your Vee Form.
7. **Knowing Side of the Vee** List the **Concepts** and new **Vocabulary Words** on your Vee Form. In the **Concept Statements** section of the Vee, use these words in sentences that define and explain them.

TECHNIQUE

Doing Side of the Vee

You will travel to seven stations to observe organisms that are representative of the invertebrate groups, or phyla. In the **Additional Records and Observations** section of the Vee Form, construct and complete a table. The headings are: Group Name, Phylum Name, Type of Symmetry, Type of Body Plan, Distinguishing Characteristics, and Animal Examples. Under "Group Name" list the name from each "Station" number. Record the answers to the following questions in the **Records** section of the Vee Form.

Station 1: Sponges

1. Describe the shape of the sponge. Why is the skeleton filled with holes or pores?
2. Since a sponge is sessile or permanently attached, how does it obtain food?

Station 2: Cnidarians

3. What body form do hydra and coral exhibit?
4. How do they paralyze their prey?
5. These organisms have a medusa body form which allows them to be free-floaters. What part of their body enables them to float?

Station 3: Worms

6. What is the shape of the planarian and sheep liver fluke? Relate this shape to their method of obtaining and distributing oxygen.
7. To what group of invertebrates do these animals belong?
8. How many body openings do these animals have?
9. Describe the outside and the shape of the body of *Ascaris*.

Station 4: Mollusks

10. Of the snail, clam, and squid, which have a visible shell?

11. How do each of these animals obtain food?

Station 5: Segmented Worms

12. Observe and compare the earthworm and the leech. What features do they have in common? To what group of worms do these two specimens belong?

Station 6: Arthropods

13. Observe and compare arthropod specimens such as crayfish, pill bug, grasshopper, scorpion, millipede, and centipede. Describe the outer covering of these animals. Are the appendages straight and stiff or are they jointed?

14. Look closely at the antennae of the grasshopper and crayfish. Speculate about their function.

15. Compare and contrast the walking legs of the millipede and the centipede.

16. Describe how each of the arthropod species shown at this station obtains oxygen.

Station 7: Echinoderms

17. Observe the sea star (starfish), sand dollar, and sea urchin specimens and compare the spiny skin of these specimens.

18. Describe the symmetry of echinoderms.

19. These animals move by a water vascular system. Look at the ventral surface of the sea star, and sketch the tube feet that make up this system.

20. In the **Procedure** section of the Vee, briefly summarize the procedure you followed.

21. Clean up your materials and wash your hands before leaving the laboratory.

INQUIRY

1. Do cnidarians exhibit cephalization? Explain your answer.

2. List the advantages and disadvantages of the exoskeleton of an arthropod.

3. Echinoderms have an endoskeleton. What advantage does an endoskeleton have over an exoskeleton?

4. Use the information on the **Knowing Side** of the Vee to interpret your results from the **Doing Side**, and then write your **Knowledge Claim**. Write a **Value Claim** for this lab.

ANALYSIS

1. Animals like *Ascaris* and leeches are parasites. Speculate why their outer body covering is a tough cuticle.

2. In terms of evolutionary development, why are the annelids placed after the mollusks when discussing invertebrates?

3. List reasons why arthropods are the most successful group of animals.

FURTHER INQUIRY

Write a **New Focus Question** that could be the point of a new investigation. The following is an example:

Some invertebrates burrow and live underground. To what extent might adaptations to different environments account for variations within a phylum?

CHAPTER 28

SIMPLE INVERTEBRATES

REVIEW
- osmosis and endocytosis (Section 3-2)
- choanocytes and filter feeding (Section 27-1)
- radial symmetry and cnidocytes (Section 27-1)
- polyps and medusae (Section 27-1)
- cephalization (Section 27-2)
- solid-bodied and pseudocoelomate worms (Section 27-2)

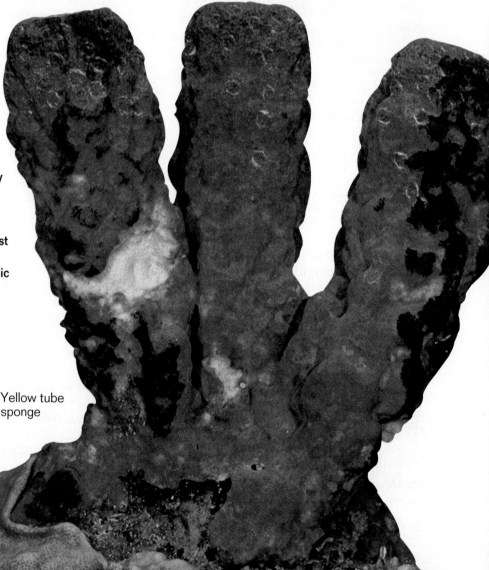

Yellow tube sponge

28-1 Sponges

Whatever their shape or size, all sponges have two characteristics in common: they have a body wall penetrated by many pores, and they are sessile. Sessile *animals are not mobile. Early in their lives, sponges attach themselves firmly to the sea bottom or some other submerged surface, like a rock or coral reef. There they remain for their entire life.*

Section Objectives

- Describe how sponge cells receive nutrients.
- Describe how a sponge's body is structurally supported.
- Compare and contrast asexual and sexual reproduction in sponges.

Sponges Are Filter Feeders

A sponge has the same basic structure as a bag—a large internal cavity with an opening at the top. As you learned in Chapter 27, facing into the internal cavity of a sponge is a layer of choanocytes, or collar cells. By beating their flagella, the collar cells draw water through the sponge's many pores and into the internal cavity of the sponge. As water circulates through a sponge, the collar cells trap food particles by functioning as sieves, as shown in Figure 28-1.

How do the other cells of the sponge, like the epithelial cells, survive if the collar cells take in all of the food? The collar cells release nutrients into the mesenchyme, the jellylike middle layer of the sponge body. Here, the nutrients are picked up by other specialized cells, called amoebocytes *(uh MEE boh seyets)*. **Amoebocytes,** cells that have irregular amoeba-like shapes, wander about the mesenchyme, supplying the rest of the sponge's cells with nutrients, and carrying away their wastes.

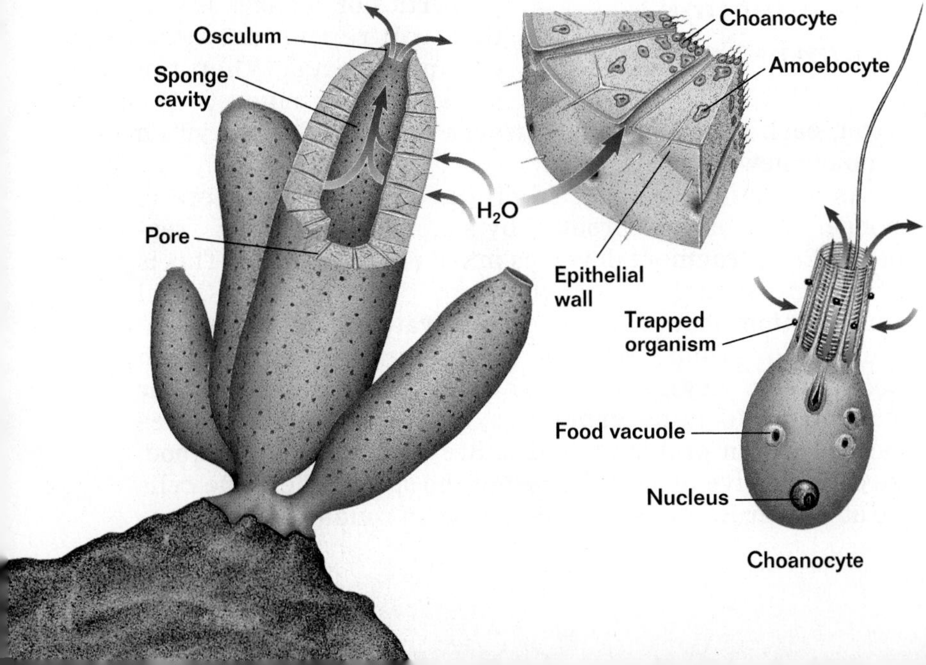

Figure 28-1 Water enters a sponge by passing through pores in the sponge's body wall, *left.* The inside of a sponge is lined by cells called choanocytes, *center,* which trap tiny organisms in the water. The collar of a choanocyte is made of small, hairlike projections resembling a picket fence, *right.* The beating of each choanocyte's flagella draws water, *blue arrows,* down through its collar. Organisms like bacteria and algae are trapped in the collar. These organisms then move toward the cell's cytoplasm, *red arrows,* where they are ingested through endocytosis.

Simple Invertebrates **639**

Figure 28-2 Calcareous sponges, *above left,* have spicules composed of calcium carbonate. In glass sponges, *above center,* the spicules are made of silica. Demosponges, *above right,* may have skeletons made of silica, spongin fibers, or both.
Eighty percent of all sponge species are demosponges.

Sponges Have Simple Skeletons of Varied Composition

For the sponge's body to function effectively, the body wall must be rigid enough to prevent the sponge from collapsing in on itself. Imagine a hollow ball made of gelatin, and you will see the point. The skeleton of a sponge is not a fixed framework like your skeleton, but rather a diffuse network of fibers or minerals. The skeletons of most sponges are made of a resilient flexible protein fiber called spongin, which gives sponges their name. A few sponges have more brittle skeletons composed of tiny hard needles of silica or calcium carbonate called spicules, often embedded within spongin. Taxonomists group sponges into three classes based on the composition of their skeletons. A representative from each class is shown in Figure 28-2.

Asexual and Sexual Reproduction Occur in Sponges

One of the most remarkable properties of sponges is that they will regenerate when cut into pieces. When the pieces are returned to the ocean, each bit of sponge, however small, will grow into a complete new sponge.

As you might suspect from this ability to regenerate, sponges frequently reproduce by simply breaking into fragments. Each fragment develops into a new individual. This is a form of asexual reproduction. Another form of asexual reproduction occurs in some freshwater sponges (a relatively small group). When living conditions become harsh (cold or very dry), some freshwater sponges ensure their survival by forming **gemmules,** clusters of amoebocytes encased within protective coats. Sealed in with ample food, the cells survive even if the rest of the sponge dies. The cells are able to grow into a new sponge when conditions improve.

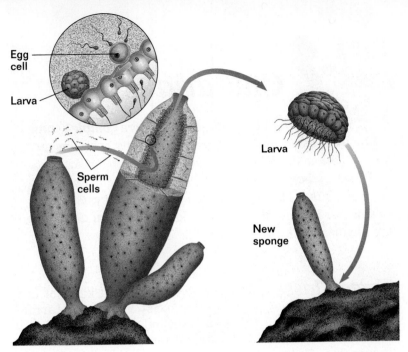

Egg
cell

Larva

Sperm
cells

Larva

New
sponge

Figure 28-3 Sperm from one sponge enter another sponge through its pores, *left*. The sperm are taken in by the receiving sponge's choanocytes and pass into the mesenchyme layer, where egg cells reside. After fertilization, larvae develop. The larvae swim out of the sponge and eventually grow into a new sponge, *right*.

Sexual reproduction among sponges, shown in Figure 28-3, is also common. Although sponges may reproduce sexually, most sponges are not exclusively male or female. Instead, each individual produces both eggs *and* sperm. An organism that produces both eggs and sperm is called a **hermaphrodite** *(huhr MAF roh deyet)*. Can eggs from a certain individual be fertilized by sperm from the same individual? No. Eggs and sperm are produced at different times, so self-fertilization is avoided. The advantage of hermaphroditism lies in the fact that the sperm of any individual can fertilize the egg of any other individual. You can see that the probability of one sponge successfully fertilizing another sponge is much greater than if only half of the population produced eggs. Many types of animals that rarely encounter members of their own species—such as sessile invertebrates and even some vertebrates like deep-sea fishes—are hermaphroditic. ❑

❑ CAPSULE SUMMARY

Sponges are simple filter-feeding animals that have a supportive skeleton composed of soft spongin fibers or hard spicules, or a combination of both. Sponges are capable of both asexual and sexual reproduction.

Section Review

1. *What are amoebocytes, and what are their functions?*
2. *Explain why not all sponges would make good bath sponges.*
3. *In what way is sponge reproduction and human reproduction similar?*

Critical Thinking
4. *What evolutionary advantage is there to a free-swimming larval stage in sponges?*

28-2 Cnidarians

The body plan of a cnidarian is more complex than that of a sponge. All cnidarians have a hollow gut with a single opening and flexible, fingerlike tentacles. Located on the tentacles are stinging cells called cnidocytes that contain harpoon-like nematocysts. Tiny freshwater hydra, jellyfish, and flowerlike coral all belong to the phylum Cnidaria.

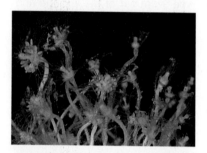

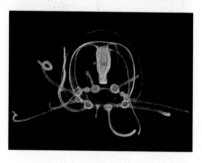

Figure 28-4 Polyps, *top,* are specialized for a sessile existence. They attach to a surface with their mouth facing upward. A medusa, *bottom,* is specialized for swimming. Its mouth faces downward, with its tentacles dangling down around it.

Three Classes of Cnidarians

Two very different kinds of body forms exist among cnidarians: vase-shaped polyps and umbrella-shaped medusae, each shown in Figure 28-4. Taxonomists divide cnidarians into three classes based on their life cycles. The most primitive cnidarians are members of the class Hydrozoa. Most hydrozoans spend part of their lives as medusae and part as polyps. The other two classes of cnidarians, which evolved from hydrozoans, emphasize one or the other body form. The jellyfish (class Scyphozoa) pass briefly through a polyp stage but spend most of their lives as medusae. The other class, corals and sea anemones (class Anthozoa), spend all of their lives as polyps and have no free-swimming medusa stage.

Hydrozoans Are Found in Fresh and Salt Water

The class Hydrozoa includes about 2,700 named species. Most species are colonial marine organisms with both polyp and medusa stages in their life cycles. But the freshwater hydrozoans, so often studied in school laboratories, are probably most familiar to us. The abundant freshwater genus *Hydra* is unique among hydrozoans because it has no medusa stage and exists only as a solitary polyp. Hydras live in quiet ponds, lakes, and streams. Hydras attach themselves to rocks or water plants by means of a sticky secretion produced by an area called the **basal disk**. Hydras can glide around by decreasing the stickiness of the material secreted by their basal disk. They can also move by tumbling, as shown in Figure 28-5. Most hydras are white or brown, although some appear green because of the algae living beneath their outer cells.

Marine hydrozoans are typically far more complex than freshwater hydrozoans. They often form colonies, with many individuals living together. While the cells of the

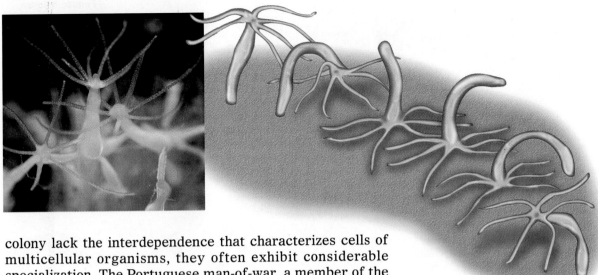

colony lack the interdependence that characterizes cells of multicellular organisms, they often exhibit considerable specialization. The Portuguese man-of-war, a member of the marine genus *Physalia*, incorporates both medusae and polyps. As shown in Figure 28-6, a gas-filled medusa of *Physalia* floats like a balloon on the surface of the water. Dangling below, nematocyst-studded tentacles that can be 15 m (50 ft.) long in large specimens stun and entangle prey. Several different kinds of polyps are attached to the tentacles, each carrying out a different function, such as feeding, sexual reproduction, and defense.

Figure 28-5 A hydra, *above,* can move by tumbling. The hydra's body bends over and touches the bottom, and then the hydra pulls its basal disk free. The basal disk end then somersaults over the mouth end and re-attaches on the other side.

Figure 28-6 As many as 1,000 individual medusae and polyps may compose a single Portuguese man-of-war colony, *far left.* Ocean currents and winds often direct Portuguese man-of-war colonies onto beaches in the southern United States, where they become a hazard to swimmers, *left.*

Reproduction in Hydrozoans

Most hydrozoans are colonial organisms whose polyps reproduce asexually by forming small buds on the outside of their body. These buds develop into miniature polyps. Eventually the offspring separate from the colony and begin living independently. Many hydrozoans are also capable of sexual reproduction. The genus *Obelia* is a typical colonial hydrozoan. The body of *Obelia* is branched like deer antlers, with various polyps attached to the branched stalks. As in the Portuguese

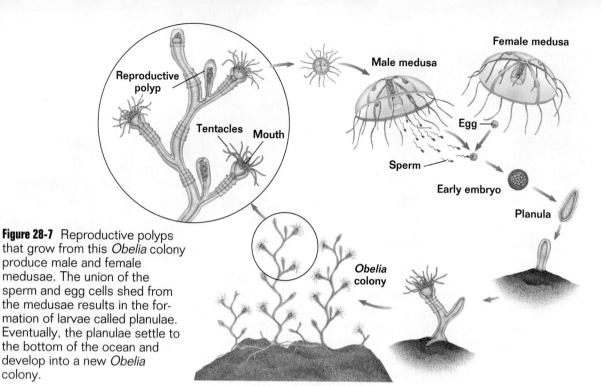

Figure 28-7 Reproductive polyps that grow from this *Obelia* colony produce male and female medusae. The union of the sperm and egg cells shed from the medusae results in the formation of larvae called planulae. Eventually, the planulae settle to the bottom of the ocean and develop into a new *Obelia* colony.

man-of-war, some of the polyps are specialized for feeding, and others are for reproduction. Reproductive polyps give rise to male and female medusae. The medusae leave the polyps and release eggs and sperm into the water. The gametes fuse and produce zygotes that develop into free-swimming larvae called **planulae** (*PLAN yoo lee*). The planulae eventually settle on the ocean bottom and develop into new polyps, shown in Figure 28-7. Sexual reproduction in *Obelia* usually occurs in the fall, when low water temperatures trigger the development of eggs and sperm in the medusae.

Like sponges, some species of *Hydra* are hermaphrodites; a single individual is capable of producing both eggs and sperm. The gametes are produced in swellings along the hydra's body wall. In other species, there are separate sexes. During sexual reproduction, sperm released into the water swim to eggs contained in a nearby hydra. Fertilization typically occurs in the fall. The resulting zygote forms a cell mass with a hard covering that protects it through the winter. In the spring the embryo bursts through the cover and develops into a new hydra.

Scyphozoans Are Marine Jellyfish

The best known cnidarians are the jellyfish, members of the class Scyphozoa (*seye fuh ZOH uh*). The name Scyphozoa is from the Greek *skyphos*, meaning "cup," and *zoia*, meaning "animal." Their name refers to the fact that members of this class spend most of their lives as medusae, which have

Figure 28-8 The diver is photographing a sea wasp jellyfish in the waters along the coast of Australia. The stings from a large sea wasp jellyfish can cause a human to die in a matter of minutes.

the shape of an inverted cup. There are about 200 species of jellyfish. Some are as small as a thimble, and others are as large as a queen-sized mattress. Jellyfish are active predators that ensnare and sting prey with their tentacles. The toxins contained within the nematocysts of some species are extremely potent. The sea wasp jellyfish, shown in Figure 28-8, lives in the sea along the tropical northern coast of Australia and can inflict severe pain and even death on humans.

Although the jellyfish that you see in the ocean are medusae, most species also go through a small, inconspicuous polyp stage at some point in their life cycle. Most jellyfish reproduce sexually, and their life cycle often includes the development of larval planulae. One of the most familiar jellyfish is the stinging nettle, *Aurelia*. The life cycle of *Aurelia*, shown in Figure 28-9, includes both medusa and polyp stages.

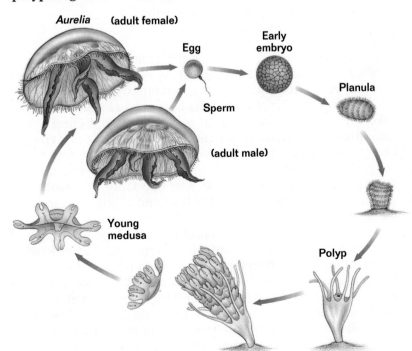

Figure 28-9 The adult male and female medusae of *Aurelia* produce gametes. The early embryos that result from the fusion of the gametes develop into planulae that settle and attach to the ocean bottom. The unattached end of the planula develops a mouth and tentacles, becoming a polyp. As the polyp grows, it forms medusae, which eventually bud off and develop into jellyfish.

Figure 28-10 The flat, lacelike branches of this sea fan from Micronesia are penetrated by many tiny holes. A polyp lives in each hole.

Figure 28-11 Once the sea star is paralyzed by the sea anemone's stinging tentacles, it will be directed through a centrally located, slitlike mouth and dissolved in the anemone's digestive cavity.

Anthozoans Are Marine Polyps

The largest class of cnidarians is the anthozoans. The class name Anthozoa is from the Greek *anthos*, meaning "flower," and *zoia*, meaning "animal." There are approximately 6,200 species in this class. The most familiar anthozoans are the brightly colored sea anemones and corals. Other members are known by such fanciful names as sea pansies, sea fans, and sea whips. A sea fan is shown in Figure 28-10. Anthozoans typically have a thick, stalklike body topped by a crown of tentacles that occur in groups of six. Nearly all of the shallow-water species contain symbiotic dinoflagellates. The anthozoans provide a place for the dinoflagellates to live in exchange for some of the food that the dinoflagellates produce. The fertilized eggs of anthozoans develop into planulae that develop into polyps. No medusae are formed by anthozoans.

Sea Anemones

Sea anemones are a large group of soft-bodied polyps found in coastal areas all over the world. Most do not grow very large, only from 5 mm (0.2 in.) to 100 mm (4.0 in.) in diameter. But one species on the northern Pacific coast of the United States may reach up to 1 m (3.3 ft.) in diameter. Sea anemones feed on fish and other marine life that happen to swim within reach of their tentacles. When they are touched, most sea anemones retract their tentacles into their body cavity and contract into a tight ball. Sea anemones are highly muscular and relatively complex animals with many divided internal cavities. Many species, such as the one shown in Figure 28-11, are quite colorful.

Corals

Most corals live in colonies. Each small polyp secretes a tough, stonelike outer skeleton. Coral reefs form when these calcium-rich skeletons become cemented to those of their neighbors. When a polyp dies, its hardened skeleton remains. Like the cement foundation of a house, the old coral skeletons provide a foundation for new coral polyps growing above them. Eventually, the reef may grow all the way to the surface. Hundreds of thousands of polyps live together on the upper part of the reef at any one time. A coral reef, shown in Figure 28-12, provides food and shelter for an enormous variety of fishes and invertebrates.

There are three kinds of coral reefs. **Fringing reefs** form close to a beach. The reefs off the coastal beaches of Florida are fringing reefs. **Barrier reefs** form in much deeper water farther out from the shore. The Great Barrier Reef, off the coast of eastern Australia, is the largest reef in the world, easily visible from outer space. Even farther out from land are **atolls,** coral islands that form far out at sea and grow into a ring shape, with a lagoon in the center.

Figure 28-12 A growing coral reef in Fiji teems with life, *left*. The many creatures that live around a coral reef interact to form complex food webs. These feeding corals, *above*, may appear to be individual organisms. Actually, the body walls of all of them are connected.

Coral reefs are usually found in tropical seas in shallow, clear water. The symbiotic algae that live inside the cells of the coral require sufficent light and warmth for photosynthesis. Given this fact about coral reef growth, naturalists were long puzzled by how an atoll could form. The deepest corals of atolls are far below the surface light. Darwin solved the puzzle over a century ago. He suggested that atolls form from coral colonies that are attached to the tops of undersea mountains or volcanoes that were formerly at sea level but have slowly sunk beneath the sea. The coral colonies continued to grow upward at the same speed that the mountains sank, keeping the coral colony's "living zone" near the surface. ☐

☐ **CAPSULE SUMMARY**

Hydras, jellyfish, corals, and sea anemones are members of the phylum Cnidaria. These soft-bodied animals have tentacles armed with stinging cells (cnidocytes). Cnidarians are grouped into the three classes Hydrozoa, Scyphozoa, and Anthozoa. Sexual reproduction among many cnidarians leads to the development of an intermediate larval stage called a planula.

Section Review

1. *Explain why a Portuguese man-of-war is not considered a jellyfish.*
2. *What is a planula, and what does it become?*
3. *Which body form of* Aurelia *dominates its life cycle?*
4. *Compare and contrast sea anemones with corals.*
5. *Describe the formation of an atoll.*

Critical Thinking
6. *Damage to coral reefs by human activity is increasing worldwide. Identify some activities that may be causing this damage, and explain how the damage could affect humans and other animals.*

28-3 Kinds of Simple Worms

*T*he remaining major phyla of simple invertebrates all have basically tubular bodies—many are what you call worms. Here, you will consider four phyla: flatworms (phylum Platyhelminthes), ribbon worms (phylum Rhynchocoela), roundworms (phylum Nematoda), and rotifers (phylum Rotifera). These organisms, although all simple in body plan, are a great deal more complex than sponges or cnidarians.

Section Objectives

- Compare and contrast turbellarians with trematodes.
- Describe the life cycle of a beef tapeworm.
- List the major characteristics of the ribbon worms.
- Contrast the life cycles of several different parasitic roundworms.
- Describe the characteristics of a rotifer.

Flatworms Have a Solid Body

Flatworms, like the one shown in Figure 28-13, are bilaterally symmetrical with simple bodies that lack both respiratory and circulatory systems. Flatworms have no need for these systems because each cell of a flatworm's body is close to the animal's exterior. A flatworm's body is solid because it has no body cavity. There are approximately 20,000 named species of flatworms, which compose three major classes: Turbellaria, Trematoda, and Cestoda.

Figure 28-13 Many of the free-living flatworms are beautiful marine species that swim with wavelike movements of their flattened body.

Turbellaria

There are over 3,000 species in the class Turbellaria, almost all of them free-living marine flatworms. Marine flatworms are rarely studied by students, however, as they are difficult to raise. Instead, students usually study a freshwater flatworm like the planarian *Dugesia*, shown in Figure 28-14 and in *Up Close: Planarian* on page 651. Like all platyhelminthes, its body is flat like a piece of tape. Dugesia's rear is tapered, while its front end is rounded like a shovel. It swims by flexing its body in a wavelike motion that somewhat resembles the butterfly stroke.

Figure 28-14 This small planarian is common in clear lakes and streams. Since planarians avoid light, they can most often be found clinging to the undersides of rocks or logs in the water.

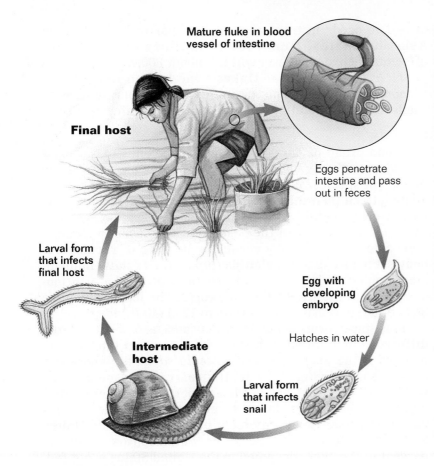

Mature fluke in blood vessel of intestine

Final host

Eggs penetrate intestine and pass out in feces

Larval form that infects final host

Egg with developing embryo

Hatches in water

Intermediate host

Larval form that infects snail

Figure 28-15 Adult blood flukes, *above,* normally live as male-female pairs in the blood vessels of the small intestine. The males are the thick-bodied worms, and females are the thinner, thread-like worms. Workers may be exposed to blood fluke larvae when they wade in fields contaminated with human feces, *left.* The larvae are able to bore through the workers' skin and enter their bloodstream.

Trematoda

Most flatworm species are parasitic. The largest flatworm class, Trematoda, consists of about 6,000 species of parasitic worms called **flukes**. Some flukes are endoparasites. **Endoparasites** live *inside* their hosts. Others are ectoparasites. **Ectoparasites** live on the outside of their hosts. To avoid being digested by their host, endoparasites often have a thick protective covering of cells called the **tegument**.

Flukes and the other parasitic flatworms have extremely simple bodies with few organs. Biologists think that parasitic flatworms evolved from free-living forms. After adopting their parasitic lifestyle, they no longer needed many of the organs necessary for independent living, like well-developed mouths or digestive systems. Other organs became modified. For example, many parasitic flatworms have one or more suckers that they use to attach themselves to their host.

Most flukes have complex life cycles involving more than one host. Blood flukes of the genus *Schistosoma* are responsible for schistosomiasis *(shihs tuh soh MEYE uh sihs),* a major public health problem in many parts of the tropics. These parasites, shown in Figure 28-15, live in the blood vessels of infected individuals and cause bleeding of the intestinal wall and decay of the liver due to blocked blood passages. The life cycle of blood flukes includes a particular species of snail as an intermediate host. Most blood fluke infestations of people occur in tropical countries, particularly those in Asia and Africa. Hundreds of millions of people are affected. Schistosomiasis will kill about 800,000 people this year! A second kind of fluke is the human liver fluke, *Clonorchis sinensis*. This parasite passes from humans to snails to fish, reinfecting humans who eat the infected fish.

Cestoda

A second group of parasitic flatworms belongs to the class Cestoda, commonly called tapeworms. There are about 1,500 tapeworm species. In contrast to flukes, tapeworms permanently attach themselves to the inner intestinal wall of their host. Tapeworms do not have mouths or digestive systems. They absorb food from the host's intestine directly through their skin. They grow by producing a string of rectangular body sections called **proglottids** *(proh GLAHT ihds)* immediately behind their head. These sections are added continuously during the life of the tapeworm. The long, ribbonlike body of a tapeworm may grow up to 12 m (40 ft.) long!

Most tapeworms occur in vertebrates, and about a dozen different kinds infect humans. One tapeworm that infects humans is the beef tapeworm, *Taenia saginata*, whose life cycle is shown in Figure 28-16. Humans can become infected if they eat beef that has not been cooked to a temperature high enough to kill the larval tapeworms that occur as cysts inside the meat. About 1 percent of the cattle in the United States are infected by tapeworms. Since approximately 20 percent of the beef consumed in the United States is not federally inspected, meat from infected cattle may reach the marketplace and be eaten. As a result, the beef tapeworm is a frequent human parasite in the United States. ◻

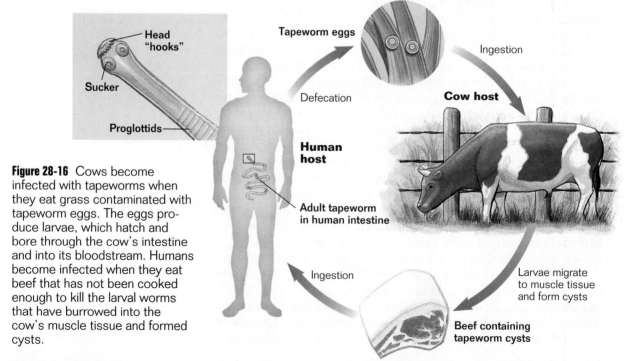

Figure 28-16 Cows become infected with tapeworms when they eat grass contaminated with tapeworm eggs. The eggs produce larvae, which hatch and bore through the cow's intestine and into its bloodstream. Humans become infected when they eat beef that has not been cooked enough to kill the larval worms that have burrowed into the cow's muscle tissue and formed cysts.

Head "hooks"

Sucker

Proglottids

Tapeworm eggs

Ingestion

Cow host

Defecation

Human host

Adult tapeworm in human intestine

Ingestion

Larvae migrate to muscle tissue and form cysts

Beef containing tapeworm cysts

UP CLOSE PLANARIAN

- **Scientific name:** *Dugesia* sp.
- **Range:** Worldwide
- **Habitat:** Cool, clear, permanent lakes and streams
- **Size:** Average length of 3–15 mm (0.1–0.6 in.)
- **Diet:** Protozoa and dead and dying animals

External Structures

Nervous System Sensory information gathered by the brain is sent to the muscles by two main nerve cords that are connected by cross branches. Planarians show an ability to learn complex tasks, such as how to find their way through an experimental maze. Interestingly, their memory of learned tasks appears to be stored chemically. Some investigators even claim that one planarian can sometimes learn to pass through a maze by eating another planarian that already knows the way!

Feeding *Dugesia*, a free-living flatworm, must extend its muscular **pharynx** out of its centrally located mouth in order to feed.

Digestion and Excretion Food drawn into the pharynx passes into a closed, branched intestine. The branching of a flatworm's intestine enables nutrients to pass close to all of the worm's tissues. The nutrients are absorbed through the intestinal wall, and undigested food is expelled through the mouth.

Brain

Nerve cord

Eye

Intestine

Mouth

Pharynx

Female reproductive system

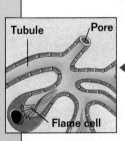

Tubule — Pore

◄ Excretory system (partial view)

Flame cell

Male reproductive system

Reproductive pore

Water Balance Because *Dugesia*'s body cells are hypertonic to fresh water, water continually enters its body by osmosis. Excess water moves into a network of tiny tubules that run the length of its body. Side branches are lined with many **flame cells,** specialized cells with beating tufts of cilia that resemble a candle flame. The beating cilia draw water through pores to the outside of the worm's body. Most flatworms, including flukes and tapeworms, have a similar water-removal system.

Reproduction
Dugesia reproduces in the summer. It reproduces asexually by attaching its posterior end to a stationary object and stretching until it tears in two. Each half then regenerates another complete animal. *Dugesia* is also capable of sexual reproduction. As **hermaphrodites**—individuals that have both male and female sex organs—two individuals simultaneously fertilize each other, each transferring sperm to the other. Protective capsules enclose groups of several fertilized eggs. The capsules are laid in bunches, and the eggs inside hatch in two to three weeks.

Figure 28-17 This beautiful ribbon worm is from Panama. Some ribbon worms burrow beneath sand. Others live under rocks or in algae.

Ribbon Worms Have a Digestive Tube With Two Openings
....

The longest worms by far are the free-living, flat-bodied, marine ribbon worms, members of the phylum Rhynchocoela *(RIHNG koh seel ah)*. There are about 650 species of marine ribbon worms. Individual worms are typically 0.31 m (1 ft.) long, but some may reach more than 30 m (100 ft). They are characterized by a long proboscis *(proh BAHS kihs)*, a muscular tube that they quickly thrust out to capture prey. Ribbon worms, like the one shown in Figure 28-17, share many characteristics with the free-living flatworms; their bodies are flattened, and most biologists consider them to be solid-bodied (acoelomate). At the same time, the body structure of a ribbon worm is far more complex than that of a flatworm. For example, they are the simplest animals that have a digestive tube open at both ends. They also have a circulatory system. The structure of ribbon worms provides early indications of important evolutionary trends that became fully developed in the more advanced animals.

Figure 28-18 This adult *Ascaris* roundworm is a human parasite that may wander into the ducts of the pancreas or gallbladder, causing a blockage. Other, related species infect dogs, cats, pigs, and other vertebrates.

Roundworms Are the Simplest Animals With a Body Cavity
....

Roundworms are members of the phylum Nematoda. There are about 12,000 known species, although it has been estimated that between 500,000 and 750,000 species may actually exist. As you learned in Chapter 27, roundworms are characterized by the presence of a pseudocoelom, a body cavity lined on the inside by endoderm and on the outside by mesoderm.

Roundworms have bodies shaped like pencils sharpened at both ends. While some roundworms grow to be a foot or more in length, most are microscopic or only a few millimeters long. The vast majority of roundworms are free-living, active hunters. They are found on land, in lakes and streams, and in all oceans.

About 50 roundworm species are plant and animal parasites that cause considerable economic damage and human suffering. One roundworm that infects humans is the intestinal roundworm *Ascaris*, shown in Figure 28-18. Because the eggs of *Ascaris* are carried to soil by way of human waste, infestation is greatest in those areas without modern plumbing. The eggs of *Ascaris* can live in the soil for years and may enter the body when aspects of personal hygiene, such as hand washing, are not properly practiced. In the intestine, the eggs develop into larvae that bore through blood vessels and enter the bloodstream. The blood carries the larvae to the lungs, where they cause respiratory distress. Eventually, they return to the intestine, where they mature and mate.

Adult *Ascaris* may grow up to 0.3 m (1 ft.) in length while in the intestine.

Another roundworm that infects people is the hookworm *Necator*. Members of this genus live mostly in the warm, moist soils of the tropics. When people step barefooted on infected soil, the hookworms penetrate the soles of their feet and attach to the sides of blood vessels. They consume blood cells, causing anemia in the infected person. Their eggs are shed in human feces and hatch in the soil, where they await another barefooted human to begin the cycle again. Hookworm infections can be prevented by wearing shoes and properly disposing of waste matter.

Trichinella, a parasitic roundworm that infects pigs, causes a disease called trichinosis *(trihk ih NOH sihs)* in humans. The disease occurs when infected, undercooked pork is consumed. Research has shown that freezing as well as proper cooking kills the parasite. Trichinosis is now rare in the United States. ☐

Most Rotifers Are Microscopic

Rotifers are complex, free-swimming hunters too tiny to see with the naked eye. There are over 1,750 named species, some marine and others living in fresh water. Rotifers, shown in Figure 28-19, are characterized by a row of cilia surrounding their mouth. The action of the cilia sweeps food into the mouth. Under a microscope, the beating of the cilia looks much like a rotating wheel, which explains why these animals are called rotifers. Rotifers feed on bacteria and protists. Like roundworms, they have a well-developed, one-way gut and a pseudocoelom.

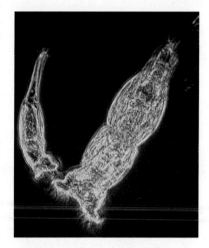

Figure 28-19 Rotifers are often found living in debris at the bottom of lakes and ponds and feed mainly on protozoans.

Section Review

1. *How does the internal anatomy of a fluke differ from that of a planarian?*

2. *Why is it important to properly cook beef?*

3. *Why might it be accurate to describe ribbon worms as a "transitional" group between flatworms and roundworms?*

4. *Compare and contrast* Ascaris *and* Necator.

Critical Thinking

5. *A student observing pond water through a microscope identifies a paramecium and a rotifer, both ciliated organisms. The student classifies both as protists. Is he correct? Explain your answer.*

CHAPTER REVIEW

Vocabulary

amoebocyte (639)
atoll (646)
barrier reef (646)
basal disk (642)
ectoparasite (649)

endoparasite (649)
fluke (649)
fringe reef (646)
gemmule (640)
hermaphrodite (641)

planula (644)
proglottid (650)
spicule (640)
spongin (640)
tegument (649)

Review

Multiple Choice

1. A protein sponge skeleton is composed of
   a. spicules.
   b. spongin.
   c. mesenchyme.
   d. amoebocytes.

2. What prevents self-fertilization among sponges?
   a. Eggs and sperm are released at different times.
   b. Few male sponges exist.
   c. Sponges are hermaphrodites.
   d. Encounters between members of the same species are rare.

3. A Portuguese man-of-war and a hydra are alike because both
   a. are colonial.
   b. contain medusae and polyps.
   c. are hydrozoans.
   d. produce planulae.

4. Which sequence reflects the life cycle of *Aurelia?*
   a. polyp → medusa → planula
   b. medusa → polyp → planula
   c. planula → medusa → polyp
   d. polyp → planula → medusa

5. Which is an anthozoan?
   a. hydra
   b. jellyfish
   c. Portuguese man-of-war
   d. sea anemone

6. Sinking volcanoes explain the existence of
   a. sponges.
   b. barrier reefs.
   c. atolls.
   d. ocean ridges.

7. The covering that protects endoparasites from the actions of digestive enzymes is called the
   a. osculum.
   b. tegument.
   c. proglottid.
   d. basal disk.

8. Which is *not* true about the life cycle of a beef tapeworm?
   a. Humans can become infected by eating raw steak.
   b. Cattle are not the only hosts of the parasite.
   c. The adult tapeworm attaches to the human liver.
   d. Eggs are deposited in human feces.

9. The most characteristic physical feature of members of the phylum Rhynchocoela is their
   a. proboscis.
   b. parasitic lifestyle.
   c. coelomate body.
   d. ciliated mouth.

10. Humans living in an area that lacks modern plumbing are likely to be parasitized by
    a. *Ascaris.*
    b. rotifers.
    c. *Dugesia.*
    d. *Trichinella.*

Completion

11. In sponges, amoebocytes are responsible for the movement of _____ from choanocytes to other sponge cells.

12. Sponges are usually _____ , producing both egg and sperm. But they are not _____ , because egg and sperm are produced at different times.

13. Some cnidarians live as free-floating _____ , some only as sessile _____ . Others pass through both in their life cycles.

14. In *Obelia*, reproductive _____ give rise to both male and female medusae. The zygotes produced from the union of the sperm and eggs released by the medusae develop into free-swimming larvae called _____ .

15. Flukes that live on the external surface of their hosts are called _____ .

16. Trichinosis in humans can be prevented by _____ or properly _____ pork.

17. Ribbon worms are more highly organized than flatworms, possessing a simple _____ system and a separate exit for the _____ tract.

18. Even though rotifers are microscopic, they are classified in kingdom _____ rather than in kingdom Protista.

Themes Review

19. **Structure and Function** Describe the structure of gemmules. How do gemmules ensure the survival of sponges?

20. **Evolution** Porifera has been called a dead-end phylum. List several reasons why no animal group evolved from the sponges.

21. **Flow of Energy** How does seasonal variation in light energy (and therefore heat) play a role in the life cycle of hydrozoans?

Critical Thinking

22. **Making Inferences** A single species of sponge may assume different appearances due to differences in substrate, availability of space, and the velocity and temperature of water currents. How might these factors make the classification of sponges confusing?

23. **Interpreting Data** Two living coral samples of equal size were placed in identically lighted, 100 gal. aquariums. The symbiotic algae were removed from one coral sample. The graph below shows the rate at which coral skeletons were deposited each year for four years. What do the data show about how the presence of the symbiotic algae affected the deposit of coral skeletons?

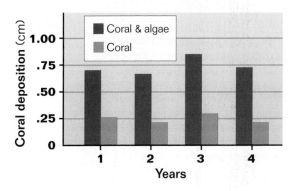

24. **Designing Experiments** Freshwater planarians reproduce sexually only during the fall. Design an experiment to determine whether day length or lower water temperature is the stimulus for sexual reproduction.

Activities and Projects

25. **Research and Writing** Look in the library for information about three of the parasitic roundworms that are not discussed in the text. Write a brief report about them, including information about their life cycles (include a diagram of each) and their economic impact.

26. **Multicultural Perspective** Compare and contrast the hydra of Greek mythology with the invertebrate hydra that you learned about in this chapter.

Hydra Behavior

OBJECTIVE

Observe hydras to determine how they respond to different stimuli and how they capture and feed on prey.

PROCESS SKILLS

• observing an animal's feeding behavior and response to stimuli
• relating structure to function

MATERIALS

• silicone culture gum
• microscope slide
• hydra culture
• 2 medicine droppers
• stereomicroscope
• filter paper cut into pennant shapes
• forceps
• concentrated beef broth
• *Daphnia* culture

BACKGROUND

1. How do animals respond to stimuli in their environment?
2. How does a sessile animal, such as a hydra, obtain food?
3. Write your own **Focus Question** on your Vee Form.
4. **Knowing Side of the Vee** List the **Concepts** and new **Vocabulary Words** on your Vee Form. In the **Concept Statements** section of the Vee, use these words in sentences that define and explain them.

TECHNIQUE

Doing Side of the Vee

Part A: Response to Stimuli

1. Using a long piece of silicone culture gum, make a circular "well" on a microscope slide, as shown in the illustration.

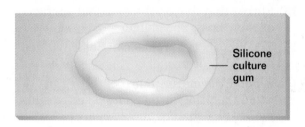

Silicone culture gum

2. With a medicine dropper, gently transfer a hydra from the culture dish to the well on the slide, making sure the hydra is in water. Allow the hydra to settle. Then examine it under the high power of a stereomicroscope. Identify and draw the hydra's body stalk, mouth, and tentacles in the **Records** section of your Vee Form.

3. In the **Records** section of your Vee Form, make a table like the one shown on the next page. Observe whether a hydra responds to a chemical stimulus, in this case a nutrient. First, hold a pennant-

Observations of Hydra

Response to filter paper	
Response to beef broth	
Response to touch	
Feeding behavior	

shaped piece of filter paper with forceps and move the long tip of the pennant near, but not touching, the hydra's tentacles. Observe and record the hydra's response to the filter paper in the **Records** section of the Vee. Next, dip the same piece of filter paper in beef broth and repeat the procedure. Record the hydra's response to the beef broth in the **Records** section.

4. Investigate how a hydra responds to touch. Using the long tip of a clean pennant-shaped piece of filter paper, touch the hydra's tentacles, mouth disk, and stalk. **CAUTION: Touch the hydra gently.** Record your observations in the **Records** section of the Vee.

Part B: Feeding Behavior

5. Hydras eat small crustaceans, such as *Daphnia*. Use a medicine dropper to transfer live *Daphnia* to the well with the hydra on the microscope slide. Observe the hydra carefully under the stereomicroscope. Watch for threadlike nematocysts shooting out from the hydra. Some nematocysts release a poison that paralyzes the prey. If the hydra does not respond after a few minutes, it may not be hungry. In that case, obtain another hydra from the culture dish.

6. Observe the way the hydra captures and ingests the *Daphnia,* and record your observations in the **Records** section of the Vee. In the **Procedure** section of the Vee, briefly summarize the procedure you followed.

7. Clean up your materials and wash your hands before leaving the lab.

INQUIRY

1. Based on your observations, how do you think a hydra behaves when threatened in its natural habitat?
2. Describe a hydra's feeding behavior.
3. What happens to food that has not been digested by a hydra?
4. What was the purpose of using the untreated filter paper in step 3?
5. Did the hydra show a feeding response or a defensive response to the beef broth? Explain.
6. Use the information on the **Knowing Side** of the Vee to interpret your results from the **Doing Side,** and then write your **Knowledge Claim.** Write a **Value Claim** for this lab.

ANALYSIS

1. How is a hydra adapted to a sedentary lifestyle?
2. How is the feeding method of a hydra different from that of a sponge?

FURTHER INQUIRY

Write a **New Focus Question** that could be the point of a new investigation. The following are examples:

How does a hydra respond to light?

What kinds of food does a hydra eat?

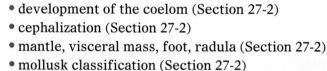

CHAPTER 29

MOLLUSKS AND ANNELIDS

REVIEW

- development of the coelom (Section 27-2)
- cephalization (Section 27-2)
- mantle, visceral mass, foot, radula (Section 27-2)
- mollusk classification (Section 27-2)
- segmentation (Section 27-3)
- annelid classification (Section 27-3)

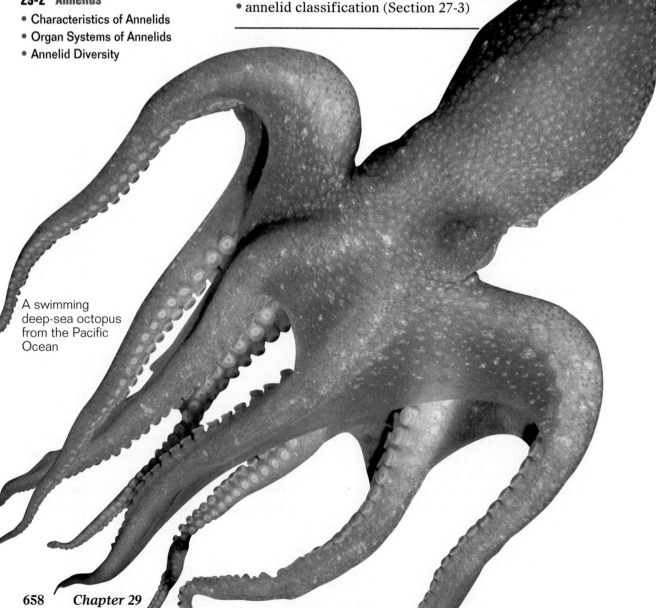

A swimming deep-sea octopus from the Pacific Ocean

29-1 Mollusks

A *snail may not seem to have much in common with an earthworm, but in fact, these two very different-looking animals are related. Mollusks and annelids were probably the first major groups of organisms to develop a true coelom. Also, the fertilized eggs of both groups develop into a distinct larval form called a trochophore (TRAHK oh fawr), shown in Figure 29-1.*

Section Objectives

■ Explain the evolutionary relationship between mollusks and annelids.

■ Describe the respiratory, circulatory, and excretory systems of mollusks.

■ Describe four classes of mollusks.

Characteristics of Mollusks

Mollusks are one of the most successful of all animal phyla. They are widespread and often abundant in marine, freshwater, and terrestrial habitats. They are the largest animal phylum, except for the arthropods, with over 100,000 named species. Members include a wide variety of animals, such as snails, oysters, octopuses, and cuttlefishes like the one shown in Figure 29-2. Despite their varied appearance, all mollusks share the following characteristics.

1. **Body cavity** All mollusks have a true coelom, although in most it is reduced to a small area immediately surrounding the heart.

2. **Symmetry** Most mollusks have bilateral symmetry, and many have one or more shells called valves.

3. **Organ systems** Mollusks have organ systems for circulation, respiration, digestion, and excretion.

4. **Three-part body plan** The body of every mollusk has three distinct parts: the muscular foot, the head, and the visceral mass.

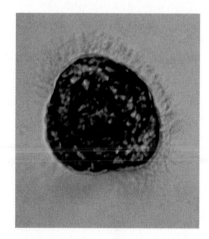

Figure 29-1 A trochophore larva has a belt of cilia that circles its body. The beating of the cilia propels the larva through the water. The cilia may also function to trap the tiny plankton that the trochophore feeds on.

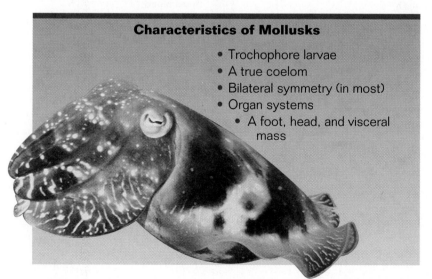

Characteristics of Mollusks

- Trochophore larvae
- A true coelom
- Bilateral symmetry (in most)
- Organ systems
  - A foot, head, and visceral mass

Figure 29-2 This cuttlefish has all of the major mollusk characteristics, although the foot has been modified into tentacles. Extremely agile swimmers, most species of cuttlefish hunt small fishes and crustaceans at night.

Organ Systems of Mollusks

As you learned in Chapter 27, the evolution of a coelom in mollusks enabled the development of complex organ systems in these animals. Mollusks are one of the earliest evolutionary lines to have developed an efficient excretory system.

Respiration

Most mollusks breathe with ciliated gills located in their mantle cavity. The **mantle cavity** is a space between the mantle and the visceral mass. The constant beating of the cilia causes a continuous stream of water to pass over the gills. Mollusk gills are very efficient and may extract 50 percent or more of the dissolved oxygen from the water that passes over them. Some terrestrial snails lack gills. Instead, their mantle cavity functions as a simple lung.

Circulation

Most mollusks have a three-chambered heart and an open circulatory system. In an **open circulatory system,** the blood leaks out of blood vessels and bathes the body's tissues directly, as shown in Figure 29-3. Two of the heart's chambers collect blood from the gills. The third chamber pumps the oxygenated blood out of the vessels and into spaces in the mollusk's tissues, where nutrients, oxygen, and carbon dioxide are exchanged between the blood and tissues. The blood then returns to the heart via the gills. Among mollusks, only octopuses and squids have a **closed circulatory system,** in which the blood never leaves the blood vessels. In a closed circulatory system, materials pass into and out of the blood by diffusing across the walls of the vessels.

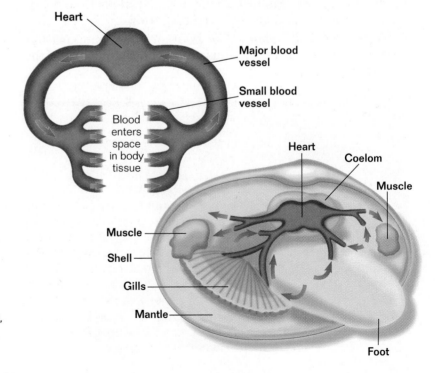

Figure 29-3 Most mollusks have an open circulatory system. The diagram, *top,* shows that in an open circulatory system blood leaks out of vessels and bathes tissue directly. In a bivalve's open circulatory system, *bottom,* blood flows out of some blood vessels and into others.

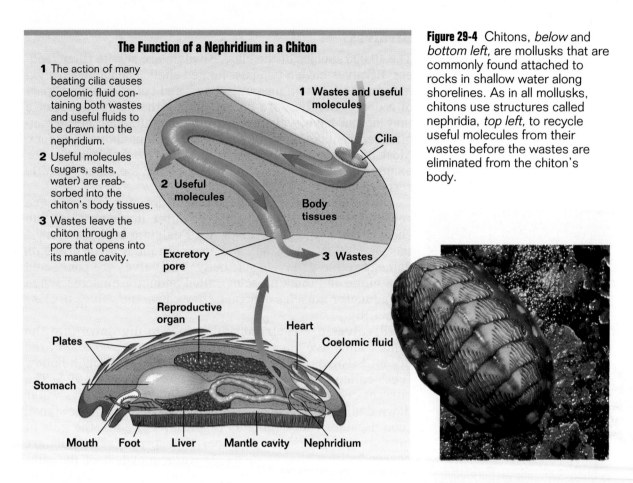

The Function of a Nephridium in a Chiton

1 The action of many beating cilia causes coelomic fluid containing both wastes and useful fluids to be drawn into the nephridium.

2 Useful molecules (sugars, salts, water) are reabsorbed into the chiton's body tissues.

3 Wastes leave the chiton through a pore that opens into its mantle cavity.

1 Wastes and useful molecules

Cilia

2 Useful molecules

Body tissues

Excretory pore

3 Wastes

Reproductive organ

Heart

Coelomic fluid

Plates

Stomach

Mouth   Foot   Liver   Mantle cavity   Nephridium

Figure 29-4 Chitons, *below* and *bottom left,* are mollusks that are commonly found attached to rocks in shallow water along shorelines. As in all mollusks, chitons use structures called nephridia, *top left,* to recycle useful molecules from their wastes before the wastes are eliminated from the chiton's body.

Excretion

Mollusks use their coelom as a refuse dump in which waste-laden body fluids are collected. Nitrogen-rich wastes are filtered from the coelomic fluid by tiny tubular structures called **nephridia** *(nee FRIHD ee ah).* Nephridia, diagrammed in Figure 29-4, are found in all coelomates except arthropods and chordates. They are a highly successful evolutionary innovation that allow useful molecules to be efficiently recovered from body fluids before wastes are discharged from the animal. ❑

❑ *CAPSULE SUMMARY*

Most mollusks breathe with gills and have an open circulatory system and nephridia. Nephridia enable mollusks to recover the useful substances from wastes before they are excreted from the animal's body.

Mollusk Diversity

There are seven classes of mollusks. By studying four of the minor classes, taxonomists have learned a great deal about the probable ancestor of this phylum. Research has indicated that the ancestor was a flattened, unsegmented, wormlike animal that glided along its ventral surface. Chitons, shown in Figure 29-4, are members of the class Polyplacophora. This class is one of the smaller groups of mollusks and still has many of the characteristics of its mollusk ancestors. As you read in Chapter 27, the three major mollusk classes are the bivalves, gastropods, and cephalopods.

Bivalves

The 10,000 species of the class Bivalvia are sessile filter feeders. Bivalves have a two-part hinged shell. The name *Bivalvia* is from the Latin *bi*, meaning "two," and *valva*, meaning "part of a door." Bivalves are unique among the mollusks because they do not have a distinct head region, although nerve ganglia (remnants of a simple brain) are present above their foot. Most bivalves have at least rudimentary sense organs. For example, sensory cells along the edge of the clam's mantle respond to light and touch.

The valves (shells) of a bivalve are layered and are secreted by the mantle. A tough outer layer protects the shell, a thick middle layer of hard calcium carbonate crystals strengthens the shell, and a smooth inner layer prevents damage to the bivalve's soft body. The valves are connected by a hinge and thick muscles called adductor muscles. When the **adductor muscles** contract, they cause the valves to close forcefully.

Bivalves use their muscular foot to dig down into the sand. Clams live buried in mud or sand on the bottom of the ocean. Bivalves feed by sucking in sea water through hollow tubes called **siphons** *(SEYE fuhns)*, shown in Figure 29-5. The current created by the cilia that cover the gills draws water down one siphon tube, over the gills, and out the other siphon tube. Besides functioning as respiratory organs, the gills of bivalves also work like flypaper to trap prey. The gills are covered with a sticky mucus. As water moves over the gills, small marine animals, protists, and organic material become trapped in the mucus. The cilia then direct the food-laden mucus to the bivalve's mouth.

Figure 29-5 Many bivalves, like this quahog clam, *above,* burrow into sand or mud on the ocean bottom. They feed by drawing sea water in one siphon and expelling it out the other. Bivalves, like all mollusks, have all of the major organ systems, *right.*

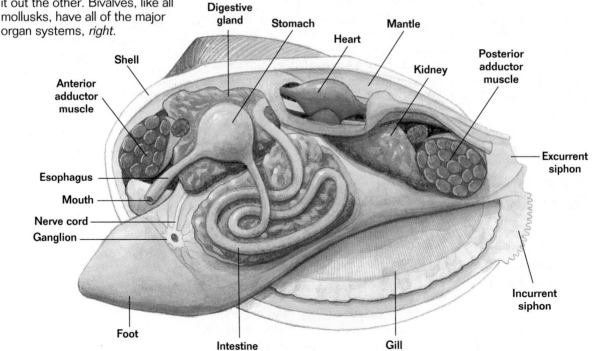

Different kinds of bivalves feed in various ways. Oysters feed in the open water with their shells permanently attached to rocks, while scallops don't attach themselves to anything. Instead, water passes over their gills as they swim. A swimming scallop looks like it might be eating, its valves rapidly opening and shutting like jaws. Actually, it is pushing itself through the water with the jets of water it expels when its valves snap shut. One bivalve does not filter-feed at all. The teredo, or shipworm, shown in Figure 29-6, digests the cellulose in wood by using symbiotic protists that live inside its intestine, much as a termite does.

Bivalves reproduce sexually. Most are either male or female, but a few species are hermaphroditic. They reproduce by shedding sperm and eggs into the water, where fertilization occurs. The fertilized eggs develop into free-swimming trochophore larvae. The trochophore larvae of marine bivalves, as well as those of marine gastropods, develop into a second free-swimming stage called a **veliger** (*VEE lah jur*). Veligers drift in ocean currents and are dispersed far and wide. In the veliger stage, the beginnings of the foot, shell, and mantle become evident. Eventually, veligers settle to the ocean bottom and grow into adults.

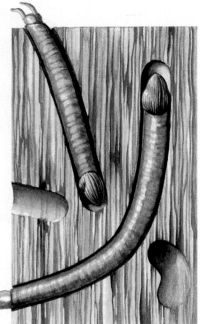

Figure 29-6 A shipworm's valves are highly reduced and function as a drill, which enables the animal to bore into wood. Shipworms can cause extensive damage to wooden ship bottoms.

Gastropods

The 80,000 species of the class Gastropoda are snails and slugs. Gastropods are primarily a marine group that has very successfully invaded freshwater and terrestrial habitats. As you learned in Chapter 27, their bodies are generally divisible into a conspicuous head, foot, and visceral mass. The foot of gastropods is adapted for locomotion. Terrestrial species secrete mucus from the base of their foot, forming a slimy path that they can glide along. Most gastropods have a pair of tentacles, on which the eyes are often located, on their head. Many gastropods have a single shell. During the evolution of slugs and nudibranchs (*NOO dih brangks*), or sea slugs, shown in Figure 29-7, the shell was lost. The shells of most marine snails can be closed by a plate that the animal pulls into place like a door; most land snails lack this plate.

Figure 29-7 Many sea slugs, *below left,* secrete toxic substances from their skin. Their bright coloration acts as a warning sign to other animals. Many terrestrial slugs, *below,* breathe air. Their mantle cavity has been modified into a simple lung.

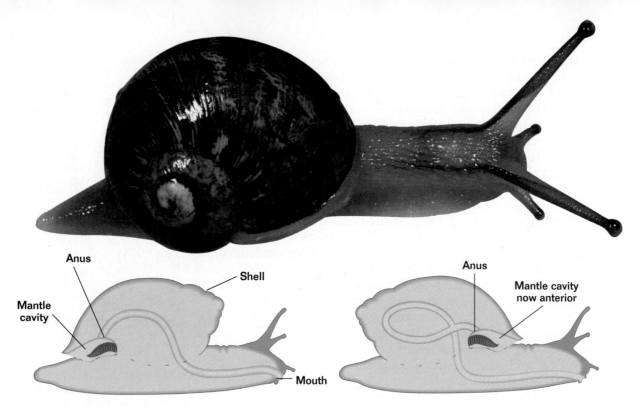

Mantle cavity

Anus

Shell

Mantle cavity

Anus

Mantle cavity now anterior

Mouth

Figure 29-8 Torsion, a 180° twisting of a gastropod's visceral mass, occurs during the gastropod's embryonic development. Before torsion, *above left,* the snail's mantle cavity is posterior. After torsion, *above right,* the mantle cavity is anterior, giving the snail a space to pull its head into when it is threatened.

☐ *CAPSULE SUMMARY*

Bivalves (oysters, mussels, and scallops) are aquatic and have hard valves (shells) made of calcium carbonate that surround and protect their soft bodies. Gastropods (snails, slugs, and nudibranchs) live in oceans, in fresh water, and on land. Gastropods are unique among mollusks because their bodies undergo torsion.

The body plan of gastropods has undergone a significant change from that of their mollusk ancestors. In gastropods, the visceral mass rotates 180° during development. This twisting, called **torsion** (*TAWR shuhn*), causes a rearrangement of organs and moves the mantle cavity from the back to the front of the animal, as shown in Figure 29-8. The spiraling of a gastropod's shell is not caused by torsion. It occurs before torsion begins.

Respiration among gastropods is carried out in a variety of ways. Aquatic snails breathe with gills that are located in the mantle cavity. Nudibranchs (marine snails) lost their gills as well as their shells through evolution—gas exchange takes place directly through their skin. Gills, which can function only when the delicate filaments of gill tissue are supported by water, have been lost in terrestrial snails as well. The empty mantle cavity of terrestrial snails acts as a primitive lung. Oxygen in the air diffuses across the thin membrane that lines the cavity. Because this membrane must be kept moist for respiration to occur, terrestrial snails are most active when air has a high moisture content, such as at night or after it rains. During dry weather, a terrestrial snail avoids water loss by creeping back into its shell and plugging the opening with a wad of mucus to keep water in. ☐

Gastropods display extremely varied feeding habits. Many are herbivores that scrape algae off rocks with their radula. Remember from Chapter 27 that a radula is a tongue-like scraping organ. Some terrestrial snails can be serious garden and agricultural pests by using their radula to saw off leaves. Nudibranchs and many other gastropods are active

predators. Whelks and oyster drills, for example, use their radula to bore holes in the shells of other mollusks. Then the tissue of their prey is sucked out. In gastropods called cone shells, shown in Figure 29-9, the radula has been modified into a kind of poison-tipped harpoon that is shot into prey.

Cephalopods

The more than 600 species of the class Cephalopoda include squids, octopuses, cuttlefishes, and nautiluses. The name *cephalopod* is from the Greek *kephalicos,* meaning "head," and *pous,* meaning "foot." Appropriately named, most of their body is a large head attached to tentacles (the foot divided into numerous parts), as shown in Figure 29-10. The

Figure 29-9 The eastern Pacific cone shell sweeps its long proboscis back and forth over the ocean bottom in search of prey, *left.* When a fish gets close, the cone shell quickly paralyzes the fish by stabbing it with the poison-tipped radula located at the tip of the proboscis. The fish is then swallowed whole, *right.*

Figure 29-10 Most octopuses are bottom dwellers that live in crevices among rocks and corals. They move around mainly by crawling, and they swim only to escape from enemies.

Figure 29-11 A nautilus, *above left,* swims backward, with its coiled shell positioned over its head. The animal lives only in the outermost compartment of its partitioned shell, *above right.* The inner chambers of the shell are filled with gas. By regulating the amount of gas in the chambers, the nautilus is able to adjust its buoyancy and thus control its depth in the water.

tentacles of cephalopods are equipped with suction cups or hooks for seizing prey. Squids have 10 tentacles, octopuses have 8, and the nautiluses have 80 to 90. All cephalopods are active marine predators. They feed on fish, mollusks, crustaceans, and worms. Once the prey has been snared by the tentacles, it is pulled to the mouth, where it is bitten by strong, beaklike jaws. The cephalopod's radula then pulls the prey into the mouth.

Like most aquatic mollusks, cephalopods draw water into their mantle cavity and expel it through a siphon. The squids and octopuses, however, have modified this system into a means of jet propulsion. When threatened, they quickly close their mantle cavity, causing water to shoot forcefully out of the siphon. Both squids and octopuses can also release a dark fluid that clouds the water and thus disguises the direction of their escape.

Although they evolved from shelled ancestors, most modern cephalopods lack an external shell. The nautilus, shown in Figure 29-11, is the only living cephalopod that has retained its outer shell.

Cephalopods are the most intelligent of all invertebrates. In contrast to the other mollusks, cephalopods have a complex nervous system that includes a well-developed brain, and they are capable of exhibiting complex behaviors. Octopuses can easily be trained to distinguish among classes of objects.

The structure of a cephalopod eye is similar to that of a vertebrate eye, and some cephalopod eyes grow quite large. A giant squid that washed up on a beach in New Zealand in 1933 had eyes that were 40 cm (about 16 in.) across, the largest eyes ever measured in an animal.

Cephalopods exhibit both sexes. In males, sperm are stored in sacs that open into the mantle cavity. During breeding, the male uses a modified tentacle to transmit a packet of sperm from its own mantle cavity into the female's mantle cavity, as shown in Figure 29-12. The eggs are fertilized as they leave her body and are attached to stones or other objects. ◻

☐ CAPSULE SUMMARY

Cephalopods (octopuses, squids, and nautiluses) have a well-developed head region and many tentacles equipped with suction cups. An effective propulsion system and a complex nervous system have enabled these mollusks to become effective predators.

Section Review

1. *What evidence suggests that mollusks and annelids share a common ancestor?*

2. *Sketch a nephridium, and describe how it functions.*

3. *Why would you expect the blood pressure inside a mollusk's blood vessels to be quite low?*

4. *Define* torsion, *and identify the mollusk group in which it occurs.*

5. *Contrast the feeding habits of cephalopods with those of gastropods.*

Critical Thinking

6. *A chemical pollutant accidentally spills into a bay. One of the effects of the chemical is that it paralyzes cilia. The next day almost all of the oysters in the bay are dead. Explain why.*

7. *Squids are the fastest swimmers of all aquatic invertebrates. Name two structural adaptations in squids that may have enhanced their swimming ability.*

29-2 *Annelids*

Section Objectives

- State the major annelid characteristic that distinguishes annelids from mollusks.
- Describe the circulatory system of an annelid.
- Describe three classes of annelids.
- Identify the internal structures of an earthworm.

The second group of early coelomates is the segmented worms, members of the phylum Annelida. Annelids that are familiar to everyone are the earthworms, shown in Figure 29-13, which burrow through the ground. Annelids also live in fresh water and are abundant in all of the world's oceans.

The great evolutionary advancement achieved by the annelids is segmentation. Many scientists believe that segmentation first evolved as an adaptation for burrowing. Segmentation enables the worm to produce strong waves of muscular contraction along the length of its body, making burrowing easier and faster.

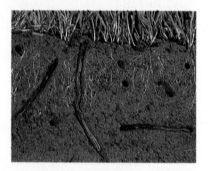

Figure 29-13 Earthworms come to the surface only at night or during heavy rains. During dry or cold weather, they burrow deep into the soil and become inactive.

Characteristics of Annelids

Annelids are easily recognized by their segments, which are visible externally as a series of ringlike structures along the length of their body. The name *annelid* is from the Latin word *annellus*, meaning "ring." At first glance the annelids that burrow beneath the ground seem to have little in common with the fierce predatory annelids of the open ocean. But in fact, all annelids, like the one shown in Figure 29-14, exhibit certain basic characteristics.

1. **Body cavity** The body cavity in annelids is a true coelom.
2. **Segmentation** In all annelids, the body is segmented, divided into many nearly identical units. Some segments fuse during development, but many segments remain separate.

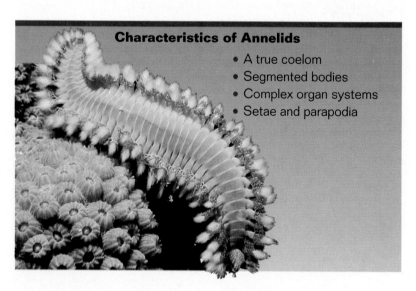

Characteristics of Annelids

- A true coelom
- Segmented bodies
- Complex organ systems
- Setae and parapodia

Figure 29-14 Notice the paired parapodia and numerous setae of this polychaete. The term *polychaete* is from the Greek *poly*, meaning "many," and *chaite*, meaning "hair."

3. **Organ systems** The major organ systems of annelids include a highly specialized gut, a closed circulatory system, and many nephridia.
4. **Bristles** Most annelids have external bristles called **setae** *(SEET ee)*. Marine annelids also have many fleshy appendages called **parapodia** *(par uh POH dee uh)*.

Organ Systems of Annelids

The organ systems of annelids display a high degree of specialization. Cephalization has led to the development of anterior sense organs and a brain. A ventral nerve cord and pairs of segmental ganglia make the coordinated movement of each body segment possible. The digestive tube is modified into specialized regions: the crop, stomach, and intestine.

Circulation and Respiration

Annelids and many other coelomates have a closed circulatory system, like the one shown in Figure 29-15. Only the two largest animal phyla, arthropods and mollusks, have an open circulatory system. Blood moves through a closed system of blood vessels faster than through an open system because the blood is under greater pressure. In several

Figure 29-15 All annelids have a closed circulatory system. The diagram, *top,* shows that in a closed circulatory system, the blood never leaves the hearts or blood vessels. The locations of the hearts and major blood vessels in an earthworm, *bottom,* are also shown.

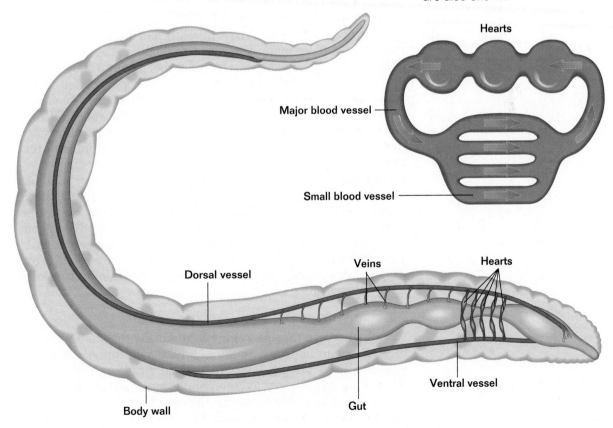

Hearts

Major blood vessel

Small blood vessel

Dorsal vessel

Veins

Hearts

Body wall

Gut

Ventral vessel

CONTENT LINK
....
You can learn about how hemo-globin functions in human blood in **Chapter 37.**

❏ **CAPSULE SUMMARY**
━━━━━━━━

Annelids are coelomate worms that have segmented bodies. Respiration in annelids occurs through their skin, and their blood is pumped through a closed circulatory system.

segments of an annelid, the blood vessels are enlarged and heavily muscled. These enlarged vessels serve as simple hearts that pump the blood. Earthworms, for example, have five pairs of hearts. Annelids do not have gills, lungs, or other specialized respiratory organs. They exchange oxygen and carbon dioxide with the environment directly through their body surfaces. Oxygen then diffuses into blood vessels that lie under the skin. An annelid's blood typically contains hemoglobin, a protein that binds to oxygen molecules. In some annelids, other respiratory molecules with a similar function are found.

Excretion

The excretory system of annelids is very similar to that of mollusks, consisting of ciliated, funnel-shaped nephridia. Each segment of an annelid has a pair of nephridia that collect waste products. The wastes are then transported out of the annelid's body through pores that open on the sides of each segment. ❏

Annelid Diversity
....

There are roughly 12,000 known species of annelids. They range in size from less than 1 mm (0.04 in.) long to more than 3 m (10 ft.) long. Annelids are classified into three classes based on the number of setae they have and the presence or absence of parapodia. The approximately 8,000 species of polychaete (PAHL ih keet) worms are all marine. Polychaetes have many setae and parapodia. The oligochaete (AHL ih goh keet) worms, which number about 3,100 species, include terrestrial earthworms as well as some related freshwater worms. Oligochaetes have only a few setae on each segment and no parapodia. The hirudinean (hihr yoo DIHN ee ahn) worms are the leeches. Most of the 600 species of leeches live in fresh water, although there are a few marine and terrestrial species. Leeches lack both setae and parapodia.

Annelids are an ancient phylum. Their fossils can be found in rock that is 530 million years old. Scientists think that annelids evolved in the sea, with the polychaetes being the ancestral group. Oligochaetes appear to have evolved from polychaetes, perhaps by way of freshwater worms. Taxonomists generally agree that leeches evolved from oligochaetes, some becoming specialized as bloodsucking external parasites.

Marine Worms

Polychaetes are marine segmented worms that live in virtually all ocean habitats. Some are free-swimming predators that use their strong jaws to feed on small animals. Some burrowing species excavate tunnels by ingesting sediment. Others feed by pumping water through their bodies or scouring the ocean bottom with their tentacles.

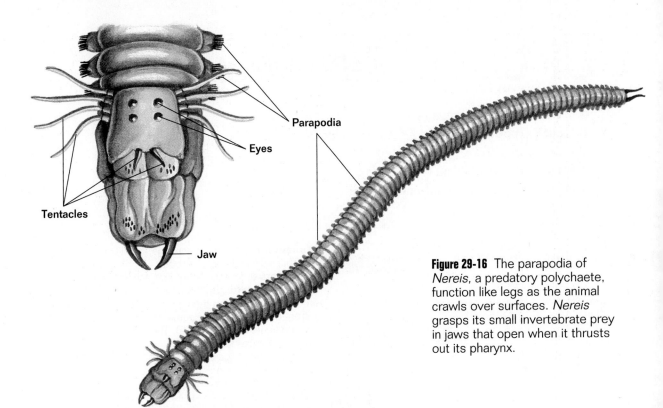

Parapodia

Eyes

Tentacles

Jaw

Figure 29-16 The parapodia of *Nereis*, a predatory polychaete, function like legs as the animal crawls over surfaces. *Nereis* grasps its small invertebrate prey in jaws that open when it thrusts out its pharynx.

Polychaetes, unlike earthworms, have a well-developed head, as shown in Figure 29-16. Many have antennae, specialized mouthparts, and sense organs, often including stalked eyes. Polychaetes are often beautiful; some have unusual forms and iridescent colors. Spectacular polychaetes called feather dusters are shown in Figure 29-17. A distinctive characteristic of polychaetes is the pair of fleshy, paddle-like flaps called parapodia that occur on most of their segments. The parapodia, which usually have setae, are used to swim, burrow, or crawl. They also greatly increase the surface area of the body, making gas exchange between the animal and the water more efficient.

Individual polychaetes are either male or female, and fertilization is usually external. There are no distinct male sex organs that produce sperm. Instead, sperm are produced from cells that line the male's coelom. Eggs are typically attached to a rock by the female and are fertilized by sperm released into the water by the male. Fertilization results in the growth of ciliated trochophore larvae. After a long period of development, the larvae begin to add segments and thus change to juvenile polychaetes that more closely resemble the adult worms.

Earthworms

Earthworms are terrestrial worms that literally eat their way through the soil. As highly specialized scavengers, earthworms take organic matter and other soil material into their mouths as they tunnel through the soil. The ingested soil moves through a long, straight digestive tube. In one

Figure 29-17 Feather dusters live in hard, narrow tubes that they make with their body secretions. They filter-feed by trapping food particles in the spiral of feather-like head structures that extend from the end of their burrow. The beating of the cilia located inside their feathery cone directs the food to their mouth.

UP CLOSE EARTHWORM

- **Scientific Name:** *Lumbricus terrestris*
- **Range:** Europe and eastern North America
- **Habitat:** Damp soil
- **Size:** Grows up to 30 cm (12 in.) long
- **Diet:** Organic matter contained in soil

Characteristics

Respiration Oxygen and carbon dioxide diffuse through the earthworm's skin, as in all annelids. This exchange can take place only if the worm's skin is kept moist.

Digestion Soil is taken into the digestive passage by the muscular throat, or **pharynx.** The soil then passes through the **esophagus** to a temporary storage area called a **crop,** and from there it passes to the **gizzard.** The thick, muscular gizzard walls contract and grind the soil, breaking up the organic matter contained in it. Food molecules pass across the walls of the **intestine** and are absorbed into the bloodstream. The blood is pumped throughout the earthworm's closed circulatory system by a series of muscular hearts.

Movement As in most annelids, circular and longitudinal muscles line the interior body wall of an earthworm. In order to crawl, an earthworm first anchors several of its segments by sinking stiff bristles, called **setae,** into the ground (1). The worm then contracts the circular muscles in front of the anchored segments. This causes the anterior segments to elongate (2). The worm then retracts the rear set of setae and grips the ground again with setae that are in front of the stretched region (3). The rear segments are then pulled forward (4).

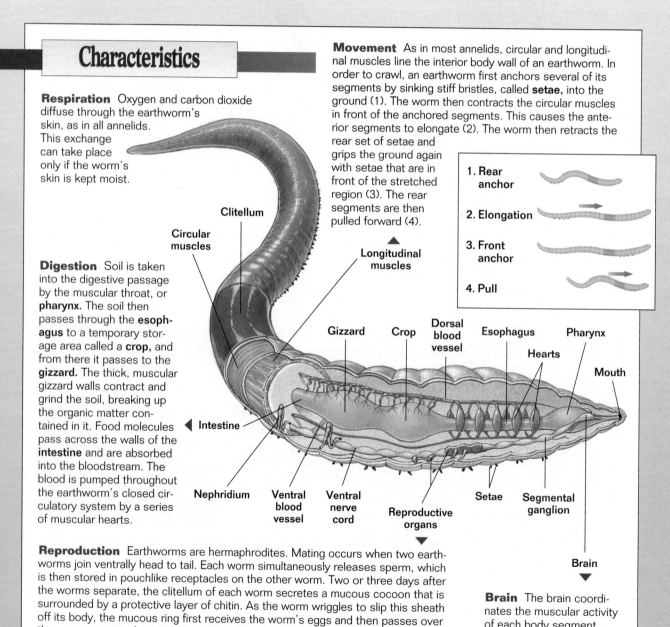

1. Rear anchor

2. Elongation

3. Front anchor

4. Pull

Clitellum

Circular muscles

Longitudinal muscles

Gizzard   Crop   Dorsal blood vessel   Esophagus   Pharynx

Hearts

Mouth

Intestine

Nephridium   Ventral blood vessel   Ventral nerve cord   Reproductive organs   Setae   Segmental ganglion

Brain

Reproduction Earthworms are hermaphrodites. Mating occurs when two earthworms join ventrally head to tail. Each worm simultaneously releases sperm, which is then stored in pouchlike receptacles on the other worm. Two or three days after the worms separate, the clitellum of each worm secretes a mucous cocoon that is surrounded by a protective layer of chitin. As the worm wriggles to slip this sheath off its body, the mucous ring first receives the worm's eggs and then passes over the sperm receptacles, acquiring sperm. When the cocoon finally passes over the end of the worm, its ends pinch together, enclosing the fertilized eggs. Young worms emerge from the cocoon after several weeks.

Brain The brain coordinates the muscular activity of each body segment, thus controlling total body movement.

modified portion of the tube, called the **gizzard**, strong muscles grind up the organic material in the soil. The material that passes all of the way through the digestive system without being absorbed exits the worm through the anus and is deposited outside its burrow in the form of characteristic castings. The tunneling activity of earthworms aerates the soil, and their castings fertilize it. An earthworm eats its own weight in soil every day. Rich, organic soil may contain thousands of earthworms per acre.

Earthworms lack the distinctive head region of polychaetes, and they have no eyes. This is not surprising when you consider an earthworm's underground lifestyle. However, earthworms do have light-sensitive and touch-sensitive organs located at each end of their body. Earthworms also have sensory cells that detect moisture. Learn more about the structure and lifestyle of earthworms in *Up Close: Earthworm* on page 672. ◻

Leeches

The body of a leech is flattened. Most species are 2.5–5.0 cm (1–2 in.) long, although one tropical species grows up to 30.5 cm (1 ft.) long. The coelom of leeches is not segmented as in other annelids. A leech, shown in Figure 29-18, has suckers at both ends of its body. Leeches move by attaching first one sucker and then the other and pulling themselves forward. Most species of leeches are predators or scavengers, but some have evolved into parasites. Parasitic leeches suck the blood from mammals and other vertebrates. A few species even suck blood from crustaceans. Many parasitic freshwater leeches remain on their hosts for extended periods. Most of the terrestrial parasitic leeches position themselves on low-growing plants, waiting to attach themselves to a suitable endothermic host. Some leeches, however, climb trees to seek out mammals and birds.

◻ **CAPSULE SUMMARY**

Polychaetes, earthworms, and leeches are kinds of annelids. Both the marine polychaetes and the terrestrial earthworms have external appendages (parapodia or setae). Leeches are both aquatic and terrestrial and lack parapodia and setae.

Figure 29-18 Well-developed, powerful muscles in the body wall of a leech enable it to carry out the complex body movements that are necessary for it to crawl along.

Section Review

1. *If you have two worms with different external characteristics, how can you tell if either is an annelid?*
2. *How are the circulatory systems of a human and an earthworm similar?*
3. *Contrast polychaetes with leeches.*

Critical Thinking
4. *A mutation results in the birth of an earthworm that lacks moisture-sensing cells in its skin. Explain why this earthworm is less likely to survive than one with the sensory cells.*

adductor muscle (662)
closed circulatory system (660)
gizzard (673)

mantle cavity (660)
nephridia (661)
open circulatory system (660)
parapodia (669)

setae (669)
siphon (662)
torsion (664)
veliger (663)

Review

Multiple Choice

1. Mollusks and annelids are alike in that they both
   a. are coelomates.
   b. undergo torsion.
   c. have veliger larvae.
   d. have a visceral mass.

2. Cephalopods have all of the following characteristics except
   a. bilateral symmetry.
   b. a three-part body plan.
   c. an open circulatory system.
   d. a true coelom.

3. Which reflects the circulatory path exhibited by the bivalves?
   a. heart→tissue sinus→gills→heart
   b. heart→gills→heart→tissue sinus
   c. heart→lungs→arteries→veins
   d. heart→tissue sinus→lungs→heart

4. Which mollusk has adductor muscles?
   a. clam   c. squid
   b. snail   d. nudibranch

5. The visceral mass of gastropods twists during the larval stage, bringing the mantle cavity toward the front of the animal. This is called
   a. spiraling.
   b. segmentation.
   c. torsion.
   d. radial symmetry.

6. Annelids are divided into three classes. This classification is based on the number of setae and the presence or absence of
   a. segments.   c. sauropodia.
   b. hearts.   d. parapodia.

7. Annelids are most easily recognized by their
   a. cephalization.   c. nephridia.
   b. segmentation.   d. body cavity.

8. Blood in the circulatory system of an annelid
   a. flows in open sinuses.
   b. moves very slowly.
   c. passes through gills.
   d. transports oxygen.

9. The nephridia of annelids and mollusks function in
   a. respiration.   c. digestion.
   b. circulation.   d. excretion.

10. Earthworm movement requires all of the following except
    a. circular muscles.
    b. secretion of a mucus layer.
    c. muscle contractions.
    d. traction provided by setae.

Completion

11. Mollusks and annelids commonly develop a(n) _____ larva, but only certain mollusks develop a second larva called a(n) _____ .

12. Oysters have a(n) _____ circulatory system in which blood leaves the vessels, but earthworms have a(n) _____ circulatory system in which blood circulates entirely within vessels.

13. All mollusks exhibit a(n) _____ body plan that includes the foot, head, and visceral mass. The head is most specialized among members of the class _____ . Torsion of the visceral mass is seen among members of the class _____ .

14. Ancestral characteristics of the phylum Mollusca are shown by the flattened, wormlike animals called _____ .

15. Soil taken in by an earthworm passes through a muscular throat called a _____ before it enters the esophagus. Upon leaving the esophagus, the soil is stored for a short time in the _____ . Then it is ground up by the contracting walls of the _____ . Food extracted from the soil is absorbed by the blood as the soil passes through the intestine.

Themes Review

16. **Homeostasis** Mollusks and annelids possess nephridia. How do nephridia function to regulate the balance of sugar, water, salt, and other useful molecules in the bodies of these animals? What other function is performed by nephridia?

17. **Evolution** The earthworm gizzard, which is used for grinding food, is lined with a cuticle and is very muscular. In comparison, the gizzards of annelids that have returned to an aquatic environment are much smaller and less muscular. How do the differences between the gizzards represent adaptations to terrestrial and aquatic feeding?

18. **Structure and Function** Most gastropods are marine and breathe with gills, but terrestrial snails are an exception. What modification to the mantle cavity enables these snails to live on land?

Critical Thinking

19. **Controlling Variables** As a science-fair project, a student decided to determine if a new, experimental paint reduces the boring by shipworms on ocean pier pilings more than other paints do. What variables must the student control for if the test of the new paint is to be a fair one?

20. **Making Inferences** Classification of organisms has traditionally been based on physical similarities. Given the physical differences of adult chitons, gastropods, bivalves, and cephalopods, how did they come to be grouped into the phylum Mollusca?

21. **Making Inferences** Examine the photograph below. Is the annelid shown in the photograph a polychaete, oligochaete, or hirudinean? What observable characteristics support your choice?

Activities and Projects

22. **Science-Technology-Society** *Nautilus* was the name given to the U.S. Navy's first nuclear submarine. Compare the shell structure of the cephalopod nautilus and the functioning of its siphuncle with the structures that enable a submarine to submerge and surface. Why do you think the Navy chose to name the submarine *Nautilus?* Support your explanation with cut-away sketches of the cephalopod and a submarine.

23. **Multicultural Perspectives** Many marine seashells are named after Greek or Roman goddesses. Using the *Audubon Society Field Guide to North American Seashells* (published by Alfred A. Knopf, Inc., 1992) and sources of Greek and Roman mythology, compile a list of 5–10 shells named after mythological figures.

Mollusks

OBJECTIVES

- Observe the behavior of a live clam.
- Examine the structure and composition of a clam shell.

PROCESS SKILLS

- identifying structures
- relating structure and function

MATERIALS

- safety goggles
- lab apron
- disposable gloves
- live clam
- small beaker or dish
- food coloring in a dropper bottle
- glass stirring rod
- clam shell
- petri dish
- scalpel
- stereomicroscope
- 10% HCl in a dropper bottle

BACKGROUND

1. What kinds of mollusks do people eat?
2. What makes a clam a bivalve?
3. What kinds of food do bivalves eat?
4. What keeps the two shells of a clam together?
5. Write your own **Focus Question** on your Vee Form.
6. **Knowing Side of the Vee** List the **Concepts** and new **Vocabulary Words** on your Vee Form. In the **Concept Statements** section of the Vee, use these words in sentences that define and explain them.

TECHNIQUE

Doing Side of the Vee

Part A: Live Clam

1. CAUTION: **Put on safety goggles, a lab apron, and disposable gloves.** Place a live clam upright in a small beaker or dish. Using a medicine dropper, apply two drops of food coloring between the two shells, as shown below. Observe what happens to the food coloring. Record your observations in the **Records** section of your Vee Form.

2. Using the stirring rod, gently touch the clam's mantle. The mantle is the internal membrane that lines the shells. Describe the clam's response in the **Records** section.

Part B: Examining a Clam Shell

3. Observe a clam shell. Look at the concentric growth rings on the shell. As a clam grows, its mantle secretes a layer of shell. Locate the knob-shaped umbo on the shell. The umbo is the oldest part of the shell. Each successive ring of shell formed beyond the umbo is younger than the one before it. In the **Records** section, record the number of growth rings on the clam shell.

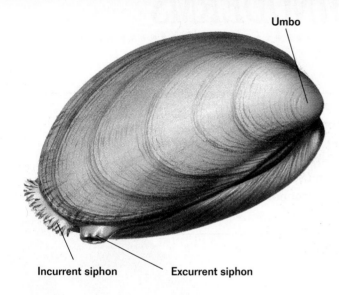

Umbo

Incurrent siphon   Excurrent siphon

INQUIRY

1. Find the incurrent and excurrent siphons of the clam in the diagram on this page. The incurrent siphon draws water into the clam, and the excurrent siphon expels the water from the clam. Using this information, explain your observations in step 1.

2. What is the purpose of a clam's shell?

3. Use the information on the **Knowing Side** of the Vee to interpret your results from the **Doing Side,** and then write your **Knowledge Claim.** Write a **Value Claim** for this lab.

ANALYSIS

1. Based on your observations, how do you think clams respond when they are touched or threatened in their natural habitat?

2. What does a clam take in from water that passes through it?

3. Water that enters a clam's incurrent siphon passes over the clam's gills. How does this help the clam to breathe?

4. How does a clam get rid of its wastes? What kinds of wastes are removed?

5. Look at the picture of the clam. Where would you expect the clam's foot to appear if it was extended?

6. Bivalve fossils are one of the most common kinds of fossils. Explain why.

FURTHER INQUIRY

Write a **New Focus Question** that could be the point of a new investigation. The following is an example:

How similar are the layers of the shells of univalves and bivalves?

4. **CAUTION: Use extreme care when working with sharp objects.** Place the clam shell in a petri dish and use a scalpel to chip away part of the shell to expose its three layers. Observe the layers under a stereomicroscope. The outermost layer of shell protects the clam from acids in the water. The innermost layer is a pearly material called mother-of-pearl. It is this material that forms a pearl inside a shell.

5. The middle layer of the shell contains crystals of calcium carbonate. **CAUTION: Hydrochloric acid can burn skin and damage clothing.** Test for the presence of this compound by placing one drop of 10% HCl on the middle layer of the shell. If calcium carbonate is present, it will form bubbles in the shell. Record your observations in the **Records** section. In the **Procedure** section of the Vee, briefly summarize the procedure you followed.

6. Clean up your materials and wash your hands before leaving the lab.

ARTHROPODS AND ECHINODERMS

REVIEW

- arthropod evolution (Section 27-3)
- jointed appendages, exoskeleton (Section 27-3)
- protostome and deuterostome embryo development (Section 27-3)
- endoskeleton (Section 27-3)
- nerve cord, notochord, pharyngeal slits (Section 27-3)

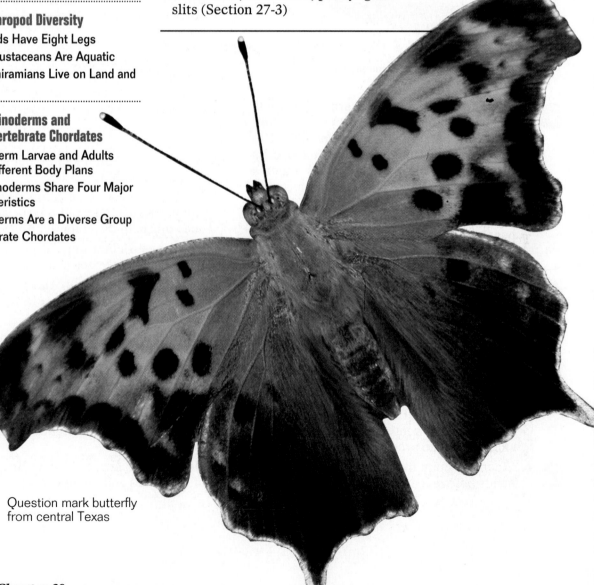

Question mark butterfly from central Texas

30-1 Features of Arthropods

There are more arthropods living on Earth than any other kind of animal, and many are of enormous importance to human life. The largest arthropod group is the class Insecta. Insects eat virtually every kind of plant, including most agricultural crops. On the other hand, insects are beneficial to agriculture. They eat the weeds that compete with the crops, and they are primary crop pollinators.

Section Objectives

- Describe the evolution of arthropods.
- Relate the characteristics of arthropods to their evolutionary success.
- Describe how growth occurs in arthropods.

Arthropods Evolved From Annelids

The ancestor of arthropods is undoubtedly some kind of annelid. Like annelids, arthropods have a coelom and a segmented body. Scientists are not quite sure which annelid gave rise to arthropods. Until recently, experts thought velvet worms, shown in Figure 30-1, might be the "missing link" between annelids and arthropods. Velvet worms appear to have a combination of both annelid and arthropod characteristics—segmented, wormlike bodies with paired legs. However, Australian researchers in 1993 analyzed the DNA of velvet worms and discovered that they are in fact true arthropods and are closely related to scorpions.

Arthropod fossils are among the oldest, best-preserved fossils of multicellular animals; some are 630 million years old. Among the most successful of the early arthropods were the trilobites—highly segmented, plate-shaped animals. Trilobites were the first animals with eyes capable of forming images. Trilobites, shown in Figure 30-2, were very abundant in the seas until they became extinct 250 million years ago. The first terrestrial arthropods were probably scorpions, whose fossils date back 425 million years.

Figure 30-1 Velvet worms live in the leaf litter of tropical and subtropical forests. This species is from the rain forests of Costa Rica. Active mainly at night, velvet worms capture insects by trapping them in a sticky substance secreted near their mouth.

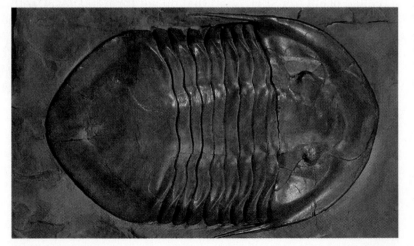

Figure 30-2 Although trilobites no longer exist, their hard exoskeletons made fossilization easy. As a result, trilobites are extremely common in the fossil record. By studying the fossils, scientists have been able to identify nearly 4,000 different trilobite species.

Figure 30-3 Notice the mandibles (jaws) of the wood borer beetle, *top*, a mandibulate arthropod. The baboon spider, *bottom,* is a chelicerate arthropod because it has chelicerae (fangs) as its mouthparts.

Figure 30-4 The compound eyes of this fruitfly, an insect, are not capable of forming images as clear as those formed by vertebrate eyes, but insect eyes are much better at detecting motion. That is why it is difficult to sneak up on a fly.

The Body Plan and Characteristics of Arthropods

As you read in Chapter 27, arthropods have many kinds of jointed appendages, including legs, antennae, and mouthparts. Living arthropods are traditionally separated into two large groups. This separation is based on the kind of mouthparts they have. Arthropods with jaws (crustaceans, insects, centipedes, and millipedes) are called **mandibulates**. The scientific word for jaw is mandible. Arthropods with fangs or pincers (spiders, mites, and scorpions) are called **chelicerates** *(kuh LIHS uhr ayts)*. The scientific name for their mouthparts is chelicerae *(kuh LIHS uhr ee)*. A mandibulate and a chelicerate arthropod are shown in Figure 30-3.

There are nine arthropod characteristics of particular importance.

1. **Jointed appendages** All arthropods have jointed legs and other jointed appendages. Joints enable an appendage to be far more flexible.

2. **Segmentation** All arthropods are segmented, although the adults of the more advanced species have many of the segments fused together to form regions called the head, thorax, and abdomen.

3. **Head** Most arthropods have a distinct head, although in some (lobsters and spiders) the head fuses with the thorax (the mid-body region) to form a body region called the **cephalothorax**. This sort of fusion is unique and does not occur in any other animal phylum.

4. **Exoskeleton** As you read in Chapter 27, the bodies of arthropods are encased in a shell-like **exoskeleton.**

5. **Compound eyes** Many arthropods have compound eyes, like those shown in Figure 30-4. A compound eye is made of thousands of individual visual units called **ommatidia** *(ahm uh TIHD ee ah),* each with its own lens and retina. An insect's brain receives the many inputs of the individual ommatidia and composes a detailed image of the object. Some arthropods also have simple, single-lens eyes called **ocelli** *(oh SEHL eye)* that do not form images. Ocelli can only distinguish light from dark.

6. **Spiracles** Crustaceans breathe with gills, and spiders have tiny lungs, but the majority of terrestrial arthropods breathe with a network of fine tubes called **tracheae.** Air from outside the animal passes into the tracheae by way of special openings called **spiracles** that are opened and closed by valves. Tracheae, shown in Figure 30-5, transmit oxygen throughout the body of the arthropod. The ability to close their spiracles and prevent water loss was a key adaptation for insects when they invaded land.

7. **Circulation** The arthropod circulatory system is open, like that of mollusks, with a heart along the top of the body that squeezes blood out into internal body spaces.

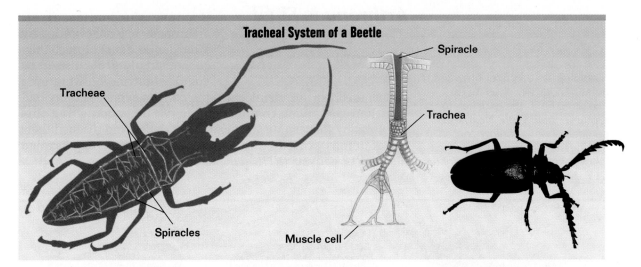

Tracheal System of a Beetle

Tracheae

Spiracles

Spiracle

Trachea

Muscle cell

8. **Malpighian tubules** Terrestrial arthropods have a unique excretory system that efficiently conserves water and eliminates metabolic wastes. This system is composed of excretory units called **Malpighian tubules.** Malpighian *(mal PIHG ee an)* tubules, shown in Figure 30-6, are slender, fingerlike structures that extend from the arthropod's gut and are bathed by the surrounding blood. The fluid component of the blood (consisting of water and small dissolved particles) moves through the tubule and into the arthropod's midgut. As this fluid flows down to the hindgut, most of the water, valuable ions, and metabolites from the fluid are reabsorbed into the arthropod's body tissues by osmosis. Metabolic wastes remain in the gut and eventually leave the body through the anus.

9. **Wings** Insects, the most abundant of the arthropods, were the first animals to evolve wings. For more than 100 million years, until flying reptiles appeared, insects were the only flying organisms in existence.

Figure 30-5 A complex series of hollow tubes called tracheae run throughout the bodies of terrestrial arthropods like this long-horned beetle, *above right.* Air enters the tracheal system through spiracles that open to the outside, and it is delivered to all of the beetle's cells, *above left.*

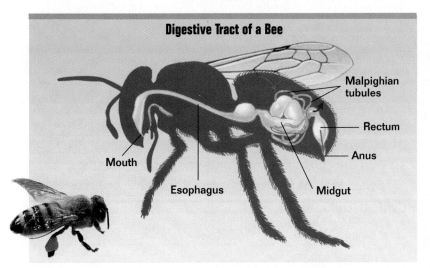

Digestive Tract of a Bee

Malpighian tubules

Rectum

Anus

Mouth

Esophagus

Midgut

Figure 30-6 Most terrestrial arthropods, like bees, have excretory organs called Malpighian tubules that extend from their gut. A mixture of both wastes and useful molecules passes into the tubules from the surrounding blood that bathes them (red arrow). As this mixture travels down the gut, the useful molecules are reabsorbed back into the blood (blue arrow). The wastes remain in the gut and leave the bee's body through its anus (green arrow).

Arthropods Shed Their Exoskeletons to Grow

Because the bodies of all arthropods are surrounded by a hardened exoskeleton, they cannot simply grow bigger, as you do. Imagine blowing up a balloon inside a soft drink can—the balloon cannot get any bigger than the space inside the can. Arthropods solve this dilemma by discarding the "can." They shed and discard their exoskeletons periodically as they grow in a process called molting, or **ecdysis** *(EHK duh sihs)*, shown in Figure 30-7.

Figure 30-7 As a lobster grows, it must periodically shed its exoskeleton. First, the exoskeleton ruptures, *below left*. Next, the carapace opens, *below center*. Finally, the lobster emerges from its old exoskeleton, *below right*.

Molting is signaled by hormones. Just prior to molting, a new exoskeleton forms underneath the old one. When the new exoskeleton is fully developed, it becomes separated from the old one by fluid. This fluid dissolves the chitin and, in a crustacean, the calcium carbonate of the old exoskeleton. The weakened exoskeleton then cracks open and is shed, and the arthropod emerges in its new, still-soft exoskeleton. Almost immediately, the new exoskeleton begins absorbing air or water and expands to fit the larger size of the animal. Within an hour or two after exposure to air or water, the new exoskeleton is hardened. Most insects molt four to eight times during their development, but some may molt as many as thirty times. ◻

◻ CAPSULE SUMMARY

Arthropods evolved from annelids and are characterized by jointed appendages, a segmented body, and an exoskeleton. An arthropod must shed its exoskeleton to grow.

Section Review

1. *Name the invertebrate phylum that gave rise to the first arthropods.*
2. *Name three important structural characteristics of arthropods.*
3. *What are compound eyes, and how do they function?*
4. *Define* ecdysis, *and explain why it occurs in arthropods.*

Critical Thinking
5. *Hypothesize why aquatic arthropods, like crayfish, evolved thicker, stronger exoskeletons than terrestrial arthropods, like insects.*

30-2 Arthropod Diversity

*T**he arthropods are the most diverse of all animal phyla. The great majority of arthropods are small, about 1 mm (0.04 in.) in length—the size of a grain of rice. The very smallest are parasitic mites only 80 μm (0.003 in.) long. The largest arthropods are gigantic crabs 3.6 m (12 ft.) across, found in the sea off the coast of Japan.*

Section Objectives

■ List the three subphyla of arthropods.

■ Describe the characteristics of arachnids and crustaceans.

■ Compare and contrast millipedes, centipedes, and insects.

■ Identify the external and internal structures of the Eastern lubber grasshopper.

■ Compare and contrast complete and incomplete metamorphosis.

Arachnids Have Eight Legs

Arthropods are traditionally grouped into three separate subphyla, each with a distinct evolutionary line, as shown in Figure 30-8. They are Chelicerata (spiders, scorpions, and their relatives), Crustacea (crabs, lobsters, and their relatives), and Uniramia (insects, millipedes, and centipedes).

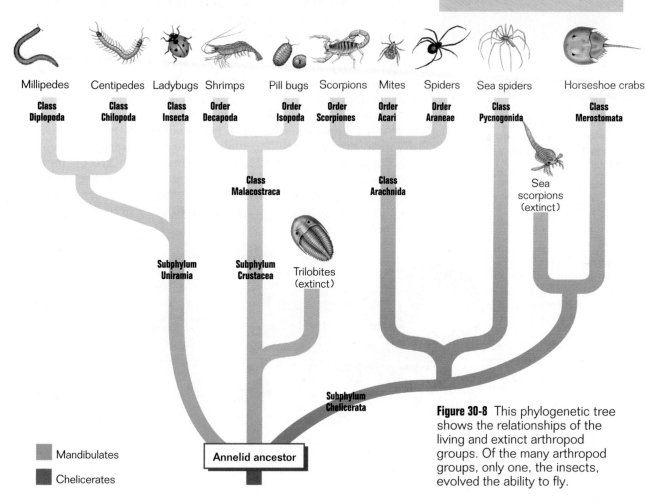

Millipedes	Centipedes	Ladybugs	Shrimps	Pill bugs	Scorpions	Mites	Spiders	Sea spiders	Horseshoe crabs
Class Diplopoda	Class Chilopoda	Class Insecta	Order Decapoda	Order Isopoda	Order Scorpiones	Order Acari	Order Araneae	Class Pycnogonida	Class Merostomata

Class Malacostraca

Class Arachnida

Sea scorpions (extinct)

Subphylum Uniramia

Subphylum Crustacea

Trilobites (extinct)

Subphylum Chelicerata

Annelid ancestor

■ Mandibulates
■ Chelicerates

Figure 30-8 This phylogenetic tree shows the relationships of the living and extinct arthropod groups. Of the many arthropod groups, only one, the insects, evolved the ability to fly.

Figure 30-9 A tarantula, *top,* does not capture its prey in a web. Instead, tarantulas stalk, lunge at, and then bite their victims. The house dust mite, *bottom,* is a major cause of allergies in humans when it is inhaled in household dust.

Arachnids, such as those in Figure 30-9, compose the largest class in the subphylum Chelicerata. Two additional minor classes of chelicerates, horseshoe crabs and sea spiders, are marine. There are over 57,000 named species of arachnids, including spiders, ticks, mites, scorpions, and daddy longlegs. Arachnids have a pair of chelicerae as their foremost appendage. A second pair of appendages, called **palps,** are used to catch and handle prey, although they are sometimes specialized for sensory or even reproductive functions. The palps are followed by four pairs of walking legs.

All arachnids, except mites, are carnivores. Arachnids have no jaws. Because they are able to ingest only liquid food, spiders and other carnivorous arachnids must digest their prey externally. They accomplish this by injecting their prey with powerful enzymes, which cause the prey's tissues to liquefy. The liquid contents are then sucked up through the arachnid's muscular pharynx.

Arachnids occur primarily in terrestrial habitats. Three major orders of arachnids are discussed below.

Scorpions

Scorpions are arachnids whose palps have evolved into large, grasping pincers. Scorpions use these pincers for handling and tearing apart their food and for reproducing, as shown in Figure 30-10. They are not used as weapons. Scorpions have long, slender, segmented abdomens that end in a venomous stinger used to stun their prey. A scorpion holds its abdomen folded forward over its body as it walks.

Scorpions are thought to have been the first animals to successfully invade land; some fossil scorpions are over 425 million years old. Scorpions evolved from eurypterids, ancient sea scorpions. The sea scorpions, all of which are extinct, breathed with gills. Some grew several feet long. There are some 1,200 species of scorpions living today, all terrestrial, and they range in size from 1 to 18 cm (0.4 to 7 in.).

Figure 30-10 In this courtship ritual, the male scorpion, *far right,* grasps the female's pincers in his and dances her around until he positions her directly over a sperm packet that he has previously deposited on the ground. Once positioned, the packet bursts open, and the sperm enter the female's reproductive opening. In this pair of scorpions, the female carries offspring, the result from a previous mating, on her back. The red spots on the male's back are mites.

Figure 30-11 This garden spider has spun what is called an orb web. The spider begins building its web by laying down a foundation of spokes that radiate from a center point. Next, it spins a continuous spiral outward from the center.

Spiders

Spiders have poison glands that secrete toxin through chelicerae that have been modified into fangs. Spiders use their fangs to bite and paralyze prey. Only two species of spiders living in the United States, the black widow and brown recluse, are dangerous to humans. There are about 35,000 named species of spiders, but many more await discovery.

Spiders are important insect predators in virtually every terrestrial ecosystem. Many, like the familiar wolf spiders, actively hunt their prey by running them down. Others, like the trapdoor spiders, construct silk-lined burrows with lids that open easily. The spider hides underneath the lid and then leaps out to seize its prey as it passes by the burrow. Many spiders trap their prey in sticky webs of remarkable diversity, as shown in Figure 30-11. The strands of the webs are made of pure silk. The silk is formed from a liquid protein forced out of specially modified appendages called spinnerets that are located at the end of the spider's abdomen.

Mites

The mites are by far the largest and most diverse group of arachnids. Few scientists study mites, so many of the members of this group are not well known. There are about 35,000 identified species, but taxonomists estimate that there may be up to a million species that exist. Some mites, including chiggers and ticks, are well known to humans because of their irritating bites.

In mites, the head, thorax, and abdomen are fused into a single, unsegmented body. Most mites pass through a series of larval phases, changing from a six-legged to an eight-legged animal. As adults, most mites are quite small, typically less than 1 mm (0.04 in.) long. A tick, shown in Figure 30-12, grows larger. Some can be up to 3 cm (1.2 in.) long. Mites are found in virtually every terrestrial, freshwater, and marine habitat. Many marine and freshwater mites are herbivores, while terrestrial mites are usually predators. Many plants have special pits on their leaves inhabited by predaceous mites that protect the plants from herbivores. ❑

Figure 30-12 Some ticks carry diseases that can be transmitted to humans when the tick bites them. This particular species, commonly called a deer tick, harbors the bacteria that cause Lyme disease.

❑ **CAPSULE SUMMARY**

Arachnids include spiders, scorpions, and mites. Arachnids are characterized by four pairs of legs, fused body regions, and mouthparts called chelicerae.

Most Crustaceans Are Aquatic

The crustaceans are a large group of primarily aquatic arthropods that includes crabs, lobsters, crayfish, shrimps, barnacles, water fleas, and pill bugs. There are eight separate crustacean orders, which include some 35,000 species. Crustaceans are often very abundant in marine and freshwater habitats and are a major food source for humans.

Like insects, crustaceans have jaws and are considered mandibulates. However, crustaceans differ from insects in a number of important respects.

1. Most appendages of crustaceans are branched at the ends, although some have become unbranched in the course of evolution. The appendages of insects are unbranched.

2. All crustaceans have a distinctive larval form, called a nauplius, that has three pairs of branched appendages, as shown in Figure 30-13.

3. Aquatic crustacean exoskeletons are hardened with calcium carbonate.

4. Crustaceans are the only arthropods with two pairs of antennae. Insects have only one pair.

5. Most crustaceans have three pairs of chewing appendages, while insects have only one pair.

6. Crustaceans, unlike insects, have legs attached to their abdomen as well as to their thorax.

7. All crustaceans breathe with gills.

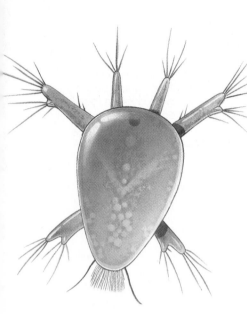

Figure 30-13 All crustaceans pass through a free-swimming larval stage called a nauplius. This is a nauplius of a tiny crustacean called a copepod.

Decapods

Large marine crustaceans such as shrimps, lobsters, and crabs, along with the freshwater crayfish shown in Figure 30-14, have five pairs of walking legs. These crustaceans are often referred to as decapods. They are all members of the order Decapoda. The term *decapod* is from the Greek *deka*, meaning "ten," and *pous*, meaning "foot." The head and thoracic body segments of decapods are fused into a single cephalothorax. The cephalothorax is covered on top by a shield called a **carapace.** In lobsters and crayfish, appendages called **swimmerets** are attached to the underside of the abdomen. The swimmerets are used in swimming and in reproduction. Flattened, paddle-like appendages, called **uropods,** are at the end of the abdomen, and many decapods have a **telson,** or tail spine. Decapods such as crayfish, lobsters, and shrimps can propel themselves through the water by forcefully flexing their abdomen.

Unlike decapods, the members of the other seven orders of crustaceans are quite small. Minute crustaceans smaller than the tip of a pencil are abundant in marine and freshwater habitats all over the world. Tiny copepods *(KOH puh pahds)* are among the most abundant multicellular organisms on Earth and are a key food source in the marine food chain. Also common are the minute ostracods, fairy shrimps, and water fleas.

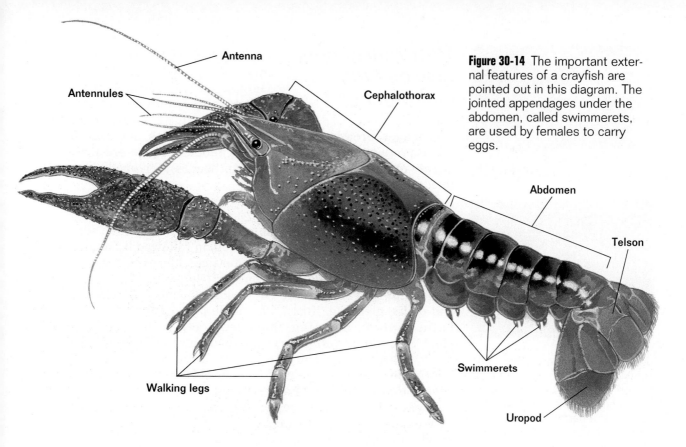

Antenna

Antennules

Cephalothorax

Figure 30-14 The important external features of a crayfish are pointed out in this diagram. The jointed appendages under the abdomen, called swimmerets, are used by females to carry eggs.

Abdomen

Telson

Walking legs

Swimmerets

Uropod

One marine crustacean, the barnacle, is sessile (permanently attached to a rock or other object) as an adult, although the larvae are free swimming. Barnacles attach themselves upside down to a rock or other object. With their bodies protected by hard, sharp plates, barnacles stir food into their mouths with their feathery legs.

Terrestrial Crustaceans

Only a few crustaceans are terrestrial. Other than a few crab species, there are two principal groups of terrestrial crustaceans. One group is composed of the pill bugs and sow bugs, which belong to a large order known as the **isopods.** A pill bug is shown in Figure 30-15. About half of the 4,500 species of isopods are terrestrial, and the rest are marine. The other group of terrestrial crustaceans is composed of the sand fleas, with several thousand terrestrial species typically found along beaches. ▢

▢ *CAPSULE SUMMARY*

Crustaceans have hardened exoskeletons and mouthparts called mandibles (jaws). Most crustaceans (crabs, lobsters, and crayfish) have greatly modified appendages and are aquatic. Other crustaceans (pill bugs and sand fleas) are terrestrial.

Figure 30-15 Although this pill bug is a terrestrial arthropod, it still breathes with gills that must be kept moist to function. Consequently, pill bugs are usually found in damp environments and are most active at night, when the humidity is high.

Most Uniramians Live on Land and Can Fly

The subphylum Uniramia is an enormous group of mostly terrestrial arthropods that differ structurally from crustaceans. Uniramians have unbranched appendages, breathe with tracheae and spiracles, and excrete waste products through Malpighian tubules. There are three classes of uniramians: Diplopoda (millipedes), Chilopoda (centipedes), and Insecta (insects).

Millipedes and Centipedes

Millipedes and centipedes, shown in Figure 30-16, have similar body designs. Each has a head region followed by numerous segments (usually 15, but sometimes more) that are all similar. Each segment bears one or two pairs of legs. The name *millipede* is from the Latin *mille*, meaning "thousand," and *pedis*, meaning "foot." The name *centipede* is from the Latin *centum*, meaning "hundred," and *pedis*, meaning "foot." Actually, millipedes have two pairs of legs on each body segment, so they typically have 60 legs in all. Centipedes have one pair of legs on each segment, so they have a total of 30 legs. Each segment of modern millipedes evolved from two segments of their ancestors. Approximately 2,500 species of centipedes and over 10,000 species of millipedes have been named.

Insect Diversity

The insects are by far the largest group of organisms on Earth, as shown in Figure 30-17. There are over 700,000 named insect species, comprising more than 50 percent of all named animal species. Most scientists agree that there may be several million insect species in existence. Most of the undiscovered species live in the tropics. There are more kinds of beetles (350,000 species) than all the kinds of non-insect animals put together! Insects are primarily a terrestrial group, and it is thought that those that are aquatic probably had terrestrial ancestors. Seven of the most common orders of insects are described in Table 30-1. You can learn more about the structure of one particular insect, the grasshopper, in *Up Close: Eastern Lubber Grasshopper* on pages 690–691.

Figure 30-16 Centipedes, *top,* are carnivorous predators that inject poison into their prey through a pair of hollow fangs. Millipedes, *bottom,* are harmless herbivores that feed on decayed vegetation. Millipedes avoid predators by releasing a foul-smelling fluid from the side of each body segment.

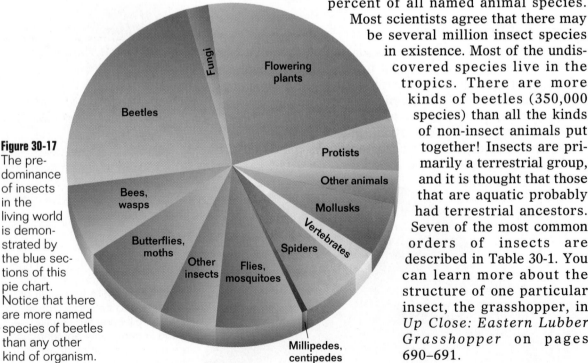

Figure 30-17 The predominance of insects in the living world is demonstrated by the blue sections of this pie chart. Notice that there are more named species of beetles than any other kind of organism.

Table 30-1 Major Orders of Insects

Order	Number of Species	Main Characteristics	Examples
Coleoptera "Shield winged"	350,000	Two pairs of wings (front pair covers transparent hind pair); heavy, armored exoskeleton; biting and chewing mouthparts; complete metamorphosis	Beetles, weevils
Diptera "Two winged"	120,000	Transparent front wings; hind wings reduced to knobby balancing organs; sucking, piercing, and lapping mouthparts; complete metamorphosis	Flies, mosquitoes
Lepidoptera "Scale winged"	120,000	Two pairs of broad, scaly wings; hairy bodies; tubelike, sucking mouthparts; complete metamorphosis	Butterflies, moths
Hymenoptera "Membrane winged"	100,000	Two pairs of transparent wings; mobile head; well-developed eyes; chewing and sucking mouthparts; stinging; many social species; complete metamorphosis	Ants, bees, wasps
Hemiptera "Half winged"	60,000	Two pairs of wings or wingless; piercing, sucking mouthparts; incomplete metamorphosis	Giant water bug, bedbug, chinch bug
Orthoptera "Straight winged"	20,000	Two pairs of wings or wingless; biting and chewing mouthparts in adults; incomplete metamorphosis	Grasshoppers, crickets, cockroaches, mantids
Odonata "Toothed"	5,000	Two pairs of transparent wings; chewing mouthparts; incomplete metamorphosis	Dragonflies, damselflies

UP CLOSE   EASTERN LUBBER GRASSHOPPER

- **Scientific name:** *Romalea microptera*
- **Range:** Eastern United States
- **Habitat:** Fields and meadows
- **Size:** 5 cm (2 in.) to 6.5 cm (2.6 in.) in length
- **Diet:** Grasses and other leafy vegetation

External Structures

Head Two antennae contain sense organs for both touch and smell. On the sides of the head are a pair of very large compound eyes, each containing hundreds of six-sided lenses. Located high on the forehead are three light-detecting ocelli. The mouthparts are composed of four kinds of appendages. The stiff upper labrum and lower labium are "lips" that hold a leaf or blade of grass. The mandibles (jaws), assisted by maxillas (graspers), are then used to tear off pieces of the plant.

Wings Like most insects, grasshoppers have two pairs of wings. A pair of leathery forewings protects the more delicate flying wings.

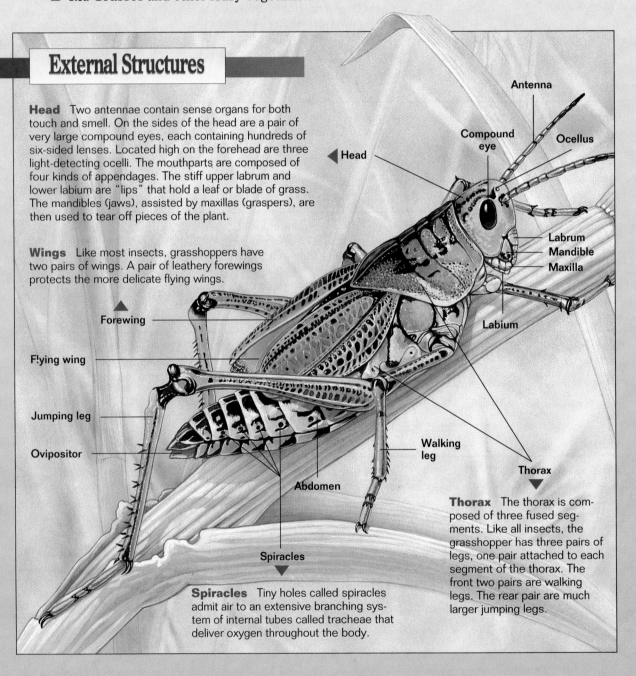

Antenna

Compound eye

Ocellus

Head

Labrum
Mandible
Maxilla

Labium

Forewing

Flying wing

Jumping leg

Ovipositor

Abdomen

Walking leg

Thorax

Spiracles

Spiracles Tiny holes called spiracles admit air to an extensive branching system of internal tubes called tracheae that deliver oxygen throughout the body.

Thorax The thorax is composed of three fused segments. Like all insects, the grasshopper has three pairs of legs, one pair attached to each segment of the thorax. The front two pairs are walking legs. The rear pair are much larger jumping legs.

Internal Structures

Reproductive system The reproductive organs (testes and ovaries) of grasshoppers are located in the abdomen. During the summer mating season each male "sings" to potential mates. A "song" is produced when he rubs the row of pegs on the inside of a jumping leg against ridges on a forewing. During mating, the female collects the male's sperm in a storage pouch called a seminal receptacle. Later that summer, the female digs a hole using two pairs of pointed organs called ovipositors. The eggs are fertilized by the stored sperm as she releases the eggs into the hole. They stay dormant over the winter and hatch the following spring.

Circulatory system The grasshopper's circulatory system is composed of a long blood vessel that runs along its back, with a series of muscular "hearts" located over the abdomen. The circulatory system is open; that is, the blood does not stay in the vessel, but is pumped out and bathes the tissues directly.

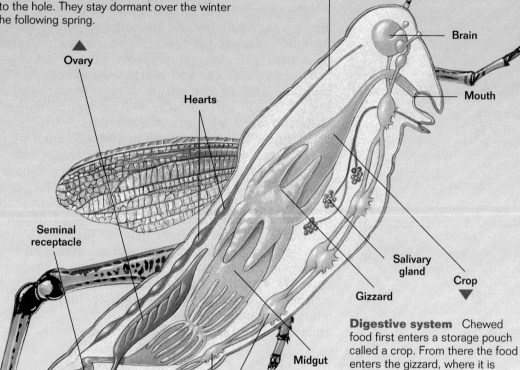

Ovary

Hearts

Dorsal blood vessel

Brain

Mouth

Seminal receptacle

Salivary gland

Gizzard

Crop

Midgut

Anus

Ganglia

Nerve cord

Digestive system Chewed food first enters a storage pouch called a crop. From there the food enters the gizzard, where it is shredded and crushed. It then passes into the midgut, which acts as a stomach and digests the food with the aid of enzymes. Food molecules then pass across the wall of the midgut into the fluid of the coelom. This fluid eventually enters the circulatory system, and the nutrients are delivered to body tissues.

Nervous system The nervous system is composed of a major ventral nerve cord with ganglia located in each body segment. Three fused ganglia in the head serve as the brain. The brain coordinates responses such as the very rapid leap that the grasshopper makes when threatened.

Figure 30-18 This stop-action photograph of a green lacewing in flight shows how its wings move. The movement is controlled by muscles inside the insect's thorax.

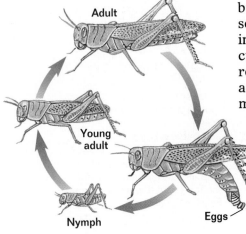

Adult

Young adult

Nymph

Eggs

Figure 30-19 The Monarch butterfly, *right*, undergoes complete metamorphosis. During complete metamorphosis, each developmental stage appears markedly different. The Alutacea bird grasshopper, *above*, undergoes incomplete metamorphosis, during which the developmental differences between stages are not nearly as great.

Insect Structure and Life Cycles

Most insects are small, only a few centimeters in length. All insects have the same general body plan: three body sections (a head, a thorax, and an abdomen); three pairs of legs, all attached to the thorax; and one pair of antennae.

As shown in Figure 30-18, flight is one of the great evolutionary achievements of insects. While some insects, like fleas, lice, and silverfish, lack wings, all other insects have one or two pairs. The insect thorax consists of three fused segments. The wings are attached to the middle and rear segments. Each wing is a solid sheet of chitin that grows from a saclike outgrowth of the body wall. The wings are strengthened by veins made of tubes of chitin.

The life cycles of most insects are complex, often requiring several molts before the adult stage is reached. In the final molt to adult, the juvenile undergoes a process of physical change called **metamorphosis**. Flies, beetles, ants, bees, wasps, butterflies, and moths—90 percent of all insect species—undergo "complete" metamorphosis, shown in Figure 30-19. In complete metamorphosis, the juvenile changes from a wingless, wormlike larva (called a caterpillar in butterflies and moths) to a winged adult by becoming a pupa enclosed within a protective capsule called a **chrysalis** (*KRIHS uh lihs*). What is the evolutionary advantage of this complicated life cycle? During development, the larvae exploit different habitats and different food sources than the adult. For example, the larvae of nectar-drinking butterflies are caterpillars that eat leaves! This ecological separation of young from adults eliminates competition, thus increasing the chance of survival for each phase of the life cycle. In a smaller number of species (grasshoppers, cockroaches, and mosquitoes), metamorphosis from juvenile to adult is much less dramatic and is described as "incomplete" metamorphosis. In incomplete metamorphosis, also shown in

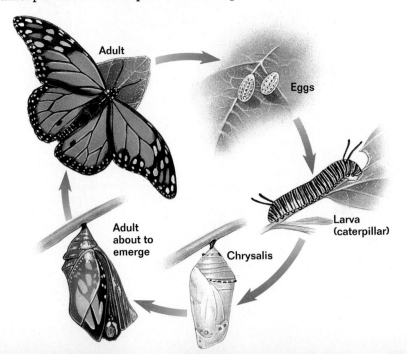

Adult

Eggs

Larva (caterpillar)

Chrysalis

Adult about to emerge

Figure 30-19, the juveniles have wings, and there is usually no pupa before the last molt. The juvenile, called a **nymph** (*NIHMF*), is essentially a smaller version of the adult.

Two orders of insects, Hymenoptera (ants, bees, and wasps) and Isoptera (termites), have evolved elaborate social systems. These insects often live in highly organized societies of genetically related individuals. The societies are characterized by marked division of labor, meaning that different kinds of individuals serve the group by performing specific functions. In the termite colony shown in Figure 30-20, for example, small, active members called workers gather the food, raise the young, and excavate tunnels. Other, larger termites, called soldiers, defend the colony with their immense jaws. Both workers and soldiers are sterile. Only the queen and king reproduce. The role played by an individual in a colony is called its **caste.** Social insects have no choice about their caste. Their genes determine their caste, just as your genes determine your sex. ◻

◻ **CAPSULE SUMMARY**

Insects, the most numerous of all animals, have jaws and three body regions: head, thorax, and abdomen. They are the only invertebrates capable of flight. Metamorphosis in insects may be either complete or incomplete. Some insects, like bees and termites, have evolved complex social systems.

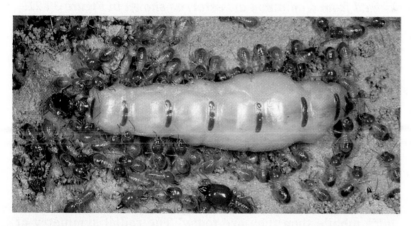

Figure 30-20 Most of the members of this termite colony are sterile (unable to reproduce). The queen, with her distended abdomen, is the egg-laying machine of the colony.

Section Review

1. *What is the major difference between chelicerate and mandibulate arthropods?*
2. *Compare the body segmentation of a crayfish with that of an insect.*
3. *Explain why a centipede is more closely related to a housefly than to a scorpion.*
4. *Describe the digestive system of the Eastern lubber grasshopper.*
5. *Describe the life cycle of a butterfly.*

Critical Thinking

6. *Arthropods first invaded land about 400 millions years ago. They have endured several mass extinctions in which many other kinds of organisms became extinct. What characteristics have enabled arthropods to thrive?*

Echinoderms and Invertebrate Chordates

Section Objectives

- Describe the major characteristics of echinoderms.
- Compare and contrast the lifestyles of the organisms in each of the five echinoderm classes.
- Describe how sea stars feed.
- Compare and contrast tunicates and lancelets.

Mollusks, annelids, and arthropods follow a protostome pattern of development. In echinoderms (sea urchins and sea stars) and chordates (tunicates, lancelets, and vertebrates), however, development is organized very differently. Their deuterostome pattern of development represents one of the most fundamental changes in body plan in the history of animal evolution. The developmental similarity between echinoderms and chordates unites these two seemingly dissimilar animal phyla and leads scientists to believe that both groups were derived from a common ancestor, as shown in Figure 30-21.

Echinoderm Larvae and Adults Have Different Body Plans

Echinoderms are familiar to all of us as "starfish." They are not really fish and are more properly called sea stars. Most people recognize the obvious radial symmetry of adult sea stars. It may be less obvious, however, that like all the advanced animal phyla, echinoderms are also *bilaterally* symmetric. In fact, it is the larvae of all echinoderms that are bilaterally symmetric. Biologists believe that the echinoderms were once far more mobile than they are today. The radial symmetry of

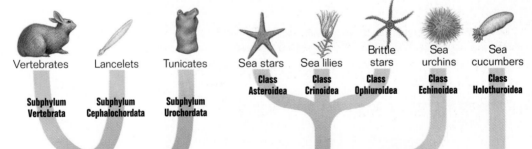

Vertebrates	Lancelets	Tunicates	Sea stars	Sea lilies	Brittle stars	Sea urchins	Sea cucumbers
			Class Asteroidea	Class Crinoidea	Class Ophiuroidea	Class Echinoidea	Class Holothuroidea
Subphylum Vertebrata	Subphylum Cephalochordata	Subphylum Urochordata					

Phylum Chordata — Phylum Echinodermata

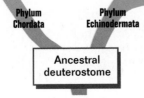

Ancestral deuterostome

Figure 30-21 This phylogenetic tree shows the evolution of chordates and echinoderms from a common ancestor, as well as the major chordate and echinoderm groups.

adults evolved later, when echinoderms adopted a more sessile style of living.

The adult echinoderm body has no head or brain. Instead, the nervous system consists of a central ring of nerves with branches extending into the arms. Although the animal is capable of complex response patterns, each arm acts more or less independently. Apparently, centralization of the nervous system is not feasible in radially symmetric bodies.

All Echinoderms Share Four Major Characteristics

The different kinds of echinoderms vary considerably in the details of their body design and in how they go about acquiring food. Despite the apparent differences among echinoderms, all share four fundamental characteristics.

1. **Endoskeleton** Echinoderms have an endoskeleton composed of individual plates called ossicles. In sea stars and many other echinoderms, a large number of these plates are fused together. The fused plates function much as an arthropod exoskeleton by providing muscle attachment sites and shell-like protection. In most echinoderms, the endoskeletal plates bear spines that project upward through their skin. The term *echinoderm* is from the Greek *echinos*, meaning "spiny," and *derma*, meaning "skin."

2. **Five-part radial symmetry** Most echinoderms have five arms extending radially from a central point.

3. **Water-vascular system** Echinoderms have a water-filled system of interconnected canals and tube feet called a water-vascular system. Echinoderms use their many tube feet to crawl across the sea floor.

4. **Coelomic circulation and respiration** The echinoderm body cavity functions as a simple circulatory system. Particles move freely throughout the large, fluid-filled coelom. In many echinoderms, respiration and waste removal are performed by **skin gills,** shown in Figure 30-22. Skin gills are small, fingerlike projections that grow between the spines. ◻

Figure 30-22 The extensions of skin on the surface of this sea star are called skin gills. Skin gills create an increased surface area through which respiratory gases (oxygen and carbon dioxide) can be exchanged. Skin gills also function as excretory organs. Wastes that accumulate in the skin gills are released into the surrounding water.

◻ CAPSULE SUMMARY

Echinoderms (sea stars, sea urchins, and sea cucumbers) are deuterostome marine invertebrates that are radially symmetric as adults and have a water-vascular system.

Echinoderms Are a Diverse Group

Echinoderms are one of the most successful of all marine phyla, although they were far more so in the past than they are now. There are more than 20 extinct classes of echinoderms and an additional 5 classes of living members. The 5 living classes are sea stars, feather stars and sea lilies, brittle stars, sea urchins and sand dollars, and sea cucumbers.

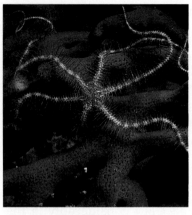

Figure 30-23 These three representative echinoderms each belong to a separate class. Feather stars, *above left,* along with sea lilies, are primitive echinoderms. Brittle stars, *above center,* are the largest of the major groups of echinoderms. Sea urchins, *above right,* often prefer to live on rocky ocean bottoms.

Figure 30-24 The tentacles of a sea cucumber are covered with sticky mucus, to which plankton adhere. Periodically, the tentacles are inserted into the mouth, cleaned of the organism-bearing mucus, and recoated with a fresh supply of mucus.

Sea stars are the echinoderms most familiar to people. All of the roughly 1,500 species of sea stars are carnivores. They are among the most important predators in many marine ecosystems. For example, the crown-of-thorns sea star consumes the cnidarians that make coral reefs. In one hour, these sea stars are able to graze along 20 m of a reef and can quickly destroy entire coral reef ecosystems. Other sea stars specialize in hunting bivalve mollusks—clams and oysters.

The feather stars and sea lilies differ from all other living echinoderms because their mouth is located on their upper surface. Sea lilies, the most primitive living echinoderms, are sessile. They are attached to the ocean floor by a stalk that can be 1 m (3.3 ft.) long. More than 600 fossil species of sea lilies are known, but there are only 80 living species. Also in this class are 520 living species of feather stars. Feather stars, shown in Figure 30-23, have stalks initially but lose them early in their development. The adult has a centrally located disk with hooklike projections that it uses to attach itself directly to the ocean bottom or a coral reef, but it is not fully sessile. It can crawl along the surface or even swim short distances.

Brittle stars have slender branched arms that they move in pairs from side to side, rowing along the ocean floor. Close relatives of the sea stars, the 2,000 species of brittle stars are sometimes grouped with them by taxonomists.

The sea urchins and sand dollars lack distinct arms but have the same basic five-part body plan as all echinoderms. Both sea urchins and sand dollars have hard endoskeletons of fused plates. There are about 900 living species of sea urchins and sand dollars.

The 1,500 species of sea cucumbers differ from other echinoderms in that their ossicles are small and not connected. Because they do not have a fused skeleton, the bodies of sea cucumbers are soft. Sea cucumbers, shown in Figure 30-24, are sluggish animals that often have a tough, leathery epidermis. They lie on their side on the ocean floor with their mouth at one end. The mouth is surrounded by several dozen tube feet that have been modified into tentacles. The tentacles capture the small organisms that the animal eats.

Invertebrate Chordates

The phylum Chordata is divided into three subphyla containing some 42,500 species. One subphylum, the vertebrates (subphylum Vertebrata), contains almost all chordate species. The other two, much smaller invertebrate subphyla are the tunicates (subphylum Urochordata) and the lancelets (subphylum Cephalochordata).

The tunicates are a group of about 1,250 species of sessile, filter-feeding marine animals. An adult tunicate, shown in Figure 30-25, exhibits few chordate characteristics. Only tunicate larvae have a body cavity, nerve cord, and notochord. All of these features are lost during development of the adult body form. The adult develops a tough sac, called a tunic, around its body. The phylum is named for this structure. Interestingly, the tunic is made of cellulose, a substance commonly found in plants and algae but rarely in animals.

Lancelets are scale-less, fishlike chordates a few centimeters long that live in shallow water throughout the oceans of the world. There are only 23 species. Lancelets spend most of their time buried in mud or sand with only their mouths protruding. They create a current with cilia that line their tentacled mouths and filter microscopic plankton from the water. They have no head or eyes. Scientists have determined that lancelets are the direct ancestors of the fishes, the first vertebrates.

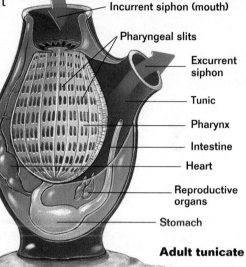

- Incurrent siphon (mouth)
- Pharyngeal slits
- Excurrent siphon
- Tunic
- Pharynx
- Intestine
- Heart
- Reproductive organs
- Stomach

Adult tunicate

Figure 30-25 Tunicates are found in shallow- and deep-water marine environments. They are often called "sea squirts," because many species shoot water out from their siphons when they are touched. The species shown is a blue sea squirt from the Philippines.

Section Review

1. List the major echinoderm characteristics.
2. Explain why some echinoderms have bodies that are softer than others.
3. Compare and contrast the feeding habits of a sea star and a sea cucumber.
4. Name the echinoderms that are completely sessile, and describe their basic structure.
5. Name two groups of invertebrate chordates, and describe how they feed.

Critical Thinking

6. A scientist collects several specimens of an animal that he has never seen before. After conducting an in-depth study of the organisms, the scientist observes that they have tube feet, an endoskeleton, and a protostome pattern of embryonic development. Why is the classification of this organism difficult?

Vocabulary

arachnid (684)
carapace (686)
caste (693)
cephalothorax (680)
chelicerate (680)
chrysalis (692)
ecdysis (682)
exoskeleton (680)

isopod (687)
Malpighian tubule (681)
mandibulate (680)
metamorphosis (692)
nymph (693)
ocelli (680)
ommatidia (680)
palp (684)

skin gill (695)
spiracle (680)
swimmeret (686)
telson (686)
trachea (680)
uropod (686)

Review

Multiple Choice

1. What evidence suggests that arthropods evolved from annelids?
   a. Arthropods and annelids have gills.
   b. Both groups have marine species.
   c. Segmentation is present in both groups.
   d. Arthropods have vestigial parapodia.

2. Lobsters, barnacles, and isopods
   a. breathe with gills.
   b. have only calcium carbonate in their exoskeleton.
   c. are scavengers.
   d. have unbranched appendages.

3. Millipedes and centipedes differ in that millipedes
   a. are terrestrial and segmented.
   b. have one pair of legs on each segment.
   c. have poisonous fangs.
   d. are herbivores.

4. The chief organ of excretion in insects is the
   a. Malpighian tubule.
   b. palp.
   c. nephridium.
   d. spiracle.

5. Which sequence shows the life cycle of an insect that undergoes incomplete metamorphosis?
   a. egg → larva → adult
   b. larva → pupa → adult
   c. egg → nymph → adult
   d. nymph → pupa → adult

6. Which of the following characteristics is typical of echinoderms?
   a. nauplius larva
   b. notochord
   c. water-vascular system
   d. exoskeleton present, endoskeleton absent

7. Which pair shows a correct match between an adult echinoderm and one of its unique characteristics?
   a. sea cucumber: leathery epidermis
   b. sand dollar: has five arms
   c. sea lily: free-swimming
   d. sea star: sessile

8. Which echinoderm group contains species that specialize in hunting bivalves?
   a. sea cucumbers   c. sea stars
   b. sand dollars   d. sea lilies

Completion

9. The compound eyes of insects form images and are made up of visual units called _____ , while some arthropods have simple eyes, called _____ , that do not form images.

10. Arthropods are classified into three subphyla: _____ , which contains spiders and scorpions; _____ , which includes crabs and lobsters; and _____ , which includes the millipedes and centipedes.

11. The nervous system of the Eastern lubber grasshopper is composed of a major _____ nerve cord with _____ located in each body segment.

12. Only the _____ stage of urochordates possesses the distinctive chordate characteristics: a nerve cord, notochord, and pharyngeal slits.

Themes Review

13. **Evolution** Included among the mandibulates, or jawed arthropods, are the crustaceans, insects, millipedes, and centipedes. There is growing acceptance among zoologists of the view that the crustacean mandibles and the mandibles of insects, millipedes, and centipedes reflect convergent evolution. What anatomical evidence suggests that the modern crustaceans are rather distantly related to other mandibulates?

14. **Structure and Function** Gas exchange is essential to crustaceans, insects, and echinoderms. Compare the manner in which gas exchange occurs in most crustaceans, insects, and echinoderms.

Critical Thinking

15. **Interpreting Data** As part of a study of North American spiders, data were collected on spider size and the number of molts undergone by spiders before reaching sexual maturity. Spider size was determined at time of sexual maturity by measuring from the tip of the spider's head to the end of its abdomen. The graph shows data collected from 100 individuals of four species. What is the relationship between spider size and number of molts that a spider undergoes before reaching sexual maturity?

16. **Designing Experiments** Two students agree that sea stars can regenerate but disagree on what body parts are required for regeneration to occur. One student claims that a complete sea star will develop from one arm. The other student insists that complete development will occur only when a part of the central disk is attached to the arm. Design an experiment to test the students' hypotheses and find out if one is correct.

Activities and Projects

17. **Cooperative Group Project** Work in teams to collect a variety of insects. Pool your team's collections and use field guides to group the insects into orders. Construct a bar graph that shows the number of individuals from each insect order collected by your team. Attach a photograph of a representative insect next to the portion of the graph that shows the number of individuals of that order. Speculate about how your team's graph might be different if insects were collected from a different location or during a different season.

18. **Multicultural Perspective** People from the Russian steppes and the plains of central Europe invaded the Indus Valley (of present day India) about 1300 B.C. Afterwards, the native people, the Dasa, were separated from the invaders, the Aryans. This separation evolved into a caste system that affects Indian culture today. What are the castes that evolved from the separation of the Dasa and Aryans? How do people who are members of one caste interact with people from a different caste? How is this caste system similar to the caste systems that exist among social insects?

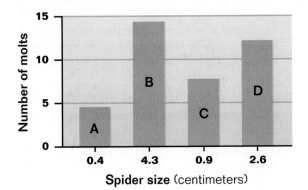

Spider size (centimeters)

Arthropods and Echinoderms 699

Arthropod Responses

OBJECTIVES

- Recognize arthropod characteristics in the pill bug, a representative arthropod.
- Observe the behavior of pill bugs on surfaces with different textures.
- Infer the type of texture pill bugs prefer.

PROCESS SKILLS

- relating structure to function
- inferring adaptive advantages of certain behaviors

MATERIALS

- 4 adult pill bugs
- 2 petri dishes
- stereomicroscope or hand lens
- blunt probe
- 4 fabrics of different texture (wool, polyester, silk, flannel, etc.)
- scissors
- transparent tape
- clock or watch with second hand

BACKGROUND

1. What are some characteristics of arthropods?
2. What is a crustacean? How do crustaceans obtain oxygen?
3. In what kind of environment do pill bugs live?
4. What kinds of stimuli do pill bugs respond to in their environment? How do they respond to the stimuli?
5. Write your own **Focus Question** on your Vee Form.
6. **Knowing Side of the Vee** List the **Concepts** and new **Vocabulary Words** on your Vee Form. In the **Concept Statements**

section of the Vee, use these words in sentences that define and explain them.

TECHNIQUE

Doing Side of the Vee

Part A: Pill Bug Characteristics

1. Place a pill bug in a petri dish. Use the stereomicroscope or hand lens to observe the pill bug. In the **Records** section of your Vee Form, list the characteristics that enable you to recognize that the pill bug is an arthropod.

2. Using the blunt probe, gently touch the pill bug. In the **Records** section of your Vee Form, describe your observations.

Part B: Pill Bug Behavior

3. On one of the fabrics, trace the outline of the bottom of the petri dish. **CAUTION: Be careful when using the scissors.** Cut out the circle from the fabric and fold it in half. Then cut along the fold to produce two half-circles.

4. Repeat step 3 using the other three fabrics. You should now have eight half-circles.

5. Tape together two half-circles, each of a different fabric.

6. Place the two-fabric circle in the bottom of a petri dish, tape side down.

7. In the **Records** section of your Vee Form, draw the circle you made in step 5. Label the two types of fabric that make up the circle.

8. Place a pill bug in the center of the circle. Observe the movement of the pill bug for five minutes. Have one member of the group keep track of the amount of time the pill bug spends on each fabric. Have another member draw in the **Records** section the path the pill bug travels in the circle. Below the circle, record the amount of time spent on each type of fabric.

9. Repeat steps 5–8 using the other two fabrics.

10. Repeat the investigation twice, each time using a different pill bug. Reuse the circles you made. Record your observations in the **Additional Records and Observations** section of the Vee. In the **Procedure** section of the Vee, briefly summarize the procedure you followed.

11. Clean up your materials and wash your hands before leaving the lab.

INQUIRY

1. Why were three pill bugs tested in this investigation instead of just one?

2. Did the three pill bugs show a similar pattern of movement? Explain.

3. Rank the fabrics according to the total amount of time spent on them by the three pill bugs.

4. Use the information on the **Knowing Side** of the Vee to interpret your results from the **Doing Side,** and then write your **Knowledge Claim.** Write a **Value Claim** for this lab.

ANALYSIS

1. In what way is the pill bug's response to disturbance advantageous?

2. In Part B, which fabric did the pill bugs prefer? Describe the texture of that fabric.

3. What stimulus did the pill bugs respond to? What was their response?

4. How is being able to detect surface texture a good adaptation for pill bugs in their natural habitat?

FURTHER INQUIRY

Write a **New Focus Question** that could be the point of a new investigation. The following are examples:

Do pill bugs prefer moist or dry environments?

Do pill bugs prefer light or dark environments?

CHAPTER 31
OVERVIEW OF VERTEBRATES

Turkey skeleton showing vertebrae

31-1 Story of Vertebrate Evolution

*V*ertebrates are chordates with a backbone. These animals are named for the individual segments, called vertebrae, that make up the backbone. Vertebrae form a central axis for muscle attachment. They also protect the dorsal nerve cord. These functions paved the way for the development of the brain and an internal skeleton that allowed vertebrates to grow larger than their invertebrate ancestors.

Section Objectives

■ Identify the major groups of vertebrates, and list the order in which they evolved.

■ Describe the key adaptations that enabled fishes to dominate the oceans.

■ Describe the adaptations of amphibians for life on land.

■ Describe the key adaptations of terrestrial vertebrates.

Figure 31-1 Like all of the earliest vertebrates, this ancient fish had no jaws.

History of Vertebrates

The first chordates probably evolved about 600 million years ago. At that time many different groups of organisms appeared in the shallow seas covering much of Earth's continents. The oldest vertebrate fossils, dating back to about 500 million years ago, are of fishes similar to the one shown in Figure 31-1. Unlike most of the fishes with which you are familiar, the earliest fishes had neither jaws nor paired fins. Many of them looked something like a flattened hot dog with a hole at one end and a fin at the other.

For more than 100 million years, fishes were the only vertebrates. They became the dominant animals in the sea, diversifying into a great variety of species. Some were just a few centimeters long; others were bigger than a car. Amphibians, which evolved from fishes, were the first vertebrates to invade the land. Frogs and salamanders are the descendants of those early amphibians. Like their ancestors, frogs and salamanders live successfully on land but must reproduce in water or very moist environments.

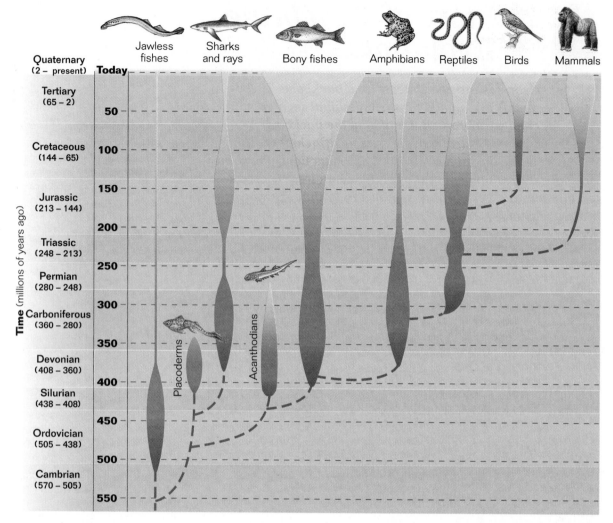

Quaternary
(2 – present)

Today

Tertiary
(65 – 2)

50

Cretaceous
(144 – 65)

100

150

Jurassic
(213 – 144)

200

Triassic
(248 – 213)

250

Permian
(280 – 248)

300

Carboniferous
(360 – 280)

350

Devonian
(408 – 360)

400

Silurian
(438 – 408)

450

Ordovician
(505 – 438)

500

Cambrian
(570 – 505)

550

Time (millions of years ago)

Jawless
fishes

Sharks
and rays

Bony fishes

Amphibians

Reptiles

Birds

Mammals

Placoderms

Acanthodians

Figure 31-2 This diagram shows a simplified history of the vertebrates and the relationships among the major groups. Dotted lines indicate the origin of one group from another. Each group shown here is a class. The width of the colored bar representing the class indicates its approximate diversity (number of species).

CONTENT LINK
....
*Curious about what caused the extinction of the dinosaurs? Part of **Chapter 33** is devoted to this puzzle.*

Amphibians, in turn, gave rise to reptiles, which were better suited to living out of water. Within 50 million years of their origin, reptiles replaced amphibians as the dominant land vertebrates. Because reptiles are **terrestrial**, able to live their whole lives on land, they were able to spread out and diversify. Many different species of reptiles evolved. Some were smaller than a chicken, while others were as big as a truck.

From these early reptiles, the two great lines of terrestrial vertebrates evolved—dinosaurs and mammals. Dinosaurs and mammals appear at about the same time in the fossil record, 220 million years ago. But the dinosaurs quickly won out in the evolutionary competition and dominated the Earth for 150 million years. Over this long period, the largest mammal was no bigger than a cat. When the dinosaurs abruptly disappeared 65 million years ago, mammals quickly took their place. The reason for the disappearance of the dinosaurs is still hotly debated, but there is no doubt that their extinction allowed mammals to become abundant and diverse. Figure 31-2 summarizes the evolutionary history of the vertebrates.

The history of vertebrates has been a series of evolutionary advances that allowed these animals to diversify first in the sea and then on the land. In this section, you will look briefly at some evolutionary changes in the vertebrate body plan that contributed to their success. You can read more about key steps in vertebrate evolution in *Highlights: Vertebrate Evolution* on pages 731–738. In the remaining sections of this chapter, you will examine the key adaptations that permitted vertebrates to invade the land. This invasion was a staggering evolutionary achievement that involved fundamental changes in many body systems.

Fishes Dominate the Sea

Jawless fishes were the first vertebrates. As they swam along the ocean bottom, these animals fed by filtering tiny invertebrates from water passing through their mouths. Their bodies were covered with thick, bony plates and smaller scales.

Jawless fishes dominated the world's oceans for about 100 million years, until they were eventually replaced by new kinds of fishes that were hunters. Only 63 species of jawless fishes exist today, the hagfishes and lampreys. Figure 31-3 shows a lamprey.

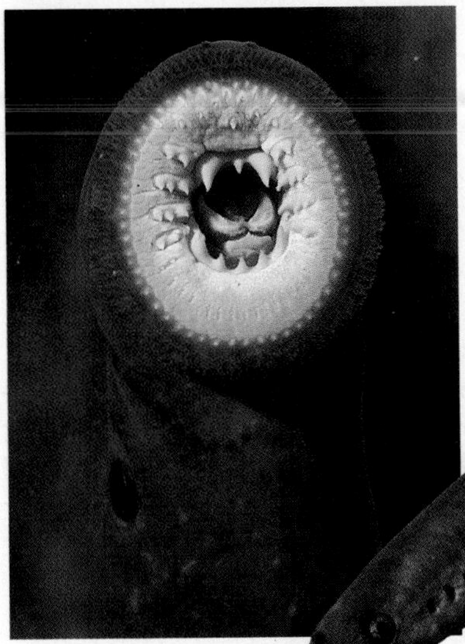

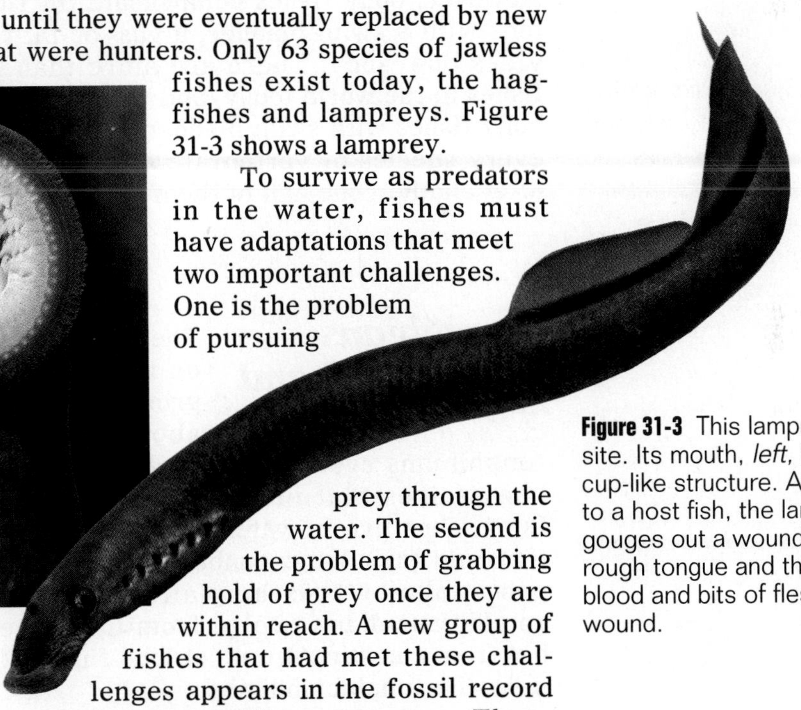

To survive as predators in the water, fishes must have adaptations that meet two important challenges. One is the problem of pursuing prey through the water. The second is the problem of grabbing hold of prey once they are within reach. A new group of fishes that had met these challenges appears in the fossil record about 430 million years ago. These fishes were the acanthodians (*uh KAN thoh dee uhns*). They had strong biting jaws with jagged bony edges that served as teeth. Such adaptations made it possible to seize prey. Chasing prey through the water was possible because of several other changes. For instance, the bodies of these early fishes had become streamlined and flattened sideways. This made it easier to move through the water. Paired fins provided better control of fast swimming. Acanthodians had small bodies covered with protective spines.

Figure 31-3 This lamprey is a parasite. Its mouth, *left,* has a suction-cup-like structure. After attaching to a host fish, the lamprey gouges out a wound with its rough tongue and then feeds on blood and bits of flesh from the wound.

Figure 31-4 Sharks were among the first vertebrates to have jaws, which allow them to bite. Sharks also have streamlined bodies and paired fins, which allow for fast swimming with better control.

About 410 million years ago, larger jawed fishes called placoderms (*PLAK uh durms*) appeared. These fishes had massive heads armored with bony plates. Placoderms and acanthodians are now extinct. Soon after the appearance of the placoderms, two other groups of jawed fishes evolved—sharks and bony fishes. Each group had somewhat different features for fast and maneuverable swimming. In sharks, such as *Hybodus* shown in Figure 31-4, the skeleton is composed of **cartilage,** a lightweight, strong, and flexible material. In bony fishes, represented by the coelacanth (*SEE luh kanth*) shown on page 731, the skeleton is made of bone, which is heavier and less flexible than cartilage. However, bony fishes compensate for their heavier skeletons with a **swim bladder,** a gas- or fat-filled sac that provides buoyancy. There are more than 20,000 species of fishes in the world today, and the vast majority of them are bony fishes with swim bladders. In fact, if you could watch every species of vertebrate alive today pass by you, one after another, one-half of them would be bony fishes. ◻

❑**CAPSULE SUMMARY**

Jawless fishes were the first vertebrates. Lampreys and hagfishes are the only surviving jawless fishes. Sharks have a lightweight skeleton of cartilage. Bony fishes have a heavier skeleton of bone and a swim bladder, which provides buoyancy.

CONTENT LINK

*See **Chapter 32** for more information on lungfishes and coelacanths.*

Amphibians Invade the Land

The first vertebrates able to live on land were amphibians. This group of animals first appeared about 370 million years ago. Amphibians evolved from lobe-finned fishes, a group of bony fishes that includes coelacanths and lungfishes. Lungfishes live in water that often has a low oxygen content, and their lungs enable them to supplement their oxygen intake by breathing air. Most scientists think that amphibians didn't evolve from lungfishes or coelacanths, but from an extinct group of lobe-finned fishes that had fins more like amphibian limbs.

Life on land is quite different from life in the water. Thus, the successful invasion of land by vertebrates involved a number of major innovations. Look at the characteristics of *Ichthyostega,* an early amphibian shown on page 732, as an example. Air is less buoyant than water, so legs were necessary to support body weight as well as to allow movement from place to place. To live on land, amphibians also needed lungs. The delicate structure of a fish's gills depends on the buoyancy of water for support. Out of water, gills stick together, reducing the surface area available for gas exchange. That is why a fish will suffocate on land, even though there is more oxygen in air than in water. Walking around on land requires a higher metabolism, which in turn uses greater amounts of oxygen. Thus, the heart also underwent change.

Some problems remained, however. For example, the eggs of amphibians are not watertight. Thus, amphibians generally seek out water or damp areas in which to reproduce and in which their young can live as they grow and mature. But evolution does not result in perfect solutions, only workable ones. The adaptations that allowed the first amphibians to climb out onto land have enabled their descendants to survive for over 350 million years. Today there are approximately 4,200 species of amphibians, including the familiar frogs, toads, and salamanders. ❑

❑ **CAPSULE SUMMARY**

Amphibians were the first vertebrates to live on land. However, they lack watertight eggs and so cannot reproduce in environments in which their eggs would dry out.

Reptiles Conquer the Land

If you think of amphibians as the "first draft" of a manuscript about survival on land, then reptiles are the finished book. Each of the adaptations that allowed amphibians to lead a terrestrial existence were refined in reptiles. Their legs, for example, were positioned to support the body more effectively. So reptiles not only were bigger, but also could run. Changes in the lungs and heart made these organs more efficient. The most significant adaptations, though, were internal fertilization and watertight eggs. Watertight eggs made reptiles the first completely terrestrial vertebrates. Today there are about 7,000 species of reptiles, mostly snakes and lizards, found in practically every habitat on Earth. The chameleon shown in Figure 31-5 is one of these species.

When reptiles first evolved, about 320 million years ago, Earth was entering a long, dry period. Early reptiles, as represented by *Dimetrodon* shown on page 733, were well suited to these conditions and quickly diversified. In particular, their ability to conserve water allowed reptiles to have large bodies in dry conditions. This was something amphibians could not do. Within 50 million years, reptiles had replaced amphibians as the large, terrestrial vertebrates. All land vertebrates larger than a chicken were reptiles.

Living reptiles are **ectothermic**. Their metabolism is too slow to produce enough heat to warm their bodies. They must absorb heat from the environment. **Endothermic** animals, such as mammals and birds, maintain a high, constant body temperature because they produce heat internally through a faster metabolism. Therapsids, an order of extinct reptiles believed to be the ancestors of mammals, were probably endotherms. They were replaced by ectothermic thecodonts and then by dinosaurs.

CONTENT LINK

More information on characteristics, classification, and anatomy of reptiles can be found in Chapter 33.

Figure 31-5 Unlike the moist skin of most present-day amphibians, reptilian skin is covered with scales that keep the body from drying out.

Evidence indicates that at least some, perhaps many, dinosaurs were endothermic. Studies of radioactive isotopes show that some bones at the tip of the tail of large dinosaurs appear to have formed at the same temperature as the rib bones in the body's center. In an ectothermic reptile, the extremities are much cooler than the body's center. Generally speaking, endothermic animals are more active and are able to live in a wider range of habitats than ectothermic animals. The distribution of dinosaur fossils provides additional evidence that some of these animals were endothermic. ◼

Mammals and Birds Adapt to Colder Climates

Sixty-five million years ago, global temperatures cooled dramatically. Many scientists think that this cooling was caused, at least in part, by huge clouds of dust thrown into the atmosphere by the impact of a huge meteorite or comet. If dinosaurs were endothermic, they may have been particularly vulnerable to the cooler temperatures because, unlike mammals and birds, dinosaurs had no insulation. No land vertebrate bigger than a cat survived the cold.

When the dinosaurs disappeared, their ecological roles as the large land vertebrates were taken over by mammals. With bodies insulated by fur, mammals were better suited to the colder climates that were typical then. Look at *Megazostrodon*, shown on page 735, and *Eozostrodon*, shown in Figure 31-6, as representative examples of early mammals.

Figure 31-6 *Eozostrodon* was an early mammal that was only about 10 cm (4 in.) long. It lived during the late Triassic period. All of the earliest known mammals were very small.

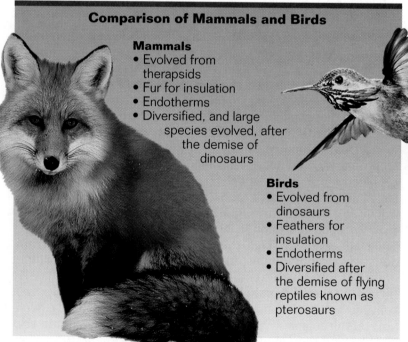

Comparison of Mammals and Birds

Mammals
• Evolved from therapsids
• Fur for insulation
• Endotherms
• Diversified, and large species evolved, after the demise of dinosaurs

Birds
• Evolved from dinosaurs
• Feathers for insulation
• Endotherms
• Diversified after the demise of flying reptiles known as pterosaurs

Figure 31-7 Feathers on birds and fur on mammals are adaptations that enabled these groups to survive in colder climates.

Birds, which evolved from dinosaurs early in their history, had feathers for insulation. Once the pterosaurs were gone, birds were able to diversify. Look at *Archaeopteryx*, shown on page 738, which is the oldest known bird, to see some characteristics of birds. Today over 8,000 species of birds fly the world's skies. Figure 31-7 contrasts birds and mammals.

Figure 31-8 The blue whale is the largest living mammal. An adult may be up to 30 m (99 ft.) long and weigh 136,000 kg (150 tons).

The mammals that evolved 220 million years ago alongside the dinosaurs would look strange to you. These early mammals were quite small and probably in some ways resembled the most primitive mammals living today—the monotremes. Monotremes, such as the duckbill platypus, lay shelled eggs and have a shoulder structure similar to that of early reptiles. Like all mammals, monotremes nurture their young with milk produced by mammary glands (hence the name *mammal*). Only three species of monotremes exist today: the duckbill platypus and two species of spiny anteaters.

Monotremes were replaced 100 million years ago by marsupials. Unlike monotremes, marsupials do not lay shelled eggs. The embryos develop within the mother's body and are born live. Upon birth, the young marsupials, which are tiny and immature, crawl into an external pouch to nurse and complete their development. About 280 species of marsupials survive today, mostly in Australia and New Guinea.

Placental mammals evolved at about the same time as marsupials and eventually replaced them throughout most of the world. Marsupials survived in Australia and New Guinea because continental drift separated these land masses from the other continents before placental mammals arrived. Placental mammals, such as the blue whale shown in Figure 31-8, invest even more time in nurturing their young than marsupials do. Embryos spend a much longer period of

Figure 31-9 *Smilodon* was a little over 1m (about 4 ft.) in length, roughly the size of a mountain lion.

development within the mother's body. This long period of growth and development is made possible by the placenta. The **placenta** is a structure that transfers nutrients from the mother's blood supply directly to the growing embryo. You will read more about its function in Section 31-5.

Today there are over 4,000 species of placental mammals, ranging in size from 1.5 g (less than 0.1 oz.) shrews to 136,000 kg (150 tons) whales. Almost half of these species are rodents (mice and their relatives), and another quarter are bats, the only flying mammals. Mammals have also invaded the seas, just as the reptiles known as plesiosaurs and ichthyosaurs did so successfully millions of years earlier. There are 79 living species of whales and dolphins. Primates, the order to which humans belong, are not a major group in terms of species diversity. There are only 233 known species. Humans evolved very recently—less than 2 million years ago. There have been at least three species of humans, but *Homo sapiens* is the only one that survives. Humans are notable among primates for lacking a complete body covering of fur. The climate in which humans evolved was so much warmer, that fur—so central to the success of mammals—was no longer essential.

Today, all very large land animals are mammals. However, the largest land mammals reached their peak during past ice ages, when the saber-toothed cat shown in Figure 31-9 lived in North America. The warming of Earth's climate in recent times has favored smaller bodies that are easier to cool. ◻

◻ **CAPSULE SUMMARY**

There are three groups of mammals. Monotromes lay eggs. Marsupials suckle their young, which are born after a short period of development, in an external pouch. Placental mammals nourish their developing embryos by means of a placenta.

Section Review

1. *From which group of vertebrates did the reptiles evolve? What groups of vertebrates evolved from reptiles?*
2. *Explain the advantages of jaws and paired fins.*
3. *Describe the characteristics that enable amphibians to survive on land.*
4. *Name two adaptations that allowed reptiles to live and reproduce completely on land.*

Critical Thinking

5. *What role have mass extinctions played in the evolution of mammals?*
6. *It is conventional to view mammals as superior to or "more advanced" than reptiles. What evidence from the history of vertebrates contradicts this view?*

31-2 Challenge of Obtaining Oxygen

One of the defining challenges faced by animals has been the need to get enough oxygen to fuel their metabolism. Sponges, cnidarians, many flatworms and roundworms, and some annelids obtain oxygen by diffusion through the body surface. The more advanced marine invertebrates—mollusks, arthropods, and echinoderms—have gills, which are specialized respiratory organs. Fishes also obtain oxygen with gills. But vertebrate gills are useless on land. In this section you will learn how gills function and how terrestrial vertebrates obtain oxygen without gills.

Section Objectives

- Describe how countercurrent flow increases oxygen absorption.
- Explain how vertebrates obtain oxygen from the air.
- Identify the key changes in the structure of the vertebrate lung.

Fishes Use Gills to Get Oxygen From Water

Have you ever looked closely at the face of a swimming fish? If so, you've probably noticed that as it swims, a fish continuously opens and closes its mouth, as if it were trying to eat the water. What looks like eating is actually breathing. The major respiratory organ of a fish is the gill. A gill is a thin sheet of tissue that increases the surface area available for the diffusion of oxygen. The gills of vertebrates, specifically those of bony fishes, are constructed in a way that makes them the most efficient oxygen-gathering organs ever to have evolved. In a fish, gills hang like a curtain between the mouth and the cheeks. At the rear of the cheek cavity is an opening called a gill slit. The whole point of a fish's "swallowing" water is to force the water from the mouth, over the gills, and out through the gill slits.

This swallowing procedure is the core of a great advance in gill design shown by fishes. To understand the importance of this process, first think about the way a mollusk gets oxygen with its gills. In a mollusk, the gills are in an internal cavity, but the water simply sloshes in and out, washing back and forth over the gills. In a fish, the swallowing process forces water to flow only in one direction. One-way flow of water, combined with a specific arrangement of gill tissue, permits countercurrent flow, which is a very efficient way of extracting oxygen.

Here is how countercurrent flow works. Each gill is made of two rows of **gill filaments**, which are fingerlike projections composed of thin membranous plates stacked on top of one another, as shown in Figure 31-10. These filaments stick out into the flow of the water. The water entering the gills is higher in oxygen content than the water leaving. Within each filament, blood flows from the back to the front,

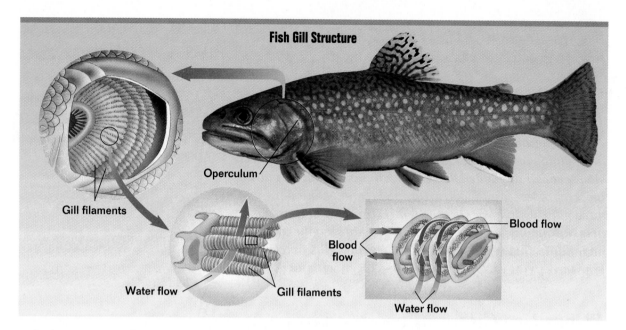

Fish Gill Structure

Operculum

Gill filaments

Water flow

Gill filaments

Blood flow

Blood flow

Water flow

Figure 31-10 The gills of a bony fish are located behind the head, underneath a flap called the operculum. In the filaments of the gills, oxygen from water diffuses into the blood vessels carrying oxygen-poor blood to be transported throughout the body.

□ CAPSULE SUMMARY

The structure of a bony fish's gills ensures that water and blood flow in opposite directions. This arrangement maximizes the amount of oxygen that can be extracted from water.

as shown in Figure 31-10. This blood is low in oxygen (indicated by blue) when it enters the gills and high in oxygen (indicated by red) when it leaves. Water, however, flows past the filaments in the opposite direction—from front to back.

Why is the countercurrent arrangement important? Countercurrent flow ensures that oxygen diffuses into the blood over the whole length of the blood vessels in the gills. If blood and water flowed in the same direction, the amount of oxygen in the blood would increase while that in the water decreased until an equilibrium was reached and diffusion stopped. The gills of bony fishes are the most efficient respiratory organs possible. They are able to extract up to 85 percent of the available oxygen from water. □

Amphibians Use Lungs to Get Oxygen From Air

Air contains about 20 times as much oxygen as sea water does. Yet all of this oxygen won't help a land-dwelling animal much if the structures for gathering oxygen don't work. And gills, which work very well in water, do not work in the air. Out of water the delicate gill membranes have no support. Therefore, they collapse into a soggy mass. The surface area for gas exchange is so dramatically reduced that a fish out of water suffocates from lack of oxygen. Thus, one of the major challenges that faced the first land vertebrates was obtaining oxygen from air.

The evolutionary solution to the problem of getting oxygen from air is the lung. A **lung** is basically a baglike respiratory organ that allows gas exchange between the air and the blood. The amount of oxygen a lung can absorb depends on

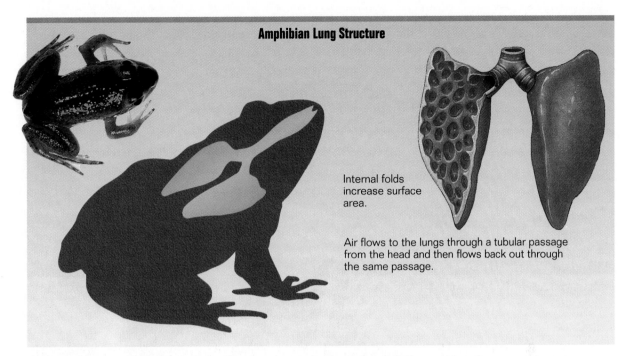

Amphibian Lung Structure

Internal folds increase surface area.

Air flows to the lungs through a tubular passage from the head and then flows back out through the same passage.

its internal surface area. The greater the surface area, the more oxygen that can be absorbed. To get as much oxygen as possible, the interior surface of the lung came to be highly folded, which greatly increased the surface area. Although amphibians could absorb oxygen through their skins, lungs eventually proved to be far more efficient. In amphibians, the lungs are hardly more than sacs with a wrinkled inner membrane, as shown in Figure 31-11. With each breath, fresh air rich in oxygen is drawn into the lungs, where it mixes with a small volume of air that has already given up its oxygen. Because the diffusion surface of the lung is exposed to a mixture of fresh and partly depleted air, the respiratory efficiency of lungs is much less than that of gills. But because there is so much more oxygen in air, amphibian lungs don't have to be as efficient as gills. This arrangement works perfectly well for amphibians, many of which also obtain some oxygen through their thin, moist skin.

Figure 31-11 The lungs of amphibians are sacs with a folded internal membrane that provides a large surface area for gas exchange.

Lung Surface Area Increases in Reptiles and Mammals

Reptiles are far more active than amphibians, so they have much greater metabolic demands for oxygen. An amphibian's lungs cannot provide that much oxygen. However, reptiles cannot rely on their skin for additional respiration in the way that many amphibians can. Remember that reptiles have dry, scaly skin to prevent water loss; they don't have a moist outer membrane for gas exchange.

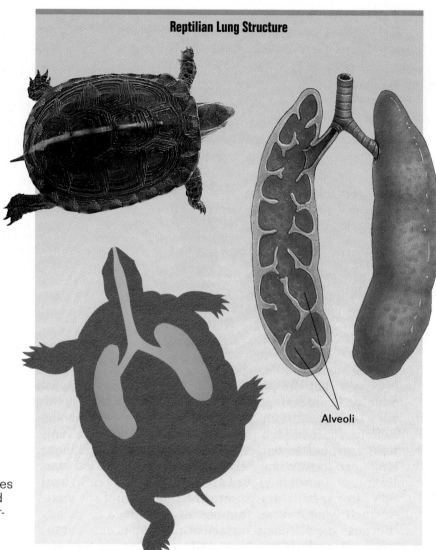

Reptilian Lung Structure

Alveoli

Figure 31-12 The lungs of reptiles contain small chambers called alveoli that create a larger surface area than do the folds in amphibian lungs.

□ CAPSULE SUMMARY

Efficiency of gas exchange in the lung increases with increasing surface area. Reptilian lungs have many alveoli that greatly increase internal surface area. Mammalian lungs have even more alveoli and so have an even larger internal surface area.

In reptiles, increased oxygen demand is met by greatly enlarged surface area of the lung available for diffusion. The inner surface of reptile lungs consists of many small chambers called **alveoli** (singular, alveolus) that are clustered together like grapes, as shown in Figure 31-12.

Mammals have an even greater demand for oxygen than reptiles do because mammals maintain a constant body temperature. The problem of harvesting more oxygen is solved by increasing the diffusion surface area within the lung even further, as shown in Figure 31-13.

Some mammals are far more active than others, but the more active ones do not have proportionally larger lungs. Instead, in active mammals the individual alveoli are smaller and more numerous, which further increases the surface area for diffusion. Throughout the long history of evolutionary refinement of the vertebrate lung, increases in efficiency have been achieved by increases in the lung's internal surface area. □

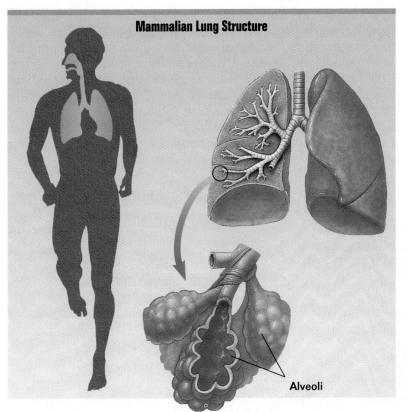

Mammalian Lung Structure

Alveoli

Figure 31-13 The lungs of mammals contain many clusters of alveoli that increase surface area even more than the alveoli in reptilian lungs. Humans have about 300 million alveoli in their lungs. In each lung, the total surface area devoted to diffusion is about 80 sq. m (860 sq. ft.)—slightly less than one-fourth the area of a full-sized basketball court.

The Lungs Are Most Efficient in Birds

There is a limit to how much efficiency can be improved by increasing the surface area of the lung. The more active mammals have, in fact, already reached this limit. Birds, however, have a respiratory demand for oxygen that exceeds the capacity of the lungs of even the most active mammal. Bird flight uses a lot of energy very quickly. Unlike bats, whose flight involves considerable gliding, most birds beat their wings rapidly as they fly. And they often fly this way for a long time. Thus, flying birds must carry out very rapid oxidative respiration within their cells to replace the ATP expended by their contracting flight muscles. The amount of oxygen needed is much greater than any mammalian lung can deliver.

If the oxygen demands of flight cannot be met by an increase in lung surface area, how do birds get the energy they need? In a sense, they do what fishes do. Just as water flows over a fish's gills in one direction, air flow in birds occurs in only one direction. One-way air flow was made possible by the evolution of a series of air sacs connected to the bird's lungs, shown in Figure 31-14.

There are two important advantages to this complicated system. First, there is no oxygen-poor air left in the lungs, as there is in mammalian lungs, so the diffusion surfaces of a bird's lungs are exposed only to fully oxygenated air. Second, the flow of blood in the lungs runs in a different

Figure 31-14 Birds can fly at high altitudes where there is limited oxygen because there is always fully oxygenated air in their lungs and because their lungs are very efficient at absorbing oxygen. There is no gas exchange in the air sacs; they act simply as holding tanks.

direction to the flow of air. The flows are not completely opposite, as in fish gills. Instead, the network of lung capillaries is arranged across the air flow at a 90-degree angle. Even though this is less efficient than in gills, blood leaving a bird's lung can still contain more oxygen than exhaled air does. No mammalian lung can do this. That is why a sparrow can fly at an altitude of 6,000 m (about 20,000 ft.), while a mouse with the same body mass would be unable to move on a mountain peak at the same height. The sparrow is simply getting more oxygen than the mouse.

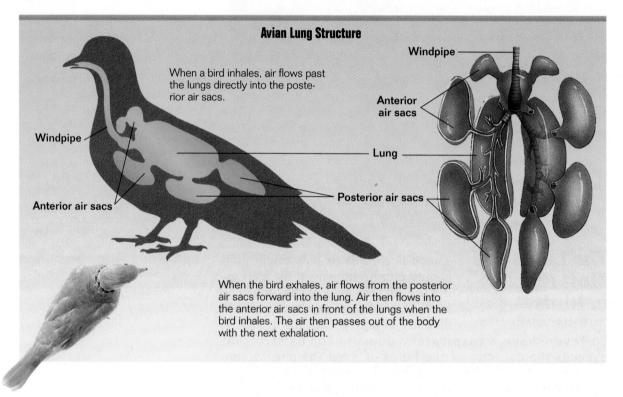

Avian Lung Structure

When a bird inhales, air flows past the lungs directly into the posterior air sacs.

Windpipe

Anterior air sacs

Lung

Posterior air sacs

Windpipe

Anterior air sacs

When the bird exhales, air flows from the posterior air sacs forward into the lung. Air then flows into the anterior air sacs in front of the lungs when the bird inhales. The air then passes out of the body with the next exhalation.

Section Review

1. *How does a fish obtain the maximum amount of oxygen from the water flowing over its gills?*
2. *What function do lungs serve in terrestrial vertebrates?*
3. *How does the structure of the lung in reptiles and mammals allow these animals to obtain sufficient oxygen?*
4. *What adaptations improved the efficiency of bird respiration?*

Critical Thinking

5. *In addition to capturing oxygen, a bird's unusual respiratory system plays a role in flight. Explain how.*

31-3 Evolution of a Better Heart

*A*long with the need to obtain oxygen, vertebrates need to deliver it efficiently to the body's tissues. Delivering oxygen to and removing carbon dioxide from the body's tissues are the functions of the vertebrate circulatory system. The circulatory systems of all vertebrates share two basic features: a network of vessels through which blood can circulate, and a heart to pump that blood. Much of the evolution of the vertebrate circulatory system involved changes in the heart.

Section Objectives

- Describe the structure of a fish's heart.
- Describe the evolution of the vertebrate heart.
- Discuss the significance of the pulmonary veins in terrestrial vertebrates.

The Chordate Heart Is a Simple Pump

Chordates that were ancestral to the vertebrates had simple tubular hearts similar to those seen in lancelets. This heart was little more than a specialized zone of the ventral (lower) artery that was more heavily muscled than the rest of the arteries. This artery-pump beat in simple peristaltic waves. A peristaltic wave starts at one end of a tube and moves progressively along toward the other end. Suppose you take a soft-drink straw and squeeze one end, and then move your squeezing fingers down the length of the straw. That's a peristaltic wave. If you squeezed a straw in the middle, any liquid in it would squirt in both directions. The same thing happens with peristaltic pumps such as a chordate heart. Blood is pushed in both directions as the heart contracts, so these organs are not very efficient.

The evolution of the vertebrate heart reflects two important transitions in vertebrate behavior. The first was the shift from filter-feeding to active prey capture that accompanied the evolution of jaws. The second was the increased activity that accompanied the invasion of land and the control of body temperature.

A Heart With Chambers Evolves in Fishes

The evolution of gills in fishes created a serious problem. As the diameter of a tube gets smaller, resistance to the flow of liquid increases dramatically. So the tiny diameters of the many small blood vessels in a fish's gills create enormous resistance to the flow of blood. This resistance is too great for any peristaltic pump to overcome. In fishes, the peristaltic pump of early chordates has been replaced with a pump of a very

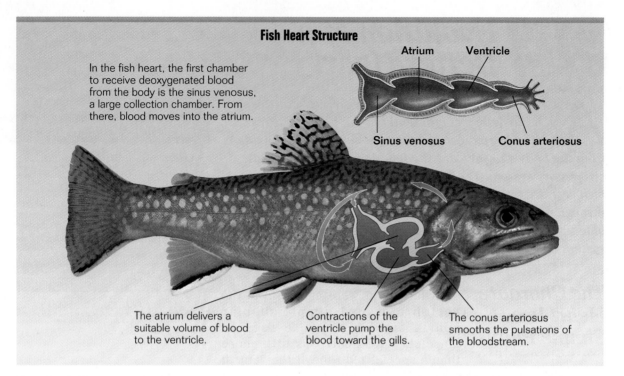

Fish Heart Structure

In the fish heart, the first chamber to receive deoxygenated blood from the body is the sinus venosus, a large collection chamber. From there, blood moves into the atrium.

Atrium   Ventricle

Sinus venosus   Conus arteriosus

The atrium delivers a suitable volume of blood to the ventricle.

Contractions of the ventricle pump the blood toward the gills.

The conus arteriosus smooths the pulsations of the bloodstream.

Figure 31-15 The heart of a fish is a tube of four connected chambers.

different nature—a chamber-pump heart, which is shown in Figure 31-15.

The heart of a gill-breathing fish can be thought of as a tube with four chambers in a row. The first two chambers are collection chambers. The second two are pumping chambers.

1. **Sinus venosus** (*SEYE nuhs vuh NOH suhs*) This chamber is a large collection chamber that acts to reduce the resistance of blood flow into the heart. A one-way valve at its entrance prevents blood from flowing backward out of the chamber when its walls contract.

2. **Atrium** Blood from the sinus venosus fills this chamber, which is large and has thin, muscular walls.

3. **Ventricle** To provide enough force to push the blood through the gills, the third chamber is a thick-walled pump with enough muscle to contract strongly.

4. **Conus arteriosus** (*KOH nuhs ahr TIHR ee oh suhs*) A second, more elongated pump, the fourth chamber smooths the pulsations and adds still more force.

You can see the ancestry of fish in how their hearts operate. As in their chordate ancestors, the sequence of contraction is a peristaltic wave starting at the rear and moving to the front. The first of the four chambers to contract is the sinus venosus, which loads the atrium. Then the atrium contracts and fills the ventricle with blood. Next the ventricle contracts, forcing blood into the conus arteriosus. This chamber contracts last, smoothing the pulsations of the bloodstream. Although the relative positions of the heart chambers have changed throughout vertebrate evolution, this heartbeat sequence has remained the same. ◼

The fish heart represents one of the great evolutionary innovations in the vertebrates. Because it pumps blood first through the gills and then to the rest of the body, the blood delivered to the tissues is fully oxygenated. This arrangement has one important limitation, however. After passing through the fine network of tiny blood vessels in the gills, the flow of blood has lost much of the force applied by the contraction of the heart. Hence, the circulation of blood from the gills to the rest of the body is sluggish. This means that oxygen cannot be quickly delivered to body muscles.

The Pulmonary Vein Evolves in Amphibians

With the evolution of amphibians, the limitations of pumping blood through the lungs to the rest of the body presented an even greater problem. The first lungs were probably quite inefficient, but to be active on land, amphibians needed even more oxygen to be delivered to their muscles. So in this group of organisms, the circulatory system underwent another change—the addition of another pair of major blood vessels in the heart. These vessels, the **pulmonary veins**, carry blood from the lungs to the heart. Therefore, after picking up oxygen in the lungs, blood returns to the heart to be pumped to the rest of the body, as shown on the right in Figure 31-16.

Not only did the path of circulation in amphibians change, but so did the heart itself. A dividing wall known as

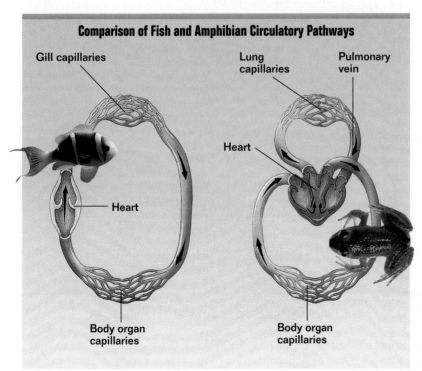

Comparison of Fish and Amphibian Circulatory Pathways

Gill capillaries

Lung capillaries

Pulmonary vein

Heart

Heart

Body organ capillaries

Body organ capillaries

Figure 31-16 Circulation in fishes differs from amphibian circulation in that amphibians have a pulmonary circulation loop. This loop moves blood from the heart to the lungs and back to the heart.

a **septum** separates the atrium into right and left halves. The pulmonary veins, carrying oxygen-rich blood from the lungs, empty into the left atrium. The sinus venosus, carrying oxygen-poor blood from the rest of the body, empties into the right atrium. Since no wall divides the ventricle, some mixing of oxygen-rich and oxygen-poor blood does occur in that chamber. A number of amphibians also have a spiral valve that divides the conus arteriosus.

In effect, the changes in the amphibian heart match the two circulation paths created by the pulmonary veins. Blood entering from the lungs tends to stay on the side of the heart that will exit toward the rest of the body. Blood entering from the rest of the body tends to stay on the other side of the heart, and it exits toward the lungs.

The advantage of this new arrangement is that oxygenated blood can be pumped to the muscles at much higher pressures. The disadvantage is that oxygenated blood from the lungs is mixed in the heart with blood that has already given up its oxygen as it circulated through the rest of the body. So the amphibian heart pumps out a mixture of oxygen-rich and oxygen-poor blood, as shown in Figure 31-17.

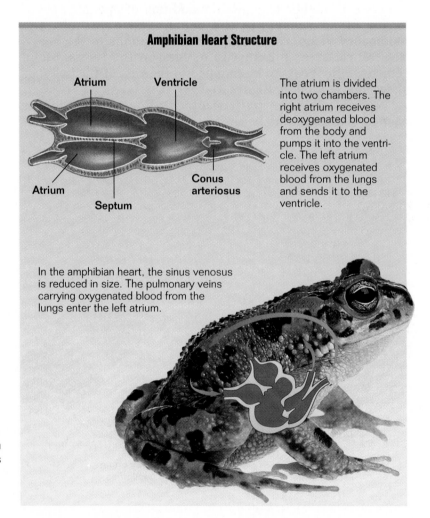

Amphibian Heart Structure

Atrium   Ventricle

Atrium

Septum

Conus arteriosus

The atrium is divided into two chambers. The right atrium receives deoxygenated blood from the body and pumps it into the ventricle. The left atrium receives oxygenated blood from the lungs and sends it to the ventricle.

In the amphibian heart, the sinus venosus is reduced in size. The pulmonary veins carrying oxygenated blood from the lungs enter the left atrium.

Figure 31-17 The amphibian atrium is divided by a septum that helps limit the mixing of oxygenated and deoxygenated blood.

The Heart Is Divided in Terrestrial Vertebrates

In the amphibian heart, oxygen-poor blood dilutes the oxygen-rich blood that has returned from the lungs. It is no accident, then, that amphibians are usually sluggish. Among reptiles, however, the septum that completely divides the atrium into right and left halves extends into the ventricle. There is therefore much better separation of oxygen-rich and oxygen-poor blood. The conus arteriosus is also divided completely in two. It is no longer a heart chamber. Instead, it forms the trunks of the two large arteries leaving the heart.

Because the reptilian heart achieves a better separation of blood, it works more efficiently than the amphibian heart at driving the unique "double" circulation that first evolved in amphibians. However, the efficiency of the system is still somewhat limited since the separation of the ventricle is not complete. Mixing of blood still occurs, as shown in Figure 31-18.

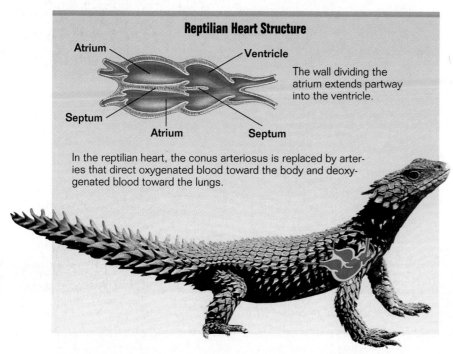

Reptilian Heart Structure

Atrium

Ventricle

The wall dividing the atrium extends partway into the ventricle.

Septum

Atrium

Septum

In the reptilian heart, the conus arteriosus is replaced by arteries that direct oxygenated blood toward the body and deoxygenated blood toward the lungs.

Figure 31-18 In all reptiles but crocodiles, the ventricle is partly divided by a septum.

Only a slight change was needed to complete the evolutionary transition of the vertebrate heart begun in amphibians. Mammals, birds, and crocodiles all have completely separated ventricles. In these animals, the septum completely divides the ventricle into two pumping chambers, as illustrated in Figure 31-19. The closing of the ventricular septum solves the problem originally created when pulmonary veins evolved in amphibians. With total separation of the lung and body circulations, blood pumped to the tissues is now fully oxygenated.

*A fish's heart only pumps deoxy-
genated blood to the gills. The
amphibian heart sends deoxygenated blood to the lungs and oxygenated blood to the body, with
some mixing of the two bloodstreams. A partial division of the
ventricle in most reptiles reduces
this mixing. No mixing occurs in the
hearts of crocodiles, birds, and
mammals because they have two
completely separate ventricles.*

What has happened to the sinus venosus over the course of
vertebrate evolution? Remember that this is a major chamber
in the fish heart. In amphibians the sinus venosus is reduced
in size, and it is even smaller in reptiles. Mammals and birds
no longer use the sinus venosus as a separate chamber. However, some tissue from it remains and plays an important role.

Throughout the evolutionary history of the heart, the
sinus venosus has had two functions. One function is to act
as a collection chamber for blood. The other function is as
the site from which the heartbeat begins. The second function is indispensable. Thus, mammals and birds have
retained some of the excitatory tissue of the sinus venosus in
the wall of the right atrium, at the point where it was located
in the fish heart. The sinus venosus has become the pacemaker in the hearts of mammals and birds. It is the point of
origin of each heartbeat. □

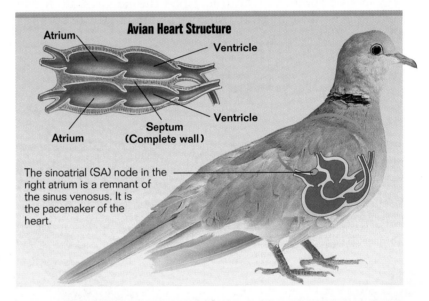

Avian Heart Structure

Atrium

Ventricle

Ventricle

Septum
(Complete wall)

Atrium

The sinoatrial (SA) node in the
right atrium is a remnant of
the sinus venosus. It is
the pacemaker of the
heart.

Figure 31-19 The avian heart
shows a complete division of the
ventricle by the septum. Oxygenated and deoxygenated blood
are kept completely separate,
meaning that oxygen is moved
through the body more efficiently.
This type of heart is also found in
mammals and crocodiles.

Section Review

1. *Trace the flow of blood through a fish's heart.*
2. *How do the hearts of fish, amphibians, and reptiles
   differ? How are they alike?*
3. *Describe how the circulatory system of a mammal is
   an improvement on the double-loop circulation of an
   amphibian?*
4. *What is the advantage of separating oxygen-rich and
   oxygen-poor blood during circulation?*

Critical Thinking

5. *In your heart, the left ventricle, which pumps blood to
   the body, is much stronger than the right ventricle,
   which pumps blood to the lungs. Explain this difference.*

31-4 Challenge of Retaining Water

Vertebrates evolved in water. Though vertebrates now also live on land, no vertebrate can do without water for long. About two-thirds of a vertebrate's body is water, and if the amount of water falls much lower than this, the animal will die. Losing water is called dehydration. Preventing dehydration has been a key evolutionary challenge facing vertebrates in all environments. Even marine fishes, which live in sea water, must cope with the problem of water loss. If this seems strange to you, remember the process of osmosis. Osmosis causes a net movement of water through membranes toward regions of higher ion concentration. Sea water is three times as salty as the tissues of a marine bony fish. As a result, water tends to leave its body by osmosis. How, then, do marine fishes avoid dehydrating in sea water?

Section Objectives

- Describe how marine and freshwater fishes maintain water and salt balance.
- Identify the key adaptations of terrestrial vertebrates that prevent water loss.
- Explain how terrestrial vertebrates are able to reproduce on land.

Fishes Use Kidneys to Balance Water and Salt

Marine bony fishes lose water continuously by osmosis to the saltier water in which they swim. To make up for this water loss, marine fishes drink a lot of water. But because this water contains high levels of salt, marine fishes have an osmotic problem. How can they avoid losing water and taking in too much salt?

Freshwater fishes have the opposite osmotic problem. Their bodies have more salt than the surrounding water. So freshwater fishes need to avoid taking in too much water and losing too much salt. The evolutionary solution for both marine and freshwater fishes is a pair of kidneys.

The **kidney**, shown in Figure 31-20, regulates salt and water balance and removes metabolic wastes from the blood. It is a complex organ made up of thousands of disposal units called nephrons. Each **nephron** has the structure of a bent tube, as shown in Figure 31-21 on the next page. Blood pressure forces the fluid in blood past a filter at the top of each nephron. This filter keeps blood cells, proteins, and other useful large molecules in the blood. However, water and the small molecules and wastes dissolved in it pass through the filter and into the bent part of the tube. As the filtered fluid passes through the nephron tube, useful sugars and ions are recovered by active transport. Some of the water is also reabsorbed. The remaining water and dissolved metabolic wastes (such as urea) that are left behind form a fluid called **urine**.

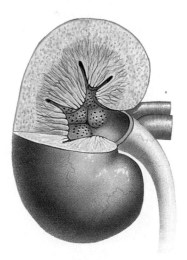

Figure 31-20 The kidney is the organ that filters waste from the vertebrate body. Kidneys differ among fish, amphibians, reptiles, birds, and mammals to address different needs in maintaining water balance. As vertebrates moved onto land, kidneys became more specialized in concentrating urine to prevent water loss.

How concentrated the urine is depends on the environment in which the animal lives. Freshwater fishes produce large amounts of dilute urine. In marine fishes, ions and nitrogenous wastes are excreted by the gills, which are their major organs for regulating ion balance. Also, the ion channels in the nephron membranes of marine fishes are reversed in orientation. This allows the nephron to actively excrete excess salts into the urine.

Sharks are an exception. These marine fishes do not actively pump ions out of their bodies through their kidneys. Instead, sharks maintain the same ion concentration as the surrounding water. So a shark does not gain or lose water by osmosis, since there is no osmotic difference between its fluids and the sea water in which it is swimming.

The vertebrate kidney evolved in fresh water. But over millions of years of evolution, it has become adapted for life in a wide range of environments.

Urine Is Concentrated in Terrestrial Vertebrates

Because they spend so much time in or near fresh water, amphibians, like freshwater fishes, produce dilute urine. Reptiles, however, face not an osmotic problem but an evaporation problem. Thus, reptiles reabsorb much more of the water in their kidneys and produce a far more concentrated urine than amphibians do. This urine, though, cannot become any more concentrated than a reptile's blood plasma. Otherwise, water from the animal's body would simply flow into its urine by osmosis while the urine was still in the kidneys.

Mammals do an even better job of retaining water than do reptiles because mammalian kidneys have evolved to remove far more water from urine. Human urine may be as

Figure 31-21 The nephron, *above*, is the collecting device in the kidney. The bend in the tube, which evolved in mammals, helps concentrate urine and further reduce water loss. The excretory systems of bony fishes, birds, and mammals, *right*, show increasing complexity in the structure of the kidney.

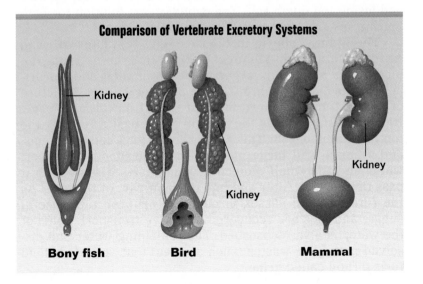

Comparison of Vertebrate Excretory Systems

Kidney

Kidney

Kidney

Kidney

Bony fish　　**Bird**　　**Mammal**

much as four times as concentrated as blood plasma. In desert mice, the urine is over 20 times as concentrated. How do mammalian kidneys avoid the osmotic trap of having body water flow back into the urine within the kidney? In a remarkably simple and powerful innovation, the nephron tube is bent back on itself, as you can see in Figure 31-21. This structure places the end segment of the nephron in a region of the kidney with a very high local concentration of ions and metabolic wastes. The high salt concentration draws water out of the tube by osmosis. Thus, a very highly concentrated urine solution is left behind.

CONTENT LINK
....
Suppose you drink a large glass of water. How will your kidneys respond? You can find out in **Chapter 38.**

Watertight Skin Evolves in Reptiles
....

On land, water tends to evaporate from surfaces into the air. So a terrestrial animal faces the problem of water loss through its skin. The first amphibians had bodies covered with the same kind of small bony scales as their fish ancestors. Most amphibians that survive today have moist skins, however. To slow water loss, many present-day amphibians secrete a slippery mucus that tends to retain water. This mucus is what makes these animals feel slimy. Even with this coating, though, the skin of amphibians is not very watertight.

Reptiles had a different evolutionary solution to the problem of water loss through the skin. Instead of bony scales, the earliest reptiles were covered with lighter, flexible scales made of protein. These scales overlap and form a watertight, dry covering that minimizes evaporation of body water through the skin. This dry skin helped to free reptiles from the necessity of living in a wet environment. Mammals and birds, which evolved from reptiles, also have skin that is dry and virtually watertight.

Watertight Eggs Evolve in Reptiles
....

Water loss is not a significant problem for the eggs of fishes and amphibians. The eggs of most species of amphibians are released by the female into water, where they are fertilized by sperm released by a male. Once fertilized, the eggs go on to develop in the water.

For a reptile living on dry land, however, reproduction presents a serious water-loss problem. Without a watery environment, both eggs and sperm will dry out. Also, sperm need water in which to swim. And finally, the fertilized egg needs a moist environment in which to develop. In the eggs of reptiles, the embryo develops surrounded by a pair of protective membranes, as shown in Figure 31-22. The **amnion**

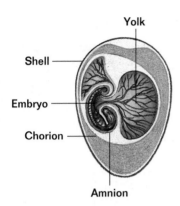

Yolk

Shell

Embryo

Chorion

Amnion

Figure 31-22 This cross section, *above*, shows the internal structure of an amniotic egg. The evolution of the amniotic egg allowed reptiles, such as these hognose snakes, *right*, to make the complete transition to land.

■ *CAPSULE*
SUMMARY

The transition from life in water to life on land required a more complex excretory system to maintain water and salt balance. It also required the evolution of protective coverings to prevent the body and eggs from losing water.

(*AM nee awn*) encloses the embryo within a watery environment. In a sense, this membrane creates a little pond that substitutes for the water in which amphibians lay their eggs. Surrounding the amnion is another membrane, the **chorion** (*KAWR ee awn*). This membrane allows oxygen to enter the egg and carbon dioxide to leave, but it is impermeable to water. This kind of watertight, fluid-filled egg is called an **amniotic egg**. Reptiles, birds, and mammals produce amniotic eggs. Most reptiles, all birds, and three species of mammals (the monotremes) lay amniotic eggs surrounded by a tough shell. ■

Section Review

1. *Describe two ways marine bony fishes maintain their salt and water balance.*
2. *How are mammals able to produce a urine that is more concentrated than that of reptiles?*
3. *How does a reptile's dry skin help conserve body water?*
4. *What are the functions of the amnion and chorion in a bird's egg?*

Critical Thinking

5. *A few species of sharks live in fresh water. Would you expect these species to have the same internal ion concentration as marine sharks? Explain your answer.*

31-5 Reproduction and Development

*I*n Section 31-4, you saw that one of the most severe problems faced by vertebrates as they began living on land was the danger of drying out. This problem was particularly severe for their small and vulnerable gametes. If released on land, the gametes of fishes would dry out and perish. As vertebrates adapted to life on land, different ways of protecting gametes and embryos from drying out evolved in each vertebrate class.

Section Objectives

- List the advantages of internal fertilization over external fertilization.
- Contrast oviparity, ovoviviparity, and viviparity.
- Contrast the parental care provided by reptiles, birds, and mammals.

Fishes Fertilize Their Eggs Externally

Most fishes fertilize their eggs like other aquatic animals, by simply releasing male and female gametes near one another in the water. This approach is called **external fertilization**. A fish egg contains only enough yolk to nourish the developing zygote for a short time. A hatchling fish has a very short larval stage during which the yolk sac of the egg is still attached. When the yolk is depleted, the growing fish must get along on its own, seeking food in the waters it lives in. Young, growing fishes are unlikely to survive because food is often hard to come by, infection by microbes is common, and predators abound. Thus, while many thousands of eggs may be fertilized in a single mating, few of them survive this harsh introduction to life and grow to maturity. As you might expect, natural selection has favored changes that make this dangerous period as short as possible. The fertilized egg develops very quickly, and the young that survive the dangers of early growth mature quickly.

Amphibians Are Still Tied to Water

As you have seen repeatedly in this chapter, amphibians, the first vertebrates to successfully invade land, did not make a full transition to terrestrial life. The reproductive cycle of most living amphibians is still dependent on the presence of free water. Among most amphibians today, fertilization is still external, just as it is among most fishes. Where do amphibians find water in which to carry out external fertilization? Many female frogs and toads lay their eggs in puddles or ponds. The male grasps the female and discharges fluid containing sperm onto the eggs as she releases them into the water, as shown in Figure 31-23.

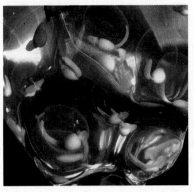

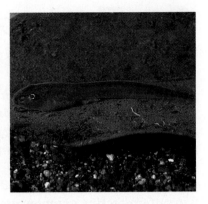

Figure 31-23 The male frog, *above left,* has grasped a female and is about to fertilize her eggs. Each egg is enclosed in a jelly-like coating, *above center.* The tadpole, *above right,* is a larval stage of the developing frog.

☐ CAPSULE SUMMARY

Fertilization of eggs is external in most fishes and amphibians. After hatching, the larval amphibian feeds and grows until it reaches a certain size, then transforms into an adult.

The body of an amphibian is far more complex than that of a fish, and, as you might expect, it takes an amphibian far longer to develop. However, amphibian eggs have about the same amount of yolk as fish eggs, so the amphibian embryo isn't protected in the egg any longer than a fish embryo. Instead, amphibian development takes place in two phases. First, the egg hatches into a larval stage, like some of the larvae found among insects. The development of the embryo into a larva is rapid because it uses yolk from the egg. The larva then functions, often for a considerable period of time, as an independent food gatherer, living in water and getting oxygen through gills. Larval amphibians typically grow quite rapidly. **Tadpoles**, which are frog larvae, can grow in a few days from the size of a grain of rice to the size of a goldfish. Only when an individual has grown to a sufficient size does it undergo the second phase of its development—**metamorphosis**, a radical developmental transformation into the terrestrial adult form. ☐

Reptiles, Birds, and Mammals Have Internal Fertilization

Reptiles were the first class of vertebrates to completely abandon aquatic habitats. Unlike the eggs of most amphibians, reptilian eggs are fertilized within the female before they are laid, a process called **internal fertilization**. The male introduces his **semen**, a fluid containing sperm and fluid secretions, directly into the female's body. In this way, fertilization takes place in a wet environment and the gametes are protected from drying out, even though the adult animals are fully terrestrial. Most vertebrates that fertilize internally use the **penis**, a tubelike organ, to inject semen into the female. Composed mostly of tissue that can become rigid and erect, the penis penetrates far into the female reproductive tract.

Most birds do not have a penis (swans are an exception). Birds achieve internal fertilization simply by the male pressing his reproductive opening against the female's reproductive opening and releasing sperm.

Many reptiles are **oviparous**, meaning their young hatch from eggs laid outside the mother's body. Other reptiles are **ovoviviparous**, meaning their young are born live from eggs that hatch within the mother's body. Reptile eggs contain a considerable amount of yolk, which provides a rich food supply for the developing embryo. The shells of most reptile eggs are leathery, but in birds the shells are hard. Because adult birds sit on the eggs to maintain the temperature needed for the embryo to develop, the shells must be strong enough to resist cracking. Both reptiles and birds produce amniotic eggs.

Young that hatch from reptile eggs are typically fully formed at birth and able to fend for themselves. Bird hatchlings, on the other hand, are not able to survive unaided, since their development is still incomplete. While most young reptiles are ignored by their parents (crocodiles and alligators are exceptions), young birds are fed and nurtured by their parents, growing to maturity only gradually. ❑

❑ **CAPSULE SUMMARY**

Fertilization is internal in birds and reptiles. Most reptiles and all birds are oviparous, meaning they lay shelled eggs

Mammals Nourish Their Young

Mammals are unique among the vertebrates in nourishing their young after birth with a nutrient-rich fluid called **milk,** which is also a rich energy source. There are three types of reproduction among mammals, as described in Figure 31-24. The first mammals to evolve, monotremes, had a reproductive strategy similar to the reptiles from which they evolved—they were oviparous. No other mammals lay eggs. All other members of this class are **viviparous**, which describes the situation in which the young are born live from egg cells that develop within the mother's body and are nourished from nutrients that pass from the mother to the embryo.

Just as among birds, the young of viviparous mammals are nourished and protected by their parents. Mammals

Figure 31-24 The duckbill platypus, *below left,* lays shelled eggs. When the young hatch, they are nursed by the mother. Marsupials, such as the kangaroo, *below center,* give birth to immature live young that continue development within the mother's pouch. Placental mammals, such as the house cat, *below right,* nourish their developing offspring through the placenta, then give birth to live young that receive milk from the mother.

other than monotremes show two patterns of development and parental care.

1. Marsupials give birth to live embryos at a very early stage of development. Born just days or weeks after fertilization, newborn marsupials are no bigger than the tip of your finger. The tiny animals crawl to and enter a pouch on the mother's body, where they attach to a nipple and continue their development for many months. They eventually emerge when they are able to function on their own.

2. Placental mammals retain their young for a much longer time within the body of the mother. To nourish the developing fetus, placental mammals have a placenta, through which nutrients are channeled to the embryo from the blood of the mother, as described in Figure 31-25.

CONTENT LINK
....
More information about the structure and function of the placenta is found in **Chapter 42**.

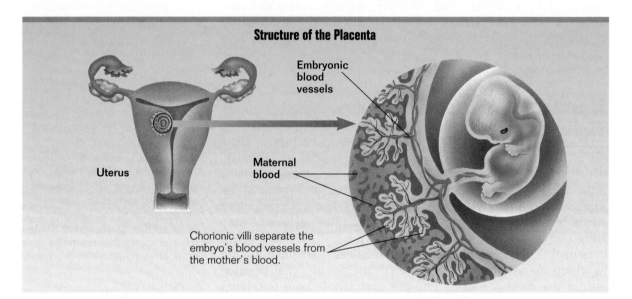

Structure of the Placenta

Embryonic blood vessels

Maternal blood

Uterus

Chorionic villi separate the embryo's blood vessels from the mother's blood.

Figure 31-25 The placenta is a composite structure containing tissue from both mother and embryo. Chorionic villi, outgrowths from the chorion, penetrate into the wall of the uterus and absorb nutrients from pools of maternal blood that form around them. As a result of this structure, the blood of the developing embryo does not mix with that of the mother.

Section Review

1. *List the advantages of internal fertilization over external fertilization.*
2. *How does viviparity increase the likelihood that offspring will survive?*
3. *How might the amount of parental care provided be related to the number of eggs laid or offspring produced? Explain your answer.*

Critical Thinking
4. *Describe two disadvantages to the parents of internal fertilization.*

Bony Fishes

Control of Buoyancy

Bony fishes, such as this lobe-finned coelacanth, evolved a **swim bladder** that gave them active control of buoyancy in the water. This control allowed them to float at any level in the water without swimming, which expends energy. Bony fishes use less energy to maintain their positions than sharks, which sink when they stop swimming. Lobe-finned fishes evolved **lungs**, which let them extract oxygen from air and enabled them to remain out of the water for long periods. Lobe-finned fishes are the direct ancestors of amphibians.

Latimeria
Length: 1.8 m (6 ft.)
Present day

Conquering the Sea

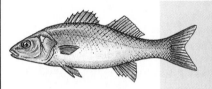

The "slimy" surface of a fish reduces water friction by more than 66 percent.

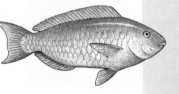

In almost all bony fishes, fertilization of eggs by sperm occurs outside the female's body.

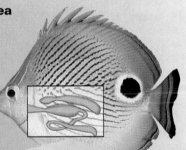

The swim bladder makes a fish more buoyant.

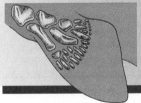

Pectoral fin — Pelvic fin

The Road to Land

Bony fishes have paired pectoral (shoulder) and pelvic (hip) fins. In most species, fins are fan shaped and supported by thin bony rays.

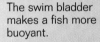

Lobe-finned fishes have fleshy fins supported by central bones, from which the limbs of amphibians are thought to have evolved.

Bony fishes are the most numerous vertebrates. There are more than 18,000 species.

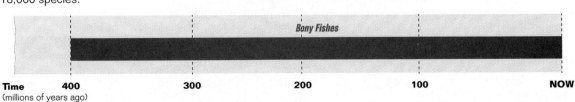

Bony Fishes

Time	400	300	200	100	NOW

(millions of years ago)

Early Amphibians

Moving on Land

Ichthyostega and other early amphibians were the first vertebrates to live largely on land, beginning about 370 million years ago. **Pectoral and pelvic legs** supported their body weight on land. A **pulmonary vein** sent oxygenated blood from the lungs back to the heart for repumping (see Figure 31-16). Early amphibians had a dry skin to avoid water loss (the moist skins of today's amphibians evolved much later). Many early amphibians had body armor and grew to be quite large. Large amphibians were the dominant land animals for 70 million years, until they were slowly replaced by reptiles. *Ichthyostega* became extinct 360 million years ago. Most other large amphibians had died out by 160 million years ago.

Ichthyostega
Length: 1 m (39 in.)
Late Devonian period

Tied to Water

Most modern amphibians have a moist, scaleless skin that aids in gas exchange but promotes water loss and limits body size.

The majority of amphibians lay their eggs in the environment. The eggs are not watertight, so amphibians must reproduce in water or moist habitats. Most amphibians have an aquatic larval phase.

And Adjusting to Life on Land

The amphibian heart is partly divided, and mixing of oxygenated and deoxygenated blood occurs within it.

Amphibians are ectotherms (their body temperature is determined by their surroundings). During winter, most hibernate in soft mud at the bottom of ponds.

Over 70 extinct families of ancient amphibians are known, compared with 37 families of today's amphibians, which fall into three groups: frogs and toads, salamanders, and caecilians.

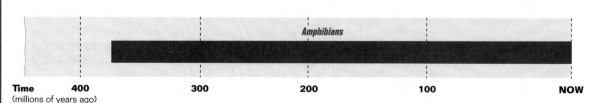

Amphibians

| Time | 400 | 300 | 200 | 100 | NOW |

(millions of years ago)

Early Reptiles

Becoming a Better Predator

Early reptiles, such as this pelycosaur called *Dimetrodon*, had powerful jaws. With large teeth and powerful jaws, pelycosaurs could devour much larger prey than their ancestors. Dominant for 50 million years, they were better adapted to life on dry land than amphibians, partly because they had **watertight eggs**. Pelycosaurs once composed 70 percent of all land vertebrates. They died out about 250 million years ago, replaced by their direct descendants, the therapsids.

Dimetrodon
Length: 3 m (10 ft.)
Early Permian period

Pelycosaur skulls have a single, large opening behind each eye socket, allowing long jaw muscles to attach to the rear of the skull. This kind of skull is called a synapsid skull. The lower pelycosaur jaw is composed of several bones.

Eating

With long, sharp, "steak knife" teeth, *Dimetrodon* was the first land vertebrate able to kill animals its own size.

Pelycosaurs appear to have been the first land vertebrates with teeth of different sizes and shapes.

Standing

Pelycosaurs stood in a sprawling stance: legs bent, knees and elbows stuck out, and feet flat on the ground, pointed outward.

The spectacular "sail" on the back of many pelycosaurs earned them the nickname "fin backs." The framework of the sail is made of spines up to 1 m (39 in.) long that are extensions of the back vertebrae. The sail may have helped control body temperature.

Pelycosaurs

Time	400	300	200	100	NOW

(millions of years ago)

Therapsids

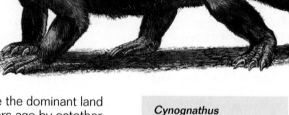

Endothermy

Like their pelycosaur ancestors, therapsids such as *Cynognathus* were synapsids with powerful jaws. But therapsids ate 10 times more frequently. This frequent eating provided the fuel to produce body heat. Therapsids probably were **endotherms,** meaning they maintained a constant, high body temperature. Endotherms could be far more active during prolonged cold periods than other vertebrates of that time. For 20 million years therapsids were the dominant land vertebrates, until largely replaced 230 million years ago by ectothermic thecodonts, ancestors of dinosaurs. The last therapsid became extinct 170 million years ago.

Cynognathus
Length: 1 m (39 in.)
Early Triassic period

Eating

The lower jaw of a therapsid is made of fewer bones than a pelycosaur's jaw. One dominant bone bears the teeth, and another occupies the rear of the jaw, forming the joint with the skull.

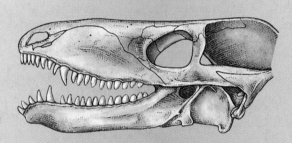

Therapsid teeth were replaced only once or twice in a lifetime, unlike reptile teeth, which are lost and replaced periodically. This long tooth life allows opposing teeth to mesh, creating a very effective chewing surface. Therapsids also had a secondary palate, a sheet of bone that separates the nasal passages from the mouth, making simultaneous breathing and chewing possible.

Improved Stance

Therapsids walked on all fours, but adopted a more upright stance than pelycosaurs, with their legs forming a "vee" (∧) rather than a horizontal "chair" (⊓).

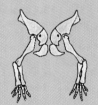

The rib cage is greatly reduced in the lower body, which is thought to reflect the closing off of the front of the body cavity (which houses the lungs and heart) by a muscular sheet of tissue, the diaphragm. The diaphragm helps ventilate the lungs.

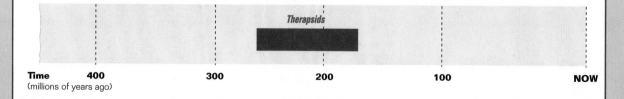

Therapsids

| Time | 400 | | 300 | | 200 | | 100 | | NOW |
(millions of years ago)

Mammals

Fur for Insulation

Mammals, such as *Megazostrodon*, evolved from therapsids at the same time that the dinosaurs first appeared. Throughout the 150-million-year reign of the dinosaurs, all mammals were small, the largest no bigger than a cat. Like their therapsid ancestors, mammals are endotherms and devote 90 percent of their food energy to making heat. Mammals are covered with insulating **fur** to retain body heat. Females produce **milk** to feed their young. When the dinosaurs disappeared and world climates turned progressively colder, mammals became the world's dominant large land animals, as they still are.

Megazostrodon
Length: 12 cm (5 in.)
Late Triassic period to early Jurassic period

Eating

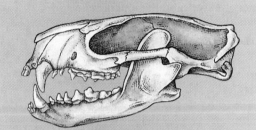

Mammals have a characteristic arrangement of four kinds of teeth: chiseling incisors in front, stabbing canines at the corners, and cutting or grinding premolars and molars on the sides.

The mammalian jaw is able to move sideways as it closes, permitting very effective grinding of food by molar teeth.

The mammalian jaw is a single bone; the other bones at the rear of the therapsid jaw evolved to become the bones of the mammalian middle ear.

Past Their Prime?

Mammals reached their maximal diversity 15 million years ago. Many large mammals flourished 2 million year ago, but most disappeared as humans became common over the last 12,000 years.

Most mammals living today are small, no larger than during the long reign of the dinosaurs. Seventy-eight percent of mammal species are rodents, bats, or insectivores, most of which are small enough to hold in your hand.

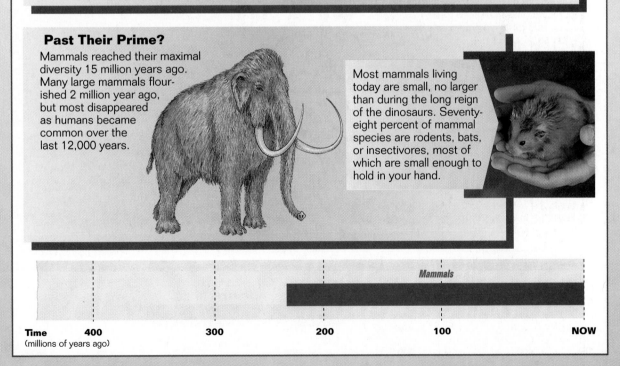

Time				
400	300	200	100	NOW

(millions of years ago)

Mammals

Dinosaurs

An Upright Stance

Dinosaurs are reptiles. They evolved from the crocodile-like reptiles known as **thecodonts** and had a significant improvement in body design—**legs** positioned **directly underneath** the body. This placed the weight of the body directly over the legs, which could bear far more weight, and allowed some dinosaurs to run with great speed and agility. Scientists now think many dinosaurs were endotherms, maintaining a high, constant body temperature in all environments. Dinosaurs were the most successful of all land vertebrates—the dominant large animals on Earth for 150 million years. They disappeared abruptly 65 million years ago. Most scientists now agree that dinosaurs became extinct as the result of the impact of a meteorite or comet. Perhaps the endothermic but uninsulated dinosaurs could not survive the intense cold produced by debris in the atmosphere.

Apatosaurus
Length: 20 m (66 ft.)
Mass: 30,000 kg (33 tons)
Late Jurassic Period

Standing Upright

The key trait that distinguishes a dinosaur from a thecodont is the presence of a hole in the side of the hip socket. Because the dinosaur leg is positioned underneath the socket, force is directed upward, not inward, so there was no need for bone on the side of the socket.

Dinosaurs can be divided into two groups based on pelvic structure. In the lizard-hipped dinosaurs, one of the bones of the pelvis, the pubis, points forward. Lizard-hipped dinosaurs include *Apatosaurus*, *Deinonychus*, and the famous predator *Tyrannosaurus rex*.

Pubis

A lizard-hipped dinosaur

Deinonychus
Length: 3-4 m (10-13 ft.)
Height: 1.8 m (6 ft.)
Early Cretaceous period

Dinosaurs

Time   400   300   200   100   NOW
(millions of years ago)

Stomach Rocks for Grinding Food

Apatosaurus ground up plant material with rocks in its stomach.

A New Food Source

When angiosperm plants evolved, dinosaurs with grinding teeth dominated the land.

The other group is the bird-hipped dinosaurs. In this group, the pubis is directed backward. Examples of bird-hipped dinosaurs include duckbill dinosaurs such as *Corythosaurus*, iguanodonts, horned dinosaurs such as *Triceratops*, and armored dinosaurs such as *Ankylosaurus*.

Pubis

A bird-hipped dinosaur

Nurturing Young

Duckbill dinosaurs such as *Corythosaurus* are thought to have cared for their young in nests.

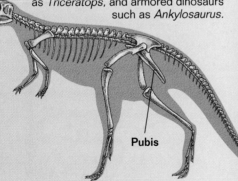

Corythosaurus
Length: 9 m (30 ft.)
Early Cretaceous period

Birds

Flying

Birds, such as *Archaeopteryx,* evolved from small bipedal dinosaurs about 150 million years ago but were not common until the flying reptiles called pterosaurs became extinct along with the dinosaurs. Unlike pterosaurs, birds are insulated with **feathers**. Birds are so structurally similar to dinosaurs in all other respects that many scientists consider birds to be simply feathered dinosaurs. Modern birds have no teeth, although early birds did.

Feathers

Feathers evolved from reptilian scales.

Archaeopteryx
Length: 35 cm (14 in.)
Late Jurassic period

Skeleton

The bones of birds are hollow and light. In all but the earliest birds, the breastbone is enlarged and has a keel down the midline, providing a solid attachment for flight muscles that make up as much as 30 percent of total body weight.

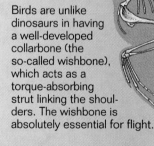

Respiration

Birds have very efficient lungs, directing air first to a series of air sacs, then to the lungs; this one-way air flow permits blood and air to flow across each other, greatly increasing the efficiency of gas exchange.

Birds are unlike dinosaurs in having a well-developed collarbone (the so-called wishbone), which acts as a torque-absorbing strut linking the shoulders. The wishbone is absolutely essential for flight.

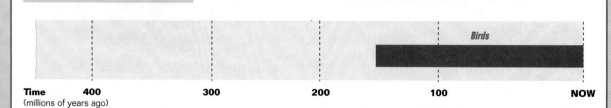

Time	400	300	200	100	NOW

(millions of years ago)

Birds

alveolus (714)
amnion (725)
amniotic egg (726)
atrium (718)
cartilage (706)
chorion (726)
conus arteriosus (718)
ectothermic (707)
endothermic (707)
external fertilization (727)
gill filament (711)

internal fertilization (728)
kidney (723)
lung (712)
metamorphosis (728)
milk (729)
nephron (723)
oviparous (729)
ovoviviparous (729)
penis (728)
placenta (710)
pulmonary vein (719)

semen (728)
septum (720)
sinus venosus (718)
swim bladder (706)
tadpole (728)
terrestrial (704)
urine (723)
ventricle (718)
viviparous (729)

Concept Mapping

Construct a concept map that shows comparisons of the circulatory and respiratory systems of the various groups of vertebrates. Use as many terms as needed from the vocabulary list. Try to include the following items in your map: pulmonary vein, lungs, gills, ventricle, atrium, air sacs, and alveoli. Use additional concepts in your map as needed.

Review

Multiple Choice

1. Which sequence reflects the order in which the major groups of vertebrates are thought to have evolved?
   a. bony fishes → reptiles → amphibians
   b. bony fishes → amphibians → reptiles
   c. amphibians → mammals → reptiles
   d. birds → mammals → reptiles

2. Mammals are thought to have survived the climatic changes that caused the extinction of dinosaurs because the bodies of mammals
   a. had little insulation.
   b. are ectothermic.
   c. maintain a constant temperature.
   d. had insulation.

3. Fish are much more efficient in getting oxygen from the water than oysters because
   a. fish gills are in an internal cavity.
   b. oxygen diffuses directly into the blood of fish.
   c. water flows in one direction over fish gills.
   d. gill slits permit fish to swallow a lot of water.

4. The heart of an amphibian
   a. has a fully divided atrium.
   b. pumps blood to the lungs through the left atrium.
   c. completely separates oxygenated and deoxygenated blood.
   d. receives oxygen-poor blood from the lungs.

5. Compared with reptilian kidneys, mammalian kidneys are more effective water conservation organs because they
   a. are paired.
   b. remove more water from urine.
   c. produce a very dilute urine solution.
   d. allow urine to diffuse into the kidney.

6. Which is *not* an adaptation of reptiles for life on land?
   a. watertight skin
   b. external fertilization
   c. amniotic egg
   d. kidneys

7. What makes marsupial reproduction unique among mammals?
   a. marsupials are viviparous
   b. the mother nourishes her young with milk
   c. the young develop in their mother's pouch
   d. the young hatch from eggs

Completion

8. The early fishes that replaced jawless fishes were successful because they had _____ to capture prey and _____ bodies to aid their movement.

9. Characteristics that make amphibians transitional land-dwellers include _____ skin, _____ fertilization, and eggs that develop in water.

10. In amphibians, the _____ veins carry _____ blood from the lungs to the heart.

11. In terrestrial vertebrates, gas exchange between the air and blood occurs in the _____ . The efficiency of this baglike structure can be improved by _____ its internal surface area.

12. Small chambers called _____ are present in the lungs of reptiles and mammals. They are where _____ diffuses into the blood.

13. The kidney is the organ in vertebrates that works to prevent water _____ . The fluid discharged from the kidney, called urine, contains dissolved metabolic wastes and _____ .

14. Reptiles and birds produce _____ eggs. Eggs of this type keep the developing embryo from _____ .

15. An advantage of _____ is that the embryos are protected and nourished within the mother's body while they develop.

Short Answer

16. How does the structure and function of the sinus venosus in fish compare with its structure and function in mammals?

17. In what ways is a snake better suited for life on land than a frog is?

18. In foot races staged between reptiles and amphibians, the reptiles won 9 out of 10 times. Explain the race results in terms of the structure of reptilian and amphibian hearts.

19. By comparing the structure of mammalian and avian lungs, predict whether a bat or a bird would be able to achieve a higher maximal altitude in flight. Explain your prediction.

20. Contrast the ways in which saltwater and freshwater fishes maintain salt and water balance.

Highlights Review

21. Explain why lobe-finned fishes, rather than other kinds of bony fishes, are thought to be the ancestors of amphibians.

22. How does the skin of modern amphibians differ from the skin of the first amphibians?

23. How is the synapsid skull related to the feeding habits of pelycosaurs?

24. The diaphragm increases the efficiency of breathing. How would this structure benefit an endothermic therapsid?

25. Describe two similarities between mammals and therapsids. What conclusion about the relationship between these two groups have scientists drawn from these similarities?

26. Scientists have discovered piles of smooth stones within the ribcages of some dinosaurs. What do these stones suggest about the feeding habits of these dinosaurs?

27. In what ways does a bird's skeleton differ from that of other vertebrates? How are these differences related to flying ability?

Themes Review

28. Structure and Function Bottom-dwelling fishes often lack a swim bladder. Explain the adaptive advantage of this condition.

29. Homeostasis Saltwater fishes drink more water and produce less urine than freshwater fishes. How do you account for this difference?

30. Evolution In what ways are the adaptations of reptiles to land similar to the adaptations of vascular plants to land?

Critical Thinking

31. Making Inferences When a female leatherback turtle comes onto a beach to lay eggs, she digs a deep hole, lays her eggs, and covers them with sand. Then she crawls about 100 m and digs another hole. This time she lays no eggs, and then covers the hole with sand. Suggest a possible explanation for this behavior.

32. Making Predictions How would having the same lung structure as birds affect the performance of a marathon runner?

33. Making Inferences Fossil evidence collected in the 1980s in Arctic Alaska suggests that some dinosaurs were year-round residents of areas that reached freezing temperatures and were in total darkness during the winter months. Does this evidence of "Arctic dinosaurs" support or contradict the hypothesis that the extinction of dinosaurs was due to a period of intense cold produced by debris in the Earth's atmosphere? Explain your answer.

Activities and Projects

34. Research and Writing Research the Alvarez hypothesis, proposed in 1980 by Water Alvarez and his father, Luis, to explain what triggered the mass extinction of dinosaurs near the end of the Mesozoic era. What evidence supports the hypothesis?

35. Research and Writing Find out how marine mammals differ structurally from fish. In your report, include structural diagrams showing similarities and differences between these animals.

36. Multicultural Perspective Research the importance of fish in the Japanese culture. Find out about the use of carp in Japan and how it differs from that in the United States.

Readings

37. Read the book *Dinosaurs Rediscovered* by Don Lessem. Describe two recent discoveries that have changed how scientists view dinosaurs.

38. Read the article "Frogs and Toads in Deserts," in *Scientific American*, March 1994, pages 82–88. Why are frogs and toads such unlikely desert inhabitants? Describe two adaptations that enable these animals to suvive or reproduce in deserts.

LABORATORY Investigation | Chapter 31

Comparing Vertebrate Characteristics

OBJECTIVE

Categorize similarities and differences among vertebrate classes by observing representative organisms.

PROCESS SKILLS

- classifying organisms
- comparing and contrasting structural features
- relating structural features to functions
- organizing observations

MATERIALS

- field guides for fishes, amphibians, reptiles, birds, and mammals
- live or preserved specimens of fishes, amphibians, reptiles, birds, and mammals
- aquarium with fish
- frog or toad
- cricket or other food for the amphibian
- bird feather
- compound light microscope
- chicken or turkey bones
- beef bones or pork bones

BACKGROUND

1. What characteristic distinguishes vertebrates from other phyla?
2. Why do all amphibians have to be near water at some times in their lives?
3. What factors allowed reptiles to live exclusively on land?
4. Why do mammals and birds have higher metabolic rates than other vertebrates?
5. Write your **Focus Question** on your Vee Form.

6. **Knowing Side of the Vee** List the **Concepts** and the new **Vocabulary** words on your Vee Form. In the **Concept Statements** section of the Vee Form, use these words in sentences that define and explain them.

TECHNIQUE

Doing Side of the Vee

You will travel to five stations to observe organisms that are representative of the most common vertebrate classes. Record your observations and answer the following questions in the **Records** section of the Vee Form.

Station 1: Fishes

1. Observe a freshwater tank containing representative bony fishes.
   a. What do you notice about the motion of their gills?
   b. Describe how they behave and eat.
   c. Describe their body coverings.
2. Using the field guide to fishes, look up two very different bony fishes and describe their similarities and differences.

Station 2: Amphibians

3. Observe a frog or toad in an aquarium. Pay particular attention to the skin of the animal.
   a. Describe the animal's behavior.
   b. How does it breathe?
4. Feed it a cricket. Describe how it eats.
5. Using the field guide to amphibians, find frogs, toads, and salamanders and compare their similarities and differences.
   a. Relate the way each moves to the structure of its legs.

Station 3: Reptiles

6. Observe a live or preserved turtle. Describe the design of the shell. What are the advantages of having a shell? What are the limitations?

7. Using the field guide to reptiles, look up snakes. What adaptations enhance the survival of snakes?

Station 4: Birds

8. Observe a live or preserved bird specimen. List two qualities birds have in common with reptiles.

9. Run your finger and thumb along a feather in one direction. Now examine the feather under the microscope. Run your finger and thumb along the feather in the opposite direction. Look at the feather under the microscope. How has it changed? Predict what could happen if a bird never groomed itself.

10. Examine the *cleaned, broken* wing and leg bones from a chicken or turkey. Describe the interior of the bones. Look at cleaned beef bones or pork bones. How do they differ from those of the birds?

11. Using the field guide to birds, find five examples of how a bird's feet relate to its feeding habits and habitat.

Station 5: Mammals

12. Observe live or preserved mammal specimens. What two characteristics distinguish mammals from other vertebrates?

13. Using the field guide to mammals, find the only true flying mammal.

14. Wash your hands thoroughly before leaving the laboratory.

INQUIRY

1. Describe the differences between snakes and amphibians.

2. How might the characteristics of the bird bones you examined in step 10 be beneficial to birds?

3. What quality do mammals share with birds?

4. Use the information on the **Knowing Side** of your Vee Form to interpret your results from the **Doing Side,** and then write your **Knowledge Claim.** Write a **Value Claim** for this lab.

ANALYSIS

1. Construct a table of vertebrate characteristics using your observations from Stations 1–5. Put the characteristics down the left column and the five common classes of vertebrates along the top of the table. Check off the characteristic in each column if it is common to that vertebrate class.

2. The echidna and crocodile are alike in that they both lay eggs, have similarly shaped pelvises, and have a single opening through which feces, urine, and reproductive products pass. Why is the echidna in a class different from the crocodile?

FURTHER INQUIRY

Write a **New Focus Question** that could be the focus of a new investigation. The following is an example:

A bird's wing is made of feathers. A bat's wing is a thin membrane of skin stretched across elongated fingers. Could a functional wing be made of hair? Why or why not?

CHAPTER 32

FISHES AND AMPHIBIANS

REVIEW

- vertebrate evolution (Sections 13-3 and 33-1)
- *Highlights: Bony Fishes* on page 731 and *Early Amphibians* on page 732
- vertebrate respiration (Section 31-2)
- vertebrate circulation (Section 31-3)
- vertebrate excretion (Section 31-4)
- vertebrate reproduction (Section 31-5)

A green frog and koi, or fancy carp

What is a fish? All of us have a pretty good idea. A fish is a vertebrate (animal with a backbone) that lives in water, breathes with gills, and swims with fins. More than one-half of all living vertebrates are fishes. The largest and most diverse vertebrate group, fishes are the evolutionary base from which amphibians, the first land vertebrates, arose. In many ways an amphibian can be thought of as a "fish out of water"—a vertebrate able to live on land but still tied to water. For this reason, fishes and amphibians will be considered together in this chapter so that the similarities between them are not lost among the host of obvious differences.

Section Objectives

- Describe the first verte-brates.
- Describe the evolution of jaws.
- Compare and contrast ostra-coderms with placoderms.
- Give examples of two mod-ern agnathans.
- Identify several important characteristics of sharks.
- Describe the evolution of bony fishes.

Jawless Fishes Were the First Vertebrates

The story of vertebrate evolution started over 500 million years ago during the Cambrian period, as indicated in Figure 32-1. It was then that the first animals with backbones, the fishes, appeared in the ancient seas. Wriggling through the water, jawless and toothless, these

Figure 32-1 This phylogenetic tree shows the evolutionary relation-ships among different groups of fishes as well as between fishes and amphibians. Living species are at the top of the tree, and extinct forms are at the bottom.

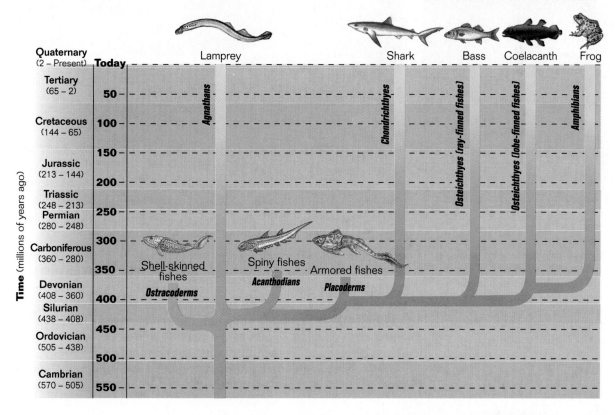

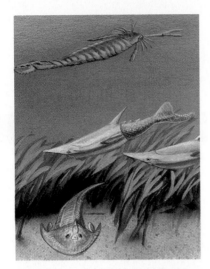

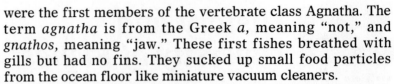

were the first members of the vertebrate class Agnatha. The term *agnatha* is from the Greek *a*, meaning "not," and *gnathos*, meaning "jaw." These first fishes breathed with gills but had no fins. They sucked up small food particles from the ocean floor like miniature vacuum cleaners.

For 50 million years, throughout the Ordovician period, simple jawless fishes were the only vertebrates that existed. Few reached more than 35 cm (1 ft.) long. By the end of this period, they had developed primitive fins that helped them swim. They had also developed massive shields of bone that may have protected them from attacks by the giant sea scorpions that were fearsome predators. These armored fishes, called ostracoderms *(AHS truh kah durms)*, are shown in Figure 32-2. The term *ostracoderm* is from the Greek *ostrakon*, meaning "shell," and *dermis*, meaning "skin."

At one time there were five ostracoderm orders containing nearly 40 families. Most had become extinct by the close of the Devonian period. The descendants of one of the ancient ostracoderm orders survive today as lampreys and hagfishes. Lampreys and hagfishes, shown in Figure 32-3, are the only remaining agnathans.

Figure 32-2 A predatory sea scorpion swims above three bottom-dwelling ostracoderms. Only the head shields of ostracoderms were made of bone. Their internal skeleton was constructed of cartilage. Cartilage weighs less than bone, but it is not as strong.

Figure 32-3 Most lamprey species are parasitic on other living fishes, *above*. Hagfishes, *above right*, are scavengers of dead and dying fishes on the ocean bottom. The slimy skins of lampreys and hagfishes have neither the plates nor scales of their ostracoderm ancestors.

The Evolution of Jaws

A key evolutionary advancement occurred in fishes 440 million years ago—the development of jaws. The transformation from cartilaginous supports called gill arches to jaws is illustrated in Figure 32-4. The top half of the jaw in early fishes was directly attached to the skull only at the rear, which indicates that the first jaws were probably not very powerful. Teeth developed from skin that lined the mouth.

The first fishes to develop jaws were called spiny fishes, members of the class Acanthodia. Spiny fishes were very

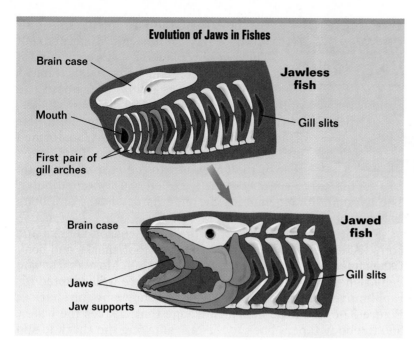

Evolution of Jaws in Fishes

Brain case

Mouth

First pair of
gill arches

**Jawless
fish**

Gill slits

Brain case

Jaws

Jaw supports

**Jawed
fish**

Gill slits

Figure 32-4 Jaws evolved from a front pair of a series of structures called gill arches. A gill arch is the cartilage support that reinforces the tissue between two gill slits (openings between each gill) and holds the slits open.

common during the early Devonian period and replaced most of the ostracoderms. By the end of the Devonian, spiny fishes were extinct. The spiny fishes had internal skeletons made of cartilage, although some fossils indicate that their skeletons also contained some bone. Their scales also contained small plates of bone. The presence of bone in the spiny fishes foreshadows the much larger role that bone would play in their descendants.

By the mid-Devonian, spiny fishes were being replaced by fishes with stronger, more efficient jaws—the heavily armored placoderms. A placoderm is shown in Figure 32-5. ◻

☐ CAPSULE SUMMARY

The first vertebrates, the jawless fishes, evolved in the sea over 500 million years ago. The only jawless fishes that survive today are lampreys and hagfishes. Jaws, derived from gill arches, first evolved in the spiny fishes.

Figure 32-5 *Dunkleosteus*, a placoderm, grew to more than 3 m (10 ft.) long. The placoderm jaw was fused to the skull, and the skull was hinged on the back, in the "shoulder" area. Only the front of the placoderm body was armored. Most lived near the sea floor because of their great weight.

The Rise of Sharks and Bony Fishes

Figure 32-6 Streamlined, with paired fins and a light, flexible, cartilaginous skeleton, sharks have evolved into extremely powerful swimmers.

☐ *CAPSULE SUMMARY*

A reinforced jaw, a streamlined body, and functional fins have made sharks and bony fishes the dominant fishes of today.

By the end of the Devonian, almost all of the early fishes had disappeared. They were replaced by more successful forms—sharks and bony fishes. Sharks and bony fishes evolved at about the same time, 400 million years ago. In sharks and bony fishes, the jaw was improved even further. The next pair of arches, behind the jaws, was transformed into a supporting strut, or prop, joining the rear of the lower jaw to the rear of the skull. This extra support strut is called the hyomandibular. The hyomandibular enables the mouths of sharks and bony fishes to open very wide.

Scientists hypothesize that the main reason sharks and bony fishes replaced primitive fishes is that sharks and bony fishes have a superior design for swimming. Most sharks and bony fishes have streamlined bodies that are well adapted for rapid movement through the water. As shown by the shark in Figure 32-6, the head acts as a wedge that cleaves the water, and the body tapers back to the tail, allowing the shark to slip through the water with minimal resistance. In addition, sharks and bony fishes have an assortment of movable fins that greatly aid their swimming. ☐

Sharks Became Top Predators

In the period following the Devonian, the Carboniferous period, sharks became the dominant vertebrates in the sea. Sharks are members of the class Chondrichthyes, the cartilaginous (*KAHRT'l aj uh nuhs*) fishes. These fishes have skeletons made completely of cartilage. Many of the early sharks died out during a worldwide mass extinction that occurred at the end of the Permian period, 250 million years ago. Those sharks that survived this extinction underwent a great burst of evolution during the age of dinosaurs. Most of the modern groups of sharks evolved during this time. Competing successfully with the marine reptiles of that time, sharks became—and remain today—major predators in the seas.

Like many earlier fishes, sharks have a skeleton made of cartilage, but the shark's skeleton is different because it is calcified. It is strengthened by the mineral calcium carbonate (the material oyster shells are made of). The calcium carbonate is deposited in the outer layers of cartilage, and a thin layer of bone covers this reinforced cartilage. The result is a very light but strong skeleton.

Sharks were among the first vertebrates to develop bony teeth, shown in Figure 32-7. In a shark's mouth, the teeth are arranged in 6 to 10 rows. The teeth in front are pointed and sharp to do the work of biting and cutting, while behind them rows of immature teeth are growing. When a functional tooth breaks or is worn down, a replacement tooth moves forward. One shark may use more than 20,000 teeth during its lifetime! This system of tooth replacement guarantees that the teeth being used are always new and sharp.

Figure 32-7 The teeth of sharks, *top*, are not set into the jaw as yours are, but rather sit atop it. Therefore, their teeth are easily lost. These shark scales, *bottom*, have been magnified 500 times. The scales give the shark's skin a rough "sandpaper" texture.

The method of reproduction among sharks is quite advanced. Shark eggs are fertilized internally. During mating, the male grasps the female with modified pelvic fins called claspers. Sperm runs from the male into the female through grooves in the claspers. A few shark species lay eggs that are enclosed in leathery cases. But the eggs of most species develop within the female's body, and the offspring (commonly called pups) are born alive.

Skates and rays, shown in Figure 32-8, are cartilaginous relatives of sharks. They evolved about 200 million years after the first sharks appeared. Today there are 275 species of sharks, apparently more kinds than existed during the Carboniferous period.

Figure 32-8 Rays, *top*, and skates, *bottom*, have flattened bodies and live on the sea floor. Most species have flattened teeth that are used to crush their prey, mainly mollusks and crustaceans.

Bony Fishes Dominate the Waters

Bony fishes are members of the class Osteichthyes. They evolved at the same time as sharks, about 400 million years ago. The early bony fishes took quite a different evolutionary road from sharks. Instead of gaining speed through lightness, as sharks did so successfully, bony fishes evolved a heavy internal skeleton made completely of bone. Such a skeleton is very strong and provides a solid base against which powerful swimming muscles can pull. The process of ossification (the replacement of cartilage by bone) is complete in the bony fishes.

Unlike sharks, bony fishes evolved in fresh water. The first bony fishes were small and had paired air sacs connected to the back of the throat. These air sacs could be inflated with gas to make the fish more buoyant in the water.

Ray-finned bony fishes, shown in Figure 32-9, comprise the vast majority of living fishes. In **ray-finned fishes** the primitive air sacs were transformed into a structure called a swim bladder. By adjusting the amount of air in its swim bladder, a ray-finned fish can easily adjust its depth in the water. A shark, however, must swim through the water or sink, because it has no swim bladder and its body is denser than water.

Teleosts (*TEL ee ahsts*) are the most advanced of the ray-finned bony fishes. Teleosts have highly mobile fins, very thin scales, and completely symmetrical tails. About 95 percent of all living fish species are teleosts.

The first teleosts evolved during the Jurassic period and included forms that were similar to modern sardines, tarpons, and eels. In the mid-Cretaceous period, a second burst of teleost evolution occurred, producing fishes with heavily muscled bodies and a second dorsal fin. Examples include salmon, trout, carp, cod, and haddock. A third, and particularly intense, burst of evolution occurred in the early Tertiary period and produced highly specialized fishes with

Figure 32-9 This large sturgeon, *top*, is a modern survivor from an early evolutionary line of ray-finned fishes. The characteristic feature of all ray-finned fishes is an internal skeleton consisting of parallel bony rays that support and stiffen each fin. There are no muscles within the fins; the fins are moved by muscles within the body. The bluegill, *bottom*, evolved much more recently than the sturgeon.

more flattened, perchlike bodies. This very diverse group includes perches, sunfishes, bass, snappers, flounders, barracudas, swordfish, and many others. To learn about a common freshwater teleost, read *Up Close: Yellow Perch* on pages 752–753. ❑

❑ CAPSULE SUMMARY

The bony fishes evolved along with sharks 400 million years ago. Powerful muscles for swimming, fins capable of subtle movements, and a fully functional swim bladder have enabled the bony fishes to become the most diverse and abundant group of fishes today.

Lobe-Finned Fishes Paved the Way to Land

The other major group of bony fishes, the **lobe-finned fishes**, evolved 390 million years ago, shortly after the first bony fishes appeared. Only seven species of lobe-finned fishes survive today. One species is the coelacanth *(SEE luh kanth)*. The other six species are called lungfishes. One species of lungfish is shown in Figure 32-10. The lobe-finned fishes have paired fins that are structurally very different from the fins of ray-finned fishes. In many lobe-finned fishes, each fin consists of a long, fleshy, muscular lobe (hence the name of the group) that is supported by a central core of bones. The bones form fully articulated joints with one another, like the joints between the bones in your hand. Bony rays are found only at the tips of each lobed fin. Muscles within each lobe can move the fin rays independently of each other, a feat no ray-finned fish can accomplish.

Although rare today, lobe-finned fishes have played a particularly important role in the evolutionary history of vertebrates. It is from lobe-finned fishes that amphibians, the first land vertebrates, almost certainly evolved.

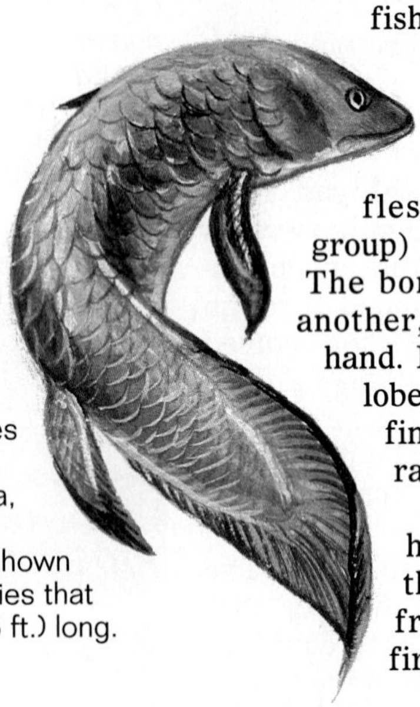

Figure 32-10 Lungfishes are found in streams and rivers in Australia, South America, and Africa. The lungfish shown is an Australian species that grows up to 1.5 m (5 ft.) long.

Section Review

1. *When and where did the first vertebrates evolve?*
2. *Explain the meaning of the following sentence: The evolution of gill arches is directly related to the evolution of predatory fishes.*
3. *The appearance of sharks probably caused the extinction of placoderms. Why?*
4. *Compare and contrast a sunfish and a coelacanth in terms of time of evolutionary appearance, classification, and structural characteristics.*

Critical Thinking

5. *How is an ostracoderm more similar to you than to a lamprey?*

32-2 Key Adaptations of Fishes

The great diversity of fishes reflects the many ways that fishes can live in the oceans and fresh waters around the world. Fishes vary in size from whale sharks, 18 m (59 ft.) giants that feed on plankton, to tiny cichlids no larger than your fingernail. Some fishes live in freezing Arctic seas, while others live in warm freshwater lakes. Mudskippers, such as the one shown in Figure 32-11, are found in Asia and Africa and spend a lot of time out of water. As you have learned, the majority of living fishes are bony fishes. In this section you will first look at the characteristics common to all fishes, and then you will learn about important adaptations of bony fishes in particular.

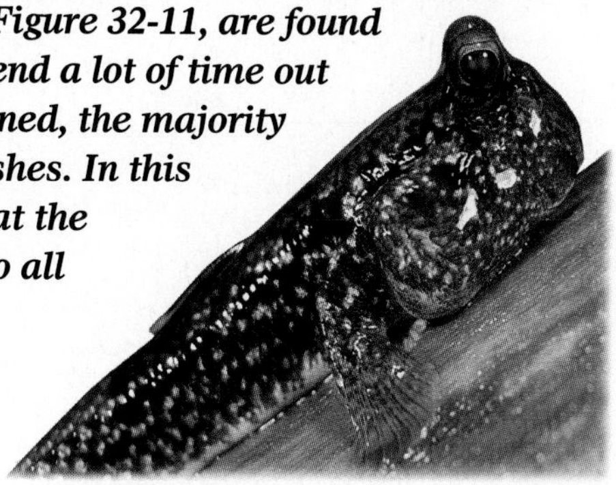

Figure 32-11 The mudskipper is found in shallow bodies of fresh water in Africa and Asia. The modified front fins of this teleost fish enable it to creep out of the water on a regular basis.

Section Objectives

- *Describe the characteristics common to all fishes.*
- *Summarize the unique features of bony fishes.*
- *Identify and describe the major external and internal characteristics of the yellow perch.*

Characteristics of Fishes

However varied their appearance, all fishes share four characteristics. For the most part, these characteristics enable fishes to live more successfully in their aquatic environments.

1. **Gills** Fishes are water-dwelling organisms and must obtain the oxygen required for metabolism from the oxygen gas dissolved in the water around them. They do this by pumping a great deal of water through their mouths. The water passes over fine filaments of tissue called gills in the back of the mouth, and then exits the body through slits in the side of the throat.
2. **Vertebral column (backbone)** All fishes have an internal skeleton made of either cartilage or bone, with a vertebral column surrounding the spinal cord. The brain is fully encased within a protective covering called the skull or cranium.
3. **Single-loop blood circulation** Blood is pumped from the heart to the gills, where the blood is oxygenated. From the gills, the oxygenated blood passes to the rest of the body and then returns to the heart.
4. **Nutritional requirements** Fishes are unable to synthesize the aromatic (ring-structured) amino acids and must consume these important protein-building amino acids in their diet. This inability to synthesize one or more aromatic amino acids has been inherited by all the vertebrate descendants of fishes, including humans.

UP CLOSE YELLOW PERCH

- **Scientific name:** *Perca flavescens*
- **Range:** Found in lakes and rivers from the Great Lakes to the Atlantic coast and as far south as South Carolina
- **Habitat:** Lives concealed among vegetation or submerged tree roots
- **Size:** Grows to about 0.30 m (1 ft.) long and up to 2.3 kg (5 lb.)
- **Diet:** Feeds on insect larvae, crustaceans, and other fishes

External Structures

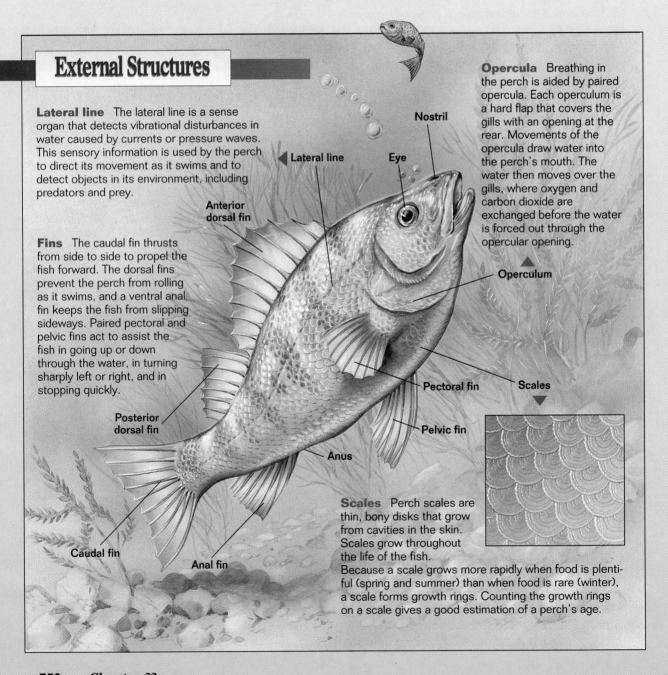

Lateral line The lateral line is a sense organ that detects vibrational disturbances in water caused by currents or pressure waves. This sensory information is used by the perch to direct its movement as it swims and to detect objects in its environment, including predators and prey.

Fins The caudal fin thrusts from side to side to propel the fish forward. The dorsal fins prevent the perch from rolling as it swims, and a ventral anal fin keeps the fish from slipping sideways. Paired pectoral and pelvic fins act to assist the fish in going up or down through the water, in turning sharply left or right, and in stopping quickly.

Opercula Breathing in the perch is aided by paired opercula. Each operculum is a hard flap that covers the gills with an opening at the rear. Movements of the opercula draw water into the perch's mouth. The water then moves over the gills, where oxygen and carbon dioxide are exchanged before the water is forced out through the opercular opening.

Scales Perch scales are thin, bony disks that grow from cavities in the skin. Scales grow throughout the life of the fish. Because a scale grows more rapidly when food is plentiful (spring and summer) than when food is rare (winter), a scale forms growth rings. Counting the growth rings on a scale gives a good estimation of a perch's age.

Labels: Nostril, Lateral line, Eye, Anterior dorsal fin, Operculum, Posterior dorsal fin, Pectoral fin, Scales, Pelvic fin, Anus, Caudal fin, Anal fin

Internal Structures

Reproductive organs Yellow perch produce gametes (sperm cells and egg cells) during their breeding season in the spring. During this period the testes of males produce enormous numbers of tiny sperm cells, and the ovaries of females become swollen with egg cells. Eggs are fertilized externally. The male deposits milt, a fluid containing the sperm, on strings of eggs laid by the female that are wound in and out of weeds and twigs in the water. The young hatch from fertilized eggs within days in warm water (in cold water it may take much longer) and grow quickly.

Brain The brain is divided into anterior (front), middle, and posterior (back) regions. Optic lobes receive sensory information from the eyes. The large size of these lobes indicates the importance of vision. The olfactory bulbs are devoted to receiving information concerning smell (another important sense in perch) from chemical-sensing cells. The cerebrum processes mainly sensory information. The cerebellum coordinates muscle activity, and the medulla oblongata controls the function of many internal organs.

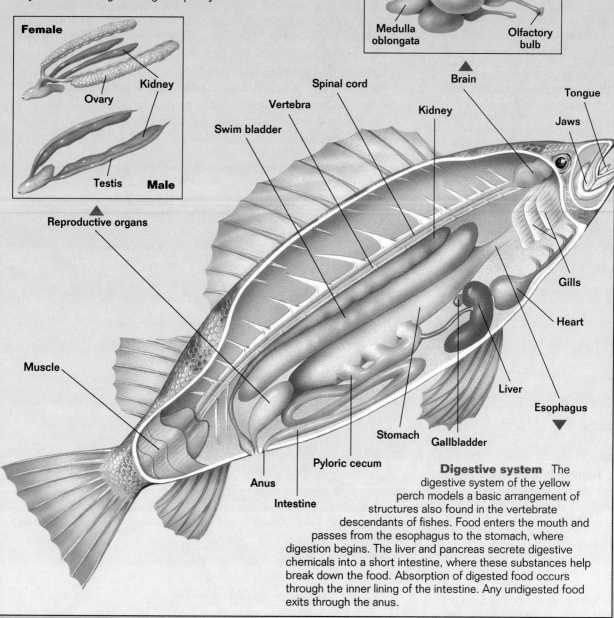

Female

Ovary

Kidney

Testis **Male**

Reproductive organs

Cerebellum   Optic lobe

Cerebrum

Medulla oblongata

Olfactory bulb

Brain

Spinal cord

Vertebra

Swim bladder

Kidney

Tongue

Jaws

Gills

Heart

Liver

Esophagus

Muscle

Anus

Intestine

Pyloric cecum

Stomach   Gallbladder

Digestive system The digestive system of the yellow perch models a basic arrangement of structures also found in the vertebrate descendants of fishes. Food enters the mouth and passes from the esophagus to the stomach, where digestion begins. The liver and pancreas secrete digestive chemicals into a short intestine, where these substances help break down the food. Absorption of digested food occurs through the inner lining of the intestine. Any undigested food exits through the anus.

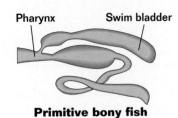

Pharynx   Swim bladder

Primitive bony fish

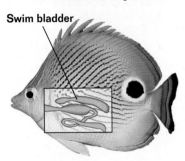

Swim bladder

Modern bony fish

Figure 32-12 In primitive bony fishes, the swim bladder is an outpocket of the pharynx behind the throat, *top*. In modern bony fishes, the swim bladder has separated from the pharynx and is an independent organ, *bottom*.

Bony Fishes Have Additional Adaptations

Although it is correct to consider jawless and cartilaginous fishes successful in their own right, they in no way approach the numbers or diversity seen in the bony fishes. The remarkable success of bony fishes is due to a series of unique structural adaptations.

1. **Swim bladder**   Early bony fishes had to gulp air to fill their swim bladder. In modern bony fishes, the swim bladder, shown in Figure 32-12, is able to secrete and absorb its own gases. The cells that make up the swim bladder generate carbon dioxide, CO_2, by carrying out cellular respiration, which fills the bladder with CO_2 gas. The filled swim bladder makes the fish rise in the water. Carbon dioxide is absorbed from the swim bladder into the fish's bloodstream where it combines with water, forming carbonate ions, H_2CO_3. The decreased amount of gas in the swim bladder causes the fish to sink.

2. **Lateral line system**   Although found to a limited degree in sharks, only bony fishes have a fully developed lateral line system. The lateral line, shown in Figure 32-13, is a specialized sensory system that extends along each side of the fish's body. Nerve impulses from ciliated sensory cells in the lateral line permit the fish to perceive its rate of movement, sensing the movement as water presses

Figure 32-13 Clusters of sensory cells are contained in pits located within a fish's lateral line system. The movement of water over these cells causes signals to be sent along nerve fibers to the fish's brain. The brain assesses this sensory information to determine the speed and direction of the water current and the position of the objects in the water.

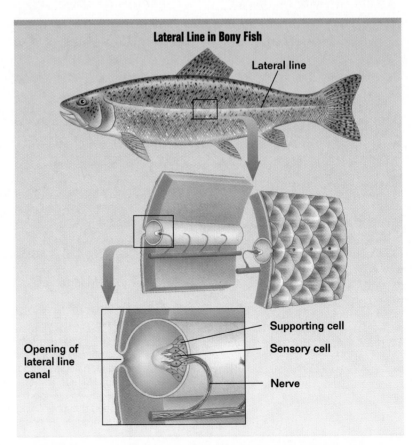

Lateral Line in Bony Fish

Lateral line

Supporting cell

Sensory cell

Opening of lateral line canal

Nerve

against its lateral line. A trout uses its lateral line system to obtain the sensory information it needs to orient itself with its head upstream.

The lateral line system also enables a fish to detect a motionless object by the movement of water reflected off that object. In a very real sense, this is equivalent to hearing. The way that a fish detects an object with its lateral line and the way that you hear a symphony with your inner ear share the same basic mechanism—cilia deflected by waves of pressure. Like fishes, you detect patterns in the waves of pressure in the medium (air, in your case) around you. The sound receptors within the ears of terrestrial vertebrates are thought to have evolved from lateral line receptors.

3. **Gill cover** What is a goldfish doing when it hovers at some depth in its fish tank, periodically taking in a mouthful of water? If it is not eating, then it is probably breathing. Most bony fishes have a hard plate called an **operculum** that covers the gills on each side of the head. Movements of the opercula, shown in Figure 32-14, permit a bony fish to pump water over the gills, enabling the fish to breathe. By using this very efficient pump, bony fishes can move water over their gills while remaining stationary in the water. A bony fish doesn't have to swim forward with its mouth open to move water over its gills as fishes without opercula, like sharks, must do. This ability to breathe without swimming enables a bony fish to conserve energy—energy that can be spent chasing after prey and escaping from predators.

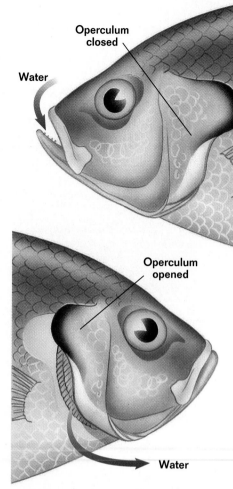

Figure 32-14 A fish breathes by moving its operculum. When the fish's mouth opens, the operculum closes over the gills, sealing off the body wall opening. This increases the volume of the mouth cavity, causing water to be drawn into the mouth, *top.* When the mouth is closed, the operculum moves away from the body wall, revealing the opening. As a result, the volume of the mouth cavity is decreased, and water is forced over the gills and out of the fish, *bottom.*

Section Review

1. *Of the characteristics that are common to all fishes, which one could affect a fish's ability to produce enzymes?*

2. *An unidentified species of fish is caught. It has rough skin, several rows of teeth, and no opercula. Based on this external description of the fish, would you expect it to have a swim bladder? Explain why or why not.*

3. *Describe the digestive system of the yellow perch.*

4. *What are the similarities between a yellow perch's lateral line and your ear?*

Critical Thinking

5. *Two identical goldfish bowls each contain one goldfish of the same type and size. A student observes that the opercula of one fish are moving at a much slower rate than the opercula of the other fish. Hypothesize about what might cause the different rates.*

32-3 History and Diversity of Amphibians

Section Objectives

- Define the term amphibian.
- Identify and describe the direct ancestors of the first amphibians.
- Discuss the major changes that have occurred in amphibians throughout their evolutionary history.

*F*rogs, toads, salamanders, and newts—these air-breathing, damp-skinned vertebrates are the direct descendants of fishes. They are the sole survivors of a very successful group, the amphibians, the first verte-brates to walk on land. Most modern amphibians are small and live unnoticed by humans. When describing amphibians in 1758, the biologist Carolus Linnaeus said, "These foul and loathsome animals are abhorrent because of their cold body, pale color, filthy skin, fierce aspect, calculating eye, offensive smell, harsh voice, squalid habitation, and terrible venom; and so their Creator has not exerted his powers to make many of them."

Many of Linnaeus's observations were inaccurate, particularly his claim that amphibians were few in number. Modern biologists have discovered that amphibians are among the most numerous of terrestrial vertebrates. There are as many amphibian species as mammalian species, if not more, and throughout the world amphibians play key roles in terrestrial as well as aquatic food chains.

Amphibians Evolved From a Type of Lobe-Finned Fish

The term *amphibian* is from the Greek *amphi*, meaning "double," and *bios*, meaning "life." The term nicely describes the essential characteristic of today's amphibians, reflecting their ability to live in two worlds—the aquatic world of their fish ancestors and the terrestrial world that their ancestors first invaded.

Fossil evidence indicates that amphibians evolved from lobe-finned fishes about 370 million years ago. For many years there was considerable disagreement about which kind of lobe-finned fish was the direct ancestor of amphibians. Some scientists argued that the coelacanths were the direct ancestors, while others insisted it was the lungfishes. A third group of scientists argued that a different group of lobe-finned fishes called rhipidistians *(RIHP uh DIST tee uhns)*, all of which are now extinct, were the true ancestors of amphibians. Good arguments can be made to support each of these possibilities. Recent analysis of the DNA of coela-canths and lungfishes indicates lungfishes are in fact far more closely related to amphibians than are coelacanths. But due to similarities in the skeletal features of amphibians and rhipidistian fishes, most paleontologists think that

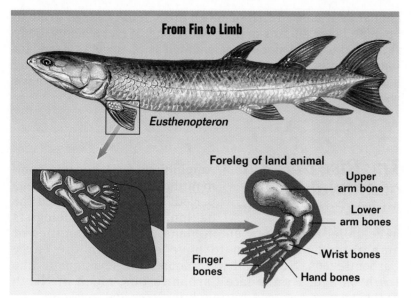

From Fin to Limb

Eusthenopteron

Foreleg of land animal

Upper arm bone

Lower arm bones

Finger bones

Wrist bones

Hand bones

Figure 32-15 The pattern of bones in an amphibian's limbs bears a remarkable resemblance to that of the fin bones of *Eusthenopteron,* a rhipidistian fish. This evidence leads most biologists to think that rhipidistian fishes are the direct ancestors of amphibians.

amphibians evolved directly from the rhipidistian fishes rather than from lungfishes. Figure 32-15 compares the fin bones of *Eusthenopteron,* an extinct rhipidistian fish, to the front leg bones of a typical land animal.

The earliest known amphibian fossil is that of an extinct amphibian called *Ichthyostega.* This fossil was found in 370-million-year-old rock in Greenland. At the time that this animal lived, Greenland was part of the North American continent and lay near the equator. Fossils from amphibians that lived during the next 100 million years have been found only in North America. Only when Asia and the southern continents merged with North America 250 million years ago to form one giant supercontinent did amphibians spread throughout the world.

Ichthyostega was a strongly built animal, with four sturdy legs well supported by hip and shoulder bones, as shown by the skeleton in Figure 32-16. Like its fish ancestors, *Ichthyostega* had bony scales, although the scales were set only in the skin of its belly and flattened tail. It also had a long fin that extended along the length of the tail. In the water the tail could be thrust from side to side to propel the

Figure 32-16 *Ichthyostega* had long, broad ribs that overlapped each other and formed a solid cage for the lungs and heart. Because its rib cage was not expandable, *Ichthyostega* probably breathed with mouth movements similar to those of a fish—lowering the floor of the mouth to draw in air and raising it to push air down the windpipe into the lungs.

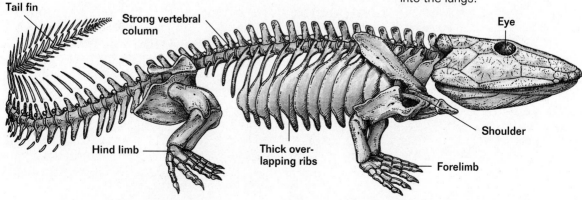

Tail fin

Strong vertebral column

Eye

Hind limb

Thick over-lapping ribs

Shoulder

Forelimb

Figure 32-17 *Eryops* grew to 1.8 m (6 ft.) long and had a heavy, flat skull bearing strong teeth. A thick, strong backbone indicates that *Eryops* was well adapted for life on land. Bony lumps in the skin formed a heavy armor. *Eryops* probably lived like a modern alligator, in and out of the water of upland streams, rivers, and lakes.

animal as it chased fish or other prey. *Ichthyostega* grew to be quite large, up to 1 m (3 ft. 3 in.) long. It had a heavy, bony skull and a short neck that enabled it to turn its head. Its four legs splayed out from the sides of its body, and each leg had a foot with five toes. On land *Ichthyostega* would have walked with an awkward sprawling gait. ■

The Rise and Fall of Amphibians

Amphibians first became common during the Carboniferous period. This period marked the beginning of a time biologists have called the age of amphibians. Fourteen families of amphibians are known to have existed in the early Carboniferous. Nearly all were aquatic or semiaquatic like *Ichthyostega*. By the late Carboniferous, much of North America was covered by low-lying tropical swamplands. Thirty-four families of amphibians thrived in this wet terrestrial environment, sharing it with pelycosaurs and other early reptiles. In the Permian period that followed, amphibians reached their greatest diversity, increasing to 40 families. In the early Permian a remarkable change occurred among amphibians—many of them began to leave the marshes for dry uplands. Many of these terrestrial amphibians had bony plates and armor covering their bodies and grew to be very large, some as big as a pony. Both their large size and shielded bodies suggest that the skin of these amphibians did not play a respiratory role, as does the skin of present-day amphibians. Rather, these early terrestrial amphibians had an impermeable leathery skin that prevented water loss. By the mid-Permian, 60 percent of all amphibian species were completely terrestrial. *Eryops*, shown in Figure 32-17, was typical of the terrestrial amphibians.

The middle Permian period marked the peak of amphibian success. By the end of this period a new kind of terrestrial reptile, the therapsid, had become common. Gradually the therapsids would be successful in replacing all terrestrial amphibians. Over half of all remaining amphibians were aquatic by the end of the Permian, and this trend continued into the Triassic period. By the end of the Triassic there were only 15 families of amphibians, including the first frogs. Almost without exception, these families were aquatic. Some contained species that grew to enormous sizes, up to 4 m (13 ft.) long. By the Jurassic period only two families of amphibians survived. One was the anurans *(uh NUR uhns)*, frogs and toads, and the other was the urodeles *(YOOR oh dehls)*, salamanders and newts. The age of amphibians was over. ❑

❑ **CAPSULE SUMMARY**

Amphibians were the dominant land vertebrates for 100 million years. The rise of reptiles marked the beginning of the decline of amphibians. Reptiles halted the expansion of amphibians into terrestrial habitats, greatly diminishing amphibian diversity and numbers.

The Success of Amphibians Today

All of today's amphibians descended from the anurans and urodeles that survived the Age of Reptiles, which lasted throughout the Mesozoic era. During the Tertiary period these remaining moist-skinned amphibians underwent a highly successful invasion of wet habitats all over the world. Today there are over 4,200 species of amphibians in 37 different families. They consist of frogs, toads, salamanders, and caecilians. While amphibians are not nearly as diverse in form or habitat as they were during the Permian period, modern amphibians have been very successful in occupying their particular niches.

Section Review

1. *Why is the term* amphibian *very appropriate for the members of this class?*

2. *Why would the capture of a rhipidistian fish today be considered a monumental scientific event?*

3. *Describe the physical appearance of* Ichthyostega, *and state when it lived.*

4. *State the scientific evidence suggesting that amphibians did not always live in aquatic or moist environments.*

5. *What kind of animal is a therapsid, and what influence did they have on early amphibians?*

Critical Thinking

6. *Offer a possible reason why amphibians evolved when they did, rather than 50 million years earlier. To answer this question, it may be helpful to refer to the timeline on pages 274–287 in Chapter 13.*

32-4 Key Adaptations of Amphibians

Section Objectives

- Describe the characteristics common to all modern amphibians.
- Contrast the three orders of modern amphibians.
- Identify and describe the major external and internal characteristics of the leopard frog.

Biologists classify today's species of amphibians into three orders. The order Anura is composed of frogs and toads. The term **Anura** is from the Greek **a**, meaning "without," and **oura**, meaning "tail." The order Urodela is composed of salamanders and newts. The term **Urodela** is from the Greek **oura**, meaning "tail," and **delos**, meaning "visible." The order Apoda is made up of wormlike, nearly blind organisms called caecilians. The term **Apoda** is from the Greek **a**, meaning "without," and **pous**, meaning "foot."

Characteristics of Modern Amphibians

Even though each amphibian order evolved at a different time during the Tertiary period, all modern amphibians have certain key characteristics in common.

1. **Legs** Frogs and salamanders have four legs and can move about on land quite well. The evolution of legs was one of the key adaptations for vertebrates that live on land. Caecilians gradually lost their legs during the evolutionary course of adapting to a burrowing existence.

2. **Lungs** Most larval amphibians have gills, but by adulthood the gills have usually disappeared and breathing is accomplished with a pair of lungs (lungless salamanders are an exception). The internal surfaces of amphibian lungs are poorly developed, with much less surface area than reptilian, bird, or mammalian lungs. Modern amphibians still breathe as *Ichthyostega* did, by lowering the floor of the mouth to suck in air and then raising it back up to force the air down into the lungs. Figure 32-18 demonstrates the breathing movements of a frog.

Figure 32-18 A frog breathes by forcing air into and out of its lungs. The steps involved in one inhalation and exhalation are numbered in sequence.

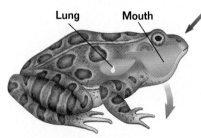

1 The floor of the mouth is lowered, causing air to be sucked in through the nostrils.

2 The nostrils close and the floor of the mouth is elevated, forcing air down into the lungs.

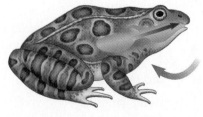

3 Contraction of muscles in the body wall forces air out through the mouth.

3. **Cutaneous respiration** Frogs, salamanders, and caecilians supplement the use of their lungs by respiring directly through their skin. The skin of amphibians is moist and provides an extensive surface area. Cutaneous respiration ("skin breathing") limits the maximum body size of amphibians because it is efficient only when there is a high ratio of skin surface area to body volume.

4. **Pulmonary veins** After blood is pumped through the lungs, two large veins called pulmonary veins return the oxygenated blood to the heart for repumping. This allows the oxygenated blood to be pumped to the tissues at a much higher pressure than it has when it leaves the lungs. Under high pressure, the blood travels quickly through the circulatory system, ensuring that the body tissues receive the oxygen they need.

5. **Partially divided heart** As you learned in Chapter 31, the amphibian heart is partitioned internally to form chambers. Because the division is incomplete (the atrium is divided into left and right sides, but the ventricle is not), a mixture of oxygenated and deoxygenated blood is delivered to the amphibian's body tissues.

Frogs and Toads Are Tailless

Frogs and toads, shown in Figure 32-19, make up the order Anura. There are 3,680 species of frogs and toads in 22 families. They live in environments ranging from deserts to mountains and ponds to puddles. All adult anurans are carnivores, eating a wide variety of insects. To learn about an anuran called the leopard frog, see *Up Close: Leopard Frog* on pages 762–763.

Figure 32-19 Frogs, like the green frog, *below left,* and the tree frog, *below center,* have smooth, moist skin, a broad body, and long hind legs that make them excellent jumpers. Most frogs live in or near water, although some tropical species live in holes in trees. Unlike frogs, toads, like the common Asiatic toad, *below right,* have dry, bumpy skin, short legs, and are well adapted to dry environments.

UP CLOSE LEOPARD FROG

- ■ **Scientific name**: *Rana pipiens*
- ■ **Range:** From northern Canada to southern New Mexico and from eastern California to the Atlantic coast
- ■ **Habitat:** Lives in the short grass of meadows and around ponds
- ■ **Size:** Body length (legs excluded) of 5–9 cm (2–3.5 in.)
- ■ **Diet:** Feeds on crickets, mosquitoes, and other insects

External Structures

Skin Numerous mucous glands embedded within the skin supply a lubricant that keeps the leopard frog's skin moist. A moist surface is necessary for respiration. Unlike many other frogs and toads, the leopard frog does not have skin glands that secrete poisonous or foul-tasting substances. Instead, the leopard frog must rely on its protective coloration and speed to evade predators, which include fishes, birds, turtles, and small mammals.

Eye Because its eyes bulge out from the head, the leopard frog can stay almost fully submerged while literally "keeping an eye out" for predators. Its eyes work equally well in or out of water. Eyelids that blink protect the eyes from dust. In addition, a transparent membrane covers each eyeball, keeping it moist and protecting it when the frog is underwater.

Tympanic membrane When sound causes the tympanic membrane (eardrum) to vibrate, a tiny bone transmits the vibrations to the middle ear. Within the middle ear are ciliated sensory cells (similar to those found in the lateral line of a fish) that are able to detect sound and help the frog maintain balance. Leopard frogs hear well in both water and air.

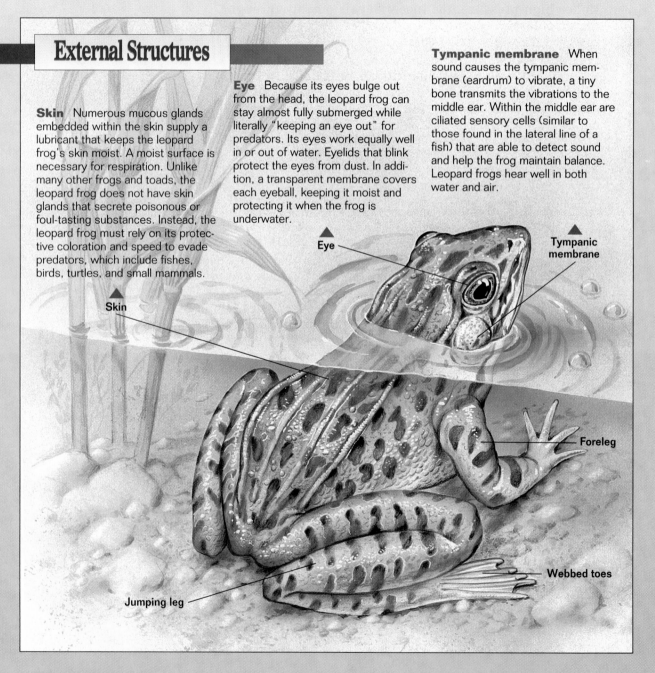

Eye

Tympanic membrane

Skin

Foreleg

Jumping leg

Webbed toes

Internal Structures

Reproductive organs Prior to breeding, the reproductive organs of male and female leopard frogs produce enormous numbers of sperm and egg cells, respectively. Next, males establish breeding territories and call to females using prominent vocal sacs that amplify their croak. When a female approaches, the male frog climbs on her back and grasps her firmly just behind the forelegs. When the female leopard frog releases a cluster of eggs into the water, the male discharges his sperm over them, fertilizing them externally.

Brain What makes the frog's brain different from the fish's brain is the degree of development of each brain component. For example, the larger, more complex cerebrum of a frog is able to process a wider assortment of sensory information than that of a fish.

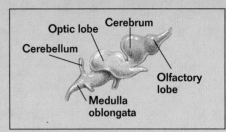

Optic lobe
Cerebrum
Cerebellum
Olfactory lobe
Medulla oblongata

Brain

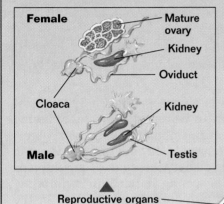

Female
Mature ovary
Kidney
Oviduct
Cloaca
Kidney
Male
Testis

Reproductive organs

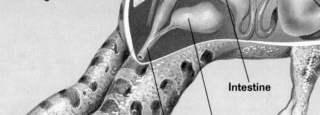

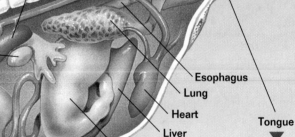

Kidney

Sacral vertebra

Urostyle

Pelvic girdle

Teeth

Esophagus
Lung
Heart
Liver

Tongue

Stomach

Intestine

Urinary bladder

Cloaca

Cloaca The excretory system of the leopard frog is very straightforward—there is only one way out. From the intestine, undigested foods are pushed into a cavity called the cloaca. Urine from the kidneys and bladder also passes into the cloaca, as do either egg cells or sperm cells from the reproductive organs. All of these materials exit the body through the cloacal opening.

Tongue The tongue flicks out at great speed, curls around the prey, then flicks back into the mouth with the insect. To prevent escape of its prey, the frog's upper jaw is lined with small, sharp teeth. In addition, two larger teeth project inward from the roof of the mouth to impale struggling prey. Food is swallowed whole.

Skeleton The skeletal system of the leopard frog (and all modern frogs) is highly reduced. It has only nine vertebrae and no ribs. From the sacral vertebrae, a long slender bone called the urostyle extends back to the center of the pelvic (hip) girdle. The bones of the frog's hind legs insert directly into sockets in the pelvic girdle. The two bones that make up the pelvic girdle extend forward to meet the sacral vertebrae. This three-strut support structure acts as a shock absorber for the long leg bones when the frog lands.

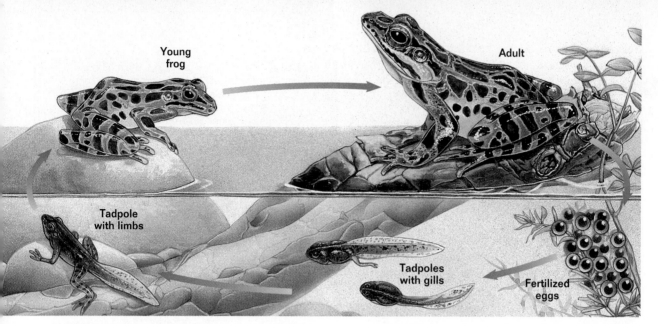

Young frog

Adult

Tadpole with limbs

Tadpoles with gills

Fertilized eggs

Figure 32-20 The transition from a larval frog (tadpole) to an adult involves a complex series of external and internal body changes. The tadpole's tail, gills, and lateral line system all disappear, and legs grow from the body. A saclike bladder in the throat (the position of the air sacs in its fish ancestors) divides into two sacs that become lungs. The pulmonary vein develops, and the heart develops its internal walls.

The life cycle of a frog is shown in Figure 32-20. Most frogs and toads must return to water to reproduce, laying their eggs directly in the water. Amphibian eggs lack watertight external membranes and would dry out quickly if not in water. Like the eggs of most fishes, the eggs of amphibians are fertilized externally. The young hatch into swimming, fishlike larval forms called tadpoles. Tadpoles live in the water and, being herbivores, feed mostly on algae. After considerable growth, the body of the tadpole abruptly changes into that of an adult frog. This process of dramatic physical change is called metamorphosis.

Figure 32-21 Most salamanders, like this marbled salamander, *left*, live in damp places, such as under stones or logs or among the leaves of plants. Some aquatic salamanders, like the Texas spring salamander, *right*, spend their entire life in water.

Salamanders Have Tails

Salamanders, shown in Figure 32-21, are members of the order Urodela. Salamanders have elongated bodies, long tails, and smooth, moist skin. There are about 369 species of salamanders in 9 families. They typically range from 10 cm to 0.3 m (4 in. to 1 ft.) in

length, although giant Asiatic salamanders of the genus *Andrias* grow to as long as 1.5 m (5 ft.) and weigh up to 41 kg (90 lb.). In general, salamanders are unable to remain away from water for long periods as toads do, although some salamander species manage to live in dry areas by remaining inactive during the day.

Salamanders lay their eggs in water or in moist places. Fertilization is usually external, although a few species practice a type of internal fertilization in which the female picks up sperm packets that have been deposited by the male. Unlike anuran larvae, salamander larvae do not undergo a dramatic metamorphosis. The young that hatch from salamander eggs are carnivorous and resemble small versions of the adults, except that the young have gills.

Caecilians Are Legless

Caecilians, members of the order Apoda, are a highly specialized group of tropical, burrowing amphibians. This order is made up of about 168 species in 6 families. These legless, wormlike animals, shown in Figure 32-22, grow to about 0.3 m (1 ft.) long, but some species can be up to 1.2 m (4 ft.) long. During breeding, the male caecilian deposits sperm directly into the female. Depending on the species, the female may bear live young or lay eggs that develop externally. ❏

Figure 32-22 This caecilian, *left,* is from Colombia, in South America. A caecilian has very small eyes and is often blind. As it burrows through the soil, it searches for small invertebrate prey. A unique group of amphibians, caecilians have small bony scales embedded in their skin.

❏ CAPSULE SUMMARY

Modern amphibians include frogs, toads, salamanders, and the legless caecilians. Most amphibian species live in moist or aquatic habitats, eat small invertebrates, and lay eggs that are fertilized externally.

Section Review

1. *Offer an explanation for the fact that there are no frogs as large as elephants.*
2. *Although most amphibians have lungs, a few salamander species are lungless. In what kinds of environments would you expect to find these lungless salamanders?*
3. *Compare and contrast the anurans and apodans.*
4. *Explain why it is difficult to "sneak up" on a leopard frog.*

Critical Thinking

5. *Would you expect the digestive system of a tadpole to look and function the same way as the digestive system of an adult frog? Explain your answer.*

lobe-finned fish (750)
operculum (755)
ray-finned fish (749)
teleost (749)

Review

Multiple Choice

1. Modern-day lampreys and hagfish evolved from an ancient order of
   a. sharks.
   b. bony fishes.
   c. ostracoderms.
   d. coelacanths.

2. The jaws of spiny fishes evolved from
   a. plates of bone.
   b. gill arches.
   c. muscles.
   d. ray fins.

3. Placoderms differed from ostracoderms in that placoderms had
   a. a jaw fused to the skull.
   b. full body armor.
   c. a swim bladder.
   d. two dorsal fins.

4. What characteristic did not help sharks become successful ocean predators?
   a. internal fertilization of eggs
   b. streamlined body
   c. powerful jaws
   d. sharp, replaceable teeth

5. Yellow perch and sharks share all of the following characteristics except
   a. gills.
   b. an internal skeleton.
   c. a single-loop circulatory system.
   d. a swim bladder.

6. The first amphibian evolved from
   a. chondrichthyes.
   b. reptiles.
   c. agnathans.
   d. rhipidistians.

7. Caecilians are different from other amphibians in that they
   a. are legless.
   b. can breathe through their skin.
   c. have a partially divided heart.
   d. mix oxygenated and deoxygenated blood.

Completion

8. The first vertebrates were jawless _____ that evolved during the _____ period.

9. Living agnathans include the parasitic _____ and scavenger _____ .

10. Sharks have a cartilaginous _____ that is covered by a thin layer of bone. Sharks also have bony _____ that are used for biting and tearing their food.

11. A fish will _____ when its swim bladder is filled with gas, but it will _____ when its swim bladder is emptied.

12. A typical amphibian lives a double life. It spends part of the time in the _____ and part of the time on _____ .

13. Amphibians reached their greatest diversity during the _____ period. The skin of the terrestrial amphibians that lived during this time was impermeable to water and was not used for cutaneous _____ .

14. During the tadpole stage, most amphibians respire by means of _____ . However, as adults they respire using simple, saclike _____ .

15. The order _____ includes frogs and toads. Amphibians that have tails are members of the order _____ .

Themes Review

16. **Evolution** Most sharks bear live young rather than laying eggs. How is bearing live young an adaptive advantage for these fishes?

17. **Structure and Function** Contrast the location and function of the swim bladder of a yellow perch with the urinary bladder of a leopard frog.

18. **Levels of Organization** What organs comprise the digestive system of the yellow perch?

19. Responding Critically A guide who leads tours of a large aquarium bases his discussion of the evolution of fishes on the idea that all bony fishes evolved from cartilaginous sharks. At the end of the tour, audience members are asked to rate and comment on various aspects of the guide's presentation. Suppose that you are a member of the audience. You are asked to rate and comment on the accuracy of the presentation. What rating on a scale of 1 to 10 (10 being the best) would you give the presentation's accuracy? What comments would you make to justify your rating?

20. Interpreting Data An ecologist collected data on the average level of the pesticide DDT contained in the tissues of insects eaten by leopard frogs in a pond. Additional data showed the number of leopard frog tadpoles captured from and released to the pond for six years. Based on the following data, what conclusions can you draw about the relationship between DDT levels and the number of tadpoles in the pond? Imagine that the average DDT level in insects is 10 parts per million in year 7. Will the ecologist likely see more or fewer tadpoles than in year 6?

Year	DDT in parts per million	Number of tadpoles
1	5	167
2	11	75
3	9	123
4	14	51
5	8	146
6	6	159

21. Multicultural Perspectives Research how dart-poison frogs are used by the native hunters of Central and South America. Find out why North American scientists are interested in studying the poison that is secreted by the skin of these frogs.

22. Research and Writing Do research to learn what caviar is and where it comes from. Also, contact a grocery store or market to find out the current price of caviar. Make a poster that presents the findings of your research in both words and pictures.

23. Cooperative Group Project Hundreds of thousands of frogs are killed and preserved each year by biological supply companies for use as specimens in laboratories. Organize a debate regarding the pros and cons of this practice. After forming debate teams, research various perspectives of the issue, including classroom demand for preserved specimens and the practices of the supply companies. Debate the issue before a class of students, following agreed-on rules of conduct. Allow the class to select a winner using predetermined criteria that are approved by your teacher.

Live Frogs

OBJECTIVES

• Examine the external features of a frog.
• Observe the behavior of a frog.
• Explain how a frog is adapted to life on land and in water.

PROCESS SKILLS

• relating structure to function
• recognizing the relationship between structure and evolutionary success

MATERIALS

• live frog in a terrarium
• aquarium half-filled with dechlorinated water
• live insects (crickets or mealworms)
• 600 mL beaker

BACKGROUND

1. What does *amphibious* mean?
2. How do amphibians live part of their life on land and part in water?
3. What are some major characteristics of amphibians?
4. Write your own **Focus Question** on your Vee Form.
5. **Knowing Side of the Vee** List the **Concepts** and new **Vocabulary Words** on your Vee Form. In the **Concept Statements** section of the Vee, use these words in sentences that define and explain them.

TECHNIQUE

Doing Side of the Vee

1. Observe a live frog in a terrarium. Closely examine the external features of the frog. Make a drawing of the frog in the **Records** section of your Vee Form.

Label the eyes, nostrils, tympanic membranes, front legs, and hind legs. The tympanic membrane, or eardrum, is a disklike membrane behind each eye.

2. Make a table like the one below in the **Records** section to note all your observations of the frog in this investigation.

Observations of Frog	
Breathing	
Eyes	
Legs	
Response to food	
Response to noise	
Skin	
Swimming behavior	

3. Watch the frog's movements as it breathes air with its lungs. Record your observations in the **Records** section of your Vee.

4. Look closely at the frog's eyes and note their location. Examine the upper and lower eyelids, as well as a third transparent eyelid called a nictitating membrane. The upper and lower eyelids do not move. The nictitating membrane moves upward over the eye. This eyelid protects the eye when the frog is underwater and keeps it moist when the frog is on land.

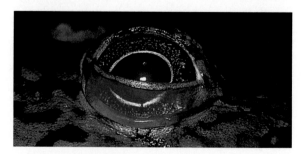

5. Study the frog's legs, noting the difference between the front and hind legs.

6. Place a live insect, such as a cricket or a mealworm, into the terrarium. Observe how the frog reacts.

7. Tap the side of the terrarium farthest from the frog and observe the frog's response.

8. **CAUTION: You will be working with a live animal. Handle it gently and follow instructions carefully.** Frogs are slippery! Do not allow the frog to injure itself by jumping from the lab bench to the floor. Place a 600 mL beaker in the terrarium. Carefully pick up the frog and examine the skin. How does it feel? The skin of a frog acts as a respiratory organ, exchanging oxygen and carbon dioxide with the air or water. A frog also takes in and loses water through its skin. Place the frog in the beaker. Cover the beaker with your hand and carry it to a freshwater aquarium. Tilt the beaker and gently submerge it beneath the surface of the water until the frog swims out of the beaker.

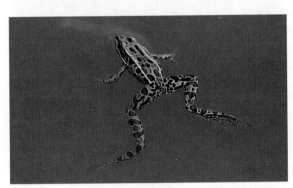

9. Closely watch the frog float and swim in the aquarium. How does the frog use its legs to swim? Notice the position of the frog's head. As the frog swims, bend down and look up into the aquarium so that you can see the underside of the frog. Then look down on the frog from above. Compare the color on the dorsal and ventral sides of the frog. When you are finished observing the frog, your teacher will remove the frog from the aquarium. In the **Procedure** section of the Vee, briefly summarize the procedure you followed.

10. Clean up your materials and wash your hands before leaving the lab.

INQUIRY

1. From the position of the frog's eyes, what can you infer about the frog's field of vision?

2. How does the position of the frog's eyes benefit the frog while it is swimming?

3. How does a frog hear?

4. How can a frog breathe air while it is swimming in water?

5. Why must a frog keep its skin moist while it is on land?

6. Use the information on the **Knowing Side** of the Vee to interpret your results from the **Doing Side,** and then write your **Knowledge Claim.** Write a **Value Claim** for this lab.

ANALYSIS

1. How are the hind legs of a frog adapted for life on land and in water?

2. What adaptive advantage do frogs have in showing different coloration on their dorsal and ventral sides?

3. What features provide evidence that an adult frog has an aquatic life and a terrestrial life?

FURTHER INQUIRY

Write a **New Focus Question** that could be the point of a new investigation. The following is an example:

How are other amphibians adapted to life on land and in water?

REPTILES AND BIRDS

REVIEW

- continental drift (Section 13-3)
- angiosperms (Section 23-2)
- main characteristics of reptiles and birds (Section 31-1)
- ectothermy and endothermy (Section 31-1)
- differences between reptilian and avian circulatory and respiratory systems (Sections 31-2 and 31-3)
- structure of the amniotic egg (Section 31-4)
- oviparous, ovoviviparous, and viviparous reproduction (Section 31-5)
- *Highlights: Therapsids* on page 734

A collared lizard, a common inhabitant of the deserts of the Southwest.

33-1 History of Today's Reptiles

More than 7,000 species of reptiles (Class Reptilia) currently live on Earth. They are a highly successful group, more diverse than even the mammals (there are three species of reptiles for every two species of mammals). Although it is traditional to think of reptiles as more primitive than mammals, the great majority of reptiles that live today belong to groups that appeared after the therapsids, the group from which mammals evolved. This section examines the history of the living reptiles.

Section Objectives

- Identify the four orders of living reptiles.
- Recognize the close relationship between crocodiles and birds.
- Describe the evolution of the reptilian skull.

Origin of Present-Day Reptiles

Figure 33-1 shows the relationship of the living reptiles to some of the extinct groups of reptiles, such as the dinosaurs, and to the groups that evolved from reptiles, the mammals and birds. Of the 16 orders of reptiles known to have existed, only 4 survive. The most ancient surviving group is the turtles (order

Figure 33-1 There are four orders of living reptiles: turtles, crocodilians, tuataras, and lizards and snakes. This phylogenetic tree shows how these four orders are related to each other and to dinosaurs, birds, and mammals.

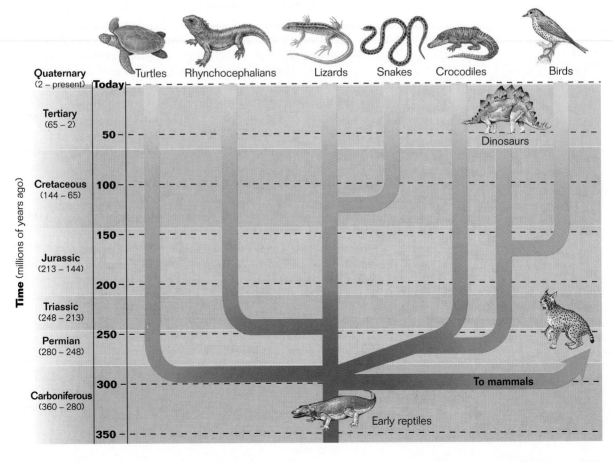

Turtles   Rhynchocephalians   Lizards   Snakes   Crocodiles   Birds

Quaternary (2 – present) — Today

Tertiary (65 – 2) — 50 — Dinosaurs

Cretaceous (144 – 65) — 100

— 150

Jurassic (213 – 144) — 200

Triassic (248 – 213) — 250

Permian (280 – 248)

— 300 — To mammals

Carboniferous (360 – 280) — Early reptiles

— 350

Time (millions of years ago)

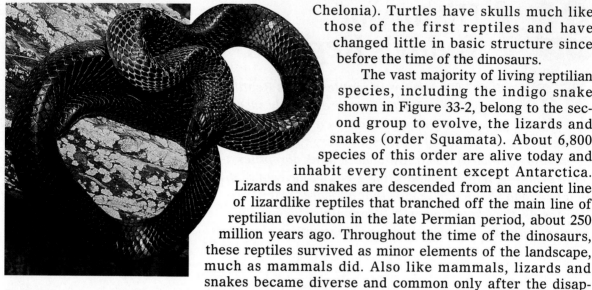

Figure 33-2 The indigo snake is one of the largest snakes in the United States, reaching a length of 262 cm (100 in.). It lives in the southeastern states, Texas, and Mexico and feeds on mammals, birds, and other snakes.

◻ *CAPSULE*
  SUMMARY

Present-day reptiles are classified into four orders: Chelonia (turtles), Rhynchocephalia (tuataras), Squamata (lizards and snakes), and Crocodilia (crocodilians).

Chelonia). Turtles have skulls much like those of the first reptiles and have changed little in basic structure since before the time of the dinosaurs.

The vast majority of living reptilian species, including the indigo snake shown in Figure 33-2, belong to the second group to evolve, the lizards and snakes (order Squamata). About 6,800 species of this order are alive today and inhabit every continent except Antarctica. Lizards and snakes are descended from an ancient line of lizardlike reptiles that branched off the main line of reptilian evolution in the late Permian period, about 250 million years ago. Throughout the time of the dinosaurs, these reptiles survived as minor elements of the landscape, much as mammals did. Also like mammals, lizards and snakes became diverse and common only after the disappearance of the dinosaurs.

The third group of surviving reptiles to evolve were members of the order Rhynchocephalia. Rhynchocephalians *(RIHN koh suh FAYL ee uhnz)* are small reptiles that appeared shortly before the dinosaurs. They lived throughout the time of the dinosaurs and were common in the Jurassic period. They began to decline in abundance during the Cretaceous period, however, possibly due to competition from lizards. Rhynchocephalians were already rare by the time the dinosaurs disappeared. Today only two very closely related species of this order survive, the tuataras *(TOO uh TAHR uhz)* of New Zealand. You will read more about tuataras in the next section.

The fourth line of living reptiles, the crocodiles (order Crocodilia), appeared on the evolutionary scene much later than the other groups of living reptiles. Crocodiles are descended from thecodonts *(THEE kuh dawntz)*, crocodile-like reptiles that gave rise to the dinosaurs, and they resemble dinosaurs in many ways. Crocodiles have changed very little in over 200 million years. Together with thecodonts and dinosaurs, crocodiles belong to a group called archosaurs, meaning "ruling reptiles."

Crocodiles resemble birds far more than they resemble other living reptiles. For instance, crocodiles are the only living reptiles that, like birds, care for their young. They are also the only living reptiles that have a four-chambered heart like that of birds. In many other points of anatomy, crocodiles differ from all other living reptiles and resemble birds. Why are crocodiles so much more like birds than are other living reptiles? Most biologists now think that birds are the direct descendants of dinosaurs. Crocodiles and birds, then, are far more closely related to dinosaurs than are lizards and snakes. That is why crocodiles and birds appear so similar; they are more closely related to each other than they are to lizards and snakes. ◻

Evolution of the Reptilian Skull

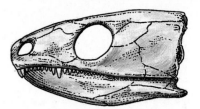

Anapsid skull

The classification of reptiles into major groups is based largely on the structure of the skull. Early reptiles, like their amphibian ancestors, had **anapsid** skulls. This type of skull has openings only for the eyes and nostrils, as illustrated in Figure 33-3. The muscles that move the jaws lie underneath this solid covering. Of today's reptiles, only turtles have anapsid skulls. Because of this characteristic, turtles are thought to be closely related to the early reptiles.

As you learned in Chapter 31, reptiles evolved openings in the skull behind the eyes. **Diapsid** skulls have two openings on each side and characterize tuataras, crocodiles, and dinosaurs. In a living animal, these openings are covered by connective tissue. In a fossil, of course, the connective tissue has long since decayed, leaving "windows" in the skull. One advantage of these openings may have been a reduction in skull weight. They may also have permitted larger, stronger jaw muscles to evolve, since these muscles would no longer be confined by the solid skull wall and could expand into the opening when contracting.

Diapsid skull

The skulls of lizards and snakes are modified versions of the diapsid skull. Lizards have lost the arch of bone below the lower of the two openings. This arrangement increases the flexibility of the skull and allows the mouth to open wider. Snakes have even more flexible jaws because they have lost the arches below both openings, resulting in one large opening.

Modified diapsid skull of a lizard

Very different skulls evolved among the reptiles that led to mammals. Therapsids had **synapsid** skulls with a single opening on each side. Mammals have the same type of skull, except that the opening is much larger and has merged with the eye socket.

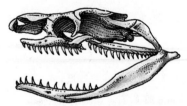

Modified diapsid skull of a snake

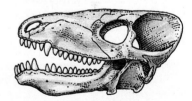

Synapsid skull

Figure 33-3 Skull structure is an important characteristic for classifying reptiles. One reason this characteristic is informative is that it is evolutionarily conservative, changing little within major groups. Characteristics such as tooth and limb structure tend to be more flexible and therefore provide less information for identifying major groups.

Section Review

1. *Which reptilian order is most closely related to the early reptiles?*
2. *Describe two similarities between crocodiles and birds.*
3. *If you found a reptilian skull, what characteristics would you look for to identify what type of reptile the skull belonged to?*

Critical Thinking

4. *Explain why a lighter skull would be particularly advantageous for a terrestrial vertebrate. In what ways would a lighter skull be disadvantageous?*

Characteristics
and Diversity of
Living Reptiles

- Describe the structure of a turtle's shell.
- Identify two differences between lizards and snakes.
- Describe the timber rattlesnake's adaptations for locating and capturing prey.
- Contrast parental care in crocodilians with that in other reptiles.

Reptiles occur worldwide except in the coldest regions, where it is impossible for ectotherms to survive. In recent times, humans have had an adverse impact on the number and distribution of reptiles. Many species of turtles, for example, are prized for food and have been hunted almost to extinction. Despite overhunting and the destruction of many of their natural habitats, reptiles remain among the most numerous and diverse of terrestrial vertebrates. In this section you will study the major characteristics of reptiles and the four surviving reptilian orders, which contain about 7,000 species.

Key Characteristics of Living Reptiles

All living reptiles share certain fundamental characteristics, features they retain from the time when they replaced amphibians as the dominant terrestrial vertebrates. Figure 33-4 summarizes these key features, several of which are described in more detail in this section.

Amniotic Eggs Make Terrestrial Reproduction Possible

Amphibians never succeeded in becoming fully terrestrial because their eggs dry out and die unless laid in water or in moist areas. In contrast, reptilian eggs can be laid on

Figure 33-4 The thorny devil, a fierce-looking but harmless lizard of arid Australia, is a representative reptile. Like North American horned lizards, the thorny devil feeds on ants.

Characteristics of Living Reptiles

- Amniotic eggs, with embryo surrounded by two membranes, the chorion and the amnion
- Internal fertilization of eggs
- Dry, watertight, scaly skin that minimizes water loss
- Partially divided ventricle of heart, reducing dilution of oxygen-rich blood by oxygen-poor blood; in crocodiles, completely divided ventricle
- Lungs have larger internal surface area than lungs of amphibians
- Ectothermic

dry land because they are watertight and contain their own supply of water. A reptilian egg contains a food source (the yolk) and a series of four membranes: the yolk sac, the amnion, the allantois (uh LAHN toh ihs), and the chorion. Each membrane plays a role in making the egg an independent life-support system. The outermost membrane of the egg is the chorion. This membrane allows oxygen to enter, but it retains water within the egg. The amnion encloses the developing embryo within a fluid-filled cavity. The yolk sac contains the yolk, which the embryo absorbs through blood vessels connected to its gut. The allantois surrounds a cavity into which waste products from the embryo are excreted. All modern reptiles show this pattern of membranes within the egg.

The majority of reptiles are oviparous. Their eggs are surrounded by a protective shell and are deposited in a suitable place in the environment. Heat from the environment incubates the eggs, which are usually left unprotected by the parents. However, several species of lizards and snakes, including rattlesnakes and some horned lizards, are ovoviviparous or viviparous and give birth to live, fully formed young.

Reptiles Are Ectothermic

The metabolism of present-day reptiles is too slow to generate enough heat to warm their bodies, so they must absorb heat from their surroundings. A reptile's body temperature is largely determined by the temperature of its environment. Many reptiles regulate their temperature by their behavior. They bask in the sun to warm up and seek shade to prevent overheating. Figure 33-5 shows that a lizard can maintain a fairly constant body temperature throughout the day by moving between sunlight and shade. You can also see why it is inaccurate to call ectotherms "cold blooded." The lizard's body temperature is higher than yours for part of the day. ▪

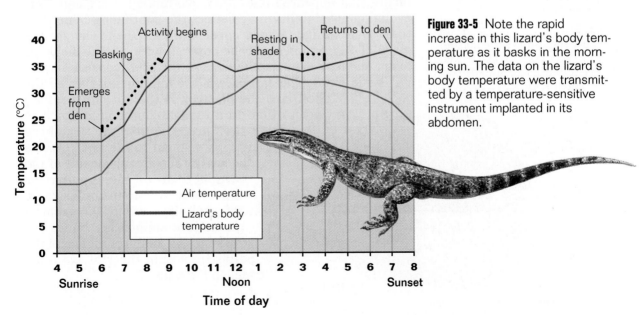

Figure 33-5 Note the rapid increase in this lizard's body temperature as it basks in the morning sun. The data on the lizard's body temperature were transmitted by a temperature-sensitive instrument implanted in its abdomen.

At very cold temperatures, reptiles become sluggish and unable to function. Intolerance of cold limits reptiles' geographical range and forces them to hibernate through the winter in temperate climates. Furthermore, the "low-power" metabolism of an ectotherm cannot support sustained, energy-demanding activity. Lizards are often swift sprinters, but they tire quickly. However, an ectotherm has one significant advantage over an endotherm, especially in environments where food is scarce. Because an endotherm uses 80 percent of its food energy to maintain a high body temperature, it must eat about 10 times more food than an ectotherm of the same size.

The thecodont ancestors of crocodiles were ectothermic, as crocodiles are today. The later dinosaurs, from which birds evolved, were probably endothermic, which is why crocodiles and birds differ in this one important aspect. This difference is a principal reason why crocodiles are classified as reptiles, while birds are not.

Figure 33-6 Like other sea turtles, this hawksbill turtle, *top*, spends virtually its entire life in the sea. Females leave the water only to lay their eggs on sandy beaches. Darwin was amazed by the diversity of shell shapes among the Galápagos tortoises, *bottom*. Each island has a different variety of tortoise with a uniquely shaped shell.

Figure 33-7 This cutaway view of a turtle's shell shows the unique arrangement of the ribs, pelvis, and pectoral girdle. The plastron has been removed.

Turtles and Tortoises

There are about 250 species of turtles (which generally live in water) and tortoises (which live on land), all classified in the order Chelonia. They differ from other reptiles in that their bodies are encased within a protective shell. Many of them can pull their head and legs into the shell for effective protection from predators. While most tortoises have a dome-shaped shell, water-dwelling turtles have a streamlined, disk-shaped shell that permits rapid turning in water. Turtles and tortoises lack teeth but have a sharp beak. Figure 33-6 shows a turtle and a tortoise.

Today's turtles and tortoises differ little from the first turtles that appeared more than 200 million years ago. This evolutionary stability may reflect the continuing benefit of their basic design, a body covered with a shell. The shell is made of fused plates of bone covered with horny shields or tough leathery skin. In either case, the shell consists of two basic parts. The **carapace** is the top (dorsal) part of the shell, and the **plastron** is the bottom (ventral) portion. In a fundamental commitment to this shell architecture, the vertebrae and ribs of most turtle and tortoise species are fused to the inside of the carapace; all of the support for muscle attachment comes from the shell. In addition, the pectoral girdle (the supporting bones for the bones of the forelimbs) and the pelvis lie within the ribs, as illustrated in Figure 33-7.

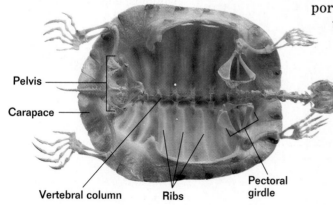

Pelvis

Carapace

Vertebral column   Ribs   Pectoral girdle

Tuataras Live Only in New Zealand

Today the order Rhynchocephalia contains only two very closely related species, the tuataras, members of the genus *Sphenodon*. *Sphenodon punctatus*, the more common and widespread species, is shown in Figure 33-8. Tuataras are lizardlike reptiles up to 70 cm (2 ft.) long. Unlike most reptiles, tuataras are most active at low temperatures. They burrow or bask in the sun during the day and feed on insects, worms, and small animals at night.

Figure 33-8 Tuataras formerly lived on several of the islands of New Zealand. When humans colonized New Zealand, they destroyed most of the forests and introduced rats, cats, dogs, weasels, and other animals that preyed on or competed with tuataras. Tuataras survived this onslaught only on a few small islands that were largely spared from human interference.

Lizards and Snakes

The order Squamata consists of about 3,800 species of lizards and about 3,000 species of snakes. The distinguishing characteristics of this order are the paired reproductive organs of males and a lower jaw that is loosely connected to the skull. This loose connection allows the mouth to open wide enough to accommodate large prey. In addition, the loss of the lower arch of bone below the lower opening in the skull of lizards makes room for large muscles to move the jaws. Most lizards and all snakes are carnivores. These improvements in jaw design have made a major contribution to their success as predators.

Lizards appeared in the late Permian period, about 250 million years ago. Common present-day lizards include iguanas, chameleons, geckos, horned lizards (often mistakenly called "horny toads"), and anoles. Several lizard species are shown in Figure 33-9. Most lizards are small, measuring less than 30 cm (1 ft.) in length. The largest lizards belong to the monitor family. The largest of these is the Komodo dragon of Indonesia, which reaches 3 m (10 ft.) in length and weighs up to 250 kg (550 lb.).

Figure 33-9 This gecko, *bottom,* like many species of lizards, has the ability to lose its tail when seized by a predator and to then grow a replacement. This individual's tail apparently did not completely detach, and a second tail has begun to grow. The Gila monster, *top right,* of the southwestern United States and northern Mexico is one of only two species of venomous lizards. Because it lacks the venom-injecting fangs found in poisonous snakes, the Gila monster is not dangerous to humans unless harassed. A common lizard throughout the western United States is the western fence lizard, *top left,* named for its habit of perching on fences.

UP CLOSE TIMBER RATTLESNAKE

- **Scientific name:** *Crotalus horridus*
- **Range:** Eastern and central United States, from northern New York to northern Florida and central Texas
- **Habitat:** Prefers areas of thick brush, dense woodland, or swamp
- **Size:** Typically 90–150 cm (36–60 in.); maximum 189 cm (74 in.)
- **Diet:** Mainly mammals

External Structures

Rattle The rattle typically consists of 5 to 7 interlocking rings made of keratin. When shaken it produces a hissing sound that serves as a warning. As the skin is shed, a new ring is added to the base of the rattle. Contrary to popular myth, the snake does not add a rattle each year; it simply adds one whenever its skin is shed. The more rapidly the snake grows, the more rattles it has (that is why the number of rattles is proportional to the size of the snake).

Rattle

Eye

Pit organ

Nostril

Pit organ Between the eye and nostril on each side of the head is a pit containing a heat-sensing organ. Using this organ, the snake can locate and strike a motionless warm animal in total darkness by sensing the heat that the animal gives off. The pit organ has an outer and an inner chamber, separated by a membrane. The organ operates by comparing the signals from temperature-sensitive nerve endings in the two chambers. The two pit organs provide stereoscopic information in much the same way that two eyes do. In fact, the information from the pit organs is processed by the same part of the brain that processes impulses from the eyes.

Internal Structures

Venom The timber rattlesnake injects its prey with toxic venom. Its upper front teeth are large hollow fangs. When the rattlesnake strikes, these hinged fangs swing forward from the roof of the mouth and inject venom deep into the prey. The venom contains hemotoxins, proteins that attack the circulatory system, destroying red blood cells and causing internal hemorrhaging. Modified salivary glands in the upper jaw produce the venom.

Venom gland

Fang

Jacobson's organs

Tongue

Trachea

Jacobson's organs
Flicking its forked tongue into the air, the rattlesnake gathers chemicals from the environment. These chemicals are transferred to two depressions in the roof of the mouth called Jacobson's organs, which act as very sensitive taste buds.

Reproductive organs This male rattlesnake produces sperm in his testes. Female timber rattlesnakes are ovoviviparous. A female carries her fertilized eggs in her body throughout development. Each egg has a thin membrane through which water and oxygen pass from the mother to the embryo, although all nourishment is provided by the egg's yolk. After the eggs hatch in the mother's body, the hatchlings are ejected to fend for themselves.

Small intestine

Testes

Large intestine

Kidneys

Pancreas

Gallbladder

Stomach

Unique internal anatomy The internal organs are elongated, matching the snake's body shape. The left lung is nonfunctional.

Liver

Cloaca

Right lung

Left lung

Esophagus

Heart

Movement without legs The rattlesnake moves by slithering—undulating gracefully and with surprising speed. This motion is made possible by the unique anatomy of its backbone and muscles. The backbone is made up of several hundred vertebrae, each with its own pair of attached ribs. These bones provide the framework for thousands of muscles. The muscles manipulate not only the skeleton but also the snake's skin, causing this fabric of overlapping scales to extend and contract. To move, the timber rattlesnake slides its head to one side, which initiates a wave of muscular contractions along the vertebrae. As this lateral motion moves down its trunk, the rattlesnake uses the sides of its body to push off against pebbles, twigs, and small irregularities of the ground, moving forward in an S-shaped path.

Figure 33-10 A South American emerald boa rests on a branch, *left.* The copperhead, *right,* is a common venomous snake in eastern and southern North America.

Figure 33-11 A snake's very flexible jaws allow it to swallow prey much larger than its head. This copperhead is swallowing a mouse.

Figure 33-10 shows a small sample of the diversity among snakes. Snakes probably evolved from lizards during the Cretaceous period. The close relationship between lizards and snakes is reflected in their many similarities. In fact, it is often difficult to distinguish between lizards and snakes. All snakes are limbless, but so are a number of lizards, such as the glass lizard of the United States. Snakes lack movable eyelids and external ears, as do several species of lizards. However, the internal anatomy of snakes is unique. No snake has any trace of a pectoral girdle, which is found even in legless lizards. Recall also that snakes have lost both arches of bone behind the eye, while lizards have lost only the lower arch. The snake jaw is also distinctive. It is very flexible because it has five points of movement (your jaw has only one movement point). One of these points is the chin, where the halves of the lower jaw are connected by a ligament and can spread apart when a large meal is being swallowed, as shown in Figure 33-11.

Very few species of snakes are poisonous. The poison is produced by modified salivary glands and is injected into the victim through grooved or hollow teeth. Snakes dangerous to humans are concentrated in four families: cobras, kraits, and coral snakes; sea snakes; adders and vipers; and rattlesnakes, water moccasins, and copperheads.

You can read more about the biology of snakes in *Up Close: Timber Rattlesnake* on pages 778–779.

Crocodiles and Alligators

The order Crocodilia is composed of 25 species of large, aquatic reptiles. In addition to crocodiles and alligators, this order also includes the alligator-like caimans and the fish-eating gavial. You can see two species of crocodilians in Figure 33-12.

All crocodilians are aggressive carnivores. They generally capture prey by stealth, often floating just beneath the water's surface near the shore. If an animal comes to the water to drink, the crocodilian explodes out of the water, seizes the prey, and hauls it back into the water to be

drowned and eaten. The bodies of crocodilians are well adapted for this form of hunting. Their eyes are high on the sides of the head, and their nostrils are on top of the snout; this enables them to see and breathe while lying nearly submerged in the water. They have an enormous mouth studded with sharp teeth and have a very strong neck. A valve in the back of the mouth prevents water from entering the air passages when crocodilians feed underwater.

Unlike other living reptiles, crocodilians care for their young after hatching. For instance, a female American alligator builds a nest for her eggs from rotting vegetation. Heat from the decaying vegetation helps to incubate the eggs. After the eggs hatch, the mother may tear open the nest to free the hatchlings. The young alligators remain under her protection for up to a year.

Figure 33-12 The American alligator, *above left,* lives throughout the southern United States from Texas to Florida. Twenty years ago, the alligator was an endangered species, but careful management has restored its numbers. These crocodiles, *above right,* are basking alongside a river in India.

Section Review

1. *Contrast the position of a turtle's shoulder with that of yours.*

2. *Why is the lack of legs not a good characteristic for distinguishing a lizard from a snake?*

3. *A blinded rattlesnake can still strike its prey. Explain how.*

4. *Explain how the parental care shown by alligators differs from that shown by most other reptiles.*

Critical Thinking

5. *Many viviparous snakes and lizards live in cold climates. Why might viviparity be advantageous in such environments?*

33-3 Dinosaurs

*O*f the major groups of terrestrial vertebrates, dinosaurs are considered the most successful. They dominated life on land for roughly 150 million years. During their long history, dinosaurs changed a great deal because the world they inhabited changed. One reason they changed was that the continents moved and radically altered the Earth's climates. Thus, you cannot study dinosaurs as if they were a particular kind of animal, with one type representing the group. Rather, you have to look at dinosaurs more as a story, a long parade of change and adaptation. This section will examine a variety of very different dinosaurs—animals that lived at different times and were adapted to very different worlds.

Figure 33-13 How do paleontologists know what dinosaurs were like, what they ate, and how they walked? Paleontologists must draw their conclusions from traces the dinosaurs left behind, such as this skeleton of a small, carnivorous dinosaur.

The Triassic Period: Origin of Dinosaurs

Dinosaurs are reptiles that first appeared in the Triassic period (248–213 million years ago). The first dinosaurs evolved from thecodonts, a group of crocodile-like, carnivorous reptiles that are now extinct. Fossils of the oldest dinosaurs for which there is clear evidence were found in Argentina in early Triassic rock some 235 million years old. Only about 30 cm (1 ft.) long, the oldest known dinosaur is a bipedal carnivore named *Eoraptor*, illustrated in Figure 33-14. Almost as old is *Herrerasaurus*, a bulky carnivore 4 m (13 ft.) long. *Herrerasaurus* had sharp, pointed teeth and a sliding jaw joint that enabled it to slice its victims as it bit into them.

These early dinosaurs were the first vertebrates to have a key improvement in body structure: their legs were positioned directly under the body, allowing them to run swiftly after prey. By the end of the Triassic period, small, carnivorous dinosaurs were very common; most of them were lightly built and bipedal.

In the late Triassic period, all of the continents were joined in a single supercontinent called **Pangaea** (*pan GEE uh*). There were few mountain ranges over this enormous stretch of land, and the interior was arid. Coastal climates were much the same all over the world—quite warm, with a dry season followed by a very wet monsoon season. By the end of the Triassic period, some 22 million years after the oldest known dinosaur fossils appear in the fossil record, dinosaurs had become common and had largely replaced the thecodonts.

There are at least three reasons why dinosaurs were so successful.

1. **Leg structure** Legs positioned directly under the body enabled the dinosaurs to be faster and more agile runners than the thecodonts.

2. **Drought resistance** In the late Triassic period, Pangaea's interior was dry, and dinosaurs were superbly adapted to arid conditions. The mammals of the late Triassic period, by contrast, were not as effective at water conservation. They lost water by sweating, their chief means of releasing excess body heat.

3. **Luck** At the end of the Triassic period, a large meteorite landed in northeastern Canada (the site, the Manicuoagan Crater, is still visible today), and it might have been responsible for the great loss of diversity that occurred at the end of the Triassic period. Thecodonts and many other species became extinct, but the dinosaurs survived.

Figure 33-14 *Eoraptor* was about the size of a chicken. Fossils of this dinosaur, the oldest species yet unearthed, were discovered in Argentina in 1989. Since no one has seen a living dinosaur, the coloration of the dinosaurs shown in this section is the artist's speculation.

The Jurassic Period: The Golden Age of Dinosaurs

At the beginning of the Jurassic period (213–144 million years ago), vast deserts still covered much of Pangaea. What was to become western North America was covered by a sea of sand during much of this time. The dominant trees were cycads, which resemble palm trees and are well adapted to arid climates.

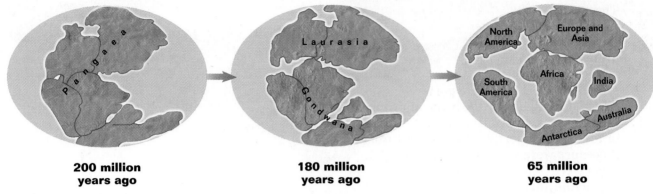

| 200 million years ago | 180 million years ago | 65 million years ago |

Figure 33-15 The continents were not in their present positions when the dinosaurs lived. In the Triassic period, *above left,* all of the continents were connected into a huge supercontinent, Pangaea. This landmass began to break up during the Jurassic period, *above center.* By the Cretaceous period, *above right,* today's continents began to be recognizable, although Australia and Antarctica did not separate until 50 million years ago, 15 million years after the dinosaurs became extinct.

The great supercontinent of Pangaea was beginning to break up, as shown in Figure 33-15. Long fingers of ocean began to separate the northern part, called Laurasia (the future continents of North America, Europe, and Asia), from the southern part, called Gondwana (India and the future continents of South America, Africa, Australia, and Antarctica). These two landmasses were fully separated by the end of the Jurassic period. World sea levels began to rise, and much of Laurasia and Gondwana were flooded by sea water, forming shallow inland seas. Because so much of the land was nearer to the oceans, conditions became progressively less arid. Also, the world's climate became even warmer.

The Jurassic period is called the golden age of dinosaurs because of the variety and abundance of dinosaurs that lived during this time, including the largest land animals of all time, the sauropods (*SAWR oh pawdz*). Sauropods, such as *Brachiosaurus* and *Diplodocus* shown in Figure 33-16, were the dominant herbivores of the Jurassic period. All sauropods had enormous barrel-shaped bodies, heavy columnlike legs, and very long necks and tails.

Figure 33-16 Sauropods were giant herbivores that were abundant in the Jurassic period. *Brachiosaurus, left,* stood 12.5 m (41 ft.) tall and was 23 m (75 ft.) in length. It weighed 81,000 kg (89 tons), more than 14 African elephants or 1,500 students. *Diplodocus, right,* was longer and slimmer, reaching 26 m (86 ft.) in length and weighing 10,000 kg (11 tons). *Apatosaurus,* which is commonly but incorrectly known as *Brontosaurus,* is closely related to *Diplodocus.*

The second largest Jurassic herbivores, about the size of a big pickup truck, were the stegosaurs *(STEHG uh sawrz)*. For example, *Stegosaurus*, illustrated in Figure 33-17, weighed about 900 kg (1 ton) and was 4.5 m (15 ft.) long. Stegosaurs fed on plants that grew close to the ground. They had a row of narrow plates along their back and sharp spikes at the tip of their tail.

By the late Jurassic period, very sophisticated carnivorous dinosaurs had evolved. These dinosaurs, known as theropods *(THEHR uh pawdz)*, were descendants of *Herrerasaurus* and preyed on the large herbivorous dinosaurs. Theropods were the dominant terrestrial predators until the dinosaurs disappeared at the end of the Cretaceous period. Figure 33-18 shows three representative theropods, all of which had the typical body structure of this group: bipedal stance, powerful legs, short arms, and a large head. This anatomy was well suited for rapid running and quick, slashing attacks.

Figure 33-17 What was the function of the plates along *Stegosaurus*'s back? Like the spikes on the tail, they may have been defensive weapons. However, the plates were interlaced with numerous blood vessels, which suggests they played a part in heating and cooling the animal.

Figure 33-18 There were three types of theropods. Coelurosaurs *(suh LUHR uh sawrz)* were small and swift with hollow bones. Birds are thought to have descended from this group, which is typified by *Composgnathus, bottom.* Raptors *(RAHP turz)*, typified by *Velociraptor, top,* had lethal, sickle-shaped claws on each foot for ripping open their prey. Most raptors were about the size of a human. The largest theropods were the carnosaurs *(KAHRN uh sawrz)*, typified by *Tyrannosaurus rex, above left.* *T. rex* was 6 m (20 ft.) tall and weighed 7,300 kg (8 tons). Each of its sharp teeth was 15 cm (6 in.) long.

The Cretaceous Period: Triumph of the Chewing Dinosaurs

The Jurassic period ended about 144 million years ago and was followed by the Cretaceous period (144–65 million years ago), a time of profound change for dinosaurs. During the Cretaceous period, Laurasia and Gondwana for the most part split into the continents we know today. Sea levels continued to rise. By the mid-Cretaceous period, sea levels had reached an all-time high, and the interior of North America was a vast inland sea. The climate of much of the world was tropical—hot and wet, like in a greenhouse. Most important, the Cretaceous period saw the rise to dominance of

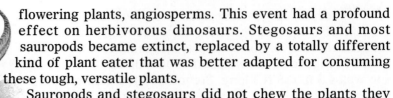

flowering plants, angiosperms. This event had a profound effect on herbivorous dinosaurs. Stegosaurs and most sauropods became extinct, replaced by a totally different kind of plant eater that was better adapted for consuming these tough, versatile plants.

Sauropods and stegosaurs did not chew the plants they ate. Their teeth had no grinding surfaces. Sauropods and stegosaurs shredded leaves and stems and swallowed the shreds whole. Rocks within their stomachs then battered the plant material to a pulp. Though a good strategy with cycads, which have soft, pulpy interiors, this approach didn't work as well with angiosperms, which are much tougher. With the rise of angiosperms during the Cretaceous period, sauropods and stegosaurs were replaced first by iguanodonts (*ih GWAHN uh dawntz*), which had chewing teeth. The jaws of iguanodonts contained enormous grinding teeth that could shred, pound, and grind even the toughest angiosperms. Even bigger than the stegosaurs, *Iguanodon*, the iguanodont illustrated in Figure 33-19, was as heavy as an elephant.

Later in the Cretaceous period, iguanodonts were replaced by three very successful groups of large chewing herbivores. The first group was the duckbill dinosaurs, typified by *Maiasaura*, illustrated in Figure 33-19. Many of these dinosaurs had bony crests on their heads. The second group was the horned dinosaurs, which had horns and bony head frills. *Triceratops*, shown in Figure 33-20, is a well-known member of this group. The third group, the armored dinosaurs, is typified by *Ankylosaurus*, also shown in Figure 33-20. Armored dinosaurs were the most diverse of all dinosaurs.

The flesh-eating theropods of the Cretaceous period were more diverse and formidable than those of the Jurassic. Among the largest was *Tyrannosaurus rex*. Only a few dozen *Tyrannosaurus rex* skeletons have been discovered, indicating that this large predator was probably never common.

Figure 33-19 Early in the Cretaceous period, iguanodonts such as *Iguanodon, top,* fed on angiosperm plants, which were becoming more abundant. Fossils of *Iguanodon* were among the first dinosaurs discovered. Later in the Cretaceous period, duckbill dinosaurs such as *Maiasaura, bottom,* replaced the iguanodonts.

Figure 33-20 Two Cretaceous herbivores were *Triceratops, left,* and *Ankylosaurus, right.*

Extinction of the Dinosaurs

Toward the end of the Cretaceous period, sea levels began to fall and the climate began to cool. Many kinds of dinosaurs became less common, and then suddenly, at the end of the Cretaceous period 65 million years ago, all dinosaurs disappeared. What caused the sudden extinction of the dinosaurs after 170 million years of existence? Most scientists now agree that the most likely cause was the impact of a gigantic meteorite 8–16 km (5–10 mi.) in diameter off the coast of the Yucatan peninsula in Mexico, as illustrated in Figure 33-21. The thin line of sediment that marks the end of the Cretaceous period in rocks is rich in iridium (a mineral rare in the Earth's crust but common in meteorites), tiny spheres of cooled molten rock, and bits of quartz shocked by a high-velocity impact. The impact created a huge crater 300 km (185 mi.) in diameter and threw massive amounts of material into the atmosphere that would have blocked out all sunlight for a considerable period of time, creating a worldwide period of low temperature. The endothermic birds and mammals, insulated with feathers or fur, survived. The ectothermic reptiles and amphibians also survived, because they could simply lower their activity levels.

No one can be sure why the dinosaurs did not live through the deep cold. Disease might have killed them, or massive volcanic eruptions might have led to their extinction. However, the most reasonable and widely accepted explanation is that the cold itself killed them. Most, if not all, Cretaceous dinosaurs appear to have been endothermic. Unlike birds and mammals, however, dinosaurs had no insulation, no way to retain body heat. Endothermy was a great contribution to the success of the dinosaurs, but it created an evolutionary dead end from which they could not emerge.

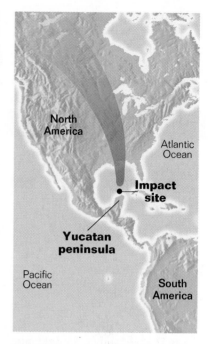

Figure 33-21 A very large impact crater lies off the coast of the Yucatan peninsula in Mexico. The age of this crater coincides with the extinction of the dinosaurs at the end of the Cretaceous period. Scientists have calculated that an impact large enough to create this crater would have thrown enough dust into the atmosphere to drastically reduce the amount of sunlight reaching the Earth's surface.

Section Review

1. Describe two of the characteristics of dinosaurs that gave them an advantage in the late Triassic period.
2. How did the breakup of Pangaea alter the global climate?
3. What effect did the rise to dominance of the angiosperms have on the evolution of dinosaurs?
4. What combination of features made the dinosaurs particularly vulnerable to cold periods?

Critical Thinking

5. Describe two pieces of evidence that would disprove the meteorite-impact hypothesis for the extinction of the dinosaurs.

33-4 History of Birds

Section Objectives

- Recognize the position of *Archaeopteryx in* avian evolution.
- *Identify two characteristics that Archaeopteryx shares with modern birds.*

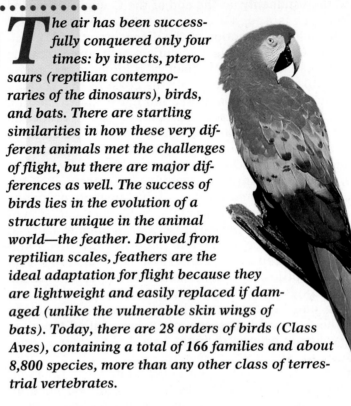

The air has been successfully conquered only four times: by insects, pterosaurs (reptilian contemporaries of the dinosaurs), birds, and bats. There are startling similarities in how these very different animals met the challenges of flight, but there are major differences as well. The success of birds lies in the evolution of a structure unique in the animal world—the feather. Derived from reptilian scales, feathers are the ideal adaptation for flight because they are lightweight and easily replaced if damaged (unlike the vulnerable skin wings of bats). Today, there are 28 orders of birds (Class Aves), containing a total of 166 families and about 8,800 species, more than any other class of terrestrial vertebrates.

Figure 33-22 Macaws, *far right,* are fruit-eating parrots of Central America and South America.

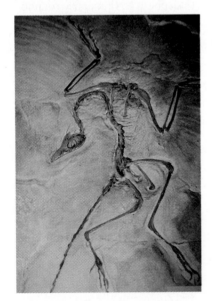

Figure 33-23 Notice the clear impressions of feathers on this specimen of *Archaeopteryx.*

Origin of Birds

The earliest known bird is *Archaeopteryx* (meaning "ancient wing"), which is shown in Figure 33-23. The first specimen of *Archaeopteryx* was found in a limestone quarry in Bavaria (in southern Germany) and is about 150 million years old. *Archaeopteryx* was about the size of a crow and shared many features with small theropod dinosaurs. For example, it had teeth and a long reptilian tail, and very few of its bones were fused to each other. These are features of dinosaurs, not of birds. Also, it had no breastbone such as modern birds have to anchor flight muscles. And unlike the hollow bones of present-day birds, its bones were solid. Finally, it had the forelimbs of a dinosaur. Because of these dinosaurian features, several *Archaeopteryx* fossils were originally classified as *Compsognathus*, a dinosaur of similar size, until impressions of feathers were discovered on the fossils. What makes *Archaeopteryx* distinctly avian is the presence of feathers on its wings and tail. It also had other avian features, notably a wishbone (dinosaurs had no wishbone).

Today almost all biologists agree that *Archaeopteryx* is very closely related to *Compsognathus*. Indeed, some biologists go so far as to classify *Archaeopteryx* and other birds as "feathered dinosaurs" and speak jokingly of "carving the dinosaur" at Thanksgiving dinner. However, most biologists continue to classify birds in a separate class, Aves, because of their key evolutionary novelties: feathers, hollow bones, and physiological mechanisms—such as super-efficient lungs—that permit sustained, powered flight. This judgment should not conceal the agreement among almost all biologists that birds are the direct descendants of theropod dinosaurs.

By the early Cretaceous period, only 15 million years after *Archaeopteryx,* a variety of birds with many of the features of modern birds had evolved. Fossils discovered within the last few years in Mongolia, Spain, and China reveal a diverse collection of toothed birds with the hollow bones and breastbones necessary for sustained flight. Other birds were highly specialized for a flightless, diving existence. The diverse birds of the Cretaceous period shared the skies with pterosaurs for 70 million years.

Since the impressions of feathers are rarely fossilized, and since modern birds have hollow, delicate bones, the fossil record of birds is incomplete. Relationships among the families of modern birds are mostly inferred from studies of the degree of DNA similarity among living birds. These studies suggest that the ostrich and its relatives belong to the oldest group of living birds. Ducks, geese, and other waterfowl arose next, in the early Cretaceous period, followed by a diverse group of woodpeckers, parrots, swifts, and owls. The largest of the bird orders, the Passeriformes, or songbirds (containing 60 percent of present-day bird species), appeared in the mid-Cretaceous period. The more specialized orders of birds, such as shorebirds, birds of prey, flamingos, and penguins, did not evolve until the late Cretaceous period. All but a few of the modern orders of birds are thought to have arisen before the disappearance of the pterosaurs and dinosaurs 65 million years ago. ◻

❏ CAPSULE SUMMARY

The oldest known bird is Archaeopteryx, *which dates from 150 million years ago. Most of the modern orders of birds are thought to have arisen before the extinction of the dinosaurs.*

Section Review

1. *In later editions of* The Origin of Species, *Darwin cited* Archaeopteryx *as an example of a transitional form. Is* Archaeopteryx *a good candidate for a transitional form between reptiles and birds? Explain your answer.*

2. *Explain why most biologists classify birds in a separate class rather than with the reptiles.*

Critical Thinking

3. *What is an adaptive advantage of toothlessness for birds?*

33-5 Characteristics and Diversity of Birds

Modern birds lack teeth and have only a vestigial tail, but they retain many other reptilian characteristics. For instance, birds lay amniotic eggs. Also, reptilian scales are present on the feet and lower legs of birds. What makes birds unique? What traits distinguish them from living reptiles? This section will answer these questions.

Section Objectives

- Describe two functions of feathers.
- Identify two avian adaptations for weight reduction.
- Describe the digestive system of the bald eagle.

Main Characteristics of Birds

Figure 33-24 lists several distinguishing features of birds. Some of these features are explained in more detail following the figure.

In addition, more information on the anatomy and habits of birds is presented in *Up Close: Bald Eagle* on pages 792–793.

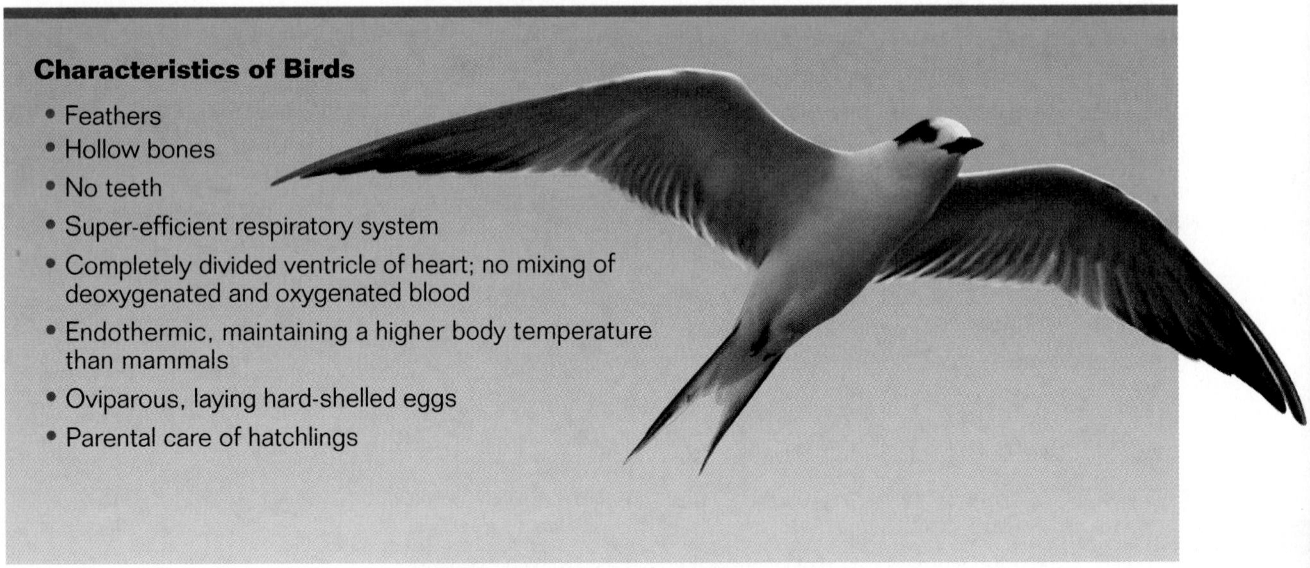

Figure 33-24 This tern is a representative bird.

Characteristics of Birds

- Feathers
- Hollow bones
- No teeth
- Super-efficient respiratory system
- Completely divided ventricle of heart; no mixing of deoxygenated and oxygenated blood
- Endothermic, maintaining a higher body temperature than mammals
- Oviparous, laying hard-shelled eggs
- Parental care of hatchlings

Feathers Insulate and Assist in Flight

Feathers are modified reptilian scales that serve two functions: providing lift for flight and conserving heat. The structure of a feather, illustrated in Figure 33-25, combines maximal flexibility and strength with minimal weight. Feathers develop from tiny pits, called follicles, in the skin. A shaft emerges from the follicle. Pairs of vanes develop from opposite sides of the shaft. At maturity each vane has many branches called barbs. The barbs in turn have many projections called barbules that are equipped with microscopic

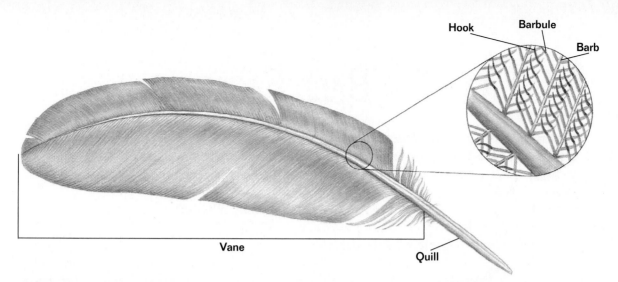

Hook   Barbule   Barb

Vane

Quill

hooks. These hooks link the barbs to one another, giving the feather a continuous surface and a sturdy but flexible shape. Like scales, feathers can be regrown.

Birds Have a Strong, Lightweight Skeleton

The bones of birds are thin and hollow. Many of the bones are fused, making a bird's skeleton more rigid than a reptile's. The fused sections of the backbone and of the pectoral girdle and pelvis form a sturdy frame that anchors muscles during flight. The power for active flight comes from large breast muscles that can make up 30 percent of a bird's total body weight. These muscles stretch from the wing to the breastbone, which is greatly enlarged and bears a prominent keel for muscle attachment, as illustrated in Figure 33-26. Muscles also attach to the fused collarbones that form the so-called wishbone. No other living vertebrates have a fused collarbone or a keeled breastbone.

Figure 33-25 The microscopic structure of a feather helps create a smooth, aerodynamic surface. Interlocking hooks and barbs form a continuous surface.

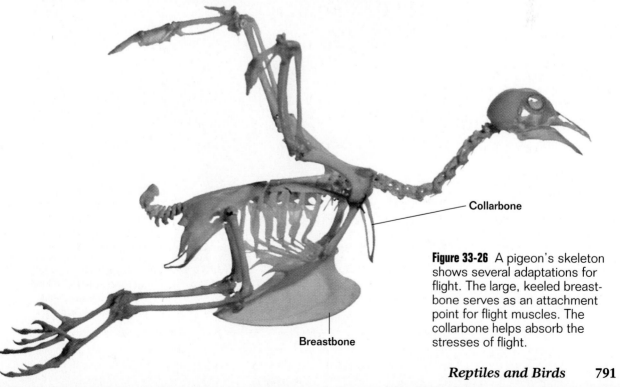

Collarbone

Breastbone

Figure 33-26 A pigeon's skeleton shows several adaptations for flight. The large, keeled breastbone serves as an attachment point for flight muscles. The collarbone helps absorb the stresses of flight.

Reptiles and Birds   **791**

UP CLOSE BALD EAGLE

- ■ **Scientific name:** *Haliaeetus leucocephalus*
- ■ **Range:** Nearly all of North America, from Florida to northern Alaska
- ■ **Habitat:** Forested areas near water that have tall trees for perching and nesting
- ■ **Size:** Wingspan is typically over 2 m (6.5 ft.), and body mass often exceeds 7 kg (15 lb.)
- ■ **Diet:** Fish, small mammals, birds, carrion

External Structures

Eye Vision is the most important sense of the bald eagle. Keen eyesight allows it to see prey at great distances. Its eyes are so large that they occupy most of the space in the head. The bald eagle's visual acuity is 3–4 times higher than yours.

Eye

Feathers

Feathers The body of the bald eagle is covered with feathers everywhere except the feet, which are bare. Both sexes develop the characteristic white head and neck at maturity.

Nostril

Beak

Grasping feet The bald eagle has large feet and talons—the hind claw may be 5 cm (2 in.) long. The talons are used to snatch fish from the water while the eagle is on the wing. When the muscles of the legs contract, the tendons in the lower legs contract, and the talons lock together around the fish.

Grasping feet

Beak The beak is massive, with an elongated, sharp, downward-curving tip. Having no teeth, bald eagles do not chew their food. Instead, they use their beak to tear their prey into manageable portions that are swallowed whole.

Internal Structures

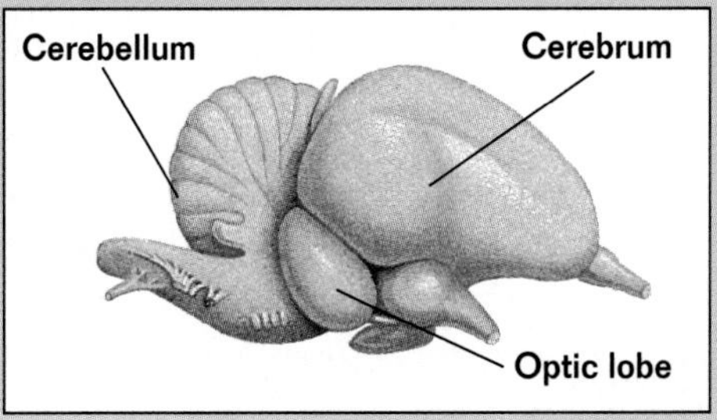

Brain In ratio of brain size to body size, birds rank second among vertebrates, behind only mammals. The large cerebellum receives and integrates information from the muscles, eyes, and inner ears, making possible the precise control of movement and balance necessary for flight. Since the optic lobe processes input from the eagle's most important sense organs—the eyes—it is large. The cerebrum performs many functions, including evaluation of sensory information, control of behavior, and learning.

Cerebellum   Cerebrum

Optic lobe

Brain

Excretory system The bald eagle's excretory system is efficient and lightweight. It does not store waste liquids in a bladder as you do. Your excretory system cannot concentrate the urea that is the byproduct of metabolism because urea is very toxic, so you must expel it in dilute form, urine. The bald eagle (and other birds and reptiles) instead converts its nitrogenous wastes to a nontoxic form called uric acid, which is concentrated into a harmless white paste. Undigested food and uric acid travel to the terminal portion of the gut, the cloaca, and are eliminated.

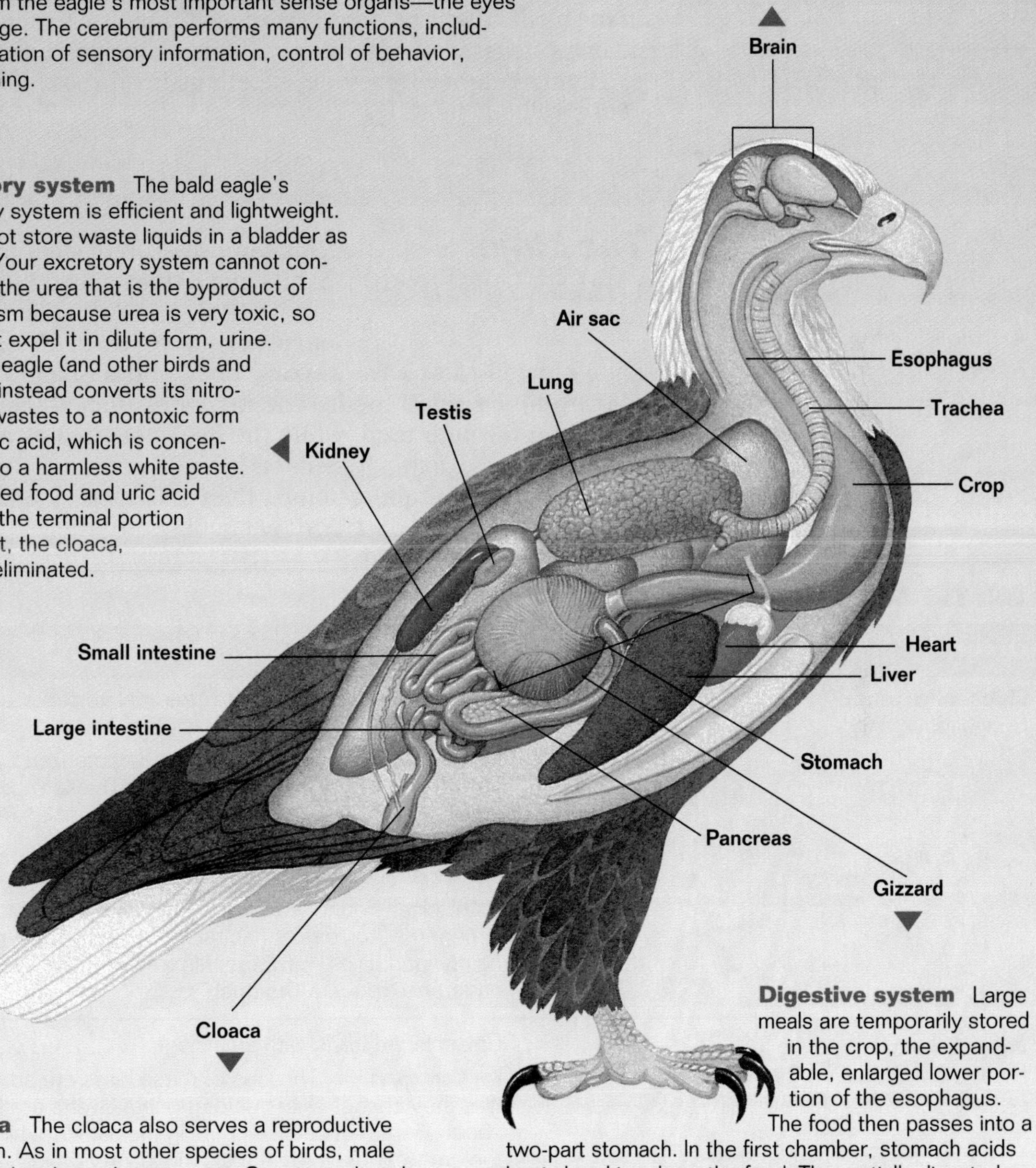

Kidney

Testis

Air sac

Lung

Esophagus

Trachea

Crop

Small intestine

Large intestine

Heart

Liver

Stomach

Pancreas

Gizzard

Cloaca

Cloaca The cloaca also serves a reproductive function. As in most other species of birds, male bald eagles do not have a penis. Sperm produced in the testes pass into the male's cloaca. During mating, the male presses his cloaca against the female's cloaca and releases sperm.

Digestive system Large meals are temporarily stored in the crop, the expandable, enlarged lower portion of the esophagus. The food then passes into a two-part stomach. In the first chamber, stomach acids begin breaking down the food. The partially digested food is then passed to a second chamber, the gizzard, where it is ground and crushed. The gizzard often contains small stones that the bird has swallowed.

Birds Are Endothermic

Birds, like mammals, are endothermic. They generate enough heat through metabolism to maintain a high body temperature. Many paleontologists think the later dinosaurs, from which birds evolved, were endothermic. Birds maintain body temperatures significantly higher than those of most mammals, ranging from 40°C to 42°C (104°F to 108°F). For comparison, your body temperature is 37°C (98°F). The high temperatures maintained by endothermy permit metabolism in the bird's flight muscles to proceed rapidly. A rapid metabolism is necessary to satisfy the large energy requirements of flight. Feathers provide excellent insulation, helping to conserve body heat. ☐

The Major Orders of Birds
....

You can tell a great deal about the habits and diet of a bird by examining its beak and feet. For instance, carnivorous birds such as hawks have curved talons for seizing prey and a sharp beak for tearing apart their meal. The beaks of ducks are flat for shoveling through mud, while the beaks of finches are short and thick for crushing seeds. Of the 28 orders of birds, the dozen orders containing more than 100 species are briefly described in Table 33-1. Pay particular attention to the feet and beaks of these birds.

Table 33-1 Major Orders of Birds

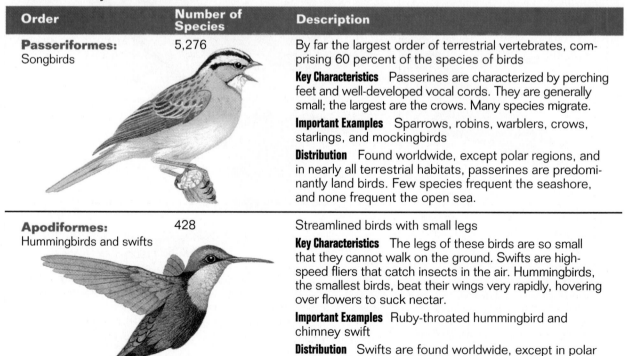

Order	Number of Species	Description
Passeriformes: Songbirds	5,276	By far the largest order of terrestrial vertebrates, comprising 60 percent of the species of birds **Key Characteristics** Passerines are characterized by perching feet and well-developed vocal cords. They are generally small; the largest are the crows. Many species migrate. **Important Examples** Sparrows, robins, warblers, crows, starlings, and mockingbirds **Distribution** Found worldwide, except polar regions, and in nearly all terrestrial habitats, passerines are predominantly land birds. Few species frequent the seashore, and none frequent the open sea.
Apodiformes: Hummingbirds and swifts	428	Streamlined birds with small legs **Key Characteristics** The legs of these birds are so small that they cannot walk on the ground. Swifts are high-speed fliers that catch insects in the air. Hummingbirds, the smallest birds, beat their wings very rapidly, hovering over flowers to suck nectar. **Important Examples** Ruby-throated hummingbird and chimney swift **Distribution** Swifts are found worldwide, except in polar regions. Hummingbirds live only in the Western Hemisphere.

Order	Number of Species	Description
Piciformes: Woodpeckers	383	Birds with highly specialized beaks

Key Characteristics These are forest birds that nest in holes in trees and have grasping feet. Many have strong, chisel-like beaks and feed on insects they dig from under bark by pounding into it with their beaks. Most species are solitary and do not migrate.

Important Examples Woodpeckers, toucans, and honeyguides

Distribution Woodpeckers occur on all continents except Australia; toucans and honeyguides are tropical.

Psittaciformes: Parrots — 340

Brightly colored vegetarians of the tropics

Key Characteristics All members of this distinctive order are similar, with short necks, compact bodies, and short, stout, hooked bills. All are vegetarians and are very fond of fruits and berries. They nest in holes in trees and are very gregarious. In captivity, some can be taught to imitate human speech.

Important Examples Parrots, parakeets, cockatoos, and cockatiels

Distribution Mainly tropical, but also New Zealand and temperate areas of Australia

---

Charadriiformes: Shorebirds — 331

A varied group of shorebirds

Key Characteristics Plovers and sandpipers have long legs and long, slender bills used to probe for marine animals in the sand. Terns and gulls have streamlined, pointed wings and fly over water in search of fish. Most shorebirds are very gregarious, and many migrate long distances.

Important Examples Terns, gulls, plovers, and sandpipers

Distribution Marshes and seashores throughout the world

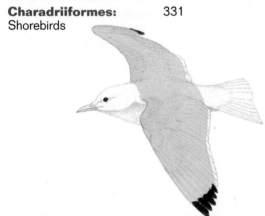

Columbiformes: Pigeons — 303

Pigeons and doves

Key Characteristics These are compact, plump birds with small heads, short beaks, perching feet, and soft, dense plumage. Most are strong fliers and gather in flocks. They feed on seeds, fruit, and berries. They vary greatly in size, from smaller than a sparrow to the size of a small turkey.

Important Examples Mourning dove and common pigeon

Distribution Worldwide, except polar regions

Order	Number of Species	Description
Falconiformes: Birds of prey	288	Day-active carnivores, the lions of the air **Key Characteristics** Strong fliers with keen vision, they have curved, pointed beaks for tearing flesh and powerful, curved talons for seizing prey. Birds of prey range in size from small, insect-eating sparrow hawks to the South American harpy eagle, which has a wingspan of 3 m (10 ft.). **Important Examples** Eagles, hawks, falcons, ospreys, and vultures **Distribution** Worldwide

Order	Number of Species	Description
Galliformes: Fowl	268	Small, rapid-running ground feeders **Key Characteristics** Sometimes called "gamebirds" because many of them are hunted for sport, these birds typically have rounded bodies and often only limited flying ability. Most fowl are browsers or grazers and live in forests and fields. **Important Examples** Quail, grouse, turkeys, partridges, pheasants, and domestic chickens **Distribution** Worldwide

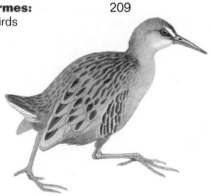

Order	Number of Species	Description
Gruiformes: Marsh birds	209	Ground-living marsh dwellers **Key Characteristics** Marsh birds have small bodies and long bills. Rails are small, have long legs, and bob their heads as they walk. Many rails are nocturnal, solitary, and rarely fly. Cranes are the exact opposite—powerful fliers that mate for life and often migrate long distances. **Important Examples** Rails, coots, and cranes **Distribution** Worldwide

Order	Number of Species	Description
Anseriformes: Waterfowl	150	Aquatic diving birds **Key Characteristics** Waterfowl have long necks and blunt, flat tails. Three of the toes are linked by webs for improved swimming. Flight feathers are molted after the breeding season, and individuals pass through a flightless period of up to a month. Three-quarters of the species are the so-called dabbling ducks—including mallards, teals, and pintails—which tip down to feed on vegetation a few feet below the water's surface. Waterfowl migrate seasonally along established flyways. **Important Examples** Ducks, geese, and swans; all domestic ducks are descended from the mallard **Distribution** Worldwide

Order	Number of Species	Description
Strigiformes: Owls	146	Nocturnal carnivores; the nighttime equivalents of birds of prey **Key Characteristics** Owls have soft plumes, short tails, and large heads with hooked bills and large eyes that are directed forward. Owls have extraordinarily good hearing and excellent night vision. They cannot see in absolute darkness, but they have a very large number of light-sensitive rods in the retinas of their eyes that provide exceptional vision in poor light, such as moonlight. **Important Examples** Barn owl, great horned owl, and snowy owl **Distribution** Worldwide

Order	Number of Species	Description
Ciconiiformes: Herons	114	Long-legged waders **Key Characteristics** Often confused with the marsh birds, the members of this order have large bodies and long necks. Herons have long, sharp bills that they use to spear or grasp fish or other prey. **Important Examples** Herons, bitterns, egrets, and flamingos **Distribution** Worldwide, except polar regions

Section Review

1. *Could a bird fly without feathers? Explain your answer.*
2. *Describe two skeletal features of birds that help reduce weight.*
3. *What is the function of the gizzard? Explain why this function is more critical for a seed-eating bird than for a carnivorous bird.*

Critical Thinking

4. *In the majority of species of birds, both parents care for the young. In contrast, only the mother provides parental care in most mammalian species. Propose a hypothesis to explain this difference. How would you test your hypothesis?*

Vocabulary

anapsid (773)
carapace (776)
diapsid (773)
Pangaea (783)
plastron (776)
synapsid (773)

Review

Multiple Choice

1. The skulls of the earliest reptiles are most similar to those of
   a. crocodiles.   **c.** snakes.
   b. turtles.   **d.** lizards.

2. Which group of living reptiles is most closely related to birds?
   a. snakes
   b. turtles
   c. rhynchocephalians
   d. crocodiles

3. Snakes are different from lizards because snakes do not have
   a. a tail.
   b. reproductive organs.
   c. a pectoral girdle.
   d. a loosely connected lower jaw.

4. Compared with other reptiles, crocodiles
   a. are harmless and tame.
   b. provide better care for their young.
   c. have weak jaw muscles.
   d. can tolerate drier conditions.

5. The oldest known dinosaur is
   a. *Brachiosaurus.*
   b. *Dimetrodon.*
   c. *Eoraptor.*
   d. *Herrerasaurus.*

6. Which factor did *not* contribute to the success of dinosaurs?
   a. the ability to live in arid conditions
   b. legs positioned directly under the dinosaur body
   c. a meteorite striking the Earth at the end of the Triassic period
   d. the break up of Pangaea

7. What evidence best supports the hypothesis that the impact of a meteorite caused the extinction of the dinosaurs?
   a. fossils
   b. iridium sediments
   c. continental drift
   d. plant evolution

8. What evidence convinced scientists that fossils once classified as *Compsognathus* should be reclassified as *Archaeopteryx*?
   a. feathers
   b. fused bones
   c. teeth
   d. hard-shelled eggs

9. The fossil record of birds is incomplete because
   a. reptiles gave rise to birds.
   b. birds have thin, hollow bones.
   c. mammals prey on birds.
   d. bird cells do not contain carbon.

10. What structure has taken over the function of teeth in birds?
    a. the beak   **c.** the gizzard
    b. the crop   **d.** the claws

11. The feathers of most birds are well adapted for
    a. swimming and repelling water.
    b. flying and insulating.
    c. flying and conducting.
    d. expelling heat and feeding.

Completion

12. Like dinosaurs, _____ evolved from thecodonts. They are aggressive carnivores and, unlike other reptiles, care for their _____ .

13. Living reptiles that have diapsid skulls include tuataras and _____ . These reptiles are thought to have appeared more recently than reptiles with _____ skulls, such as turtles.

14. Turtles and tortoises are members of the order _____ . Their bodies are covered with a shell that consists of two parts: the carapace and the _____ .

15. Herbivores of the Cretaceous period include the horned _____ and the duck-billed *Maiasaura*. A very large carnivore of the same period is _____ .

16. Due to their _____ rate of metabolism, birds probably eat a _____ amount of food for their size.

17. The feathers and _____ of birds are strong yet _____ . These adaptations are important for flight.

Themes Review

18. **Evolution** Coevolution occurs when interacting groups of organisms evolve together. How is the evolution of modern angiosperms and herbivorous dinosaurs an example of coevolution?

19. **Homeostasis** The plates along *Stegosaurus*'s back contained many blood vessels. How did these blood vessels help *Stegosaurus* maintain a constant body temperature?

20. **Structure and Function** The young of birds and reptiles develop in an amniotic egg. Name the four membranes contained in an amniotic egg. Describe how they help make the egg an independent life-support system.

Critical Thinking

21. **Designing an Experiment** Rattlesnakes and other pit vipers have special sensory pits on each side of the head near the nostril. Scientists think that these sensory pits help snakes locate and bite objects that have a higher temperature than the surrounding environment. Design an experiment to test this hypothesis.

22. **Making Predictions** A hummingbird has a four-chambered heart. How would having a three-chambered heart, like that of most reptiles, affect a hummingbird in flight?

23. **Interpreting Data** The graph below shows the body temperature range of the members of the major groups of reptiles.

    a. Which two groups show a much narrower range of temperatures than the other groups? What might account for this? (Hint: Consider the number of different species in each group.)

    b. Which groups do you think could best tolerate the temperature extremes found in deserts?

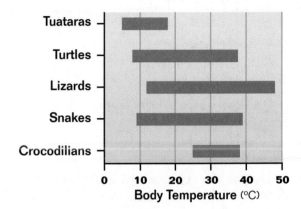

Activities and Projects

24. **Cooperative Group Project** Organize a bird-watching trip. Have different members of your group visit different ecosystems in your area, such as woods, lakes, apartment houses, and fields. Using a field guide to birds, identify as many birds in your ecosystem as you can. Make a list of the different species of birds you see, and record the number of times you see each species. Share your findings with your class.

25. **Multicultural Perspective** Different cultures have different attitudes toward snakes. Research the important cultural roles of snakes in India and Africa. Write a report that summarizes what you have learned.

Chicken Anatomy

OBJECTIVE

Analyze how the muscular and skeletal systems of a bird are adapted for flight.

PROCESS SKILLS

- observing the skeletal and muscular structure of an organism
- identifying anatomical features

MATERIALS

- disposable gloves
- whole, fresh chicken
- dissecting pan
- scalpel, scissors, blunt probe

BACKGROUND

1. How is the structure of a bird typical of all vertebrates?
2. How are hollow bones and fused bones adaptations for flight?
3. What muscles of a bird are especially large, allowing flight?
4. Write your own **Focus Question** on your Vee Form.
5. **Knowing Side of the Vee** List the **Concepts** and new **Vocabulary Words** on your Vee Form. In the **Concept Statements** section of the Vee, use these words in sentences that define and explain them.

TECHNIQUE

Doing Side of the Vee

1. Put on disposable gloves. **CAUTION: Chicken is sometimes contaminated with *Salmonella* bacteria. Wear gloves throughout this investigation and keep your hands away from your face and mouth.**

2. **CAUTION: Be careful not to cut yourself with the scalpel or scissors.** Use the scalpel and scissors to remove the skin from the chicken, exposing as many muscles as possible. Record your observations in the **Records** section of your Vee Form.

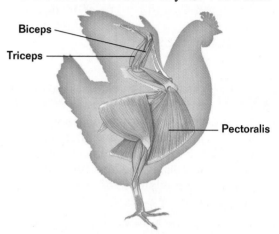

3. Compare the size and color of the muscles of the wing with those of the leg. How do they differ?
4. Compare the muscles on the back and on the chest. How do they differ?
5. The main muscle of the chest is the pectoralis. This muscle originates on the breastbone and the fused clavicle bones, or "wishbone," and attaches to the underside of the humerus at some distance from the shoulder joint. When this muscle contracts, it pulls the wing down. Use the blunt probe to pull on the pectoralis.
6. Cut the pectoralis from the humerus and peel back the muscle. This will expose the supracoracoideus muscle. It originates on the coracoid and the breastbone and attaches on the upper side of the humerus. Pull on this muscle to find out how it functions.
7. Identify the biceps and triceps of the bird. How do these muscles function?
8. Use the scissors and scalpel to remove muscles from the breast, back, and thigh of the chicken.

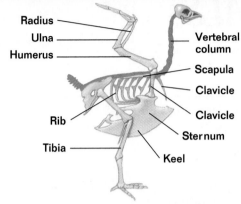

Radius
Ulna
Humerus
Vertebral column
Scapula
Clavicle
Clavicle
Sternum
Rib
Tibia
Keel

9. Examine the vertebral column of the chicken, which is shown above in pink. Unlike most other vertebrates, birds have a rigid backbone. This provides support for flight. In contrast, the relatively long neck is flexible, allowing a bird to use its beak as a tool.

10. The ribs come together in a large shield-shaped sternum, or breastbone. The backbone, ribs, and sternum together form a flexible but strong box that houses the heart, lungs, and visceral organs. Find the sternum and the keel that extends from it. The pectoralis and supracoracoideus attach to the keel.

11. Follow the sternum up to the coracoid. The coracoid, scapula (shoulder blade), and clavicle (collarbone) form the shoulder. The clavicles, one from each shoulder, join together in the front to form the "wishbone." Find these bones.

12. Next find the femur, or thighbone. Break the thighbone and examine the broken ends. Notice that it is not solid, but is instead filled with air pockets. In the lower leg, the tibia is the main bone and the fibula is reduced to a splint. Because they have only one bone in this part of the leg, birds cannot twist their legs or step sideways; they can only walk forward or backward.

13. In the **Procedure** section of the Vee, briefly summarize the procedure you followed. In the **Additional Records**

section of your Vee, sketch the bones and muscles you have located in this lab.

14. Follow your teacher's instructions to dispose of the chicken and clean up your materials. Before leaving the lab, use soap and water to thoroughly wash your hands and any surfaces touched by the chicken.

INQUIRY

1. How does the color of the wing muscles differ from that of the leg muscles?

2. What is the chief function of the chest muscles?

3. Use the information on the **Knowing Side** of the Vee to interpret your results from the **Doing Side,** and then write your **Knowledge Claim.** Write a **Value Claim** for this lab.

ANALYSIS

1. What adaptations for flight did you find in the chicken skeleton and musculature?

2. Muscles that have a large blood supply are darker than those that have a smaller amount of blood flow. Generally, those muscles that are the most active have the greatest supply of blood. Relate these facts to your observations during this investigation.

3. The keel increases the area avavilable for the attachment of flight muscles. Compared with that of other birds, the keel of the chicken is small. Why?

FURTHER INQUIRY

Write a **New Focus Question** that could be the point of a new investigation. The following is an example:

How do the structures of flightless birds and songbirds differ?

CHAPTER
34

MAMMALS

REVIEW

- continental drift (Sections 13-3 and 33-3)
- cladogram (Section 19-4)
- mammalian characteristics (Section 31-1)
- endothermy (Section 31-1)
- differences among marsupials, monotremes, and placental mammals (Section 31-1)
- structure of mammalian lungs (Section 31-2)
- structure of mammalian heart (Section 31-3)
- viviparous and oviparous reproduction (Section 31-5)
- *Highlights: Therapsids* on page 734

Giraffes in Africa

34-1 History of Mammals

*T*oday almost all large, land-dwelling vertebrates are mammals, and they therefore tend to dominate terrestrial communities. When you look out over an African plain, for example, you see the big mammals—the lions, elephants, antelopes, and zebras. Your eye does not as readily pick out the many birds, lizards, snakes, and frogs that also live there. But only within the last 65 million years have mammals become dominant. For most of their history, mammals were small animals living in the shadow of the dinosaurs. In this section you will learn about the history of mammals (class Mammalia).

Section Objectives

- Recognize the role of pelycosaurs in mammalian evolution.
- Describe three similarities between therapsids and mammals.
- Identify three characteristics of the earliest mammals.

Ancestors of Mammals

The first step in the evolutionary journey leading to mammals occurred in the warm, moist tropical forests that covered most of the globe 300 million years ago. This step was the evolution of reptiles known as pelycosaurs *(PEHL uh kuh SAWRZ)*, including *Archaeothyris*, shown in Figure 34-1. Early pelycosaurs, which were small and lizardlike, seem very different from mammals. However, both groups have the same type of skull, the synapsid skull. Recall from Chapter 33 that a synapsid skull has a single opening behind each eye socket. This arrangement enables larger, stronger muscles to attach to the jaw, increasing biting power. In their powerful jaws, pelycosaurs had teeth differing in size and shape, another characteristic shared with mammals. The degree of difference between teeth in pelycosaurs is much less than in mammals, however. For 40 million years, over 70 percent of the land vertebrates were pelycosaurs. Some were large herbivores; others, such as *Dimetrodon*, which you saw in Chapter 31, were ferocious predators.

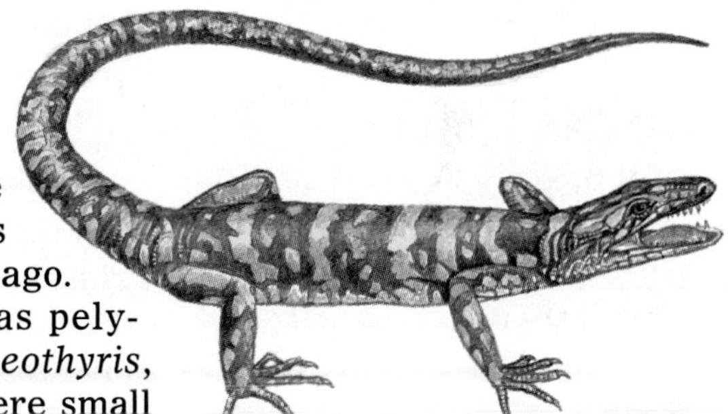

About 260 million years ago, the therapsids *(thuh RAP sihdz)* evolved from and replaced the pelycosaurs. As you learned in Chapter 31, therapsids, such as *Cynognathus* illustrated in Figure 34-2, are the direct ancestors of mammals. Therapsids had

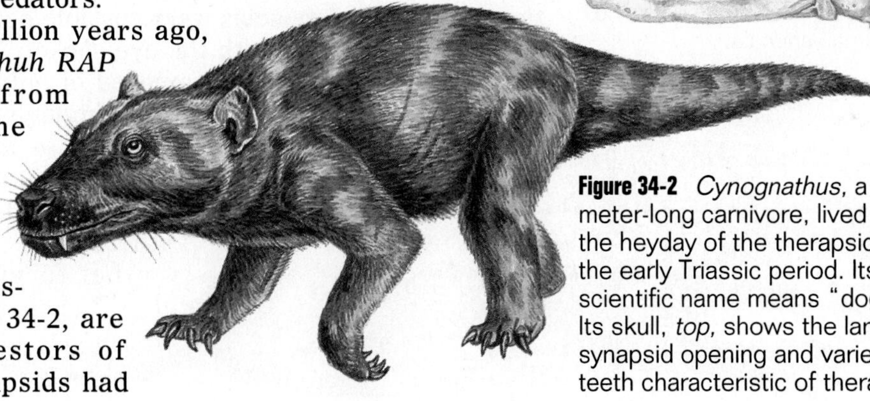

Figure 34-1 *Archaeothyris* is the oldest known pelycosaur. About 50 cm (20 in.) from nose to tail, it preyed on other early reptiles.

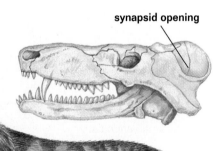

synapsid opening

Figure 34-2 *Cynognathus,* a meter-long carnivore, lived during the heyday of the therapsids in the early Triassic period. Its scientific name means "dog jaw." Its skull, *top*, shows the large synapsid opening and varied teeth characteristic of therapsids.

several adaptations for more efficient feeding, as evidenced by the skull of *Cynognathus*. Therapsid teeth, for instance, are much more diverse than the teeth of pelycosaurs. In addition to sharp, stabbing teeth, some therapsids had teeth with complex surfaces for crushing, grinding, or tearing food. The secondary palate, a sheet of bone that separates the mouth cavity from the nasal passages, is another adaptation that appears in therapsids. An animal with a secondary palate can breathe and chew simultaneously. (Your secondary palate is the roof of your mouth.) In therapsid skulls, the synapsid opening is much larger, making room for even bigger jaw muscles. These three adaptations—complex teeth, secondary palate, and large synapsid openings—are also characteristic of mammals.

Therapsids were active, and some were probably endothermic. How is it possible to know the metabolic state of these animals without examining a living specimen? For one thing, the therapsid adaptations for efficient feeding suggest animals with a high demand for energy. In addition, therapsid fossils have been found in areas that had cold winters in the Permian period. Evidence also suggests that at least some therapsids had fur to insulate their endothermic bodies.

Therapsids were the dominant terrestrial animals from the middle of the Permian period until the middle of the Triassic period. As climates warmed in the mid-Triassic period, however, therapsids entered a period of decline, perhaps because of competition from thecodonts. It was during this decline that the mammals evolved. Therapsids became extinct about 150 million years ago. ▫

▫ CAPSULE SUMMARY

Therapsids were the direct ancestors of mammals. Like mammals, they had diverse teeth and a secondary palate. At least some therapsids may have been endothermic.

Origin of Mammals

The first mammals appeared about 220 million years ago, just as the first dinosaurs were evolving from thecodonts. These early mammals, such as *Eozostrodon* illustrated in Figure 34-3, were small (about the size of mice), shrewlike, insectivorous tree dwellers. Their relatively large eye sockets indicate they were active at night. The early mammals had a number of improvements on the therapsid feeding adaptations. Larger jaw muscles were possible because the synapsid opening had merged with the eye socket. The jaw itself was stronger because it was composed of only one bone, rather than several as in therapsids. The single jawbone is one of the defining features of mammals. Paleontologists use it to distinguish between fossils of mammals and therapsids. Two of the bones that formed the therapsid

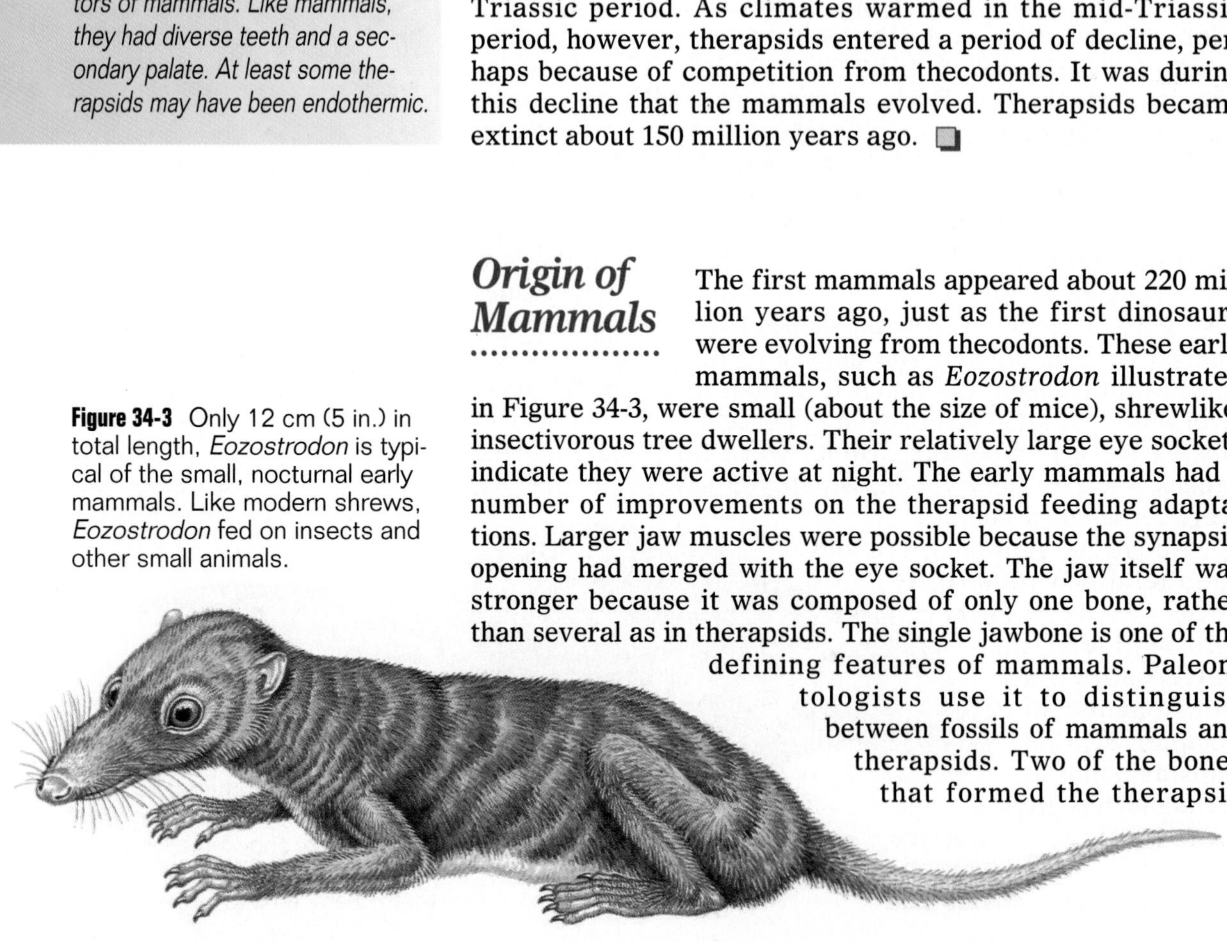

Figure 34-3 Only 12 cm (5 in.) in total length, *Eozostrodon* is typical of the small, nocturnal early mammals. Like modern shrews, *Eozostrodon* fed on insects and other small animals.

jaw joint became part of the chain of tiny bones that transmits sound in the middle ear of mammals.

For 155 million years, while the dinosaurs flourished, mammals were a minor group that changed little. Only five orders of mammals arose in that time, and their fossils are scarce, indicating that mammals were not abundant. However, the two groups to which present-day mammals belong did appear. The most primitive mammals, direct descendants of therapsids, were members of the subclass Prototheria *(PROH toh THIHR ee uh)*. Most prototherians were small and resembled modern shrews. All prototherians laid eggs, as did their reptilian ancestors. The only prototherians surviving today are the monotremes—the duckbill platypus and the echidnas, or spiny anteaters. The other major mammalian group is the subclass Theria *(THIHR ee uh)*. All of the mammals you are familiar with, including humans, are therians. Therians are viviparous. The two major therian groups are marsupials, or pouched mammals, and placental mammals. Kangaroos, opossums, and koalas are marsupials. Dogs, cats, humans, horses, and most other mammals are placentals. You will learn more about these groups later in the chapter.

The Age of Mammals

At the end of the Cretaceous period, 65 million years ago, the dinosaurs and numerous other land and marine animals became extinct, but mammals survived. In the Tertiary period (lasting from 65 million years ago to 2 million years ago), mammals rapidly diversified, taking over many of the ecological roles once dominated by dinosaurs. Mammals reached their maximal diversity late in the Tertiary period, about 15 million years ago. At that time, tropical conditions existed over much of the world. During the last 15 million years, world climates have deteriorated, and the area covered by tropical habitats has decreased, causing a decline in the total number of mammalian species.

Section Review

1. *Identify two mammalian features shown by pelycosaurs.*
2. *Name two features you share with therapsids.*
3. *How could you distinguish between a therapsid fossil and a fossil of an early mammal?*

Critical Thinking
4. *How might mammalian evolution have been different if the dinosaurs had not become extinct?*

34-2 Key Mammalian Adaptations

34-2

*O*ne hundred million years ago dinosaurs were the large land animals, and mammals were small, insectivorous, and nocturnal. No mammal was larger than a cat until after the dinosaurs became extinct. Today, mammals have succeeded dinosaurs as the dominant land animals. Dinosaurs were successful in diverse habitats because they had adaptations such as horns, claws, expandable jaws, and specialized kinds of teeth. In this section you will explore a few of the adaptations that contributed to the success of mammals.

Section Objectives

■ Identify three functions of hair.

■ Contrast parental care in reptiles and mammals.

■ Describe two hunting adaptations of the Siberian tiger.

■ Compare patterns of reproduction in monotremes, marsupials, and placental mammals.

Hair Is a Unique Mammalian Feature

Mammals are hairy. Even whales and dolphins, which appear to be hairless, have a few sensitive bristles on the snout. No other living animals have hair. A hair is a filament composed mainly of dead cells filled with the protein keratin. Each hair is anchored in and produced by a bulb-shaped structure, the hair follicle, which lies beneath the surface of the skin. The evolutionary origin of hair is unknown, but it is probably not derived from reptilian scales.

The primary function of hair is insulation. Mammals, such as the polar bear in Figure 34-4, tend to lose body heat because they typically maintain body temperatures higher than the temperature of their surroundings. Most mammals are covered with a dense coat of hair that reduces the amount of body heat escaping into the environment. We humans need clothes in most climates because our hair is too sparse to be adequate insulation.

Hair has many functions besides insulation. The coloration and pattern of a mammal's coat often make very effective camouflage. A little brown mouse is practically invisible against the brown leaf litter of the forest floor, and the orange and black stripes of a Bengal tiger cause it to blend in with the tall, orange-brown grass in which it hunts. A mammal's coat may also be a conspicuous signal. The black and white fur of a skunk, for instance, warns would-be predators to stay away.

In some mammals, hairs serve a sensory function. The whiskers of cats and dogs are stiff hairs that are very sensitive to touch. Mammals that are active at night or that live underground often rely on their whiskers for information about the environment. Hair can also be a defensive weapon. Porcupines and hedgehogs are protected by long, sharp, stiff hairs called quills.

Figure 34-4 A thick coat of fur, coupled with a layer of blubber beneath the skin, insulates this polar bear in its frigid arctic habitat. The coloration of the coat helps conceal the bear as it stalks seals, its main prey.

Nursing and Caring for the Young

Female mammals have mammary glands, the unique feature for which the class Mammalia was named. The word *mammary* is derived from the Latin word *mamma*, meaning "breast." These glands, located on the chest or abdomen, produce milk. Newborn mammals grow rapidly, and milk provides the nutrition to support this growth. Milk is rich in protein, carbohydrates (chiefly the sugar lactose), and fat, which accounts for 50 percent of the energy in milk. It also contains water to prevent dehydration and contains minerals, such as calcium, that are critical to early growth. Young mammals are nourished on milk from birth (or hatching, in the case of monotremes) until weaning, when the mother stops nursing them.

Figure 34-5 By watching their mother stalk and capture prey, these lion cubs will learn to hunt. Though weaned at the age of one year, the cubs will not be skilled enough to survive on their own until they are about three years old.

Reptiles seldom provide parental care to their offspring. Typically, a female reptile buries and leaves her eggs. The hatchlings must fend for themselves. In contrast, mammalian young are dependent on parental care for a relatively long period, receiving milk and other food, protection, and shelter. Like the lion cubs shown in Figure 34-5, young mammals often learn necessary skills, such as hunting, during their period of dependence. Learning is especially important for primates, which have particularly long periods of dependence.

You can read more about the habits and anatomy of mammals in *Up Close: Siberian Tiger* on pages 808–809.

UP CLOSE SIBERIAN TIGER

- **Scientific name:** *Panthera tigris altaica*
- **Range:** Far northeastern China and southeastern Russia
- **Habitat:** Undisturbed forests with ample cover; fewer than 200 Siberian tigers remain in the wild
- **Size:** Up to 300 kg (660 lb.) and 3 m (10 ft.) from nose to tip of tail
- **Diet:** Large mammals such as wild boars and red deer

External Structures

Fur A Siberian tiger has dense fur colored reddish orange with dark vertical stripes; the underparts are creamy white. The striped pattern acts as effective camouflage in high grass or dense, dry forest.

Senses Tigers hunt primarily at night. They do not have as acute a sense of smell as dogs and other carnivores do. To locate prey, tigers instead rely on excellent vision—focused on detecting motion—and sensitive hearing.

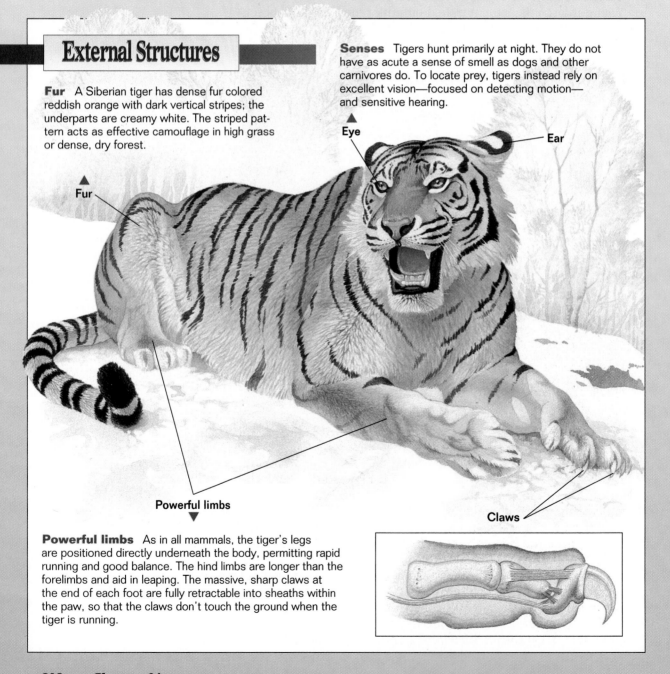

Fur

Eye

Ear

Powerful limbs

Claws

Powerful limbs As in all mammals, the tiger's legs are positioned directly underneath the body, permitting rapid running and good balance. The hind limbs are longer than the forelimbs and aid in leaping. The massive, sharp claws at the end of each foot are fully retractable into sheaths within the paw, so that the claws don't touch the ground when the tiger is running.

Internal Structures

Reproductive system Like all placental mammals, Siberian tigers have internal fertilization and nourish their embryos through the placenta. Among wild Siberian tigers, reproduction occurs only every 4–6 years. Litters are small, only 2 or 3 cubs, each weighing only about 1 kg (2 lb.) at birth. Born blind and helpless, the cubs open their eyes after 6–12 days and nurse for about 6 months. The mother guards the cubs as they grow and teaches them to hunt. The cubs will hunt with their mother for over 2 years before striking off on their own.

Skull The skull is particularly heavy and anchors massive jaw muscles. The tiger's head is large and foreshortened, providing leverage for its powerful jaws. In the lower jaw there is a gap between the canines and the premolars, allowing the long, sharp upper canines to pierce deeply into prey during a bite. Tigers usually kill their prey by biting its upper neck and then using their powerful muscles to twist and snap its neck.

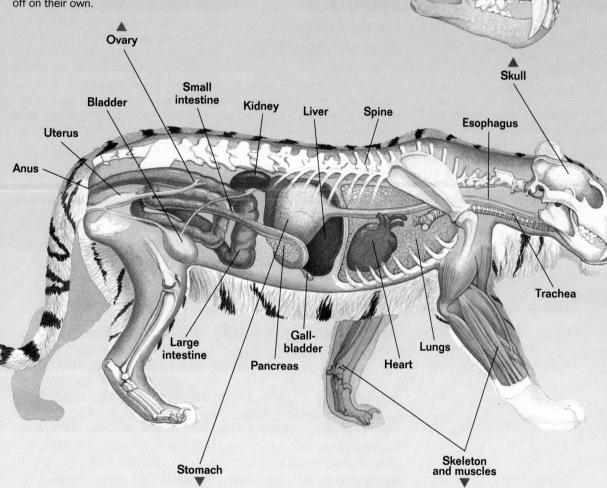

Ovary

Bladder

Uterus

Anus

Small intestine

Kidney

Liver

Spine

Skull

Esophagus

Trachea

Large intestine

Pancreas

Gall-bladder

Heart

Lungs

Skeleton and muscles

Stomach

Digestive system A tiger is pure carnivore. Because a tiger's largely protein diet is more easily digested than grass, its digestive system is much shorter than that of a herbivore. A tiger in a zoo will eat 6 kg (13 lb.) of meat per day. In the wild, tigers eat far less frequently but eat more—up to 40 kg (90 lb.)—in each meal.

Skeleton and muscles The skeletal system of a tiger, like that of all mammals, is made of bone. A tiger's body is compact and has the most powerful muscular system of any cat.

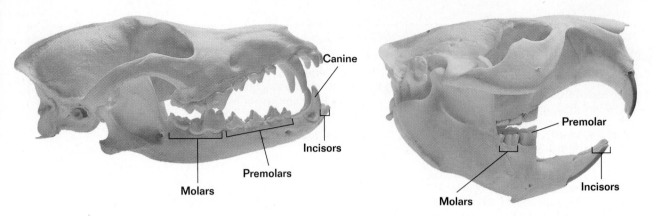

Figures labeled: Canine, Incisors, Premolars, Molars (coyote skull, left); Premolar, Incisors, Molars (beaver skull, right)

Figure 34-6 The coyote's skull, *above left,* reveals it to be a predator. Long canine teeth enable it to bite and hold prey; it shears off chunks of flesh with its sharp, triangular premolars and molars. The beaver's skull, *above right,* shows that it is a herbivore that gnaws with its incisors. Gnawing causes the back surface of the incisor to wear away faster than the front surface, maintaining the tooth's sharp, chisel-like edge. The incisors continue to grow throughout the beaver's life. The surfaces of its premolars and molars show a complex pattern of ridges and grooves that forms a grinding surface.

Mammals Have Diverse, Specialized Teeth

You need only look at your teeth in a mirror to confirm that mammals have teeth of different sizes and shapes. In most mammals, four types of teeth can be recognized: incisors, canines, premolars, and molars. Each type of tooth performs a different function in eating. Incisors, the front teeth, are for biting and cutting. Behind them are canines for stabbing and holding. Along the cheeks are the premolars and molars, which grind, crush, or cut. Unlike dinosaurs, whose teeth were constantly being lost and replaced, an adult mammal keeps its teeth all its life.

The teeth of a mammal are highly specialized to match the food it eats. In fact, it is usually possible to determine a mammal's diet just by examining its teeth. For example, notice the differences between the teeth of a coyote (a carnivore) and those of a beaver (a herbivore), shown in Figure 34-6.

Maintaining a High Body Temperature

Like birds, therapsids, and some dinosaurs, mammals are endothermic. Recall from Chapter 31 that endothermic animals keep their body temperature high and nearly constant by producing large amounts of heat through metabolism. Because their body temperature remains steady regardless of the temperature of the surroundings, mammals can be active at any time of day or night and can inhabit many severe environments not habitable by ectothermic reptiles and amphibians. In addition, endotherms are capable of sustained activity, such as running or flying long distances.

To maintain the high metabolic rate necessary for endothermy, a mammal must eat about 10 times as much food as an ectotherm of similar size. Endothermy also requires rapid delivery of oxygen to metabolizing tissues. The mammalian respiratory and circulatory systems are very efficient at acquiring and distributing oxygen.

In Chapter 31 you learned that mammals have a four-chambered heart in which oxygenated and deoxygenated blood never mix. Only oxygen-rich blood is delivered to the tissues. Recall also that mammalian lungs, because of their large internal surface area, are much more effective at absorbing oxygen than are reptilian and amphibian lungs. Efficient respiration in mammals is aided by the **diaphragm,** a sheet of muscle at the bottom of the rib cage. Contraction of the diaphragm helps draw air into the lungs.

Three Styles of Reproduction
....

In the previous section you learned that present-day mammals are divided into three groups: mono-tremes, marsupials, and placental mammals. These three groups differ in their manner of reproduction. Monotremes are oviparous, laying shelled eggs in which yolk nourishes the embryo. After hatching, a young monotreme feeds on its mother's milk for several months.

Placental mammals are viviparous. The key feature of embryonic development in placental mammals is the placenta, which you studied in Chapter 31. Recall that the placenta brings the bloodstreams of the mother and offspring into close proximity. Nutrients and oxygen pass from mother to offspring; wastes pass in the opposite direction and are eliminated by the mother's excretory system. Nourished by nutrients from the placenta, the offspring remains in the uterus until development is essentially complete. The length of time between fertilization and birth is the **gestation period** (jeh STAY shuhn). Compared with marsupials, placental mammals have long gestation periods.

Marsupials, such as the Virginia opossum shown in Figure 34-7, are also viviparous, but only a few species have a placenta. Instead, much of the embryonic marsupial's nutrition is provided by a rich fluid secreted by the uterus. The gestation period in marsupials is very short—as little as eight days in some species. A newborn marsupial is tiny, hairless, and

Figure 34-7 Young Virginia opossums, *left*, ride on their mother's back. A newborn opossum, *above*, is smaller than your thumb. It must crawl through the mother's fur to reach her pouch, where the nipples are located.

CONTENT LINK

Chapter 42 further describes the structure and function of the placenta.

incompletely developed, as you can see in Figure 34-7. Without any help from its mother, the newborn struggles to the nipples, which are usually located in a pouch on the mother's abdomen. It attaches to a nipple and continues its growth and development, which may take several months. Table 34-1 summarizes some of the similarities and differences among monotremes, marsupials, and placental mammals.

Table 34-1 Comparison of Monotremes, Marsupials, and Placental Mammals

Group	Mode of Reproduction	Placenta Present?	Distribution	Number of Species
Monotremes	Oviparity	No	Confined to Australia and New Guinea	3
Marsupials	Viviparity	Only in a few species	Concentrated in Australia and New Guinea, but also in nearby Asian islands and in the Americas	280
Placental Mammals	Viviparity	Yes, in all species	Worldwide	4,100+

Section Review

1. *Explain two ways a cat would be hampered by the removal of all its hair.*
2. *How does the parental care shown by mammals and crocodilians differ?*
3. *What adaptations enable the Siberian tiger to hunt at night?*
4. *Name two differences between marsupials and placental mammals.*

Critical Thinking

5. *Reptiles swallow their food without chewing. Mammals usually chew their food. Why is chewing an advantage for an endotherm?*
6. *How is the ability to chew related to the replacement pattern of mammalian teeth?*

34-3 Mammalian Diversity

There are about 4,400 living species of mammals. In number of species, mammals are not the most successful vertebrate class; there are more species of birds, reptiles, and bony fishes. In diversity of anatomy and habitat, however, mammals far surpass the other vertebrate groups. For instance, consider the differences between a bat and a whale, some of which are listed in Figure 34-8. In this section, you will get an idea of the diversity within the class Mammalia by studying the 21 mammalian orders recognized by most biologists.

Section Objectives

- *Recognize the limited distribution of monotremes and marsupials.*
- *Recognize the diversity of placental mammals.*

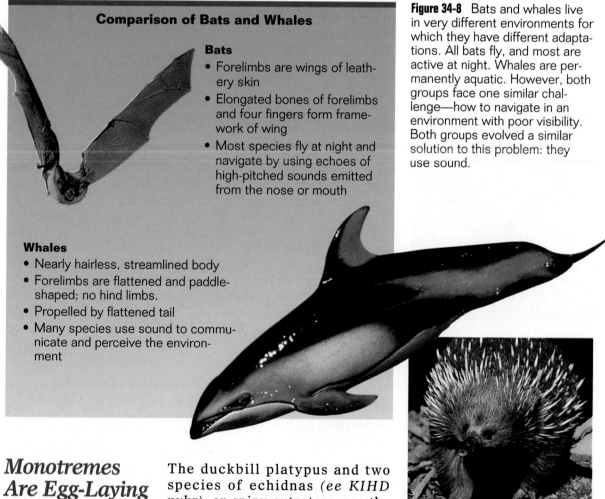

Comparison of Bats and Whales

Bats

- Forelimbs are wings of leathery skin
- Elongated bones of forelimbs and four fingers form framework of wing
- Most species fly at night and navigate by using echoes of high-pitched sounds emitted from the nose or mouth

Whales

- Nearly hairless, streamlined body
- Forelimbs are flattened and paddle-shaped; no hind limbs.
- Propelled by flattened tail
- Many species use sound to communicate and perceive the environment

Figure 34-8 Bats and whales live in very different environments for which they have different adaptations. All bats fly, and most are active at night. Whales are permanently aquatic. However, both groups face one similar challenge—how to navigate in an environment with poor visibility. Both groups evolved a similar solution to this problem: they use sound.

Monotremes Are Egg-Laying Mammals

The duckbill platypus and two species of echidnas *(ee KIHD nuhz)*, or spiny anteaters, are the only living monotremes (order Monotremata). The Australian echidna is shown in Figure 34-9. Recall from Chapter 31 that mammals evolved from reptiles. Because monotremes

Figure 34-9 Termites and ants are the main foods of echidnas.

share several characteristics with reptiles, biologists think monotremes are more closely related to the early mammals than are any other living mammals. For instance, monotremes, like reptiles, lay shelled eggs. The monotreme shoulders and forelimbs are also quite reptilian. Monotremes, like reptiles and birds, have a cloaca, a common passageway for the digestive, reproductive, and urinary systems. Despite their reptilian features, monotremes are definitely mammals; they have hair, and the females produce milk. All monotremes live in Australia and New Guinea. This limited distribution resulted from continental drift, which you studied in Chapter 33. Australia and New Guinea moved away from the other landmasses about 50 million years ago, before placental mammals had a chance to reach them.

The platypus inhabits lakes and streams in eastern and southern Australia. With its soft, sensitive bill, it searches through mud and gravel for worms and other soft-bodied animals. Echidnas are terrestrial, and they have very strong, sharp claws, which they use for burrowing and for digging out insects.

Figure 34-10 The koala is one of nearly 100 species of marsupials found in Australia. Its cuddly, "teddy bear" appearance belies a testy, pugnacious disposition. Koalas have a very specialized diet, feeding only on the leaves of 17 species of eucalyptus trees.

Marsupials Are Pouched Mammals

The order Marsupialia contains 280 species of kangaroos, opossums, wombats, and a variety of other animals, including the koala shown in Figure 34-10. As you learned in Section 34-2, marsupials differ from placental mammals in their pattern of development. The gestation period is very short, and the young are born incompletely formed. They crawl to the pouch, attach to nipples, and complete their development. Only one species of marsupial, the Virginia opossum, lives in North America today. Australia, New Guinea, and a few nearby islands contain the majority of marsupial species. Most mammalian species in Australia and New Guinea are marsupials.

Most Mammals Are Placental Mammals

Nineteen orders of mammals, comprising more than 90 percent of the species, contain only placental mammals. Recall that the offspring of placental mammals remain in the uterus through a relatively long gestation period, during which they receive nourishment through the placenta. At birth, placental young are further along in development than are marsupial young. Terrestrial placental mammals inhabit all continents except Antarctica, and aquatic placental mammals inhabit all oceans. Table 34-2 summarizes the orders of placental mammals.

Table 34-2 Orders of Placental Mammals

Order	Number of Species	Main Characteristics
Rodentia: Gnawing mammals 	1,814	Over 40 percent of mammalian species are rodents. **Key Characteristics** While most mammals have four incisors (front biting teeth) in each jaw, a rodent has only two, which continue to grow throughout its lifetime. Famous for their reproductive powers, rodents can give birth to a new litter every three weeks. **Important Examples** The families Muridae (mice and rats), Sciuridae (squirrels and chipmunks), and Cricetidae (voles, lemmings, deer mice, and hamsters) **Distribution** Worldwide, in almost all habitats
Chiroptera: Flying mammals 	986	Bats are the only mammals capable of powered flight. **Key Characteristics** The wing of a bat is a leathery membrane stretching across the forelimb and four elongated fingers. Most bats feed on night-flying insects and navigate in the dark with a sonar system that employs very high-pitched sounds. **Important Examples** Little brown bat (*Myotis lucifugus*), freetail bats (*Tadarida*), and flying foxes (*Pteropus*) **Distribution** Worldwide, except in polar environments
Insectivora: Insect eaters 	390	Insectivores are small and have a high metabolic rate. They are the mammals most similar to the ancestors of the placental mammals. **Key Characteristics** Most have long pointed snouts that they use to root for insects and worms. Their sharp teeth are adapted for grabbing and piercing prey. **Important Examples** Shrews, among the smallest of mammals, feed above ground by sweeping invertebrates into their mouths with clawed paws. Moles, which burrow, have small eyes and no external ears. **Distribution** All continents except Australia
Carnivora: Flesh eaters	240	The large predators of the animal kingdom, the carnivores include some of the strongest and most intelligent of animals. **Key Characteristics** Carnivores have long canine teeth, strong jaws, and clawed toes that aid in seizing and holding prey. Most have keen senses of sight and smell and can run quickly. **Important Examples** The families Canidae (domestic dogs, wolves, foxes, and coyotes), Felidae (the cats, including tigers, lions, leopards, and domestic cats), Mustelidae (weasels, minks, otters, and skunks), Procyonidae (raccoons), and Ursidae (bears, many of which are omnivores) **Distribution** Native to all continents except Australia

Order	Number of Species	Main Characteristics

Primates:
Large-brained mammals

233

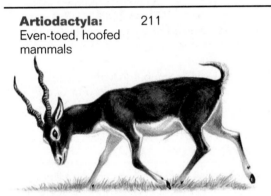

Primates have the largest brains for their body size of any animals.

Key Characteristics The only mammals with binocular vision and opposable thumbs, primates have five digits on all four limbs. None have horns or hoofs, and most have flat fingernails rather than claws. Hair covers the bodies of all primates except humans.

Important Examples Prosimians (lemurs, lorises, and tarsiers), monkeys (baboons, squirrel monkeys, and howler monkeys), and hominoids (apes and humans). The hominoids include the families Hylobatidae (gibbons), Pongidae (orangutans, chimpanzees, and gorillas), and Hominidae (australopithecines and humans).

Distribution Humans live on all continents; most other species live in tropical regions of Africa, Asia, and the Western Hemisphere.

Artiodactyla:
Even-toed, hoofed mammals

211

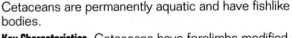

Fast-running antelopes and other grazers and browsers

Key Characteristics On each foot, artiodactyls have an even number of toes, each encased in a horny hoof. All are herbivores, with large flat molars for grinding plant material. Many have stomachs with a storage chamber called the rumen, where bacteria break down cellulose.

Important Examples There are nine families, grouped into three suborders: Suina (pigs and hippopotamuses), Tylopoda (camels), and Ruminantia (deer, antelope, cattle, sheep, and giraffes).

Distribution All continents except Australia

Cetacea:
Fully marine mammals

79

Cetaceans are permanently aquatic and have fishlike bodies.

Key Characteristics Cetaceans have forelimbs modified as flippers, no hind limbs, and broad, flat tails for swimming. They breathe through a nostril, or blowhole, on top of the head. A thick layer of blubber beneath the skin serves as insulation. Except for a few bristles on the muzzle, cetaceans are hairless.

Important Examples Cetaceans are divided into two groups: the predatory toothed whales (sperm whales, killer whales, dolphins, and porpoises) and the filter-feeding baleen whales (blue whales, gray whales, and humpback whales). The blue whale is the largest animal that ever lived.

Distribution All oceans; some species of dolphins live in freshwater rivers and lakes.

Lagomorpha:
Rabbits

69

Small, swiftly running herbivores with powerful hind legs

Key Characteristics Like rodents, they have one pair of long, continually growing incisors; unlike rodents, they have an additional pair of peglike incisors growing behind the front pair.

Important Examples Rabbits and hares

Distribution Native to all continents except Australia

Order	Number of Species	Main Characteristics
Pinnipedia: Marine hunters	34	Partially marine carnivores that feed in the sea but mate and sleep on land **Key Characteristics** All four limbs have been modified as flippers for swimming. Their bodies are streamlined for rapid movement through the water. **Important Examples** Seals, sea lions, and walruses **Distribution** Temperate and polar oceans

| **Edentata:** Toothless mammals | 30 | Mammals with poorly developed or no teeth
Key Characteristics The name of this order means "without teeth," and its members are toothless or have simple, peglike, rootless molars without enamel.
Important Examples Anteaters use a long sticky tongue to lap up insects; sloths are sluggish, tree-dwelling herbivores; armadillos feed on any kind of meat, including carrion.
Distribution Western Hemisphere only, primarily South America and Central America |

| **Macroscelidea:** Elephant shrews | 19 | Hopping shrews
Key Characteristics Elephant shrews are insect eaters that have a long, flexible snout. They resemble true shrews but hop about somewhat like small kangaroos.
Distribution Africa |

| **Perissodactyla:** Odd-toed, hoofed mammals | 17 | Horses and rhinos
Key Characteristics Hoofed herbivores with an odd number of toes per foot. They have a cecum, a pouch branching from the intestine in which bacteria digest cellulose.
Important Examples Members of the family Equidae (horses, asses, and zebras) have one toe per foot. Rhinoceroses have three toes per foot.
Distribution Most species are native to Asia and Africa. |

Order	Number of Species	Main Characteristics
Scandentia: Tree shrews	16	Small, squirrel-like mammals of the rain forest **Key Characteristics** They have a shrewlike body, a long snout, and sharp teeth. Despite their name, tree shrews live mainly on the ground. They feed on animals and fruit. **Distribution** Tropical rain forests of southern Asia

Pholidota: Pangolins	7	Armored anteaters **Key Characteristics** The word *pholia,* the root word of the name for this order, means "horny scale." The bodies of pangolins are covered with overlapping scales formed from fused bundles of hair. Pangolins have no teeth and feed on termites and ants they capture with a very long tongue. **Distribution** Tropical Asia and Africa

Hyracoidea: Hyraxes	7	Rabbitlike desert dwellers **Key Characteristics** Hyraxes have rabbitlike bodies, short ears, and hoofs. They have four hoofed toes on the front feet and three on the back. **Distribution** Africa and the Middle East

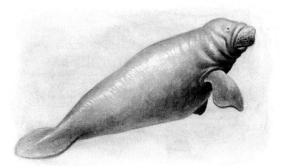

Sirenia: Manatees	4	Tubby, clumsy-looking aquatic herbivores **Key Characteristics** Like whales, sirenians have front limbs modified as flippers, no hind limbs, and a flattened tail used for propulsion. **Important Examples** Manatees, dugongs, and sea cows **Distribution** Tropical coastlines and estuaries of rivers

Order	Number of Species	Main Characteristics
Proboscidea: Elephants	2	The largest land animals alive today, elephants can weigh up to 5,400 kg (6 tons). **Key Characteristics** Elephants have a long, boneless, trunked nose. Their two upper incisors are long, curved tusks. **Important Examples** The African (large-eared) elephant and the much smaller Asian (small-eared) elephant **Distribution** Africa and Asia

Dermoptera: Flying lemurs	2	Large, squirrel-like gliders **Key Characteristics** Flying lemurs are misnamed on both counts: they cannot fly, and they are not lemurs (which are primates). Like flying squirrels, they glide on a sheet of skin stretching between their forelegs and hind legs. **Distribution** Southeastern Asia

Tubulidentata: Aardvark	1	The most unusual mammalian order **Key Characteristics** The name *aardvark* means "earth pig" in Afrikaans. The nocturnal aardvark has a pig-like body, big ears, and a long snout. Like pangolins, they feed on ants and termites. **Distribution** Southern Africa

Section Review

1. *To which country would you have to travel to see a monotreme? to see a marsupial?*

2. *Like reptiles in the Mesozoic era, mammals have taken to the air and live in the ocean. Describe one adaptation that enables mammals to live in each habitat.*

Critical Thinking

3. *Few species of placental mammals have been able to reach Australia without human help. Most that did are rodents and bats. Why do you think these mammals were able to make this journey, while members of other orders could not?*

34 CHAPTER REVIEW

Vocabulary

diaphragm (811)
gestation period (811)

Review

Multiple Choice

1. What shared characteristics suggest that therapsids are the ancestors of mammals?
   a. secondary palate and milk production
   b. complex teeth and a large synapsid opening
   c. endothermic body and primary palate
   d. fur and large eye sockets

2. Evidence that therapsids were efficient feeders and lived in cold climates indicates that they may have been
   a. covered with scales.
   b. ectothermic.
   c. endothermic.
   d. carnivores.

3. A paleontologist finds a fossilized jaw believed to be that of a mammal. If it is that of a mammal, how many bones will compose the jaw?
   a. 12   c. 2
   b. 4   d. 1

4. After the dinosaurs became extinct, mammalian species flourished because
   a. many new habitats became available to them.
   b. the climate became warmer and dryer.
   c. they have four-chambered hearts.
   d. young mammals can learn necessary skills.

5. A dolphin is classified as a mammal because it
   a. has a bony jaw with teeth.
   b. breathes air.
   c. has a four-chambered heart.
   d. feeds milk to its offspring.

6. Sheep are sheared annually for their hair, called wool. The function served by this hair is
   a. camouflage.   c. defense.
   b. insulation.   d. sensory.

7. The care given by mammals to their young
   a. begins after the young are weaned.
   b. is similar to that given by reptiles to their young.
   c. involves nursing and teaching survival skills.
   d. ends soon after the young are born or hatched.

8. Monotremes differ from marsupials in that monotremes
   a. lay eggs.
   b. are viviparous.
   c. nourish unborn young via the placenta.
   d. do not have mammary glands.

9. Members of the order Rodentia include
   a. polar bears and bats.
   b. chipmunks and gerbils.
   c. deer mice and wolves.
   d. deer and cats.

Completion

10. Scientific evidence suggests that mammals evolved from reptiles. Reptiles called _____ are considered to be the ancestors of therapsids, which are the direct ancestors of mammals.

11. The class Mammalia is named for the _____ glands found on female mammals. These glands produce _____ .

12. The spiny anteater and duckbill platypus belong to the order _____ . They share several characteristics with _____ , including forelimb structure and a cloaca.

13. Monotremes are _____ , meaning they lay eggs. Marsupials and placental mammals are _____ .

14. The hippopotamus and camel belong to the order _____ .

15. Not all placental mammals are active solely on land. For example, members of the order Chiroptera are active primarily in _____ , while members of the order Cetacea are active primarily in _____ .

Themes Review

16. **Evolution** Fewer species of mammals live today than did during the Tertiary period. What contributed to the decline in mammalian diversity?

17. **Structure and Function** The photograph shows a polar bear in its habitat. What visible characteristics enable the polar bear to survive in this habitat?

18. **Homeostasis** How do hair and a high rate of metabolism help mammals maintain homeostasis?

19. **Structure and Function** How does the shape of a beaver's teeth relate to its diet?

20. **Structure and Function** Explain why a tiger has a shorter digestive system than a herbivore of the same size.

Critical Thinking

21. **Making Inferences** The fossil record indicates that early mammals had large eye sockets. What does this evidence suggest about how early mammals survived competition with dinosaurs?

22. **Making Predictions** The gestation period of a mouse is about 21 days, whereas the gestation period of a moose is about 8 months. If you were a scientist looking for a laboratory animal for experiments dealing with mammalian development and heredity, which animal would you select for your experiments? Explain your choice.

23. **Making Comparisons** Some mammal species must care for their young for many years until the young reach maturity and can survive on their own. This pattern of rearing young has advantages over the way reptiles rear their young. Longer periods of parental care offer the young food, protection, and time to learn necessary survival skills. What are some disadvantages of this pattern of rearing young, compared with the reptilian pattern?

Activities and Projects

24. **Research and Writing** Find out about the theory of continental drift and how the theory can explain the distribution of mammals on Earth. In your report, describe how the breakup of Pangaea led to the predominance of marsupials in Australia and the predominance of placental mammals on other continents.

25. **Cooperative Group Project** Obtain a live female mammal, such as a mouse, hamster, or gerbil, that has just given birth. Get instructions for the care and feeding of the mammal. Then observe the mother and her babies for several weeks, or more if possible. Take notes on the feeding, sleeping, and interaction patterns of the mother and her babies. Share what you learn with your class.

26. **Multicultural Perspective** Research the importance of bison to the Native Americans that lived on the Great Plains. Write a report that summarizes what you have learned.

Mammalian Behavior: Human Postural Signals

OBJECTIVE

Observe human postural cues, a type of nonverbal communication.

PROCESS SKILL

• observing and identifying behaviors

MATERIALS

• stopwatch or watch with second hand
• paper and pencil

BACKGROUND

1. What are some types of communication among mammals?
2. What is the most common way in which humans communicate?
3. What are some types of nonverbal communication in humans?
4. Write your own **Focus Question** on your Vee Form.
5. **Knowing Side of the Vee** List the **Concepts** and new **Vocabulary Words** on your Vee Form. In the **Concept Statements** section of the Vee, use these words in sentences that define and explain them.

TECHNIQUE

Doing Side of the Vee

1. Work in a group of two or three to observe pairs of people in conversation. It is important that your subjects are unaware of being observed.
2. You will observe at least three conversations. Each conversation must last between 45 seconds and 5 minutes, and it must involve only two people. Data for conversations that last less than 45 seconds

will not give significant results for this investigation.

3. People who study behavior give specific definitions to certain actions. An interaction between two people is called a *dyadic exchange*. The position of the body while standing is called a *stance*. In an *equal stance*, the body weight is supported equally by both legs. And in an *unequal stance* more weight is supported by one leg than by the other.
4. For each dyadic exchange you will record the actions of only one person. Another member of your team will be the timekeeper and will alert you to 15-second intervals. Within each 15-second interval, you will make note of all the changes of stance by the person you are observing.
5. How often do you think people change their stance while they are in conversation? How might the gender of the people in conversation influence the number of stance shifts? How might the timing of the conversation—beginning, middle, or end—affect the number of stance shifts? Record your expectations in the **Records** section of your Vee.
6. In the **Records** section, set up a table for your observations. You may choose to use a table like the one shown below. Record the sexes of both people in the dyad: male/female, male/male, or female/female. Then record the sex of the person you are observing (*s.o.*). Each of the numbered boxes represents a 15-second interval. Notice that the table

Observations of Dyadic Exchanges

	sexes	s.o.	1	2	3	...	18	19	20
Dyad 1									
Dyad 2									
Dyad 3									

is reduced here and that you will need to allow extra space for additional 15-second intervals on your own table.

7. Record every time your subject shifts from an equal stance to an unequal one, or vice versa. When the subject assumes an unequal stance, record the number of discernible weight shifts from one foot to the other. To record this simply, you may write *E* or *U* to identify an equal or unequal stance, and *W* to identify a weight shift. Record all of the changes within a single 15-second block until the timekeeper signals you to go on to the next block.

8. At the end of the conversation, write down whether the pair departed together or separately. Record your observations in the **Records** section of your Vee.

9. After you have completed each observation, tally the total number of shifts within each 15-second block.

   IMPORTANT! Only retain data for conversations that last at least 45 seconds. If you have data for 30 seconds and the pair depart, you must observe a new dyadic exchange.

10. After you have completed your observations, share your data with other teams in your class. Combine and analyze the data from all the teams. Find the most common stance during the first 15 seconds of an exchange, the middle intervals, and the last 15 seconds.

11. Average the number of weight shifts for the beginning, middle, and end intervals. Record these numbers in the **Records** section of your Vee.

12. Analyze the data by the gender of the subject you observed and by the gender of the partner. Compare these with the averages not related to gender, and record any differences in the **Records** section.

13. Construct bar graphs to express data in the **Records** section of your Vee Form. In the **Procedure** section of the Vee, briefly summarize the procedure you followed.

INQUIRY

1. Which is the most common stance in a dyadic exchange?

2. Which is seen more often as the members of the dyad prepare to depart: stance change or weight shift?

3. What happens when a member of the dyad prepares to leave?

4. Use the information on the **Knowing Side** of the Vee to interpret your results from the **Doing Side,** and then write your **Knowledge Claim**. Write a **Value Claim** for this lab.

ANALYSIS

1. Were your observations what you had expected? If not, how were they different?

2. Are there differences in the departure signals of males and females? If so, what are they?

3. What do you think might be an adaptive significance of a departure signal?

4. What other types of behaviors might serve as departure signals for humans?

5. What other nonverbal communication did you observe during this investigation?

FURTHER INQUIRY

Write a **New Focus Question** that could be the point of a new investigation. The following is an example:

Do people use body language even when their actions cannot be seen? For example, does a person speaking at a public pay phone shift stances as the conversation comes to a close?

CHAPTER 35

OVERVIEW OF THE HUMAN BODY

REVIEW

- homeostasis (Section 1-3)
- cytoskeleton (Section 2-4)
- tissues (Section 19-3)
- mesoderm (Section 27-1)
- coelom (Section 27-2)
- exoskeleton and endoskeleton (Section 27-3)
- endothermy (Section 31-1)
- vertebrate characteristics (Section 31-1)

A rock climber

35-1 The Human Body Plan

I n this unit you will study the human body and how it functions. You will begin your study by exploring the general nature of the human body plan. To understand how the human body functions, you must first look at its overall architecture. Only then can you properly appreciate how the various parts function. Imagine trying to explain an automatic transmission to someone who doesn't know what a car is. In this section you will examine six general functions of the human body plan. Just as in the discussion of the animal body plan in Chapter 27, the use of the term plan in no way implies a conscious design. Rather, it refers to the organization of the human body that has been programmed into its genes by thousands of centuries of evolution.

Section Objectives

- Describe the role of joints in the skeleton.
- Contrast the functions of the central and peripheral nervous systems.
- Identify two advantages of maintaining a constant body temperature.

Supporting the Body

Perhaps the most important feature of vertebrates is their endoskeleton, or internal skeleton. Instead of a rigid exterior skeleton like that of arthropods, vertebrates have soft, flexible skin that stretches to accommodate the body's movement. When you bend your arm, the skin covering the joint of your elbow stretches; if it didn't, it would tear. The human endoskeleton, shown in Figure 35-1, is composed of bone, a strong and nonbrittle material. There are 206 individual bones in the skeleton.

Like insects, vertebrates have jointed appendages. In the human body, the jointed appendages are the arms, hands, legs, and feet. The movement of an appendage such as the arm occurs when its bones move relative to the rest of the skeleton. The movement of the skeleton occurs at **joints,** where one bone meets another. The freely movable joints of the skeleton are cushioned by pads of cartilage and held together by bands of elastic tissue called **ligaments.** The pads of cartilage at the tips of bones do not touch

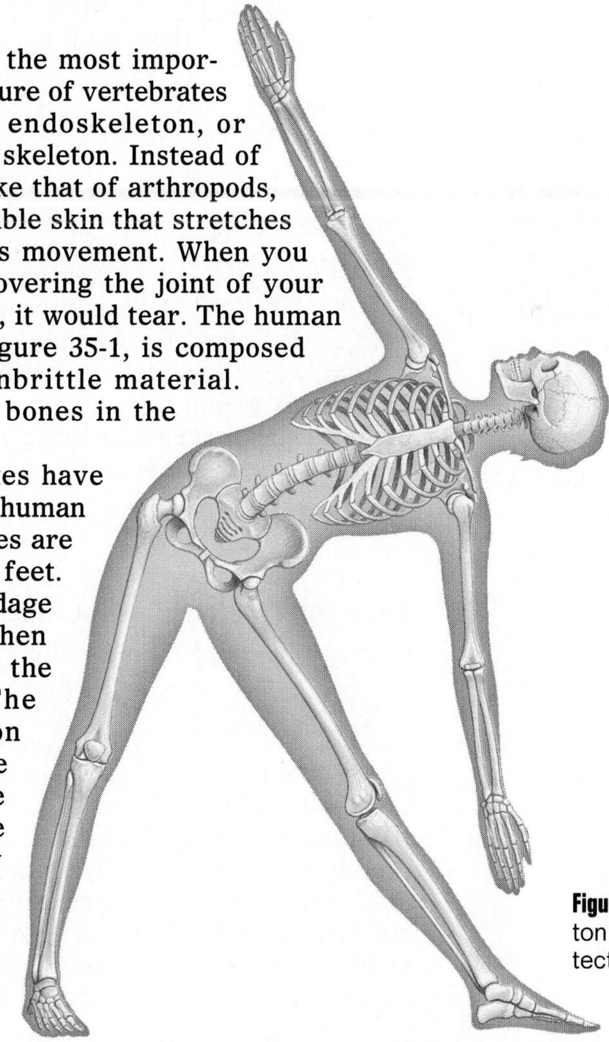

Figure 35-1 Your skeleton supports and protects your body.

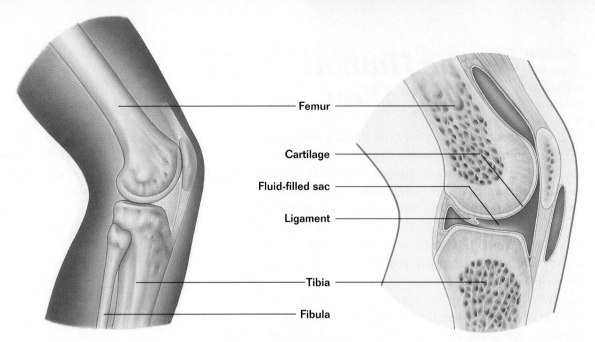

Femur

Cartilage

Fluid-filled sac

Ligament

Tibia

Fibula

Figure 35-2 Movement in the skeleton is possible at joints, where bones meet. This diagram shows the structure of the knee.

at the joints. As shown in Figure 35-2, membranes filled with fluid occur where the bones meet, and this fluid lubricates the movement of these pads across one another.

CONTENT LINK
. . . .
Rheumatoid arthritis is a very painful disease in which, among other things, the cartilage of the joints is destroyed. To find out more about this disease, turn to **Chapter 36.**

Collecting and Evaluating Information

Humans, like all vertebrates, have a brain that coordinates and directs the activities of the body. The human nervous system is composed of two contrasting functional groups. The first is the central processing region, the **central nervous system,** which includes the brain and spinal cord. The second functional group, called the **peripheral nervous system,** includes the nerves that bring information to the brain and transmit commands from it.

Transporting Oxygen and Nutrients

Most animals have a circulatory system that carries oxygen and nutrients to the cells of the body and removes carbon dioxide and water. Humans and other vertebrates have a closed circulatory system. Blood is enclosed within blood vessels that separate it from the rest of the body's fluids and prevent it from mixing freely with them. Materials pass into and out of the blood by diffusion. The great advantage of a closed circulatory system is that the body can maintain different circulation rates in different organs by changing the diameter of blood vessels. A second advantage of a closed circulatory system is that it allows the blood to be pumped under pressure. The heart pushes the blood out through the body's arteries, sending the blood through the body much faster than would be possible in an open system.

Protecting the Heart, Lungs, and Digestive Organs

Like all vertebrates, humans are coelomates. Every human body contains a large body cavity, the coelom, that develops entirely within mesoderm tissue in the embryo. As you can see in Figure 35-3, within this cavity are found the large organs of the body, which are suspended in fluid that supports their weight and prevents them from being deformed by body movements. If your heart were compressed every time you turned at the waist or bent over, it could not function correctly. Why doesn't it get deformed by body movements? Imagine a balloon full of water that is floating within another, larger balloon, also full of water. Pushing your thumb into the outer balloon doesn't deform the inner one. The human coelom protects the body's organs in much the same way.

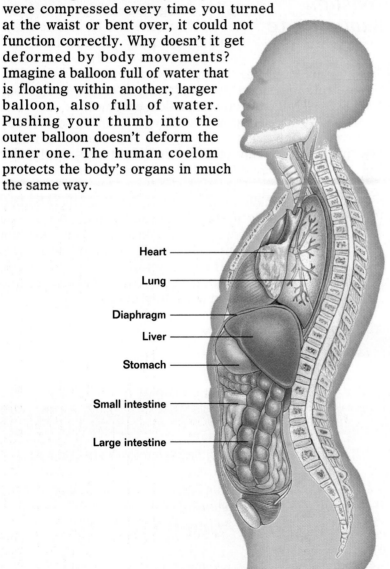

Heart

Lung

Diaphragm

Liver

Stomach

Small intestine

Large intestine

Figure 35-3 The heart, lungs, and digestive organs lie within the coelom. This fluid-filled cavity protects and supports the organs. The diaphragm separates the body cavity into the thoracic cavity, which houses the lungs and heart, and the abdominal cavity, which houses the digestive organs.

CONTENT LINK

To find out what role each organ of the digestive system plays in extracting nutrients from the food you eat, turn to **Chapter 38**.

Obtaining Nutrients From Food

The human body, like that of nearly all animals, is organized to carry out extracellular digestion within the gut. As in most animals, the flow of materials through the gut proceeds in one direction. Food enters the body through the mouth and passes through the digestive tract and out the anus. In humans, the digestive tract is a long tube that extends from mouth to anus and is suspended within the coelom.

Maintaining a Relatively Constant Temperature

Like all mammals, human beings are endothermic, maintaining a fairly constant internal temperature of 37°C (98°F). A large percentage of the calories you consume are devoted to maintaining your body temperature, and you cannot live if it falls far below normal for long. Very high temperatures, such as fevers sometimes induced by infections, are also dangerous because they can inactivate critical enzymes.

Why invest so much energy keeping your body warm? One great advantage of endothermy is that it permits the body to maintain its activity at all times and in many different places, regardless of the temperature of the surroundings. Mountain butterflies, which use the sun's rays to heat their bodies, cannot fly if the sun doesn't shine, just as you could not walk on a cold day if you were not endothermic. In addition, endothermy permits you to sustain strenuous activity, while ectotherms such as lizards and frogs are limited to short periods of exertion.

Figure 35-4 Running a marathon requires more than two hours of exertion. Strenuous exercise of this duration, though beyond the capabilities of most people, depends on the large amount of energy supplied by an endothermic metabolism. No ectotherm could run for even a tiny fraction of this time.

Section Review

1. *Explain how the joints of the skeleton both permit and limit movement.*
2. *Describe the relationship between the central and peripheral nervous systems.*
3. *Why are very high body temperatures often fatal?*

Critical Thinking
4. *Explain why your body temperature usually rises slightly after you eat.*

35-2 How the Body Is Organized

*T*he human body is composed of over a hundred different kinds of cells. As illustrated in Figure 35-5, cells
within the body are organized into larger functional
units. Recall from Chapter 19 that a tissue is a group of similar cells that work together to perform a common function.
The diverse cell types of the human body are traditionally
grouped by function into four basic types of tissues: epithelial
(**ehp uh THEE lee uhl**), *connective, nerve, and muscle tissues. These four kinds of tissues are the building blocks of
the human body. Each* **organ** *of the human body is composed
of a mixture of these tissues organized in various ways. An*
organ system *is a group of organs that function together to
carry out a major activity of the body.*

Section Objectives

■ Contrast the characteristics
and functions of the four
basic types of tissue.

■ Contrast smooth, skeletal,
and cardiac muscle.

■ Describe the function of each
organ system.

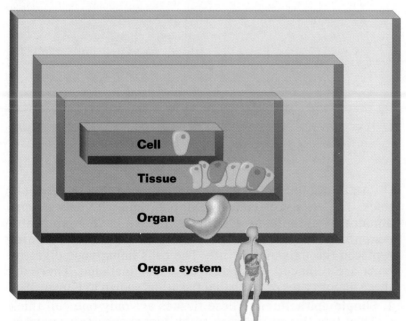

Cell

Tissue

Organ

Organ system

Figure 35-5 Within your body, cells
belong to functional units called
tissues. Tissues are organized
into organs, and organs belong to
groups of organs called organ
systems.

Epithelial Tissue Is Protective Tissue

Epithelial cells are the
guards of the body. Epithelial
tissue, or **epithelium,** covers
the body's surfaces, protect-
ing the tissues beneath from dehydration and physical damage. Because epithelium covers the body's outer surface and
lines the gut and lungs, every substance that enters or leaves
the body must cross an epithelial layer. Therefore, epithelial
cells control which substances enter and leave the cells of the
body. Many secretory glands are derived from epithelium, as
are many of the body's sensory organs.

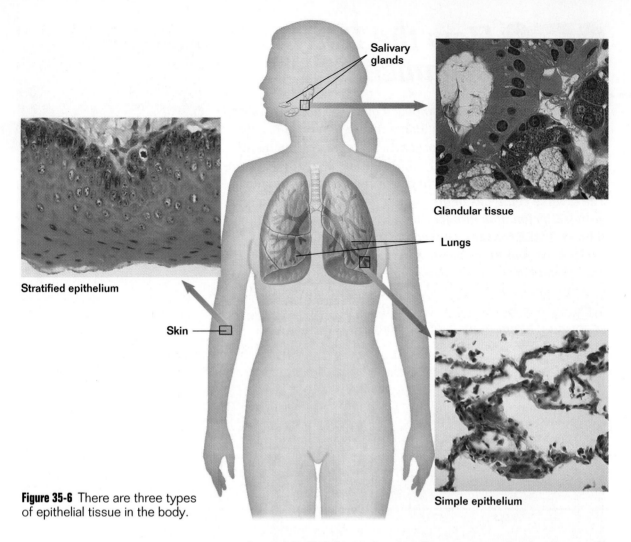

Figure 35-6 There are three types of epithelial tissue in the body.

Salivary glands

Glandular tissue

Lungs

Stratified epithelium

Skin

Simple epithelium

An epithelial layer is usually no more than a few cells thick. The cells are typically flat and thin, with only a small amount of cytoplasm. Epithelial layers have remarkable regenerative powers—their cells are constantly being replaced throughout your life. The cells lining your digestive tract are replaced every few days, for instance. There are three major types of epithelial tissue, as shown in Figure 35-6.

1. **Simple epithelium**   These tissues are only one cell thick. They line the respiratory tract, lungs, and major cavities of the body.

2. **Stratified epithelium**   These tissues are several cell layers thick. Stratified epithelium is most abundant in the skin, where it forms the outer layer. A characteristic property of stratified epithelial cells is their ability to produce keratin, a very strong, fibrous protein. The calluses that occur on your hands are composed of keratin, and so is hair.

3. **Glandular tissue**   The human body contains many glands, all composed of stratified epithelia. These glands produce sweat, milk, saliva, digestive enzymes, hormones, and many other substances.

Connective Tissue Supports and Defends the Body

The cells of connective tissue provide the body with its structural building blocks and its most potent defenses. Connective tissue is the most diverse of the four principal types of tissue. Some connective tissue cells, such as those in bone, are densely packed together. Others, such as blood cells, are spaced well apart from each other. Connective tissue cells fall into three functional categories: cells of the immune system, cells of the skeleton, and cells that accumulate and store molecules.

Cells of the Immune System

The immune system is composed of cells that defend the body from infection and cancer. Immune system cells are typically small, and they roam the bloodstream searching for invaders. Some immune system cells, called **macrophages** (*MAK roh FAY juhz*), are mobile cells that engulf and digest bacteria and other microbes. Other cells, called **lymphocytes** (*LIM foh seyets*), make antibodies or attack virus-infected cells and cancer cells. Both types of cells are shown in Figure 35-7.

Cells of the Skeletal System

The human skeleton is composed of three kinds of tissue, shown in Figure 35-8, that are distinguished by the nature of the material that is laid down between the cells.

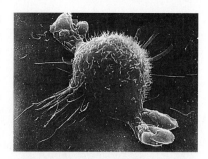
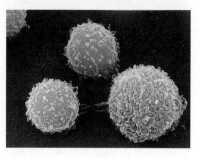

Figure 35-7 Macrophages, *top*, are immune system cells that consume invaders. B cells, *bottom*, are lymphocytes that produce defensive proteins.

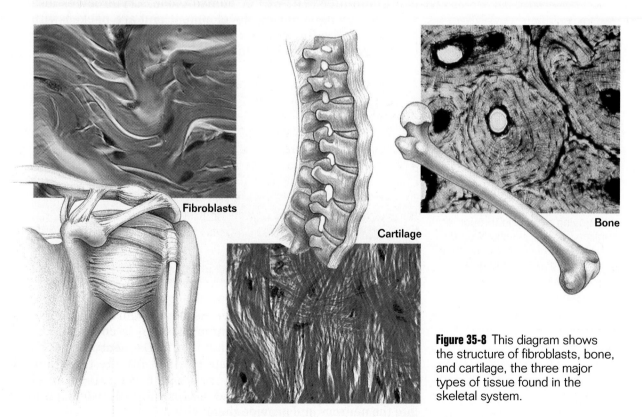

Fibroblasts

Cartilage

Bone

Figure 35-8 This diagram shows the structure of fibroblasts, bone, and cartilage, the three major types of tissue found in the skeletal system.

1. **Fibroblasts** *(FEYE broh blasts)*   These cells are the most common connective tissue cells. They are flat, irregularly shaped cells that secrete the strong structural protein **collagen** into the spaces between the cells. Collagen makes up one-fourth of the protein in the body. Fibroblasts also produce scar tissue.
2. **Cartilage**   This material is firm, flexible, and very strong. It is formed at positions of mechanical stress by fibroblasts, which lay down long parallel strands of collagen fibers along the lines of stress. Cartilage also covers the ends of bones at joints.
3. **Bone**   In bone, the collagen fibers between cells are coated with a calcium phosphate salt, making bone more rigid than collagen, but not brittle. The human endoskeleton is composed mainly of bone.

Cells That Accumulate and Transport Molecules

Many of the body's connective tissues are specialized for storing and transporting substances; examples include pigment-containing cells and the fat-storing cells of adipose (fat) tissue. The most important storage and transportation cells are red blood cells, called **erythrocytes** *(eh REE throh seyets)*, which are shown in Figure 35-9. Erythrocytes circulate in the blood, carrying oxygen from the lungs to the body's cells and transporting carbon dioxide from the cells back to the lungs. There are about 5 billion erythrocytes in every milliliter (0.03 oz.) of human blood. Erythrocytes lose their cell nucleus during development and are packed with the oxygen-binding protein hemoglobin. Each erythrocyte contains about 300 million molecules of hemoglobin.

CONTENT LINK

An erythrocyte lives for only about 120 days. In **Chapter 37** you will learn how and where new erythrocytes are produced.

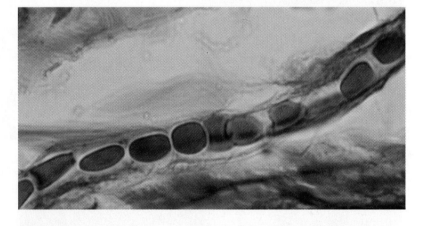

Figure 35-9 These red blood cells are moving single file through a capillary, a very small blood vessel. Red blood cells release their load of oxygen and nutrients while in capillaries.

Nerve Tissue Conducts Signals Rapidly

Nerve tissue, the third major class of human tissue, is composed of two kinds of cells: **neurons,** which are specialized for the transmission of nerve impulses, and **supporting cells,** also called **glial** *(GLEE uhl)* **cells,** which insulate the neurons and provide them with nutrients.

Neurons have a central cell body, which contains the nucleus, and two kinds of cytoplasmic extensions, as you can see in Figure 35-10. The first kind of extension consists of threadlike protrusions called **dendrites,** which act as antennae for the reception of nerve impulses from other cells or sensory systems. A single neuron may have many dendrites. The second kind of cytoplasmic extension that projects from the cell body of a neuron is the **axon.** An axon is a long, tubular extension of the neuron that carries the nerve impulse away from the cell body, often for considerable distances. A single axon connects your heel with the base of your spine, a distance of more than 1 m (39 in.). Another neuron runs from your spine all the way across your shoulder to your thumb. Single neurons more than 1 m long are common. Nerves in the human body appear as fine white threads, but they actually are composed of hundred of axons bunched together like the wires of a telephone cable.

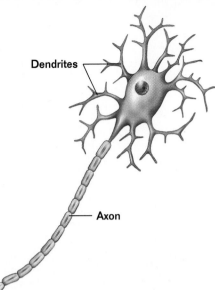

Dendrites

Axon

Figure 35-10 The small, hairlike projections on this neuron are dendrites. The larger projection is an axon.

Muscle Tissue Permits Movement

Muscle cells, formed from mesoderm early in development, are the workhorses of the human body. The key property of muscle cells is the relative abundance of the protein filaments **actin** and **myosin,** which enable the muscle cell to contract. These protein filaments are present as part of the cytoskeleton in all eukaryotic cells, but they are far more abundant in muscle cells. The three kinds of muscle cells in the human body are illustrated in Figure 35-11.

1. **Smooth muscle** Smooth muscle cells are long, spindle shaped, and packed with myosin and actin filaments. The individual filaments are not aligned with each other. Smooth muscle cells are typically organized into sheets of cells. Some smooth muscles, such as those lining the walls of blood vessels, contract only when stimulated by nerves or hormones. Other smooth muscles, such as those lining the wall of the gut, contract spontaneously, leading to a slow, steady contraction of the tissue.

2. **Skeletal muscle** Skeletal muscles (also called striated muscles) move the endoskeleton. Each muscle is a tissue made of many individual muscle cells acting together. The cells fuse end to end during development to form a long muscle fiber with a central cable of filaments and many nuclei pushed out to the edges. Because the filaments are aligned, skeletal muscles are much stronger than smooth muscles.

3. **Cardiac muscle** The muscles of the human heart, called cardiac muscles, are organized very differently from skeletal muscles. Instead of long cells with many nuclei, cardiac muscle is composed of chains of single cells, each cell having its own nucleus. These chains are organized

CONTENT LINK

The interaction between actin and myosin creates the force of muscular contraction. Read how the body controls this interaction in **Chapter 36.**

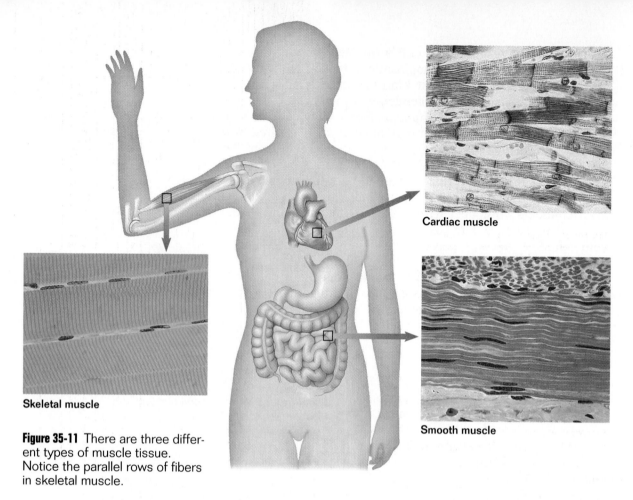

Skeletal muscle

Cardiac muscle

Smooth muscle

Figure 35-11 There are three different types of muscle tissue. Notice the parallel rows of fibers in skeletal muscle.

into fibers that branch and interconnect, forming a latticework. When two fibers touch in the lattice, they make an electrical junction. When the heart starts to contract, the electrical wave that initiates contraction rapidly spreads over the whole lattice. Thus, a group of neighboring cells tends to contract all at once, rather than individually.

Organs and Organ Systems

All of the organs in your body are composed of combinations of the four types of tissue just described. The heart, for example, is composed of cardiac muscle tissue and connective tissue, and is connected with many nerves. All of these tissues work together to pump blood through your body. Each organ, in turn, belongs to an organ system that carries out a critical function. For example, the digestive system is composed of individual organs that break up and digest food, absorb nutrients, and compact and expel the unabsorbed residue. The human body contains 11 principal organ systems that are briefly described in Table 35-1. *Highlights: Organ Systems* on pages 841-848 further examines the structure and function of these organ systems. ◻

❏ CAPSULE SUMMARY

Cells in the body are organized into tissues. There are four kinds of tissue: epithelial, connective, nervous, and muscular. Tissues are organized into organs, which, in turn, belong to organ systems.

Table 35-1 Major Organ Systems of the Human Body

System	Major Components	Functions
Systems That Cover, Support, or Produce Movement		
Integumentary	Skin and associated structures	Protects against injury, infection, and fluid loss
Muscular	Skeletal and smooth muscles	Movement
Skeletal	Bones of the skeleton	Protects and supports the body
Systems That Regulate Body Functions		
Endocrine	Various glands, including pituitary gland, thyroid, adrenal glands, and pancreas	Regulates and coordinates the body's functions
Nervous	Brain, spinal cord, nerves, sense organs	Collects and evaluates information; issues commands
Reproductive	Testes, penis, and associated ducts (males); ovaries, uterus (females)	Produces gametes; delivers gametes (males); nourishes and shelters fertilized egg (females)
Systems That Transport or Protect		
Circulatory	Heart, blood vessels, lymphatic vessels	Transports oxygen, carbon dioxide, nutrients, and cells
Immune	White blood cells	Defends against pathogens and cancer
Systems Involved in Metabolism or Excretion		
Respiratory	Lungs, air passages	Obtains oxygen; releases carbon dioxide
Digestive	Mouth, esophagus, stomach, small intestine, large intestine, liver, pancreas	Extracts and absorbs nutrients from food
Urinary	Kidneys, urinary bladder, ureters, urethra	Removes wastes from blood; regulates concentration of the body's fluids

Section Review

1. *Why is it advantageous to be able to rapidly replace epithelial cells?*
2. *Describe two differences between cardiac and skeletal muscle.*
3. *Complex machines often have redundant systems so that if one system fails, another can take over its function. Are there any redundant organ systems in the human body? Explain your answer.*

Critical Thinking

4. *Some people living at high elevation have more erythrocytes in their blood than do people living at lower elevations. Propose a hypothesis to account for this observation.*

35-3 *Homeostasis*

Section Objectives

- Recognize why it is essential for the body to maintain homeostasis.
- Contrast positive feedback with negative feedback.
- Contrast chemical signals with electrical signals.

Every cell in your body is continuously bathed in fluid that supplies it with nutrients and carries away its wastes. No human cell can live for long without this supply line; a brain cell will die in minutes if its supply of oxygen and nutrients is cut off. There is nothing unusual about this. All living cells, whether single-celled paramecia or your own cells, face this same stringent requirement. In your body, however, there are over 100 trillion cells. Each cell must get nutrients from and release wastes into the same 15 L (4 gal.) of body fluid.

Why Homeostasis Is Critical

The body fluid not inside your cells is called extracellular fluid. It occupies the spaces between cells and forms the plasma, the fluid portion of the blood. Because so much exchange occurs across the membranes of cells, any extreme change in the composition or volume of the extracellular fluid can have very serious effects on the activities inside cells, as shown in Figure 35-12. The pH level, the concentration of key ions such as potassium, sodium, and calcium, and the level of sugar in the extracellular fluid are all critical to the body's functions. If any of these factors vary outside the range that individual cells require, death may result. A stable fluid environment is one of the key achievements of multicellular organisms, and it is maintained by a complex set of physiological mechanisms. As you learned in Chapter 1, maintaining a relatively constant internal environment is called homeostasis.

Figure 35-12 Red blood cells normally are flat and disk-shaped, *left*. When placed in a solution with lower-than-normal salt concentration, *right*, the cells take up water, swell, and may burst.

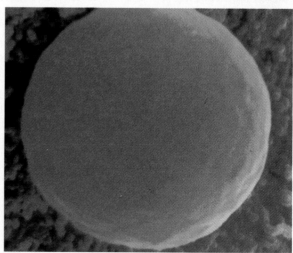

The Feedback of Information

To maintain homeostasis, your body must keep track of what is going on in all of its organs and tissues. In particular, the concentrations of ions and other chemicals in the extracellular fluid must be monitored. To do this, the central nervous system gathers information about the body's condition, evaluates it, and issues commands to counteract any deviation from normal conditions. Such a process of surveillance and response is called a **feedback loop.** Homeostasis is maintained by feedback loops.

Many of the most important homeostatic feedback loops involve **negative feedback.** Negative feedback prevents a variable such as pH or temperature from deviating from its normal value, called its **set point.** A good everyday example of negative feedback is driving a car. The variable the driver wants to keep at a constant value is the position of the car in its lane; the set point is the center of the lane. The eyes of the driver serve as sensors that feed information to the driver's brain about the car's position and constantly compare the car's actual position with the center of the lane. Deviations from the set point, caused by bumps or curves in the road, are recognized by the brain, which issues signals to the driver's muscles to use the car's steering system to correct the deviations. Figure 35-13 illustrates a generalized negative feedback loop.

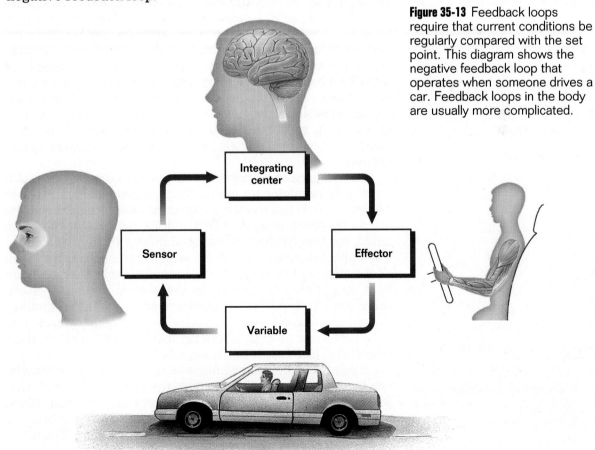

Figure 35-13 Feedback loops require that current conditions be regularly compared with the set point. This diagram shows the negative feedback loop that operates when someone drives a car. Feedback loops in the body are usually more complicated.

Integrating center

Sensor

Effector

Variable

In the negative feedback controls of the human body, the values of variables such as body temperature, blood pressure, and pH are continuously compared with their set point values. Any changes that increase the difference between the variable and its set point initiate responses that tend to oppose the changes and restore the variable to its set point. That is why it is called negative feedback—it *reduces* the difference between variable and set point. However, the human body may change its physiological set points from time to time. For example, body temperature decreases during sleep.

Most physiological feedback is negative, but there are exceptions. **Positive feedback** refers to the condition in which a change in a variable causes the body to drive the variable even farther from the initial value. Positive feedback systems tend to be highly unstable—an explosion is an example of positive feedback. In humans, positive feedback plays an important role in childbirth; the pressure of the baby's head on the lower part of the uterus increases the frequency and intensity of uterine contractions. ◻

Homeostasis in Action

Using sensors scattered throughout the body, the central nervous system constantly monitors temperature, pH, blood pressure, ion concentrations, and many other factors. Much of this surveillance is centered in a small, marble-sized region of the brain called the **hypothalamus.** When the hypothalamus detects a disturbance in the body's condition, it issues orders to correct the disturbance. Sometimes these orders alter body functions, such as how rapidly you breathe. At other times, they call for production of chemical signals, which you will study later in this section.

An example of homeostasis in action is the way in which your body regulates its temperature, a process illustrated in Figure 35-14. When the temperature of your blood exceeds its set point, which is usually 37°C (98°F), neurons in the brain detect the temperature change and inform the hypothalamus. The hypothalamus responds by triggering mechanisms, such as sweating and the expansion of blood vessels, that dissipate heat. In addition to neurons in the central nervous system, there are two types of temperature-sensitive nerve endings in your skin. One type is sensitive to low temperatures, and the other type to high temperatures. These "surface thermometers" also report to the hypothalamus.

Other parts of the brain are also important in maintaining homeostasis. The brain stem, the part of the brain that joins to the spinal cord, keeps blood pressure constant. Blood pressure is measured at sites in the major arteries where the wall of the artery is thin and contains a highly branched

CONTENT LINK

*In **Chapter 41** you will learn more about the hypothalamus's regulatory functions.*

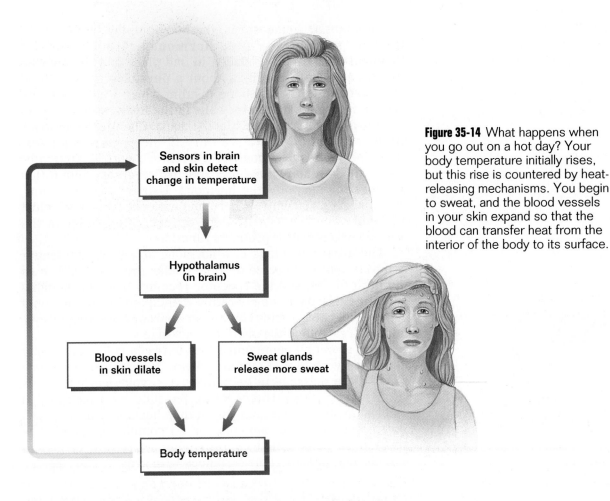

Figure 35-14 What happens when you go out on a hot day? Your body temperature initially rises, but this rise is countered by heat-releasing mechanisms. You begin to sweat, and the blood vessels in your skin expand so that the blood can transfer heat from the interior of the body to its surface.

Sensors in brain and skin detect change in temperature

Hypothalamus (in brain)

Blood vessels in skin dilate

Sweat glands release more sweat

Body temperature

network of nerve endings. When blood pressure increases, the thin wall of the artery bulges out, stretching the nerve endings, which relay signals to the brain stem. The brain stem responds by adjusting the rate at which the heart beats, the force of its contractions, and the diameter of some blood vessels. This negative feedback loop moves blood pressure back toward its set point.

The Body's Signaling System

Maintaining homeostasis requires a great deal of signaling back and forth between the central nervous system and the rest of the body. There are two types of signals. Some are rapidly transmitted **electrical signals,** which are carried by neurons from one place in the body to another much as wires carry telephone messages. An electrical signal's effects usually last only for brief periods. **Chemical signals** are more slowly transmitted and are carried by the bloodstream from one place in the body to another. Unlike the effects of electrical signals, the effects of chemical signals tend to last for a long time.

CONTENT LINK
. . . .
How does a hormone change the
functions of its target tissue?
You will study this topic in
Chapter 41.

❑ *CAPSULE*
SUMMARY

*The hypothalamus, a region of the
brain, monitors and regulates the
body's condition. Information is
transmitted from one part of the
body to another by the nervous
system and by hormones.*

To maintain homeostasis, the body employs a battery of specific molecules called **hormones.** A hormone is a chemical messenger produced in one place and transported to another, where it brings about a physiological response. One example is the hormone insulin, which is produced by the pancreas and stimulates cells in the liver and muscles to absorb glucose from the bloodstream. Chemical messages obviously can be effective only if they are recognized. How does a tissue recognize a hormone with a particular shape? It does so by having a specific receptor protein that matches the shape of the hormone. Using such a receptor, the tissue can recognize a hormone with precision and select one molecule from billions of others. Specificity is the essence of a receptor protein's function.

The great advantage of a hormone as a body messenger is that it can be directed at a particular protein receptor on the target cells, and the receptor recognizes only this molecule. In every case the operating principle is the same: only cells whose membranes contain an appropriate receptor protein will respond to the chemical message.

Why aren't all communications between cells in the body handled by chemical signals? Their transmission is very slow. If the message to be delivered to your leg muscles is "Contract quickly; we are being pursued by a leopard," a quicker means of communication than hormones is required. To answer this need, humans and all complex animals have specialized cells called neurons, which transmit electrical signals over considerable distances with great speed. ❑

Section Review

1. *Why is homeostasis crucial to the body?*
2. *Explain why most body functions are regulated by negative, rather than positive, feedback loops.*
3. *Identify one advantage and one disadvantage of chemical signals, compared with electrical signals.*

Critical Thinking

4. *A thermostat plays the same role in regulating the temperature of a room as your brain does in regulating the temperature of your body. Which situation would produce a more pleasant room temperature: a thermostat controlling a positive feedback loop or one controlling a negative feedback loop? Explain your answer.*
5. *Why is it advantageous for uterine contractions during birth to be regulated by a positive feedback loop?*

Protecting and Supporting the Body

The bones, cartilage, and ligaments of the skeletal system provide a strong, rigid internal framework against which the body's muscles can pull. The skeletal system also protects the body by encasing the central nervous system in a shell of bone and shielding the thoracic cavity with a framework of ribs. A flexible endoskeleton of bone allows the body a wide range of motion while supporting a body of considerable size. There are 206 individual bones in the human body.

The Axial Skeleton

The bones of the **axial** (*AX ee uhl*) **skeleton** support and protect the trunk. Twenty-eight of these bones form the skull, which protects the brain; 26 bones compose the backbone, which protects the spinal cord; and 24 bones form the rib cage (12 pairs of ribs attached to the breastbone and the backbone), which protects the heart and lungs.

The Appendicular Skeleton

The bones of the **appendicular** (*ap ehn DIHK yoo luhr*) **skeleton** form the limbs and the points where the limbs attach to the axial skeleton. The 30 bones of each arm and hand are attached to the axial skeleton at the shoulder, and the 30 bones of each leg and foot are attached to the axial skeleton at the hip.

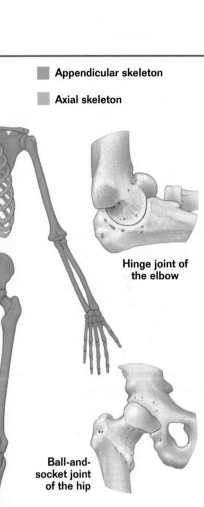

■ Appendicular skeleton

■ Axial skeleton

Hinge joint of the elbow

Ball-and-socket joint of the hip

Bone

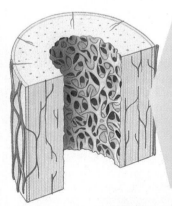

Bone is a dynamic living tissue composed of many kinds of cells. Blood vessels passing through channels in the bone bring oxygen and nutrients to these cells. Bone is hard because some of its cells form crystals of calcium phosphate.

Ligaments and Tendons

Strong fibers called **ligaments** hold bones together. They act as cables to hold bones flexibly within their sockets. Muscles are attached to bones by straps of tough cartilage called **tendons.**

Joints

A joint is the place at which two bones meet. Some joints, such as those in the skull, are locked together. Others allow your trunk to twist or your limbs to move. In movable joints, the tips of the bones are lubricated and covered with a layer of smooth cartilage, allowing them to slide easily against each other. The elbow and knee are hinge joints. They can swing up and down, but not side to side. The hip and shoulder are ball-and-socket joints. In this type of joint, a long bone that ends in a ball fits into a hollow socket. The long bone can move in almost any direction by swiveling within its socket.

Muscular and Integumentary Systems

Producing Movement

The **muscular system** is composed of

1. smooth muscles that line the blood vessels and hollow internal organs, such as those of the digestive tract, and

2. skeletal muscles that move the head, chest, and appendicular skeleton.

Kinds of Smooth Muscles

Smooth muscles are organized into sheets of tissue. In some smooth muscles, such as those that line the walls of many blood vessels, the cells contract simultaneously when the sheet is stimulated by a nerve or hormone. In other smooth muscles, such as those that line the wall of the gut, individual cells contract spontaneously, producing a slow, steady contraction.

Kinds of Skeletal Muscles

Skeletal muscles have different functions. Muscles of the fingers and eyes must contract rapidly, and they fatigue with continued use. The muscles that control posture, such as the long muscles of the back, contract slowly and do not fatigue. Skeletal muscles can function in these different ways because they have different kinds of fibers. Muscles that contract rapidly have a fast but inefficient metabolism, while muscles that contract slowly have a slower but more efficient metabolism.

Covering the Body

The **integumentary system** covers and protects the body. Human skin is from 10 to 30 cells thick, about as thick as this page; the outer layer of the skin is epithelial tissue.

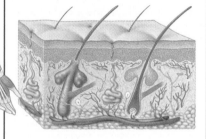

Maintaining the Skin

Cells from the outer layer are continually being injured and worn; they also lose moisture and dry out. The integumentary system deals with this damage not by repairing cells but by replacing them. Cells from the outer layer are shed continuously and replaced by new cells that have been produced deeper within the skin. A cell normally lives on the surface of the skin for about a month before it is shed.

Motor Units

Each cell of a skeletal muscle is called a **muscle fiber.** Nerve cells called motor neurons control the contraction of muscle fibers. Motor neurons usually branch at their tips and contact several muscle fibers. All of the fibers controlled by one motor neuron, called a **motor unit,** contract together when the motor neuron fires. Large muscles, such as those of the leg, have more than 100 muscle fibers per motor unit, while small muscles, such as those of the eye, have less than 10.

Motor Unit

Muscle fibers   Neuron

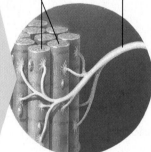

How Much Force?

The total amount of force developed by a muscle depends on two things:

1. how often the motor neuron leading to the muscle fires—repetitive firing produces stronger contraction of individual muscle fibers, an effect called **summation;** and

2. the number of muscle fibers involved in the contraction— stronger contractions occur when additional motor neurons fire, activating additional groups of muscle fibers, an effect called **recruitment.**

Circulatory and Immune Systems

The Body's Transportation System

The **circulatory system** contains the network of blood vessels that connects the organs and muscles of the body, the muscular heart that pumps blood through the vessels, and the lymphatic vessels that collect excess fluid. The circulatory system transports food and nutrients, oxygen-bearing red blood cells, and defensive immune system cells. It also helps maintain a uniform body temperature by transferring heat from one part of the body to another.

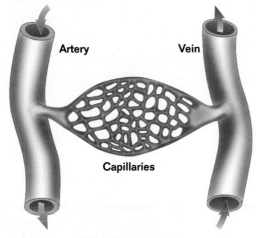

Arteries and Veins

Arteries carry blood away from the heart. Because they have elastic walls, arteries can expand their volume when the heart contracts. Their walls also contain a layer of smooth muscle that allows the body to control the amount of blood flowing to particular tissues. Veins return blood to the heart. The walls of veins are much thinner than those of arteries—when a vein is empty, its walls collapse.

Capillaries

Capillaries connect arteries to veins. They are very short and narrow. The average capillary is only 1 mm (0.04 in.) long with an internal diameter of 8 μm (slightly larger than the diameter of a red blood cell). However, there are many capillaries in the body. Laid end to end, the capillaries in your body would extend across the United States. Oxygen and food are transferred from the blood to the body's cells, and wastes are collected in capillaries. No cell of the human body is more than 100 μm from a capillary.

The Lymphatic System

Because some liquid is always being lost from the blood supply by diffusion through the walls of the smallest blood vessels, it is necessary to constantly re-collect this fluid. This recycling task is carried out by the lymphatic system, an open series of vessels that collects fluid and returns it to the blood vessels.

Defending the Body Against Infection and Cancer

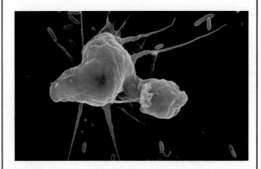

The **immune system** is composed of defensive white blood cells that circulate in the bloodstream and line the lymphatic system. The cells of the immune system constantly monitor the bloodstream and tissues to detect invading bacteria, fungi, protists, or viruses; to identify body cells infected with viruses; and to screen for cancer cells. When an infection occurs or an infected or cancerous cell is identified, the cells of the immune system marshal an attack and destroy the invading microbes or abnormal cell. No one can live for long without the defense they provide.

Blood

Blood is a protein-rich fluid in which many kinds of cells are suspended—cells are 40 percent of the total volume. Each milliliter of blood contains about 5 million oxygen-carrying **erythrocytes** (red blood cells) and 10,000 body-defending **leukocytes** (white blood cells), as well as cells and cell fragments involved in blood clotting and waste removal.

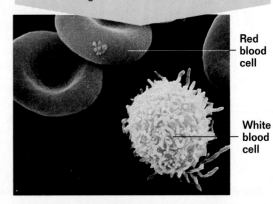

Red blood cell

White blood cell

Exchanging Gases

The respiratory system captures oxygen and releases carbon dioxide. This system is composed of a pair of **lungs,** which hang free within the fluid-filled chest cavity, and the passages that link the lungs to the nose and mouth.

Trachea

Air travels from the nose or mouth to the lungs through a long tube called the **trachea,** which branches near the lungs into the left and right **bronchi** (singular, bronchus). At the lung, each bronchus branches into many bronchioles, each connected to a large number of alveoli. Mucus produced by cells lining the trachea traps dust particles before they reach the lungs. Cilia on cells lining the trachea sweep the mucus upward so that it can be swallowed.

Alveoli

Each lung consists of some 300 million tiny air-filled sacs called **alveoli.** Each alveolus is surrounded by a network of about 100 tiny blood vessels. Gases diffuse across the short distance from alveolus to blood and back. The key to the structure of the lung is the enormous surface area that this arrangement provides for gas exchange—60 to 80 sq. m (650 to 860 sq. ft.) per lung. The total inner surface area of your lungs is about one-half the area of a tennis court.

Trachea
Lungs
Bronchus
Bronchiole

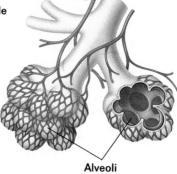

Alveoli

Diaphragm

Moving Air Into and Out of the Lungs

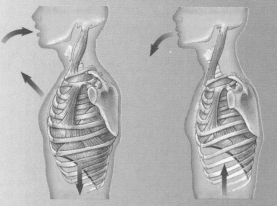

Inhalation

Exhalation

Diaphragm contracts

Diaphragm relaxes

A thick layer of muscle called the **diaphragm** separates the thoracic cavity, containing the lungs, from the abdominal cavity, containing the digestive organs. The diaphragm is dome-shaped when relaxed. During inhalation, the diaphragm contracts, causing it to flatten. Muscles on the ribs raise and expand the rib cage. The actions of these muscles enlarge the thoracic cavity and reduce the pressure in the lungs, allowing air to rush inward. During exhalation, the diaphragm relaxes and the muscles on the ribs lower the rib cage, compressing the thoracic cavity and driving air outward.

Digestive and Urinary Systems

Obtaining Nutrients From Food

The **digestive system** is a series of organs involved in breaking down food and absorbing nutrients. The digestive system includes the organs through which food passes and also includes the liver and pancreas, which manufacture many digestive enzymes and hormones.

The function of the digestive system is to break up the large molecules in food (proteins, carbohydrates, and fats) into small subunits (amino acids, sugars, glycerol, and fatty acids). These molecules are then absorbed into the bloodstream and carried to the cells of the body.

Regulating the Body's Salt and Water Levels

The **urinary system** filters metabolic wastes from the bloodstream and controls the ionic composition of the blood. This system is composed of the kidneys, urinary bladder, and associated ducts.

Kidneys

The kidneys are about the size of a small fist and are located in the lower back region. The kidneys play a crucial role in waste removal and water retention. Your blood passes through the kidneys for cleansing about once every four minutes; in the course of a full day, about 2,000 L (530 gal.) of fluid pass out of the bloodstream and into the kidneys, and all but a very small portion of that fluid is reabsorbed.

Urinary Bladder

Urine, the residual unabsorbed fluid that contains the body's wastes, is stored in a hollow, muscular sac called the bladder. The bladder can hold over 0.5 L (about 0.5 qt.) when full.

Mouth
Esophagus

Liver
Stomach
Kidney
Pancreas
Small intestine
Large intestine
Urinary Bladder

■ Functional Area #1: Processing and Initiation of Digestion

The **mouth** grinds and lubricates food, and enzymes in saliva begin to break down its large molecules. The **esophagus** carries food to the stomach.

■ Functional Area #2: Acidic and Enzymatic Digestion

The **stomach** receives the food and gives it an acid bath. This unfolds proteins. The stomach also releases digestive enzymes.

■ Functional Area #3: Completion of Digestion and Absorption

From the stomach, the food passes to the **small intestine,** where most digestion takes place. Carbohydrates, proteins, and fats are broken down into sugars, amino acids, glycerol, and fatty acids. Once these small molecules have been produced, they pass through the wall of the small intestine and into the bloodstream.

■ Functional Area #4: Concentration of Solids

Undigested solids pass into the **large intestine,** where they are compacted and much of the water they contain is reabsorbed.

Regulating the Body's Functions

The endocrine system uses chemical signals called **hormones** to coordinate and integrate the activities of the body. This system is composed of ductless glands called endocrine glands, such as the pituitary, thyroid, and adrenal glands. Hormones regulate growth and sexual development and maintain physiological conditions within narrow bounds. The proper balance of sodium, potassium, and calcium ions is under careful hormonal control, as is the level of glucose in the blood.

Hypothalamus

Pituitary

Thyroid

Parathyroids

Adrenal glands

Pancreas

Hypothalamus

A small area in the lower portion of the brain called the hypothalamus directs much of the endocrine system. Special brain hormones diffuse from the hypothalamus to a nearby endocrine gland called the **pituitary gland,** which responds by producing one of eight pituitary hormones.

Parathyroids

Regulate calcium metabolism and bone building

Thyroid

Regulates metabolism

Adrenal Glands

The interior of the adrenal gland, the **adrenal medulla,** releases adrenaline, a hormone that enables you to act in an emergency. Among adrenaline's effects are increased heart rate, heightened awareness, expansion of the airways leading to the lungs, and an increase in metabolic rate. Secretions of the outer layer of the adrenal gland, the **adrenal cortex,** maintain the levels of sodium and potassium in the body.

Pancreas

Controls levels of glucose in the blood

Hormones

Hormones are chemical messenger molecules that are manufactured in a small quantity in one part of the body and then transported through the bloodstream to another location, where they bring about a physiological response. Unlike the effects of chemicals used as messengers in the nervous system, the effects of hormones tend to persist for a long time.

Pituitary Hormones

Pituitary hormones are action orders. They pass to endocrine glands located in other parts of the body and cause these glands to begin production of particular hormones. The hypothalamus controls the commands that the pituitary issues to the endocrine system like an army general controls the orders that a captain issues to the troops.

Nervous System

Receiving and Interpreting Information

The nervous system is composed of

1. the body's **sense organs**, which provide information about the body and its environment,
2. **sensory nerves**, which relay the sensory information to the central nervous system,
3. the **brain** and **spinal cord**, which process and integrate sensory information and issue commands in the form of nerve impulses, and
4. **motor nerves**, which transmit commands to the body's muscles, organs, and glands.

Cerebrum

Cerebellum

Brain stem

Spinal cord

Cerebrum

The cerebrum is responsible for learning, memory, and thought. It evaluates sensory information and issues commands. The cerebrum is the largest part of the brain. The cerebrum is so large that it envelops much of the rest of the brain. A human brain weighs about 1.5 kg (3 lb.), and over 1 kg (2 lb.) of it is the cerebrum.

Cerebral Cortex

Much of the learning activity that occurs in the cerebrum takes place within a thin layer on its outer surface called the cerebral cortex. The human cerebral cortex contains over 10 billion nerve cells, amounting to roughly 10 percent of all the neurons in the brain.

Nerve Impulses

A nerve impulse is a short-lived electrical disturbance that passes rapidly along a neuron. Every nerve impulse in the nervous system is the same, involving the same magnitude of electrical change in the neuron cell membrane. The frequency (pattern) and the point of origin of nerve impulses are what determine what information is carried from one part of the body to another.

Cerebellum

The cerebellum receives information from the sense organs, muscles, joints, and inner ears. It integrates this information to maintain balance and coordination.

Brain Stem

The brain stem controls functions such as heart rate, blood pressure, and breathing.

Spinal Cord

The spinal cord carries input from sensory nerves to the brain and transmits commands from the brain to the motor nerves.

Producing Gametes

The female and male reproductive systems act to join **egg cells** from a female with **sperm cells** from a male. As in all mammals, the egg is fertilized within the female. The fertilized egg develops into a mature fetus in the uterus.

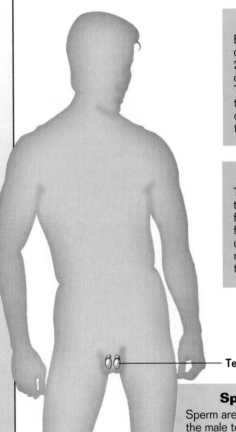

Ovaries

Eggs are formed in the ovaries and develop from cells called **oocytes.** At birth, a female has over 2 million oocytes, all she will ever produce. These oocytes have begun, but not completed, meiosis. They remain in a dormant state until puberty, when they resume meiosis and begin to mature. Usually only one egg matures and is available for fertilization each month.

Fallopian Tubes

The **fallopian tubes** connect the ovaries to the uterus. Mature egg cells are released from the ovary into the fallopian tubes. For fertilization to occur, a sperm must travel up a fallopian tube and encounter a mature egg soon after its release from the ovary.

Fallopian tube

Ovary

Uterus

 Testes

Sperm

Sperm are formed in the male testes, develop over a period of about two months, and are delivered in incredible numbers by ejaculation from the penis. A typical male produces several million sperm every day of his adult life.

Uterus

A successfully fertilized egg journeys down one of the fallopian tubes to the uterus. There it embeds in the lining and starts the long developmental journey that eventually leads to birth.

Vocabulary

actin (833)
axon (833)
central nervous system (826)
chemical signal (839)
collagen (832)
dendrite (833)
electrical signal (839)
epithelium (829)
erythrocyte (832)

feedback loop (837)
glial cell (832)
hormone (840)
hypothalamus (838)
joint (825)
ligament (825)
lymphocyte (831)
macrophage (831)
myosin (833)

negative feedback (837)
neuron (832)
organ (829)
organ system (829)
peripheral nervous system (826)
positive feedback (838)
set point (837)
supporting cell (832)

Concept Mapping

Construct a concept map that shows how the human body is organized. Use as many terms as needed from the vocabulary list. Try to include the following items in your map: cells, tissues, organs, organ systems, epithelial tissue, connective tissue, nerve tissue, muscle tissue, endocrine system, nervous system, reproductive system, circulatory system, immune system, respiratory system, and digestive system. Include additional terms in your map as needed.

Review

Multiple Choice

1. Movement of the skeleton is made possible by
   a. bones.
   c. cartilage.
   b. joints.
   d. skin.

2. What enables chemical reactions occurring in your body to proceed at a constant rate?
   a. endothermy
   b. fat content of muscles
   c. body movement
   d. the coelom

3. What tissue type functions to cover and protect the body?
   a. connective
   c. muscle
   b. nerve
   d. epithelial

4. Cells of the nervous system that surround and insulate neurons are called
   a. axons.
   c. glial cells.
   b. dendrites.
   d. myosin.

5. What muscle type lines the walls of the blood vessels and gut?
   a. cardiac muscle
   c. smooth muscle
   b. skeletal muscle
   d. rough muscle

6. Which organ system concentrates and removes waste from the body?
   a. the urinary system
   b. the immune system
   c. the skeletal system
   d. the endocrine system

7. Which medical test is least likely to reveal information about homeostasis in the human body?
   a. chest X-ray
   b. urinalysis
   c. blood test
   d. measurement of body temperature

8. Which is an example of positive feedback?
   a. the operation of a home thermostat
   b. an explosion
   c. upward and downward trends in the stock market
   d. fluctuations in pH of body fluids

Completion

9. The site at which two bones meet is called a(n) _____ . Tough bands of connective tissue called _____ hold the bones in place.

10. The brain and spinal cord make up the _____ nervous system, while the _____ nervous system consists of nerves that connect the brain and the spinal cord to the rest of the body.

11. The lungs are lined by _____ epithelium that is one cell thick. The skin is partly composed of _____ epithelium that is several cells thick.

12. Organ systems are made up of _____ that, in turn, are composed of different types of tissues.

13. The _____ system defends the body against infection. The functional units of this system are _____ cells that make antibodies or attack infected cells or pathogens.

14. Homeostasis of the body is regulated by a region of the brain called the _____, which detects and initiates changes in the body's condition.

15. Maintaining homeostasis in the human body involves the operation of _____ signals carried by neurons and _____ signals transmitted by hormones.

Short Answer

16. List two benefits of endothermy. What is the major disadvantage of endothermy?

17. What type of muscle tissue is shown in the photograph below? How does the structure of this kind of tissue reflect its function?

18. Contrast the functions and characteristics of each kind of connective tissue.

19. Explain how your body responds to an increase in its temperature. What role does the hypothalamus play in the response?

Highlights Review

20. The human skeleton is composed of two major parts: the axial and the appendicular skeletons. What bones make up the axial skeleton? What bones make up the appendicular skeleton? What are the functions of these two parts of the human skeleton?

21. The lymphatic system is sometimes called a backup circulatory system. What does the lymphatic system do that has earned it this title?

22. The main organs of the respiratory system are the lungs. Air is drawn into and forced out of the lungs, but they do not have muscles. How is air moved into and out of the lungs?

23. Serious diarrhea is associated with failure of the large intestine to reabsorb water. How can serious diarrhea affect water balance in the body?

Themes Review

24. **Levels of Organization** What is the relationship between an organ and its organ system?

25. **Structure and Function** Blood vessels have one layer of smooth muscle that encircles the vessel wall. The stomach and intestine have one layer of circular smooth muscle and one layer of longitudinal (running lengthwise) smooth muscle. How does this muscle arrangement reflect the function of each structure?

26. **Homeostasis** How does a feedback loop that involves negative feedback help the body to maintain homeostasis?

27. Structure and Function Compare and contrast the structures of smooth muscle, skeletal muscle, and cardiac muscle. How is the structure of each muscle type related to its function?

Critical Thinking

28. Interpreting Data The graph below shows the relationship between body temperature and environmental temperature for two animals. Which line, A or B, best represents the human condition? Explain your answer.

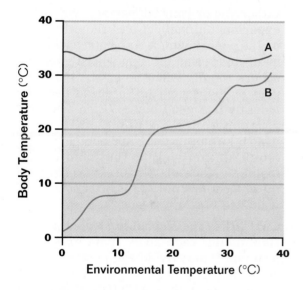

29. Making Inferences Polio is a disease that paralyzes muscles by affecting the nerves that make them move. Before polio vaccines were developed, many people were stricken by polio, and they sometimes died because they could not breathe. From what you know about the respiratory system, why might this happen?

30. Making Inferences The hormone insulin travels throughout the body in the bloodstream, but only certain types of cells respond to its presence by taking in glucose. How can this selective influence be explained?

Activities and Projects

31. Research and Writing Do library research and talk to medical doctors to find out about arthroscopic surgery. If possible, view a videotape of a surgical procedure to gain additional information. Write a report including a description of the nature of the surgical procedure, the types of injuries best suited for arthroscopic surgery, and the benefits associated with this type of surgery.

32. Science-Technology-Society Medical care for AIDS patients is very expensive. For this reason, insurance companies routinely require blood tests for persons applying for medical insurance and deny insurance to applicants carrying HIV, the virus that causes AIDS. Is this practice ethical? Construct an answer to this question by surveying students in your school, insurance agents, and medical workers who care for AIDS patients. Share your findings with your class.

Readings

33. Read the article "A vaccine for AIDS," in *The American Biology Teacher*, April 1993, pages 198–202. Describe one AIDS vaccine in development, and explain its disadvantages. What problems hamper the development of all AIDS vaccines?

34. Read the article "The Inside Dope on Runner's High," in *Runner's World*, August 1994, pages 60–64. What triggers runner's high and what physical sensations are associated with it? How far must a person run to experience runner's high?

35. Read the article "On the trail of lethal DNA," in *Newsweek*, May 17, 1993, page 63. About how many people die of colon cancer each year in the United States? What have new discoveries revealed about the cause of colon cancer?

LABORATORY Investigation | Chapter 35

Thermoregulation

OBJECTIVES

- Demonstrate the distribution of sweat glands in skin.
- Determine the sweat gland activity with a change of environmental temperature and body temperature.

PROCESS SKILLS

- organizing data in tables
- comparing effects of different experimental conditions

MATERIALS

- safety gloves
- pre-moistened alcohol pads
- antibacterial soap (with iodine)
- cotton swabs
- 2 in. squares of erasable bond paper (no substitution)
- heat source lamp (no sun lamps)
- towel
- metric ruler

BACKGROUND

1. What is homeostasis?
2. How is internal body temperature regulated in humans?
3. How does the internal body temperature of mammals and birds compare to that of reptiles and amphibians?
4. What are the sensory organs that communicate with the brain's temperature control center in the hypothalamus?
5. Write your own **Focus Question** on your Vee Form.
6. **Knowing Side of the Vee** List the **Concepts** and new **Vocabulary Words** on your

Vee Form. In the **Concept Statements** section of the Vee, use these words in sentences that define and explain them.

TECHNIQUE

Doing Side of the Vee

CAUTION: Wear safety goggles during this investigation. Students should be careful when handling any solutions they use to swab the skin, especially on the face.

Part A: Heat and Sweat Gland Activity

1. Either you or your lab partner will be the experimental subject and the other will be the experimenter. Ask your partner to name an area on the skin where sweating occurs rapidly when he or she gets nervous or too warm (the forehead near the hairline is a good place). Also ask your partner to name a place on the skin where sweating does not occur rapidly (such as the back of the hand or the inside surface of the forearm). Record these locations in the **Records** section of your Vee.

2. **CAUTION: The alcohol in alcohol pads is flammable; keep away from excessive heat and flame. Avoid eye contact.** Using an alcohol pad, wipe a 3 in. square area on each location. **CAUTION: Soap with iodine can stain clothing. Avoid eye contact. In case of eye contact, call your teacher and flush eye with water for 15 minutes.** Using the soap with iodine and a cotton swab, carefully cover a 2 in. square in each of those areas. Blot them dry. After they have dried, gently hold a 2 in. square of the bond paper on each area for 30 seconds.

3. Take the paper from the skin and observe it closely for any blue dots. The bond paper contains starch, which reacts with the iodine to form a blue-black color.

4. Find the area of the bond paper with the greatest concentration of dots and mark off a 1 cm square. In the **Records** section of your Vee, make a table similar to the one below. Record the number of dots in your table under the column labeled "No stimulus."

Sweat Gland Activity

Number of dots

	No stimulus	After heat	After exercise
Forehead			
Back of hand			

5. Wipe the areas of the skin with a fresh alcohol pad and reswab them with iodine using a fresh cotton swab. Cover those areas with a towel or other cloth to prevent heat from being directed onto bare skin.

6. After you turn on the lamp, your lab partner should tell you the instant the heat is actually felt. DO NOT bring the lamp closer than 6 in. from the skin. DO NOT touch the lamp. Gently hold a 2 in. square of bond paper on each area, asking the subject to help hold it in place. From that point, time the heat application for 10 minutes. Record the resulting number of dots in the column labeled "After heat" in your table in the **Records** section of your Vee.

Part B: Exercise and Sweat Gland Activity

7. Repeat steps 5 and 6 with exercise rather than heat as the variable. Have the subject jog in place or do jumping jacks, push-ups, sit-ups, or another form of moderate exercise for 10 minutes, allowing brief rest breaks as needed. Record the number of dots under the column labeled "After exercise" in your table in the **Records** section. In the **Procedure** section of the Vee, briefly summarize the procedure you followed.

8. Clean up your materials and wash treated areas with soap and water before leaving the lab.

INQUIRY

1. Where do you hypothesize that the greatest number of sweat glands are located: on the back of the hand or forearm or on the forehead?

2. What do the dots on the bond paper represent?

3. Were there more dots on the paper before or after the heat exposure?

4. Use the information on the **Knowing Side** of the Vee to interpret your results from the **Doing Side,** and then write your **Knowledge Claim.** Write a **Value Claim** for this lab.

ANALYSIS

1. What does sweating accomplish for the body?

2. How is sweating an adaptation that maintains homeostasis?

3. As your subject reacted to heat and exercise, what changes were observed in the sweat glands?

4. How do the sweat glands fit into the analogy of the body as a human thermostat?

FURTHER INQUIRY

Write a **New Focus Question** that could be the point of a new investigation. The following is an example:

Will the placement of cold compresses near a skin area reduce the activity of sweat glands there?

CHAPTER 36

SKELETON, MUSCLES, AND SKIN

REVIEW

Olympic gold medalist Gail Devers

36-1 The Skeletal System

An internal framework of calcium-hardened bones shapes and supports the human body. This endoskeleton is one of the hallmarks of the vertebrate body. All of the skeletons discussed in previous chapters have a serious limitation. Chitin, a carbohydrate found in arthropod exoskeletons, is brittle and is not very strong. Calcium alone is hard and strong but is also brittle and likely to fracture. The calcium-hardened bone endoskeleton of vertebrates, however, has enabled them to become the most mobile animals. The human skeleton provides protection for internal organs and acts as a versatile system of levers and joints. The body's muscles pull against these levers and joints, enabling the arms and legs to move.

Section Objectives

- Describe the structure of a bone.
- Differentiate between the axial skeleton and the appendicular skeleton.
- Name the three main types of joints, and explain how an example of each works.
- Discuss causes and effects of osteoporosis, and explain how it can be prevented.

The Human Skeleton Is Mainly Bone

Your skeleton is made mostly of bone, a dynamic living tissue that is constantly formed and replaced as long as you live. Bone has a great advantage as a structural material. Its calcium-hardened collagen fibers make it strong but flexible, unlike the brittle chitin of the arthropod exoskeleton. As Figure 36-1 shows, bone tissue occurs in two forms. **Compact bone** is a dense, almost solid tissue that provides a great deal of support. **Spongy bone,**

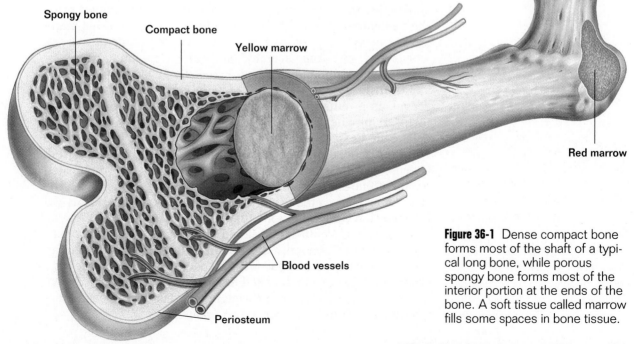

Spongy bone

Compact bone

Yellow marrow

Red marrow

Blood vessels

Periosteum

Figure 36-1 Dense compact bone forms most of the shaft of a typical long bone, while porous spongy bone forms most of the interior portion at the ends of the bone. A soft tissue called marrow fills some spaces in bone tissue.

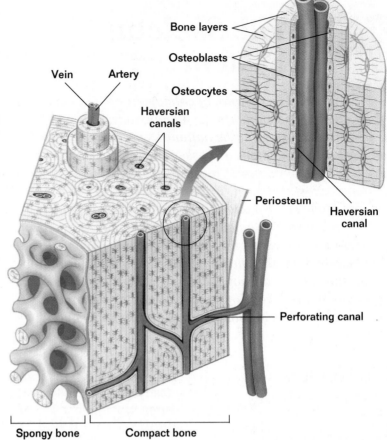

Figure 36-2 A scanning electron micrograph (SEM) of a section of compact bone from which all organic tissue was removed, *above,* shows the concentric rings of bone that surround a Haversian canal. The diagrams, *right,* show the structure of compact and spongy bone, including the many living cells and blood vessels that are an important part of bone tissue.

Labels in figure:
Bone layers
Osteoblasts
Osteocytes
Vein   Artery
Haversian canals
Periosteum
Haversian canal
Perforating canal
Spongy bone   Compact bone

which provides lightweight support, consists of hardened fibers interspersed with many spaces. Some of these spaces are filled with a soft tissue called **red marrow,** which produces red blood cells in some bones. The hollow interiors of the arm and leg bones are filled with **yellow marrow,** which stores fat. Individual bones are surrounded by a membrane, the **periosteum** *(pehr ee AHS tee uhm),* which contains many blood vessels. The word *periosteum* comes from the Greek words *peri,* meaning "around," and *osteon,* meaning "bone."

When the human body first takes shape as an embryo, the skeleton is made of cartilage. This cartilaginous skeleton serves as a template for bone formation. As you grow, bone forms along lines of stress. New bone is formed by young bone cells called **osteoblasts** *(AHS tee oh blasts),* which secrete the collagen fibers on which calcium phosphate is deposited. Osteoblasts build bone by laying down thin layers of collagen on top of one another, like layers of paint. In compact bone, the first layer is secreted around a narrow, hollow channel, called a **Haversian canal,** that extends the length of the bone. As you can see in Figure 36-2, additional layers are laid down, creating a series of concentric tubes of bone. Eventually, young bone cells become trapped within spaces in the bone they lay down, and they are then called **osteocytes** *(AHS tee oh seyets).* The blood vessels that run through each Haversian canal supply the osteocytes with oxygen and nourishment. ❑

Osteoporosis Results From Bone Loss

Bone tissue, which also serves as a storehouse for minerals needed by the body, is continuously broken down and replaced. In babies, new bone is added more rapidly than it is broken down, and bones grow larger and denser. In young adults, bone tissue is broken down and replaced at the same rate, so the density of bone usually remains relatively constant. However, during middle age, bone replacement gradually becomes less efficient. As a result, bones become less dense and store fewer minerals. When bone loss is severe, as seen in Figure 36-3, the condition is called **osteoporosis** *(ahst ee oh puh ROH sihs)*, which means "porous bone." Bones that are affected by osteoporosis become brittle and are easily broken. In the United States, more than 600,000 fractures result from osteoporosis each year.

Figure 36-3 In the photos of bone tissue, *below left,* compare the density of bone tissue in a normal bone, *left,* with that in a bone weakened by osteoporosis, *right.* An active lifestyle, *below,* and regular exercise help prevent or delay osteoporosis.

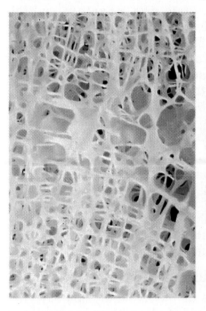
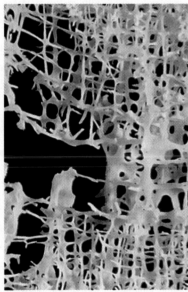

Although both women and men lose bone as they age, many more women are affected by osteoporosis. Because their bones are smaller, women cannot afford to lose as much bone tissue as men. Women also lose a lot of calcium each month during the menstrual cycle. Moreover, the production of sex hormones, which help maintain bone density, declines rapidly during menopause, resulting in an increased rate of bone loss in women.

To avoid the consequences of osteoporosis, you must act at an early age. Bone density can be increased only during the teens and twenties with regular exercise and a calcium-rich diet that includes dairy products such as milk and yogurt. Although regular exercise throughout your life will help maintain bone density, strengthening your bones now will make you less likely to be affected by osteoporosis later in life. ❑

❑ CAPSULE SUMMARY

Osteoporosis, which affects many more women than men, is a loss of bone density that causes bones to become brittle and to break easily. Osteoporosis can be delayed or prevented by a calcium-rich diet and regular exercise starting at an early age.

The Human Skeleton Has Two Main Parts
....

The human skeleton is made up of 206 individual, calcium-hardened bones. Of these, 80 bones form the main body axis and are called the **axial skeleton.** The other 126 bones form the arms and legs and are called the **appendicular** *(ap ehn DIHK yoo luhr)* **skeleton.** The word *appendicular* is derived from the Latin word *appendere,* meaning "to hang." The bones of the axial and appendicular skeletons can be seen in Figure 36-4.

The Axial Skeleton

The skull, backbone, and rib cage compose the axial skeleton. The most complex element of the axial skeleton is the skull. Of the 29 bones in the skull, 8 bones form the **cranium,** which encases the brain. The skull also has 14 facial bones, 6 middle-ear bones, and a single bone that supports the base of the tongue. The skull is attached to the top of the backbone, or **spine,** which is a flexible stack of 26 vertebrae. The spinal cord passes through all but the lowermost vertebrae. Curving forward from the middle vertebrae are 12 pairs of ribs. At the front of the body, most of the ribs are attached to the breastbone, or **sternum,** forming a protective **rib cage** around the heart and lungs.

The Appendicular Skeleton

The 126 bones of the appendicular skeleton form the human body's appendages—the arms and legs. The arms and legs, each of which contains 30 bones, are attached to the axial skeleton at the shoulders and hips, respectively. The shoulder attachment is called the **pectoral girdle** and is composed of two large, flat shoulder blades (scapulas) and two slender, curved collarbones (clavicles). The collarbones, which connect the shoulder blades to the sternum, hold the shoulders apart. This arrangement permits the full rotation of the arms about the shoulder joints. When you fall on an outstretched arm, much of the force is transmitted to the collarbones—the most frequently broken bones in the human body. The hip attachment is called the **pelvic girdle** and is composed of two large pelvic bones that form a bowl. The pelvic bones transmit the weight of the body squarely down the legs. ◻

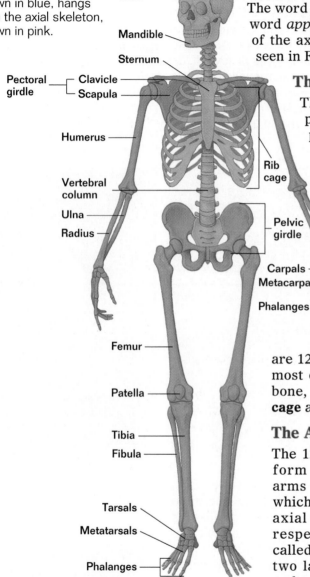

Figure 36-4 Some major bones of the human skeleton are identified here. Notice that the appendicular skeleton, shown in blue, hangs from the axial skeleton, shown in pink.

Skull
Mandible
Sternum
Pectoral girdle — Clavicle — Scapula
Humerus
Vertebral column
Ulna
Radius
Rib cage
Pelvic girdle
Carpals
Metacarpals
Phalanges
Femur
Patella
Tibia
Fibula
Tarsals
Metatarsals
Phalanges

◻ CAPSULE SUMMARY

The skull, spine, and rib cage make up the axial skeleton. The appendicular skeleton consists of the bones of the arms and legs and of the pectoral and pelvic girdles.

Joints Fasten Bones Together

The junction of two bones is called a **joint.** Most joints permit movement of the bones they join. These movements are aided by the structure of the joint. For example, pads of cartilage cushion the ends of both bones in a joint. The axial and appendicular skeletons contain three different kinds of joints that enable varying degrees of movement. Examples of the three basic types of joints can be seen in Figure 36-5.

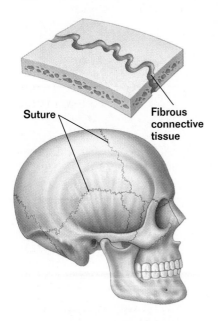

Suture

Fibrous connective tissue

Body of vertebra

Cartilage   Disc

Immovable joint

Slightly movable joint

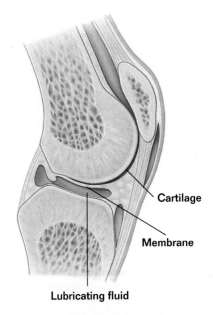

Cartilage

Membrane

Lubricating fluid

Freely movable joint

1. **Immovable joints**   Very tight joints that permit little or no movement of the bones they join are called **immovable joints.** The cranial bones of the skull are joined by sutures, a type of immovable joint in which the bones have only a thin layer of connective tissue between them.

2. **Slightly movable joints**   Joints that permit limited movement of the bones they join are called **slightly movable joints.** The vertebrae in the spine are joined by cartilaginous joints, a type of slightly movable joint in which a bridge of cartilage joins the two bones. The pads of cartilage, or disks, that separate the vertebrae also act as shock absorbers.

3. **Freely movable joints**   The joints that permit the greatest degree of movement are called **freely movable joints.** Recall that ligaments hold the bones of a freely movable joint together and that a membrane surrounding the joint contains a lubricating fluid that separates the ends of the bones. **Rheumatoid arthritis,** a very painful degeneration of movable joints, occurs when cells of the immune system attack these membranes, weaken the cartilage, and deposit bone in its place. The hinge joints in elbows and

Figure 36-5 Individual bones in the skull are joined by sutures, *left,* a type of immovable joint. The vertebrae in the spine are joined by cartilaginous joints, *center,* a type of slightly movable joint. The knee, *right,* is an example of a hinge joint, a type of freely movable joint.

knees are perhaps the most familiar freely movable joints. Table 36-1 lists some other freely movable joints found in the appendicular skeleton.

Table 36-1 Types of Freely Movable Joints

Name	Type of Movement	Examples
Ball-and-socket joint		Shoulders and hips
Pivot joint		Elbows
Plane joint		Carpals of hands and tarsals of feet
Saddle joint		Thumbs

Section Review

1. *How does compact bone differ from spongy bone?*
2. *Where are the axial and appendicular skeletons attached to each other?*
3. *Which type of freely movable joint allows the widest range of motion? Give an example of this kind of joint.*
4. *What are two reasons that women are more likely than men to develop osteoporosis?*

Critical Thinking

5. *The bones of a newborn baby are made mostly of cartilage. Why is this is an advantage?*

36-2 The Muscular System

While smooth muscle lines your organs and blood vessels, your body is composed largely of skeletal muscle that is devoted to moving the parts of your body. If you have ever tried to lift a heavy object, you are familiar with a basic problem of movement—force must be applied to overcome the forces that tend to hold the object in place. By splitting ATP molecules into ADP and inorganic phosphate, muscle cells utilize the body's store of chemical energy to contract (shorten their length). Thus, muscle cells exert force by pulling on surrounding tissue. When a lot of muscle cells contract at one time, they are able to exert a great deal of force.

Section Objectives

- Describe the action of flexors and extensors in moving the parts of the body.
- Explain how muscle contractions are produced.
- Discuss the two factors that determine the strength of a muscle contraction.
- Discuss the importance of exercise in increasing endurance and maintaining muscle strength.

Muscles Make the Skeleton Move

For muscles to move the parts of the body, they must be attached to something they can pull against. The bones of the skeleton provide these points of attachment. Most skeletal muscles are attached to bones by strips of dense connective tissue called **tendons.** One end of the muscle, the **origin,** is attached to a bone that remains stationary during a muscle contraction. This gives the muscle something to pull against. The other end of the muscle, the **insertion,** is attached to a bone that moves when the muscle contracts. A muscle's insertion always moves towards its origin.

Skeletal muscles are attached to the bones of the appendicular skeleton in opposing pairs called flexors and extensors. As Figure 36-6 illustrates, **flexors** cause the limbs to

Figure 36-6 A flexor and an extensor work together to enable you to bend your arm. Contraction of the biceps muscle, a flexor, bends the arm at the elbow. Contraction of the triceps muscle, an extensor, straightens the arm.

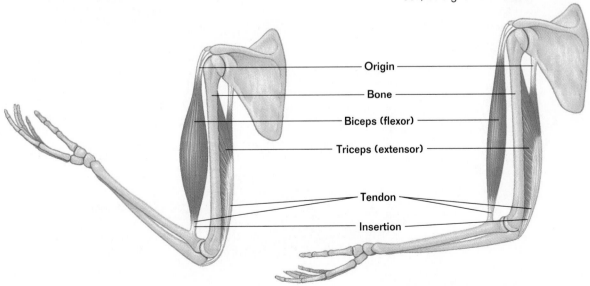

- Origin
- Bone
- Biceps (flexor)
- Triceps (extensor)
- Tendon
- Insertion

Skeleton, Muscles, and Skin   **861**

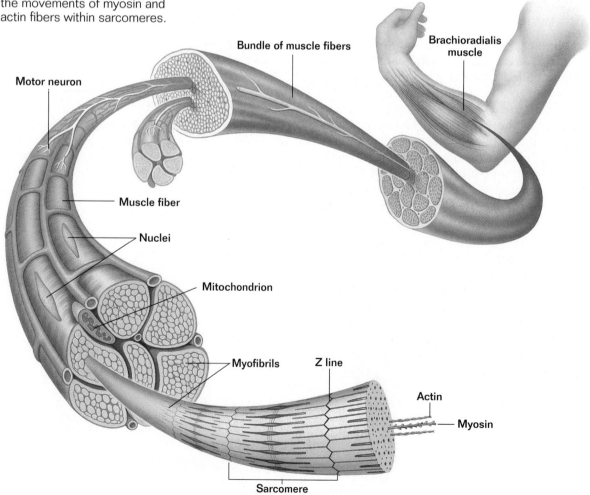

CAPSULE SUMMARY

Most muscles are attached to the bones they move by tendons. Muscles work in pairs to move the parts of the skeleton. A flexor pulls a bone in one direction, and an extensor pulls the bone back.

bend at a joint, while **extensors** cause them to straighten. Thus, one muscle in a pair pulls a bone in one direction, and the other pulls it back. When you move, the command to contract goes from the nervous system to one muscle or the other, but not to both muscles at the same time. If both muscles were commanded to contract, they would simply pull against each other, and the limb would not move. Thus, to produce movement, muscle contraction and relaxation are carefully coordinated and controlled by the nervous system.

Myosin and Actin Cause Muscle Fibers to Shorten

The structure of a muscle, such as that of the skeletal muscle seen in Figure 36-7, enables it to contract. Recall that skeletal muscles are made of many muscle fibers, which are strands of muscle cells joined end to end. Within each muscle fiber are many bundles of smaller

Figure 36-7 Bundles of muscle fibers (muscle cells) make up a skeletal muscle. Each muscle fiber is a bundle of many smaller fibers (myofibrils) that are bundles of protein (myosin and actin) fibers. Muscle contractions begin with the movements of myosin and actin fibers within sarcomeres.

Motor neuron

Bundle of muscle fibers

Brachioradialis muscle

Muscle fiber

Nuclei

Mitochondrion

Myofibrils

Z line

Actin

Myosin

Sarcomere

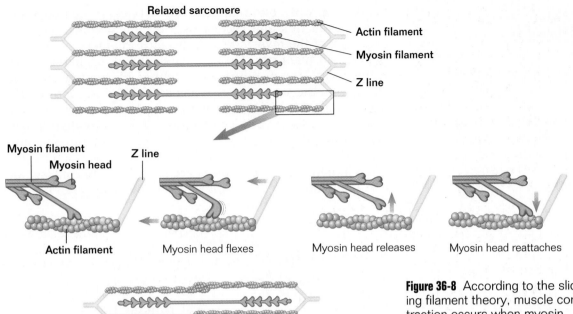

Relaxed sarcomere

Actin filament

Myosin filament

Z line

Myosin filament
Myosin head
Z line

Actin filament

Myosin head flexes

Myosin head releases

Myosin head reattaches

Contracted sarcomere

Figure 36-8 According to the sliding filament theory, muscle contraction occurs when myosin fibers within a sarcomere "walk" along adjacent actin fibers. This movement pulls the Z lines (to which actin is attached) of a sarcomere together, shortening the sarcomere. The shortening of the sarcomeres along a myofibril causes the myofibril to contract.

fibers called **myofibrils,** which in turn contain many filaments of the proteins myosin and actin. Notice that each myosin filament consists of a cluster of myosin molecules, each of which looks like a long rod with a large "head" at one end. The thinner actin filaments resemble two strings of beads twisted together. One end of each actin filament is anchored at a **Z line,** which is the location of a protein that crosses a myofibril. Myosin and actin filaments lie parallel to one another and are arranged in **sarcomeres** *(SAHR koh mihrz),* repeating units that are bounded by two Z lines. The characteristic light and dark bands of striated muscle result from the orderly and slightly overlapping arrangement of the myosin and actin filaments in the sarcomeres.

Sarcomeres are the functional units of contraction in a myofibril. In a relaxed muscle fiber, the ends of the myosin and actin filaments in each sarcomere barely overlap. When a muscle fiber contracts, adjacent myosin and actin filaments slide along one another so that they overlap a great deal and shorten each sarcomere. This explanation of muscle contraction is called the **sliding filament theory.**

How do actin filaments slide along myosin filaments, and how does this lead to the contraction of a muscle? As illustrated in Figure 36-8, myosin and actin filaments are so close to each other that the myosin heads can touch the adjacent actin filaments. Muscles contract when the myosin heads attach to the adjacent actin filaments and bend inward, pulling the actin filaments along with them. Each myosin head then releases the actin filament, "recocks" (flexes back

Skeletal muscles consist of bundles of fibers that contain bundles of smaller fibers called myofibrils. Muscles contract when myosin and actin fibers arranged in repeating units called sarcomeres slide together and shorten the sarcomeres along a myofibril.

to where it was), and reattaches to it, ready to bend and pull again. Each flex and recock uses a molecule of ATP. With each flex of the myosin heads, the actin filaments slide along the myosin filaments. The actin filaments drag their attached Z lines with them, pulling the Z lines closer together and shortening the sarcomeres. When stimulated to contract, this shortening of the sarcomeres occurs along the entire length of a muscle fiber. The contraction of an individual muscle fiber is an all-or-nothing response. Thus, the total amount of force that a muscle can exert depends on two factors: how often individual muscle fibers are stimulated to contract and how many muscle fibers contract in a given muscle. ◻

Exercise Increases Muscle Size and Efficiency

The ATP used to fuel muscle contraction is usually supplied by the aerobic pathway of cellular respiration. However, in rapidly contracting muscles, the oxygen supply soon becomes inadequate, and anaerobic processes take over. In the absence of oxygen, fermentation follows glycolysis, and the muscle has only glycolysis as a source of ATP. The lactic acid produced during fermentation causes muscle fatigue (the more acidic conditions lower the activity of glycolytic enzymes and interfere with the action of the myosin heads). Because the production of ATP by glycolysis is so inefficient, the body begins using the glycogen stored in muscles as a source of glucose. When the glycogen is depleted, the body must begin using fat as its only source of energy. However, energy production from fat occurs at only about half of the rate of energy production

Figure 36-9 Aerobic exercises, such as jogging, *below*, improve your body's ability to supply oxygen to its muscles and increase the efficiency of ATP production. Resistance exercises, such as chin-ups, *below right*, increase muscle size and strength.

from glucose, so the depletion of glycogen stores is marked by a substantial decrease in muscle performance. Long-distance runners refer to this as "hitting the wall."

With aerobic training and exercise, chest muscles can be strengthened so that more air (and oxygen) enters the body with each breath. Such exercises also make the heart pump more efficiently and increase the number of blood vessels in the muscles. These changes expand your body's ability to supply oxygen to its muscles and thus make ATP production more efficient. The increase in muscle efficiency results in greater endurance (ability to continue exercising).

Although exercise cannot increase the number of muscle cells, resistance exercises such as the chin-ups shown in Figure 36-9 can increase muscle size and strength. However, because these exercises do not significantly improve your body's ability to deliver oxygen to your muscles, they do not increase your endurance. Some athletes are tempted to use drugs called **anabolic steroids** to increase the size (and thus the strength) of their muscles. Anabolic steroids are powerful synthetic chemicals that resemble the male sex hormone testosterone and that trick the muscles into growing larger. Unfortunately, their use has many serious, often irreversible side effects, including cancer, heart disease, and altered sexual development. Anabolic steroids are dangerous drugs, and their use is illegal in both amateur and professional sports.

Overusing your muscles by exercising more than you usually do can lead to muscle injury. A **muscle strain,** commonly called a "pulled muscle," is the overstretching or tearing of a muscle. Muscle strain occurs when a muscle is overused or when strenuous exercise is done without warm-up exercises. A **sprain** is a torn or overly stretched muscle, ligament, or tendon. If excessive stress causes the tendons that attach the muscles to bone to become inflamed, a painful condition called **tendinitis** results.

CONTENT LINK

Steroids are described in greater detail in Chapter 41.

❏ CAPSULE SUMMARY

Exercise can increase the efficiency of your muscles by increasing your body's ability to supply oxygen to its muscle cells. Exercise can also increase the size and strength of your muscles.

Section Review

1. *How do flexors and extensors work together to move the limbs of the body?*

2. *How are the Z lines of a myofibril pulled closer together?*

3. *How can the force that a muscle exerts be increased?*

4. *What type of exercise helps to increase muscle size and strength?*

Critical Thinking

5. *After looking back at Figure 5-16 on page 111, explain why an increase in muscle efficiency increases your endurance.*

36-3 The Skin

Section Objectives

- Name the three layers of the skin, and describe the main functions of each.
- Explain how the epidermis, hair, and nails are formed.
- Explain how the dermis helps regulate body temperature.
- Identify the causes of skin cancer and acne, and describe how you can minimize your chances of getting skin cancer and how you can minimize the symptoms of acne.

*T*he largest organ of the human body—accounting for about 15 percent of your total body weight—is the skin. Many kinds of specialized structures are found in the skin, which forms the integumentary system. Among the stratified epithelial cells in 1 sq. cm of your skin (an area about the size of a dime) are about 200 nerve endings, 10 hairs and muscles, 100 sweat glands, 15 oil glands, 3 blood vessels, 12 heat sensors, 2 cold sensors, and 25 pressure sensors.

The Skin Is Composed of Three Layers

Skin is the outermost layer of the human body. It protects the body from injury, provides the first line of defense against disease-causing microbes, helps regulate body temperature, and prevents the body from drying out through evaporation. Human skin is a very complex organ that is composed of three layers: an outer epidermis, a lower dermis, and an underlying layer of subcutaneous tissue. The structure of the skin is seen in Figure 36-10.

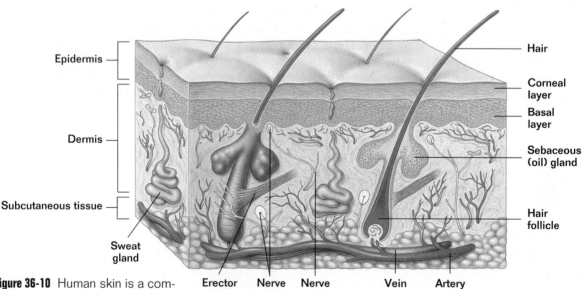

Figure 36-10 Human skin is a complex organ that consists of three layers as well as many blood vessels, nerve endings, muscles, hairs, and glands.

Epidermis

The **epidermis,** which is from 10 to 30 cells thick, is the outermost layer of the skin. About as thick as this page, the epidermis consists of several layers of stratified epithelial cells. The part of the epidermis that you see when you look in a mirror is the **corneal layer,** a thin layer of flattened, mostly dead cells that are filled with the protein keratin, an excellent waterproofing agent. The cells of the corneal layer are continuously damaged by encounters with the outside world. They are scraped, ripped, worn away by friction, and dried out by moisture loss. Your body deals with this damage not by repairing the cells but by replacing them.

The outermost cells of the epidermis are continuously shed and replaced by cells from the **basal layer,** a layer of actively dividing cells that lies at the base of the epidermis. As new skin cells form, they migrate upward and produce large amounts of keratin, the protein that makes the skin tough. After reaching the surface, a skin cell normally lives in the corneal layer for about a month before it is shed. The basal layer also contains **melanocytes** *(muh LAN uh seyets),* which are cells that produce the brown pigment melanin *(MEHL uh nihn).* Human skin color results from melanin, which provides the skin with protection from the sun's ultraviolet rays. The more these cells are exposed to the sun, the more melanin they produce. That is why skin "tans."

Hair and nails are also produced by the cells of the epidermis. Specialized epidermal structures called **hair follicles** produce individual hairs. As Figure 36-11 shows, hair is composed of dead, keratin-filled cells stacked on top of one another like a pile of roof shingles. Each hair on your head grows for several years; then the follicle enters a resting phase for several months, and the hair is shed. Nails are produced by specialized epidermal cells located in the light "half-moon" area at their base. These cells fill with tough keratin as they are pushed outward by the production of new cells.

Dermis

The **dermis** is the chief framework of the skin. It is 15 to 40 times thicker than the epidermis. The leather used to make belts and shoes is made from very thick animal dermis. The dermis serves as a structural support and as a matrix for the many nerve endings, blood vessels, and specialized cells of the skin. For example, the human sense of touch originates with the nerve endings in the dermis. These nerve endings, which are sometimes coupled to simple sensory receptors, enable you to sense pressure, pain, and temperature.

The dermis has tiny muscles that are attached to the hair follicles in your skin. When you are cold or afraid, these muscles contract and pull the hairs upright. Similar muscles cause a cat's fur to stand up when it is frightened (making it look bigger and more dangerous) or cold (trapping air around its body to insulate it from heat loss). In humans, the muscles that pull hairs upright also cause goose bumps.

The dermis is crisscrossed by a network of blood vessels that provide nourishment to the living cells of the skin. These blood vessels also help regulate body temperature either by radiating heat into the air or by helping to insulate the body. If your body gets too hot, muscles around the skin's blood vessels relax. When the muscles relax, the blood vessels enlarge

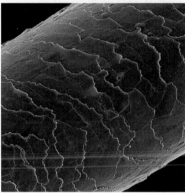

Figure 36-11 Human hair, *top,* grows from hair follicles in the skin. Hair color results from melanin pigments produced by melanocytes in the hair follicles. A scanning electron micrograph, *bottom,* shows the overlapping, flattened epidermal cells that cover the shaft of a hair.

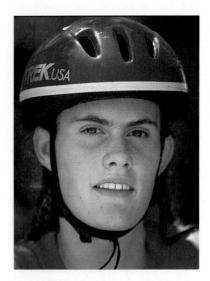

Figure 36-12 The flushed (reddened) skin of this cyclist is a sign that her body is trying to cool itself.

☐ CAPSULE SUMMARY

Human skin, the largest organ of the body, consists of three layers: the epidermis, dermis, and subcutaneous layer. The skin and its specialized structures help protect the body from disease and help maintain the body's homeostasis.

so that more blood flows near the skin's surface and more heat is dissipated. This is why the skin of people with light complexions may become reddish during strenuous exercise, as seen in Figure 36-12. If your body gets too cold, the muscles around the the skin's blood vessels contract, making the blood vessels constrict. This constriction keeps the blood deeper in the skin and helps insulate the body to reduce heat loss.

The dermis has another way to remove excess body heat. It contains about 100 sweat glands per square centimeter. The evaporation of sweat from the surface of your skin removes heat much more efficiently than by dissipation through the skin from the blood. Most sweat is about 99 percent water and 1 percent dissolved salts and acids. Certain sweat glands located in body areas with dense hair, such as the armpits, also secrete proteins and fatty acids. Because these substances provide a rich food source for bacteria, stale sweat often has the rank odor of bacterial waste products.

Subcutaneous Tissue

The **subcutaneous tissue** is a layer of fat-rich cells lying just beneath the dermis. These cells act as shock absorbers, provide additional insulation to conserve body heat, and store energy and fat-soluble vitamins. Different parts of the body have very different thicknesses of subcutaneous tissue—from the eyelids, which have none, to the buttocks and thighs, which may have a lot. The pads of subcutaneous tissue in the soles of your feet may be one-fourth of an inch thick or more. ☐

Skin Disorders Are Common

The skin is the most exposed part of the body and is therefore continually exposed to damaging agents such as insects, microorganisms, and sunlight. Injuries such as insect bites, scrapes, and blisters are often minor and usually heal rapidly without permanent scarring. Burns, however, may be very serious and may result in permanent scarring or even death. Some skin disorders are the result of changes that occur within your body.

Skin cancer may result from mutations caused by years of exposure to the sun's ultraviolet rays. The most common types of skin cancer are **carcinomas,** which originate in the non-pigment-producing cells of the epidermis. If detected early, carcinomas have a very high cure rate. About 1 percent of skin cancers, however, result from mutations that occur in the pigment-producing melanocytes. These cancers, called **malignant melanomas,** grow very fast and spread easily to other parts of the body. People with malignant melanomas have a very low survival rate. A carcinoma and a melanoma can be seen in Figure 36-13. The best way to minimize your risk of skin cancer is to avoid overexposure to sunlight.

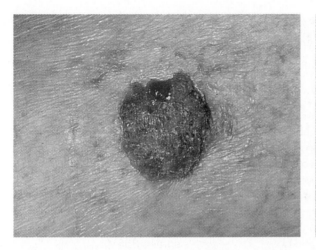
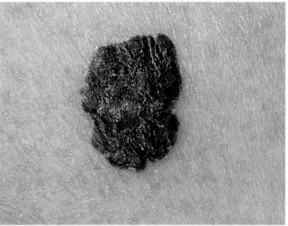

In many teens, high levels of sex hormones produced during adolescence increase the activity of the skin's oil glands. Like sweat glands, oil glands are **exocrine glands,** which release their products through ducts. If the oil ducts become clogged, **acne** results. In the first stage of acne, the buildup of oil in a clogged oil duct causes a swelling called a whitehead. When exposed to air, a whitehead becomes a blackhead. If the oil buildup continues until the gland bursts, the surrounding skin becomes red and inflamed. The swollen pimple that forms is the result of your body's attempts to fight both the tissue damage and the bacteria that infect the damaged area. Although acne cannot be prevented, it can usually be controlled with proper skin care.

Psoriasis *(suh REYE uh sihs)*, which affects about 4 million Americans, is an inherited skin disorder in which new epidermal cells are produced rapidly in the basal layer and reach the skin's surface every three or four days—about eight times faster than normal. The cause of this rapid cell production is still unknown. Though the raised red patches and scaly white skin caused by psoriasis may look unpleasant, the health of most people with this condition is not affected. ◻

Figure 36-13 In its early stages, a carcinoma such as a squamous cell carcinoma, *above left,* may look like a wart. If what you think is merely a wart grows in size and begins to bleed, you should consult a doctor because it could be a carcinoma. A malignant melanoma, *above,* often resembles a mole. If you have a mole that changes in size, shape, or color, you should immediately consult a doctor because it could be a malignant melanoma.

◻ CAPSULE SUMMARY

Carcinomas and melanomas are skin cancers, which result from mutations in skin cells. Acne is a common skin condition that results from clogged oil ducts.

Section Review

1. *What is the function of the skin's subcutaneous layer?*
2. *Where does hair form, and how does it grow in length?*
3. *Describe the two main ways that the dermis helps regulate body temperature.*
4. *What is the most common cause of skin cancer, and how can skin cancer best be prevented?*

Critical Thinking

5. *Why is a third-degree burn, which destroys the epidermis and dermis of the skin, such a serious injury?*

CHAPTER REVIEW

Vocabulary

acne (869)
anabolic steroid (865)
appendicular skeleton (858)
axial skeleton (858)
basal layer (867)
carcinoma (868)
compact bone (856)
corneal layer (866)
cranium (858)
dermis (867)
epidermis (866)
exocrine gland (869)
extensor (862)
flexor (861)
freely movable joint (859)
hair follicle (867)

Haversian canal (856)
immovable joint (859)
insertion (861)
joint (859)
malignant melanoma (868)
melanocyte (867)
muscle strain (865)
myofibril (863)
origin (861)
osteoblast (856)
osteocyte (856)
osteoporosis (857)
pectoral girdle (858)
pelvic girdle (858)
periosteum (856)
psoriasis (869)

red marrow (856)
rheumatoid arthritis (859)
rib cage (858)
sarcomere (863)
sliding filament theory (863)
slightly movable joint (859)
spine (858)
spongy bone (855)
sprain (865)
sternum (858)
subcutaneous tissue (868)
tendinitis (865)
tendon (861)
yellow marrow (856)
Z line (863)

Review

Multiple Choice

1. Which of the following is *not* a role of the skeletal system?
a. support **c.** movement
b. protection **d.** homeostasis

2. The outer membrane of the bone that contains blood vessels is called the
a. marrow. **c.** periosteum.
b. spongy bone. **d.** Haversian canal.

3. The bones of the arms and legs form the
a. axial skeleton.
b. appendicular skeleton.
c. compact bone system.
d. spongy bone system.

4. The end of a muscle that does not move during muscle contraction is called the
a. tendon. **c.** origin.
b. insertion. **d.** flexor.

5. Synchronized shortening of sarcomeres along the length of a muscle fiber causes
a. muscles to contract.
b. Z lines to move apart.
c. muscles to relax.
d. myofibrils to slide.

6. The factor that affects the amount of force exerted when a muscle contracts is the
a. mass of the object being moved.
b. rate of the muscle fiber contractions.
c. distance between the Z lines.
d. length of the sarcomeres.

7. Resistance exercises
a. decrease endurance.
b. decrease steroid uptake.
c. increase muscle size.
d. increase muscle cell number.

8. Which one of the following layers of skin provides structural support?
a. epidermis **c.** subcutaneous
b. dermis **d.** corneal

9. The dermis helps to regulate body temperature by producing
a. sweat. **c.** oil.
b. goose bumps. **d.** melanocytes.

10. The risk of developing skin cancer is greatly increased by
a. eating oily foods.
b. using suntan lotion.
c. exercising.
d. sunbathing.

Completion

11. Living bone cells called _____ get trapped between concentric layers of bone that surround a tube called a(n) _____ .

12. The sutures of your skull are _____ joints. A painful condition that affects movable joints is called _____ .

13. _____ are muscles that cause limbs to bend at joints, and _____ are muscles that cause limbs to straighten.

14. The strength of a muscle contraction depends on the number of _____ that contract and on how often they _____ .

15. _____ exercises increase strength, and _____ exercises increase endurance.

16. Cells of the _____ layer of the epidermis are filled with keratin. These cells are constantly shed and replaced by cells from the _____ layer, which also makes cells for the _____ and _____ .

17. A _____ is a very serious type of skin cancer that results from mutations in pigment-producing cells. _____ are the most common type of skin cancer.

Themes Review

18. **Structure and Function** Give an example of a ball-and-socket joint and a hinge joint. What type of movement is permitted by each of these joints?

19. **Levels of Organization** How are the actions of myosin and actin fibers translated into the contractions of muscles attached to the bones of the arm?

20. **Evolution** How do the tiny muscles in the dermis benefit dogs and cats during cold weather? What vestige (sign) of this benefit is observed in humans?

Critical Thinking

21. **Communicating Effectively** Write a letter about the consequences of osteoporosis to a mature female friend or relative, and persuade her to start a diet and exercise program to delay osteoporosis.

22. **Making Inferences** Look at the X rays below of two different hands. Which is an adult's hand? Which is a child's hand? Explain your decision.

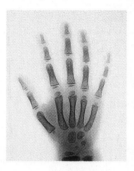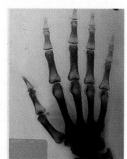

23. **Making Predictions** Leg muscles have a rich supply of blood and energy stored in the form of fat and glycogen. In contrast, arm muscles have a smaller supply of blood and energy, which comes from stored glycogen and blood sugar. Which muscles will tire more quickly? Explain.

Activities and Projects

24. **Cooperative Group Project** Visit a gym that has exercise machines. Test several of the machines, and talk to exercise consultants to determine which muscles are worked by each machine. Document the visit with photographs or a videotape. Identify other activities that work the same muscles as the machines tested. Present your findings to your class.

25. **Multicultural Perspective** Research the incidence of skin cancer in several cultures from different parts of the world. Look for differences and similarities in skin cancer type, prevalence, and occurrence in males and females among several ethnic groups. Relate your findings in a written or oral report.

26. **Research and Writing** Research the link between Accutane® and birth defects. Find out how its effectiveness compares with that of other prescription and over-the-counter acne treatments. Relate your findings in a written report.

Muscle Contraction

OBJECTIVES

- Demonstrate muscle tone.
- Determine muscle fatigue.

PROCESS SKILLS

- measuring changes in muscle size
- organizing data in tables

MATERIALS

- tape measure
- 5–7 lb. weight
- used tennis ball or hollow rubber bouncing ball
- stopwatch or second hand

BACKGROUND

1. What are the two protein filaments that are abundant in muscle cells and allow them to contract?

2. What is the structure of skeletal muscle, and how does it function?

3. How do smooth and cardiac muscle differ from skeletal muscle?

4. Write your own **Focus Question** on your Vee Form.

5. **Knowing Side of the Vee** List the **Concepts** and new **Vocabulary Words** on your Vee Form. In the **Concept Statements** section of the Vee, use these words in sentences that define and explain them.

TECHNIQUE

Doing Side of the Vee

Part A: Muscle Tone

1. Each pair of students in your class will take turns performing the following tasks. In each case, note the degree of

biceps and triceps contraction by feeling each of these muscles to check for hardening and/or tightening.

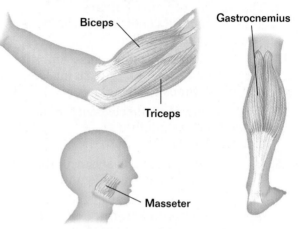

a. Sit comfortably with your forearms on the table, palms down.
b. Sit comfortably in the same position, but this time with palms up. Note differences, if any.
c. Sit comfortably, grasp the weight and lift it with your palms facing up. This is called a weight curl.
d. Stand up straight and let both arms hang down at your sides.
e. Stand up straight and hold the weight with one hand at the side of the body.

Record all observations in the **Records** section of your Vee Form.

2. Take turns doing the following tasks and note the degree of contraction of the gastrocnemius muscle, recording your observations in the **Additional Records** section of your Vee Form.

a. Sit on the edge of a chair with both feet on the floor.
b. Sit on the edge of a table with feet dangling.
c. Stand up straight and lock knees for 5 seconds.
d. Stand up straight with legs relaxed.
e. Stand on toes, extend arms overhead, and reach for the ceiling.

3. Place your hands on the masseter muscles of your lab partner's face and notice what happens when your partner relaxes and then tightens these muscles. Record your observations in the **Additional Records** section of your Vee Form.

Part B: Muscle Fatigue

Each member of each team may do Part B in turn.

4. Hold the ball in your right hand and squeeze the ball repeatedly as quickly as you can. Count the number of complete contractions during each 10 second interval for a total period of 3 minutes or until fatigue is complete. Record the time of fatigue as well as the number of contractions during each 10 second interval the in a table similar to the one below in the **Additional Records** section in your Vee Form.

Muscle Contraction During Ten-Second Intervals			
Seconds	Contractions	Seconds	Contractions
10		100	
20		110	
30		120	
40		130	
50		140	
60		150	
70		160	
80		170	
90		180	

5. Relax for 1 minute and repeat step 4. Record the number of contractions and the time of complete fatigue in the **Additional Records** section.

6. Wait 5 minutes, repeat step 4 again, and record your data.

7. Now repeat steps 4–6 with your left hand. Record all data in the **Additional Records** section. Graph your data as well.

8. Each lab partner should keep records of his or her personal data. In the **Procedure** section of the Vee, briefly summarize the procedure you followed.

9. Clean up your materials and wash your hands before leaving the laboratory.

INQUIRY

1. Compare the muscle tone and contractions of the muscles studied in the lab.

2. Which tasks were harder to complete than the others? Why?

3. Were differences in muscle fatigue found between the right and left hands?

4. Use the information on the **Knowing Side** of the Vee to interpret your results from the **Doing Side,** and then write your **Knowledge Claim.** Write a **Value Claim** for this lab.

ANALYSIS

1. In which tasks was the muscle tension evident?

2. What variables might have influenced the differences between the left and right arm?

3. What is muscle fatigue?

FURTHER INQUIRY

Write a **New Focus Question** that could be the point of a new investigation. The following is an example:

What is the relationship of muscle tone to fatigue among people who exercise regularly compared with those who do not?

CIRCULATORY AND RESPIRATORY SYSTEMS

REVIEW

- surface-area-to-volume ratio (Section 2-1)
- glands and hormones (Sections 3-1, 3-2, 3-3, and 35-3)
- diffusion and osmosis (Section 3-2)
- oxidative respiration (Section 5-3)
- hemophilia (Section 7-30)
- vertebrate heart (Section 27-3)
- closed circulatory systems (Sections 29-2 and 35-1)
- white and red blood cells (Section 35-2)
- cardiac muscle (Section 35-2)
- homeostasis (Section 35-3)

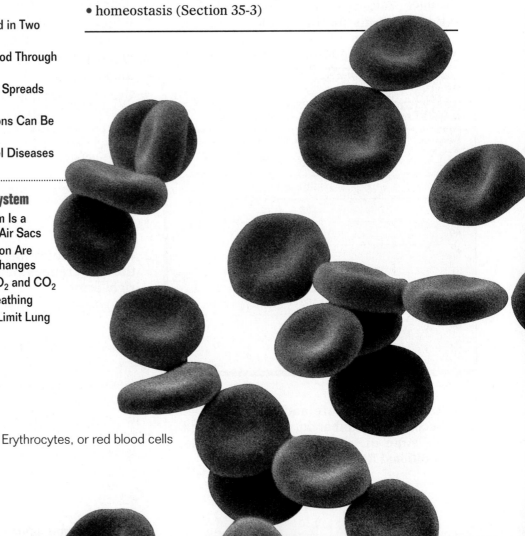

Erythrocytes, or red blood cells

37-1 The Circulatory System

Humans, like all vertebrates, are mobile animals. We are able to walk, run, and swim because of an extensive muscular system and an elaborate internal skeleton. All of this muscle and bone would be of little use, however, if there were no way to provide it with food and no way to take away its wastes. This need is met in humans by a highly efficient closed circulatory system, a network of vessels that extends to every tissue of the body.

The Circulatory System Transports Materials and Distributes Heat

The human circulatory system functions like a network of highways. It connects the various muscles and organs of the body with one another. Four kinds of traffic move along this highway, as shown in Figure 37-1.

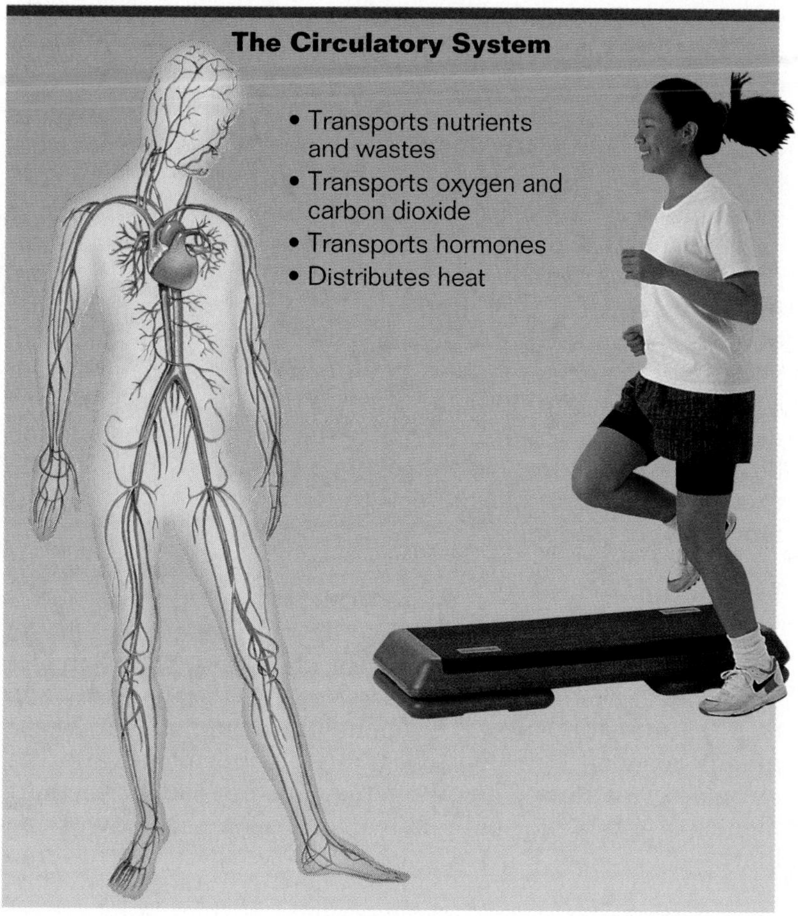

The Circulatory System

- Transports nutrients and wastes
- Transports oxygen and carbon dioxide
- Transports hormones
- Distributes heat

Figure 37-1 The circulatory system transports a variety of materials throughout the body. The circulatory system also helps maintain a constant body temperature by releasing heat into the environment when you are hot and retaining heat when you are cold. The girl doing step aerobics releases her excess body heat through blood vessels under her skin.

Nutrients and Wastes

Food molecules are the nutrients that fuel muscle contraction and other cell activities. After nutrients pass from the small intestine into the bloodstream, they are transported to the cells of the body by the circulatory system.

Cells that metabolize (break down) proteins dump the resulting nitrogen-containing wastes into the circulatory system. The circulatory system carries wastes to excretory organs called kidneys, where the wastes are filtered from the blood.

Oxygen and Carbon Dioxide

Our cells must have a way of getting oxygen and disposing of carbon dioxide. Both of these gases are transported to and from cells by the circulatory system. In a process known as gas exchange, oxygen is carried to cells, and carbon dioxide is carried away from cells. The transportation provided by the circulatory system is vital. It has been estimated that it would take a molecule of oxygen three years to travel from your lung to your toe if the transport depended only on diffusion rather than the circulatory system.

Heat

The body maintains a relatively constant temperature by using metabolic energy to make heat. The circulatory system distributes this heat more or less uniformly to all parts of the body.

But environmental (outside) temperature also affects body temperature. Passing just beneath the skin, the circulatory system absorbs heat from a hot environment and gives off heat to a cold one. Distribution of heat by the circulatory system thus tends to adjust the entire body's temperature toward that of the environment. However, your body has several mechanisms that limit how much your body temperature can fluctuate. When it is cold outside, your body attempts to conserve heat by constricting the blood vessels that lie near the surface so that less heat escapes into the environment. In the opposite manner, the blood vessels under your skin dilate when it is hot outside, thereby delivering more blood to the surface and allowing more heat to escape.

Hormones

As you learned in Chapter 35, the body coordinates the activities of its many organs with hormones, the chemical messengers of the endocrine system. The brain sends signals that activate special hormone-producing glands. These glands respond by releasing particular hormones into the circulatory system. The circulatory system carries the hormones to the target cells at various sites throughout the body.

CONTENT LINK

You can learn more about glands and the hormones they release into the bloodstream in **Chapter 41**.

Components of the Circulatory System

There are four major components of the human circulatory system. Together, they form the transportation network that delivers essential materials to every cell in your body.

1. **Blood vessels** Blood vessels are a network of tubes through which blood moves.
2. **Lymphatic vessels** Lymphatic vessels intertwine with blood vessels. They recover fluid that leaks out of the blood vessels.
3. **Blood** Blood is a complex mixture of specialized cells and fluid.
4. **Heart** The heart is a muscular pump that propels blood through the blood vessels of the circulatory system.

Circulation Pathway

The general path of blood through the circulatory system is shown in Figure 37-2. Blood leaves the heart through vessels known as **arteries.** From the arteries, the blood passes into a large network of **arterioles** *(ahr TIHR ee uhls)*, or small arteries. Eventually, the blood is pushed through the **capillaries,** an extensive network of very narrow tubes. The term *capillary* is from the Latin *capillaris,* meaning "hair." It is while the blood is passing through the capillaries that gases and metabolites are exchanged with the cells of the body. After leaving the capillaries, the blood flows into **venules** *(VEHN yools)*, or small veins. The network of venules empties into larger **veins,** which carry the blood back to the heart.

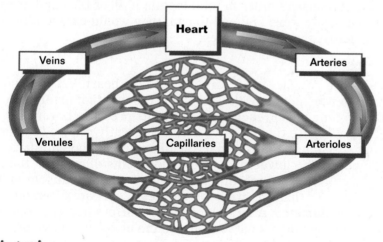

Figure 37-2 Blood leaves the heart through blood vessels called arteries, which branch into many smaller arterioles. The smallest and most numerous blood vessels are capillaries. Blood returns to the heart through venules and veins.

Arteries

Arteries are the tubes that carry blood away from the heart, but they must be more than just tubes to do their job. Blood leaves the heart not in a smooth flow but rather in pulses as the heart forcefully ejects it with contractions. To accommodate each forceful pulse of blood, an artery's wall must be able to expand.

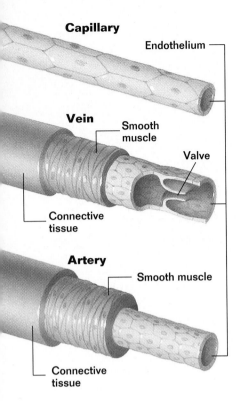

Capillary

Endothelium

Vein

Smooth muscle

Valve

Connective tissue

Artery

Smooth muscle

Connective tissue

Figure 37-3 The thick, muscular wall of an artery, *bottom,* enables the artery to withstand the high blood pressure that develops inside it every time the heart contracts. Veins, *center,* have thinner walls than arteries since the blood pressure inside veins is much lower. Capillaries, *top,* have the thinnest walls, allowing materials like oxygen and carbon dioxide to easily pass through.

How is an artery able to expand? The wall of an artery is made up of three layers of tissue, as shown in Figure 37-3. The innermost layer is a thin skin of endothelial cells. Surrounding this is a layer of elastic, smooth muscle tissue. Finally, a protective layer of connective tissue wraps around the smooth muscle tissue. Just as a balloon expands when you blow more air into it, so the elastic artery expands when the blood is pumped into it each time the heart contracts.

Capillaries

In capillaries, food and oxygen molecules are transferred from the blood to the body's cells, and carbon dioxide and wastes are picked up. This extensive back-and-forth traffic is possible only because of two key properties of capillaries.

1. **Thin walls**   Built like a soft-drink straw, capillaries are simple tubes with walls that are only one cell thick. The walls of capillaries are so thin that food and gas molecules easily pass through them.

2. **Narrow diameter**   All capillaries are very narrow, with an internal diameter of about 8 µm (0.0003 in.). Because this is only slightly larger than the diameter of a red blood cell, blood cells passing through a capillary slide along the capillary's inner wall. This tight fit makes it easy for oxygen to move through the wall as it diffuses from the blood cells to body cells.

No cell in your body is more than about 100 µm (0.004 in.) away from a capillary. At any one moment, about 5 percent of your blood is in capillaries. The network of capillaries is several thousand miles in total length; if all of the capillaries of your body were laid end to end, they would extend all the way across the United States!

Veins

Veins are vessels that return blood to the heart. Veins do not have to accommodate the pulsing pressures that arteries do. By the time blood reaches the veins, its pressure is very low. This is because the pressure has been greatly reduced by the high resistance that the narrow diameter of the capillaries provides. The lower pressure inside veins allows for their walls to have a much thinner layer of smooth muscle and elastic fiber than the walls of arteries. Veins are often quite large in diameter. A large blood vessel offers less resistance to blood flow than a narrow one, so the blood can move more quickly through large veins. The largest vein in the human body is the vena cava (*VEE nuh KAYE vuh*), the vein that leads into the heart. The vena cava is about 3 cm (1.2 in.) in diameter—you could easily slide your thumb into it. Most veins have internal, one-way valves spaced at regular intervals. These valves prevent the blood from flowing backward during its trip to the heart. ◼

Lymphatic Vessels

The heart and the blood vessels, collectively, are called the cardiovascular system. The cardiovascular system is very leaky. Fluids are forced out of the thin walls of the capillaries by the pressure generated every time the heart pumps. This loss of fluid is unavoidable because the capillaries could not function without very thin walls. But the loss must be made up or the cardiovascular system would soon dry up. About 3 L (3.2 qt.) of fluid leak out of your cardiovascular system each day—more than half of your total supply of about 5.6 L (5.9 qt.) of blood. To collect and recycle this fluid, the body utilizes another circulatory system called the **lymphatic** *(lihm FAT ihk)* **system,** shown in Figure 37-4. The fluid that leaks out of the capillaries accumulates in the spaces around the body's cells and diffuses into vessels called lymphatic capillaries. Once the fluid enters the lymphatic capillaries, it is called **lymph.** The recovered fluid passes through a series of progressively larger lymphatic vessels into two large lymphatic ducts that drain into veins in the lower part of the neck.

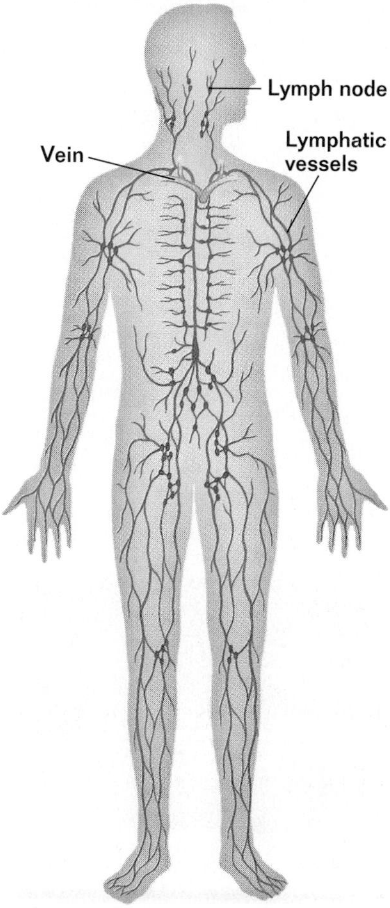

Blood Has Liquid and Solid Components

The blood that circulates through the cardiovascular system is composed of water, a variety of molecules dissolved in the water, and three kinds of cells, as shown in Table 37-1.

Blood Plasma

The noncellular portion of blood is called **plasma.** Plasma is a complex solution of 90 percent water and 10 percent solute. Three very different kinds of substances compose the solute component of plasma.

Figure 37-4 The lymphatic system, in green, returns the fluid that collects around tissues to veins in the neck. The fluid, called lymph, is driven through lymphatic vessels as the vessels are squeezed by the movements of the body's muscles. Structures called lymph nodes filter foreign substances from the lymph.

Table 37-1 Composition of Blood

Components of Blood	Function
Plasma portion (60% of total blood volume)	
Water	Acts as solvent
Metabolites and wastes, salts and ions, proteins	Play diverse roles (nourish cells, catalyze chemical reactions, act as chemical messengers, maintain blood volume, fight infection, etc.)
Cellular portion (40% of total blood volume)	
Red blood cells	O_2 + CO_2 transport
White blood cells	Produce antibodies, ingest foreign materials
Platelets	Aid in clotting blood

1. **Metabolites and wastes** Dissolved within the plasma are glucose and other food molecules, as well as vitamins, hormones, and nitrogen-containing wastes.

2. **Salts and ions** The chief plasma ions are sodium, chloride, and bicarbonate. Plasma also contains trace amounts of calcium, magnesium, and metallic ions such as copper, potassium, and zinc.

3. **Proteins** If not for the high concentration of protein in the plasma, the cells of your body would soak up much of the water in plasma. Remember from Chapter 3 that water tends to move from a region of lower solute concentration toward a region of higher solute concentration in a process called osmosis. Plasma avoids losing its water to cells because it is rich in dissolved proteins. In fact, the total amount of protein inside cells and in plasma is the same, making cytoplasm and plasma essentially isotonic to each other.

Blood Cells

About 40 percent of the total volume of blood is not plasma; it consists of cells and cell fragments that are suspended in the plasma. There are three principal types of cells in human blood: red blood cells, white blood cells, and pieces of cells called platelets.

1. **Red blood cells** Each milliliter of your blood contains about 5 million oxygen-carrying red blood cells, also called **erythrocytes** *(eh RIHTH roh seyets)*. Each red blood cell has the shape of a flat disk with a collapsed center, as shown in Figure 37-5. Almost all of the interior of a red blood cell is packed with hemoglobin, a protein that binds to oxygen in the lungs and transports it to the tissues of the body. A red blood cell is simply a hemoglobin container; it has no nucleus, and it cannot make proteins or repair itself. As a result, red blood cells have a short life span—about four months. New red blood cells are produced constantly by stem cells, specialized cells in bone marrow. If the production of new red blood cells slows, the blood's population of them soon decreases, and a condition called **anemia** develops.

2. **White blood cells** A small proportion of the cells in your blood are white blood cells, or leukocytes *(LOO koh seyets)*. There are only one or two white blood cells for every 1,000 red blood cells. White blood cells are larger than red blood cells and contain no hemoglobin. **Leukocytes** are the primary cells of the immune system, your body's defense against disease.

There are different kinds of white blood cells. Each kind has a different function in the immune system. Some are called lymphocytes *(LIM foh seyets)*, many of which produce antibodies. Others, called macrophages *(MAK roh fayj ehs)*, are unique among blood cells in that they do not always stay within the vessels of the cardiovascular system. They are mobile soldiers that migrate into the fluid

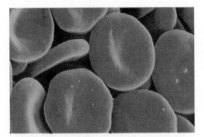
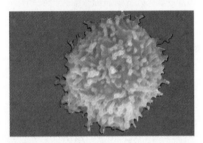
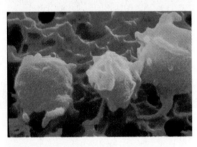

Figure 37-5 Red blood cells, *top,* get their red color from the hemoglobin molecules inside them. Lymphocytes, *center,* are specific kinds of white blood cells. Lymphocytes, slightly larger than red blood cells, have large nuclei with very little cytoplasm. Platelets, *bottom,* are essentially membrane-bound "enzyme packets." The enzymes are used for blood clotting.

surrounding the body's cells. Sometimes cancer occurs among the cells that give rise to white blood cells. As a result, too many white blood cells are produced, and a disease called **leukemia** *(loo KEE mee uh)* develops. Leukemia is often fatal.

3. **Platelets** Certain large cells in bone marrow called megakaryocytes *(mehg uh KAR ee oh seyets)* regularly pinch off bits of their cytoplasm. These unnucleated cell fragments, called **platelets** *(PLAYT lihts)*, play a key role in blood clotting. Remember that humans, like other vertebrates, have a closed circulatory system. If a blood vessel gets a hole in its wall, the hole must be plugged quickly. If the hole is not plugged, all of the blood will leak out of the system, and death will occur.

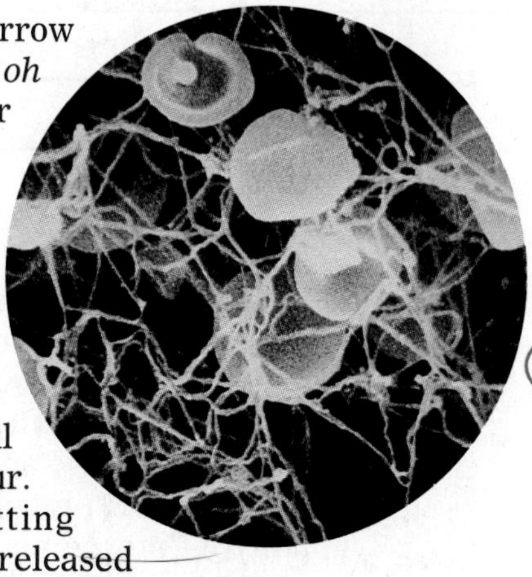

Circulating platelets start the clotting process when they encounter chemicals released by damaged blood vessel cells. The platelets then release a clotting protein into the blood that initiates a series of chemical reactions resulting in the formation of a protein called **fibrin.** The fibrin begins to form a netlike covering over the damaged site, as shown in Figure 37-6. Very quickly, the gluey mesh of fibrin and platelets develops into a mass, or clot, that plugs the hole in the blood vessel. Because of its initially gluey nature, the clot fits itself to the shape of the rupture in the blood vessel and provides a tight, strong seal. As you learned in Chapter 7, the lack of one of the clotting proteins causes hemophilia. ◻

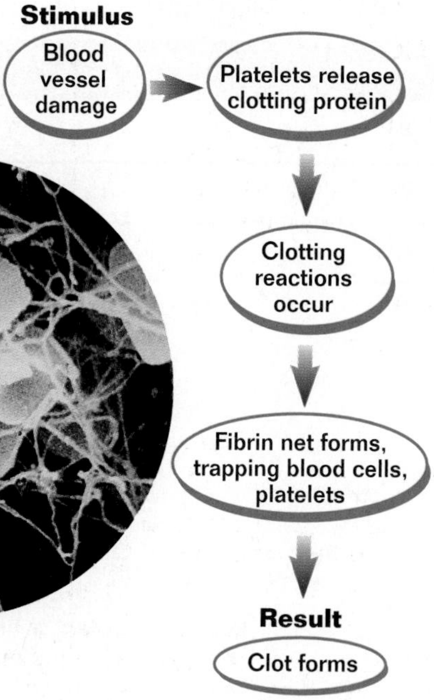

Stimulus

Blood vessel damage → Platelets release clotting protein

↓

Clotting reactions occur

↓

Fibrin net forms, trapping blood cells, platelets

↓

Result

Clot forms

Figure 37-6 The release of enzymes from platelets at the site of a damaged blood vessel initiates a "clotting cascade." During this cascade, a series of chemical reactions occurs, resulting in the formation of a fibrin net in which blood cells are trapped, *above left.* The net eventually forms into a clot in the hole of the damaged vessel.

Surface Proteins on Red Blood Cells Determine Blood Type

Before a blood transfusion can be performed, it is important to know the blood type of the donor and recipient. If the blood of each is not compatible, a life-threatening situation can develop. Blood compatibility is determined by proteins called antigens that coat the outside of red blood cells.

The system used to classify human blood is called the **A-B-O system.** Your red blood cells have either A antigens, B antigens, both A and B antigens, or no antigens on their surface. The kinds of antigens present on red blood cells determine your blood type. The danger of mixing bloods of different blood types comes from the presence of other proteins, called antibodies, that are present in the blood plasma. Antibodies seek out specific foreign antigens and attack them. For example, type B blood contains antibodies that will attack A antigens. If a person with type B blood receives type A or AB blood, the recipient's antibodies will react with

Table 37-2 Blood Types

Type	Antigen on the RBC	Antibodies in Plasma	Can Receive Blood From	Can Donate Blood to
A	A	B	O, A	A, AB
B	B	A	O, B	B, AB
AB	A, B	None	O, A, B, AB	AB
O	None	A, B	O	O, A, B, AB

the donor's blood and cause the blood to clump. Table 37-2 lists the four different blood types and the antigens and antibodies found in each.

Another important antigen on the surface of red blood cells is called **Rh factor.** People who have this protein are said to be Rh+, and those who lack it are Rh–. The greatest danger of Rh incompatibility arises in pregnancies in which an Rh– mother is carrying an Rh+ child. Throughout the pregnancy, the mother's blood remains separate from the child's, but during delivery some mixing usually occurs. As a result of this mixing, the mother's immune system responds to the Rh antigens in the child's blood by producing Rh antibodies. Because it takes some time for the mother's body to produce the antibodies, the first child is usually born unharmed. But if the next child that the mother carries is also Rh+, the antibodies that remain in the mother's blood will attack this child's blood cells, causing severe complications for the child. Fortunately, a drug can be given to the mother shortly after delivery of her first Rh+ child that will destroy any Rh antigens in her blood before her body has a chance to produce antibodies to them, thus making it safe for the next Rh+ child. ◻

□ **CAPSULE SUMMARY**

Blood type (A, B, AB, O) is determined by the presence or absence of A and B antigens located on the surface of red blood cells. Another important antigen found on red blood cells is called Rh factor. Rh+ blood has Rh antigens, and Rh- blood does not.

Section Review

1. Why might a person's face become flushed and red when the weather is hot?

2. Why is a cut artery usually considered more life threatening than a cut vein?

3. How does the total volume of blood in the cardiovascular system remain constant?

4. How is oxygen transported in the blood?

5. A person with type O blood is considered a universal blood donor. Explain why.

Critical Thinking

6. Standing in one place for a long period of time often results in swollen ankles and feet. Explain why this happens. How could the swelling be relieved?

37-2 The Heart

The human heart is a **double** *pump. One side of the heart powers an oxygen-acquiring circulatory phase, while the other side powers an oxygen-delivering circulatory phase. But oxygen is not the only gas carried by the circulatory system. When oxygen is acquired or delivered, it is exchanged with carbon dioxide.*

Section Objectives

- *Describe pulmonary and systemic circulation.*
- *Identify the chambers, vessels, and valves of the heart.*
- *Trace the path of blood through the heart.*
- *Describe two ways that medical technicians can monitor cardiovascular performance.*
- *Describe two different blood vessel diseases.*

The Heart Pumps Blood in Two Separate Loops

The pattern of circulation in the human body includes two separate circulatory loops, as shown in Figure 37-7. The right side of the heart is responsible for driving the pulmonary circulation loop, which pumps oxygen-poor blood through pulmonary arteries to the lungs. In the lungs, gas exchange occurs: carbon dioxide is released, and the blood receives oxygen. This oxygenated blood is then returned to the left side of the heart through pulmonary veins. The left side of the heart is responsible for driving the systemic circulation loop, which pumps oxygen-rich blood through a network of arteries to the tissues of the body. The deoxygenated blood is then returned to the right side of the heart by veins.

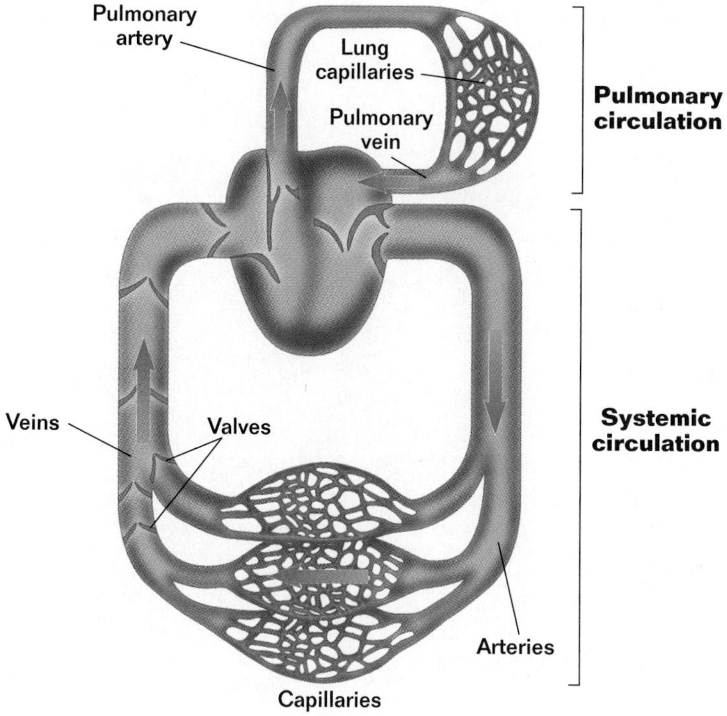

Figure 37-7 The pulmonary circulatory loop transports blood from the right side of the heart to the lungs and then to the left side of the heart. The systemic circulatory loop transports blood from the left side of the heart to all the body's tissues and then to the right side of the heart. The designation of the left and right sides of the heart in the diagram is based on the left and right sides of an intact heart that is still in a body.

Tracing the Path of Blood Through the Heart

Let's follow the journey of blood through the human heart, shown in Figure 37-8, starting with the entry of oxygen-rich blood from the lungs into the heart. This oxygenated blood enters the left side of the heart through the **pulmonary veins,** emptying directly into the heart's collection chamber, the **left atrium.** From the left atrium, the blood flows down into the **left ventricle.** Most of this flow occurs before a contraction starts. When the heart starts to contract, the atrium contracts first, pushing its remaining blood into the ventricle.

After a slight delay that permits the atrium to empty fully, the ventricle contracts. The walls of the ventricle are far more muscular than those of the atrium, so the ventricle's contraction is much more forceful than that of the atrium. As the left ventricle contracts, the blood is prevented from going back into the left atrium by a one-way valve called the **mitral** *(MEYE truhl)* **valve.** All four valves in the heart are one-way valves. Each has flaps that act as doors that open in only one direction, ensuring that the blood that passes through them will not flow back.

Prevented from reentering the atrium, the blood within the contracting left ventricle enters the largest artery of the body, the **aorta** *(ay AWR tuh).* Once inside the aorta, the blood is prevented from reentering the left ventricle by another large

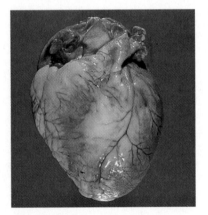

Figure 37-8 The ventricles (lower chambers) of the human heart have thicker walls than the atria (upper chambers) because the ventricles must pump the blood inside them a long distance. The arrows indicate the pathway that the blood takes as it travels into, through, and out of the heart. Oxygen-rich blood from the lungs, red arrows, enters the left side of the heart through pulmonary veins. Oxygen-poor blood from the body, blue arrows, enters the right side of the heart through two large veins, the superior vena cava and the inferior vena cava.

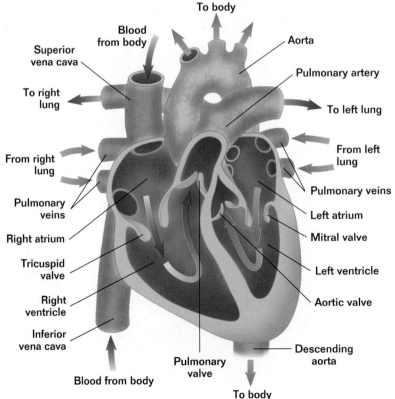

valve, the **aortic valve.** Many major arteries branch from the aorta and carry oxygen-rich blood to all parts of the body.

The first arteries to branch from the aorta are the **coronary** (*KAWR uh neh ree*) **arteries,** which carry freshly oxygenated blood to the heart muscle itself. Many other arteries also branch from the aorta. For example, two renal (*REEN uhl*) arteries leave the aorta and carry blood to the kidneys, where nitrogen wastes are filtered from the blood.

After delivering its cargo of oxygen to the cells of the body, the blood makes its way back to the heart through the body's many veins. Two large veins collect all the oxygen-poor blood from the systemic circulatory system. The **superior vena cava** (*VEE nuh KAY vuh*) drains blood from the upper body, while the **inferior vena cava** drains blood from the lower body. These two veins empty directly into the **right atrium** of the heart. Blood passes from the right atrium into the right ventricle through the **tricuspid** (*treye KUHS pihd*) **valve.** As the right ventricle contracts, it sends the blood through the **pulmonary valve** and into the pulmonary arteries, which carry the blood to the lungs. The blood then returns from the lungs to the left atrium with a new cargo of oxygen, and the cycle begins again. ◻

◻ **CAPSULE SUMMARY**

Oxygen-rich blood from the lungs enters the left atrium and is pumped from the left ventricle to the body's tissues. After returning to the right atrium, the oxygen-poor blood is pumped from the right ventricle to the lungs. Blood traveling through the heart does not flow backward because the heart contains one-way valves.

A Wave of Contraction Spreads Over the Heart

Contraction of the heart is initiated by a small cluster of cardiac muscle cells embedded in the upper wall of the right atrium, as shown in Figure 37-9. Called the **sinoatrial** (*SEYE noh ay tree ahl*) **node,** or SA node for short, these cells act as the pacemaker of the heart, spontaneously starting contractions with a regular rhythm. Each contraction initiated at the SA node travels quickly in a wave that causes both the right and the left atria to contract almost simultaneously.

The wave of contraction does not immediately spread to the ventricles, however. Almost one-tenth of a second passes before the lower half of the heart starts to contract. The delay is critical to proper functioning of the heart, as it permits the atria to finish emptying blood into the ventricles before the ventricles contract. The reason for the delay is that the two upper chambers of the heart are separated from the two lower chambers by connective tissue. Connective tissue cannot transmit the contraction. The wave of contraction would not pass to the ventricles at all except for a slender bridge of cardiac muscle cells that connects the upper chambers of the heart to the lower chambers. This bridge is called the atrioventricular node, or AV node. From the AV node, the wave of contraction is conducted rapidly over both ventricles by a network of fibers called the Bundle of His. Both ventricles contract almost simultaneously.

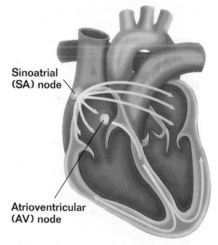

Sinoatrial (SA) node

Atrioventricular (AV) node

Figure 37-9 The sinoatrial node, or pacemaker, initiates each heart contraction. After initiation, the wave of contraction spreads across both atria, delays for an instant at the atrioventricular node, travels to the bottom of the heart, and then moves back up and across both ventricles. On average, heart contractions are initiated at a rate of about 72 times per minute. During sleep the rate decreases, and during exercise it increases.

Cardiovascular Functions Can Be Monitored

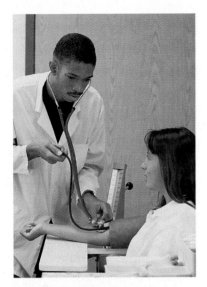

Figure 37-10 Blood pressure is quickly and easily measured with a sphygmomanometer, or blood pressure cuff, and a stethoscope. The blood pressure readings obtained from females are generally slightly lower than the readings from males.

Figure 37-11 The electrical changes that occur in the heart each time it contracts are measured with a device called an electrocardiograph, *below right*. Each time the heart beats, a recording pen graphs characteristic up-and-down waves, *below*. These waves are analyzed by medical professionals to assess the health of the heart.

As you have seen, the contraction of the heart is not simply a continuous squeeze-release cycle, but rather a series of events that occurs in a predictable order. Medical technicians can monitor these events with special instruments.

Blood pressure is measured with a device called a sphygmomanometer (*sfihg moh muh NAHM uht uhr*), shown in Figure 37-10. During the first part of the heartbeat, the atria are filling, so the pressure in the arteries leading out to the tissues of the body falls slightly. This low pressure that occurs during relaxation of the heart is called the **diastolic** (*DEYE uh stahl ihk*) **pressure.** Then, with the contraction of the ventricles, a pulse of blood is forced into the systemic arterial system, immediately raising the blood pressure. This higher blood pressure is called the **systolic** (*sihs TAHL ihk*) **pressure.** A blood pressure reading is usually reported as the systolic pressure written over the diastolic pressure. Normal blood pressure values are from 100 to 130 for systolic and from 70 to 90 for diastolic, so an example of a normal reading would be written 120/80. Many Americans suffer from a condition called high blood pressure, or hypertension. People with this condition have elevated systolic and diastolic blood pressures. Left untreated, hypertension can lead to heart damage, a stroke (rupture of a blood vessel in the brain), or kidney failure.

Another way to monitor the heartbeat is to measure the tiny electrical impulses produced by the heart muscle when it contracts. Because the human body is composed mostly of water, it conducts electrical currents rather well. A wave of muscle contraction passing over the surface of the heart generates an electrical current that passes in a wave through the entire body. Although the magnitude of this electrical pulse is tiny, it can be detected with special sensors that are placed on the skin. A recording of these impulses is called an electrocardiogram (*ee LEHK troh kahr dee ah graym*). In one normal heartbeat, three successive electrical impulses are recorded, as shown in Figure 37-11.

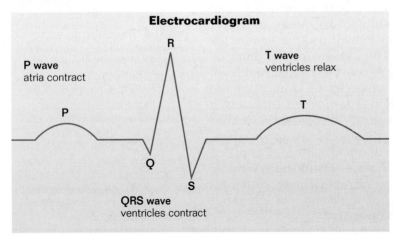

Electrocardiogram

P wave
atria contract

R

T wave
ventricles relax

P

T

Q

S

QRS wave
ventricles contract

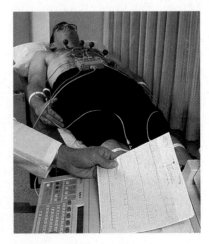

Heart and Blood Vessel Diseases Kill Many Americans

Diseases of the heart and blood vessels are the leading cause of death in the United States. Over 16 percent of the population, or about 42 million Americans, are affected by some form of cardiovascular disease.

Heart attacks are the main cause of death due to cardiovascular disease. They result when an area of heart muscle does not receive enough blood (and consequently, oxygen), leading to the death of cardiac muscle cells. Heart attacks are sometimes caused by a blood clot that has formed somewhere else in the body. The clot can be carried through the heart and into the coronary arteries, where it blocks the passage of blood to part of the heart. Heart attacks may also occur if the coronary arteries become blocked by deposits of fatty materials on their inner walls.

Atherosclerosis *(ath uhr oh skluh ROH sihs)* is a buildup of fatty deposits on the inner walls of arteries, as shown in Figure 37-12. The deposits are composed of cholesterol and cellular debris of various kinds. When this condition is severe, blood flow through the artery becomes greatly restricted. Exercise, along with diets low in cholesterol and saturated fats are prescribed to help prevent atherosclerosis. **Arteriosclerosis** *(ahr tihr ee oh skluh ROH sihs)*, or hardening of the arteries, occurs when calcium is deposited in the fatty buildup caused by atherosclerosis. Hardened arteries cannot expand to accommodate the volume of blood that enters them every time the heart contracts. As a result, pressure builds up in the artery and feeds back to the heart, causing it to work harder. ◻

Figure 37-12 A normal artery, *top*, has smooth interior walls and a large diameter through which blood can flow. An atherosclerotic artery, *bottom*, has accumulated deposits on its interior walls. As a result, blood flow through the artery is greatly diminished.

◻ **CAPSULE SUMMARY**

Monitoring blood pressure and the electrical activity of the heart enables medical professionals to assess the condition of the heart. Cardiovascular diseases, like atherosclerosis and arteriosclerosis, are the leading cause of death in America.

Section Review

1. *Is the blood that your heart pumps to your stomach part of the systemic or pulmonary circulatory loop?*

2. *A red blood cell loaded with oxygen must be located in which side of the heart?*

3. *Beginning with the excitation of the SA node, list the sequence of events that results in atrial and ventricular contraction.*

4. *If a person has severe atherosclerosis, would you expect their blood pressure to be low, normal, or high? Explain your answer.*

Critical Thinking

5. *The pulmonary artery is the only artery in the body that carries oxygen-poor blood. Explain why this blood vessel cannot be a vein.*

37-3 The Respiratory System

Section Objectives

- Name the structures of the respiratory system.
- Describe the body changes associated with a single inhalation and exhalation.
- Explain how oxygen and carbon dioxide are transported in the blood.
- Explain how breathing is regulated.
- Describe three respiratory diseases.

O *ne of the major tasks of the circulatory system is to transport oxygen to the cells of the body. Air is about 21 percent oxygen gas. When you breathe, air moves into the large, saclike organs in your chest called lungs. The oxygen in the air then moves into lung capillaries and is carried to the heart by the pulmonary veins. The simultaneous uptake of oxygen and release of carbon dioxide by your lungs is called* **respiration.**

The Respiratory System Is a Network of Tubes and Air Sacs

The human respiratory system, shown in Figure 37-13, is composed of the respiratory passages, the lungs, and the thoracic cavity. Together, these structures function to provide your body with a constant supply of oxygen.

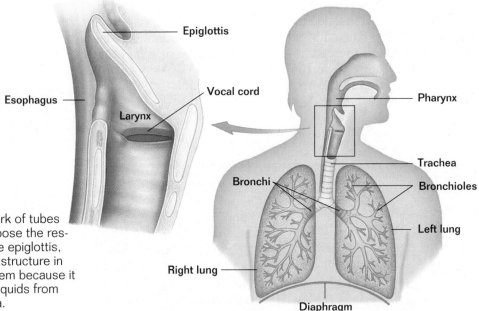

Figure 37-13 A network of tubes and two lungs compose the respiratory system. The epiglottis, *left,* is an important structure in the respiratory system because it prevents food and liquids from entering the trachea.

Epiglottis
Vocal cord
Esophagus
Larynx
Pharynx
Trachea
Bronchi
Bronchioles
Left lung
Right lung
Diaphragm

Respiratory Passages

Air normally enters your respiratory system through the nostrils. Hairs in your nose filter dust and other particles out of the air, and the epithelial tissue that lines the nasal cavity moistens and warms it.

The air then passes through the **pharynx** *(FAR ingks),* or upper throat, and the **larynx** *(LAR ingks),* or voice box. Next, the air enters the long, straight **trachea** *(TRAY kee uh),* the tube that carries air down to the lungs. A flap of tissue called the **epiglottis** *(ehp uh GLAHT ihs)* covers the opening to the trachea when you swallow. Without an epiglottis, food and liquids could easily pass into your lungs. Air moves down the trachea and then through two branches called **bronchi** *(BRAHNG keye)*. Each bronchus enters a lung. Ciliated cells in the bronchi and trachea secrete mucus that traps any foreign particles that may remain in the air. This mucus is directed upward by the beating of the cilia. When it reaches the epiglottis, it is swallowed and digested. Bacteria and other microbes are destroyed by the strong acids and enzymes in the stomach.

Lungs

The lungs are suspended in the **thoracic cavity,** bounded on the sides by the ribs and on the bottom by the **diaphragm** *(DEYE uh fram)*. A protective double membrane, called the pleural membrane, surrounds both lungs. The outermost membrane is attached to the the wall of the thoracic cavity, and the inner membrane is attached to the surface of the lungs. Between both membranes is a small space called the pleural cavity. Your lungs are among the largest organs in your body. The interior of each lung is not an open cavity like a bag. Instead, each lung is subdivided into about 300 million small chambers called **alveoli** *(al VEE uh leye)*.

The alveoli, shown in Figure 37-14, are clustered together in groups, like bunches of grapes. Alveoli increase the surface area of your lungs to as much as 80 sq. m, or 42 times the surface area of your body. The alveoli are surrounded by many capillaries, enabling blood to flow over them in a nearly continuous sheet.

The alveoli are connected to the bronchi by a network of tiny tubes called **bronchioles** *(BRAHNG kee ohls)* within each lung. Most of the bronchioles have smooth muscle in their walls. ▫

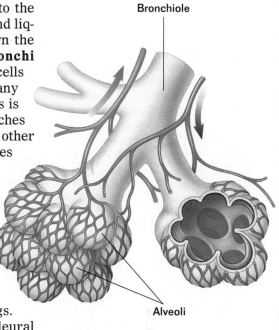

Bronchiole

Alveoli

Figure 37-14 The walls of the alveoli inside the lungs are extremely thin, enabling oxygen and carbon dioxide to easily pass through. Oxygen diffuses from the alveoli into the surrounding capillaries, and carbon dioxide diffuses from the capillaries into the alveoli.

▫ **CAPSULE SUMMARY**

Air is carried to the lungs by way of a branched system of tubes that looks much like an upside-down tree. The ends of the smallest branches inside the lungs are connected to microscopic air sacs called alveoli.

Inhalation and Exhalation Are Caused by Pressure Changes

Breathing is the result of pressure changes that occur inside the pleural cavity and lungs. Air moves into the lungs when the air pressure inside them is reduced by the expanding walls of the chest cavity, which contains a great deal of skeletal muscle. During inhalation, the diaphragm contracts and moves downward, and the rib cage moves upward and outward. In effect, this enlarges the volume of the chest cavity as it is pulled in all directions. The air pressure inside the pleural cavity and lungs decreases, causing air to rush into the lungs from outside the body.

During exhalation, the diaphragm and ribs return to their original resting position. The compression exerted by the ribs and diaphragm forces air out of the lungs. The air pressure changes that occur in the pleural cavity and lungs during inhalation and exhalation are described in Figure 37-15.

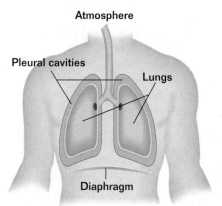

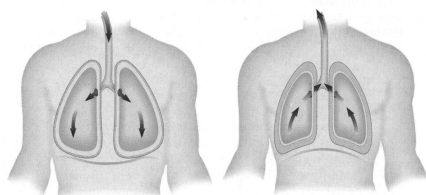

Figure 37-15 Immediately before inhalation, *left,* the pressure inside the lungs is equal to atmospheric pressure. Once the chest expands, *center,* the pressure inside the pleural cavity decreases, causing the lungs to push outward. As a result, the pressure inside the lungs drops below atmospheric pressure, and air enters. During exhalation, *right,* the pressure inside the pleural cavity increases, compressing the lungs. Air is forced from the lungs until the pressure inside the lungs equals atmospheric pressure.

Breathing Exchanges O_2 and CO_2

When oxygen molecules diffuse from the air into the cells of your alveoli, their journey has just begun, as shown in Figure 37-16. Passing into the plasma of the bloodstream, the oxygen is picked up by red blood cells that contain an oxygen-carrying protein, hemoglobin.

Oxygen Transport

Each hemoglobin molecule contains an atom of iron that binds reversibly with oxygen. Therefore, at the appropriate time, the oxygen can be released elsewhere in the body and be taken up by the cells that need it. Hemoglobin is manufactured within red blood cells, giving them their red color. Oxygen is simply loaded and unloaded from the hemoglobin inside red blood cells.

Hemoglobin molecules act like little sponges, soaking up the oxygen that diffuses into the red blood cells and causing more oxygen to diffuse in from the blood plasma. At the high oxygen levels that occur in the blood inside the lungs, most hemoglobin molecules carry a full load of oxygen. Later, in the tissues, oxygen levels are much lower, causing the hemoglobin to release its bound oxygen.

In tissues, the presence of carbon dioxide produced by cellular respiration makes the blood more acidic and causes the hemoglobin molecules to assume a different shape, one that gives up oxygen more easily. This speeds the unloading of oxygen from hemoglobin even more. The effect of carbon dioxide on oxygen unloading is called the **Bohr effect.** The

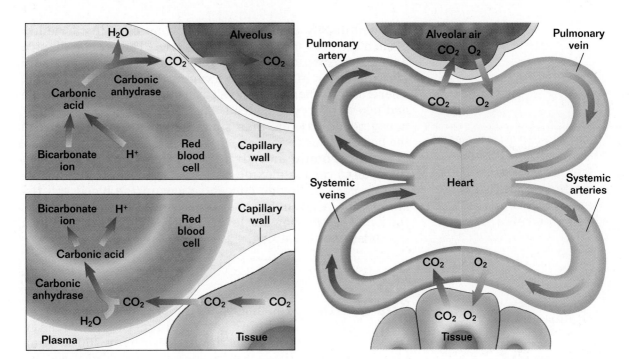

Bohr effect is of great importance because it means that blood unloads oxygen more readily within those tissues that undergo a high rate of cellular respiration. These are the very tissues that need oxygen the most.

Carbon Dioxide Transport

At the same time that the red blood cells are unloading oxygen to tissues, they are also absorbing carbon dioxide. Only a tiny fraction of the carbon dioxide that the blood carries is dissolved in plasma (about 7 percent). About 23 percent is carried by the hemoglobin molecules inside red blood cells. The remaining 70 percent is carried within the cytoplasm of red blood cells.

How do the red blood cells manage to keep all of this carbon dioxide inside their cytoplasm? Why doesn't it simply diffuse back into the plasma, where carbon dioxide levels are lower? An enzyme within the red blood cells, called **carbonic anhydrase,** combines carbon dioxide molecules with water to form carbonic acid, H_2CO_3. The carbonic acid then dissociates (breaks apart) to form bicarbonate ions, HCO_3^-.

$$H_2O + CO_2 \longrightarrow H_2CO_3 \longrightarrow HCO_3^- + H^+$$

Bicarbonate ions are unable to pass through the membranes of red blood cells. The red blood cells carry their cargo of bicarbonate ions back to the lungs, where the low carbon dioxide concentration of the air causes the carbonic anhydrase reaction to proceed in the reverse direction.

$$HCO_3^- + H^+ \longrightarrow H_2CO_3 \longrightarrow H_2O + CO_2$$

As a result, gaseous carbon dioxide is released and diffuses from the blood into the alveoli. The carbon dioxide is then

Figure 37-16 This overview of respiration, *above,* shows that in the lungs, oxygen diffuses into the blood and carbon dioxide diffuses into alveoli. In the tissues, oxygen diffuses into the tissue's cells and carbon dioxide diffuses into the blood. Most carbon dioxide is transported in the form of bicarbonate ions inside red blood cells. Bicarbonate ions are chemically produced from carbon dioxide that has diffused from the tissue, through a capillary wall, and into a red blood cell, *bottom left.* In the lung, the chemical reaction is reversed. The carbon dioxide that is produced diffuses out of the red blood cell, through a capillary wall, and into an alveolus, *top left.*

expelled during exhalation. The carbon dioxide carried in plasma diffuses into the alveoli as well. The carbon dioxide bound to hemoglobin is also released because hemoglobin has a greater affinity for oxygen than carbon dioxide. Therefore, the hemoglobin releases its bound carbon dioxide and takes up oxygen instead. The red blood cells, with their newly bound oxygen, then start their next journey back to the body's tissues.

The carbonic anhydrase reaction is critical to the removal of carbon dioxide from tissues because the difference in carbon dioxide concentrations of the blood and tissues is not large (only about 5 percent). We will see in the following section how the level of carbon dioxide in the blood regulates your breathing. ☐

☐ CAPSULE SUMMARY

Oxygen is transported to tissues by combining with hemoglobin molecules inside red blood cells. Most carbon dioxide is transported to the lungs as bicarbonate ions inside the cytoplasm of red blood cells.

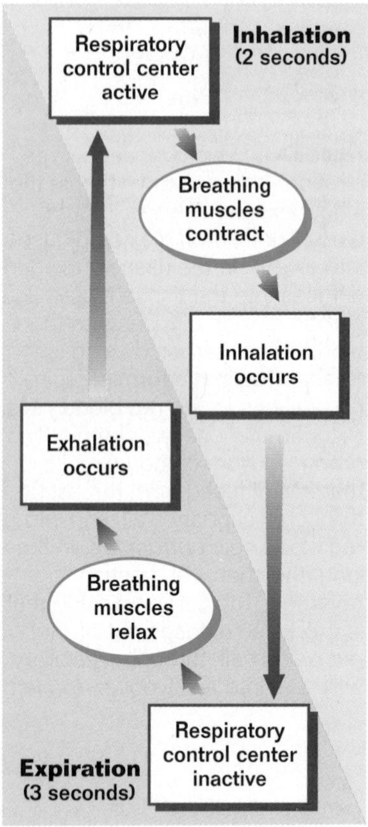

Figure 37-17 The respiratory control center in the brain initiates each breath. When the center sends out a nerve impulse, inhalation occurs. When the center is inhibited, exhalation occurs.

The Brain Controls Breathing

You took your first breath within moments of being born. Since then, you have repeated the process over 200 million times. Every one of these breaths was initiated by the respiratory control center of the brain, as is shown in Figure 37-17.

When the body is at rest, the respiratory control center sends nerve signals to the diaphragm to initiate inhalations. During vigorous breathing, such as during exercise, the level of carbon dioxide in the blood increases, making the blood more acidic. The respiratory control center responds to this increased acidity by sending signals to the muscles between your ribs as well as to the diaphragm. The contraction of the rib muscles causes the chest cavity to expand even further, enabling you to take in more air. Right after inhalation occurs, other neurons from the respiratory center inhibit the stimulation of the diaphragm and the muscles between the ribs so that they relax, and the body exhales.

The respiratory control center also regulates breathing rate (how many breaths you take per minute). Again, blood pH is the controlling factor. When you are sleeping, your muscles are not being used very much, so they don't produce much carbon dioxide. As a result, blood pH remains relatively constant, and you breathe at a slow rate. When you run, intense muscle activity increases the level of carbon dioxide in the blood, and blood pH drops. In response, you breathe more rapidly.

The signals that travel from the breathing center of the brain are not subject to voluntary control. You cannot simply decide to stop breathing. You can hold your breath for a while, but ultimately your respiratory control center will take over and force your body to breathe.

Respiratory Diseases Limit Lung Function

Respiratory diseases affect millions of Americans. **Asthma** is a respiratory disease in which the bronchioles of the lungs become constricted (narrowed) because of their sensitivity to certain stimuli in the air. The narrowing of the airways restricts air flow, making it difficult to exhale air from the alveoli. Left untreated, asthma can be dangerous. In severe asthma attacks, the alveoli may swell enough to rupture. Fortunately, inhalant medicines can counteract an asthma attack by dilating (expanding) the bronchioles.

Cigarette smoking has been linked to emphysema *(ehm fuh SEE muh)* and lung cancer, two respiratory diseases that claim millions of lives annually. Over 2 million American teenagers smoke, a pattern of behavior that may someday kill them. Almost no Americans begin smoking as adults. In people who develop **emphysema,** the lungs' alveoli lose their elasticity, making it difficult for the alveoli to release their air during exhalation. There is also a great reduction in the efficiency of gas exchange. Severely affected individuals must breathe from tanks of pure oxygen in order to live. Carcinogens present in cigarette smoke can also cause **lung cancer.** As you read in Chapter 9, cancer is a disease characterized by abnormal cell growth. Once the lung cancer, shown in Figure 37-18, is detected, the affected lung is usually removed surgically. Even with such drastic measures, fewer than 10 percent of lung cancer victims live more than five years after diagnosis.

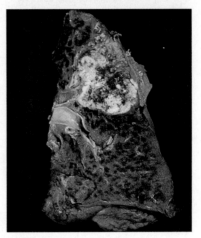

Figure 37-18 A healthy lung, *top,* contains normal cells and tissues. A cancerous lung, *bottom,* contains cells that divide uncontrollably. Scientific studies indicate that between 85 and 90 percent of all lung cancers are the direct result of cigarette smoking.

Section Review

1. *Draw a diagram of the respiratory system.*

2. *After an exhalation is completed, in what direction will the diaphragm move next?*

3. *Explain the role that carbonic anhydrase plays in transporting carbon dioxide in the blood.*

4. *If the nerves going to the diaphragm were severed, would breathing be affected? Explain your answer.*

5. *Explain why a person with emphysema would have trouble climbing stairs.*

Critical Thinking

6. *Researchers have determined that cigarette smoke paralyzes the cilia in the trachea. Knowing this, explain why smokers often suffer from severe bouts of coughing.*

Vocabulary

A-B-O system (881)
alveoli (889)
anemia (880)
aorta (884)
aortic valve (884)
arteriole (877)
arteriosclerosis (887)
artery (877)
asthma (893)
atherosclerosis (887)
Bohr effect (890)
bronchi (889)
bronchiole (889)
capillary (877)
carbonic anhydrase (891)
coronary artery (885)

diaphragm (889)
diastolic pressure (886)
emphysema (893)
epiglottis (889)
erythrocyte (880)
fibrin (881)
inferior vena cava (885)
larynx (888)
left atrium (884)
left ventricle (884)
leukemia (881)
leukocyte (880)
lung cancer (893)
lymph (879)
lymphatic system (879)
mitral valve (884)

pharynx (888)
plasma (879)
platelet (881)
pulmonary valve (885)
pulmonary vein (884)
respiration (888)
Rh factor (882)
right atrium (885)
sinoatrial node (885)
superior vena cava (885)
systolic pressure (886)
thoracic cavity (889)
trachea (888)
tricuspid valve (885)
vein (877)
venule (877)

Review

Multiple Choice

1. Which is not a function of the circulatory system?
   a. gas exchange
   b. temperature regulation
   c. hormone transport
   d. action potential transport

2. Lymphatic vessels
   a. transport blood.
   b. return fluid to the blood.
   c. produce antibodies.
   d. control blood clotting.

3. Blood that contains A antibodies (but not B antibodies) in the plasma and lacks Rh antigens is labeled
   a. AB negative.   c. B negative.
   b. A positive.   d. O positive.

4. Blood traveling through the pulmonary veins is
   a. oxygenated.   c. oxygen-poor.
   b. iron-poor.   d. calcium-rich.

5. A sphygmomanometer is used to measure
   a. heart rate.   c. blood pressure.
   b. lung volume.   d. breathing rate.

6. The diaphragm contracts and the pressure in the pleural cavity decreases during
   a. inhalation.   c. respiration.
   b. exhalation.   d. asthma attacks.

7. Breathing rate will automatically increase when
   a. blood pH is high.
   b. the amount of carbon dioxide in the blood decreases.
   c. blood acidity increases.
   d. hemoglobin is unloaded.

8. A disease in which the alveoli lose their elasticity is called
   a. atherosclerosis.   c. emphysema.
   b. arteriosclerosis. d. asthma.

Completion

9. Blood flows from arteries to _____ and then capillaries. From capillaries, blood flows through venules and into _____ before returning to the heart.

10. In each side of the heart, blood flows from the _____ to the ventricle. Blood flows into the _____ from the left ventricle and into the _____ from the right ventricle.

11. During inhalation, air flows through the trachea, into the right and left _____ , and then into the bronchioles before reaching the _____ of the lungs.

Themes Review

12. **Homeostasis** How is body temperature regulated by blood vessel constriction and dilation?

13. **Levels of Organization** The site of gas exchange in the human body is the lungs. How do alveoli function to increase the overall efficiency of the lungs?

Critical Thinking

14. **Interpreting Data** Salt intake and blood pressure data were collected from people representing more than 20 cultures, ranging from factory workers in France to natives of Colombia. The data are displayed as a best fit curve on the graph below. What is the relationship between salt intake and blood pressure? If an American businessman averages a daily salt intake of 26 g, what would you predict his systolic pressure to be? Knowing that high blood pressure can lead to a stroke or kidney failure, what advice would you give this businessman?

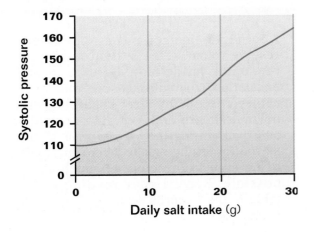

15. **Making Inferences** The frequency of blood clots and heart attacks is much lower among the Inuit, the nomadic hunters of the North American Arctic, than it is among other North Americans and Europeans. This difference is due to fish oils in the Inuit diet that cause their platelets to be more slippery. How do you think the clotting ability of the Inuit's blood is affected by the slippery platelets?

16. **Making Predictions** As altitude increases, the gases that make up the atmosphere, including oxygen, become more scarce. When a runner who trained at sea level competes at a location 500 m above sea level, how will his performance at this altitude compare with his performance during training?

Activities and Projects

17. **Cooperative Group Project** With a small group of classmates, obtain training in first aid procedures from the American Red Cross. Be sure your training includes the Heimlich maneuver and cardiopulmonary resuscitation (CPR). After receiving first aid training, organize training sessions with Red Cross officials for students and teachers at your school. Build interest in the first aid training by constructing and displaying posters that describe or picture medical emergencies and ask if the reader is prepared to help save a life.

18. **Multicultural Perspective** Dr. Charles Drew was an African American physician who lived during the first half of the twentieth century. Research the life of Dr. Drew and his many accomplishments, and how his work with blood helped save many lives.

LABORATORY Investigation

Lung Capacity and Carbon Dioxide Production

OBJECTIVES

- Indirectly measure lung capacity.
- Measure the amount of carbon dioxide produced before and after exercise.

PROCESS SKILLS

- observing and comparing differences
- charting effects of experimental results

MATERIALS

- spirometer
- safety goggles
- marker
- 250 mL flasks
- 0.1% methyl red in dropper bottle
- plastic wrap
- drinking straws
- 10 mL and 100 mL graduated cylinders
- graduated pipettes with rubber bulbs
- 0.04% sodium hydroxide in dropper bottle

BACKGROUND

1. Where does the carbon dioxide we exhale come from?
2. What is lung capacity?
3. How does exercise influence the amount of carbon dioxide you produce?
4. Write your own **Focus Question** on your Vee Form.
5. **Knowing Side of the Vee** List the **Concepts** and new **Vocabulary Words** on your Vee Form. In the **Concept Statements** section of the Vee, use these words in sentences that define and explain them.

TECHNIQUE

Doing Side of the Vee

Part A: Determining Lung Capacity

CAUTION: Do not share spirometer mouthpieces; change them as your teacher directs.

1. Tidal volume is one of several measurements that can be made by a spirometer. Tidal volume is the amount of air inhaled or exhaled in a normal breath. Remember that your lung capacity is influenced by many factors (age, diaphragm and chest muscle strength, disease, gender, and body position while testing).

2. With the tube of the spirometer near your mouth, inhale a normal breath. Then exhale a normal breath into the spirometer tube and take your reading. In the **Records** section of your Vee Form, record your results in a table similar to the one below.

Measurements of Lung Capacity		
	Avg. for young adult male	Your readings
Tidal volume	500 mL	
Expiratory reserve volume	100 mL	
Vital capacity	4600 mL	

3. Another measurement is expiratory reserve volume. This is the amount of air remaining in the lungs after a normal exhalation. You will measure your expiratory reserve by first breathing in a normal breath and exhaling normally and then putting the tube to your mouth as you forcefully exhale whatever is left, being sure to force out as much air as possible. Record your reading in the table in the **Records** section of your Vee Form.

4. Another measurement determined with a spirometer is vital capacity. This is the maximum amount of air that you can possibly inhale or exhale. Inhale the biggest breath you can, and then exhale all of the air possible into the spirometer tube. Record your reading in the table in the **Records** section.

5. The table on the previous page includes the values for young adult males. The average volumes for young adult females are 20–25% lower. Athletes can have volumes 30–40% greater than the average for their gender.

Part B: Carbon Dioxide Production

6. **CAUTION: Wear safety goggles at all times during this procedure.** Label and mark three flasks as 1, 2, and "Control."

7. **CAUTION: Avoid skin and eye contact with methyl red and sodium hydroxide.** Add 100 mL of tap water into each flask. Add 10 drops of methyl red to each and swirl to mix. Cover the mouth of each flask with plastic wrap.

8. Remove the plastic wrap from flask 1. Blow gently through one straw into flask 1 for exactly 2 minutes, exhaling slowly so that the solution does not bubble up. Be careful not to inhale the solution.

9. Using your pipette with a bulb, add NaOH 1 mL at a time, swirling with each addition and counting the milliliters you add to get the same pink color as in the control flask. In the **Records** section of your Vee Form, record the number of milliliters of NaOH added.

10. Exercise by jogging in place or doing jumping jacks for 2 minutes. Immediately blow gently through a new straw into flask 2 for exactly 2 minutes.

11. Repeat step 9 on flask 2 until you get the same pink color as in the control flask.

Record the number of milliliters of NaOH added to flask 2 in the **Records** section. In the **Procedure** section of the Vee, briefly summarize the procedure you followed.

12. Clean up your materials and wash your hands before leaving the lab.

INQUIRY

1. How did your tidal volume compare with that of your classmates?

2. What color change was observed in Part B when you blew into the flask?

3. What part of respiration is measured indirectly in Part B?

4. Use the information on the **Knowing Side** of the Vee to interpret your results from the **Doing Side,** and then write your **Knowledge Claim.** Write a **Value Claim** for this lab.

ANALYSIS

1. Why would males and athletes have greater vital capacities than females?

2. If a person's vital capacity was extremely low, how could it be increased?

3. How do you know if you produced more carbon dioxide before or after you exercised? What is your evidence from this lab?

4. Why were the flasks covered with plastic wrap?

FURTHER INQUIRY

Write a **New Focus Question** that could be the point of a new investigation. The following is an example:

How do asthma and emphysema reduce the efficiency of the respiratory system as measured by a spirometer?

DIGESTIVE AND EXCRETORY SYSTEMS

REVIEW

- pH scale (Section 2-2)
- carbohydrates, fats, and proteins (Section 2-3)
- diffusion and osmosis (Section 3-2)
- homeostasis (Sections 3-2 and 35-3)
- cellular respiration (Sections 5-1 and 5-3)
- water and salt balance in vertebrates (Section 31-4)
- negative feedback (Section 35-3)
- antigens (Section 37-1)
- carbon dioxide transport (Section 37-3)
- *Highlights: Digestive and Urinary Systems* (p. 845)

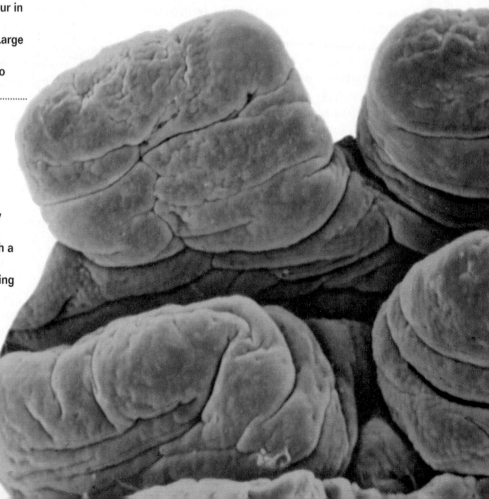

Villi from the small intestine

38-1 Diet: What We Need to Eat and Why

You obtain energy from the foods you eat to fuel your every activity. Your body uses energy to move, to grow, and to do countless other things—like read these words. The energy in food is stored in its chemical bonds. As you learned in Chapter 5, your cells break the chemical bonds of food molecules and harvest their energy to make ATP.

In addition to providing energy, the food you eat provides raw materials that your body uses to manufacture molecules for its own use. For example, calcium is used to make bone, and amino acids are used to make proteins. Every molecule in your body, every hair and bone and cell, is built from raw materials that came from food you ate.

Section Objectives

■ Describe how the body uses carbohydrates, fats, and proteins.

■ Explain what essential amino acids are.

■ Compare and contrast vitamins and trace elements.

Energy-Rich Compounds Are Used to Make ATP

Much of your need for food is the result of your body's constant use of and need for energy. Everything that you eat—meals and snacks and soda at the movies—is your diet, a record of what you consume. An optimal diet contains more carbohydrates than fats and also contains a significant amount of protein. To help educate consumers about the food choices they make, the United States government has set new standards for the nutrition labels that appear on food products. Figure 38-1 shows you one of these labels and explains how to interpret it.

Figure 38-1 The Food and Drug Administration's new system for labeling foods closely reflects the dietary guidelines developed by the American Heart Association. The label is not as complex as it looks. Study this sample label from a frozen dinner to learn how to interpret the new system.

1 Serving Size Serving sizes formerly were unrealistically small. New serving sizes set by the FDA are closer to the amounts that people actually eat.

2 Calories From Fat Beware of any food that gets more than one-third of its Calories from fat.

3 % Daily Value (DV) This tells you what percentage of the daily requirement for a nutrient you are getting.

4 Total Fat Keep a close eye on saturated fats; they cause clogged arteries.

5 Cholesterol High levels of fatlike cholesterol molecules in the blood can also lead to clogged arteries. Junk foods made from vegetables, like corn chips, may be advertised as cholesterol-free, but they are often high in other fats.

Nutrition Facts

1 Serving Size 1 Package (258 g)
Servings Per Container 1

Amount per Serving	
2 **Calories** 270	From Fat 70
3	**% Daily Value**
4 **Total Fat** 8 g	**12%**
Saturated Fat 3.5 g	**15%**
Polyunsaturated Fat .5 g	
Monosaturated Fat 1.5 g	
5 **Cholesterol** 30 mg	**9%**
6 **Sodium** 500 mg	**20%**
7 **Total Carbohydrate** 28 mg	**9%**
Dietary Fiber 3 g	**13%**
Sugars 5 g	
8 **Protein** 21 g	
9 Vitamin A 8% • Vitamin C 20%	
Calcium 35% • Iron 6%	

6 Sodium Sodium is abundant in table salt and many food products. Too much sodium can cause high blood pressure.

7 Total Carbohydrate Getting enough carbohydrates, especially complex ones, like starches, is important. Carbohydrates are used by your cells as a source of energy.

8 Protein Although the government has not set a percent DV for proteins, it is generally agreed they should compose no more than about 15 percent of your total daily Calories.

9 Vitamins and Minerals A DV of about 10 percent indicates that a food is a good source of these essential nutrients.

Figure 38-2 Carbohydrate-rich foods, *left,* should compose a major portion of your diet (50 to 55 percent). The majority of these foods should be foods such as breads, pastas, fruits, and vegetables, which contain complex carbohydrates and natural sugars. Foods containing refined sugars like candies and soft drinks, should be consumed sparingly. Protein-rich foods, *center,* should compose only about 15 percent of your diet, and fatty foods, *right,* no more than 30 percent of your diet. Fats contain more than twice as many Calories per gram as carbohydrates or proteins, so fat-rich foods should be avoided if you are trying to lose weight.

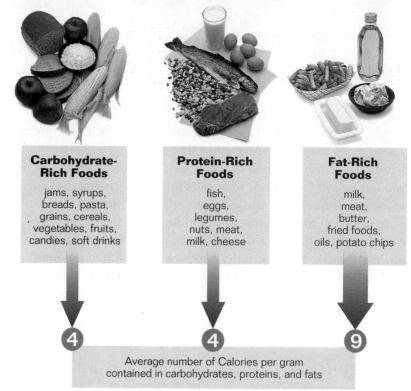

Carbohydrate-Rich Foods	Protein-Rich Foods	Fat-Rich Foods
jams, syrups, breads, pasta, grains, cereals, vegetables, fruits, candies, soft drinks	fish, eggs, legumes, nuts, meat, milk, cheese	milk, meat, butter, fried foods, oils, potato chips
4	4	9

Average number of Calories per gram contained in carbohydrates, proteins, and fats

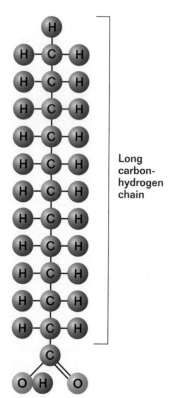

Long carbon-hydrogen chain

Figure 38-3 Fats are composed of molecules called fatty acids. The many carbon-hydrogen bonds of a fatty acid molecule store a great deal of energy.

Carbohydrates are a major energy source for your cells. The carbohydrates in a typical human diet are obtained primarily from cereals, grains, breads, fruits, and vegetables like the ones shown in Figure 38-2. On average, carbohydrates contain 4 Calories per gram. A **Calorie** is a unit of energy that indicates the amount of energy contained in food. Specifically, a Calorie is the amount of heat energy required to raise the temperature of 1 kg (2.2 lb.) of water 1°C (1.8°F). The greater the number of Calories in a quantity of food, the more energy it contains.

Proteins, like carbohydrates, contain 4 Calories per gram. Proteins are obtained from dairy products, poultry, fish, meat, and grains. Although cells can use proteins as an energy source, your body uses them mainly as building materials for cell structures, enzymes, hormones, muscles, and bones.

Fats are obtained from oils, margarine, and butter. They are abundant in fried foods, meats, and processed snack foods such as potato chips and crackers. Your body uses fats to construct cell membranes and other cell structures, to insulate nervous tissue, and as an energy source. Fats also contain certain fat-soluble vitamins that are essential for proper health.

A fat molecule with its long fatty acids, like the one shown in Figure 38-3, contains far more energy-rich carbon-hydrogen chemical bonds than a carbohydrate or protein molecule does. Therefore, fats contain much more energy per gram than carbohydrates or proteins do, about 9 Calories per gram. Although fats may be used as an energy source by cells, fats are also a very efficient way to store energy in the body.

Calories that are not used by your body are converted into fat tissue. When you eat a meal, your body can do only two things with the energy it gets from the food. The energy is either used by the muscles and other cells of the body, or it is converted into fat and stored in fat cells. This relationship can be expressed by the following simple equation.

$$\begin{array}{r} \text{energy obtained from food} \\ - \quad \text{energy used by body's cells} \\ \hline = \quad \text{energy stored in fat} \end{array}$$

In wealthy countries like ours, obesity, being more than 20 percent heavier than the average person of the same sex and height, is common. Obesity usually results from habitually overeating an unbalanced diet that is too high in fat. In the United States, about 30 percent of middle-aged women and 15 percent of middle-aged men are classified as obese. Obesity significantly increases an individual's risk of diabetes, coronary heart disease, and many other disorders. ▢

▢ CAPSULE SUMMARY

Carbohydrates, proteins, and fats are energy-rich compounds present in food. Your cells usually use carbohydrates as an energy source, although they are also capable of using proteins and fats.

Other Compounds Serve as Raw Materials

Over the course of evolution, many animals have lost the ability to manufacture some of the substances they need. These substances often play critical roles in metabolism. Typically, this ability is lost because a substance has become plentiful in the animal's diet. For example, mosquitoes and many other bloodsucking insects cannot manufacture cholesterol (a critical component of cell membranes). But that does not matter because the blood they consume is rich in cholesterol. There are essential substances that humans can get only from the foods they eat. These substances include certain amino acids, vitamins, and trace elements.

Amino Acids

Humans are unable to manufacture 8 of the 20 amino acids used to make proteins: lysine, tryptophan, threonine, methionine, phenylalanine, leucine, isoleucine, and valine. These eight amino acids are called essential amino acids because without them in your diet, your cells would be unable to manufacture the many proteins they need. For this reason, it is important to eat complete proteins, those containing all of the essential amino acids.

Vitamins

Vitamins are essential organic substances that the body requires in tiny amounts for normal growth and activity. Just as a computer uses only a tiny amount of grease (to lubricate the spinning hard disk) but cannot function without it, so our bodies use minute traces of vitamins to maintain good health.

Many vitamins function as integral parts of cellular enzymes. Other vitamins are essential components of coenzymes, organic molecules that take part in metabolic reactions. In Chapter 5, you learned that FAD accepts hydrogen atoms during the Krebs cycle. FAD is a coenzyme that has riboflavin, vitamin B_2, as an essential component. Vitamins must be obtained from the foods we eat because we are unable to manufacture them.

Some vitamins are soluble in water, while others are soluble in fat. Ingesting too much of a fat-soluble vitamin can be dangerous. Excessive amounts accumulate in the body's fatty tissues and may reach toxic levels. Table 38-1 lists several important water-soluble and fat-soluble vitamins and describes their food sources and biological roles.

Table 38-1 Vitamins

Vitamin	Food Sources	Role	Effects of Deficiency
Water-Soluble			
Vitamin B_1 (thiamin)	Most vegetables, nuts, organ meats	Carbohydrate metabolism, helps nerves and heart to function properly	Digestive disturbances, impaired senses
Vitamin B_2 (riboflavin)	Fish, poultry, cheese, yeast, green vegetables	Needed for healthy skin and tissue repair, carbohydrate metabolism	Blurred vision, cataracts, cracking of skin, lesions of intestinal lining
Vitamin B_3 (niacin)	Whole grains, fish, poultry, liver, tomatoes, legumes, potatoes	Keeps skin healthy, carbohydrate metabolism	Mental disorders, diarrhea, inflamed skin
Vitamin B_{12} (cobalamin)	Meat, poultry, green vegetables, milk, dairy products	Needed for formation of red blood cells	Reduced number of red blood cells
Vitamin C (ascorbic acid)	Citrus fruits, strawberries, potatoes	Needed for wound healing, healthy gums and teeth	Swollen and bleeding gums, loose teeth, slow-healing wounds
Fat-Soluble			
Vitamin A (retinol)	Carrots, green leafy vegetables, butter, eggs, liver, sweet potatoes	Keeps eyes and skin healthy, needed for strong bones and teeth	Infections of urinary and digestive systems, night blindness
Vitamin D (cholecalciferol)	Salmon, tuna, fish liver oils, fortified milk	Calcium and phosphorus metabolism, needed for strong bones and teeth	Bone deformities in children, loss of muscle tone
Vitamin E (tocopherol)	Many foods, especially wheat germ oil and olives	Protects cell membranes from damage by reactive oxygen compounds	Reduced number of red blood cells; nerve tissue damage in infants
Vitamin K	Leafy green vegetables, liver, cauliflower	Necessary for normal blood clotting	Bleeding caused by a prolonged clotting time

Trace Elements

Trace elements are minerals that are required by your body. Unlike calcium, sodium, and other inorganic elements that are present in your body in large quantities, trace elements are present in only minute amounts. Most trace elements are also essential for plant growth. Therefore, humans usually obtain adequate amounts of the required trace elements directly from the plants we eat or indirectly from animals that have eaten plants. Some important trace elements, their food sources, and their biological roles are presented in Table 38-2. ▢

Table 38-2 Trace Elements

Trace Element	Best Sources	Role
Iodine	Iodized salt, seafood, plants grown in high-iodine soil	Synthesis of thyroid hormone
Cobalt	Leafy vegetables, liver, kidney	Synthesis of vitamin B_{12}
Zinc	Meat, shellfish, dairy products	Synthesis of digestive enzymes
Molybdenum	Legumes, cereals, milk	Protein synthesis
Manganese	Whole grains, nuts, legumes	Hemoglobin synthesis, urea formation
Selenium	Meat, seafood, cereal grains	Preventing chromosome breakage

Section Review

1. State several reasons why you need more carbohydrates than fats in your diet.
2. Why would a diet without any protein be dangerous to your health?
3. Explain why zinc and manganese are not considered vitamins.

Critical Thinking

4. Each week for a month, an overweight person used more Calories exercising than were obtained from the food he or she had eaten that week. Contrast the amount of fat tissue in the individual's body at the beginning of the month with the amount present at the end of the month, and explain any difference.
5. Explain why a physician would prescribe an injection of several B vitamins to a person whose major symptoms are persistent fatigue and lack of energy.

38-2 Digestion

Your cells obtain the energy they need by extracting it from sugars, fatty acids, and amino acids during cellular respiration. Eating a plant or an animal provides you with a rich source of complex starches, fats, and proteins. These molecules occur as long chains composed of individual sugars, fatty acids, or amino acids, like strings of beads. Your cells cannot extract energy from these large chains. First, these large molecules must be broken down into their individual components during a process called digestion.

Section Objectives

- Trace the path of food through the digestive system.
- Name the enzymes that digest carbohydrates, fats, and proteins.
- Describe the major digestive processes that occur in the stomach, small intestine, and large intestine.
- Describe the role of the pancreas and liver in digestion.
- Explain how nutrients are absorbed into the bloodstream from the digestive system.

Food Molecules Are Broken Down by Enzymes

Each of the three types of large food molecules—starch, fat, and protein—presents a different digestive challenge and requires a unique strategy, as shown in Figure 38-4. For example, starches, which are chains of sugar molecules, are readily broken down into sugars by enzymes called **amylases** (*AM uh lays uhs*).

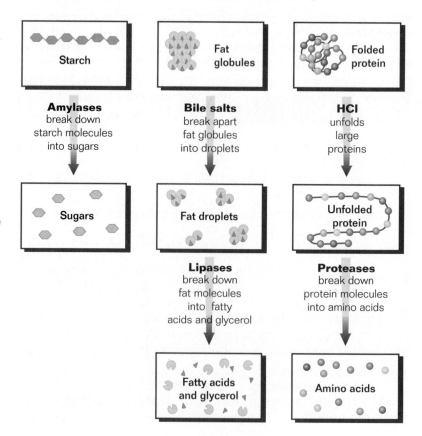

Figure 38-4 Starches, *top left,* are chemically digested into sugars by enzymes called amylases. Globules of fat, *top center,* must be broken apart by bile salts before the individual fat molecules can be digested into fatty acids (yellow) and glycerol (orange) by enzymes called lipases. Large, folded proteins, *top right,* must be unfolded by hydrochloric acid (HCl) before the proteins can be broken into amino acids by enzymes called proteases.

Fats are insoluble in water and tend to aggregate into large globules consisting of many fat molecules. As a result, many of the individual fat molecules are not easily attacked by enzymes. Before fats can be digested, they must first be treated with a detergent to make them water soluble, a process that is called emulsification. **Emulsification** (*ee MUHL suh fih kay shuhn*) breaks up fat globules into many tiny fat droplets, thus exposing many more fat molecules to enzymes. Your liver produces detergent molecules called **bile salts,** which emulsify the fats contained in the foods you eat. After these fats have been emulsified, enzymes called **lipases** (*LEYE pays uhs*) break the fat molecules into fatty acids and glycerol.

Proteins, which are chains of amino acids, present a particularly tough challenge to the digestive system. Almost all proteins are either folded into tight balls or wound together into tough fibers. Enzymes cannot break down these balls and fibers because they cannot get at the individual protein chains. The human body solves this problem by carrying out protein digestion in two steps. First, hydrochloric (*HEYE droh klawr ihk*) acid, HCl, in the stomach is used to unfold large proteins into single polypeptide strands; then enzymes called **proteases** (*PROHT ee ays uhs*) attack the strands, cutting them into smaller fragments.

Most digestive enzymes cannot tolerate extremely acidic conditions, so human digestion is carried out in phases, each of which occurs in a different part of the digestive system. The digestion of proteins by acid, as well as some enzymatic digestion of proteins, takes place in the stomach. Next, the food moves to the small intestine, where the acid from the stomach is neutralized, thus allowing the intestinal enzymes to function. In the small intestine fats are emulsified, and carbohydrates, fats, and proteins are completely dismantled by amylases, lipases, and proteases. The products of this digestion—amino acids, fatty acids, glycerol, and sugars—then pass through the wall of the small intestine and enter the bloodstream. ❏

❏ **CAPSULE SUMMARY**

Amylases, lipases, and proteases are three classes of digestive enzymes. Amylases break down starches into sugars. After fat globules have been emulsified by bile salts, lipases break down individual fat molecules into fatty acids and glycerol. After compacted proteins have been unfolded by HCl, proteases break down the proteins into polypeptides and amino acids.

Figure 38-5 Like most mammals, humans have teeth that are differentiated. Canine and incisor teeth, used for cutting and tearing food, are located toward the front of both the upper, *top,* and lower, *bottom,* jaws. The molars, located toward the back of the jaws, are used to grind food.

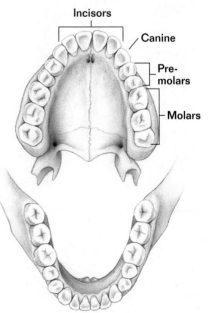

Incisors

Canine

Pre-molars

Molars

The Digestive Journey Begins in the Mouth

Humans regularly eat a wide variety of plants and animals, and the structure of our teeth, shown in Figure 38-5, reflects our omnivorous (*ahm NIHV uh ruhs*) diet. The front teeth of humans are structurally similar to the pointed, cutting, and ripping teeth characteristic of meat eaters (carnivores). Our back teeth resemble the flat, grinding teeth characteristic of plant eaters (herbivores).

After food has been ripped or chewed into shreds by the teeth, the tongue mixes it with a watery solution called saliva. Saliva is secreted into your mouth by three pairs of salivary

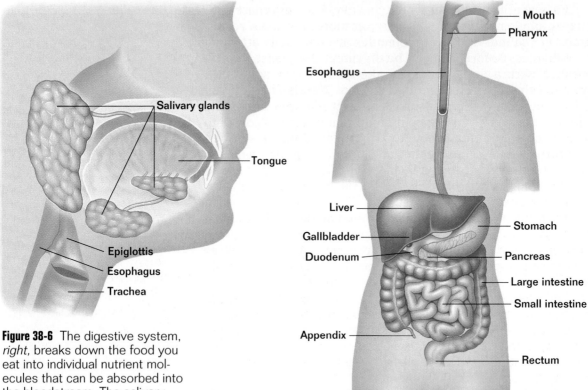

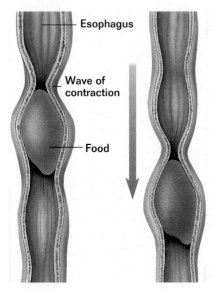

Figure 38-6 The digestive system, *right*, breaks down the food you eat into individual nutrient molecules that can be absorbed into the bloodstream. The salivary glands, *left*, secrete saliva through ducts that open into the mouth.

Figure 38-7 Food enters the esophagus from the pharynx. The food is pushed down the esophagus and toward the stomach by a wave of muscular contraction that occurs in the wall of the esophagus.

glands, shown in Figure 38-6, located above the upper jaw and below the lower jaw. Saliva moistens and lubricates the food so that it can be swallowed more easily. Saliva also contains amylase, which begins the breakdown of starch.

After passing through the region in the back of the throat called the **pharynx** *(FAIR ihnks)*, the food enters the esophagus. The **esophagus** *(ih SAHF uh guhs)* is a long tube that connects the mouth to the stomach. No digestion takes place in the esophagus. Its role is to act as a kind of descending escalator, moving food down to the stomach. Your esophagus is about 25 cm (10 in.) long. The lower two-thirds of the esophagus is wrapped in sheets of smooth muscle. Food doesn't just fall down into the stomach—it is "pushed" down, as shown in Figure 38-7. Successive rhythmic waves of contraction of the smooth muscle in the wall of the esophagus, called **peristaltic** *(pehr uh STAHL tihk)* **contractions,** move the food along toward the stomach.

Proteins Are Dismantled in the Stomach

Food exits the esophagus and passes into the stomach through a muscular valve called a sphincter *(SFIHNGK tuhr)*. The sphincter prevents acid-soaked food in the stomach from making its way back into the esophagus. The stomach is a saclike organ located just beneath the diaphragm. Besides temporarily storing food, the stomach also mechanically breaks down food and chemically unravels and breaks down proteins.

When food enters the stomach, "gastric juice," a combination of HCl and acid-stable proteases, is secreted by the stomach's epithelial lining. In the upper portion of the stomach's inner wall are deep depressions called gastric pits. These pits contain cells that secrete HCl. The proteases contained in gastric juice are secreted by other cells in the gastric pits. The stomach mixes its contents with a churning action caused by the contraction and relaxation of its muscular wall. The mixture of food, HCl, and enzymes eventually forms a soupy, semisolid material called **chyme** (KEYEM).

Your stomach secretes about 2 L (2.11 qt.) of HCl every day, creating a very concentrated acid solution—about 3 million times more acidic than your bloodstream. HCl unfolds proteins because its low pH (between 1.5 and 2.5) breaks the molecular attractions that hold the polypeptide chains of proteins together. Once the chains have been unfolded, the proteases cut them into smaller polypeptide fragments.

It is important that your stomach not produce *too* much acid, for if it did, it would be impossible for your body to neutralize the acid later in the small intestine. Acid production is controlled by hormones produced in endocrine cells that are scattered within the walls of the stomach. The hormone **gastrin** regulates the synthesis of HCl, permitting it to be made only when the pH in the stomach is higher than about 1.5. This negative feedback loop is shown in Figure 38-8. Some people oversecrete gastrin, causing their stomachs to produce too much HCl. The excessive acid may eventually eat holes in the walls of the stomach and small intestine. These holes are called ulcers (UHL suhrs).

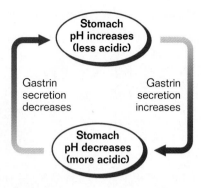

Figure 38-8 The pH in the stomach remains relatively constant (about 2) due to a hormonal negative feedback loop. If the pH in the stomach begins to rise too high, the stomach secretes the hormone gastrin, which stimulates the stomach to secrete more HCl, thus decreasing the pH inside the stomach. If the pH begins to drop too low, gastrin secretion decreases, less HCl is secreted, and the pH increases.

Figure 38-9 The pancreas and liver are often called accessory digestive organs because the food within the digestive tract never actually enters them. Instead, bile from the liver and digestive enzymes from the pancreas pass through ducts into the first part of the small intestine, the duodenum.

Digestion and Absorption Occur in the Small Intestine

From the stomach, food passes into the small intestine. Only small portions of acidic chyme are introduced into the small intestine, thereby ensuring that the acid will be neutralized and that the intestinal enzymes will be able to function.

Some of the digestive enzymes in the small intestine are secreted by the cells of the intestinal wall. Most of the digestive enzymes, however, are produced in a gland called the pancreas, shown in Figure 38-9. The **pancreas** (PAN kree uhs) sends enzymes through a duct into the first part of the small intestine, called the **duodenum** (doo oh DEE nuhm). Your small intestine is approximately 6 m (19.8 ft.) long. Unwound and

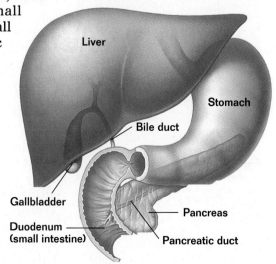

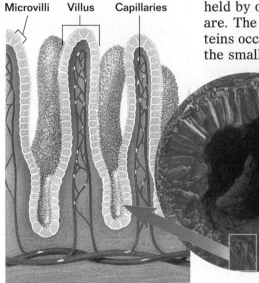

Microvilli Villus Capillaries

Figure 38-10 This cross section of the small intestine, *right,* has been magnified about 30 times. The fingerlike projections called villi, *left,* are the sites where food molecules are absorbed from the intestine into the bloodstream. Hairlike extensions called microvilli on the surface of the villi greatly increase the internal surface area of the small intestine.

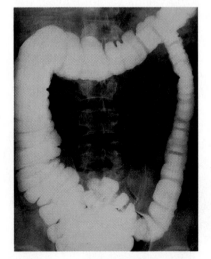

Figure 38-11 This X ray of the large intestine shows its location in the abdomen. The diameter of the large intestine is about three times that of the small intestine.

held by one end, it would be over three times taller than you are. The complete digestion of carbohydrates, fats, and proteins occurs in the duodenum, about the first 25 cm (10 in.) of the small intestine. The rest of the small intestine is devoted to absorbing water and the products of digestion into the bloodstream.

The lining of the small intestine is covered with fine fingerlike projections called **villi,** shown in Figure 38-10, each too small to see with the naked eye. In turn, the cells covering each villus have cytoplasmic projections on their outer surface called **microvilli.** The villi and microvilli greatly increase the absorptive surface area of the small intestine. The average surface area of an adult human's small intestine is about 300 sq. m (3,229 sq. ft.)—greater than the surface area of many large swimming pools.

The volume of material passing through the small intestine is surprisingly large. An average human consumes about 0.85 L (0.90 qt.) of solid food and 1.2 L (1.27 qt.) of water a day, for a total volume of about 2 L (2.11 qt.). To this amount are added secretions from the salivary glands, stomach, liver, pancreas, and small intestine, bringing the total up to a remarkable 9 L (9.51 qt.)—more than 10 percent of the total volume of your body!

Although the volume of material that flows through the digestive system is great, the amount that leaves the body as waste is small. This is because almost all of these fluids and solids are absorbed during their passage through the small intestine. Only a small quantity is absorbed through the wall of the large intestine. Of the 9 L (9.51 qt.) of material that enter the digestive tract each day, only about 0.05 L (0.06 qt.) of solid and 0.1 L (0.11 qt.) of liquid leave the body as wastes. The fluid absorption efficiency of the intestine approaches 99 percent.

Solids Are Compacted in the Large Intestine

Food passes from the small intestine to the large intestine, shown in Figure 38-11. The large intestine, or **colon,** is much shorter than the small intestine, about 1 m (3.3 ft.) long. The large intestine is not coiled up like the small intestine. Instead, it is in three relatively straight segments. No digestion takes place within the large intestine.

The inner surface of the large intestine does not have villi. Consequently, the large intestine has only one-thirtieth of the absorptive surface area of the small intestine. Although a small amount of fluid, along with sodium and vitamin K, are absorbed through its walls, the primary function of the large intestine is to act as a trash compactor. Within it, undigested material, including large amounts of plant fiber and cellulose, is compacted and stored. Many bacteria live and actively divide within the large intestine, where they play a role in processing undigested material into the final excretory product, feces.

The final segment of the digestive tract is a short extension of the large intestine called the **rectum.** Compacted solids within the colon pass into the rectum as a result of the peristaltic contractions of the muscles in the wall of the large intestine. From the rectum, the solid feces pass out of the body through the anus. ▪

How Nutrients Are Delivered to Cells

Blood that leaves the small intestine, rich with the products of digestion, is collected in the portal vein and carried to the liver.

Your **liver,** a very large organ, is the size of a football and weighs over 1.4 kg (3.1 lb.). The liver performs a wide variety of metabolic functions. For example, the liver supplies quick energy to the body by releasing sugar into the blood, and it builds complex carbohydrates from sugar. It also makes proteins from amino acids, stores vitamins and minerals, produces bile salts, regulates blood clotting, monitors the production of cholesterol, and detoxifies alcohol and poisons.

After flowing through the many fine passages of the liver, the blood is collected into the hepatic vein, which carries the blood back toward the heart to be circulated to the rest of the body.

❏ CAPSULE SUMMARY

Carbohydrate digestion begins in the mouth, and protein digestion begins in the stomach. The pancreas secretes amylases, lipases, and proteases into the small intestine, where carbohydrates, fats, and proteins are completely digested. The end products of digestion are absorbed through intestinal villi and pass into the bloodstream. The waste products of digestion are stored in the large intestine before passing out of the body.

CONTENT LINK

You can learn about the hormones that cause the liver to store and release sugar in **Chapter 41.**

Section Review

1. *Why does protein digestion begin in the stomach?*
2. *Define the term* emulsification, *and explain why it occurs during digestion.*
3. *When during the digestive process does food enter the liver and pancreas?*
4. *Describe digestion in the small intestine, and explain how food molecules enter the bloodstream.*
5. *What is the function of the large intestine?*

Critical Thinking
6. *A person has a small intestine that has villi but lacks microvilli. Would you expect this person to be underweight or overweight? Explain your answer.*

38-3 Excretion

If you live to be 70 years old, you will, in the course of your life, eat some 45,000 lb. of food and drink over 7,250 gal. of fluid—enough to fill a tanker truck. Did you ever wonder what happens to all of this food and fluid? Every atom of it still exists, if not in your body, then in the wastes that have been eliminated from it. Humans eliminate wastes in a process called excretion.

Section Objectives

- Identify the major wastes produced by humans, and describe how they are eliminated from the body.
- Draw a simple sketch of a nephron.
- Explain how a nephron forms urine.
- Trace the flow of urine through the human urinary system.
- Compare and contrast kidney dialysis with kidney transplants.

Carbon Dioxide, Nitrogen Compounds, and Water Are Excreted

Excretion rids the body of toxic chemicals, excess water, salts, and carbon dioxide. Among the many kinds of substances you excrete, the most important are carbon dioxide, nitrogen wastes, and water, as shown in Figure 38-12.

Almost all of the energy in the food you eat is contained in the food's carbon-hydrogen bonds. During cellular respiration this energy is extracted from the food molecules. The carbon and hydrogen atoms that are left behind combine with oxygen to form carbon dioxide and water. The carbon dioxide is transported to your lungs by the circulatory system and excreted from your body every time you exhale.

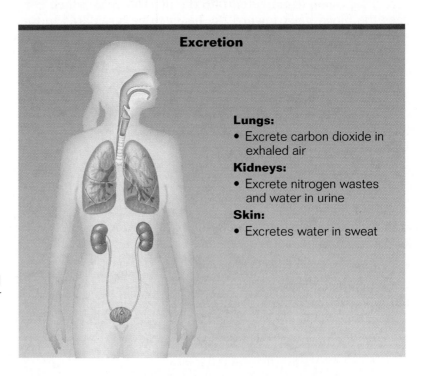

Excretion

Lungs:
- Excrete carbon dioxide in exhaled air

Kidneys:
- Excrete nitrogen wastes and water in urine

Skin:
- Excretes water in sweat

Figure 38-12 The lungs, the kidneys (along with their associated system of tubes), and the skin all function as excretory organs. For example, every time you exhale, you release carbon dioxide, a waste product of cellular respiration, into the atmosphere.

The removal of the amino groups from the amino acids of proteins produces ammonia, NH_3. Ammonia is extremely toxic to cells. Getting rid of this metabolic nitrogen waste product is a problem that every organism must solve. In aquatic animals like fishes, the ammonia is excreted through gills, directly into the surrounding water. Land animals, like humans, must solve the problem of ammonia accumulation a different way. In the liver, a complex series of chemical reactions combines pairs of amino groups with carbon dioxide to form a much less toxic nitrogen waste called **urea** (*yoo REE uh*). Urea is a principal component of urine, the fluid formed by the urinary system.

The average person drinks over 1 L (1.1 qt.) of fluid every day. You would swell up like a balloon if your body did not lose an equal amount of water through sweat and urine. As you learned in Chapter 35, sweating plays an important role in controlling body temperature, while urine is a vehicle for discharging urea and excess salts from the body.

Your body eliminates urea and excess water by means of a pair of bean-shaped, reddish brown kidneys located in the lower back. Kidneys, shown in Figure 38-13, are the size of a small fist. Urea is carried by the bloodstream to the kidneys, where it is removed from the blood. The kidneys receive a flow of about 2,000 L (2,114 qt.) of blood each day—more than the volume of a car! Since your body holds only 5.6 L (5.9 qt.) of blood, you can see that your blood must circulate through the kidneys many times during the day (about 350 times per day, or once every four minutes).

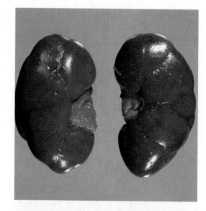

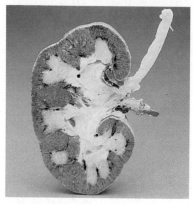

Figure 38-13 Human kidneys, *top,* filter wastes (mostly urea and excess water) from the blood that circulates through them. The blood supply to a kidney is enormous, as can be seen in this dissected kidney, *bottom.* This kidney has been injected with a dye that stains arteries red and veins blue.

The Kidneys Regulate Blood Plasma Composition

The kidneys are extremely important organs because of the role they play in homeostasis—maintaining a balanced state within the body. As you learned in Chapter 35, the chemical composition of extracellular fluid (which includes blood plasma) cannot vary to any great degree without causing serious harm to the body's cells and organ systems. The kidneys play a key role in regulating the amount of water and salt contained in blood plasma.

If you were to drink a large amount of water, your blood would become diluted and your blood pressure would sky-rocket unless you somehow got rid of the extra water. Your body uses the kidneys to remove the excess water by increasing urine production. Besides removing excess water from the blood, your kidneys also conserve water. Being a terrestrial organism, your body has a much higher concentration of water than the surrounding air does, so you constantly lose water to the air by evaporation. Water is also lost as water vapor in the air that you exhale. In order to minimize water loss, your kidneys tend to concentrate the urine they produce.

Your body also monitors the levels of sodium (salt) in your blood. When the number of sodium ions falls below normal, your kidneys direct the adrenal glands to send out hormones that cause the kidneys to remove more sodium ions from the urine forming within them. This sodium then returns to the blood. On the other hand, if levels of sodium in the bloodstream rise too high, hormone levels are decreased so that more sodium is excreted in the urine. ■

Figure 38-14 A kidney, *left*, contains microscopic filtering units called nephrons. Each nephron, *center*, is composed of a hollow tubule system with blood vessels wrapped around it. The processes of filtration and reabsorption that occur in the nephron lead to the formation of urine, *right*. During filtration, urea, water, sodium, and glucose from the blood pass into a Bowman's capsule. During reabsorption, all of the glucose, and most of the sodium and water, pass into the capillaries that surround the loop of Henle. Urine is composed of the urea and excess water and sodium that remain in the collecting duct.

Kidneys Are Composed of Tiny Blood Filters

The kidney is a complex organ composed of roughly 1 million microscopic blood-filtering units called **nephrons** (*NEHF rahns*), shown in Figure 38-14. Each nephron produces urine. A nephron is composed of three elements. A different phase of urine production occurs in each element.

1. **Filtration** The filtration device at the top of each nephron is called a **Bowman's capsule.** Within each Bowman's capsule, an arteriole enters and splits into a fine network of capillaries called a **glomerulus** (*gloo MEHR yoo luhs*). The glomerulus acts as a filtration device. The blood pressure inside the glomerulus forces fluid through the capillary walls. The walls prevent blood cells, proteins, and other large molecules from leaving the blood but allow water

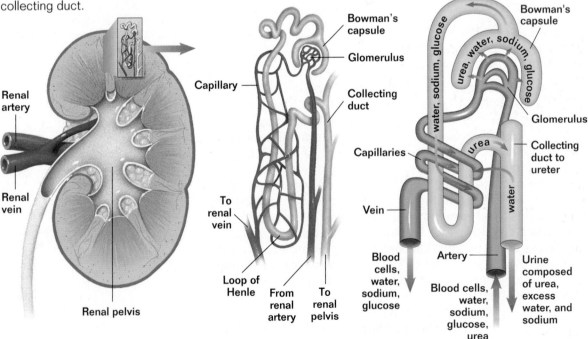

and small molecules such as urea, sodium ions, and glucose to pass through and enter the hollow interior of a Bowman's capsule. The fluid inside a Bowman's capsule is called the **filtrate.**

2. **Reabsorption** A Bowman's capsule is connected to a long narrow tube called the **renal tubule.** The renal tubule bends back on itself in its center, forming what is called the **loop of Henle.** The renal tubule is a reabsorption device. It extracts a variety of useful molecules, including glucose, ions, and some water, from the filtrate that passes through it. These substances reenter the bloodstream by passing into capillaries that wrap around the tubule. If it were not for the reabsorption that occurs in this portion of the nephron, these molecules would be eliminated from the body in the urine.

3. **Urine formation** The renal tubule empties into a larger tube called a **collecting duct.** The collecting duct removes much of the water from the filtrate that passes through it. As a result, human urine is four times more concentrated than blood plasma. Your kidneys achieve this remarkable degree of water reabsorption by a simple but superbly efficient mechanism. Because the collecting duct is slightly permeable to urea, some urea diffuses out and accumulates in the tissue around the duct. In a similar manner, some sodium also diffuses out from the portion of the tubule that is directly after the loop of Henle. Because of the high concentration of solute that exists around the collecting duct, an osmotic gradient is created, and water is drawn out of the collecting duct into the surrounding tissue, as is shown in Figure 38-15. The tissue absorbs water from the filtrate and passes it on to blood vessels that carry it out of the kidneys and back into the bloodstream. The filtrate that remains after salts, nutrients, and water have been removed is called urine. ◻

◻ CAPSULE SUMMARY

The microscopic blood-filtering units of the kidneys are called nephrons. Water, urea, sugars, amino acids, and salts enter one end of a nephron. As these materials move through the nephron's tubule and collecting duct, most of the water, some of the salts, and all of the sugar and amino acids are reabsorbed into the bloodstream. The mixture of water, urea, and salts that remains in the nephron is called urine.

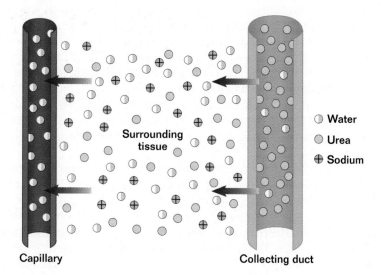

Water
Urea
Sodium

Capillary   Surrounding tissue   Collecting duct

Figure 38-15 The high concentration of urea and sodium that surrounds the collecting duct of a nephron creates a hypertonic environment, which draws water out of the collecting duct. The water then diffuses into capillaries and is returned to the body. The capillaries are impermeable to urea. Although some sodium enters the capillaries, equal quantities of sodium exit. Thus, the hypertonic nature of the surrounding tissue is maintained.

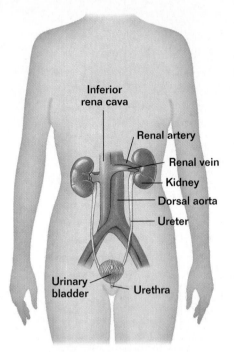

Inferior
rena cava

Renal artery

Renal vein

Kidney

Dorsal aorta

Ureter

Urinary
bladder

Urethra

Figure 38-16 Urine exits the kidneys by way of two ureters that empty into a storage organ called the urinary bladder. Urine exits the body through the urethra.

CONTENT LINK

You can learn more about the structural relationship between the urinary and reproductive systems in **Chapter 42.**

Urine Leaves the Body Through a System of Tubes

Urine produced in the kidneys passes into the urinary bladder through tubes called **ureters** *(yoo REET uhrs),* shown in Figure 38-16. The ureters have smooth muscle in their walls. The slow, rhythmic contractions of this muscle move the urine through the ureters.

The ureters direct the urine into the **urinary bladder,** a hollow, muscular sac that gradually expands as it fills. The bladder can hold up to 0.6 L (0.63 qt.) when full. Urine leaves the bladder and exits the body through a tube called the **urethra** *(yoo REE thruh).* In males the urethra passes through the penis. In females the urethra lies in front of the vagina and is only about 2.5 cm (1 in.) long. Such a short length makes it easy for bacteria and other pathogens to invade the urinary system, which explains why females are more prone to urinary infections than males are. There is no connection between the urethra and the genital (reproductive) system in females. In males, both sperm and urine exit the body through the urethra.

The elimination of urine from the body through the urethra is called urination. When the bladder fills with urine, stretch receptors in its wall send nerve impulses to the spinal cord. In response, the spinal cord returns impulses to the bladder, causing its muscular walls to contract and the tight rings of muscle closing off the urethra to relax. The bladder then empties its contents through the urethra. In older children and adults, the brain controls this urination reflex, delaying the release of urine until a convenient time.

Kidney Failure Is Life Threatening

Because of the important role that the kidneys play in maintaining homeostasis of body fluids, kidney failure can quickly become life threatening. The most common causes of kidney failure are infection, long-term diabetes, untreated long-term high blood pressure, and damage to the kidneys by the body's own immune system (autoimmune kidney disease). When kidneys stop working, toxic waste materials such as urea accumulate in the blood plasma. In addition, blood plasma ion levels rapidly depart from their normal values. If the kidneys fail, there are only two treatment options, dialysis *(deye AL uh sihs)* or transplant.

In **hemodialysis**, tubes called catheters are surgically inserted into an artery and a vein, usually on the lower arm. These catheters are equipped with valves that can be opened and closed. Every few days the individual must go to a clinic where the catheters are connected to a dialysis machine, shown in Figure 38-17. Blood passes from the patient's artery into the machine and then back into the vein. Inside the dialysis machine, the blood travels through a unit of many hollow tubes. The tubes are surrounded by a thin permeable membrane. Waste materials and ions that have accumulated in the person's blood plasma diffuse through the membrane into a fluid that bathes the outside of the membrane. The fluid has the same composition as normal blood plasma and is free of wastes. Dialysis patients must carefully manage their salt and water intake because the dialysis machine, unlike the kidney, cannot regulate blood volume and sodium levels.

Dialysis is not a permanent solution to kidney failure. A single healthy kidney can meet all of the homeostatic needs of the body, but no dialysis machine can. A more permanent solution to kidney failure is transplantation of a kidney from a healthy donor.

One great problem with kidney transplants is common to all organ transplants—rejection of the transplanted organ by the recipient's immune system. All of the cells of your body have "self-markers" on their surfaces to identify them so that they are not attacked by your body's immune system. These markers are called **histocompatibility** *(hihs toh kehm pat uh BIHL uh tee)* **antigens.** The combination of these antigens displayed on your body's cells is as unique as your fingerprints. Only identical twins have the same histocompatibility antigens. The more closely related two individuals are to one another, the more likely they are to have some common histocompatibility antigens. This is why tissue transplants are more likely to succeed if the donor and recipient are closely related. But even in close matches, there is some chance of transplant rejection. To reduce chances of rejection, the recipient is treated with drugs like cyclosporin, which suppresses the activity of the immune system. However, when such drugs are given, there is an increased risk of infection.

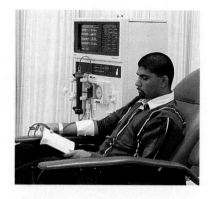

Figure 38-17 Hemodialysis has prolonged the lives of many people with damaged or diseased kidneys. The dialysis machine functions like a kidney in that it filters urea and excess ions from the blood.

CONTENT LINK
You will learn more about why the body rejects foreign tissues in **Chapter 39.**

Section Review

1. *How is the carbon dioxide in your body produced and excreted?*
2. *Name the liquid that would be found inside the collecting duct of a nephron.*
3. *How is urine stored and eliminated from the body?*
4. *What are the benefits and risks involved in kidney transplants?*

Critical Thinking
5. *Suppose that a urine sample contains a high concentration of protein. Explain why this may indicate damaged kidneys.*

Vocabulary

amylase (904)
bile salt (904)
Bowman's capsule (912)
Calorie (900)
carbohydrate (900)
chyme (907)
collecting duct (913)
colon (908)
duodenum (907)
emulsification (904)
esophagus (906)
excretion (910)
fat (900)

filtrate (913)
gastrin (907)
glomerulus (912)
hemodialysis (914)
hepatic vein (909)
histocompatibility antigen (915)
lipase (905)
liver (909)
loop of Henle (913)
microvilli (908)
nephron (912)
pancreas (907)
peristaltic contraction (906)

pharynx (906)
portal vein (909)
protease (905)
protein (900)
rectum (909)
renal tubule (913)
ulcer (907)
urea (911)
ureter (914)
urethra (914)
urinary bladder (914)
villi (908)

Review

Multiple Choice

1. Which compound is used by the body mainly to build muscle and bone?
   a. fat
   b. carbohydrate
   c. protein
   d. iodine

2. Select the sequence that most accurately reflects the order of movement and digestive activity as food passes through the digestive tract.
   a. duodenum, pancreas, stomach, small intestine, large intestine
   b. stomach, small intestine + pancreas, large intestine
   c. duodenum, esophagus, small intestine + pancreas, large intestine
   d. esophagus, small intestine, large intestine, pancreas, colon

3. Which pair matches the enzymes with the food molecules they digest?
   a. lipases: starches
   b. amylases: fats
   c. lipases: fats
   d. proteases: starches

4. Nutrients in the small intestine enter the bloodstream by passing through structures called
   a. villi.
   b. macrovilli.
   c. gastric pits.
   d. both a and b.

5. Which substance is *not* a waste eliminated from the body through the kidneys?
   a. urea
   b. water
   c. salts
   d. oxygen

6. The fine capillary network of the nephron system is called the
   a. collecting duct.
   b. glomerulus.
   c. loop of Henle.
   d. Bowman's capsule.

7. A kidney transplant is likely to be most successful if the kidney donor is the recipient's
   a. friend.
   b. uncle.
   c. wife.
   d. sister.

Completion

8. Studies have shown that 1 g of carbohydrate yields about _____ Calories; 1 g of protein yields about _____ Calories; and 1 g of fat yields approximately _____ Calories.

9. Of the 20 amino acids used to make proteins in the human body, _____ must be obtained through food.

10. The large intestine, or _____ , is much shorter than the small intestine. It functions to compact undigested materials and reabsorb vitamin _____ .

11. Urine is produced in the _____ , each of which drains into the urinary bladder through a tube called the _____ .

12. **Homeostasis** How would liver failure affect the concentration of ammonia within the body? What is the relationship of ammonia to urea?

13. **Evolution/Structure and Function** The loop of Henle functions to conserve water by reabsorbing it. Its length varies among mammal species. Would you expect the loop of Henle of an animal like the beaver, which lives in a watery environment, to be longer or shorter than that found in humans? Explain.

14. **Interpreting Data** The table below shows nutrition data for a single serving of two breakfast cereals. If you were most concerned with preventing colon cancer, keeping your blood pressure low, and supplying your body's cells with enough energy, which cereal, A or B, would you choose for breakfast? Explain.

Cereal A		Cereal B	
Serving Size 1 box (19 g)		Serving Size 1 box (32 g)	
Amount/serving		Amount/serving	
Calories	70	Calories	100
Fat Calories	0	Fat Calories	5
	% DV		% DV
Total Fat 0 g	0%	Total Fat 0.5 g	1%
Saturated Fat 0 g	0%	Saturated Fat 0 g	0%
Cholesterol 0 mg	0%	Cholesterol 0 mg	0%
Sodium 210 mg	9%	Sodium 170 mg	7%
Total Carbohydrates 17 g	6%	Total Carbohydrates 25 g	8%
Fiber 1 g	4%	Fiber 4 g	16%
Sugars 1 g		Sugars 10 g	
Protein 1 g		Protein 2 g	

15. **Making Inferences** Pancreatic fluid contains sodium bicarbonate, which changes the pH of the chyme from acid to base as it moves into the duodenum. How would the absence of sodium bicarbonate in the pancreatic fluid affect the functioning of the many enzymes also carried in this fluid?

16. **Research and Writing** Write an article for the health section of your school or local newspaper that discusses diuretics. You may wish to address these questions in your article: What is a diuretic and how does it affect kidney function? Why is thirst often intensified by a diuretic? How do diuretic drugs function to lower blood pressure? To capture your reader's interest, why not begin your article by identifying diuretics that most people have heard of, such as the caffeine in coffee and soft drinks.

17. **Cooperative Group Project** Survey students in your school to find out what they know about vitamins. Ask about vitamin use in general and about specific vitamins, such as vitamin A and vitamin C. Organize your survey questions into four sections: (1) general vitamin facts, (2) sources of specific vitamins, (3) benefits of specific vitamins, and (4) deficiency diseases and symptoms associated with specific vitamins. Compile the survey results and present them on a poster. Display the poster where it can be seen by students and teachers.

18. **Multicultural Perspectives** Look at your library's reference books that describe the cultures and customs of other nations. Find information relating to ideas about ideal body weight and eating habits for at least five different countries. How do these ideas differ from those in America?

Lactose Digestion

- Identify the types of large molecules present in milk products and other foods.
- Observe the action of enzymes.
- Recognize the relationship between enzymes and the digestion of food molecules.

PROCESS SKILLS

- designing and performing an experiment
- collecting data

MATERIALS

- milk-treatment product (liquid)
- toothpicks
- depression slides
- droppers
- whole milk
- glucose
- glucose test strips

BACKGROUND

1. What are the three types of large molecules found in foods?
2. How does the human body break down food molecules?
3. Describe the chemical composition of carbohydrates.
4. What is milk sugar?
5. What is milk intolerance? What happens to people who drink milk but cannot digest it?
6. Write your own **Focus Question** on your Vee Form.
7. **Knowing Side of the Vee** List the **Concepts** and new **Vocabulary Words** on your Vee Form. In the **Concept Statements** section of the Vee, use these words in sentences that define and explain them.

TECHNIQUE

Doing Side of the Vee

1. Working with a lab partner, prepare a list of at least 15 foods that contain milk. List the number of times each week that you and your lab partner are likely to eat these foods.

2. Read the information sheet about the milk-treatment product. Discuss with your lab partner what the product is and what it does. Summarize your discussion in the **Records** section of your Vee Form.

3. Using the materials listed, design a control experiment that will test the effectiveness of the milk-treatment product. In the **Records** section of your Vee, describe the experiment that you have designed. Be sure to include one or more data tables.

4. Show your experimental design to your teacher. You must receive approval from your teacher before you can perform the actual experiment.

100% Lactose Reduced
Nonfat Milk

5. After your teacher has approved your procedure, change the measures from the ones given by the product to the following: quart of milk = drop of milk in a depression slide; teaspoon of milk-treatment product = drop of milk-treatment product.

6. To use the glucose test strips, follow the instructions printed on their container.

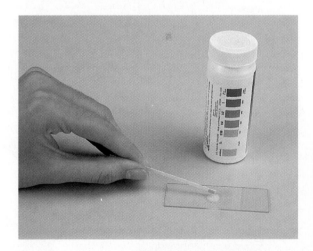

7. Fill in your self-designed data table in the **Records** section of your Vee Form.

8. Clean up your materials and wash your hands before leaving the lab.

INQUIRY

1. What are the ingredients of the milk-treatment product?

2. What does the product do when added to milk?

3. Why was it important for your experiment to include glucose?

4. Use the information on the **Knowing Side** of the Vee to interpret your results from the **Doing Side,** and then write your **Knowledge Claim.** Write a **Value Claim** for this lab.

ANALYSIS

1. What is the definition of milk intolerance?

2. What are lactose and lactase, and how do they interact with each other?

3. What do the results of this lab imply about treatments for other medical problems resulting from enzyme deficiencies?

4. As a person grows older, do you think the likelihood of that individual exhibiting milk intolerance increases or decreases? Explain your answer.

5. Do you think milk intolerance might be inherited? Why or why not?

FURTHER INQUIRY

Write a **New Focus Question** that could be the point of a new investigation. The following is an example:

In a hospital study of people with milk intolerance, how would blood glucose levels in human subjects receiving treated milk compare with those of subjects receiving untreated milk?

CHAPTER 39

THE BODY'S DEFENSES

REVIEW

- receptor proteins (Section 3-1)
- pathogens (Section 8-1)
- antibodies (Section 8-3)
- cancer (Section 9-3)
- structure of viruses (Section 20-1)
- AIDS and HIV (Section 20-1)
- macrophages (Section 35-2)
- *Highlights: Circulatory and Immune Systems* (page 843)
- structure of skin (Section 36-3)
- white blood cells (Section 37-1)
- lymphatic system (Section 37-1)

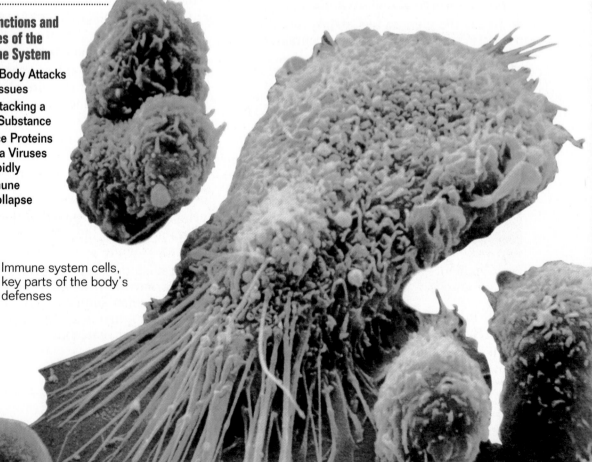

Immune system cells, key parts of the body's defenses

39-1 Defending Against Infection

When you think of how animals defend themselves, it is natural to think of armor, such as the hard shells of turtles, clams, and armadillos. However, even armor would offer no protection against the most dangerous enemies that the human body faces: pathogenic bacteria, viruses, fungi, and protists. The world is filled with pathogens, and you could not live for long unprotected. You survive because your body has a variety of very effective defenses against this constant attack, as summarized in Figure 39-1.

Section Objectives

■ Describe two ways that skin and mucous membranes defend the body.

■ Explain the roles of macrophages, neutrophils, and natural killer cells in combating pathogens.

■ Contrast the inflammatory response with the temperature response.

Skin and Mucous Membranes: The First Line of Defense

The outermost layer of the human body, the skin, provides the first defense against invasion by pathogens. Skin acts as a wall that keeps pathogens out of the body. Cells of the outer layer of skin, the epidermis, are continually worn away. These cells are quickly replaced by new cells that move up from the lower layers, where cell division occurs. In only 40 minutes, your body loses and replaces approximately 1 million skin cells.

Your skin defends your body not only by providing a nearly impermeable barrier, but also by reinforcing this defense with chemical weapons. Secretions of the oil and sweat glands within the dermis make the skin's surface very acidic, inhibiting the growth of many pathogens. Sweat also contains the enzyme lysozyme, which attacks and digests the cell walls of many bacteria.

Internal surfaces of the body through which pathogens could pass are covered by **mucous membranes**, epithelial layers that produce the sticky fluid called mucus. Mucous membranes line the digestive system, nasal passages, lungs, respiratory passages, and reproductive tract. Like the skin, mucous membranes not only serve as a barrier to pathogens, but also produce chemical defenses. Cells lining the bronchi and bronchioles in the respiratory tract secrete a layer of sticky mucus that traps microorganisms before they can

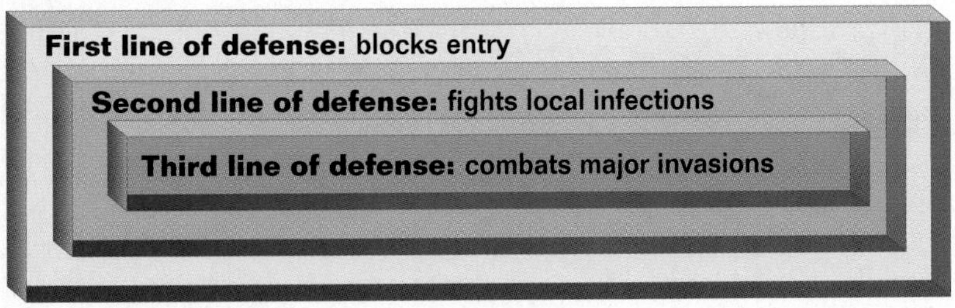

First line of defense: blocks entry

Second line of defense: fights local infections

Third line of defense: combats major invasions

Figure 39-1 Pathogens that attack your body are opposed by three lines of defense. The first defense consists of the skin and the epithelial layers that line the body's interior surfaces. This defensive line prevents pathogens from entering the body. The cells, proteins, and physiological responses of the second line of defense generally deal with minor infections, such as those resulting from small breaks in the skin. The third line of defense, the immune system, is activated only when your body is facing a major invasion by pathogens that have broken through the first two lines.

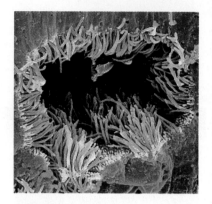

Figure 39-2 Cells lining the respiratory tract are covered by beating cilia. Movement of the cilia carries pathogens and foreign objects up to the esophagus to be swallowed. This cross section of a respiratory passage in the lung shows the hairlike cilia.

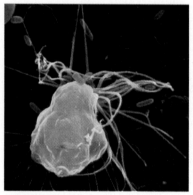

Figure 39-3 The extensions of this macrophage (shown here in yellow) are capturing bacteria.

Figure 39-4 The large cell with two dark blotches is a neutrophil. Neutrophils are white blood cells that release chemicals that are lethal to pathogens—and themselves.

reach the warm, moist lungs, which are ideal breeding grounds for microbes. Cilia on the cells of the respiratory tract, shown in Figure 39-2, continually sweep the mucus upward so that it can be swallowed, sending potential invaders to the stomach, where they are destroyed by acid.

The surface defenses of your body are very effective, but they are occasionally penetrated. You can take pathogens into your body when you eat and breathe. They can also enter through wounds or open sores. When invaders reach deeper tissue, a second line of defense comes into play.

Counterattacks: The Second Line of Defense

When the body's interior is invaded, a host of cellular and chemical defenses swing into action. Four are of particular importance: (1) cells that kill invading microbes; (2) proteins that kill invading microbes; (3) the inflammatory response, which sends defensive cells to the point of infection; and (4) the temperature response, which elevates body temperature to slow the growth of invading bacteria.

Cells That Kill Invading Microbes

The most important of the counterattacks against infection are those carried out by cells that attack invading microbes. These cells patrol the bloodstream and wait within the tissues for invaders. There are three kinds of these cells. Each kind kills invading microbes differently.

1. **Macrophages** White blood cells called macrophages, illustrated in Figure 39-3, kill bacteria one at a time by ingesting them. Although some macrophages stay within particular organs, particularly the spleen, most of the body's macrophages travel throughout the body in the blood, lymph, and fluid between cells.

2. **Neutrophils** White blood cells called **neutrophils** (*NOO truh fihlz*), shown in Figure 39-4, sacrifice themselves to defend the body. Neutrophils release chemicals that are

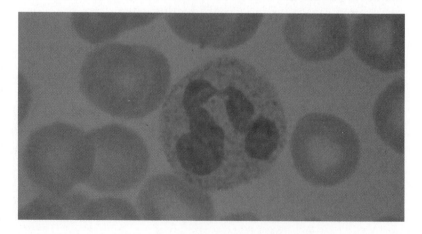

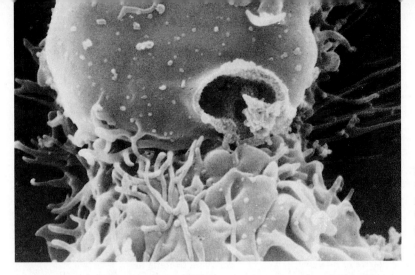

Figure 39-5 Natural killer cells monitor the body for the presence of cancer cells. The cell membrane of this cancer cell, *top,* has been punctured by a natural killer cell, *bottom;* the wound is fatal to the cell.

identical to household bleach, killing any nearby bacteria—and themselves in the process.

3. **Natural killer cells** Unlike macrophages, **natural killer cells** attack only cells that have been infected by microbes, not the microbes themselves. Natural killer cells are particularly effective at detecting and attacking body cells infected with viruses. Natural killer cells destroy a cell by puncturing its membrane, as shown in Figure 39-5. This allows water to rush into the cell, causing it to burst. Natural killer cells are also able to detect cancer cells, which they often kill before the cancer cells can develop into a tumor. Immune surveillance by natural killer cells is one of the body's most potent defenses against cancer.

Proteins That Kill Invading Microbes

Your body also uses a very effective chemical defense to assist its cellular defenses. This defense, called **complement,** consists of proteins that circulate in the blood plasma. Their defensive activity is triggered when they encounter the cell walls of bacteria or fungi. Then the complement proteins interact to form a membrane attack complex (MAC). The MAC inserts itself into the pathogen's cell membrane, creating a hole.

The Inflammatory Response

An injury or local infection causes an **inflammatory response,** a series of events that suppresses infection and speeds healing. For example, imagine that you have cut your finger, creating an entrance for pathogens. Infected or injured cells in your finger release chemical alarm signals that cause expansion of local blood vessels, increasing the flow of blood to the site of infection. Increased blood flow produces the redness and swelling so often associated with infections. It also promotes the migration of macrophages and neutrophils to the infection site, where they can attack invading microbes. Neutrophils arrive first, spilling out chemicals that kill the invading microbes and the tissue cells

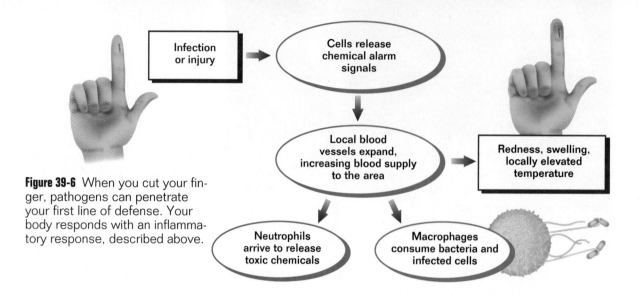

Figure 39-6 When you cut your finger, pathogens can penetrate your first line of defense. Your body responds with an inflammatory response, described above.

Infection or injury → Cells release chemical alarm signals

Local blood vessels expand, increasing blood supply to the area

Redness, swelling, locally elevated temperature

Neutrophils arrive to release toxic chemicals

Macrophages consume bacteria and infected cells

in the immediate area. After the neutrophils come macrophages, which clean up the remains of the dead cells. The pus associated with some infections is a mixture of dead or dying neutrophils, broken-down tissue cells, and dead pathogens. Figure 39-6 summarizes the inflammatory response.

The Temperature Response

When macrophages initiate their counterattack, they increase the odds in their favor by sending a message to the brain to raise the body's temperature several degrees above the normal value of 37°C (98°F). This higher temperature is called a fever. Fever is helpful because bacteria that cause human diseases do not grow well at high temperatures. Although fever inhibits microbial growth, very high fevers are dangerous because excessive heat can destroy important cellular enzymes. In general, temperatures greater than 39°C (103°F) are considered dangerous, and those greater than 41°C (105°F) are often fatal.

☐ **CAPSULE SUMMARY**

Skin and mucous membranes help exclude pathogens from the body. Any pathogens that do enter may be attacked by white blood cells, such as macrophages and neutrophils. Pathogens may also stimulate an inflammatory response or an increase in body temperature.

Section Review

1. *Describe the skin's chemical defenses.*
2. *Why are neutrophils used to combat only local infections and not body-wide diseases?*
3. *Aspirin reduces fever. Explain why taking aspirin for a mild fever might slow rather than hasten your recovery from a bacterial infection.*

Critical Thinking

4. *Would you expect the temperature response to be as effective against a viral infection as against a bacterial infection? Explain your answer.*

39-2 The Immune System

O nly occasionally do bacteria or viruses overwhelm your body's second line of defense. When this happens, the invaders face a third line of defense, the immune system, which is more difficult to evade than the first two lines of defense. Your immune system is not localized in any one place in the body, nor is it controlled by any one organ, such as the brain. Rather, it is an army of individual cells that rush to the site of an infection to combat invading microorganisms.

The Key Players in the Immune System

White blood cells, which are produced in the bone marrow and circulate in blood and lymph, constitute the immune system. These cells are very numerous; of the 100 trillion cells in your body, 2 trillion are white blood cells. Four kinds of white blood cells participate in the immune system's attack on pathogens: macrophages, killer T cells, helper T cells, and B cells. Each kind of cell performs a different function. **Killer T cells** attack and kill infected cells. **B cells** label invaders for later destruction by macrophages. **Helper T cells** activate killer T cells and B cells. And, as you have already seen, macrophages consume pathogens and infected cells. These four kinds of white blood cells exchange information and act in concert as a functional, integrated system.

To understand how this third line of defense works, imagine that you have just come down with the flu. Influenza viruses have entered your body in small water droplets inhaled into your respiratory system. They were not ensnared in the mucus covering the respiratory membranes (first line of defense), they slipped past patrolling macrophages (second line of defense), and they have begun to infect and kill mucous membrane cells. You feel sick because large numbers of the cells lining your respiratory tract are dying.

At this point, macrophages initiate the immune defense by releasing an alarm signal, the protein interleukin-1 *(IHN tuhr LOO kihn)*. This protein activates helper T cells. However, helper T cells do not attack pathogens. Instead, they serve as the "generals" of the immune system. Helper T cells respond to the alarm broadcast from macrophages by simultaneously activating two different types of immune system cells: killer T cells, which attack infected cells, and B cells, which produce defensive proteins.

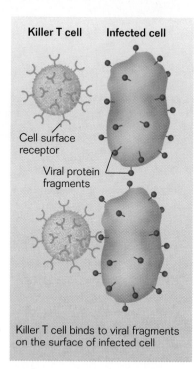

Killer T cell   Infected cell

Cell surface receptor

Viral protein fragments

Killer T cell binds to viral fragments on the surface of infected cell

Figure 39-7 How does a killer T cell recognize an infected cell? The infected cell will have some of the pathogen's proteins on its surface, and the killer T cell has receptors that detect these foreign proteins.

Killer T Cells Attack Infected Body Cells

Stimulation by the interleukin-1 signal causes helper T cells to immediately unleash a very potent attack against the virus, which now is located mainly inside the cells of your respiratory tract. Using a second chemical signal called interleukin-2, the helper T cells call into action killer T cells, which recognize and destroy body cells that have been infected with the virus. This attack is illustrated in Figure 39-7. Infected body cells display little bits of viral protein on their surface, and it is these telltale traces that the killer T cells recognize. Any of your body's cells that bear traces of viral infection are destroyed. The method used by killer T cells to kill infected cells is similar to that used by natural killer cells and complement—they puncture the cell membrane of the infected cell.

How do killer T cells recognize viral proteins? As explained in Figure 39-7, T cells have receptor proteins scattered over their cell membranes. Your body makes millions of different types of T cells. Each type of T cell bears a unique kind of receptor protein on its cell membrane, and this receptor is able to bind to a particular viral or bacterial protein. When you were still a fetus, any cells with receptors that recognized your own proteins were eliminated. ◻

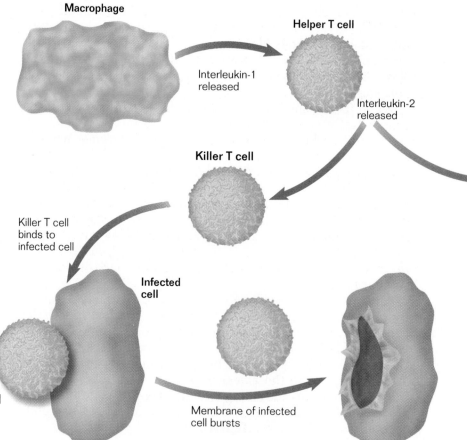

Figure 39-8 When stimulated by macrophages, helper T cells activate the two arms of the immune system: killer T cells and B cells. For simplicity, the receptors on the killer T cell and the viral fragments on the infected cell are not shown.

Macrophage

Helper T cell

Interleukin-1 released

Interleukin-2 released

Killer T cell

Killer T cell binds to infected cell

Infected cell

Membrane of infected cell bursts

B Cells Attack Invading Microbes

The interleukin-2 released by helper T cells simultaneously activates the second kind of defensive white blood cells, B cells. Like killer T cells, B cells have unique receptor proteins on their surface. These receptor proteins are called **antibodies.** B cells can release copies of their antibodies into the bloodstream or attach them directly to pathogens.

B cells do not directly attack pathogens or infected cells. Instead, they mark the pathogen for destruction by macrophages and natural killer cells. When a B cell encounters a foreign microbe with a surface protein that matches the shape of its antibodies, it simply sticks an antibody onto the microbe. The antibody acts as a flag to attract macrophages and natural killer cells. In addition, the B cell is stimulated by its encounter to divide repeatedly, forming a large population (known as a "clone") of identical B cells. Then all these B cells begin to make large amounts of antibodies, which they secrete into the bloodstream. Antibodies attach to any invading pathogens that might be present, marking them for destruction. Figure 39-8 summarizes the roles B cells and killer T cells play in fighting pathogens.

The B-cell defense is very powerful because it amplifies the reaction to an initial pathogen encounter a millionfold. It

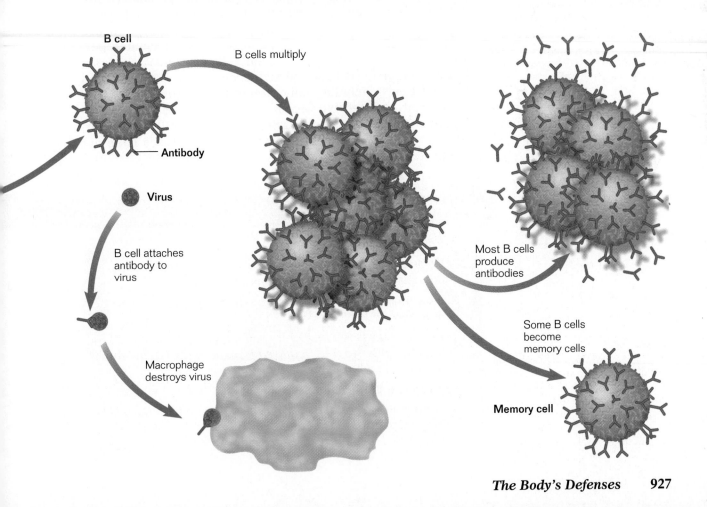

B cell

B cells multiply

Antibody

Virus

B cell attaches antibody to virus

Macrophage destroys virus

Most B cells produce antibodies

Some B cells become memory cells

Memory cell

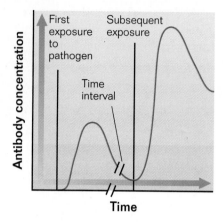

Figure 39-9 The first time you are exposed to a pathogen, your immune system responds slowly. At the second exposure, the response is faster and results in more antibodies.

is also a very long lasting defense. A few of the B cells resulting from a pathogen encounter do not become antibody producers. Instead, they become specialized memory cells that continue to patrol your body's tissues, circulating through your blood and lymph for a long time—sometimes for the rest of your life. If the pathogen that they are specialized to recognize ever appears again, those memory cells start the production of a new generation of antibody-producing cells directed against that particular pathogen. Your body does not wait until an infection is well underway to begin to fight back. The new generation of B cells produces large amounts of antibodies, as shown in Figure 39-9, and macrophages destroy the pathogen *before* you become ill. You are not even aware of the battle going on in your body and are said to be "immune" to the pathogen. ☐

Vaccination Prepares the Immune System

Vaccination is the introduction into your body of a dead or disabled pathogen or of a harmless microbe with the protein of a pathogen on its surface. Vaccination triggers an immune system response against the pathogen without an infection ever occurring. Afterward, the bloodstream of the vaccinated person contains memory cells that are directed against that specific pathogen. The vaccinated person is said to be "immunized" against the disease. Vaccination has dramatically reduced the incidence of many bacterial and viral diseases, including polio, tetanus, and diphtheria. More significant, an intensive vaccination program led to the elimination of the deadly disease smallpox in the 1970s. However, as you will see in the next section, some diseases have not responded so favorably.

Section Review

1. *Contrast B cells with killer T cells.*
2. *Could the immune system respond to an infection without helper T cells? Explain your answer.*
3. *How do macrophages assist B cells in fighting pathogens?*
4. *In what way is the immunity you acquire from a vaccination similar to that acquired from actually having the disease? In what way is it different?*

Critical Thinking
5. *What would have happened if the killer T cells that recognized your tissues hadn't been weeded out before birth?*

39-3 Malfunctions and Failures of the Immune System

*T*hough the immune system is one of the most sophisticated systems of the human body, it is not perfect. Many of today's major diseases, and other health concerns, are the result of failures or malfunctions of the immune system.

When the Body Attacks Its Own Tissues

The ability of killer T cells and B cells to distinguish cells of your own body from foreign cells is crucial to the fight against pathogens. In **autoimmune diseases,** this ability breaks down, causing the body to attack its own cells. Multiple sclerosis is an autoimmune disease that usually strikes people between the ages of 20 and 40. In multiple sclerosis, the immune system attacks and destroys the sheath of myelin that insulates motor nerves, which carry commands from the brain to the muscles and organs. Degeneration of the myelin sheath interferes with the transmission of nerve impulses, ultimately stopping their transmission altogether. Voluntary functions, such as the movement of limbs, and involuntary functions, such as bladder control, are lost, leading to paralysis and, eventually, death. Scientists do not know what stimulates the immune system to attack myelin. Table 39-1 describes several other autoimmune diseases and the tissues they affect.

Figure 39-10 Multiple sclerosis is a slow, degenerative disease in which transmission of nerve impulses is slowed and blocked. This teacher and multiple sclerosis patient requires a cane for support.

Table 39-1 Some Autoimmune Diseases

Disease	Areas affected	Symptoms
Systemic lupus erythematosus	Connective tissue, joints, kidney	Facial skin rash, painful joints, fever, fatigue, kidney problems, weight loss
Type 1 diabetes	Insulin-producing cells in pancreas	Excessive urine production, blurred vision, weight loss, fatigue, irritability
Graves' disease	Thyroid	Weakness, irritability, heat intolerance, increased sweating, weight loss, insomnia
Rheumatoid arthritis	Joints	Crippling inflammation of the joints

Figure 39-11 If your eyes water and your nose runs when you encounter dust, blame this microscopic mite that lives in the dust. You are allergic to proteins in its feces.

Allergy: Attacking a Harmless Substance

Although your immune system provides very effective protection against bacteria, viruses, fungi, and protists, it sometimes does its job too well and mounts a major defense against a harmless substance. Such an immune system response is called an **allergy**. Hay fever, the sensitivity that many people experience to even tiny amounts of plant pollen, is a familiar example of an allergy. Many people are allergic to proteins in the feces of the tiny mite shown in Figure 39-11, which lives on grains of house dust.

What makes an allergic reaction uncomfortable, and sometimes dangerous, is the involvement of antibodies that are attached to a kind of white blood cell called a mast cell. When mast cells encounter something that matches their antibodies, they initiate an inflammatory response. The inflammatory response begins when mast cells release **histamines** (*HIHST uh meens*) and other chemicals that cause capillaries to swell. Histamines also increase mucus production by the mucous membranes, resulting in the runny nose and nasal congestion that are symptoms of hay fever. Most allergy medicines relieve these symptoms with antihistamines, chemicals that block the action of histamines.

Asthma is a form of allergic response that takes place in the lungs. In addition to the reactions already described, histamines cause the narrowing of air passages in the lungs of people who have asthma. These people have trouble breathing when exposed to substances to which they are allergic. ❏

The Surface Proteins of Influenza Viruses Evolve Rapidly

If the activities of memory cells provide such an effective defense against future infection, why can you catch some diseases, such as the flu, more than once? Even if you are immunized against influenza, vaccination does not provide long-term protection against this disease. The reason you don't remain immune to flu is that the influenza virus has evolved a way to evade the human immune system. The virus changes its surface, the part that white blood cells recognize. How does it do this? The genes encoding the surface proteins of the influenza virus mutate very rapidly. Thus, the influenza virus continually makes new varieties of its surface proteins, and the immune system does not recognize these new varieties as belonging to a pathogen it has previously encountered. For each new variety of influenza virus, therefore, your body needs to mount an entirely new defense.

AIDS: Immune System Collapse

AIDS was first recognized as a disease in 1981. In Figure 39-12, you can see how rapidly the number of AIDS cases has increased since then. The World Health Organization estimates that 40 million people throughout the world will be infected with HIV (human immunodeficiency virus), the virus that causes AIDS, by the year 2000. By 1995, more than 250,000 Americans had died of AIDS, and more than 1.5 million Americans were thought to be infected with HIV. This destructive virus apparently evolved from a very similar virus that infects chimpanzees in Africa. A mutation arose in the chimpanzee virus that allowed it to recognize a human cell surface receptor called CD4. This receptor is present in the human body on certain immune system cells, notably macrophages and helper T cells.

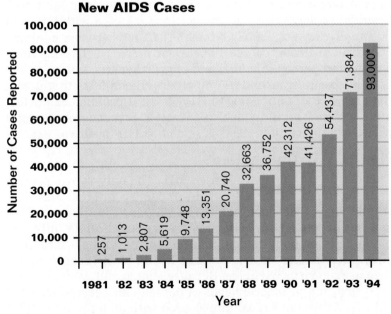

New AIDS Cases

Centers for Disease Control and Prevention Estimate

Figure 39-12 The number of cases of AIDS in the United States has grown rapidly since the disease was first described in 1981.

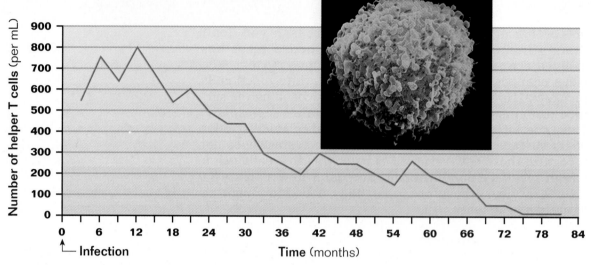

Figure 39-13 HIV destroys helper T cells. The small, green spheres, *above right,* are particles of HIV emerging from a helper T cell. These particles will infect other helper T cells, resulting in a rapid decrease in the number of these cells in the infected person. The graph shows the decline in helper T cells in the blood of an AIDs patient.

☐ CAPSULE SUMMARY

AIDS is caused by the human immunodeficiency virus (HIV). HIV destroys helper T cells in large numbers, thereby disabling the immune system. Without a functional immune system, AIDS patients cannot ward off infections and cancers.

How HIV Disables the Immune System

HIV attacks and cripples the immune system. The virus invades macrophages, which carry it throughout the body. It also attacks helper T cells and causes a drastic reduction in their numbers as the infection progresses, as shown in Figure 39-13. When the number of helper T cells in the blood of an infected person has fallen below 200/mL (a normal count is 800–1,000/mL) the person is said to have AIDS. Without enough helper T cells to activate and direct B cells and killer T cells, the immune system cannot respond to infections and cancer. As a result, the body is soon overwhelmed by pathogens and cancers that it would normally detect and destroy. Pneumocystic pneumonia, a common cause of death in AIDS patients, is an infection caused by a fungus that healthy immune systems easily defeat. Kaposi's sarcoma, another major cause of death in AIDS patients, is a form of cancer that is rare (except in very old people) because normal immune surveillance easily detects and eliminates cancer cells. Scientists believe that almost everyone infected with HIV will eventually develop AIDS, although the time between infection and onset of illness can be 10 years or more. During this time, an infected person may feel and appear healthy, but he or she is fully able to transmit HIV to others. Blood tests that detect the presence of antibodies to HIV or viral proteins can determine whether an individual is infected. Someone receiving a positive result on these tests is said to be **HIV positive.** ☐

How HIV Is Transmitted

There is no cure or vaccine for AIDS. The only way to protect yourself from AIDS is to avoid exposure to HIV. HIV is a fragile virus that cannot exist for long outside the body. You can be infected with HIV only by taking the HIV-infected blood cells or body fluids of infected individuals into your body. Most of the virus is present within macrophages rather than as free virus particles. Because semen and vaginal secretions are rich in macrophages, a person can become infected with HIV through sexual intercourse with an infected person. Worldwide, most HIV infections are spread this way. The virus can be transmitted by men and women during sexual intercourse (the worldwide frequency of AIDS is about the same in men and women). Vaginal, anal, and oral sex can transmit the virus, because macrophages can easily cross any mucous membrane. Use of a latex condom during sex greatly reduces, but does not eliminate, the risk of getting HIV.

Because blood contains a great many macrophages, it provides a vehicle for transmitting AIDS. People who inject intravenous drugs with needles or hypodermic syringes that have become contaminated with HIV-containing blood are at high risk of becoming infected with HIV. The majority of infections in the United States in the late 1980s were transmitted in this manner. In the late 1970s and early 1980s, many people were infected after receiving blood transfusions or injections of blood products, such as the Factor VIII used by hemophiliacs to help their blood to clot. The introduction of tests for HIV has almost completely eliminated this route of infection in the industrialized countries—but not in the developing countries, where such tests are often unavailable or too expensive.

HIV is not transmitted through the air, on toilet seats, or by any other medium where a macrophage would not survive. HIV cannot be contracted through shaking hands, sharing food, or drinking from a water fountain used by an infected person, for the simple reason that macrophages cannot be transmitted by casual contact. Although HIV is found in saliva, tears, and urine, these fluids contain too few virus particles to easily initiate an infection. This is why there are no known instances of anyone being infected with HIV from the small amount of saliva exchanged while kissing. Biting insects such as mosquitoes and ticks do not transmit HIV because they do not transmit macrophages. AIDS is not an easy disease to contract. There are only two common routes of infection: exposure to HIV-contaminated syringes or hypodermic needles and unprotected sexual contact with infected individuals.

Figure 39-14 The impact of AIDS is global. This clinic in Kenya counsels people on how they can avoid being infected with HIV. Africa has been hardest hit by AIDS of any continent. The World Health Organization estimates that over 6 million Africans are already infected by HIV.

Section Review

1. *Describe the differences and similarities between allergies and autoimmune diseases.*

2. *Would HIV's effects on the immune system be as damaging if it invaded only killer T cells? Explain your answer.*

3. *Describe two ways that HIV can be transmitted and two ways that it cannot.*

Critical Thinking

4. *When a pathogen initially invades a new host—such as when HIV first infected humans—it is usually deadly. After some time, however, the pathogen usually evolves to become less deadly. Explain why natural selection would favor a less-harmful pathogen.*

allergy (930)
antibody (927)
autoimmune disease (929)
B cell (925)
complement (923)

helper T cell (925)
histamine (930)
HIV positive (932)
inflammatory response
   (923)

killer T cell (925)
mucous membrane (921)
natural killer cell (923)
neutrophil (922)

Review

Multiple Choice

1. A mucous membrane defends against infection by
   a. ejecting lysozyme, which digests bacterial cell walls.
   b. secreting a sticky fluid that traps pathogens.
   c. preventing blood clots.
   d. engulfing and destroying bacteria.

2. An increased body temperature helps the body fight bacterial infection because it
   a. inhibits bacterial growth.
   b. stimulates antibody production.
   c. causes a membrane attack complex to form.
   d. produces redness and swelling.

3. B cells and killer T cells are stimulated to act by a substance called interleukin-2, which is released by
   a. macrophages.   c. helper T cells.
   b. neutrophils.   d. natural killer cells.

4. A flu shot must be taken each year to be even marginally effective because
   a. the influenza virus mutates rapidly.
   b. flu is caused by a retrovirus.
   c. very few memory cells are produced.
   d. macrophages cannot engulf the flu virus.

5. Rheumatoid arthritis, a disease that involves an immune system attack on the membranes around the joints, is an example of a(n)
   a. allergy.
   b. autoimmune disease.
   c. AIDS-related infection.
   d. passive immunity.

6. HIV disables the immune system by
   a. blocking the action of macrophages.
   b. destroying helper T cells.
   c. activating the production of B cells.
   d. triggering genetic mutations in CD5 surface receptors.

7. HIV can be transmitted by
   a. sexual intercourse.
   b. mosquito bites.
   c. shaking hands.
   d. vaccination only.

Completion

8. Your body's second line of defense against invading pathogens involves both _____ , which consume pathogens, and _____ , which destroy pathogens and themselves with chemicals.

9. The _____ response is recognized by the redness and swelling of affected tissue, while evidence of the _____ response is fever.

10. The immune system is activated by _____ that secrete the protein interleukin-1. This protein activates _____ T cells.

11. Infected cells are attacked directly by _____ T cells, while microbes are labeled by _____ cells for later attack by macrophages and natural killer cells.

12. Vaccines may be prepared from a weakened or _____ pathogen. Once vaccinated, a person's blood contains _____ cells that provide protection from the disease caused by that pathogen.

13. Destructive responses of the immune system to cells of the person's own body are called _____ diseases. One such disease is _____ .

14. The runny nose and nasal congestion associated with hay fever are due to the release of _____ . These hay fever symptoms can be relieved by using medicines that contain _____ .

15. AIDS patients die from infections and _____ because their immune systems have been crippled by HIV.

Themes Review

16. **Structure and Function** What are the outward signs of the inflammatory response? How do the changes that accompany the outward signs of the inflammatory response help to destroy pathogens?

17. **Structure and Function** When a B cell encounters a recognizable pathogen protein, it divides rapidly and produces clones that make large amounts of antibodies. These antibody-producing cells contain many Golgi apparatuses and large amounts of endoplasmic reticulum. How is the presence of these components related to the function of these cells?

18. **Evolution** Scientists have been successful in developing vaccines that provide long-term protection against a number of diseases, including polio and smallpox. Yet, they have been unsuccessful in developing a vaccine that protects against the influenza virus responsible for flu. Why has developing a flu vaccine been so difficult?

Critical Thinking

19. **Making Inferences** People who receive severe burns often die from infection. Given what you know about the body's defenses, explain how this is possible.

20. **Making Inferences** A government agency is reviewing two proposals for HIV research, but can fund only one. Suppose you are asked to provide input into the decision. Which proposal would you recommend that the agency fund? You should consider not only the likely effectiveness of the treatment but also possible side effects. Explain how you made your choice.

    Proposal 1: Develop a drug that interferes with protein synthesis.

    Proposal 2: Develop a substance that binds to the CD4 receptors on macrophages and helper T cells.

Activities and Projects

21. **Research and Writing** Do library research and talk to doctors and nurses to find out how the drug cyclosporin has affected the success of organ transplants. In your report include information about the source of the drug and how it works.

22. **History** The German physician Robert Koch was the first person to develop a step-by-step procedure for identifying the pathogen responsible for a given disease. What are the steps of Koch's procedure and what was the first pathogen identified using the procedure? What is the advantage of using Koch's procedure to identify the pathogen responsible for a given disease?

23. **Science-Technology-Society** Several AIDS vaccines are under development. The final test of these vaccines will be to inject them into healthy people and see if they develop AIDS or begin to build up antibodies to fight HIV, the virus that causes AIDS. Organize your group into two teams to debate the pros and cons of this practice.

LABORATORY Investigation | Chapter 39

Simulating Disease Transmission

OBJECTIVE

Simulate the transmission of a disease.

PROCESS SKILLS

- testing unknown solutions with chemical indicator
- recording data in tables

MATERIALS

- lab apron
- safety goggles
- disposable gloves
- dropper bottle of unknown solution
- large test tube
- indophenol indicator

BACKGROUND

1. How does a cold or flu spread from person to person?
2. How does the body fight invading viruses?
3. Why has the transmission of AIDS become a great concern worldwide?
4. Why is a person with HIV less able to combat infections?
5. Write your own **Focus Question** on your Vee Form.
6. **Knowing Side of the Vee** List the **Concepts** and new **Vocabulary Words** on your Vee Form. In the **Concept Statements** section of the Vee, use these words in sentences that define and explain them.

TECHNIQUE

Doing Side of the Vee

1. **CAUTION: Do not allow any solutions to touch your skin or clothing. Put on a lab**

apron, goggles, and disposable gloves. This investigation will involve the class in a simulation of disease transmission. After the simulation, you will try to identify the original infected person in the closed class population.

2. You have been given a dropper bottle of unknown solution and a clean test tube. Handle the unknown solutions with care because they are not simply water. When your teacher says to begin, transfer 3 dropperfuls of your solution to your clean test tube.

3. Randomly select one person to be your partner. Let one partner pour the contents of his or her test tube into the other partner's test tube. Then pour half of the solution back into the first test tube. You and your partner now share pathogens of any possible transmittable disease that either of you might have had. Record the name of your first partner (Round 1) in the **Records** section of your Vee Form.

4. For Round 2, wait for your teacher's signal and then find a different student partner and exchange solutions in the same manner as in step 3. Record the name of your second partner (Round 2) in the **Records** section of your Vee Form. Do not exchange solutions with the same person more than once. Repeat again for Round 3.

5. After all rounds are finished, your instructor will ask you to add one dropperful of indophenol indicator to your test tube to see if the fluids in your test tube have become infected. Infected solutions will be colorless or light pink in color. All uninfected solutions will appear blue. Record the outcome of your test in the **Records** section of your Vee.

6. If you are an "infected" person, give your name to your teacher. As names of "infected" people are written on the chalkboard or on the overhead projector, record them in the **Additional Records**

Path of Disease Transmission

Infected Person	Names of Partners		
	Round 1	Round 2	Round 3

section of your Vee Form in a table similar to the one above.

7. Try to trace the original source of the infection, then determine the transmission route of the disease. In your table, cross out the names of all the uninfected partners in Rounds 1, 2, or 3. There should be only two people in Round 1 who were infected. One of these was the original carrier. Devise a method to determine the route of transmission by constructing a diagram similar to the one below. Draw your diagram on the back of your Vee Form. Insert the names of the two people in Round 1 who were infected and the names of their partners in Round 2 and Round 3.

Transmission Route

Round 1	Round 2	Round 3

8. To test which person was the original disease carrier, pour a sample from his/her dropper bottle into a clean test tube and test it with indophenol indicator.

9. Clean up your materials and wash your hands before leaving the lab. In the **Procedure** section of the Vee, briefly summarize the procedure you followed.

1. What might the clear fluid in each student's dropper bottle represent?
2. Does the simulated "disease" have any apparent "symptoms"?
3. What chemical is added to the test tubes when the rounds are completed?
4. What color indicates a positive result?
5. What color indicates a negative result?
6. Who was the original disease carrier?
7. Use the information on the **Knowing Side** of the Vee to interpret your results from the **Doing Side,** and then write your **Knowledge Claim.** Write a **Value Claim** for this lab.

ANALYSIS

1. After the three rounds, how many students were "infected"? Express this as a percentage of the number of students in the class.
2. If there were an epidemic in your community, how might public health officials work to stop the spread of the disease?

FURTHER INQUIRY

Write a **New Focus Question** that could be the point of a new investigation. The following is an example:

> How can a public health official determine whether a disease has been caused by the passing of pathogens from person to person or by environmental conditions?

NERVOUS SYSTEM

REVIEW

- sodium-potassium pump (Section 3-2)
- gated ion channels (Section 3-3)
- binocular vision (Section 14-1)
- neurons (Section 35-1)

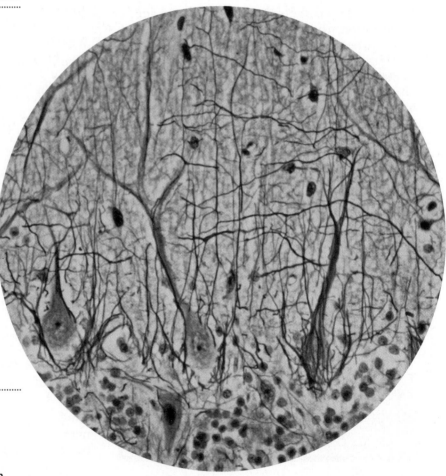

Micrograph of human nerve cells from the cerebral cortex

40-1 Neurons and Nerve Impulses

Cells in the human body could communicate using only chemical signals, except for one problem: circulating chemicals is a very slow form of communication. A quicker means of communication is needed, especially if your brain has an urgent message for the muscles in your leg, such as "Contract quickly, a speeding car is headed this way!" To solve this problem, humans and most other animals have neurons, specialized cells that can quickly transmit messages throughout the body.

Neurons Transmit Electrical Signals

Recall from Chapter 35 that a neuron is a cell specialized for transmitting electrical signals called nerve impulses. A neuron has an electrical charge because sodium-potassium pumps transport ions across the plasma membrane. This activity creates a difference in electrical charge on both sides of the membrane, which is said to be polarized. Loss of this charge difference is called **depolarization**. A neuron transmits a nerve impulse as a wave of depolarization that travels along its membrane like a rapidly burning fuse.

The human body contains many different types of neurons. Some neurons are tiny with few projections. Others are bushy with many projections, like the neuron shown in Figure 40-1, and still others have a single projection several meters long. Despite these differences, all neurons have the same functional architecture. Cytoplasmic extensions called **dendrites** (*DEHN dryts*) extend from the body of the neuron. Dendrites are the antennae of the neuron, enabling it to receive information simultaneously from many different sources. The surface of the **cell body** collects the information arriving from the many different dendrites. The information then travels as a nerve impulse from the cell body

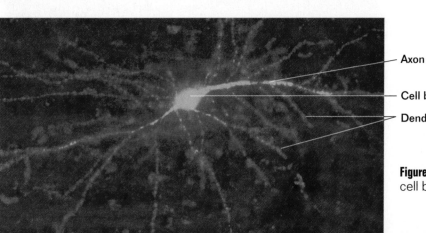

Axon

Cell body

Dendrites

Figure 40-1 A neuron consists of a cell body, dendrites and an axon.

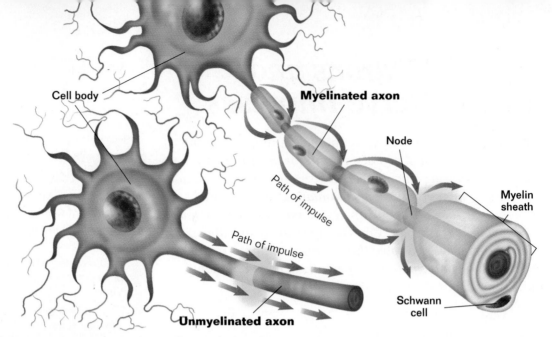

Cell body

Myelinated axon

Node

Path of impulse

Myelin sheath

Path of impulse

Schwann cell

Unmyelinated axon

Figure 40-2 Myelin sheaths cover the axons of nerves that carry information to and away from the brain. A Schwann cell wraps its cell membrane around the axon to form the insulating myelin sheath. Impulses hop from node to node, which is faster than traveling the full length of an unmyelinated axon.

☐ CAPSULE SUMMARY

A neuron consists of a cell body, dendrites that carry impulses to the cell body, and an axon that transmits impulses to other cells. Many nerves have myelin sheaths, which expedite impulses.

along an **axon.** Most neurons have only a single axon, which may be quite long.

Bundles of neurons are called **nerves,** which appear as fine, white threads when viewed with the naked eye. Like a telephone line, nerves contain a large number of independent communications channels. In addition, they are composed of many supporting cells that form the structure of nervous tissue and assist neurons. In some neurons, a form of supporting cell called a **Schwann cell** wraps around the axon, forming a fatty, insulating covering called a **myelin** *(MY uh lihn)* **sheath.** As you can see in Figure 40-2, the myelin sheath is interrupted at intervals, leaving exposed gaps called nodes. A nerve impulse traveling down a myelinated axon jumps from node to node, which is much faster than traveling along the full length of a bare, unmyelinated axon. Destruction of large patches of myelin characterizes a disease called **multiple sclerosis.** In multiple sclerosis, small, hardened scars appear throughout the myelin sheath and interfere with the transmission of impulses. Normal nerve function is impaired, and symptoms such as double vision, muscular weakness, loss of memory, and paralysis result. ☐

A Reversal in Voltage Triggers an Impulse
....

When a neuron is not transmitting an impulse, the sodium-potassium pumps in its cell membrane are transporting sodium ions (Na^+) out of the cell and potassium ions (K^+) into it. Once they are pumped out, sodium ions cannot easily move back into the cell; therefore, the concentration of sodium ions is higher outside the cell. At the same time, potassium ions build up inside the cell. However, potassium ions are able to diffuse out through open channels. As a result, the outside of the neuron has a higher positive charge

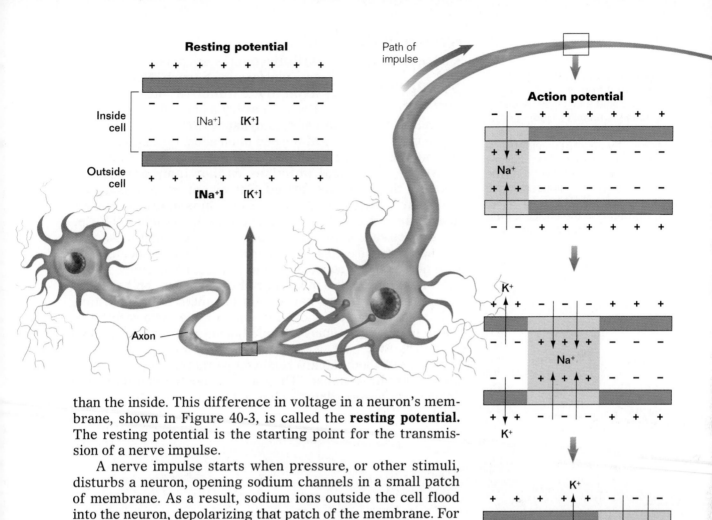

Resting potential

Inside cell

[Na+]   [K+]

Outside cell

[Na+]   [K+]

Axon

Path of impulse

Action potential

Na+

K+

Na+

K+

K+

Na+

K+

than the inside. This difference in voltage in a neuron's membrane, shown in Figure 40-3, is called the **resting potential.** The resting potential is the starting point for the transmission of a nerve impulse.

A nerve impulse starts when pressure, or other stimuli, disturbs a neuron, opening sodium channels in a small patch of membrane. As a result, sodium ions outside the cell flood into the neuron, depolarizing that patch of the membrane. For a brief moment the inside of that section of membrane becomes more positively charged than the outside. This sudden local reversal of voltage across the neuron membrane, shown in Figure 40-3, is called an **action potential.**

The sodium channels in the small patch of the membrane with the action potential remain open for only about half of a millisecond. The change in voltage causes other nearby voltage-gated sodium channels to open, reversing the voltage in that patch of membrane. The action potential that results causes more channels to open. The action potential moves down the neuron like a chain of falling dominoes as the reversal of voltage causes adjacent gated channels to open.

When the action potential has passed, the voltage-gated sodium channels snap shut, and the resting potential is restored. Why does the potential return to normal? The depolarization briefly opens voltage-gated potassium channels and allows potassium ions to leave the cell, making the inside of the cell less positively charged again. The full potential difference is then re-created by the sodium-potassium pump, and the resting potential is restored. Restoration of the resting potential is called **repolarization.** The neuron cannot transmit another signal for a short period of time until this recovery is complete.

Figure 40-3 At rest, sodium-potassium pumps in a neuron's membrane keep a higher concentration of sodium ions outside the cell and a higher concentration of potassium ions inside, creating a voltage difference called the resting potential, *above left.* When an impulse moves down an axon, sodium ions rush into the cell, creating a reversal in voltage called an action potential, *above right.*

Figure 40-4 Most neurons are separated from each other by a gap called a synapse. A nerve impulse is unable to jump across a synapse. Instead, it is carried across by neurotransmitters, chemicals released from tiny sacs in the end of the axon. The neurotransmitter diffuses across the synapse and binds to receptor proteins in the adjacent neuron, triggering an impulse.

If you examine a nerve closely, you will discover a surprising fact—no two neurons ever touch. Nor does an axon ever touch a muscle cell. Between every axon and its neighboring cell is a tiny gap called a **synapse** *(SIHN aps)*. When a nerve impulse arrives at the end of an axon, the action potential must cross the synapse for the message to continue. However, electrical impulses cannot jump across the synapse; they must be carried. The end of an axon contains tiny sacs filled with chemical messengers called **neurotransmitters.** When a nerve impulse reaches the tip, the sacs release neurotransmitters into the synapse. As you can see in Figure 40-4, the neurotransmitter molecules diffuse across the synapse and bind to receptor proteins in the cell membrane on the other side of the synapse. This binding of neurotransmitters causes ion channels to open, which results in voltage changes in the membrane. As a result, the nerve impulse continues in the adjacent neuron, passing the signal to that cell. Neurotransmitter molecules in the synapse are then quickly removed so that another message can cross the synapse. They are either broken down by enzymes or absorbed into the axon for recycling.

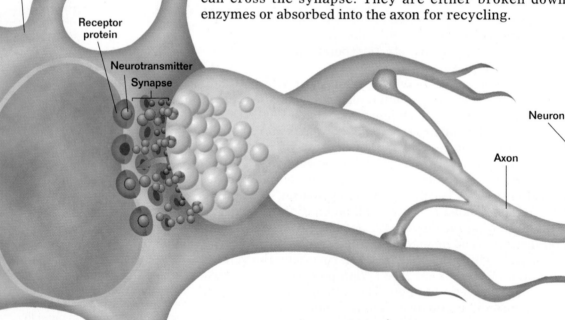

Neuron

Receptor protein

Neurotransmitter

Synapse

Neuron

Axon

Why Are Synapses Advantageous?

Electrical signals travel along an axon much faster than neurotransmitters cross a synapse. If impulses travel faster along axons, what is the evolutionary advantage of synapses? Why don't neurons have continuous electrical contact? The great advantage of a chemical junction like the synapse is that the nature of the chemical signal can be different in different junctions, permitting different kinds of responses. Over 60 different chemicals have been identified that either act as specific neurotransmitters or modify the activity of neurotransmitters. ◻

◻ *CAPSULE SUMMARY*

When a nerve impulse reaches the end of an axon, neurotransmitters are released into the synapse. They cross the synapse and bind to receptors in the neighboring cell membrane.

Neurons Form an Integrated Circuit

Synapses fall into two general classes depending on what happens when their cell membrane receptors bind neurotransmitters. In an excitatory synapse, the receptor is a gated sodium channel that is closed while at rest. When a neurotransmitter binds to it, the sodium channel opens, allowing sodium ions to flood inward and depolarize the membrane. If enough sodium channels are opened by neurotransmitters, an action potential that initiates a new nerve impulse in the neuron is created. In an inhibitory synapse, the receptor is a gated potassium channel. When a neurotransmitter binds to it, the potassium channel opens, allowing positively charged potassium ions to leave the cell, giving it a more negatively charged interior. This change in voltage inhibits nerve impulses because even more sodium channels must be opened to initiate an action potential.

An individual cell can have both excitatory synapses and inhibitory synapses. When signals from both excitatory and inhibitory synapses reach the neuron cell body, the excitatory and inhibitory effects interact with one another. The result is a process of **integration** in which the various excitatory and inhibitory electrical effects tend to cancel or reinforce one another. Individual neurons often receive many inputs. A single motor neuron in the spinal cord may have as many as 50,000 synapses. The combined influences of all of the inputs determine whether the cell membrane will be sufficiently depolarized to initiate a nerve impulse along the axon. Your every thought and feeling result from such integration.

Section Review

1. Diagram and label a neuron. Next to each part, list its function.
2. What is a myelin sheath? What is its function?
3. Define the following terms: resting potential, action potential, depolarization, and repolarization.
4. Outline the major steps in the transmission of a nerve impulse.
5. What events are involved in the transmission of a nerve impulse across a synapse?

Critical Thinking

6. How are nerve impulses inhibited? Of what advantage is this to the body?
7. How are synapses involved in integration?

40-2 Architecture of the Nervous System

Section Objectives

- Explain the structural and functional divisions of the nervous system.
- Name the major regions of the brain, and explain their functions.
- Describe the peripheral nervous system, and distinguish between sensory and motor nerves.
- Explain the action of a reflex.
- Compare the functions of the two divisions of the autonomic nervous system.

*H*umans, like all other animals except sponges, use a network of nerve cells to gather and integrate information about the body's internal and external environments and to send messages to the body's muscles and glands. Nerves connect every part of your body to its command-and-control center, the brain and spinal cord. Your body is run like a submarine: information about what is happening inside and outside the body flows into the control center, which analyzes the data and issues commands.

The Nervous System Has Two Divisions

As you learned in Chapter 35, the human nervous system is divided into two principal parts—the central nervous system and the peripheral nervous system. The central nervous system, composed of the brain and the spinal cord, is the body's main processing center. It receives and sends information through the branches of nerves that make up the peripheral nervous system.

The peripheral nervous system consists of two types of nerves. **Sensory nerves** gather information about your environment and your body's condition and deliver it to the central nervous system. **Motor nerves** transmit commands from the central nervous system to muscles and glands all over your body. Motor nerves that conduct impulses to skeletal muscles under our conscious control make up the **somatic nervous system;** other motor nerves that regulate the activity of cardiac muscles, smooth muscles, and glands make up the **autonomic nervous system.** Figure 40-5 shows how the different parts of the nervous system are integrated.

Figure 40-5 The nervous system is divided into the central nervous system and the peripheral nervous system, which consists of two other divisions of nervous systems.

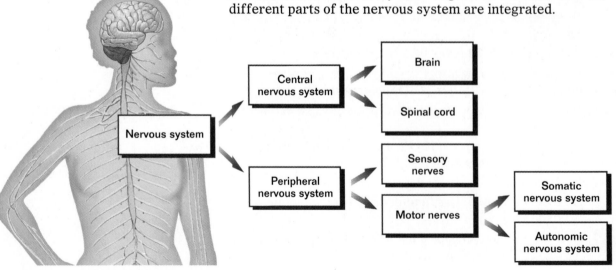

The Central Nervous System Directs and Coordinates Information

In an average adult, the brain is one of the largest organs of the body, weighing about 1,500 g (approximately 3 lb.). The most conspicuous part of the brain is its rounded, wrinkled outer layer, the **cerebrum,** as shown in Figure 40-6. A long, narrow cleft, or fissure, divides the cerebrum into right and left halves, or hemispheres, which communicate with each other through a connecting tract of fibers. Much of the activity in the cerebrum occurs in the **cerebral cortex,** an outer layer only 2–4 mm thick. The cerebral cortex contains thick layers of unmyelinated neurons that appear gray and are therefore referred to as gray matter. Beneath the cerebral cortex lies a region of myelinated neurons (white matter) that shuttles information between the cortex and the rest of the brain.

Like a hand covering a fist, the cerebrum surrounds the rest of the brain. As shown in Figure 40-6, directly beneath the cerebrum are the thalamus and hypothalamus, important centers of information processing. The egg-shaped **thalamus** is the main site of sensory processing in the brain. Most of

Figure 40-6 A midsagittal cut through a brain divides it into two mirror image halves. This midsagittal view of a human brain shows one hemisphere in which the major structures of the brain can be seen.

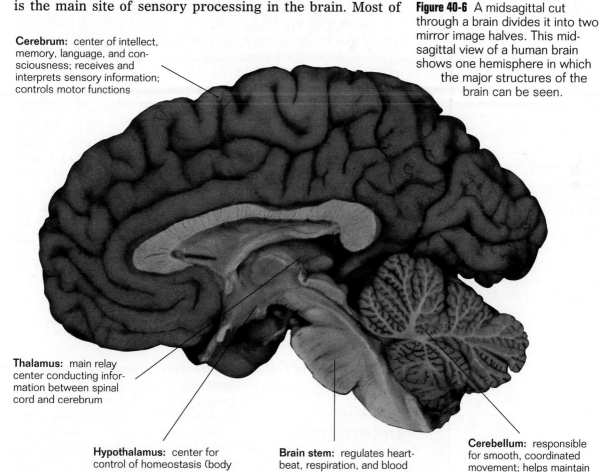

Cerebrum: center of intellect, memory, language, and consciousness; receives and interprets sensory information; controls motor functions

Thalamus: main relay center conducting information between spinal cord and cerebrum

Hypothalamus: center for control of homeostasis (body temperature, appetite, water balance); links nervous and endocrine systems

Brain stem: regulates heartbeat, respiration, and blood pressure; helps control swallowing, coughing, sneezing and vomiting

Cerebellum: responsible for smooth, coordinated movement; helps maintain posture, muscle tone, and equilibrium

the sensory nerves from all parts of the body converge on the thalamus, which sorts and relays information to appropriate areas in the cerebral cortex. Below the thalamus is the **hypothalamus,** a slender thread of tissue. Despite its small size, the hypothalamus controls many body activities related to homeostasis, such as body temperature, blood pressure, respiration, and heart rate. It also directs the secretions of the brain's major hormone-producing gland, the pituitary gland. The hypothalamus is linked to some areas of the cerebral cortex by an extensive network of neurons. This network, along with the hypothalamus, is called the **limbic system.** The limbic system, sometimes referred to as the emotional brain, is responsible for many basic drives and emotions, such as pain, pleasure, anger, sex, hunger, and thirst.

Extending back from the base of the brain, as shown in Figure 40-6, is a region known as the **cerebellum.** The cerebellum controls balance, posture, and voluntary muscle contractions. This small cauliflower-shaped structure, while well developed in humans and other mammals, is even more developed in birds. Birds perform more complicated feats of balance than do humans. Imagine the kind of balance and coordination needed for a bird to land on a branch.

The base of the brain, called the **brain stem,** connects the rest of the brain to the spinal cord. This stalklike structure contains nerves that control breathing, swallowing, digestive processes, heartbeat, and the diameter of blood vessels. A network of nerves called the **reticular formation** runs through the brain stem and connects to other parts of the brain. Their widespread connections make these nerves essential to consciousness, awareness, and sleep. One part of the reticular formation filters sensory input, enabling you to sleep through repetitive noises such as traffic, yet awaken instantly when a telephone rings.

The **spinal cord** is a cable of nerve tissue that extends from the base of the brain through the backbone to the level just below the ribs. Messages from the brain and the rest of the body run up and down the spinal cord, making it the "information highway" of the central nervous system. The center of the spinal cord consists of a column of gray matter covered by a sheath of white matter. ▢

CONTENT LINK

In **Chapter 41,** you will learn why the pituitary gland is sometimes called "the master gland."

❏ **CAPSULE SUMMARY**

The central nervous system is made up of the brain and spinal cord, which work together to collect and process information.

The Peripheral Nervous System Shuttles Information

If you were to see a car speeding toward you, your brain would send messages through motor neurons to glands that secrete the hormone adrenaline. Adrenaline would increase your heart rate and breathing rate. Your brain would also send messages through motor neurons to muscles in your legs, which would quickly contract and get you out of the way fast. Without these nerve pathways, your brain would be useless because it would never be able to receive or send messages.

As you can see in Figure 40-7, peripheral nerves arise from either the brain or the spine. Twelve pairs of nerves from the brain are associated with motor and sensory functions. Thirty-one pairs of spinal nerves supply the communication links between the central nervous system and the neck, trunk, arms, and legs. This is why injuries to the spinal cord often paralyze the lower part of the body.

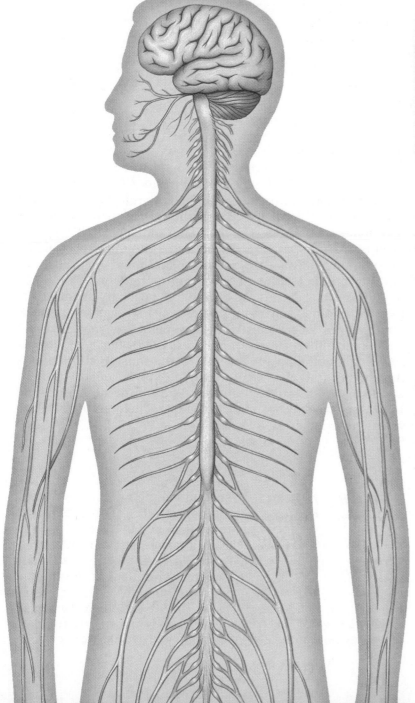

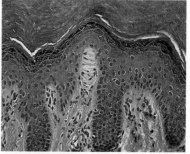

Sensory neuron

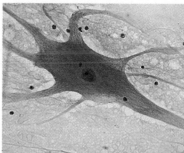

Motor neuron

Figure 40-7 The peripheral nervous system, *left,* has two main types of neurons: sensory neurons and motor neurons. Sensory neurons send messages to the brain; motor neurons deliver messages to muscles and glands.

Figure 40-8 In the patellar reflex, tapping the tendon below the kneecap excites muscle fibers in the quadriceps muscle. Nerve impulses travel along a sensory neuron, (*red*), to the spinal cord. There, it synapses with motor neurons, (*blue*), which send an impulse back to the muscle telling it to contract. The contraction causes extension of the knee and a forward movement of the foot.

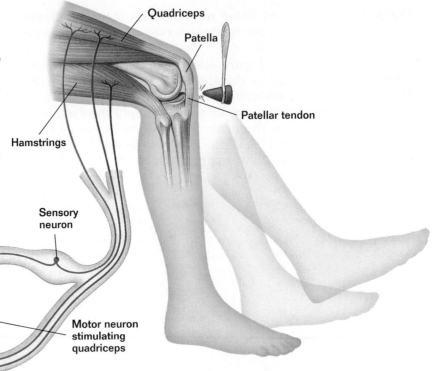

Quadriceps

Patella

Patellar tendon

Hamstrings

Sensory neuron

Motor neuron stimulating quadriceps

Spinal cord

Motor neuron stimulating hamstrings

Many motor nerves in your peripheral nervous system are wired to give your body the ability to respond quickly and involuntarily to a stimulus. This sudden, rapid response is called a **reflex.** A reflex produces a rapid motor response to a stimulus because the sensory neuron carrying the information connects directly to a motor neuron. The nerve impulse never reaches the brain; it travels only as far as the spinal cord and then comes right back as a motor response. For example, when you strike a muscle below your knee with a reflex hammer, as shown in Figure 40-8, your leg kicks out for a moment. The sudden jolt to the muscle stimulates a nerve impulse that travels to the spinal cord where it synapses with motor neurons. The motor neurons send impulses back to the muscle, causing it to contract. As a result, your leg kicks forward. ▢

The Autonomic Nervous System Keeps the Body Functioning
....

The autonomic nervous system is the command network that the central nervous system uses to maintain the body's homeostasis. By using it, the central nervous system regulates heart rate and controls muscle contractions in the walls of your blood vessels and digestive, urinary, and reproductive tracts. It also carries messages that stimulate glands to secrete tears, mucus, and digestive enzymes.

Some motor neurons in the autonomic nervous system are active all the time, even when you are asleep. These neurons carry messages from the central nervous system to keep the body functioning even when it is not active. For example, they direct the muscles that control blood pressure, breathing, and the movement of food through the digestive system.

The division of the autonomic nervous system that is referred to as the **sympathetic division** dominates in times of stress. It controls the "fight or flight" reaction—the feeling you experience when you find yourself in an emergency or a dangerous situation, such as being in the path of a speeding car. Its activity increases blood pressure, heart rate, breathing rate, and blood flow to the muscles—things your body needs in order to deal with a crisis. Another division of the autonomic nervous system, called the **parasympathetic division,** has the opposite effect, one that you feel when you relax after eating a heavy meal. The parasympathetic division conserves energy by slowing the heartbeat and breathing rate and by promoting digestion and elimination. Most glands, smooth muscles, and cardiac muscles receive information from *both* the sympathetic and parasympathetic divisions of the body.

Although the autonomic nervous system can carry out its tasks automatically, it is not completely independent of voluntary control. For instance, it enables you to breathe involuntarily, but you can decide to stop breathing for a short time. However, any voluntary control of the autonomic nervous system that endangers life will disturb the homeostasis of the brain tissue and cause unconsciousness. Then the autonomic nervous system takes over and restores normal functioning, which explains why you cannot hold your breath indefinitely. ❑

❑ CAPSULE SUMMARY

The autonomic nervous system is the division of the peripheral nervous system that controls activities involved with maintaining homeostasis.

Section Review

1. *What are the principal components of the nervous system? What are their general functions?*

2. *What are six major parts of the brain? Describe the function of each part.*

3. *How do sensory nerves and motor nerves differ?*

4. *What is a reflex? Why is a reflex quicker than other responses?*

5. *How does the autonomic nervous system affect your body when you become frightened?*

Critical Thinking

6. *Basic reflexes are unlearned and said to be built into neural anatomy. What is the evolutionary advantage of this?*

40-3 *The Sense Organs*

O ver a dozen kinds of sensory cells carry impulses to the central nervous system. These specialized cells, called *sensory receptors*, are capable of detecting various forms of stimuli such as changes in blood pressure, strain on ligaments, and odors in the air. Organs that contain sensory receptors, such as your eyes and ears, are composed of many cell and tissue types and are called *sensory organs*.

- Describe the structure of the inner ear, and explain how it functions to maintain balance and to sense sound waves.

- Trace the pathway of light through the eye to the retina, and describe the events involved in sensing light waves.

- List four kinds of sensory receptors that help maintain homeostasis.

- Describe how taste and smell are sensed.

Figure 40-9 A view of the anatomy of the ear shows the structures that are responsible for sensing balance, *top arrows,* and the structures that enable you to hear, *bottom arrows.*

Receptors in the Ear Sense Equilibrium and Sound

Although you may think of ears as flattened cups of flesh on the sides of your head, the ear is actually an intricate sense organ, responsible for sensing equilibrium and hearing. The sensory cells for equilibrium are located in chambers within the inner ear. As you can see in Figure 40-9, clusters of hairs in these chambers respond to changes in head position with respect to gravity. The hairs project into a jellylike fluid that contains tiny pieces of calcium carbonate called otoliths, meaning ear stones. Gravity always pulls downward on the otoliths and hairs, sending a steady signal to the brain. When the head moves, the fluid slides over the hairs and bends them in the opposite direction, changing the signal to the brain. The brain interprets the changes in impulses to determine the position of the head. In a similar manner, the three semicircular canals in the inner ear detect rotation of the head.

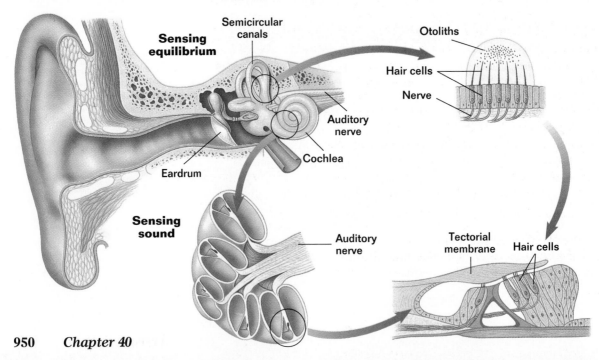

Sensing equilibrium — Semicircular canals — Otoliths — Hair cells — Nerve — Auditory nerve — Cochlea — Eardrum — Sensing sound — Auditory nerve — Tectorial membrane — Hair cells

Responding to Sound Waves

Look again at the structure of the human ear shown in Figure 40-9. When a sound wave enters the ear canal, it strikes the eardrum and causes it to vibrate. Behind the eardrum, three small bones called ossicles transfer the vibration to a fluid-filled chamber within the inner ear. This chamber, called the **cochlea** *(KAHK lee uh)*, is shaped like a tightly coiled snail shell. The sound receptors within the cochlea are hair cells that rest on a membrane separating the chamber into two halves. The hair cells do not project into the fluid filling the cochlea; instead, they are covered by another membrane, called the tectorial membrane. When a sound enters the cochlea, the sound waves cause this membrane "sandwich" to vibrate, bending the hairs pressed against the outer membrane and causing them to send nerve impulses to sensory neurons that travel along the auditory nerve to the brain.

Sounds of different frequencies cause different parts of the membrane to vibrate and thus fire different sensory neurons. Our ability to hear depends upon the flexibility of the membranes within the cochlea. ◻

◻ *CAPSULE SUMMARY*

The receptors for equilibrium and hearing are located in the inner ear.

Photoreceptors in the Eye Sense Light

No other stimulus provides as much detailed information about the environment as light. Vision depends on a special sensory apparatus called an eye. All of the sensory receptors described so far respond to chemical or mechanical stimuli. Eyes contain sensory receptors that respond to photons of light. The light energy is absorbed by pigments that trigger nerve impulses in sensory neurons. Humans have extremely good eyesight. We see in color and can distinguish fine details and movement. Birds are the only animals with better eyesight than humans.

The human eye can be compared to a camera. Light first passes through a transparent, protective covering called the **cornea,** which begins to focus the light onto the back of the eye. The amount of light entering the eye is controlled by a shutter called the **iris,** which consists of tiny muscles arranged in a ring. The black hole in the center of the iris is the **pupil,** the area through which light enters the eye. The pupil gets larger in dim light and smaller in bright light. The beam of light then passes through the **lens,** which completes the focusing. The lens is a fat disk that resembles a flattened balloon. It is attached to ciliary muscles by suspended ligaments. When these muscles contract, they change the shape of the lens and thus the point of focus on the rear of the eye.

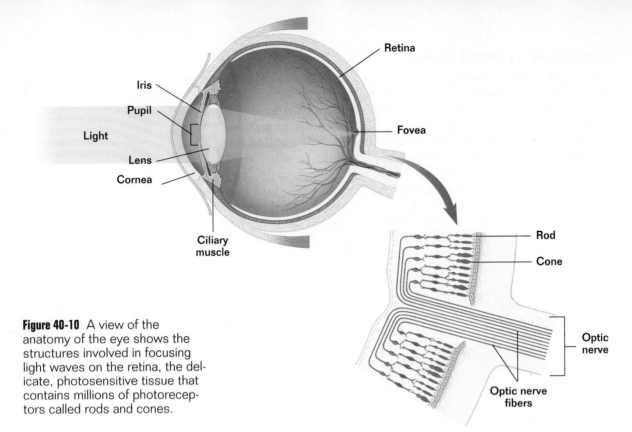

Figure 40-10 A view of the anatomy of the eye shows the structures involved in focusing light waves on the retina, the delicate, photosensitive tissue that contains millions of photoreceptors called rods and cones.

As light enters the eye, as shown in Figure 40-10, it passes through several transparent materials that focus it onto a delicate layer of tissue called the **retina**. The retina is the light-sensing portion of the eye. It contains about 1 billion light-sensitive receptor cells that, when stimulated by light, generate nerve impulses. **Rods** are receptor cells that are extremely sensitive to light and can detect various shades of gray even in dim light. However, they cannot distinguish colors, and because they do not detect edges well, they produce poorly defined images. **Cones** are receptor cells that detect color and are sensitive to edges, so they produce sharp images. The center of the retina contains a tiny pit densely packed with some 3 million cones. This area, called the fovea, produces the sharpest image, which is why we tend to move our eyes so that the image of an object we want to see clearly falls on this area.

A photoreceptor in your eye is able to detect a single photon of light. The primary sensing event of vision is the absorption of a photon of light by pigment complexes called rhodopsin *(roh DAHP sihn)* in rods and photopsins in cones. When photoreceptors absorb a photon of light, the pigment changes shape and initiates a chain of events that leads to the generation of an action potential. The nerve impulse travels along a short, thick nerve pathway called the optic nerve to the brain, which translates the information into a meaningful image.

In humans, the field of vision of the two eyes overlaps; each eye sees about one-third of what the other sees. The image that each eye sees of the same object, however, is

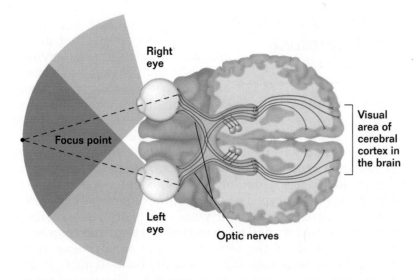

Right eye

Focus point

Left eye

Optic nerves

Visual area of cerebral cortex in the brain

Figure 40-11 Binocular vision results from the overlap of the visual fields from both eyes. In addition, approximately half the optic nerves from each eye cross over and provide each visual area of the cerebral cortex with information from each eye. Binocular vision provides a small visual field, but it enables us to perceive depth.

slightly different because the two eyes view the object from different angles, as shown in Figure 40-11. This slight displacement of images permits the brain to figure out how far away an object is. Interestingly, we are not born with this ability to perceive depth and distance. We learn it as babies by trial and error as the brain explores and remembers how the physical distance to an object compares with the difference between the images. ◻

◻ **CAPSULE SUMMARY**

The receptors for sensing light are rods and cones, which are located in the retina in the back of the eye.

Receptors Throughout the Body Sense Its Internal Condition

Sensory receptors inform the central nervous system about the internal conditions of your body. Much of this information passes to the hypothalamus to help maintain homeostasis— that is, to keep the body's internal environment constant. Table 40-1 lists several stimuli and describes the body's receptors for each and how they work.

Table 40-1 Sensory Receptors

Stimulus	Receptor	Location	Process
Temperature	Heat receptors and cold receptors	Skin, hypothalamus	Changes in temperature alter activity of ion channels
Pain	Nociceptors	Throughout all tissues and organs except the brain	Changes in temperature and pressure open membrane channels
Touch	Mechanoreceptors	Skin surface	Changes in pressure deform nerve
Muscle contraction	Stretch receptors	Wrapped around muscle fibers	Motion of muscle fiber changes nerve

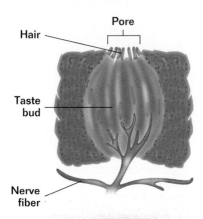

Hair

Pore

Taste
bud

Nerve
fiber

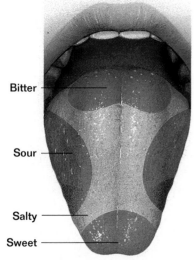

Bitter

Sour

Salty

Sweet

Figure 40-12 A taste bud, *above,* is a cluster of cells surrounding a pore. Several hairs protrude from the tips of the cells through the pore to the surface of the taste bud, where they are bathed in saliva. Taste buds are located all over the tongue, but they tend to be grouped into regions according to which "flavor" they sense, *below.*

Taste and Smell Are Chemical Senses

The senses of taste and smell depend on receptors that detect specific chemicals in the environment. In the presence of chemicals, these sensory neurons transmit information to an area of the brain where the information is processed and analyzed. These neurons are far more sensitive in many vertebrates than they are in humans.

Embedded within the surface of the tongue are over 10,000 taste receptors, or **taste buds.** A taste bud, shown in Figure 40-12, is a globular cluster of cells specialized to detect four basic types of chemicals: sugars (sweet), acids (sour), alkaloids (bitter), and metal ions (salty). Sensitivity to these chemicals varies on different parts of the tongue. In general, the tip of the tongue is most sensitive to sweet tastes, the sides to sour tastes, and the back to bitter tastes. Areas sensitive to salty tastes seem to be distributed evenly over the tongue. A taste bud is stimulated when a chemical dissolved in saliva binds to small hairs that protrude from the tip of the taste bud. The binding generates an action potential that travels along associated sensory neurons to the brain, where taste is perceived. The "hot" sensation of foods such as chili peppers is detected by pain receptors, not chemical receptors.

The receptors that sense smell are in a patch of membrane located in the roof of the nasal passage. There are approximately 5 million receptor cells in this area that respond to chemicals in a gaseous state. The sense of smell, in addition to the sense of taste, is very important in telling us about our food. That is why when you have a bad cold and your nose is stuffed up, your food seems to have little taste.

Section Review

1. *What are two functions of the inner ear?*
2. *Describe the mechanism in the inner ear responsible for equilibrium.*
3. *How are sound vibrations transmitted through the ear?*
4. *Trace the path of light as it passes through the eye. What roles do rods and cones play in the perception of light?*
5. *How are taste buds stimulated?*

Critical Thinking

6. *Describe a possible cause for colorblindness, a condition in which an individual cannot distinguish some colors from others.*
7. *Why is the brain an integral part of any sensation?*

40-4 Drugs and the Nervous System

We live in a drug-oriented society. Television commercials and other forms of advertising tell you about pain relievers, antacids, cough syrups, and other medications you can buy to help you feel better. Many drugs can prevent, treat, or cure illnesses that used to be deadly. However, when drugs are taken for nonmedical reasons drug abuse can occur. Few social problems in this country have had a greater impact on people's lives than the spreading abuse of addictive drugs.

Psychoactive Drugs Affect the Nervous System

In the broadest sense, a drug is a chemical that can alter biological functions and structures. Drugs most commonly abused are **psychoactive drugs,** substances that produce their effects on nervous system tissue and are often addictive. Scientists have recognized an important fact: addiction to psychoactive drugs is a physiological response that involves drug molecules and receptors in neuron membranes. Addiction is the body's attempt to cope with the chemical disruption a drug inflicts on a neuron's signaling systems.

To understand the nature of drug addiction, keep in mind how nerves communicate with one another because psychoactive drugs interfere with this process. Recall that neurotransmitters are released from the end of an axon, cross a synapse, and bind to receptor proteins on the adjacent neuron, triggering an impulse. Most psychoactive drugs produce their effects in the synapse. For example, some drugs have chemical structures similar to those of neurotransmitters. When a molecule of such a drug reaches the cell membrane of a neuron, it might be able to fit into a receptor protein for a particular neurotransmitter, thus causing the neuron to react as if a neurotransmitter were present. This process occurs with a number of drugs, such as morphine and heroin. Other drugs work by preventing neurotransmitters from being destroyed or recycled in the synapse, as shown in Figure 40-13.

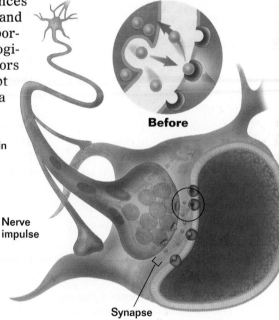

Serotonin

Nerve impulse

After

Before

Synapse

Figure 40-13 After the neurotransmitter serotonin carries a signal across a synapse, it is reabsorbed or broken down, *before.* When a drug that blocks the removal of a neurotransmitter such as Prozac® is taken, the neurotransmitter serotonin remains in the synapse, asserting its effects longer, *after.* This drug is helpful in treating depression, which can result from a shortage of serotonin.

Addiction Is Due to a Change in Receptor Protein Numbers

When a cell is exposed to a chemical signal for a prolonged period of time, it tends to lose its ability to respond to the stimulus with its original intensity. You are familiar with this loss of sensitivity—when you put on a wristwatch, how long are you aware you are wearing it? Neurons are particularly affected by this loss of sensitivity. If receptor proteins within synapses are exposed to high levels of neurotransmitter molecules for prolonged periods of time, a neuron will often respond by building and inserting fewer receptor proteins into the membrane. This feedback process is a normal part of the functions of all neurons. It is a mechanism that has evolved to adjust the number of receptor proteins in a cell to a certain level of neurotransmitter, enabling a cell to use its energy more efficiently.

Figure 40-14 shows an example of this mechanism, in which a kind of drug that prevents neurotransmitter uptake is present in a synapse. If this kind of drug is taken, large amounts of a neurotransmitter will remain in the synapses for a long time. In response to this surplus, a neuron will produce fewer receptor proteins for that neurotransmitter. In effect, the decreased number of receptor proteins creates a less sensitive neuron. If it didn't, the overabundance of neurotransmitter would cause a neuron to receive information over and over.

The decrease in production of receptor proteins results in addiction because a neuron cannot function normally unless the drug is present. When the drug is removed from the synapses, the neurotransmitter can once again be removed from the synapses. Suddenly there is no longer enough neurotransmitter to enable a neuron to receive information. The neuron cannot function normally until its number of receptor

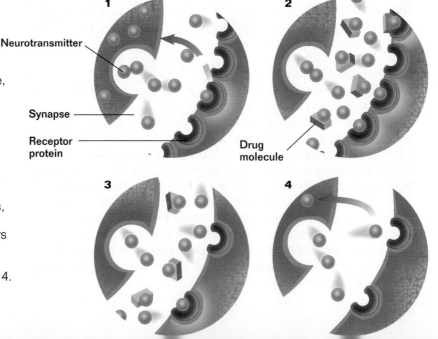

Figure 40-14 In a normal synapse, neurotransmitters are rapidly reabsorbed, 1. When a drug blocks the removal of a neurotransmitter, receptors across the synapse are flooded with excess neurotransmitters, 2. The receiving neuron responds to this surplus by lowering the number of its receptor proteins, 3. When the drug is removed, the surplus of neurotransmitters can be removed, leaving too few in the synapse to fire the reduced numbers of receptors, 4.

proteins have adjusted to the decrease in neurotransmitter. Sometimes in drug treatment programs, a drug is withdrawn slowly to allow the number of receptor proteins to increase gradually, minimizing withdrawal symptoms such as anxiety, depression, and cravings for the drug. As you can see, addiction is not simply a psychological state to be overcome by willpower; addiction is a physiological dependence caused by a change in a neuron's receptor proteins. ◻

◻ **CAPSULE SUMMARY**

Drugs can change the levels of neurotransmitter in a synapse. Addiction occurs when the number of receptor proteins in a neuron's membrane adjusts in response to these changes.

Narcotics and Cocaine Quickly Produce Addiction

Narcotics are powerful drugs that are used to relieve pain and induce sleep. They are **depressants,** substances that decrease the activity of the central nervous system. Many of the most potent narcotics are derived from chemicals extracted from *Papaver somniferum*, a species of the poppy plant, shown in Figure 40-15. The sap that oozes from the cut seed pod forms a thick, gummy substance called opium. Derivatives of opium are called opiates. Throughout history, the remarkable pain-relieving property of opium made it seem helpful for many illnesses. However, the price for this property is high—opiates produce addiction.

The major active ingredient in opium is morphine, a derivative that is about 10 times more potent than opium and highly addictive. Morphine is one of the most effective pain-relieving drugs known. It was widely used during the American Civil War to treat severely wounded soldiers, a practice that produced an addiction so common that it was often called "soldier's disease." A relatively simple chemical modification of morphine produces heroin, a powerful drug that is even more likely than morphine to produce addiction. Heroin abuse is among the most serious drug problems in society today. To understand how narcotics affect the nervous system and cause addiction, you first need to understand how the body perceives pain.

Figure 40-15 Opium is derived from the poppy *Papaver somniferum*. When scored with a knife, the seed pod oozes a milky white sap. After drying, the sap forms thick, gummy, brown opium.

Narcotics Mimic Natural Painkillers

As uncomfortable as it may feel, pain plays a very important role in the body. It notifies you of damaged tissue and other injury. Imagine how your body would look and function today if you did not have the ability to sense pain. Pain begins as a signal at damaged nerve endings and travels up the spinal cord to the brain. After reaching the brain, a pain signal is shut off when the central nervous system secretes a class of neurotransmitter called **enkephalins** (*ihn KEHF uh lihnz*). Enkephalins are one of several kinds of natural pain relievers your body releases in response to pain and stress. When enkephalins bind to receptor proteins in spinal neurons, potassium channels open, inhibiting pain messages from traveling to the brain.

Narcotics function by imitating enkephalins. Their similar molecular structure makes it easy for them to bind to the receptor proteins for enkephalins. These enkephalin receptor proteins are called opiate receptors because scientists observed opiates binding to them before enkephalins were ever discovered. Morphine and heroin are potent pain-blocking drugs because they act to block pain signals traveling up the spine to the brain. Narcotics also interact with the brain's limbic system, the "pleasure center," producing a perception of intense well-being. With prolonged use, neurons adjust their internal chemistry to make themselves less sensitive to stimulation. The neurons become so desensitized that the number of opiate receptors may actually increase.

What happens when an addict stops taking a narcotic? Removing the inhibitory action of the narcotic causes the neuron to become extremely sensitive. Unpleasant symptoms result, including extreme anxiety, tremors, and heightened sensitivity to pain. Sensitivity to pain results because the changes in potassium ion levels caused by enkephalins are no longer enough to inhibit the spinal nerves. As a result, pain signals travel freely up the spine, making the body very receptive to pain. The narcotic is required in significant amounts just to make the person's body feel normal. ◻

Cocaine Is a Stimulant

Unlike narcotics, cocaine is a **stimulant,** a substance that excites the central nervous system and speeds up body processes. Cocaine is found in the leaves of coca plants that grow at high altitudes in the mountains of South America. Even though many South Americans had chewed the coca leaf for centuries, the process for extracting cocaine was not developed until the mid-1800s. Many physicians at first con-

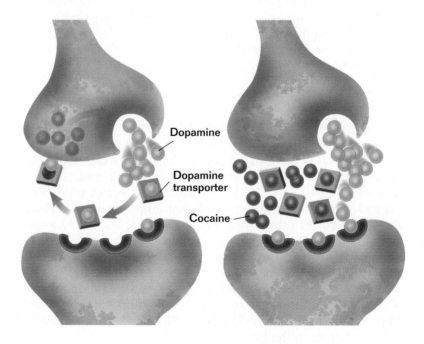

Figure 40-16 Dopamine is a neurotransmitter that helps send pleasure messages to the brain. Cocaine binds with dopamine transporter to block the reabsorption of dopamine so that it remains in the synapse longer.

sidered it a miracle drug and prescribed it for all sorts of physical and mental ailments; it was even added to soft drinks. However, cocaine is highly addictive. Today, laws in the United States forbid the importation, manufacture, and use of cocaine for nonmedical purposes, and medical use is extremely limited.

Despite being illegal and highly addictive, cocaine is still abused by many people. They risk becoming addicted because of the changes cocaine makes in the brain. As shown in Figure 40-16, cocaine works by preventing the reabsorption of dopamine, a neurotransmitter that stimulates the limbic system. The trapped dopamine repeatedly stimulates neurons, producing an intense feeling of well-being. Over a short period of time, neurons adjust to the presence of cocaine by decreasing their number of dopamine receptors. As a result, the neurons in the limbic system become less sensitive. As more time passes, more and more cocaine is needed to stimulate the limbic system. As you can see, cocaine addiction is a result of changes in the number of receptor proteins as neurons desensitize in response to the drug. Even more potent and addictive than cocaine is crack, a smokable form of the drug. When crack is smoked, it is absorbed very rapidly in the lungs and produces addiction very quickly. □

□ CAPSULE SUMMARY

Cocaine blocks the reabsorption of dopamine, a neurotransmitter that stimulates the pleasure center in the brain.

Dangerous Social Drugs
....

Not all dangerous drugs are illegal. More deaths and suffering are caused by familiar, legal drugs such as nicotine and alcohol. **Nicotine** is a highly addictive stimulant found in the leaves of the tobacco plant. Tobacco has many other constituents, but nicotine has the broadest and most immediate effects in the body. The other substances in tobacco, mostly tars, are highly mutagenic, causing changes in DNA that can lead to cancer.

Nicotine is extremely toxic; only 60 mg is lethal in humans. When a smoker inhales a cigarette, the nicotine in the tobacco reaches the brain within 10 seconds. Nicotine binds to specific receptors in the brain that are similar to those for acetylcholine *(as ee tihl KOH leen)*, a neurotransmitter that stimulates skeletal muscles. The binding of these receptors produces a mild, short-lived pleasurable sensation. Nicotine addiction is similar to cocaine addiction. Neurons in the brain adjust to prolonged exposure to nicotine by making fewer receptor proteins, and eventually the body requires nicotine to maintain a "normal" feeling.

You do not have to smoke tobacco to feel the effects of nicotine. Nicotine is also absorbed by the body from smokeless tobacco products, such as chewing tobacco and snuff. These tobacco products are just as addictive as cigarettes. In addition, the tars in chewing tobacco and snuff lead to increased risk of mouth and throat cancer. Chewing tobacco instead of smoking it simply trades one form of deadly cancer for another.

It was once thought that cigarette smokers were only harming themselves. However, scientists now know that just being in the same vicinity of people smoking will introduce nicotine, carbon monoxide, and other elements of tobacco smoke into your body. A study of smokers and nonsmokers confined in airplanes revealed that blood samples from nonsmokers contained high levels of nicotine. And a 1992 study of women who died in accidents revealed that almost all of those married to smokers had precancerous lesions in their lungs. Those women married to nonsmokers did not. For these reasons and others, smoking is banned from many public places. □

Alcohol

Of all the psychoactive drugs, alcohol (ethanol) is one of the most widely used and abused. Consumed for centuries as wine from fermented grapes or as beer from fermented grain, alcohol reduces inhibitions and produces a sense of well-being. Unfortunately, consuming alcohol can alter both judgment and reaction time, a condition known as being "drunk." Every year, many high school students die in automobile accidents resulting from drunken driving, making alcohol a truly dangerous drug.

Alcohol is an unusual psychoactive drug. Unlike narcotics and cocaine, it has no receptor. Instead, alcohol is able to alter the structure of the membrane bilayer, producing changes in the shape of receptor proteins. An altered receptor protein may become more sensitive to a stimulus or may be, in effect, switched off. As you can imagine, each change in a receptor protein may have complex effects on normal brain functions. Alcohol also inhibits the nerves that repress the limbic system, causing feelings of pleasure and inhibition.

Addiction to alcohol, or alcoholism, is the major drug abuse problem in the United States. People who drink excessive amounts of alcohol over long periods of time develop serious health problems. For example, many alcoholics suffer from a lack of vitamins because they do not eat properly when drinking heavily. This can lead to malnutrition, abnormalities in the circulatory system, and inflammation in the stomach lining. In addition, alcohol is readily converted to energy, preventing the body from breaking down other nutrients, such as sugars, amino acids, and fatty acids. These nutrients are stored as fat in the liver. After several years of drinking, liver cells are filled with fat and begin to die. If heavy drinking continues, a liver condition called **cirrhosis** may develop. In cirrhosis, liver cells are replaced with useless scar tissue, and the liver gradually shrinks into a small, hard mass. In this form, the liver can no longer eliminate body wastes, produce blood clotting, or carry out its other functions. ▪

□ *CAPSULE SUMMARY*

Nicotine acts like a stimulant because it binds to receptor proteins similar to those for acetylcholine, a neurotransmitter that stimulates skeletal muscles.

□ *CAPSULE SUMMARY*

Alcohol alters neuron membrane structure, which changes the shape of receptor proteins.

Psychoactive drugs include a variety of other substances. The caffeine found in coffee and soft drinks is a psychoactive drug, as is the drug in marijuana. Table 40-2 lists and describes additional psychoactive drugs that are commonly abused and that can lead to many serious health problems.

Table 40-2 Additional Psychoactive Drugs

Drug	Examples	Effects	Risks Associated With Abuse
Depressants	Barbiturates, tranquilizers, methaqualone, phencyclidine hydrochloride (PCP)	Slow down the action of the central nervous system	Drowsiness, depression, emotional instability
Stimulants	Amphetamines, caffeine	Speed up the central nervous system, metabolism, blood pressure, heartbeat, and respiratory rate	Irregular heartbeat, high blood pressure, headaches, stomach disorders, exhaustion, violent behavior
Inhalants	Nitrous oxide, ether, paint thinners, glue, cleaning fluids, correction fluids	Disorientation, confusion, memory loss	Hallucinations; permanent damage to brain, kidneys, liver; death
Hallucinogens	Lysergic acid diethylamide (LSD), mescaline, peyote	Distortion in the way the brain translates impulses	Dangerous hallucinations, unpredictable behavior
Marijuana	Derived from dried leaves, flowers, and stems of *Cannabis sativum*	Wide range of effects; short-term memory loss, disorientation, impaired judgment	Lung damage, loss of motivation

Section Review

1. *What is a psychoactive drug?*
2. *What role do enkephalins play in the body?*
3. *Why are narcotics so addictive?*
4. *How does cocaine addiction differ from heroin addiction?*
5. *How does nicotine affect the nervous system?*
6. *How does alcohol differ from other drugs in its effect on the nervous system?*

Critical Thinking
7. *Why is addiction defined as a physiological response?*

action potential (941)
autonomic nervous
   system (944)
axon (939)
brain stem (946)
cell body (939)
cerebellum (946)
cerebral cortex (945)
cerebrum (945)
cirrhosis (960)
cochlea (951)
cone (952)
cornea (951)
dendrite (939)
depolarization (939)
depressant (957)
enkephalin (957)

hypothalamus (946)
integration (943)
iris (951)
lens (951)
limbic system (946)
motor nerve (944)
multiple sclerosis (940)
myelin sheath (940)
nerve (940)
neurotransmitter (942)
nicotine (959)
parasympathetic division
   (949)
psychoactive drug (955)
pupil (951)
reflex (948)
repolarization (941)

resting potential (941)
retina (952)
rods (952)
Schwann cell (940)
sensory nerve (944)
sensory organ (950)
sensory receptor (950)
somatic nervous system
   (944)
spinal cord (946)
stimulant (958)
sympathetic division (949)
synapse (942)
taste bud (954)
thalamus (945)

Review

Multiple Choice

1. Compared with the rate at which nerve impulses travel down a myelinated axon, the rate of travel down an unmyelinated axon is
   a. faster.
   b. much faster.
   c. slower.
   d. no different.

2. Which of the following describes the condition in which there is both a high concentration of sodium ions outside the membrane of the neuron and a high concentration of potassium ions inside the membrane?
   a. resting potential
   b. action state
   c. synapse
   d. neurotransmission

3. Which of the following describes an action potential?
   a. sodium ions flood into a neuron
   b. sodium ions diffuse out of a neuron
   c. potassium ions flood into a neuron
   d. potassium ions reach their resting potential

4. Which of the following might result from damage to the cerebellum?
   a. insatiable thirst
   b. loss of balance
   c. indigestion
   d. dilation of blood vessels

5. A reflex is a very rapid response because the nerve impulse
   a. is passed along an unmyelinated axon.
   b. never reaches the brain.
   c. increases heart rate.
   d. involves only motor neurons.

6. Identify the pathway of light from the environment to the retina.
   a. → cornea → lens → receptor cells
   b. → receptor cells → cornea → lens
   c. → lens → receptor cells → cornea
   d. → lens → cornea → receptor cells

7. Addiction to most psychoactive drugs is considered to be a physiological response because the drugs
   a. can be purchased illegally.
   b. cause long-term suffering among addicts, family, and friends.
   c. affect the operation of neurotransmitters in the synapse.
   d. cause hallucinations, depression, and brain damage.

8. What is the function of enkephalins?
   a. pain relief
   b. increasing the flow of blood to infected areas
   c. restoration of equilibrium
   d. alleviate depression

Completion

9. The small space between an axon of one neuron and a dendrite of another neuron is the _____ . Impulses are carried across this space by chemical messengers called _____ .

10. The _____ nervous system is subdivided into the sympathetic and the parasympathetic nervous system. The sympathetic system dominates in times of _____ , and the parasympathic system dominates in times of rest.

11. Photoreceptor cells of the eye that detect color and produce sharp images are called _____ . Photoreceptor cells that are especially light-sensitive while producing poorly defined images are called _____ .

12. Two legal drugs considered potentially dangerous are _____ and _____ . One of these is found in tobacco and interacts with the _____ receptors on neurons. The other drug inhibits the nerves that repress the _____ system.

Themes Review

13. **Evolution** Many of the axons of humans are wrapped with myelin sheaths. Why are myelinated axons considered an evolutionary advancement?

14. **Structure and Function** Useful information travels in only one direction along a neuron: toward the central nervous system in sensory neurons and away from the central nervous system in motor neurons. What structures of the neuron ensure that this pattern is always followed?

15. **Levels of Organization** The nervous system is divided into various subunits. The two motor subunits that make up the peripheral nervous system are the somatic system and the autonomic system. How do the functions performed by the motor nerves of these two systems compare?

Critical Thinking

16. **Making Inferences** For some people too little potassium in the bloodstream is a serious medical problem. Symptoms of this condition include an increased heart rate and headaches. How might you explain these symptoms?

17. **Interpreting Data** Look at the diagram of the brain of a goose. Using what you know about the human brain, what would you assume the size of the portions of this brain indicate about their relative importance?

Cerebellum   Cerebrum

Optic lobe

Activities and Projects

18. **Research and Writing** A number of potentially addictive substances, including opiates and cocaine, have now been linked to specific receptors in the brain. Find out how the work of neurochemist Solomon H. Snyder led to a greater understanding of how potentially addictive substances affect the operation of receptors in the brain. In your report, be sure to describe the evidence that led Snyder to infer the existence of endorphins, the naturally occurring opiate-like substances in brain tissue.

Neuron Model

OBJECTIVES

- Construct a model to simulate the electrical communication between nerve cells.
- Observe the distribution of ions in a model neuron.

PROCESS SKILLS

- assembling a model
- measuring voltage

MATERIALS

- dialysis tubing
- scissors
- 150 mL beakers (2)
- 15 cm pieces of nickel-chromium alloy wire (2)
- glass stirring rods (2)
- 40 cm pieces of copper wire (2)
- alligator clips (2)
- screwdriver
- DC millivoltmeter
- safety goggles
- 3 M sodium chloride solution
- 3 M potassium chloride solution
- string
- elastic band

BACKGROUND

1. How do nerve cells communicate with each other?
2. Where does the electricity in nerve cells come from?
3. How is the resting potential of nerve cells like a live battery?
4. What is an action potential?
5. Write your own **Focus Question** on your Vee Form.

6. **Knowing Side of the Vee** List the **Concepts** and new **Vocabulary Words** on your Vee Form. In the **Concept Statements** section of the Vee, use these words in sentences that define and explain them.

TECHNIQUE

Doing Side of the Vee

1. **CAUTION: Be careful when using scissors.** Cut a 20 cm strip of dialysis tubing and place it in a beaker of water.
2. Coil the 15 cm strip of nickel-chromium wire tightly around one end of the glass stirring rod, with about 3 cm of the wire uncoiled and laying straight against the rest of the glass rod. Prepare the second glass rod the same way.
3. Pull the copper wire through the hollow end of the alligator clip, loop it around the screw, and tighten the screw with a screwdriver. Repeat with the second clip and copper wire.
4. Clip each alligator clip to each stirring rod with the clip touching the uncoiled part of the nickel-chromium wire, as shown below.

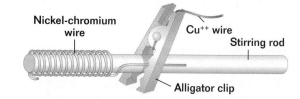
Nickel-chromium wire • Cu++ wire • Stirring rod • Alligator clip

5. Attach the copper wire of one clip to the positive terminal of the meter and the copper wire of the other clip to the negative terminal of the meter by twisting the end of the copper wire around the terminal post and tightening the nut.
6. **CAUTION: Sodium chloride and potassium chloride solutions are strong eye irritants; avoid eye contact and wear safety goggles for the remainder of the**

procedure. Fill a 150 mL beaker with 75 mL of 3 M sodium chloride (NaCl) solution.

7. Remove the dialysis tubing from the beaker and tie off one end of it with string to make a sack. Fill this sack halfway with 3 M potassium chloride (KCl) solution. This sack is the model of a neuron.

8. Place the stirring rod of the apparatus attached to the negative terminal into the potassium chloride solution in the sack and use the elastic band to close the open end of the tubing around the stirring rod as you see in the drawing below. **DO NOT** let any wires touch each other.

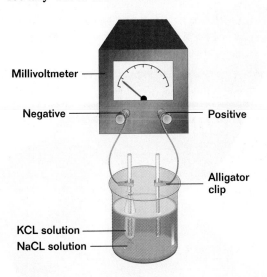

Millivoltmeter

Negative — Positive

Alligator clip

KCL solution
NaCL solution

9. The sodium chloride solution represents the fluid surrounding a neuron when it is not transmitting an impulse. Certain ions can pass through the dialysis tubing (neuron cell membrane) much more easily than others, so a difference in net charge develops across the membrane. This difference in charge, or voltage, is measured in volts or millivolts. Observe the millivoltmeter reading and record it in the **Records** section of your Vee Form. This is the resting potential of your model neuron. In the **Procedure** section of the Vee, briefly summarize the procedure you followed.

10. Clean up your materials and wash your hands before leaving the lab.

INQUIRY

1. What reading was recorded from the voltmeter?

2. Compare your reading with the other lab groups. Were they the same?

3. Use the information on the **Knowing Side** of the Vee to interpret your results from the **Doing Side,** and then write your **Knowledge Claim**. Write a **Value Claim** for this lab.

ANALYSIS

1. Because the millivoltmeter indicator moves in the direction of the flowing electrons, which way with respect to the potassium chloride solution are electrons flowing in the model?

2. How does the charge on the inside of the nerve cell model compare to the charge on the outside?

3. The potassium (K^+), chloride (Cl^-), and sodium (Na^+) ions are all moving, but perhaps with different ease through the dialysis tubing. Which must be moving more easily to create a net negative charge inside the tubing (neuron)?

4. How is this set up like an unused battery?

5. Why are the sodium and potassium ions rather than the chloride ions responsible for the difference in charge?

FURTHER INQUIRY

Write a **New Focus Question** that could be the point of a new investigation. The following is an example:

How do the concentrations of the sodium chloride and potassium chloride solutions affect the amount of voltage made by the model? Is this important to the study of a functioning neuron?

CHAPTER
41

HORMONES AND THE ENDOCRINE SYSTEM

REVIEW

- action of enzymes (Section 2-3)
- role of DNA and mRNA in protein synthesis (Section 9-1)
- negative feedback (Section 35-3)
- *Highlights: Endocrine System* (p. 846)
- nature of nerve messages (Section 40-1)
- hypothalamus (Section 40-2)

Speed skater ready to race

41-1 Hormones

The tissues and organs of your body carry out a multitude of activities. Activities such as those in Figure 41-1 must be coordinated to prevent them from conflicting with one another. This coordination, plus a smoothly functioning set of organs, is what truly makes an organism alive. As you learned in Chapter 35, the central nervous system coordinates the body's activities. Fast-acting but short-lived nerve messages are sent by electrical signals that travel along neurons, much as telephone messages travel along wires. For a longer-lasting effect, the central nervous system signals the release of hormones—chemical messengers that are produced in one place and transported to another by the bloodstream.

Section Objectives

- Describe several advantages that hormones have over the electrical signals sent through nerves.
- Compare and contrast peptide hormones with steroid hormones.
- Compare and contrast hormones with nonendocrine chemical signals.
- Explain how the hypothalamus regulates the activities of hormones.

Figure 41-1 Combining two activities, such as walking and eating an apple, requires the coordination of many body processes. Such coordination is maintained by hormones, the messengers of the endocrine system.

Hormones Have Advantages Over Nerve Signals

The hormones produced by the endocrine system are the body's chief means of regulating its ongoing activities. Although hormones are slow-acting, their effects tend to persist for a long time. In addition to their longer-lasting effects, hormones have several other distinct advantages over the electrical signals transmitted by nerves.

1. **Scope** For an electrical signal to be effective, it must be transmitted directly to an individual cell through a neuron. To maintain the homeostasis of a group of tissues, organs, or organ systems, electrical signals would have to reach all the cells involved. This would require an enormous number of nerves—far too many to be practical. By contrast, hormones readily spread to all cells and tissues via the blood.

2. **Specificity**   Like letters and postcards, hormones are sent to specific addresses. Each kind of hormone molecule has a shape unlike any other, much as every human face is unique. A hormone's shape matches a particular receptor protein on its target cells. Thus, a hormone will bind only to cells that have a particular receptor protein, ignoring all other cells. In this way, the body ensures that a hormone signal will affect only certain cells.

3. **Flexibility**   A particular receptor protein can be present on different kinds of cells in different organs. Therefore, the same hormone can be used by the body to achieve different effects in different tissues. Like the blowing of an official's whistle in a sporting event, what happens after the signal is released depends on the context. Figure 41-2 illustrates the advantages of using hormones to regulate ongoing body processes. □

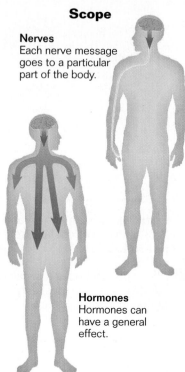

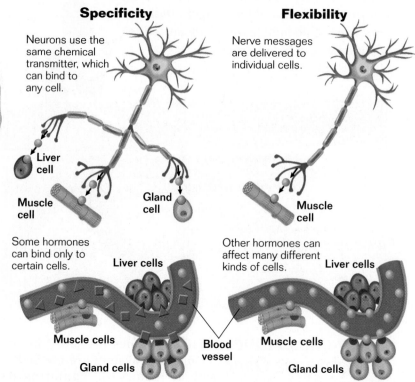

Scope

Nerves
Each nerve message goes to a particular part of the body.

Hormones
Hormones can have a general effect.

Specificity

Neurons use the same chemical transmitter, which can bind to any cell.

Liver cell

Muscle cell

Gland cell

Some hormones can bind only to certain cells.

Liver cells

Muscle cells

Gland cells

Blood vessel

Flexibility

Nerve messages are delivered to individual cells.

Muscle cell

Other hormones can affect many different kinds of cells.

Liver cells

Muscle cells

Gland cells

Figure 41-2 While nerve messages are faster-acting, messages carried in the bloodstream by hormones have a greater scope and more flexibility, and they affect only those cells with receptors that match specific hormones.

Endocrine Glands Produce Most Hormones

Throughout the body, a variety of tissues produce hormones and other similar chemical messengers. Most hormones are produced by **endocrine glands,** ductless glands that release their products directly into the bloodstream. The thyroid gland and the adrenal glands are examples of endocrine glands. In contrast to endocrine glands, exocrine glands, such as the salivary glands, deliver their products directly to where they are needed through ducts. The pancreas is both an exocrine gland and an endocrine gland. Acting as an exocrine gland, the pancreas produces digestive enzymes and delivers them to the small intestine through ducts. Clusters of cells within the pancreas secrete hormones into the bloodstream, making the pancreas an endocrine gland. In addition to the body's endocrine glands, several other body organs also contain cells that secrete hormones. These organs include the brain, stomach, small intestine, and heart.

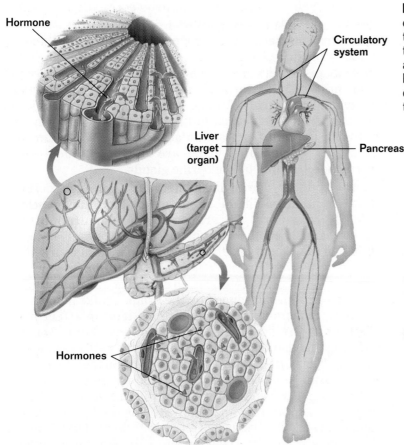

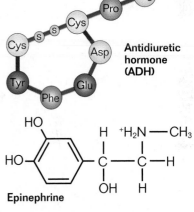

Peptide hormones

Antidiuretic hormone (ADH)

Epinephrine

Steroid hormone

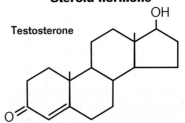

Testosterone

Figure 41-3 The liver is a target organ of hormones produced in the pancreas. Even though the two organs are very near one another, the hormones secreted by endocrine cells within the pancreas travel to the liver through the bloodstream.

Once they are in the bloodstream, hormones travel to a **target cell** where they produce their effect, as illustrated in Figure 41-3. Because they must often be transported in their active form over long distances in the body, hormones are very stable molecules. Most hormone molecules are either peptides or steroids. **Peptide hormones,** which consist of chains of amino acids, are water-soluble. Insulin is an example of a peptide hormone. A few important hormones, such as adrenaline (epinephrine) and norepinephrine, are derived from a single amino acid. **Steroid hormones,** which are lipid molecules similar to cholesterol, are fat-soluble. The sex hormones, estrogen and testosterone, are examples of steroid hormones. Examples of peptide and steroid hormones are illustrated in Figure 41-4.

Figure 41-4 These are the structures of some representative hormones. ADH, *top,* is a peptide hormone. Epinephrine (also called adrenaline), *center,* contains a single amino acid. Testosterone, *bottom,* is a steroid hormone.

Nonendocrine Chemical Signals Also Regulate Cell Activities

The human body also employs many nonendocrine chemical signals to regulate the activities of cells. These substances affect cells in a manner similar to hormones but are not usually referred to as hormones because they act locally instead of at a distance. Nonendocrine chemical signals include many chemicals secreted by the brain and nerves, as well as chemicals that are secreted by cells in a variety of the body's other tissues.

1. **Neurotransmitters**   One group of nonendocrine chemical signals produced by the brain and nervous tissue is the neurotransmitters. As you learned in Chapter 40, neurotransmitters take nerve signals across the gaps between individual nerve cells. Unlike hormones, neurotransmitters act only from one neuron to the next. Most neurotransmitters are derived from amino acids. Some neurotransmitters, such as norepinephrine, are chemically identical to hormones secreted by endocrine glands.

2. **Neuropeptides**   The brain and nervous tissue also secrete a large number of chemical signals called **neuropeptides.** There are several different groups of neuropeptides. Enkephalins, which were discussed in Chapter 40, are a group of neuropeptides that inhibit pain messages traveling toward the brain. **Endorphins** *(ehn DAWR fihnz)*, which are thought to regulate emotions, are another important group of neuropeptides produced by the brain. Unlike neurotransmitters, enkephalins and endorphins tend to affect many cells near the nerve cells that produce them. Because enkephalins and endorphins often alter a cell's response to a neurotransmitter, they are called **neuromodulators.** Many of the brain's neuropeptides are true hormones. The peptide hormones produced in the brain are secreted by the hypothalamus and are then delivered by the bloodstream to other parts of the body such as the pituitary gland. You will learn more about the relationship between the hypothalamus and the pituitary gland later in this chapter.

3. **Prostaglandins**   Among the most important nonendocrine chemical signals are the **prostaglandins** *(prahs tuh GLAN dihnz)*, modified lipids that are made from phospholipids by virtually all cells. Instead of circulating in the blood as hormones do, prostaglandins tend to accumulate in areas where tissue is disturbed or injured. Dozens of different prostaglandins produce a variety of effects. Some prostaglandins stimulate smooth muscle contractions that cause the constriction of blood vessels. The constricted blood vessels in turn affect blood pressure and body temperature. Other prostaglandins cause blood vessels to dilate, which causes inflammation. A headache may result when blood vessels swell and their walls press against nerves in the brain. Aspirin relieves headaches and reduces fever and inflammation by inhibiting prostaglandin production. ◻

◻ CAPSULE SUMMARY

The ductless glands called endocrine glands produce most of the body's hormones. Many nonendocrine chemical signals are produced in the brain and in many cells and tissues throughout the body.

The Hypothalamus Initiates Most Chemical Signals

A part of the brain controls most of the glands of the endocrine system. In fact, the endocrine system and the nervous system work together so closely that they can be thought of as one system—the **neuroendocrine system.** Until recently, however, one of the great mysteries of medicine was how the brain regulates the endocrine system. Scientists understood that the hypothalamus of the brain issues commands to the **pituitary gland**, an endocrine gland that is located at the base of the brain. They also knew that the pituitary gland in turn sends chemical signals to the body's other endocrine glands. But the way that the hypothalamus issues its commands was not clear. The pituitary gland is very close to the hypothalamus—actually suspended from it by a short stalk, as Figure 41-5 shows. However, no nerves connect the pituitary gland to the hypothalamus or to any other part of the brain.

In the 1930s, researchers discovered a network of tiny blood vessels that spans the short distance between the hypothalamus and the pituitary. This discovery suggested to researchers that perhaps chemical messages carried by hormones passed from the hypothalamus to the pituitary. In 1969, thyrotropin-releasing hormone (TRH), the first of several hormones produced by the hypothalamus to be isolated, was isolated from the brains of pigs. The release of TRH from the hypothalamus triggers the pituitary to release thyrotropin. This hormone then travels to the thyroid gland and causes the release of the thyroid hormones. Six other hypothalamic regulatory hormones that govern the pituitary have since been isolated.

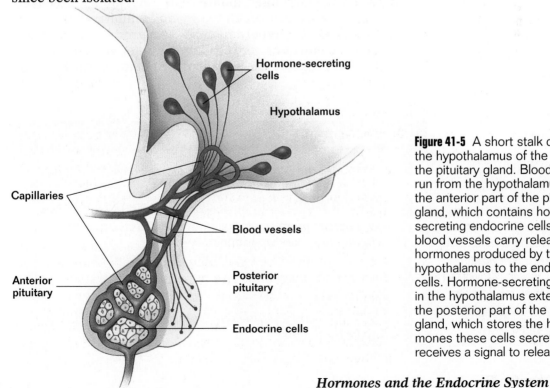

Figure 41-5 A short stalk connects the hypothalamus of the brain to the pituitary gland. Blood vessels run from the hypothalamus to the anterior part of the pituitary gland, which contains hormone-secreting endocrine cells. These blood vessels carry releasing hormones produced by the hypothalamus to the endocrine cells. Hormone-secreting cells in the hypothalamus extend into the posterior part of the pituitary gland, which stores the hormones these cells secrete until it receives a signal to release them.

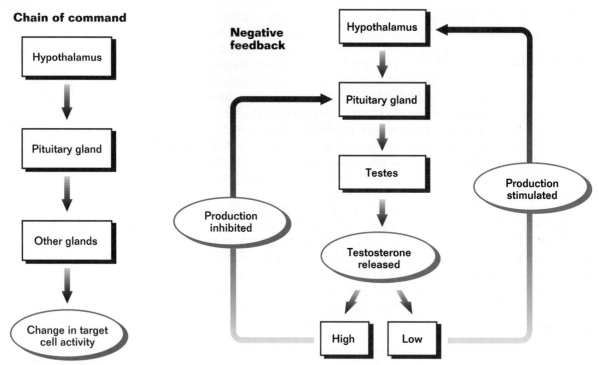

Chain of command

Hypothalamus

Pituitary gland

Other glands

Change in target cell activity

Negative feedback

Hypothalamus

Pituitary gland

Testes

Testosterone released

High

Low

Production inhibited

Production stimulated

Figure 41-6 The hypothalamus regulates many of the body's activities through a chain of command, *above,* in which the pituitary is ordered to release hormones that in turn stimulate other endocrine glands to release their hormones. A negative-feedback system that involves both the pituitary gland and hypothalamus, *above right,* inhibits the secretion of a hormone when a rise in its concentration is detected, or stimulates the hormone's secretion when a drop in its concentration is detected.

Scientists have learned through intensive research that the central nervous system regulates the body's hormones through a chain of command, as illustrated in Figure 41-6. Each of the "releasing" hormones made by the hypothalamus causes the pituitary to synthesize a corresponding pituitary hormone. This hormone then travels to a distant endocrine gland and causes that gland to begin producing its particular hormone or hormones. Some of the hormones released by the pituitary are regulated by an "inhibiting" hormone produced by the hypothalamus. The production of most hormones, however, is regulated by negative-feedback mechanisms like the one seen on the right in Figure 41-6. ◻

Section Review

1. *Compare the scope of a chemical messenger with that of a nerve impulse.*
2. *How do peptide hormones differ from steroid hormones?*
3. *Where are neuropeptides produced? What are some functions of neuropeptides?*
4. *What role does the hypothalamus play in the transmission of chemical messages by hormones?*

Critical Thinking
5. *Why are nonendocrine chemical signals not considered to be hormones?*

41-2 How Hormones Work

After hormones are produced, they travel through the bloodstream to their target cells, where they cause a change in the activity of the cells. Hormones cause an effect in a cell by attaching to a receptor protein in one of three ways. Some hormones enter the cells they affect, while others do not. One hormone goes directly to the nucleus of a cell. The location of the receptor proteins for a hormone determines its mode of action. In any case, the hormone's message must cross a plasma membrane in order to cause a response, such as the one shown in Figure 41-7. How does the binding of a hormone to a receptor cause a change in a cell's activity? The main way that hormones affect cells is by altering the activity or amounts of certain enzymes, which in turn alters chemical reactions that occur in the cells.

Section Objectives

- Describe how a peptide hormone produces a response.
- Describe how a steroid hormone produces a response.
- Explain how the action of thyroxine differs from that of other hormones.

Figure 41-7 A cat's response to danger is triggered by a hormone that causes changes to occur inside of cells. In order for these changes to occur, the hormone must send a message across the plasma membrane.

Peptide Hormones Remain Outside of Cells

A peptide hormone produces its effect from outside a cell. Because peptide hormones are not fat-soluble, they cannot pass through the plasma membrane of a cell. The receptor proteins for peptide hormones are embedded in the plasma membranes of target cells. When a peptide hormone binds to a receptor on the surface of a target cell, a change in the receptor protein's shape is triggered. This change unleashes a series of events in the cell's cytoplasm. One of these events is the production of a **second messenger**, which is a chemical that is produced in response to the binding of a chemical signal on the outside of a cell. **Cyclic AMP**, which is produced by the removal of two phosphate groups from ATP, is a common second messenger.

A second messenger activates or deactivates certain enzymes, altering the chemical activity of a peptide hormone's target cells. Because many molecules of the second messenger are produced by the binding of a single hormone molecule, second messengers amplify the effect of a peptide hormone. The effect of a peptide hormone may be further amplified because a single second-messenger molecule may stimulate the activation or production of many molecules of a particular enzyme. Each of these molecules may in turn activate many molecules of another enzyme, in a cascade of reactions that greatly amplifies the original signal. For discovering how this pathway works, Alfred Gilman and Martin Rodbell won the Nobel Prize for medicine in 1994.

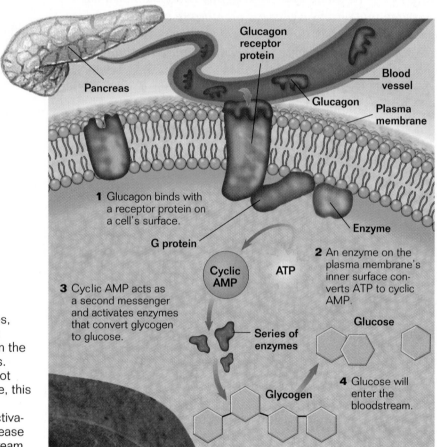

Glucagon receptor protein

Blood vessel

Pancreas

Glucagon

Plasma membrane

1 Glucagon binds with a receptor protein on a cell's surface.

G protein

Enzyme

2 An enzyme on the plasma membrane's inner surface converts ATP to cyclic AMP.

Cyclic AMP

ATP

3 Cyclic AMP acts as a second messenger and activates enzymes that convert glycogen to glucose.

Glucose

Series of enzymes

Glycogen

4 Glucose will enter the bloodstream.

Figure 41-8 Peptide hormones, such as glucagon, bind with receptor proteins located on the surfaces of their target cells. Even though glucagon cannot cross the plasma membrane, this binding initiates a series of events that results in the activation of enzymes and the release of glucose into the bloodstream.

The mechanism by which peptide hormones produce a response is summarized in Figure 41-8. **Glucagon** (*GLOO kuh gahn*) is a peptide hormone that is produced in the pancreas. It travels through the bloodstream and binds to receptor proteins located on the surfaces of liver cells. When glucagon molecules bind to their receptor proteins, the receptor proteins change shape and cause the production of cyclic AMP. Acting as a second messenger, the cyclic AMP molecules alter the activity of liver cells, causing them to activate a series of enzymes that converts glycogen into glucose. ☐

Steroid Hormones Work Inside of Cells

Because steroid hormones are fat-soluble, they can pass readily through the plasma membranes of their target cells and produce a response from inside the cells. A steroid hormone binds to a receptor protein located in a target cell's cytoplasm, which produces a **hormone-receptor complex.** This complex then enters the nucleus of the cell and binds to DNA, causing a change in a gene's activity and thus a change in the cell's activity. Some steroid hormones stimulate genes to synthesize certain proteins, and some repress the synthesis of other proteins.

The mechanism by which steroid hormones produce a response is summarized in Figure 41-9. **Cortisol** (*KAWRT uh sahl*) is a steroid hormone that is produced by the adrenal glands in response to stress. When cortisol molecules reach a target cell, the molecules diffuse through the plasma membrane and bind to receptor proteins located in the cell's cytoplasm. The hormone-receptor complex then enters the cell's nucleus, binds to DNA, and activates the genes that produce certain enzymes. These enzymes catalyze reactions that break down fats and proteins into fatty acids and amino acids, respectively. Other enzymes produced as a result of cortisol's action catalyze the conversion of fatty acids and amino acids to glucose. These changes help you deal with stress by providing extra sources of energy and building blocks for repairing damaged tissues and for making enzymes. ◻

◻ CAPSULE SUMMARY

Steroid hormones work by passing through a cell's plasma membrane and binding to protein receptors located in the cytoplasm. The hormone-receptor complex then enters the nucleus and attaches to DNA, altering the cell's production of enzymes and thus changing the cell's activity.

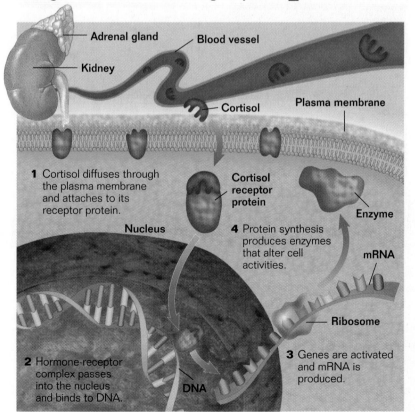

1 Cortisol diffuses through the plasma membrane and attaches to its receptor protein.

Adrenal gland

Blood vessel

Kidney

Cortisol

Plasma membrane

Cortisol receptor protein

Enzyme

Nucleus

4 Protein synthesis produces enzymes that alter cell activities.

mRNA

Ribosome

2 Hormone-receptor complex passes into the nucleus and binds to DNA.

DNA

3 Genes are activated and mRNA is produced.

Figure 41-9 Steroid hormones, such as cortisol, enter their target cells and bind to receptor proteins in the cytoplasm. This binding activates certain genes, which produce enzymes that alter cell activity.

One Hormone Works Inside the Nucleus

The receptor proteins for one important hormone are located not on the plasma membrane or in the cytoplasm, but inside the nucleus, as seen in Figure 41-10. **Thyroxine** *(theye RAHKS ihn)*, which is produced by the thyroid gland, consists of an amino acid and four attached iodine atoms. Thyroxine molecules are small enough to pass directly through the plasma membranes of their target cells. After diffusing through the cytoplasm, they enter the nucleus and bind to a receptor protein that is attached to a DNA molecule. The binding of a thyroxine molecule to its receptor protein initiates the production of mRNA molecules. These mRNA molecules cause the production of specific enzymes that stimulate cell metabolism and promote growth. ◻

◻ **CAPSULE SUMMARY**

Thyroxine passes through the plasma membrane and cytoplasm of a cell and enters the nucleus, where it attaches to a receptor protein on a DNA molecule and initiates the production of enzymes that change the cell's activity.

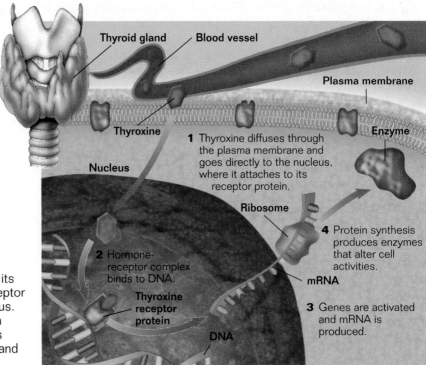

Figure 41-10 Thyroxine enters its target cells and binds to receptor proteins located in the nucleus. This binding activates certain genes that produce enzymes which stimulate metabolism and growth.

Thyroid gland · **Blood vessel**

Plasma membrane

Thyroxine

Nucleus

1 Thyroxine diffuses through the plasma membrane and goes directly to the nucleus, where it attaches to its receptor protein.

Ribosome

Enzyme

4 Protein synthesis produces enzymes that alter cell activities.

2 Hormone-receptor complex binds to DNA.

Thyroxine receptor protein

mRNA

3 Genes are activated and mRNA is produced.

DNA

Section Review

1. *How does a peptide hormone produce its response?*
2. *How does a steroid hormone produce its response?*
3. *How does the action of thyroxine differ from that of either a peptide hormone or a steroid hormone?*

Critical Thinking

4. *The excessive use of steroid hormones has been linked to cancer. Why are these hormones likely to cause cancer?*

41-3 The Body's Endocrine Glands

About a dozen major endocrine glands collectively make up your endocrine system, which is illustrated in Figure 41-11. Table 41-2 on pages 984–985 lists the major endocrine glands, the hormones they release, the target tissues for each hormone, and the effects that these hormones produce.

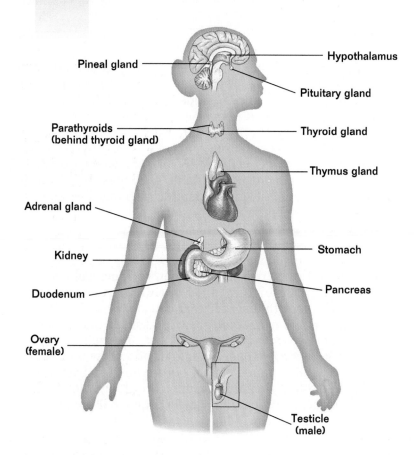

Pineal gland

Hypothalamus

Pituitary gland

Parathyroids (behind thyroid gland)

Thyroid gland

Thymus gland

Adrenal gland

Kidney

Stomach

Duodenum

Pancreas

Ovary (female)

Testicle (male)

Figure 41-11 Major endocrine glands are located throughout the human body. As research continues, more and more of the body's other organs are being added to the list of endocrine glands because they contain cells that secrete hormones.

The Pituitary Gland Regulates Other Endocrine Glands

The pituitary gland is a small endocrine gland that is located just beneath the hypothalamus within a bony recess at the base of the skull. One of the most important endocrine glands in the body, the pituitary releases nine major hormones. Because many of these hormones act principally to influence other endocrine glands, it was fashionable until recently to regard the pituitary as the "master gland" of the endocrine system. However, that role actually belongs to the hypothalamus, because it controls the pituitary.

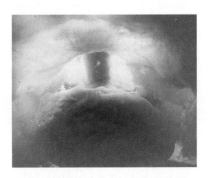

Figure 41-12 Despite its tiny size, the pituitary gland, *above,* consists of two lobes, *right,* which release different hormones.

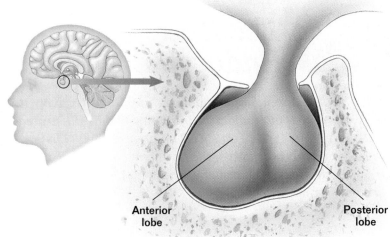

Anterior lobe

Posterior lobe

◻ **CAPSULE SUMMARY**

The pituitary gland, which is located in the brain, controls many of the body's other endocrine glands. It consists of an anterior lobe, which produces several hormones in response to signals from the hypothalamus, and a posterior lobe, which stores and releases hormones produced by the hypothalamus.

As Figure 41-12 illustrates, the pituitary is actually two glands. The front portion of the gland, known as the **anterior lobe,** produces seven major peptide hormones. Each of these hormones is released in response to a particular releasing signal secreted by the hypothalamus. For example, the release of thyroid-stimulating hormone (TSH) by the anterior pituitary is triggered by thyroid-releasing hormone (TRH). The back portion of the gland, known as the **posterior lobe,** stores and releases other peptide hormones, which are produced by the hypothalamus.

Many of the roles of the endocrine glands were first discovered by studies of medical conditions caused by malfunctions of the glands. For example, one key role of the anterior pituitary was discovered in 1909 when the surgical removal of a pituitary tumor cured a South Dakota farmer of **acromegaly,** a growth disorder that causes facial features to thicken. **Gigantism** is another growth disorder that is almost always associated with pituitary tumors. The tallest human being ever recorded—Robert Wadlow, who grew to a height of 8 ft. 11 in. and weighed 475 lb. before he died from infection at age 22—had a pituitary tumor. Pituitary tumors produce giants because such tumor cells produce large amounts of somatotropin, or growth hormone (GH). ◻

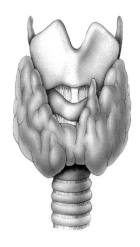

Figure 41-13 The thyroid gland, located in the neck, is wrapped around the windpipe.

The Thyroid Gland Regulates Metabolic Rate

The thyroid gland, which is shaped like a shield, is located just below the Adam's apple in the front of the neck, as seen in Figure 41-13. The name *thyroid* comes from the Greek word *thyros,* which means "shield." The thyroid gland produces several hormones, but the most important is thyroxine.

Thyroxine increases the body's metabolic rate and promotes the normal growth of the brain, bones, and muscles during childhood. Neither a peptide nor a steroid, thyroxine

Figure 41-14 A goiter, *left,* is a swelling in the throat that results from a lack of iodide in the diet. A diet that includes seafood and iodized salt, *right,* provides the iodide necessary for proper thyroid function.

is produced by the addition of iodide to the amino acid thyronine. If iodide salts are lacking in the diet, the thyroid gland becomes greatly enlarged as a result of futile attempts to make more thyroxine and forms a **goiter.** Once common in the United States, goiters, like the one seen in Figure 41-14, are now rare because of the addition of iodide to salt. The underproduction of thyroxine is known as **hypothyroidism.** If it occurs early in childhood, hypothyroidism may cause stunted growth or mental retardation, or both. In adults, hypothyroidism tends to produce a lack of energy, dry skin, and weight gain. Overproduction of thyroid hormones, or **hyperthyroidism,** causes nervousness, sleep disorders, an irregular heart rate, and weight loss.

Specialized cells within the thyroid gland produce another important hormone, **calcitonin** *(kal sih TOH nihn),* which plays a key role in maintaining a proper calcium level in the body. A high level of calcium in the blood stimulates calcitonin production, which in turn stimulates the deposition of calcium in bone tissue and lowers the blood-calcium level. ◻

CAPSULE SUMMARY

The thyroid gland, which is located in the neck, releases thyroxine and calcitonin. Thyroxine regulates growth and metabolism. Calcitonin helps regulate calcium in the body by stimulating the deposition of calcium in bone tissue.

Parathyroid Glands Regulate Blood Calcium

Attached to the thyroid gland, as seen in Figure 41-15, are four **parathyroid glands,** which also help regulate calcium in the body. Small and unobtrusive, they were ignored by researchers until well into this century. The parathyroid glands produce **parathyroid hormone** (PTH). PTH is essential for survival because it helps maintain an adequate supply of calcium in the blood. Recall from Chapter 40 that nerve impulses cause muscles to contract by initiating the release of calcium ions. You cannot live without the muscles that pump your heart, and these muscles cannot function if the blood-calcium level is not kept within narrow limits.

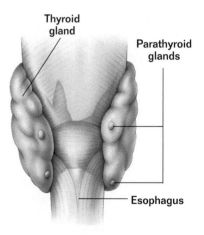

Figure 41-15 The parathyroid glands are located on the back of the thyroid gland.

Figure 41-16 This graphic organizer shows three ways that parathyroid hormone (PTH) causes the blood-calcium level to rise.

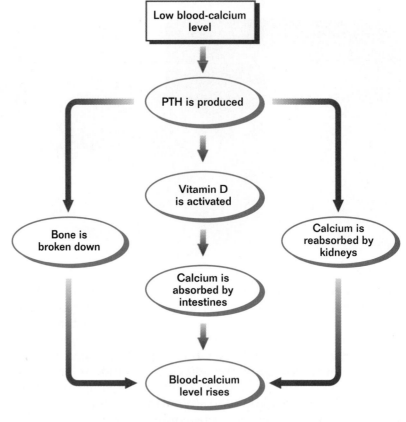

Low blood-calcium level

↓

PTH is produced

Vitamin D is activated

Bone is broken down

Calcium is reabsorbed by kidneys

Calcium is absorbed by intestines

Blood-calcium level rises

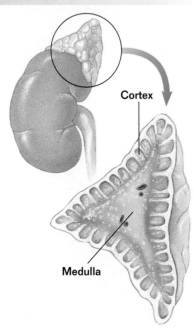

Cortex

Medulla

Figure 41-17 The adrenal glands, located on top of each kidney, consist of an inner medulla and an outer cortex.

PTH, which is synthesized in response to a falling level of calcium in the blood, affects the blood-calcium level in three ways, as diagramed in Figure 41-16. First, it ensures that the blood-calcium level never falls too low. When released into the bloodstream, PTH is absorbed by the bones, where it stimulates the osteocytes to dismantle bone tissue and release calcium into the bloodstream. Second, PTH acts on the kidneys, causing them to reabsorb calcium ions from urine. Third, PTH leads to activation of vitamin D, which is necessary for calcium absorption by the intestine. ■

The Adrenals Are Two Glands in One

Your body has two adrenal glands, one located just above each kidney. Each almond-sized adrenal gland is actually two glands in one, as seen in Figure 41-17. An inner core called the **adrenal medulla** produces the peptide hormones epinephrine *(ehp uh NEF rihn)*, also called adrenaline *(uh DRIHN uh lihn)*, and norepinephrine. Unlike other endocrine glands, the adrenal medulla is stimulated to release its hormones by nerves that come from the hypothalamus. An outer shell called the **adrenal cortex** produces the steroid hormones cortisol and aldosterone.

The adrenal medulla acts as an emergency warning system in times of stress by releasing **adrenaline** (or epinephrine)

Figure 41-18 The "fight-or-flight" hormones help this stockbroker handle the stress of his high-pressure job. Perspiration, rapid breathing, and an elevated pulse are outward signs that the "fight-or-flight" hormones are at work. Inside, extra blood glucose and oxygen enable his cells to supply the energy needed to keep going.

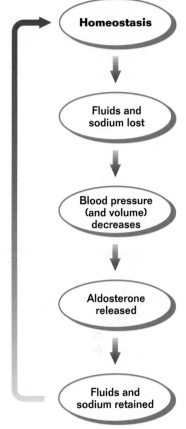

Figure 41-19 Aldosterone helps maintain the body's fluid and sodium balance by the mechanism shown here.

and **norepinephrine,** the "fight-or-flight" hormones. The effects of these hormones, which prepare the body for action in emergencies, are identical to the effects of the sympathetic nervous system but are longer-lasting. In stressful situations, such as the one in Figure 41-18, the fight-or-flight hormones accelerate heartbeat and increase blood pressure, blood-sugar level, and blood flow to the heart and lungs. These hormones can thus be thought of as extensions of the sympathetic nervous system, which was discussed in Chapter 40.

The adrenal cortex produces hormones that affect your metabolic health and maintain the proper amount of salt in your body. Cortisol (also called hydrocortisone) is produced in response to adrenocorticotropic hormone (ACTH) from the pituitary. It acts on many different cells in the body to maintain nutritional well-being. Cortisol stimulates carbohydrate metabolism and acts to reduce inflammation. Derivatives of this hormone, such as **prednisone** *(PREHD nih sohn),* are widely used as anti-inflammatory agents.

Aldosterone *(al DAHS tuh rohn)* affects primarily the kidneys by promoting the uptake of sodium and other salts from body fluids, as summarized in Figure 41-19. Recall that sodium ions play crucial roles in nerve conduction and many other bodily functions, such as maintaining blood pressure. Without aldosterone, sodium ions are not retrieved from the fluids removed by the kidneys and are lost in the urine. The resulting loss of salt from the blood causes water to leave the bloodstream and enter cells, and blood pressure falls. Aldosterone affects potassium in the opposite way. If the potassium level rises, aldosterone stimulates the kidneys to secrete potassium ions into the urine. When the aldosterone level is too low, the potassium level in the blood may rise to a dangerous level. Thus, aldosterone, as well as PTH, is essential for survival. ◻

◻ CAPSULE SUMMARY

The adrenal glands are located on top of each kidney and consist of an outer cortex, which produces steroid hormones, and an inner medulla, which produces the "fight-or-flight" hormones.

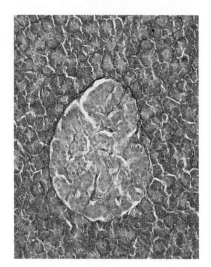

Figure 41-20 Islets of Langerhans are clusters of hormone-secreting cells that are easily recognized among the other cells of the pancreas. Two types of islet cells secrete hormones. The A cells, stained red in this light micrograph, produce glucagon. The B cells, stained purple, produce insulin. Notice that many blood vessels, shown in blue and white, pass through islets of Langerhans.

❏ *CAPSULE*
SUMMARY

The islets of Langerhans, which secrete insulin, are hormone-producing cells found in the pancreas. Either an abnormally low level of insulin in the body or a lack of insulin receptors makes an individual's cells unable to absorb glucose from the blood and causes a condition called diabetes mellitus.

The Pancreas Regulates Blood Sugar

Scattered throughout the pancreas, once thought to be solely an exocrine gland, are clusters of specialized cells that act as endocrine glands. These clusters, which are seen in Figure 41-20 and are called **islets of Langerhans,** produce two peptide hormones that interact to govern the level of glucose in the blood. **Insulin,** a storage hormone, puts away nutrients for leaner times by promoting the accumulation of glycogen in the liver. Glucagon causes liver cells to release glucose that was stored in glycogen. The two hormones thus work together to keep the blood-glucose level within narrow bounds.

Diabetes mellitus is a serious disorder in which the cells of affected individuals are unable to obtain glucose from the blood, causing the blood-glucose level to become very high. Affected individuals may lose weight, suffer brain damage, and ultimately starve to death. About 12 million Americans, and over 100 million people worldwide, have diabetes mellitus, which is the seventh-leading cause of death in the United States and is the leading cause of blindness among adults. It also accounts for one-third of all kidney failures.

There are actually *two* kinds of diabetes mellitus. The warning signs of each are listed in Table 41-1. About 10 percent of affected individuals suffer from **Type I diabetes,** a hereditary autoimmune disease in which the immune system attacks the islets of Langerhans, causing insulin secretion to be abnormally low. Also called juvenile-onset diabetes, Type I diabetes usually develops before age 20. Type I diabetes can be treated with daily injections of insulin. Active research on the possibility of transplanting islets of Langerhans holds the promise of a lasting treatment for Type I diabetes. However, researchers must find a way to prevent the immune system from attacking the pancreas. People with **Type II diabetes** often have an abnormally low number of insulin receptors, while the level of insulin in their blood is often higher than normal. Type II, or adult-onset, diabetes usually develops in people over 40, almost always as a consequence of obesity. In the United States, 90 percent of Type II diabetics are obese. Type II diabetes is usually treated with diet and exercise. Most Type II diabetics do not need daily injections of insulin. ◼

Table 41-1 Warning Signs of Diabetes Mellitus

Type	Symptoms
Type I	Frequent urination, unusual thirst, extreme hunger, unusual weight loss, extreme fatigue, irritability
Type II	Any of the Type I symptoms; frequent infections; blurred vision; cuts or bruises that are slow to heal; tingling or numbness in the hands or feet; recurring skin, gum, or bladder infections

Other Organs and Glands Produce Hormones

In addition to the organs and glands mentioned so far, several other organs and glands produce hormones. Among these are the female and male reproductive organs, the ovaries and the testes. These organs, which also produce gametes, secrete hormones that regulate reproduction. Ovaries secrete **estrogen** and **progesterone,** and testes produce **testosterone.** These hormones not only affect the formation of gametes but also control the development of the secondary sex characteristics, such as breasts in females and coarse facial hair in males.

CONTENT LINK

*The reproductive roles of the ovaries and testes and the hormones they produce are described in **Chapter 42.***

Daily Human Body Temperature Variation

Body temperature (°F) — Awake period — Asleep period

6 A.M. — 6 P.M. — 6 A.M.

Time of day

Figure 41-21 Sleep, *above left,* is an example of a daily biorhythm. A graph, *above,* shows how body temperature fluctuates daily, another type of biorhythm.

The **pineal** *(PIHN ee uhl)* **gland** is a pea-sized gland located in the top of the brain and named for its resemblance to a pine cone. The pineal gland secretes **melatonin,** which is a modified form of the amino acid tryptophan. In hamsters, melatonin regulates reproduction, and in frogs it influences pigmentation. However, the function of the pineal gland in humans is not yet known. Although the pineal gland is not connected directly to the central nervous system, it is connected to the eyes through the sympathetic nervous system. Also, melatonin seems to be released by the human pineal gland as a response to darkness. Therefore, the pineal gland is thought to be involved in establishing daily biorhythms, such as those seen in Figure 41-21. The pineal gland has also been implicated in mood disorders such as winter depression, also called seasonal affective disorder syndrome (SADS), and in a variety of aspects of sexual development.

Other organs that produce hormones and thus act as endocrine glands include the stomach, small intestine, and heart. As discussed in Chapter 38, the stomach and small intestine secrete hormones, such as gastrin, that regulate the release of acids and digestive enzymes and play a key role in digestion. The heart secretes atrial *(AY tree uhl)* natriuretic *(na tree yoo REHT ihk)* factor (ANF). Cells in the blood vessels, kidneys, and adrenal glands have receptors for ANF, which apparently helps regulate blood pressure. ❑

❑ CAPSULE SUMMARY

The pineal gland, which is located in the brain, and several other body organs, such as the ovaries, testes, stomach, and heart, produce hormones and thus act as endocrine glands.

Table 41-2 Major Endocrine Glands and Hormones

Gland	Hormone	Target Tissue	Effects
Pituitary gland	**Anterior lobe**		
	Adrenocorticotropic hormone (ACTH)	Adrenal glands	Stimulates the production of steroid hormones
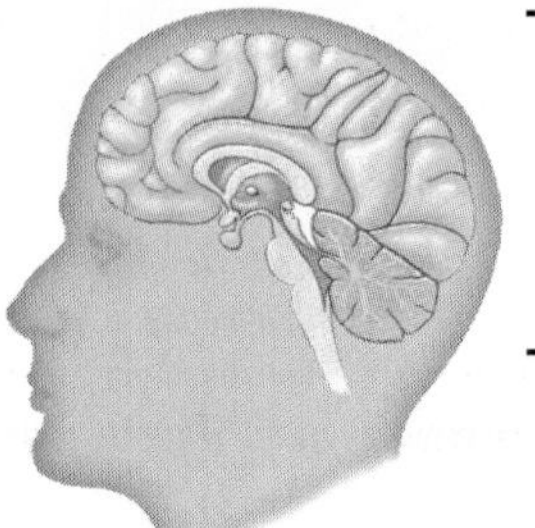	Follicle-stimulating hormone (FSH)	Ovaries and testes	Regulates the development of male and female gametes; stimulates the production of testosterone (male sex hormone) in males
	Luteinizing hormone (LH)	Ovaries and testes	Stimulates the release of an egg (ovulation) from an ovary; stimulates testosterone production by the testes
	Prolactin	Mammary glands	Stimulates milk production in breasts
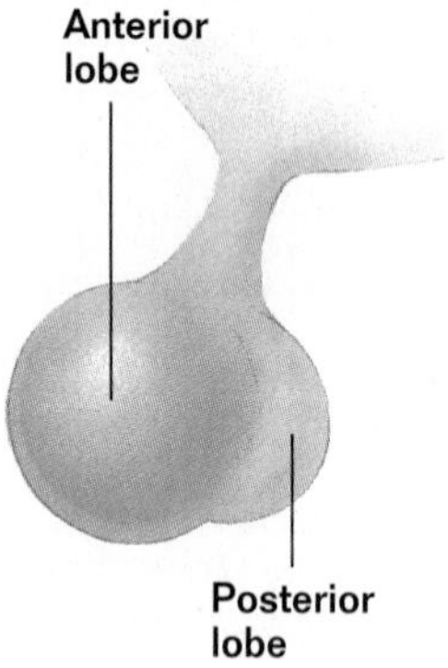	Somatotropin, or growth hormone (GH)	All tissues	Promotes protein synthesis; stimulates growth of muscles and bones
	Thyroid-stimulating hormone (TSH)	Thyroid gland	Stimulates production of thyroxine by the thyroid gland
	Posterior lobe		
	Antidiuretic hormone (ADH)	Kidneys, blood vessels	Stimulates reabsorption of water; constricts blood vessels
	Oxytocin	Mammary glands, uterus	Stimulates uterine contractions and milk secretion
Adrenal glands	**Cortex**		
	Aldosterone	All tissues	Controls salt (sodium and potassium) and water balance
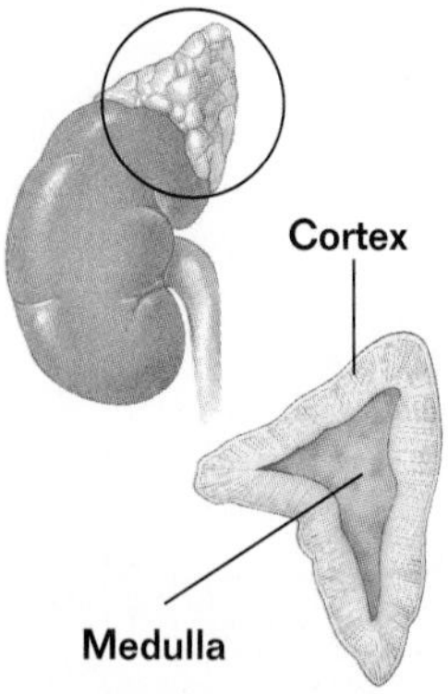	Cortisol	Kidneys	Stimulates metabolism of carbohydrates, lipids, and proteins; raises blood sugar
	Medulla		
	Epinephrine (adrenaline) and norepinephrine	Skeletal and cardiac muscle, blood vessels	Initiates the response to stress; increases metabolic rate, heart rate, and blood pressure; dilates blood vessels; raises blood sugar
Islets of Langerhans (Pancreas)	Glucagon	Liver, fatty tissues	Stimulates conversion of glycogen to glucose; raises blood sugar
	Insulin	All tissues	Stimulates conversion of glucose to glycogen; lowers blood sugar

Anterior lobe

Posterior lobe

Cortex

Medulla

Gland	Hormone	Target Tissue	Effects
Parathyroids	Parathyroid hormone	Bone tissue, digestive tract, kidneys	Stimulates breakdown of bone tissue and absorption of calcium by kidneys; raises blood calcium; activates vitamin D
Pineal	Melatonin	Uncertain, possibly ovaries and testes	May regulate biorhythms and moods; may affect the onset of puberty
Thyroid	Calcitonin	Bone tissue	Inhibits loss of calcium from bone; lowers blood calcium
	Thyroxine	All tissues	Raises metabolic rate; necessary for normal growth and development
Ovaries	Estrogen	All tissues, female reproductive structures	Controls development of secondary female sex characteristics and sex organs; initiates preparation of the uterus for pregnancy
	Progesterone	Uterus, breasts	Completes preparation of the uterus for pregnancy; stimulates breast development
Testes	Testosterone	All tissues, male reproductive structures	Controls development of secondary male sex characteristics and sex organs; stimulates sperm formation

Section Review

1. Which endocrine glands are actually two glands in one? Which of these glands produces the greatest number of hormones?
2. What is diabetes mellitus? How is it treated?
3. What are the consequences of an underproduction of thyroxine during childhood? How are adults affected by underproduction of thyroxine?
4. How is parathyroid hormone essential to your health?

Critical Thinking
5. Why is the pituitary gland often referred to as the master gland?

acromegaly (978)
adrenal cortex (980)
adrenal medulla (980)
adrenaline (980)
aldosterone (981)
anterior lobe (978)
calcitonin (979)
cortisol (975)
cyclic AMP (973)
diabetes mellitus (982)
endocrine gland (968)
endorphin (970)
estrogen (983)
gigantism (978)

glucagon (974)
goiter (979)
hormone-receptor complex (975)
hyperthyroidism (979)
hypothyroidism (979)
insulin (982)
islets of Langerhans (982)
melatonin (983)
neuroendocrine system (971)
neuromodulator (970)
neuropeptide (970)
norepinephrine (981)
parathyroid gland (979)
parathyroid hormone (979)

peptide hormone (969)
pineal gland (983)
pituitary gland (971)
posterior lobe (978)
prednisone (981)
progesterone (983)
prostaglandin (970)
second messenger (973)
steroid hormone (969)
target cell (969)
testosterone (983)
thyroxine (976)
Type I diabetes (982)
Type II diabetes (982)

Review

Multiple Choice

1. The chemical messengers of the endocrine system are
   a. neurons.
   b. blood cells.
   c. hormones.
   d. pheromones.

2. The nonendocrine chemical signals secreted by the brain
   a. are never derived from amino acids.
   b. regulate high blood pressure.
   c. include neuropeptides and neurotransmitters.
   d. do not travel through the bloodstream.

3. The production of hormones by glands of the endocrine system is initiated by the
   a. hypothalamus.
   b. nonendocrine glands.
   c. hormone-receptor complex.
   d. medulla of the adrenal glands.

4. Which of the following describes how a steroid hormone produces a response?
   a. The hormone binds to a receptor on the cell membrane.
   b. The hormone first binds to mRNA.
   c. A hormone-receptor complex binds to DNA.
   d. The hormone passes directly into the nucleus.

5. Peptide hormones may use cyclic AMP as a
   a. receptor.
   b. second messenger.
   c. target cell.
   d. coenzyme.

6. The receptor protein for thyroxine is located
   a. in the cytoplasm.
   b. on mRNA.
   c. on the cell membrane.
   d. in the nucleus.

7. Insulin promotes glucose uptake by cells, which leads to
   a. higher blood sugar.
   b. lower blood sugar.
   c. release of additional insulin.
   d. glycogen breakdown.

8. A pituitary tumor in a child could lead to
   a. hyperthyroidism.   c. Type I diabetes.
   b. gigantism.   d. hypothyroidism.

9. What adrenal cortex hormone acts to reduce inflammation?
   a. calcitonin   c. prostaglandin
   b. aldosterone   d. cortisol

10. Which of the following endocrine glands secretes melatonin and is believed to be involved in establishing biorhythms?
    a. pituitary gland   c. pineal gland
    b. thyroid gland   d. adrenal gland

Completion

11. Responses caused by _____ are slower but last longer than those caused by _____ .

12. _____ hormones are fat-soluble, and _____ hormones are water-soluble.

13. _____ hormones bind to receptors in the cytoplasm, while _____ hormones remain outside of cells.

14. The _____ pituitary releases hormones that are produced by the hypothalamus. Hormones secreted by the _____ pituitary induce other endocrine glands to secrete particular hormones.

15. Low blood sugar stimulates the release of _____ , while high blood sugar stimulates the release of _____ .

16. A treatment for Type _____ diabetes is proper diet and exercise, while treatment for Type _____ diabetes usually involves insulin injections.

17. Parathyroid hormone regulates _____ level in the blood, and aldosterone regulates _____ level in the blood.

Themes Review

18. **Homeostasis** Explain how your hypothalamus and endocrine glands work together to maintain homeostasis.

19. **Structure and Function** Describe the structure of an adrenal gland, and explain how it acts as two glands in one.

Critical Thinking

20. **Identifying Variables** Suppose that a friend tells you that he or she has recently experienced some of the warning signs of diabetes mellitus. What else could cause symptoms similar to diabetes mellitus. What questions could you ask your friend to determine whether there may be another cause for these symptoms?

21. **Making Inferences** Before iodide was added to table salt, goiters were common among people living in inland regions but were rare among people living in coastal areas. Why do you think this was so?

22. **Making Inferences** The graph below shows the blood glucose levels of three experimental rats, measured over time. At time T_1, two rats received injections of a saline solution plus one hormone, and the control rat was injected with a saline solution only. Which rat (A, B, or C) received insulin? Which rat received glucagon? Which rat was the control?

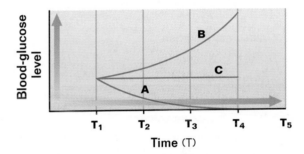

Activities and Projects

23. **Research and Writing** Urine tests that detect steroid use by athletes are being advocated by many high school coaches. Interview several coaches to determine their attitudes toward steroid testing. Write an article for your school newspaper that discusses your findings and explains how steroids affect the body and why doctors oppose their use.

24. **Cooperative Group Project** Using materials such as sturdy cardboard, paint, batteries, flashlight bulbs, wire, and brads, build a model that shows the location of the endocrine glands in the human body. Electrical circuits can be used to "light up" a light at a particular gland when a wire is touched to its name on a key. Display the model in your classroom where other students can use it as a study guide.

Effects of Thyroxine on Frog Metamorphosis

OBJECTIVE

Determine the effect of the hormone thyroxine on the development of tadpoles.

PROCESS SKILLS

- measuring and comparing anatomical features
- organizing data using tables and graphs

MATERIALS

- glass-marking pen
- 600 mL beakers (6)
- pond water
- 10 mL graduated cylinder
- 0.01% thyroxine solution
- strained spinach
- graph paper marked in 1 mm squares
- petri dish
- small fish net
- 9 tadpoles with hind legs just beginning to show
- 3 pencils in different colors

BACKGROUND

1. What is a hormone?
2. What is metamorphosis?
3. Describe the stages of frog development.
4. What is the importance of thyroxine in human development? What gland secretes thyroxine?
5. Write your **Focus Question** on your Vee Form.
6. **Knowing Side of the Vee** List the **Concepts** and new **Vocabulary Words** on your Vee Form. In the **Concepts Statements** section of the Vee, use these words in sentences that define and explain them.

TECHNIQUE

Doing Side of the Vee

1. Use a glass-marking pen to label three beakers A, B, and C. Add your initials to each.
2. Add 500 mL of pond water to each beaker.
3. Use a graduated cylinder to measure 10 mL of thyroxine solution, and add the solution to beaker A. Add 5 mL of thyroxine solution to beaker B. Add nothing to beaker C.
4. Add approximately 1 cu. cm of strained spinach to each beaker.
5. Place a sheet of graph paper, ruled side up, underneath a petri dish.
6. **CAUTION: You will be working with live animals. Be sure to treat them gently and follow directions carefully.** Catch a tadpole with a fish net and place the tadpole in the petri dish. Measure the tadpole's total length, tail length, and body length in millimeters by counting the number of squares it covers on the graph paper. Place the tadpole in beaker A.
7. Repeat step 6 with two more tadpoles. Average the total length, tail length, and body length of the three tadpoles. In the **Records** section, record these averages in a table similar to the table shown on the next page.
8. Repeat step 6 with six more tadpoles, placing three tadpoles in beaker B and three in beaker C.
9. Feed the tadpoles about 1 cu. cm of spinach every other day. Be careful to avoid overfeeding. Change the water every four days, adding thyroxine solution to beakers A and B in the original amounts.
10. Measure the tadpoles once a week for three weeks, and average the lengths of the tadpoles in each beaker. Record the average lengths in the table you

Measurement of Tadpole Growth

	Beaker A			Beaker B			Beaker C		
	Avg. total length	Avg. tail length	Avg. body length	Avg. total length	Avg. tail length	Avg. body length	Avg. total length	Avg. tail length	Avg. body length
Initial									
End of week 1									
End of week 2									
End of week 3									
Growth in 1st week									
Growth in 2nd week									
Growth in 3rd week									

constructed in the **Records** section of the Vee.

11. Calculate the average growth per week for each group of tadpoles. For example, the average growth in total length during the second week is equal to the average total length at the end of week 2 minus the average total length at the end of week 1. Record these values in the appropriate spaces in the table in the **Records** section of your Vee.

12. Graph your data using different colored pencils for the tadpoles in beaker A, beaker B, and beaker C. In the **Procedure** section of the Vee, briefly summarize the procedure you followed.

 13. Clean up your materials and wash your hands before leaving the lab.

INQUIRY

1. What is the purpose of putting some tadpoles in water without the thyroxine solution?

2. Why are three tadpoles used for each solution, rather than just one?

3. Use the information on the **Knowing Side** of the Vee to interpret your results from the **Doing Side,** and then write your **Knowledge Claim.** Write a **Value Claim** for this lab.

ANALYSIS

1. What is the effect of thyroxine on tadpole metamorphosis?

2. Which concentration of thyroxine caused the greatest visible change in the tadpoles?

3. How do average body length and tail length change during metamorphosis?

FURTHER INQUIRY

Write a **New Focus Question** that could be the point of a new investigation. The following is an example:

What is the effect of iodine on frog development?

BREAST CANCER

What's Really Behind the Epidemic?

The growth of cancer cells in the body may be related to exposure to certain chemicals, like those routinely applied as pesticides.

BY TRACEY COHEN

The following article presents one person's editorial viewpoint regarding possible conflicts of interest that can influence scientific research priorities. Read this article critically to determine whether the author provides sufficient factual evidence to support her opinion.

About 182,000 women in the United States will be found to have breast cancer this year. Nearly 2 million women have been diagnosed with the disease and perhaps another 1 million have it but do not yet know it. Standard medical treatments have had little impact on the long-range outcome of the disease. The death rate from metastatic breast cancer—disease that has spread to other parts of the body—has remained unchanged for over 40 years. Twenty five percent of women with metastatic breast cancer die within 5 years of their diagnosis and 40 percent die within 10 years. Unfortunately, what has changed is a woman's chance of developing breast cancer at some time in her life. In 1940, that chance was 1 in 20. Today the National Cancer Institute estimates that by the time a woman is 85 years old, her chance is 1 in 8.

What Causes Breast Cancer?

Scientists have identified several factors that appear to increase a woman's chance of getting breast cancer. Having a close relative like a mother or sister with the disease is one. Another risk factor is increased lifetime exposure to natural or synthetic estrogen. For instance, early puberty, late menopause, having a first child after age 30, or not having children at all increases a woman's exposure to estrogen. A diet high in fats may also put women at risk, but the actual role of dietary fat is still controversial.

Altogether known risk factors account for just 20 to 30 percent of all breast cancers. The majority of women with the disease have none of these. The lack of progress in treatment and the limited number of cases explained by known risk factors is both frightening and frustrating. As a result, some researchers now think that exposure to pollutants and other chemicals may be responsible for the breast cancer epidemic.

Is Breast Cancer an Environmental Disease?

In the growing body of evidence linking breast cancer to environmental causes, one group of compounds—organochlorines—is highly suspect. Organochlorines include DDT and other pesticides; PCBs, which are used in electrical transformers; and PVC, used to make many kinds of plastic. About 11,000 different organochlorines are used by industries. Hundreds of other organochlorines are formed as byproducts of industrial processes like bleaching paper pulp and disinfecting wastewater. Burning garbage containing plastics and other chlorinated materials also creates organochlorines.

Organochlorines are highly toxic and slow to degrade. Studies have found 177 different organochlorines in samples of fat, blood, semen, and mother's milk taken from people living in the United States and Canada.

Research shows that organochlorines can have a number of harmful effects on the body. Some organochlorines cause genetic mutations that lead to cancer. Some suppress the immune system, which may also lead to cancer. Others interfere with sex hormones, acting like synthetic estrogen. The International Agency for Research on Cancer reports over 100 organochlorine compounds cause cancer in humans or other animals.

One study supporting the link between breast cancer and environmental pollutants comes from Israel. Before 1976, the breast cancer death rate among Israeli women under age 44 was unusually high. At the same time, high concentrations of three pesticides—DDT, BHC, and lindane—were found in women's breast milk and in commercial milk and dairy products. In 1976, use of the three pesticides was banned. By 1978, DDT levels in breast milk dropped 43 percent, lindane levels dropped 90 percent, and BHC levels dropped 98 percent. Less than a decade later, the breast cancer rate for Israeli women under age 44 had fallen 30 percent. In the time period studied, Israel was the only one of 28 countries observed in which the death rate from breast cancer actually declined.

Pesticides in current use have also been linked to cancer, either in humans or in other animals. Atrazine, used by farmers to kill weeds, has been shown to cause mammary cancers in rodents. It also appears to increase the risk of ovarian cancer in women. Farmers in the United States apply between 70 million and 90 million pounds of atrazine to their fields each year.

Federal law requires that the National Cancer Institute expand its research program on preventing cancers caused by exposures to environmental or workplace carcinogens. But at a congressional budget hearing in October of 1993, NCI Director Samuel Broder testified that the NCI spent only 1% of its almost $2 million budget on environmental cancer studies, and he gave no indications that the NCI

planned to shift its priorities. NCI research will continue to emphasize the treatment and cure of cancer. Of the part of the budget given to researching causes of the disease, the two areas to receive the most emphasis will remain the links to diet and smoking.

Many health specialists have criticized this approach. They want additional funds devoted to research into identifying carcinogens and studying the link between exposure and the incidence of cancer. Some critics have speculated that the ties between the chemical companies and the organizations devoted to cancer research are too close. These critics point to the interlocking relationships among cancer researchers, policy makers, and industry representatives. They question appointments such as that in the 1980s of Armand Hammer, then president of Occidental Petroleum, to a position as chair of the President's Cancer Advisory Board.

The fear often expressed by critics is that lawmakers may be unduly influenced by lobbyists for the chemical industry. Critics also worry that chemical companies have a vested interest in protecting their own industry. Will chemical companies be willing to support research that may identify their products as probable causes of cancer?

Industry spokespeople, on the other hand, claim a role in helping to identify and solve the problem of carcinogens. They argue that the funds donated by the chemical companies for research are essential to finding both the causes and treatment of cancer. The chemical companies view their involvement as stemming from a natural, communal interest—not as a conflict of interest. From this viewpoint, everyone involved is focused on the problem but working from a different perspective.

The issue is complicated, but it is clear that the value of scientific research is its ability to confirm or deny the link between a substance and the incidence of cancer. The problem of how this research will be funded and directed is the issue. To be effective, a researcher needs a degree of independence, and critics of the current policy worry that researchers will be reluctant to "bite the hand that feeds them." As long as chemical companies are directly involved, many health specialists worry that the objectivity of the research may be compromised.

Tracey Cohen is a freelance writer specializing in science and environmental issues.

Analyzing the Issue

1. **Detecting Bias** Do you think the argument presented in the article is convincing? Why or why not? Where does the writer use opinions rather than facts to make a point?

2. **Get the Facts** Go to the library and find at least five books or articles on this subject. Make a list of facts you can use to challenge the writer's argument. Then write a counterargument

supporting a different point of view.

3. **Examine Social Consequences** Find out how many chemicals are made each year. How does the U. S. Environmental Protection Agency determine their safety or risk? Is this method adequate to protect public health and the environment? Explain.

CHAPTER 4

REPRODUCTION AND DEVELOPMENT

REVIEW

- meiosis (Section 6-3)
- mutation (Section 9-3)
- development (Section 19-3)
- ectoderm, mesoderm, and endoderm (Section 27-2)
- notochord and hollow dorsal nerve cord (Section 27-3)
- *Highlights: Reproductive System* (p. 848)
- HIV and AIDS (Section 39-3)
- hypothalamus (Sections 40-2 and 41-2)
- pituitary gland, estrogen, progesterone, and testosterone (Section 41-3)

42-1 The Male Reproductive System

*T*he male reproductive system produces the male gametes, or sperm, and prepares them for delivery to a female's body so that fertilization, or conception, can occur. Sperm are highly specialized carriers of genetic information. Formed after meiosis, sperm cells are haploid, which means that they have only 23 chromosomes instead of the 46 (diploid number) found in most cells of the human body. As you can see in Figure 42-1, sperm cells consist of a head with very little cytoplasm and a long tail. Digestive enzymes located in the head enable a sperm cell to penetrate an egg. Mitochondria located between the head and tail supply the ATP energy that a sperm cell needs to swim through a female's reproductive system to an egg.

Section Objectives

- Describe the structure of a sperm cell, and explain how it is adapted for its function.
- Discuss the two functions of the testes.
- Trace the path taken by sperm from the testes to the outside of the body.
- Explain how the structure of the penis enables it to deposit sperm inside the female reproductive tract.

Head   Mitochondria   Tail

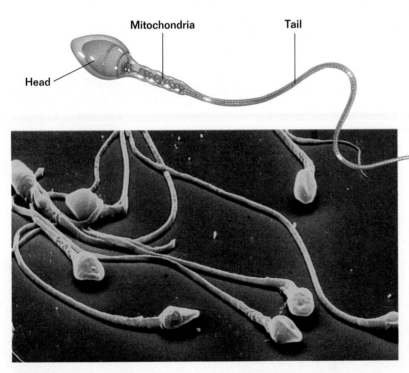

Figure 42-1 A diagram of a sperm cell, *left,* shows that it is a highly specialized cell that is able to move quickly through a female's reproductive system. The scanning electron micrograph, *bottom left,* shows human sperm at a high magnification.

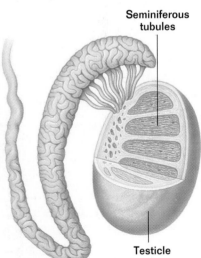

Seminiferous tubules

Testicle

Figure 42-2 A cutaway view of a testicle shows that the testes are filled with tiny coiled tubes, the sperm-producing seminiferous tubules.

Testes Are the Male Reproductive Organs

Two spherical **testes** (*TEHS teez*), or testicles, are the main organs of the male reproductive system. Each testicle contains hundreds of compartments that are packed with large numbers of tightly coiled tubes, as you can see in Figure 42-2. Sperm cells are produced within these tubes, called **seminiferous** (*sehm uh NIHF uhr uhs*) **tubules.** As you learned in

Chapter 41, the testes also secrete testosterone, the male sex hormone. Testosterone is secreted by cells that are scattered among the connective tissues between the seminiferous tubules. Two pituitary hormones regulate the functioning of the testes. Luteinizing hormone (LH) stimulates the interstitial cells to release testosterone. Follicle stimulating hormone (FSH) stimulates the seminiferous tubules to produce sperm.

The structure of the male reproductive system is illustrated in Figure 42-3. The testes are located outside the body cavity in the **scrotum** *(SKROHT uhm)*, a sac that hangs between the legs. First formed inside the body cavity, the testes descend into the scrotum either before or shortly after birth. Unlike other human body cells, sperm cannot successfully complete their development at the normal human body temperature of 37°C (98°F). In the scrotum, however, the temperature is about three degrees Celsius cooler than in the rest of the body.

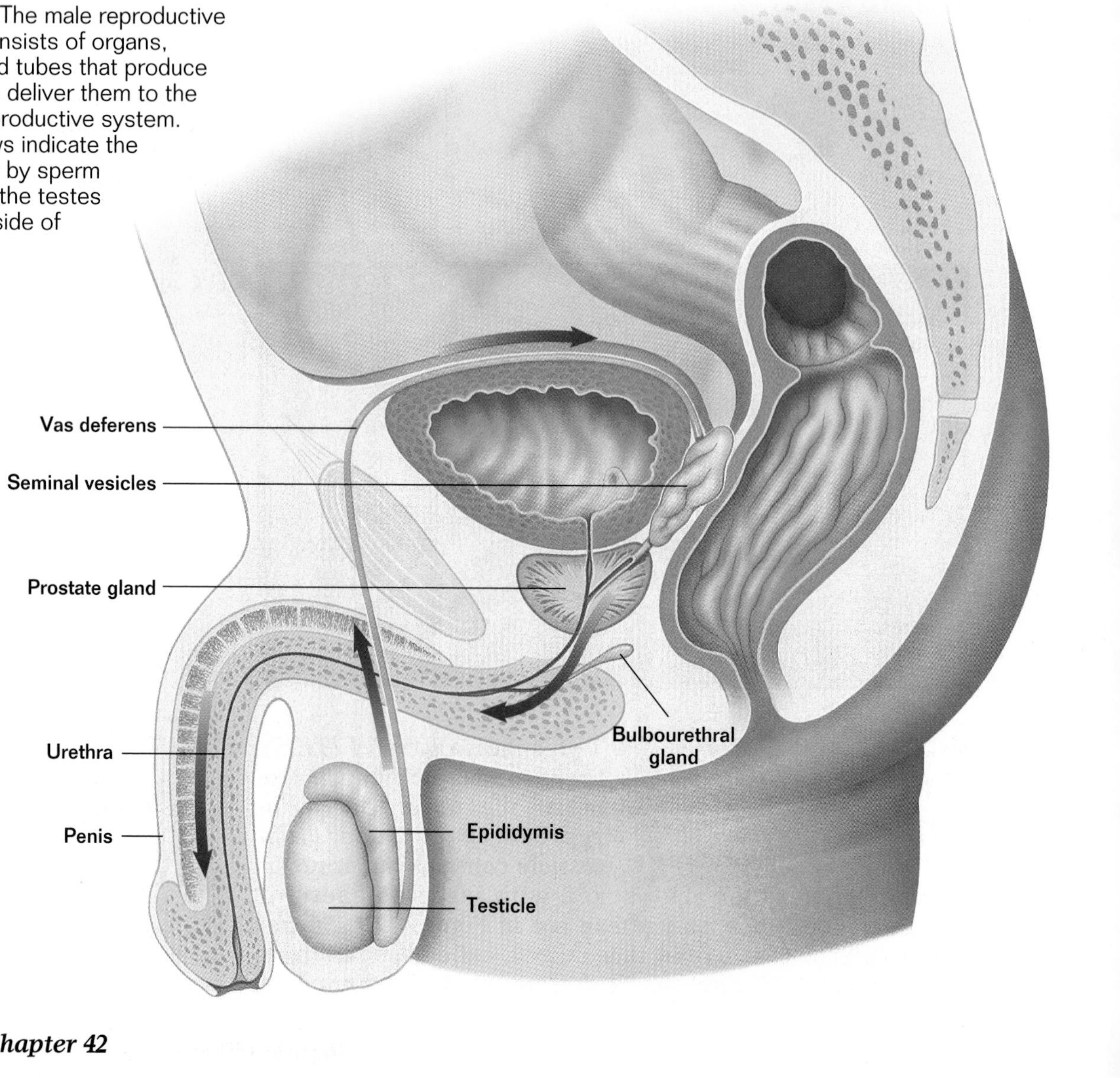

Figure 42-3 The male reproductive system consists of organs, glands, and tubes that produce sperm and deliver them to the female reproductive system. Blue arrows indicate the path taken by sperm cells from the testes to the outside of the body.

Vas deferens

Seminal vesicles

Prostate gland

Urethra

Penis

Bulbourethral gland

Epididymis

Testicle

Sperm Are Mixed With Fluids to Make Semen

A typical adult male produces several hundred million sperm each day of his life. After being produced in the seminiferous tubules, the sperm mature as they travel through a series of long tubes. During this journey, indicated by the arrows in Figure 42-3, fluids secreted by several glands are added to the sperm. These fluids nourish the sperm and lubricate the tubes through which they pass on their journey out of the body. Sperm that do not leave the body are broken down, and their materials are absorbed by the body to be recycled.

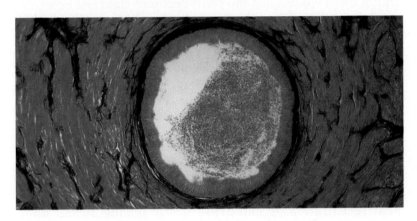

Figure 42-4 This light micrograph shows thousands of sperm cells within a human epididymis.

The full process of sperm production and maturation takes about two and a half months. After the sperm cells are manufactured within a testicle, they begin their long journey out of the body. First, the sperm are delivered to a long coiled tube called the **epididymis** *(ehp uh DIHD ih mihs)*, where they mature and develop the ability to become motile. Sperm do not actually become motile until after they leave the male body. Figure 42-4 shows sperm cells within the epididymis. From the epididymis, the sperm are delivered to another long tube, the **vas deferens** *(vas DEHF uh rehnz)*, where they are stored until they leave the body. Cutting or tying each vas deferens, an operation called a vasectomy *(va SEHK tuh mee)*, prevents sperm from leaving the male body. Sperm leave the body by passing through a tube that leads from the vas deferens to the urethra, where the reproductive and urinary tracts join.

As the sperm pass through the vas deferens and urethra to the outside of the body, the fluid secretions of several glands are added to them. **Seminal vesicles,** which lie between the bladder and the rectum, produce a fluid that is rich in sugars that nourish the sperm cells. The **prostate** *(PRAHS tayt)* **gland,** which is located just below the bladder, secretes an alkaline fluid that will counteract the acids produced by the female reproductive tract. The bulbourethral gland also secretes an alkaline fluid that is added to the mixture. **Semen** is a mixture of these secretions and sperm. ❑

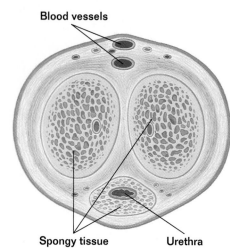

Blood vessels

Spongy tissue   Urethra

Figure 42-5 A cross section of a penis shows that it contains three cylinders of spongy tissue that run the length of the penis. When the spaces in the spongy tissue of these cylinders fill with blood, the penis becomes rigid and erect.

☐ CAPSULE SUMMARY

During ejaculation, the penis delivers sperm to the female reproductive tract. Fertilization, which usually requires a high sperm count, can be prevented by abstaining from sexual intercourse or by blocking the path of sperm as they leave the male body.

Sperm Are Delivered by the Penis

. . . .

The **penis** deposits sperm in the female reproductive tract during sexual intercourse. The penis must harden, or become erect, before it can enter the female reproductive tract. The structure of the penis enables this to happen. As you can see in Figure 42-5, the penis contains two cylinders of spongy tissue that lie side by side. Below and between them is a third cylinder of spongy tissue that surrounds the urethra. Small spaces separate the cells of these spongy tissues. When nerve impulses from the central nervous system cause the small arteries that lead into this tissue to expand, blood collects within these spaces, and the penis becomes rigid and erect.

Sperm exit the penis through **ejaculation,** which is the forceful expulsion of semen. During ejaculation, muscles encircling each vas deferens contract, moving the sperm they contain into the urethra. Eventually, the contractions of muscles at the base of the penis force semen out of the urethra. Only semen passes through the urethra at this time. After they are deposited in a female's body, sperm travel through the female reproductive tract toward the point where fertilization occurs. If, however, sperm are unable to reach an egg, fertilization cannot occur. One way to prevent fertilization is to block the path of the sperm. Covering the penis with a thin rubber sheath called a condom helps prevent fertilization by capturing sperm during sexual intercourse. Abstaining from sexual intercourse is the surest way to prevent fertilization.

About 3.5 mL of semen is expelled during ejaculation. This amount of semen normally contains several hundred million sperm. Still, the odds against any one sperm cell successfully completing the long journey to an egg and fertilizing it are extraordinarily high. Therefore, fertilization requires a high sperm count. Males with less than 20 million sperm per milliliter are generally considered sterile. ☐

Section Review

1. *Describe the structure of a sperm cell.*
2. *What are the two functions of the testes?*
3. *What path do mature sperm follow when they leave the vas deferens?*
4. *How does the structure of a penis enable it to become rigid?*

Critical Thinking

5. *If a male's left vas deferens were to become blocked, would you expect his sperm count to be normal, below normal, or zero? Explain your answer.*

42-2 The Female Reproductive System

*I*n addition to producing and maturing the female gametes, or eggs, the female reproductive system plays other important roles in reproduction and development. Fertilization occurs within specialized structures of the female reproductive system. This union of two haploid cells, as seen in Figure 42-6, restores the normal diploid number of 46 chromosomes. Following fertilization, the female reproductive system houses and nourishes a developing embryo (later called a fetus) through the period known as pregnancy.

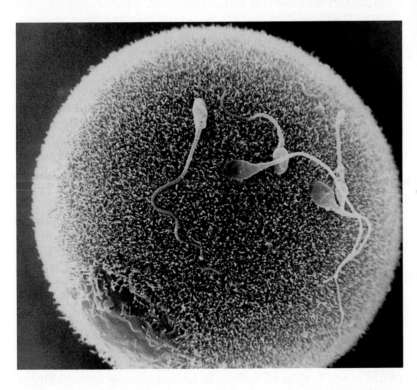

Figure 42-6 This photomicrograph shows five human sperm cells on a human ovum. During fertilization, an ovum and a sperm cell, both haploid (*n*) cells, fuse to form a zygote, which is a diploid (2*n*) cell.

Ovaries Produce Female Gametes

Two egg-shaped **ovaries** are the gamete-producing organs of the female reproductive system. Unlike testes, ovaries are located within the abdominal cavity. And while new sperm cells are constantly formed in males who have reached sexual maturity, females are born with all of the egg cells they will ever make. At birth, a female's ovaries contain some 2 million immature egg cells that have begun the first division of meiosis. Each of these cells has the ability to develop further but does not continue with meiosis. Instead, it waits to receive the proper signal, arrested in prophase of the first meiotic division.

After a female reaches **puberty,** or reproductive maturity, the first division of meiosis is able to resume. However, only one immature egg cell completes its development each month. The others remain in a developmental holding pattern. When they mature, the egg cells are called **ova** (singular, **ovum**). The Latin word *ovum* means "egg." Only about 400 of the immature egg cells that a woman has at birth will mature. The long period during which immature egg cells exist is one reason that developmental abnormalities occur with increasing frequency in the pregnancies of women over 35 years old. Immature egg cells are exposed to mutation-causing agents throughout a woman's life. After age 35, the odds that a harmful mutation will have occurred become high enough to significantly increase the incidence of abnormalities. There is evidence that aging in men also plays a role in developmental abnormalities, such as Down syndrome, that result from chromosomal defects. ■

□ *CAPSULE SUMMARY*

The ovaries, which are located in the abdominal cavity, produce the female gametes, or ova. At birth, a female already has all of the immature egg cells she will ever produce. At puberty, immature egg cells begin to mature one at a time.

Fertilization Occurs as an Ovum Journeys to the Uterus

When an ovum is released, it begins a journey that may end in pregnancy. The ovum's journey takes it through the organs of the female reproductive system, seen in Figure 42-7. First, cilia sweep the ovum into one of the **fallopian** *(fuh LOH pee uhn)* **tubes,** which lead from the ovaries to the uterus. The **uterus** *(YOO tuh ruhs)* is a hollow, muscular, pear-shaped organ about the size of a small fist. The inner wall of the uterus, which is called the **endometrium** *(ehn doh MEE tree uhm),* has two layers. The outer layer thickens and is shed monthly, while the inner layer generates another outer layer. Smooth muscles lining the fallopian tubes contract rhythmically. These contractions move the ovum down the tube to the uterus in much the same way that food moves through your intestines. An ovum's journey through a fallopian tube is slow, taking from three to four days to complete. If the ovum is not fertilized within 24 hours, it loses its capacity to develop.

To successfully fertilize an egg, a sperm must make its way far up a fallopian tube, a long journey that only a few complete. Sperm are first deposited within the **vagina** *(vuh JEYE nuh),* a muscular tube about 7 cm long that leads to the uterus. The entrance to the uterus from the vagina is called the **cervix** *(SUR vihks).* A soft rubber cap called a diaphragm *(DEYE uh fram)* can be used to cover the cervix and help prevent fertilization by blocking the passage of sperm into the uterus. A diaphragm is more effective when used with a sperm-killing chemical, or spermicide. After swimming through the vagina, the sperm pass through the cervix and uterus and then swim up the fallopian tubes. This journey requires the sperm to

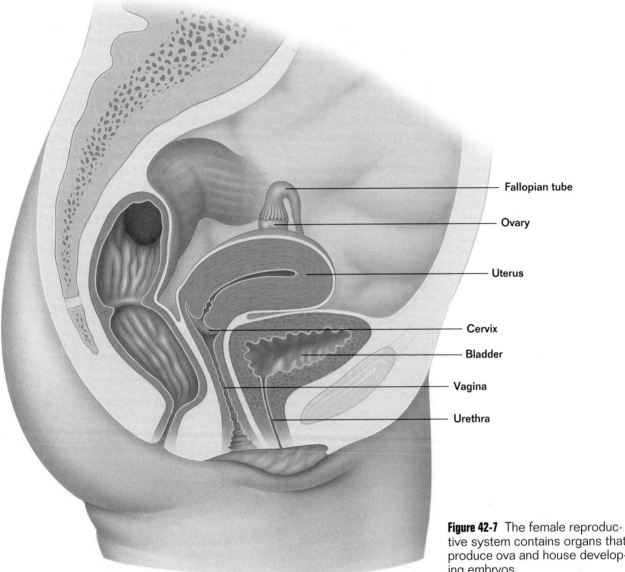

Figure 42-7 The female reproductive system contains organs that produce ova and house developing embryos.

Fallopian tube

Ovary

Uterus

Cervix

Bladder

Vagina

Urethra

expend a great deal of energy because they are constantly swimming against the current that carries the ovum to the uterus. Cutting or tying the fallopian tubes, an operation called a tubal ligation *(leye GAY shun)*, prevents fertilization by blocking the sperm's path to the ovum.

When a sperm succeeds in fertilizing an ovum high in the fallopian tube, the fertilized ovum begins to divide and continues on its journey to the uterus. After reaching the uterus, the fertilized ovum (now a hollow ball of cells) burrows into the thick, outer lining of the uterus, an event called **implantation.** Implantation, which marks the beginning of pregnancy, occurs about six days after fertilization. Following implantation, the fertilized ovum starts the long developmental process that eventually leads to the birth of a child. If fertilization does not occur, the outer layer of the endometrium and the unfertilized ovum are shed. ☐

☐ *CAPSULE SUMMARY*

Fertilization occurs in the fallopian tubes. Blocking the path of sperm to the fallopian tubes prevents fertilization. When a fertilized egg enters the uterus, it implants in the uterine wall.

Eggs Mature in the Ovarian Cycle

The ovaries prepare and release a mature ovum in a series of events called the **ovarian cycle,** which takes an average of 28 days to complete. The events of the ovarian cycle are under the control of hormones released by the pituitary gland, which is in turn regulated by hormones released from the hypothalamus. The ovarian cycle is composed of two distinct phases: the follicular phase and the luteal phase. The events of the ovarian cycle are summarized in Figure 42-8.

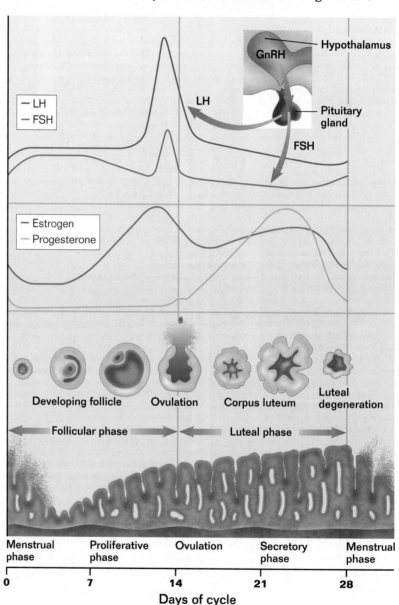

Figure 42-8 This chart summarizes the human ovarian and menstrual cycles. The ovarian cycle, *top,* is regulated by hormones produced by the hypothalamus and pituitary gland. The menstrual cycle, *bottom,* is regulated by the hormones estrogen and progesterone, which are produced by the follicle and corpus luteum.

Triggering the Maturation of an Egg Cell

During the follicular phase of the ovarian cycle, hormones produced by the pituitary gland regulate the completion of an ovum's development. Ova develop within **follicles,** which are clusters of cells that surround the immature egg cells in an ovary. The anterior lobe of the pituitary, after receiving a chemical signal (gonadotropin releasing hormone, or GnRH) from the hypothalamus, starts the cycle by secreting FSH and LH. FSH binds to receptors on the surface of the cells in an immature follicle and triggers the development of the follicle. Together, FSH and LH cause the follicle to begin producing estrogen, which aids in the maturation of the follicle and initiates the thickening of the uterine lining. Normally, only one follicle at a time is able to respond immediately to FSH. The FSH level falls before other immature follicles and egg cells can mature, thus usually allowing only one ovum to mature in every cycle.

The decrease in FSH level is achieved by a negative-feedback mechanism. Neither FSH nor LH carries this signal. Instead, estrogen acts as the messenger. The rising level of estrogen in the bloodstream is detected by the hypothalamus, which responds by commanding the pituitary to cut off the further production of FSH. Thus, the hypothalamus "shuts the door" on further egg development and ordinarily ensures that only one egg cell reaches the final stage of development each month.

A continued rise in the estrogen level initiates the thickening of the uterine lining and signals the end of the follicular phase of the ovarian cycle. The anterior pituitary responds to a high level of estrogen by greatly increasing its secretion of LH. The high level of LH causes the maturation of a follicle and its ovum to be completed and causes the wall of the follicle to burst. When the follicle bursts, the mature ovum within it is released in a process called **ovulation.** Ovulation, seen in Figure 42-9, sends the mature ovum on a journey toward possible fertilization.

Preparing the Body for Fertilization

The luteal phase of the ovarian cycle follows smoothly from the follicular phase. After ovulation, LH causes the ruptured follicle to fill in and become yellowish. The repaired follicle is called a **corpus luteum,** which comes from the Latin words *corpus,* meaning "body," and *luteum,* meaning "yellow." The corpus luteum soon begins to secrete progesterone, which then inhibits FSH and LH (a backup for estrogen in preventing further ovulation). Prescription pills containing synthetic estrogen and progesterone disrupt the ovarian cycle and prevent ovulation. Progesterone is the body's signal to prepare itself for fertilization. If fertilization occurs, the corpus luteum continues to produce progesterone for several weeks. Rising levels of progesterone initiate many physiological changes associated with pregnancy. If, however, fertilization does *not* occur soon after ovulation, production of progesterone slows and eventually ceases, marking the end of the luteal phase. ❑

Figure 42-9 In this series of photos showing ovulation, a mature ovum emerges from an ovary, *top* and *center,* and retreats toward a fallopian tube, *bottom.*

❑ *CAPSULE SUMMARY*

During the ovarian cycle, FSH from the pituitary gland causes an ovum to mature inside a follicle. After ovulation, LH converts a ruptured follicle to a corpus luteum.

The Menstrual Cycle Prepares the Uterus for Pregnancy

While changes occur in the ovaries during the ovarian cycle, changes also occur in the uterus. Look back at Figure 42-8 to see how the uterus changes during the ovarian cycle. These changes, which prepare the uterine lining for a possible pregnancy each month, are called the **menstrual cycle.** The term *menstrual* comes from the Latin word *mensis*, meaning "month." Like the ovarian cycle, the menstrual cycle is, on the average, a 28-day cycle. During the period before ovulation, increasing levels of estrogen initiate the thickening of the endometrium. After ovulation, high levels of estrogen and progesterone cause further thickening of the endometrium and maintain this thickened uterine lining. If pregnancy does not occur, however, the levels of estrogen and progesterone decrease. The decreasing levels of these hormones mark the end of the ovarian cycle and cause the thickened outer layer of the endometrium to be shed.

As the outer layer of the endometrium is shed, blood vessels are broken and bleeding results. A mixture of blood and discarded endometrial cells, the menstrual fluid, then leaves the uterus through the vagina. This process, called **menstruation,** usually occurs about 14 days after ovulation. Thus, the end of the menstrual cycle also marks the end of the luteal phase of the ovarian cycle. At the end of the female reproductive cycles, neither estrogen nor progesterone is being produced. In their absence, the pituitary is again stimulated to produce FSH and LH, thus starting another ovarian and menstrual cycle. Between the ages of 45 and 55, menstruation usually ceases. After this event, which is called **menopause,** a woman is no longer able to conceive. ∎

❑ CAPSULE SUMMARY

During the menstrual cycle, which occurs along with the ovarian cycle, estrogen and progesterone produced by the follicle and corpus luteum prepare the uterus for possible pregnancy. If pregnancy does not occur, estrogen and progesterone production stops, and the thickened uterine lining is shed during menstruation.

Section Review

1. *What are three functions of the female reproductive system?*
2. *How does egg production in females differ from sperm production in males, and how is this difference significant?*
3. *What roles do the hypothalamus and pituitary gland play in regulating the ovarian cycle?*
4. *What causes the lining of the uterus to thicken and then be shed during the menstrual cycle?*

Critical Thinking
5. *How could the maturation of an egg be blocked during the ovarian cycle?*

42-3 Human Development

*H*uman development takes an average of nine months from fertilization to birth. The nine months of pregnancy are often divided into three **trimesters,** or three-month periods. Much of the first trimester is devoted to the formation of a fetus, like the one seen in Figure 42-10. The development of the basic human body plan is essentially completed by the end of the first trimester. The second and third trimesters are devoted mainly to the growth of the fetus. As you study the course of this long developmental period, notice that the most crucial events of development occur very early in pregnancy, many of them before a woman knows that she is pregnant.

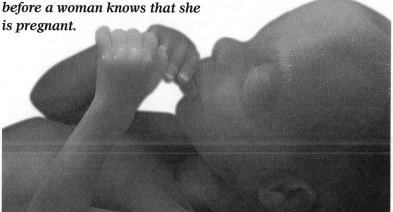

Figure 42-10 At 12 weeks old, this fetus, shown here in the uterus, already looks distinctly human.

The Embryo Forms During Early Development

Embryonic development in humans is a complex and dynamic process—a symphony of cell movement and change that starts with a single diploid cell. From this single cell, 100 trillion others will arise as the individual develops and grows to adulthood. After fertilization, the developmental journey begins with a series of rapid cell divisions that results in a ball of cells. Some of these cells then migrate to form a structure composed of the three primary tissue layers. The development of the specific tissues of the human body follows.

Week One: Cleavage

The first major event in the development of a human embryo is the rapid division of the zygote into a large number of cells—first two cells, then four, then eight, and so on. During this period of division, called **cleavage,** the overall size of the embryo does not increase; the cells simply become smaller. While cleavage is occurring, the embryo continues its

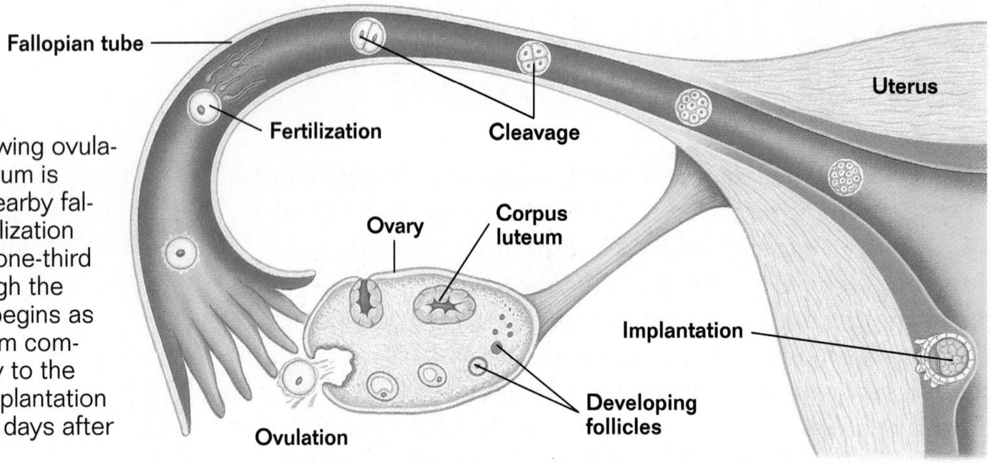

Figure 42-11 Following ovulation, a mature ovum is swept into the nearby fallopian tube. Fertilization occurs at about one-third of the way through the tube. Cleavage begins as the fertilized ovum completes its journey to the uterus, where implantation occurs about six days after fertilization.

Fallopian tube

Uterus

Fertilization

Cleavage

Ovary

Corpus luteum

Implantation

Developing follicles

Ovulation

journey down the fallopian tube, as seen in Figure 42-11. By the time the embryo reaches the uterus, it is a hollow ball of about 100 cells, called a **blastocyst.** On about the sixth day, the embryo begins to implant in the tissue of the uterine lining.

Week Two: Gastrulation

About eight days after fertilization, shortly after implantation, the embryo begins to grow rapidly, and the membranes that will later protect and nourish it begin to form. One of these membranes, the **amnion,** will enclose the developing embryo. Another embryonic membrane, the **chorion,** will interact with uterine tissue to form the **placenta,** which will nourish the growing embryo. To learn more about the placenta, look at Figure 42-12. As the placenta forms, the inner cells of the blastocyst begin a carefully orchestrated migration called **gastrulation.** This migration largely determines the future development of the embryo. By the end of gastrulation, the distribution of cells into the three primary tissue types—endoderm, mesoderm, and ectoderm—has been completed. The ectoderm is destined to form the epidermis and nerve tissue. The mesoderm is destined to form the connective tissue,

Figure 42-12 An embryo (later called a fetus) develops within the fluid-filled amniotic sac and is nourished by its mother through the placenta, *below*. Shaped like a pancake, the placenta forms where chorionic villi (extensions of the chorion containing fetal blood vessels) penetrate the endometrium. The umbilical cord connects the developing embryo (fetus) to the placenta. Food and oxygen pass from the mother's blood into the embryo's blood while the embryo's wastes pass into the mother's blood, but the mother's blood and embryo's blood remain separate.

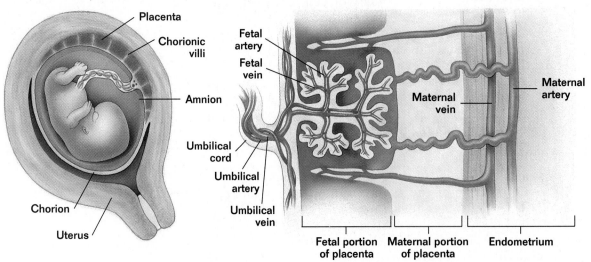

Placenta

Chorionic villi

Amnion

Chorion

Uterus

Fetal artery

Fetal vein

Maternal vein

Maternal artery

Umbilical cord

Umbilical artery

Umbilical vein

Fetal portion of placenta

Maternal portion of placenta

Endometrium

muscle, and vascular elements. The endoderm will form the lining of the organs in the gut. One of the characteristic features of all vertebrates—the notochord—also develops from mesoderm during gastrulation.

Week Three: Neurulation

In the third week, the three primary cell types begin to differentiate into the tissues and organs of the body. As in all vertebrates, differentiation begins with the formation of another characteristic feature, the hollow dorsal nerve cord. The process by which the hollow dorsal nerve cord forms from ectoderm is called **neurulation.** While neurulation is occurring, the rest of the basic architecture of the human body is being rapidly determined by changes in the mesoderm and endoderm. By the end of the third week, blood vessels and the gut have begun to develop. At this point, the embryo is about 2 mm (less than 0.1 in.) long.

Week Four: Organogenesis

In the fourth week, the body organs begin to form in a process called **organogenesis.** The developing heart begins a rhythmic beating that stops only at death. At an average of 70 beats per minute, the heart is destined to beat more than 2.5 billion times during a lifetime of about 70 years. The arm and leg buds also begin to form in the fourth week, as you can see in Figure 42-13. The embryo more than doubles in length during this week to about 5 mm. By the end of the fourth week of development, all of the major organs of the human body have begun to form. Although development is now far advanced, many women are not aware that they are pregnant at this stage.

Figure 42-13 A four-week-old human embryo, *above,* has a tail, arm and leg buds, eyes, and a heart that has begun to beat. The placenta and umbilical cord have also formed.

The Second Month: Morphogenesis

During the second month, the final stage in the formation of a human embryo takes place. During **morphogenesis,** the miniature limbs of the embryo assume their adult shapes. Within the body cavity, the major internal organs are evident, including the liver and pancreas. By the end of the second month, the embryo is about 22 mm (less than 1 in.) in length, weighs about 1 g, and is beginning to look distinctly human.

Because so much is happening and the proper course of events can be easily interrupted, early pregnancy is a very crucial time in development. Most miscarriages occur during this period. Alcohol use by pregnant women, especially during early pregnancy, is a leading cause of birth defects, such as the one seen in Figure 42-14. **Fetal alcohol syndrome,** or **FAS,** is a common birth defect (affecting 1 in 750 newborns in the United States), in which a baby is born with a deformed face and often severe mental and physical retardation. While alcohol should be avoided throughout pregnancy, it is particularly important for women to avoid alcohol if they are planning to become pregnant. Organogenesis can also be upset during the first months of pregnancy if the mother contracts rubella (German measles).

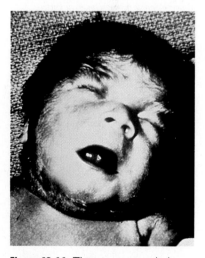

Figure 42-14 This prematurely born infant has several of the signs of fetal alcohol syndrome (FAS), including an abnormally small head and widely spaced eyes.

The Fetus Grows Rapidly Until Birth

The body plan of a human embryo is essentially complete at eight weeks. From this point on, a developing human is referred to as a **fetus.** By the end of the first trimester, all of the major body organs have differentiated, and the sex of the fetus has been established. During the second and third trimesters, the fetus grows rapidly as its organs are completed and become functional. Table 42-1 describes some of the major events of fetal development.

Table 42-1 Events of Human Fetal Development

Stage	Major Events
First Trimester (0 to 3 months)	Fertilization, cleavage, implantation, gastrulation, neurulation, and organogenesis occur as the embryo becomes a fetus; all major organ systems are formed; the fetus begins to move, but movements cannot be felt
Second Trimester (4 to 6 months)	Skin and hair grow; eyes blink; fetal movements can be felt; arms and legs reach final proportions; heartbeat can be heard
Third Trimester (7 to 9 months)	Substantial increase in size; skin is red and wrinkled; development of the lungs is completed; fingernails and toenails grow; fetus can survive if born during this stage

By the end of the third trimester, a fetus is able to exist outside of its mother's body. Yet the neurological development of the fetus is far from complete. At the end of the third trimester, the fetus is about as large as it can safely be delivered without damage to the mother or child. Physical growth and neurological development continue after birth. ◼

☐ CAPSULE SUMMARY

During the first eight weeks after fertilization, a developing human is called an embryo. After that, the embryo is called a fetus. Human development is most easily disturbed during the first trimester of pregnancy. Growth and neurological development continue after birth, which occurs about nine months after fertilization.

Section Review

1. *What events in human development occur during the first month after fertilization?*
2. *What is the placenta, and what is its function?*
3. *When does an embryo become a fetus?*
4. *At what stage of development can a fetus survive if it is born?*

Critical Thinking
5. *Why can certain drugs so profoundly affect an embryo's development when they are ingested or injected by a woman during early pregnancy?*

42-4 Sexually Transmitted Diseases

Disease-causing agents travel from one host to another in many ways. When certain disease-causing microbes are present in human body fluids, they can be passed from one person to another when these fluids are exchanged during sexual activity. Diseases that are spread by sexual contact are called **sexually transmitted diseases**, or **STDs**. Both viruses and bacteria can cause STDs.

Viral STDs Cannot Be Cured With Antibiotics

STDs that are caused by viruses are called viral STDs. Because viruses are not affected by antibiotics, viral STDs cannot be treated and cured with antibiotics. Two viral diseases that can be transmitted by sexual activity are AIDS and genital herpes.

Acquired Immune Deficiency Syndrome

Acquired immune deficiency syndrome (AIDS) is a fatal disease caused by the human immunodeficiency virus (HIV). As you learned in Chapter 39, transmission during sexual intercourse is the most common way that people get AIDS. However, HIV can be transmitted in many other ways, and AIDS, therefore, is not exclusively a sexually transmitted disease. HIV destroys the immune system of infected individuals. AIDS patients die from infections and cancers that a healthy immune system normally defeats. The graph in Figure 42-15 shows how AIDS cases among young adults have increased over the last decade. AIDS is now the leading killer of American men between the ages of 25 and 44. In addition to the known AIDS cases and HIV-positive individuals, some 1.5 million more people in the United States are thought to be infected—most unknowingly—with HIV.

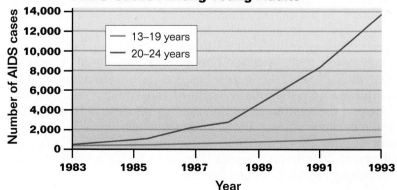

Figure 42-15 As this graph shows, the number of AIDS cases among teenagers and young adults rose sharply from 1983 to 1993. Although the greatest increase in AIDS cases occurred among 20- to 24-year-old individuals, you should keep in mind that most of these individuals contracted HIV during their teens.

Genital Herpes

Genital herpes is caused by a herpes simplex virus (HSV). The symptoms of genital herpes include periodic outbreaks of painful blisters in the genital area, as seen in Figure 42-16, and flulike aches and fever. Two types of HSV can cause genital herpes. About 80 percent of genital herpes infections are caused by HSV-2. The rest are caused by HSV-1, which more commonly causes cold sores, or fever blisters, around and in the mouth. Antiviral drugs can temporarily eliminate the blisters, but they cannot eliminate HSV from the body. Although genital herpes is not life threatening, it can have serious consequences. Women with genital herpes have a greater risk of developing cervical cancer. Also, like HIV, HSV can be passed from mother to fetus during pregnancy or birth. A baby infected with HSV may suffer severe damage to its nervous system or even die as a result of the infection.

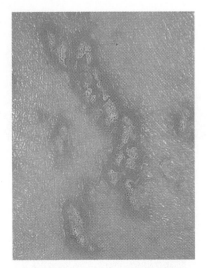

Figure 42-16 Painful blisters such as these are a symptom of genital herpes. In males, genital herpes blisters may appear on the penis, scrotum, or skin near the genitals. In females, genital herpes blisters may develop on the labia (external parts of the genital area), within the vagina, or on the skin near the genital area.

Bacterial STDs Can Be Difficult to Detect

STDs that are caused by bacteria are called bacterial STDs. Unlike viral STDs, bacterial STDs can be successfully treated and cured with antibiotics. Unfortunately, the early symptoms of most bacterial STDs are very mild and often go undetected. Routine laboratory tests that detect bacterial STDs are available at hospitals and clinics. Early detection and treatment are necessary to prevent serious consequences that can result from infection. For example, untreated bacterial STDs can cause sterility in both men and women. One of the most common causes of infertility (inability to become pregnant) in women is **pelvic inflammatory disease,** or **PID.** PID is a severe inflammation of the uterus, ovaries, fallopian tubes, or pelvic cavity that is usually caused by a bacterial STD. Three major bacterial STDs are syphilis, gonorrhea, and chlamydia.

Syphilis

Syphilis (*SIHF uh lihs*) is a serious bacterial STD that usually begins with the appearance of a small, painless ulcer called a **chancre** (*SHAHNG kuhr*) about two to three weeks after infection. In males, the chancre usually appears on the penis. In females, the chancre may form inside the vagina or on the cervix, making it difficult to detect. If syphilis is not treated, it may cause fever, swollen lymph glands, or a rash, like that seen in Figure 42-17, a few weeks after infection. These symptoms also disappear without treatment. Years later, however, the syphilis infection may cause destructive lesions on the nervous system, blood vessels, bones, and skin. A pregnant woman infected with syphilis can also transmit it to her unborn child, who as a result may be stillborn or suffer serious damage to major organ systems.

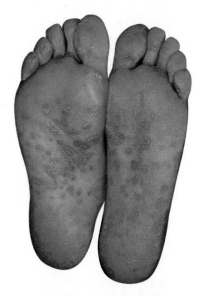

Figure 42-17 A rash such as the one on these feet is a symptom of the second stage of syphilis. A person with syphilis may also develop a fever and swollen lymph glands. Even at this stage, syphilis can easily be cured by treatment with antibiotics.

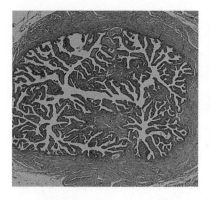 

Figure 42-18 A normal fallopian tube, *left,* has a highly folded lining and many spaces through which gametes can pass. In a fallopian tube scarred by PID, *right,* many of these spaces have become filled with tissue, blocking the path of gametes and decreasing the likelihood of fertilization.

Gonorrhea

Gonorrhea *(gahn uh REE uh)* is a bacterial STD that causes painful urination and pus discharge from the penis in males. In females, it sometimes causes a vaginal discharge but more often has no symptoms. In males, untreated gonorrhea can spread to the vas deferens, epididymis, or testes. In females, it can spread to the fallopian tubes and cause scarring that results in infertility. Some strains of gonorrhea are resistant to the more commonly used antibiotics, such as penicillin.

Chlamydia

Chlamydia is the most common bacterial STD in the United States. Between 3 million and 10 million new cases occur each year. The symptoms of chlamydia are similar to those of a mild case of gonorrhea: painful urination in males and vaginal discharge in females. Like gonorrhea, chlamydia often produces no symptoms and often is not diagnosed. Chlamydia, even more so than gonorrhea, can cause scar tissue to form on infected fallopian tubes and close them, leading to infertility. Most cases of PID are the result of gonorrhea or chlamydia infections. Figure 42-18 shows the damage that can be caused by pelvic inflammatory disease. ❏

❏ CAPSULE SUMMARY

Diseases that can be spread by sexual contact are called sexually transmitted diseases (STDs). STDs caused by viruses cannot be cured with antibiotic treatment. STDs caused by bacteria are difficult to detect but can be cured with antibiotic treatment.

Section Review

1. *What are two sexually transmitted diseases caused by viruses?*

2. *Name three symptoms of genital herpes.*

3. *What are three common STDs caused by bacteria? Why is early detection of these diseases important?*

4. *What is one main difference between viral STDs and bacterial STDs?*

Critical Thinking

5. *How can you best protect yourself from contracting an STD?*

amnion (1004)
blastocyst (1004)
cervix (998)
chancre (1008)
chlamydia (1009)
chorion (1004)
cleavage (1003)
corpus luteum (1001)
ejaculation (996)
endometrium (998)
epididymis (995)
fallopian tube (998)
fetal alcohol syndrome, FAS (1005)
fetus (1006)
follicle (1001)

gastrulation (1004)
genital herpes (1008)
gonorrhea (1009)
implantation (999)
menopause (1002)
menstrual cycle (1002)
menstruation (1002)
morphogenesis (1005)
neurulation (1005)
organogenesis (1005)
ova, ovum (998)
ovarian cycle (1000)
ovary (997)
ovulation (1001)
pelvic inflammatory disease, PID (1008)

penis (996)
placenta (1004)
prostate gland (995)
puberty (998)
scrotum (994)
semen (995)
seminal vesicles (995)
seminiferous tubules (993)
sexually transmitted disease, STD (1007)
syphilis (1008)
testes (993)
trimester (1003)
uterus (998)
vagina (998)
vas deferens (995)

Review

Multiple Choice

1. The correct pathway of sperm is
   a. testes to vas deferens to epididymis.
   b. epididymis to urethra to vas deferens.
   c. testes to epididymis to vas deferens.
   d. urethra to vas deferens to testes.

2. Which of the following is *not* a function of the female reproductive system?
   a. production of gametes
   b. nourishment of the fetus
   c. maturation of eggs
   d. secretion of FSH

3. Fertilization normally takes place in the
   a. fallopian tubes.   c. epididymis.
   b. cervix.   d. vas deferens.

4. The maturation of an ovum is ultimately controlled by GnRH released by the
   a. hypothalamus.   c. fallopian tubes.
   b. corpus luteum.   d. follicles.

5. The follicular phase of the ovarian cycle
   a. occurs when LH levels drop to zero.
   b. starts when fertilization occurs.
   c. causes estrogen levels to drop to zero.
   d. ends when ovulation occurs.

6. An embryo develops endoderm, mesoderm, and ectoderm during
   a. cleavage.   c. neurulation.
   b. gastrulation.   d. organogenesis.

7. Which of the following is *not* true for human development?
   a. Alcohol and drugs taken at any time during pregnancy may damage the embryo or fetus.
   b. Most miscarriages occur during the first trimester of pregnancy.
   c. Birth defects are never the result of drugs and alcohol taken by parents prior to pregnancy.
   d. Normal development may be affected by viral diseases.

8. Morphogenesis occurs during the
   a. first week of development.
   b. second month of development.
   c. third trimester.
   d. fourth week of development.

9. Which of the following sexually transmitted diseases cannot be treated with antibiotics?
   a. genital herpes   c. gonorrhea
   b. syphilis   d. chlamydia

10. A symptom associated with the earliest stage of syphilis is
   a. painful urination.
   b. blisters in the genital area.
   c. fever blisters and cold sores.
   d. a painless chancre.

Completion

11. The testes contain seminiferous tubules, in which the _____ are produced, and cells that secrete the hormone _____ .

12. A male produces _____ throughout his life, but all a female's _____ are produced during fetal development.

13. In the human female, timely reductions in the FSH level will ensure that one _____ will mature in each cycle.

14. The discharge of endometrial cells mixed with blood is called _____ .

15. Organogenesis begins in week _____ following fertilization.

16. AIDS is a disease of the _____ system, apparently caused by a _____ .

17. Two bacterial STDs that cause sterility in females are _____ and _____ .

Themes Review

18. Structure and Function Human sperm cells have a tail that whips back and forth and a head that contains digestive enzymes. How do these structures enable a sperm to fertilize an ovum?

19. Homeostasis FSH controls final egg development. How is the FSH level regulated by estrogen in the female body? What could happen if the FSH level increases and remains high?

20. Levels of Organization During the first four weeks of development, the human embryo undergoes many changes. Describe the changes that occur during each of the first four weeks of development, and identify the names associated with these weekly changes.

Critical Thinking

21. Making Inferences A man interested in fathering children has a sperm count of over 60 million for a 3.5 mL sample of semen. As his physician, what would you tell him about the results of the test?

22. Making Predictions Data on levels of pituitary and ovarian hormones collected from two women are presented in the table below. Based on the data, which woman could have conceived two or three days after the day the data were collected? Explain.

Hormone	Woman 1	Woman 2
FSH	high	low
LH	high	high
Estrogen	high	moderate
Progesterone	low	high

Activities and Projects

23. Research and Writing Research the causes of infertility, and find out about different biotechnological methods that assist conception. Relate your findings in a written report.

24. Research and Writing Some scientists do not believe that AIDS is caused by HIV. Look up and read articles by Peter Duesberg and others on this subject. In a written report, evaluate the arguments presented on both sides of this controversial issue.

25. Multicultural Perspective Almost every culture in the world celebrates puberty, a child's transition into adolescence. Research the puberty rites of two cultures from different parts of the world. What similarities and differences do they have? Relate your findings in a written report.

Embryonic Development

Doing Side of the Vee

1. Most members of the animal kingdom (including both sea stars and humans) begin life as a single cell — the fertilized egg, or zygote. The early stages of development in different species are quite similar. The zygote divides many times during cleavage, the new cells begin to specialize as they become part of specific tissues, and complex structures are formed as the embryo grows into a fully developed organism. Similarities and differences in these early stages of development reflect evolutionary relationships among species.

2. Obtain a set of prepared slides showing sea star eggs at different stages of development. Choose slides labeled unfertilized egg, zygote, 2-cell stage, 4-cell stage, 8-cell stage, 16-cell stage, 32-cell stage, 64-cell stage, blastula, early gastrula, middle gastrula, late gastrula, and young sea star larva. (Note: A blastula is the general term for the embryonic stage resulting from cleavage. A blastocyst in mammals is a modified form of the blastula.) Examine each slide under a low-power compound microscope. For each slide, focus on one good example of the developmental stage listed on the label. Then switch to high power.

3. Draw a diagram of each developmental stage you examine (in chronological order) in the **Records** and **Additional Records** sections of your Vee Form. Label each diagram with the name of the stage it represents and at what magnification it was observed. Record your observations as soon as they are made. Do not recopy your diagrams. Draw what you see. Lab drawings do not need to be artistic or elaborate. They should be well organized and include specific detail.

OBJECTIVES

- Identify the stages of early animal development.
- Describe the changes that occur during early development.

PROCESS SKILLS

- to observe prepared slides of sea star embryonic development
- to communicate microscopic observations through drawings

MATERIALS

- **prepared slides of sea star development**
- **compound microscope**
- **paper and pencil**

BACKGROUND

1. How does a fertilized egg become a completely developed organism?

2. What are the early stages (first four weeks) in the development of a human embryo?

3. Describe what happens during cleavage.

4. What are the three primary tissue types resulting from gastrulation, and what will they form in the completely developed individual?

5. Write your own **Focus Question** on your Vee Form.

6. **Knowing Side of the Vee** List the **Concepts** and new **Vocabulary Words** on your Vee Form. In the **Concept Statements** section of the Vee, use these words in sentences that define and explain them.

4. Compare your diagrams to the diagrams of human embryonic stages shown below. In the **Procedure** section of the Vee, briefly summarize the procedure you followed.

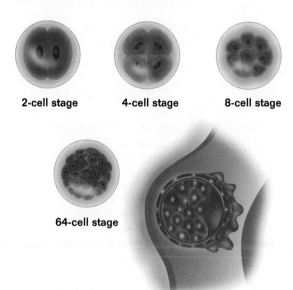

2-cell stage 4-cell stage 8-cell stage

64-cell stage

Blastocyst

 5. Clean up your materials and wash your hands before leaving the lab.

INQUIRY

1. Compare the size of the zygote with that of the blastula. At what stage does the embryo become bigger than the zygote?

2. By what stage do all of the cells in the embryo not look exactly like each other?

3. How do cell shape and size change during successive stages of development?

4. Are the cell nuclei the same size, larger, or smaller as the stages progress?

5. Use the information on the **Knowing Side** of the Vee to interpret your results from the **Doing Side,** and then write your **Knowledge Claim.** Write a **Value Claim** for this lab.

ANALYSIS

1. Why are sea star eggs a good choice for the study of embryonic development?

2. Compare the number of chromosomes in a fertilized sea star egg with the number of chromosomes in one cell of each of the following: 2-cell stage, blastula, gastrula, and adult organism.

3. From your observations of changes in cellular organization, why do you think the blastocoel (the space in the center of the hollow sphere of cells of a blastula) is important during embryonic development?

4. Label the endoderm and ectoderm in your drawing of the late gastrula stage. Describe the developmental fates of these two tissue types.

5. How is the symmetry of a sea star embryo and larva different from the symmetry of an adult sea star? Would you expect to see a similar change in human development?

6. In what ways can sea star embryos be used to study early human development?

7. Describe one way in which the cleavage of echinoderms and mammals is identical.

FURTHER INQUIRY

Write a **New Focus Question** that could be the point of a new investigation. The following is an example:

How is embryonic development in other organisms, such as birds, similar to or different from embryonic development in sea stars?

Careers

The following list provides the names and addresses of organizations that can supply information on a broad range of careers that can be pursued with a background in biology.

Agricultural Scientist
American Society of Agronomy
Crop Science Society of America
Soil Science Society of America
677 South Segoe Road
Madison, WI 53711

AIDS Information Specialist
New York Department of Health
Office of Public Health,
  AIDS Institute
Empire State Plaza
Albany, New York 12237-0684

Biochemist
American Society for Biochem-
  istry and Molecular Biology
9650 Rockville Pike
Bethesda, MD 20814

Biological Scientist
American Institute of
  Biological Sciences
730 11th Street NW
Washington, DC 20001

Biomedical Engineer
Biomedical Engineering Society
PO Box 2399
Culver City, CA 90231

Biomedical Researcher
Federation of American
  Societies for
  Experimental Biology
9650 Rockville Pike
Bethesda, MD 20814

Botanist
Botanical Society of America
Department of Genetics
Ohio State University
1735 Neil Avenue
Columbus, OH 43210

Cardiovascular Technologist
Society of Vascular Technology
1101 Connecticut Ave NW,
  Suite 700
Washington, DC 20036-4303

Dentist/Dental Assistant
American Dental Association
211 East Chicago Avenue
Chicago, IL 60611

Diagnostic Medical Sonographer
Society of Diagnostic
  Medical Sonographers
12770 Coit Road, Suite 508
Dallas, TX 75251

Ecologist
Ecological Society of America
Center for Environmental
  Studies
Arizona State University
Tempe, AZ 85287-3211

Emergency Medical Technician/Paramedic
National Association of
  Emergency Medical
  Technicians
9140 Ward Parkway
Kansas City, MO 64114

Entomological Inspector
Entomological Society
  of America
9301 Annapolis Road
Lanham, MD 20706

Environmental Engineer
American Society
  of Sanitary Engineering
PO Box 40362
Bay Village, OH 44140

Epidemiologist
Epidemiology Program Office
MS C08, Centers for Disease
  Control and Prevention
Atlanta, GA 30333

Forester
Society of American Foresters
5400 Grosvenor Lane
Bethesda, MD 20814

Forensics Expert
American Academy of
  Forensic Sciences
218 East Cache La Poudre
Colorado Springs, CO 80901-0669

Genetic Counselor
National Society of Genetic
  Counselors
233 Canterbury Drive
Wallingford, PA 19086

Geneticist
Genetics Society of America
9650 Rockville Pike
Bethesda, MD 20814

Health Inspector
Public Health Service
Department of Health and
  Human Services
200 Independence Avenue SW
Washington, DC 20201

Horticultural Therapist
American Horticultural
  Therapy Association
9200 Wightman Road, Suite 400
Gaithersburg, MD 20879

Landscaper
National Landscape Association
1250 I Street NW, Suite 5000
Washington, DC 20005

Medical Technologist
American Society for
  Medical Technology
2021 L Street NW, Suite 400
Washington, DC 20036

Medical Illustrator
Association of Medical
  Illustrators
1819 Peachtree Street NE,
  Suite 560
Atlanta, GA 30309

Medical Record Administrator
American Health Information
  Management Association
919 N. Michigan Ave., Suite 1400
Chicago, IL 60611-1683

Nurse
American Nurses Association
2420 Pershing Road
Kansas City, MO 64108

Occupational Therapist
American Occupational Therapy
  Association
1383 Piccard Drive
PO Box 1725
Rockville, MD 20849-1725

Oceanographer
Scripps Institute of
  Oceanography
A-033 University of California,
  San Diego
La Jolla, CA 92093

Optometrist
American Optometric
  Association
243 North Lindbergh Boulevard
St. Louis, MO 63141

Parasitologist
American Society of
  Parasitologists
1041 New Hampshire Street
Lawrence, KS 66044

Park Ranger
National Recreation and
  Park Association
2775 South Quincy Street,
  Suite 300
Arlington, VA 22206

Pharmacologist
American Society for
  Pharmacology
9650 Rockville Pike
Bethesda, MD 20814

Physical Therapist
American Physical Therapy
  Association
1111 North Fairfax Street
Alexandria, VA 22314

Physician
American Medical Association
515 North State Street
Chicago, IL 60610

Physiologist
American Physiological Society
9650 Rockville Pike
Bethesda, MD 20814

Pollution-Control Technician
Environmental Protection
  Agency
401 M Street SW
Washington, DC 20005

Public Health Microbiologist
American Society for
  Microbiology
1325 Massachusetts Avenue NW
Washington, DC 20005

Radiation Therapy Technologist
American Society of
  Radiological Technologists
15000 Central Avenue SE
Albuquerque, NM 87123

Respiratory Therapist
American Association for
  Respiratory Care
11030 Ables Lane
Dallas, TX 75229

Science Writer and Editor
National Association of Science
  Writers and Editors
PO Box 294
Greenlawn, NY 11740

Science Teacher
National Science Teachers
  Association
1742 Connecticut Avenue NW
Washington, DC 20009

Soil Scientist
U.S. Department of Agriculture
Soil Conservation Office
PO Box 2890
Washington, DC 20013

Surgical Technologist
Association of Surgical
  Technologists
8307 Shaffer Parkway
Littleton, CO 80127

Toxicologist
Society of Toxicology
1101 14th Street NW, Suite 1100
Washington, DC 20005

Veterinarian
American Veterinary Medical
  Association
930 North Meacham Rd.
Schaumburg, IL 60196

Virologist
American Society for
  Microbiology
1325 Massachusetts Ave. NW
Washington, DC 20005

Wildlife Manager
U.S. Department of Interior
Fish and Wildlife Service
Mail Stop 100 Arlington Square
Washington, DC 20240

Wood Scientist
Society of Wood Science and
  Technology
One Gifford Pinchot Drive
Madison, WI 53705

Zoo Curator
American Association of
  Zoological Parks and
  Aquariums
Oglebay Park, Route 88
Wheeling, WV 26003

Zoologist
American Society of Zoologists
Box 2739
California Lutheran University
Thousand Oaks, CA 91360

Measurement

All measurements in this book are expressed in metric units. Scientists throughout the world use the metric system, and you will always use metric units when you make measurements in the laboratory. The official name of the measurement system is the Système International d'Unités, or International System of Measurements. It is usually referred to simply as SI.

The metric system is a decimal system, that is, all relationships between units of measurement are based on powers of 10. Most units have a prefix that indicates the relationship of that unit to the base unit. For example, a meter equals 100 cm, or 1,000 mm. The lists below show the most commonly used prefixes as well as the main units used for each type of measurement.

Metric Prefixes

Prefix	Symbol	Factor of Base Unit
giga	G	1,000,000,000
mega	M	1,000,000
kilo	k	1,000
hecto	h	100
deka	da	10
deci	d	0.1
centi	c	0.01
milli	m	0.001
micro	µ	0.000001
nano	n	0.000000001

Length

1 kilometer (km) = 1,000 meters

1 meter (m) = base unit of length

1 centimeter = 0.01 meter

1 millimeter (mm) = 0.001 meter

1 micrometer (µm) = 0.000001 meter

1 nanometer (nm) = 0.000000001 meter

Temperature

In the metric system, temperature is measured on the Celsius (C) scale. On the Celsius scale, 0° is the freezing point of water, and 100° is the boiling point of water. Thus 1°C equals 0.01 of the difference between the freezing point and boiling point of water. You can use the scale shown to convert between the Celsius scale and the Fahrenheit scale, which is commonly used in the United States.

Area

square kilometer (sq.km) = 100 hectares

1 hectare (ha) = 10,000 square meters

1 square meter (sq.m) = 10,000 square centimeters

1 square centimeter (sq.cm) = 100 square millimeters

Mass

1 kilogram (kg) = 1,000 grams

1 gram (g) = derived from kg (base unit of mass)

1 milligram (mg) = 0.001 gram

1 microgram (µg) = 0.000001 gram

Liquid Volume

1 kiloliter (kL) = 1,000 liters

1 liter (L) = base unit of liquid volume

1 milliliter (mL) = 0.001 liter

Note: When measuring liquid volume in a graduated cylinder, be sure to read the measurement at the bottom of the meniscus, or curve.

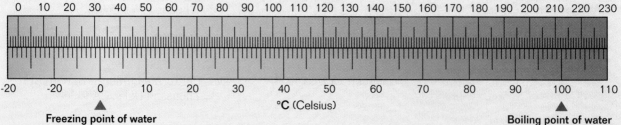

°F (Fahrenheit)
0 10 20 30 40 50 60 70 80 90 100 110 120 130 140 150 160 170 180 190 200 210 220 230

-20 -20 0 10 20 30 40 50 60 70 80 90 100 110
°C (Celsius)

▲ Freezing point of water

▲ Boiling point of water

Safety

You will avoid accidents in the biology laboratory by following directions, handling materials carefully, and taking your work seriously. Before you begin working, be sure that you are familiar with safety procedures and know the location of fire extinguishers and first aid supplies. Be aware of your classmates' safety as well as your own. Never attempt any laboratory procedure without an instructor's direction, and never work in the laboratory by yourself.

Read the text that follows to familiarize yourself with the safety symbols used in your text and the guidelines that you should follow when you see these symbols.

ANIMAL CARE
You are both legally and ethically required to treat animals as humanely as possible.

ANIMAL SAFETY
Wear leather or thick gloves when handling laboratory animals, especially rodents. When working in the field, be aware of poisonous or dangerous animals in the area. Do not touch or approach wild animals.

CAUSTIC SUBSTANCE
Use extreme care when handling caustic substances such as hydrochloric acid. These substances can injure the skin. If your skin comes in contact with a caustic substance, wash thoroughly with water and notify your teacher immediately.

CHEMICAL SAFETY
Use poisonous chemicals with extreme caution. Two harmless chemicals may become poisonous when combined. Never mix chemicals without teacher direction. Never put chemicals in your mouth, and avoid chemical contact with eyes or skin.

CLOTHING PROTECTION
Wear a laboratory apron in the classroom. Confine loose clothing.

ELECTRICAL SAFETY
Watch for loose plugs or worn electric cords. Be sure that cords are not placed where they could cause a fall. Do not use electrical equipment near water or with wet hands.

EXPLOSION DANGER
Many chemicals are explosive when combined, and some will explode when jarred, heated, or exposed to air. When heating chemicals, always point the test tube away from people.

EYE SAFETY
Wear approved safety goggles when working near an open flame or handling chemicals. If any chemical gets into your eyes, flush it out thoroughly with water and notify your instructor immediately.

FIRE SAFETY
Keep combustible materials away from sources of fire. Shield any open flame with an asbestos-protected screen.

GAS PRECAUTION
Do not inhale fumes directly. When instructed to smell a substance, wave fumes toward your nose and inhale gently. Use flammable liquids only in small amounts and in a well-ventilated room or under a fume hood. Always use a fume hood with toxic or flammable fumes. Do not breathe pure gases such as hydrogen, argon, helium, nitrogen, or high concentrations of carbon dioxide.

GLASSWARE SAFETY
Handle glassware carefully. Never attempt to clean up broken glass. Notify your teacher immediately.

HAND SAFETY
Dissect specimens in dissecting pans—never in your hand. Never use chipped or cracked glassware. Wear gloves when working with an open flame or with caustic chemicals.

HEATING SAFETY
Use proper procedures when lighting Bunsen burners. Heat flasks or beakers on a ring stand with a wire gauze between glass and flame. Use only heat-resistant glassware for heating materials or storing hot liquids. Turn off hot plates and open flames when not in use.

HYGIENIC CARE
Always wash your hands after the lab. Keep hands away from your face and mouth. Use sterile technique when transferring bacteria or other microorganisms between cultures or to a microscope slide. Do not open a petri dish to observe or count bacterial colonies.

PLANT SAFETY
Some plants cause ill effects when touched or eaten. Use a good field guide when collecting specimens; never eat any part of an unknown plant.

PROPER WASTE DISPOSAL
Clean up the laboratory after you finish. Follow your teacher's directions for waste disposal, especially for chemicals and microbes. Place broken glass in a special container.

WATER SAFETY
When working near water, always work with a partner or adult. Always wear a life jacket. Do not work near water during stormy weather.

Study Skill: Concept Mapping

To remember information longer and to be able to use it more effectively, you need to move that information into your long-term memory. Concept mapping can help you do this.

Identifying Concepts

Concept mapping helps you understand ideas by showing you their connections to other ideas. It is different from taking notes or making an outline. A concept map not only identifies the major ideas of interest from a chapter or your class notes, but also shows the relationships among the ideas, much as a road map illustrates how highways and other roads link cities. The figure below shows what a concept map looks like.

Suppose you have just finished reading a section of a chapter and wish to make a concept map as a study aid. How do you begin? First you need to identify the concepts in that section. Concepts usually form a picture in your mind. For instance, examine the following words: pool, biking, grass, sky, tree. All of these words are concepts because they create images in your mind. Now read this series of words: the, to, be, with, is, can. Are they concepts? No. They don't form a picture in your mind. They are linking words that connect ideas or concepts. Linking words play an important role in concept mapping because you use them to connect concepts in your map. See if you can identify the concepts and linking words in the figure below.

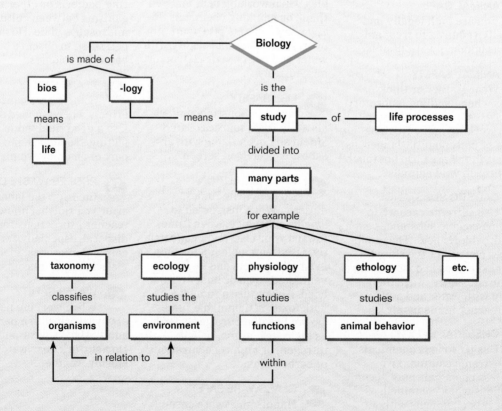

Sorting and Ranking Ideas

Another feature of a concept map is that some concepts or ideas are more general and include other concepts. Such general concepts will be the main ideas in your map. What is the main idea for this list of concepts: Tokyo, Mexico City, Seoul, New York, Bombay? Cities, of course. What is the main idea for this list: car, bus, train, bicycle, truck? Each of these concepts is an example of a vehicle.

Every concept map should have at its top the most general and all-inclusive concept. The smaller, more specific concepts and examples should go below. It might help you to capitalize the first letter of the main concept or idea. You can write the other, more specific, concepts using all lowercase letters. Study the concept map on page 1018 before you try one yourself. Then read the following paragraph and make a list of any important words you think should be learned.

What is life? What is the difference between living and nonliving things? If you were in a wilderness area, it would be easy for you to pick out the living and nonliving things. The animals and plants are the living organisms. Organisms are made up of many substances organized into living systems. The rocks, air, water, and soil you see are nonliving. They contribute substances to the living organisms.

Connecting Ideas

Two general ideas prevail in the preceding paragraph: living and nonliving. You could either make two separate maps, the chief concept of one being "living" and of the other being "nonliving," or you could make one map and use "natural things" as the main idea. Capitalize the first letter and put it at the top of the map. "Natural things" did not actually appear in the paragraph, but frequently the main idea does.

Natural things can be living or nonliving. Therefore, the ideas of "living" and "nonliving" are parts of the main idea and should be placed below it in the map. In the paragraph, you learned that living things are organisms, such as plants and animals, and that they are made up of substances that are organized into living systems. The paragraph also explained that living things are different from but related to nonliving things. So you should make the connection between living and nonliving things near the top of the map. And the rest of the map should explain how living and nonliving things are related. Nonliving things such as air, soil, water, and rocks all contribute substances that are organized into living systems.

Now you have the entire map shown on page 1020. If you had a choice between studying all those sentences or looking at this map, you would probably agree that the map shows the concepts more clearly. This map gives you the main idea more quickly, and it's easier to understand all the ideas because their relationships to other ideas are shown.

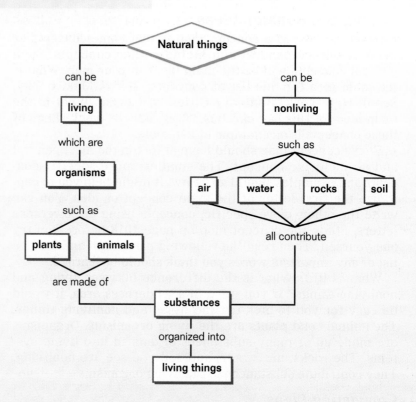

Features of Good Concept Maps

Remember, practice is the key to good concept mapping. You'll get better as you go along. Here are some things to remember that will help.

- A concept map does not have to be symmetrical. It can have more concepts on one side than the other.
- There are no perfectly correct concept maps, only maps that come closer to the meanings you have for those concepts. As the mapmaker, you must make it correct for you.
- Do not put more than three words in a concept box.
- Do not string out more than three boxes in a row or line without branching out.
- Write linkage words connecting every two concepts. Use as few words as it takes to make the connection between the concepts clear.

If the relationships you have made between any two concepts are wrong, your teacher will help you sort out your misconception. Even if you are absolutely correct in your relationships, maps made by your fellow students may be different. These maps could be equally correct, even though they may look nothing like yours. Everyone thinks a little bit differently, and, as a result, other people may see different relationships between certain concepts.

As you practice making concept maps, your teacher will examine your linkage statements more closely. Since these lines represent the relationships between concepts, whatever you write on the line will tell you if you really understand how those two concepts are connected. After you have practiced, your maps should always

- be two dimensional—not just a list of concepts connected by lines.
- show which concepts are more important by their placement on the map and by what concepts branch off them.
- have many branches with no more than three concept boxes in a row and no more than three or four words in a concept box.
- have only concepts in the boxes and only linkage words on the lines.

Evaluating Your Skills

For the first map you make on your own, think about something you know very well. Do you play a team sport or an individual sport? Do you have a hobby? Do you enjoy a particular kind of music? Whatever you choose, this will be the major concept for your next concept map. This will be more fun and easier since you know this topic so well.

Laboratory Skill: Using a Compound Light Microscope

Parts of the Compound Light Microscope

- The **eyepiece** magnifies the image, usually 10×.
- The **low-power objective** magnifies the image even more, such as 4×.
- The **high-power objectives** magnify the image even more, such as 10× and 43×.
- The **nosepiece** holds the objectives and can be turned to change from one magnification to another.
- The **body tube** maintains the correct distance between the eyepiece and the objectives. This is usually about 25 cm (10 in.), the normal distance for reading and viewing objects with the naked eye.
- The **coarse adjustment** moves the body tube up and down in large increments to allow gross positioning and focusing of the objective lens.

- The **fine adjustment** moves the body tube slightly to bring the image into sharp focus.
- The **stage** supports a slide.
- The **stage clips** secure the slide in position for viewing.
- The **diaphragm** (not labeled) controls the amount of light allowed to pass through the object being viewed.
- The **light source** provides light for viewing the image. It can either be a light reflected with a mirror or an incandescent light from a small lamp. NEVER use reflected direct sunlight as a light source.
- The **arm** supports the body tube.
- The **base** supports the microscope.

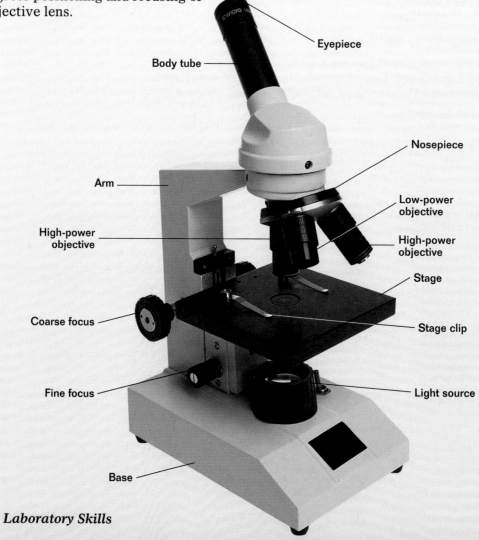

Eyepiece

Body tube

Arm

High-power objective

Coarse focus

Fine focus

Base

Nosepiece

Low-power objective

High-power objective

Stage

Stage clip

Light source

Proper Handling and Use of the Compound Light Microscope

1. Carry the microscope to your lab table using both hands, one beneath the base and the other holding the arm of the microscope. Hold the microscope close to your body.

2. Place the microscope on the lab table, at least 5 cm (2 in.) in from the edge of the table.

3. Check to see what type of light source the microscope has. If the microscope has a lamp, plug it in, making sure that the cord is out of the way. If the microscope has a mirror, adjust it to reflect light through the hole in the stage.

CAUTION: If your microscope has a mirror, do not use direct sunlight as a light source. Direct sunlight can damage your eyes.

4. Adjust the revolving nosepiece so that the low-power objective is in line with the body tube.

5. Place a prepared slide over the hole in the stage, and secure the slide with stage clips.

6. Look through the eyepiece and move the diaphragm to adjust the amount of light coming through the specimen.

7. Now look at the stage from eye level, and slowly turn the coarse adjustment to lower the objective until it almost touches the slide. Do not allow the objective to touch the slide.

8. While looking through the eyepiece, turn the coarse adjustment to raise the objective until the image is in focus. Never focus objectives downward. Use the fine adjustment to achieve a sharply focused image. Keep both eyes open while viewing a slide.

9. Make sure that the image is exactly in the center of your field of vision. Then switch to the high-power objective. Focus the image with the fine adjustment. Never use the coarse adjustment at high power.

10. When you are finished using the microscope, remove the slide. Clean the eyepiece and objectives with lens paper, and return the microscope to its storage area.

Making a Wet Mount

1. Use lens paper to clean a glass slide and coverslip.

2. Place the specimen you wish to observe in the center of the slide.

3. Using a medicine dropper, place one drop of water on the specimen.

4. Position the coverslip so that it is at the edge of the drop of water and at a 45° angle to the slide. Make sure that the water runs along the edge of the coverslip.

5. Lower the coverslip slowly to avoid trapping air bubbles.

6. If a stain or solution is to be added to a wet mount, place a drop of the staining solution on the microscope slide along one side of the coverslip. Place a small piece of paper towel on the opposite side of the coverslip.

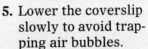

7. As the water evaporates from the slide, add another drop of water by placing the tip of the medicine dropper next to the edge of the coverslip, just as you would if adding stains or solutions to a wet mount. If you have added too much water, remove the excess by using the corner of a paper towel as a blotter. Do not lift the coverslip to add or remove water.

A Six-Kingdom System for the Classification of Organisms

This is a classification of organisms based on the six-kingdom system explained in Chapter 19. Not all groups are shown. Numbers of species are approximate. Also, how the Eubacteria and Archaebacteria should be divided into phyla is controversial. Here, only broad, generally recognized groups are presented; these groups may not correspond to phyla.

KINGDOM EUBACTERIA

Typically unicellular; prokaryotic; without membrane-bound organelles; nutrition mainly by absorption, but some are photosynthetic or chemosynthetic; reproduction usually by fission or budding; about 5,000 species have been recognized, but there are undoubtedly many times that number

Cyanobacteria
Photosynthetic; surrounded by a gooey, deeply pigmented covering; common on land and in the ocean; probably ancestors of chloroplasts in some kinds of protists: *Anabaena, Oscillatoria, Spirulina*

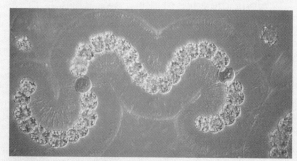

Anabaena

Chemoautotrophs
An ancient group of bacteria that can grow without sunlight or other organisms; derive energy from reduced gases—ammonia (NH_3), methane (CH_4), hydrogen sulfide (H_2S); play critical roles in the Earth's nitrogen cycles: nitrobacteria, sulfur bacteria (*Nitrosomonas, Nitrobacter*)

Enterobacteria
Typically rigid, rod-shaped, heterotrophic bacteria; usually aerobic; have flagella; responsible for many serious diseases of plants and humans, including cholera (*Vibrio cholerae*): *Escherichia coli, Salmonella typhimurium*

Pseudomonads
Straight or curved rods with flagella at one end; very common in soil; many are serious plant pathogens: *Pseudomonas aeruginosa*

Spirochaetes
Long, spiral cells; flagella originating at each end; responsible for several serious diseases, including syphilis (*Treponema pallidum*) and Lyme disease (*Borrelia burgdorferi*)

Actinomycetes
Filamentous bacteria that are often mistaken for fungi; spore-producing; sources of antibiotics including streptomycin, tetracycline, and chloramphenicol; cause dental plaque, leprosy, and tuberculosis: *Mycobacterium tuberculosis*

Rickettsias
Obligate parasites within the cells of vertebrates and arthropods; responsible for Rocky Mountain spotted fever: *Rickettsia rickettsii*

Gliding and budding bacteria
Long, rod-shaped cells that secrete slimy polysaccharides; often aggregate into gliding masses; live mainly in soil: *Myxobacteria*

KINGDOM ARCHAEBACTERIA

Anaerobic and aerobic bacteria adapted to environments with extreme temperatures, acidity, or salt content; prokaryotic; differ from eubacteria in structure of cell membrane and cell wall; RNA polymerase and a ribosomal protein similar to those in eukaryotes, suggesting that archaebacteria are more closely related to eukaryotes than to eubacteria; asexual reproduction only; fewer than 100 named species, divided into three broad groups

Methanogens
Anaerobic methane producers; most species use carbon dioxide as a carbon source; found in the soil, swamps, and the digestive tracts of animals, particularly grazing mammals such as cattle; produce nearly 2 trillion kg (2 billion tons) of methane gas annually

Thermoacidophiles

Inhabit very hot environments that are often very acidic; some species can tolerate temperatures of 110°C (230°F); require sulfur; nearly all are anaerobes: *Sulfolobus*

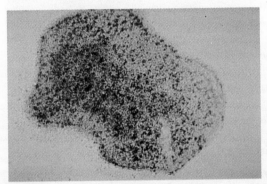

Sulfolobus

Extreme halophiles

Live in environments with very high salt content, including the Dead Sea and the Great Salt Lake; nearly all are aerobic; all are gramnegative

KINGDOM PROTISTA

A catchall kingdom for the eukaryotes that are not plants, fungi, or animals; the most structurally diverse kingdom; unicellular and multicellular representatives; all have a membranebound nucleus; nearly all have chromosomes, mitochondria, and internal compartments; many have chloroplasts; most have cell walls; reproduce sexually and asexually; aquatic or parasitic; about 43,000 species

Phylum Rhizopoda

Unicellular and heterotrophic; amorphously shaped cells that move using cytoplasmic extensions called pseudopods: amoebas (*Amoeba, Entamoeba*); about 300 species

Phylum Foraminifera

Unicellular and heterotrophic; marine; have shells of organic material with pores through which many cytoplasmic threads project: forams; about 300 species

Phylum Oomycota

Heterotrophic; unicellular parasites or decomposers; cell walls composed of cellulose, not chitin as in fungi: water molds, white rusts, downy mildews (*Phytophthora*); about 580 species

Phylum Ciliophora

Very complex single cells; heterotrophic; have rows of cilia and two types of cell nuclei: ciliates (*Didinium, Paramecium, Stentor, Vorticella*); about 8,000 species

Phylum Zoomastigina

Mostly unicellular; heterotrophic; all have at least one flagellum: zoomastigotes (*Giardia, Leishmania, Trichonympha, Trypanosoma*); about 3,000 species

Phylum Sporozoa

Unicellular; heterotrophic; nonmotile; sporeforming parasites of animals; have complex life cycles; asexual and sexual reproduction; *Plasmodium* is responsible for malaria, which kills more than 1 million people each year: sporozoans (*Plasmodium, Toxoplasma*); about 3,900 species

Phylum Myxomycota

Heterotrophic; individuals stream along as a multinucleate mass of cytoplasm; when dry or starving, can give rise to spores that start a new individual in a more favorable environment: plasmodial slime molds (*Physarum*); about 500 species

Physarum

A Six-Kingdom System continued

Phylum Acrasiomycota

Heterotrophic; amoeba-shaped cells that aggregate into a moving mass called a slug when deprived of food; cells within the slug retain their membranes and do not fuse; a slug produces spores that form new amoebas elsewhere: cellular slime molds *(Dictyostelium)*; about 70 species

Phylum Caryoblastea

Unicellular; lacks mitochondria and chloroplasts; possibly an early stage in the evolution of eukaryotes; lives in low-oxygen conditions in the mud on pond bottoms: *Pelomyxa palustris* is the only species

Phylum Chlorophyta

Unicellular, colonial, and multicellular species; all are photosynthetic; have chlorophylls *a* and *b*; contain chloroplasts very similar to those of plants; most scientists think that plants descended from this group: green algae *(Chlamydomonas, Chlorella, Oedogonium, Spyrogyra, Ulva, Volvox)*; about 7,000 species

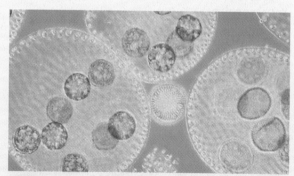

Volvox

Phylum Phaeophyta

Multicellular and photosynthetic; nearly all are marine; contain chlorophylls *a* and *c* and fucoxanthin, the source of their brownish color: brown algae *(Fucus, Laminaria, Postelsia, Sargassum)*; about 1,500 species

Phylum Rhodophyta

Almost all are multicellular; all are photosynthetic; most are marine; contain chlorophyll *a* and phycobilins; chloroplasts probably evolved from symbiotic cyanobacteria: red algae *(Porphyra)*; about 4,000 species

Red Algae

Phylum Bacillariophyta

Unicellular and photosynthetic; secrete a unique shell made of opaline silica that resembles a box with a lid; chloroplasts resemble those of brown algae; contain chlorophylls *a* and *c* and fucoxanthin: diatoms; more than 11,500 species

Phylum Dinoflagellata

Unicellular; heterotrophic and autotrophic species; mostly marine; body enclosed within two cellulose plates; contain chlorophylls *a* and *c* and carotenoids: dinoflagellates *(Gonyaulax, Noctiluca)*; more than 2,100 species

Phylum Euglenophyta

Unicellular; both photosynthetic and heterotrophic species; asexual; most live in fresh water; chloroplasts are similar to those of green algae and are thought to have evolved from the same symbiotic bacteria: euglenoids *(Euglena)*; about 1,000 species

KINGDOM FUNGI

Eukaryotic heterotrophs with nutrition by absorption; all but yeasts are multicellular; nearly all are terrestrial; body is typically composed of filaments (called hyphae) and is multinucleate, with incomplete divisions (called septae) between cells; cell walls made of chitin; reproduction asexual or sexual; about 77,000 species

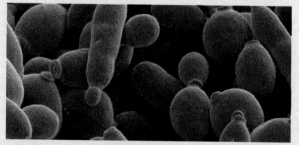

Yeasts

Phylum Ascomycota

Hyphae usually have perforated septae; fusion of hyphae leads to formation of densely interwoven mass that contains characteristic microscopic reproductive structures called asci (singular, ascus); terrestrial, marine, and freshwater species: bread molds, morels, truffles, *Neurospora, Saccharomyces*; about 30,000 species

Phylum Basidiomycota

Hyphae usually have incomplete septae; reproduction is typically sexual; fusion of hyphae leads to the formation of densely interwoven reproductive structure (mushroom) with characteristic microscopic, club-shaped structures called basidia (singular, basidium); terrestrial: mushrooms, toadstools, shelf fungi, rusts, smuts; about 16,000 species

A basidiomycete

Phylum Deuteromycota

(also called Fungi Imperfecti); Sexual stages of life cycle not observed; most are probably ascomycetes that have lost the ability to reproduce sexually; includes the molds that produce penicillin (*Penicillium*) and athlete's foot fungus (*Trichophyton*); about 17,000 species

Phylum Zygomycota

Usually lack septae; fusion of hyphae leads directly to formation of a zygote, which divides by meiosis when it germinates; terrestrial or parasitic: bread molds (*Pilobolus, Rhizopus*); about 660 species

FUNGAL ASSOCIATIONS

Fungi form symbiotic associations with plants, green algae, and cyanobacteria.

Lichens

Mutualistic relationships between fungi (almost always ascomycetes) and cyanobacteria, green algae, or both; the photosynthetic partners actually live among the hyphae of the fungus; the fungus derives energy from its photosynthetic partners and cannot survive without them; about 15,000 species

Mycorrhizae

Mutualistic relationships between fungi and the roots of plants; 80 percent of all plants have mycorrhizae associated with their roots; the plant provides sugars to the fungi; in return, the fungi serve as accessory roots, greatly increasing the surface area available for the absorption of nutrients; about 5,000 species

KINGDOM PLANTAE

Multicellular, eukaryotic, autotrophic, terrestrial organisms having tissues and organs; cell walls contain cellulose; chlorophylls *a* and *b* present and localized in plastids; all have alternation of generations; about 265,000 species

Phylum Bryophyta

Small and usually found in moist environments; most have simple vascular tissues; gametophyte is the dominant generation; gametophyte lacks roots, stems, and leaves; mosses (*Sphagnum*); about 10,000 species

Phylum Hepatophyta

Gametophyte is dominant, usually small, and grows close to the ground; gametophyte lacks stomata, vascular tissue, roots, stems, and leaves; sporophyte inconspicuous and parasitic on the gametophyte: liverworts (*Marchantia*); about 6,000 species

A Six-Kingdom System *continued*

Marchantia

Phylum Anthocerophyta

Gametophyte is dominant generation; usually small and flat; stomata present on sporophyte; lacks vascular tissue, roots, stems, and leaves: hornworts *(Anthoceros):* about 100 species

Phylum Psilotophyta

Vascular system present; seedless; no roots or stems; gametophyte is small and independent; sporophyte is dominant and has small leaves: whisk ferns *(Psilotum);* 21 species

Phylum Lycophyta

Vascular system present; sporophyte dominant and mosslike; has roots, stems, and leaves; gametophyte is small and independent; seedless: club mosses *(Lycopodium, Selaginella);* about 1,000 species

Phylum Sphenophyta

Vascular system present; gametophyte is small and independent; sporophyte dominant; roots present; seedless: horsetails *(Equisetum);* 15 species

Phylum Pterophyta

Vascular system present; gametophyte is small and independent; sporophyte generation dominant; has roots, stems, and leaves; seedless: ferns *(Salvinia);* about 12,000 species

Phylum Cycadophyta

Palmlike gymnosperms; vascular system present; male and female cones produced on different trees; naked seeds; sporophyte dominant; tropical and subtropical: cycads *(Cycas);* about 100 species

Phylum Coniferophyta

Gymnosperms that produce cones; vascular system lacks vessels; leaves usually needles or scales; typically evergreen; sporophyte dominant; ovules exposed at time of pollination; pollen is dispersed by wind: pines, spruces, firs, larches, yews *(Pinus, Taxus);* about 550 species

Lodgepole pine

Phylum Gnetophyta

Specialized gymnosperms; vascular system contains water-conducting vessels; seeds naked; sporophyte dominant; gnetophytes *(Ephedra, Welwitschia);* about 70 species

Phylum Ginkgophyta

Deciduous, gymnosperm tree; vascular system present; has fanlike leaves; sporophyte dominant; produces conelike male reproductive structures and uncovered seeds on different individuals: the ginkgo, *Ginkgo biloba,* is the only species

Phylum Anthophyta

Angiosperms; vascular system present; sporophyte dominant; ovules are fully enclosed by ovary; after fertilization, ovary and seed mature to become fruit; flowers are reproductive structures: flowering plants *(Aster, Prunus, Quercus, Zea);* about 250,000 species

Class Monocotyledones

Embryo has one cotyledon; flower parts in threes; leaf veins parallel; vascular bundles scattered through stem tissue: grasses, sedges, lilies, irises, palms, orchids; around 70,000 species

Widow's tears

Class Dicotyledones

Embryo has two cotyledons; flower parts in fours or fives; leaves with netlike veins; vascular bundles in orderly arrangement in stems: roses, maples, elms; about 170,000 species

KINGDOM ANIMALIA

Multicellular, eukaryotic, heterotrophic organisms; nutrition mainly by ingestion; most have specialized tissues, and many have complex organs and organ systems; no cell walls or chloroplasts; sexual reproduction predominates; both aquatic and terrestrial forms; about 1 million species

Phylum Porifera

Asymmetrical; lack tissues and organs; body wall consists of two cell layers, penetrated by numerous pores; internal cavity is lined with unique food-filtering cells called chaonocytes; mostly marine: sponges; about 9,000 species

Purple tube sponges

Phylum Cnidaria

Radially symmetrical and gelatinous; most have distinct tissues; baglike body of two cell layers; marine and freshwater species: hydras, jellyfish, corals, sea anemones (*Hydra, Obelia*); about 10,000 species

Class Hydrozoa

Most have both polyp and medusa stages in life cycle: hydras, Portuguese man-of-war; about 2,700 species

Class Scyphozoa

Exclusively marine; medusa stage dominant: jellyfish; about 200 species

Class Anthozoa

Marine; solitary or colonial; medusa stage absent: sea anemones, corals, sea fans; about 6,200 species

Phylum Ctenophora

Transparent, gelatinous marine animals resembling jellyfish; radially symmetrical: comb jellies; about 100 species

Phylum Platyhelminthes

Body flat and ribbonlike, without true segments; bilaterally symmetrical acoelomates; organs present; three germ layers: flatworms (*Dugesia, Planaria, Schistosoma*); about 20,000 species

Marine flatworm

Class Turbellaria

Mostly free-living aquatic or terrestrial forms: planarians (*Dugesia*)

Class Trematoda

Internal parasites, with mouth at anterior end; often have complex life cycle with alternation of hosts: human blood flukes (*Schistosoma*), human liver fluke (*Chlonorchis sinensis*)

Class Cestoda
Extremely specialized internal parasites; hooked scolex for attaching to host: tapeworms

Phylum Rotifera
Small, wormlike or spherical animals; bilaterally symmetrical; pseudocoelomates; almost all live in fresh water: rotifers; about 1,750 species

Phylum Nematoda
Typically tiny, parasitic, unsegmented worms; body slender and elongated; pseudocoelomates; includes important human parasites such as *Ascaris*, pinworms, hookworms, *Trichinella*, and *Wuchereria*: roundworms; more than 12,000 species

Phylum Loricifera
Minute, bilaterally symmetrical pseudocoelomates; live in spaces between grains of sand; loriciferans (*Nanaloricus mysticus*); 6 species

Phylum Mollusca
Soft-bodied animals with a true coelom; three-part body plan consisting of foot, visceral mass, and mantle; protostomes; most have a unique rasping tongue called a radula; terrestrial, freshwater, and marine: clams, snails, octopuses, squid, mussels, slugs; about 110,000 species

Class Cephalopoda
Foot modified into tentacles: squids, octopuses, nautilus; over 600 species

Brief squid

Class Bivalvia
Two shells connected by a hinge; no radula; large, wedge-shaped foot: clams, oysters, scallops; about 10,000 species

Class Gastropoda
Visceral mass twisted during development; head, distinct eyes, and tentacles usually present: snails, slugs, whelks; about 80,000 species

Class Polyplacophora
Elongated body and reduced head: chitons; about 600 species

Phylum Rhynchocoela
Bilaterally symmetrical acoelomates; long, typically ribbonlike body: ribbon worms; about 650 species

Phylum Annelida
Serially segmented worms; bilaterally symmetrical; protostomes: segmented worms; about 12,000 species

Class Polychaeta
Fleshy outgrowths called parapodia extend from segments; marine; many bristles: sandworms; about 8,000 species

Fireworm

Class Oligochaeta
Head not well developed; no parapodia; terrestrial and freshwater forms: earthworms; about 3,100 species

Class Hirudinea
Body flattened; no parapodia; usually suckers at both ends; many are external parasites: leeches; about 600 species

Phylum Onychophora
Protostomes; chitinous exoskeletons; wormlike: velvet worms (*Peripatus*); about 70 species

Phylum Pogonophora
Long, deep-sea worms that live within chitinous tubes on the ocean floor: tube worms, or beard worms; about 100 species

Phylum Arthropoda

Segmented bodies with paired, jointed appendages; bilaterally symmetrical; chitinous exoskeleton; protostomes; aerial, terrestrial, and aquatic forms: arthropods; about 1 million species

Subphylum Chelicerata

Distinguished by absence of antennae and presence of chelicerae; all appendages unbranched; four pairs of walking legs; body has two regions (cephalothorax and abdomen); predominantly terrestrial

Class Merostomata

Cephalothorax covered by protective "shell"; sharp spike on tail: horseshoe crabs (*Limulus*); 5 species

Class Pycnogonida

Small, marine predators or parasites; usually four pairs of legs, but sometimes five or six pairs: sea spiders; about 1,000 species

Class Arachnida

Terrestrial; use book lungs and tracheae for respiration: spiders, scorpions, ticks, mites; about 57,000 species

Subphylum Crustacea

Two pairs of antennae, mandibles, and appendages with two branches; predominantly aquatic

Class Malacostraca

Typically five pairs of legs; two pairs of antennae; most are aquatic: crayfish, lobsters, crabs, shrimp, sow bugs; about 20,000 species

Subphylum Uniramia

Have antennae, mandibles, and unbranched appendages

Class Chilopoda

Body flattened and consisting of 15 to 170 or more segments; one pair of legs attached to each segment: centipedes; about 2,500 species

Class Diplopoda

Elongated body of 15 to 200 segments; two pairs of legs per segment; primarily herbivorous: millipedes; about 10,000 species

Class Insecta

Longhorned beetle

Body has three regions—head, thorax, and abdomen; three pairs of legs, all attached to thorax; usually two pairs of wings: insects; about 750,000 described species, but millions more may exist

- **Order Thysanura:** silverfish
- **Order Ephemeroptera:** mayflies
- **Order Odonata:** dragonflies, damsel flies
- **Order Orthoptera:** grasshoppers, cockroaches, walking sticks, praying mantises, crickets
- **Order Isoptera:** termites
- **Order Dermaptera:** earwigs
- **Order Mallophaga:** chicken lice
- **Order Anoplura:** human body louse
- **Order Hemiptera:** true bugs—water striders, water boatmen, back swimmers, bedbugs, squash bugs, stink bugs, assassin bugs
- **Order Homoptera:** cicadas, aphids, leaf hoppers, scale insects
- **Order Neuroptera:** ant lions, lacewings
- **Order Coleoptera:** beetles—ladybugs, fireflies, boll weevil
- **Order Lepidoptera:** butterflies and moths
- **Order Diptera:** flies, mosquitoes, gnats, midges
- **Order Siphonaptera:** fleas
- **Order Hymenoptera:** bees, ants, wasps, hornets, ichneumon fly

Phylum Echinodermata

Deuterostomes; adults radially symmetrical with five-part body plan; most forms have water vascular system with tube feet for locomotion; marine: echindoderms; about 6,000 species

Class Crinoidea

Mouth faces upward and is surrounded by many arms: sea lilies, feather stars; About 600 species

Class Asteroidea

Body usually with five arms and double rows of tube feet on each arm; mouth directed downward: sea stars; about 1,500 species

Blacktip shark

Class Ophiuroidea

Usually with five slender, delicate arms or rays: brittle stars, basket star; about 2,000 species

Class Echinoidea

Body spherical, oval, or disk-shaped; arms lacking but five-part body plan still apparent: sea urchins, sand dollars; about 900 species

Class Holothuroidea

Elongated, thickened body with tentacles around the mouth: sea cucumbers; about 1,500 species

Phylum Hemichordata

Wormlike chordates; deuterostomes; body in three regions: acorn worms; about 90 species

Phylum Chordata

Bilaterally symmetrical; deuterostomes; coelom present; have a notochord, a dorsal nerve cord, pharyngeal slits, and a tail at some stage of life; aquatic and terrestrial; about 42,500 species

Subphylum Urochordata

Saclike covering, or tunic, in adults; larvae are free-swimming and have nerve cord and notochord; marine: tunicates; about 1,250 species

Subphylum Cephalochordata

Small and fishlike with a permanent notochord; filter feeders: lancelets (*Branchiostoma*); 21 species

Subphylum Vertebrata

Most of the notochord is replaced by a spinal column composed of vertebrae that protect the dorsal nerve cord; recognizable head containing a brain: vertebrates; about 40,000 species

Class Agnatha

Freshwater or marine eel-like fishes without true jaws, scales, or paired fins; cartilaginous skeleton: lampreys, hagfish; 63 species

Class Chondrichthyes

Fishes with jaws and paired fins; gills present; no swim bladder; cartilaginous skeleton: sharks, rays, skates; about 850 species

Class Osteichthyes

Freshwater and marine fishes with gills free and attached to gill arch; jaws and paired fins; bony skeleton; most have swim bladder: bony fishes; about 18,000 species

Class Amphibia

Freshwater or terrestrial; gills present at some stage; skin often slimy and lacking scales; eggs typically laid in water and fertilized externally: amphibians; about 4,200 species

Order Apoda: caecilians
Order Urodela: salamanders and newts
Order Anura: frogs and toads

Class Reptilia

Terrestrial or semiaquatic vertebrates; breathe by lungs at all stages; body covered by scales; most species lay amniotic eggs covered with a protective shell; fertilization internal: reptiles; about 7,000 species

Order Rhynchocephalia: tuataras
Order Chelonia: turtles and tortoises
Order Crocodilia: alligators, crocodiles, gavials, caimans
Order Squamata: lizards and snakes

Gila monster

Class Aves
Body covered with feathers; forelimbs modified into wings; four-chambered heart; endothermic; lay shelled, amniotic eggs: birds; about 9,000 species

Order Gaviiformes: loons

Order Pelecaniformes: pelicans, cormorants, gannets

Order Ciconiiformes: herons, bitterns, egrets, storks, spoonbills, ibises

Order Anseriformes: ducks, geese, swans

Order Falconiformes: hawks, falcons, eagles, kites, vultures

Order Galliformes: pheasants, turkeys, quails, partridges, grouse

Order Gruiformes: cranes, coots, gallinules, rails

Order Charadriiformes: snipes, sandpipers, plovers, gulls, terns, auks, puffins, ibises

Order Procellariiformes: albatrosses, petrels

Order Columbiformes: pigeons and doves

Order Psittaciformes: parrots, parakeets, macaws, cockatoos

Order Cuculiformes: cuckoos, roadrunners

Macaw

Order Strigiformes: owls

Order Caprimulgiformes: goatsuckers, whippoorwills, nighthawks

Order Apodiformes: swifts, hummingbirds

Order Coraciiformes: kingfishers

Order Sphenisciformes: penguins

Order Piciformes: woodpeckers, sapsuckers, flickers, toucans, honeyguides

Order Passeriformes: robins, bluebirds, sparrows, warblers, thrushes

Order Struthioniformes: ostrich

Order Apterygiformes: kiwis

Class Mammalia
Hair on at least part of body; young nourished with milk secreted by mammary glands; endothermic; breathe with lungs; mammals; about 4,000 species

Order Monotremata: duckbill platypus and spiny anteaters

Order Marsupialia: opossums, kangaroos, koalas, wallabies

Order Insectivora: moles and shrews

Order Chiroptera: bats

Order Edentata: armadillos, sloths, anteaters

Order Pholidota: pangolins

Deer mouse

Order Rodentia: squirrels, woodchucks, mice, rats, muskrats, beavers

Order Lagomorpha: rabbits, hares, pikas

Order Carnivora: bears, weasels, mink, otters, skunks, lions, tigers, wolves

Order Cetacea: whales, porpoises, dolphins

Order Sirenia: sea cows, dugongs, manatees

Order Proboscidea: elephants

Order Pinnipedia: seals, sea lions, walruses

Order Perissodactyla: tapirs, rhinoceroses, horses, zebras

Order Artiodactyla: hippopotamuses, camels, llamas, deer, giraffes, cattle, sheep, goats

Order Primates: monkeys, lemurs, gibbons, orangutans, gorillas, chimpanzees, humans

Order Macroscelidea: elephant shrews

Order Scandentia: tree shrews

Order Hyracoidea: hyraxes

Order Dermoptera: flying lemurs

Order Tubulidentata: aardvark

Glossary

Pronunciation Key

Sound	As In	Phonetic Respelling
ahy	bat	(BAT)
ay	face	(FAYS)
ah	lock argue	(LAHK) (AHR gyoo)
ow	out	(OWT)
ch	chapel	(CHAP uhl)
eh	test	(TEHST)
ai	rare	(RAIR)
ee	eat feet ski	(EET) (FEET) (SKEE)
ih	bit	(BIHT)
eye	idea	(eye DEE uh)
y	ripe	(RYP)

Sound	As In	Phonetic Respelling
ihng	going	(GOH ihng)
k	card kite	(KAHRD) (KYT)
ng	anger	(ANG guhr)
oh	over	(OH vuhr)
aw	dog horn	(DAWG) (HAWRN)
oy	foil	(FOYL)
u	pull	(PUL)
oo	pool	(POOL)
s	cell sit	(SEHL) (SIHT)
sh	sheep	(SHEEP)

Sound	As In	Phonetic Respelling
th	that thin	(THAT) (THIHN)
uh	cut	(CUHT)
ur	fern	(FURN)
y	yes	(YEHS)
yoo	globule	(GLAHB yool)
yu	cure	(KYUR)
z	bags	(BAGZ)
zh	treasure	(TREHZH uhr)
uh	medal pencil onion	(MEHD uhl) (PEHN suhl) (UHN yuhn)
uhr	paper	(PAY puhr)

A

A-B-O system the system used to classify human blood by antigens (881)

acetylcholine neurotransmitter that stimulates skeletal muscles (67)

acetyl-Co A molecule derived from glucose and fatty acid metabolism; compound that enters the Krebs cycle (109)

acid compound that is a proton donor (33)

acid rain precipitation with below-normal pH, often the result of industrial pollution and automobile exhaust (6, 390)

acne a skin condition resulting from clogged oil ducts (869)

acoelomate an animal that lacks a coelom, or body cavity (618)

acromegaly disorder that results from overproduction of growth hormone and causes facial figures to thicken (978)

actin *See* actin filament

actin filament a protein found in a muscle cell that functions in contraction (46)

action potential sudden reversal of voltage across the neuron membrane (941)

activation energy amount of energy required to start a chemical reaction (81)

active site site on an enzyme that attaches to a substrate (83)

active transport movement of a particle through a membrane against a concentration gradient with the use of energy from ATP (62)

adaptation process of becoming adapted to an environment; an anatomical structure, physiological process, or behavioral trait that improves an organism's likelihood of survival and reproduction (249)

adductor muscle attachment between the two valves of a mollusk that causes the shell to open and close (662)

adenine a purine base; a component of nucleotides (169)

adrenal cortex outer shell of the adrenal gland that produces the steroid hormones cortisol and aldosterone (980)

adrenaline peptide hormone released by the adrenal medulla in times of stress; also called epinephrine (980)

adrenal medulla part of the adrenal gland that produces the peptide hormone adrenaline, or epinephrine (980)

adventitious roots roots that grow from aboveground parts of a plant, such as stems and leaves (559)

aerobic term for processes that require oxygen (107)

aggregation a temporary collection of cells that come together for a period of time and then separate (427)

aldosterone hormone that affects the kidneys by promoting the uptake of sodium and other salts from body fluids (981)

allele an alternative form of a gene (146)

allele frequency the relative abundance of an allele of a gene within a population, expressed as a percentage (322)

allergy a major defense mounted by the body's immune system against a harmless substance (930)

allosteric enzyme an enzyme whose shape can be altered by the binding of a signal molecule to its surface (88)

allosteric site the site where the signal molecule binds to an allosteric enzyme's surface (88)

alternation of generations life cycle in which a haploid individual alternates with a diploid individual (524)

alveolus microscopic air sac in the lung where oxygen and carbon dioxide are exchanged (714, 889)

amino acid organic molecule that is the building block of protein (36)

amniocentesis procedure in which a sample of amniotic fluid is withdrawn and tested for genetic abnormalities in a fetus (124)

amnion membrane enclosing the embryo (725, 1004)

amniotic egg watertight, fluid-filled egg in birds, reptiles, and mammals (726)

amoeba a protist that moves using flexible, cytoplasmic extensions (485)

amoebocyte amoeba-like cell that supplies nutrition and removes wastes from sponge body cells (638)

amylase enzyme that breaks down starches into sugars (904)

anabolic steroids synthetic chemicals that resemble testosterone and aid in increasing muscle size and strength (863)

anaerobic term for processes not requiring oxygen (107)

analogous characters similar features of organisms that evolve independently (438)

anaphase stage of cell division in which chromosome copies separate (131)

anapsid type of vertebrate skull that has openings only for the eyes and nostrils (773)

androecium part of a flower that produces male gametes, or pollen grains (536)

anemia condition in which red blood cell production slows down (880)

angiosperm seed plant that reproduces via flowers, which produce seeds within an ovary (531)

annual plant plant that completes its life cycle during one growing season (577)

anterior front end of bilaterally symmetric animal (617)

anterior lobe front portion of the pituitary gland from which seven major hormones are secreted (978)

anther the sac at the tip of the stamen in which pollen grains are formed (536)

antheridium reproductive structures in seedless plants that produce sperm by mitosis (525)

antibiotic substance used as a drug to kill bacteria (469)

antibody defensive protein released by B cells in response to a foreign substance in the body (211, 458, 927)

anticoagulant protein that prevents blood from clotting (211)

anticodon a three-nucleotide sequence on tRNA that recognizes a codon on mRNA (137)

aorta main artery in the body; receives blood from the left ventricle (884)

aortic valve one-way valve that prevents blood in the aorta from flowing back into the left ventricle (884)

apical dominance the inhibition of lateral bud growth by auxin produced in the terminal bud (582)

apical meristem meristem in the tips of stems and roots (573)

appendicular skeleton bones that form the arms and legs (858)

aquifer ground water trapped within porous rock (397)

arachnid any of a large class of arthropods, usually with four pairs of legs, including spiders, scorpions, mites, and ticks (684)

archaebacteria one of the two kingdoms of prokaryotes; represented today by a few groups of bacteria inhabiting extreme environments (274)

archegonium reproductive structure in seedless plants that produces eggs by mitosis (525)

artery vessel that carries blood away from the heart to the body's organs (877)

arteriole a branch of an artery that gives rise to capillaries (877)

arteriosclerosis a cardiovascular disease caused by the formation of hard plaques in the walls of arteries, which leads to decreased elasticity (887)

arthropod members of the phylum Arthropoda, which includes invertebrate animals such as insects, crustaceans, and arachnids (281)

ascocarp compact mass of hyphae constituting the reproductive structure in Ascomycota (506)

ascus sac that forms on the surface of an asocarp in which haploid spores are formed (506)

asexual reproduction reproduction that does not involve a union of gametes (422)

A site a binding site on the ribosome filled by a tRNA molecule carrying an amino acid (187)

aspirin compound originally derived from the white willow tree; used as a pain reliever (601)

asthma a chronic respiratory disorder caused by constricted air passages in the lungs (893)

atherosclerosis a cardiovascular condition caused by a buildup of fat deposits on artery walls (887)

atoll a ring-shaped coral island with a lagoon in the center (646)

atom smallest unit of matter that cannot be broken down by chemical means (29)

ATP-synthetase protein channel in a membrane through which protons are pumped to produce ATP (102)

atrioventricular node a bundle of cardiac cells in the right atrium that helps conduct impulses through the heart (885)

atrium a chamber that receives blood returning to the heart (718)

autoimmune disease disease in which the body loses the ability to distinguish its own cells from foreign cells and as a result, attacks itself (929)

autonomic nervous system the network of motor nerves that regulate cardiac muscle, smooth muscle, and gland activity (944)

autosomes a chromosome that is not directly involved in determining sex (124)

autotroph organism that obtains energy from sunlight or chemicals (78)

auxin a class of plant hormones that stimulates cell elongation, secondary growth, and leaf and fruit development and inhibits lateral bud growth (582)

axial skeleton bones that form the main body axis (858)

axon the elongated extension of a neuron that carries impulses away from the cell body (833, 940)

Glossary

bacillus rod-shaped bacterial cell (462)

bacteriophage virus that infects bacteria (167, 457)

balancing selection type of natural selection in which the homozygous genotypes are selected against but the heterozygous genotype is selected for, maintaining both alleles in a population (263)

barrier reef coral reef that serves as a barrier between waves and beaches (646)

basal disk an small area on hydra that secretes a sticky substance, which enables them to adhere to rocks or plants in the water (642)

basal layer a layer of the epidermis that continuously replaces lost cells (867)

base compound that is a proton acceptor (33)

base-pairing rules Chargaff's rules that state the amount of adenine equals the amount of thymine and the amount of guanine equals the amount of cytosine (170)

basidium club-shaped sexual reproductive structure that forms on the gills of basidiomycetes (508)

B cell white blood cell that produces antibodies (925)

biennial plant plant that completes its life cycle within two years (577)

bilaterally symmetric animal configuration with left and right halves that mirror each other (617)

bile salts fat-emulsifying molecules produced by the liver (905)

binary fission form of asexual reproduction that produces identical offspring (126)

binomial nomenclature a system for giving each organism a two-word scientific name that consists of the genus followed by the species (432)

biochemical pathway series of enzyme-catalyzed reactions that perform a specific function (85)

biological species concept group of actually or potentially interbreeding natural populations, reproductively isolated from other such groups (436)

biology science of life (12)

biomass the dry weight of tissue and other organic matter used to determine the amount of energy present in trophic levels (348)

biome major type of terrestrial ecological community such as a grassland and desert (379)

bipedal term used to denote the ability to walk erect on two feet (299)

blastocyst hollow-ball embryo that consists of about 700 cells when it arrives at the uterus (1004)

blood type the antigenic characteristic of the blood of an individual (881)

Bohr effect effect of carbon dioxide on oxygen unloading (890)

brain stem the part of the brain that connects to the spinal cord; contains nerves controlling breathing, swallowing, digestive processes, heartbeat, and blood vessel diameter (946)

bran papery outer coat, or husk, of a grain (592)

bronchus one of the two branches of the trachea that leads to the lungs (889)

bronchiole tiny air passages that connect alveoli to the bronchi (889)

bundle of His network of fibers over which a wave of contraction of the heart is conducted (885)

C_3 plant plant that fixes carbon using the Calvin cycle (556)

C_4 plant plant that fixes carbon using an alternative pathway in which the first detectable product is a four-carbon compound (556)

calcitonin hormone produced in the thyroid gland that helps maintain proper calcium levels in the body (979)

Calorie amount of energy needed to raise the temperature of 1 gram of water 1°C. The Calorie used to indicate the energy content of food is a kilocalorie. (900)

Calvin cycle the second major pathway in photosynthesis involving carbon fixation and carbohydrate formation (105)

calyx outermost whorl of a flower; the sepals (536)

cancer a disease characterized by abnormal cell growth (17, 195)

capillaries tiny blood vessels that allow exchange between blood and cells in tissue (877)

capsid a protein sheath that surrounds the nucleic acid core in a virus (456)

capsule a gelatinous outer layer enclosing many bacterial cell walls (463)

carapace shieldlike plate covering the cephalothorax of decapods; dorsal part of a turtle's shell (686, 776)

carbohydrate organic compound composed of carbon, hydrogen, and oxygen; used by living things as an energy source (34, 900)

carbon fixation process by which carbon dioxide is incorporated into organic compounds (105)

carbonic anhydrase enzyme within red blood cells that combines carbon dioxide with water to form carbonic acid (891)

carcinogen cancer-causing substance (195)

carcinoma common type of skin cancer originating in the non-pigment producing cells of the epidermis (868)

carnivore flesh-eating organism (344)

carotenoids yellow and orange plant pigments that aid in photosynthesis (101)

carrying capacity population size that an environment can sustain (318)

cartilage lightweight, strong, and flexible connective tissue (706)

caste role played by individual insect in a colony (693)

catalysis process of increasing a chemical reaction rate through the action of a catalyst (82)

catalyst material that speeds up a chemical reaction without being used itself (82)

cell smallest unit that can perform all the life processes (12)

cell body the part of a neuron that processes incoming information (939)

cell cycle repeating five-phase sequence of eukaryotic cell growth and division (127)

cell membrane bilipid layer that encloses the cytoplasm, essential to the cell's cytoplasm; also called the plasma membrane (26)

cell specialization ability of different cells to perform specific functions in a multicellular organism, such as protection, mobility, and reproduction (429)

cell surface marker membrane protein of a cell that distinguishes it from other cells and foreign matter (57)

cell surface protein protein within a cell's plasma membrane (56)

cell theory principle that states all organisms are made of one or more cells that are derived from other cells (25)

cellular respiration process by which living things obtain energy from the bonds of food molecules (78)

cellulose carbohydrate that is the main structural component of cell walls (34)

central nervous system system composed of the brain and spinal cord (826)

central vacuole membrane-bound cavity in plant cells used for storage (46)

centriole rod-shaped organelle that help move chromosomes during cell division (46)

centromere region joining two chromatids (120)

cephalization process of head development in bilaterally symmetric animals (617)

cephalothorax the mid-body region in arthropods; consists of a head fused with a thorax (680)

cereal grass grown as food for humans and livestock (592)

cerebellum region of the brain that controls coordination and balance (946)

cerebral cortex thin outer layer of the cerebrum (945)

cerebral ganglion the brain contained in the anterior segment of an annelid (625)

cerebrum rounded, wrinkled anterior portion of the brain; center for memory, learning, emotions, and other highly complex functions(945)

cervix portion of the uterus that joins with the vagina (998)

chancre initial sore caused by syphilis (1008)

channel cell surface proteins that loop back and forth through membrane bilayer (57)

character displacement situation in which two potentially competing species differ more where their ranges overlap (368)

chelicera one of the first pair of appendages of arachnids and their relatives (680)

chelicerate arthropods with fangs (680)

chemically gated description of a channel that opens and closes due to the binding of a chemical (67)

chemical reaction process by which the atoms of one or more molecules are rearranged to form molecules of one or more new substances (80)

chemical signal a molecular signal carried by the blood stream (839)

chemiosmosis process by which cells pump protons across a plasma membrane and use the resulting proton gradient to produce ATP (62)

chemosynthesis production of organic compounds using energy contained in inorganic molecules (78)

chitin tough carbohydrate found in many fungi and in the exoskeletons of all arthropods (501)

chlamydia sexually transmitted disease marked by discharge, burning, and pain, caused by bacteria (1009)

chloroflurocarbon any of a group of compounds that contain carbon, chlorine, and florine, often used as coolants, propellants, or foaming agents (392)

chlorophyll the green pigment molecule responsible for trapping light energy in photosynthesis (101)

choanocyte unique flagellated cell surrounded by a collar of microvilli found inside sponges (613)

chorion outer membrane surrounding embryos of birds, reptiles, and mammals that contributes to the development of the placenta (726, 1004)

chorionic villi sampling fetal-testing procedure in which pieces of the chorionic villi in a mother's uterus are removed and examined for genetic abnormalities (124)

chromatid one of a pair of strands of DNA that make up a chromosome during meiosis or mitosis (120)

chromosome cellular structure on which genes are located (44, 119)

chromosome puff material accumulating at a site on a chromosome that is undergoing rapid transcription (192)

chrysalis in insects, protective capsule enclosing the transforming larva (692)

chyme semisolid material in the stomach and duodenum composed of food, HCl, and enzymes (907)

cilia in cells, tightly packed rows of short flagella used for movement (46)

circulatory system network of vessels that carries nutrients and oxygen through the body (621)

cirrhosis degenerative condition of the liver in which cells are replaced with scar tissue and the organ shrinks into a hard mass (960)

cladistics phylogenetic method in which relationships are inferred based on presence of derived characters (438)

cladogram diagram based on patterns of shared, derived traits that shows the evolutionary relationships among groups of organisms (438)

class taxonomic category containing orders with common characteristics (434)

cleavage in development, the rapid, mitotic division of the zygote (1003)

cloning process of growing a large number of genetically identical cells from a single cell (204)

closed circulatory system system in which the blood does not leave the blood vessels and materials pass in and out by diffusing across the walls of the vessels (660)

closed system in thermodynamics, a system that allows no matter or energy to enter or leave (76)

cnidocyte stinging cell used by cnidarians to stun prey (614)

coccus spherical-shaped bacterial cell (462)

cochlea fluid-filled chamber of the inner ear that is involved in the perception of sound (950)

codominance condition in which both alleles for a gene are expressed when present(152)

codon a three-nucleotide sequence in DNA or RNA that encodes an amino acid or signifies a stop signal (186)

coelom fluid-filled body cavity that forms within the mesoderm (621)

coenzyme nonprotein molecule that assists an enzyme in carrying out a reaction (86)

Glossary

coevolution reciprocal evolutionary adjustments between interacting members of an ecosystem (361)

collagen fibrous structural protein present in all multicellular organisms, particularly in connective tissue (35, 832)

collecting duct in the urinary system, the tube into which the renal tubule empties (913)

colon organ that compacts waste for excretion; also called the large intestine (908)

colonial organism collection of cells that are permanently associated but in which little or no integration of cell activities occurs (427)

commensalism ecological interaction in which one species benefits and the other is neither harmed nor helped (363)

common ancestor species from which two or more species diverged (254)

community the many different species that live together in a habitat (338)

compact bone dense, almost solid tissue within a bone (855)

companion cell of plant phloem, cell alongside a sieve tube member that carries out metabolic functions for it (554)

competition ecological interaction between two or more species that use the same scarce resource such as food, light, and water (363)

competitive exclusion local extinction of one species due to competition (369)

complement defensive proteins that circulate in the bloodstream (923)

complementary characteristic of nucleic acids in which the sequence of bases on one strand determines the sequence of bases on the other (171)

complete flower flower that has all four whorls of appendages—sepals, petals, stamens, and pistils (537)

complete protein protein that contains all the essential amino acids needed by the body (595)

compound substance that is made up of more than one kind of atom (30)

cone photoreceptor of the retina of the eye that can detect color (952)

conifer gymnosperm that produces seeds in cones (532)

conjugation temporary union of two protists to exchange nuclear material (482)

consumer organism that must obtain energy to build its molecules by consuming other organisms; heterotroph (97, 343)

continental drift movement of the continents over geologic time (287)

continuous variation a genetic phenomenon in which a trait is controlled by several genes and therefore exhibits a variety of phenotypes (153)

control factor that is held constant throughout an experiment to test a hypothesis (10)

conus arteriosus elongated heart chamber that smooths heart pulsations (718)

convergent evolution process by which unrelated species become similar as they adapt to similar environments (438)

cork cambium lateral meristem of woody plants that produces cork cells of the outer bark (574)

cork cells cells that replace the epidermis of woody stems; part of the bark (558)

cornea transparent outer covering of the eye (950)

corneal layer the outermost layer of the epidermis, consisting mostly of dead cells containing keratin (866)

corolla whorl of flower that consists of the petals (536)

coronary arteries arteries that branch from the aorta and carry oxygenated blood to the heart muscle (885)

corpus luteum structure that forms from the ruptured follicle in the ovary after ovulation and releases hormones (1001)

cortex the outer layer of ground tissue in roots and stems of plants (557)

cortisol steroid compound produced by the adrenal glands in response to stress (975)

cortisone drug used in the treatment of inflammation and allergies, and in birth control pills (602)

cotyledon leaflike structure of a plant embryo in which food is stored (539)

coupled channel form of channel that carries into the cell sodium ions that accumulate outside the cell, as with nerve cells (62)

coupled reaction endergonic reaction that is driven by the splitting of ATP molecules (86)

cranium portion of the skull that encases the brain (858)

crop plant cultivated for use by humans (96)

crossing over the exchange of reciprocal segments of DNA by chromosomes at the beginning of meiosis; source of genetic recombination (133)

cross-pollination transfer of pollen from the male structures of one plant to the female structures of a different plant of the same species (143)

cuticle waxy, watertight outer covering of aboveground parts of a plant (520)

cyanobacteria group of photosynthetic eubacteria (273)

cyclic AMP common second messenger in cells (973)

cystic fibrosis genetic disorder in which excessive amounts of mucus are secreted, blocking intestinal and bronchial ducts and causing difficulty in breathing (18)

cytokinesis division of the cytoplasm (127)

cytokinin plant hormone produced in root tips that stimulate cell division (583)

cytoplasm the material between the cell membrane and the nuclear membrane (26)

cytosine nitrogenous base of the pyrimidine class; component of RNA and DNA (169)

cytoskeleton cytoplasmic network of protein filaments that plays an essential role in cell movement, shape, and division (46)

cytosol liquid portion of the cytoplasm (26)

D

day-neutral plant plant whose flowering cycle is not affected by day length (585)

deciduous describes trees, shrubs, and woody vines that drop all of their leaves at the end of each growing season (577)

decomposer organism that causes decay (344)

deletion mutation in which a nucleotide or segment of DNA is lost (124, 195)

demography statistical study of populations (316)

dendrite cytoplasmic extension from the body of a neuron (833, 939)

deoxyribonucleic acid. *See* DNA

deoxyribose five-carbon sugar that is a component of DNA nucleotides (169)

depolarization loss of the electrical charge across the membrane of a neuron (939)

depressant drug that decreases the activity of the central nervous system (957)

derived trait unique characteristic of a particular group of organisms (438)

dermal tissue system the outer protective layer of dermal tissues of vascular plants (551)

dermis thick layer of skin beneath the epidermis (867)

detritivore organism that obtains its energy by feeding on dead organisms or wastes (344)

deuterostome an animal whose mouth forms from an opening other than the blastopore (629)

diabetes mellitus serious disorder in which cells are unable to obtain glucose from the blood (982). *See also* Type I diabetes; Type II diabetes

diaphragm in mammals, sheet of muscle at the bottom of the rib cage that aids in respiration (811, 889)

diapsid term used to identify vertebrate skull that has two openings on each side (773)

diastolic pressure the low blood pressure that occurs during relaxation of the heart (886)

diatom photosynthetic unicellular protist of the phylum Bacillariophyta (486)

dicot dicotyledon; angiosperm that has seeds with two cotyledons (539)

differentiation process in which the cells of a multicellular individual become specialized during development (430, 573)

diffusion movement of particles from an area of high concentration to an area of low concentration (58)

digitalis drug used in the treatment of cardiac disorders (602)

dihybrid cross a cross that involves two pairs of contrasting traits (150)

dikaryon a cell that contains a pair of nuclei that stem from different parent cells (504)

dikaryotic in fungi, condition in which two nuclei are present in each cell (504)

dinoflagellate unicellular, photosynthetic protist of the phylum Dinoflagellata (488)

diploid term used to indicate cell containing two homologues of each chromosome (120)

directional selection natural selection that causes the frequency of a particular allele to move in one direction (263, 329)

dispersion the pattern of distribution of organisms in a population (317)

disruptive selection natural selection in which individuals with extreme forms of a trait have an advantage (330)

diurnal term describing animals that are active during the day and sleep at night (296)

divergence accumulation of differences between groups; can lead to the formation of new species (264)

diversity. *See* species diversity

division in taxonomy, an alternative term for phylum (434)

DNA (deoxyribonucleic acid) nucleic acid that stores hereditary information (37)

DNA fingerprint the pattern of bands that result when DNA fragments are separated by gel electrophoresis (208)

DNA polymerase enzyme that catalyzes the replication of DNA (172)

dominant trait trait that is expressed when its allele is homozygous or heterozygous (145)

dormancy condition in which a seed or plant remains inactive for a period of time (585)

dorsal top surface of a bilaterally symmetrical animal (617)

double fertilization process by which two sperm fuse with cells of the megagametophyte, producing both a zygote and an endosperm (539)

double helix spiral-staircase structure characteristic of the DNA molecule (170)

Down syndrome a syndrome of congenital defects, especially mental retardation, resulting from an additional copy of chromosome 21 (122)

duodenum first part of the small intestine (907)

duplication form of mutation in which a chromosome contains an extra copy of a segment of DNA (124)

ecdysis process of shedding and discarding the exoskeleton; also called molting (682)

ecological race population of a species that differs genetically because of adaptations to different living conditions (265)

ecology the study of the interactions of living organisms with one another and with their environment (337)

ecosystem ecological system encompassing a community and all the physical aspects of its habitat (338)

ectoderm in animals, the outer layer of embryonic tissue from which the skin and nervous system develop (614)

ectomycorrhizae mycorrhizae that do not physically penetrate plant roots but wrap around them (513)

ectoparasites parasites that live outside their host (649)

ectothermic referring to an animal whose body temperature is determined by the temperature of the environment (707)

ejaculation the forceful expulsion of semen from the penis (996)

electrical signal a signal carried from one place in the body to another by neurons (839)

electromagnetic spectrum total range of all electromagnetic waves (100)

electron elementary particle with negative electric charge (29)

electron transport chain series of molecules in an inner cell membrane through which high-energy electrons are passed to pump protons across the membrane and generate ATP by chemiosmosis (102)

element substance composed of a single type of atom (29)

emphysema respiratory condition in which the alveoli of the lungs lose their elasticity, making it difficult for them to release air during exhalation (893)

Glossary

emulsification process in which fat globules are broken down into droplets and exposed to enzymes (905)

endergonic reaction reaction that absorbs free energy (81)

endocrine glands ductless glands in the body that release their products directly into the bloodstream (968)

endocytosis process by which extracellular matter is taken up by a cell (64)

endoderm in animals, inner layer of embryonic tissue from which the digestive organs develop (614)

endometrium lining of the inner wall of the uterus (998)

endomycorrhizae mycorrhizae in which the fungal hyphae penetrate the outer cells of the roots (512)

endoparasite parasite that lives inside its host (649)

endoplasmic reticulum (ER) cell membranes in the cytoplasm that transport substances made by the cell (42)

endorphins chemical signals secreted by the brain and nervous tissue and thought to regulate emotions (970)

endoskeleton an internal skeleton (629)

endosperm highly nutritious tissue developed by the seeds of angiosperms (538)

endospore dormant cell enclosed by a tough coating that is highly resistant to environmental stress (463)

endothermic refers to an animal that generates its own body heat (707)

energy capacity for doing work, often described as the ability to make things move or change (75)

enhancer region preceding a eukaryotic gene that must be activated prior to gene expression (193)

enkephalins a class of neurotransmitters secreted by the central nervous system that act as pain relievers (957)

entropy amount of disorder in a system; amount of unavailable energy in a system (77)

envelope outer layer covering the capsid of many kinds of viruses (456)

enzyme protein that catalyzes a chemical reaction (35, 82)

epidermis outermost layer of tissue, consisting of several layers of cells (866)

epididymis long, coiled tube on the surface of the testicle where sperm mature (995)

epiglottis flap of tissue that covers the trachea during swallowing (889)

epithelium protective tissue that covers the body's interior surfaces (829)

equilibrium state in a chemical reaction when the rates of the forward and reverse reactions are equal (58)

erythrocyte oxygen-carrying red blood cell (832, 880)

esophagus tube that connects the mouth to the stomach (906)

essential amino acids amino acids not synthesized by the human body and must be obtained from food (595)

estrogen hormone secreted by the ovaries that regulates maturation of an ovum in reproduction and regulates the development of female characteristics (983)

ethylene plant hormone that causes fruit ripening (583)

eubacteria one of the two kingdoms of prokaryotes (274)

euglenoids members of the phylum Euglenophyta (489)

eukaryotic cell complex cell that has nucleus enclosed by a membrane (26)

eumetazoans animals with both tissues and symmetry (614)

evergreens trees, shrubs, and woody vines that drop only a few leaves at a time throughout the year (577)

evolution change in the genetic makeup of a population or species over time (48)

evolutionary systematics method of constructing phylogenies that involves weighting characters by their presumed evolutionary significance (442)

excretion the process of eliminating wastes (910)

exergonic reaction a reaction that releases free energy (81)

exhalation expulsion of air from the lungs (890)

exocrine glands glands that release their products through ducts (869)

exocytosis releasing materials outside a cell by discharge from waste vacuoles (64)

exon sequence of nucleotides that gets translated and transcribed (174)

exoskeleton hard external covering of some invertebrates (628, 680)

experiment test of a hypothesis in which data are gathered under controlled conditions (9)

exponential growth curve J-shaped curve showing the rapid increase in an exponentially growing population (318)

extensor skeletal muscle that causes a limb to straighten at a joint (862)

external fertilization union of egg and sperm occurring outside the body of either parent (727)

extinct term used to indicate species that have disappeared permanently (251)

extraterrestrial origin origin (of life) elsewhere in the universe than Earth (225)

eyespot organelle containing light-sensitive pigments (481)

F

F_1 generation the offspring from a cross of two varieties (144)

F_2 generation the offspring from crosses among individuals of the F_1 generation (144)

facilitated diffusion transport of substances through a cell membrane along a concentration gradient with the aid of carrier molecules (61)

fallopian tubes tubelike organs of the female reproductive system that lead from the ovaries to the uterus (998)

family taxonomic category containing genera with similar properties (434)

family tree a diagram that shows the evolutionary relationships among a set of organisms (254)

fat class of organic compound containing carbon, hydrogen, and oxygen and used to construct membranes and to store energy (900)

feedback inhibition negative feedback mechanism in which the end product of a metabolic pathway inhibits an enzyme that catalyzes a reaction previous in the pathway (88)

feedback loop process of surveillance of internal conditions and response used to maintain homeostasis (837)

fermentation anaerobic pathway of cellular respiration that converts pyruvate to either lactic acid or ethyl alcohol and carbon dioxide (111)

fetal alcohol syndrome (FAS) birth defect resulting from alcohol use by the mother during pregnancy (1005)

fetus the developing human from the age of eight weeks until birth (1006)

fibrin clot-forming protein in the blood (881)

fibrous root system root system consisting of roots that are about the same size (559)

filament stalk of a stamen (536)

filtrate in the kidneys, fluid that has passed from the bloodstream into Bowman's capsule (913)

first law of thermodynamics law that states energy cannot be created or destroyed, but only converted from one form to another (76)

flagellum whiplike structure that grows out of a cell and enables it to move (46)

flexor skeletal muscle that causes a limb to bend at a joint (861)

fluke parasitic flatworm of the class Trematoda (649)

follicle cluster of cells that surround the immature egg cells in an ovary (1001)

food chain linear pathway of energy transfer in an ecosystem (97, 344)

food web a network of feeding relationships in an ecosystem (344)

foram marine protist of the phylum Foraminifera (485)

fossil preserved or mineralized remains or traces of an organism that lived long ago (252)

free energy amount of energy available for work, e.g., to drive cell activities (81)

freely movable joint joint that permits the greatest movement of bones (859)

fringing reef coral reef that forms close to the beach (646)

frond long, highly divided leaf of ferns (530)

fruit a mature ovary that contains one or more seeds (540)

fundamental niche the entire range of conditions an organism can tolerate (367)

Fungi Imperfecti group of fungi in which sexual reproduction has not been observed; also known as Deuteromycota (510)

G

gametangium reproductive structure in which gametes form (504)

gamete haploid cell that participates in fertilization by fusing with another haploid cell (120)

gametic meiosis sexual life cycle in which gametes are the only haploid cells; all of the other cells of the individuals in the life cycle are diploid (424)

gametocyte third stage of the *Plasmodium* life cycle; undergoes a sexual phase in the bloodstreams of infected humans (494)

gametophyte in the life cycle of a plant, the haploid phase that produces gametes (425)

gastrin hormone that regulates the synthesis of HCl by the stomach (907)

gastrulation inward migration of the outer cells of the blastula, resulting in the distribution of cells into three primary tissue layers (1004)

gated channel a channel that opens and closes like a gate in a fence (67)

gel electrophoresis technique that uses an electrical field passed through a gel to separate molecules in a mixture (207)

gemmules food-filled buds produced by freshwater sponges in cold or dry weather (640)

gene section of chromosome that codes for a protein or RNA molecule (119)

gene expression two-stage processing of information from DNA to proteins (183)

gene flow movement of alleles into or out of a population due to the migration of individuals to or from the population (324)

genetic code sequence of nucleotides that specifies the amino acid sequence of a protein (186)

genetic disorder harmful effect, such as sickle cell anemia, produced by mutated genes (154)

genetic drift random change in allele frequency in a population (325)

genetic engineering process of isolating a gene from DNA of one organism and transferring it to the DNA of another organism (203)

genetic polymorphism genetic variation in a population that results from a more than one allele for a gene (327)

genetic recombination rearrangement of genetic material (134)

genetics study of heredity (141)

genital herpes sexually transmitted disease caused by the herpes simplex virus (1008)

genotype genetic constitution of an organism as indicated by its set of alleles (146)

genus taxonomic category containing similar species (431)

germination resumption of growth by a plant embryo (572)

gestation period length of time between fertilization and birth (811)

gibberellin type of plant hormone that regulates growth, especially stem elongation (583)

gigantism condition of excess growth; almost always associated with pituitary tumors (978)

gill of fishes, structure located in the pharynx that is the site of gas exchange (623)

gill filament fingerlike projection from a gill in which gases enter and leave the blood (711)

gizzard a portion of the digestive tube of earthworms where strong muscles grind up the organic material in ingested soil (673)

glial cell *See* supporting cell

global warming increase in global temperatures as a result of increased concentration of greenhouse gases in the atmosphere (394)

glomerulus in the kidney, a cluster of capillaries that receives blood from the renal artery and that serves as a filter (912)

glucagon a peptide hormone produced by the pancreas that causes liver cells to release glucose stored in glycogen (974, 982)

glycogen polymer of glucose used for short-term energy storage (34)

glycolysis biochemical pathway that breaks down glucose into pyruvate (108)

Glossary

glycoprotein protein with carbohydrate molecules attached (456)

goiter enlargement of all or part of the thyroid gland (979)

Golgi apparatus cell organelle of a eukaryotic cell that consists of flattened sacs and collects, packages, and distributes molecules produced by the cell (43)

gonorrhea sexually transmitted disease caused by bacteria that cause inflammation of the mucous membranes in the urinary and reproductive tracts (1009)

gradualism model of evolution in which gradual change over a long period of time leads to macroevolution (258)

grain edible dry fruit of a cereal grass (592)

gram-negative designates a bacterium that does not retain the Gram stain (463)

gram-positive designates a bacterium that retains the Gram stain (463)

granum stack of thylakoids in a chloroplast (99)

gravitropism growth response to gravity (583)

greenhouse effect atmospheric warming resulting from heat trapped by gases such as carbon dioxide (393)

ground tissue system all the tissues of a vascular plant except the dermal tissue and vascular tissue (551)

ground water water found beneath Earth's surface (349)

guanine nitrogen base of the purine class; component of DNA and RNA nucleotides (169)

guard cells pair of specialized cells that border a stoma (521)

gymnosperm seed plant that produces seeds that do not develop within a fruit (531)

gynoecium part of a flower that houses the female gametophytes; the pistils, collectively (537)

H

habitat place where an organism lives (338)

hair follicle specialized dermal structure that produces hair (867)

half-life the period of time that it takes for one-half of a radioisotope to decay (233)

haploid having only one set of chromosomes (120)

Hardy-Weinberg principle principle stating that the frequency of alleles in a population does not change unless evolutionary forces such as selection act on the population (321)

Haversian canal hollow channel surrounded by concentric rings of bone and through which blood vessels and nerve pass (856)

heartwood dark, nonconducting wood in the center of a log (557)

helicase enzyme that unwinds a DNA molecule's double helix before replication (172)

helper T cell white blood cell that activates killer T cells and B cells (925)

hemodialysis treatment that cleanses the blood of patients with kidney failure by machine outside the body (914)

hemoglobin component of red blood cells that binds with and carries oxygen through the body (154)

hemophilia genetic disorder that impairs the blood's ability to clot and can cause excessive bleeding (155)

hepatic vein blood vessel that carries blood collected from the liver toward the heart (909)

herbaceous stem flexible, usually green (nonwoody) stem (557)

herbivore organism that eats only plants or algae (343)

heredity transmission of genetic traits from parent to offspring (14)

hermaphrodite organism that produces both eggs and sperm (641)

heterocyst thick-walled, cyanobacterial cell with enzymes that fix nitrogen gas (N_2) into ammonia (NH_3) (467)

heterotroph organism that cannot make its own food (78)

heterozygous refers to a pair of genes, or an individual, with two different alleles for a trait (146)

histamine chemical released by mast cells in an inflammatory response (930)

histocompatibility antigen marker on a body cell that protects it from attack by the body's immune system (915)

HIV human immunodeficiency virus that causes AIDS (17)

HIV positive condition of an individual whose blood tests detect the presence of antibodies to HIV or of viral proteins (932)

homeostasis maintenance of the internal stability of a cell, organism, or population in its environment (13)

hominid member of the family Hominidae of the order Primates; characterized by opposable thumbs, no tail, longer lower limbs, and erect bipedalism (299)

homologous chromosomes chromosomes that are similar in shape, size, and the genes they carry (120)

homologous structures structures that share a common ancestry (256)

homozygous refers to a pair of genes, or an individual, with two identical alleles for a trait (146)

hormone chemical produced in one part of an organism and then transported to another part of the organism, where it causes a response (57, 582, 840)

hormone-receptor complex complex formed by the binding of a hormone to its receptor protein (975)

Human Genome Project research effort to identify and locate the entire collection of genes in a human cell (209)

hybrid offspring of individuals from two different species (436)

hydrogen bond weak chemical bond in which a slightly positive hydrogen atom in a polar bond of one molecule is attracted to a slightly negative atom (usually oxygen or nitrogen) in a polar bond of another molecule (31)

hyperthyroidism condition resulting from overproduction of thyroxine by the thyroid gland, characterized by nervousness, sleep disorders, irregular heartbeat, and weight loss (979)

hypertonic describes a solution with a higher concentration of solute molecules than the solution across a selectively permeable membrane (60)

hypha slender filament that is part of the body of a multicellular fungus (502)

hypothalamus region of the brain located below the thalamus that produces several hormones and that controls many body activities related to homeostasis (838, 946)

hypothesis proposed explanation (9)

hypothyroidism condition resulting from underproduction of thyroxine by the thyroid gland, characterized in childhood by stunted growth and mental retardation and characterized in adulthood by lack of energy and weight gain (979)

hypotonic describes a solution with a lower concentration of solute molecules than the solution across a selectively permeable membrane (60)

immovable joint tight joint that permits little or no movement of the bones (859)

imperfect flower flower that lacks either a gynoecium or an androecium (537)

implantation burrowing of a blastocycst into the thick outer lining of the uterus (999)

incomplete dominance condition in which a trait is intermediate between two parents (152)

incomplete flower flower that lacks any one of the floral whorls: calyx, corolla, androecium, or gynoecium (537)

incomplete protein food that lack one or more of the essential amino acids (595)

inducer molecule that enables transcription to resume (191)

industrial melanism darkening of populations of organisms over time in response to industrial pollution (261)

inferior vena cava large vein that delivers blood from the lower portion of the body back to the heart (885)

inflammatory response series of events initiated by an injury or local infection that suppress infection and promote healing by removing disease-causing agents and dead cells (923)

in-group in cladograms, closely-related organisms (439)

inhalation part of breathing in which the diaphragm contracts and air moves into the lungs (889)

inhibitory synapse chemical junction where the receptor is a gated potassium channel that is closed at rest (943)

insertion in point mutations, addition of one or more nucleotides to a gene (195)

insertion in the muscular system, end of a muscle attached to a bone that moves when the muscle contracts (861)

insulin peptide hormone produced by the islets of Langerhans in the pancreas that stores excess glucose by promoting the accumulation of glycogen in the liver (982)

integration sum of the interactions among the signals from all the excitatory and inhibitory synapses of a neuron, which tend to cancel or reinforce one another (943)

intercellular coordination adjustment of a cell's activity in response to what other cells are doing (430)

intermediate filament long cytoplasmic protein filament found in the cytoskeletons of many eukaryotic cells (46)

internal fertilization fertilization that occurs within the body of the female parent (728)

internode area between two nodes of a plant stem (557)

interphase period of growth between two mitotic or meiotic divisions of a eukaryotic cell (128)

intron segment of mRNA transcribed from eukaryotic DNA but removed before translation of mRNA into a protein (174)

inversion mutation in which a chromosome fragment rejoins its original chromosome with its nucleotides reversed (124)

ion electrically charged atom or molecule (29)

ionic bond chemical bond joining positive and negative ions (31)

iris ring of tiny muscles that controls the amount of light entering the eye (951)

islets of Langerhans clusters of endocrine cells in the pancreas that secrete the hormone insulin (982)

isolation condition in which two populations of a species are separated so that they cannot interbreed (251)

isopod crustacean with its first thoracic segment fused to the head, seven pairs of similarly-sized walking legs, and a dorsoventrally flattened body (687)

isotonic describes solutions with equal solute concentrations on either side of a selectively permeable membrane (60)

joint junction of two bones (825, 859)

karyotype array of the chromosomes found in an individual's cells arranged in order of size and shape (122)

kidney organ that removes metabolic wastes from the blood and regulates salt and water balance (723)

killer T cell white blood cell that attacks and kills foreign cells and body cells infected by pathogens (925)

kinetic energy energy of an object due to its motion (76)

kinetochore disk of protein on a chromosome's centromere to which microtubules attach during mitosis and meiosis (129)

kingdom taxonomic category that contains phyla with similar characteristics (434)

Krebs cycle cyclic biochemical pathway of cellular respiration that uses pyruvate from glycolysis, releases CO_2, and produces ATP, NADH, and $FADH_2$ (109)

K-strategist species characterized by slow maturation, low fertility, slow population growth, and high competitive ability (319)

***lac* operon** gene system with an operator gene and three structural genes that control lactose metabolism in *E. coli* (190)

larva independent, immature stage in animal development that emerges from an egg (626)

larynx voice box; structure at the upper end of the trachea containing the vocal cords (888)

lateral bud bud located at a node of a stem that grows into a branch of the stem (557)

lateral line system row of pressure- and vibration-sensing organs running the length of both sides of a fish's body (754)

Glossary

lateral meristem meristem that produces secondary growth in woody plants; vascular cambium or cork cambium (574)

latex milky white sap of certain plants, such as those of the genus *Hevea*, that is used to make natural rubber (603)

law of independent assortment law stating that pairs of genes separate independently of one another in meiosis (147)

law of segregation law stating that pairs of genes separate in meiosis and each gamete receives one gene of a pair (147)

leaflet individual segment of a compound leaf (555)

legume member of the Fabaceae (pea) family of plants; type of fruit produced by members of the pea family (598)

lens part of the eye that focuses light passing through it (951)

lenticle area of loosely-packed cork cells on a woody stem that enables gas exchange (558)

leukemia disease that results in the overproduction of leucocytes (881)

leukocyte white blood cell; the primary cell of the immune system (880)

lichen symbiotic association between a fungus and an alga or cyanobacterium (511)

ligament band of connective tissue that holds together the bones in a joint (825)

light microscope microscope that uses a beam of light passing through one or more lenses (39)

limbic system network of neurons linked to the cerebral cortex that is responsible for many drives and emotions (946)

lipase enzyme that breaks down fat molecules into fatty acids and glycerol (905)

lipid bilayer basic structure of a plasma membrane; composed of two layers of phospholipids (55)

liver large organ that secretes bile and performs a wide variety of functions such as detoxification of poisons and storage and metabolism of carbohydrates, fats, and proteins (909)

lobe-finned fish bony fish with paired fins consisting of long, fleshy, muscular lobes (750)

logistic model model of population growth that assumes finite resource levels limit population growth (318)

long-day plant plant that produces flowers when days become longer than a critical length (585)

loop of Henle U-shaped section of a renal tubule in a kidney (913)

lumber wood from trees that have been cut down and sawed into boards, beams, or planks (600)

lung spongelike respiratory organ of a vertebrate that enables gas exchange between the air and the blood (712)

lung cancer malignant growth of cells of the lungs (893)

lymph fluid found in the intracellular spaces and lymphatic vessels of vertebrates (879)

lymphatic system system of the body that collects and recycles fluids leaked from the cardiovascular system (879)

lymphocyte type of white blood cell that matures in the organs of the lymphatic system (831, 880)

lysosome cell organelle of a eukaryotic cell containing hydrolytic, digestive enzymes (45)

M

macroevolution change that occurs among species over time as new species evolve and old species become extinct (251)

macrophage large white blood cell that engulfs pathogens (831, 880)

magnification enlargement or enlarging of an image (39)

malignant melanoma cancer derived from the pigment-producing melanocytes of the skin (868)

Malpighian tubule slender, fingerlike organ of excretion that opens into the gut of certain arthropods (681)

mandibulate arthropod with jaws (680)

mantle heavy fold of tissue that surrounds the visceral mass of mollusks (623)

mantle cavity space between the mantle and the visceral mass of mollusks (660)

mass extinction episode during which large numbers of species become extinct (278)

medusa free-swimming, bell-shaped, mouth-down body plan of a cnidarian; jellyfish (616)

megagametophyte female gametophyte of seed plants (531)

megakeryocyte giant bone-marrow cell characterized by a large, irregularly lobed nucleus; precursor to platelets (881)

megaspore spore that grows into a megagametophyte (531)

meiosis process in which the nucleus of a cell completes two successive divisions that produce four nuclei, each with a chromosome number that has been reduced by half (132)

melanocyte type of cell in the basal layer of the skin's epidermis that produces the dark pigment melanin (867)

melatonin hormone secreted by the pineal gland (983)

menopause time when a woman stops menstruating, usually between the ages of 45 and 55, and is no longer able to conceive (1002)

menstrual cycle series of hormone-induced changes that prepare the uterine lining for a possible pregnancy each month (1002)

menstruation periodic flow of blood and tissue shed from the outer layer of the endometrium of a woman's uterus that occurs approximately every 28 days (1002)

meristem region (or zone) of actively-dividing undifferentiated plant cells that are capable of developing into specialized plant tissues (528)

merozoite second stage of the life cycle of *Plasmodium* in which the protist divides rapidly and produces millions of cells in the liver of infected humans (494)

mesoderm middle layer of embryonic tissue in animals from which the skeleton and muscles develop (614)

mesophyll ground tissue of a leaf (556)

messenger RNA RNA copy of a gene used as a blueprint for the making of a protein during translation (185)

metabolism sum of all chemical processes occurring in an organism (13, 85)

metamorphosis process of change through which an immature organism passes as it grows to adulthood (692, 728)

metaphase stage of mitosis and meiosis when chromosomes move to and line up at center of a cell (131)

metastasis spread of malignant cells beyond their original site (196)

Index

Index

Krebs cycle, 109, **109**
krill, **344–345**
K-strategist, 319–320

Index

Index

Index

Credits

Credits *continued*

Timeline of the History of Life on Earth

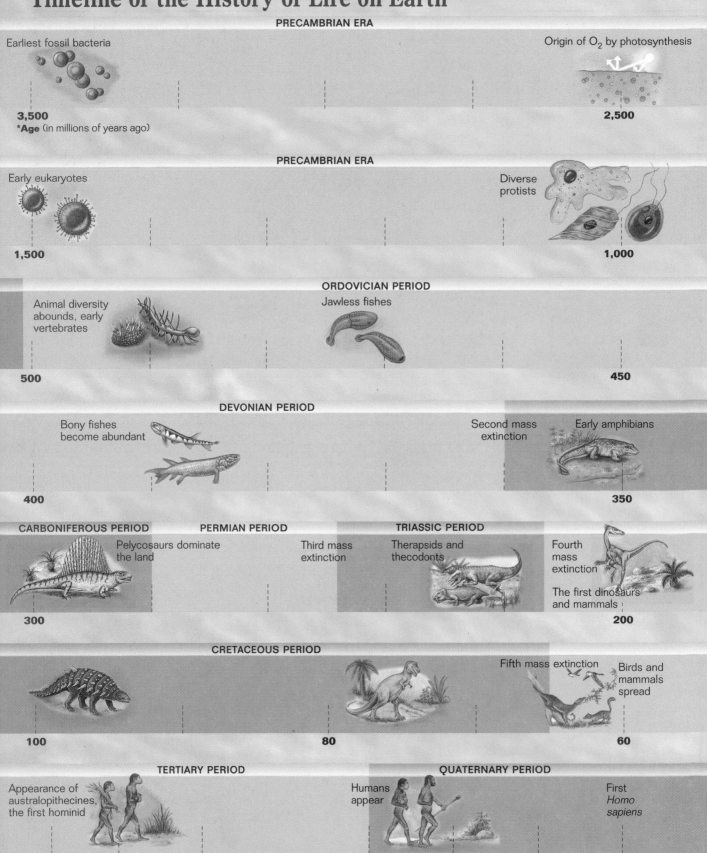

PRECAMBRIAN ERA

Earliest fossil bacteria

Origin of O₂ by photosynthesis

3,500
*Age (in millions of years ago)

2,500

PRECAMBRIAN ERA

Early eukaryotes

Diverse protists

1,500

1,000

ORDOVICIAN PERIOD

Animal diversity abounds, early vertebrates

Jawless fishes

500

450

DEVONIAN PERIOD

Bony fishes become abundant

Second mass extinction

Early amphibians

400

350

CARBONIFEROUS PERIOD   **PERMIAN PERIOD**   **TRIASSIC PERIOD**

Pelycosaurs dominate the land

Third mass extinction

Therapsids and thecodonts

Fourth mass extinction

The first dinosaurs and mammals

300

200

CRETACEOUS PERIOD

Fifth mass extinction

Birds and mammals spread

100

80

60

TERTIARY PERIOD   **QUATERNARY PERIOD**

Appearance of australopithecines, the first hominid

Humans appear

First *Homo sapiens*

3.9

3

2

1

.5